Artificial Neural Network-based Optimized Design of Reinforced Concrete Structures

Artificial Neural Network-based Optimized Design of Reinforced Concrete Structures introduces AI-based Lagrange optimization techniques that can enable more rational engineering decisions for concrete structures while conforming to codes of practice. It shows how objective functions including cost, CO_2 emissions, and structural weight of concrete structures are optimized either separately or simultaneously while satisfying constraining design conditions using an ANN-based Lagrange algorithm. Any design target can be adopted as an objective function. Many optimized design examples are verified by both conventional structural calculations and big datasets.

- Uniquely applies the new powerful tools of AI to concrete structural design and optimization
- Multi-objective functions of concrete structures optimized either separately or simultaneously
- Design requirements imposed by codes are automatically satisfied by constraining conditions
- Heavily illustrated in color with practical design examples

The book suits undergraduate and graduate students who have an understanding of college-level calculus and will be especially beneficial to engineers and contractors who seek to optimize concrete structures.

Artificial Neural Network-based Optimized Design of Reinforced Concrete Structures

Won-Kee Hong

CRC Press
Taylor & Francis Group
Boca Raton London New York

CRC Press is an imprint of the
Taylor & Francis Group, an **informa** business

First edition published 2023
by CRC Press
6000 Broken Sound Parkway NW, Suite 300, Boca Raton, FL 33487-2742

and by CRC Press
4 Park Square, Milton Park, Abingdon, Oxon, OX14 4RN

CRC Press is an imprint of Taylor & Francis Group, LLC

Library of Congress Cataloging-in-Publication Data
Names: Hong, Won-Kee (Professor of architectural engineering), author.
Title: Artificial neural network-based optimized design of reinforced
concrete structures / Won-Kee Hong.
Description: Boca Raton : CRC Press, 2023 | Includes bibliographical references and index.
Identifiers: LCCN 2022029378 | ISBN 9781032323688 (hardback) |
ISBN 9781032323695 (paperback) | ISBN 9781003314684 (ebook)
Subjects: LCSH: Reinforced concrete construction—Mathematical models |
Structural optimization | Lagrangian functions | Neural networks (Computer science)
Classification: LCC TA683.2 .H65 2023 | DDC 624.1/8341—dc23/eng/20221011
LC record available at https://lccn.loc.gov/2022029378

ISBN: 978-1-032-32368-8 (hbk)
ISBN: 978-1-032-32369-5 (pbk)
ISBN: 978-1-003-31468-4 (ebk)

DOI: 10.1201/9781003314684

Typeset in Sabon
by codeMantra

Contents

Preface

Structural engineers face several code-restricted design decisions. Codes impose many conditions and requirements to the designs of structural frames, such as columns and beams. Engineers commonly make design decisions based on empirical observations, however, it is difficult to intuitively find optimized solutions, while satisfying all code requirements simultaneously. Mathematician Joseph-Louis Lagrange proposed a method of Lagrange multipliers to identify local maxima and minima of objective functions subject to one or more equality constraints. However, this great idea faces some issues today because the explicit formulation of objective and constraining functions is complex when it is applied to structural engineering areas where design targets called objective functions are optimized while satisfying constraints imposed by code requirements and engineer's interests.

In this book, an artificial neural network (ANN)-based Lagrange algorithm provides a breakthrough in optimized structural designs. A novel way to perform an optimization of structural designs including reinforced concrete structures is suggested in this book. A conventional optimization of structural designs for reinforced concrete structures is not simple because deriving explicit objective functions is not easy.

An optimization application of the structural designs has been overlooked, being missed to enhance design balances and related economy. Hong et al. published several papers, highlighting the importance of optimizations for structural designs in general, and reinforced concrete structures in particular. This book titled "Artificial Neural Network-based Optimized Design of Reinforced Concrete Structures" is to introduce an optimized design of reinforced concrete structures to readers. Design conditions constrained by equality and inequality equations are based on material properties, member dimensions, and design code-based requirement such as ACI. Readers will have opportunities to learn how an ANN-based structural design can be implemented in optimizing designated objectives and targets.

Chapter 1 encapsulates Lagrange multiplier method that are taught in calculus course, which is then extended to an optimal design of concrete structures. Fundamental information provided in this chapter may help understand how constraining conditions including equality and inequality conditions are formulated to identify stationary points under KKT optimality conditions when inequality constraints are imposed. Constrained optimization problems are transformed into unconstrained problems in which gradients of objective functions and constraints are aligned on one line based on Lagrange multipliers when formulating Lagrange functions. In the previous book published by the author and by Daega, cost, CO_2 emissions, and structural weight of columns and beams were also calculated using ANN-based iterations, but not optimized.

Chapter 2 introduces an ANN-based optimization of objective functions subject to constraining functions. When optimizing designs, one of the biggest obstacles is to find explicit objective functions describing target functions, for example, a cost index, CO_2 emissions, and structural weight of RC beams and columns that need to be minimized. Chapter 2 explains how objective and constraint functions are described based on ANNs, replacing complex explicit functions which might be inaccurate when they are too much simplified with respect to design variables.

Examples provided will help readers derive their own ANN-based objective and constraining functions for their own optimizations. Jacobian and Hessian matrices of the ANN-based functions are also formulated to implement Newton-Raphson iteration to find stationary points of the Lagrange functions. Jacobian and Hessian matrices are a fundamental tool to optimize RC beams and columns as demonstrated in Chapters 3 and 4.

In Chapters 3 and 4, objective functions including cost index, CO_2 emissions, and structural weight for both doubly RC beams and columns are minimized as specified by the American concrete institute (ACI) using ANN-based Lagrange algorithm. Design parameters such as rebar ratios and concrete volumes for RC beams and columns are also calculated based on design criteria. Explicit objective and constraint functions with respect to design variables for both RC beams and columns are replaced by ANN-based functions obtained using the proposed method. An optimization using ANN-based objective functions trained by large datasets accompanies design parameters for best practices, resulting in designs that can meet various code restrictions at the same time. Optimized results are verified by conventional structural calculations and big datasets.

Minimizing multi-objective functions, such as costs, CO_2 emissions and structural weight, simultaneously is a challenging task. Five-steps for optimizing multi-objective functions are introduced in Chapter 5 for structural engineers and decision-makers who are not familiar with optimization algorithms. A large system of differential equations representing Jacobi equations of the Lagrange function for Unified Functions of Objective (UFO) is solved based on the five steps to optimize a design of a doubly reinforced concrete columns and beams. The design conditions of UFO are imposed by equality and equality constraints, yielding optimized designs. A Pareto frontier calculates tradeoff ratios contributed by each objective function. The results are then verified. A design example would offer a preliminary optimization guide for engineers. He also runs a YouTube channel (Deep learning for beginners; Won-Kee Hong, Kyung Hee University) on how AI-based structural designs and optimizations are performed based on this book.

The author cordially hopes his book serves as a stepping-stone for the next step in structural analysis and design. This book can be extended further for an AI-based Data-centric Engineering and Science. The author would like to thank for the support of the National Research Foundation of Korea (NRF) grant funded by the Korean government [MSIT 2019R1A2C2004965]. The author would like to appreciate his students, Nguyen Dinh Han, Tien Dat Pham, Thuc-Anh Le, Manh Cuong Nguyen, and Van Tien Nguyen for their contribution to the birth of this book. Finally, the author could not have published this book without the unflagging spiritual support of his wife Debbie, his son David, and daughter-in-law Sharon. His parents and parents-in-law also have been with him during the tough time of the preparation of this book. Call to me and I will answer you and tell you great and unsearchable things you do not know (Jeremiah 33:3).

Author

Won-Kee Hong is a professor of architectural engineering at Kyung Hee University, South Korea. He earned his bachelor's degree from Yonsei (Korea), master's and PhD degrees from UCLA (USA). He has worked for Englekirk and Hart, Inc. (USA), Nihhon Sekkei (Japan), and the Samsung Engineering and Construction Company (Korea). Dr. Hong has more than 35 years of professional experience in structural and construction engineering. He has been both an inventor and researcher in the field of modularized composite structures and is the author of more than 100 technical papers and over 100 patents in both Korea and The United States. He also published "Hybrid Composite Precast Systems: Numerical Investigation to Construction", in the Woodhead Publishing Series in Civil and Structural Engineering, published by Elsevier in December 2019, which has had an enthusiastic reception from readers. In this book, the author presented the basic concepts of artificial neural networks (ANNs) in Chapter 10 which contains an introduction to the implementation of AI-based neural networks applied to the design of reinforced concrete beams. More than 15 papers on the area of AI-based structural designs have been published or accepted in Journals published by Taylors & Francis and Elsevier. Dr. Hong has received several requests to write a book related to the AI-based structural optimizations in the areas of civil and architectural engineering. Designs of steel structures will be introduced in follow-on volumes. He also is preparing a book introducing an optimized auto-design of tall buildings.

He cordially hopes his publications contribute to the advent of new structural engineering with aids by Artificial Intelligence. Dr. Hong is a registered professional engineer in the United States and structural engineer in Korea. He also runs a YouTube channel (Deep learning for beginners; Won-Kee Hong, Kyung Hee University) on AI-based structural designs and optimizations.

Chapter 1

Introduction to Lagrange optimization for engineering applications

1.1 SIGNIFICANCE OF THIS CHAPTER

This chapter encapsulates Lagrange multiplier method that is taught in a calculus course, which is then extended for an optimal design of concrete structures. The method of Lagrange multipliers is proposed by the mathematician Joseph-Louis Lagrange to identify the local maxima and minima of objective functions subject to one or more equality constraints (Hoffmann and Bradley, 2004). The targets called objective functions are optimized while satisfying constraints as specified by constraining conditions. Constrained optimization problems are transformed into unconstrained problems in which derivatives are still implemented. Gradients of objective functions and gradients of constraints are aligned on one line based on Lagrange multipliers when formulating Lagrange functions (Beavis and Dobbs, 1990). KKT optimality conditions are also introduced to identify stationary points when inequality conditions are imposed. Fundamental information provided in this chapter may help understand how constraining conditions including equality and inequality conditions are formulated into Lagrange functions. In later chapters, objective functions, equality, and inequality conditions are derived using ANNs because these functions are difficult to be explicitly derived.

Optimization of structural designs such as reinforced concrete structures is not simple because multiple design parameters including material properties, member dimensions, and design code-based requirements, such as ACI and EC2, are constrained by equality and inequality conditions. Hong et al. published several papers which use Artificial Neural Networks (ANNs) to optimize structural designs in general, and reinforced concrete structures in particular, while an application of the optimization of this area has been overlooked, being missed to enhance design balances and related economy. The greatest advantage of Lagrange multipliers is to convert a constrained optimization problem into an unconstrained optimization problem while adding Lagrange multipliers for both equality (λ_c) and inequality (λ_v) constraints to the original constrained optimization. Mathematical discovery by Lagrange (Lagrange, 1804) states that gradients vectors of objective functions and those of constraints of original problems are to be shared (Beavis and Dobbs, 1990), such that gradients vectors based on partial differential equations for constrained problems can be solved. Lagrange function unifies gradients vectors in terms of the partial derivatives of an objective and constraint functions with respect to input variables [refer to Equations (1.2.4.3-2) and (1.2.4.3-3)] and Lagrange multipliers [refer to Equation (1.2.4.3-4)] together. However, KKT optimality equations must be considered with Lagrange optimization to identify stationary points of objective functions when inequality constraints are considered.

DOI: 10.1201/9781003314684-1

1.2 AN OPTIMALITY FORMULATION BASED ON EQUALITY CONSTRAINTS

1.2.1 Formulation of Lagrange functions

Objective functions can be established based on diverse reasons. Some of them are cost index (CI_c), CO_2 emission, and weight (W_c) for optimizing a design of reinforced concrete structures. Cost (CI_c) is an interest of structural engineers, whereas CO_2 emission, and weight (W_c) are the interests of governments and contractors, respectively. A multivariate objective function shown in Equation (1.2.1.1) is subjected to optimization. m equality constraining function $g_i(x_i)$ of Equation (1.2.1.2) should be set as zero so that it can be added or subtracted to an objective function for being formulated as a Lagrange function with multi-variables shown in Equation (1.2.1.3). A Lagrange Multiplier Method (LMM) identifies stationary points (stationary points including a maximum or minimum) of a Lagrange function when a Lagrange function takes the following form shown in Equation (1.2.1.3). It is noted that Lagrange formulation shown in Equations (1.2.1.1)–(1.2.1.3) of Chapter 1 does not include inequality constraints, whereas Section 1.2.3.2 consider inequality constraints which should be solved using KKT equations. Section 1.2.4 shows examples having inequality constraints that must be solved using KKT equations to find stationary points of objective functions.

$$y = f(x_1, x_2, ..., x_n) \tag{1.2.1.1}$$

$$g_i(x_1, x_2, ..., x_n) = 0, \quad i = 1,2,..., m \tag{1.2.1.2}$$

$$\mathcal{L}(x_1, x_2, ..., x_n, \lambda_1, \lambda_2, ..., \lambda_m) = f(x_1, x_2, ..., x_n) + or - \sum_{i=1\ i}^{m} \lambda_i g_i(x_1, x_2, ..., x_n) \tag{1.2.1.3}$$

1.2.2 Formulation of gradient vectors

Constrained minimum or maximum points of objective functions occur at constraining variables (x_i, y_i) where a constraint function $g(x, y)$ is tangent to an objective function $f(x, y)$ as shown in Figure 1.2.2.1. These stationary points can be found by proving level

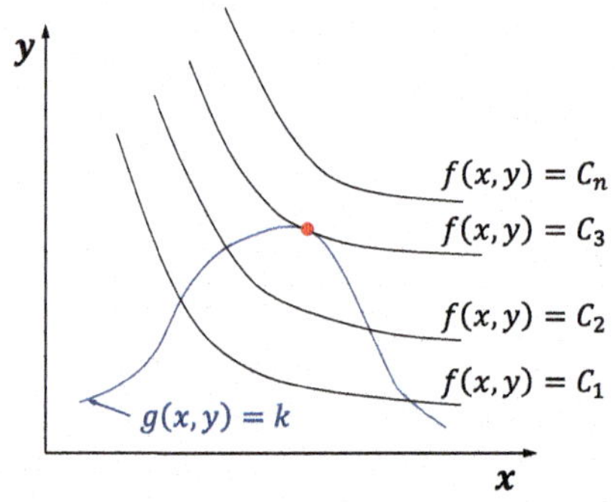

Figure 1.2.2.1 Constrained minimum or maximum points of objective functions occurring at constraining variables.

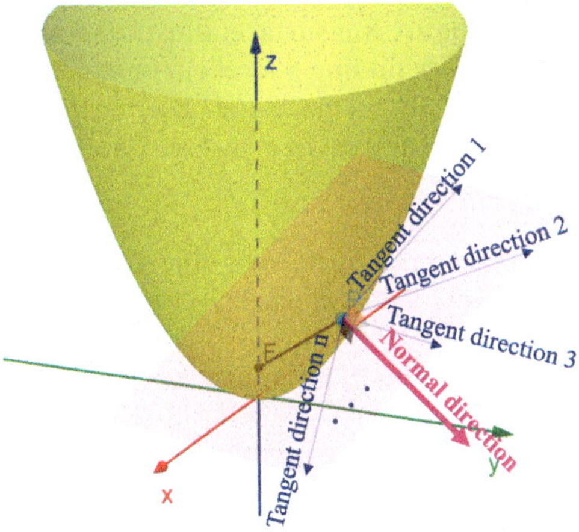

Figure 1.2.2.2 Definition of gradient and tangential vector on objective function $f(x, y)$ and on constraining function $g(x, y)$.

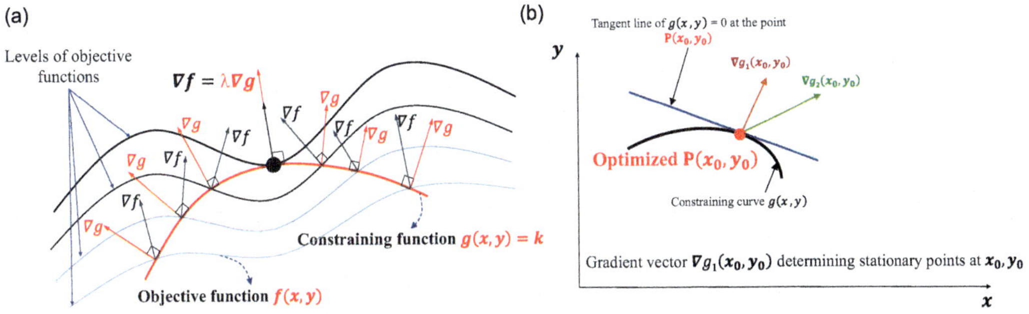

Figure 1.2.2.3 Gradient vector $\nabla g_1(x_0, y_0)$ showing stationary point at x_0, y_0. (a) $f(x, y)$ and $g(x, y)$ are tangent. (b) Two vectors $\nabla g_1(x_0, y_0), \nabla g_2(x_0, y_0)$ pointing different directions from $g(x)$ at x_0, y_0.

curves of $f(x, y)$ and $g(x, y)$ are tangent at constraining variables (x_i, y_i) (Osborne, 2009). However, tangents on functions are not uniquely determined at one point as shown in Figure 1.2.2.2. This is why gradients need to be introduced to prove $f(x, y)$ and $g(x, y)$ are tangent as shown in Figure 1.2.2.3. A gradient vector of a function $f(x, y)$ is expressed as $\nabla f(x, y)$ which is defined as a partial derivative of $f(x, y)$ with respect to corresponding variables (x, y) as shown in Equation (1.2.2.1-1). Similarly, a gradient vector of a function $g(x, y)$ is expressed as $\nabla g(x, y)$ as shown in Equation (1.2.2.1-2).

$$\nabla f(x, y) = \left(\frac{\partial f}{\partial x} + \frac{\partial f}{\partial y} \right) \tag{1.2.2.1-1}$$

$$\nabla g(x, y) = \left(\frac{\partial g}{\partial x} + \frac{\partial g}{\partial y} \right) \tag{1.2.2.1-2}$$

The objective function in this figure is not a line.

The objective function in this figure is a surface, illustrated by four black level curves.

The aiming of this optimization is finding a level curve (thick black curve) in a surface of objective function *f* which contacts to a constraint function *g* (a red line). The intersection point of that level curve (thick black curve) in a surface f and a constraint function *g* (a red line) has three prosperities:

1. The intersection point located in a surface of objective function *f*
2. The intersection point located in a line of a constraint function g, satisfying constrained defined by *g*
3. The intersection point is a stationary point in a surface of objective function *f* => minimum value of objective function *f*.

   $f(x,y)$ is level curves of the objective function
   Many level curves $f(x,y)$ create a surface present the objective function
   $g(x,y) = k$ is a curve defines constraints.

A gradient of $f(x, y)$ is always perpendicular to a level curve, $f(x, y)$, at a given pair of variables (x, y). To prove this, consider that $f(x,y)$ is written as Equation (1.2.2.2) (Dan's Blog, 2015).

$$f(x+\Delta x, y+\Delta y) \approx f(x,y) + \frac{\partial f}{\partial x}\Delta x + \frac{\partial f}{\partial y}\Delta y \tag{1.2.2.2}$$

when $(\Delta x, \Delta y)$ is a small amount by which a level curve, $f(x, y)$, changes from a given location (x, y), infinitesimally. The relationship $f(x+\Delta x, y+\Delta y) \approx f(x,y)$ holds when $\frac{\partial f}{\partial x}\Delta x + \frac{\partial f}{\partial y}\Delta y = 0$ when $(\Delta x, \Delta y)$ is infinitesimally small, leading to a dot product of $\nabla f(x)$ and $(\Delta x, \Delta y)$ is zero. A level curve, $f(x, y)$, and gradient vectors are, therefore, perpendicular to each other when $(\Delta x, \Delta y)$ is in the direction of a level curve, $f(x, y)$.

1.2.3 Optimality conditions of objective functions constrained by equality functions based on gradient vectors; finding stationary points based on gradients vectors

Two functions, $f(x, y)$ and $g(x, y)$, share gradient vectors, allowing two functions to be tangent each other when their gradients are scalar multiples of each other, having gradients point in the same direction (or in opposite directions). Stationary values of Lagrange functions shown in Equation (1.2.3.1) with a scalar λ are, thus, found at constraining variables (x_i, y_i) when $\nabla f(x) = \lambda \nabla g(x)$ is satisfied as shown in Equation (1.2.3.2-1) with multivariables (Dan's Blog, 2015; Dawkins, 2018). Gradient vectors constrained by constraining conditions will pass through constraining variables (x_i, y_i) when constraining function $g(x, y)$ is satisfied. System of equations shown in Equations (1.2.3.2-2) and (1.2.3.2-3), are obtained to solve for constraining variables (x_i, y_i) and Lagrange multiplier λ which will also satisfy Equation (1.2.3.2-1). Section 1.3.3 presents optimality example based on equality conditions.

In Figures 1.2.2.3a and b and Equation (1.2.3.2-1), two functions $f(x)$ and $g(x)$ share the same tangents at constraining variables (x_i, y_i) when gradients of $f(x)$ and $g(x)$ are scalar multiples of each other at variables (x_i, y_i), having gradients point in the same direction (or in opposite directions. Stationary points of objective function shown in Equation

(1.2.1.1) subject to constraining function $g_i(x_i)$ of Equation (1.2.1.2) can, then, be identified by solving Equation (1.2.3.2-1) based on Equations (1.2.3.2-2) and (1.2.3.2-3). Stationary points of objective function can also be found by equating the first derivative of a Lagrange function to 0 as shown in Equations (1.2.3.2-2) and (1.2.3.2-3) for multi-variables. The relationship $\nabla f(x) = \lambda \nabla g(x)$ shown in Figure 1.2.2.3 holds at constraining variables (x_i, y_i) for a scalar called Lagrange multiplier λ. Lagrange multiplier λ can be positive or negative to adjust a direction and length of a gradient vector. Thus, mathematically, stationary points of Lagrange functions are identified by equating gradient vectors for two functions λ_c and $g(x)$ with scalar multiples (Lagrange multiplier λ). This is equivalent to equating first-order partial differentiations of Lagrange function with respect to x_k and λ to zero as shown in shown in Equations (1.2.3.2-2) and (1.2.3.2-3). It is noted that two functions $f(x)$ and $g(x)$ shown in Figure 1.2.2.3a do not share the same tangents at constraining variables (x_i, y_i) when gradients $\nabla f(x)$, $\nabla g(x)$ are not scalar multiples of each other at constraining variables (x_i, y_i). Gradients $(\nabla f(x), \nabla g(x))$ of $f(x)$ and $g(x)$ will point in the different direction.

$$\mathcal{L}(x, \lambda) = f(x) + \lambda g(x) \tag{1.2.3.1}$$

$$\nabla f(x) = \lambda \nabla g(x) \text{ or } \nabla f(x) - \lambda \nabla g(x) = 0 \tag{1.2.3.2-1}$$

$$\frac{d\mathcal{L}}{dx_k} = \frac{df}{dx_k} + \sum_{i=1}^{m} \lambda_i \frac{dg_i}{dx_k} = 0, \quad k = 1,2,\ldots,n \tag{1.2.3.2-2}$$

$$\frac{\partial \mathcal{L}}{\partial \lambda} = g_i = 0, \quad i = 1,2,\ldots,m \tag{1.2.3.2-3}$$

1.2.4 Optimizations of an objective function constrained by equality functions

This section optimizes an objective function $f(x, y)$ shown in Equation (1.2.4.1-1) with equality constraint $g(x, y)$ shown in Equation (1.2.4.1-2) (Engineer2009Ali, 2017). Stationary points of Lagrange functions shown in Equation (1.2.4.2) can be identified when two functions $f(x)$ shown in Equation (1.2.4.1-1) and $g(x)$ shown in Equation (1.2.4.1-2) share the same gradients shown in Equation (1.2.4.3-1) with scalar multiples of each other, having gradients point in the same direction (or in opposite directions). This can be achieved by equating the two gradient vectors identical with scalar multiples of each other, as shown in Equation (1.2.4.3-1). Alternatively, equating the first derivative of a Lagrange function to 0 which is shown in Equations (1.2.4.3-2)–(1.2.4.3-4) will yield stationary points of Lagrange functions. Gradient vectors normal to both an objective function $f(x)$ and a constraining function $g(x)$ are formulated in terms of 3×3 simultaneous differential equations shown in Equations (1.2.4.3-2)–(1.2.4.3-4) by differentiating the Lagrange function shown in Equation (1.2.4.2) with respect to x, y, and Lagrange multiplier λ. The 3×3 simultaneous differential equations yield $x=0$, $y=+1$ or -1, and corresponding Lagrange multiplier $\lambda=+1$ or -1. Objective function $f(x, y)$ is, then, calculated as 2 or -2, respectively, when $y=+1$ or -1 as shown in Equations (1.2.4.4-1) and (1.2.4.4-2), resulting in minimized value as -2. Stationary point (0,-1) will also satisfy Equation (1.2.4.3-1). However, finding stationary points of Lagrange functions can be difficult when complex constraints such as inequality conditions are to be considered, and hence, KKT conditions based on trial error technique must be implemented not to miss any constraint including inequality conditions. There were no inequality constraints in this example, and hence finding stationary points of Lagrange functions is rather straightforward, without the use of KKT conditions. KKT conditions are introduced in Section 1.3 where inequality constraints are imposed.

$$\text{Minimize}: f(x, y) = 2y + x \tag{1.2.4.1-1}$$

$$\text{Subject to}: g(x, y) = y^2 + xy - 1 = 0 \tag{1.2.4.1-2}$$

$$\mathcal{L}(x,y,\lambda) = 2y + x + \lambda\left(y^2 + xy - 1\right) \tag{1.2.4.2}$$

$$\nabla f(x) = \lambda \nabla g(x) \text{ or } \nabla f(x) - \lambda \nabla g(x) = 0 \tag{1.2.4.3-1}$$

$$\mathcal{L}_x = \frac{\partial \mathcal{L}}{\partial x} = 1 + \lambda y = 0 \Rightarrow y = -\frac{1}{\lambda} \tag{1.2.4.3-2}$$

$$\mathcal{L}_y = \frac{\partial \mathcal{L}}{\partial y} = 2 + 2\lambda y + \lambda x = 0 \Rightarrow 2 - 2\lambda\frac{1}{\lambda} + \lambda x = 0$$
$$\Rightarrow 2 - 2 + \lambda x = 0 \Rightarrow \lambda x = 0 \Rightarrow x = 0 \tag{1.2.4.3-3}$$

$$\mathcal{L}_\lambda = \frac{\partial \mathcal{L}}{\partial \lambda} = g(x,y) = y^2 + xy - 1 = 0 \Rightarrow y^2 + (0)y - 1 = 0$$
$$\Rightarrow y^2 - 1 = 0 \Rightarrow y = \pm 1 \tag{1.2.4.3-4}$$

$$(1.2.4.3\text{-}1) \Rightarrow \lambda = -\frac{1}{y} = \lambda \pm 1$$

$$f(0, 1) = 2(1) + (0) = 2 \tag{1.2.4.4-1}$$

$$f(0, -1) = 2(-1) + (0) = -2 \text{ (solution)} \tag{1.2.4.4-2}$$

1.3 AN OPTIMALITY FORMULATION BASED ON INEQUALITY CONSTRAINTS

1.3.1 KKT (Karush-Kuhn-Tucker Conditions) optimality conditions

A general constrained optimization problem is described in Equation (1.3.1.1-1), in which $f(x)$ shown in Equation (1.3.1.1-1) is an objective function with an input vector $x=[x_1, x_2,..., x_n]^T$ subjected to equality constraining equations $c(x)=[c_1(x), c_2(x),...,c_m(x)]^T$ shown in Equation (1.3.1.1-2) and inequality constraining equations $v(x)=[v_1(x), v_2(x), ..., v_l(x)]^T \geq 0$ shown in Equation (1.3.1.1-3). Any optimal solution of objective function shown in Equation (1.3.1.1-1) belongs to a feasible set $\Gamma=\{x \in X : c_i(x)=0,\ i = 1,...,m,\ v_j(x)\geq 0,\ j = 1,...,l\}$.

$$\text{Minimize / Maximize}: f(x) \tag{1.3.1.1-1}$$

$$\text{Subject to}: c(x)=[c_1(x),...,c_m(x)]^T = 0 \tag{1.3.1.1-2}$$

$$v(x)=[v_1(x), v_2(x), ..., v_l(x)]^T \geq 0 \tag{1.3.1.1-3}$$

In Equation (1.3.1.2), Lagrange Multipliers Method (LMM) associates an objective function $f(x)$ with constrained conditions by applying Lagrange multipliers $\boldsymbol{\lambda}_c=[\lambda_1, \lambda_2,..., \lambda_m]^T$ and $\boldsymbol{\lambda}_v=[\lambda_1, \lambda_2,..., \lambda_l]^T$ for equality and inequality constraints. These multipliers formulate a multivariate objective function $f(x)$ subjected to various constraints in a non-boundary Lagrange function that allows conventional optimization algorithms to be implemented. A gradient of $f(x)$ and gradient of constraints functions $c(x)$, $v(x)$ are aligned by "factors" $\boldsymbol{\lambda}_c^*$ and $\boldsymbol{\lambda}_v^*$ at critical points x^* as shown in Equation (1.3.1.3). The solution $(x^*, \boldsymbol{\lambda}_c^*, \boldsymbol{\lambda}_v^*)$ of Lagrange optimization function must satisfy first-order necessary conditions for optimality known as Karush-Kuhn-Tucker (KKT) conditions (Peel and Moon, 2020) when inequality conditions are imposed as shown in Equations (1.3.1.4-1)–(1.3.1.4-3).

$$\mathcal{L}(x, \boldsymbol{\lambda}_c, \boldsymbol{\lambda}_v)=f(x)-\boldsymbol{\lambda}_c^T c(x)-\boldsymbol{\lambda}_v^T Sv(x) \tag{1.3.1.2}$$

$$\text{Case 1 Stationarity}: \nabla f(x^*)-\nabla\boldsymbol{\lambda}_c^T c(x^*)-\nabla\boldsymbol{\lambda}_v^T Sv(x^*) = 0 \tag{1.3.1.3}$$

$$\text{Case 2 Primal feasibility}: \boldsymbol{v}(x^*)\geq 0, \boldsymbol{c}(x^*)=0 \tag{1.3.1.4-1}$$

$$\text{Case 3 Dual feasibility}: \boldsymbol{\lambda}_v \geq \boldsymbol{0} \text{ for active constraints} \tag{1.3.1.4-2}$$

$$\text{Case 4 Hybrid KKT condition} \tag{1.3.1.4-3}$$

A KKT condition of primal feasibility shown in Equation (1.3.1.4-1) indicates that each of its inequality constraint $v_j(x^*)$ must be satisfied for given optimality conditions $v_j(x^*)>0$ if x^* is a solution or a critical stationary point. In Equation (1.3.1.4-2), a dual feasibility condition aligns a gradient vector of an objective function $f(x)$ to that of an inequality constraining function, ensuring they are aligned in a correct direction with a positive Lagrange multiplier $\lambda_{v_j} \geq 0$. Dual feasibility implies two gradient vectors, one for a gradient vector of an objective function and one for a gradient vector of an inequality constraint, point the same direction (the same or opposite gradient vectors). Dual feasibility holds for equality conditions when gradients are clearly defined.

In Equation (1.3.1.4-3), the condition expresses that Lagrange multipliers take up a hybrid KKT condition which indicates two possible cases when implementing the KKT condition. One case is $v_j(x^*)=0$; $\lambda_{v_j}>0$ expressing an "active" inequality constraint v_j, making a critical point x^* lying on a boundary line of an inequality constraint v_i. This indicates that inequality constraint is "active" ($\lambda_{v_j}(x)=0$; $\lambda_{v_j}>0$) which can be treated similarly to equality constraints by binding to transfer inequality conditions to equality conditions. Another case representing an inactive inequality constraint v_j, which is also called a slack condition, ensures $v_j(x^*)>0$, whereas $\lambda_{v_j}=0$ or $\lambda_v^T v(x^*)=0$. This optimality condition ignores inactive inequality constraint v_j when the Lagrange function shown in Equation (1.3.1.2) is formulated, however, a slack condition $v_j(x^*)>0$ must be verified at a critical point x^*. A critical stationary point x^*, in a slack condition, is not located on a constraint's boundary line. It is noteworthy that a condition of $v_j(x^*)<0$ is excluded from solution sets since it conflicts inequality primal feasibility condition $v_j(x^*)>0$ as shown in Case 2 shown in Equation (1.3.1.4-1). All inequality conditions must be arranged with $v_j(x^*)>0$, not $v_j(x^*)<0$ when formulating slack conditions in this study. A stationary point x^* cannot be a solution of an optimality when a condition $v_j(x^*)<0$ occurs in slack condition, because it contradicts to primal feasibility condition $v_j(x^*)>0$. However, in general, inequality conditions are specified in either $v_j(x^*)>0$, *or* $v_j(x^*)<0$.

1.3.2 Formulation of KKT optimality conditions (active and inactive), their implications on economy and structural engineering

Let's consider an objective (utility) function $f(x, y, z)$ shown in Equation (1.3.2.1-1) constrained by two functions $g_1(x, y, z)$, representing cash constraint and $g_2(x, y, z)$, representing coupon constraint shown in Equations (1.3.2.1-2) and (1.3.2.1-3), respectively. Constraints do not have to be strictly equal to certain values (equality constraints), but they can be greater than or less than certain values (inequality constraints) (Leslie Major, 2020). KKT conditions are introduced to bind inequality constraints to inequality constraints for finding optimality solutions based on Lagrange multipliers method. Lagrange optimization can be very complex when one or more inequality constraints are considered because gradient vectors cannot be defined at one location (x, y).

KKT conditions are a useful tool to solve for Lagrange multipliers method having more than one inequality constraint, such as when two constraining functions $g_1(x, y, z)$ and $g_2(x, y, z)$ shown in Equations (1.3.2.1-2) and (1.3.2.1-3) are imposed to identify constraining variables (x_i, y_i) maximizing an objective function $f(x, y, z)$ shown in Equation (1.3.2.1-1). Four KKT conditions for solving Lagrange optimization problem are assumed to identify stationary points of Lagrange function shown in Equations (1.3.2.2) constrained by Equations (1.3.2.1-2) and (1.3.2.1-3). KKT optimality equations based on trial-and-error technique are implemented to find a set of candidate points for stationary values of an objective function among which a sufficiency with constraint qualification must be checked to identify true a set of points (x, y). KKT conditions assume inequality constraints as either active or inactive conditions. Inequality conditions are less or greater than certain values (different from equality constraints) when complementary slack (or inactive) conditions are assumed. Complementary slack conditions or inactive conditions are ignored during Lagrange optimization, leading to inactive Lagrange multipliers λ_{v_j} which are 0 ($\boldsymbol{\lambda}_{v_j} = \boldsymbol{0}$) during optimization, or Lagrange multipliers times the slack conditions must be equal to 0 ($\boldsymbol{\lambda}_v^T v\left(\boldsymbol{x}^*\right) = 0$), whereas inactive constraining conditions with complementary slack conditions should be verified at the end. Active KKT conditions bind inequality KKT condition $v_j\left(\boldsymbol{x}^*\right) > 0$ to equality condition $v_j\left(\boldsymbol{x}^*\right) = 0$.

A slack variable is introduced to ignore inequality constraints, formulating Lagrange function in terms of only equality constraints in an optimization problem (Wikipedia, 2021). In economics, any resources related to an inactive condition are not exhausted economically, neither imposing any cost nor incurring any opportunity cost with 0 shadow price. This is the reason why the inactive condition is referred to as slack condition. In structural engineering, any requirement including structural, architectural, and financial conditions imposed by inactive constraints are ignored until the last moment when the final design parameters are verified to meet the ignored requirements. However, active conditions transform to bind inequality constraints to equality constraints, treating Lagrange optimization as a regular one with equality constraints as shown in an example of Section 1.3.3. KKT conditions with active constraints are imposed to cost opportunities which exhaust resources if capacities were to be increased. Active constraints are operated at capacity when a constraint is not slack, but active or binding, leading to slack as 0 (Mathematics for Economists, 2017). For example, stationary points minimizing or maximizing objective (utility) functions should be found based on reflecting given active constraining conditions while active inequality constraints are regarded as equality conditions. Lagrange multipliers λ_{v_j} must be strictly positive for active or binding KKT conditions while 0 or negative Lagrange multipliers are found for inactive or non-binding conditions which can also be referred to as slack conditions.

Four KKT conditions show a set of candidate points at which KKT conditions evaluate stationary points of an objective function when Lagrange functions are constrained by two constraining functions including one inequality condition in which constraints do not have to be strictly equal to 0 but can be less than or equal to 0. They can be greater than or equal to certain values (Mathematics for Economists, 2017). A sufficiency with constraints qualification must be verified to identify a stationary point.

$$f(x, y, z) \tag{1.3.2.1-1}$$

$$g_1(x, y, z) \tag{1.3.2.1-2}$$

$$g_2(x, y, z) \tag{1.3.2.1-3}$$

$$\mathcal{L} = f(x,y,z) + \lambda_1 g_1(x,y,z) + \lambda_2 g_2(x,y,z) \tag{1.3.2.2}$$

1.3.3 Optimality examples with inequality conditions

1.3.3.1 Example #1

An objective function $f(x_1,x_2)$ shown in Equation (1.3.3.1-1) is minimized subject to constraining functions $g_1(x_1,x_2)$ and $g_2(x_1,x_2)$ as shown in Equations (1.3.3.1-2) and (1.3.3.1-3), respectively (Engineer2009Ali, 2017). Four KKT equations are explored to identify stationary values of an objective function when inequality conditions are imposed.

$$\text{Minimize}: f(x_1, x_2) = x_1^2 + x_2^2 - 14x_1 - 6x_2 \tag{1.3.3.1-1}$$

$$\text{Subject to}: g_1(x_1, x_2) = x_1 + x_2 - 2 \le 0 \tag{1.3.3.1-2}$$

$$g_2(x_1, x_2) = x_1 + 2x_2 - 3 \le 0 \tag{1.3.3.1-3}$$

1. KKT condition #1
   Assuming both inequality constraining functions $g_1(x_1,x_2)$ and $g_2(x_1,x_2)$ are active where Lagrange multipliers λ_1 and λ_2 should be positive. And hence, both Lagrange multipliers λ_1 and λ_2 appear in Lagrange function as shown in Equation (1.3.3.2) whereas KKT optimality condition transform inequality constraints of $g_1(x_1,x_2)$ and $g_2(x_1,x_2)$ to equality constraints. Gradient vectors are normal to both an objective function $f(x)$ and constraining functions $g_1(x_1,x_2)$ and $g_2(x_1,x_2)$ are formulated in terms of 4×4 simultaneous differential equations shown in Equations (1.3.3.3-5)–(1.3.3.3-9) based on Equations(1.3.3.3-1)–(1.3.3.3-4) obtained by differentiating Lagrange function shown in Equation (1.3.3.2) with respect to x_1 and x_2 with Lagrange multipliers λ_1 and λ_2.

$$\mathcal{L}(x_1, x_2, \lambda_1, \lambda_2) = x_1^2 + x_2^2 - 14x_1 - 6x_2 + \lambda_1(x_1 + x_2 - 2) + \lambda_2(x_1 + 2x_2 - 3) \tag{1.3.3.2}$$

$$\frac{\partial \mathcal{L}}{\partial x_1} = 2x_1 - 14 + \lambda_1 + \lambda_2 = 0 \tag{1.3.3.3-1}$$

$$\frac{\partial \mathcal{L}}{\partial x_2} = 2x_2 - 6 + \lambda_1 + 2\lambda_2 = 0 \tag{1.3.3.3-2}$$

$$\frac{\partial \mathcal{L}}{\partial_1} = x_1 + x_2 - 2 = 0 \tag{1.3.3.3-3}$$

$$\frac{\partial \mathcal{L}}{\partial_2} = x_1 + 2x_2 - 3 = 0 \tag{1.3.3.3-4}$$

$$(1.3.3.3-1) \Rightarrow 2x_1 + 0x_2 + \lambda_1 + \lambda_2 = 14 \tag{1.3.3.3-5}$$

$$(1.3.3.3-2) \Rightarrow 0x_1 + 2x_2 + \lambda_1 + 2\lambda_2 = 6 \tag{1.3.3.3-6}$$

$$(1.3.3.3-3) \Rightarrow x_1 + x_2 + 0_1 + 0_2 = 2 \tag{1.3.3.3-7}$$

$$(1.3.3.3-4) \Rightarrow x_1 + 2x_2 + 0\lambda_1 + 0\lambda_2 = 3 \tag{1.3.3.3-8}$$

$$\text{In matrix form}: \begin{bmatrix} 2 & 0 & 1 & 1 \\ 0 & 2 & 1 & 2 \\ 1 & 1 & 0 & 0 \\ 1 & 2 & 0 & 0 \end{bmatrix} \begin{bmatrix} x_1 \\ x_2 \\ \lambda_1 \\ \lambda_2 \end{bmatrix} = \begin{bmatrix} 14 \\ 6 \\ 2 \\ 3 \end{bmatrix} \tag{1.3.3.3-9}$$

Four equations for four unknowns are solved when equality conditions are adopted for $g_1(x_1,x_2)$ and $g_2(x_1,x_2)$. Solving for 4×4 simultaneous equations yields constraining input variables $x_1 = 1$ and $x_2 = 1$, yielding Lagrange multipliers $\lambda_1 = 20$ and $\lambda_2 = -8$. However, negative Lagrange multiplier λ_2 leads to a false KKT optimality condition. Lagrange multiplier λ_2 should be greater than 0 for active KKT condition. The solution of First KKT condition fail to yield optimal objective function even if the KKT condition binds inequality constraints $g_1(x_1,x_2)$ and $g_2(x_1,x_2)$ to equality constraint as shown in Equations (1.3.3.3-10)–(1.3.3.3-12).

$$f(x_1,x_2) = x_1^2 + x_2^2 - 14x_1 - 6x_2$$
$$\Rightarrow f(1,1) = 1^2 + 1^2 - 14(1) - 6(1) = -18 \tag{1.3.3.3-10}$$

$$g_1(x_1,x_2) = x_1 + x_2 - 2 \Rightarrow \; g_1(1,1) = 1 + 1 - 2 = 0 \tag{1.3.3.3-11}$$

$$g_2(x_1,x_2) = x_1 + 2x_2 - 3 \Rightarrow \; g_2(1,1) = 1 + 2(1) - 3 = 0 \tag{1.3.3.3-12}$$

2. KKT condition #2
   Another KKT condition is assumed with both constraining functions $g_1(x_1,x_2)$ and $g_2(x_1,x_2)$ are not active. Lagrange function shown in Equation (1.3.3.4) does not include any of constraining functions $g_1(x_1,x_2)$ and $g_2(x_1,x_2)$ because both Lagrange multipliers λ_1 and λ_2 are regarded as 0 when being assumed as inactive. This condition is also referred to as a non-binding or slack condition, in which both Lagrange multipliers λ_1 and λ_2 disappear in Lagrange function as shown in Equation (1.3.3.4). Solving for 2×2 simultaneous equations shown in Equations (1.3.3.5-1) and (1.3.3.5-2) identifies constraining input variables x_1 and x_2 which yield positive values for constraining functions $g_1(x_1,x_2) = 8$ and $g_2(x_1,x_2) = 10$ as shown in Equations (1.3.3.6-1) and (1.3.3.6-2). However, this result also conflicts constraining functions $g_1(x_1,x_2)$ and $g_2(x_1,x_2)$ which should be less than or equal to 0 as constrained by Equations (1.3.3.1-2) and (1.3.3.1-3).

$$\mathcal{L}(x_1,x_2) = x_1^2 + x_2^2 - 14x_1 - 6x_2 \tag{1.3.3.4}$$

$$\frac{\partial \mathcal{L}}{\partial x_1} = 2x_1 - 14 = 0 \Rightarrow x_1 = 7 \tag{1.3.3.5-1}$$

$$\frac{\partial \mathcal{L}}{\partial x_2} = 2x_2 - 6 = 0 \Rightarrow x_2 = 3 \tag{1.3.3.5-2}$$

$$g_1(x_1,x_2) = g_1(7,3) = 7 + 3 - 2 = 8 \tag{1.3.3.6-1}$$

$$g_2(x_1,x_2) = g_2(7,3) = 7 + 2(3) - 3 = 10 \tag{1.3.3.6-2}$$

3. KKT condition #3
   Next KKT condition #3 assumes that a constraining function $g_1(x_1,x_2)$ is active while a constraining function $g_2(x_1,x_2)$ is inactive. In Equation (1.3.3.7), Lagrange function

does not include Lagrange multiplier λ_2 for a constraining function $g_2(x_1,x_2)$ because Lagrange multiplier λ_2 is regarded as 0 while including Lagrange multiplier λ_1. Gradient vectors normal to both an objective function $f(x_1,x_2)$ and a constraining function $g_1(x_1,x_2)$ are formulated in terms of 3×3 simultaneous differential equations shown in Equations (1.3.3.9-1)–(1.3.3.9-4) obtained from Equations (1.3.3.8-1)–(1.3.3.8-3) by differentiating Lagrange function shown in Equation (1.3.3.7) with respect to x_1 and x_2 and Lagrange multiplier λ_1. Solving 3 × 3 simultaneous equations gives constraining input variables $x_1 = 3$ and $x_2 = -1$, yielding constraining conditions for $g_1(x_1,x_2)=0$ and $g_2(x_1,x_2)=-2$ which are less than or equal to 0 as shown in Equations (1.3.3.10-1)–(1.3.3.10-2). Lagrange multiplier $_1$ was calculated as 8 which is greater than 0. An objective function was minimized as –26 as shown in Equation (1.3.3.10-3). It is noted that $g_1(x_1,x_2)=0$ is found because inequality constraining function $g_1(x_1,x_2)$ is bound as inequality constraining function when function $g_1(x_1,x_2)$ is assumed active.

$$\mathcal{L}(x_1,x_2,\lambda_1)=x_1^2+x_2^2-14x_1-6x_2+\lambda_1(x_1+x_2-2) \quad (1.3.3.7)$$

$$\frac{\partial\mathcal{L}}{\partial x_1}=2x_1-14+\lambda_1=0 \quad (1.3.3.8\text{-}1)$$

$$\frac{\partial\mathcal{L}}{\partial x_2}=2x_2-6+\lambda_1=0 \quad (1.3.3.8\text{-}2)$$

$$\frac{\partial\mathcal{L}}{\partial \lambda_1}=x_1+x_2-2=0 \quad (1.3.3.8\text{-}3)$$

$$(1.3.3.8\text{-}1)\Rightarrow 2x_1+0x_2+\lambda_1=14 \quad (1.3.3.9\text{-}1)$$

$$(1.3.3.8\text{-}2)\Rightarrow 0x_1+2x_2+\lambda_1=6 \quad (1.3.3.9\text{-}2)$$

$$(1.3.3.8\text{-}3)\Rightarrow x_1+x_2+0\lambda_1=2 \quad (1.3.3.9\text{-}3)$$

$$\text{In matrix form}: \begin{bmatrix} 2 & 0 & 1 \\ 0 & 2 & 1 \\ 1 & 1 & 0 \end{bmatrix}\begin{bmatrix} x_1 \\ x_2 \\ \lambda_1 \end{bmatrix}=\begin{bmatrix} 14 \\ 6 \\ 2 \end{bmatrix} \quad (1.3.3.9\text{-}4)$$

$$\Rightarrow x_1=3, \quad x_2=-1, \quad \lambda_1=8$$

$$g_1(x_1,x_2)=g_1(3,-1)=3+(-1)-2=0 \quad (1.3.3.10\text{-}1)$$

$$g_2(x_1,x_2)=g_2(3,-1)=3+2(-1)-3=-2 \quad (1.3.3.10\text{-}2)$$

$$f(x_1,x_2)=f(3,-1)=3^2+(-1)^2-14(3)-6(-1)=-26 \quad (1.3.3.10\text{-}3)$$

4. KKT condition #4

KKT condition #4 is tested if another stationary point is available. Assuming function $g_2(x_1,x_2)$ is active while $g_1(x_1,x_2)$ is inactive. In Equation (1.3.3.11), Lagrange function does not include constraining function $g_1(x_1,x_2)$ because Lagrange multiplier λ_1 is regarded as 0. Gradient vectors normal to both an objective function $f(x_1,x_2)$ and a constraining function $g_2(x_1,x_2)$ are formulated in terms of 3×3 simultaneous differential equations shown in Equations (1.3.3.12-4)–(1.3.3.12-7) obtained from Equations (1.3.3.12-1)–(1.3.3.12-3) by differentiating Lagrange function shown in Equation (1.3.3.11) with respect to x_1 and x_2 and Lagrange multiplier λ_2. Solving 3x3 simultaneous equations yields constraining input variables x_1 and x_2, leading to constraining functions $g_1(x_1,x_2)=2$ and function $g_2(x_1,x_2)=0$ as shown in Equations (1.3.3.13-1) and (1.3.3.13-2). This KKT condition with constraining function $g_1(x_1,x_2)=2$ violates the constraining conditions specified by Equations (1.3.3.1-2) and (1.3.3.1-3) even if Lagrange multiplier λ_2 is calculated as 4 which is greater than 0. This result, thus, conflicts a constraining function $g_1(x_1,x_2)$ which should be less than or equal to 0, failing to provide an optimal solution of an objective function. It is noted that $g_2(x_1,x_2)=0$ is found because inequality constraining function $g_2(x_1,x_2)$ is transformed to inequality constraining function when function $g_2(x_1,x_2)$ is assumed to be active. Objective function $f(x_1,x_2)$ was minimized as –28 in this KKT condition as shown in Equation (1.3.3.13-3).

$$\mathcal{L}(x_1,x_2,\lambda_2)=x_1^2+x_2^2-14x_1-6x_2+\lambda_2(x_1+2x_2-3) \quad (1.3.3.11)$$

$$\frac{\partial\mathcal{L}}{\partial x_1}=2x_1-14+\lambda_2=0 \quad (1.3.3.12\text{-}1)$$

$$\frac{\partial\mathcal{L}}{\partial x_2}=2x_2-6+\lambda_2=0 \quad (1.3.3.12\text{-}2)$$

$$\frac{\partial\mathcal{L}}{\partial \lambda_2}=x_1+2x_2-3=0 \quad (1.3.3.12\text{-}3)$$

$$(1.3.3.12\text{-}1)\Rightarrow 2x_1+0x_2+\lambda_2=14 \quad (1.3.3.12\text{-}4)$$

$$(1.3.3.12\text{-}2)\Rightarrow 0x_1+2x_2+\lambda_2=6 \quad (1.3.3.12\text{-}5)$$

$$(1.3.3.12-3)\Rightarrow x_1+2x_2+0\lambda_2=3 \quad (1.3.3.12\text{-}6)$$

$$\text{In matrix form}: \begin{bmatrix} 2 & 0 & 1 \\ 0 & 2 & 1 \\ 1 & 2 & 0 \end{bmatrix}\begin{bmatrix} x_1 \\ x_2 \\ \lambda_2 \end{bmatrix}=\begin{bmatrix} 14 \\ 6 \\ 3 \end{bmatrix} \quad (1.3.3.12\text{-}7)$$

$$\Rightarrow x_1=5, \quad x_2=-1, \quad \lambda_1=4$$

$$g_1(x_1,x_2) = g_1(5,-1) = 5+(-1)-2 = 2 \tag{1.3.3.13-1}$$

$$g_2(x_1,x_2) = g_2(5,-1) = 5+2(-1)-3 = 0 \tag{1.3.3.13-2}$$

$$f(x_1,x_2) = f(5,-1) = 5^2+(-1)^2-14(5)-6(-1) = -38 \tag{1.3.3.13-3}$$

SUMMARY

An objective function $f(x_1,x_2)$ shown in Equation (1.3.3.1-1) is minimized subject to constraining functions $g_1(x_1,x_2)$ and $g_2(x_1,x_2)$ shown in Equations (1.3.3.1-2) and (1.3.3.1-3), respectively. KKT condition #3 among four KKT optimality conditions satisfies the constraining conditions shown by $g_1(x_1,x_2)=0$ and $g_2(x_1,x_2)=-2$ (less than or equal to 0) as shown in Equations (1.3.3.10-1) and (1.3.3.10-2). In Equation (1.3.3.10-3), stationary points based on KKT condition #3 let an objective function be minimized as –26 when function $g_1(x_1,x_2)$ is active while $g_2(x_1,x_2)$ is inactive. In Chapters 3 and 4, KKT equations are implemented in an optimization of structures in general and reinforce concrete structures in particular in which equality and inequality constraints consisting of material properties, structural dimensions, and design code such as ACI and EC2. Computation time can be lengthy when a combination of inequality constraints for KKT equations is many.

1.3.3.2 Example #2

An objective function $f(x,y)$ shown in Equation (1.3.3.14-1) is minimized subject to constraining functions $g_1(x,y)$ governing cash budget and $g_2(x,y)$ governing coupon budget as shown in Equations (1.3.3.14-2) and (1.3.3.14-3), respectively (Leslie Major, 2020), when x and y are butter and bread, respectively, which are constrained by cash and coupon budget. This example is to minimize total budget $x \times y$ constrained by functions $g_1(x,y)$ and $g_2(x,y)$. Four KKT equations are explored to identify stationary points (x,y) of an objective function when inequality conditions are imposed.

$$\text{Maximize}: f(x,y) = xy \tag{1.3.3.14-1}$$

$$\text{Subject to}: g_1(x,y) = x+y \le 50 \Rightarrow 50-x-y \ge 0 \tag{1.3.3.14-2}$$

$$g_2(x,y) = 2x+y \le 60 \Rightarrow 60-2x-y \ge 0 \tag{1.3.3.14-3}$$

1. KKT condition 1; assuming $\lambda_2 = 0$ (constraint $g_1(x,y)$ is active but $g_2(x,y)$ is not binding):
   First, KKT condition assumes that an inequality constraining function $g_1(x,y)$ is active, leading to Lagrange multiplier λ_1 greater than 0 while $g_2(x,y)$ is inactive, leading to Lagrange multiplier λ_2 which is ignored. Lagrange function does not include constraining function $g_2(x,y)$ because Lagrange multiplier λ_2 is regarded as 0. Coupon

is enough when it is assumed that coupon constraint $g_2(x,y)$ is satisfied, leading constraint $g_2(x,y)$ to non-binding or inactive constraint for slack condition. Slack constraint $g_2(x,y)$ is ignored when KKT conditions are considered; however, slack constraint $g_2(x,y)$ should be verified to satisfy at the end of optimization. No opportunity cost incurs with 0 shadow price when there is a slack on constraint $g_2(x,y)$, or when Lagrange multipliers times the slack must be equal to 0. Economically any resources are not exhausted, imposing that this constraint does not cost anything (APMonitor.com, 2016). However, Lagrange multiplier of constraint $g_1(x,y)$ should be greater than 0 for active KKT condition, indicating that money is not enough, exhausting resources, when constraint $g_1(x,y)$ is active, binding inequality constraint $g_1(x,y)$ to equality constraint during optimization when complementary slack conditions are 0.

First KKT condition indicates that we do not have enough cash budget, constraining inequality condition shown in Equation (1.3.3.14-2) as equality condition $x + y = 50$, while we do have enough coupons, ignoring inequality condition constraining coupon shown in Equation (1.3.3.14-3) during KKT optimality. Therefore, it is assumed that coupon constraint $g_2(x,y)$ is satisfied, having enough coupons, while solving for the remaining Lagrange problems as shown in Lagrange function of Equation (1.3.3.15) where λ_2 is 0.

Gradient vectors normal to both an objective function $f(x,y)$ and a constraining function $g_1(x,y)$ are formulated in terms of 3×3 simultaneous differential equations shown in Equations (1.3.3.16-4)–(1.3.3.16-6) obtained from Equations (1.3.3.16-1)–(1.3.3.16-3) by differentiating Lagrange function shown in Equation (1.3.3.15) with respect to x, y, and Lagrange multipliers λ_1, λ_2.

As demonstrated in Example #1, solving 3×3 simultaneous equations shown in Equations (1.3.3.16-1)–(1.3.3.16-3) yields stationary points $x = 25$ and $y = 25$ shown in Equation (1.3.3.16-7), indicating that Lagrange multiplier λ_1 is also calculated as 25 which is greater than 0. However, this solution violates non-binding slack constraining function $g_2(x\ ,y) = 2x + y \le 60$ or $60 - 2x - y \ge 0$ as shown in Equation (1.3.3.14-3) because Equation (1.3.3.17-2) does not meet non-binding slack constraining function $g_2(x\ ,y)$.

This KKT condition does not provide an optimality solution for an objective function, failing to find stationary values when assuming function $g_1(x\ ,y)$ is active (λ_1 is assumed greater than 0) while $g_2(x\ ,y)$ are inactive or not binding with λ_2 being assumed as 0.

$$\mathcal{L}(x,y,\lambda_1,\lambda_2) = xy + \lambda_1(50 - x - y) \tag{1.3.3.15}$$

$$\frac{\partial \mathcal{L}}{\partial x} = y + \lambda_1(0 - 1 - 0) = 0 \tag{1.3.3.16-1}$$

$$\frac{\partial \mathcal{L}}{\partial y} = x + \lambda_1(0 - 0 - 1) = 0 \tag{1.3.3.16-2}$$

$$\frac{\partial \mathcal{L}}{\partial \lambda_1} = 50 - x - y = 0 \tag{1.3.3.16-3}$$

$$(1.3.3.16\text{-}1) \Rightarrow\ 0x\ + y - \lambda_1 = 0 \tag{1.3.3.16-4}$$

$$(1.3.3.16\text{-}2) \Rightarrow x + 0y - \lambda_1 = 0 \tag{1.3.3.16-5}$$

$$(1.3.3.16\text{-}3) \Rightarrow x + y + 0\lambda_1 = 50 \tag{1.3.3.16-6}$$

$$\text{In matrix form:} \begin{bmatrix} 0 & 1 & -1 \\ 1 & 0 & -1 \\ 1 & 1 & 0 \end{bmatrix} \begin{bmatrix} x \\ y \\ \lambda_1 \end{bmatrix} = \begin{bmatrix} 0 \\ 0 \\ 50 \end{bmatrix} \tag{1.3.3.16-7}$$

$$\Rightarrow x = y = \lambda_1 = 25$$

$$g_1(x, y) = g_1(25, 25) = 50 - 25 - 25 = 0 \tag{1.3.3.17-1}$$

$$g_2(x, y) = g_2(25, 25) = 60 - 2 \times 25 - 25 = -15 \tag{1.3.3.17-2}$$

2. KKT condition 2; assuming function $g_2(x, y)$ is active while $g_1(x, y)$ is inactive. Lagrange function does not include constraining function $g_1(x, y)$, and hence, Lagrange multiplier λ_1 regarded as 0 does not appear in Equation (1.3.3.18). Solving 3×3 simultaneous equations shown in Equations (1.3.3.19-1)–(1.3.3.19-3) yields constraining input variables $x = 15$ and $y = 30$, indicating Lagrange multiplier λ_2 is calculated as 15 which is greater than 0 in Equation (1.3.3.19-4). Slack constraint $g_1(x, y)$ is ignored when KKT conditions are considered, however, slack constraint $g_1(x, y)$ should be verified to satisfy at the end of optimization. Inactive constraint $g_1(x, y)$ must be verified to ensure that the KKT optimality conditions are properly adopted. This solution verifies inactive constraining function $g_1(x, y) = x + y = 45$ which is less than 50. This KKT condition satisfies the inactive constraining condition for $g_1(x, y)$ shown in Equation (1.3.3.14-2) with Lagrange multiplier λ_2 calculated as 15. It is noted that $g_2(x, y)$ is calculated as $2x + y = 60$ because inequality constraint $g_2(x, y)$ was regarded as equality constraint for an active KKT condition.

   No opportunity cost incurs with 0 shadow price when there is a slack on the constraint $g_1(x, y)$, or when Lagrange multipliers times the slack must be equal to 0. Economically any resources are not exhausted, imposing that this constraint does not cost anything, and implying that the cash budget is enough. However, Lagrange multiplier should be greater than 0 for active KKT condition for constraint $g_2(x, y)$, implying that the coupon budget is not enough which exhausts resources when constraint $g_2(x, y)$ is active. This case binds inequality constraint $g_2(x, y)$ to equality constraint during optimization when complementary slack conditions are 0. The minimized budget $x \times y$ is calculated as $15 \times 30 = 450$ unit based on a utility function shown in Equation (1.3.3.14-1). Cash budget is enough when cash budget constraint $g_1(x, y)$ is satisfied as non-binding or inactive constraint for slack condition. The rest of KKT conditions must be performed until stationary points are identified.

   Lagrange function is shown in Equation (1.3.3.18) when function $g_2(x, y)$ is active while $g_1(x, y)$ is inactive, assuming λ_1 as 0. Calculations similar to Equations (1.3.3.15)–(1.3.3.17) are shown in Equations (1.3.3.18)–(1.3.3.20).

KKT#2 satisfies inequality conditions and gives the maximized $f(x,y) = f(15, 30) = 15 \times 30 = 450$ as shown in Equation (1.3.3.20-3).

$$\mathcal{L}(x,y,\lambda_1,\lambda_2) = xy + \lambda_2(60 - 2x - y) \tag{1.3.3.18}$$

$$\frac{\partial \mathcal{L}}{\partial x} = y - 2\lambda_2 = 0 \tag{1.3.3.19-1}$$

$$\frac{\partial \mathcal{L}}{\partial y} = x - \lambda_2 = 0 \tag{1.3.3.19-2}$$

$$\frac{\partial \mathcal{L}}{\partial \lambda_2} = 60 - 2x - y = 0 \tag{1.3.3.19-3}$$

$$(1.3.3.19\text{-}1) \Rightarrow 0x + y - 2\lambda_2 = 0 \tag{1.3.3.19-4}$$

$$(1.3.3.19\text{-}2) \Rightarrow x + 0y - \lambda_2 = 0 \tag{1.3.3.19-5}$$

$$(1.3.3.19\text{-}3) \Rightarrow 2x + y + 0\lambda_2 = 60 \tag{1.3.3.19-6}$$

$$\text{In matrix form}: \begin{bmatrix} 0 & 1 & -2 \\ 1 & 0 & -1 \\ 2 & 1 & 0 \end{bmatrix} \begin{bmatrix} x \\ y \\ \lambda_2 \end{bmatrix} = \begin{bmatrix} 0 \\ 0 \\ 60 \end{bmatrix} \tag{1.3.3.19-7}$$

$$\Rightarrow x = 15, \quad y = 30, \quad \lambda_2 = 15$$

$$g_1(x,y) = g_1(15, 30) = 50 - 15 - 30 = 5 \tag{1.3.3.20-1}$$

$$g_2(x,y) = g_2(15, 30) = 60 - 2 \times 15 - 30 = 0 \tag{1.3.3.20-2}$$

$$f(x,y) = f(15, 30) = 15 \times 30 = 450 \tag{1.3.3.20-3}$$

3. KKT condition 3; assuming both inequality constraining functions $g_1(x_1,x_2)$ and $g_2(x_1,x_2)$ are active.

   Lagrange function is shown in Equation (1.3.3.21) when both $g_1(x_1,x_2)$ and $g_2(x_1,x_2)$ are active. Calculations similar to Equations (1.3.3.15)–(1.3.3.17) are shown in Equations (1.3.3.21)–(1.3.3.22).

   KKT#3 does not satisfy inequality conditions as shown in Equation (1.3.3.22-9) because active Lagrange multiplier should be positive.

$$\mathcal{L}(x,y,\lambda_1,\lambda_2) = xy + \lambda_1(50 - x - y) + \lambda_2(60 - 2x - y) \tag{1.3.3.21}$$

$$\frac{\partial \mathcal{L}}{\partial x} = y - \lambda_1 - 2\lambda_2 = 0 \quad (1.3.3.22\text{-}1)$$

$$\frac{\partial \mathcal{L}}{\partial y} = x - \lambda_1 - \lambda_2 = 0 \quad (1.3.3.22\text{-}2)$$

$$\frac{\partial \mathcal{L}}{\partial \lambda_1} = 50 - x - y = 0 \quad (1.3.3.22\text{-}3)$$

$$\frac{\partial \mathcal{L}}{\partial \lambda_2} = 60 - 2x - y = 0 \quad (1.3.3.22\text{-}4)$$

$$(1.3.3.19\text{-}1) \Rightarrow 0x + y - \lambda_1 - 2\lambda_2 = 0 \quad (1.3.3.22\text{-}5)$$

$$(1.3.3.19\text{-}2) \Rightarrow x + 0y - \lambda_1 - \lambda_2 = 0 \quad (1.3.3.22\text{-}6)$$

$$(1.3.3.19\text{-}3) \Rightarrow x + y + 0\lambda_1 + 0\lambda_2 = 50 \quad (1.3.3.22\text{-}7)$$

$$(1.3.3.19\text{-}4) \Rightarrow 2x + y + 0\lambda_1 + 0\lambda_2 = 60 \quad (1.3.3.22\text{-}8)$$

$$\text{In matrix form}: \begin{bmatrix} 0 & 1 & -1 & -2 \\ 1 & 0 & -1 & -1 \\ 1 & 1 & 0 & 0 \\ 2 & 1 & 0 & 0 \end{bmatrix} \begin{bmatrix} x \\ y \\ \lambda_1 \\ \lambda_2 \end{bmatrix} = \begin{bmatrix} 0 \\ 0 \\ 50 \\ 60 \end{bmatrix} \Rightarrow \begin{cases} x = 10 \\ y = 40 \\ \lambda_1 = -20 < 0 \\ \lambda_2 = 30 \end{cases} \quad (1.3.3.22\text{-}9)$$

4. KKT condition 4; assuming both inequality constraining functions $g_1(x_1,x_2)$ and $g_2(x_1,x_2)$ are inactive.

   Lagrange function is shown in Equation (1.3.3.23) when both $g_1(x_1,x_2)$ and $g_2(x_1,x_2)$ are inactive. Calculations similar to Equations (1.3.3.15)–(1.3.3.17) are shown in Equations (1.3.3.23)–(1.3.3.25).

   KKT#4 satisfies inequality conditions when $(x,y) = f(0, 0)$. However, but $f(x,y) = f(0, 0) = 0$ is less than $f(x,y) = f(15, 30) = 15 \times 30 = 450$ optimized by KKT#2 as shown in Equation (1.3.3.20-3), indicating that KKT#4 does not maximize Lagrange function.

$$\mathcal{L}(x, y) = xy \quad (1.3.3.23)$$

$$\frac{\partial \mathcal{L}}{\partial x} = y = 0 \quad (1.3.3.24\text{-}1)$$

$$\frac{\partial \mathcal{L}}{\partial y} = x = 0 \quad (1.3.3.24\text{-}2)$$

$$g_1(x, y) = g_1(0, 0) = 50 - 0 - 0 = 50 \quad (1.3.3.25\text{-}1)$$

$$g_2(x, y) = g_2(0, 0) = 60 - 2 \times 0 - 0 = 60 \quad (1.3.3.25\text{-}2)$$

$$f(x, y) = f(0, 0) = 0 \times 0 = 0 \quad (1.3.3.25\text{-}3)$$

1.3.3.3 Example #3

A standard form of an optimum problem using KKT equations is shown in Equations (1.3.3.26-1)–(1.3.3.26-3) (APMonitor.com, 2016, Mathematics for Economists, 2017).

$$\text{Minimize}: f(x) \quad (1.3.3.26\text{-}1)$$

$$\text{Subject to}: \ g_i(x) - b_i \geq 0, \ i = 1, \ldots, k \quad (1.3.3.26\text{-}2)$$

$$g_i(x) - b_i = 0, \ i = k+1, \ldots, m \quad (1.3.3.26\text{-}3)$$

A stationary point minimizing an objective function shown in Equation (1.3.3.27-1) is sought in this example. KKT equations should be implemented with inequality constraints shown in Equations (1.3.3.27-2) and (1.3.3.27-3).

$$\text{Minimize}: f(x_1, x_2, x_3) = x_1^2 + 2x_2^2 + 3x_3^2 \quad (1.3.3.27\text{-}1)$$

$$\text{Subject to}: \ g_1(x_1, x_2, x_3) = - \ 5x_1 + x_2 + 3x_3 \leq -3 \quad (1.3.3.27\text{-}2)$$

$$g_2(x_1, x_2, x_3) = 2x_1 + x_2 + 2x_3 \geq 6 \quad (1.3.3.27\text{-}3)$$

Inequality constraining functions can be rewritten in Equations (1.3.3.28-1) and (1.3.3.28-2) based on a standard form [refer to Equations (1.3.3.26-2) and (1.3.3.26-3)] of an optimum problem using KKT equations.

$$g_1(x_1, x_2, x_3) = 5x_1 - x_2 - 3x_3 - 3 \geq 0 \quad (1.3.3.28\text{-}1)$$

$$g_2(x_1, x_2, x_3) = 2x_1 + x_2 + 2x_3 - 6 \geq 0 \quad (1.3.3.28\text{-}2)$$

1. KKT condition #1; both inequality constraints are assumed as binding
   In Equations (1.3.3.29-1) and (1.3.3.29-2), both inequality constraints are assumed as binding, treating them like equality constraints which will be verified if this is a good

assumption. Signs of Lagrange multipliers should be positive when constraints are regarded as active or binding.

$$g_1(x_1, x_2, x_3) = 5x_1 - x_2 - 3x_3 = 3 \tag{1.3.3.29-1}$$

$$g_2(x_1, x_2, x_3) = 2x_1 + x_2 + 2x_3 = 6 \tag{1.3.3.29-2}$$

The first KKT conditions can be written as shown in Equations (1.3.3.30) when both of slack conditions shown in Equations (1.3.3.29-1) and (1.3.3.29-2) are 0 which is active. Gradient vectors normal to both an objective function $f(x_1, x_2, x_3)$ and inequality constraining functions $g_1(x_1, x_2, x_3)$ and $g_2(x_1, x_2, x_3)$ are formulated in terms of 5×5 simultaneous differential equations as shown in Equations (1.3.3.31-1)–(1.3.3.31-5) by differentiating Lagrange function shown in Equation (1.3.3.30) with respect to x_1, x_2, and x_3 and Lagrange multipliers (λ_1, λ_2). And hence, five equations to solve for five unknowns (input variables x_1, x_2, and x_3, and Lagrange multipliers (λ_1, λ_2) are obtained when inequality conditions $g_1(x_1, x_2, x_3)$andg$_2(x_1, x_2, x_3)$ are transformed to equality conditions. Solving 5×5 simultaneous equations yields stationary points x_1=1.450, x_2=0.800, and x_3=1.150, and Lagrange multipliers λ_1, λ_2. However, Lagrange multiplier λ_1 was calculated as −0.5 whereas λ_2 was calculated as 2.7, as shown in Equation (1.3.3.31-11), which cannot be accepted by KKT optimality conditions because Lagrange multiplier λ_1 should be greater than 0 for an active KKT condition. The KKT equations fail to yield optimal solutions when one of Lagrange multipliers is negative for an active KKT condition. Therefore, a test to find KKT conditions should move on until stationary points minimizing an objective function are identified with correct KKT conditions. Next KKT conditions must be tried with different assumptions, such that both of constraints are assumed as inactive or not binding. An inequality constraining function $g_1(x_1,x_2,x_3)$ is also assumed to be active while another inequality constraining $g_2(x_1,x_2,x_3)$ is assumed to be inactive, expecting Lagrange multiplier λ_1 being greater than 0 for active constraining function $g_1(x_1,x_2,x_3)$. Alternatively, inequality constraining function $g_2(x_1,x_2,x_3)$ is assumed to be active while another inequality $g_1(x_1,x_2,x_3)$ is assumed to be inactive.

Lagrange function is shown in Equation (1.3.3.30) when both $g_1(x_1,x_2,x_3)$ and $g_2(x_1,x_2,x_3)$ are active. Calculations similar to Equations (1.3.3.15)–(1.3.3.17) are shown in Equations (1.3.3.30) and (1.3.3.31).

$$\begin{aligned}\mathcal{L}(x_1,x_2,x_3,\lambda_1,\lambda_2) &= x_1^2 + 2x_2^2 + 3x_3^2 + \lambda_1(5x_1 - x_2 - 3x_3 - 3) \\ &\quad + \lambda_2(2x_1 + x_2 + 2x_3 - 6)\end{aligned} \tag{1.3.3.30}$$

$$\frac{\partial \mathcal{L}}{\partial x_1} = \frac{\partial f}{\partial x_1} - \lambda_1 \frac{\partial g_1}{\partial x_1} - \lambda_2 \frac{\partial g_2}{\partial x_1} = 0 \Rightarrow\ 2x_1 - \lambda_1(5) - \lambda_2(2) = 0 \tag{1.3.3.31-1}$$

$$\frac{\partial \mathcal{L}}{\partial x_2} = \frac{\partial f}{\partial x_2} - \lambda_1 \frac{\partial g_1}{\partial x_2} - \lambda_2 \frac{\partial g_2}{\partial x_2} = 0 \Rightarrow\ 4x_2 - \lambda_1(-1) - \lambda_2(1) = 0 \tag{1.3.3.31-2}$$

$$\frac{\partial \mathcal{L}}{\partial x_3} = \frac{\partial f}{\partial x_3} - \lambda_1 \frac{\partial g_1}{\partial x_3} - \lambda_2 \frac{\partial g_2}{\partial x_3} = 0 \Rightarrow 6x_3 - \lambda_1(-3) - \lambda_2(2) = 0 \tag{1.3.3.31-3}$$

$$\frac{\partial \mathcal{L}}{\partial \lambda_1} = g_1(x_1, x_2, x_3) = 5x_1 - x_2 - 3x_3 - 3 = 0 \tag{1.3.3.31-4}$$

$$\frac{\partial \mathcal{L}}{\partial \lambda_2} = g_2(x_1, x_2, x_3) = 2x_1 + x_2 + 2x_3 - 6 = 0 \tag{1.3.3.31-5}$$

$$(1.3.3.31\text{-}1) \Rightarrow 2x_1 + 0x_2 + 0x_3 - 5\lambda_1 - 2\lambda_2 = 0 \tag{1.3.3.31-6}$$

$$(1.3.3.31\text{-}2) \Rightarrow 0x_1 + 4x_2 + 0x_3 + \lambda_1 - \lambda_2 = 0 \tag{1.3.3.31-7}$$

$$(1.3.3.31\text{-}3) \Rightarrow 0x_1 + 0x_2 + 6x_3 + 3\lambda_1 - 2\lambda_2 = 0 \tag{1.3.3.31-8}$$

$$(1.3.3.31\text{-}4) \Rightarrow 5x_1 - x_2 - 3x_3 + 0\lambda_1 + 0\lambda_2 = 3 \tag{1.3.3.31-9}$$

$$(1.3.3.31\text{-}5) \Rightarrow 2x_1 + x_2 + 2x_3 + 0\lambda_1 + 0\lambda_2 = 6 \tag{1.3.3.31-10}$$

$$\text{In matrix form}: \begin{bmatrix} 2 & 0 & 0 & -5 & -2 \\ 0 & 4 & 0 & 1 & -1 \\ 0 & 0 & 6 & 3 & -2 \\ 5 & -1 & -3 & 0 & 0 \\ 2 & 1 & 2 & 0 & 0 \end{bmatrix} \begin{bmatrix} x_1 \\ x_2 \\ x_3 \\ \lambda_1 \\ \lambda_2 \end{bmatrix}$$

$$= \begin{bmatrix} 0 \\ 0 \\ 0 \\ 3 \\ 6 \end{bmatrix} \Rightarrow \begin{Bmatrix} x_1 = 1.450 \\ x_2 = 0.800 \\ x_3 = 1.150 \\ \lambda_1 = -0.500 \\ \lambda_2 = 2.700 \end{Bmatrix} \tag{1.3.3.31-11}$$

2. KKT condition #2; assuming function $g_1(x_1, x_2, x_3)$ is active while $g_2(x_1, x_2, x_3)$ is inactive.

   Lagrange function is shown in Equation (1.3.3.32) when function $g_1(x_1, x_2, x_3)$ is active while $g_2(x_1, x_2, x_3)$ is inactive. Calculations similar to Equations (1.3.3.15)–(1.3.3.17) are shown in Equations (1.3.3.32)–(1.3.3.34). KKT condition #2 does not satisfy inequality constraining function shown in Equation (1.3.3.27-3) or (1.3.3.28-2), noting that Equation (1.3.3.34) shows $g_2(x_1, x_2, x_3) < 0$.

$$\mathcal{L}(x_1,x_2,x_3,\lambda_2)=x_1^2+2x_2^2+3x_3^2+\lambda_1(5x_1-x_2-3x_3-3) \tag{1.3.3.32}$$

$$\frac{\partial \mathcal{L}}{\partial x_1}=\frac{\partial f}{\partial x_1}-\lambda_1\frac{\partial g_1}{\partial x_1}=2x_1-5\lambda_1=0 \tag{1.3.3.33-1}$$

$$\frac{\partial \mathcal{L}}{\partial x_2}=\frac{\partial f}{\partial x_2}-\lambda_1\frac{\partial g_1}{\partial x_2}=4x_2-\lambda_1(-1)=0 \tag{1.3.3.33-2}$$

$$\frac{\partial \mathcal{L}}{\partial x_3}=\frac{\partial f}{\partial x_3}-\lambda_1\frac{\partial g_1}{\partial x_3}=6x_3-\lambda_1(-3)=0 \tag{1.3.3.33-3}$$

$$\frac{\partial \mathcal{L}}{\partial \lambda_1}=g_1(x_1,x_2,x_3)=5x_1-x_2-3x_3-3=0 \tag{1.3.3.33-4}$$

$$(1.3.3.33\text{-}1)\Rightarrow 2x_1+0x_2+0x_3-5\lambda_1=0 \tag{1.3.3.33-5}$$

$$(1.3.3.33\text{-}2)\Rightarrow 0x_1+4x_2+0x_3+\lambda_1=0 \tag{1.3.3.33-6}$$

$$(1.3.3.33\text{-}3)\Rightarrow 0x_1+0x_2+6x_3+3\lambda_1=0 \tag{1.3.3.33-7}$$

$$(1.3.3.33\text{-}4)\Rightarrow 5x_1-x_2-3x_3+0\lambda_1=3 \tag{1.3.3.33-8}$$

$$\text{In matrix form:}\begin{bmatrix}2&0&0&-5\\0&4&0&1\\0&0&6&3\\5&-1&-3&0\end{bmatrix}\begin{bmatrix}x_1\\x_2\\x_3\\\lambda_1\end{bmatrix}=\begin{bmatrix}0\\0\\0\\3\end{bmatrix}\Rightarrow\left\{\begin{matrix}x_1=0.526\\x_2=-0.053\\x_3=-0.105\\\lambda_1=0.211\end{matrix}\right\} \tag{1.3.3.33-9}$$

$$\begin{aligned}g_2(x_1,x_2,x_3)&=g_2(0.526,-0.053,-0.105)=2\times0.526+(-0.053)+2\times(-0.105)-6\\&=-5.211<0\end{aligned} \tag{1.3.3.34}$$

3. KKT condition #3; assuming function $g_2(x_1,x_2,x_3)$ is active while $g_1(x_1,x_2,x_3)$ is inactive.
   Lagrange function is shown in Equation (1.3.3.35-1) when function $g_2(x_1,x_2,x_3)$ is active while $g_1(x_1,x_2,x_3)$ is inactive. Calculations similar to Equations (1.3.3.15)–(1.3.3.17) are shown in Equations (1.3.3.35)–(1.3.3.36) which yield $(x_1,x_2,x_3)=(2.057, 0.514, 0.686)$. g_1 $(x_1, x_2, x_3)>0$ imposed by Equation (1.3.3.28-1) is satisfied as shown in Equation (1.3.3.36-1) whereas g_2 $(x_1, x_2, x_3) = 0$ is obtained as shown in Equation (1.3.3.36-2) because g_2 (x_1, x_2, x_3) is assumed active. The candidate solutions of KKT condition #3 are, then, determined as (x_1, x_2, x_3) = (2.057, 0.514, 0.686).

$$\mathcal{L}(x_1,x_2,x_3,\lambda_2)=x_1^2+2x_2^2+3x_3^2+\lambda_2(2x_1+x_2+2x_3-6) \tag{1.3.3.35-1}$$

$$\frac{\partial \mathcal{L}}{\partial x_1} = \frac{\partial f}{\partial x_1} - \lambda_2 \frac{\partial g_1}{\partial x_1} = 2x_1 - 2\lambda_2 = 0 \quad (1.3.3.35\text{-}2)$$

$$\frac{\partial \mathcal{L}}{\partial x_2} = \frac{\partial f}{\partial x_2} - \lambda_2 \frac{\partial g_1}{\partial x_2} = 4x_2 - \lambda_2 = 0 \quad (1.3.3.35\text{-}3)$$

$$\frac{\partial \mathcal{L}}{\partial x_3} = \frac{\partial f}{\partial x_2} - \lambda_2 \frac{\partial g_1}{\partial x_2} = 6x_3 - 2\lambda_2 = 0 \quad (1.3.3.35\text{-}4)$$

$$\frac{\partial \mathcal{L}}{\partial \lambda_2} = g_2(x_1, x_2, x_3) = 2x_1 + x_2 + 2x_3 - 6 = 0 \quad (1.3.3.35\text{-}5)$$

$$(1.3.3.35\text{-}1) \Rightarrow 2x_1 + 0x_2 + 0x_3 - 2\lambda_2 = 0 \quad (1.3.3.35\text{-}6)$$

$$(1.3.3.35\text{-}2) \Rightarrow 0x_1 + 4x_2 + 0x_3 - \lambda_2 = 0 \quad (1.3.3.35\text{-}7)$$

$$(1.3.3.35\text{-}3) \Rightarrow 0x_1 + 0x_2 + 6x_3 - 2\lambda_2 = 0 \quad (1.3.3.35\text{-}8)$$

$$(1.3.3.35\text{-}4) \Rightarrow 2x_1 + x_2 + 2x_3 + 0\lambda_2 = 6 \quad (1.3.3.35\text{-}9)$$

$$\text{In matrix form}: \begin{bmatrix} 2 & 0 & 0 & -2 \\ 0 & 4 & 0 & -1 \\ 0 & 0 & 6 & -2 \\ 2 & 1 & 2 & 0 \end{bmatrix} \begin{bmatrix} x_1 \\ x_2 \\ x_3 \\ \lambda_2 \end{bmatrix} = \begin{bmatrix} 0 \\ 0 \\ 0 \\ 6 \end{bmatrix} \Rightarrow \left\{ \begin{array}{l} x_1 = 2.057 \\ x_2 = 0.514 \\ x_3 = 0.686 \\ \lambda_2 = 2.057 \end{array} \right\} \quad (1.3.3.35\text{-}10)$$

$$\begin{aligned} g_1(x_1, x_2, x_3) &= g_1(2.057,\ 0.514,\ 0.686) \\ &= 5 \times 2.057 - 0.514 - 3 \times 0.686 - 3 = 4.713 \geq 0 \end{aligned} \quad (1.3.3.36\text{-}1)$$

$$g_2(x_1, x_2, x_3) = g_2(2.057,\ 0.514,\ 0.686) = 2 \times 2.057 + 0.514 + 2 \times 0.686 - 6 = 0 \quad (1.3.3.36\text{-}2)$$

$$\begin{aligned} f(x_1, x_2, x_3) &= f(2.057,\ 0.514,\ 0.686) \\ &= (2.057)^2 + 2 \times (0.514)^2 + 3 \times (0.686)^2 = 6.171 \end{aligned} \quad (1.3.3.36\text{-}3)$$

4. KKT condition #4; assuming both inequality functions $g_1(x_1, x_2, x_3)$ and $g_2(x_1, x_2, x_3)$ are inactive.
   Lagrange function is shown in Equation (1.3.3.37) when both inequality functions $g_1(x_1, x_2, x_3)$ and $g_2(x_1, x_2, x_3)$ are inactive. Calculations similar to Equations (1.3.3.15)–(1.3.3.17) are shown in Equations (1.3.3.37)–(1.3.3.39). KKT condition #4 does not satisfy inequality constraining functions shown in Equations (1.3.3.27-2) and (1.3.3.27-3) because $g_1(x_1, x_2, x_3)$ and $g_2(x_1, x_2, x_3)$ obtained as shown in Equations (1.3.3.39-1) and (1.3.3.39-2), respectively, do not satisfy $g_1(x_1, x_2, x_3)>0$ and $g_2(x_1, x_2, x_3)>0$ imposed by inequality Equations (1.3.3.28-1) and (1.3.3.28-2) for KKT condition #4.

$$\mathcal{L}(x_1, x_2, x_3) = x_1^2 + 2x_2^2 + 3x_3^2 \quad (1.3.3.37)$$

$$\frac{\partial \mathcal{L}}{\partial x_1} = 2x_1 = 0 \quad (1.3.3.38\text{-}1)$$

$$\frac{\partial \mathcal{L}}{\partial x_2} = 4x_2 = 0 \quad (1.3.3.38\text{-}2)$$

$$\frac{\partial \mathcal{L}}{\partial x_3} = 6x_3 = 0 \quad (1.3.3.38\text{-}3)$$

$$(1.3.3.33\text{-}1) \Rightarrow x_1 = 0 \quad (1.3.3.38\text{-}4)$$

$$(1.3.3.33\text{-}2) \Rightarrow x_2 = 0 \quad (1.3.3.38\text{-}5)$$

$$(1.3.3.33\text{-}3) \Rightarrow x_3 = 0 \quad (1.3.3.38\text{-}6)$$

$$g_1(x_1, x_2, x_3) = g_1(0, 0, 0) = 5 \times 0 - 1 \times 0 - 3 \times 0 = 0 < 3 \quad (1.3.3.39\text{-}1)$$

$$g_2(x_1, x_2, x_3) = g_2(0, 0, 0) = 2 \times 0 + 1 \times 0 + 2 \times 0 = 0 < 6 \quad (1.3.3.39\text{-}2)$$

1.4 HOW MANY KKT CONDITIONS (KUHN AND TUCKER, 1951; KUHN AND TUCKER, 2014) MUST BE CONSIDERED?

Considering an optimization problem constrained by n inequality conditions, in which an inequality constraint is assumed either active or inactive in each KKT condition, and hence, 2^n possible combinations for defining n inequality constraints exist.

Figure 1.4.1 shows that a number of KKT conditions are estimated as $2^2 = 4$ when two inequality constraints are imposed as presented in Equations (1.4.1) and (1.4.2).

$$\text{Minimize}: f(x_1, x_2) = x_1^2 + x_2^2 - 14x_1 - 6x_2 \quad (1.4.1)$$

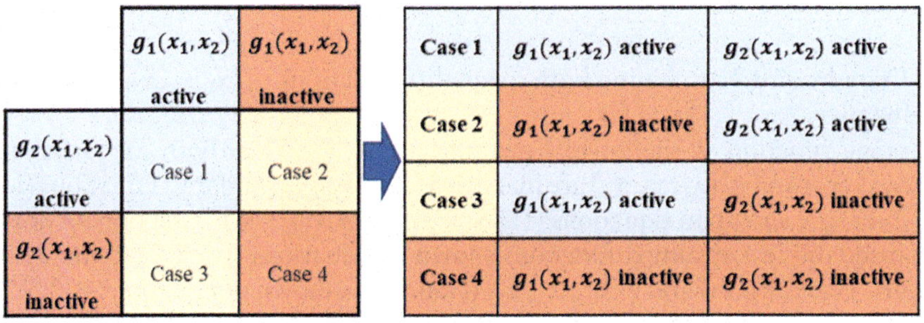

Figure 1.4.1 A number of KKT conditions with two inequality constraints.

$$\text{Subject to}: g_1(x_1, x_2) = x_1 + x_2 - 2 \le 0 \quad (1.4.2\text{-}1)$$

$$g_2(x_1, x_2) = x_1 + 2x_2 - 3 \le 0 \quad (1.4.2\text{-}2)$$

A number of KKT conditions are estimated as $2^{20} = 1{,}048{,}576$ for possible combinations to define 20 inequality constraints whereas possible combinations of KKT conditions are $2^{60} = 1.15 \times 10^{18}$ for 60 inequality constraints. However, all these combinations do not always exist because some combinations cannot occur due to constraining conflicts.

1.5 CONCLUSIONS

Gradient vectors normal to both an objective function $f(x)$ and a constraining function $g(x)$ share the same gradients with scalar multiples of each other, having gradients point in the same direction (or in opposite directions). Such points will be candidates for constrained minima (or maxima) based on KKT optimality conditions when an objective function is minimized subject to inequality constraining functions. Computation time can be lengthy for structural designs described in Chapters 3–5 when a combination of equality and inequality constraints consisting of material properties, structural dimensions, and design code such as ACI and EC2 are considered as KKT equations where an optimization of structures in general and reinforce concrete structures, in particular, is performed.

REFERENCES

APMonitor.com. (2016). KKT Conditions with Inequality Constraints. YouTube. https://www.youtube.com/watch?v=JTTiELgMyuM

Beavis, B., & Dobbs, I. (1990). *Optimization and Stability Theory for Economic Analysis* (p. 40). Cambridge University Press, Cambridge, United Kingdom.

Dan's Blog. (2015). Understanding Lagrange Multipliers. https://danstronger.wordpress.com/2015/08/08/lagrange-multipliers/amp/

Dawkins, P. (2018). Calculus IIII - Lagrange Multipliers. https://tutorial.math.lamar.edu/classes/calciii/lagrangemultipliers.aspx

Engineer2009Ali. (2017). Lagrange Multipliers with equality and inequality constraints (KKT conditions). YouTube. https://www.youtube.com/watch?v=eREvLgRJWrE

Hoffmann, L.D., & Bradley, G.L. (2004). *Calculus for Business, Economics, and the Social and Life Sciences* (8th ed., pp. 575–588). McGraw-Hill, Avenue of the Americas, New York, USA.

Kuhn, H. W.; Tucker, A. W. (1951). "Nonlinear programming". Proceedings of 2nd Berkeley Symposium. Berkeley: University of California Press. pp. 481– 492.

Kuhn, H. W.; Tucker, A. W.(2014). Nonlinear programming. In: Traces and emergence of nonlinear programming. Birkhäuser, Basel,pp. 247-258.

Lagrange, J. L. (1804). *Leçons sur le calcul des fonctions*. Imperiale, Paris, France.

Leslie Major, (2020). How to solve a basic Kuhn Tucker problem with 2 constraints (using the Lagrange Multiplier Method). YouTube. https://www.youtube.com/watch?v=eREvLgRJWrE

Mathematics for Economists (2017). Examples for optimization subject to inequality constraints, Kuhn-Tucker. YouTube. https://www.youtube.com/watch?v=TqN-8fxYUYY

Osborne, J.M. (2009). Lagrange Multipliers. Department of Mathematics & Statistics Boston University. http://math.bu.edu/people/josborne/MA225and230/MA230/notes/ConstrainedOptimizationNotes.pdf

Peel, C., & Moon, T. K. (2020). Algorithms for optimization. *IEEE Control Systems*, *40*(2), 92–94. https://doi.org/10.1109/MCS.2019.2961589

Wikipedia. (2021). Slack variable. https://wiki2.org/en/Slack_variable

Chapter 2

Lagrange optimization using artificial neural network-based generalized functions

2.1 IMPORTANCE OF AN OPTIMIZATION FOR ENGINEERING DESIGNS

2.1.1 Significance of ANN-based optimization

Joseph–Louis Lagrange (Lagrange, 1804)-optimized objective functions with constraints, identifying stationary points of Lagrange function. Lagrange functions are formed as a function of constraining input variables and Lagrange multiplier λ (Protter and Morrey, 1985). Maxima or minima of Lagrange function subjected to inequality and equality constraints points can be identified by solving systems of nonlinear differential equations (Hoffmann, 2004) based on Hessian matrix (which are differentiated twice) (Silberberg and Wing, 2001).

In a previous book published by the author (Hong, 2021), design parameters such as rebar ratios and concrete volumes for RC beams and columns subject to design criteria were determined based on ANN-based reverse analysis. Cost, CO_2 emissions, and structural weight of columns and beams were also calculated using ANN-based iteration. This chapter deals with an ANN-based Lagrange optimization of objective functions subject to constraining functions. When optimizing designs, one of the biggest obstacles is to find explicit objective functions describing target functions, for example, a cost index, CO_2 emissions, and structural weight of a RC beam and column that need to be minimized. The use of ANN-based optimization can overcome the need to explicitly describe complex objective functions and constraints with respect to design variables which are difficult to analytically derive.

In this chapter, objective functions are described based on ANNs, replacing complex analytical objective functions which can be inaccurate when they are too simplified with respect to design variables. In Chapters 3 and 4, objective functions including cost index, CO_2 emissions, and structural weight for both doubly RC beams and columns are minimized as specified by the American concrete institute (ACI) using ANN-based Lagrange algorithm. Optimized results are verified by conventional structural calculations. Optimization using ANN-based objective functions trained by large datasets can effectively aid a selection of design parameters for a holistic design is concrete structures, resulting in designs that can meet various code restrictions at the same time. Readers can refer to examples of ANN-based optimization including an optimization of a truss design presented in Section 2.4. This chapter uses ANNs to derive generalized objectives, constraint functions, and other output parameters with respect to design variables to replace complex explicit objective functions and constraints. Jacobian and Hessian matrices of the ANN-based functions are also formulated to implement Newton–Raphson iteration to find stationary points of the Lagrange functions.

Lagrange multipliers with equality and inequality constraints are used to identify stationary points which make specific design targets minimum. Two types of forward and

DOI: 10.1201/9781003314684-2

reverse ANNs are developed in this chapter; a reverse Lagrange Multiplier Method (LMM) is convenient and has fast computing speed; however, it may pose input conflicts among variables when unrealistic inputs are preassigned. Conversely, a forward LMM seeks optimized solutions in outputs using feed-forward networks; therefore, it does not pose any cross-relationship conflicts; however, a computational speed is lower than that of a reverse LMM. This chapter aims to present objective functions obtained based on both reverse and forward artificial neural networks.

2.1.2 Why ANN-based generalized functions?

Engineers commonly make design decisions based on empirical observations. Optimization techniques can be employed to help add efficiencies to engineering decisions, which results in designs that can meet various code restrictions simultaneously. However, only in a limited circumstance, it is possible to explicitly describe the complex analytical objective functions with respect to design parameters, targeting design goals such as a cost of RC beams and columns. It is not common to derive analytical objective functions representing a complex behavior of structural components such as columns and beams. Shariat et al. (2018) derived an objective function representing an optimum reinforcement ratio and optimum effective depth of RC beams based on BS (Standard, 1985), ACI (2014), and ICS (Tahouni, 2005) regulations, which are presented in Tables 2–4 of their paper, respectively. The procedure for calculating Lagrange multipliers according to the ACI code for singly and doubly reinforcement beams is also given in Equations (18)–(21) of their paper. These studies, however, were based on analytically derived objective functions which were simplified. It is difficult to express objective functions analytically directly in terms of design variables to use derivative methods, such as LMM. This chapter presents ANN-based Lagrange optimization implementing functions obtained based on large datasets to replace analytically obtained functions. ANN-based approach was provided to optimize structural designs in the level that conventional approach is unable to achieve. An optimization and sensitivity analysis using a LMM were performed based on both forward and reverse ANNs to optimize a design of RC beams and columns.

2.2 ANN-BASED LAGRANGE FORMULATION CONSTRAINED BY INEQUALITY FUNCTIONS

Optimization having inequality constraints is difficult to achieve directly because it is not possible to define gradient vectors for inequality functions, which makes it difficult to identify contact points for an objective and inequality functions. In Karush–Kuhn–Tucker (KKT) conditions, inequality constraints are ignored in Lagrange functions when inequality constraints are assumed as inactive conditions before solving for candidate solutions, among which one satisfying inequality conditions is an optimized solution. However, when inequality constraints are assumed as active conditions, inequality constraints are bound to transform into equality constraints. In Chapters 3 and 4, KKT conditions are led to designs of RC structures that meet various design requirements, including code restrictions and/or architectural criteria, while optimizing objective targets such as cost index, CO_2 emissions, or structural weight. Unified Function of Optimization (UFO) is also derived in Chapter 5 to optimize multiple objective functions, such that a cost of materials, manufacture, CO_2 emissions, and a beam weight are optimized simultaneously for RC structures.

2.3 ANN-BASED GENERALIZABLE OBJECTIVE AND CONSTRAINING FUNCTIONS

2.3.1 A limitation of an analytical function-based objective and inequality functions

Structural engineers face to meet several code-restricted requirements during their designs. Codes impose many conditions and requirements for designs; however, it is difficult to explicitly derive the analytical objective functions to optimize complex RC structures that meet all code requirements simultaneously. ANN-based Lagrange optimization techniques with constraints introduced in this chapter are not based on explicit mathematical formulations, but are based on solving ANN-based non-linear optimization problems under strict constraints imposed by design codes and interests of engineers. In LMM, design parameters $\left(\mathbf{x}^{(k)}\right)$ and Lagrange multipliers $\left(\lambda_c^{(k)},\ \lambda_v^{(k)}\right)$ are calculated, at which the first derivative of Lagrange function leading to Jacobi matrix (gradient vector) shown in Equations (2.3.2.4-1) and (2.3.2.4-2) converges to zero. Newton–Raphson iteration is used when it is difficult to find $\left(\mathbf{x}^{(k)},\ \lambda_c^{(k)},\ \lambda_v^{(k)}\right)$ analytically. A first derivative Jacobi and second derivative Hessian matrices of objective functions are obtained to optimize objective functions based on ANNs shown in Equations (2.3.3.1-1)–(2.3.3.1-3), not based on analytical functions. Generalized functions derived by training ANNs on large datasets enhance prediction. This chapter elicits how ANN and Large datasets-based objective, and constraining functions are derived using LMM. Section 2.3.3 formulates ANN-based objective and inequality functions.

2.3.2 Formulation of ANN-based Lagrange functions and KKT condition

In Chapters 3–5, Karush–Kuhn–Tucker conditions (Kuhn and Tucker, 2014) are adopted to account for inequality conditions constraining, designs of reinforced concrete structures, while meeting various design requirements including code restrictions and/or architectural criteria. In addition, objective functions such as cost index, CO_2 emissions, or structural weight are also optimized. ANN-based Lagrange optimization techniques without explicit parameterizations for an objective and constraining functions are implemented to optimize objective functions. Nonlinear differential equations obtained by KKT conditions are solved under tight constraints provided by design codes.

Design parameters defined as stationary points over a domain of optimization are sought by optimizing objective functions constrained by design requirements. An objective function $f(\mathbf{x})$ for a column design is given, in general, material and erection cost of a concrete column is given in Equation (2.3.2.1), in particular, as a function of seven input design parameters $\left(b, h, \rho_s, f_c', f_y, P_u, M_u\right)$. m equality constraints and l inequality constraints are also given in Equations (2.3.2.2-1) and (2.3.2.2-2). Lagrange function is then formulated in Equation (2.3.2.3) by converting constrained equations into unconstrained equations by introducing Lagrange multipliers.

A first derivative of Lagrange function leading to Jacobi matrix (gradient vector) is given in Equation (2.3.2.4-1) which is rearranged in Equation (2.3.2.4-2) according to similar terms. Candidate solutions are determined from Equations (2.3.2.4-1) and (2.3.2.4-2) based on Newton–Raphson iteration (Newton, 1687) method which is adopted to find stationary points based on linearized Lagrange functions, as shown in Equation (2.3.4.1-1) when Equation (2.3.2.4-1) which is rearranged in Equation (2.3.2.4-2) is not solved analytically. It is noted that Lagrange formulation shown in Equations (1.2.1.1)–(1.2.1.3) of Chapter 1 does not include inequality constraints, whereas Equations (2.3.2.1)–(2.3.2.4) in this chapter consider inequality constraints which should be solved using KKT equations. A first-order KKT conditions is formulated in Equation (2.3.2.4-2) (Villarrubia, 2018), by ignoring inequality constraints of $\mathbf{v}(\mathbf{x})$ or binding it to equality constraints of $\mathbf{c}(\mathbf{x})$ when inequality constraints of $\mathbf{v}(\mathbf{x})$ shown in Equation (2.3.2.2-2) are imposed. KKT theorem is sometimes referred to as a stationary-point theorem (Hoffmann, 2004). A number of KKT conditions depend on a number of inequality constraints shown in Equation (2.3.2.2-2). In Section 2.4, objective functions with inequality constraints are optimized based on KKT equations to find stationary points. Section 2.4 presents examples including minimizing fourth-order polynomial, a truss frame weights, and maximizing distance of projectiles based on both analytical and ANN-based Lagrange optimizations.

$$\begin{aligned}&\text{Objective function}: f(\mathbf{x}) = CI_c = f_{CIc}\left(b,\ h,\ \rho_s,\ f_c^{'},\ f_y,\ P_u,\ M_u\right)\\&\text{Input variables}: \mathbf{x} = \left[x_1,\ x_2,\ldots,\ x_n\right]\end{aligned} \tag{2.3.2.1}$$

$$\text{Equality constraints}: \mathbf{c}(\mathbf{x}) = \left[c_1(\mathbf{x}), c_2(\mathbf{x}), \ldots, c_m(\mathbf{x})\right]^T = 0 \tag{2.3.2.2-1}$$

$$\text{Inequality constraint}: \mathbf{v}(\mathbf{x}) = \left[v_1(\mathbf{x}), v_2(\mathbf{x}), \ldots, v_l(\mathbf{x})\right]^T \geq 0 \tag{2.3.2.2-2}$$

Alternatively,

$$\begin{aligned}&\text{Equality constraints}: c_j(\mathbf{x}) = 0, \qquad j = 1,\ldots,m\\&\text{Inequality constraint}: v_j(\mathbf{x}) \geq 0, \qquad k = 1,\ldots,l\\&\text{Lagrange function}: \mathcal{L}(\mathbf{x},\ \lambda_c,\ \lambda_v) = f(\mathbf{x}) - \boldsymbol{\lambda}_c^T\mathbf{c}(\mathbf{x}) - \boldsymbol{\lambda}_v^T\mathbf{S}\mathbf{v}(\mathbf{x})\end{aligned} \tag{2.3.2.3}$$

First-order gradient vectors presenting first-order KKT conditions:

$$\nabla\mathcal{L}\left(\mathbf{x}^{(k)},\ \boldsymbol{\lambda}_c^{(k)},\ \boldsymbol{\lambda}_v^{(k)}\right) = \nabla f(\mathbf{x}) - \boldsymbol{\lambda}_c^T\nabla\mathbf{c}(\mathbf{x}) - \boldsymbol{\lambda}_v^T\mathbf{S}\nabla\mathbf{v}(\mathbf{x}) = 0 \tag{2.3.2.4-1}$$

$$\begin{cases}\nabla_{\mathbf{x}}\mathcal{L} = 0 \rightarrow \dfrac{\partial\mathcal{L}}{\partial x_i} = 0, & i = 1,\ldots,n\\[2ex]\nabla_{\lambda_c}\mathcal{L} = 0 \rightarrow \dfrac{\partial\mathcal{L}}{\partial\lambda_{c,j}} = 0, & j = 1,\ldots,m\\[2ex]\nabla_{\lambda_v}\mathcal{L} = 0 \rightarrow \dfrac{\partial\mathcal{L}}{\partial\lambda_{v,k}} = 0, & k = 1,\ldots,l\end{cases} \tag{2.3.2.4-2}$$

2.3.3 Formulation of ANN-based objective and inequality functions

Objective functions for column cost (CI_c,) shown in Equation (2.3.2.1) are presented in Equation (2.3.3.1-1) based on ANN. Objective functions are extended to CO_2 emissions, and weight W_c in Equations (2.3.3.1-2) and (2.3.3.1-3) for columns as a function of seven design variables $\left(b, h, \rho_s, f_c^{'}, f_y, P_u, M_u\right)$. ANN-based objective functions for beams (CI_b,CO_2 emissions, and W_b) can be obtained similarly in Equations (2.3.3.1-4)–(2.3.3.1-6).

In Equations (2.3.3.1-1)–(2.3.3.1-6) (Krenker et al., 2011), $\mathbf{x}$ is an initial trial inputs consisting of design parameters for ANNs; L is a number of layers including hidden and output layers; $\mathbf{W}^l$ a weight matrix between layer $l-1$ and layer l; $\mathbf{b}^l$ is a bias matrix of layer l; and g^N and g^D are a normalization and denormalization functions, respectively. Activation functions (*tansig*, *tanh*) shown in Figure 2.3.3.1, f_t^l at layer l, were implemented to formulate nonlinear relationships between networks, whereas a linear activation function f_{lin}^L was selected for an output layer because output values are not squashed by an activation function. Objective functions are subject to both equality constraints $\mathbf{c}(\mathbf{x})$ shown in Equation (2.3.2.2-1) and inequality constraints $\mathbf{v}(\mathbf{x})$ shown in Equation (2.3.2.2-2) which represent diverse design goals and code requirement, i.e., inequality constraints for minimum and maximum rebar ratio are established based on ACI design requirements. Stationary points based on Lagrange multiplier equations are identified under constraints with one or more equalities and inequalities. Lagrange function shown in Equations (2.3.2.3) and (2.3.2.4) is derived with equality and equality constraints. Equality constraints are established with design parameters such as, P_u=1,000 kN or P_u − 1,000 kN = 0, M_u = 3,000 kNm or M_u − 3,000 kNm = 0, f_c' = 40 MPa or f_c' − 40 MPa = 0, f_y = 500 MPa or f_y − 500 MPa = 0.

$$\underbrace{CI_c}_{[1\times1]} = g_{CI_c}^D\left(f_{\text{lin}}^L\left(\underbrace{\mathbf{W}_{CI_c}^L}_{[1\times80]} f_t^{L-1}\left(\underbrace{\mathbf{W}_{CI_c}^{L-1}}_{[80\times80]}\cdots f_t^1\left(\underbrace{\mathbf{W}_{CI_c}^1}_{[80\times7]}\underbrace{g^N(\mathbf{x})}_{[7\times1]}+\underbrace{\mathbf{b}_{CI_c}^1}_{[80\times1]}\right)\cdots+\underbrace{\mathbf{b}_{CI_c}^{L-1}}_{[80\times1]}\right)+\underbrace{b_{CI_c}^L}_{[1\times1]}\right)\right) \quad (2.3.3.1\text{-}1)$$

$$\underbrace{CO_2}_{[1\times1]} = g_{CO_2}^D\left(f_{\text{lin}}^L\left(\underbrace{\mathbf{W}_{CO_2}^L}_{[1\times80]} f_t^{L-1}\left(\underbrace{\mathbf{W}_{CO_2}^{L-1}}_{[80\times80]}\cdots f_t^1\left(\underbrace{\mathbf{W}_{CO_2}^1}_{[80\times7]}\underbrace{g^N(\mathbf{x})}_{[7\times1]}+\underbrace{\mathbf{b}_{CO_2}^1}_{[80\times1]}\right)\cdots+\underbrace{\mathbf{b}_{CO_2}^{L-1}}_{[80\times1]}\right)+\underbrace{b_{CO_2}^L}_{[1\times1]}\right)\right) \quad (2.3.3.1\text{-}2)$$

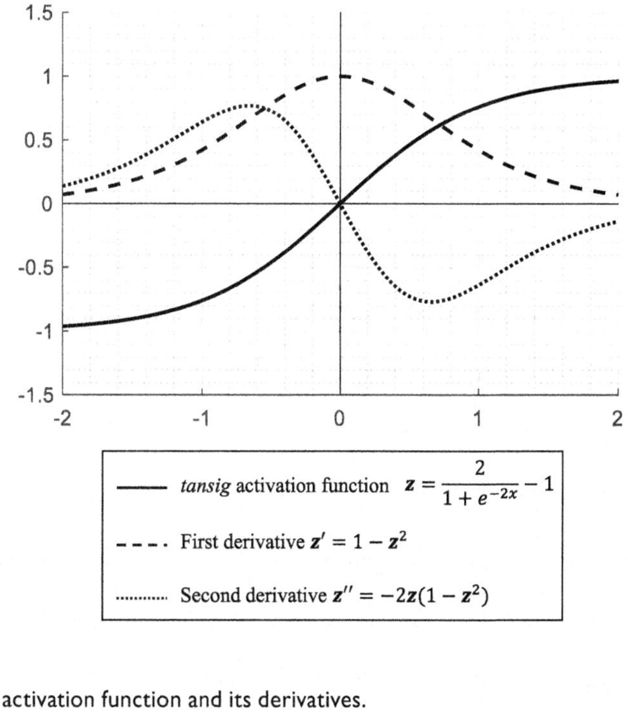

Figure 2.3.3.1 *tansig* activation function and its derivatives.

$$\underbrace{W_c}_{[1\times1]} = g^N_{W_c}\left(f^L_{\text{lin}}\left(\underbrace{\mathbf{W}^L_{W_c}}_{[1\times80]} f_t^{L-1}\left(\underbrace{\mathbf{W}^{L-1}_{W_c}}_{[80\times80]} \ldots f_t^1\left(\underbrace{\mathbf{W}^1_{W_c}}_{[80\times7]} \underbrace{g^N(\mathbf{x})}_{[7\times1]} + \underbrace{\mathbf{b}^1_{W_c}}_{[80\times1]}\right) \ldots + \underbrace{\mathbf{b}^{L-1}_{W_c}}_{[80\times1]}\right) + \underbrace{b^L_{W_c}}_{[1\times1]}\right)\right) \tag{2.3.3.1-3}$$

$$\underbrace{CI_b}_{[1\times1]} = g^D_{CI_b}\left(f^L_{\text{lin}}\left(\underbrace{\mathbf{W}^L_{CI_b}}_{[1\times30]} f_t^{L-1}\left(\underbrace{\mathbf{W}^{L-1}_{CI_b}}_{[30\times30]} \ldots f_t^1\left(\underbrace{\mathbf{W}^1_{CI_b}}_{[30\times7]} \underbrace{g^N(\mathbf{x})}_{[9\times1]} + \underbrace{\mathbf{b}^1_{CI_b}}_{[30\times1]}\right) \ldots + \underbrace{\mathbf{b}^{L-1}_{CI_b}}_{[30\times1]}\right) + \underbrace{b^L_{CI_b}}_{[1\times1]}\right)\right) \tag{2.3.3.1-4}$$

$$\underbrace{CO_2}_{[1\times1]} = g^D_{CO_2}\left(f^L_{\text{lin}}\left(\underbrace{\mathbf{W}^L_{CO_2}}_{[1\times80]} f_t^{L-1}\left(\underbrace{\mathbf{W}^{L-1}_{CO_2}}_{[80\times80]} \ldots f_t^1\left(\underbrace{\mathbf{W}^1_{CO_2}}_{[80\times7]} \underbrace{g^N(\mathbf{x})}_{[7\times1]} + \underbrace{\mathbf{b}^1_{CO_2}}_{[80\times1]}\right) \ldots + \underbrace{\mathbf{b}^{L-1}_{CO_2}}_{[80\times1]}\right) + \underbrace{b^L_{CO_2}}_{[1\times1]}\right)\right) \tag{2.3.3.1-5}$$

$$\underbrace{W_c}_{[1\times1]} = g^N_{W_b}\left(f^L_{\text{lin}}\left(\underbrace{\mathbf{W}^L_{W_b}}_{[1\times80]} f_t^{L-1}\left(\underbrace{\mathbf{W}^{L-1}_{W_b}}_{[80\times80]} \ldots f_t^1\left(\underbrace{\mathbf{W}^1_{W_b}}_{[80\times7]} \underbrace{g^N(\mathbf{x})}_{[7\times1]} + \underbrace{\mathbf{b}^1_{W_b}}_{[80\times1]}\right) \ldots + \underbrace{\mathbf{b}^{L-1}_{W_b}}_{[80\times1]}\right) + \underbrace{b^L_{W_b}}_{[1\times1]}\right)\right) \tag{2.3.3.1-6}$$

2.3.4 Linear approximation of a first derivative (Jacobi) of Lagrange functions

2.3.4.1 Optimization based on linearized Lagrange functions based on first-order (Jacobian matrix $\nabla\mathcal{L}(\mathbf{x}^{(k)}, \lambda_c^{(k)}, \lambda_v^{(k)})$) using Newton–Raphson iteration

Stationary points of a Lagrange function $\mathcal{L}(\mathbf{x}, \lambda_c, \lambda_v)$ with respect to input design parameters ($\mathbf{x}$) indicated by red color shown in Equation (2.3.2.3), Lagrange multipliers of equality constraints (λ_c), and inequality constraints (λ_v) shown in Equation (2.3.2.4) are found from gradient vectors $\nabla\mathcal{L}(\mathbf{x}, \lambda_c, \lambda_v)$ of Lagrange function $\mathcal{L}(\mathbf{x}, \lambda_c, \lambda_v)$ and constraining function which are perpendicular to their tangents.

As shown in Equation (2.3.2.4-1), Gradient vectors of an objective function, $\nabla f(\mathbf{x})$, and constraining functions, $\nabla\mathbf{c}(\mathbf{x})$ and $\nabla\mathbf{v}(\mathbf{x})$, should point in the same direction (or in opposite directions) if they are a scalar multiple or Lagrange multipliers (λ_c, λ_v) of the other (refer to Sections 1.2.3 for gradient directions and 1.3.1 for KKT conditions). This means that gradient vector $\nabla\mathcal{L}(\mathbf{x}, \lambda_c, \lambda_v)$ of a Lagrange function $\mathcal{L}(\mathbf{x}, \lambda_c, \lambda_v)$ shown in Equations (2.3.2.4-1) and (2.3.2.4-2) is set to zero to find stationary points of a Lagrange function $\mathcal{L}(\mathbf{x}, \lambda_c, \lambda_v)$. Lagrange functions are derived based on combinations two cases depending on active and inactive conditions.

2.3.4.1.1 Linear approximation of a first derivative (Jacobi) of Lagrange functions $\nabla\mathcal{L}(\mathbf{x}, \lambda_c, \lambda_v)$

However, it is difficult to directly find stationary points of a Lagrange function $\mathcal{L}(\mathbf{x}, \lambda_c, \lambda_v)$ when Lagrange functions are complex such as those found in structural designs. This is why Newton–Raphson iteration method is adopted to find stationary points by linearizing Lagrange functions as shown in Equation (2.3.4.1-1). A first-order approximation of $\nabla\mathcal{L}(\mathbf{x}, \lambda_c, \lambda_v)$ obtained by expanding $\nabla\mathcal{L}(\mathbf{x}, \lambda_c, \lambda_v)$ with respect to stationary points $\mathbf{x}$ and Lagrange multipliers λ_c, λ_v is implemented to find stationary points of Lagrange functions.

Gradient vectors $\nabla\mathcal{L}(\mathbf{x}, \lambda_c, \lambda_v)$ expressed with a set of partially differential equations is to be linearized before being solved by Newton–Raphson technique. Jacobi of Lagrange function consisting of objective and constraining functions are linearized at a specific space point. In Equation (2.3.4.1-1), a gradient vector of a Lagrange function is linearized in terms of higher derivatives at $\mathbf{x}, \lambda_c, \lambda_v$ including Hessian matrices. Gradient vectors $\nabla\mathcal{L}(\mathbf{x}, \lambda_c, \lambda_v)$ are linearized by differentiating Lagrange functions with respect to $\mathbf{x}$, λ_c, and λ_v.

The first derivative (Jacobi) of Lagrange functions is tested at the distance moved infinitesimally by $\Delta\mathbf{x}$, $\Delta\lambda_c$, $\Delta\lambda_v$ if a first derivative of Lagrange functions doesn't converge to zero at initial input parameters $\mathbf{x}^{(0)}, \lambda_c^{(0)}, \lambda_v^{(0)}$, where $(\Delta\mathbf{x}, \Delta\lambda_c, \Delta\lambda_v)$ is a small amount by which first derivative of Lagrange functions changes from a given location $(\mathbf{x}^{(0)}, \lambda_c^{(0)}, \lambda_v^{(0)})$, infinitesimally. An amount moved infinitesimally from the $\mathbf{x}^{(0)}, \lambda_c^{(0)}, \lambda_v^{(0)}$ is calculated by multiplying the slop of a first derivative (Jacobi) of Lagrange functions by an infinitesimal distance $(\Delta\mathbf{x}, \Delta\lambda_c, \Delta\lambda_v)$ as shown by $\left[\boldsymbol{H}_{\mathrm{L}}\left(\mathbf{x}^{(0)}, \lambda_c^{(0)}, \lambda_v^{(0)}\right)\right]\begin{bmatrix}\Delta\mathbf{x}\\ \Delta\lambda_c\\ \Delta\lambda_v\end{bmatrix}$ in Equation (2.3.4.1).

A slop of the first derivative (Jacobi) of Lagrange functions is called Hessian and they are derived in Sections 2.3.5 and 2.3.6. Newton-Raphson method is repeated until the solution converges using MATLAB toolbox.

2.3.4.1.2 Derivation of a first derivative (Jacobi) of Lagrange functions $\nabla\mathcal{L}(\mathbf{x}, \lambda_c, \lambda_v)$

A first-order derivative of Lagrange function $\mathcal{L}(\mathbf{x}, \lambda_c, \lambda_v)$ denoted as $\nabla\mathcal{L}(\mathbf{x}, \lambda_c, \lambda_v)$ is referred to as a Jacobian matrix. Jacobian matrix which is differentiated one more time to be expanded in terms of second derivatives with respect to $\mathbf{x}, \lambda_c, \lambda_v$ is referred to as a Hessian matrix of Lagrange function as shown in Equation (2.3.4.1-1). Newton–Raphson technique is adopted to reach a convergence of an iteration shown in Equations (2.3.4.1-1)–(2.3.4.1-5) consisting of Jacobian and Hessian matrices.

In Equation (2.3.4.1-1), a gradient vector of Lagrange functions $\nabla\mathcal{L}(\mathbf{x}, \lambda_c, \lambda_v)$ at $\mathbf{x}^0 + \Delta\mathbf{x}$ which is very close to $\mathbf{x}^{(0)}$ can be obtained shown as $\nabla\mathcal{L}\left(\mathbf{x}^{(0)} + \Delta\mathbf{x}, \lambda_c^{(0)} + \Delta\lambda_c, \lambda_v^{(0)} + \Delta\lambda_v\right)$ by adding a second term of Equation (2.3.4.1-1) $\left[\boldsymbol{H}_{\mathcal{L}}\left(\mathbf{x}^{(0)}, \lambda_c^{(0)}, \lambda_v^{(0)}\right)\right]\begin{bmatrix}\Delta\mathbf{x}\\ \Delta\lambda_c\\ \Delta\lambda_v\end{bmatrix}$ to $\nabla\mathcal{L}\left(\mathbf{x}^{(0)}, \lambda_c^{(0)}, \lambda_v^{(0)}\right)$. A second term of Equation (2.3.4.1-1) is a Hessian matrix of $\mathcal{L}\left(\mathbf{x}^{(0)}, \lambda_c^{(0)}, \lambda_v^{(0)}\right)$, which is a second order of partial differential equations denoted as $\mathbf{H}_{\mathcal{L}}\left(\mathbf{x}^{(0)}, \lambda_c^{(0)}, \lambda_v^{(0)}\right)$. Both Jacobian matrix and Hessian matrix of Lagrange function should be differentiable at $\mathbf{x}^{(0)}$. Now, Newton–Raphson method is ready to be implemented in the first-order approximation of $\nabla\mathcal{L}(\mathbf{x}, \lambda_c, \lambda_v)$, leading to finding stationary points of Lagrange function $\mathcal{L}(\mathbf{x}, \lambda_c, \lambda_v)$. Newton–Raphson iterations repeat until solutions of partial differential equations (gradient vectors normal to Lagrange functions) shown in Equations (2.3.2.4-1) and (2.3.4.1-1) converge. The solutions are the candidates for stationary points with assumed KKT optimality conditions. Readers are referred to Sections 1.2 and 1.3 for their review of Lagrange function, which is formulated based on objective functions, regular equality constraints (Sections 1.2), and those obtained with active and inactive inequality conditions (Sections 1.3). Lagrange multipliers λ_c and λ_v adopted with active conditions should be greater than 0.

2.3.4.1.3 Updating input design parameters based on Newton–Raphson iteration

In Equation (2.3.4.1-1), a linear approximation of a gradient vector of Lagrange functions $\nabla\mathcal{L}(\mathbf{x}, \lambda_c, \lambda_v)$ can be obtained as $\nabla\mathcal{L}\left(\mathbf{x}^{(0)} + \Delta\mathbf{x}, \lambda_c^{(0)} + \Delta\lambda_c, \lambda_v^{(0)} + \Delta\lambda_v\right)$ at $\mathbf{x}^0 + \Delta\mathbf{x}$ which is very close to $\mathbf{x}^{(0)}$

$$\nabla\mathcal{L}\left(\mathbf{x}^{(0)} + \Delta\mathbf{x}, \lambda_c^{(0)} + \Delta\lambda_c, \lambda_v^{(0)} + \Delta\lambda_v\right) \approx \nabla\mathcal{L}\left(\mathbf{x}^{(0)}, \lambda_c^{(0)}, \lambda_v^{(0)}\right) + \left[\mathbf{H}_\mathcal{L}\left(\mathbf{x}^{(0)}, \lambda_c^{(0)}, \lambda_v^{(0)}\right)\right]\begin{bmatrix}\Delta\mathbf{x}\\ \Delta\lambda_c\\ \Delta\lambda_v\end{bmatrix} \tag{2.3.4.1-1}$$

where $\left[\mathbf{H}_\mathcal{L}\left(\mathbf{x}^{(0)}, \lambda_c^{(0)}, \lambda_v^{(0)}\right)\right]$ is a Hessian matrix of Lagrange function $\mathcal{L}$ at $\mathbf{x}^{(0)}, \lambda_c^{(0)}, \lambda_v^{(0)}$ with Iteration 0. Incremental $\begin{bmatrix}\Delta\mathbf{x} & \Delta\lambda_c & \Delta\lambda_v\end{bmatrix}^T$ shown in Equation (2.3.4.1-1) can be computed as shown in Equation (2.3.4.1-2) when Equation (2.3.4.1-1) is set at as 0, since seeking stationary points $\mathbf{x}, \lambda_c, \lambda_v$ at which $\nabla\mathcal{L}\left(\mathbf{x}^{(0)} + \Delta\mathbf{x}, \lambda_c^{(0)} + \Delta\lambda_c, \lambda_v^{(0)} + \Delta\lambda_v\right)$ converges to 0.

$$\begin{bmatrix}\Delta\mathbf{x}\\ \Delta\lambda_c\\ \Delta\lambda_v\end{bmatrix} \approx -\left[\mathbf{H}_\mathcal{L}\left(\mathbf{x}^{(0)}, \lambda_c^{(0)}, \lambda_v^{(0)}\right)\right]^{-1}\nabla\mathcal{L}\left(\mathbf{x}^{(0)}, \lambda_c^{(0)}, \lambda_v^{(0)}\right) \tag{2.3.4.1-2}$$

In general, input design variables $\mathbf{x}^{(k)}$ for stationary points and Lagrange multipliers $\lambda_c^{(k)}, \lambda_v^{(k)}$ at Iteration k can be updated to $\mathbf{x}^{(k+1)}, \lambda_c^{(k+1)}, \lambda_v^{k+1}$ at Iteration $k+1$ based on Newton–Raphson technique as derived in Equations (2.3.4.1-3)–(2.3.4.1-5). Newton–Raphson iteration is repeated until identifying stationary points $\mathbf{x}^{(k+1)}, \lambda_c^{(k+1)}, \lambda_v^{k+1}$ to achieve a convergence indicated by $\nabla\mathcal{L}(\mathbf{x}, \lambda_c, \lambda_v) = 0$ shown in Equation (2.3.4.1-1). This iteration is illustrated in Figures 2.3.4.1 and 2.3.4.2.

$$\begin{bmatrix}\Delta\mathbf{x}^{(k)}\\ \Delta\lambda_c^{(k)}\\ \Delta\lambda_v^{(k)}\end{bmatrix} = -\left[\mathbf{H}_\mathcal{L}\left(\mathbf{x}^{(k)}, \lambda_c^{(k)}, \lambda_v^{(k)}\right)\right]^{-1}\nabla\mathcal{L}\left(\mathbf{x}^{(k)}, \lambda_c^{(k)}, \lambda_v^{(k)}\right) \tag{2.3.4.1-3}$$

$$\begin{bmatrix}\mathbf{x}^{(k+1)}\\ \lambda_c^{(k+1)}\\ \lambda_v^{(k+1)}\end{bmatrix} = \begin{bmatrix}\mathbf{x}^{(k)}\\ \lambda_c^{(k)}\\ \lambda_v^{(k)}\end{bmatrix} + \begin{bmatrix}\Delta\mathbf{x}^{(k)}\\ \Delta\lambda_c^{(k)}\\ \Delta\lambda_v^{(k)}\end{bmatrix} \tag{2.3.4.1-4}$$

$$\begin{bmatrix}\mathbf{x}^{(k+1)}\\ \lambda_c^{(k+1)}\\ \lambda_v^{(k+1)}\end{bmatrix} = \begin{bmatrix}\mathbf{x}^{(k)}\\ \lambda_c^{(k)}\\ \lambda_v^{(k)}\end{bmatrix} - \left[\mathbf{H}_\mathcal{L}\left(\mathbf{x}^{(k)}, \lambda_c^{(k)}, \lambda_v^{(k)}\right)\right]^{-1}\nabla\mathcal{L}\left(\mathbf{x}^{(k)}, \lambda_c^{(k)}, \lambda_v^{(k)}\right) \tag{2.3.4.1-5}$$

Equations (2.3.4.1-1)–(2.3.4.1-5) are repeated to provide candidates for stationary points minimizing a Lagrange function with assumed KKT optimality. It is noted from Equation (2.3.4.1-5) that input design variables at Iteration $k+1$ are obtained based on $\left[\mathbf{H}_\mathcal{L}\left(\mathbf{x}^{(k)}, \lambda_c^{(k)}, \lambda_v^{(k)}\right)\right]^{-1}$ and $\nabla\mathcal{L}\left(\mathbf{x}^{(k)}, \lambda_c^{(k)}, \lambda_v^{(k)}\right)$ of previous step. A Jacobi matrix

$\nabla\mathcal{L}\left(\mathbf{x}^{(k)}, \lambda_c^{(k)}, \lambda_v^{(k)}\right)$ and a Hessian matrix $\mathbf{H}_{\mathcal{L}}\left(\mathbf{x}^{(k)}, \lambda_c^{(k)}, \lambda_v^{(k)}\right)$ as functions of $\mathbf{x}^{(k)}$, $\lambda_c^{(k)}$, and $\lambda_v^{(k)}$ are derived in Sections 2.3.5 and 2.3.6.

2.3.4.2 Formulation of generalized Jacobian and Hessian matrices

First derivative $\nabla\mathcal{L}\left(\mathbf{x}^{(k)}, \lambda_c^{(k)}, \lambda_v^{(k)}\right)$ called Jacobian matrix can be obtained from Equations (2.3.4.2-1a) and (2.3.4.2-1b) for analytical use and ANN use, respectively, while, in Equations (2.3.4.3-1)–(2.3.4.3-8), Hessian matrix $\mathbf{H}_{\mathcal{L}}\left(\mathbf{x}^{(k)}, \lambda_c^{(k)}, \lambda_v^{(k)}\right)$ of second-order partial derivative of Lagrange function $\mathcal{L}(\mathbf{x}, \lambda_c, \lambda_v) = f(\mathbf{x}) - \lambda_c^T \mathbf{c}(\mathbf{x}) - \lambda_v^T \mathbf{S}\mathbf{v}(\mathbf{x})$ is derived upon substituting Equations (2.3.4.3-3)–(2.3.4.3-8) into Equations (2.3.4.3-1) and (2.3.4.3-2).

Jacobian matrices $\nabla f_f(\mathbf{x})$, $\mathbf{J}_c\left(\mathbf{x}^{(k)}\right)$, and $\mathbf{J}_v\left(\mathbf{x}^{(k)}\right)$ of an objective, equality and inequality functions are shown in Equations (2.3.4.2-2a)–(2.3.4.2-2c), respectively, for two input variables $\mathbf{x}^{(k)} = \left[x_1^{(k)}, x_2^{(k)}\right]$ From an example shown in Section 2.4.3, Jacobi matrices $\nabla f_W(x_A, x_h)$ for objective function and $\mathbf{J}_v(x_A, x_h)$ for inequality functions are derived in Equations (2.4.3.6-2) and (2.4.3.6-3) of Section 2.4.3 in advance based on Equations (2.3.4.2-2a) and (2.3.4.2-2c), respectively, when two input variables are considered.

A global Hessian matrix $\mathbf{H}_{\mathcal{L}}\left(\mathbf{x}^{(k)}, \lambda_c^{(k)}, \lambda_v^{(k)}\right)$ shown in Equations (2.3.4.3-1) and (2.3.4.3-2) comprises subsets of global Hessian matrix which are $\mathbf{H}_{\mathcal{L}}(\mathbf{x})$, $\mathbf{H}_f(\mathbf{x})$, $\mathbf{H}_{c_i}(\mathbf{x})$, and $\mathbf{H}_{v_i}(\mathbf{x})$ which are Hessian matrices of Lagrange, objective, inequality, and inequality functions, respectively, as shown in Equation (2.3.4.3-3). Hessian matrix $\mathbf{H}_{\mathcal{L}}\left(\mathbf{x}^{(k)}, \lambda_c^{(k)}, \lambda_v^{(k)}\right)$ are obtained from Equations (2.3.4.3-2a) and (2.3.4.3-2b) for analytical use and ANN use, respectively. $\mathbf{H}_{\mathcal{L}}\left(\mathbf{x}^{(k)}, \lambda_c^{(k)}, \lambda_v^{(k)}\right)$ are a Hessian matrix of Lagrange function as functions of $\mathbf{x}^{(k)}$, $\lambda_c^{(k)}$, and $\lambda_v^{(k)}$, whereas $\mathbf{H}_{\mathcal{L}}\left(\mathbf{x}^{(k)}\right)$ is a Hessian matrix of Lagrange function as a function of initial trial input parameter $\mathbf{x}^{(k)} = \left[x_1^{(k)}, x_2^{(k)},\ldots,x_n^{(k)}\right]$ only, indicating that $\mathbf{H}_{\mathcal{L}}\left(\mathbf{x}^{(k)}\right)$ is not differentiated by $\lambda_c^{(k)}$ and $\lambda_v^{(k)}$. $\mathbf{H}_{\mathcal{L}}(\mathbf{x})$, $\mathbf{H}_f(\mathbf{x})$, $\mathbf{H}_{c_i}(\mathbf{x})$, and $\mathbf{H}_{v_i}(\mathbf{x})$ required to calculate a global Hessian matrix $\mathbf{H}_{\mathcal{L}}\left(\mathbf{x}^{(k)}, \lambda_c^{(k)}, \lambda_v^{(k)}\right)$ shown in Equation (2.3.4.3-3) are also functions of $\mathbf{x}^{(k)} = \left[x_1^{(k)}, x_2^{(k)},\ldots, x_n^{(k)}\right]$ only. $\mathbf{H}_{\mathcal{L}}\left(\lambda_c^{(k)}\right)$ and $\mathbf{H}_{\mathcal{L}}\left(\lambda_v^{(k)}\right)$ shown in Equations (2.3.4.3-6) and (2.3.4.3-7) are Hessian matrices of Lagrange function as functions of $\lambda_c^{(k)}$ and $\lambda_v^{(k)}$, respectively, in which $\mathbf{H}_{\mathcal{L}}\left(\lambda_c^{(k)}\right)$ and $\mathbf{H}_{\mathcal{L}}\left(\lambda_v^{(k)}\right)$ are not differentiated by $\mathbf{x}^{(k)}$.

Section 2.3.6 describes the formulation of generalizable functions based on Jacobian and Hessian matrices.

$$\nabla\mathcal{L}(\mathbf{x}, \lambda_c, \lambda_v) = \begin{bmatrix} \nabla_{\mathbf{x}}\mathcal{L}(\mathbf{x}, \lambda_c, \lambda_v) \\ \nabla_{\lambda_c}\mathcal{L}(\mathbf{x}, \lambda_c, \lambda_v) \\ \nabla_{\lambda_v}\mathcal{L}(\mathbf{x}, \lambda_c, \lambda_v) \end{bmatrix} = \begin{bmatrix} \nabla f(\mathbf{x}) - \mathbf{J}_c(\mathbf{x})^T \lambda_c - \mathbf{J}_v(\mathbf{x})^T \mathbf{S}\lambda_v \\ -\mathbf{c}(\mathbf{x}) \\ -\mathbf{S}\mathbf{v}(\mathbf{x}) \end{bmatrix} \tag{2.3.4.2-1a}$$

$$\nabla\mathcal{L}(\mathbf{x}, \lambda_c, \lambda_v) = \begin{bmatrix} \left[\mathbf{J}_f^{(D)}(\mathbf{x})\right]^T - \left[\mathbf{J}_c^{(D)}(\mathbf{x})\right]^T \lambda_c - \left[\mathbf{J}_v^{(D)}(\mathbf{x})\right]^T \boldsymbol{S}\lambda_v \\ -\mathbf{c}(\mathbf{x}) \\ -\mathbf{S}\mathbf{v}(\mathbf{x}) \end{bmatrix} = 0 \tag{2.3.4.2-1b}$$

$$\nabla f_f(\mathbf{x}) = \begin{bmatrix} \dfrac{\partial f_f(x_1, x_2)}{\partial x_1} \\ \dfrac{\partial f_f(x_1, x_2)}{\partial x_2} \end{bmatrix} \begin{bmatrix} \dfrac{\partial v_1(x_A, x_h)}{\partial x_A} & \dfrac{\partial v_1(x_A, x_h)}{\partial x_h} \\ \dfrac{\partial v_2(x_A, x_h)}{\partial x_A} & \dfrac{\partial v_2(x_A, x_h)}{\partial x_h} \end{bmatrix} \tag{2.3.4.2-2a}$$

For two equality constraints,

$$\mathbf{J}_c(\mathbf{x}) = \begin{bmatrix} \mathbf{J}_{c1}(\mathbf{x}) \\ \mathbf{J}_{c2}(\mathbf{x}) \\ \vdots \\ \mathbf{J}_{cm}(\mathbf{x}) \end{bmatrix} = \begin{bmatrix} \dfrac{\partial c_1(x_1, x_2)}{\partial x_1} & \dfrac{\partial c_1(x_1, x_2)}{\partial x_2} \\ \dfrac{\partial c_2(x_1, x_2)}{\partial x_1} & \dfrac{\partial c_2(x_1, x_2)}{\partial x_2} \end{bmatrix} \tag{2.3.4.2-2b}$$

For two inequality constraints,

$$\mathbf{J}_v(\mathbf{x}) = \begin{bmatrix} \mathbf{J}_{v1}(\mathbf{x}) \\ \mathbf{J}_{v2}(\mathbf{x}) \\ \vdots \\ \mathbf{J}_{vl}(\mathbf{x}) \end{bmatrix} = \begin{bmatrix} \dfrac{\partial v_1(x_1, x_2)}{\partial x_1} & \dfrac{\partial v_1(x_1, x_2)}{\partial x_2} \\ \dfrac{\partial v_2(x_1, x_2)}{\partial x_1} & \dfrac{\partial v_2(x_1, x_2)}{\partial x_2} \end{bmatrix} \tag{2.3.4.2-2c}$$

$$\nabla f_W(x_A, x_h) = \begin{bmatrix} \dfrac{\partial f_W(x_A, x_h)}{\partial x_A} \\ \dfrac{\partial f_W(x_A, x_h)}{\partial x_h} \end{bmatrix} = \begin{bmatrix} 0.016 + 0.008\sqrt{x_h^2 + 4} \\ \dfrac{0.008 x_A x_h}{\sqrt{x_h^2 + 4}} \end{bmatrix} \tag{2.4.3.6-2}$$

$$\mathbf{J}_v(x_A, x_h) = \begin{bmatrix} \dfrac{\partial v_1(x_A, x_h)}{\partial x_A} & \dfrac{\partial v_1(x_A, x_h)}{\partial x_h} \\ \dfrac{\partial v_2(x_A, x_h)}{\partial x_A} & \dfrac{\partial v_2(x_A, x_h)}{\partial x_h} \end{bmatrix}$$

$$= \begin{bmatrix} \dfrac{\sigma_1}{|\sigma_1|}\left(\dfrac{55}{x_A^2} - \dfrac{200}{x_A^2 x_h}\right) & \left(-\dfrac{\sigma_1}{|\sigma_1|}\dfrac{200}{x_h^2 x_A}\right) \\ \left(\dfrac{100\sqrt{x_h^2 + 4}}{x_A^2 |x_h|}\right) & \left(\dfrac{100 x_h \sqrt{x_h^2 + 4}}{x_A |x_h|^3} - \dfrac{100 x_h}{x_A |x_h| \sqrt{x_h^2 + 4}}\right) \end{bmatrix} \tag{2.4.3.6-3}$$

A global Hessian matrix $\mathbf{H}_{\mathcal{L}}\left(\mathbf{x}^{(k)}, \lambda_c^{(k)}, \lambda_v^{(k)}\right)$ shown in Equations (2.3.4.3-1) and (2.3.4.3-2) comprises subsets of global Hessian matrix which are $\mathbf{H}_{\mathcal{L}}(\mathbf{x})$, $\mathbf{H}_f(\mathbf{x})$, $\mathbf{H}_{c_i}(\mathbf{x})$, and $\mathbf{H}_{v_i}(\mathbf{x})$ which are Hessian matrices of Lagrange, objective, inequality, and inequality functions, respectively, as shown in Equation (2.3.4.3-3). Global Hessian matrix $\mathbf{H}_{\mathcal{L}}\left(\mathbf{x}^{(k)}, \lambda_c^{(k)}, \lambda_v^{(k)}\right)$ are obtained from Equations (2.3.4.3-2a) and (2.3.4.3-2b) for analytical use and ANN use, respectively. $\mathbf{H}_{\mathcal{L}}\left(\mathbf{x}^{(k)}, \lambda_c^{(k)}, \lambda_v^{(k)}\right)$ are a Global Hessian matrix of Lagrange function as functions of $\mathbf{x}^{(k)}$, $\lambda_c^{(k)}$, and $\lambda_v^{(k)}$, whereas $\mathbf{H}_{\mathcal{L}}\left(\mathbf{x}^{(k)}\right)$ is a Hessian matrix of Lagrange function as a function of input design parameters $\mathbf{x}^{(k)} = \left[x_1^{(k)}, x_2^{(k)}, \ldots, x_n^{(k)}\right]$ only, indicating that $\mathbf{H}_{\mathcal{L}}\left(\mathbf{x}^{(k)}\right)$ is not differentiated by $\lambda_c^{(k)}$ and $\lambda_v^{(k)}$. $\mathbf{H}_{\mathcal{L}}(\mathbf{x})$, $\mathbf{H}_f(\mathbf{x})$, $\mathbf{H}_{c_i}(\mathbf{x})$, and $\mathbf{H}_{v_i}(\mathbf{x})$ shown in Equation (2.3.4.3-3) required to calculate a global Hessian matrix $\mathbf{H}_{\mathcal{L}}\left(\mathbf{x}^{(k)}, \lambda_c^{(k)}, \lambda_v^{(k)}\right)$ are also a function of $\mathbf{x}^{(k)} = \left[x_1^{(k)}, x_2^{(k)}, \ldots, x_n^{(k)}\right]$ only. $\mathbf{H}_{\mathcal{L}}\left(\lambda_c^{(k)}\right)$ and $\mathbf{H}_{\mathcal{L}}\left(\lambda_v^{(k)}\right)$ shown in Equations (2.3.4.3-6) and (2.3.4.3-7) are Hessian matrices of Lagrange function as functions of $\lambda_c^{(k)}$ and $\lambda_v^{(k)}$, respectively, in which $\mathbf{H}_{\mathcal{L}}\left(\lambda_c^{(k)}\right)$ and $\mathbf{H}_{\mathcal{L}}\left(\lambda_v^{(k)}\right)$ are not differentiated by $\mathbf{x}^{(k)}$. Section 2.3.6 describes the formulation of generalizable functions based on Jacobian and Hessian matrices.

$$\left[\mathbf{H}_{\mathcal{L}}\left(\mathbf{x}^{(k)}, \lambda_c^{(k)}, \lambda_v^{(k)}\right)\right] = \begin{bmatrix} \dfrac{\partial^2 \mathcal{L}}{\partial \mathbf{x}^2} & \dfrac{\partial^2 \mathcal{L}}{\partial \mathbf{x}\,\partial \lambda_c} & \dfrac{\partial^2 \mathcal{L}}{\partial \mathbf{x}\,\partial \lambda_v} \\ \dfrac{\partial^2 \mathcal{L}}{\partial \lambda_c\,\partial \mathbf{x}} & \dfrac{\partial^2 \mathcal{L}}{\partial \lambda_c^{\,2}} & \dfrac{\partial^2 \mathcal{L}}{\partial \lambda_c\,\partial \lambda_v} \\ \dfrac{\partial^2 \mathcal{L}}{\partial \lambda_v\,\partial \mathbf{x}} & \dfrac{\partial^2 \mathcal{L}}{\partial \lambda_v\,\partial \lambda_c} & \dfrac{\partial^2 \mathcal{L}}{\partial \lambda_v^{\,2}} \end{bmatrix} \tag{2.3.4.3-1}$$

$$\left[\mathbf{H}_{\mathcal{L}}\left(\mathbf{x}^{(k)}, \lambda_c^{(k)}, \lambda_v^{(k)}\right)\right] = \begin{bmatrix} \mathbf{H}_{\mathcal{L}}\left(\mathbf{x}^{(k)}\right) & -\left[\mathbf{J}_c\left(\mathbf{x}^{(k)}\right)\right]^T & -\left[\mathbf{S}\mathbf{J}_v\left(\mathbf{x}^{(k)}\right)\right]^T \\ -\mathbf{J}_c\left(\mathbf{x}^{(k)}\right) & 0 & 0 \\ -\mathbf{S}\mathbf{J}_v\left(\mathbf{x}^{(k)}\right) & 0 & 0 \end{bmatrix} \tag{2.3.4.3-2a}$$

$$\left[\mathbf{H}_{\mathcal{L}}\left(\mathbf{x}^{(k)}, \lambda_c^{(k)}, \lambda_v^{(k)}\right)\right] = \begin{bmatrix} H_{\mathcal{L}}\left(\mathbf{x}^{(k)}\right) & -\left[\mathbf{J}_c^{(D)}\left(\mathbf{x}^{(k)}\right)\right]^T & -\left[\mathbf{S}\mathbf{J}_v^{(D)}\left(\mathbf{x}^{(k)}\right)\right]^T \\ -\mathbf{J}_c^{(D)}\left(\mathbf{x}^{(k)}\right) & 0 & 0 \\ -\mathbf{S}\mathbf{J}_v^{(D)}\left(\mathbf{x}^{(k)}\right) & 0 & 0 \end{bmatrix} \tag{2.3.4.3-2b}$$

where

$$\frac{\partial^2 \mathcal{L}}{\partial \mathbf{x}^2} = \mathbf{H}_{\mathcal{L}}(\mathbf{x}) = \frac{\partial^2 f(\mathbf{x})}{\partial \mathbf{x}^2} - \frac{\partial^2\left(\lambda_c^T \mathbf{c}(\mathbf{x})\right)}{\partial \mathbf{x}^2} - \frac{\partial^2\left(\lambda_v^T \mathbf{S}\mathbf{v}(\mathbf{x})\right)}{\partial \mathbf{x}^2}$$

$$\rightarrow \mathbf{H}_{\mathcal{L}}(\mathbf{x}) = \mathbf{H}_f(\mathbf{x}) - \sum_{i=1}^{m} \lambda_{c_i} \mathbf{H}_{c_i}(\mathbf{x}) - \sum_{i=1}^{l} s_i \lambda_{v_i} \mathbf{H}_{v_i}(\mathbf{x}) \tag{2.3.4.3-3}$$

$$\frac{\partial^2 \mathcal{L}}{\partial \lambda_c\,\partial \mathbf{x}} = \frac{\partial^2 f(\mathbf{x})}{\partial \lambda_c\,\partial \mathbf{x}} - \frac{\partial^2\left(\lambda_c^T \mathbf{c}(\mathbf{x})\right)}{\partial \lambda_c\,\partial \mathbf{x}} - \frac{\partial^2\left(\lambda_v^T \mathbf{S}\mathbf{v}(\mathbf{x})\right)}{\partial \lambda_c\,\partial \mathbf{x}} = 0 - \mathbf{J}_c(\mathbf{x}) - 0 \tag{2.3.4.3-4}$$

$$\frac{\partial^2 \mathcal{L}}{\partial \lambda_v\,\partial \mathbf{x}} = \frac{\partial^2 f(\mathbf{x})}{\partial \lambda_c\,\partial \mathbf{x}} - \frac{\partial^2\left(\lambda_c^T \mathbf{c}(\mathbf{x})\right)}{\partial \lambda_c\,\partial \mathbf{x}} - \frac{\partial^2\left(\lambda_v^T \mathbf{S}\mathbf{v}(\mathbf{x})\right)}{\partial \lambda_c\,\partial \mathbf{x}} = 0 - 0 - \mathbf{S}\mathbf{J}_v(\mathbf{x}) \tag{2.3.4.3-5}$$

$$\frac{\partial^2 \mathcal{L}}{\partial \lambda_c^{\,2}} = \mathbf{H}_{\mathcal{L}}(\lambda_c) = \frac{\partial^2 f(\mathbf{x})}{\partial \lambda_c^{\,2}} - \frac{\partial^2\left(\lambda_c^T \mathbf{c}(\mathbf{x})\right)}{\partial \lambda_c^{\,2}} - \frac{\partial^2\left(\lambda_v^T \mathbf{S}\mathbf{v}(\mathbf{x})\right)}{\partial \lambda_c^{\,2}} = 0 \tag{2.3.4.3-6}$$

$$\frac{\partial^2 \mathcal{L}}{\partial \lambda_v^{\,2}} = \mathbf{H}_{\mathcal{L}}(\lambda_v) = \frac{\partial^2 f(\mathbf{x})}{\partial \lambda_v^{\,2}} - \frac{\partial^2\left(\lambda_c^T \mathbf{c}(\mathbf{x})\right)}{\partial \lambda_v^{\,2}} - \frac{\partial^2\left(\lambda_v^T \mathbf{S}\mathbf{v}(\mathbf{x})\right)}{\partial \lambda_v^{\,2}} = 0 \tag{2.3.4.3-7}$$

$$\frac{\partial^2 \mathcal{L}}{\partial \lambda_v \partial \lambda_c} = \frac{\partial^2 f(\mathbf{x})}{\partial \lambda_v \partial \lambda_c} - \frac{\partial^2\left(\lambda_c^T \mathbf{c}(\mathbf{x})\right)}{\partial \lambda_v \partial \lambda_c} - \frac{\partial^2\left(\lambda_v^T \mathbf{S}\mathbf{v}(\mathbf{x})\right)}{\partial \lambda_v \partial \lambda_c} = 0 \tag{2.3.4.3-8}$$

Hessian matrices $\mathbf{H}_f(\mathbf{x})$, $\mathbf{H}_{c_i}(\mathbf{x})$, and $\mathbf{H}_{v_i}(\mathbf{x})$ of an objective, equality, and inequality functions are shown in Equations (2.3.4.4a)–(2.3.4.4c) for two input variables $\mathbf{x}^{(k)} = \left[x_1^{(k)},\ x_2^{(k)}\right]$. From an example shown in Section 2.4.3, Hessian matrix $\mathbf{H}_{fw}(x_A, x_h)$ for objective function is shown in Equations (2.4.3.8-3) based on Equations (2.3.4.4a) whereas $\mathbf{H}_{v_1}(x_A, x_h)$ and $\mathbf{H}_{v_2}(x_A, x_h)$ for inequality functions are shown Equations (2.4.3.8-4), and (2.4.3.8-5), respectively, based on Equations (2.3.4.4c) when two input variables are considered in functions. Section 2.4.3 is referred for more explanations.

$$\mathbf{H}_f(\mathbf{x}) = \begin{bmatrix} \dfrac{\partial^2 f_f(\mathbf{x})}{\partial x_1^2} & \dfrac{\partial^2 f_f(\mathbf{x})}{\partial x_1 \partial x_2} \\ \dfrac{\partial^2 f_f(\mathbf{x})}{\partial x_2 \partial x_1} & \dfrac{\partial^2 f_f(\mathbf{x})}{\partial x_2^2} \end{bmatrix} \tag{2.3.4.4a}$$

For two equality constraints,

$$\mathbf{H}_{c_i}(\mathbf{x}) = \begin{bmatrix} \dfrac{\partial^2 c_1(\mathbf{x})}{\partial x_1^2} & \dfrac{\partial^2 c_1(\mathbf{x})}{\partial x_1 \partial x_2} \\ \dfrac{\partial^2 c_1(\mathbf{x})}{\partial x_2 \partial x_1} & \dfrac{\partial^2 c_1(\mathbf{x})}{\partial x_2^2} \end{bmatrix} \tag{2.3.4.4b}$$

For two inequality constraints,

$$\mathbf{H}_{v_i}(\mathbf{x}) = \begin{bmatrix} \dfrac{\partial^2 v_1(\mathbf{x})}{\partial x_1^2} & \dfrac{\partial^2 v_1(\mathbf{x})}{\partial x_1 \partial x_2} \\ \dfrac{\partial^2 v_1(\mathbf{x})}{\partial x_2 \partial x_1} & \dfrac{\partial^2 v_1(\mathbf{x})}{\partial x_2^2} \end{bmatrix} \tag{2.3.4.4c}$$

$$\mathbf{H}_{fw}(x_A, x_h) = \begin{bmatrix} \dfrac{\partial^2 f_W(\mathbf{x})}{\partial x_A^2} & \dfrac{\partial^2 f_W(\mathbf{x})}{\partial x_A \partial x_h} \\ \dfrac{\partial^2 f_W(\mathbf{x})}{\partial x_h \partial x_A} & \dfrac{\partial^2 f_W(\mathbf{x})}{\partial x_h^2} \end{bmatrix} = \begin{bmatrix} 0 & \dfrac{0.008x_h}{\sqrt{x_h^2+4}} \\ \dfrac{0.008x_h}{\sqrt{x_h^2+4}} & \dfrac{0.032x_A}{\left(x_h^2+4\right)^{1.5}} \end{bmatrix} \tag{2.4.3.8-3}$$

$$\mathbf{H}_{v_1}(x_A, x_b) = \begin{bmatrix} \dfrac{\partial^2 v_1(\mathbf{x})}{\partial x_A^2} & \dfrac{\partial^2 v_1(\mathbf{x})}{\partial x_A \, \partial x_b} \\ \dfrac{\partial^2 v_1(\mathbf{x})}{\partial x_b \, \partial x_A} & \dfrac{\partial^2 v_1(\mathbf{x})}{\partial x_b^2} \end{bmatrix} = \begin{bmatrix} \dfrac{\sigma_1}{|\sigma_1|}\left(\dfrac{400}{x_A^3 x_b} - \dfrac{110}{x_A^3}\right) & \dfrac{\sigma_1}{|\sigma_1|}\left(\dfrac{200}{x_b^2 x_A^2}\right) \\ \dfrac{\sigma_1}{|\sigma_1|}\left(\dfrac{200}{x_b^2 x_A^2}\right) & \dfrac{\sigma_1}{|\sigma_1|}\left(\dfrac{400}{x_b^3 x_A}\right) \end{bmatrix} \tag{2.4.3.8-4}$$

$$\mathbf{H}_{v_2}(x_A, x_b) = \begin{bmatrix} \dfrac{\partial^2 v_2(\mathbf{x})}{\partial x_A^2} & \dfrac{\partial^2 v_2(\mathbf{x})}{\partial x_A \, \partial x_b} \\ \dfrac{\partial^2 v_2(\mathbf{x})}{\partial x_b \, \partial x_A} & \dfrac{\partial^2 v_2(\mathbf{x})}{\partial x_b^2} \end{bmatrix}$$

$$= \begin{bmatrix} \left(-\dfrac{200\sqrt{x_b^2+4}}{x_A^3 |x_b|}\right) & \left(\dfrac{100 x_b}{x_A^2 |x_b| \sqrt{x_b^2+4}} - \dfrac{100 x_b \sqrt{x_b^2+4}}{x_A^2 |x_b|^3}\right) \\ \left(\dfrac{100 x_b}{x_A^2 |x_b| \sqrt{x_b^2+4}} - \dfrac{100 x_b \sqrt{x_b^2+4}}{x_A^2 |x_b|^3}\right) & \left(\dfrac{100 |x_b|}{x_A \left(x_b^2+4\right)^{1.5}} - \dfrac{200\sqrt{x_b^2+4}}{x_A |x_b|^3} + \dfrac{100 x_b}{x_A |x_b| \sqrt{x_b^2+4}}\right) \end{bmatrix} \tag{2.4.3.8-5}$$

2.3.4.3 Formulation of KKT non-linear equations based on Newton–Raphson iteration

In Equation (2.3.2.4-1), a Jacobi matrix $\nabla \mathcal{L}(\mathbf{x}, \boldsymbol{\lambda}_c, \boldsymbol{\lambda}_v)$ of Lagrange functions $\mathcal{L}(\mathbf{x}, \boldsymbol{\lambda}_c, \boldsymbol{\lambda}_v)$ shown in Equation (2.3.2.3) is derived to solve for stationary points of Lagrange functions $\mathcal{L}(\mathbf{x}, \boldsymbol{\lambda}_c, \boldsymbol{\lambda}_v)$ based on KKT nonlinear equations shown in Equation (2.3.2.4-2) subject to constraining conditions shown in Equation (2.3.2.2-1) and (2.3.2.2-2). Stationary points $\mathbf{x}$, $\boldsymbol{\lambda}_c$, $\boldsymbol{\lambda}_v$ are updated based on Equation (2.3.4.1-5) and Newton–Rapson method until gradient vector $\nabla \mathcal{L}(\mathbf{x}, \boldsymbol{\lambda}_c, \boldsymbol{\lambda}_v)$ of Lagrange functions $\mathcal{L}(\mathbf{x}, \boldsymbol{\lambda}_c, \boldsymbol{\lambda}_v)$ shown in Equation (2.3.4.1-1) converges to 0. Jacobian and Hessian matrices calculated based on Section 2.3.6 are used during this iteration. Iterations for convergence to seek an optimized Lagrange function is illustrated in Figures 2.3.4.1 and 2.3.4.2.

2.3.5 Stationary points of Lagrange functions based on gradient vectors

Equations (2.3.2.4-1) are gradient vectors normal to tangential lines of objective functions $f(\mathbf{x})$ and constraints $\mathbf{c}(\mathbf{x})$ and $\mathbf{v}(\mathbf{x})$. As explained in Sections 1.2 and 1.3 of Chapter 1, stationary points of Lagrange functions shown in Equation (2.3.2.3) can be identified by equating Equation (2.3.2.4-1) to 0 with respect to $\mathbf{x}$, $\boldsymbol{\lambda}_c$, and $\boldsymbol{\lambda}_v$, leading to simultaneous equations shown in Equation (2.3.2.4-2). However, finding $\mathbf{x}$, $\boldsymbol{\lambda}_c$, and $\boldsymbol{\lambda}_v$ to locate stationary points of Lagrange function $\boldsymbol{\mathcal{L}}$ directly is difficult for complex Lagrange functions. Hence, Equation (2.3.2.4-1) is linearly expanded as shown in Equations (2.3.4.1-1) with higher order terms including Hessian matrix shown in Equations (2.3.4.3-1) and (2.3.4.3-2) before Newton–Raphson method is implemented to find stationary points of Lagrange function. Gradient vectors of $f(\mathbf{x})$, $\mathbf{c}(\mathbf{x})$, and $\mathbf{v}(\mathbf{x})$ which are normal to these functions are presented

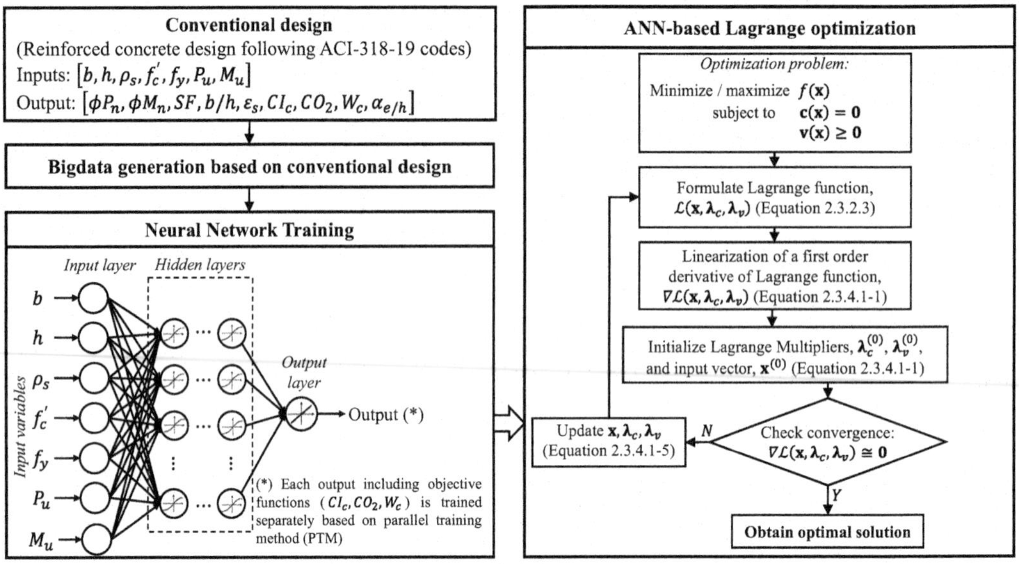

Figure 2.3.4.1 Iterations for convergence to seek an optimized Lagrange.

in Equations (2.3.2.4-1) and (2.3.4.2-1b). $\mathbf{J}_c(\mathbf{x})$ and $\mathbf{J}_v(\mathbf{x})$ shown in Equations (2.3.4.2-2b) and (2.3.4.2-2c), respectively, are Jacobi matrices of equality and inequality functions with respect to input design parameters $\mathbf{x}$. In Equation (2.3.4.3-3), Hessian matrices $\mathbf{H}_{\mathcal{L}}(\mathbf{x})$, $\mathbf{H}_f(\mathbf{x})$, $\mathbf{H}_{c_i}(\mathbf{x})$, and $\mathbf{H}_{v_i}(\mathbf{x})$ are obtained by differentiating a Lagrange, objective, equality, and inequality functions twice with respect to input design parameters $\mathbf{x}$, whereas Hessian matrices $\mathbf{H}_{\mathcal{L}}(\boldsymbol{\lambda}_c)$ and $\mathbf{H}_{\mathcal{L}}(\boldsymbol{\lambda}_v)$ are obtained by differentiating a Lagrange function twice with respect to Lagrange multipliers $\boldsymbol{\lambda}_c$ and $\boldsymbol{\lambda}_v$ as shown in Equations (2.3.4.3-6) and (2.3.4.3-7), respectively. Initial trial input parameters for obtaining design parameters are updated based on Equations (2.3.4.1-1) and (2.3.4.1-5) until $\nabla\mathcal{L}(\mathbf{x}, \boldsymbol{\lambda}_c, \boldsymbol{\lambda}_v)$ converges within tolerated error ranges, finding stationary points of Lagrange functions. Iterations for convergence to seek an optimized Lagrange function is illustrated in Figures 2.3.4.1 and 2.3.4.2.

2.3.6 ANN-based generalized functions replacing analytical functions

2.3.6.1 Formulation of Jacobian and Hessian matrices

Objective functions $f(\mathbf{x})$ and constraining functions $\mathbf{c}(\mathbf{x})$ and $\mathbf{v}(\mathbf{x})$ appearing in Lagrange optimization are sometimes complex or impossible to explicitly derive twice-differentiable analytical functions in order to calculate Jacobian and Hessian matrices of Lagrange function. Formulation of Jacobian and Hessian matrices of Lagrange functions shown in Equations (2.3.4.2-1) and (2.3.4.3-1) are required when obtaining candidate solutions with KKT conditions. Jacobian $\nabla f(\mathbf{x})$, $\mathbf{J}_c(\mathbf{x})$, and $\mathbf{J}_v(\mathbf{x})$ for objective, equality, and inequality functions are shown in Equation (2.3.4.2-2a), Equations (2.3.4.2-2b) and (2.3.4.2-2c), respectively, whereas Hessian $\mathbf{H}_{c_i}(\mathbf{x})$ and $\mathbf{H}_{v_i}(\mathbf{x})$ matrices of constraining functions also shown in Equation (2.3.4.3-3). ANNs for objective and constraining functions used to optimize RC structures are shown in Equation (2.3.6.1), which replace analytical objective and constraining functions. ANNs shown in Equation (2.3.6.1) can be also used to calculate

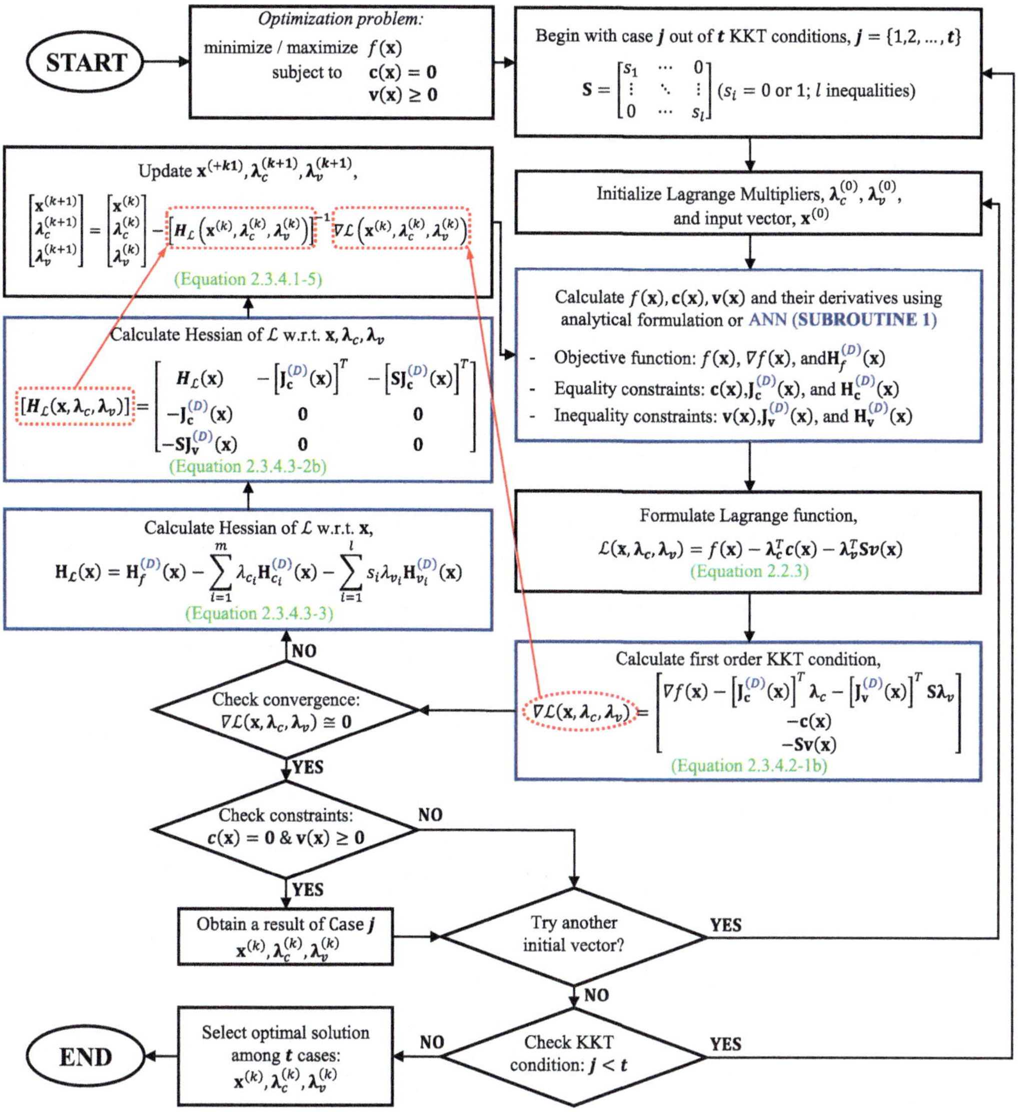

Figure 2.3.4.2 A flow of equations adopted in the Newton–Raphson iteration.

Jacobian and Hessian matrices. Variable **x** in Equation (2.3.6.1) represents input design parameters.

$$f(\mathbf{x}) = g^D\left(f_{lin}^L\left(\mathbf{W}^L f_t^{L-1}\left(\mathbf{W}^{L-1}\ldots f_t^1\left(\mathbf{W}^1 g_{\mathbf{x}}^N(\mathbf{x})+\mathbf{b}^1\right)\ldots+\mathbf{b}^{L-1}\right)+b^L\right)\right) \tag{2.3.6.1}$$

Min-max normalization function shown in Equations (2.3.6.2-1) and (2.3.6.2-2) is adopted in this formulation.

$$\bar{x} = g^N(x) = \alpha_x\left(x - x_{\min}\right) + \bar{x}_{\min} \tag{2.3.6.2-1}$$

$$\alpha_x = \frac{\overline{x}_{\max} - \overline{x}_{\min}}{x_{\max} - x_{\min}} \tag{2.3.6.2-2}$$

where $\overline{x}$ represents a specified minimum and maximum range for datasets, for example, $\overline{x}_{min} = -1$ and $\overline{x}_{\max} = 1$ when datasets are normalized between $\overline{x}_{\min} = -1$ and $\overline{x}_{max} = 1$. A coefficient α_x is a ratio used to normalize input datasets between $\overline{x}_{\max}$ and $\overline{x}_{\min}$ over the ranges of large datasets $(x_{\max} - x_{\min})$ as shown in Equation (2.3.6.2-2). Similarly, de-normalization function, a reverse of α_x, is expressed in Equation (2.3.6.2–3) to take output parameters back to the original scale.

$$x = g^D(\overline{x}) = \frac{1}{\alpha_x}(x - \overline{x}_{\min}) + x_{\min} \tag{2.3.6.2-3}$$

Activation functions (*tansig*, *tanh*, f_t^l) at layer l shown in Figure 2.3.3.1 are implemented to formulate nonlinear relationships of a network, whereas a linear activation function (f_{lin}^L) is selected for an output layer because output parameters are not squashed, allowing them to take linear behavior at the output layer. Objective and constraining functions extracted from Equation (2.3.6.1) are de-normalized at the output layer, which is denoted by L.

2.3.6.2 Formulation of Jacobian matrix based on ANN

2.3.6.2.1 Generalization of functions based on ANN

ANN shown in Equation (2.3.6.1) to derive objective and constraining functions is also described in Equation (2.3.6.3-1) using as a series of composite mathematical operations, e.g., operation sequence is shown in $f(g(\mathbf{x})) = \boldsymbol{f} \circ g \circ \boldsymbol{x})$ where weight $\mathbf{W}$ and bias $\mathbf{b}$ matrices are used to connect hidden layers to derive objective and constraining functions. ANN-based objective functions presented in Equations (2.3.3.1-1)–(2.3.3.1-6) and Equation (2.3.6.1) can be also expressed as a series of composite mathematical operations as shown in Equation (2.3.6.3-1).

In ANN, an output at each hidden layer is relayed to neurons of successive layers fully connected using weights at each neuron and bias in each hidden layer as described in Equation (2.3.6.1). Layers are then summed at an output layer as a function generalized using ANN. Equation (2.3.6.3-1) represents a function based on ANN consisting of multiple hidden layers and neurons for given input design parameters. Weights and bias matrices obtained during training ANNs denoted as red in Equations (2.3.6.3-1) and (2.3.6.3-2) are used to generalize functions. Equation (2.3.6.3-2) encapsulates how functions are derived based on ANN from an initial to the last layer. Readers are also referred to Section 1.4 of the publication by the author (Hong, 2021). Neural networks shown in Equation (2.3.6.3-1) are trained on large datasets to yield weight and bias matrices which can be used to generalize functions and their derivatives based on ANN. Equation (2.3.6.3-2) presents how functions are sequentially generalized to yield Equation (2.3.6.3-2f) using weight and bias matrices obtained from Equation (2.3.6.3-1). De-normalized Jacobi matrices are derived in Equations (2.3.6.5)–(2.3.6.6) whereas de-normalized Hessian matrices are obtained in Equations (2.3.6.7)–(2.3.6.17) based on functions derived at Equation (2.3.6.3-2f). ANN-based Jacobi and Hessian matrices are necessary to reach convergence of a first derivatives (Jacobi) of Lagrange functions shown in Equations (2.3.4.1) and (2.3.4.2). One of the main goals of this chapter is to show Equations (2.3.4.1) and (2.3.4.2) can converge. Almost any function can be generalized by ANN to present their derivative such as Jacobi and Hessian when large datasets of good quality are provided, enabling Lagrange-based optimization while explicit analytical functions are replaced. ANN-based functions derived in Equation (2.3.6.3-2) based on Equation (2.3.6.3-1) is applied to Equations (2.4.3.26-5a)–(2.4.3.26-37b) for truss frame optimization. Section 2.4 also presents optimization examples of fourth order of polynomial and distance of projectile movement using function derived in Equations (2.3.6.3-1) based on Equation (2.3.6.3-2). $\mathbf{z}^{(N)}$ shown in Equation (2.3.6.3-2a) is a normalized input parameter based on normalizing function $g^N(x)$ as shown in Equations (2.3.6.2-1) and (2.3.6.3-2a). The notation $\odot$ denotes the Hadamard (element-wise) product operation (Horn, 1990). The following example shows the Hadamard product operation ($\odot$)for3×3matrixAwith3×3matrixB. $\begin{bmatrix} a_{11} & a_{12} & a_{13} \\ a_{21} & a_{22} & a_{23} \\ a_{31} & a_{32} & a_{33} \end{bmatrix} \circ \begin{bmatrix} b_{11} & b_{12} & b_{13} \\ b_{21} & b_{22} & b_{23} \\ b_{31} & b_{32} & b_{33} \end{bmatrix} = \begin{bmatrix} a_{11}b_{11} & a_{12}b_{12} & a_{13}b_{13} \\ a_{21}b_{21} & a_{22}b_{22} & a_{23}b_{23} \\ a_{31}b_{31} & a_{32}b_{32} & a_{33}b_{33} \end{bmatrix}$

In Equation (2.3.6.3-2a), both $\boldsymbol{\alpha}_x$ and x are vectors, requiring using $\odot$ to multiply an element of $\boldsymbol{\alpha}_x$ to corresponding element of $(\boldsymbol{x} - \boldsymbol{x}_{\text{min}})$.

In Equation (2.3.6.3-1), $\mathbf{x}$ is input design parameters and L is a number of layers including hidden layers and output layer. $\mathbf{z}^{(1)}$ shown in Equation (2.3.6.3-2b) is a neural value to be used in the first layer (1) in which $\mathbf{z}^{(N)}$ is used as normalized neural inputs for $\mathbf{z}^{(1)}$ with weights $\mathbf{W}^{(1)}$ and biases $\mathbf{b}^{(1)}$. The weights $\mathbf{W}^{(1)}$ and biases $\mathbf{b}^{(1)}$ are obtained by training training Equation (2.3.6.3-1) or (2.3.3.1-1) to (2.3.3.1-6) at the first layer. But all $\mathbf{W}^{(i)}$ and $\mathbf{b}^{(i)}$ should be obtained by training ANNs shown in Equation (2.3.6.3-1) or (2.3.3.1-1) to (2.3.3.1-6). $\mathbf{z}^l$ shown in Equation (2.3.6.3-2c) is a normalized neural value at layer l, whereas $\mathbf{z}^{(L=\text{output})}$ shown in Equation (2.3.6.3-2e) is a normalized neural value at the output layer L before returning to the original scale by de-normalizing function $g^D(\overline{x})$ as shown in Equation (2.3.6.3-2f). $\mathbf{W}^l$ is weight matrix between layer $l-1$ and layer l; $\mathbf{b}^l$ is bias matrix of layer l which are also obtained by Equation (2.3.6.3-1). Normalizing and de-normalizing functions g^N, g^D are shown in Equations (2.3.6.2-1) and (2.3.6.2-3), respectively. The calculation procedure using Equation 2.3.6.3-2 is also summarized in Figure 2.3.6.2.

$$y = f(\mathbf{x}) = \mathbf{z}^{(D)} \circ \mathbf{z}^{(L)} \circ \mathbf{z}^{(L-1)} \circ \ldots \circ \mathbf{z}^{(l)} \circ \ldots \circ \mathbf{z}^{(1)} \circ \mathbf{z}^{(N)} \quad (2.3.6.3\text{-}1)$$

$$\begin{aligned}
&\mathbf{z}^{(N)} = \mathbf{g}^{(N)}(\mathbf{x}) = \alpha_{\mathbf{x}} \odot (\mathbf{x} - \mathbf{x}_{\text{min}}) + \overline{\mathbf{x}}_{\text{min}} && \text{(a)} \\
&\mathbf{z}^{(1)} = f_t^{(1)}\left(\mathbf{W}^{(1)}\mathbf{z}^{(N)} + \mathbf{b}^{(1)}\right) && \text{(b)} \\
&\ldots && \text{(c)} \\
&\mathbf{z}^{(l)} = f_t^{(l)}\left(\mathbf{W}^{(l)}\mathbf{z}^{(l-1)} + \mathbf{b}^{(l)}\right) && \\
&\ldots && \text{(d)} \\
&\mathbf{z}^{(L-1)} = f_t^{(L-1)}\left(\mathbf{W}^{(L-1)}\mathbf{z}^{(L-2)} + \mathbf{b}^{(L-1)}\right) && \\
&\mathbf{z}^{(L=\text{output})} = f_{\text{lin}}^{(L)}\left(\mathbf{W}^{(L)}\mathbf{z}^{(L-1)} + b^{(L)}\right) && \text{(e)} \\
&y = \mathbf{z}^{(D)} = g^{(D)}(y) = \frac{1}{\alpha_y}\left(\mathbf{z}^{(L)} - \overline{\mathbf{z}_{\text{min}}^{(L)}}\right) + \mathbf{z}_{\text{min}}^{(L)} && \text{(f)}
\end{aligned} \quad (2.3.6.3\text{-}2)$$

where $\alpha_{\mathbf{x}} = \left[\alpha_{x_1}, \alpha_{x_2}, \ldots, \alpha_{x_n}\right]^T$ shown in Equation (2.3.6.3-2a) are ratios used to normalize input datasets between $\overline{x}_{\text{max}}$ and $\overline{x}_{\text{min}}$ over the ranges of large datasets $(x_{\text{max}} - x_{\text{min}})$. For example, each α_{x_i} is calculated by dividing $(\overline{x}_{i,\text{max}} - \overline{x}_{i,\text{min}})$ for each input design parameter by ranges of large datasets $(x_{i,\text{max}} - x_{i,\text{min}})$ corresponding to the input design parameter. For example, if α_{x_1} and α_{x_2} are ratios of beam width (b) and beam height (h), ratios of beam width (b) α_{x_1} and beam height (h) α_{x_2} are calculated by dividing $(\overline{x}_{b,\text{max}} - \overline{x}_{b,\text{min}})$ for beam width (b) and $(\overline{x}_{d,\text{max}} - \overline{x}_{d,\text{min}})$ beam height (h) by ranges of large datasets $(x_{b,\text{max}} - x_{b,\text{min}})$ and $(x_{d,\text{max}} - x_{d,\text{min}})$ corresponding to the beam width (b) and the beam height

(h), respectively. $\mathbf{x}_{\min} = \left[x_{1,\min}, x_{2,\min}, \ldots, x_{n,\min}\right]^T$ and $\mathbf{x}_{\max} = \left[x_{1,\max}, x_{2,\max}, \ldots, x_{n,\max}\right]^T$ are minimum and minimum ranges of input design parameter x in large datasets, respectively. $\overline{\mathbf{x}}_{\min} = \left[\overline{x}_{1,\min}, \overline{x}_{2,\min}, \ldots, \overline{x}_{n,\min}\right]^T$ and $\overline{\mathbf{x}}_{\max} = \left[\overline{x}_{1,\max}, \overline{x}_{2,\max}, \ldots, \overline{x}_{n,\max}\right]^T$ are a minimum and maximum range of a normalized input datasets, respectively, e.g., a range beam width(b) is normalized between –1 and 1 if $\overline{\mathbf{x}}_{\text{width,min}}=-1$ and $\overline{\mathbf{x}}_{\text{width, max}}=1$ are set. y shown in Equation 2.3.6.3-2(f) is de-normalized by using de-normalizing function $g^D(\overline{x})$ is $\mathbf{z}^{(D)}$, returning to the original scale. α_y, $\mathbf{z}^{(L)}$, $\mathbf{z}_{\min}^{(L)}(\overline{y}_{\min})$, and $\mathbf{z}_{\min}^{(L)}(y_{\min})$ are the parameters to de-normalize $\mathbf{z}^{(L)}$. An activation function (*tansig, tanh*) shown in Equation (2.3.6.4-1) is differentiated once and twice, yielding convenient forms as shown in Equations (2.3.6.4-2) and (2.3.6.4-3).

$$f_t(x) = \frac{2}{1+e^{-2x}} - 1 \tag{2.3.6.4-1}$$

$$f_t'(x) = 1 - f_t(x)^2 \tag{2.3.6.4-2}$$

$$f_t''(x) = -2f_t(x)\left(1 - f_t(x)^2\right) \tag{2.3.6.4-3}$$

A derivation of ANN-based function begins with $\mathbf{z}^{(N)}$ shown in Equation (2.3.6.3-2a) and ends with $\mathbf{z}^{(D)}$ shown in Equation (2.3.6.3-2f). A first derivative Jacobi of a function $\mathbf{z}$ is also derived from $\mathbf{z}^{(N)}$ shown in Equation (2.3.6.6-1) and ends in terms of $\mathbf{z}^{(D)}$ shown in Equation (2.3.6.6-9), i.e., Jacobi $\mathbf{J}^{(N)}$ is derived in Equation (2.3.6.6-1) using $\mathbf{z}^{(N)}$ and $\mathbf{J}^{(D)}$ is obtained in Equation (2.3.6.6-9) using $\mathbf{z}^{(D)}$. A second derivative Hessian of a function $\mathbf{z}$ is also derived in terms of $\mathbf{z}^{(D)}$ shown in Equations (2.3.4.3-2b) and (2.3.6.16). A derivation of $\mathbf{z}^{(D)}$ shown in Equation (2.3.6.3-2) presents, therefore, one of the most important equations with ANN-based Lagrange optimizations.

2.3.6.2.2 A derivation of ANN-based of Jacobi matrix

Chain rule is employed to obtain derivatives of composite functions shown in Equation (2.3.6.3-1). As shown in Equation (2.3.6.5), Jacobian matrix of $\mathbf{J}^{(l)}$ with respect to $\mathbf{x}$ can be derived as Jacobian matrix of $\mathbf{J}^{(l-1)}$ by multiplying a derivative of $\mathbf{z}^{(l)}$ over $\mathbf{z}^{(l-1)}$ with respect to $\mathbf{x}$. Jacobian matrix for $\mathbf{J}^{(L)}$ at the output layer is then computed by a feed-forward network as the following procedure.

Jacobian matrix with a normalized initial trial input parameter is obtained in Equation (2.3.6.6-1) by substituting (2.3.6.3-2a) into Equation (2.3.6.5).

$$\mathbf{J}^{(l)} = \frac{\partial \mathbf{z}^{(l)}}{\partial \boldsymbol{x}} = \frac{\partial \mathbf{z}^{(l)}}{\partial \mathbf{z}^{(l-1)}} \frac{\partial \mathbf{z}^{(l-1)}}{\partial \boldsymbol{x}} = \frac{\partial \mathbf{z}^{(l)}}{\partial \mathbf{z}^{(l-1)}} \mathbf{J}^{(l-1)} \tag{2.3.6.5}$$

$$\mathbf{J}^{(N)} = \frac{\partial \mathbf{z}^{(N)}}{\partial \mathbf{x}} = \frac{\partial\left(\alpha_{\mathbf{x}} \odot (\mathbf{x} - \mathbf{x}_{\min}) + \overline{\mathbf{x}}_{\min}\right)}{\partial \mathbf{x}} = I_n \odot \alpha_{\mathbf{x}} \tag{2.3.6.6-1}$$

where I_n is an identity matrix with a size of $n \times n$ used for transforming vertical vector of α_x to a diagonal matrix $\begin{bmatrix} \alpha_{x_1} & \cdots & 0 \\ \vdots & \ddots & \vdots \\ 0 & \cdots & \alpha_{x_n} \end{bmatrix}$. Secondly, all input design variables such as beam width (b) and beam height (h) shown in Equation (2.3.6.3-2a) indicated by N are normalized prior to being fed to a first layer shown in Equation (2.3.6.3-2b), where a neural value of Jacobian matrix at the first hidden layer (1) is obtained based on a chain rule as shown in Equation (2.3.6.6-2).

$$\mathbf{J}^{(1)} = \frac{\partial \mathbf{z}^{(1)}}{\partial \mathbf{x}} = \frac{\partial \mathbf{z}^{(1)}}{\partial \mathbf{z}^{(N)}}\mathbf{J}^{(N)} = \frac{\partial\left(f_t^{(1)}\left(\mathbf{W}^{(1)}\mathbf{z}^{(N)} + \mathbf{b}^{(1)}\right)\right)}{\partial\left(\mathbf{W}^{(1)}\mathbf{z}^{(N)} + \mathbf{b}^{(1)}\right)} \frac{\partial\left(\mathbf{W}^{(1)}\mathbf{z}^{(N)} + \mathbf{b}^{(1)}\right)}{\partial \mathbf{z}^{(N)}}\mathbf{J}^{(N)} \tag{2.3.6.6-2}$$

$\mathbf{J}^{(1)}$ is computed in Equation (2.3.6.6-3) based on Equation (2.3.6.6-2) when using the first derivative of *tanh* or *tansig* activation function shown in Equation (2.3.6.4-2). The last term of Equation (2.3.6.6-3) $\mathbf{W}^{(1)}\mathbf{J}^{(N)}$ is also obtained by differentiating $\partial\left(\mathbf{W}^{(1)}\mathbf{z}^{(N)} + \mathbf{b}^{(1)}\right)$ by $\partial \mathbf{z}^{(N)}$.

$$\mathbf{J}^{(1)} = \frac{\partial \mathbf{z}^{(1)}}{\partial \mathbf{x}} = \left(1 - \left(\mathbf{z}^{(1)}\right)^2\right) \odot \mathbf{W}^{(1)}\mathbf{J}^{(N)} \tag{2.3.6.6-3}$$

Similarly, neural values of Jacobian at the next layers are computed in Equations (2.3.6.6-4)–(2.3.6.6-9).

$$\mathbf{J}^{(2)} = \frac{\partial \mathbf{z}^{(2)}}{\partial \mathbf{x}} = \frac{\partial \mathbf{z}^{(2)}}{\partial \mathbf{z}^{(1)}}\mathbf{J}^{(1)} \tag{2.3.6.6-4}$$

$$\mathbf{J}^{(2)} = \frac{\partial\left(f_t^{(2)}\left(\mathbf{W}^{(2)}\mathbf{z}^{(1)} + \mathbf{b}^{(2)}\right)\right)}{\partial\left(\mathbf{W}^{(2)}\mathbf{z}^{(1)} + \mathbf{b}^{(2)}\right)} \frac{\partial\left(\mathbf{W}^{(2)}\mathbf{z}^{(1)} + \mathbf{b}^{(2)}\right)}{\partial \mathbf{z}^{(1)}}\mathbf{J}^{(1)} \tag{2.3.6.6-5}$$

$$\mathbf{J}^{(2)} = \left(1 - \left(\mathbf{z}^{(2)}\right)^2\right) \odot \mathbf{W}^{(2)}\mathbf{J}^{(1)} \tag{2.3.6.6-6}$$

...

$$\mathbf{J}^{(l)} = \frac{\partial \mathbf{z}^{(l)}}{\partial \mathbf{x}} = \frac{\partial \mathbf{z}^{(l)}}{\partial \mathbf{z}^{(l-1)}}\mathbf{J}^{(l-1)} = \left(1 - \left(\mathbf{z}^{(l)}\right)^{(2)}\right) \odot \mathbf{W}^{(l)}\mathbf{J}^{(l-1)} \tag{2.3.6.6-7}$$

$$\mathbf{J}^{(L)} = \frac{\partial \mathbf{z}^{(L)}}{\partial \mathbf{x}} = \frac{\partial \mathbf{z}^{(L)}}{\partial \mathbf{z}^{(L-1)}}\mathbf{J}^{(L-1)} = \frac{\partial\left(\mathbf{W}^{(L)}\mathbf{z}^{(L-1)} + b^{(L)}\right)}{\partial \mathbf{z}^{(L-1)}}\mathbf{J}^{(L-1)} = \mathbf{W}^{(L)}\mathbf{J}^{(L-1)} \tag{2.3.6.6-8}$$

$$\mathbf{J}^{(D)} = \frac{\partial \mathbf{z}^{(D)}}{\partial \mathbf{x}} = \frac{\partial \mathbf{z}^{(D)}}{\partial \mathbf{z}^{(L)}}\mathbf{J}^{(L)} = \frac{\partial\left(\frac{1}{\alpha_y}\left(\mathbf{z}^{(L)} - \overline{y}_{\min}\right) + y_{\min}\right)}{\partial \mathbf{z}^{(L)}}\mathbf{J}^{(L)} = \frac{1}{\alpha_y}\mathbf{J}^{(L)} \tag{2.3.6.6-9}$$

Activation functions (*tansig*, *tanh*, $f_{\text{lin}}^{L=\text{output}}$) are not implemented in an output layer because output design parameters at output layer are allowed to take linear behavior with a linear activation function. Jacobian matrix at the last layer shown in Equation (2.3.6.6-8) are de-normalized, being transformed to the original scale as shown in Equation (2.3.6.6-9). Neural values at the last output layer (L) are calculated for all input design variables, such as beam width (b) and beam height (h), etc. prior to being de-normalized at layer D.

2.3.6.3 Formulation of universally generalizable Hessian matrix based on ANN

This section documents derivations of ANN-based Hessian matrix which was suggested by MathWorks Technical Support Department (MATLAB, 2022a).

2.3.6.3.1 *Definition of a slice* $\mathbf{H}_n^D$ *and global* $\mathbf{H}^D$ *of the Hessian matrix*

At a hidden layer l, Jacobian matrix has one dimension with $(1\times n)$ for each neuron such as shown in blue box of Figure 2.3.6.1b when n input design variables $(\mathbf{x}=[x_1,\ x_2,\ldots,\ x_n]^T)$ are considered while Hessian matrix has two dimensions with $(n\times n)$ for each neuron such as shown in blue box of Figure 2.3.6.1c when n input design variables $(\mathbf{x}=[x_1,\ x_2,\ldots,\ x_n]^T)$ are considered. It is noted that Hessian matrix of two dimensions with $(n\times n)$ at the first neuron is seen downwards as a blue box of Figure 2.3.6.1c when having when n input design variables $(\mathbf{x}=[x_1,\ x_2,\ldots,\ x_n]^T)$. Jacobian matrix having m neurons with respect to n input design variables $(\mathbf{x}=[x_1,\ x_2,\ldots,\ x_n]^T)$ at hidden layer (l) has a matrix of two dimensions $m\times n$ as shown in Figure 2.3.6.1b. Hessian matrix needs to be differentiated one more time, resulting in one more dimension as three dimensions of $m\times n\times n$ when m neurons and n input design variables are considered as shown in Figure 2.3.6.1c. Hessian matrix is third-order tensor with m neurons and n input variables.

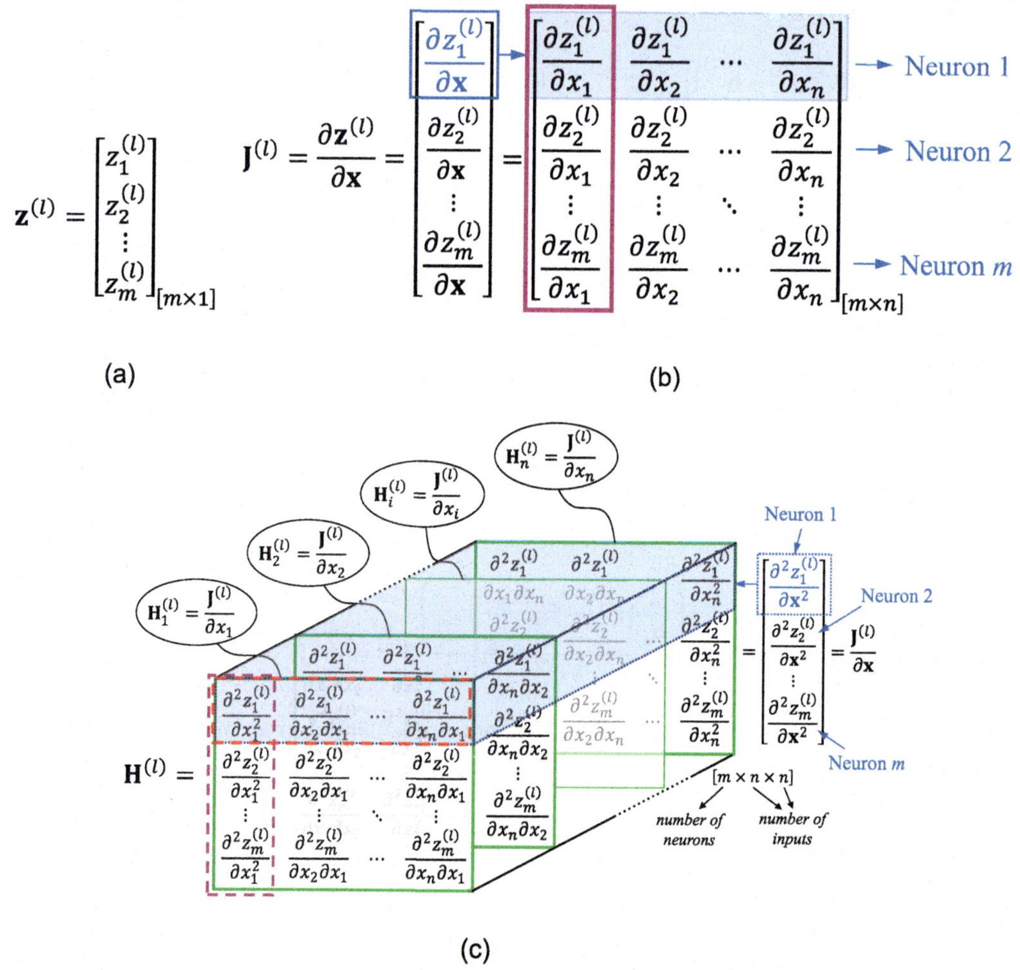

Figure 2.3.6.1 Dimensions of output vector, Jacobian, and Hessian matrices with m neurons at Hidden layer l. (a) Output vector of Hidden layer l - $\mathbf{z}^{(l)}$: Vector of size $m\times 1$. (b) Jacobian of Hidden layer l - $\mathbf{J}^{(l)}$: Matrix of size $m\times n$. (c) Hessian of hidden layer l - $\mathbf{H}^{(l)}$: third order tensor of $m\times n\times n$.

Using Hessian matrix with three dimensions of $m \times n \times n$ having m neurons and n input design variables is difficult to obtain. Thus, three-dimensional Hessian tensor $\mathbf{H}^{(l)}$ which is a third-order tensor of size $m^{(l)} \times n \times n$ is replaced by an explicit slice with a second derivative for each input design variable. Such a slice of the Hessian $\mathbf{H}_i^{(l)}$ is obtained by $\mathbf{H}_i^{(l)} = \partial \mathbf{J}^{(l)} / \partial x_i$ with respect to one of input design variables x_i among $i = 1$ to n as shown in Figure 2.3.6.1c and Equation (2.3.6.17). A slice with respect to each of n input design variables ($x = [x_1, x_2, ..., x_n]^T$) is shown as a plane parallel with paper surface shown with a green box as displayed in Figure 2.3.6.1c where a $m \times n$ matrix represents m neurons and n input design variables. A slice $\mathbf{H}_i^{(L=\text{output})}$ of three-dimensional global Hessian matrix $\mathbf{H}^{(D)}$ at an output layer shown in Equation (2.3.6.15-1) is de-normalized as $\mathbf{H}_i^{(D)}$ shown in Equation (2.3.6.16-1) which is, then, combined to form global Hessian matrix $\mathbf{H}^{(D)}$ shown in Equation (2.3.6.17). Finally, a full Hessian $\mathbf{H}^{(D)}$ of de-normalized output is obtained by adding each slice $\mathbf{H}_i^{(D)}$ $\mathbf{H}_i^{(D)} = \dfrac{\partial \mathbf{J}^{(D)}}{\partial x_i}$ into an appropriate position as shown in Equation (2.3.6.17) (MATLAB, 2022b). As shown in Figure 2.3.6.1c, full $\mathbf{H}^{(L = \text{output})}$ and $\mathbf{H}^{(D)}$ are matrices with $(m \times n \times n)$ because an output layer (L=output) and (D) have m neurons with n input design variables ($\mathbf{x} = [x_1, x_2, ..., x_n]^T$).

Hessian slices are of size 1 x n for one neuron with input design variables x_i for i =1 to n as shown with a red dotted box, whereas a slice is shown with a pink dotted box of Figure 2.3.6.1c where $m \times 1$ matrix represents m neurons when one input design variable, i.e., $\boldsymbol{x} = [x_1]$ is considered. A Jacobi matrix is also shown with a pink dotted box of Figure 2.3.6.1b where $m \times 1$ matrix represents m neurons when one input design variable, i.e., $x = [x_1]$ is considered. A slice of the Hessian $\mathbf{H}_i^{(l)}$, a derivative of Jacobian $\mathbf{J}^{(l)}$ with respect to each of input design parameters x_i between $i = 1$ to n (e.g., beam width or beam height, etc.), is obtained as shown in Equation (2.3.6.7-1). A calculation procedure is presented in detail.

2.3.6.3.2 Derivation of a slice $\mathbf{H}_n^D$ and global $\mathbf{H}^D$ of the Hessian matrix

A slice matrix $\mathbf{H}_i^{(l)}$ is obtained by differentiating Jacobi $\mathbf{J}^{(l)}$ with respect to one (x_1) of input design variables among $\mathbf{x}$ ($x_1, x_2, ..., x_n$), such as beam width or beam height, etc. A slice matrix is derived for one input variable x_1. Equations (2.3.6.7-2) and (2.3.6.7-3) are obtained by substituting $\mathbf{J}^{(l)} = {\partial \mathbf{z}^{(l)}} / {\partial \mathbf{z}^{(l-1)}} \, \mathbf{J}^{(l-1)}$ shown in Equation (2.3.6.5) into Equation (2.3.6.7-1) and applying the product rule.

$$\mathbf{H}_i^{(l)} = \frac{\partial \mathbf{J}^{(l)}}{\partial x_i} \tag{2.3.6.7-1}$$

$$\mathbf{H}_i^{(l)} = \frac{\partial\left(\left(\dfrac{\partial \mathbf{z}^{(l)}}{\partial \mathbf{z}^{(l-1)}}\right)\mathbf{J}^{(l-1)}\right)}{\partial x_i} = \frac{\partial\left(\dfrac{\partial \mathbf{z}^{(l)}}{\partial \mathbf{z}^{(l-1)}}\right)}{\partial x_i}\mathbf{J}^{(l-1)} + \frac{\partial \mathbf{z}^{(l)}}{\partial \mathbf{z}^{(l-1)}}\frac{\partial \mathbf{J}^{(l-1)}}{\partial x_i} \tag{2.3.6.7-2}$$

$$\mathbf{H}_i^{(l)} = \frac{\partial^2 \mathbf{z}^{(l)}}{\partial x_i \, \partial \mathbf{z}^{(l-1)}}\mathbf{J}^{(l-1)} + \frac{\partial \mathbf{z}^{(l)}}{\partial \mathbf{z}^{(l-1)}}\mathbf{H}_i^{(l-1)} \tag{2.3.6.7-3}$$

where ${\partial^2 \mathbf{z}^{(l)}} / {\partial x_i \, \partial \mathbf{z}^{(l-1)}}$ can be obtained by applying a chain rule as shown in Equation (2.3.6.8) taken according to (MATLAB, 2022b, 2022c).

$$\frac{\partial^2 \mathbf{z}^{(l)}}{\partial x_i\, \partial \mathbf{z}^{(l-1)}} = \frac{\partial\left(\dfrac{\partial \mathbf{z}^{(l)}}{\partial \mathbf{z}^{(l-1)}}\right)}{\partial \mathbf{z}^{(l-1)}} \frac{\partial \mathbf{z}^{(l-1)}}{\partial x_i} = \frac{\partial^2 \mathbf{z}^{(l)}}{\partial\left(\mathbf{z}^{(l-1)}\right)^2} \odot \frac{\partial \mathbf{z}^{(l-1)}}{\partial x_i} \tag{2.3.6.8}$$

The expression $\partial \mathbf{z}^{(l-1)} / \partial x_i$ can be written as $\mathbf{i}_i^{(l-1)}$, which is an ith column of Jacobian $\mathbf{J}^{(l-1)}$ for an ith input design parameter as shown in pink dotted box of Figure 2.3.6.1b and Equation (2.3.6.9-1) for $x_i = x_1$. Equation (2.3.6.9-2) is obtained by substituting Equations (2.3.6.8) and (2.3.6.9-1) into Equation (2.3.6.7-3). In Equations (2.3.6.7-3) and (2.3.6.9), a slice "i" of Hessian indicated by a green part of matrix shown in Figure 2.3.6.1c is calculated for each input variable ($i = 1, 2,\ldots, n$), i.e., for x_i; beam width (b) and depth (d). Hessian matrix is also shown as a plane parallel with a paper surface with a green box shown in Figure 2.3.6.1c where a $m \times n$ matrix represents m neurons and n input design variables. Hessian matrix is derived at Hidden layer (l) based on an ANN with a forward network as shown in Equation (2.3.6.9-2). Hessian matrix $\mathbf{H}_i^{(N)}$ is derived in Equation (2.3.6.10) and Figure 2.3.6.1c based on Jacobi $\mathbf{J}^{(N)}$ of Equation (2.3.6.6-1) calculated using normalized $\mathbf{z}^{(N)}$. Hessian matrix $\mathbf{H}_i^{(N)}$ is calculated by differentiating Jacobi $\mathbf{J}^{(N)}$ with respect to one of input variables, i.e., beam width (b) and depth (d). $\mathbf{z}^{(N)}$ shown in Equation (2.3.6.3-2(a)) is linear function, and hence, its second derivative with respect to one (x_1) of the input parameters (x_i) yields zero value as expressed in Equation (2.3.6.10). A sequence of derivation of Hessian matrices, such as $\mathbf{z}^{(N)} \Rightarrow \mathbf{J}^{(N)} \Rightarrow \mathbf{H}^{(N)}$, $\mathbf{z}^{(l)} \Rightarrow \mathbf{J}^{(l)} \Rightarrow \mathbf{H}^{(l)}$, $\mathbf{z}^{(L)} \Rightarrow \mathbf{J}^{(L)} \Rightarrow \mathbf{H}^{(L)}$, $\mathbf{z}^{(D)} \Rightarrow \mathbf{J}^{(D)} \Rightarrow \mathbf{H}^{(D)}$, is summarized in Figure 2.3.6.2.

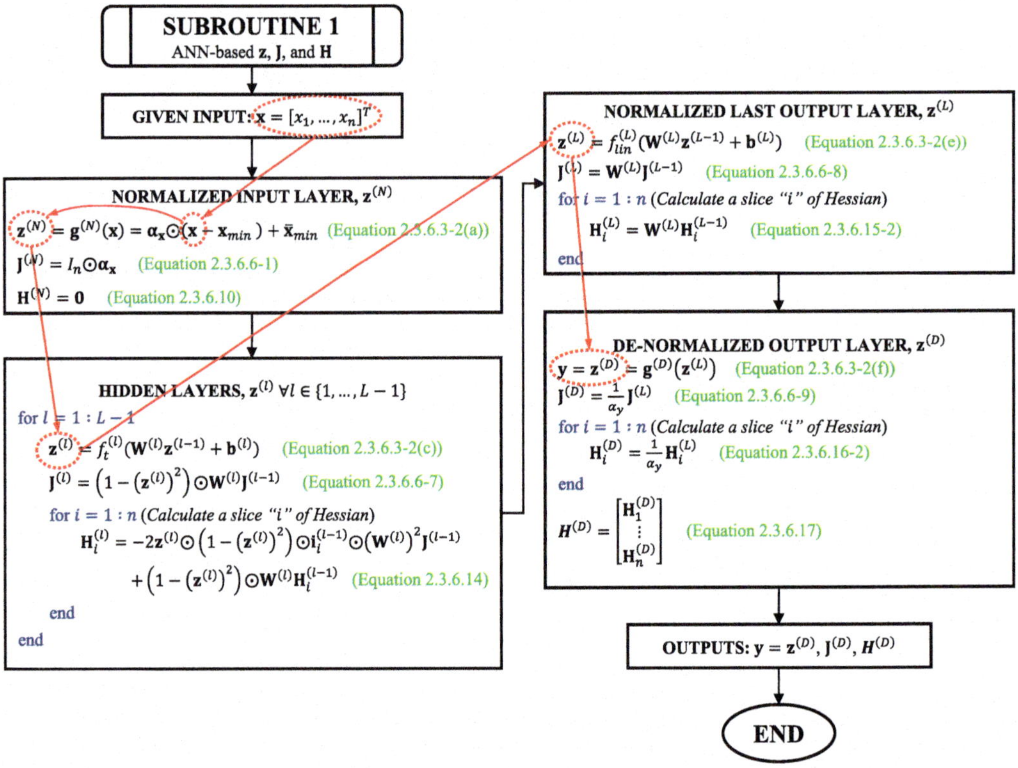

Figure 2.3.6.2 A flow of equations to obtain functions including objective, equality, inequality functions, and their Jacobian and Hessian matrices using obtained weight and bias matrices.

$$\partial \mathbf{z}^{(l-1)} \Big/ \partial x_i = \mathbf{i}_i^{(l-1)} \tag{2.3.6.9-1}$$

$$\mathbf{H}_i^{(l)} = \frac{\partial^2 \mathbf{z}^{(l)}}{\partial \left(\mathbf{z}^{(l-1)}\right)^2} \odot \mathbf{i}_i^{(l-1)} \odot \mathbf{J}^{(l-1)} + \frac{\partial \mathbf{z}^{(l)}}{\partial \mathbf{z}^{(l-1)}} \mathbf{H}_i^{(l-1)} \tag{2.3.6.9-2}$$

$$\mathbf{H}_i^{(N)} = \frac{\partial^2 \mathbf{z}^{(N)}}{\partial x_i \, \partial \mathbf{x}} = 0 \tag{2.3.6.10}$$

A slice of Hessian matrix $\mathbf{H}_i^{(1)}$ at Hidden layer (1) is obtained in Equation (2.3.6.11) by differentiating Jacobian $\mathbf{z}^{(1)}$ with respect to one of the input parameters (x_i; e.g., beam width (b) and depth (d), etc.).

$$\mathbf{H}_i^{(1)} = \frac{\partial^2 \mathbf{z}^{(1)}}{\partial x_i \, \partial \mathbf{x}} \tag{2.3.6.11-1}$$

Note that $\mathbf{z}^{(1)}$ is not a linear function, yielding non-zero values when its second derivative with respect to one of the input parameters (x_i) is performed. Based on Equation (2.3.6.9-2), $\mathbf{H}_i^{(1)}$ can be decomposed as shown in Equation (2.3.6.11-2). Current layer (l) is (1), and hence, a layer ($l-1 = 0$) is N. $\mathbf{z}^{(l-1)}$ and $\mathbf{H}_i^{(l-1)}$ are a normalized function and its Hessian matrix of starting input layer. Substituting $\mathbf{H}_i^{(N)}=0$ shown in Equation (2.3.6.10) into Equation (2.3.6.11-2) yield Equation (2.3.6.11-3) when applying chain rules twice.

$$\mathbf{H}_i^{(1)} = \frac{\partial^2 \mathbf{z}^{(1)}}{\partial \left(\mathbf{z}^{(N)}\right)^2} \odot \mathbf{i}_i^{(N)} \odot \mathbf{J}^{(N)} + \frac{\partial \mathbf{z}^{(1)}}{\partial \mathbf{z}^{(N)}} \mathbf{H}_i^{(N)} \tag{2.3.6.11-2}$$

$$\mathbf{H}_i^{(1)} = \frac{\partial^2 \mathbf{z}^{(1)}}{\partial \left(\mathbf{W}^{(1)}\mathbf{z}^{(N)} + \mathbf{b}^{(1)}\right)^2} \left(\frac{\partial \left(\mathbf{W}^{(1)}\mathbf{z}^{(N)} + \mathbf{b}^{(1)}\right)}{\partial \mathbf{z}^{(N)}} \right)^2 \odot \mathbf{i}_i^{(N)} \odot \mathbf{J}^{(N)} + 0 \tag{2.3.6.11-3}$$

By applying *tansig* activation function shown in Equation (2.3.6.4-1) and the second derivative of *tansig* activation function shown in Equation (2.3.6.4-3), $\partial^2 \mathbf{z}^{(1)} \Big/ \partial\left(\mathbf{W}^{(1)}\mathbf{z}^{(N)} + \mathbf{b}^{(1)}\right)^2$ is computed as $-2\mathbf{z}^{(1)} \odot \left(1 - \left(\mathbf{z}^{(1)}\right)^2\right)$, whereas $\left(\partial\left(\mathbf{W}^{(1)}\mathbf{z}^{(N)} + \mathbf{b}^{(1)}\right) \Big/ \partial \mathbf{z}^{(N)} \right)^2 = \left(\mathbf{W}^{(1)}\right)^2$, leading to Equation (2.3.6.11-4).

$$\mathbf{H}_i^{1} = -2\mathbf{z}^{(1)} \odot \left(1 - \left(\mathbf{z}^{(1)}\right)^2\right) \odot \mathbf{i}_i^{(N)} \odot \left(\mathbf{W}^{(1)}\right)^2 \mathbf{J}^{N} \tag{2.3.6.11-4}$$

As shown in Equation (2.3.6.7-1), a slice of Hessian matrix $\mathbf{H}_i^{l}$ of $\mathbf{z}^{l}$ at Hidden layer l is obtained by differentiating Jacobian $\mathbf{J}^{(l)}$ with respect to one of the input parameters [x_i; e.g., beam width (b) and depth (d), etc.]. $\mathbf{H}_i^{l}$ is, then, obtained when substituting Equations (2.3.6.12) and (2.3.6.13) into Equation (2.3.6.9-2), resulting in Equation (2.3.6.14).

$$\frac{\partial \mathbf{z}^{(l)}}{\partial \mathbf{z}^{(l-1)}} = \frac{\partial\left(f_t^{(l)}\left(\mathbf{W}^{(l)}\mathbf{z}^{(l-1)} + \mathbf{b}^{(l)}\right)\right)}{\partial\left(\mathbf{W}^{(l)}\mathbf{z}^{(l-1)} + \mathbf{b}^{(l)}\right)} \frac{\partial\left(\mathbf{W}^{(l)}\mathbf{z}^{(l-1)} + \mathbf{b}^{(l)}\right)}{\partial \mathbf{z}^{(l-1)}} = \left(1 - \left(\mathbf{z}^{l}\right)^2\right) \odot \mathbf{W}^{(l)} \tag{2.3.6.12}$$

$$\frac{\partial^2 \mathbf{z}^{(l)}}{\partial\left(\mathbf{z}^{(l-1)}\right)^2} = \frac{\partial^2\left(f_t^{(l)}\left(\mathbf{W}^{(l)}\mathbf{z}^{(l-1)} + \mathbf{b}^{(l)}\right)\right)}{\partial\left(\mathbf{W}^{(l)}\mathbf{z}^{(l-1)} + \mathbf{b}^{(l)}\right)^2} \left(\frac{\partial\left(\mathbf{W}^{(l)}\mathbf{z}^{(l-1)} + \mathbf{b}^{(l)}\right)}{\partial \mathbf{z}^{(l-1)}}\right)^2 = -\,2\mathbf{z}^{(l)} \odot \left(1 - \left(\mathbf{z}^{(l)}\right)^2\right) \odot \left(\mathbf{W}^{(l)}\right)^2 \tag{2.3.6.13}$$

$$\mathbf{H}_i^{(l)} = -2\mathbf{z}^{(l)} \odot \left(1 - \left(\mathbf{z}^{(l)}\right)^2\right) \odot \mathbf{i}_i^{(l-1)} \odot \left(\mathbf{W}^{(l)}\right)^2 \mathbf{J}^{(l-1)} + \left(1 - \left(\mathbf{z}^{(l)}\right)^2\right) \odot \mathbf{W}^{(l)}\mathbf{H}_i^{(l-1)} \tag{2.3.6.14}$$

$\mathbf{H}_i^{(L)}$ similar to $\mathbf{H}_i^{(l)}$ except for implementing a linear activation function is calculated when L indicates the output layer. L is linear because the last output remains without nonlinear activation, not being squashed by a nonlinear activation function. Equation (2.3.6.15-1) is obtained from Equation (2.3.6.9) when l becomes L. A second derivative of a linear activation function $\left(\frac{\partial^2 \mathbf{z}^{(L)}}{\partial\left(\mathbf{z}^{(L-1)}\right)^2}\right)$ which appears from Equation (2.3.6.4-3) is zero at an output layer where a linear activation function L is implemented. Equation (2.3.6.15-1), then, becomes Equation (2.3.6.15-2). $\mathbf{z}^{(L=\text{output})}$ is shown in Equation (2.3.6.3-2e) in which a linear activation function is applied.

$$\mathbf{H}_i^{(L)} = \frac{\partial^2 \mathbf{z}^{(L)}}{\partial\left(\mathbf{z}^{(L-1)}\right)^2} \odot \mathbf{i}_i^{(L-1)} \odot \mathbf{J}^{(L-1)} + \frac{\partial \mathbf{z}^{(L)}}{\partial \mathbf{z}^{(L-1)}} \mathbf{H}_i^{(L-1)} \tag{2.3.6.15-1}$$

$$\mathbf{H}_i^{(L)} = 0 \odot \mathbf{i}_i^{(L-1)} \odot \mathbf{J}^{(L-1)} + \frac{\partial\left(\mathbf{W}^{(L)}\mathbf{z}^{(L-1)} + b^{(L)}\right)}{\partial \mathbf{z}^{(L-1)}} \mathbf{H}_i^{(L-1)} = \mathbf{W}^{(L)}\mathbf{H}_i^{(L-1)} \tag{2.3.6.15-2}$$

Equation (2.3.6.16-1) is obtained from Equation (2.3.6.15-1) at the last output layer when L-1 and L become L and D, respectively. Equation (2.3.6.16-1) computes a de-normalized slice of the Hessian matrix $\mathbf{H}_i^{(D)}$ of de-normalized output $\mathbf{z}^{(D)}$. Similar to $\mathbf{z}^{(N)}$, $\mathbf{z}^{(D)}$ is linear function as shown in Equation (2.3.6.3-2(f)), and hence, its second derivative $\left(\frac{\partial^2 \mathbf{z}^{(D)}}{\partial\left(\mathbf{z}^{(L)}\right)^2}\right)$ with respect to $\mathbf{z}^{(L)}$ shown in Equation (2.3.6.16-1) becomes zero, presenting Equation (2.3.6.16-2) by also substituting $y = \mathbf{z}^{(D)}$ shown in Equation(2.3.6.3-2f) into Equation (2.3.6.16-1).

$$\mathbf{H}_i^{(D)} = \frac{\partial^2 \mathbf{z}^{(D)}}{\partial\left(\mathbf{z}^{(L)}\right)^2} \odot \mathbf{i}_i^{(L)} \odot \mathbf{J}^{(L)} + \frac{\partial \mathbf{z}^{(D)}}{\partial \mathbf{z}^{(L)}} \mathbf{H}_i^{(L)} \tag{2.3.6.16-1}$$

$$\mathbf{H}_i^{(D)} = 0 \odot \mathbf{i}_i^{(L)} \odot \mathbf{J}^{(L)} + \frac{\partial\left(\frac{1}{\alpha_y}\left(\mathbf{z}^{(L)} - \overline{\mathbf{z}^{(L)}}_{\min}\right) + \mathbf{z}^{(L)}_{\min}\right)}{\partial \mathbf{z}^{(L)}} \mathbf{H}_i^{(L)} = \frac{1}{\alpha_y} \mathbf{H}_i^{(L)} \tag{2.3.6.16-2}$$

A number of input design variables is indicated by n shown in Figure 2.3.6.1c. Hessian matrix $\mathbf{H}^D$ shown in Equation (2.3.6.17) is derived if Figure 2.3.6.1c is seen downwards at any neuron between 1 and m of the layer D. A size $(n \times n)$ of $\mathbf{H}^D$ shown in Equation (2.3.6.17) at a de-normalized output layer $l=D$ is a size of input design variables ($\mathbf{x}=[x_1,\ x_2,\ldots,\ x_n]^T$) which is combined with a size of Jacobian $\left(\mathbf{J}^{(D)}=\left[\frac{\partial z^{(D)}}{\partial x_1},\frac{\partial z^{(D)}}{\partial x_2},\ldots,\frac{\partial z^{(D)}}{\partial x_n}\right]^T\right)$. Jacobian $\mathbf{J}^{(D)}$ is a derivative of de-normalized $\mathbf{z}^{(D)}$ with respect to input design variables ($\mathbf{x}=[x_1,\ x_2,\ldots,\ x_n]^T$) which can be obtained based on y of Equation (2.3.6.3-2(f)).

Full Hessian matrix $\mathbf{H}^D$ at all neurons between 1 and m is obtained in terms of sliced Hessian matrices obtained by differentiating Jacobian matrices $\mathbf{J}^{(D)}$ with respect to all input design variables x_i (x_i; e.g., beam width (b) and depth (d), etc. at de-normalized output layer D. For example, Hessian slice $\mathbf{H}_2^{(D)}$ having a size $1\times n$ shown in the blue box of Equation (2.3.6.17) is obtained with input design variable $i=2$ at all neurons between 1 and m of a de-normalized output layer $l = D$. Finally, a total Hessian matrix at all neurons between 1 and m of a de-normalized output layer $l = D$ (single last layer) is obtained by regrouping n Hessian slices of $\mathbf{H}_i^{(D)}$ having a size $1\times n$ obtained for all input design variables ($\mathbf{x}=[x_1,\ x_2,\ldots,\ x_n]^T$) into $n\times n$ $\mathbf{H}^{(D)}$ matrix as presented in Equation (2.3.6.17) and Figure 2.3.6-1c. As shown in Figure 2.3.6-1c, de-normalized total Hessian matrix $\mathbf{H}^{(D)}$ for all neurons is finally derived by folding slice Hessian matrices $\mathbf{H}_i^{(D)}$ with two dimensions $(n\times n)$ indicated by green matrix, leading to global Hessian matrix derived for entire n input design variables with three dimensions ($m \times n \times n$). Section 2.4 shows how Hessian matrices are implemented in solving ANN-based Lagrange optimization.

$$H^D=\begin{bmatrix}\mathbf{H}_1^{(D)}\\ \mathbf{H}_2^{(D)}\\ \vdots\\ \mathbf{H}_n^{(D)}\end{bmatrix}_{(n\times n)}=\begin{bmatrix}\frac{\partial \mathbf{J}^{(D)}}{\partial x_1}\\ \frac{\partial \mathbf{J}^{(D)}}{\partial x_2}\\ \vdots\\ \frac{\partial \mathbf{J}^{(D)}}{\partial x_n}\end{bmatrix}_{(n\times n)}$$

$$=\begin{bmatrix}\frac{\partial^2 \mathbf{z}^{(D)=1\text{ to }m}}{\partial x_1^2} & \frac{\partial^2 \mathbf{z}^{(D)=1\text{ to }m}}{\partial x_1\,\partial x_2} & \cdots & \frac{\partial^2 \mathbf{z}^{(D)=1\text{ to }m}}{\partial x_1\,\partial x_n}\\ \frac{\partial^2 \mathbf{z}^{(D)=1\text{ to }m}}{\partial x_2\,\partial x_1} & \frac{\partial^2 \mathbf{z}^{(D)=1\text{ to }m}}{\partial x_2^2} & \cdots & \frac{\partial^2 \mathbf{z}^{(D)=1\text{ to }m}}{\partial x_2\,\partial x_n}\\ \vdots & \vdots & \ddots & \vdots\\ \frac{\partial^2 \mathbf{z}^{(D)=1\text{ to }m}}{\partial x_n\,\partial x_1} & \frac{\partial^2 \mathbf{z}^{(D)=1\text{ to }m}}{\partial x_n\,\partial x_2} & \cdots & \frac{\partial^2 \mathbf{z}^{(D)=1\text{ to }m}}{\partial x_n^2}\end{bmatrix} \quad (2.3.6.17)$$

2.3.6.4 Flow chart for Lagrange-based optimization

Figure 2.3.4.2 presents a flow of equations which are adapted for use in an iteration based on Newton–Raphson method to find solutions of the first derivative (Jacobi matrix) of Lagrange functions shown in Equation (2.3.4.2). Diagram shown in Figure 2.3.6.2 summarizes a flow of equations to obtain functions including objective, equality, equality functions, and their

Jacobian and Hessian matrices. Figure 2.3.6.2 shows that how de-normalized objective, equality, equality functions, and their Jacobian, Hessian matrix obtained in Equations (2.3.6.3-2f), (2.3.6.6-9), and (2.3.6.16), respectively, are inter-related with weight and bias matrices connecting hidden layers derived by training ANN shown in Figure 2.3.4.1 and Equations (2.3.6.1), (2.3.6.3-1) on large datasets. These flow charts help readers understand how to use weight and bias matrices. Figure 2.3.6.2 must be understood well to effectively use Figure 2.3.4.2. Examples are provided to further explain Figures 2.3.4.2 and 2.3.6.2 in Section 2.4.

2.3.6.5 Summary

ANN-based functions $\mathbf{z}$ are derived in Equations (2.3.6.1), (2.3.6.3-1), and (2.3.6.3-2). Jacobi matrices are derived in Equations (2.3.6.6-1)–(2.3.6.6-9) whereas Hessian matrices are derived in Equations (2.3.6.7)–(2.3.6.17). All these functions are using weight and bias matrices connecting each hidden layer derived by Figure 2.3.4.1, Equations (2.3.6.1), and (2.3.6.3-1). Figure 2.3.6.2 summarizes a derivation procedure for Jacobi and Hessian matrices based on ANN-based functions $\mathbf{z}$. Figure 2.3.6.2 also illustrates how weight, bias matrices are used to calculate ANN-based functions $\mathbf{z}$, Jacobian, and Hessian matrices.

Lagrange optimization is greatly influenced by the accuracies of weight and bias matrices which determine the accuracies of ANN-based functions $\mathbf{z}$ and their derivatives. Figures 2.3.4.1 and 2.3.4.2 also summarize an iteration to find solutions of the first derivative (Jacobi matrix) of Lagrange functions shown in Equation (2.3.4.2). ANN-based functions $\mathbf{z}$, Jacobian, and Hessian matrices shown in Figures 2.3.4.1, 2.3.4.2, and 2.3.6.2 are used for Lagrange optimization examples presented in Section 2.4. ANN-based Lagrange optimization method is summarized following three steps.

- **Step 1:** Structural large datasets are generated from structural design software. A proper number of large datasets needed for training ANNs should be selected discreetly based on a level of complexity for a considered problem as shown in Chapters 3–5.
- **Step 2:** ANN-based objective functions are derived based on ANNs trained on large datasets obtained from Step 1. ANN-based objective functions are derived to replace analytical functions which are difficult to explicitly obtain. Accuracies of an ANN-based functions are considerably affected by not only a number of large datasets but also training parameters of ANNs, such as a number of hidden layers, neurons, and required epochs, etc. Appropriate training parameters of ANNs should be adopted to obtain good training results.
- **Step 3:** ANN-based objective functions which are twice differentiable, being capable of deriving Jacobian and Hessian matrices, are used with Lagrange multipliers to optimize objective functions under KKT conditions based on Newton–Raphson method.

2.4 EXAMPLES OF OPTIMIZING LAGRANGE FUNCTIONS USING ANN-BASED OBJECTIVE AND CONSTRAINING FUNCTIONS WITH KKT CONDITIONS

2.4.1 Purpose of examples

This section presents examples including minimizing fourth-order polynomial, a truss frame weights, and maximizing the distance of projectiles using both analytical and ANN-based Lagrange optimizations. These examples aim to show that stationary points of objective functions can be obtained more conveniently and rapidly using ANN-based objective and constraining functions than using those based on analytical functions. These examples will also help readers follow ANN-based Lagrange optimizations using Jacobian and Hessian matrices.

2.4.2 Optimization of a fourth-order polynomial with KKT conditions

2.4.2.1 *Optimization of a fourth-order polynomial considering inequality constraints based on analytical objective and constraining functions*

2.4.2.1.1 Establishing KKT conditions

In this chapter, an optimization of a fourth-order polynomial based on analytical objective and constraining functions is demonstrated. Objective function shown in Equations (2.4.2.1) and (2.4.2.2) is optimized while being constrained by an inequality condition shown in Equation (2.4.2.3) based on the conventional Lagrange method. Lagrange function is obtained after two KKT conditions are established.

Lagrange function is formulated in Equations (2.4.2.4-1) and (2.4.2.4-2) when a KKT condition with inactive inequality constraint is applied, whereas Lagrange function is derived in Equations (2.4.2.5-1) and (2.4.2.5-2) when a KKT condition with active inequality constraint is implemented. Finding stationary points of Lagrange function with KKT conditions with both inactive and active conditions is even more difficult and is complex when optimizing reinforced concrete structures as shown in Chapter 3. This section explains why a first derivative (Jacobi or gradient vectors) of Lagrange function is useful to be linearized in terms of Jacobian and Hessian matrices as shown in Equation (2.3.4.1-1) for finding stationary points of Lagrange function $\mathcal{L}\left(x_{(k)}, \lambda_{v,(k)}\right)$ under first-order KKT equations. A fourth-order polynomial using Lagrange multiplier is minimized in this example based on conventional Lagrange multipliers to demonstrate how analytical Jacobian and Hessian matrices are implemented.

Minimizing using Lagrange multiplier

$$\text{minimize} \quad f(x) = x^4 - x^3 - 4x^2 + 5x + 5 \tag{2.4.2.1}$$

$$\text{with respect to} \quad x \in \mathbb{R} \tag{2.4.2.2}$$

$$\text{subjected to} \quad v(x) = x^2 - 4 \geq 0; \tag{2.4.2.3}$$

The Lagrange function for this problem becomes,

$$\mathcal{L}(x, \lambda_v) = x^4 - x^3 - 4x^2 + 5x + 5 - \lambda_v \underset{(=0)}{S} \left(x^2 - 4\right) \tag{2.4.2.4-1}$$

$$\mathcal{L}(x) = x^4 - x^3 - 4x^2 + 5x + 5 \tag{2.4.2.4-2}$$

$$\mathcal{L}(x, \lambda_v) = f(x) - \lambda_v^T S v(x) \tag{2.4.2.5-1}$$

$$\mathcal{L}(x, \lambda_v) = x^4 - x^3 - 4x^2 + 5x + 5 - \lambda_v S\left(x^2 - 4\right) \tag{2.4.2.5-2}$$

where λ_v is Lagrange multiplier applied to inequality constraint $v(x) = x^2 - 4 \geq 0$ and $\boldsymbol{S}$ is a diagonal matrix designating an inequality function either as an inactive or active condition. In this case, S is scalar because it designates only one condition. S determines inequality status as either inactive or active by specifying $S = 1$ or $S = 0$ as shown in Table 2.4.2.1.

Table 2.4.2.1 KKT Conditions Based on Inequality Constraint

Case	S	*Comment*
Case 1	0	Inequality $v(x)$ is inactive
Case 2	1	Inequality $v(x)$ is active

2.4.2.1.2 *Case 1 of inactive KKT condition*

2.4.2.1.2.1 FINDING STATIONARY POINTS OF LAGRANGE FUNCTIONS BASED ON NEWTON–RAPHSON ITERATION

Diagonal scalar matrix S becomes 0 to inactivate inequality constraint in which inequality constraint are not bound to equality constraint but considered as slack condition. Inequality $v(x)$ is assumed not to constrain an objective function $f(x)$ when reaching a minimum of $f(x)$ in a case that inequality $v(x)$ is inactive. The Lagrange multiplier λ_v^T of Equations (2.4.2.4-1) and (2.4.2.4-2) is equal to zero to ignore inactive inequality $v(x)$, resulting in an objective function which is equal to Lagrange function. However, an unconstrained minimum of an objective function $f(x)$ should be verified to exist within an inequality constraint set, $v(x)=x^2-4\geq 0$. Necessary Case 1 of KKT condition simply, then, yields first-order differential equation (Jacobi) of Lagrange function as shown in Equations (2.4.2.6-1) and (2.4.2.6-2).

$$\nabla f(x)=0 \tag{2.4.2.6-1}$$

$$\nabla f(x)=4x^3-3x^2-8x+5=0 \tag{2.4.2.6-2}$$

In Case 1 KKT condition, $\nabla\mathcal{L}\left(x_{(k)},\, \lambda_{v,(k)}\right)$ becomes $\nabla f(x)$ as shown in Equations (2.4.2.6-1) and (2.4.2.6-2) because inequality constraint $v(x)=x^2-4\geq 0$ is ignored due to non-binding inequality constraint. However, it is difficult to directly find stationary points of a Lagrange function $\nabla\mathcal{L}\left(x_{(k)},\, \lambda_{v,(k)}\right)=\nabla f(x)$ when Jacobi of Lagrange functions are higher order of polynomials as shown in Equations (2.4.2.6-2). This is why Lagrange functions are linearized, as shown in Equation (2.3.4.1-1), after which Newton–Raphson method can be adopted to find stationary points iteratively. Jacobi shown in Equation (2.4.2.6-1) and Hessian shown in Equation (2.4.2.7-2) are included in linearized Lagrange function $\nabla\mathcal{L}\left(x_{(k)},\, \lambda_{v,(k)}\right)=\nabla f(x)$, in which recursion shown in Equations (2.3.4.1-3)–(2.3.4.1-5) is also implemented in Equations (2.4.2.7-1) and (2.4.2.7-2) to have initial trial input parameters $x_{(k)}$, $\lambda_{v,(k)}$ converge, resulting in $\nabla f(x)=0$ shown in Equation (2.4.2.6-1).

$$x_{(k+1)}=x_{(k)}-H_f\left(x_k\right)^{-1}\nabla f\left(x_k\right) \tag{2.4.2.7-1}$$

where

$$H_f(x)=\frac{\partial^2 f}{\partial x^2}=12x^2-6x-8 \tag{2.4.2.7-2}$$

Stationary points minimizing the Lagrange function shown in Equation (2.4.2.4-1) are identified by letting Jacobi of Lagrange function shown in Equation (2.4.2.6-2) be 0.

Table 2.4.2.2 Newton–Raphson Iteration Based on Initial Trial Input Parameters for an Inactive Case I KKT Condition

	Initial Guess $x_{(0)} = -3$			Initial Guess $x_{(0)} = 0$			Initial Guess $x_{(0)} = 3$		
Iteration	$x_{(k)}$	$\nabla f(x_k)$	$f(x_k)$	$x_{(k)}$	$\nabla f(x_k)$	$f(x_k)$	$x_{(k)}$	$\nabla f(x_k)$	$f(x_k)$
0	−3	−106	62	0	5	5	3	62	38
1	−2.102	−28.572	5.617	0.625	−0.195	6.471	2.244	17.137	10.133
2	−1.606	−6.452	−2.554	0.597	0.003	6.474	1.804	4.289	5.723
3	−1.408	−0.842	−3.249	0.598	8.8E-07	6.474	1.592	0.800	5.211
4	−1.373	−0.024	−3.264	0.598	6.0E-14	6.474	1.530	0.061	5.185
5	−1.372	−2.1E−05	−3.264	0.598	0	6.474	1.524	4.9E-04	5.184

Equation (2.4.2.6-2) should be linearized as a first-order approximation based on Jacobian and Hessian matrices using Equation (2.3.4.1-1). Initial trial input parameters $\left[x_{(0)}, \lambda_{v(0)}\right]$ should be assumed because solutions are unknown at this time. Newton–Raphson iterations based on Equations (2.3.4.1-3)–(2.3.4.1-5) and Equations (2.4.2.7-1) and (2.4.2.7-2) are implemented in finding solutions by converging initial trial input parameters $\left[x_{(0)}, \lambda_{v(0)}\right]$ to stationary points.

It is well known that Newton–Raphson iteration relies on an appropriate initial guess to expedite a convergence and to enhance accuracy. Thus, three initial guesses $x_{(0)}$ of −3, 0, and 3 are adopted to find stationary points for Case 1 KKT as shown in Table 2.4.2.2 which is obtained based on Hessian shown in Equations (2.4.2.7-1) and (2.4.2.7-2). Table 2.4.2.2 summarizes iteration process for finding stationary points of the first-order derivative (Jacobi) of Lagrange function shown in Equation (2.4.2.6). For initial trial input parameter $x_{(0)}=0$, $f\left(x_{(0)}\right)$ and $\nabla f\left(x_{(0)}\right)$ are obtained both as 5 from Equation (2.4.2.1) and Equation (2.4.2.6), respectively, thus not being able to converge to 0 at Iteration 0. Initial trial input parameter must be updated to next $x_{(1)}$ to verify if the first-order derivative (Jacobi) of Lagrange function converges to 0 at Iteration 1. Hessian which is obtained as 8 with $x_{(0)}=0$ from Equation (2.4.2.7-2) is substituted into Equation (2.4.2.7-1) to find updated $x_{(1)} = x_{(0)} - H_f\left(x_{(0)}\right)^{-1} \nabla f\left(x_{(0)}\right) = 0 - \frac{1}{8} \times 5 = 0.625$. For next initial trial input parameter $x_{(1)}=0.625$ at Iteration 1, $f\left(x_{(1)}\right)$ and $\nabla f\left(x_{(1)}\right)$ are updated as 6.471 and −0.195 from Equations (2.4.2.1) and (2.4.2.6), respectively, not being able to converge to 0 yet.

2.4.2.1.2.2 VERIFICATION OF KKT CONDITION

Table 2.4.2.2 summarizes five iterations for three initial trial input parameters. A stationary point at $x_{(3)} = 0.598$ is found at the third iteration (Iteration 3) for initial trial input parameter 0 as presented in Table 2.4.2.2 where $\nabla f\left(x_{(3)}\right)$ and $f\left(x_{(3)}\right)$ converge to 8.8E-07 and 6.474, respectively. Similarly, $\nabla f\left(x_{(5)}\right)$ converges to 0 for initial trial input parameters −3 and 3 at the fifth iteration (Iteration 5). Figure 2.4.2.1 shows that three initial trial input parameters $x_{(0)}=(-3, 0, 3)$ yielded $\nabla f\left(x_0\right)$ of −106, 5, and 62, respectively at Iteration 0, which did not converge. The stationary points and the minimized functions of $x_{(5)}$ and $f\left(x_5\right)$ for the three initial trial input parameters $x_{(0)}=(-3, 0, 3)$ are finally found to be (−1.372, −3.264(minimum)), (0.598, 6.474(maximum)), and (1.524, 5.184(minimum)), respectively, at fifth iteration (Iteration 5). However, all initial trial input parameters converge to $x_{(5)} = -1.372$, 0.598, and 1.524, which violates an inequality constraint shown in Equation (2.4.2.3),

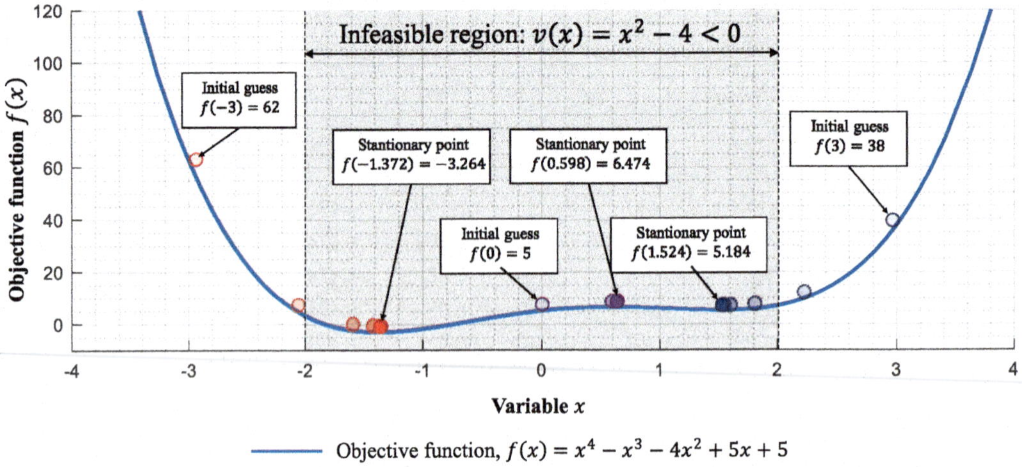

Figure 2.4.2.1 Verification of optimized results based on an inequality constraint for an inactive Case I KKT condition.

requiring $x_{(5)}$ should converge in the region, $x_{(5)} \geq 2$ and $x_{(5)} \leq -2$, showing that $x_{(5)}$ converged between $2 \leq x_{(5)} \leq 2$ as shown in Figure 2.4.2.1 where a minimum of an objective function of Case 1 KKT condition is plotted graphically.

2.4.2.1.3 Case 2 of active KKT condition

2.4.2.1.3.1 FINDING STATIONARY POINTS OF LAGRANGE FUNCTIONS BASED ON NEWTON–RAPHSON ITERATION

In this case of an active condition, inequality $v(x)$ is assumed to constrain an objective function $f(x)$ when reaching a minimum of $f(x)$. The Lagrange multiplier λ_v^T of Equations (2.4.2.5) and (2.4.2.8) should be greater than 0. Lagrange function and necessary Case 2 KKT condition are, thus, obtained in Equations (2.4.2.8-1) and (2.4.2.8-2). Equation (2.4.2.8-2) is solved based on the initial trial input parameter, leading to Equation (2.4.2.8-3) in which a recursion shown in Equations (2.3.4.1-3)–(2.3.4.1-5) applies to obtain Equations (2.4.2.9-1) and (2.4.2.9-2). Diagonal scalar matrix S becomes 1 to activate inequality constraint in which inequality constraints are bound to equality constraint as shown in Equation (2.4.2.8-1) with a Lagrange multiplier greater than 0 when deriving a Lagrange function. Stationary points minimizing the Lagrange function shown in Equation (2.4.2.8-1) are identified by letting Jacobi of Lagrange function shown in Equation (2.4.2.8-2) be 0. Equation (2.4.2.8-2) should be linearized as a first-order approximation based on Jacobian and Hessian matrices using Equation (2.3.4.1-1). Initial trial input parameters $\left[x_{(0)}, \lambda_{v(0)}\right]$ should be assumed because solutions are unknown at this time. Newton–Raphson iterations based on Equations (2.3.4.1-3)–(2.3.4.1-5) are implemented in finding solutions by converging initial trial input parameters $\left[x_{(0)}, \lambda_{v(0)}\right]$ to stationary points shown in Equation (2.4.2.8-2).

Table 2.4.3.3 traces the convergence of solutions based on initial trial input parameters (−3, 1) and (1.5, 7). $f\left(x_{(0)}\right)$ and $\mathcal{L}\left(x_{(0)}, \lambda_{v(0)}\right)$ are obtained as 62 and 57 from Equations (2.4.2.1) and (2.4.2.8-1), respectively, for an initial trial input parameter $\left[x_{(0)}, \lambda_{v(0)}\right] = [-3, 1]$,

Table 2.4.2.3 Converges of Stationary Points Based on Newton–Raphson Iteration Using Initial Trial Input Parameters for Active Case 2 KKT Condition

(a): With Two Trial Vectors

	Initial Guess $x_{(0)} = -3, \lambda_{v,(0)} = 1$					Initial Guess $x_{(0)} = 1.5, \lambda_{v,(0)} = 7$				
Iteration	$x_{(k)}$	$\lambda_{v,(k)}$	*MSE*	$\mathcal{L}$	$f(x_k)$	$x_{(k)}$	$\lambda_{v,(k)}$	*MSE*	$\mathcal{L}$	$f(x_k)$
0	−3	1	5012.5	57	62	1.5	7	227.312	17.438	5.188
1	−2.167	1.556	330.34	6.518	7.598	2.083	−0.86	113.602	8.144	7.851
2	−2.006	5.332	1.880	3.012	3.149	2.002	2.103	0.198	7.001	7.015
3	−2.000	5.748	2.1E-05	3	3	2	2.25	1.5E-07	7	7
4	−2	5.750	5.7E-16	3	3	2	2.25	1.9E-20	7	7

(b): With Four Trial Vectors

	Initial guess $x_{(0)} = -3, \lambda_{v,(0)} = 1$					Initial guess $x_{(0)} = 1.5, \lambda_{v,(0)} = 7$				
Iteration	$x_{(k)}$	$\lambda_{v,(k)}$	*MSE*	$\mathcal{L}$	$f(x_k)$	$x_{(k)}$	$\lambda_{v,(k)}$	*MSE*	$\mathcal{L}$	$f(x_k)$
0	−3	1	5012.5	57	62	1.5	7	227.312	17.438	5.188
1	−2.167	1.556	330.34	6.518	7.598	2.083	−0.86	113.602	8.144	7.851
2	−2.006	5.332	1.880	3.012	3.149	2.002	2.103	0.198	7.001	7.015
3	−2.000	5.748	2.1E-05	3	3	2	2.25	1.5E-07	7	7
4	−2	5.750	5.7E-16	3	3	2	2.25	1.9E-20	7	7
	Initial Guess $x_{(0)} = -1.5, \lambda_{v,(0)} = 8$					Initial Guess $x_{(0)} = 3, \lambda_{v,(0)} = 0$				
Iteration	$x_{(k)}$	$\lambda_{v,(k)}$	*MSE*	$\mathcal{L}$	$f(x_k)$	$x_{(k)}$	$\lambda_{v,(k)}$	*MSE*	$\mathcal{L}$	$f(x_k)$
0	−1.5	8	216.8	10.94	−3.06	3	0	1934.5	38	38
1	−2.083	3.417	88.331	3.940	5.103	2.167	−1.06	177.763	9.655	8.922
2	−2.002	5.631	0.149	3.001	3.038	2.006	1.986	1.162	7.009	7.058
3	−2.000	5.750	1.1E-07	3.000	3.000	2.000	2.243	1.4E-05	7.000	7.000
4	−2	5.75	1.4E-20	3	3	2	2.25	3.8E-16	7	7

whereas the first derivative of Lagrange function is obtained from Equation (2.4.2.8-2). Mean squared error (MSE) of $\nabla\mathcal{L}\left(x_{(0)}, \lambda_{v(0)}\right)$ with respect to an initial value $\left[x_{(0)}, \lambda_{v(0)}\right]$ of $[-3, 1]$ is obtained as $\frac{1}{2}\left((-100)^2 + (-5)^2\right) = 5012.5$ in Equations (2.4.2.8-2) and (2.4.2.8-3) when compared with $\nabla\mathcal{L}\left(x_{(k)}, \lambda_{v,(k)}\right) = 0$. MSE did not converge to $\nabla\mathcal{L}\left(x_{(k)}, \lambda_{v(k)}\right) = 0$. Initial trial input parameter $\left[x_{(0)}, \lambda_{v(0)}\right]$ should be updated to $\left[x_{(1)}, \lambda_{v(1)}\right]$ in Equation (2.4.2.9-4) by substituting Hessian of Lagrange function $H_{\mathcal{L}}\left(x_{(0)}, \lambda_{v(0)}\right)$ with respect to $\left[x_{(0)}, \lambda_{v(0)}\right]$ obtained from Equations (2.3.4.2-2a), (2.4.2.9-2), and (2.4.2.9-3) into Equation (2.4.2.9-1). Initial trial input parameter $\left[x_{(0)}, \lambda_{v(0)}\right] = [-3, 1]$ is indicated by red and blue color in Equations (2.4.2.8-2)–(2.4.2.9-4).

$$\mathcal{L}(x, \lambda_v) = x^4 - x^3 - 4x^2 + 5x + 5 - \lambda_v\left(x^2 - 4\right) \tag{2.4.2.8-1}$$

$$\nabla\mathcal{L}\left(x_{(k)},\ \lambda_{v(k)}\right)=\begin{bmatrix}\nabla_x\mathcal{L}(x,\ \lambda_v)\\ \nabla_{\lambda_v}\mathcal{L}(x,\ \lambda_v)\end{bmatrix}=\begin{bmatrix}4x^3-3x^2-8x+5-2\lambda_v x\\ 4-x^2\end{bmatrix}=0 \quad (2.4.2.8\text{-}2)$$

$$\nabla\mathcal{L}\left(x_{(0)},\ \lambda_{v(0)}\right)=\begin{bmatrix}4\times(-3)^3-3\times(-3)^2-8\times(-3)+5-2\times1\times(-3)\\ 4-(-3)^2\end{bmatrix}=\begin{bmatrix}-100\\ -5\end{bmatrix} \quad (2.4.2.8\text{-}3)$$

$$\begin{bmatrix}x_{(k+1)}\\ \lambda_{v(k+1)}\end{bmatrix}=\begin{bmatrix}x_{(k)}\\ \lambda_{v(k)}\end{bmatrix}-H_{\mathcal{L}}\left(x_{(k)},\ \lambda_{v(k)}\right)^{-1}\nabla\mathcal{L}\left(x_{(k)},\ \lambda_{v(k)}\right) \quad (2.4.2.9\text{-}1)$$

$$H_{\mathcal{L}}\left(x_{(k)},\ \lambda_{v(k)}\right)=\begin{bmatrix}\dfrac{\partial^2\mathcal{L}}{\partial x^2} & \dfrac{\partial^2\mathcal{L}}{\partial x\,\partial\lambda_v}\\ \dfrac{\partial^2\mathcal{L}}{\partial x\,\partial\lambda_v} & \dfrac{\partial^2\mathcal{L}}{\partial\lambda_v^2}\end{bmatrix}=\begin{bmatrix}12x^2-6x-8-2\lambda_v & -2x\\ -2x & 0\end{bmatrix} \quad (2.4.2.9\text{-}2)$$

$$H_{\mathcal{L}}\left(x_{(0)},\ \lambda_{v(0)}\right)=\begin{bmatrix}12\times(-3)^2-6\times(-3)-8-2(1) & -2\times(-3)\\ -2\times(-3) & 0\end{bmatrix}=\begin{bmatrix}116 & 6\\ 6 & 0\end{bmatrix} \quad (2.4.2.9\text{-}3)$$

$$\begin{bmatrix}x_{(1)}\\ \lambda_{v(1)}\end{bmatrix}=\begin{bmatrix}x_{(0)}\\ \lambda_{v(0)}\end{bmatrix}-H_{\mathcal{L}}\left(x_{(0)},\lambda_{v(0)}\right)^{-1}\nabla\mathcal{L}\left(x_{(0)},\lambda_{v(0)}\right)$$

$$=\begin{bmatrix}-3\\ 1\end{bmatrix}-\begin{bmatrix}116 & 6\\ 6 & 0\end{bmatrix}^{-1}\begin{bmatrix}-100\\ -5\end{bmatrix}\rightarrow\begin{bmatrix}x_{(1)}\\ \lambda_{v(1)}\end{bmatrix}=\begin{bmatrix}-2.167\\ 1.556\end{bmatrix}. \quad (2.4.2.9\text{-}4)$$

First derivative (Jacobi) of Lagrange function updated at the first iteration with respect to $\left[x_{(1)},\ \lambda_{v(1)}\right]=[-2.167, 1.556]$ is obtained in Equation (2.4.2.10) by substituting $\left[x_{(1)},\ \lambda_{v(1)}\right]=[-2.167, 1.556]$ into Equation (2.4.2.8-2). Updated input parameter $\left[x_{(1)},\lambda_{v(1)}\right]=[-2.167, 1.556]$ is indicated by red and blue color in Equation (2.4.2.10).

$$\nabla\mathcal{L}\left(x_{(1)},\ \lambda_{v(1)}\right)=\begin{bmatrix}4\times(-2.167)^3-3\times(-2.167)^2-8\times(-2.167)+5-2\times1.556\times(-2.167)\\ 4-(-2.167)^2\end{bmatrix}$$

$$=\begin{bmatrix}-25.694\\ -0.694\end{bmatrix} \quad (2.4.2.10)$$

The MSE decreased significantly to $\frac{1}{2}\left((-25.694)^2+(-0.694)^2\right)=330.343$ when an initial trial input parameter is updated to $\left[x_{(1)},\ \lambda_{v(1)}\right]=[-2.167,\ 1.556]$, compared to MSE 5012.5 shown in Equation (2.4.2.8-3) with an initial trial input parameter $\left[x_{(0)},\ \lambda_{v(0)}\right]$ of $[-3,\ 1]$.

Two pairs of initial values $\left[x_{(0)},\ \lambda_{v,(0)}\right]$ of [−3, 1] and [1.5, 7] are used to find stationary points as shown in Table 2.4.2.3(a) to optimize Equation (2.4.2.8-1) for Case 2 KKT condition. The MSE decreased to 5.7E-16 at fourth iteration (Iteration 4) with $\left[x_{(4)},\ \lambda_{v(4)}\right]=[-2,\ 5.75]$ based on the initial value $\left[x_{(0)},\ \lambda_{v,(0)}\right]$ of [−3, 1] as shown in Table 2.4.2.3(a), converging to 0. Another local stationary point is obtained at $\left[x_{(4)},\ \lambda_{v(4)}\right]=[2,\ 2.25]$ with decreased MSE to 1.4E-20 at fourth iteration (Iteration 4) based on another initial trial input parameter $\left[x_{(0)},\ \lambda_{v(0)}\right]$ of $[1.5,\ 7]$ as shown in Table 2.4.2.3(a).

Convergence based on four initial trial input parameters $[-3,\ 1]$, $[1.5,\ 7]$, $[1.5,\ 8]$, and $[3,\ 0]$ similar to Table 2.4.2.3(a) is also presented in Table 2.4.2.3(b). The MSE decreased to 5.7E-16 and 1.4E-20 at fourth iteration (Iteration 4) with $\left[x_{(4)},\ \lambda_{v(4)}\right]=[-2,\ 5.75]$ based on initial value $\left[x_{(0)},\ \lambda_{v,(0)}\right]$ of [−3, 1] and $[1.5,\ 8]$, respectively, as shown in Table 2.4.2.3(b), converging to 0. Another local stationary point is obtained at $\left[x_{(4)},\ \lambda_{v(4)}\right]=[2,\ 2.25]$ based on another initial trial input parameters $\left[x_{(0)},\ \lambda_{v(0)}\right]$ of $[1.5,\ 7]$ and $[3,\ 0]$ as shown in Table 2.4.2.3(b).

2.4.2.1.3.2 A GLOBAL OPTIMAL SOLUTION MINIMIZING LAGRANGE FUNCTIONS

A constrained optimization always occurs at stationary points of Lagrange functions $\mathcal{L}\left(x_{(k)},\ \lambda_{v(k)}\right)$. Initial trial input parameter $\left[x_{(0)},\ \lambda_{v(0)}\right]=[-3,\ 1]$ converged to local stationary point $\left[x_{(4)},\ \lambda_{v(4)}\right]=[-2,\ 5.75]$, yielding $\mathcal{L}\left(x_{(k)},\ \lambda_{v,(k)}\right)=3$ and $f_{min}(x)=3$ whereas another initial trial input parameter $\left[x_{(0)},\ \lambda_{v(0)}\right]=[1.5,\ 7]$ converged to another local stationary point $\left[x_{(4)},\ \lambda_{v(4)}\right]=[2,\ 2.25]$, yielding $\mathcal{L}\left(x_{(k)},\ \lambda_{v,(k)}\right)=7$ and $f_{min}(x)=7$ as shown in Table 2.4.2.3(a) and Figure 2.4.2.2c. A global optimal solution is, thus, found at $\left[x_{(4)},\ \lambda_{v(4)}\right]=[-2,\ 5.75]$ which yields $\mathcal{L}\left(x_{(k)},\ \lambda_{v,(k)}\right)=3$ and $f_{min}(x)=3$ when Case 2 KKT condition is constrained by an active inequality condition $v(x)$. Observation similar to one shown in Figure 2.4.2.2(c) is shown in Figure 2.4.2.2(d).

In case 2, $f(x)$ is function of x only, whereas $\mathcal{L}\left(x_{(k)},\ \lambda_{v,(k)}\right)$ is function of x and λ_v as shown in Equation (2.4.2.8-1). An optimal solution $f_{min}(x)$ of $f(x)$ constrained by an active inequality $v(x)$ is obtained as to $\mathcal{L}=3$. It is noted that Lagrange function $\mathcal{L}(x,\ \lambda_v)$ is identical to an objective function $f(x)$ when an inequality constraint is active as shown in Equation(2.4.2.8-1), when inequality constraint $v(x)=x^2-4=0$, however, they are not similar for the first few iterations, but eventually they become identical when they converge at the final iteration as shown in Table 2.4.2.3 and Figure 2.4.2.2c and d. Stationary point $\left[x_{(4)},\ \lambda_{v(4)}\right]=[-2,\ 5.75]$ verifies that an objective function $\mathcal{L}\left(x_{(k)},\ \lambda_{v,(k)}\right)$ is globally minimized to 3 in the range $x_{(4)}\geq 2$ and $x_{(4)}\leq -2$ as defined by inequality constraint $v(x)=x^2-4\geq 0$ shown in Equation (2.4.2.3).

2.4.2.1.3.3 EFFICIENT ESTABLISHMENT OF INITIAL TRIAL INPUT PARAMETERS

Objective functions can be plotted to estimate a number of stationary points graphically prior to a selection of initial trial input parameters. Contours of Lagrange function can be plotted by varying Lagrange multiplier λ_v and input variable **x** prior to solving Lagrange optimization as shown in Figure 2.4.2.2a. Intersections of contour lines can be where stationary points are located. However, these contours are difficult to be visualized when there are parameters of more than two variables including Lagrange multiplier λ_v.

As in Figure 2.4.2.2a, knowing Lagrange multipliers λ_v of 5.75 and 2.25 for local stationary points $x=-2$ and $x=+2$, respectively, helps to locate a selection of initial trial input variables shown in Table 2.4.2.3. Two initial trial input variables $\left[x_{(0)}, \lambda_{v(0)}\right] = [-3, 1]$ and $[1.5, 7]$ shown in Figure 2.4.2.2c and d are assumed, demonstrating that the initial trial input variable

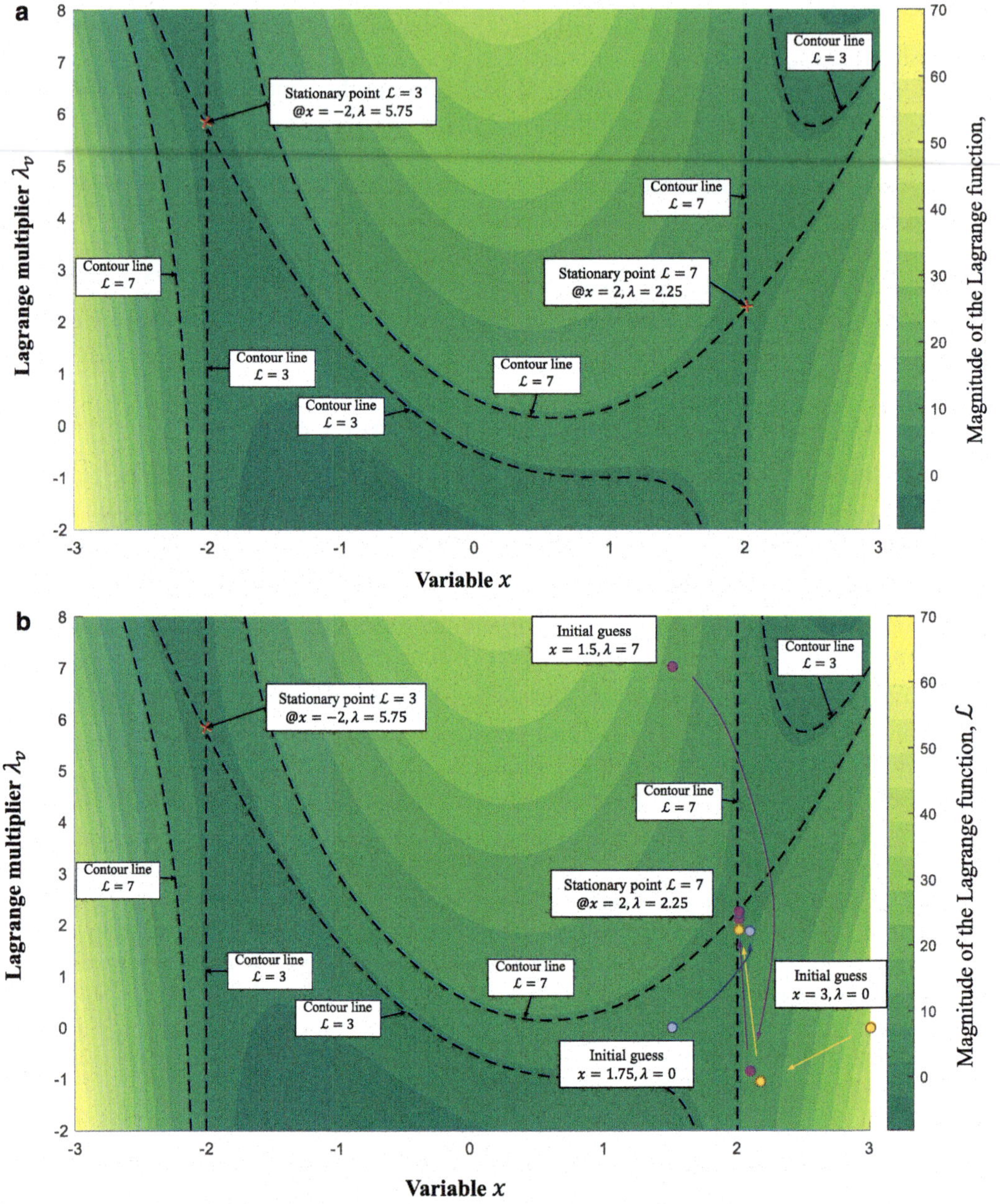

Figure 2.4.2.2 Verification of positive Lagrange multipliers $\lambda_{v,(k)}$ for active Case 2 KKT condition minimizing objective function $f(x)$ at $x=-2$, $\lambda_v=5.75$, L = 3. (a) Contours of Lagrange function based on Lagrange multiplier λ_v and input variable. (b) With one initial vector leading to an incorrect stationary point. *(Continued)*

$\left[x_{(0)}, \lambda_{\upsilon(0)}\right] = [-3, 1]$ converges to correct stationary point $\left[x_{(4)}, \lambda_{\upsilon(4)}\right] = [-2,\ 5.75]$, whereas $\left[x_{(0)}, \lambda_{\upsilon(0)}\right] = [1.5, 7]$ did not converges to correct stationary point $\left[x_{(4)}, \lambda_{\upsilon(4)}\right] = [2, 2.25]$. In Figure 2.4.2.2c and d, and Table 2.4.2.3, two and four initial trial input parameters are used to start Newton–Raphson iteration. Both of them yield the same solutions. However, using initial trial input parameter of bad choices such as $\left[x_{(0)}, \lambda_{\upsilon(0)}\right] = [3, 0]$ instead of $\left[x_{(0)}, \lambda_{\upsilon(0)}\right] = [-3, 1]$ shown in Figure 2.4.2.2b may lead to an incorrect stationary point, being unable to identify minimized Lagrange function.

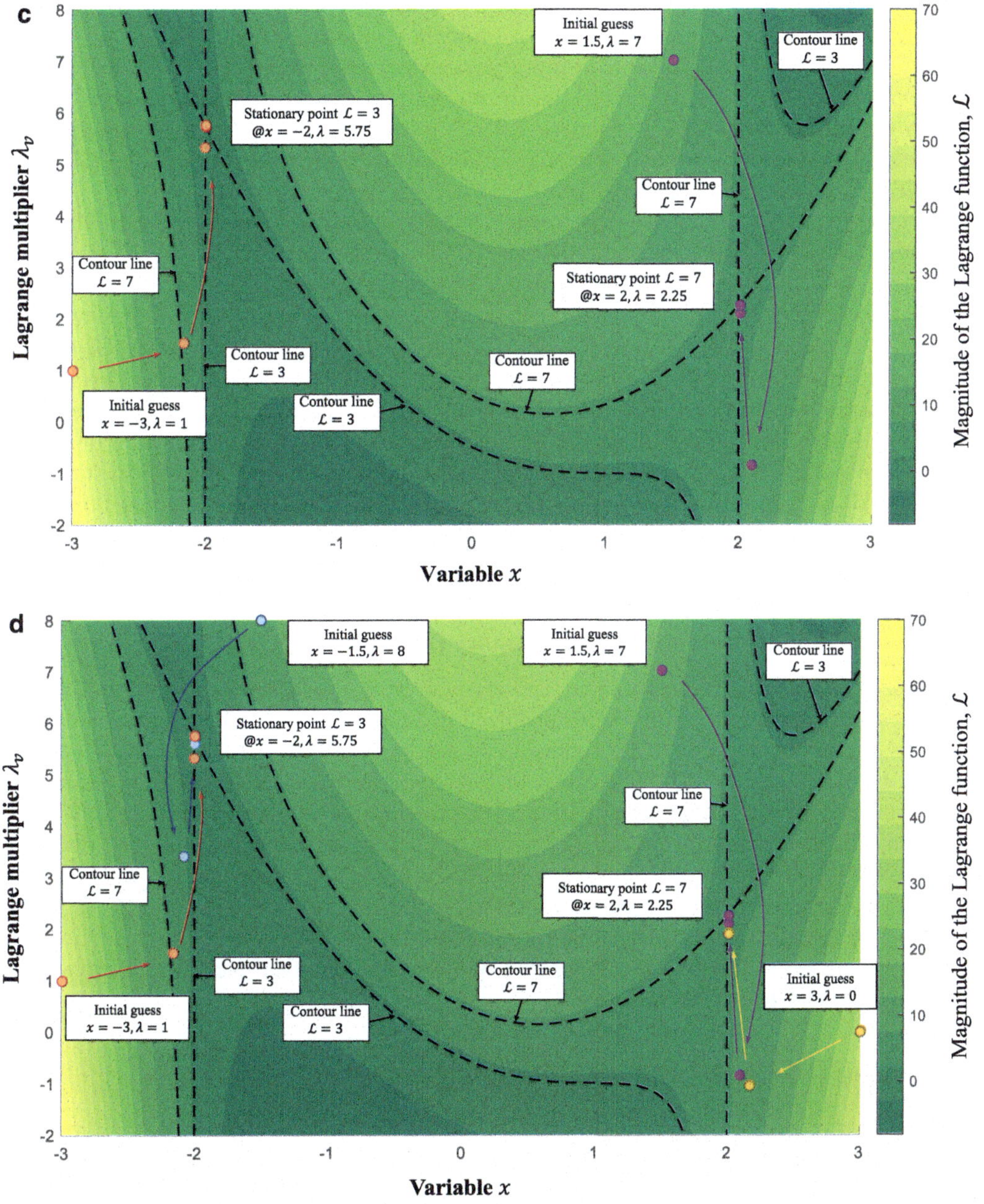

Figure 2.4.2.2 (Continued) (c) With two initial vectors. (d) With four initial vectors.

Choosing an appropriate number of initial trial input parameters covering entire solution domains is very important. A number of initial trial input parameters is normally selected intuitively if a number of stationary points are not known. However, there is no rule of thumb for selecting good initial trial input parameters and their locations. Each trial parameter should be distributed evenly but ensure that initial trial input parameters are to be separated far enough from each other not to converge to the same stationary point. Initial trial input parameters which are well distributed covering areas sufficiently wide will move to real stationary that exist to identify minimized Lagrange function. The more initial trial input parameters are selected; the better results will result in not missing any stationary point when a number of stationary points are not known. However, too many selections of initial trial input parameters may require significant computing time. Readers are referred to Section 3.3.1 where 25 initial trial input parameters are selected to solve for Lagrange-based optimization for columns. Trial parameters can be selected based on dividing a range of each input variable shown in Figures 2.4.2.2 into five, for example, selecting 25 trial parameters to capture all possible stationary points when two unknown input variables (for example, beam width b and beam depth d) are to be sought.

There exist three local stationary points in Case 1 as shown in Figure 2.4.2.1, so three initial trial input parameters are enough for each local stationary point. Figure 2.4.2.1 shows three local stationary points for the Lagrange function $f(x) = x^4 - x^3 - 4x^2 + 5x + 5$ shown in Equation(2.4.2.4-2) with an inactive inequality whereas Figure 2.4.2.2 shows two local stationary points for the Lagrange function $\mathcal{L}(x, \lambda_v) = x^4 - x^3 - 4x^2 + 5x + 5 - \lambda_v(x^2 - 4)$ shown in Equation(2.4.2.8-1) with an active inequality. Three local initial trial input parameters are assumed in Tables 2.4.2.2 whereas two and four local initial parameters are assumed in Table2.4.2.3 to find stationary points. However, a number of initial trial input parameters should be selected intuitively covering wider domains to find all local stationary points by converging initial trial input parameters accurately. Code for Lagrange optimization based on analytical functions is presented in Appendix B1.

2.4.2.2 *ANN-based optimization of a fourth-order polynomial constrained by inequality functions*

In this section, ANNs are formulated and trained on large datasets to replace analytical objectives and constraining functions optimized in Section 2.4.2.1. This section shows that any analytical function shown in Equations (2.4.2.11-1) and (2.4.2.11-2) can be replaced by that derived based on ANN which is at least twice differentiable, yielding Jacobi and Hessian matrices without almost any limit. The same fourth-order polynomial is optimized based on ANN, yielding faster and easier results than those optimized analytically. Step-by-step Newton-Raphson procedures for the ANN-based Lagrange method are presented to find stationary points of Lagrange functions using functions derived based on ANNs. ANN-based design equations that constrain any design condition including member sizes, design requirements, and material properties will replace analytical constraining conditions which are difficult to explicitly derive.

2.4.2.2.1 Step 1: Generation of large datasets for training ANN

$$\text{Minimize} \quad f(x) = x^4 - x^3 - 4x^2 + 5x + 5 \tag{2.4.2.11-1}$$

$$\text{with respect to} \quad x \in \mathbb{R}$$

subjected to $v(x) = x^2 - 4 \geq 0;$ (2.4.2.11-2)

In this problem, both an objective function $f(x) = x^4 - x^3 - 4x^2 + 5x + 5$ shown in Equation (2.4.2.11-1) and inequality constraint $v(x) = x^2 - 4$ shown in Equation (2.4.2.11-2) are non-linear functions, which will be replaced by ANN-based functions. A proper number of large datasets to train an ANN over a reasonable range should be selected carefully based on a level of a complexity of nonlinear objective and inequality constraining function. As shown in Table 2.4.2.4, ten thousand datasets with a range of input variable x between −4 and 4 are generated for training an ANN. The large datasets are not normalized in Table 2.4.2.4(a) whereas they are normalized between −1 and +1 as shown in Table 2.4.2.4(b). Minimum and maximum ranges of a non-normalized and normalized data are shown below the table. Objective function $f(x) = x^4 - x^3 - 4x^2 + 5x + 5$ shown in Equation (2.4.2.11-1) and constraining function $v(x) = x^2 - 4$ shown in Equation (2.4.2.11-2) are obtained based on ANN shown in Equations (2.4.2.12-1) and (2.4.2.12-2), respectively. Equations (2.4.2.12-1) and (2.4.2.12-2) are trained by large datasets to obtain weight and bias matrices. Readers are referred to Section 2.3.3 for a review.

Table 2.4.2.4 Generation of Large Datasets

(a): Non-Normalized				*(b): Normalized*			
10,000 Datasets (Non-Normalized)				**10,000 Datasets (normalized)**			
Data	*x*	*f(x)*	*v(x)*	Data	*x*	*f(x)*	*v(x)*
1	0.919	6.154	−3.156	1	0.184	−0.970	−0.932
2	−0.489	1.769	−3.760	2	−0.098	−0.984	−0.981
3	2.637	20.386	2.954	3	0.527	−0.925	−0.444
4	4.910	395.846	20.105	4	0.982	0.262	0.929
5	−3.872	208.509	10.993	5	−0.775	−0.330	0.200
6	1.019	5.961	−2.961	6	0.204	−0.971	−0.917
7	−2.319	13.300	1.379	7	−0.464	−0.948	−0.570
8	−1.062	−2.352	−2.872	8	−0.213	−0.997	−0.910
9	1.499	5.188	−1.753	9	0.300	−0.973	−0.820
10	0.752	6.392	−3.434	10	0.150	−0.969	−0.955
11	1.139	5.712	−2.703	11	0.228	−0.972	−0.896
12	3.070	42.554	5.426	12	0.614	−0.855	−0.246
13	0.987	6.025	−3.026	13	0.197	−0.971	−0.922
14	1.448	5.214	−1.905	14	0.289	−0.973	−0.832
15	2.331	11.778	1.433	15	0.466	−0.952	−0.565
16	−3.527	136.226	8.440	16	−0.706	−0.559	−0.005
17	−2.647	31.370	3.006	17	−0.530	−0.891	−0.439
18	−3.190	84.335	6.175	18	−0.638	−0.723	−0.186
...	...	...	...	...	...	...	...
10,000	−2.756	39.494	3.598	10,000	−0.551	−0.865	−0.392
Max	3.996	240.920	11.998	Max	1.000	1.000	1.000
Min	−4.000	−3.264	−4.000	Min	−1.000	−1.000	−1.000
Mean	0.026	34.324	1.285	Mean	0.007	−0.692	−0.339

$$\underbrace{f(x)}_{[1\times1]} = g_f^D\left(f_{\text{lin}}^{(L=3)}\left(\underbrace{\mathbf{W}_f^{(L=3)}}_{[1\times2]} f_t^{(2)}\left(\underbrace{\mathbf{W}_f^{(2)}}_{[2\times2]} f_t^{(1)}\left(\underbrace{\mathbf{W}_f^{(1)}}_{[2\times1]} \underbrace{g_x^N(x)}_{[1\times1]} + \underbrace{\mathbf{b}_f^{(1)}}_{[2\times1]} \right) + \underbrace{\mathbf{b}_f^{(2)}}_{[2\times1]} \right) + \underbrace{b_f^{(L=3)}}_{[1\times1]} \right)\right) \quad (2.4.2.12\text{-}1)$$

$$\underbrace{v(x)}_{[1\times1]} = g_v^D\left(f_{lin}^{(L=3)}\left(\underbrace{\mathbf{W}_v^{(L=3)}}_{[1\times2]} f_t^{(2)}\left(\underbrace{\mathbf{W}_v^{(2)}}_{[2\times2]} f_t^{(1)}\left(\underbrace{\mathbf{W}_v^{(1)}}_{[2\times1]} \underbrace{g_x^N(x)}_{[1\times1]} + \underbrace{\mathbf{b}_v^{(1)}}_{[2\times1]} \right) + \underbrace{\mathbf{b}_v^{(2)}}_{[2\times1]} \right) + \underbrace{b_v^{(L=3)}}_{[1\times1]} \right)\right) \quad (2.4.2.12\text{-}2)$$

2.4.2.2.2 Step 2: ANN trained on large datasets

An ANN with proper training parameters including a number of hidden layers, neurons, suggest epochs, validation checks, etc. should be adopted to obtain good training results and weight and bias matrices.

An ANN-based objective function $f(x) = x^4 - x^3 - 4x^2 + 5x + 5$ shown in Equation (2.4.2.11-1) is formulated with two hidden layers and two neurons using Equations (2.4.2.12-2), and (2.4.2.17) based on Figure 2.4.2.3a. ANN-based inequality function $v(x) = x^2 - 4$ shown in Equation (2.4.2.11-2) is formulated with two hidden layers and two neurons using Equations (2.4.2.12-2), and (2.4.2.17) based on Figure 2.4.2.3b.

Weight and bias matrices for objective $f(x) = x^4 - x^3 - 4x^2 + 5x + 5$ and inequality functions $v(x) = x^2 - 4$ obtained by training ANNs shown in Equations (2.4.2.12-1) and (2.4.2.12-2) based on Figure 2.4.2.3a and b, respectively, are essential parts in performing

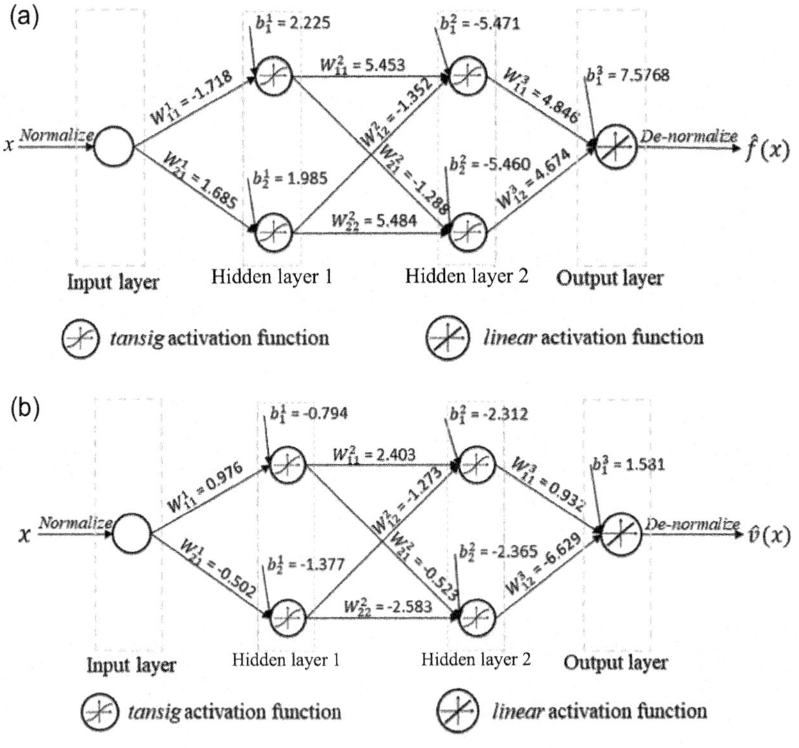

Figure 2.4.2.3 ANN network deriving objective and constraining functions based on two layers with two neurons. (a) ANN-based objective function $f(x) = x^4 - x^3 - 4x^2 + 5x + 5$ [Equation (2.4.2.12-1)]. (b) ANN-based constraint $v(x) = x^2 - 4$ [Equation (2.4.2.12-2)].

ANN-based Lagrange optimization. Readers are referred to Equations (2.3.6.3-1) and (2.3.6.3-2) for a derivation of ANN-based functions.

In this example, Equations (2.3.6.15) and (2.3.6.18) are used to derive ANN-based objective and inequality functions using weight and bias matrices shown in Figures 2.4.2.3(a) and 2.4.2.3(b). A good training accuracy of ANN is obtained in Figure 2.4.2.4 whereas Figure 2.4.2.5 compares ANN-based objective and constraining functions with analytical functions, demonstrating ANN-based functions and explicit analytical functions are close when they are obtained based on Equations (2.4.2.12-1), (2.4.2.12-2) for objective function and Equations (2.3.6.15), (2.3.6.18) for constraining function. It is noted that, in Figure 2.4.2.5, ANN-based objective and constraining functions demonstrate an error in the region outside trained ranges, i.e., region smaller than –4 and greater than +4. Weights and biases are calculated in Figure 2.4.2.3 when training is completed.

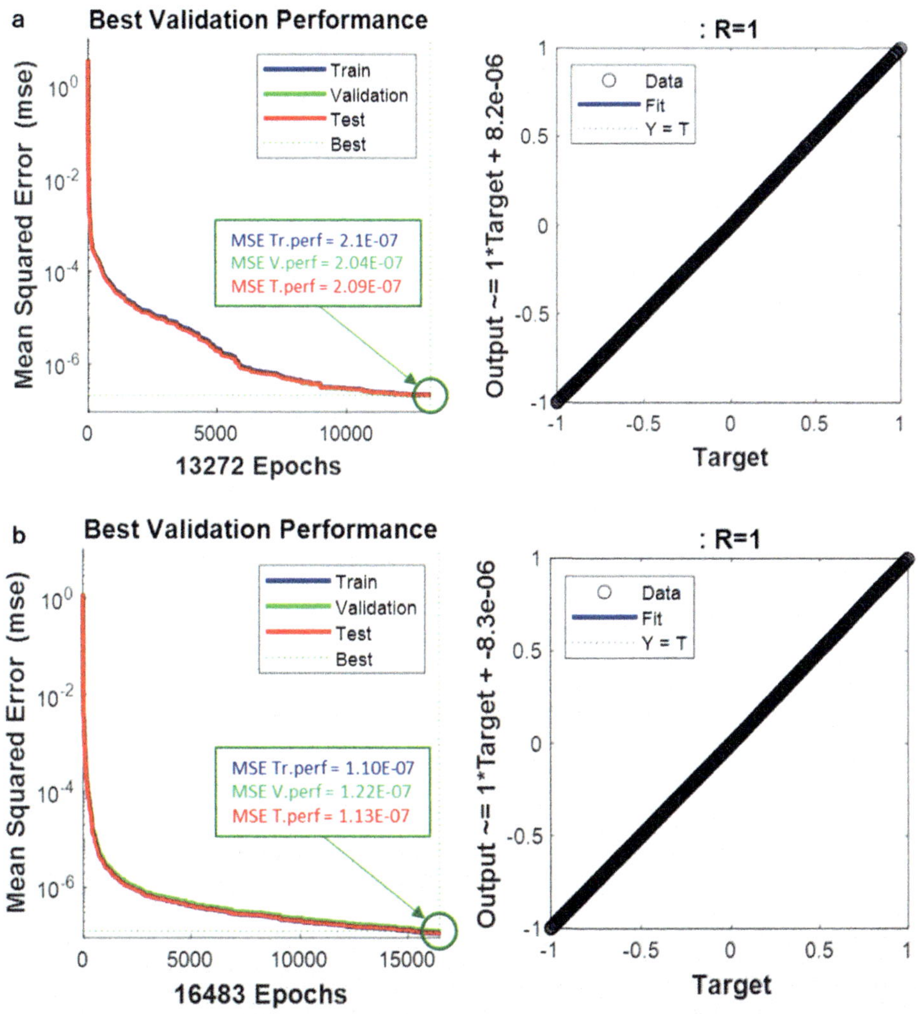

Figure 2.4.2.4 Training performance of an ANN. (a) Training performance of ANN for an objective function $f(x)$ [Equation (2.4.2.12-1)]. (b) Training performance of ANN for an inequality constraint $v(x)$ [Equation (2.4.2.12-2)].

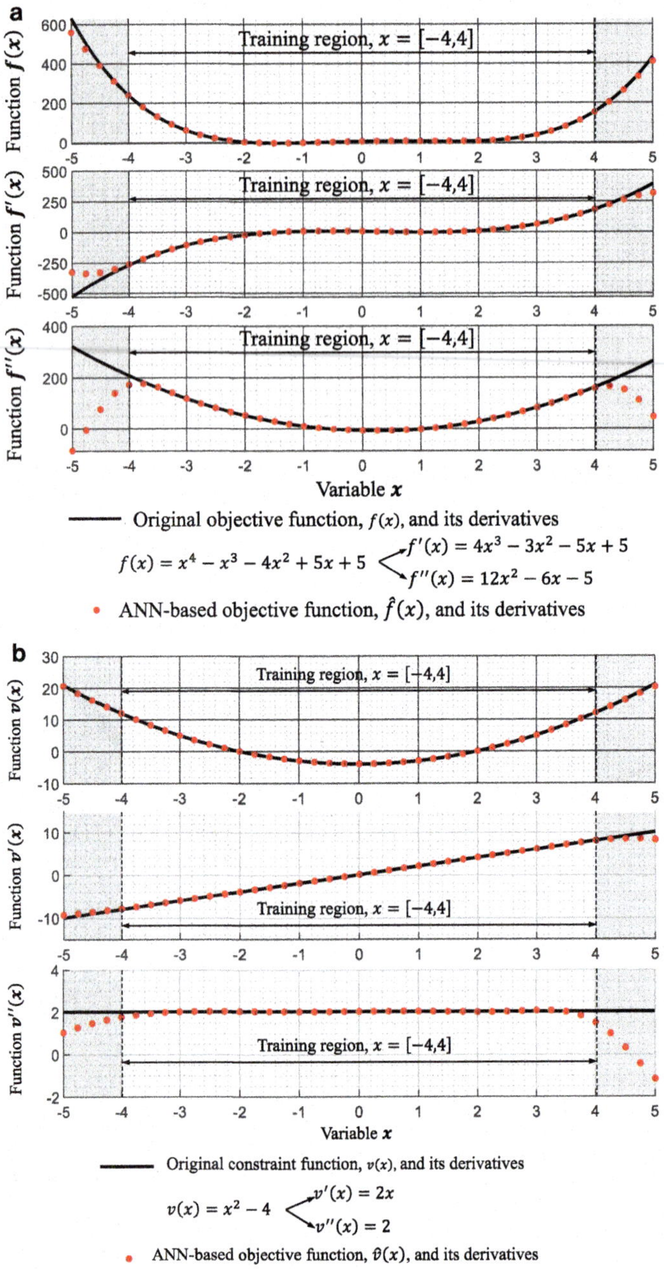

Figure 2.4.2.5 ANN-based objective and constraining functions compared with analytical functions. (a) Objective function $f(x)$ [Equation (2.4.2.12-1)]. (b) Constraining function $v(x)$ [Equation (2.4.2.12-2)].

2.4.2.2.3 Step 3: Formulation of ANN-based Lagrange function

An aim of obtaining ANN-based objective functions is to replace analytical functions while calculating Jacobian and Hessian matrices to solve KKT conditions using Newton–Raphson iteration. ANN-based Lagrange function is derived in Equation (2.4.2.13-1), similarly to conventional Lagrange function using an analytical objective function. Equation (2.4.2.13-2) is an analytical Jacobi of Lagrange function whereas Equation (2.4.2.13-3) is a Jacobi of ANN-based Lagrange function. ANN-based objective function derived in Equations (2.3.6.3-1) and (2.3.6.3-2), or in Equations (2.4.2.14-1), (2.4.2.14-2), and (2.4.2.15) becomes $\mathbf{z}_f^{(D)}$ at de-normalization layer shown in Equation (2.3.6.3-2f) or in Equation (2.4.2.15-5) which will replace an analytical objective function $f(x)$ shown in Equation (2.4.2.11-1). Similarly, the ANN-based constraining function also becomes $\mathbf{z}_v^{(D)}$ at de-normalization layer shown in Equation (2.4.2.18-5) which will replace an analytical constraining function $v(x)$ shown in Equation (2.4.2.11-2). As shown in Equation (2.4.2.13-3), Jacobi of objective function $\nabla f\left(x^{(k)}\right)$ becomes $\left[\mathbf{J}_f^{(D)}(x)\right]$ whereas Jacobi of inequality constraining function $\nabla v(x)$ also becomes $\mathbf{J}_v(x) = \left[\mathbf{J}_v^{(D)}(x)\right]$ at de-normalization layer. A Jacobi of Lagrange function show in Equation (2.4.2.13-3) is derived in terms of $\left[\mathbf{J}_f^{(D)}(x)\right]$ and $\left[\mathbf{J}_v^{(D)}(x)\right]$ which are generalized in Equations (2.3.6.6-1)–(2.3.6.6-9) based on ANN.

$$\mathcal{L}(x, \lambda_v) = f(x) - \lambda_v^T S v(x) \tag{2.4.2.13-1}$$

$$\nabla\mathcal{L}(x, \lambda_v) = \begin{bmatrix} \nabla_x \mathcal{L}(x, \lambda_v) \\ \nabla_{\lambda_v} \mathcal{L}(x, \lambda_v) \end{bmatrix} = \begin{bmatrix} \nabla f(x) - \mathbf{J}_v(x)^T S\lambda_v \\ -v(x) \end{bmatrix} \tag{2.4.2.13-2}$$

$$\nabla\mathcal{L}(x, \lambda_v) = \begin{bmatrix} \nabla_x \mathcal{L}(x, \lambda_v) \\ \nabla_{\lambda_v} \mathcal{L}(x, \lambda_v) \end{bmatrix} = \begin{bmatrix} \left[\mathbf{J}_f^{(D)}(x)\right]^T - \left[\mathbf{J}_v^{(D)}(x)\right]^T S\lambda_v \\ -Sv(x) \end{bmatrix} = 0 \tag{2.4.2.13-3}$$

Equation (2.3.6.3-1) is rewritten in Equation (2.4.2.14-1), in which the ANN-based objective function presented in Equation (2.4.2.14-1) and Figure 2.4.2.3a can be expressed as a series of composite mathematical operations with two hidden and output layers having two neurons as shown in Equation (2.4.2.14-2).

$$f(\mathbf{x}) = \mathbf{z}^{(D)} \circ \mathbf{z}^{(L)} \circ \mathbf{z}^{(L-1)} \circ \ldots \circ \mathbf{z}^{(1)} \circ \mathbf{z}^{(N)} \tag{2.4.2.14-1}$$

$$f(x) = \mathbf{z}_f^{(D)} \circ \mathbf{z}_f^{(L=3)} \circ \mathbf{z}_f^{(2)} \circ \mathbf{z}_f^{(1)} \circ \mathbf{z}_f^{(N)} \tag{2.4.2.14-2}$$

$\mathbf{z}^{(N)}$ shown in Equations (2.4.2.14-2) and (2.4.2.15-1) is normalized input parameter based on normalizing function $g^{(N)}(x)$ as shown in Equations (2.3.6.2-1) and (2.3.6.3-2a). $\mathbf{z}_f^{(l)}$ is a neural value of an objective function $f(x)$ at layer l which can be calculated using Equation (2.3.6.3-2c), whereas $\mathbf{z}^{(3)}$ shown in (2.4.2.15-4) is a normalized output value at an output layer before returning to the original scale by de-normalizing function $g^{(D)}(\bar{x})$ as shown in Equation (2.4.2.15-5). $\mathbf{W}^l$ is weight matrix between layer $l-1$ and layer l; $\mathbf{b}^l$ is bias matrix of layer l; and g^N, g^D are normalization and de-normalization function, respectively. An objective function is derived using Equations (2.3.6.3-2) and (2.4.2.15) when two hidden and output layers are implemented based on Figure 2.4.2.3a.

$$\begin{aligned}
&\mathbf{z}_f^{(N)} = \alpha_x \odot (x - x_{\min}) + \bar{x}_{\min} && (1)\\
&\mathbf{z}_f^{(1)} = f_t^{(1)}\left(\mathbf{W}_f^{(1)}\mathbf{z}_f^{(N)} + \mathbf{b}_f^{(1)}\right) && (2)\\
&\mathbf{z}_f^{(2)} = f_t^{(2)}\left(\mathbf{W}_f^{(2)}\mathbf{z}_f^{(1)} + \mathbf{b}_f^{(2)}\right) && (3)\\
&\mathbf{z}_f^{(L=3=\text{output})} = f_{\text{lin}}^{(L)}\left(\mathbf{W}_f^{(L)}\mathbf{z}_f^{(2)} + b_f^{(L)}\right) && (4)\\
&f(x) = \mathbf{z}_f^{(D)} = \frac{1}{\alpha_{f(x)}}\left(\mathbf{z}_f^{(L)} - \overline{f(x)}_{\min}\right) + f(x)_{\min} && (5)
\end{aligned} \qquad (2.4.2.15)$$

Ratios α_x used to normalize input datasets between $\bar{x}_{\max}$ and $\bar{x}_{\min}$ over the ranges of large datasets $(x_{\max} - x_{\min})$ are calculated by dividing $(\bar{x}_{\max} - \bar{x}_{\min})$ by the ranges of large datasets $(x_{\max} - x_{\min})$ as shown in Equation (2.4.2.16-1). Ratios $\alpha_{f(x)}$ shown in Equation (2.4.2.16-2) are calculated by dividing $(\overline{f(x)}_{\max} - \overline{f(x)}_{\min})$ by the ranges $(f(x)_{\max} - f(x)_{\min})$ of $f(x)$ to normalize neural outputs of an objective function between $\overline{f(x)}_{\max}$ and $\overline{f(x)}_{\min}$. Reverse of ratio $\alpha_{f(x)}$ shown in Equation (2.4.2.15-5) is also required to de-normalize normalized neural outputs of an objective function. As shown in Equation (2.4.2.15-5), $f(x)$ becomes $\mathbf{z}_f^{(D)}$ when $f(x)$ is de-normalized by using de-normalizing function $g^D(\bar{x})$, returning to the original scale. $\alpha_{f(x)}$, $\mathbf{z}_f^{(L)}$, $\overline{f(x)}_{\min}$, and $f(x)_{\min}$ are parameters to de-normalize $\mathbf{z}_f^{(L)}$. Ranges for normalizations used in Equations (2.4.2.16-1) and (2.4.2.16-2) are found in Table 2.4.2.4.

$$\alpha_x = \frac{\bar{x}_{\max} - \bar{x}_{\min}}{x_{\max} - x_{\min}} = \frac{1-(-1)}{3.996-(-4)} = 0.2501 \qquad (2.4.2.16\text{-}1)$$

$$\alpha_{f(x)} = \frac{\overline{f(x)}_{\max} - \overline{f(x)}_{\min}}{f(x)_{\max} - f(x)_{\min}} = \frac{1-(-1)}{240.92-(-3.264)} = 0.00819 \qquad (2.4.2.16\text{-}2)$$

Finally, the weight and bias matrices of an objective function, $\mathbf{W}_f^{(l)}$ and $\mathbf{b}_f^{(l)}$, can be obtained in Figure 2.4.2.3a and Table 2.4.2.5(a).

ANN-based inequality constraint $v(x)$ shown in (2.4.2.17) is expressed as a series of composite mathematical operations with two hidden and output layers, similarly to an objective function $f(x)$ shown in Equation (2.4.2.14-2). ANN-based inequality constraint $v(x)$ is also shown in Equation (2.4.2.12-2) and Figure 2.4.2.3b with two hidden and output layers having two neurons. $\mathbf{z}_v^l$ is ANN-based inequality constraint $v(x)$ at layer (l) calculated in Equation (2.4.2.18) when $L=l$.

$$v(x) = \mathbf{z}_v^{(D)} \circ \mathbf{z}_v^{(L=3)} \circ \mathbf{z}_v^{(2)} \circ \mathbf{z}_v^{(1)} \circ \mathbf{z}_v^{(N)} \qquad (2.4.2.17)$$

$$\begin{aligned}
&\mathbf{z}_v^{(N)} = \alpha_x \odot (x - x_{\min}) + \bar{x}_{\min} && (1)\\
&\mathbf{z}_v^{(1)} = f_t^{(1)}\left(\mathbf{W}_v^{(1)}\mathbf{z}_v^{(N)} + \mathbf{b}_v^{(1)}\right) && (2)\\
&\mathbf{z}_v^{(2)} = f_t^{(2)}\left(\mathbf{W}_v^{(2)}\mathbf{z}_v^{(1)} + \mathbf{b}_v^{(2)}\right) && (3)\\
&\mathbf{z}_v^{(L=3=\text{output})} = \mathbf{z}_v^{(3)} = f_{\text{lin}}^{(3)}\left(\mathbf{W}_v^{(3)}\mathbf{z}_v^{(2)} + b_v^{(3)}\right) && (4)\\
&v(x) = \mathbf{z}_v^{(D)} = \frac{1}{\alpha_{v(x)}}\left(\mathbf{z}_v^{(L=3)} - \overline{v(x)}_{\min}\right) + v(x)_{\min} && (5)
\end{aligned} \qquad (2.4.2.18)$$

Table 2.4.2.5 ANN –Based Weight and Bias Matrix

	(a): Objective Function $f(x)$	
	Weight Matrix	Bias Matrix
Hidden layer 1	$W_f^{(1)} = \begin{bmatrix} -1.718 \\ 1.685 \end{bmatrix}$	$b_f^{(1)} = \begin{bmatrix} 2.225 \\ 1.985 \end{bmatrix}$
Hidden layer 2	$W_f^{(2)} = \begin{bmatrix} 5.453 & -1.352 \\ -1.288 & 5.484 \end{bmatrix}$	$b_f^{(2)} = \begin{bmatrix} -5.471 \\ -5.460 \end{bmatrix}$
Output layer	$W_f^{(3=L)} = \begin{bmatrix} 4.846 & 4.674 \end{bmatrix}$	$b_f^{(L)} = 7.577$
	(b): Constraint $v(x)$	
	Weight Matrix	Bias Matrix
Hidden layer 1	$\mathbf{W}_v^{(1)} = \begin{bmatrix} 0.976 \\ -0.502 \end{bmatrix}$	$\mathbf{b}_v^{(1)} = \begin{bmatrix} -0.794 \\ -1.377 \end{bmatrix}$
Hidden layer 2	$W_v^{(2)} = \begin{bmatrix} 2.403 & -1.273 \\ -0.523 & -2.583 \end{bmatrix}$	$b_v^{(2)} = \begin{bmatrix} -2.312 \\ -2.365 \end{bmatrix}$
Output layer	$W_v^{(3=L)} = \begin{bmatrix} 0.932 & -6.629 \end{bmatrix}$	$b_v^{(L)} = 1.581$

Table 2.4.2.6 KKT Conditions of Active and Inactive Inequalities

Case	*S*	*Comment*
Case 1	0	Inequality $v(x)$ is inactive
Case 2	1	Inequality $v(x)$ is active

Weight and bias matrices of inequality constraint $v(x)$, $\mathbf{W}_v^{(l)}$ and $\mathbf{b}_v^{(l)}$, are obtained in Figure 2.4.2.3b and Table 2.4.2.5(b). $\alpha_{v(x)}$ is a ratio of $(\overline{v(x)}_{\max} - \overline{v(x)}_{\min})$ to the ranges of large datasets, $(v(x)_{\max} - v(x)_{\min})$, as shown in Equation (2.4.2.9).

$$\alpha_{v(x)} = \frac{\overline{v(x)}_{\max} - \overline{v(x)}_{\min}}{v(x)_{\max} - v(x)_{\min}} = \frac{1-(-1)}{11.998-(-4)} = 0.125 \tag{2.4.2.19}$$

As shown in Table 2.4.2.6, there are two cases of ANN-based KKT conditions; inactive Inequality $v(x)$ and active Inequality $v(x)$, similar to a conventional optimization based on analytical functions presented in Section 2.4.2.1.

2.4.2.2.4 *Step 4: Identification of stationary points based on KKT conditions using Jacobian and Hessian matrices*

A first-order approximation of $\nabla\mathcal{L}(\mathbf{x}, \lambda_c, \lambda_v)$ shown in Equation (2.4.2.13-2) is set to 0 to find stationary points of Lagrange functions. However, finding stationary points of $\nabla\mathcal{L}(\mathbf{x}, \lambda_c, \lambda_v)$ can be difficult, and thus, $\nabla\mathcal{L}(\mathbf{x}, \lambda_c, \lambda_v)$ should be linearized as a first-order approximation as shown in Equation (2.3.4.1-1) based on Jacobian and Hessian matrices.

2.4.2.2.4.1 CASE 1 FOR INACTIVE INEQUALITY $v(x)$

Initial trial input parameter

In this KKT condition, inequality $v(x)$ is assumed to be inactive which does not constrain an objective function $f(x)$ when reaching a minimum of $f(x)$. The Lagrange function shown in Equation (2.4.2.20-1) is simply equal to an objective function, ignoring inequality $v(x)$ when Lagrange multiplier λ_v is 0. Lagrange function is also obtained as a series of composite mathematical operations shown in Equation (2.4.2.21). Jacobi of Lagrange function is presented in Equation (2.4.2.20-2) based on analytical functions whereas Equation (2.4.2.20-3) is based on ANN-based functions. An unconstrained minimum should be verified to exist within an inequality constraint set, $v(x) \geq 0$ as shown in Equations (2.4.2.11-2) or (2.4.2.12-2).

$$\mathcal{L}(x,\lambda_v) = f(x) - \lambda_v v(x) \text{ where } \lambda v = 0 \tag{2.4.2.20-1}$$

$$\nabla\mathcal{L}(x,\lambda_v) = \begin{bmatrix} \nabla_x \mathcal{L}(x,\lambda_v) \\ \nabla_{\lambda_v}\mathcal{L}(x,\lambda_v) \end{bmatrix} = \begin{bmatrix} \nabla f(x) - J_v(x)^T \times 0 \\ -v(x) \times 0 \end{bmatrix} \tag{2.4.2.20-2}$$

$$\nabla\mathcal{L}(x,\lambda_v) = \begin{bmatrix} \left[\mathbf{J}_f^{(D)}(x)\right]^T - \left[\mathbf{J}_v^{(D)}(x)\right]^T \times 0 \\ -v(x) \times 0 \end{bmatrix} = 0 \tag{2.4.2.20-3}$$

$$\mathcal{L}(x) = f(x) = \mathbf{z}_f^{(D)} \circ \mathbf{z}_f^{(L=3=\text{output})} \circ \mathbf{z}_f^{(2)} \circ \mathbf{z}_f^{(1)} \circ \mathbf{z}_f^{(N)} \tag{2.4.2.21}$$

Three initial trial inputs $x^{(0)}$ of –3, 0, and 3 are assumed to evaluate all local stationary points. The initial trial inputs $x^{(0)}$ of –3, 0, and 3 are used to begin to obtain the three local stationary points corresponding to the three initial trial inputs for Case 1 as shown in Table 2.4.2.7. Initial trial inputs are to be updated until Equation (2.4.2.23-7) is satisfied. For initial trial input $x^{(0)}$ of –3, iteration begins by substituting initial trial input –3 into $\mathbf{z}_f^{(N)}$ shown in Equation (2.4.2.22-1) to obtain –0.75. Normalized input –0.75 is, then, continued to be substituted into $\mathbf{z}_f^{(1)}$ shown in Equation (2.4.2.22-2). Normalized $\mathbf{z}_f^{(L=3=\text{output})}$ of –0.465 at $x^{(0)}$=–3 is calculated as shown in Equations (2.4.2.22-1)–(2.4.2.22-4) using weight matrix $\mathbf{W}_f^{(1)}$, $\mathbf{W}_f^{(2)}$, and $\mathbf{W}_f^{(L=3=\text{output})}$ shown in Table 2.4.2.5(a). A de-normalized

Table 2.4.2.7 Newton–Raphson Iteration Based on Initial Trial Inputs for Inactive Case 1 KKT Condition

	Initial Guess $x_{(0)} = -3$			*Initial Guess* $x_{(0)} = 0$			*Initial Guess* $x_{(0)} = 3$		
Iteration	$x_{(k)}$	$\nabla f(x_k)$	$f(x_k)$	$x_{(k)}$	$\nabla f(x_k)$	$f(x_k)$	$x_{(k)}$	$\nabla f(x_k)$	$f(x_k)$
0	−3	−105.96	61.99	0	4.829	4.946	3	62.148	38.018
1	−2.104	−28.988	5.658	0.637	−0.103	6.418	2.238	16.784	9.981
2	−1.607	−6.339	−2.614	0.6215	9.5E-04	6.419	1.811	4.218	5.783
3	−1.417	−0.820	−3.269	0.6216	7.7E-08	6.419	1.602	0.815	5.285
4	−1.384	−0.023	−3.282	0.6216	2.4E-14	6.419	1.537	0.068	5.258
5	−1.384	−2.0E-05	−3.282	0.6216	−2.4E-14	6.419	1.531	6.7E-04	5.257

objective function $f\left(x^{(0)}\right) = \mathbf{z}_f^{(D)} = 61.99$ based on initial trial input parameter $x^{(0)} = -3$ is calculated in Equation (2.4.2.22-5) using Equations (2.3.6.3-2f) or (2.4.2.15-5). α_x and $\alpha_{f(x)}$ are calculated using Equations (2.4.2.16-1) and (2.4.2.16-2). Calculations similar to those for $f\left(x^{(0)}\right) = \mathbf{z}_f^{(D)}$ yield de-normalized inequality function $v\left(x^{(0)}\right) = \mathbf{z}_v^{(D)}$ at $x^{(0)} = -3$.

Jacobian matrices of an objective function are also obtained in Equations (2.4.2.23-1)–(2.4.2.23-5) based on Equations (2.3.6.6-1)–(2.3.6.6-9). A normalized Jacobian matrices of objective function $\mathbf{J}_f^{(N)}$ is calculated in Equation (2.4.2.23-1) whereas Jacobi $\mathbf{J}_f^{(D)}$ of ANN-based objective function $f(x)$ is obtained at a de-normalized layer D as obtained in Equation (2.4.2.23-5) based on Equation (2.3.6.6-9). A first derivative of Lagrange function $\nabla\mathcal{L}\left(x^{(0)}, \lambda_v\right) = \nabla f\left(x^{(0)}\right) = \left[\mathbf{J}_f^{(D)}\right]^T$ with initial trial input parameter, −3, calculated in Equations (2.4.2.23-6) did not converge to 0.

An inequality constraint $v(x)$ is not considered with $S = 0$ in a derivation of Lagrange functions, resulting in a Lagrange function $\mathcal{L}(x, \lambda_v)$ identical to the objective function $f(x)$ as shown in Equations (2.4.2.21) and (2.4.2.23-6) when an inactive KKT condition is assumed. First-order differential equation $\nabla\mathcal{L}(\mathbf{x}, \lambda_v)$ of a Lagrange function, then, becomes Jacobi $\nabla f(x)$ or gradient vector of an objective function for Case 1. However, first derivative of Lagrange function $\nabla\mathcal{L}\left(x^{(0)}, \lambda_v\right) = \nabla f\left(x^{(0)}\right)$ which is Jacobian of $f\left(x^{(0)}\right)$ shown in Equation (2.4.2.23-6) is calculated as −105.596 which did not converge to 0 when the first trial initial value of $x^{(0)} = -3$ is used. This indicates that a trial input of $x^{(0)} = -3$ does not lead to a root of Case 1 KKT solution shown in Table 2.4.2.6, failing to find stationary points minimizing an objective function as shown in Equation (2.4.2.23-6). And hence, KKT condition for Case 1 based on $x^{(0)} = -3$ is not verified yet, requiring trial inputs to be updated based on Newton–Raphson iteration until Equation (2.4.2.23-7) is reached.

$$\mathbf{z}_f^{(N)} = \alpha_x \odot \left(x^{(0)} - x_{\min}\right) + \bar{x}_{\min} = 0.2501(-3 - (-4.000)) + (-1) = -0.75 \qquad (1)$$

$$\mathbf{z}_f^{(1)} = f_t^{(1)}\left(\mathbf{W}_f^{(1)}\mathbf{z}_f^{(N)} + \mathbf{b}_f^{(1)}\right) = f_t^{(1)}\left(\begin{bmatrix} -1.718 \\ 1.685 \end{bmatrix}(-0.75) + \begin{bmatrix} 2.225 \\ 1.985 \end{bmatrix}\right) = \begin{bmatrix} 0.998 \\ 0.618 \end{bmatrix} \qquad (2)$$

$$\mathbf{z}_f^{(2)} = f_t^{(2)}\left(\mathbf{W}_f^{(2)}\mathbf{z}_f^{(1)} + \mathbf{b}_f^{(2)}\right) = f_t^{(2)}\left(\begin{bmatrix} 5.453 & -1.352 \\ -1.288 & 5.484 \end{bmatrix}\begin{bmatrix} 0.998 \\ 0.618 \end{bmatrix} + \begin{bmatrix} -5.471 \\ -5.460 \end{bmatrix}\right)$$

$$= \begin{bmatrix} -0.698 \\ -0.998 \end{bmatrix} \qquad (3)$$

$$\mathbf{z}_f^{(L=3=\text{output})} = f_{\text{lin}}^{(3)}\left(\mathbf{W}_f^{(3)}\mathbf{z}_f^{(2)} + b_f^{(3)}\right) = f_{\text{lin}}^{(3)}\left(\begin{bmatrix} 4.846 & 4.674 \end{bmatrix}\begin{bmatrix} -0.698 \\ -0.998 \end{bmatrix} + 7.577\right)$$

$$= -0.465 \qquad (4)$$

$$f\left(x^{(0)}\right) = \mathbf{z}_f^{(D)} = \frac{1}{\alpha_{f(x)}}\left(\mathbf{z}_f^{(L=3=\text{output})} - \overline{f(x)}_{\min}\right) + f(x)_{\min}$$

$$= \frac{1}{0.00819}(-0.465 - (-1)) + (-3.264) = 61.99 \qquad (5)$$

(2.4.2.22)

$$\mathbf{J}_f^{(N)} = \frac{\partial \mathbf{z}^{(N)}}{\partial \mathbf{x}} = \frac{\partial\left(\boldsymbol{\alpha}_{\mathbf{x}} \odot \left(\mathbf{x} - \mathbf{x}_{\min}\right) + \bar{\mathbf{x}}_{\min}\right)}{\partial \mathbf{x}} = I_n \odot \alpha_x = \alpha_x = 0.2501 \tag{2.4.2.23-1}$$

$$\mathbf{J}_f^{(1)} = \left(1 - \left(\mathbf{z}_f^{(1)}\right)^2\right) \odot \mathbf{W}_f^{(1)} \mathbf{J}_f^{(N)} = \left(1 - \begin{bmatrix} 0.998 \\ 0.618 \end{bmatrix} \odot \begin{bmatrix} 0.998 \\ 0.618 \end{bmatrix}\right) \odot \begin{bmatrix} -1.718 \\ 1.685 \end{bmatrix} 0.2501$$

$$= \begin{bmatrix} -0.0015 \\ 0.2607 \end{bmatrix} \tag{2.4.2.23-2}$$

$$\mathbf{J}_f^{(2)} = \left(1 - \left(\mathbf{z}_f^{(2)}\right)^2\right) \odot \mathbf{W}_f^{(2)} \mathbf{J}_f^{(1)}$$

$$= \left(- \begin{bmatrix} -0.698 \\ -0.998 \end{bmatrix} \odot \begin{bmatrix} -0.698 \\ -0.998 \end{bmatrix}\right) \odot \begin{bmatrix} 5.453 & -1.352 \\ -1.288 & 5.484 \end{bmatrix} \begin{bmatrix} -0.0015 \\ 0.2607 \end{bmatrix}$$

$$= \begin{bmatrix} -0.1852 \\ 0.0069 \end{bmatrix} \tag{2.4.2.23-3}$$

$$\mathbf{J}_f^{(L=3=)} = \mathbf{W}^{(L)} \mathbf{J}^{(2)} = \begin{bmatrix} 4.846 & 4.674 \end{bmatrix} \begin{bmatrix} -0.1852 \\ 0.0069 \end{bmatrix} = -0.8649 \tag{2.4.2.23-4}$$

$$\mathbf{J}_f^{(D)} = \frac{1}{\alpha_{f(x)}} \mathbf{J}_f^{(L)} \tag{2.4.2.23-5}$$

$$\nabla \mathcal{L}\left(x^{(0)}, \lambda_v\right) = \nabla f\left(x^{(0)}\right) = \left[\mathbf{J}_f^{(D)}\right]^T = \left[\frac{1}{\alpha_{f(x)}} \mathbf{J}_f^{(L)}\right]^T = \frac{-0.8649}{0.00819}$$

$$= -105.596 \tag{2.4.2.23-6}$$

$$\nabla f(x) = \left[\mathbf{J}_f^{(D)}\right]^T = 0 \tag{2.4.2.23-7}$$

Iteration to update initial trial input parameter

The first trial input of $x^{(0)} = -3$ is now updated to $x^{(1)}$. Hessian matrix of an objective function is used when the initial trial input is updated to $x^{(1)}$ as shown in Equation (2.3.4.1-5). Hessian matrix is calculated using Equations (2.4.2.24-1)–(2.4.2.24-5) based on Equations (2.3.6.10)–(2.3.6.16). A calculation of Hessian matrix begins by substituting $\mathbf{z}_f^{(1)}$ obtained in Equation (2.4.2.22-2) and $\mathbf{J}_f^{(N)}$ obtained in Equation (2.4.2.23-1) into Equation (2.4.2.24-2) to calculate $\mathbf{H}_f^{(1)}$. Purple weight matrix $\mathbf{W}_f^{(1)}$ of objective function was calculated in Table 2.4.2.5(a). And hence, Hessian matrix $\mathbf{H}_f^{(D)}$ of an objective function is finally obtained in Equations (2.4.2.24-1)–(2.4.2.24-5) based on Equations (2.3.6.16-2). Finally, an initial trial input parameter $x^{(0)}$ is updated to $x^{(1)}$ by substituting $\nabla f\left(x^{(0)}\right)$ calculated in Equation

(2.4.2.23-6) and $\mathbf{H}_f^{(D)}$ calculated in Equation (2.4.2.24-5) into Equation (2.3.4.1-5), yielding an updated input $x^{(1)}$ shown in Equations (2.4.2.25-1) and (2.4.2.25-2).

$$\mathbf{H}_f^{(N)} = 0 \quad (2.4.2.24\text{-}1,\ 2.3.6.10)$$

$$\mathbf{H}_f^{(1)} = -2\mathbf{z}_f^{(1)} \odot \left(1-\left(\mathbf{z}_f^{(1)}\right)^2\right) \odot \mathbf{J}_f^{(N)} \odot \left(\mathbf{W}_f^{(1)}\right)^2 \mathbf{J}_f^{(N)}$$

$$= -2\begin{bmatrix} 0.998 \\ 0.618 \end{bmatrix} \odot \left(1-\begin{bmatrix} 0.998 \\ 0.618 \end{bmatrix} \odot \begin{bmatrix} 0.998 \\ 0.618 \end{bmatrix}\right)$$

$$\odot\, 0.2501 \begin{bmatrix} -1.718 \\ 1.685 \end{bmatrix} \odot \begin{bmatrix} -1.718 \\ 1.685 \end{bmatrix} 0.2501 = \begin{bmatrix} -0.0013 \\ -0.1357 \end{bmatrix}$$

(2.4.2.24-2, 2.3.6.11-4)

$$\mathbf{H}_f^{(2)} = -2\mathbf{z}_f^{(2)} \odot \left(1-\left(\mathbf{z}_f^{(2)}\right)^2\right) \odot \mathbf{J}_f^{(1)} \odot \left(\mathbf{W}_f^{(2)}\right)^2 \mathbf{J}_f^{(1)} + \left(1-\left(\mathbf{z}_f^{(2)}\right)^2\right) \odot \mathbf{W}_f^{(2)} \mathbf{H}_f^{(1)} = \begin{bmatrix} -0.1837 \\ -0.0162 \end{bmatrix}$$

(2.4.2.24-3, 2.3.6.14)

$$\mathbf{H}_f^{(L)} = \mathbf{W}_f^{(L)} \mathbf{H}_f^{(2)} = \begin{bmatrix} 4.846 & 4.674 \end{bmatrix} \begin{bmatrix} -0.1837 \\ -0.0162 \end{bmatrix} = 0.9656 \quad (2.4.2.24\text{-}4,\ 2.3.6.15\text{-}2)$$

$$\mathbf{H}_f^{(D)} = \frac{\mathbf{H}_f^{(L)}}{\alpha_{f(x)}} = 117.89 \quad (2.4.2.24\text{–}5,\ 2.3.6.16\text{-}2)$$

$$x^{(1)} = x^{(0)} - \mathbf{H}_f\left(x^{(0)}\right)^{-1} \nabla f\left(x^{(0)}\right) \quad (2.4.2.25\text{-}1)$$

$$x^{(1)} = -3 - \frac{(-105.596)}{117.89} = -2.104 \quad (2.4.2.25\text{-}2)$$

Updated input parameter $x^{(1)}$ =-2.104 shown in Equation (2.4.2.25-2) is substituted into Equation (2.4.2.22-1) to repeat calculation for Iteration 1. The first derivative of Lagrange function $\nabla f\left(x^{(1)}\right)$ is reduced to –28.988 at Iteration 1 as shown in Table 2.4.2.7, which is closer to 0 than that of –105.596 at Iteration 0. The iteration converged at $x = -1.3827$, producing $f(x) = -3.2824$ at Iteration 5. The Newton-Raphson iteration is summarized in Figure 2.3.6.3.

In Figure 2.4.2.6, three corresponding initial trial inputs $x_{(0)}$=(–3, 0, 3) showing initial errors of $\nabla f(x_0)$ =–105.96, 4.829, 62.148, respectively, converged to three stationary points of (–1.384, –3.282), (0.6216, 6.419), and (1.531, 5.257) for $x_{(5)}$ and $f(x_5)$, respectively, at fifth iteration (Iteration 5). However, all initial trial inputs converged to $x_{(5)}$ = –1.384, 0.6216, and 1.531 violated an inequality constraint of $v(x) = \mathbf{z}_v^{(D)} \circ \mathbf{z}_v^{(L=3=\text{output})} \circ \mathbf{z}_v^{(2)} \circ \mathbf{z}_v^{(1)} \circ \mathbf{z}_v^{(N)} \geq 0$ shown in Equations(2.4.2.3) and (2.4.2.17). $x_{(5)}$ should converge at $x_{(5)} \geq 2$ and $x_{(5)} \leq -2$, however, $x_{(5)}$ converged to 1.384, 0.6216, and 1.5 which are between $2 < x_{(5)} < 2$ as shown in Figure 2.4.2.6, Equation (2.4.2.11–2) or (2.4.2.12-2). And hence, inequality constraint $v(x)$ are not satisfied for Case I KKT condition, thus, objective function was not minimized in a domain defined by $v(x)$ shown in Equation (2.4.2.18-5).

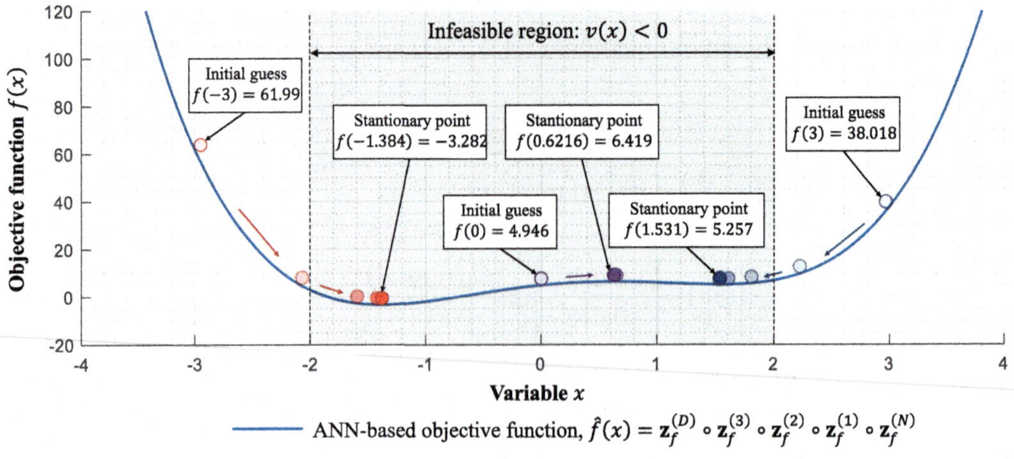

Figure 2.4.2.6 Inactive Case I for the first KKT condition, failing to meet an inequality constraint of $v(x) = z_v^{(D)} \circ z_v^{(3)} \circ z_v^{(2)} \circ z_v^{(l)} \circ z_v^{(N)} \geq 0$, converging at $2 \leq x_{(5)} \leq 2$.

2.4.2.2.4.2 CASE 2 FOR ACTIVE INEQUALITY $v(x)$

Diagonal scalar matrix S becomes 1 to activate inequality constraint in which inequality constraint are bound to equality constraint as shown in Equation (2.4.2.8-1) when deriving a Lagrange function. In this KKT condition, inequality $v(x)$ is assumed to constrain an objective function $f(x)$ while reaching a stationary point of $f(x)$. The Lagrange multiplier λ_v^T of Equation (2.4.2.26-1) should be greater than 0. ANN-based Lagrange function and necessary KKT condition for Case 2 are obtained in Equations (2.4.2.26-1). Jacobi of Lagrange function is presented in Equation (2.4.2.26-2) based on analytical functions whereas Equation (2.4.2.26-3) is based on ANN-based functions. $\mathbf{J}_f^{(D)}$ and $\mathbf{J}_v^{(D)}$ are de-normalized Jacobian matrices of ANN-based objective function $f(x)$ and inequality constrain $v(x)$ at layer D, respectively. Calculation procedures of $\mathbf{J}^{(D)}$ are discussed in Equations (2.3.6.6-1)–(2.3.6.6-9) in detail. ANN-based Lagrange function $\mathcal{L}\left(x^{(k)}, \lambda_v^{(k)}\right)$ presented in Figure 2.4.2.7b proves a good agreement with an analytical Lagrange function $\mathcal{L}\left(x^{(k)}, \lambda_v^{(k)}\right)$ shown in Figure 2.4.2.7a which is identical to Figure 2.4.2.2, where contours of magnitude of Lagrange function $\mathcal{L}\left(x^{(k)}, \lambda_v^{(k)}\right)$ are illustrated as functions of Lagrange multiplier λ_v and input parameter x. Equation (2.4.2.26-3) is solved based on initial trial inputs which is updated in Equation (2.4.2.27) until initial trial inputs converge to stationary points. A recursion shown in Equations (2.3.4.1-3)–(2.3.4.1-5) applies using Equation (2.4.2.28).

$$\mathcal{L}(x, \lambda_v) = f(x) - \lambda_v v(x) \tag{2.4.2.26-1}$$

$$\nabla\mathcal{L}(x, \lambda_v) = \begin{bmatrix} \nabla_x \mathcal{L}(x, \lambda_v) \\ \nabla_{\lambda_v} \mathcal{L}(x, \lambda_v) \end{bmatrix} = \begin{bmatrix} \nabla f(x) - [\mathbf{J}_v(x)]^T S\lambda_v \\ -Sv(x) \end{bmatrix} \tag{2.4.2.26-2}$$

$$\nabla\mathcal{L}(x, \lambda_v) = \begin{bmatrix} \left[\mathbf{J}_f^{(D)}(x)\right]^T - \left[\mathbf{J}_v^{(D)}(x)\right]^T S\lambda_v \\ -Sv(x) \end{bmatrix} = 0 \tag{2.4.2.26-3}$$

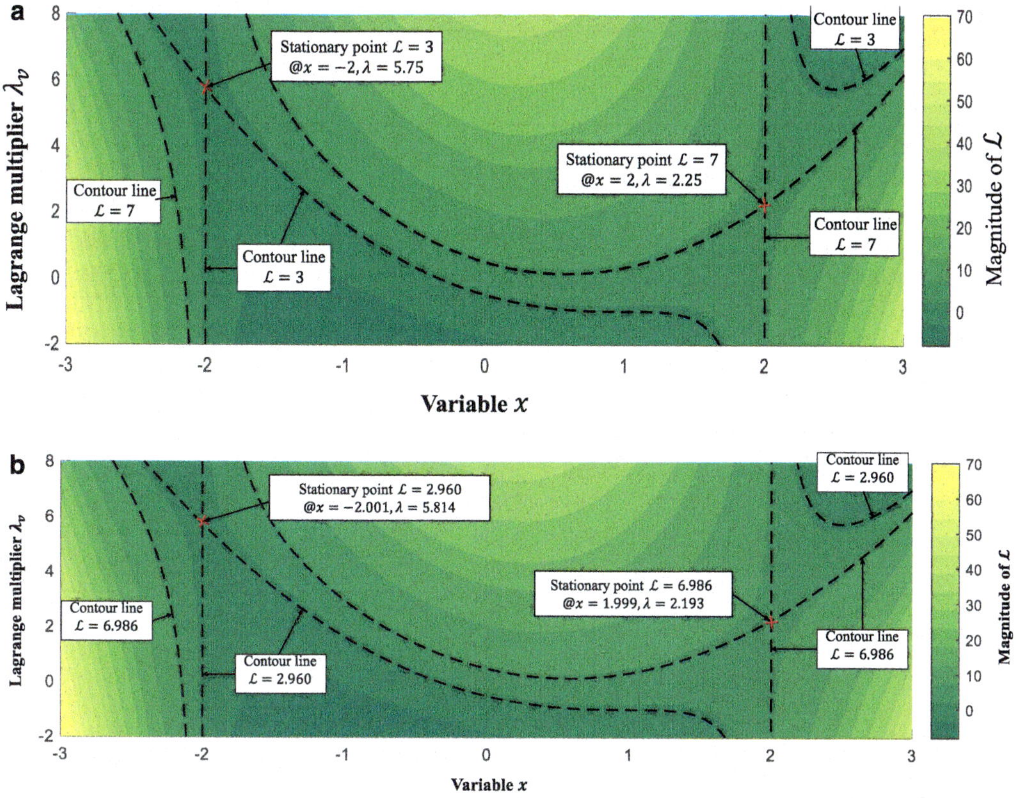

Figure 2.4.2.7 ANN-based Lagrange function compared well with an analytical function for active Case 2 KKT. (a) Analytical Lagrange function, $\mathcal{L}(x, \lambda_v)$. (b) ANN-based Lagrange function, $\hat{\mathcal{L}}(x, \lambda_v)$

Equations (2.3.4.3-1)–(2.3.4.3-8) are used to calculate Hessian matrix of Lagrange function $\mathbf{H}_{\mathcal{L}}\left(\mathbf{x}^{(k)}, \lambda_c^{(k)}, \lambda_v^{(k)}\right)$. ANN-based Hessian of Lagrange function shown in Equation (2.3.4.3-2b) becomes one shown in Equation (2.4.2.28-1) when equality constraint $\mathbf{c}(\mathbf{x})$ is not considered. $\mathbf{H}_{\mathcal{L}}(x)$ of Equation (2.3.4.3-3) which is substituted in Equation (2.4.2.28-1) also becomes one shown in Equation (2.4.2.28-2) when equality constraint $\mathbf{c}(\mathbf{x})$ is not considered. $\mathbf{H}_{\mathcal{L}}\left(x^{(k)}, \lambda_v^{(k)}\right)$ shown in Equation(2.4.2.28-1) is functions of $x^{(k)}$, $\lambda_v^{(k)}$ whereas $\mathbf{H}_f\left(x^{(k)}\right)$ and $\mathbf{H}_v\left(x^{(k)}\right)$ shown in Equation (2.4.2.28-2) are Hessian matrices of ANN-based objective function $f(x)$ and inequality constraint $v(x)$ with respect to $x^{(k)}$ only. It is noted that calculating Hessian matrices is complicated when multi-input parameters are considered, so that slices of Hessian matrices $\mathbf{H}_i^{(l)}$ are obtained using Equations (2.3.6.7-1)–(2.3.6.17), after which Newton–Raphson's iteration is implemented to find stationary points of a first derivative of Lagrange function based on Equation (2.4.2.27) using Equations (2.3.4.1-3)–(2.3.4.1-5). However, slice calculation is not necessary in this example because only one input parameter x is used.

$$\left[x^{(k+1)}, \lambda_v^{(k+1)}\right] = \begin{bmatrix} x^{(k)} \\ \lambda_v^{(k)} \end{bmatrix} - \mathbf{H}_{\mathcal{L}}\left(x^{(k)}, \lambda_v^{(k)}\right)^{-1} \nabla\mathcal{L}\left(x^{(k)}, \lambda_v^{(k)}\right) \tag{2.4.2.27}$$

Table 2.4.2.8 Newton–Raphson Iteration Based on Initial Trial Inputs for Active Case 2 KKT Condition

	Initial Guess $x_{(0)} = -3, \lambda_{v,(0)} = 1$					Initial Guess $x_{(0)} = 1.5, \lambda_{v,(0)} = 7$				
Iteration	$x_{(k)}$	$\lambda_{v,(k)}$	MSE	$\mathcal{L}$	$f(x_k)$	$x_{(k)}$	$\lambda_{v,(k)}$	MSE	$\mathcal{L}$	$f(x_k)$
0	−3	1	4971.0	56.99	61.99	1.5	7	229.36	17.508	5.262
1	−2.168	1.528	343.285	6.564	7.629	2.082	−0.99	118.36	8.145	7.810
2	−2.006	5.396	1.892	2.972	3.113	2.001	2.044	0.202	6.987	6.999
3	−2.001	5.813	2.1E-05	2.960	2.960	1.999	2.193	1.4E-07	6.986	6.986
4	−2.001	5.814	6.4E-16	2.960	2.960	1.999	2.193	1.69E-20	6.986	6.986

	Initial Guess $x_{(0)} = -1.5, \lambda_{v,(0)} = 8$					Initial Guess $x_{(0)} = 3, \lambda_{v,(0)} = 0$				
Iteration	$x_{(k)}$	$\lambda_{v,(k)}$	MSE	$\mathcal{L}$	$f(x_k)$	$x_{(k)}$	$\lambda_{v,(k)}$	MSE	$\mathcal{L}$	$f(x_k)$
0	−1.5	8	220.04	10.89	−3.11	3	0	1934.5	38.02	38.02
1	−2.085	3.454	90.73	3.918	5.10	2.167	−0.96	167.59	9.560	8.888
2	−2.003	5.695	0.151	2.961	3.000	2.006	1.856	1.1355	6.995	7.043
3	−2.001	5.814	1.2E-07	2.960	2.960	1.999	2.192	1.3E-05	6.986	6.986
4	−2.001	5.814	1.6E-20	2.960	2.960	1.999	2.193	3.7E-16	6.986	6.986

$$\mathbf{H}_{\mathcal{L}}\left(x^{(k)}, \lambda_v^{(k)}\right) = \begin{bmatrix} \mathbf{H}_{\mathcal{L}}\left(x^{(k)}\right) & -\left[\mathcal{S}\mathbf{J}_v^{(D)}\left(x^{(k)}\right)\right]^T \\ -\mathcal{S}\mathbf{J}_v^{(D)}\left(x^{(k)}\right) & 0 \end{bmatrix} = \begin{bmatrix} \mathbf{H}_f\left(x^{(k)}\right) - \lambda_v^{(k)}\mathbf{H}_v\left(x^{(k)}\right) & -\left[\mathcal{S}\mathbf{J}_v^{(D)}\left(x^{(k)}\right)\right]^T \\ -\mathcal{S}\mathbf{J}_v^{(D)}\left(x^{(k)}\right) & 0 \end{bmatrix} \tag{2.4.2.28-1}$$

$$\mathbf{H}_{\mathcal{L}}\left(x^{(k)}\right) = \mathbf{H}_f\left(x^{(k)}\right) - \lambda_v^{(k)}\mathbf{H}_v\left(x^{(k)}\right) \tag{2.4.2.28-2}$$

As shown in Table 2.4.2.8, four initial trial input parameters of $\left[x^{(0)}, \lambda_v^{(0)}\right]$= [−3, 1], [1.5, 7], [−1.5, 8], and [3, 0] are used to find all local stationary points of ANN-based Lagrange function shown in Equation (2.4.2.26-2) based on Newton–Raphson iteration shown in Equation (2.4.2.27). Two initial trial inputs of $\left[x^{(0)}, \lambda_v^{(0)}\right] = [-3, 1]$ and $[-1.5, 8]$ converge at $\left[x^{(4)}, \lambda_v^{(4)}\right] = [-2.001, 5.814]$ which is a local stationary point to minimize $\mathcal{L}\left(x^{(4)}, \lambda_v^{(4)}\right) = 2.960$ at Iteration 4. A stationary point $\left[x^{(4)}, \lambda_v^{(4)}\right] = [-2.001, 5.814]$ verifies that an objective function is minimized when $x^{(4)}$ = −2.001 is obtained close to activated inequality function $v(x) = x^4 - 4 \geq 0$ for Case 2 with active inequality $v(x)$. Another local stationary point of $\left[x^{(4)}, \lambda_v^{(4)}\right] = [1.999, 2.193]$ which minimizes $\mathcal{L}\left(x^{(4)}, \lambda_v^{(4)}\right) = 6.986$ is converged by two other initial trial inputs of $\left[x^{(0)}, \lambda_v^{(0)}\right] = [1.5, 7]$ and $[3, 0]$ at Iteration 4. It is also noted that $x^{(4)}$ = 1.999 is obtained close to activated inequality function $v(x)=x^{(4)}-4=0$ for Case 2 with active inequality $v(x)$. By comparing two local stationary points found in Table 2.4.2.8 and Figure 2.4.2.8, a global stationary point in Case 2 KKT condition is obtained at $\left[x^{(4)}, \lambda_v^{(4)}\right] = [-2.001, 5.814]$, minimizing $f\left(x^{(4)}\right) = \mathcal{L}\left(x^{(4)}, \lambda_v^{(4)}\right)$ to 2.960 which is less than 6.986. Figure 2.4.2.8 traces convergence of initial trial Inputs 1, 2, 3, and 4,

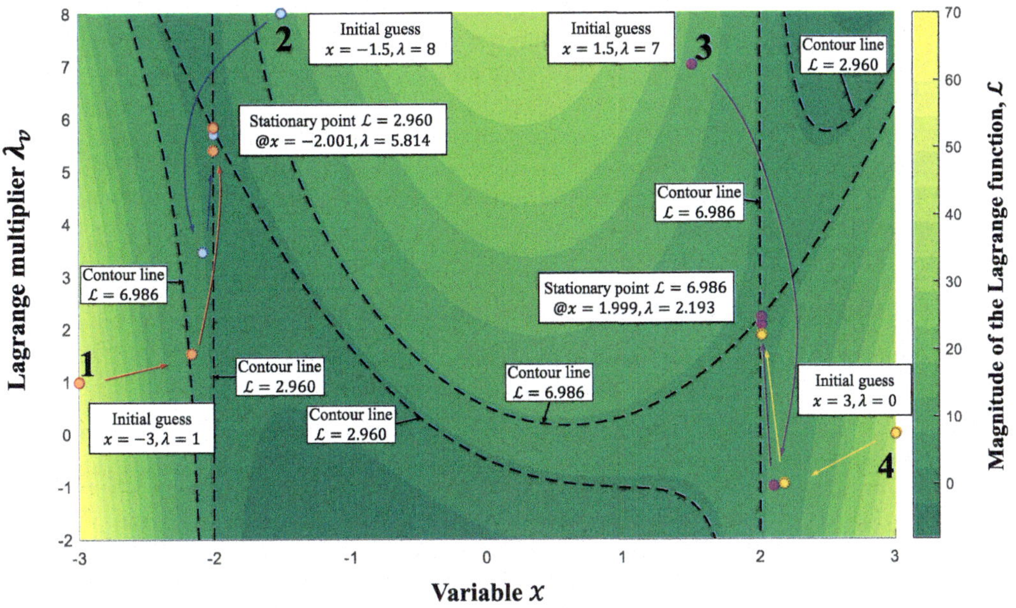

Figure 2.4.2.8 Convergence trace of initial trial inputs for active Case 2 KKT condition based on Newton–Raphson iteration.

Table 2.4.2.9 A Good Agreement between ANN-Based Lagrange Optimization and a Convention Lagrange Optimization Shown in Tables 2.4.2.8 and 2.4.2.3

	Conventional Lagrange	*ANN-Based Lagrange*	*Difference (%)*
x	−2	−2.001	0.05
$f(x)$	3	2.960	−1.33

implying that local stationary points of Lagrange function can be missed when a choice of initial trial inputs is not widely distributed. Figures 2.4.2.2(c) and (d) show convergence sequences indicated by arrows. It is noted that KKT 2 condition converges to [-2.001,5.814] and [1.999,2.193] which are transformed for active inequalities. Readers are recommended to try another initial trial inputs than ones used in this example to understand an importance of a selection of initial trial inputs. Figure 2.4.2.8 compares a minimized solution based on ANN-based functions well with one calculated by an analytical function shown in Figure 2.4.2.2. Table 2.4.2.9 presents a good agreement between ANN-based Lagrange optimization shown in Table 2.4.2.8 and a conventional Lagrange optimization shown in Table 2.4.2.3, demonstrating –1.33% difference only.

2.4.2.3 Conclusions

In some cases, a conventional Lagrange optimization based on an analytical objective and constraining functions poses significant difficulties to perform. The explicit formulation of analytical objectives and constraining functions can be difficult when they are complex, i.e., when a doubly RC beam is to be optimized for a cost with multiple input parameters such as $(L,h,b,f_y,f_c^{'},\rho_{rt},\rho_{rc},M_D,M_L)$. This chapter provides fundamental theories for both conventional Lagrange optimization and ANN-based Lagrange optimization before diving into reinforced concrete designs in Chapters 3, 4, and 5. Lagrange multiplier method using equality (λ_c) and inequality (λ_v) performed for conventional Lagrange optimization and ANN-based Lagrange optimization is compared in Table 2.4.2.9. A good agreement is observed, presenting ANN-based Lagrange optimization that can replace a conventional Lagrange optimization based on an explicit analytical objective and constraining functions. Further examples which minimize truss frame and maximize trajectory of projectile are provided in Sections 2.4.3 and 2.4.4.

2.4.3 A design of a truss frame based on Lagrange optimization

2.4.3.1 *Lagrange optimization of a truss frame based on analytical objective and constraining functions*

2.4.3.1.1 Formulation of Lagrange function

A truss frame under vertical force $P_1 = 100$ kN and horizontal force $P_2 = 55$ kN is shown in Figure 2.4.3.1. Given material is steel with a mass density of $\rho = 0.008$ g/mm^3 and yield strength of $f_y = 200$ MPa. This example will implement Lagrange multipliers to optimize truss height x_h and member area x_A, minimizing a total weight W shown in Equation (2.4.3.1) while constraining the stress in each member (σ_1, σ_2) to 200 MPa as shown in Equations (2.4.3.2-1) and (2.4.3.2-2).

$$\text{Minimize} \quad f_W(\mathbf{x}) \tag{2.4.3.1}$$

$$\text{with respect to} \quad \mathbf{x} = [x_A, x_h]$$

$$\text{subject to} \quad v_1(\mathbf{x}) = -|\sigma_1| + f_y \geq 0 \tag{2.4.3.2-1}$$

$$v_2(\mathbf{x}) = -|\sigma_2| + f_y \geq 0 \tag{2.4.3.2-2}$$

In the first step, an analytical objective function $f_W(\mathbf{x})$ representing a total weight of a truss is shown in Equations (2.4.3.3-1)–(2.4.3.3-3) and $|\sigma_1|$, $|\sigma_1|$ representing stress in each

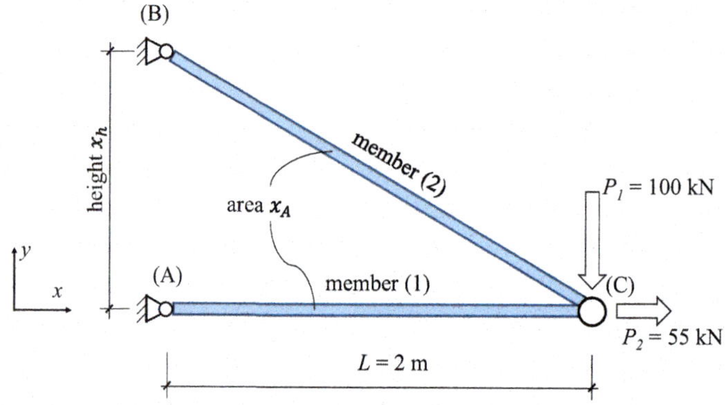

Figure 2.4.3.1 A truss frame designed by Lagrange optimization based on analytical functions.

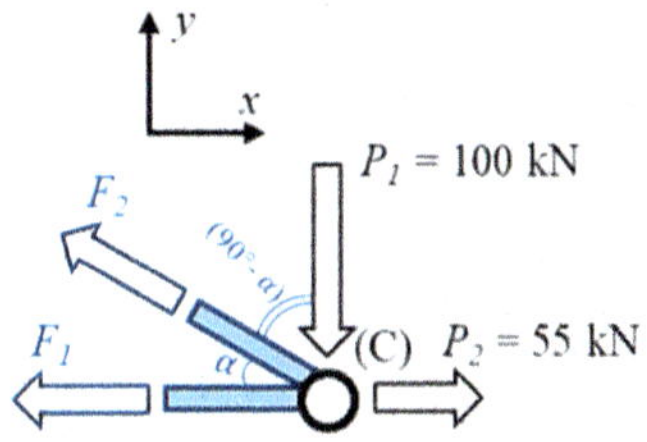

Figure 2.4.3.2 Equilibrium at Joint C.

member (σ_1, σ_2) shown in Equations (2.4.3.4-8) and (2.4.3.4-9) are formulated in order to derive a Lagrange function $\mathcal{L}(\mathbf{x}, \boldsymbol{\lambda}_v)$ as shown in Equation (2.4.3.5-1).

$$f_W(\mathbf{x}) = \rho(x_A L_1 + x_A L_2) \tag{2.4.3.3-1}$$

$$f_W(\mathbf{x}) = \rho x_A \left(L + \sqrt{x_h^2 + L^2}\right) \tag{2.4.3.3-2}$$

$$f_W(x_A, x_h) = 0.008 x_A \left(2 + \sqrt{x_h^2 + 4}\right) (\text{unit : kgf}) \tag{2.4.3.3-3}$$

where $L_1 = L$ and $L_2 = \sqrt{x_h^2 + L^2}$ are the length of members 1 and 2, respectively. Two unknown member forces F_1, F_2 acting at Joint C shown in Figure 2.4.3.2 are determined from the static equilibrium of forces at Joint C.

Equilibrium force at Joint C along the y-direction,

$$\Sigma F_y = F_2 \cos(90° - \alpha) - P_1 = 0 \rightarrow F_2 = \frac{P_1}{\sin\alpha} \tag{2.4.3.4-1}$$

Equilibrium force at Joint C along the x-direction,

$$\Sigma F_x = -F_1 - F_2 \cos\alpha + P_2 = 0 \rightarrow F_1 = -F_2 \cos\alpha + P_2 \tag{2.4.3.4-2}$$

Substituting $F_2 = P_1/\sin\alpha$ from Equation (2.4.3.4-1) into Equation (2.4.3.4-2), internal force F_1 is calculated as shown in Equation (2.4.3.4-3)

$$F_1 = -P_1 \cot\alpha + P_2 \tag{2.4.3.4-3}$$

$\sin\alpha$ and $\cot\alpha$ can be computed in Equations (2.4.3.4-4) and (2.4.3.4-5) based on a configuration of a truss shown in Figure 2.4.3.1.

$$\sin\alpha = \frac{x_h}{L_2} = \frac{x_h}{\sqrt{x_h^2 + L^2}} \tag{2.4.3.4-4}$$

$$\cot\alpha = \frac{L}{x_h} \tag{2.4.3.4-5}$$

Two unknown member forces F_1 and F_2 can be computed as a function of x_h as shown in Equations (2.4.3.4-6) and (2.4.3.4-7) by substituting Equations (2.4.3.4-4) and (2.4.3.4-5) into Equation (2.4.3.4-1) and (2.4.3.4-3) with $P_1 = 100$ kN and $P_2 = 55$ kN.

$$F_1 = \frac{-P_1 L}{x_h} + P_2 = \frac{-200}{x_h} + 55 \left(\text{unit : kN}\right) \tag{2.4.3.4-6}$$

$$F_2 = \frac{-P_1\sqrt{x_h^2 + L^2}}{x_h} = \frac{-100\sqrt{x_h^2 + L^2}}{x_h} \left(\text{unit : kN}\right) \tag{2.4.3.4-7}$$

Absolute stresses in each member (σ_1, σ_2) are then calculated accordingly in Equations (2.4.3.4-8) and (2.4.3.4-9).

$$|\sigma_1| = \frac{|F_1|}{x_A} = \left|\frac{-200}{x_h x_A} + \frac{55}{x_A}\right| \left(\text{unit : kN / mm}^2\right) \tag{2.4.3.4-8}$$

$$|\sigma_2| = \frac{|F_2|}{x_A} = \frac{100\sqrt{x_h^2 + 4}}{x_A |x_h|} \left(\text{unit : kN / mm}^2\right) \tag{2.4.3.4-9}$$

2.4.3.1.2 *Formulation of objective, inequality constraining functions and their Jacobi, Hessian matrices*

A conventional Lagrange multiplier method is implemented to optimize a total weight of truss shown in Figure 2.4.3.1 by minimizing an analytical objective function $f_W(\mathbf{x}) = f_W(x_A, x_h)$ derived in Equation (2.4.3.3-3) which is constrained by stresses in each member (σ_1, σ_2) derived in Equations (2.4.3.4-8) and (2.4.3.4-9). The Lagrange function for this problem becomes Equation (2.4.3.5-1) where λ_{v1} and λ_{v2} are Lagrange multipliers applied to inequality constraints $v_1(x) = v_1(x_A, x_h) = |\sigma_1| - f_y \geq 0$ and $v_2(x) = v_2(x_A, x_h) = |\sigma_2| - f_y \geq 0$ shown in Equations (2.4.3.5-2) and (2.4.3.5-3), respectively. **S** is a diagonal matrix which activates or inactivates an inequality constraint. First derivative (gradient vector or Jacobi) of Lagrange function is obtained in Equation (2.4.3.6-1). A Newton–Raphson iteration is employed to find roots (x_A, x_h, λ_{v1}, λ_{v2}) of a system of a first-order partial differential equations $\nabla\mathcal{L}(\mathbf{x}, \boldsymbol{\lambda}_v) = \nabla\mathcal{L}(x_A, x_h, \lambda_{v1}, \lambda_{v2}) = 0$ which is obtained from Equation (2.3.4.2-1a) when equality constraining function $\mathbf{c}(\mathbf{x})$ shown in Equation (2.3.4.2-2b) is not considered. A solution $[\mathbf{x}^{(k)}, \boldsymbol{\lambda}_v^{(k)}] = [x_A^{(k)}, x_h^{(k)}, \lambda_{v1}^{(k)}, \lambda_{v2}^{(k)}]$ of Equation (2.4.3.6-1) represents stationary points of Lagrange function, $\nabla\mathcal{L}(\mathbf{x}, \boldsymbol{\lambda}_v) = \nabla\mathcal{L}(x_A, x_h, \lambda_{v1}, \lambda_{v2}) = 0$. Newton–Raphson iteration is based on a first-order approximation, requiring initial trial inputs of $[\mathbf{x}^{(0)}, \boldsymbol{\lambda}_v^{(0)}] = [x_A^{(0)}, x_h^{(0)}, \lambda_{v1}^{(0)}, \lambda_{v2}^{(0)}]$ to be updated at every iteration step using Equation (2.4.3.7). Jacobi of Lagrange function $\nabla\mathcal{L}(x_A, x_h, \lambda_{v1}, \lambda_{v2})$ shown in Equation (2.4.3.6-1) based on Equation (2.3.4.2-1a) is calculated using Jacobi of an objective function $\nabla f_W(X) = \nabla f_W(x_A, x_h) = \mathbf{J}_{fW}^{(D)}$ shown in Equation (2.4.3.6-2) based on Equation (2.3.4.2-2a) and Jacobi of inequality function $J_v(x_A, x_h)$ shown in Equation (2.4.3.6-3) based on Equation (2.3.4.2-2c). $\nabla f_W(x_A, x_h)$ is calculated using Equation (2.4.3.6-2) based on Equation (2.3.4.2-2a). $J_v(x_A, x_h)$ are partial differential equations of inequality functions $v_1(x_A, x_h)$ and $v_2(x_A, x_h)$ obtained in Equation (2.4.3.6-3) based on Equation (2.3.4.2-2c).

$$\mathcal{L}(\mathbf{x},\boldsymbol{\lambda}_v)=\mathcal{L}(x_A,\, x_h,\, \lambda_{v1},\ \lambda_{v2})=f_W(\mathbf{x})-\boldsymbol{\lambda}_v^T\mathbf{S}\mathbf{v}(x)$$

$$=f_W(x_A,\, x_h)-\begin{bmatrix}\lambda_{v1} & \lambda_{v2}\end{bmatrix}\begin{bmatrix}s_1 & 0\\ 0 & s_2\end{bmatrix}\begin{bmatrix}-\sigma_1+f_y\\ -\sigma_2+f_y\end{bmatrix} \quad (2.4.3.5\text{-}1)$$

$$v_1(x_A,\, x_h)=|\sigma_1|-f_y\geq 0 \quad (2.4.3.5\text{-}2)$$

$$v_2(x_A,\, x_h)=|\sigma_2|-f_y\geq 0 \quad (2.4.3.5\text{-}3)$$

$$\nabla\mathcal{L}(x_A,x_h,\lambda_{v1},\ \lambda_{v2})=\begin{bmatrix}\nabla f_W(x_A,x_h)-\boldsymbol{J}_\mathrm{v}(x_A,x_h)^T\,\mathbf{S}\lambda_v\\ -\mathbf{S}\mathbf{v}(x_A,x_h)\end{bmatrix} \quad (2.4.3.6\text{-}1 \text{ and } 2.3.4.2\text{-}1\text{a})$$

$$\nabla f_W(x_A,\, x_h)=\begin{bmatrix}\dfrac{\partial f_W(x_A,\, x_h)}{\partial x_A}\\[2mm] \dfrac{\partial f_W(x_A,\, x_h)}{\partial x_h}\end{bmatrix}=\begin{bmatrix}0.016+0.008\sqrt{x_h^2+4}\\[2mm] \dfrac{0.008x_Ax_h}{\sqrt{x_h^2+4}}\end{bmatrix} \quad (2.4.3.6\text{-}2,\ 2.3.4.2\text{-}2\text{a})$$

$$\mathbf{J}_\mathrm{v}(x_A,\, x_h)=\begin{bmatrix}\dfrac{\partial v_1(x_A,\, x_h)}{\partial x_A} & \dfrac{\partial v_1(x_A,x_h)}{\partial x_h}\\[2mm] \dfrac{\partial v_2(x_A,x_h)}{\partial x_A} & \dfrac{\partial v_2(x_A,\, x_h)}{\partial x_h}\end{bmatrix}$$

$$=\begin{bmatrix}\dfrac{\sigma_1}{|\sigma_1|}\left(\dfrac{55}{x_A^2}-\dfrac{200}{x_A^2x_h}\right) & \left(-\dfrac{\sigma_1}{|\sigma_1|}\dfrac{200}{x_h^2x_A}\right)\\[3mm] \left(\dfrac{100\sqrt{x_h^2+4}}{x_A^2|x_h|}\right) & \left(\dfrac{100x_h\sqrt{x_h^2+4}}{x_A|x_h|^3}-\dfrac{100x_h}{x_A|x_h|\sqrt{x_h^2+4}}\right)\end{bmatrix} \quad (2.4.3.6\text{-}3,\ 2.3.4.2\text{-}2\text{c})$$

In summary, stationary points $[\,\mathbf{x}^{(k)},\ \boldsymbol{\lambda}_v^{(k)}\,]=[\,x_A^{(k)},\, x_h^{(k)},\, \lambda_{v1}^{(k)},\ \lambda_{v2}^{(k)}\,]$ optimizing Lagrange function $\mathcal{L}(x_A,\, x_h,\, \lambda_{v1},\ \lambda_{v2})$ presented in Equation (2.4.3.5-1) are obtained by iterating initial trial inputs $[\ \mathbf{x}^{(0)},\, \boldsymbol{\lambda}_v^{(0)}\ ]$ based on Equation(2.4.3.7) until Equation (2.4.3.6-1) converges, requiring a calculation of $\nabla\mathcal{L}\left(\mathbf{x}^{(k)},\, \boldsymbol{\lambda}_v^{(k)}\right)$ shown in Equation (2.4.3.6-1) and Hessian matrices $\left[\boldsymbol{H}_\mathcal{L}\left(\mathbf{x}^{(k)},\, \boldsymbol{\lambda}_v^{(k)}\right)\right]^{-1}$ shown in Equations (2.4.3.8-1)–(2.4.3.8-5).

$$\begin{bmatrix}\mathbf{x}^{(k+1)}\\ \boldsymbol{\lambda}_v^{(k+1)}\end{bmatrix}=\begin{bmatrix}\mathbf{x}^{(k)}\\ \boldsymbol{\lambda}_v^{(k)}\end{bmatrix}-\left[\boldsymbol{H}_\mathcal{L}\left(\mathbf{x}^{(k)},\, \boldsymbol{\lambda}_v^{(k)}\right)\right]^{-1}\nabla\mathcal{L}\left(\mathbf{x}^{(k)},\, \boldsymbol{\lambda}_v^{(k)}\right) \quad (2.4.3.7)$$

$$\mathbf{H}_\mathcal{L}(x_A,\, x_h,\, \lambda_{v1},\ \lambda_{v2})=\begin{bmatrix}\mathbf{H}_\mathcal{L}(x_A,\, x_h) & -\left(\mathbf{S}\boldsymbol{J}_\mathrm{v}(x_A,\, x_h)\right)^T\\ -\mathbf{S}\boldsymbol{J}_\mathrm{v}(x_A,\, x_h) & 0\end{bmatrix} \quad (2.4.3.8\text{-}1,\ 2.3.4.3\text{-}2\text{a})$$

$$H_{\mathcal{L}}(x_A, x_h) = H_{fW}(x_A, x_h) - \sum_{i=1}^{2} s_i \lambda_{v_i} H_{v_i}(x_A, x_h) \qquad (2.4.3.8\text{-}2,\ 2.3.4.3\text{-}3)$$

$$H_{fW}(x_A, x_h) = \begin{bmatrix} \dfrac{\partial^2 f_W(\mathbf{x})}{\partial x_A^2} & \dfrac{\partial^2 f_W(\mathbf{x})}{\partial x_A \partial x_h} \\ \dfrac{\partial^2 f_W(\mathbf{x})}{\partial x_h \partial x_A} & \dfrac{\partial^2 f_W(\mathbf{x})}{\partial x_h^2} \end{bmatrix} = \begin{bmatrix} 0 & \dfrac{0.008x_h}{\sqrt{x_h^2+4}} \\ \dfrac{0.008x_h}{\sqrt{x_h^2+4}} & \dfrac{0.032x_A}{\left(x_h^2+4\right)^{1.5}} \end{bmatrix} \qquad (2.4.3.8\text{-}3)$$

$$H_{v_1}(x_A, x_h) = \begin{bmatrix} \dfrac{\partial^2 v_1(\mathbf{x})}{\partial x_A^2} & \dfrac{\partial^2 v_1(\mathbf{x})}{\partial x_A \partial x_h} \\ \dfrac{\partial^2 v_1(\mathbf{x})}{\partial x_h \partial x_A} & \dfrac{\partial^2 v_1(\mathbf{x})}{\partial x_h^2} \end{bmatrix} = \begin{bmatrix} \dfrac{\sigma_1}{|\sigma_1|}\left(\dfrac{400}{x_A^3 x_h} - \dfrac{110}{x_A^3}\right) & \dfrac{\sigma_1}{|\sigma_1|}\left(\dfrac{200}{x_h^2 x_A^2}\right) \\ \dfrac{\sigma_1}{|\sigma_1|}\left(\dfrac{200}{x_h^2 x_A^2}\right) & \dfrac{\sigma_1}{|\sigma_1|}\left(\dfrac{400}{x_h^3 x_A}\right) \end{bmatrix}$$

(2.4.3.8-4)

$$H_{v_2}(x_A, x_h) = \begin{bmatrix} \dfrac{\partial^2 v_2(\mathbf{x})}{\partial x_A^2} & \dfrac{\partial^2 v_2(\mathbf{x})}{\partial x_A \partial x_h} \\ \dfrac{\partial^2 v_2(\mathbf{x})}{\partial x_h \partial x_A} & \dfrac{\partial^2 v_2(\mathbf{x})}{\partial x_h^2} \end{bmatrix}$$

$$= \begin{bmatrix} \left(-\dfrac{200\sqrt{x_h^2+4}}{x_A^3|x_h|}\right) & \left(\dfrac{100x_h}{x_A^2|x_h|\sqrt{x_h^2+4}} - \dfrac{100x_h\sqrt{x_h^2+4}}{x_A^2|x_h|^3}\right) \\ \left(\dfrac{100x_h}{x_A^2|x_h|\sqrt{x_h^2+4}} - \dfrac{100x_h\sqrt{x_h^2+4}}{x_A^2|x_h|^3}\right) & \left(\dfrac{100|x_h|}{x_A\left(x_h^2+4\right)^{1.5}} - \dfrac{200\sqrt{x_h^2+4}}{x_A|x_h|^3} + \dfrac{100x_h}{x_A|x_h|\sqrt{x_h^2+4}}\right) \end{bmatrix}$$

(2.4.3.8-5)

In Table 2.4.3.1, there are four cases of KKT conditions based on active and inactive inequalities.

Table 2.4.3.1 KKT Conditions of Active and Inactive Equalities

Case	S	*Comment*
Case 1	$\begin{bmatrix} 1 & 0 \\ 0 & 0 \end{bmatrix}$	Inequality $v_1(x)$ is active Inequality $v_2(x)$ is inactive
Case 2	$\begin{bmatrix} 0 & 0 \\ 0 & 1 \end{bmatrix}$	Inequality $v_1(x)$ is inactive Inequality $v_2(x)$ is active
Case 3	$\begin{bmatrix} 1 & 0 \\ 0 & 1 \end{bmatrix}$	Both inequalities are active
Case 4	$\begin{bmatrix} 0 & 0 \\ 0 & 0 \end{bmatrix}$	Both inequalities are inactive

2.4.3.1.3 Case I KKT condition based on active inequality $v_1(x)$ and inactive inequality $v_2(x)$

In this case, active inequality $v_1(\mathbf{x}) = -|\sigma_1| + f_y \geq 0$ is assumed to constrain an objective function $f_W(\mathbf{x})$ when reaching a minimum of $f_W(\mathbf{x})$ whereas inactive inequality $v_2(\mathbf{x})$ is ignored when calculating the Lagrange function $\mathcal{L}(\mathbf{x}, \boldsymbol{\lambda}_v)$ shown in Equation (2.4.3.9). However, a minimized objective condition of Case 1 should be verified to exist within an inequality constraint $v_2(\mathbf{x}) = -|\sigma_2| + f_y \geq 0$, although it is set as inactive status. The Lagrange function $\mathcal{L}(\mathbf{x}, \boldsymbol{\lambda}_v)$ becomes Equation (2.4.3.9) where only active inequality $v_1(x)$ is implemented.

2.4.3.1.3.1 CALCULATION OF JACOBI AND HESSIAN MATRIX FOR OBJECTIVE FUNCTION

Initial trial inputs of $\mathbf{x}^{(0)} = \left[x_A^{(0)},\ x_h^{(0)}\right]^T = [800, -7]^T$ and initial trial Lagrange multipliers of $\boldsymbol{\lambda}_v^{(0)} = \left[\lambda_{v1}^{(0)},\ \lambda_{v2}^{(0)}\right] = [0, 0]$ are used to find stationary points $[x_A, x_h, \lambda_{v1}, \lambda_{v2}]$, letting Equation (2.4.3.6-1) be 0 under KKT conditions using Newton–Raphson iteration. Full initial trial inputs become $\left[\mathbf{x}^{(0)}, \boldsymbol{\lambda}_v^{(0)}\right]^T = [800, -7, 0, 0]^T$. An objective function $f_W(\mathbf{x}) = f_W\left(x_A^{(0)},\ x_h^{(0)}\right)$, Jacobian $\nabla f_W\left(x_A^{(0)},\ x_h^{(0)}\right) = \mathbf{J}_{fW}^{(D)}$, and Hessian $\boldsymbol{H}_{fW}\left(x_A^{(0)},\ x_h^{(0)}\right)$ matrices of objective function with respect to initial trial inputs $\left[x_A^{(0)},\ x_h^{(0)}\right]^T$ are computed in Equations (2.4.3.10)–(2.4.3.12) using Equations (2.4.3.3-3), (2.4.3.6-2), and (2.4.3.8-3), respectively, where initial trial inputs are indicated in red.

$$\mathcal{L}(x_A, x_h, \lambda_{v1}, \lambda_{v2}) = f_W(\mathbf{x}) - \boldsymbol{\lambda}_v^T \boldsymbol{Sv}(x)$$

$$= f_W(x_A, x_h) - \begin{bmatrix} \lambda_{v1} & \lambda_{v2} \end{bmatrix} \begin{bmatrix} 1 & 0 \\ 0 & 0 \end{bmatrix} \begin{bmatrix} -\sigma_1 + f_y \\ -\sigma_2 + f_y \end{bmatrix} \tag{2.4.3.9}$$

$$f_W\left(x_A^{(0)},\ x_h^{(0)}\right) = 0.008 \times 800\left(2 + \sqrt{(-7)^2 + 4}\right) = 59.393\ (\text{unit}: \text{kgf}) \tag{2.4.3.10}$$

$$\nabla f_W\left(x_A^{(0)},\ x_h^{(0)}\right) = \begin{bmatrix} 0.016 + 0.008\sqrt{(-7)^2 + 4} \\ \dfrac{0.008 \times 800 \times (-7)}{\sqrt{(-7)^2 + 4}} \end{bmatrix} = \begin{bmatrix} 0.0742 \\ -6.154 \end{bmatrix} \tag{2.4.3.11}$$

$$\boldsymbol{H}_{fW}\left(x_A^{(0)},\ x_h^{(0)}\right) = \begin{bmatrix} 0 & \dfrac{0.008 \times (-7)}{\sqrt{(-7)^2 + 4}} \\ \dfrac{0.008 \times (-7)}{\sqrt{(-7)^2 + 4}} & \dfrac{0.032 \times 800}{\left((-7)^2 + 4\right)^{1.5}} \end{bmatrix} = \begin{bmatrix} 0 & -0.0077 \\ -0.0077 & 0.0663 \end{bmatrix} \tag{2.4.3.12}$$

Stresses in each truss member are calculated in Equations (2.4.3.13-1) and (2.4.3.13-2) using Equations (2.4.3.4-8) and (2.4.3.4-9).

$$\left|\sigma_1^{(0)}\right| = \left|\frac{-200}{(-7)\times 800} + \frac{55}{800}\right| = 0.1045\ \left(\text{unit}: \text{kN}/\text{mm}^2\right) \quad (2.4.3.13\text{-}1)$$

$$\left|\sigma_2^{(0)}\right| = \frac{100\sqrt{(-7)^2+4}}{800\times|-7|} = 0.1300\ \left(\text{unit}: \text{kN}/\text{mm}^2\right) \quad (2.4.3.13\text{-}2)$$

2.4.3.1.3.2 CALCULATION OF JACOBI AND HESSIAN MATRIX FOR INEQUALITY CONSTRAINING FUNCTION

Then, constraints $v_1(\mathbf{x})$, $v_2(\mathbf{x})$, and their Jacobian and Hessian matrices are computed in Equations (2.4.3.14-1)–(2.4.3.14-5) using Equations (2.4.3.5-2), (2.4.3.5-3), (2.4.3.6-3), (2.4.3.8-4), and (2.4.3.8-5), respectively.

$$v_1\left(x_A^{(0)},\ x_h^{(0)}\right) = -\left|\sigma_1^{(0)}\right| + f_y = 0.0955 \quad (2.4.3.14\text{-}1)$$

$$v_2\left(x_A^{(0)},\ x_h^{(0)}\right) = -\left|\sigma_2^{(0)}\right| + f_y = 0.0700 \quad (2.4.3.14\text{-}2)$$

$$\boldsymbol{J}_{\mathrm{v}}\left(x_A^{(0)},\ x_h^{(0)}\right) = \begin{bmatrix} 1.306\times 10^{-4} & -0.0051 \\ 1.625\times 10^{-4} & -0.0014 \end{bmatrix} \quad (2.4.3.14\text{-}3)$$

$$\boldsymbol{H}_{v_1}\left(x_A^{(0)},\ x_h^{(0)}\right) = \begin{bmatrix} -3.2645\times 10^{-7} & 6.3776\times 10^{-6} \\ 6.3776\times 10^{-6} & -0.0015 \end{bmatrix} \quad (2.4.3.14\text{-}4)$$

$$\boldsymbol{H}_{v_2}\left(x_A^{(0)},\ x_h^{(0)}\right) = \begin{bmatrix} -4.0626\times 10^{-7} & 1.752\times 10^{-6} \\ 1.752\times 10^{-6} & -5.856\times 10^{-4} \end{bmatrix} \quad (2.4.3.14\text{-}5)$$

2.4.3.1.3.3 CALCULATION OF JACOBI (FIRST DERIVATIVE, GRADIENT VECTOR) OF LAGRANGE FUNCTION BASED ON INITIAL TRIAL INPUT PARAMETER

Jacobi (first derivative, gradient vector) of Lagrange function based on initial trial inputs $\left[x_A^{(0)},\ x_h^{(0)}, \lambda_{v1}^{(0)},\ \lambda_{v2}^{(0)}\right]^T = [800,\ -7,\ 0,\ 0]^T$ is obtained in Equations (2.4.3.15-1) and (2.4.3.15-2) by substituting $\boldsymbol{S} = \begin{bmatrix} 1 & 0 \\ 0 & 0 \end{bmatrix}$ with $\nabla f_W\left(\mathbf{x}^{(0)}\right)$ shown in Equation (2.4.3.11), $\mathbf{v}\left(\mathbf{x}^{(0)}\right) = \left[v_1\left(x_A^{(0)},\ x_h^{(0)}\right),\ v_2\left(x_A^{(0)},\ x_h^{(0)}\right)\right]$ shown in Equations (2.4.3.14-1), and (2.4.3.14-2), $\boldsymbol{J}_{\mathrm{v}}\left(\mathbf{x}^{(0)}\right)$ shown in Equation (2.4.3.14-3) into Equation (2.4.3.6-1) or Equation (2.4.3.15-1).

$$\nabla\mathcal{L}\left(x_A^{(0)},\ x_h^{(0)}, \lambda_{v1}^{(0)},\ \lambda_{v2}^{(0)}\right) = \begin{bmatrix} \nabla f_W\left(\mathbf{x}^{(0)}\right) - \boldsymbol{J}_{\mathrm{v}}\left(\mathbf{x}^{(0)}\right)^T \mathbf{S}\lambda_v^{(0)} \\ -\boldsymbol{S}\mathbf{v}\left(\mathbf{x}^{(0)}\right) \end{bmatrix} \quad (2.4.3.15\text{-}1)$$

$$\nabla\mathcal{L}\left(x_A^{(0)},\ x_h^{(0)},\ \lambda_{v1}^{(0)},\ \lambda_{v2}^{(0)}\right)$$

$$=\begin{bmatrix} \begin{bmatrix} 0.0742 \\ -6.154 \end{bmatrix} - \begin{bmatrix} 1.306\times10^{-4} & -0.0051 \\ 1.625\times10^{-4} & -0.0014 \end{bmatrix}\begin{bmatrix} 1 & 0 \\ 0 & 0 \end{bmatrix}\begin{bmatrix} 0 \\ 0 \end{bmatrix} \\ -\begin{bmatrix} 1 & 0 \\ 0 & 0 \end{bmatrix}\begin{bmatrix} 0.0955 \\ 0.0700 \end{bmatrix} \end{bmatrix}$$

$$\rightarrow \nabla\mathcal{L}\left(x_A^{(0)},\ x_h^{(0)},\ \lambda_{v1}^{(0)},\ \lambda_{v2}^{(0)}\right)=\begin{bmatrix} 0.0742 \\ -6.154 \\ -0.0955 \\ 0 \end{bmatrix}\neq 0 \qquad (2.4.3.15\text{-}2)$$

As shown in Equations (2.4.3.14-1) and (2.4.3.14-2), inequality constraints $\left|\sigma_1^{(0)}\right| = v_1\left(x_A^{(0)},\ x_h^{(0)}\right)$ and $\left|\sigma_2^{(0)}\right| = v_2\left(x_A^{(0)},\ x_h^{(0)}\right)$ are obtained as 0.0955 $\frac{\text{kN}}{\text{mm}^2}$ and 0.0700 $\frac{\text{kN}}{\text{mm}^2}$ which is less than $f_y = 0.2\frac{\text{kN}}{\text{mm}^2}$, satisfying Case 1 KKT condition shown in Table 2.4.3.1. This indicates that a truss with the configuration of $x_A^{(0)} = 800$ and $x_h^{(0)} = -7$ can resist given load condition. However, as shown in Equation (2.4.3.15-2), this KKT condition based on $\left[x_A^{(0)},\ x_h^{(0)},\ \lambda_{v1}^{(0)},\ \lambda_{v2}^{(0)}\right]$ does not converge to yield a stationary point, failing to obtain a root of the KKT condition to calculate an optimal solution. And hence, initial trial inputs $\left[x_A^{(0)},\ x_h^{(0)},\ \lambda_{v1}^{(0)},\ \lambda_{v2}^{(0)}\right]$ should be updated in the next iteration shown in Equation (2.4.3.16-1) until Equation (2.4.3.15-2) converges to 0.

2.4.3.1.3.4 CALCULATION OF HESSIAN MATRIX OF LAGRANGE FUNCTION TO UPDATE INITIAL TRIAL INPUTS (ITERATION 0)

Initial trial inputs $\left[\mathbf{x}^{(0)},\ \boldsymbol{\lambda}_v^{(0)}\right] = \left[x_A^{(0)},\ x_h^{(0)},\ \lambda_{v1}^{(0)},\ \lambda_{v2}^{(0)}\right]$ is now updated to the next iteration (Iteration 0) $\left[x_A^{(1)},\ x_h^{(1)},\ \lambda_{v1}^{(1)},\ \lambda_{v2}^{(1)}\right]$ using Equation(2.4.3.16-1) based on the Hessian matrix of Lagrange function $\mathrm{H}_{\mathcal{L}}\left(\mathbf{x}^{(0)},\ \boldsymbol{\lambda}_v^{(0)}\right) = \mathrm{H}_{\mathcal{L}}\left(x_A^{(0)},\ x_h^{(0)},\ \lambda_{v1}^{(0)},\ \lambda_{v2}^{(0)}\right)$ shown in Equation (2.4.3.16-2). Hessian matrix of Lagrange function $\mathbf{H}_{\mathcal{L}}\left(\mathbf{x}^{(0)},\ \boldsymbol{\lambda}_v^{(0)}\right) = \mathbf{H}_{\mathcal{L}}\left(x_A^{(0)},\ x_h^{(0)},\ \lambda_{v1}^{(0)},\ \lambda_{v2}^{(0)}\right)$ with respect to initial trial inputs $\left[\mathbf{x}^{(0)},\ \boldsymbol{\lambda}_v^{(0)}\right] = \left[x_A^{(0)},\ x_h^{(0)},\ \lambda_{v1}^{(0)},\ \lambda_{v2}^{(0)}\right]$ is obtained in Equation (2.4.3.16-2).

$$\begin{bmatrix} \mathbf{x}^{(1)} \\ \boldsymbol{\lambda}_v^{(1)} \end{bmatrix} = \begin{bmatrix} \mathbf{x}^{(0)} \\ \boldsymbol{\lambda}_v^{(0)} \end{bmatrix} - \left[\boldsymbol{H}_{\mathcal{L}}\left(\mathbf{x}^{(0)},\ \boldsymbol{\lambda}_v^{(0)}\right)\right]^{-1}\nabla\mathcal{L}\left(\mathbf{x}^{(0)},\ \boldsymbol{\lambda}_v^{(0)}\right) \qquad (2.4.3.16\text{-}1)$$

alternatively,

$$\begin{bmatrix} x_A^{(1)},\ x_h^{(1)} \\ \lambda_{v1}^{(1)},\ \lambda_{v2}^{(1)} \end{bmatrix} = \begin{bmatrix} x_A^{(0)},\ x_h^{(0)} \\ \lambda_{v1}^{(0)},\ \lambda_{v2}^{(0)} \end{bmatrix} - \left[\boldsymbol{H}_{\mathcal{L}}\left(x_A^{(0)},\ x_h^{(0)},\ \lambda_{v1}^{(0)},\ \lambda_{v2}^{(0)}\right)\right]^{-1}\nabla\mathcal{L}\left(x_A^{(0)},\ x_h^{(0)},\ \lambda_{v1}^{(0)},\ \lambda_{v2}^{(0)}\right)$$

alternatively,

$$\begin{bmatrix} x_A^{(1)} \\ x_h^{(1)} \\ \lambda_{v1}^{(1)} \\ \lambda_{v2}^{(1)} \end{bmatrix} = \begin{bmatrix} x_A^{(0)} \\ x_h^{(0)} \\ \lambda_{v1}^{(0)} \\ \lambda_{v2}^{(0)} \end{bmatrix} - \left[\boldsymbol{H}_{\mathcal{L}}\left(x_A^{(0)},\ x_h^{(0)},\ \lambda_{v1}^{(0)},\ \lambda_{v2}^{(0)}\right)\right]^{-1} \nabla\mathcal{L}\left(x_A^{(0)},\ x_h^{(0)},\ \lambda_{v1}^{(0)},\ \lambda_{v2}^{(0)}\right)$$

where,

$$\mathbf{H}_{\mathcal{L}}\left(\mathbf{x}^{(0)},\ \boldsymbol{\lambda}_v^{(0)}\right) = \mathbf{H}_{\mathcal{L}}\left(x_A^{(0)},\ x_h^{(0)},\ \lambda_{v1}^{(0)},\ \lambda_{v2}^{(0)}\right),$$

$$\nabla\mathcal{L}\left(\mathbf{x}^{(0)},\ \boldsymbol{\lambda}_v^{(0)}\right) = \nabla\mathcal{L}\left(x_A^{(0)},\ x_h^{(0)},\ \lambda_{v1}^{(0)},\ \lambda_{v2}^{(0)}\right)$$

$$\mathbf{H}_{\mathcal{L}}\left(x_A^{(0)},\ x_h^{(0)},\ \lambda_{v1}^{(0)},\ \lambda_{v2}^{(0)}\right) = \begin{bmatrix} \mathbf{H}_{\mathcal{L}}\left(\mathbf{x}^{(0)}\right) & -\left(\boldsymbol{SJ}_{\mathrm{v}}\left(\mathbf{x}^{(0)}\right)\right)^T \\ -\boldsymbol{SJ}_{\mathrm{v}}\left(\mathbf{x}^{(0)}\right) & 0 \end{bmatrix} \qquad (2.4.3.16\text{-}2)$$

$\mathbf{H}_{\mathcal{L}}\left(\mathbf{x}^{(0)}\right) = \boldsymbol{H}_{\mathcal{L}}\left(x_A^{(0)},\ x_h^{(0)}\right)$ are to be calculated in advance by substituting $\boldsymbol{H}_{fw}\left(x_A^{(0)},\ x_h^{(0)}\right)$, $\boldsymbol{H}_{v_1}\left(x_A^{(0)},\ x_h^{(0)}\right)$, and $\boldsymbol{H}_{v_2}\left(x_A^{(0)},\ x_h^{(0)}\right)$ of Equations (2.4.3.12), (2.4.3.14-4), and (2.4.3.14-5), respectively, into Equation (2.4.3.16-3) to obtain Equation (2.4.3.16-4). Readers can follow the arrows to understand the calculation OF Hessian matrix $\mathbf{H}_{\mathcal{L}}\left(\mathbf{x}^{(0)}\right) = \boldsymbol{H}_{\mathcal{L}}\left(x_A^{(0)},\ x_h^{(0)}\right)$.

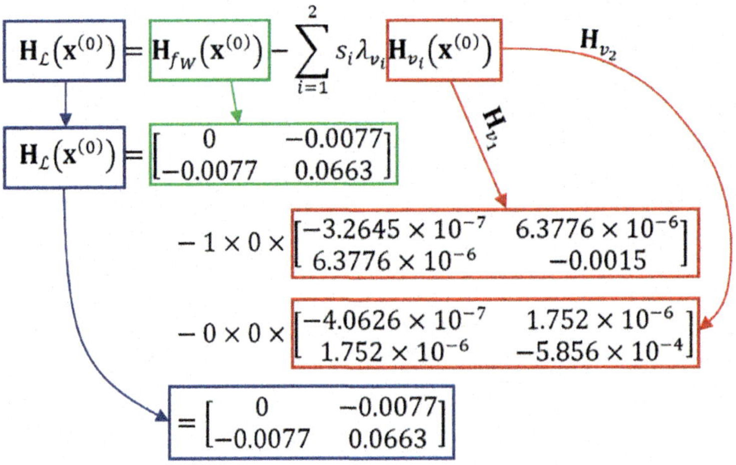

Then, global Hessian matrix $\mathbf{H}_{\mathcal{L}}\left(\mathbf{x}^{(0)},\ \boldsymbol{\lambda}_v^{(0)}\right) = \mathbf{H}_{\mathcal{L}}\left(x_A^{(0)},\ x_h^{(0)},\ \lambda_{v1}^{(0)},\ \lambda_{v2}^{(0)}\right)$ is obtained from Equations (2.4.3.16-5) and (2.4.3.16-6) by substituting $\mathbf{S} = \begin{bmatrix} 1 & 0 \\ 0 & 0 \end{bmatrix}$, $\boldsymbol{H}_{\mathcal{L}}\left(x_A^{(0)},\ x_h^{(0)}\right)$ of Equation (2.4.3.16-4) and $\boldsymbol{J}_{\mathrm{v}}\left(\mathbf{x}\right)$ of Equation (2.4.3.14-3) into Equation (2.4.3.16-2).

$$\mathbf{H}_{\mathcal{L}}\left(x_A^{(0)},\ x_h^{(0)},\ \lambda_{v1}^{(0)},\ \lambda_{v2}^{(0)}\right)$$

$$=\begin{bmatrix} \begin{bmatrix} 0 & -0.0077 \\ -0.0077 & 0.0663 \end{bmatrix} & -\left(\begin{bmatrix} 1 & 0 \\ 0 & 0 \end{bmatrix}\begin{bmatrix} 1.306\times10^{-4} & -0.0051 \\ 1.625\times10^{-4} & -0.0014 \end{bmatrix}\right)^T \\ -\begin{bmatrix} 1 & 0 \\ 0 & 0 \end{bmatrix}\begin{bmatrix} 1.306\times10^{-4} & -0.0051 \\ 1.625\times10^{-4} & -0.0014 \end{bmatrix} & 0 \end{bmatrix} \tag{2.4.3.16-5}$$

$$\mathbf{H}_{\mathcal{L}}\left(x_A^{(0)},\ x_h^{(0)},\ \lambda_{v1}^{(0)},\ \lambda_{v2}^{(0)}\right)=\begin{bmatrix} 0 & -0.0077 & 1.036\times10^{-4} & 0 \\ -0.0077 & 0.0663 & -0.0051 & 0 \\ 1.036\times10^{-4} & -0.0051 & 0 & 0 \\ 0 & 0 & 0 & 0 \end{bmatrix} \tag{2.4.3.16-6}$$

In this case, $\mathbf{H}_{\mathcal{L}}\left(x_A,\ x_h,\ \lambda_{v1},\ \lambda_{v2}\right)$ of Equation (2.4.3.16-6) is singular because the fourth row/column have zero values, indicating the fourth input (λ_{v2}) does not affect Lagrange function $\nabla\mathcal{L}\left(x_A,\ x_h,\ \lambda_{v1},\ \lambda_{v2}\right)$, because $v_2\left(x_A,\ x_h\right)=|\sigma_2|-f_y\geq0$ was assumed as an inactive KKT condition and ignored. This statement allows the fourth row/column to be excluded from the Hessian matrix when inverting $\mathbf{H}_{\mathcal{L}}\left(x_A^{(0)},\ x_h^{(0)},\ \lambda_{v1}^{(0)},\ \lambda_{v2}^{(0)}\right)$ because Inequality $v_2\left(\mathbf{x}\right)$ is inactive in Case 1 KKT condition. Thus, Hessian matrix $\mathbf{H}_{\mathcal{L}}\left(x_A^{(0)},\ x_h^{(0)},\ \lambda_{v1}^{(0)},\ \lambda_{v2}^{(0)}\right)$ with a (3×3) dimension is inverted in Equations (2.4.3.16-7) and (2.4.3.16-8).

$$\mathbf{H}_{\mathcal{L}}\left(x_A^{(0)},\ x_h^{(0)},\ \lambda_{v1}^{(0)},\ \lambda_{v2}^{(0)}\right)^{-1}$$

$$=\begin{bmatrix} \begin{bmatrix} 0 & -0.0077 & 1.036\times10^{-4} \\ -0.0077 & 0.0663 & -0.0051 \\ 1.036\times10^{-4} & -0.0051 & 0 \end{bmatrix}^{-1} & \begin{matrix} 0 \\ 0 \\ 0 \end{matrix} \\ \begin{matrix} 0 & 0 & 0 \end{matrix} & 0 \end{bmatrix} \tag{2.4.3.16-7}$$

$$\mathbf{H}_{\mathcal{L}}\left(x_A^{(0)},\ x_h^{(0)},\ \lambda_{v1}^{(0)},\ \lambda_{v2}^{(0)}\right)^{-1}=\begin{bmatrix} -2,854.8 & -73.066 & -3,354.0 & 0 \\ -73.066 & -1.870 & 110.159 & 0 \\ -3,354.0 & 110.159 & -6,489.2 & 0 \\ 0 & 0 & 0 & 0 \end{bmatrix} \tag{2.4.3.16-8}$$

2.4.3.1.3.5 CALCULATION OF JACOBI (FIRST DERIVATIVE, GRADIENT VECTOR) OF LAGRANGE FUNCTION AT ITERATION 1

Substituting $\nabla\mathcal{L}\left(x_A^{(0)},\ x_h^{(0)},\ \lambda_{v1}^{(0)},\ \lambda_{v2}^{(0)}\right)$ shown in Equation (2.4.3.15-2) and $\mathbf{H}_{\mathcal{L}}\left(\mathbf{x}^{(0)},\ \lambda_v^{(0)}\right)^{-1}=\mathbf{H}_{\mathcal{L}}\left(x_A^{(0)},\ x_h^{(0)},\ \lambda_{v1}^{(0)},\ \lambda_{v2}^{(0)}\right)^{-1}$ shown in Equation (2.4.3.16-8) into Equation (2.4.3.16-1) yields updated trial inputs $x_A^{(0)},\ x_h^{(0)},\ \lambda_{v1}^{(0)},\ \lambda_{v2}^{(0)}$ of the first iteration (Iteration 1) as shown in Equation (2.4.3.17).

$$\begin{bmatrix} x_A^{(1)} \\ x_h^{(1)} \\ \lambda_{v1}^{(1)} \\ \lambda_{v2}^{(1)} \end{bmatrix} = \begin{bmatrix} 800 \\ -7 \\ 0 \\ 0 \end{bmatrix} - \begin{bmatrix} -2,854.8 & -73.066 & -3,354.0 & 0 \\ -73.066 & -1.870 & 110.159 & 0 \\ -3354.0 & 110.159 & -6,489.2 & 0 \\ 0 & 0 & 0 & 0 \end{bmatrix} \begin{bmatrix} 0.0742 \\ -6.154 \\ -0.0955 \\ 0 \end{bmatrix}$$

$$\rightarrow \begin{bmatrix} x_A^{(1)} \\ x_h^{(1)} \\ \lambda_{v1}^{(1)} \\ \lambda_{v2}^{(1)} \end{bmatrix} = \begin{bmatrix} 241.891 \\ -2.559 \\ 306.942 \\ 0 \end{bmatrix} \tag{2.4.3.17}$$

Updated Jacobi (first derivative, gradient vector) of Lagrange function $\nabla\mathcal{L}\left(x_A^{(1)}, x_h^{(1)}, \lambda_{v1}^{(1)}, \lambda_{v2}^{(1)}\right)$ with respect to $\left[x_A^{(1)}, x_h^{(1)}, \lambda_{v1}^{(1)}, \lambda_{v2}^{(1)}\right] = [241.891, -2.559,\ 306.942,\ 0]$ at Iteration 1 is obtained in Equation (2.4.3.18) by substituting $\nabla f_W\left(x_A^{(1)}, x_h^{(1)}\right)$, $-\mathbf{J}_v\left(x_A^{(1)}, x_h^{(1)}\right)^T$, and $-\mathbf{v}\left(x_A^{(1)}, x_h^{(1)}\right)$ into Equation (2.3.4.2-1a). For updated $\left[x_A^{(1)}, x_h^{(1)}\right] = [241.891,\ -2.559]$, $\mathbf{v}\left(x_A^{(1)}, x_h^{(1)}\right)$ is calculated by Equations (2.4.3.4-8), (2.4.3.4-9) and Equations (2.4.3.5-2), (2.4.3.5-3). $\nabla f_W\left(x_A^{(1)}, x_h^{(1)}\right)$ and $\boldsymbol{J}_v\left(x_A^{(1)}, x_h^{(1)}\right)$ are calculated by Equations (2.4.3.6-2) and (2.4.3.6-3), respectively. However, updated Jacobi of Lagrange function still does not converge to yield a root of KKT condition as shown in Equation (2.4.3.18).

$$\nabla\mathcal{L}\left(x_A^{(1)}, x_h^{(1)}, \boldsymbol{\lambda}_v^{(1)}\right) = \begin{bmatrix} \nabla f_W\left(x_A^{(1)}, x_h^{(1)}\right) - \boldsymbol{J}_v\left(x_A^{(1)}, x_h^{(1)}\right)^T \mathbf{S}\boldsymbol{\lambda}_v^{(1)} \\ -\mathbf{S}\mathbf{v}\left(x_A^{(1)}, x_h^{(1)}\right) \end{bmatrix}$$

$$= \begin{bmatrix} -0.6565 \\ 37.2272 \\ -0.3505 \\ 0 \end{bmatrix} \neq 0 \tag{2.4.3.18}$$

2.4.3.1.3.6 ITERATION CONVERGED AND ANALYSIS OF THE RESULTS

A root of KKT condition converges at ninth iteration (Iteration 9), yielding a solution of $\nabla\mathcal{L}\left(x_A^{(9)}, x_h^{(9)}, \lambda_{v1}^{(9)}, \lambda_{v2}^{(9)}\right) = 0$ at $\left[x_A^{(9)}, x_h^{(9)}, \lambda_{v1}^{(9)}, \lambda_{v2}^{(9)}\right] = [558.12,\ -3.532,\ 135.\ 266,\ 0]$, resulting in an minimized weight of $f_W\left(x_A^{(9)}, x_h^{(9)}\right) = 27.05\text{kgf}$ as shown in Figure 2.4.3.3. In the last step, a solution found as $\left[x_A^{(9)}, x_h^{(9)}, \lambda_{v1}^{(9)}, \lambda_{v2}^{(9)}\right] = [558.12, -3.532, 135.266, 0]$ should be verified by inactive Inequality $v_2\left(x_A^{(9)}, x_h^{(9)}\right)$ shown in Equation (2.4.3.2-2). However, inactive Inequality $v_2\left(x_A^{(9)}, x_h^{(9)}\right)$ shown in Equation (2.4.3.19) was not met with $\left[x_A^{(9)}, x_h^{(9)}, \lambda_{v1}^{(9)}, \lambda_{v2}^{(9)}\right] = [558.12,\ -3.532,\ 135.\ 266,\ 0]$ indicated by red, failing to find stationary points of the objective function under Case 1 KKT condition based on active inequality $v_1(x)$ and inactive inequality $v_2(x)$.

$$|\sigma_2| = \frac{100\sqrt{(-3.532)^2 + 4}}{558.12 \times |-3.532|} = 0.2059 \geq f_y \ \left(\text{unit} : \text{kN/mm}^2\right) \tag{2.4.3.19}$$

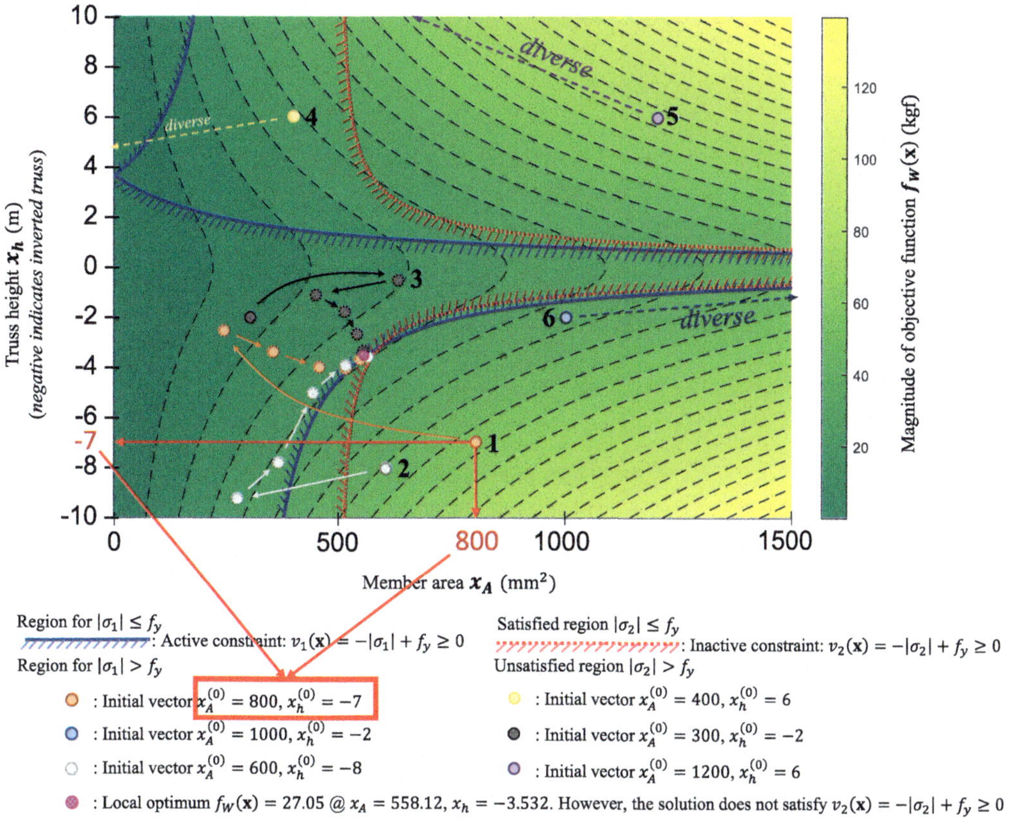

Figure 2.4.3.3 Minimized Lagrange function at $\left[x_A^{(9)}, x_h^{(9)}, \lambda_{v1}^{(9)}, \lambda_{v2}^{(9)}\right] = [558.12, -3.532, 135.266, 0]$, not in feasible region for inequality constraint $v_2(x) = -|\sigma_2| + f_y \ge 0$ based on Case 1 KKT condition.

Five more initial trial inputs of $x_A^{(0)}, x_h^{(0)}$ indicated in Figure 2.4.3.3 are implemented to find stationary points for an optimal solution with Case 1 KKT condition. It is noted that an initial trial Lagrange multiplier $\lambda_v^{(0)} = \left[\lambda_{v1}^{(0)}, \ \lambda_{v2}^{(0)}\right]$ is selected as $[0, 0]$ with all initial trial inputs. Among a total of six initial trial inputs, three initial trial inputs (indicated 1, 2, 3 of Figure 2.4.3.3) of $\left[x_A^{(0)}, x_h^{(0)}, \lambda_{v1}^{(0)}, \lambda_{v2}^{(0)}\right] = [800, -7, 0, 0]$, $[600, -8, 0, 0]$, and $[300, -2, 0, 0]$ converge to the same local stationary point $\left[x_A^{(9)}, x_h^{(9)}\right] = [558.12, -3.532]$, leading to an minimized weight of $f_W(\mathbf{x}) = 27.05kgf$. However, this solution does not satisfy inactive Inequality $v_2\left(x_A^{(9)}, x_h^{(9)}\right) = -|\sigma_2| + f_y \ge 0$ as shown in Equation (2.4.3.19). Three other initial trial inputs (indicated 4, 5, 6 of Figure 2.4.3.3) of $\left[x_A^{(0)}, x_h^{(0)}, \lambda_{v1}^{(0)}, \lambda_{v2}^{(0)}\right] = [1{,}000, -2, 0, 0]$, $[400, 6, 0, 0]$, and $[1{,}200, 6, 0, 0]$ diverge, yielding no solution for $\nabla\mathcal{L}\left(x_A^{(k)}, x_h^{(k)}, \lambda_{v1}^{(k)}, \lambda_{v2}^{(k)}\right) = 0$ since Newton–Raphson iteration diverge due to bad starting trial inputs.

Figure 2.4.3.3 graphically traces convergences for six initial trial inputs with respect to truss height (x_h) and member area (x_A) where blue and red curves represent active and inactive constraints $v_1\left(x_A^{(9)}, x_h^{(9)}\right)$ and $v_2\left(x_A^{(9)}, x_h^{(9)}\right)$, respectively. Three initial trial inputs indicated by 1, 2, and 3 converge to $\left[x_A^{(9)}, x_h^{(9)}, \lambda_{v1}^{(9)}, \lambda_{v2}^{(9)}\right] = [558.12, -3.532, 135.266, 0]$ indicated by pink color which is a local stationary point that should be located on blue line because an inequality condition $v_1\left(x_A^{(9)}, x_h^{(9)}\right) = -|\sigma_1| + f_y \ge 0$ is bound to an equality KKT condition for Case 1

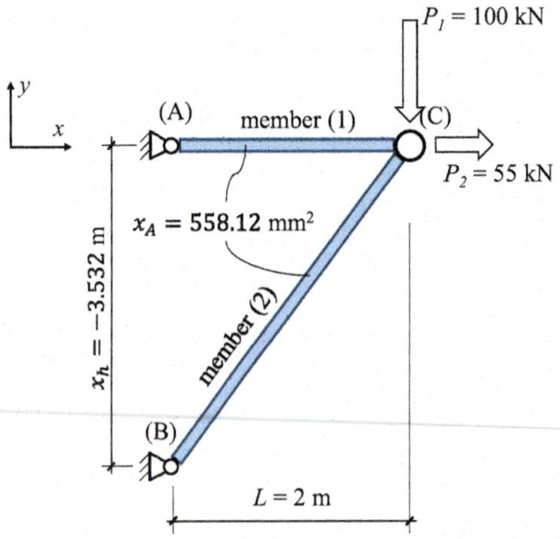

Figure 2.4.3.4 Truss topology with Case 1 KKT condition satisfying inactive Inequality $v_1(x) = -|\sigma_1| + f_y = 0$, but not satisfying inactive Inequality $v_2(x) = -|\sigma_2| + f_y \geq 0$.

KKT condition, being transformed to $|\sigma_1| = 0.2\frac{\text{kN}}{\text{mm}^2} = f_y$. The converged initial trial inputs, however, do not prove inactive Inequality $v_2\left(x_A^{(9)}, x_h^{(9)}\right) = -|\sigma_2| + f_y \geq 0$ as shown to be located within a red shaded area, which indicates that inactive Inequality $v_2\left(x_A^{(9)}, x_h^{(9)}\right) = -|\sigma_2| + f_y \geq 0$ is not met. It is also noted that $\lambda_{v1}^{(9)}$ is calculated positive whereas $\lambda_{v2}^{(9)}$ is zero under Case 1 KKT condition which is based on active inequality $v_1(x)$ and inactive inequality $v_2(x)$. Initial trial inputs 4, 5, and 6, however, diverge from stationary points.

Minimized truss weight of $f_W\left(x_A^{(9)}, x_h^{(9)}\right) = 27.05kgf$ can be read with an ordinate on right side presenting magnitude of an objective function $f_w(x_A, x_h)$. Initial trial Lagrange multipliers of $\lambda_v^{(9)} = \left[x_A^{(9)}, x_h^{(9)}\right]$ converge to positive $[135.266, 0]$ at ninth iteration (Iteration 9) when Inequality $v_1\left(x_A^{(9)}, x_h^{(9)}\right)$ is active. Case 1 KKT condition where Inequality $v_1\left(x_A^{(9)}, x_h^{(9)}\right) = -|\sigma_1| + f_y \geq 0$ is active presents $|\sigma_1| = 0.2 \text{ kN/mm}^2 = f_y$ because $v_1\left(x_A^{(9)}, x_h^{(9)}\right) = -|\sigma_1| + f_y \geq 0$ is transformed to an equality KKT condition. Truss frame shown in Figure 2.4.3.4 provides one root of the KKT condition, which, however, does not satisfy inactive Inequality $v_2\left(x_A^{(9)}, x_h^{(9)}\right) = -|\sigma_2| + f_y \geq 0$. Thus, truss weight of $f_W\left(x_A^{(9)}, x_h^{(9)}\right)$ cannot be minimized with the truss configuration shown in Figure 2.4.3.4.

2.4.3.1.4 Case 2 KKT condition based on inactive inequality $v_1(x)$ and active inequality $v_2(x)$

2.4.3.1.4.1 ITERATION CONVERGED AND ANALYSIS OF THE RESULTS

Similar to Case 1, six initial trial inputs of $\mathbf{x}^{(0)} = [x_A^{(0)}, x_h^{(0)}]$ combining with initial trial Lagrange multipliers of $\lambda_v^{(0)} = \left[\lambda_{v1}^{(0)}, \lambda_{v2}^{(0)}\right] = [0, 0]$ are implemented as illustrated in Figure 2.4.3.5. The initial trial inputs are used to begin Newton–Raphson iteration to identify stationary points minimizing Lagrange function by equating a first derivative of Lagrange function

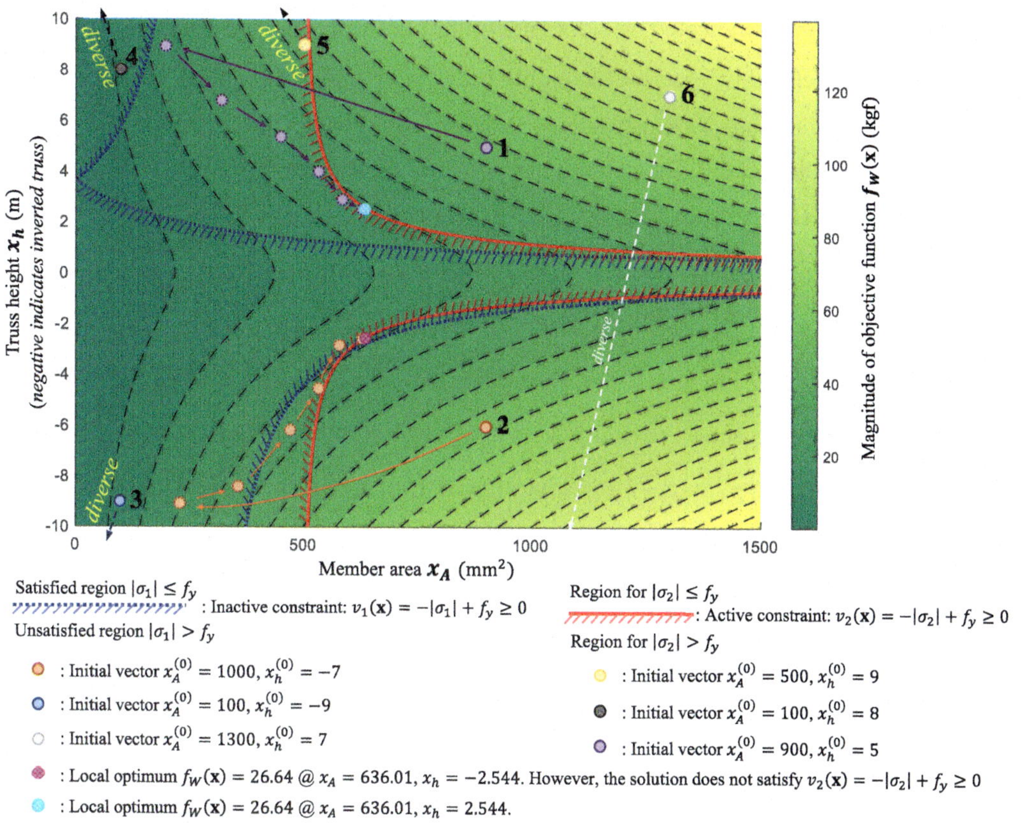

Figure 2.4.3.5 Minimized Lagrange function at $[x_A, x_h, \lambda_{v1}, \lambda_{v2}]$ based on Case 2 KKT condition; must satisfy inactive Inequality constraints $v_1(x) = -|\sigma_1| + f_y \ge 0$, $v_2(x) = -|\sigma_2| + f_y = 0$.

$\nabla\mathcal{L}(\mathbf{x},\boldsymbol{\lambda}_v) = \nabla\mathcal{L}\left(x_A^{(k)},\ x_h^{(k)},\ \lambda_{v1}^{(k)},\ \lambda_{v2}^{(k)}\right)$ to under Case 2 KKT condition. These six initial trial inputs of $\mathbf{x}^{(0)}=[\ x_A^{(0)},\ x_h^{(0)}]$ should be verified if they converge to yield minimized Lagrange function within the ranges defined by inactive Inequality $v_1\left(x_A^{(10)},\ x_h^{(10)}\right) = -|\sigma_1| + f_y \ge 0$. Entire initial trial inputs including trial Lagrange multipliers are assumed as $\left[x_A^{(0)},\ x_h^{(0)},\ \lambda_{v1}^{(0)},\ \lambda_{v2}^{(0)}\right]^T = \left[x_A^{(0)},\ x_h^{(0)},\ 0,\ 0\right]^T$, similarly to Case 1. Two initial trial inputs $\left[x_A^{(0)},\ x_h^{(0)},\ \lambda_{v1}^{(0)},\ \lambda_{v2}^{(0)}\right]^T = [900,\ 5,\ 0,\ 0]^T$ and $[1{,}000,\ -7,\ 0,\ 0]^T$ indicated by 1 and 2 converge to the same local stationary points at $\left[x_A^{(k)},\ x_h^{(k)},\ \lambda_{v1}^{(k)},\ \lambda_{v2}^{(k)}\right] = [636.01,\ 2.544,\ 0,\ 133.21]$ at tenth iteration (Iteration 10) and $[636.01,\ -2.544,\ 0,\ 133.21]$ at 11th iteration (Iteration 11), respectively, resulting in the same minimized weight of $f_W(\mathbf{x}) = f_W\left(x_A^{(k)},\ x_h^{(k)}\right) = 26.64$ kgf as illustrated by the same color contour shown in Figure 2.4.3.5. Both stationary points indicated 1 and 2 shown in Figure 2.4.3.5 are located on red line representing an active inequality condition $v_2\left(x_A^{(k)},\ x_h^{(k)}\right)$ because an active inequality condition $v_2\left(x_A^{(k)},\ x_h^{(k)}\right) = -|\sigma_2| + f_y \ge 0$ is bound to an equality constraint for Case 2 KKT condition. One initial trial input indicated by 1 satisfies inactive Inequality $v_1\left(x_A^{(10)},\ x_h^{(10)}\right) = -|\sigma_1| + f_y \ge 0$ as shown to be located the other side a blue shaded area, which indicates that the weight of $f_W(\mathbf{x}) = f_W\left(x_A^{(k)},\ x_h^{(k)}\right) = 26.64$ kgf is locally minimized within the ranges defined by inactive Inequality $v_1\left(x_A^{(10)},\ x_h^{(10)}\right) = -|\sigma_1| + f_y \ge 0$. However, another initial trial input indicated by 2 does not minimize the weight of $f_W(\mathbf{x}) = f_W\left(x_A^{(k)}, x_h^{(k)}\right)$ within the ranges defined by inactive Inequality $v_1\left(x_A^{(11)}, x_h^{(11)}\right) = -|\sigma_1| + f_y \geq 0$ as shown to be located within a blue shaded area. In Figure 2.4.3.5, both initial trial inputs indicated by 1 (indicated by sky color) and 2 (indicated by pink color) are located on red curve, because an inequality condition $v_2\left(x_A^{(k)}, x_h^{(k)}\right) = -|\sigma_2| + f_y \geq 0$ is bound to an equality KKT condition for Case 2, being transformed to $|\sigma_2| = 0.2\dfrac{\text{kN}}{\text{mm}^2} = f_y$ when Inequality $v_2\left(x_A^{(k)}, x_h^{(k)}\right) = -|\sigma_2| + f_y \geq 0$ is active. An ordinate on the right side of Figure 2.4.3.5 represents magnitude of an objective function $f_W\left(x_A^{(k)}, x_h^{(k)}\right)$ in which minimized truss weight of $f_W\left(x_A^{(10)}, x_h^{(10)}\right) = 26.64$ (kgf) is shown. Figure 2.4.3.6a shows truss topology satisfying $v_1\left(x_A^{(10)}, x_h^{(10)}\right) = -|\sigma_1| + f_y \geq 0$ when inequality constraint $v_2\left(x_A^{(10)}, x_h^{(10)}\right)$ is active. Initial trial input parameters 3, 4, 5, and 6 diverge from stationary points.

2.4.3.1.4.2 TRUSS TOPOLOGY FOR CASE 2 KKT CONDITION

Two solutions are available Figures 2.4.3.5 and 2.4.3.6 when Inequality $v_2\left(x_A^{(k)}, x_h^{(k)}\right) = -|\sigma_2| + f_y \geq 0$ is active whereas inactive Inequality $v_1\left(x_A^{(k)}, x_h^{(k)}\right) = -|\sigma_1| + f_y \geq 0$ should be verified. Case 2 KKT condition provides two roots of KKT conditions, one of which satisfies inactive Inequality $v_1\left(x_A^{(k)}, x_h^{(k)}\right) = -|\sigma_1| + f_y \geq 0$, successfully identifying stationary points for the Lagrange function. As shown in Figure 2.4.3.6a, stress ($|\sigma_1| = 0.037\dfrac{\text{kN}}{\text{mm}^2} < f_y$) of Member 1 provided by first solution is satisfactory, producing a weight of $f_W\left(x_A^{(10)}, x_h^{(10)}\right) = 26.64$ kgf. Stress of $|\sigma_1|$ in Member 1 provided by second solution

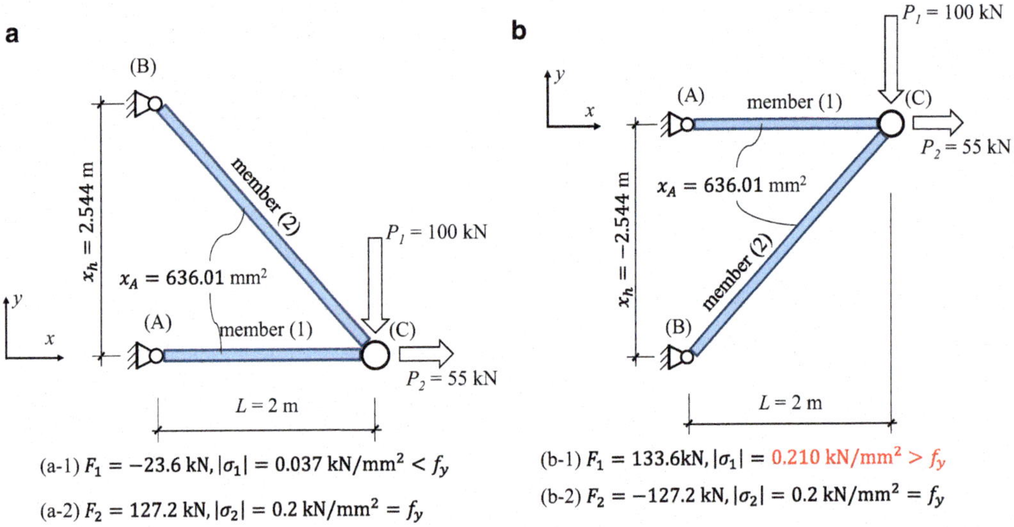

Figure 2.4.3.6 Truss topology with Case 2 KKT condition; must satisfy inactive Inequality constraints $v_1(x) = -|\sigma_1| + f_y \geq 0$, $v_2(x) = -|\sigma_2| + f_y = 0$.

is unsatisfactory because stress calculated as $210\,\frac{\text{kN}}{\text{mm}^2}$ is greater than the yield strength of $f_y = 0.2\,\frac{\text{kN}}{\text{mm}^2}$, making trusses presented in Figure 2.4.3.6b are infeasible. Case 2 also presents $|\sigma_2| = 0.2\,\frac{\text{kN}}{\text{mm}^2} = f_y$ when Inequality $v_2\left(x_A^{(k)}, x_h^{(k)}\right) = -|\sigma_2| + f_y \geq 0$ is active, because $v_2\left(x_A^{(k)}, x_h^{(k)}\right) = -|\sigma_2| + f_y \geq 0$ is bound to an equality KKT condition. It is noted that KKT condition presented in Figure 2.4.3.6(a) yields the least truss weight $f_W\left(x_A^{(10)}, x_h^{(10)}\right) = 26.64$ kgf. Trusses presented in Figure 2.4.3.6a show an optimal shape for Case 2 to minimize the objective function, satisfying all design requirements in term of KKT inequality conditions, $v_1\left(x_A^{(10)}, x_h^{(10)}\right) = -|\sigma_1| + f_y \geq 0$ and $v_2\left(x_A^{(10)}, x_h^{(10)}\right) = -|\sigma_2| + f_y = 0$. Two solutions share the same member area of $x_A = 636.01\ \text{mm}^2$ but different truss height of $|x_h| = 2.544$ m, and $= -2.544$ m, leading to truss members which are differently placed as illustrated in Figure 2.4.3.6a and b. It is noted, in Figure 2.4.3.6a, that truss weight is optimized not only based on an area of truss section, but on truss shape.

2.4.3.1.5 Case 3 KKT condition based on both inequality constraints, $v_1(x)$ and $v_2(x)$, are active

Similar to both Cases 1 and 2, six initial trial inputs of $\mathbf{x}^{(0)} = [\,x_A^{(0)}, x_h^{(0)}]$ combined with an initial Lagrange multiplier of $\lambda_v^{(0)} = \left[\lambda_{v1}^{(0)}, \lambda_{v2}^{(0)}\right] = [0, 0]$ are assumed as shown in Figure 2.4.3.7 under a Case 3 KKT condition. The initial trial inputs are used to

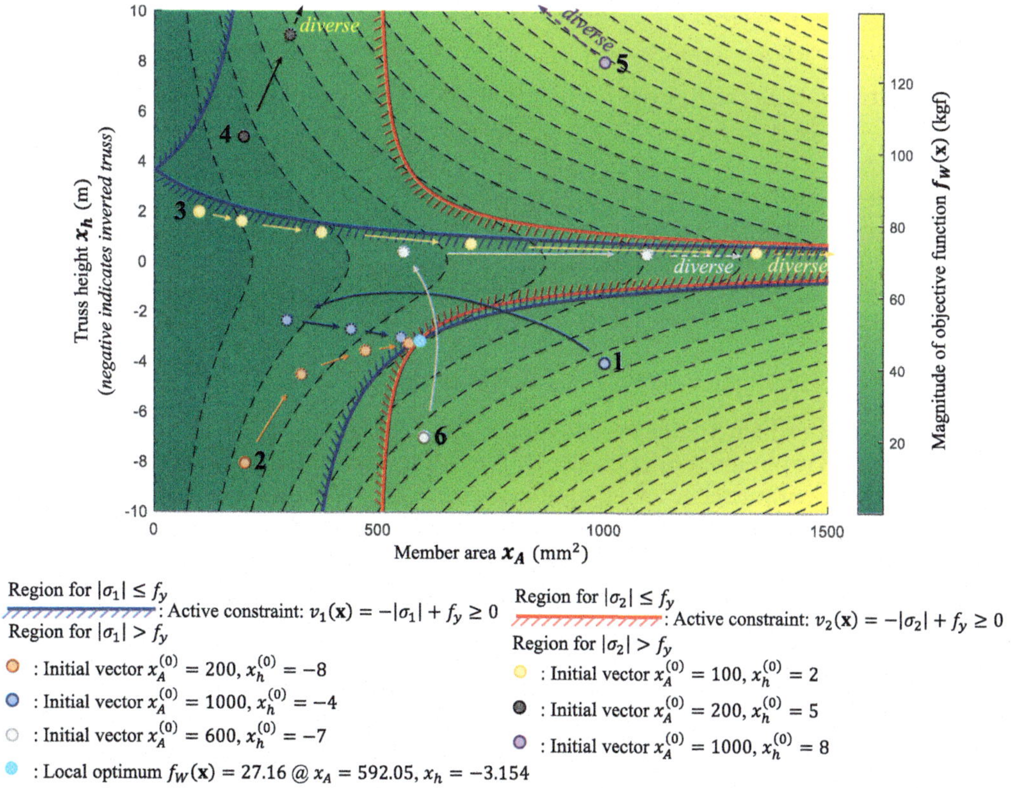

Figure 2.4.3.7 Minimized Lagrange function at $[x_A, x_h, \lambda_{v1}, \lambda_{v2}]$ based on Case 3 KKT condition satisfying active Inequality constraint $v_1(x) = -|\sigma_1| + f_y = 0$, $v_2(x) = -|\sigma_2| + f_y = 0$.

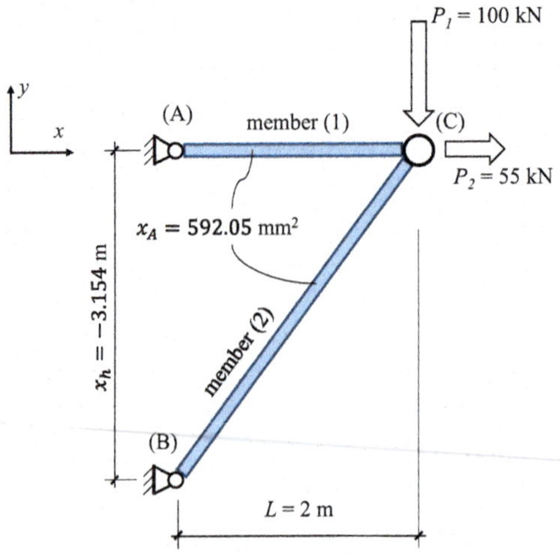

Figure 2.4.3.8 Truss topology with Case 3 KKT condition satisfying active Inequality constraint $v_1(x) = -|\sigma_1| + f_y = 0$, $v_2(x) = -|\sigma_2| + f_y = 0$.

begin Newton–Raphson iteration to identify stationary points minimizing Lagrange function by solving $\nabla\mathcal{L}(\mathbf{x},\boldsymbol{\lambda}_v) = \nabla\mathcal{L}\left(x_A^{(k)}, x_h^{(k)}, \lambda_{v1}^{(k)}, \lambda_{v2}^{(k)}\right) = 0$. Two initial trial inputs $\left[x_A^{(0)}, x_h^{(0)}, \lambda_{v1}^{(0)}, \lambda_{v2}^{(0)}\right]^T = [1000, -4, 0, 0]^T$ and $[200, -8, 0, 0]^T$ indicated by 1 and 2 of in Figure 2.4.3.7 converge to the same local $\left[x_A^{(k)}, x_h^{(k)}, \lambda_{v1}^{(k)}, \lambda_{v2}^{(k)}\right] = [592.05, -3.154, 97.03, 38.77]$ at eighth (Iteration 8) and seventh (Iteration 7), respectively. Two initial trial inputs indicated by 1 and 2 in Figure 2.4.3.7 satisfy active Inequality $v_1(x) = -|\sigma_1| + f_y \geq 0$ and $v_2(x) = -|\sigma_2| + f_y \geq 0$ as local stationary points are located on both a blue and red curves, because both inequalities, $v_1(x) = -|\sigma_1| + f_y \geq 0$ and $v_2(x) = -|\sigma_2| + f_y \geq 0$, are active transforming them to equality conditions. It is also noted that both active inequalities are calculated as $f_y = 0.2\text{kN}/\text{mm}^2$, indicating stresses in all members are equal to $f_y = 0.2\ \text{kN}/\text{mm}^2$ as shown in Figure 2.4.3.7. A root (local stationary points and Lagrange multipliers) of Case 3 KKT condition found at $\left[x_A^{(k)}, x_h^{(k)}, \lambda_{v1}^{(k)}, \lambda_{v2}^{(k)}\right] = [592.05, -3.154, 97.03, 38.77]$ resulted in positive Lagrange multipliers with an optimal weight of $f_W\left(\lambda_{v1}^{(k)}, \lambda_{v2}^{(k)}\right) = 27.16\text{kgf}$ shown in contours of an ordinate on right side Figure 2.4.3.7. However, $f_W\left(\lambda_{v1}^{(k)}, \lambda_{v2}^{(k)}\right) = 27.16\text{kgf}$ is greater than that obtained for Case 2 KKT condition, failing to minimize truss weight. Initial trial input parameters 3, 4, 5, and 6 diverge from stationary points. Figure 2.4.3.8 illustrates resulted truss topology based on a Case 3 KKT condition.

2.4.3.1.6 *Case 4 KKT condition based on none of inequality constraints are active*

In Case 4 where none of inequality constraints is active, inequality constraints are ignored in a formulation of Lagrange function $\mathcal{L}(\mathbf{x}, \boldsymbol{\lambda}_v) = \mathcal{L}(x_A, x_h, \lambda_{v1}, \lambda_{v2})$ and first derivative (Jacobi, gradient vector) $\nabla\mathcal{L}(\mathbf{x},\boldsymbol{\lambda}_v) = \nabla\mathcal{L}(x_A, x_h, \lambda_{v1}, \lambda_{v2})$, adopting only objective function $f_W(\mathbf{x}) = f_W(x_A, x_h)$ as shown in Equation (2.4.3.20-1), thus, its first derivative (Jacobi, gradient vector) $\nabla f_W(\mathbf{x}) = \nabla f_W(x_A, x_h)$ is optimized as shown in Equation (2.4.3.20–2).

$$\mathcal{L}(x_A, x_h, \lambda_{v1}, \lambda_{v2}) = f_W(x_A, x_h) = 0.008x_A\left(2 + \sqrt{x_h^2 + 4}\right)\ (\text{unit: kgf}) \qquad (2.4.3.20\text{-}1)$$

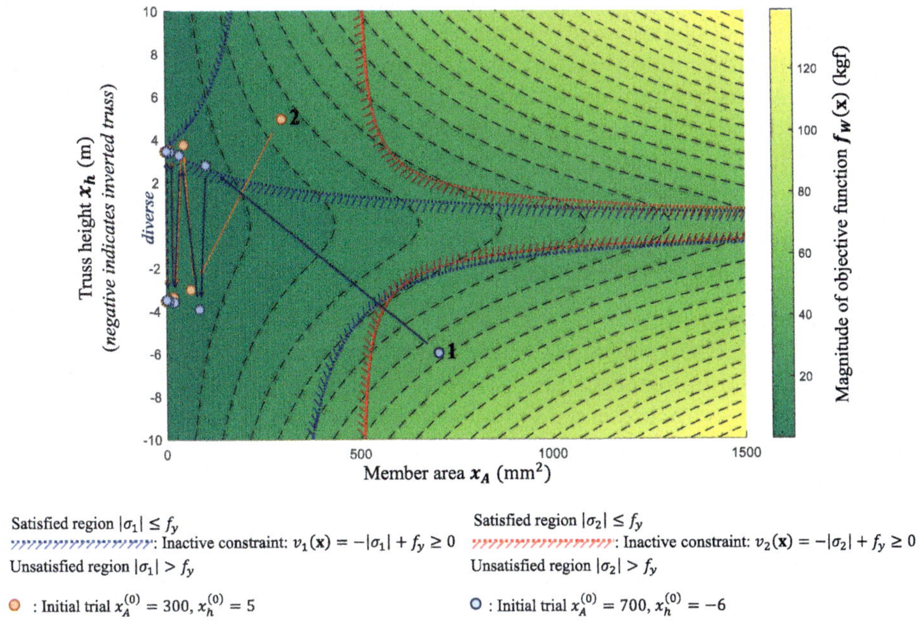

Figure 2.4.3.9 Minimized Lagrange function at $\left[x_A, x_h, \lambda_{v1}, \lambda_{v2}\right]$ based on Case 4 KKT condition; must satisfy $v_1(x) = -|\sigma_1| + f_y \ge 0$, $v_2(x) = -|\sigma_2| + f_y \ge 0$.

$$\nabla \mathcal{L}(x_A, x_h, \lambda_{v1}, \lambda_{v2}) = \nabla f_W(x_A, x_h) = \begin{bmatrix} 0.016 + 0.008\sqrt{x_h^2 + 4} \\ \dfrac{0.008 x_A x_h}{\sqrt{x_h^2 + 4}} \end{bmatrix} \tag{2.4.3.20-2}$$

Newton–Raphson iteration is implemented to find stationary points when Equation (2.4.3.20-2) are set to 0. Two initial trial inputs of $\left[\mathbf{x}^{(0)}, \boldsymbol{\lambda}_v\right] = \left[x_A^{(0)}, x_h^{(0)}, \lambda_{v1}^{(0)}, \lambda_{v2}^{(0)}\right] = [300, 5, 0, 0]$ and $[700, -6, 0, 0]$ indicated by 1 and 2 in Figure 2.4.3.9 are used to begin Newton–Raphson iteration to find roots. Both trials do not converge, but diverge, failing to calculate truss height x_h and area x_A. Mathematically, either $x_A = 0$ or $x_h = 0$ makes truss weight $0.008 x_A x_h \big/ \sqrt{x_h^2 + 4} = 0$ minimum, but these are trivial solutions to minimize truss weight.

2.4.3.1.7 Section summary

Minimized truss weight based on conventional Lagrange optimization is found as 26.64 kgf based on the Case 2 KKT condition, as shown in Table 2.4.3.2. Truss topology is

Table 2.4.3.2 Minimized Truss Weight Based on Conventional Lagrange Optimization

	Conventional Lagrange Optimization				
	Parameters	*Case 1*	*Case 2*	*Case 3*	*Case 4*
1	x_A (mm)	N/A	**636.01**	592.05	N/A
2	x_h (m)	N/A	**2.544**	−3.154	N/A
3	$\lvert\sigma_1\rvert$ (kN/mm^2)	N/A	**0.037**	0.2	N/A
4	$\lvert\sigma_2\rvert$ (kN/mm^2)	N/A	**0.2**	0.2	N/A
Objective f_W (kgf)		N/A	**26.64**	27.16	N/A

Case 1, Inequality $v_1(x)$ is active; Case 2, Inequality $v_2(x)$ is active; Case 3, Both inequalities are active; Case 4, None of inequalities are active.

depicted in Figure 2.4.3.6a which minimizes truss weight $f_W(\mathbf{x})$ equivalent to 26.64 kgf. It is noteworthy that finding analytically optimized solutions based on Lagrange optimization is lengthy and difficult to follow as demonstrated by a simple truss example. The following section uses ANN-based Lagrange optimization to minimize truss weight in a quick and accurate manner. Code for Lagrange optimization based on analytical functions is presented in Appendix B2.

2.4.3.2 Lagrange optimization of a truss frame based on ANN-based object and constraining functions

This section minimizes a weight of truss frame based on ANN-based Lagrange optimization. In this example, the same truss frame minimized based on analytical functions in Section 2.4.3 is also minimized based on ANN-based functions and compared. Step-by-step solution procedures are presented to find stationary points of Lagrange functions using ANN-based Jacobian and Hessian matrices.

2.4.3.2.1 Step 1: Generation of large datasets for training ANN

A nonlinear objective function shown in Equation (2.4.3.21) and inequality constraints shown in Equations (2.4.3.22-1) and (2.4.3.22-2) will be replaced by ANN-based functions. A proper number of generated large datasets and data ranges to train ANNs should be selected carefully based on a level of a complexity of objective and inequality constraining functions to obtain accurate weight and bias matrices. As shown in Figure 2.4.3.10 and Table 2.4.3.3, 20,000 datasets with x_A which varies from 100 to 3,000 mm^2 and x_h from −30 to 30 m are generated to train ANNs. Datasets are not normalized in Table 2.4.3.3(a), whereas they are normalized between −1 and +1 as shown in Table 2.4.3.3(b). Minimum and maximum ranges of a non-normalized and normalized data are shown below the table. ANN-based functions are derived to minimize truss weight $f_W(x_A, x_h)$ shown in Equation (2.4.3.21) constrained by truss stresses $|\sigma_1|$ and $|\sigma_2|$ shown in Equations (2.4.3.22-1) and (2.4.3.22-2), respectively. Lagrange function is derived in Equation (2.4.3.23-1) with only inequality constraints whereas the first derivative (Jacobi, gradient vector) is derived in Equation (2.4.3.24-1) based on analytical function and Equation (2.4.3.24-2) based on ANNs.

Minimize $$f_W(\mathbf{x}) = f_W(x_A, x_h) \tag{2.4.3.21}$$

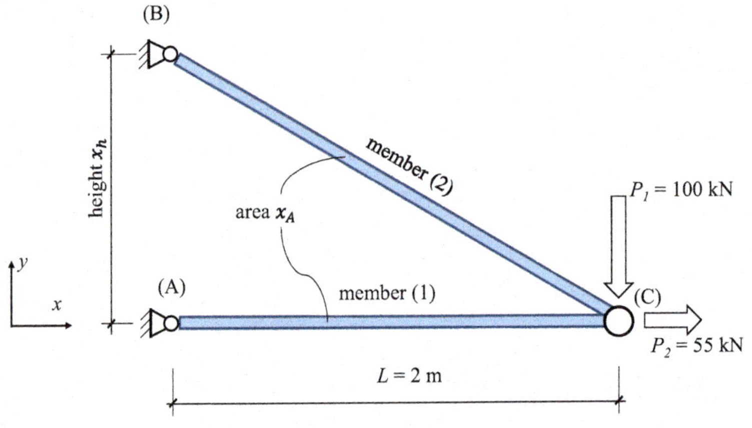

Figure 2.4.3.10 Truss frame optimized by ANN-based Lagrange method.

with respect to $\mathbf{x} = [x_A,\ x_h]$

subject to $v_1(\mathbf{x}) = v_1(x_A,\ x_h) = -|\sigma_1| + f_y \geq 0$ (2.4.3.22-1)

$$v_2(\mathbf{x}) = v_2(x_A,\ x_h) = -|\sigma_2| + f_y \geq 0 \quad (2.4.3.22\text{-}2)$$

$$\mathcal{L}(\mathbf{x}, \boldsymbol{\lambda}_v) = \mathcal{L}(x_A, x_h, \lambda_{v1},\ \lambda_{v2}) = f_W(\mathbf{x}) - \boldsymbol{\lambda}_v^T \mathbf{S}\mathbf{v}(\boldsymbol{x}) \quad (2.4.3.23\text{-}1)$$

$$\mathcal{L}(\mathbf{x}, \boldsymbol{\lambda}_v) = f_W(x_A, x_h) - \begin{bmatrix} \lambda_{v1} & \lambda_{v2} \end{bmatrix} \begin{bmatrix} s_1 & 0 \\ 0 & s_2 \end{bmatrix} \begin{bmatrix} -|\sigma_1| + f_y \\ -|\sigma_2| + f_y \end{bmatrix} \quad (2.4.3.23\text{-}2)$$

$$\nabla\mathcal{L}(\mathbf{x}, \boldsymbol{\lambda}_v) = \nabla\mathcal{L}(x_A, x_h, \lambda_{v1},\ \lambda_{v2}) = \begin{bmatrix} \nabla f_W(x_A, x_h) - \boldsymbol{J}_v(x_A, x_h)^T \mathbf{S}\boldsymbol{\lambda}_v \\ -\mathbf{S}\mathbf{v}(x_A, x_h) \end{bmatrix} \quad (2.4.3.24\text{-}1)$$

Table 2.4.3.3 Generation of Large Datasets

(a): Non-Normalized

20,000 Datasets (Non-Normalized)

Data	x_A (mm²)	x_h (m)	$\lvert\sigma_1\rvert$ (kN/mm²)	$\lvert\sigma_2\rvert$ (kN/mm²)	f_W (g)
1	206.4	13.85	0.196	0.489	26.40
2	2728.7	19.38	0.016	0.037	468.95
3	1726.5	−3.08	0.069	0.069	78.40
4	1842.1	−10.68	0.040	0.055	189.57
5	970.3	13.62	0.042	0.104	122.36
6	583.0	19.30	0.077	0.172	99.80
7	1546.3	6.47	0.016	0.068	108.48
8	1945.2	13.90	0.021	0.052	249.68
9	681.2	7.93	0.044	0.151	55.44
10	630.1	−25.10	0.100	0.159	137.00
11	500.6	29.90	0.097	0.200	128.02
12	1539.4	−26.00	0.041	0.065	345.78
13	1399.6	−24.27	0.045	0.072	295.09
14	1823.5	17.82	0.024	0.055	290.80
15	2347.3	11.15	0.016	0.043	250.34
16	2454.9	−6.34	0.035	0.043	169.79
17	1637.7	−19.54	0.040	0.061	283.58
18	1017.4	12.00	0.038	0.100	115.28
...	...	...	...	...	...
20,000	2340.2	8.56	0.014	0.044	201.97
x_{max} (y_{max})	2999.9	30.00	6.905	7.445	765.82
x_{min} (y_{min})	100.2	−30.00	0.000	0.033	3.26
x_{mean} (y_{mean})	1552.9	−0.04	0.080	0.133	216.52

(Continued)

Table 2.4.3.3 (Continued) Generation of Large Datasets

(b): Normalized

20,000 Datasets (Normalized)					
Data	x_A *(mm²)*	x_h *(m)*	$\lvert\sigma_1\rvert$ *(kN/mm²)*	$\lvert\sigma_2\rvert$ *(kN/mm²)*	f_W *(g)*
1	−0.927	0.462	−0.943	−0.877	−0.939
2	0.813	0.646	−0.995	−0.999	0.221
3	0.122	−0.103	−0.980	−0.990	−0.803
4	0.201	−0.356	−0.988	−0.994	−0.511
5	−0.400	0.454	−0.988	−0.981	−0.688
6	−0.667	0.643	−0.978	−0.962	−0.747
7	−0.003	0.216	−0.995	−0.991	−0.724
8	0.273	0.463	−0.994	−0.995	−0.354
9	−0.599	0.264	−0.987	−0.968	−0.863
10	−0.635	−0.837	−0.971	−0.966	−0.649
11	−0.724	0.997	−0.972	−0.955	−0.673
12	−0.007	−0.867	−0.988	−0.991	−0.102
13	−0.104	−0.809	−0.987	−0.990	−0.235
14	0.189	0.594	−0.993	−0.994	−0.246
15	0.550	0.372	−0.995	−0.997	−0.352
16	0.624	−0.211	−0.990	−0.997	−0.563
17	0.060	−0.652	−0.988	−0.992	−0.265
18	−0.367	0.400	−0.989	−0.982	−0.706
...	...	...	...	...	...
20,000	0.545	0.285	−0.996	−0.997	−0.479
$\bar{x}_{max}$ ($\bar{y}_{max}$)	1.000	1.000	1.000	1.000	1.000
$\bar{x}_{min}$ ($\bar{y}_{min}$)	−1.000	−1.000	−1.000	−1.000	−1.000
$\bar{x}_{mean}$ ($\bar{y}_{mean}$)	0.002	−0.001	−0.977	−0.973	−0.441
α_x (α_y)	0.0007	0.0333	0.2896	0.2699	0.0026

$$= \begin{bmatrix} \left[\mathbf{J}_{fW}^{(D)}(x_A, x_h)\right]^T - \left[\mathbf{J}_{\mathrm{v}}^{(D)}(x_A, x_h)\right]^T \mathbf{S}\lambda_v \\ -\mathbf{S}\mathrm{v}(x_A, x_h) \end{bmatrix} \quad (2.4.3.24\text{-}2)$$

2.4.3.2.2 Step 2: Training ANNs on large datasets

In this step, ANNs with proper training parameters including a number of hidden layers, neurons, suggested epochs, validation checks, etc. should be used to obtain accurate weight and bias matrices. As shown in Equation (2.4.3.25) and Figure 2.4.3.11, an ANN-based objective function $f_W(\mathbf{x})$ for weight and constraining functions for stresses of truss members $|\sigma_1|$ and $|\sigma_2|$ with five hidden layers and ten neurons are trained to minimize truss weight. Figure 2.4.3.11a shows an ANN with five hidden layers and ten neurons, which yields one output parameter for given multiple input parameters based on PTM. The ANN-based objective and constraining functions are to replace analytical functions. Training accuracies represented by mean square errors of a test subset (MSE.Tperf) for $|\sigma_1|$, $|\sigma_2|$, and $f_W(\mathbf{x})$, are obtained as 1.56E-7, 1.07E-7, and 2.27E-8, respectively, as shown in Figure 2.4.3.11b–d.

2.4.3.2.3 Step 3: Application of Lagrange multipliers to minimize an ANN-based objective function.

ANNs for objective functions for $f_W(\mathbf{x})$ shown in Equation (2.4.3.21) and inequality constraining functions $v_1(x_A, x_h)$, $v_2(x_A, x_h)$ for $|\sigma_1|$, $|\sigma_2|$ shown in Equations (2.4.3.22-1) and (2.4.3.22-2) are presented in Equations (2.4.3.25-1)–(2.4.3.25-3), respectively. In this example,

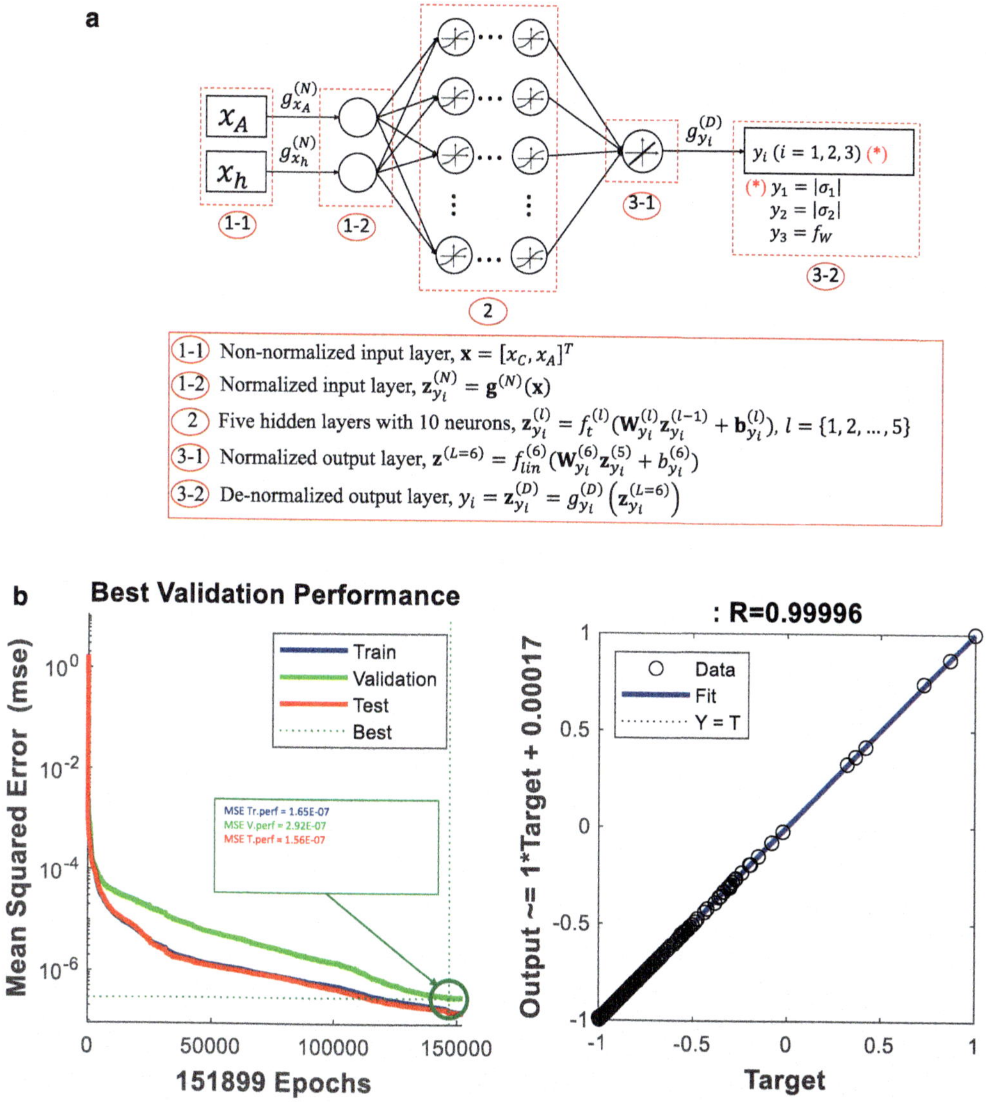

Figure 2.4.3.11 Formulation of ANNs using five hidden layers and ten neurons to minimize truss weight to obtain accurate weight and bias matrices. (a) ANN based on PTM yielding one output parameters. (b) Training performance of ANN for stress of Member 1 $|\sigma_1|$. (c) Training performance of ANN for stress of Member 2 $|\sigma_2|$. (d) Training performance of ANN for objective function $f_W(x)$ minimizing weight.

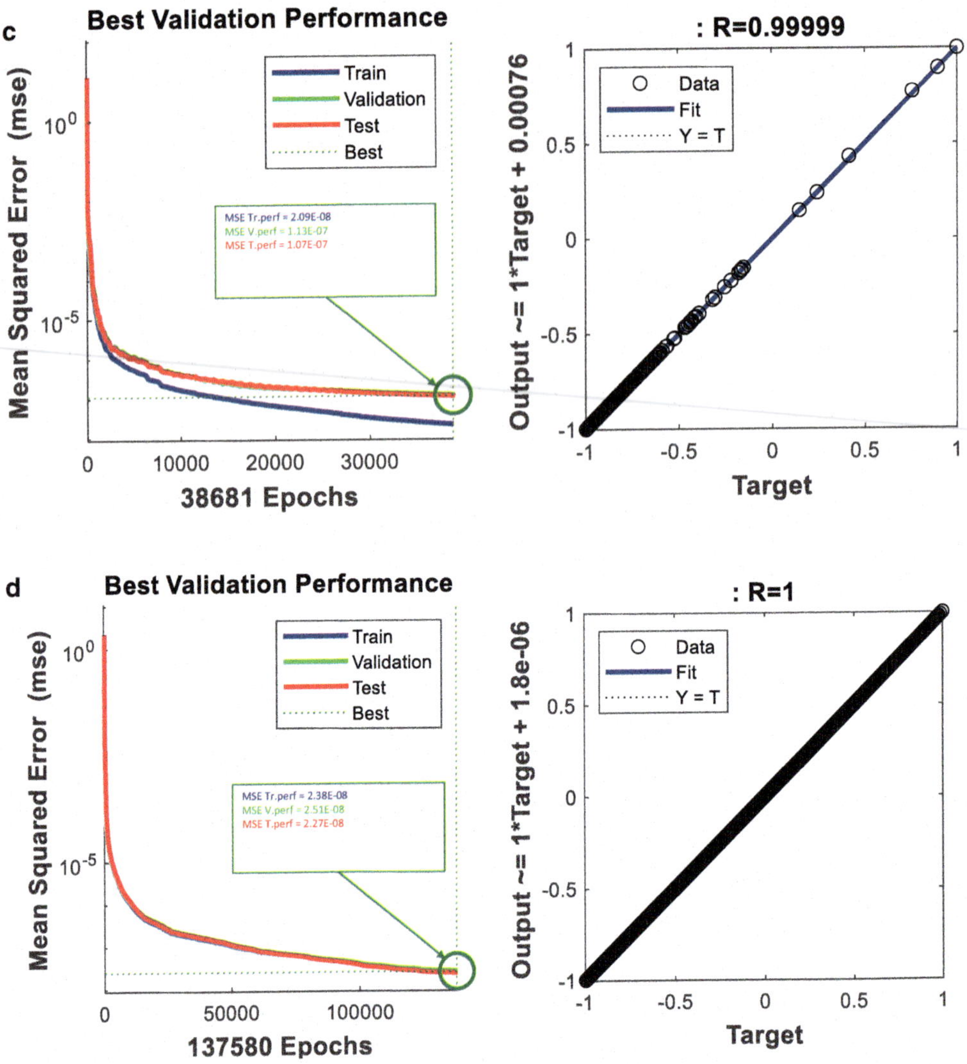

Figure 2.4.3.11 *(Continued)* Formulation of ANNs using five hidden layers and ten neurons to minimize truss weight. (a) ANN based on PTM yielding one output parameters. (b) Training performance of ANN for stress of Member 1 $|\sigma_1|$. (c) Training performance of ANN for stress of Member 2 $|\sigma_2|$. (d) Training performance of ANN for objective function $f_W(x)$ minimizing weight.

an objective function is subject to only inequality constraints $v(x)$ representing $v_1(x_A, x_h)$ and $v_2(x_A, x_h)$ whereas equality constraints $c(\mathbf{x})$ do not exist. In Equations (2.4.3.25-1)–(2.4.3.25-3), $\mathbf{x}$ is input design parameter representing area and height of truss section x_A, x_h; L is a number of layers including hidden and output layers; $\mathbf{W}^l$ a weight matrix between layer $l-1$ and layer l; $\mathbf{b}^l$ is a bias matrix of layer l; and g^N and g^D are a normalization and denormalization functions, respectively. An activation function f_t^l at layer l (*tansig*, *tanh*) shown in Figure 2.3.3.1 is implemented to formulate nonlinear relationships between networks, whereas a linear activation function f_{lin}^L is selected for an output layer because output values are not squashed by activation function. ANN-based objective $(\underbrace{W}_{[1\times1]})$ and inequality constraining $(\underbrace{|\sigma_1|}_{[1\times1]}, \underbrace{|\sigma_2|}_{[1\times1]})$ functions of this example are obtained by substituting initial input design parameters $\mathbf{x}^{(0)} = \left[x_A^{(0)},\ x_b^{(0)}\right]^T = [800,\ -7]^T$ represented by $[2 \times 1]$ red matrix into Equation (2.4.3.25-1).

$$\underbrace{W}_{[1\times1]} = \mathbf{g}_W^D\left(f_l^L\left(\underbrace{\mathbf{W}_W^L}_{[1\times10]} f_t^{L-1}\left(\underbrace{\mathbf{W}_W^{L-1}}_{[10\times10]} \ldots f_t^1\left(\underbrace{\mathbf{W}_W^1}_{[10\times2]} \underbrace{\mathbf{g}_\mathbf{x}^N(\mathbf{x})}_{[2\times1]} + \underbrace{\mathbf{b}_W^1}_{[10\times1]}\right)\ldots + \underbrace{\mathbf{b}_W^{L-1}}_{[10\times1]}\right) + \underbrace{b_W^L}_{[1\times1]}\right)\right) \quad (2.4.3.25\text{-}1)$$

$$\underbrace{|\sigma_1|}_{[1\times1]} = \mathbf{g}_{|\sigma_1|}^D\left(f_l^L\left(\underbrace{\mathbf{W}_{|\sigma_1|}^L}_{[1\times10]} f_t^{L-1}\left(\underbrace{\mathbf{W}_{|\sigma_1|}^{L-1}}_{[10\times10]} \ldots f_t^1\left(\underbrace{\mathbf{W}_{|\sigma_1|}^1}_{[10\times2]} \underbrace{\mathbf{g}_\mathbf{x}^N(\mathbf{x})}_{[2\times1]} + \underbrace{\mathbf{b}_{|\sigma_1|}^1}_{[10\times1]}\right)\ldots + \underbrace{\mathbf{b}_{|\sigma_1|}^{L-1}}_{[10\times1]}\right) + \underbrace{b_{|\sigma_1|}^L}_{[1\times1]}\right)\right) \quad (2.4.3.25\text{-}2)$$

$$\underbrace{|\sigma_2|}_{[1\times1]} = \mathbf{g}_{|\sigma_2|}^D\left(f_l^L\left(\underbrace{\mathbf{W}_{|\sigma_2|}^L}_{[1\times10]} f_t^{L-1}\left(\underbrace{\mathbf{W}_{|\sigma_2|}^{L-1}}_{[10\times10]} \ldots f_t^1\left(\underbrace{\mathbf{W}_{|\sigma_2|}^1}_{[10\times2]} \underbrace{\mathbf{g}_\mathbf{x}^N(\mathbf{x})}_{[2\times1]} + \underbrace{\mathbf{b}_{|\sigma_2|}^1}_{[10\times1]}\right)\ldots + \underbrace{\mathbf{b}_{|\sigma_2|}^{L-1}}_{[10\times1]}\right) + \underbrace{b_{|\sigma_2|}^L}_{[1\times1]}\right)\right) \quad (2.4.3.25\text{-}3)$$

ANN is implemented to derive objective and constraining functions using Equations (2.4.3.25-1)–(2.4.3.25-3) to calculate Jacobian and Hessian matrices, which must be calculated to minimize truss weight $f_W(\mathbf{x})$ constrained by stresses in truss $|\sigma_1|$ and $|\sigma_2|$ under KKT conditions based on Newton–Raphson iteration. Figures 2.4.3.12–2.4.3.14 compare ANN-based objective function $f_W(\mathbf{x})$ and inequality constraining functions $|\sigma_1|$, $|\sigma_2|$ with those based on analytical functions, indicating they match well, and demonstrating that analytical functions can be replaced by ANN-based functions. Analytical functions for truss weight and truss stress are derived in Equations (2.4.3.3-3), (2.4.3.4-8), and (2.4.3.4-9).

2.4.3.2.4 Case 1 KKT condition based on active inequality constraints $v_1(x)$ and inactive inequality constraints $v_2(x)$

2.4.3.2.4.1 DERIVATION OF ANN

In this KKT condition, inequality $v_1(x) = -|\sigma_1| + f_y \geq 0$ is assumed to constrain an objective function $f_W(\mathbf{x})$ by binding Inequality $v_1(x)$ to equality constraint, whereas inequality $v_2(x) = -|\sigma_2| + f_y \geq 0$ assumed to be inactive does not constrain an objective function $f_W(\mathbf{x})$ during optimization. However, stationary points of Lagrange function of Case 1 KKT condition should be verified to exist within the ranges defined by an inequality constraint set, $v_2(x) = -|\sigma_2| + f_y \geq 0$. Full initial input design parameters becomes $\left[\mathbf{x}^{(0)},\ \boldsymbol{\lambda}_v^{(0)}\right]^T = \left[x_A^{(0)},\ x_b^{(0)},\ \lambda_{v1}^{(0)},\ \lambda_{v2}^{(0)}\right]^T = [800, -7,\ 0,\ 0]^T$, consisting of initial trial input of $\mathbf{x}^{(0)} = \left[x_A^{(0)},\ x_b^{(0)}\right]^T = [800, -7]^T$ and initial trial Lagrange multipliers of $\boldsymbol{\lambda}_v^{(0)} = \left[\lambda_{v1}^{(0)},\ \lambda_{v2}^{(0)}\right]^T = [0, 0]^T$. Stationary points are found by setting the first derivative (Jacobian, gradient vector) of Lagrange function shown in Equation (2.4.3.24-2) equal to zero using Newton–Raphson iteration. Based on ANNs shown in Equation (2.3.6.3-1), objective $f_W(\mathbf{x})$ and inequality constraining $\mathbf{v}(\mathbf{x})$ functions shown in Equations (2.4.3.25-1)–(2.4.3.25-3) can be also expressed in Equations (2.4.3.26-1)–(2.4.3.26-3) as a series of composite mathematical operations. This expression is used to represent functions based on ANN with a last hidden layer (L=5) and output layer (L = 6). The ANNs consist of five hidden, one output layer and ten neurons for given initial input design parameters. ANN-based objective function $f_W(\mathbf{x})$ shown in Equation (2.4.3.26-1) and inequality constraining

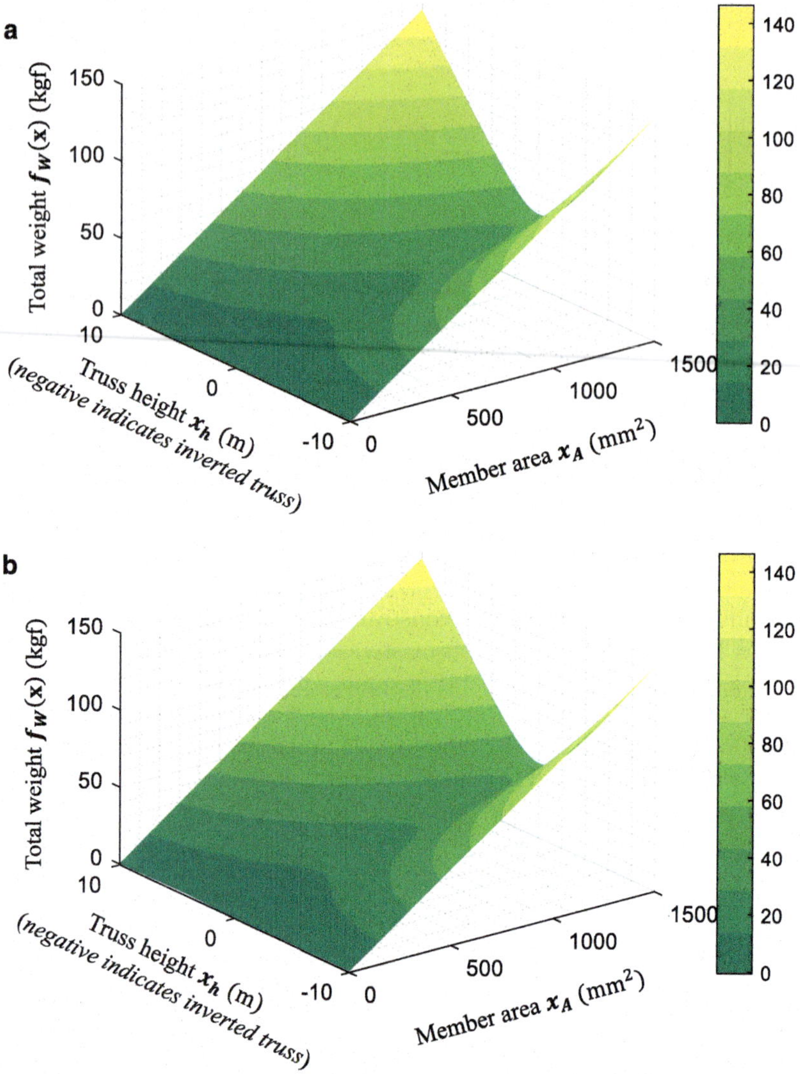

Figure 2.4.3.12 Comparison between analytical and ANN-based functions for $f_W(x)$. (a) Analytical objective function minimizing truss weight $f_W(x)$. (b) ANN-based objective function minimizing truss weight $f_W(x)$

functions $\mathbf{v}(\mathbf{x}) = v_1(x),\ v_2(x)$ shown in Equations (2.4.3.26-1)–(2.4.3.26-3) are calculated the de-normalized Layer D as 59.3664, 0.0957, and 0.0701, respectively.

$$f_W\left(x_A^{(0)},\ x_h^{(0)}\right) = \mathbf{z}_{fW}^{(D)} \circ \mathbf{z}_{fW}^{(L=6=\text{output})} \circ \mathbf{z}_{fW}^{(5)} \circ \ldots \circ \mathbf{z}_{fW}^{(1)} \circ \mathbf{z}_{fW}^{(N)} = 59.3664 \quad (2.4.3.26\text{-}1)$$

$$\begin{aligned} v_1\left(x_A^{(0)},\ x_h^{(0)}\right) &= -\left|\sigma_1\left(x_A^{(0)},\ x_h^{(0)}\right)\right| + f_y \\ &= -\mathbf{z}_{|\sigma_1|}^{(D)} \circ \mathbf{z}_{|\sigma_1|}^{(L=6=\text{output})} \circ \mathbf{z}_{|\sigma_1|}^{(5)} \circ \ldots \circ \mathbf{z}_{|\sigma_1|}^{(1)} \circ \mathbf{z}_{|\sigma_1|}^{(N)} + 0.2 \\ &= 0.0957 \end{aligned} \quad (2.4.3.26\text{-}2)$$

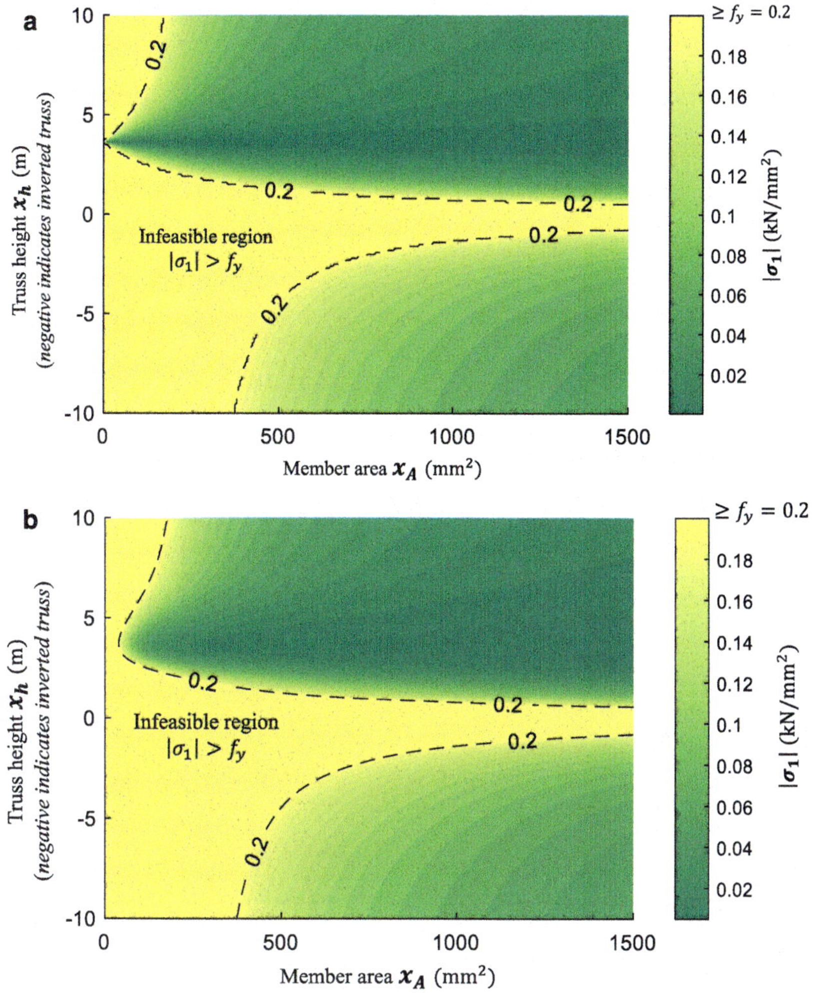

Figure 2.4.3.13 Comparison between analytical and ANN-based functions for $|\sigma_1|$. (a) Analytical function for stress $|\sigma_1|$. (b) ANN-based function for stress $|\sigma_1|$.

$$
\begin{aligned}
v_2\left(x_A^{(0)},\ x_h^{(0)}\right) &= -\left|\sigma_2\left(x_A^{(0)},\ x_h^{(0)}\right)\right| + f_y \\
&= -\mathbf{z}_{|\sigma_2|}^{(D)} \circ \mathbf{z}_{|\sigma_2|}^{(L=6=\text{output})} \circ \mathbf{z}_{|\sigma_2|}^{(5)} \circ \ldots \circ \mathbf{z}_{|\sigma_2|}^{(1)} \circ \mathbf{z}_{|\sigma_2|}^{(N)} + 0.2 \\
&= 0.0701
\end{aligned}
\tag{2.4.3.26-3}
$$

$$
\mathbf{v}\left(x_A^{(0)},\ x_h^{(0)}\right) = \begin{bmatrix} v_1\left(x_A^{(0)},\ x_h^{(0)}\right) \\ v_2\left(x_A^{(0)},\ x_h^{(0)}\right) \end{bmatrix} = \begin{bmatrix} 0.0957 \\ 0.0701 \end{bmatrix}
\tag{2.4.3.26-4}
$$

Objective $f_W(\mathbf{x})$, equality $\mathbf{c}(\mathbf{x})$ and inequality constraining $\mathbf{v}(\mathbf{x})$ functions derived in an output layer are de-normalized. ANN consists of one normalizing step (N), five normalized hidden layers, one normalized output layer, and one de-normalizing step (D). Objective

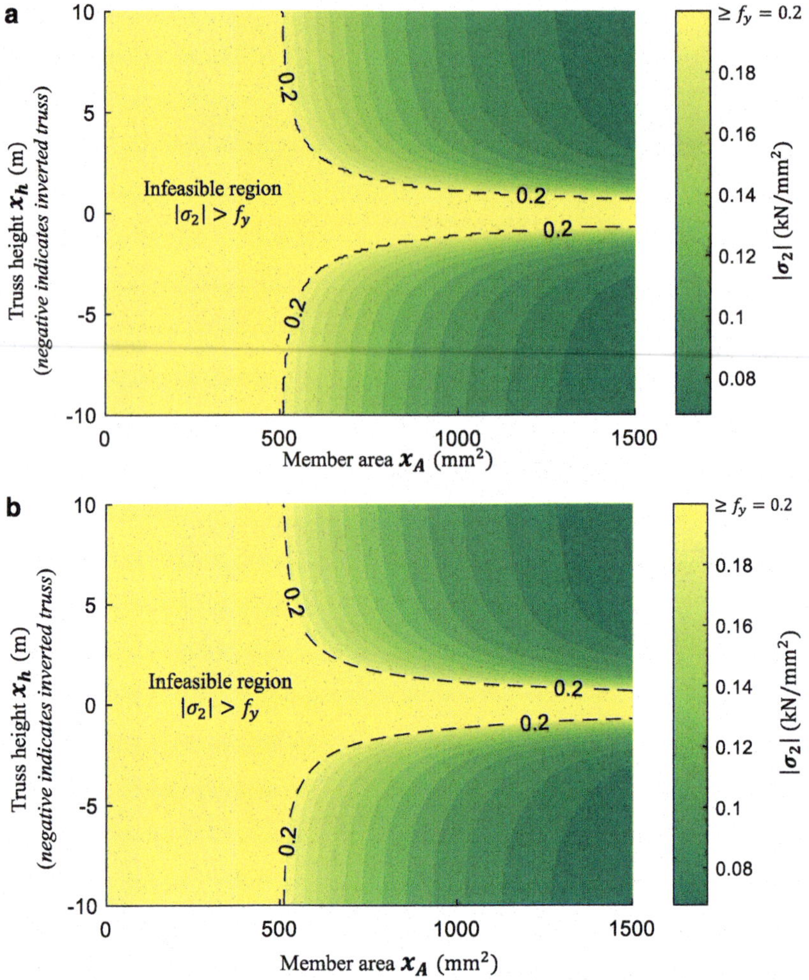

Figure 2.4.3.14 Comparison between analytical and ANN-based functions for $|\sigma_2|$. (a) Analytical function for stress $|\sigma_2|$. (b) ANN-based function for stress $|\sigma_2|$.

function $f_W(\mathbf{x})$ shown in Equation (2.4.3.26-1) and inequality constraining functions $v_1(x_A, x_h)$, $v_2(x_A, x_h)$ for truss stress $|\sigma_1|$, $|\sigma_2|$ shown in Equations (2.4.3.26-2) and (2.4.3.26-3) are trained on large datasets shown in Table 2.4.3.3 using ANNs shown in Equations (2.4.3.25-1)–(2.4.3.25-3) and Table 2.4.3.3. Figure 2.4.3.11 yields weight and bias matrices at each hidden and output layer which are presented in Table 2.4.3.4. Jacobi and Hessian matrices are derived from these matrices. Inequality constraining functions $v_1(x_A, x_h)$ and $v_2(x_A, x_h)$ are based on Equations (2.4.3.5-2) and (2.4.3.5-3), respectively.

2.4.3.2.4.2 ITERATION 0 BASED ON INITIAL INPUT DESIGN PARAMETERS

In this section, objective function $y_{fw} = f_W\left(x_A^{(0)}, x_h^{(0)}\right) = z_{fw}^{(D)}$, Jacobi matrix $\nabla f_W(\mathrm{x}) = \mathrm{J}_{fw}^{(D)}\left(x_A^{(0)}, x_h^{(0)}\right)$, $\mathrm{J}_v\left(x_A^{(0)}, x_h^{(0)}\right)$, Hessian matrix $\mathrm{H}_{fw}^{(D)}$, $\mathrm{H}_{v1}^{(D)}\left(x_A^{(0)}, x_h^{(0)}\right)$, $\mathrm{H}_{v2}^{(D)}\left(x_A^{(0)}, x_h^{(0)}\right)$ are derived using weight matrix W and bias matrix b shown in Table 2.4.3.4. Weight matrix **W** and bias matrix **b** for Hidden layers 1–5 and Output layer 6 are derived from ANNs

Table 2.4.3.4 Weight and Bias Matrices Derived from ANN Based on Equations (2.4.3.26-1)–(2.4.3.26-3)

(a): Inequality Constraining Function v_1; Stress $|\sigma_1|$; Hidden Layers ($\mathbf{L}=1\sim5$), Output Layer ($\mathbf{L}=6$)

(a-1): Hidden Layer #1

$\mathbf{W}^{(1)}_{|\sigma_1|}$		$\mathbf{b}^{(1)}_{|\sigma_1|}$
3.171	−2.368	−4.649
3.142	2.788	−3.701
−1.255	−3.906	2.248
−3.739	−1.534	1.623
−1.563	0.341	−0.357
1.900	3.271	0.373
−0.063	6.069	−0.041
2.277	−5.321	2.863
3.740	0.436	4.403
−2.581	2.857	−4.334
[10×2]		[10×1]

(a-2): Hidden Layer #2

$\mathbf{W}^{(2)}_{|\sigma_1|}$										$\mathbf{b}^{(2)}_{|\sigma_1|}$
−0.365	−0.030	0.081	−0.286	−0.611	−0.275	0.030	0.043	−1.072	−0.101	2.022
−0.389	−0.312	0.375	−0.610	0.992	0.407	−0.026	0.428	0.712	−0.836	1.500
−0.599	−1.016	0.426	−0.425	−0.529	0.096	1.346	0.441	0.948	−0.749	1.377
0.818	0.152	0.127	0.384	0.716	−0.976	−0.322	−0.286	−0.414	1.430	−0.850
0.067	−0.052	0.021	0.128	0.483	0.132	0.645	−0.096	−1.149	0.326	0.385
−0.195	−0.211	−0.289	0.079	−0.694	−0.647	−4.522	−1.354	0.931	0.037	−0.444
0.244	1.149	−0.015	0.548	0.705	0.207	−1.142	0.591	−0.492	0.696	−1.262
−0.262	0.130	−0.557	−0.025	0.481	0.488	3.318	0.914	−0.118	−0.555	−1.175
0.101	−0.175	−0.300	0.553	−1.371	0.668	1.478	−0.796	1.466	−1.210	1.798
0.441	0.229	0.612	0.043	0.396	−0.360	−0.193	0.124	0.081	−0.449	2.054
[10×10]										[10×1]

(a-3): Hidden Layer #3

$\mathbf{W}^{(3)}_{|\sigma_1|}$										$\mathbf{b}^{(3)}_{|\sigma_1|}$
−0.186	0.247	0.844	−0.122	0.321	0.255	−0.806	−0.679	−0.231	−0.755	1.949
0.423	0.829	−0.337	0.506	−1.040	0.631	0.003	0.726	0.350	0.536	−1.087
−0.774	−0.645	−0.732	0.459	−1.325	−1.913	−0.102	−0.928	0.900	0.130	0.311
0.053	−0.509	0.897	−0.049	1.178	−0.482	−0.022	0.217	0.709	1.006	−0.367
−1.093	−0.435	0.383	0.790	−1.244	−2.663	−0.087	−2.965	−1.617	−0.116	0.000
0.219	0.552	−0.438	−0.766	−0.438	−0.999	−0.495	1.635	−0.810	−0.235	−0.339
−0.901	0.312	0.761	0.599	−0.377	0.234	0.492	0.347	−0.977	−0.323	−0.987
−0.065	0.411	−0.203	0.321	−0.782	0.228	0.942	−0.416	0.798	−0.754	−1.005
−0.201	−0.398	−1.265	0.265	−0.690	1.214	1.366	−0.539	−1.452	−0.636	1.039
0.861	−0.367	0.958	1.429	1.336	−0.444	−0.759	−0.783	−0.782	−0.010	−1.273
[10×10]										[10×1]

(Continued)

Table 2.4.3.4 (Continued) Weight and Bias Matrices Derived from ANN Based on Equations (2.4.3.26-1)–(2.4.3.26-3)

(a-4): Hidden Layer #4

$\mathbf{W}^{(4)}_{\lvert\sigma_1\rvert}$										$\mathbf{b}^{(4)}_{\lvert\sigma_1\rvert}$
0.777	−0.406	0.226	−0.328	2.719	−0.363	−0.651	0.434	−0.305	−0.200	−1.739
−1.209	0.219	−1.290	0.832	2.729	−0.520	0.347	−0.188	−1.299	1.084	1.208
0.115	0.777	−0.807	0.089	0.277	0.185	−0.657	0.051	1.177	−0.545	−1.122
−0.565	0.919	−0.734	−0.287	1.239	1.504	−0.422	−0.506	0.503	1.005	0.585
0.682	−0.666	0.443	0.607	−1.092	0.530	−0.172	0.149	0.323	−0.731	−0.102
0.586	0.297	−0.661	−0.302	0.055	−0.181	−0.654	0.876	−0.203	−1.465	0.133
0.687	−0.470	−0.411	0.134	0.112	−0.205	−0.830	−0.897	−0.588	−0.641	0.638
−0.410	0.537	−0.274	0.773	−1.917	0.076	0.447	−0.828	−0.772	0.236	−0.785
−0.224	0.574	−0.744	−0.230	−1.242	−0.846	0.019	0.426	−1.089	−0.439	−1.573
−0.826	−0.297	0.193	0.636	−1.492	−0.669	−0.125	−0.836	0.938	−0.642	−1.789
[10×10]										[10×1]

(a-5): Hidden Layer #5

$\mathbf{W}^{(5)}_{\lvert\sigma_1\rvert}$										$\mathbf{b}^{(5)}_{\lvert\sigma_1\rvert}$
−0.857	−0.180	−0.949	1.002	−0.058	0.457	−0.226	0.361	0.097	−0.172	1.795
0.718	1.234	−0.399	0.178	0.598	−0.475	−0.642	−1.325	−0.500	−1.208	−1.291
−1.083	−0.029	1.149	−0.389	0.702	−1.271	0.439	−0.445	0.275	0.177	0.808
0.195	0.454	−0.744	0.952	0.620	0.139	0.463	0.934	−0.117	−0.024	−0.678
0.021	1.102	0.325	0.599	0.149	−0.587	0.292	0.190	−0.042	−1.447	0.016
0.168	−1.000	−0.028	1.610	0.453	−0.635	−0.663	0.228	0.228	−0.503	0.302
1.780	2.364	−0.176	0.831	0.505	0.256	−0.331	−1.104	1.932	−1.502	0.736
−0.239	−0.560	−0.279	−0.075	0.163	−0.615	0.612	−0.835	0.582	−0.280	−1.209
1.096	−0.458	−1.174	−0.155	−0.385	−0.656	0.372	1.244	−0.544	−1.185	1.190
−0.078	0.393	−0.007	−0.297	−0.818	−0.868	0.999	0.048	0.519	0.299	−1.776
[10×10]										[10×1]

(a-6): Normalized Output Layer

$\mathbf{W}^{(6)}_{\lvert\sigma_1\rvert}$										$\mathbf{b}^{(6)}_{\lvert\sigma_1\rvert}$
−0.289	1.001	1.623	0.909	0.882	−0.440	2.337	−0.184	1.894	0.319	0.037
[1×10]										[1×1]

(b): Inequality Constraining Function v_2; Stress $\lvert\sigma_2\rvert$; Hidden Layers ($\mathbf{L}=1\sim5$), Output Layer ($\mathbf{L}=6$)

(b-1): Hidden Layer #1

$\mathbf{W}^{(1)}_{\lvert\sigma_2\rvert}$		$\mathbf{b}^{(1)}_{\lvert\sigma_2\rvert}$
−3.540	1.762	4.836
−2.252	3.362	3.860
−2.530	−2.066	1.711
2.728	−2.515	−1.441
−0.034	−4.876	−0.176
2.585	−2.420	0.768
2.506	−2.581	2.232

(Continued)

Table 2.4.3.4 (Continued) Weight and Bias Matrices Derived from ANN Based on Equations (2.4.3.26-1)–(2.4.3.26-3)

2.063	4.362	2.395
3.132	0.021	3.188
2.093	3.691	4.266
[10×2]		[10×1]

(b-2): Hidden Layer #2

$\mathbf{W}^{(2)}_{\|\sigma_2\|}$										$\mathbf{b}^{(2)}_{\|\sigma_2\|}$
0.067	0.056	−0.413	−0.256	0.785	0.857	−0.359	−0.536	0.243	−0.461	−1.846
0.074	0.769	−0.355	−0.398	−2.159	−0.160	−0.163	0.880	0.020	−0.309	−1.336
0.111	0.273	−0.118	−0.266	1.872	0.083	−0.256	−0.842	0.944	1.503	−1.207
0.722	0.828	−0.533	0.412	0.222	0.667	0.454	0.732	0.625	0.428	−0.198
−0.457	0.320	0.887	−0.297	0.850	−0.569	−0.819	−0.167	0.424	−0.439	0.022
0.384	0.678	0.639	−0.112	−0.764	−0.062	0.004	−0.483	−0.447	0.356	0.096
−0.152	0.116	0.463	0.562	1.118	0.435	0.049	0.909	0.616	−0.279	−0.550
−0.060	0.040	−0.279	−0.153	0.607	−0.220	−0.223	0.144	−1.695	0.924	−0.906
0.617	−0.360	0.549	0.209	0.572	−0.214	−0.188	0.697	−0.690	−1.077	1.308
−0.796	0.292	0.647	−0.618	−0.183	−0.312	0.014	−0.610	0.792	0.424	−1.757
[10×10]										[10×1]

(b-3): Hidden Layer #3

$\mathbf{W}^{(3)}_{\|\sigma_2\|}$										$\mathbf{b}^{(3)}_{\|\sigma_2\|}$
0.567	0.742	−0.473	0.607	−0.034	−0.733	0.662	0.710	0.124	−0.118	−1.873
−0.370	−1.116	0.352	0.297	0.123	−0.398	0.331	1.049	0.432	0.762	1.389
−0.831	−0.239	−0.570	−0.020	0.089	0.957	0.735	−0.078	−0.156	0.479	1.248
−0.259	0.275	−0.606	0.789	−0.520	0.141	1.007	0.254	0.594	0.719	0.762
−0.231	−0.762	−0.149	1.001	−0.423	0.565	0.168	0.979	0.437	−0.627	0.137
−0.466	1.718	−1.267	0.363	−1.041	0.178	−0.670	0.183	0.325	0.631	−0.317
0.129	−0.129	0.922	1.351	0.321	−0.723	0.493	0.330	0.902	−0.347	−0.832
−0.481	−0.830	1.021	−0.071	0.823	−0.147	1.110	0.321	−0.206	−0.657	−1.102
−0.095	0.327	0.738	0.581	−0.075	−0.785	0.458	−0.016	−0.598	0.014	1.290
0.903	−0.244	−0.619	−0.318	−0.086	0.292	0.068	0.671	−0.006	−0.997	1.925
[10×10]										[10×1]

(b-4): Hidden Layer #4

$\mathbf{W}^{(4)}_{\|\sigma_2\|}$										$\mathbf{b}^{(4)}_{\|\sigma_2\|}$
0.399	−0.407	−0.472	0.657	−0.537	0.857	0.409	−0.788	0.371	0.218	−1.843
0.448	−0.038	−0.563	−1.147	−0.352	−0.241	−1.035	0.492	−0.677	0.869	−1.309
−0.011	−0.705	−0.157	−0.429	−0.410	−0.032	−1.091	0.616	0.080	−0.981	−0.961
0.454	0.361	−0.928	0.014	1.434	0.058	0.301	−0.168	−1.885	−0.586	0.052
0.855	0.270	−0.262	−0.395	−0.427	0.550	0.439	0.472	1.008	−0.402	−0.290
0.673	0.286	0.719	−1.273	0.150	−0.400	0.615	−0.277	−0.343	−0.747	0.353
−0.762	−0.365	−0.186	0.307	−0.575	0.876	0.596	0.249	0.137	−0.546	−0.657
0.778	1.118	−0.540	−1.198	0.895	−1.686	0.959	1.485	1.031	−0.401	0.910
−0.244	−0.106	0.595	0.169	0.937	−0.760	−0.208	0.508	0.372	−0.506	−1.437
0.317	1.012	−0.779	0.496	−0.212	0.175	−0.275	−0.125	0.721	0.760	1.707
[10×10]										[10×1]

(Continued)

Table 2.4.3.4 (Continued) Weight and Bias Matrices Derived from ANN Based on Equations (2.4.3.26-1)–(2.4.3.26-3)

(b-5): Hidden Layer #5

$\mathbf{W}^{(5)}_{\vert\sigma_2\vert}$										$\mathbf{b}^{(5)}_{\vert\sigma_2\vert}$
0.350	−0.009	−0.122	−0.390	−0.015	−0.735	0.585	−0.568	0.913	−0.875	−1.795
0.930	0.131	−0.828	0.193	0.022	0.869	−0.131	0.179	−0.518	−0.623	−1.431
0.759	0.802	−0.318	0.594	0.814	0.003	−0.551	−0.136	0.149	−0.652	−0.981
0.940	0.325	0.111	−1.467	−0.403	−0.827	−0.216	−1.252	−0.826	−0.629	−0.565
−0.669	−0.086	−0.233	0.048	−0.735	0.963	0.506	0.432	−0.084	0.707	0.121
0.250	0.551	−0.766	1.628	−0.466	−0.471	−0.120	−1.259	−0.674	0.439	0.404
−0.757	−0.070	−0.111	0.078	0.776	−1.066	0.773	−0.963	−0.150	−0.295	−0.576
−0.114	0.094	−0.397	0.163	0.818	0.144	−0.524	−0.017	−0.969	−0.894	−0.999
−0.489	0.811	−0.278	−0.509	−0.509	0.493	0.207	0.774	−1.046	−0.803	−1.398
−0.137	−0.536	0.827	−0.424	1.027	−0.113	0.613	0.234	0.090	−0.481	−1.954
[10×10]										[10×1]

(b-6): Normalized Output Layer

$\mathbf{W}^{(6)}_{\vert\sigma_2\vert}$										$\mathbf{b}^{(6)}_{\vert\sigma_2\vert}$
−0.138	−0.118	−0.252	−1.716	0.497	2.246	−0.446	0.003	0.611	0.949	−0.388
[1×10]										[1×1]

Objective Function $f_W(\mathbf{x})$; Truss Weight, Hidden Layers ($\mathbf{L}=1\sim5$), Output Layer ($\mathbf{L}=6$)

(c-1): Hidden Layer #1

$\mathbf{W}^{(1)}_{f_w}$		$\mathbf{b}^{(1)}_{f_w}$
3.171	−2.368	−4.649
3.142	2.788	−3.701
−1.255	−3.906	2.248
−3.739	−1.534	1.623
−1.563	0.341	−0.357
1.900	3.271	0.373
−0.063	6.069	−0.041
2.277	−5.321	2.863
3.740	0.436	4.403
−2.581	2.857	−4.334
[10×2]		[10×1]

(Continued)

Table 2.4.3.4 (Continued) Weight and Bias Matrices Derived from ANN Based on Equations (2.4.3.26-1)–(2.4.3.26-3)

(c-2): Hidden Layer #2

$\mathbf{W}_{fw}^{(2)}$										$\mathbf{b}_{fw}^{(2)}$
−0.365	−0.030	0.081	−0.286	−0.611	−0.275	0.030	0.043	−1.072	−0.101	2.022
−0.389	−0.312	0.375	−0.610	0.992	0.407	−0.026	0.428	0.712	−0.836	1.500
−0.599	−1.016	0.426	−0.425	−0.529	0.096	1.346	0.441	0.948	−0.749	1.377
0.818	0.152	0.127	0.384	0.716	−0.976	−0.322	−0.286	−0.414	1.430	−0.850
0.067	−0.052	0.021	0.128	0.483	0.132	0.645	−0.096	−1.149	0.326	0.385
−0.195	−0.211	−0.289	0.079	−0.694	−0.647	−4.522	−1.354	0.931	0.037	−0.444
0.244	1.149	−0.015	0.548	0.705	0.207	−1.142	0.591	−0.492	0.696	−1.262
−0.262	0.130	−0.557	−0.025	0.481	0.488	3.318	0.914	−0.118	−0.555	−1.175
0.101	−0.175	−0.300	0.553	−1.371	0.668	1.478	−0.796	1.466	−1.210	1.798
0.441	0.229	0.612	0.043	0.396	−0.360	−0.193	0.124	0.081	−0.449	2.054
[10×10]										[10×1]

(c-3): Hidden Layer #3

$\mathbf{W}_{fw}^{(3)}$										$\mathbf{b}_{fw}^{(3)}$
−0.186	0.247	0.844	−0.122	0.321	0.255	−0.806	−0.679	−0.231	−0.755	1.949
0.423	0.829	−0.337	0.506	−1.040	0.631	0.003	0.726	0.350	0.536	−1.087
−0.774	−0.645	−0.732	0.459	−1.325	−1.913	−0.102	−0.928	0.900	0.130	0.311
0.053	−0.509	0.897	−0.049	1.178	−0.482	−0.022	0.217	0.709	1.006	−0.367
−1.093	−0.435	0.383	0.790	−1.244	−2.663	−0.087	−2.965	−1.617	−0.116	0.000
0.219	0.552	−0.438	−0.766	−0.438	−0.999	−0.495	1.635	−0.810	−0.235	−0.339
−0.901	0.312	0.761	0.599	−0.377	0.234	0.492	0.347	−0.977	−0.323	−0.987
−0.065	0.411	−0.203	0.321	−0.782	0.228	0.942	−0.416	0.798	−0.754	−1.005
−0.201	−0.398	−1.265	0.265	−0.690	1.214	1.366	−0.539	−1.452	−0.636	1.039
0.861	−0.367	0.958	1.429	1.336	−0.444	−0.759	−0.783	−0.782	−0.010	−1.273
[10×10]										[10×1]

(c-4): Hidden Layer #4

$\mathbf{W}_{fw}^{(4)}$										$\mathbf{b}_{fw}^{(4)}$
0.777	−0.406	0.226	−0.328	2.719	−0.363	−0.651	0.434	−0.305	−0.200	−1.739
−1.209	0.219	−1.290	0.832	2.729	−0.520	0.347	−0.188	−1.299	1.084	1.208
0.115	0.777	−0.807	0.089	0.277	0.185	−0.657	0.051	1.177	−0.545	−1.122
−0.565	0.919	−0.734	−0.287	1.239	1.504	−0.422	−0.506	0.503	1.005	0.585
0.682	−0.666	0.443	0.607	−1.092	0.530	−0.172	0.149	0.323	−0.731	−0.102
0.586	0.297	−0.661	−0.302	0.055	−0.181	−0.654	0.876	−0.203	−1.465	0.133
0.687	−0.470	−0.411	0.134	0.112	−0.205	−0.830	−0.897	−0.588	−0.641	0.638
−0.410	0.537	−0.274	0.773	−1.917	0.076	0.447	−0.828	−0.772	0.236	−0.785
−0.224	0.574	−0.744	−0.230	−1.242	−0.846	0.019	0.426	−1.089	−0.439	−1.573
−0.826	−0.297	0.193	0.636	−1.492	−0.669	−0.125	−0.836	0.938	−0.642	−1.789
[10×10]										[10×1]

(Continued)

Table 2.4.3.4 (Continued) Weight and Bias Matrices Derived from ANN Based on Equations (2.4.3.26-1)–(2.4.3.26-3)

(c-5): Hidden Layer #5

$\mathbf{W}_{fw}^{(5)}$										$\mathbf{b}_{fw}^{(5)}$
−0.857	−0.180	−0.949	1.002	−0.058	0.457	−0.226	0.361	0.097	−0.172	1.795
0.718	1.234	−0.399	0.178	0.598	−0.475	−0.642	−1.325	−0.500	−1.208	−1.291
−1.083	−0.029	1.149	−0.389	0.702	−1.271	0.439	−0.445	0.275	0.177	0.808
0.195	0.454	−0.744	0.952	0.620	0.139	0.463	0.934	−0.117	−0.024	−0.678
0.021	1.102	0.325	0.599	0.149	−0.587	0.292	0.190	−0.042	−1.447	0.016
0.168	−1.000	−0.028	1.610	0.453	−0.635	−0.663	0.228	0.228	−0.503	0.302
1.780	2.364	−0.176	0.831	0.505	0.256	−0.331	−1.104	1.932	−1.502	0.736
−0.239	−0.560	−0.279	−0.075	0.163	−0.615	0.612	−0.835	0.582	−0.280	−1.209
1.096	−0.458	−1.174	−0.155	−0.385	−0.656	0.372	1.244	−0.544	−1.185	1.190
−0.078	0.393	−0.007	−0.297	−0.818	−0.868	0.999	0.048	0.519	0.299	−1.776
[10×10]										[10×1]

(c-6): Normalized Output Layer

$\mathbf{W}_{fw}^{(6)}$										$\mathbf{b}_{fw}^{(6)}$
−0.289	1.001	1.623	0.909	0.882	−0.440	2.337	−0.184	1.894	0.319	0.037
[1×10]										[1×1]

for an objective function $f_W(\mathbf{x})$ shown in Equation (2.4.3.26-1), inequality constraining functions $v_1(x_A, x_b)$, and $v_2(x_A, x_b)$ for truss stress $|\sigma_1|, |\sigma_2|$ shown in Equations (2.4.3.26-2) and (2.4.3.26-3). Super scripts (N) and (1) indicate normalizing step and Hidden layer 1, respectively, whereas super script D represents de-normalizing step. Objective function $y_{fW} = f_W\left(x_A^{(0)},\ x_b^{(0)}\right) = z_{fW}^{(D)}$ and inequality constraining functions $v_1\left(x_A^{(0)},\ x_b^{(0)}\right), v_2\left(x_A^{(0)},\ x_b^{(0)}\right)$ at Iteration 0 are obtained using weight matrix **W** and bias matrix **b** shown in Table 2.4.3.4 calculated based on Equations (2.4.3.25-1) - (2.4.3.25-3) or (2.4.3.26-1)–(2.4.3.26-3). Jacobi matrix $\nabla f_W(\mathbf{x}) = \mathrm{J}_{fW}^{(D)}\left(x_A^{(0)},\ x_b^{(0)}\right), \mathrm{J}_{v1}^{(D)}\left(x_A^{(0)},\ x_b^{(0)}\right),\ \mathrm{J}_{v2}^{(D)}\left(x_A^{(0)},\ x_b^{(0)}\right)$ and Hessian matrix $\mathrm{H}_{fW}^{(D)}, \mathrm{H}_{v1}^{(D)}\left(x_A^{(0)}, x_b^{(0)}\right), \mathrm{H}_{v2}^{(D)}\left(x_A^{(0)},\ x_b^{(0)}\right)$ at Iteration 0 are, then, calculated based on Equations (2.4.3.26-5a)–(2.4.3.26-14c) which show how weight matrix **W** and bias matrix **b** obtained from Hidden layer 1 to an output layer shown in Table 2.4.3.4 are used. Step N shown in Equation (2.3.6.6-1) to Step D shown in Equation (2.3.6.6-9) is followed to calculate Jacobi and Hessian matrices shown in Equations (2.4.3.26-5a)–(2.4.3.26-14c) at Iteration 0.

For Iteration 0, Equations (2.4.3.26-5a)–(2.4.3.26-12e) are only derived for an inequality constraining function $|\sigma_1|$ using weight matrix **W** and bias matrix **b** shown in Table 2.4.3.4(a). Readers are recommended to derive the same ANNs for objective and inequality constraining function $|\sigma_2|$. At Iteration 1 shown in Equation (2.4.3.26-25a)–(2.4.3.26-31d), an inequality constraining function $|\sigma_2|$, Jacobi $\mathrm{J}_{v2}^{(D)}\left(x_A^{(0)},\ x_b^{(0)}\right)$, and Hessian matrices $\mathrm{H}_{v2}^{(D)}\left(x_A^{(0)}, x_b^{(0)}\right)$ for an inequality constraining function $|\sigma_2|$ are only obtained using weight matrix **W** and bias matrix **b** shown in Table 2.4.3.4(b) similarly.

a. Iteration 0; Step (N) for normalizing input, $\mathbf{z}_{|\sigma_1|}^{(N)}$

$\mathbf{z}_{|\sigma_1|}^{(N)}$, $\mathbf{J}_{|\sigma_1|}^{(N)}$, and $\mathbf{H}_{|\sigma_1|}^{(N)}$ are calculated using Equations (2.4.3.26-5a), (2.4.3.26-5b), and (2.4.3.26-5c) based on Equations (2.3.6.3-2a), (2.3.6.6-1), (2.3.6.10). Normalization begins by substituting initial input design parameters $\mathbf{x}^{(0)} = \left[x_A^{(0)},\ x_h^{(0)}\right]^T = [800, -7]^T$ indicated by red color into Equation (2.4.3.26-5a) to calculate normalized function $\mathbf{z}_{|\sigma_1|}^{(N)}$.

$$\mathbf{z}_{|\sigma_1|}^{(N)} = \mathbf{g}^{(N)}\left(\mathbf{x}^{(0)}\right) = \boldsymbol{\alpha}_{\mathrm{x}} \odot \left(\mathbf{x} - \mathbf{x}_{\min}\right) + \bar{\mathbf{x}}_{\min}$$

$$= \begin{bmatrix} 0.0007 \\ 0.0333 \end{bmatrix} \odot \left(\begin{bmatrix} 800 \\ -7 \end{bmatrix} - \begin{bmatrix} 100.2 \\ -30 \end{bmatrix} \right) + \begin{bmatrix} -1 \\ -1 \end{bmatrix}$$

$$= \begin{bmatrix} -0.5173 \\ -0.2334 \end{bmatrix} \qquad (2.4.3.26\text{-}5a,\ 2.3.6.3\text{-}2a)$$

$$\mathbf{J}_{|\sigma_1|}^{(N)} = I_2 \odot \boldsymbol{\alpha}_{\mathrm{x}} = \begin{bmatrix} 0.0007 & 0 \\ 0 & 0.0333 \end{bmatrix} \qquad (2.4.3.26\text{-}5b,\ 2.3.6.6\text{-}1)$$

$$\mathbf{H}_{|\sigma_1|}^{(N)} = \begin{bmatrix} 0 & 0 \\ 0 & 0 \end{bmatrix} \qquad (2.4.3.26\text{-}5c,\ 2.3.6.10)$$

b. Iteration 0; calculations of functions, Jacobi and Hessian matrices at normalized hidden layers, $\mathbf{z}_{|\sigma_1|}^{(l)}\ \forall l \in \{1, 2, 3, 4, 5\}$

a. Iteration 0; normalized hidden layer, $\mathbf{z}_{|\sigma_1|}^{(1)}$

Readers are referred to Section 2.3.6.3(2) for the calculation sequence for Jacobian and Hessian matrices, $\mathbf{z}^{(N)} \Rightarrow \mathbf{J}^{(N)} \Rightarrow \mathbf{H}^{(N)}$, $\mathbf{z}^{(l)} \Rightarrow \mathbf{J}^{(l)} \Rightarrow \mathbf{H}^{(l)}$, $\mathbf{z}^{(L)} \Rightarrow \mathbf{J}^{(L)} \Rightarrow \mathbf{H}^{(L)}$, $\mathbf{z}^{(D)} \Rightarrow \mathbf{J}^{(D)} \Rightarrow \mathbf{H}^{(D)}$. Dimensions of matrices for an objective function are shown in Equation (2.4.3.26-6a) for Hidden layer (1). Weight matrix $\mathbf{W}_{|\sigma_1|}^{(1)}$, $\mathbf{z}_{|\sigma_1|}^{(N)}$, and bias matrix $\mathbf{b}_{|\sigma_1|}^{(1)}$ are $[10\times 2]$, $[2\times 1]$, and $[10\times 1]$ matrices, respectively, resulting in $\mathbf{z}_{|\sigma_1|}^{(1)}$ which is $[10\times 1]$. Similarly, dimensions of matrices for a Jacobi matrix of an objective function are shown in Equation (2.4.3.26-6b) in which $\mathbf{z}_{|\sigma_1|}^{(1)}$, weight matrix $\mathbf{W}_{|\sigma_1|}^{(1)}$, and $\mathbf{J}_{|\sigma_1|}^{(N)}$ are $[10\times 1]$, $[10\times 2]$, and $[2\times 2]$ matrices, respectively, resulting in $\mathbf{J}_{|\sigma_1}^{(1)}$ which is $[10\times 2]$. Similarly, dimensions of matrices for a slice (1) Hessian matrix of an objective function are shown in Equation (2.4.3.26-6c) in which $\mathbf{z}_{|\sigma_1|}^{(1)}$, $\boldsymbol{i}_1^{(N)}$, weight matrix $\mathbf{W}_{|\sigma_1|}^{(1)}$, $\mathbf{J}_{|\sigma_1|}^{(N)}$, and $\mathbf{H}_{|\sigma_1|,1}^{(N)}$ are $[10\times 1]$, $[2\times 1]$, $[10\times 2]$, $[2\times 2]$, and $[2\times 2]$ matrices, respectively, resulting in $\mathbf{H}_{|\sigma_1|,1}^{(1)}$ which is $[10\times 2]$. Similarly, dimensions of matrices for a slice (2) Hessian matrix of an objective function are shown in Equation (2.4.3.26-6d) in which $\mathbf{z}_{|\sigma_1|}^{(1)}$, $\boldsymbol{i}_2^{(N)}$, weight matrix $\mathbf{W}_{|\sigma_1|}^{(1)}$, $\mathbf{J}_{|\sigma_1|}^{(N)}$, and $\mathbf{H}_{|\sigma_1|,2}^{(N)}$ are $[10\times 1]$, $[2\times 1]$, $[10\times 2]$, $[2\times 2]$, and $[2\times 2]$ matrices, respectively, resulting in $\mathbf{H}_{|\sigma_1|,2}^{(1)}$ which is $[10\times 2]$.

$$\mathbf{z}^{(1)}_{|\sigma_1|} = f_t^{(1)}\left(\underbrace{\mathbf{W}^{(1)}_{|\sigma_1|}}_{[10\times2]}\underbrace{\mathbf{z}^{(N)}_{|\sigma_1|}}_{[2\times1]} + \underbrace{\mathbf{b}^{(1)}_{|\sigma_1|}}_{[10\times1]}\right) = \begin{bmatrix} -1.0000 \\ -1.0000 \\ 0.9990 \\ 0.9992 \\ 0.3552 \\ -0.8794 \\ -0.8907 \\ 0.9943 \\ 0.8926 \\ -0.9987 \end{bmatrix}_{[10\times1]} \quad (2.4.3.26\text{-}6a,\ 2.3.6.3\text{-}2b)$$

$$\mathbf{J}^{(1)}_{|\sigma_1|} = \left(1-\underbrace{\left(\mathbf{z}^{(1)}_{|\sigma_1|}\right)^2}_{[10\times1]}\right) \odot \underbrace{\mathbf{W}^{(1)}_{|\sigma_1|}}_{[10\times2]}\underbrace{\mathbf{J}^{(N)}_{|\sigma_1|}}_{[2\times2]} = \begin{bmatrix} 0.0000 & -0.0000 \\ 0.0000 & 0.0000 \\ -0.0000 & -0.0003 \\ -0.0000 & -0.0001 \\ 0.0009 & 0.0099 \\ 0.0003 & 0.0247 \\ -0.0000 & 0.0418 \\ 0.0000 & -0.0020 \\ 0.0001 & 0.0005 \\ -0.0000 & 0.0002 \end{bmatrix}_{[10\times2]} \quad (2.4.3.26\text{-}6b,\ 2.3.6.6\text{-}3)$$

$$\mathbf{H}^{(1)}_{|\sigma_1|,\,1} = -2\underbrace{\mathbf{z}^{(1)}_{|\sigma_1|}}_{[10\times1]} \odot \left(1-\underbrace{\left(\mathbf{z}^{(1)}_{|\sigma_1|}\right)^2}_{[10\times1]}\right) \odot \underbrace{\boldsymbol{i}^{(N)}_1}_{[2\times1]} \odot \underbrace{\left(\mathbf{W}^{(1)}_{|\sigma_1|}\right)^2}_{[10\times2]}\underbrace{\mathbf{J}^{(N)}_{|\sigma_1|}}_{[2\times2]} + \left(1-\underbrace{\left(\mathbf{z}^{(1)}_{|\sigma_1|}\right)^2}_{[10\times1]}\right) \odot \underbrace{\mathbf{W}^{(1)}_{|\sigma_1|}}_{[10\times2]}\underbrace{\mathbf{H}^{(N)}_{|\sigma_1|,\,1}}_{[2\times2]}$$

$$= \begin{bmatrix} 0.0000 & -0.0001 \\ 0.0000 & 0.0001 \\ -0.0000 & -0.0044 \\ -0.0002 & -0.0042 \\ 0.0072 & 0.0762 \\ 0.0068 & 0.5697 \\ 0.0000 & -0.0322 \\ -0.0006 & 0.0633 \\ -0.0045 & -0.0255 \\ 0.0002 & -0.0089 \end{bmatrix}_{[10\times2]} \times 10^{-4}$$

(2.4.3.26-6c, 2.3.6.14)

$$\mathbf{H}^{(1)}_{|\sigma_1|,\,2} = -2\underbrace{\mathbf{z}^{(1)}_{|\sigma_1|}}_{[10\times1]} \odot \left(1-\underbrace{\left(\mathbf{z}^{(1)}_{|\sigma_1|}\right)^2}_{[10\times1]}\right) \odot \underbrace{\boldsymbol{i}_2^{(N)}}_{[2\times1]} \odot \underbrace{\left(\mathbf{W}^{(1)}_{|\sigma_1|}\right)^2}_{[10\times2]} \underbrace{\mathbf{J}^{(N)}_{|\sigma_1|}}_{[2\times2]} + \left(1-\underbrace{\left(\mathbf{z}^{(1)}_{|\sigma_1|}\right)^2}_{[10\times1]}\right) \odot \underbrace{\mathbf{W}^{(1)}_{|\sigma_1|}}_{[10\times2]} \underbrace{\mathbf{H}^{(N)}_{|\sigma_1|,\,2}}_{[2\times2]}$$

$$= \begin{bmatrix} -0.0000 & 0.0000 \\ 0.0000 & -0.0001 \\ -0.0000 & -0.0000 \\ -0.0000 & -0.0001 \\ 0.0000 & 0.0047 \\ 0.0001 & 0.0047 \\ -0.0000 & 0.0151 \\ 0.0000 & -0.0007 \\ -0.0000 & -0.0000 \\ -0.0000 & 0.0000 \end{bmatrix}_{[10\times2]}$$

(2.4.3.26-6d, 2.3.6.14)

b. Iteration 0; normalized hidden layer, $\mathbf{z}^{(2)}_{|\sigma_1|}$

For Hidden layer (2), objective function, a Jacobi matrix of an objective function, and a slice ($1 = \sigma_1$, $2 = \sigma_2$) Hessian matrix of an objective function are shown in Equations (2.4.3.26-7a)–(2.4.3.26-7d), respectively.

$$\mathbf{z}^{(2)}_{|\sigma_1|} = f_t^{(2)}\left(\mathbf{W}^{(2)}_{|\sigma_1|}\mathbf{z}^{(1)}_{|\sigma_1|} + \mathbf{b}^{(2)}_{|\sigma_1|}\right) = \begin{bmatrix} 0.8620 \\ 0.9993 \\ 0.9986 \\ -0.9661 \\ -0.9140 \\ 0.9986 \\ -0.9255 \\ -0.9981 \\ 0.9202 \\ 0.9974 \end{bmatrix} \quad (2.4.3.26\text{-}7a)$$

$$\mathbf{J}^{(2)}_{|\sigma_1|} = \left(1-\left(\mathbf{z}^{(2)}_{|\sigma_1|}\right)^2\right) \odot \mathbf{W}^{(2)}_{|\sigma_1|}\mathbf{J}^{(1)}_{|\sigma_1|} = \begin{bmatrix} 0.0001 & -0.0032 \\ -0.0000 & 0.0000 \\ 0.0000 & 0.0001 \\ -0.0001 & -0.0020 \\ -0.0001 & 0.0057 \\ 0.0000 & -0.0006 \\ -0.0001 & -0.0053 \\ -0.0000 & 0.0006 \\ 0.0002 & 0.0102 \\ -0.0000 & -0.0001 \end{bmatrix} \quad (2.4.3.26\text{-}7b)$$

$$\mathbf{H}^{(2)}_{|\sigma_1|,\,1} = -2\mathbf{z}^{(2)}_{|\sigma_1|} \odot \left(1-\left(\mathbf{z}^{(2)}_{|\sigma_1|}\right)^2\right) \odot i^{(1)}_1 \odot \left(\mathbf{W}^{(2)}_{|\sigma_1|}\right)^2 \mathbf{J}^{(1)}_{|\sigma_1|} + \left(1-\left(\mathbf{z}^{(2)}_{|\sigma_1|}\right)^2\right) \odot \mathbf{W}^{(2)}_{|\sigma_1|}\mathbf{H}^{(1)}_{|\sigma_1|,\,1}$$

$$= \begin{bmatrix} 0.0012 & -0.0226 \\ -0.0000 & 0.0009 \\ -0.0000 & -0.0018 \\ -0.0000 & 0.0046 \\ 0.0007 & -0.0368 \\ 0.0013 & 0.0054 \\ -0.0000 & 0.0978 \\ 0.0008 & -0.0027 \\ -0.0060 & -0.2774 \\ -0.0001 & -0.0015 \end{bmatrix} \times 10^{-4} \qquad (2.4.3.26\text{-}7c)$$

$$\mathbf{H}^{(2)}_{|\sigma_1|,\,2} = -2\mathbf{z}^{(2)}_{|\sigma_1|} \odot \left(1-\left(\mathbf{z}^{(2)}_{|\sigma_1|}\right)^2\right) \odot i^{(1)}_2 \odot \left(\mathbf{W}^{(2)}_{|\sigma_1|}\right)^2 \mathbf{J}^{(1)}_{|\sigma_1|} + \left(1-\left(\mathbf{z}^{(2)}_{|\sigma_1|}\right)^2\right) \odot \mathbf{W}^{(2)}_{|\sigma_1|}\mathbf{H}^{(1)}_{|\sigma_1|,\,2}$$

$$= \begin{bmatrix} -0.0000 & -0.0003 \\ 0.0000 & 0.0000 \\ -0.0000 & 0.0000 \\ 0.0000 & -0.0005 \\ -0.0000 & 0.0021 \\ 0.0000 & -0.0004 \\ 0.0000 & -0.0020 \\ -0.0000 & 0.0004 \\ -0.0000 & 0.0027 \\ -0.0000 & -0.0000 \end{bmatrix} \qquad (2.4.3.26\text{-}7d)$$

c. Iteration 0; normalized hidden layer, $\mathbf{z}^{(3)}_{|\sigma_1|}$

For Hidden layer (3), objective function, a Jacobi matrix of an objective function, and a slice ($1 = \sigma_1$, $2 = \sigma_2$) Hessian matrix of an objective function are shown in Equations (2.4.3.26-8a)–(2.4.3.26-8d), respectively.

$$\mathbf{z}^{(3)}_{|\sigma_1|} = f_t^{(3)}\left(\underbrace{\mathbf{W}^{(3)}_{|\sigma_1|}}_{[10\times10]}\underbrace{\mathbf{z}^{(2)}_{|\sigma_1|}}_{[10\times1]} + \underbrace{\mathbf{b}^{(3)}_{|\sigma_1|}}_{[10\times1]}\right) = \begin{bmatrix} 0.9978 \\ 0.7567 \\ -0.7149 \\ 0.0162 \\ -0.9511 \\ -0.9672 \\ -0.9913 \\ -0.6004 \\ -0.9568 \\ -0.9772 \end{bmatrix}_{[10\times1]} \qquad (2.4.3.26\text{-}8a)$$

$$\mathbf{J}^{(3)}_{|\sigma_1|} = \left(1 - \Big(\underbrace{\mathbf{z}^{(3)}_{|\sigma_1|}}_{[10\times1]}\Big)^2\right) \odot \underbrace{\mathbf{W}^{(3)}_{|\sigma_1|}}_{[10\times10]} \underbrace{\mathbf{J}^{(2)}_{|\sigma_1|}}_{[10\times2]} = \begin{bmatrix} -0.0000 & 0.0000 \\ 0.0001 & -0.0020 \\ 0.0001 & 0.0020 \\ 0.0001 & 0.0145 \\ -0.0000 & -0.0020 \\ -0.0000 & -0.0004 \\ -0.0000 & -0.0002 \\ 0.0001 & -0.0013 \\ -0.0000 & -0.0023 \\ -0.0000 & -0.0001 \end{bmatrix}_{[10\times2]} \quad (2.4.3.26\text{-}8b)$$

$$\mathbf{H}^{(3)}_{|\sigma_1|,\,1} = -2\,\underbrace{\mathbf{z}^{(3)}_{|\sigma_1|}}_{[10\times1]} \odot \left(1 - \Big(\underbrace{\mathbf{z}^{(3)}_{|\sigma_1|}}_{[10\times1]}\Big)^2\right) \odot \underbrace{\boldsymbol{i}^{(2)}_1}_{[10\times1]} \odot \underbrace{\Big(\mathbf{W}^{(3)}_{|\sigma_1|}\Big)^2}_{[10\times10]} \underbrace{\mathbf{J}^{(2)}_{|\sigma_1|}}_{[10\times2]} + \left(1 - \Big(\underbrace{\mathbf{z}^{(3)}_{|\sigma_1|}}_{[10\times1]}\Big)^2\right) \odot \underbrace{\mathbf{W}^{(3)}_{|\sigma_1|}}_{[10\times10]} \underbrace{\mathbf{H}^{(2)}_{|\sigma_1|,\,1}}_{[10\times2]}$$

$$= \begin{bmatrix} 0.0000 & -0.0001 \\ -0.0013 & -0.0216 \\ -0.0034 & -0.0904 \\ -0.0028 & -0.2509 \\ 0.0011 & 0.0656 \\ 0.0002 & 0.0117 \\ 0.0001 & 0.0079 \\ -0.0029 & -0.0620 \\ 0.0011 & 0.0689 \\ 0.0004 & 0.0039 \end{bmatrix}_{[10\times2]} \times 10^{-4} \quad (2.4.3.26\text{-}8c)$$

$$\mathbf{H}^{(3)}_{|\sigma_1|,\,2} = -2\,\underbrace{\mathbf{z}^{(3)}_{|\sigma_1|}}_{[10\times1]} \odot \left(1 - \Big(\underbrace{\mathbf{z}^{(3)}_{|\sigma_1|}}_{[10\times1]}\Big)^2\right) \odot \underbrace{\boldsymbol{i}^{(2)}_2}_{[10\times1]} \odot \underbrace{\Big(\mathbf{W}^{(3)}_{|\sigma_1|}\Big)^2}_{[10\times10]} \underbrace{\mathbf{J}^{(2)}_{|\sigma_1|}}_{[10\times2]} + \left(1 - \Big(\underbrace{\mathbf{z}^{(3)}_{|\sigma_1|}}_{[10\times1]}\Big)^2\right) \odot \underbrace{\mathbf{W}^{(3)}_{|\sigma_1|}}_{[10\times10]} \underbrace{\mathbf{H}^{(2)}_{|\sigma_1|,\,2}}_{[10\times2]}$$

$$= \begin{bmatrix} -0.0000 & 0.0000 \\ -0.0000 & -0.0007 \\ -0.0000 & 0.0002 \\ -0.0000 & 0.0047 \\ 0.0000 & -0.0006 \\ 0.0000 & -0.0000 \\ 0.0000 & -0.0001 \\ -0.0000 & -0.0011 \\ 0.0000 & -0.0006 \\ 0.0000 & 0.0001 \end{bmatrix}_{[10\times2]} \quad (2.4.3.26\text{-}8d)$$

d. Iteration 0; normalized hidden layer, $\mathrm{z}^{(4)}_{|\sigma_1|}$

For Hidden layer (3), objective function, a Jacobi matrix of an objective function, and a slice $(1=\sigma_1, 2=\sigma_2)$ Hessian matrix of an objective function are shown in Equations (2.4.3.26-9a)–(2.4.3.26-9d), respectively.

$$\mathbf{z}_{|\sigma_1|}^{(4)} = f_t^{(4)}\left(\mathbf{W}_{|\sigma_1|}^{(4)}\mathbf{z}_{|\sigma_1|}^{(3)} + \mathbf{b}_{|\sigma_1|}^{(4)}\right) = \begin{bmatrix} -0.9927 \\ 0.7766 \\ -0.2509 \\ -0.9725 \\ 0.6527 \\ 0.9972 \\ 0.9992 \\ 0.9393 \\ 0.9824 \\ -0.4958 \end{bmatrix} \tag{2.4.3.26-9a}$$

$$\mathbf{J}_{|\sigma_1|}^{(4)} = \left(1-\left(\mathbf{z}_{|\sigma_1|}^{(4)}\right)^2\right) \odot \mathbf{W}_{|\sigma_1|}^{(4)}\mathbf{J}_{|\sigma_1|}^{(3)} = \begin{bmatrix} -0.0000 & -0.0001 \\ -0.0001 & 0.0026 \\ -0.0001 & -0.0048 \\ -0.0000 & -0.0006 \\ 0.0001 & 0.0070 \\ 0.0000 & -0.0000 \\ -0.0000 & 0.0000 \\ 0.0000 & 0.0019 \\ 0.0000 & -0.0000 \\ 0.0000 & 0.0094 \end{bmatrix} \tag{2.4.3.26-9b}$$

$$\mathbf{H}_{|\sigma_1|,1}^{(4)} = -2\mathbf{z}_{|\sigma_1|}^{(4)} \odot \left(1-\left(\mathbf{z}_{|\sigma_1|}^{(4)}\right)^2\right) \odot \boldsymbol{i}_1^{(3)} \odot \left(\mathbf{W}_{|\sigma_1|}^{(4)}\right)^2 \mathbf{J}_{|\sigma_1|}^{(3)} + \left(1-\left(\mathbf{z}_{|\sigma_1|}^{(4)}\right)^2\right) \odot \mathbf{W}_{|\sigma_1|}^{(4)}\mathbf{H}_{|\sigma_1|,1}^{(3)}$$

$$= \begin{bmatrix} 0.0000 & 0.0030 \\ 0.0018 & -0.0044 \\ 0.0025 & 0.1186 \\ 0.0004 & 0.0173 \\ -0.0022 & -0.1441 \\ -0.0000 & 0.0003 \\ 0.0000 & 0.0001 \\ -0.0003 & -0.0392 \\ -0.0001 & -0.0028 \\ -0.0004 & -0.1225 \end{bmatrix} \times 10^{-4} \tag{2.4.3.26-9c}$$

$$\mathbf{H}^{(4)}_{|\sigma_1|,2} = -2\mathbf{z}^{(4)}_{|\sigma_1|} \odot \left(1-\left(\mathbf{z}^{(4)}_{|\sigma_1|}\right)^2\right) \odot \boldsymbol{i}^{(3)}_2 \odot \left(\mathbf{W}^{(4)}_{|\sigma_1|}\right)^2 \mathbf{J}^{(3)}_{|\sigma_1|} + \left(1-\left(\mathbf{z}^{(4)}_{|\sigma_1|}\right)^2\right) \odot \mathbf{W}^{(4)}_{|\sigma_1|}\mathbf{H}^{(3)}_{|\sigma_1|,2}$$

$$= \begin{bmatrix} 0.0000 & -0.0000 \\ -0.0000 & 0.0012 \\ 0.0000 & -0.0012 \\ 0.0000 & -0.0001 \\ -0.0000 & 0.0020 \\ 0.0000 & -0.0000 \\ 0.0000 & 0.0000 \\ -0.0000 & 0.0006 \\ -0.0000 & -0.0000 \\ -0.0000 & 0.0035 \end{bmatrix} \tag{2.4.3.26-9d}$$

Iteration 0; normalized hidden layer, $\mathrm{z}^{(5)}_{|\sigma_1|}$

e. For Hidden layer (3), objective function, a Jacobi matrix of an objective function, and a slice $(1 = \sigma_1, 2 = \sigma_2)$ Hessian matrix of an objective function are shown in Equations (2.4.3.26-10a)–(2.4.3.26-10d), respectively.

$$\mathbf{z}^{(5)}_{|\sigma_1|} = f_t^{(5)}\left(\mathbf{W}^{(5)}_{|\sigma_1|}\mathbf{z}^{(4)}_{|\sigma_1|} + \mathbf{b}^{(5)}_{|\sigma_1|}\right) = \begin{bmatrix} 0.9920 \\ -0.9999 \\ 0.8829 \\ -0.1811 \\ -0.6993 \\ -0.7437 \\ -0.9436 \\ -0.3470 \\ 0.9203 \\ -0.9364 \end{bmatrix} \tag{2.4.3.26-10a}$$

$$\mathbf{J}^{(5)}_{|\sigma_1|} = \left(1-\left(\mathbf{z}^{(5)}_{|\sigma_1|}\right)^2\right) \odot \mathbf{W}^{(5)}_{|\sigma_1|}\mathbf{J}^{(4)}_{|\sigma_1|} = \begin{bmatrix} 0.0000 & 0.0000 \\ -0.0000 & -0.0000 \\ -0.0000 & 0.0001 \\ 0.0001 & 0.0098 \\ -0.0000 & -0.0057 \\ 0.0000 & -0.0021 \\ -0.0000 & -0.0007 \\ 0.0000 & -0.0028 \\ 0.0000 & -0.0011 \\ -0.0000 & -0.0002 \end{bmatrix} \tag{2.4.3.26-10b}$$

$$\mathbf{H}^{(5)}_{|\sigma_1|,1} = -2\mathbf{z}^{(5)}_{|\sigma_1|} \odot \left(1-\left(\mathbf{z}^{(5)}_{|\sigma_1|}\right)^2\right) \odot i^{(4)}_1 \odot \left(\mathbf{W}^{(5)}_{|\sigma_1|}\right)^2 \mathbf{J}^{(4)}_{|\sigma_1|} + \left(1-\left(\mathbf{z}^{(5)}_{|\sigma_1|}\right)^2\right) \odot \mathbf{W}^{(5)}_{|\sigma_1|}\mathbf{H}^{(4)}_{|\sigma_1|,1}$$

$$= \begin{bmatrix} -0.0000 & -0.0013 \\ 0.0000 & 0.0000 \\ 0.0002 & 0.0044 \\ -0.0022 & -0.1873 \\ 0.0017 & 0.1056 \\ -0.0009 & 0.0048 \\ 0.0005 & 0.0169 \\ -0.0015 & 0.0070 \\ -0.0004 & 0.0042 \\ 0.0003 & 0.0088 \end{bmatrix} \times 10^{-4} \tag{2.4.3.26-10c}$$

$$\mathbf{H}^{(5)}_{|\sigma_1|,2} = -2\mathbf{z}^{(5)}_{|\sigma_1|} \odot \left(1-\left(\mathbf{z}^{(5)}_{|\sigma_1|}\right)^2\right) \odot i^{(4)}_2 \odot \left(\mathbf{W}^{(5)}_{|\sigma_1|}\right)^2 \mathbf{J}^{(4)}_{|\sigma_1|} + \left(1-\left(\mathbf{z}^{(5)}_{|\sigma_1|}\right)^2\right) \odot \mathbf{W}^{(5)}_{|\sigma_1|}\mathbf{H}^{(4)}_{|\sigma_1|,2}$$

$$= \begin{bmatrix} -0.0000 & 0.0000 \\ 0.0000 & -0.0000 \\ 0.0000 & 0.0001 \\ -0.0000 & 0.0029 \\ 0.0000 & -0.0018 \\ 0.0000 & -0.0009 \\ 0.0000 & -0.0002 \\ 0.0000 & -0.0013 \\ 0.0000 & -0.0005 \\ 0.0000 & -0.0000 \end{bmatrix} \tag{2.4.3.26-10d}$$

c. Iteration 0; normalized output layer, $\mathbf{z}^{(L=6=\text{output})}_{\sigma_1}$

For Hidden layer (3), objective function, a Jacobi matrix of an objective function, and a slice ($1=\sigma_1$, $2=\sigma_2$) Hessian matrix of an objective function are shown in Equations (2.4.3.26-11a)–(2.4.3.26-11d), respectively.

$$z^{(L=6=\text{output})}_{|\sigma_1|} = f^{(6)}_{\text{lin}}\left(\underbrace{\mathbf{W}^{(6)}_{|\sigma_1|}}_{[1\times10]}\underbrace{\mathbf{z}^{(5)}_{|\sigma_1|}}_{[10\times1]} + \underbrace{b^{(6)}_{|\sigma_1|}}_{[1\times1]}\right) = -0.9698 \tag{2.4.3.26-11a}$$

$$\mathbf{J}^{(L=6=\text{output})}_{|\sigma_1|} = \underbrace{\mathbf{W}^{(6)}_{|\sigma_1|}}_{[1\times10]}\underbrace{\mathbf{J}^{(5)}_{|\sigma_1|}}_{[10\times2]} = \begin{bmatrix} -0.0000 & 0.0016 \end{bmatrix}_{[1\times2]} \tag{2.4.3.26-11b}$$

$$\mathbf{H}^{(L=6=\text{output})}_{|\sigma_1|,1} = \underbrace{\mathbf{W}^{(6)}_{|\sigma_1|}}_{[1\times10]}\underbrace{\mathbf{H}^{(5)}_{|\sigma_1|,1}}_{[10\times2]} = \begin{bmatrix} -0.0095 & -0.2270 \end{bmatrix}_{[1\times2]} \times 10^{-5} \tag{2.4.3.26-11c}$$

$$\mathbf{H}^{(L=6=\text{output})}_{|\sigma_1|,2} = \underbrace{\mathbf{W}^{(6)}_{|\sigma_1|}}_{[1\times10]}\underbrace{\mathbf{H}^{(5)}_{|\sigma_1|,2}}_{[10\times2]} = \begin{bmatrix} -0.0023 & 0.4716 \end{bmatrix}_{[10\times2]} \times 10^{-3} \tag{2.4.3.26-11d}$$

A shown in Table 2.4.3.4(a-6), weight matrix $\mathbf{W}_{|\sigma_1|}^{(L=6=\text{output})}$ at an output layer shown in Equation (2.4.3.26-11) is (1×10) matrix which has one row and ten columns, resulting in (1×1) shown in Equation (2.4.3.26-11a) or (1×2) shown in Equations (2.4.3.26-11b)–(2.4.3.26-11d) scalar output parameter. An ANN *with* ten neurons shown in Figure 2.4.3.11a is trained for truss optimization based on PTM which yields one output parameter per train. A bias matrix is (1×1) matrix.

d. Iteration 0; de-normalized output layer (D, 6=L)

Equations (2.4.3.26-12c) and (2.4.3.26-12d) are slice Hessian matrices for σ_1 obtained using weight matrix **W** and bias matrix **b** shown in Tables 2.4.3.4(a) whereas Equation (2.4.3.26-12e) is global Hessian matrix. Objective function $y_{fw} = f_W\left(x_A^{(0)},\ x_h^{(0)}\right) = z_{fw}^{(D)}$, its Jacobi $\nabla f_W(\mathrm{x}) = \mathrm{J}_{fw}^{(D)}\left(x_A^{(0)},\ x_h^{(0)}\right)$, and Hessian $\mathrm{H}_{fw}^{(D)}$ are obtained in Equations (2.4.3.26-13a)–(2.4.3.26-13c) for $f_W(\mathbf{x})$ using weight matrix **W** and bias matrix **b** shown in Table 2.4.3.4(c). Inequality constraining function $|\sigma_2| = z_{|\sigma_2|}^{(D)}$, its Jacobi $\mathrm{J}_{v2}^{(D)}\left(x_A^{(0)},\ x_h^{(0)}\right)$, and Hessian $\mathrm{H}_{fw}^{(D)}$, $\mathrm{H}_{v2}^{(D)}\left(x_A^{(0)},\ x_h^{(0)}\right)$ are also obtained in Equations (2.4.3.26-14a)–(2.4.3.26-14c) for σ_2 using weight matrix **W** and bias matrix **b** shown in Table 2.4.3.4(b).

$$|\sigma_1| = z_{|\sigma_1|}^{(D)} = g_{|\sigma_1|}^{(D)}\left(z_{|\sigma_1|}^{(6)}\right) = 0.1043 \qquad (2.4.3.26\text{-}12a,\ 2.3.6.3\text{-}2f)$$

$$\mathbf{J}_{|\sigma_1|}^{(D)} = \frac{1}{\alpha_{|\sigma_1|}}\mathbf{J}_{|\sigma_1|}^{(6)} = \begin{bmatrix} -1.282\times10^{-4} & 0.0056 \end{bmatrix} \qquad (2.4.3.26\text{-}12b,\ 2.3.6.6\text{-}9)$$

$$\mathbf{H}_{|\sigma_1|,\,1}^{(D)} = \frac{1}{\alpha_{|\sigma_1|}}\mathbf{H}_{|\sigma_1|,\,1}^{(6)} = \begin{bmatrix} 3.28\times10^{-7} & -7.84\times10^{-6} \end{bmatrix} \qquad (2.4.3.26\text{-}12c,\ 2.3.6.16\text{-}2)$$

$$\mathbf{H}_{|\sigma_1|,\,2}^{(D)} = \frac{1}{\alpha_{|\sigma_1|}}\mathbf{H}_{|\sigma_1|,\,2}^{(6)} = \begin{bmatrix} -7.84\times10^{-6} & 0.0016 \end{bmatrix} \qquad (2.4.3.26\text{-}12d,\ 2.3.6.16\text{-}2)$$

$$\mathbf{H}_{|\sigma_1|}^{(D)} = \begin{bmatrix} \mathbf{H}_{|\sigma_1|,\,1}^{(D)} \\ \mathbf{H}_{|\sigma_1|,\,2}^{(D)} \end{bmatrix} = \begin{bmatrix} 3.28\times10^{-7} & -7.84\times10^{-6} \\ -7.84\times10^{-6} & 0.0016 \end{bmatrix} \qquad (2.4.3.26\text{-}12e,\ 2.3.6.17)$$

$$y_{fw} = z_{fw}^{(D)} = g_{fw}^{(D)}\left(z_{fw}^{(6)}\right) = 59.366 \qquad (2.4.3.26\text{-}13a,\ 2.3.6.3\text{-}2f)$$

$$\mathbf{J}_{fw}^{(D)} = \frac{1}{\alpha_{fw}}\mathbf{J}_{fw}^{(6)} = \begin{bmatrix} 0.0742 & -6.1504 \end{bmatrix} \qquad (2.4.3.26\text{-}13b,\ 2.3.6.6\text{-}9)$$

$$\mathbf{H}_{fw}^{(D)} = \begin{bmatrix} \mathbf{H}_{fw,\,1}^{(D)} \\ \mathbf{H}_{fw,\,2}^{(D)} \end{bmatrix} = \begin{bmatrix} -2.00\times10^{-6} & -0.0080 \\ -0.0080 & 0.1243 \end{bmatrix} \qquad (2.4.3.26\text{-}13c,\ 2.3.6.17)$$

$$|\sigma_2| = z_{|\sigma_2|}^{(D)} = g_{|\sigma_2|}^{(D)}\left(z_{|\sigma_2|}^{(6)}\right) = 0.1299 \qquad (2.4.3.26\text{-}14a,\ 2.3.6.3\text{-}2f)$$

$$\mathbf{J}_{|\sigma_2|}^{(D)} = \frac{1}{\alpha_{|\sigma_2|}}\mathbf{J}_{|\sigma_2|}^{(6)} = \begin{bmatrix} -1.598\times10^{-4} & 0.0013 \end{bmatrix} \qquad (2.4.3.26\text{-}14b,\ 2.3.6.6\text{-}9)$$

$$\mathbf{H}^{(D)}_{|\sigma_2|} = \begin{bmatrix} \mathbf{H}^{(D)}_{|\sigma_2|,1} \\ \mathbf{H}^{(D)}_{|\sigma_2|,2} \end{bmatrix} = \begin{bmatrix} 3.60\times10^{-7} & -8.36\times10^{-7} \\ -8.36\times10^{-7} & 9.11\times10^{-4} \end{bmatrix} \quad (2.4.3.26\text{-}14c,\ 2.3.6.17)$$

e. Iteration 0; convergence verification of Lagrange function based on Jacobi, Hessian matrix

Initial input design parameters $\left[x_A^{(0)}, x_h^{(0)}, \lambda_{v1}^{(0)}, \lambda_{v2}^{(0)}\right] = [800, -7, 0, 0]$ are substituted into a first derivative (Jacobian) of the Lagrange function shown in Equation (2.4.3.26-18) to begin iterations until it converges.

Inequality constraining functions $\mathbf{v}(\mathbf{x}) = \mathbf{v}\left(x_A^{(0)}, x_h^{(0)}\right)$ are calculated in Equations (2.4.3.26-15a)–(2.4.3.26-15c) using Equations (2.4.3.26-12a) and (2.4.3.26-14a). Jacobi $\mathbf{J}_{v1}^{(D)}\left(x_A^{(0)}, x_h^{(0)}\right)$ and $\mathbf{J}_{v2}^{(D)}\left(x_A^{(0)}, x_h^{(0)}\right)$ of inequality constraining function $\mathbf{v}(\mathbf{x}) = [v_1(x), v_2(x)]^T$ are obtained in Equations (2.4.3.26-16a)–(2.4.3.26-16c) based on Equations (2.4.3.26-12b) and (2.4.3.26-14b). Hessian $\mathbf{H}_{v1}^{(D)}\left(x_A^{(0)}, x_h^{(0)}\right)$ and $\mathbf{H}_{v2}^{(D)}\left(x_A^{(0)}, x_h^{(0)}\right)$ of inequality constraining function $\mathbf{v}(\mathbf{x}) = [v_1(x), v_2(x)]^T$ are also obtained in Equations (2.4.3.26-17a) and (2.4.3.26-17b) based on Equations (2.4.3.26-12e) and (2.4.3.26-14c).

$$v_1\left(x_A^{(0)}, x_h^{(0)}\right) = -\left|\sigma_1\left(x_A^{(0)}, x_h^{(0)}\right)\right| + f_y = -|\sigma_1| + 0.2 = 0.0957 \quad (2.4.3.26\text{-}15a)$$

$$v_2\left(x_A^{(0)}, x_h^{(0)}\right) = -\left|\sigma_2\left(x_A^{(0)}, x_h^{(0)}\right)\right| + f_y = -|\sigma_2| + 0.2 = 0.0701 \quad (2.4.3.26\text{-}15b)$$

$$\mathbf{v}\left(x_A^{(0)}, x_h^{(0)}\right) = \begin{bmatrix} v_1\left(x_A^{(0)}, x_h^{(0)}\right) \\ v_2\left(x_A^{(0)}, x_h^{(0)}\right) \end{bmatrix} = \begin{bmatrix} 0.0957 \\ 0.0701 \end{bmatrix} \quad (2.4.3.26\text{-}15c)$$

$$\mathbf{J}_{v1}^{(D)}\left(x_A^{(0)}, x_h^{(0)}\right) = \mathbf{J}^{(D)}_{(-|\sigma_1|+f_y)}\left(x_A^{(0)}, x_h^{(0)}\right) = -\mathbf{J}^{(D)}_{|\sigma_1|}\left(x_A^{(0)}, x_h^{(0)}\right)$$
$$= \begin{bmatrix} 1.282\times10^{-4} & -0.0056 \end{bmatrix} \quad (2.4.3.26\text{-}16a)$$

$$\mathbf{J}_{v2}^{(D)}\left(x_A^{(0)}, x_h^{(0)}\right) = \mathbf{J}^{(D)}_{(-|\sigma_2|+f_y)}\left(x_A^{(0)}, x_h^{(0)}\right) = -\mathbf{J}^{(D)}_{|\sigma_2|}\left(x_A^{(0)}, x_h^{(0)}\right)$$
$$= \begin{bmatrix} 1.598\times10^{-4} & -0.0013 \end{bmatrix} \quad (2.4.3.26\text{-}16b)$$

$$\mathbf{J}_{\mathbf{v}}\left(x_A^{(0)}, x_h^{(0)}\right) = \begin{bmatrix} \mathbf{J}_{v1}^{(D)}\left(x_A^{(0)}, x_h^{(0)}\right) \\ \mathbf{J}_{v2}^{(D)}\left(x_A^{(0)}, x_h^{(0)}\right) \end{bmatrix}$$
$$= \begin{bmatrix} 1.282\times10^{-4} & -0.0056 \\ 1.598\times10^{-4} & -0.0013 \end{bmatrix} \quad (2.4.3.26\text{-}16c)$$

$$\mathbf{H}_{v1}^{(D)}\left(x_A^{(0)},\ x_b^{(0)}\right)=\mathbf{H}_{(-|\sigma_1|+f_y)}^{(D)}\left(x_A^{(0)},\ x_b^{(0)}\right)=-\mathbf{H}_{|\sigma_1|}^{(D)}\left(x_A^{(0)},\ x_b^{(0)}\right)$$

$$=\begin{bmatrix} -3.28\times10^{-7} & 7.84\times10^{-6} \\ 7.84\times10^{-6} & -0.0016 \end{bmatrix} \quad (2.4.3.26\text{-}17a)$$

$$\mathbf{H}_{v2}^{(D)}\left(x_A^{(0)},\ x_b^{(0)}\right)=\mathbf{H}_{(-|\sigma_2|+f_y)}^{(D)}\left(x_A^{(0)},\ x_b^{(0)}\right)=-\mathbf{H}_{|\sigma_2|}^{(D)}\left(x_A^{(0)},\ x_b^{(0)}\right)$$

$$=\begin{bmatrix} -3.60\times10^{-7} & 8.36\times10^{-7} \\ 8.36\times10^{-7} & -9.11\times10^{-4} \end{bmatrix} \quad (2.4.3.26\text{-}17b)$$

First derivative (Jacobi, gradient vector) of Lagrange function at initial input design parameters $\mathbf{x}^{(0)}=\left[x_A^{(0)},\ x_b^{(0)}\right]^T=[800,-7]^T$ is obtained by substituting $\mathbf{J}_{fw}^{(D)}$ shown in Equation (2.4.3.26-13b), $\mathbf{v}\left(x_A^{(0)},\ x_b^{(0)}\right)$ shown in Equation (2.4.3.26-15c), $\mathbf{J}_{\mathrm{v}}\left(x_A^{(0)},\ x_b^{(0)}\right)$ shown in Equation (2.4.3.26-16c), $\boldsymbol{\lambda}_v^{(0)}=\left[\lambda_{v1}^{(0)},\ \lambda_{v2}^{(0)}\right]^T=[0,\ 0]^T$, and $\boldsymbol{S}=\begin{bmatrix} 1 & 0 \\ 0 & 0 \end{bmatrix}$ into Equations (2.4.3.24-1) or (2.4.3.26-18).

$$\nabla\mathcal{L}\left(x_A^{(0)},\ x_b^{(0)},\ \lambda_{v1}^{(0)},\ \lambda_{v2}^{(0)}\right)$$

$$=\begin{bmatrix} \left[\mathbf{J}_{fw}^{(D)}\left(x_A^{(0)},x_b^{(0)}\right)\right]^T-\mathbf{J}_{\mathrm{v}}\left(x_A^{(0)},x_b^{(0)}\right)^T\mathbf{S}\boldsymbol{\lambda}_v \\ -\mathbf{S}\mathbf{v}\left(x_A^{(0)},x_b^{(0)}\right) \end{bmatrix}$$

$$=\begin{bmatrix} \begin{bmatrix} 0.0742 \\ -6.1504 \end{bmatrix}-\begin{bmatrix} 1.282\times10^{-4} & 1.598\times10^{-4} \\ -0.0056 & -0.0013 \end{bmatrix}\begin{bmatrix} 1 & 0 \\ 0 & 0 \end{bmatrix}\begin{bmatrix} 0 \\ 0 \end{bmatrix} \\ -\begin{bmatrix} 1 & 0 \\ 0 & 0 \end{bmatrix}\begin{bmatrix} 0.0957 \\ 0.0701 \end{bmatrix} \end{bmatrix}$$

$$=\begin{bmatrix} 0.0742 \\ -6.1504 \\ -0.0957 \\ 0 \end{bmatrix}\neq 0 \quad (2.4.3.26\text{-}18)$$

With respect to $\left[x_A^{(0)},\ x_b^{(0)},\ \lambda_{v1}^{(0)},\ \lambda_{v2}^{(0)}\right]^T=[800,-7,0,0]^T$, ANN-based KKT condition $\nabla\mathcal{L}\left(x_A^{(0)},\ x_b^{(0)},\ \lambda_{v1}^{(0)},\ \lambda_{v2}^{(0)}\right)$ shown in Equation (2.4.3.26-18) yields a similar result compared to analytical KKT condition presented in Equation (2.4.3.15-2). However, this KKT condition with initial input design parameters $\left[x_A^{(0)},\ x_b^{(0)},\ \lambda_{v1}^{(0)},\ \lambda_{v2}^{(0)}\right]^T=[800,-7,0,0]^T$ does not converge to yield a stationary point, failing to obtain a root of the KKT condition to yield a minimized truss weight. A root of the KKT condition should be updated based on Iteration 1 as shown in

Equation (2.4.3.26-19) using $\nabla\mathcal{L}\left(x_A^{(0)},\ x_h^{(0)},\ \lambda_{v1}^{(0)},\ \lambda_{v2}^{(0)}\right)$ shown in Equation (2.4.3.26-18) and $\left[\mathbf{H}_{\mathcal{L}}\left(x_A^{(0)},\ x_h^{(0)},\ \lambda_{v1}^{(0)},\ \lambda_{v2}^{(0)}\right)\right]^{-1}$ shown in Equation (2.4.3.26-23).

$$\begin{bmatrix} x_A^{(1)} \\ x_h^{(1)} \\ \lambda_{v1}^{(1)} \\ \lambda_{v2}^{(1)} \end{bmatrix} = \begin{bmatrix} x_A^{(0)} \\ x_h^{(0)} \\ \lambda_{v1}^{(0)} \\ \lambda_{v2}^{(0)} \end{bmatrix} - \left[\mathbf{H}_{\mathcal{L}}\left(x_A^{(0)},\ x_h^{(0)},\ \lambda_{v1}^{(0)},\ \lambda_{v2}^{(0)}\right)\right]^{-1} \nabla\mathcal{L}\left(x_A^{(0)},\ x_h^{(0)},\ \lambda_{v1}^{(0)},\ \lambda_{v2}^{(0)}\right) \tag{2.4.3.26-19}$$

Hessian matrix $\mathbf{H}_{\mathcal{L}}\left(x_A^{(0)},\ x_h^{(0)}\right)$ of Lagrange function should be obtained in advance by substituting $\mathbf{H}_{v_1}^{(D)}\left(x_A^{(0)},\ x_h^{(0)}\right)$ shown in (2.4.3.26-13c), $\mathbf{H}_{v_1}^{(D)}\left(x_A^{(0)},\ x_h^{(0)}\right)$ shown in (2.4.3.26-17a), $\mathbf{H}_{v_2}^{(D)}\left(x_A^{(0)},\ x_h^{(0)}\right)$ shown in Equation (2.4.3.26-17b) into Equation (2.4.3.26-21a), resulting in Equation (2.4.3.26-21b). Hessian matrix $\mathbf{H}_{\mathcal{L}}\left(x_A^{(0)},\ x_h^{(0)},\ \lambda_{v1}^{(0)},\ \lambda_{v2}^{(0)}\right)$ is, then, obtained in Equation (2.4.3.26-22) by substituting $\mathbf{H}_{\mathcal{L}}\left(x_A^{(0)},\ x_h^{(0)}\right)$ obtained in Equation (2.4.3.26-21b) with $J_v\left(x_A^{(0)},\ x_h^{(0)}\right)$ shown in Equation (2.4.3.26-16c) and $\mathrm{S} = \begin{bmatrix} 1 & 0 \\ 0 & 0 \end{bmatrix}$ into Equation (2.4.3.26-20b).

$$\mathbf{H}_{\mathcal{L}}\left(x_A^{(0)},\ x_h^{(0)},\ \lambda_{v1}^{(0)},\ \lambda_{v2}^{(0)}\right) = \begin{bmatrix} \mathbf{H}_{\mathcal{L}}\left(x_A^{(0)},\ x_h^{(0)}\right) & -\left(SJ_v\left(x_A^{(0)},\ x_h^{(0)}\right)\right)^T \\ -SJ_v\left(x_A^{(0)},\ x_h^{(0)}\right) & 0 \end{bmatrix} \tag{2.4.3.26-20a}$$

$$\mathbf{H}_{\mathcal{L}}\left(x_A^{(0)},\ x_h^{(0)},\lambda_{v1}^{(0)},\ \lambda_{v2}^{(0)}\right) = \begin{bmatrix} \mathbf{H}_{\mathcal{L}}\left(x_A^{(0)},\ x_h^{(0)}\right) & -\left[S\mathbf{J}_v^{(D)}\left(x_A^{(0)},\ x_h^{(0)}\right)\right]^T \\ -S\mathbf{J}_v^{(D)}\left(x_A^{(0)},\ x_h^{(0)}\right) & 0 \end{bmatrix} \tag{2.4.3.26-20b}$$

$$H_{\mathcal{L}}\left(x_A^{(0)},\ x_h^{(0)}\right) = H_{fw}\left(x_A^{(0)},\ x_h^{(0)}\right) - \sum_{i=1}^{2} s_i \lambda_{v_i} H_{v_i}\left(x_A^{(0)},\ x_h^{(0)}\right) \tag{2.4.3.26-21a}$$

$$\begin{aligned} H_{\mathcal{L}}\left(x_A^{(0)},\ x_h^{(0)}\right) &= \begin{bmatrix} -2.00\times10^{-6} & -0.0080 \\ -0.0080 & 0.1243 \end{bmatrix} - 1\times0\times\begin{bmatrix} -3.28\times10^{-7} & 7.84\times10^{-6} \\ 7.84\times10^{-6} & -0.0016 \end{bmatrix} \\ &\quad - 0\times0\times\begin{bmatrix} -3.60\times10^{-7} & 8.36\times10^{-7} \\ 8.36\times10^{-7} & -9.11\times10^{-4} \end{bmatrix} \\ &= \begin{bmatrix} -2.00\times10^{-6} & -0.0080 \\ -0.0080 & 0.1243 \end{bmatrix} \end{aligned} \tag{2.4.3.26-21b}$$

$$\mathbf{H}_{\mathcal{L}}\left(x_A^{(0)}, x_h^{(0)}, \lambda_{v1}^{(0)}, \lambda_{v2}^{(0)}\right)$$

$$= \begin{bmatrix} \begin{bmatrix} -2.00\times10^{-6} & -0.0080 \\ -0.0080 & 0.1243 \end{bmatrix} & -\left(\begin{bmatrix} 1 & 0 \\ 0 & 0 \end{bmatrix}\begin{bmatrix} 1.282\times10^{-4} & -0.0056 \\ 1.598\times10^{-4} & -0.0013 \end{bmatrix}\right)^T \\ -\begin{bmatrix} 1 & 0 \\ 0 & 0 \end{bmatrix}\begin{bmatrix} 1.282\times10^{-4} & -0.0056 \\ 1.598\times10^{-4} & -0.0013 \end{bmatrix} & 0 \end{bmatrix} \quad (2.4.3.26\text{-}22a)$$

$$\mathbf{H}_{\mathcal{L}}\left(x_A^{(0)}, x_h^{(0)}, \lambda_{v1}^{(0)}, \lambda_{v2}^{(0)}\right)$$

$$= \begin{bmatrix} -2.00\times10^{-6} & -0.0080 & 1.282\times10^{-4} & 0 \\ -0.0080 & 0.1243 & -0.0056 & 0 \\ 1.282\times10^{-4} & -0.0056 & 0 & 0 \\ 0 & 0 & 0 & 0 \end{bmatrix} \quad (2.4.3.26\text{-}22b)$$

In this case, $\mathbf{H}_{\mathcal{L}}\left(x_A^{(0)}, x_h^{(0)}, \lambda_{v1}^{(0)}, \lambda_{v2}^{(0)}\right)$ of Equation (2.4.3.26-22b) is singular because the fourth row/column have zero values, indicating the fourth input (λ_{v2}) does not affect Lagrange function $\nabla\mathcal{L}(x_A, x_h, \lambda_{v1}, \lambda_{v2})$, because $v_2(x_A, x_h) = -|\sigma_2| + f_y \geq 0$ was assumed as an inactive Case 1 KKT condition by equating Lagrange multiplier λ_{v2} as 0. The fourth row/column is, then, excluded when inverting $\mathbf{H}_{\mathcal{L}}\left(x_A^{(0)}, x_h^{(0)}, \lambda_{v1}^{(0)}, \lambda_{v2}^{(0)}\right)$ which becomes (3×3) matrix dimension. Hessian matrix $\mathbf{H}_{\mathcal{L}}\left(x_A^{(0)}, x_h^{(0)}, \lambda_{v1}^{(0)}, \lambda_{v2}^{(0)}\right)^{-1}$ is obtained in Equation (2.4.3.26-23b).

$$\mathbf{H}_{\mathcal{L}}\left(x_A^{(0)}, x_h^{(0)}, \lambda_{v1}^{(0)}, \lambda_{v2}^{(0)}\right)^{-1}$$

$$= \begin{bmatrix} \begin{bmatrix} -2.00\times10^{-6} & -0.0080 & 1.282\times10^{-4} \\ -0.0080 & 0.1243 & -0.0056 \\ 1.282\times10^{-4} & -0.0056 & 0 \end{bmatrix}^{-1} & \begin{matrix} 0 \\ 0 \\ 0 \end{matrix} \\ \begin{matrix} 0 & 0 & 0 \end{matrix} & 0 \end{bmatrix} \quad (2.4.3.26\text{-}23a)$$

$$\mathbf{H}_{\mathcal{L}}\left(x_A^{(0)}, x_h^{(0)}, \lambda_{v1}^{(0)}, \lambda_{v2}^{(0)}\right)^{-1} = \begin{bmatrix} -3{,}328.3 & -75.751 & -3{,}039.9 & 0 \\ -75.751 & -1.724 & 108.415 & 0 \\ -3{,}039.9 & 108.415 & -6{,}695.7 & 0 \\ 0 & 0 & 0 & 0 \end{bmatrix} \quad (2.4.3.26\text{-}23b)$$

Substituting Equations (2.4.3.26-18) and (2.4.3.26-23b) into Equation (2.4.3.26-19) updates initial input design parameters to $\mathbf{x}^{(1)} = \left[x_A^{(1)}, x_h^{(1)}\right]^T = [290.063, -1.607]^T$ shown in Equation (2.4.3.26-24) for use in Iteration 1. $\mathbf{H}_{\mathcal{L}}\left(x_A^{(0)}, x_h^{(0)}, \lambda_{v1}^{(0)}, \lambda_{v2}^{(0)}\right)^{-1}$ of Equation (2.4.3.26-23b) is calculated based on Equations (2.4.3.26-5a)–(2.4.3.26-14c) of Iteration 0.

$$\begin{bmatrix} x_A^{(1)} \\ x_h^{(1)} \\ \lambda_{v1}^{(1)} \\ \lambda_{v2}^{(1)} \end{bmatrix} = \begin{bmatrix} 800 \\ -7 \\ 0 \\ 0 \end{bmatrix} - \begin{bmatrix} -3328.3 & -75.751 & -3039.9 & 0 \\ -75.751 & -1.724 & 108.415 & 0 \\ -3039.9 & 108.415 & -6695.7 & 0 \\ 0 & 0 & 0 & 0 \end{bmatrix} \begin{bmatrix} 0.0742 \\ -6.1504 \\ -0.0957 \\ 0 \end{bmatrix}$$

$$\begin{bmatrix} x_A^{(1)} \\ x_h^{(1)} \\ \lambda_{v1}^{(1)} \\ \lambda_{v2}^{(1)} \end{bmatrix} = \begin{bmatrix} 290.063 \\ -1.607 \\ 251.444 \\ 0 \end{bmatrix} \tag{2.4.3.26-24}$$

2.4.3.2.4.3 ITERATION 1 BASED ON INITIAL INPUT DESIGN PARAMETERS

In this section, objective function $y_{fw} = f_W\left(x_A^{(1)}, x_h^{(1)}\right) = z_{fw}^{(D)}$, Jacobi matrix $\nabla f_W(\mathbf{x}) = \mathrm{J}_{fw}^{(D)}\left(x_A^{(1)}, x_h^{(1)}\right)$, $\mathrm{J}_v\left(x_A^{(1)}, x_h^{(1)}\right)$, Hessian matrix $\mathrm{H}_{fw}^{(D)}$, $\mathrm{H}_{v1}^{(D)}\left(x_A^{(1)}, x_h^{(1)}\right)$, $\mathrm{H}_{v2}^{(D)}\left(x_A^{(1)}, x_h^{(1)}\right)$ are derived using weight matrix W and bias matrix b shown in Table 2.4.3.4. Weight matrix $\mathbf{W}$ and bias matrix $\mathbf{b}$ for Hidden layers 1–5 and Output layer 6 are derived from ANNs for an objective function $f_W(\mathbf{x})$ shown in Equation (2.4.3.26-1), inequality constraining functions $v_1(x_A, x_h)$, and $v_2(x_A, x_h)$ for truss stress $|\sigma_1|$, $|\sigma_2|$ shown in Equations (2.4.3.26-2) and (2.4.3.26-3). Super scripts (N) and (1) indicate normalizing step and Hidden layer 1, respectively,

First derivative (Jacobi, gradient vector) of Lagrange function at initial input design parameters $\mathbf{x}^{(0)} = \left[x_A^{(0)}, x_h^{(0)}\right]^T = [800, -7]^T$ obtained based on Equations (2.4.3.26-5a)–(2.4.3.26-14c) does not converge to zero, and hence, initial input design parameters $\mathbf{x}^{(0)} = \left[x_A^{(0)}, x_h^{(0)}\right]^T = [800, -7]^T$ were updated to $\mathbf{x}^{(1)} = \left[x_A^{(1)}, x_h^{(1)}\right]^T = [290.063, -1.607]^T$ shown in Equation (2.4.3.26-24) for use in Iteration 1. Input design parameters $\mathbf{x}^{(1)} = \left[x_A^{(1)}, x_h^{(1)}\right]^T$ at Iteration 1 should be also updated to $\mathbf{x}^{(2)} = \left[x_A^{(2)}, x_h^{(2)}\right]^T$ for use in Iteration 2 using Equation (2.4.3.26–44) if first derivative (Jacobi, gradient vector) of Lagrange function at $\mathbf{x}^{(1)} = \left[x_A^{(1)}, x_h^{(1)}\right]^T = [290.063, -1.607]^T$ at Iteration 1 does not converge to zero.

Normalization begins at normalization step (N) by substituting updated input design parameters $\mathbf{x}^{(1)} = \left[x_A^{(1)}, x_h^{(1)}\right]^T = [290.063, -1.607]^T$ indicated by red color as shown in Equation (2.4.3.26-25a). Normalized $\mathbf{z}_{|\sigma_2|}^{(N)}$ is, then, calculated in Equation (2.4.3.26-25a). σ_2 is de-normalized at de-normalization step (D) as shown in Equation (2.4.3.26-32a). At Iteration 1 shown in Equation (2.4.3.26-25a)–(2.4.3.26-31d), an inequality constraining function $|\sigma_2|$, Jacobi $\mathbf{J}_{v2}^{(D)}\left(x_A^{(0)}, x_h^{(0)}\right)$, and Hessian matrices $\mathrm{H}_{v2}^{(D)}\left(x_A^{(0)}, x_h^{(0)}\right)$ are obtained using weight matrix $\mathbf{W}$ and bias matrix $\mathbf{b}$ shown in Table 2.4.3.4(b). Readers are recommended to derive the same equations for objective function, inequality constraining function $|\sigma_1|$, and their Jacobi and Hessian matrix at Iteration 1, referring to Section 2.3.6 for review.

a. Iteration 1; Step (N) for normalizing input, $\mathbf{z}_{|\sigma 2|}^{(N)}$

$$\mathbf{z}_{|\sigma 2|}^{(N)} = \mathbf{g}^{(N)}\left(\mathbf{x}^{(0)}\right) = \boldsymbol{\alpha}_{\mathrm{x}} \odot \left(\mathbf{x} - \mathbf{x}_{\min}\right) + \bar{\mathbf{x}}_{\min}$$

$$= \begin{bmatrix} 0.0007 \\ 0.0333 \end{bmatrix} \odot \left(\begin{bmatrix} 290.063 \\ -1.607 \end{bmatrix} - \begin{bmatrix} 100.2 \\ -30 \end{bmatrix} \right) + \begin{bmatrix} -1 \\ -1 \end{bmatrix}$$

$$= \begin{bmatrix} -0.8691 \\ -0.0536 \end{bmatrix} \tag{2.4.3.26-25a}$$

$$\mathbf{J}_{|\sigma 2|}^{(N)} = I_2 \odot \boldsymbol{\alpha}_{\mathrm{x}} = \begin{bmatrix} 0.0007 & 0 \\ 0 & 0.0333 \end{bmatrix} \tag{2.4.3.26-25b}$$

$$\mathbf{H}_{|\sigma 2|}^{(N)} = \begin{bmatrix} 0 & 0 \\ 0 & 0 \end{bmatrix} \tag{2.4.3.26-25c}$$

b. Iteration 1; calculations of output functions, Jacobi and Hessian matrices at normalized hidden layers, $\mathbf{z}_{|\sigma 2|}^{(l)}\ \ \forall l \in \{1, 2, 3, 4, 5\}$

a. Iteration 1; normalized hidden layer, $\mathbf{z}_{|\sigma 2|}^{(1)}$

$$\mathbf{z}_{|\sigma 2|}^{(1)} = f_t^{(1)}\left(\mathbf{W}_{|\sigma 2|}^{(1)} \mathbf{z}_{|\sigma 2|}^{(N)} + \mathbf{b}_{|\sigma 2|}^{(1)}\right) = \begin{bmatrix} 1.0000 \\ 1.0000 \\ 0.9994 \\ -0.9987 \\ 0.1141 \\ -0.8739 \\ 0.1901 \\ 0.3528 \\ 0.4347 \\ 0.9780 \end{bmatrix} \tag{2.4.3.26-26a}$$

$$\mathbf{J}_{|\sigma 2|}^{(1)} = \left(1 - \left(\mathbf{z}_{|\sigma 2|}^{(1)}\right)^2\right) \odot \mathbf{W}_{|\sigma 2|}^{(1)} \mathbf{J}_{|\sigma 2|}^{(N)} = \begin{bmatrix} -0.0000 & 0.0000 \\ -0.0000 & 0.0000 \\ -0.0000 & -0.0001 \\ 0.0000 & -0.0002 \\ -0.0000 & -0.1604 \\ 0.0004 & -0.0191 \\ 0.0017 & -0.0829 \\ 0.0012 & 0.1273 \\ 0.0018 & 0.0006 \\ 0.0001 & 0.0053 \end{bmatrix} \tag{2.4.3.26-26b}$$

$$\mathbf{H}^{(1)}_{|\sigma 2|,\,1} = -2\mathbf{z}^{(1)}_{|\sigma 2|} \odot \left(1-\left(\mathbf{z}^{(1)}_{|\sigma 2|}\right)^2\right) \odot i_1^{(N)} \odot \left(\mathbf{W}^{(1)}_{|\sigma 2|}\right)^2 \mathbf{J}^{(N)}_{|\sigma 2|} + \left(1-\left(\mathbf{z}^{(1)}_{|\sigma 2|}\right)^2\right) \odot \mathbf{W}^{(1)}_{|\sigma 2|} \mathbf{H}^{(N)}_{|\sigma 2|,\,1}$$

$$= \begin{bmatrix} -0.0000 & 0.0000 \\ -0.0000 & 0.0000 \\ -0.0000 & -0.0003 \\ 0.0000 & -0.0008 \\ -0.0000 & -0.0009 \\ 0.0013 & -0.0544 \\ -0.0011 & 0.0545 \\ -0.0013 & -0.1278 \\ -0.0033 & -0.0011 \\ -0.0002 & -0.0151 \end{bmatrix} \times 10^{-3} \qquad (2.4.3.26\text{-}26c)$$

$$\mathbf{H}^{(1)}_{|\sigma 2|,2} = -2\mathbf{z}^{(1)}_{|\sigma 2|} \odot \left(1-\left(\mathbf{z}^{(1)}_{|\sigma 2|}\right)^2\right) \odot i_2^{(N)} \odot \left(\mathbf{W}^{(1)}_{|\sigma 2|}\right)^2 \mathbf{J}^{(N)}_{|\sigma 2|} + \left(1-\left(\mathbf{z}^{(1)}_{|\sigma 2|}\right)^2\right) \odot \mathbf{W}^{(1)}_{|\sigma 2|} \mathbf{H}^{(N)}_{|\sigma 2|,2}$$

$$= \begin{bmatrix} 0.0000 & -0.0000 \\ 0.0000 & -0.0000 \\ -0.0000 & -0.0000 \\ -0.0000 & 0.0000 \\ -0.0000 & -0.0060 \\ -0.0001 & 0.0027 \\ 0.0001 & -0.0027 \\ -0.0001 & -0.0131 \\ -0.0000 & -0.0000 \\ -0.0000 & -0.0013 \end{bmatrix} \qquad (2.4.3.26\text{-}26d)$$

b. Iteration 1; normalized hidden layer, $\mathbf{z}^{(2)}_{|\sigma 2|}$

$$\mathbf{z}^{(2)}_{|\sigma 2|} = f_t^{(2)}\left(\mathbf{W}^{(2)}_{|\sigma 2|}\mathbf{z}^{(1)}_{|\sigma 2|} + \mathbf{b}^{(2)}_{|\sigma 2|}\right) = \begin{bmatrix} -0.9963 \\ -0.5153 \\ 0.7616 \\ 0.7092 \\ 0.8341 \\ 0.9528 \\ -0.5458 \\ -0.5477 \\ 0.7676 \\ -0.1976 \end{bmatrix} \qquad (2.4.3.26\text{-}27a)$$

$$\mathbf{J}_{|\sigma_2|}^{(2)} = \left(1 - \left(\mathbf{z}_{|\sigma_2|}^{(2)}\right)^2\right) \odot \mathbf{W}_{|\sigma_2|}^{(2)} \mathbf{J}_{|\sigma_2|}^{(1)} = \begin{bmatrix} -0.0000 & -0.0014 \\ 0.0006 & 0.3476 \\ 0.0001 & -0.1593 \\ 0.0015 & 0.0049 \\ -0.0003 & -0.0246 \\ -0.0001 & 0.0059 \\ 0.0017 & -0.0542 \\ -0.0022 & -0.0366 \\ -0.0003 & 0.0043 \\ 0.0005 & -0.0391 \end{bmatrix} \quad (2.4.3.26\text{-}27b)$$

$$\mathbf{H}_{|\sigma_2|,\ 1}^{(2)} = -2\mathbf{z}_{|\sigma_2|}^{(2)} \odot \left(1 - \left(\mathbf{z}_{|\sigma_2|}^{(2)}\right)^2\right) \odot \boldsymbol{i}_1^{(1)} \odot \left(\mathbf{W}_{|\sigma_2|}^{(2)}\right)^2 \mathbf{J}_{|\sigma_2|}^{(1)} + \left(1 - \left(\mathbf{z}_{|\sigma_2|}^{(2)}\right)^2\right) \odot \mathbf{W}_{|\sigma_2|}^{(2)} \mathbf{H}_{|\sigma_2|,\ 1}^{(1)}$$

$$= \begin{bmatrix} 0.0000 & 0.0015 \\ -0.0003 & 0.2171 \\ -0.0009 & 0.0907 \\ -0.0080 & -0.0788 \\ -0.0009 & -0.0411 \\ -0.0002 & 0.0209 \\ 0.0027 & -0.2406 \\ 0.0115 & 0.1077 \\ 0.0002 & -0.0236 \\ -0.0022 & 0.0785 \end{bmatrix} \times 10^{-3} \quad (2.4.3.26\text{-}27c)$$

$$\mathbf{H}_{|\sigma_2|,\ 2}^{(2)} = -2\mathbf{z}_{|\sigma_2|}^{(2)} \odot \left(1 - \left(\mathbf{z}_{|\sigma_2|}^{(2)}\right)^2\right) \odot \boldsymbol{i}_2^{(1)} \odot \left(\mathbf{W}_{|\sigma_2|}^{(2)}\right)^2 \mathbf{J}_{|\sigma_2|}^{(1)} + \left(1 - \left(\mathbf{z}_{|\sigma_2|}^{(2)}\right)^2\right) \odot \mathbf{W}_{|\sigma_2|}^{(2)} \mathbf{H}_{|\sigma_2|,\ 2}^{(1)}$$

$$= \begin{bmatrix} 0.0000 & 0.0005 \\ 0.0002 & 0.1709 \\ 0.0001 & -0.0925 \\ -0.0001 & -0.0055 \\ -0.0000 & -0.0038 \\ 0.0000 & 0.0002 \\ -0.0002 & -0.0074 \\ 0.0001 & -0.0026 \\ -0.0000 & -0.0047 \\ 0.0001 & 0.0079 \end{bmatrix} \quad (2.4.3.26\text{-}27d)$$

c. Iteration 1; normalized hidden layer, $\mathbf{z}^{(3)}_{|\sigma 2|}$

$$\mathbf{z}^{(3)}_{|\sigma 2|} = f_t^{(3)}\left(\mathbf{W}^{(3)}_{|\sigma 2|}\mathbf{z}^{(2)}_{|\sigma 2|} + \mathbf{b}^{(3)}_{|\sigma 2|}\right) = \begin{bmatrix} -0.9995 \\ 0.9612 \\ 0.9740 \\ 0.2928 \\ 0.8793 \\ -0.9418 \\ 0.5762 \\ 0.2618 \\ 0.5891 \\ 0.4206 \end{bmatrix} \quad (2.4.3.26\text{-}28a)$$

$$\mathbf{J}^{(3)}_{|\sigma 2|} = \left(1 - \left(\mathbf{z}^{(3)}_{|\sigma 2|}\right)^2\right) \odot \mathbf{W}^{(3)}_{|\sigma 2|}\mathbf{J}^{(2)}_{|\sigma 2|} = \begin{bmatrix} 0.0000 & 0.0003 \\ -0.0001 & -0.0404 \\ 0.0001 & -0.0023 \\ 0.0025 & 0.1101 \\ -0.0003 & -0.0547 \\ 0.0001 & 0.0943 \\ 0.0011 & -0.1461 \\ 0.0001 & -0.4836 \\ 0.0015 & -0.0203 \\ -0.0022 & 0.0210 \end{bmatrix} \quad (2.4.3.26\text{-}28b)$$

$$\mathbf{H}^{(3)}_{|\sigma 2|,1} = -2\mathbf{z}^{(3)}_{|\sigma 2|} \odot \left(1 - \left(\mathbf{z}^{(3)}_{|\sigma 2|}\right)^2\right) \odot \boldsymbol{i}_1^{(2)} \odot \left(\mathbf{W}^{(3)}_{|\sigma 2|}\right)^2 \mathbf{J}^{(2)}_{|\sigma 2|} + \left(1 - \left(\mathbf{z}^{(3)}_{|\sigma 2|}\right)^2\right) \odot \mathbf{W}^{(3)}_{|\sigma 2|}\mathbf{H}^{(2)}_{|\sigma 2|,1}$$

$$= \begin{bmatrix} 0.0000 & 0.0005 \\ 0.0003 & -0.1453 \\ -0.0002 & -0.0059 \\ -0.0052 & -0.3681 \\ 0.0007 & -0.1721 \\ -0.0002 & 0.1358 \\ -0.0059 & 0.1423 \\ 0.0068 & -0.3347 \\ -0.0069 & 0.0436 \\ 0.0061 & -0.0320 \end{bmatrix} \times 10^{-3} \quad (2.4.3.26\text{–}28c)$$

$$\mathbf{H}_{|\sigma_2|,2}^{(3)} = -2\mathbf{z}_{|\sigma_2|}^{(3)} \odot \left(1-\left(\mathbf{z}_{|\sigma_2|}^{(3)}\right)^2\right) \odot i_2^{(2)} \odot \left(\mathbf{W}_{|\sigma_2|}^{(3)}\right)^2 \mathbf{J}_{|\sigma_2|}^{(2)} + \left(1-\left(\mathbf{z}_{|\sigma_2|}^{(3)}\right)^2\right) \odot \mathbf{W}_{|\sigma_2|}^{(3)} \mathbf{H}_{|\sigma_2|,2}^{(2)}$$

$$= \begin{bmatrix} 0.0000 & 0.0003 \\ -0.0001 & -0.0586 \\ -0.0000 & 0.0003 \\ -0.0004 & 0.0794 \\ -0.0002 & -0.0529 \\ 0.0001 & 0.1957 \\ 0.0001 & -0.1219 \\ -0.0003 & -0.3668 \\ 0.0000 & -0.0112 \\ -0.0000 & 0.0062 \end{bmatrix} \quad (2.4.3.26\text{–}28d)$$

d. Iteration 1; normalized hidden layer, $\mathbf{z}_{|\sigma_2|}^{(4)}$

$$\mathbf{z}_{|\sigma_2|}^{(4)} = f_t^{(4)}\left(\mathbf{W}_{|\sigma_2|}^{(4)} \mathbf{z}_{|\sigma_2|}^{(3)} + \mathbf{b}_{|\sigma_2|}^{(4)}\right) = \begin{bmatrix} -0.9991 \\ -0.9971 \\ -0.9957 \\ -0.7512 \\ -0.8736 \\ 0.5061 \\ -0.8871 \\ 0.9994 \\ 0.7133 \\ 0.9604 \end{bmatrix} \quad (2.4.3.26\text{-}29a)$$

$$\mathbf{J}_{|\sigma_2|}^{(4)} = \left(1-\left(\mathbf{z}_{|\sigma_2|}^{(4)}\right)^2\right) \odot \mathbf{W}_{|\sigma_2|}^{(4)} \mathbf{J}_{|\sigma_2|}^{(3)} = \begin{bmatrix} 0.0000 & 0.0010 \\ -0.0000 & -0.0011 \\ 0.0000 & -0.0014 \\ -0.0008 & -0.0089 \\ 0.0005 & -0.0710 \\ -0.0012 & -0.1217 \\ 0.0007 & -0.0125 \\ 0.0000 & -0.0014 \\ 0.0008 & -0.1647 \\ 0.0000 & 0.0113 \end{bmatrix} \quad (2.4.3.26\text{-}29b)$$

$$\mathbf{H}^{(4)}_{|\sigma 2|,1} = -2\mathbf{z}^{(4)}_{|\sigma 2|} \odot \left(1-\left(\mathbf{z}^{(4)}_{|\sigma 2|}\right)^2\right) \odot \boldsymbol{i}^{(3)}_1 \odot \left(\mathbf{W}^{(4)}_{|\sigma 2|}\right)^2 \mathbf{J}^{(3)}_{|\sigma 2|} + \left(1-\left(\mathbf{z}^{(4)}_{|\sigma 2|}\right)^2\right) \odot \mathbf{W}^{(4)}_{|\sigma 2|}\mathbf{H}^{(3)}_{|\sigma 2|,1}$$

$$= \begin{bmatrix} -0.0000 & 0.0052 \\ 0.0007 & 0.0151 \\ 0.0001 & -0.0005 \\ 0.0054 & -0.0876 \\ 0.0000 & -0.2020 \\ -0.0025 & 0.2047 \\ 0.0019 & -0.0303 \\ 0.0000 & -0.0001 \\ -0.0025 & 0.1553 \\ -0.0002 & -0.0257 \end{bmatrix} \times 10^{-3} \quad (2.4.3.26\text{-}29c)$$

$$\mathbf{H}^{(4)}_{|\sigma 2|,2} = -2\mathbf{z}^{(4)}_{|\sigma 2|} \odot \left(1-\left(\mathbf{z}^{(4)}_{|\sigma 2|}\right)^2\right) \odot \boldsymbol{i}^{(3)}_2 \odot \left(\mathbf{W}^{(4)}_{|\sigma 2|}\right)^2 \mathbf{J}^{(3)}_{|\sigma 2|} + \left(1-\left(\mathbf{z}^{(4)}_{|\sigma 2|}\right)^2\right) \odot \mathbf{W}^{(4)}_{|\sigma 2|}\mathbf{H}^{(3)}_{|\sigma 2|,2}$$

$$= \begin{bmatrix} 0.0000 & 0.0019 \\ 0.0000 & 0.0005 \\ -0.0000 & -0.0002 \\ -0.0001 & -0.0181 \\ -0.0002 & -0.0000 \\ 0.0002 & -0.1524 \\ -0.0000 & 0.0180 \\ -0.0000 & -0.0050 \\ 0.0002 & -0.2492 \\ -0.0000 & 0.0047 \end{bmatrix} \quad (2.4.3.26\text{–}29d)$$

e. Iteration 1; normalized hidden layer, $\mathbf{z}^{(5)}_{|\sigma 2|}$

$$\mathbf{z}^{(5)}_{|\sigma 2|} = f_t^{(5)}\left(\mathbf{W}^{(5)}_{|\sigma 2|}\mathbf{z}^{(4)}_{|\sigma 2|} + \mathbf{b}^{(5)}_{|\sigma 2|}\right) = \begin{bmatrix} -0.9976 \\ -0.9683 \\ -0.9983 \\ -0.9964 \\ 0.9927 \\ -0.9560 \\ -0.9946 \\ -0.8592 \\ -0.8595 \\ -0.9980 \end{bmatrix} \quad (2.4.3.26\text{-}30a)$$

$$\mathbf{J}_{|\sigma_2|}^{(5)} = \left(1-\left(\mathbf{z}_{|\sigma_2|}^{(5)}\right)^2\right) \odot \mathbf{W}_{|\sigma_2|}^{(5)} \mathbf{J}_{|\sigma_2|}^{(4)} = \begin{bmatrix} 0.0000 & -0.0004 \\ -0.0001 & -0.0017 \\ -0.0000 & -0.0003 \\ 0.0000 & 0.0020 \\ -0.0000 & -0.0007 \\ -0.0001 & 0.0169 \\ 0.0000 & 0.0009 \\ -0.0000 & 0.0023 \\ -0.0003 & 0.0364 \\ -0.0000 & -0.0003 \end{bmatrix} \tag{2.4.3.26-30b}$$

$$\mathbf{H}_{|\sigma_2|,1}^{(5)} = -2\mathbf{z}_{|\sigma_2|}^{(5)} \odot \left(1-\left(\mathbf{z}_{|\sigma_2|}^{(5)}\right)^2\right) \odot i_1^{(4)} \odot \left(\mathbf{W}_{|\sigma_2|}^{(5)}\right)^2 \mathbf{J}_{|\sigma_2|}^{(4)} + \left(1-\left(\mathbf{z}_{|\sigma_2|}^{(5)}\right)^2\right) \odot \mathbf{W}_{|\sigma_2|}^{(5)} \mathbf{H}_{|\sigma_2|,1}^{(4)}$$

$$= \begin{bmatrix} 0.0004 & -0.0141 \\ 0.0034 & 0.1197 \\ 0.0001 & -0.0027 \\ -0.0001 & 0.0372 \\ 0.0006 & 0.0244 \\ 0.0141 & -0.7199 \\ 0.0013 & -0.0101 \\ 0.0013 & -0.1256 \\ 0.0057 & -0.4618 \\ 0.0001 & 0.0177 \end{bmatrix} \times 10^{-4} \tag{2.4.3.26-30c}$$

$$\mathbf{H}_{|\sigma_2|,2}^{(5)} = -2\mathbf{z}_{|\sigma_2|}^{(5)} \odot \left(1-\left(\mathbf{z}_{|\sigma_2|}^{(5)}\right)^2\right) \odot i_2^{(4)} \odot \left(\mathbf{W}_{|\sigma_2|}^{(5)}\right)^2 \mathbf{J}_{|\sigma_2|}^{(4)} + \left(1-\left(\mathbf{z}_{|\sigma_2|}^{(5)}\right)^2\right) \odot \mathbf{W}_{|\sigma_2|}^{(5)} \mathbf{H}_{|\sigma_2|,2}^{(4)}$$

$$= \begin{bmatrix} -0.0000 & -0.0004 \\ 0.0000 & -0.0006 \\ -0.0000 & -0.0001 \\ 0.0000 & 0.0037 \\ 0.0000 & -0.0018 \\ -0.0001 & 0.0250 \\ 0.0000 & 0.0025 \\ -0.0000 & 0.0063 \\ -0.0000 & 0.0582 \\ -0.0000 & -0.000 \end{bmatrix} \tag{2.4.3.26-30d}$$

c. Iteration 1; normalized output layer, $\mathbf{z}_{\sigma_2}^{(L=6=\text{output})}$

$$z_{|\sigma_2|}^{(L=6=\text{output})} = f_{lin}^{(6)}\left(\mathbf{W}_{|\sigma_2|}^{(6)} \mathbf{z}_{|\sigma_2|}^{(5)} + b_{|\sigma_2|}^{(6)}\right) = -0.8602 \tag{2.4.3.26-31a}$$

$$\mathbf{J}_{|\sigma_2|}^{(L=6=\text{output})} = \mathbf{W}_{|\sigma_2|}^{(6)}\mathbf{J}_{|\sigma_2|}^{(5)} = \begin{bmatrix} -0.0005 & 0.0559 \end{bmatrix} \tag{2.4.3.26-31b}$$

$$\mathbf{H}_{|\sigma_2|,\,1}^{(L=6=\text{output})} = \mathbf{W}_{|\sigma_2|}^{(6)}\mathbf{H}_{|\sigma_2|,1}^{(5)} = \begin{bmatrix} 3.41\times10^{-6} & -1.98\times10^{-4} \end{bmatrix}\times10^{-3} \tag{2.4.3.26-31c}$$

$$\mathbf{H}_{|\sigma_2|,\,2}^{(L=6=\text{output})} = \mathbf{W}_{|\sigma_2|}^{(6)}\mathbf{H}_{|\sigma_2|,2}^{(5)} = \begin{bmatrix} -1.98\times10^{-4} & 0.0836 \end{bmatrix} \tag{2.4.3.26-31d}$$

d. Iteration 1; de-normalized output layer (D, $6=L$)

Inequality constraining function $|\sigma_2| = z_{|\sigma_2|}^{(D)}$, its Jacobi $\mathbf{J}_{v_2}^{(D)}\left(x_A^{(1)},\ x_h^{(1)}\right)$, Hessian $\mathbf{H}_{v_2}^{(D)}\left(x_A^{(1)},\ x_h^{(1)}\right)$ are also obtained in Equations (2.4.3.26-32a)–(2.4.3.26-32e) which are calculated using weight matrix **W** and bias matrix **b** shown in Table 2.4.3.4(b) for $f_W(\mathbf{x})$ and σ_2. Equations (2.4.3.26-32c) and (2.4.3.26-32d) are slice Hessian matrices whereas Equation (2.4.3.26-32e) is global Hessian matrix. Similarly, objective function $y_{fw} = f_W\left(x_A^{(0)},\ x_h^{(0)}\right) = z_{fw}^{(D)}$, its Jacobi $\nabla f_W(\mathbf{x}) = \mathbf{J}_{fw}^{(D)}\left(x_A^{(1)},\ x_h^{(1)}\right)$, Hessian $\mathbf{H}_{fw}^{(D)}\left(x_A^{(1)},\ x_h^{(1)}\right)$, and inequality constraining function $|\sigma_1| = z_{|\sigma_1|}^{(D)}$, its Jacobi $\mathbf{J}_{v_1}^{(D)}\left(x_A^{(1)},\ x_h^{(1)}\right)$, Hessian $\mathbf{H}_{v_1}^{(D)}\left(x_A^{(1)},\ x_h^{(1)}\right)$ are also obtained in Equations (2.4.3.26-33a) to (2.4.3.26-34c) which are calculated using weight matrix **W** and bias matrix **b** shown in Table 2.4.3.4(c) and (a) for $f_W(\mathbf{x})$ and σ_1, respectively.

$$|\sigma_2| = z_{|\sigma_2|}^{(D)} = g_{|\sigma_2|}^{(D)}\left(z_{|\sigma_2|}^{(6)}\right) = 0.5513 \tag{2.4.3.26-32a, 2.3.6.3-2f}$$

$$\mathbf{J}_{|\sigma_2|}^{(D)} = \frac{1}{\alpha_{|\sigma_2|}}\mathbf{J}_{|\sigma_2|}^{(6)} = \begin{bmatrix} -0.0019 & 0.2071 \end{bmatrix} \tag{2.4.3.26-32b, 2.3.6.6-9}$$

$$\mathbf{H}_{|\sigma_2|,\,1}^{(D)} = \frac{1}{\alpha_{|\sigma_2|}}\mathbf{H}_{|\sigma_2|,\,1}^{(6)} = \begin{bmatrix} 1.26\times10^{-5} & -7.32\times10^{-4} \end{bmatrix} \tag{2.4.3.26-32c, 2.3.6.16-2}$$

$$\mathbf{H}_{|\sigma_2|,\,2}^{(D)} = \frac{1}{\alpha_{|\sigma_2|}}\mathbf{H}_{|\sigma_2|,\,2}^{(6)} = \begin{bmatrix} -7.32\times10^{-4} & 0.3098 \end{bmatrix} \tag{2.4.3.26-32d, 2.3.6.16-2}$$

$$\mathbf{H}_{|\sigma_2|}^{(D)} = \begin{bmatrix} \mathbf{H}_{|\sigma_2|,\,1}^{(D)} \\ \mathbf{H}_{|\sigma_2|,\,2}^{(D)} \end{bmatrix} = \begin{bmatrix} 1.26\times10^{-5} & -7.32\times10^{-4} \\ -7.32\times10^{-4} & 0.3098 \end{bmatrix} \tag{2.4.3.26-32e, 2.3.6.17}$$

$$y_{fw} = z_{fw}^{(D)} = g_{fw}^{(D)}\left(z_{fw}^{(6)}\right) = 10.577 \tag{2.4.3.26-33a, 2.3.6.3-2f}$$

$$\mathbf{J}_{fw}^{(D)} = \frac{1}{\alpha_{fw}}\mathbf{J}_{fw}^{(6)} = \begin{bmatrix} 0.0360 & -1.4038 \end{bmatrix} \tag{2.4.3.26-33b, 2.3.6.6-9}$$

$$\mathbf{H}_{fw}^{(D)} = \begin{bmatrix} \mathbf{H}_{fw,\,1}^{(D)} \\ \mathbf{H}_{fw,\,2}^{(D)} \end{bmatrix} = \begin{bmatrix} -7.36\times10^{-7} & -0.0049 \\ -0.0049 & 0.6294 \end{bmatrix} \tag{2.4.3.26-33c, 2.3.6.17}$$

$$|\sigma_1| = z_{|\sigma_1|}^{(D)} = g_{|\sigma_1|}^{(D)}\left(z_{|\sigma_1|}^{(6)}\right) = 0.6168 \qquad (2.4.3.26\text{-}34a,\ 2.3.6.3\text{-}2f)$$

$$\mathbf{J}_{|\sigma_1|}^{(D)} = \frac{1}{\alpha_{|\sigma_1|}}\mathbf{J}_{|\sigma_1|}^{(6)} = \begin{bmatrix} 0.0021 & 0.2734 \end{bmatrix} \qquad (2.4.3.26\text{-}34b,\ 2.3.6.6\text{-}9)$$

$$\mathbf{H}_{|\sigma_1|}^{(D)} = \begin{bmatrix} \mathbf{H}_{|\sigma_1|,\,1}^{(D)} \\ \mathbf{H}_{|\sigma_1|,\,2}^{(D)} \end{bmatrix} = \begin{bmatrix} 1.51\times10^{-5} & -8.95\times10^{-4} \\ -8.95\times10^{-4} & 0.3502 \end{bmatrix} \qquad (2.4.3.26\text{-}34c,\ 2.3.6.17)$$

e. Iteration 1; convergence verification of Lagrange function based on Jacobi, Hessian matrix

Updated input design parameters $\mathbf{x}^{(1)} = \left[x_A^{(1)},\ x_h^{(1)}\right]^T = [290.063,\ -1.607]^T$ at Iteration 1 is substituted into a first derivative (Jacobian) of Lagrange function $\nabla\mathcal{L}\left(x_A^{(1)},\ x_h^{(1)},\ \lambda_{v1}^{(1)},\ \lambda_{v2}^{(1)}\right)$ shown in Equation (2.4.3.26-38) to see if it converges to stationary points. Inequality constraining functions $\mathbf{v}(\mathbf{x}) = \mathbf{v}\left(x_A^{(1)},\ x_h^{(1)}\right)$ are calculated in Equations (2.4.3.26-35a)–(2.4.3.26-35c), using Equations (2.4.3.26-34a) and (2.4.3.26-32a). Jacobi $\mathbf{J}_{v_1}^{(D)}\left(x_A^{(1)},\ x_h^{(1)}\right)$ and $\mathbf{J}_{v_2}^{(D)}\left(x_A^{(1)},\ x_h^{(1)}\right)$ of inequality constraining function $\mathbf{v}(\mathbf{x}) = \left[v_1(x),\ v_2(x)\right]^T$ are obtained in Equations (2.4.3.26-36a)–(2.4.3.26-36c) based on Equations (2.4.3.26–34b) and (2.4.3.26–32b). Hessian $\mathbf{H}_{v_1}^{(D)}\left(x_A^{(1)},\ x_h^{(1)}\right)$ and $\mathbf{H}_{v_2}^{(D)}\left(x_A^{(1)},\ x_h^{(1)}\right)$ of inequality constraining function $\mathbf{v}(\mathbf{x}) = \left[v_1(x),\ v_2(x)\right]^T$ are also obtained in Equations (2.4.3.26-37a), and (2.4.3.26-37b) based on Equations (2.4.3.26-34c) and (2.4.3.26-32e).

$$v_1\left(x_A^{(1)},\ x_h^{(1)}\right) = -\left|\sigma_1\left(x_A^{(1)},\ x_h^{(1)}\right)\right| + f_y = -|\sigma_1| + 0.2 = -0.4168 \qquad (2.4.3.26\text{-}35a)$$

$$v_2\left(x_A^{(1)},\ x_h^{(1)}\right) = -\left|\sigma_2\left(x_A^{(1)},\ x_h^{(1)}\right)\right| + f_y = -|\sigma_2| + 0.2 = -0.3513 \qquad (2.4.3.26\text{-}35b)$$

$$\mathbf{v}\left(x_A^{(1)},\ x_h^{(1)}\right) = \begin{bmatrix} v_1\left(x_A^{(1)},\ x_h^{(1)}\right) \\ v_2\left(x_A^{(1)},\ x_h^{(1)}\right) \end{bmatrix} = \begin{bmatrix} -0.4168 \\ -0.3513 \end{bmatrix} \qquad (2.4.3.26\text{-}35c)$$

$$\mathbf{J}_{v_1}^{(D)}\left(x_A^{(1)},\ x_h^{(1)}\right) = \mathbf{J}_{(-|\sigma_1|+f_y)}^{(D)}\left(x_A^{(1)},\ x_h^{(1)}\right) = -\mathbf{J}_{|\sigma_1|}^{(D)}\left(x_A^{(1)},\ x_h^{(1)}\right)$$
$$= \begin{bmatrix} 0.0021 & 0.2734 \end{bmatrix} \qquad (2.4.3.26\text{-}36a)$$

$$\mathbf{J}_{v_2}^{(D)}\left(x_A^{(1)},\ x_h^{(1)}\right) = \mathbf{J}_{(-|\sigma_2|+f_y)}^{(D)}\left(x_A^{(1)},\ x_h^{(1)}\right) = -\mathbf{J}_{|\sigma_2|}^{(D)}\left(x_A^{(1)},\ x_h^{(1)}\right)$$
$$= \begin{bmatrix} -0.0019 & 0.2071 \end{bmatrix} \qquad (2.4.3.26\text{-}36b)$$

$$\mathbf{J}_{\mathbf{v}}\left(x_A^{(1)},\ x_h^{(1)}\right) = \begin{bmatrix} \mathbf{J}_{v_1}^{(D)}\left(x_A^{(1)},\ x_h^{(1)}\right) \\ \mathbf{J}_{v_2}^{(D)}\left(x_A^{(1)},\ x_h^{(1)}\right) \end{bmatrix} = \begin{bmatrix} 0.0021 & 0.2734 \\ -0.0019 & 0.2071 \end{bmatrix} \qquad (2.4.3.26\text{-}36c)$$

$$\mathbf{H}_{v1}^{(D)}\left(x_A^{(1)},\ x_h^{(1)}\right)=\mathbf{H}_{(-|\sigma_1|+f_y)}^{(D)}\left(x_A^{(1)},\ x_h^{(1)}\right)=-\mathbf{H}_{|\sigma_1|}^{(D)}\left(x_A^{(1)},\ x_h^{(1)}\right)$$

$$=\begin{bmatrix} 1.51\times10^{-5} & -8.95\times10^{-4} \\ -8.95\times10^{-4} & 0.3502 \end{bmatrix} \quad (2.4.3.26\text{-}37a)$$

$$\mathbf{H}_{v2}^{(D)}\left(x_A^{(1)},\ x_h^{(1)}\right)=\mathbf{H}_{(-|\sigma_2|+f_y)}^{(D)}\left(x_A^{(1)},\ x_h^{(1)}\right)=-\mathbf{H}_{|\sigma_2|}^{(D)}\left(x_A^{(1)},\ x_h^{(1)}\right)$$

$$=\begin{bmatrix} 1.26\times10^{-5} & -7.32\times10^{-4} \\ -7.32\times10^{-4} & 0.3098 \end{bmatrix} \quad (2.4.3.26\text{-}37b)$$

First derivative (Jacobi, gradient vector) of Lagrange function at updated input parameter $\mathbf{x}^{(1)}=\left[x_A^{(1)},\ x_h^{(1)}\right]^T=[290.063, -1.607]^T$ is obtained by substituting $\mathbf{J}_{fw}^{(D)}$ shown in Equation (2.4.3.26-33b), $\mathbf{v}\left(x_A^{(1)},\ x_h^{(1)}\right)$ shown in Equation (2.4.3.26-35c), $\mathbf{J}_{\mathbf{v}}\left(x_A^{(1)}, x_h^{(1)}\right)$ shown in Equation (2.4.3.26-36c), $\boldsymbol{\lambda}_v^{(1)}=\left[\lambda_{v1}^{(1)},\ \lambda_{v2}^{(1)}\right]^T=[251.444, 0]^T$ shown in Equation (2.4.3.26-24), and $\boldsymbol{S}=\begin{bmatrix} 1 & 0 \\ 0 & 0 \end{bmatrix}$ into Equations (2.4.3.24-1) or (2.4.3.26-38).

$$\nabla\mathcal{L}\left(x_A^{(1)},\ x_h^{(1)}, \lambda_{v1}^{(1)},\ \lambda_{v2}^{(1)}\right)=\begin{bmatrix} \left[\mathbf{J}_{fw}^{(D)}\left(x_A^{(1)},\ x_h^{(1)}\right)\right]^T-\mathbf{J}_{\mathbf{v}}\left(x_A^{(1)},\ x_h^{(1)}\right)^T\mathbf{S}\boldsymbol{\lambda}_v \\ -\mathbf{S}\mathbf{v}\left(x_A^{(1)},\ x_h^{(1)}\right) \end{bmatrix}$$

$$=\begin{bmatrix} \begin{bmatrix} 0.0360 \\ -1.4038 \end{bmatrix}-\begin{bmatrix} 0.0021 & 0.2734 \\ -0.0019 & 0.2071 \end{bmatrix}\begin{bmatrix} 1 & 0 \\ 0 & 0 \end{bmatrix}\begin{bmatrix} 251.444 \\ 0 \end{bmatrix} \\ -\begin{bmatrix} 1 & 0 \\ 0 & 0 \end{bmatrix}\begin{bmatrix} -0.4168 \\ -0.3513 \end{bmatrix} \end{bmatrix}$$

$$=\begin{bmatrix} -0.5018 \\ 67.347 \\ 0.4168 \\ 0 \end{bmatrix}\neq 0 \quad (2.4.3.26\text{-}38)$$

With respect to $\left[x_A^{(1)},\ x_h^{(1)}, \lambda_{v1}^{(1)}, \lambda_{v2}^{(1)}\right]^T=[290.063, -1.607, 251.444, 0]^T$, ANN-based KKT condition $\nabla\mathcal{L}\left(x_A^{(1)},\ x_h^{(1)}, \lambda_{v1}^{(1)},\ \lambda_{v2}^{(1)}\right)$ shown in Equation (2.4.3.26-38) does not converge to yield a stationary point, failing to obtain a root of KKT condition to minimize truss weight. A root of KKT condition should be updated for used in Iteration 2 as shown in Equation (2.4.3.26-39) using $\nabla\mathcal{L}\left(x_A^{(1)},\ x_h^{(1)},\ \lambda_{v1}^{(1)},\ \lambda_{v2}^{(1)}\right)$ shown in Equation (2.4.3.26-38) and $\left[\mathbf{H}_{\mathcal{L}}\left(x_A^{(1)},\ x_h^{(1)},\ \lambda_{v1}^{(1)},\ \lambda_{v2}^{(1)}\right)\right]^{-1}$ shown in Equation (2.4.3.26-43).

$\mathbf{H}_{\mathcal{L}}\left(x_A^{(1)},\ x_h^{(1)},\ \lambda_{v1}^{(1)},\ \lambda_{v2}^{(1)}\right)$ shown in Equation (2.4.3.26-40b) is calculated using Hessian matrix $\mathbf{H}_{\mathcal{L}}\left(x_A^{(1)},\ x_h^{(1)}\right)$ of Lagrange function which is obtained by substituting $\mathbf{H}_{fw}^{(D)}\left(x_A^{(1)},\ x_h^{(1)}\right)$ shown in (2.4.3.26-33c), $\mathbf{H}_{v1}^{(D)}\left(x_A^{(1)}, x_h^{(1)}\right)$ shown in (2.4.3.26-37a), and $\mathbf{H}_{v2}^{(D)}\left(x_A^{(1)},\ x_h^{(1)}\right)$ shown in Equation (2.4.3.26-37b) into Equation (2.4.3.26-41a), resulting in Equation (2.4.3.26-41b). Hessian matrix $\mathbf{H}_{\mathcal{L}}\left(x_A^{(1)},\ x_h^{(1)},\ \lambda_{v1}^{(1)},\ \lambda_{v2}^{(1)}\right)$ is, then, obtained in Equation (2.4.3.26-42) by substituting $\mathbf{H}_{\mathcal{L}}\left(x_A^{(1)},\ x_h^{(1)}\right)$ obtained in Equation (2.4.3.26-41b) with $J_v\left(x_A^{(1)},\ x_h^{(1)}\right)$ shown in Equation (2.4.3.26-36c) and $\mathbf{S}=\begin{bmatrix} 1 & 0 \\ 0 & 0 \end{bmatrix}$ into Equation (2.4.3.26-40b).

$$\begin{bmatrix} x_A^{(2)} \\ x_h^{(2)} \\ \lambda_{v1}^{(2)} \\ \lambda_{v2}^{(2)} \end{bmatrix} = \begin{bmatrix} x_A^{(1)} \\ x_h^{(1)} \\ \lambda_{v1}^{(1)} \\ \lambda_{v2}^{(1)} \end{bmatrix} - \left[\mathbf{H}_{\mathcal{L}}\left(x_A^{(1)},\ x_h^{(1)},\ \lambda_{v1}^{(1)},\ \lambda_{v2}^{(1)}\right)\right]^{-1} \nabla\mathcal{L}\left(x_A^{(1)},\ x_h^{(1)},\ \lambda_{v1}^{(1)},\ \lambda_{v2}^{(1)}\right) \quad (2.4.3.26\text{-}39)$$

$$\mathbf{H}_{\mathcal{L}}\left(x_A^{(1)},\ x_h^{(1)},\ \lambda_{v1}^{(1)},\ \lambda_{v2}^{(1)}\right) = \begin{bmatrix} \mathbf{H}_{\mathcal{L}}\left(x_A^{(1)},\ x_h^{(1)}\right) & -\left(\boldsymbol{S}\boldsymbol{J}_v\left(x_A^{(1)},\ x_h^{(1)}\right)\right)^T \\ -\boldsymbol{S}\boldsymbol{J}_v\left(x_A^{(1)},\ x_h^{(1)}\right) & 0 \end{bmatrix} \quad (2.4.3.26\text{-}40a)$$

$$\mathbf{H}_{\mathcal{L}}\left(x_A^{(1)},\ x_h^{(1)},\ \lambda_{v1}^{(1)},\ \lambda_{v2}^{(1)}\right) = \begin{bmatrix} \mathbf{H}_{\mathcal{L}}\left(x_A^{(1)},\ x_h^{(1)}\right) & -\left[\boldsymbol{S}\mathbf{J}_v^{(D)}\left(x_A^{(1)},\ x_h^{(1)}\right)\right]^T \\ -\boldsymbol{S}\mathbf{J}_v^{(D)}\left(x_A^{(1)},\ x_h^{(1)}\right) & 0 \end{bmatrix} \quad (2.4.3.26\text{-}40b)$$

$$H_{\mathcal{L}}\left(x_A^{(1)},\ x_h^{(1)}\right) = H_{fw}\left(x_A^{(1)},\ x_h^{(1)}\right) - \sum_{i=1}^{2} s_i \lambda_{v_i} H_{v_i}\left(x_A^{(1)},\ x_h^{(1)}\right) \quad (2.4.3.26\text{-}41a)$$

$$H_{\mathcal{L}}\left(x_A^{(1)},\ x_h^{(1)}\right) = \begin{bmatrix} -7.36\times10^{-7} & -0.0049 \\ -0.0049 & 0.6294 \end{bmatrix} - 1\times 251.444 \times \begin{bmatrix} 1.51\times10^{-5} & -8.95\times10^{-4} \\ -8.95\times10^{-4} & 0.3502 \end{bmatrix} - 0\times 0 \times \begin{bmatrix} 1.26\times10^{-5} & -7.32\times10^{-4} \\ -7.32\times10^{-4} & 0.3098 \end{bmatrix} = \begin{bmatrix} 0.0038 & -0.2300 \\ -0.2300 & 88.691 \end{bmatrix} \quad (2.4.3.26\text{-}41b)$$

$$\mathbf{H}_{\mathcal{L}}\left(x_A^{(1)},\ x_h^{(1)},\ \lambda_{v1}^{(1)},\ \lambda_{v2}^{(1)}\right) = \begin{bmatrix} \begin{bmatrix} 0.0038 & -0.2300 \\ -0.2300 & 88.691 \end{bmatrix} & -\left(\begin{bmatrix} 1 & 0 \\ 0 & 0 \end{bmatrix}\begin{bmatrix} 0.0021 & 0.2734 \\ -0.0019 & 0.2071 \end{bmatrix}\right)^T \\ -\begin{bmatrix} 1 & 0 \\ 0 & 0 \end{bmatrix}\begin{bmatrix} 0.0021 & 0.2734 \\ -0.0019 & 0.2071 \end{bmatrix} & 0 \end{bmatrix} \quad (2.4.3.26\text{-}42a)$$

$$\mathbf{H}_{\mathcal{L}}\left(x_A^{(1)},\ x_h^{(1)},\ \lambda_{v1}^{(1)},\ \lambda_{v2}^{(1)}\right)=\begin{bmatrix} 0.0038 & -0.2300 & -0.0021 & 0 \\ -0.2300 & 88.691 & 0.2734 & 0 \\ -0.0021 & 0.2734 & 0 & 0 \\ 0 & 0 & 0 & 0 \end{bmatrix} \tag{2.4.3.26-42b}$$

In this case, $\mathbf{H}_{\mathcal{L}}\left(x_A^{(1)},\ x_h^{(1)},\ \lambda_{v1}^{(1)},\ \lambda_{v2}^{(1)}\right)$ of Equation (2.4.3.26-42b) is singular because the fourth row/column have zero values, indicating the fourth input (λ_{v2}) does not affect Lagrange function $\nabla\mathcal{L}(x_A, x_h, \lambda_{v1}, \lambda_{v2})$, because $v_2(x_A, x_h) = -|\sigma_2| + f_y \geq 0$ was assumed as an inactive KKT condition by equating Lagrange multiplier λ_{v2} as 0. The fourth row/column is excluded because Inequality $v_2(x)$ is inactive in Case 1 when inverting $\mathbf{H}_{\mathcal{L}}\left(x_A^{(0)},\ x_h^{(0)},\ \lambda_{v1}^{(0)},\ \lambda_{v2}^{(0)}\right)$ which becomes (3×3) matrix dimension. Hessian matrix $\mathbf{H}_{\mathcal{L}}\left(x_A^{(1)},\ x_h^{(1)},\ \lambda_{v1}^{(1)},\ \lambda_{v2}^{(1)}\right)^{-1}$ is obtained in Equation (2.4.3.26-43).

$$\mathbf{H}_{\mathcal{L}}\left(x_A^{(1)},\ x_h^{(1)},\ \lambda_{v1}^{(1)},\ \lambda_{v2}^{(1)}\right)^{-1}$$

$$=\begin{bmatrix} \begin{bmatrix} 0.0038 & -0.2300 & -0.0021 \\ -0.2300 & 88.691 & 0.2734 \\ -0.0021 & 0.2734 & 0 \end{bmatrix}^{-1} & \begin{matrix} 0 \\ 0 \\ 0 \end{matrix} \\ \begin{matrix} 0 & \quad 0 & \quad 0 \end{matrix} & 0 \end{bmatrix} \tag{2.4.3.26-43a}$$

$$\mathbf{H}_{\mathcal{L}}\left(x_A^{(1)},\ x_h^{(1)},\ \lambda_{v1}^{(1)},\ \lambda_{v2}^{(1)}\right)^{-1}=\begin{bmatrix} 178.03 & 1.3927 & -302.00 & 0 \\ 1.3927 & 0.0109 & 1.2950 & 0 \\ -302.00 & 1.2950 & -674.06 & 0 \\ 0 & 0 & 0 & 0 \end{bmatrix} \tag{2.4.3.26-43b}$$

Substituting Equations (2.4.3.26-38) and (2.4.3.26-43b) into Equations (2.3.4.1-5) or (2.4.3.26-44) yields updated input parameter $\left[x_A^{(2)},\ x_h^{(2)},\ \lambda_{v1}^{(2)},\ \lambda_{v2}^{(2)}\right]$ for use in Iteration 2.

$$\begin{bmatrix} x_A^{(2)} \\ x_h^{(2)} \\ \lambda_{v1}^{(2)} \\ \lambda_{v2}^{(2)} \end{bmatrix}=\begin{bmatrix} 290.063 \\ -1.607 \\ 251.444 \\ 0 \end{bmatrix}-\begin{bmatrix} 178.03 & 1.3927 & -302.00 & 0 \\ 1.3927 & 0.0109 & 1.2950 & 0 \\ -302.00 & 1.2950 & -674.06 & 0 \\ 0 & 0 & 0 & 0 \end{bmatrix}\begin{bmatrix} -0.5018 \\ 67.347 \\ 0.4168 \\ 0 \end{bmatrix}$$

$$\rightarrow\begin{bmatrix} x_A^{(2)} \\ x_h^{(2)} \\ \lambda_{v1}^{(2)} \\ \lambda_{v2}^{(2)} \end{bmatrix}=\begin{bmatrix} 411.489 \\ -2.181 \\ 293.632 \\ 0 \end{bmatrix} \tag{2.4.3.26–44}$$

The updated $\mathbf{J}_{fw}^{(D)}\left(x_A^{(2)}, x_h^{(2)}\right)$, $\mathbf{v}\left(x_A^{(2)}, x_h^{(2)}\right)$, $\mathbf{J}_{\mathrm{v}}\left(x_A^{(2)}, x_h^{(2)}\right)$,$\boldsymbol{\lambda}_v^{(2)} = \left[\lambda_{v1}^{(2)}, \lambda_{v2}^{(2)}\right]^T$, and $\mathcal{S} = \begin{bmatrix} 1 & 0 \\ 0 & 0 \end{bmatrix}$ with respect to $\mathbf{x}^{(2)} = \left[x_A^{(2)}, x_h^{(2)}, \lambda_{v1}^{(2)}, \lambda_{v2}^{(2)}\right] = [411.489, -2.181, 293.632, 0]$ based on Equations (2.4.3.26-25a)–(2.4.3.26-34c) are substituted in Equation (2.4.3.26-45) to obtain updated first derivative (Jacobi) of Lagrange function with respect to updated input data $\left[x_A^{(2)}, x_h^{(2)}, \lambda_{v1}^{(2)}, \lambda_{v2}^{(2)}\right] = [411.489, -2.181, 293.632, 0]$ for Step 2. This still does not converge to yield a root of KKT condition, requiring further iterations.

$$\nabla\mathcal{L}\left(x_A^{(2)}, x_h^{(2)}, \lambda_{v1}^{(2)}, \lambda_{v2}^{(2)}\right) = \begin{bmatrix} \nabla f_{\mathrm{W}}\left(x_A^{(2)}, x_h^{(2)}\right) - J_{\mathrm{v}}\left(x_A^{(2)}, x_h^{(2)}\right)^T \mathbf{S}\boldsymbol{\lambda}_v^{(1)} \\ -\mathbf{S}\mathbf{v}\left(x_A^{(2)}, x_h^{(2)}\right) \end{bmatrix}$$

$$= \begin{bmatrix} -0.2112 \\ 27.4112 \\ 0.1545 \\ 0 \end{bmatrix} \neq 0 \qquad (2.4.3.26\text{-}45)$$

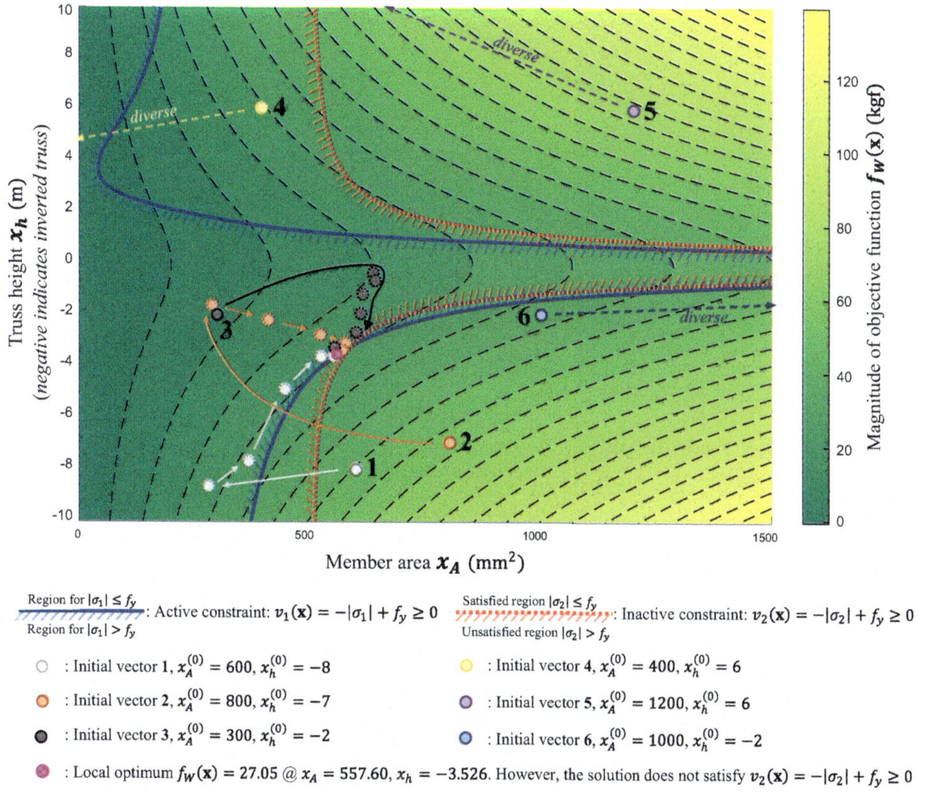

Figure 2.4.3.15 Minimized Lagrange function at $\left[x_A, x_h, \lambda_{v1}, \lambda_{v2}\right]$ based on Case I KKT condition; not satisfying inactive Inequality constraint $v_2(x) = -|\sigma_2| + f_y \ge 0$.

Table 2.4.3.5 Verification by Structural Calculation for Case I KKT Condition

Parameter		AI Results	Check (*Analytical*)	Error
		Optimimal $f_W(\mathbf{x})$ Design (Based on Five Layers with 10 Neurons)		
1	x_A (mm)	557.59	557.59	0.00%
2	x_h (m)	−3.526	−3.526	0.00%
3	$\lvert\sigma_1\rvert$ (kN/mm²)	$0.2000 = f_y$	$0.2004 = f_y$	−0.18%
4	$\lvert\sigma_2\rvert$ (kN/mm²)	$0.2067 > f_y$	$0.2062 > f_y$	0.25%
5	f_W (kgf)	27.05	27.00	0.17%

Note: (dotted box) Two inputs for AI design.
(dashed box) Two inputs for Analytical calculation.

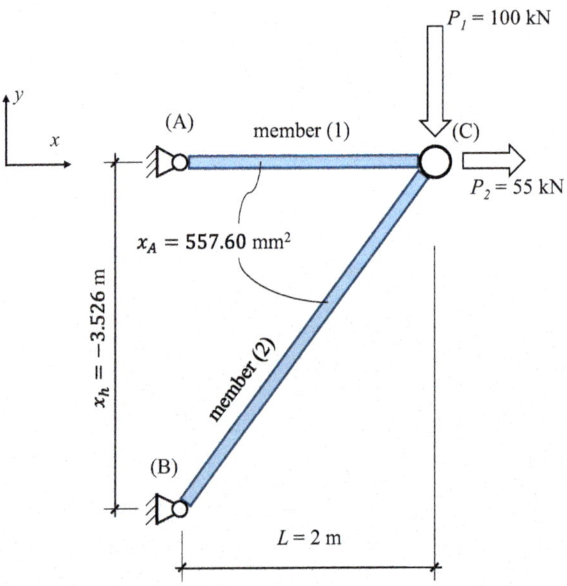

Figure 2.4.3.16 Truss topology with Case I KKT condition; not satisfying inactive Inequality constraint $v_2(x) = -\lvert\sigma_2\rvert + f_y \geq 0$.

A root of KKT condition converges at Iteration 9, yielding a stationary point having $\nabla\mathcal{L}\left(x_A^{(9)}, x_h^{(9)}, \lambda_{v1}^{(9)}, \lambda_{v2}^{(9)}\right) = 0$ at $\left[x_A^{(9)}, x_h^{(9)}, \lambda_{v1}^{(9)}, \lambda_{v2}^{(9)}\right] = [557.595, -3.526, 139.851, 0]$, resulting in a minimized weight of $f_W\left(x_A^{(9)}, x_h^{(9)}\right) = 27.05$ kgf as shown in Figure 2.4.3.15. In Iteration 9, the solution found as $\left[x_A^{(9)}, x_h^{(9)}, \lambda_{v1}^{(9)}, \lambda_{v2}^{(9)}\right] = [557.595, -3.526, 139.851, 0]$ should be verified by inactive Inequality $v_2\left(x_A^{(9)}, x_h^{(9)}\right)$, however, objective function $f_W\left(x_A^{(9)}, x_h^{(9)}\right) = 27.05$ kgf did not converge within a range defined by inactive Inequality $v_2\left(x_A^{(9)}, x_h^{(9)}\right)$ as shown in Figure 2.4.3.15 and Table 2.4.3.5. Six input design parameters assumed in Figures 2.4.3.3, 2.4.3.5, 2.4.3.7, and 2.4.3.9 of Section 2.4.3 for the calculation of Lagrange function $\nabla\mathcal{L}(\mathbf{x}, \boldsymbol{\lambda}_v) = \nabla\mathcal{L}(x_A, x_h, \lambda_{v1}, \lambda_{v2})$ with analytical functions were also used as trial input design parameters based on Newton–Raphson iteration when solving for Jacobi of Lagrange function $\nabla\mathcal{L}(\mathbf{x}, \boldsymbol{\lambda}_v) = \nabla\mathcal{L}(x_A, x_h, \lambda_{v1}, \lambda_{v2}) = 0$.

A root converged at $\left[\mathbf{x}, \lambda_v\right] = \left[x_A, x_h, \lambda_{v1}, \lambda_{v2}\right] = [557.60, -3.526, 139.850, 0]$ by trial input design parameters indicated by 1, 2, and 3 optimized weight of $f_W(\mathbf{x}) = f_W(x_A, x_h) = 27.05$ kgf as shown in Figures 2.4.3.15 and 2.4.3.16. Roots indicated by a pink circle in Figure 2.4.3.15 lie on the blue line indicating active Inequality $v_1(x) = v_1(x_A, x_h)$ is satisfied to be 0 because Inequality constraint $v_1(x)$ is bound to equality constraint for the Case 1 KKT condition. However, roots indicated by a pink circle in Figure 2.4.3.15 are unsatisfactory because it lies in infeasible region shaded by red curve. Inactive inequality constraint stress of $|\sigma_2|$ (0.2067 kN / mm^2) is greater than f_y in which $v_2(x) = v_2(x_A, x_h) = -|\sigma_2| + f_y \geq 0$ is violated even if design accuracies of optimized results are verified by structural calculation shown in Table 2.4.3.5. Case 1 KKT condition cannot be adopted as a solution even if Table 2.4.3.5 shows optimized weight of truss $f_W(\mathbf{x}) = 27.05$ kgf, demonstrating 0.17% compared with a conventional structural calculation. Initial input design parameters denoted by 4, 5, and 6 of Figure 2.4.3.15 diverged.

2.4.3.2.5 Case 2 based on inactive Inequality constraints $v_1(x)$ and active Inequality $v_2(x)$

Similarly, six initial input design parameters assumed for analytical function-based Lagrange optimization used in Figures 2.4.3.3, 2.4.3.5, 2.4.3.7, and 2.4.3.9 of Section 2.4.3 were used when minimizing truss weight using ANN-based Lagrange optimization, $\nabla\mathcal{L}(\mathbf{x}, \lambda_v) = \nabla\mathcal{L}(x_A, x_h, \lambda_{v1}, \lambda_{v2}) = 0$, for Case 2 KKT condition. Two roots at $\left[x_A^{(10)}, x_h^{(10)}, \lambda_{v1}^{(10)}, \lambda_{v2}^{(10)}\right] = [642.86, -2.466, 0, 130.44]$ indicated by 1 in Figure 2.4.3.17

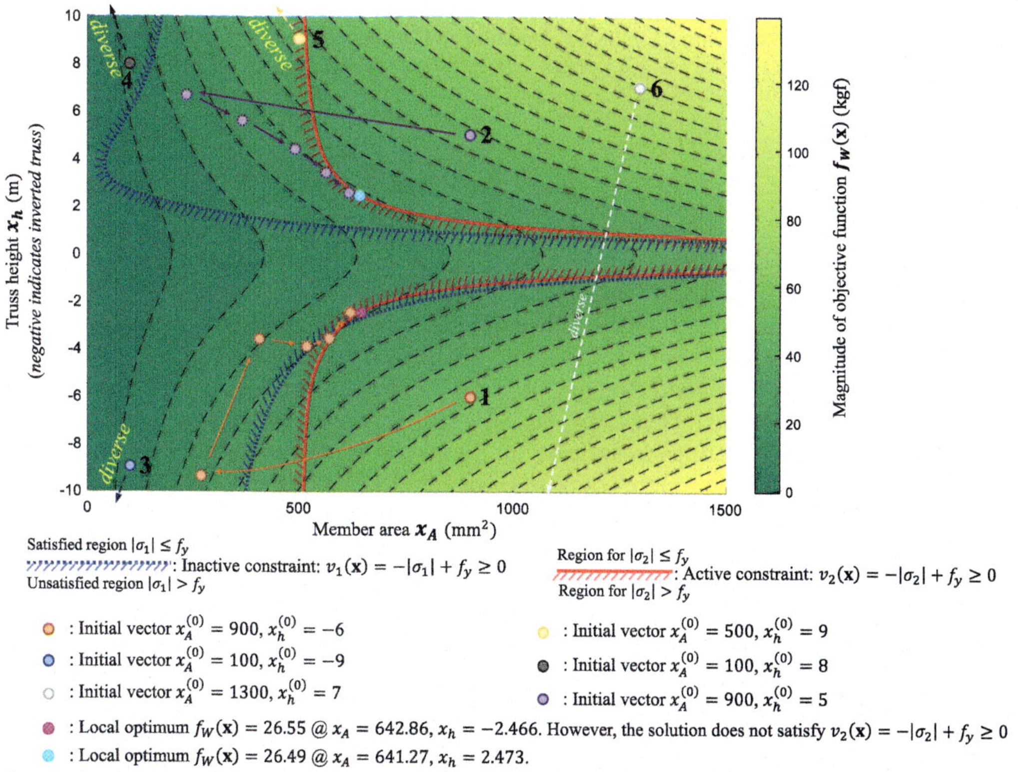

Figure 2.4.3.17 Minimized Lagrange function at $\left[x_A, x_h, \lambda_{v1}, \lambda_{v2}\right]$ based on Case 2 KKT condition; must satisfy $v_1(x) = -|\sigma_1| + f_y \geq 0$, $v_2(x) = -|\sigma_2| + f_y = 0$.

a

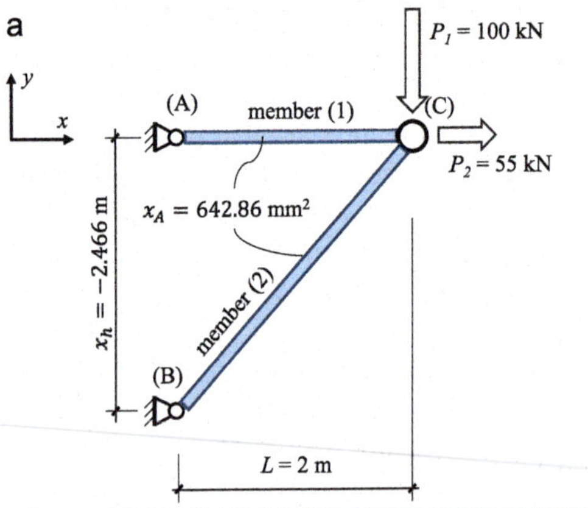

OPTIMIMAL $f_W(\mathbf{x})$ DESIGN
(based on 5 layer with 10 neurons)

	Parameter	AI results	Check (Analytical)	Error
1	x_A (mm)	642.86	642.86	0.00%
2	x_h (m)	-2.466	-2.466	0.00%
3	$\lvert\sigma_1\rvert$ (kN/mm²)	0.2124 > f_y	0.2117 > f_y	0.32%
4	$\lvert\sigma_2\rvert$ (kN/mm²)	0.2000 = f_y	0.2003 ≈ f_y	-0.13%
5	f_W (kgf)	26.55	26.62	-0.25%

Note: (dotted box) 2 inputs for AI design
(dashed box) 2 inputs for Analytical calculation

b

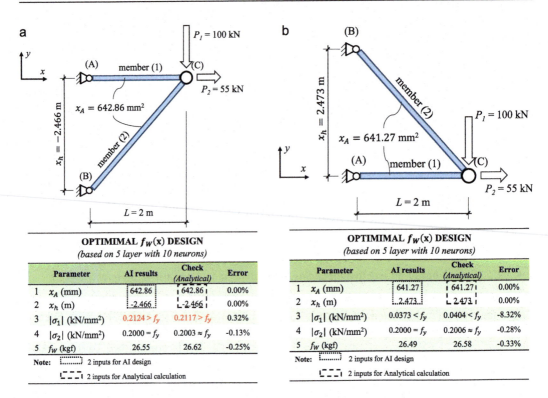

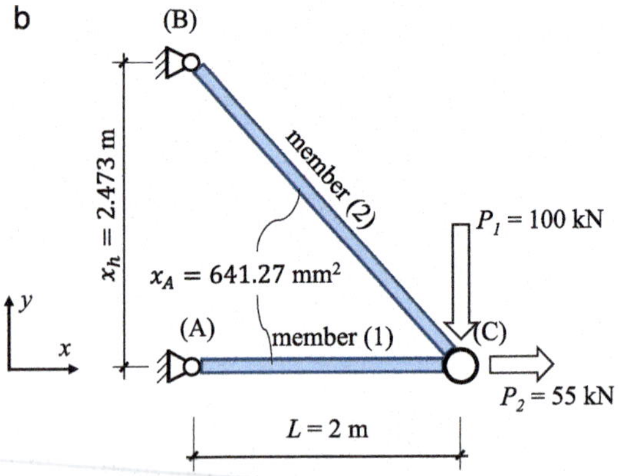

OPTIMIMAL $f_W(\mathbf{x})$ DESIGN
(based on 5 layer with 10 neurons)

	Parameter	AI results	Check (Analytical)	Error
1	x_A (mm)	641.27	641.27	0.00%
2	x_h (m)	2.473	2.473	0.00%
3	$\lvert\sigma_1\rvert$ (kN/mm²)	0.0373 < f_y	0.0404 < f_y	-8.32%
4	$\lvert\sigma_2\rvert$ (kN/mm²)	0.2000 = f_y	0.2006 ≈ f_y	-0.28%
5	f_W (kgf)	26.49	26.58	-0.33%

Note: (dotted box) 2 inputs for AI design
(dashed box) 2 inputs for Analytical calculation

Figure 2.4.3.18 Truss topology with Case 2KKT condition and verification by a structural calculation; must satisfy $v_1(x) = -|\sigma_1| + f_y \geq 0$, $v_2(x) = -|\sigma_2| + f_y = 0$. (a) Violating KKT condition. (b) Satisfying KKT condition.

and $\left[x_A^{(11)}, x_h^{(11)}, \lambda_{v1}^{(11)}, \lambda_{v2}^{(11)}\right] = [641.27, 2.473, 0, 129.18]$ indicated by 2 in Figure 2.4.3.17 converge to the local stationary points, leading to a total weights of $f_W(\mathbf{x})$ of 26.55 and 26.49 kgf, respectively, as shown in Figures 2.4.3.17 and 2.4.3.18. As shown in Figure 2.4.3.17, input design parameters denoted by 3, 4, 5, and 6 diverge. For a total weight of $f_W(\mathbf{x})$ of 26.55 kgf minimized by the first initial input design parameter, stress $\sigma_1|$ obtained at pink circle indicated by Point 1 is infeasible because stress of $\sigma_1|$ in Member 1 (0.2124 kN / mm²) is greater than a yield strength of $f_y = 0.2$ kN/mm². It lies in an infeasible region restricted by inactive Inequality $v_1(x) = -|\sigma_2| + f_y \geq 0$. It is also shown that pink circle lies on blue shaded area, indicating stress of $|\sigma_1|$ in Member 1 is greater than a yield strength of $f_y = 0.2$ kN/mm². For a total weights of $f_W(\mathbf{x}) =$ of 26.49 kgf minimized by the second input design parameter, stress $|\sigma_1|$ obtained at cyan circle indicated by Point 2 shown in Figure 2.4.3.17 is feasible because stress of $|\sigma_1|$ (0.0373 kN / mm²) in Member 1 is less than a yield strength of $f_y = 0.2$ kN/mm². It lies in the feasible region of other side the blue shaded area. Roots indicated by both pink and cyan circles in Figure 2.4.3.17 lie on the red line indicating active Inequality $v_2(x)$ is satisfied to be 0 because Inequality constraint $v_2(x)$ is bound to equality constraint for the Case 2 KKT condition. Figure 2.4.3.18 verifies that the minimized total weights of truss $f_W(\mathbf{x})$ of 26.49 kgf is well compared with 26.58 kgf based on a structural calculation, demonstrating an error of −0.33%. The minimized total weights of truss $f_W(\mathbf{x})$ of 26.49 kgf and stress $|\sigma_1|$ in Member 1 of 0.0373 kN/mm² based on ANN-based functions are also well agreed with 26.64 kgf and 0.0404 kN/mm² obtained based analytical functions as shown in Table 2.4.3.2.

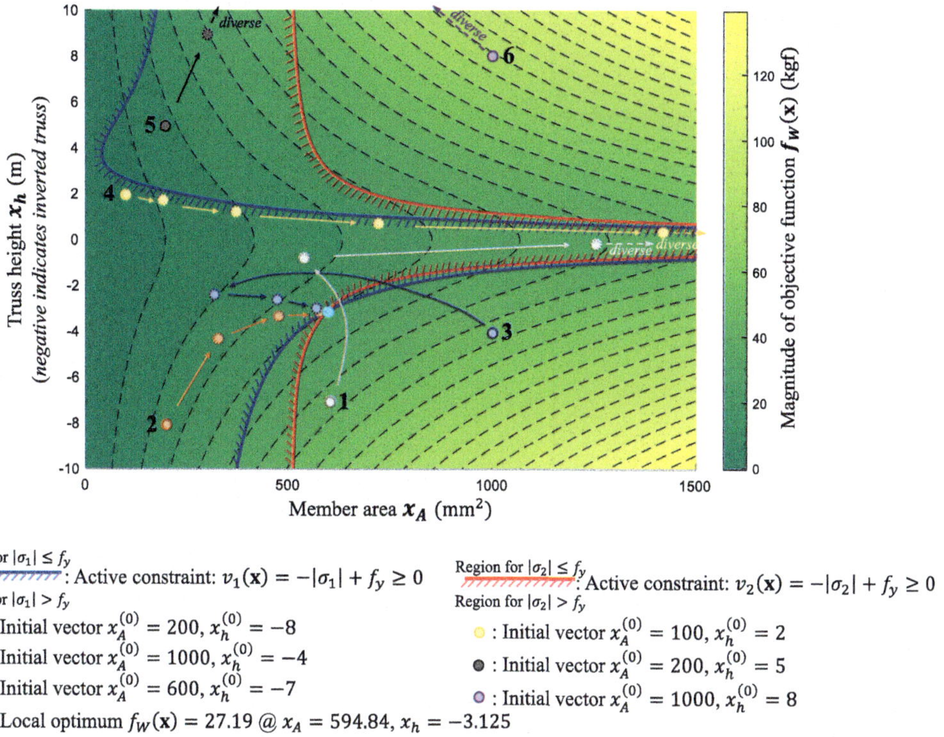

Figure 2.4.3.19 Minimized Lagrange function at $[x_A, x_h, \lambda_{v1}, \lambda_{v2}]$ based on Case 3 KKT condition; must satisfy $v_1(x) = -|\sigma_1| + f_y = 0$, $v_2(x) = -|\sigma_2| + f_y = 0$.

2.4.3.2.6 Case 3 based on both inequalities is active

Similarly to those assumed in previous KKT conditions described in Figures 2.4.3.3, 2.4.3.5, 2.4.3.7, and 2.4.3.9 of Section 2.4.3, six input design parameters $\left[x_A^{(0)}, x_h^{(0)}, \lambda_{v1}^{(0)}, \lambda_{v2}^{(0)}\right]^T = \left[x_A^{(0)}, x_h^{(0)}, 0, 0\right]^T$ were also implemented when minimizing truss weight using ANN-based Lagrange optimization, $\nabla\mathcal{L}(\mathbf{x}, \boldsymbol{\lambda}_v) = \nabla\mathcal{L}(x_A, x_h, \lambda_{v1}, \lambda_{v2}) = 0$, based on Newton–Raphson iteration. One stationary point $\left[x_A^{(k)}, x_h^{(k)}, \lambda_{v1}^{(k)}, \lambda_{v2}^{(k)}\right] = [594.84, -3.125, 97.75, 40.97]$ was converged by two input design parameters denoted by 2 and 3 at Iteration 8 and Iteration 7, respectively, resulting in an minimized weight of $f_W(\mathbf{x}) = 27.19$ (kgf) as shown in Figure 2.4.3.19. Initial trial input design parameters denoted by 1, 4, 5, and 6 of Figure 2.4.3.19 diverge. Cyan circle converged by two trial design parameters denoted by 2 and 3 shown in Figure 2.4.3.19 lies on both blue and red line where Inequalities $v_1(x)$ and $v_2(x)$ are 0, indicating that both Inequalities $v_1(x)$ and $v_2(x)$ are active, binding them to equality constraint for the Case 3 KKT condition. Figure 2.4.3.20 demonstrates that the minimized weights of truss $f_W(\mathbf{x})$ of 27.19 kgf is well compared with 27.17 kgf based on a structural calculation, demonstrating an error of 0.05%. However, the truss weight $f_W(\mathbf{x})$ of 27.19 kgf minimized by Case 2 KKT condition is greater than that minimized by Case 2 KKT condition, failing to yield an optimized truss weight.

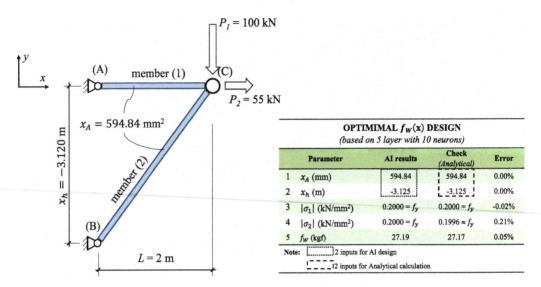

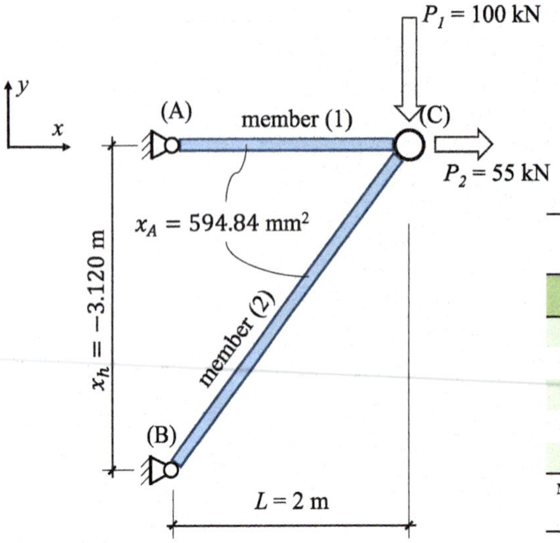

OPTIMIMAL $f_W(\mathbf{x})$ DESIGN
(based on 5 layer with 10 neurons)

	Parameter	AI results	Check (Analytical)	Error
1	x_A (mm)	594.84	594.84	0.00%
2	x_h (m)	-3.125	-3.125	0.00%
3	$\lvert\sigma_1\rvert$ (kN/mm²)	$0.2000 = f_y$	$0.2000 = f_y$	-0.02%
4	$\lvert\sigma_2\rvert$ (kN/mm²)	$0.2000 = f_y$	$0.1996 \approx f_y$	0.21%
5	f_W (kgf)	27.19	27.17	0.05%

Note: (dotted box) 2 inputs for AI design; (dashed box) 2 inputs for Analytical calculation

Figure 2.4.3.20 Truss topology with Case 3 KKT condition and verification by a structural calculation; must satisfy $v_1(x) = -|\sigma_1| + f_y = 0$, $v_2(x) = -|\sigma_2| + f_y = 0$.

2.4.3.2.7 Case 4 based on none of inequalities are active

In Case 4 where none of inequality constraints is active, inequality constraints are ignored in a formulation of Lagrange function $\mathcal{L}(\mathbf{x}, \boldsymbol{\lambda}_v)$ and gradient vector $\nabla\mathcal{L}(\mathbf{x}, \boldsymbol{\lambda}_v)$ which is equivalent to objective function $f_W(\mathbf{x})$, and hence, first derivative $\nabla f_W(\mathbf{x})$ is only minimized. Newton–Raphson iterations are performed to find roots of KKT conditions which minimize Lagrange function by equating $\nabla\mathcal{L}(\mathbf{x}, \boldsymbol{\lambda}_v)$ to 0. Two input design parameters of $\left[\mathbf{x}^{(0)}, \boldsymbol{\lambda}_v\right] = [300, 5]$ and $[700, -6]$ shown in Figure 2.4.3.21 are used to find stationary points by setting a first derivative of Lagrange function equal to 0. Both trial input design parameters diverge as shown in Figure 2.4.3.21 providing no solution for Case 4 KKT condition.

2.4.3.2.8 Section summary

As summarized in Table 2.4.3.6, the best solution for this example is found in Case 2 KKT condition, similarly to Table 2.4.3.2 which is obtained based on analytical functions. Topology of an optimal truss is presented in Figure 2.4.3.18b, minimizing truss weight of $f_W(\mathbf{x})$ to 26.49 kgf at truss area x_A = 641.27 mm and truss height x_h = 2.473 m. A minimized weight of $f_W(\mathbf{x}) = 26.64$ kgf at truss area x_A = 636.01 mm and truss height x_h = 2.544 m were also obtained in Table 2.4.3.2 which was calculated based on analytical functions. Figure 2.4.3.22 verifies ANN-based optimized designs for truss frame based on 100,000 large datasets.

2.4.3.3 Conclusions

Lower bound of a total truss weight generated by 100,000 large datasets is accurately predicted by the optimization based on ANN-based functions. ANN-based optimization for truss designs is much simpler, faster, and easy to use while yielding an accuracy as close

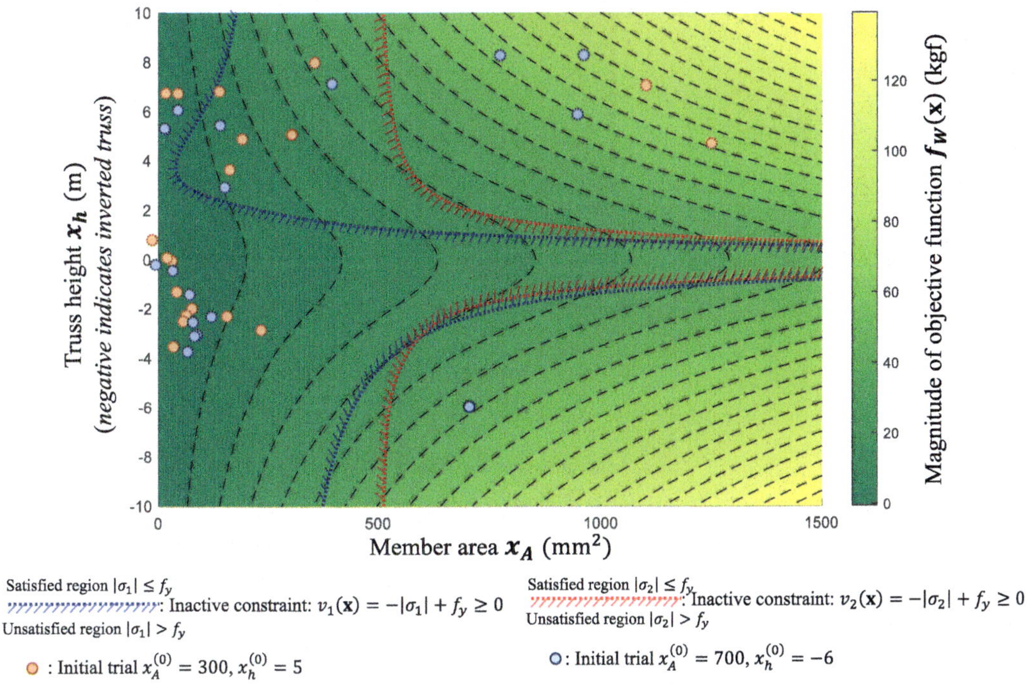

Figure 2.4.3.21 Minimized Lagrange function at $\left[x_A, x_h, \lambda_{v1}, \lambda_{v2}\right]$ based on Case 4 KKT condition; must satisfy $v_1(x) = -|\sigma_1| + f_y \geq 0, v_2(x) = -|\sigma_2| + f_y \geq 0$.

Table 2.4.3.6 ANN-based Design Summary

		ANN-based Lagrange Optimization			
Parameters		Case 1	Case 2	Case 3	Case 4
1	x_A (mm)	N/A	**641.27**	594.84	N/A
2	x_h (m)	N/A	**26.49**	−3.125	N/A
3	$\lvert\sigma_1\rvert$ (kN/mm^2)	N/A	**0.0373**	0.2	N/A
4	$\lvert\sigma_2\rvert$ (kN/mm^2)	N/A	**0.200**	0.2	N/A
Objective f_W (kgf)		N/A	**26.49**	27.17	N/A

Case 1, Inequality $v_1(x)$ is active; Case 2, Inequality $v_2(x)$ is active; Case 3, Both inequalities are active; Case 4, None of inequalities are active.

as to that provided by a Lagrange optimization based on an analytical objective function. Especially, the proposed method is efficient when the objective and constraining functions are difficult to be explicitly derived for the formulation of Lagrange functions, thus, recommending the proposed method using ANN-based functions in such cases. ANN-based functions which describe objective functions for a cost and CO_2 emissions of RC structures constrained by restricted conditions can replace analytical functions. In Chapters 3–5, an optimized design of RC structures on objective functions including a cost and CO_2 emissions is performed with respect to effective parameters such as rebar ratios and failure criteria.

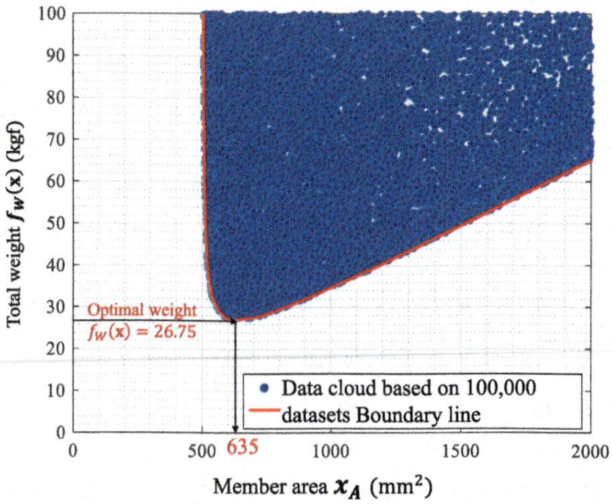

Figure 2.4.3.22 Verification of ANN-based optimization for truss designs based on 100,000 large datasets.

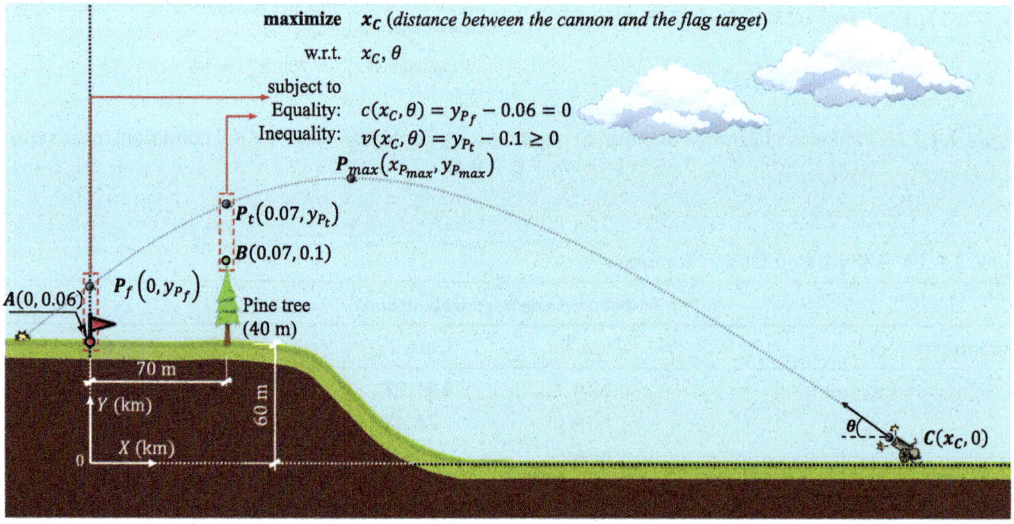

Figure 2.4.4.1 Lagrange optimization of flying distance based on analytical functions.

2.4.4 Maximizing flying distance of a projectile based on Lagrange optimization

2.4.4.1 Analytical function-based Lagrange optimization

2.4.4.1.1 Derivation of an equation of motion for projectile trajectory

This example is to find a launching angle θ of a projectile to maximize the flying distance of the projectile based on Lagrange optimization. A cannon ball is launched from ground level ($y_C = 0$) with an initial velocity of $V_0 = 0.08$ km/s at an angle of θ towards a flag target located on a hill ($y_A = 0.06$ km) as shown in Figure 2.4.4.1. Standard gravity is $g = 9.81$ m/s^2 and air resistance is neglected. A pine tree with a height of 40 m is in front of a flag target, hindering the projectile from hitting a flag target.

Optimization goals that should be achieved are defined as objective functions whereas equality and inequality functions are presented to constrain objective functions. Objective function $f_{xC}(x_C)$ should be the flying distance (x_C) of a projectile which is maximized based on Lagrange optimization. Flying height y_{P_f} of the cannon at the flag should be 0.06 km because a cannon should hit a $y_A = 0.06$ km-high flag as shown in Equation (2.4.4.2). However, flying height y_{P_f} of the cannon at the pine tree should be higher than or equal to a height of the pine tree $y_B = 0.1$ km because a cannon should not be hindered by a 0.1 km-high pine tree (obstacle) as shown in Equation (2.4.4.3). Equality constraint $c(x_C, \theta)$ and inequality constraint $v(x_C, \theta)$ are functions of a flying distance x_C and an angle of θ as shown in Equations (2.4.4.2) and (2.4.4.3), respectively, whereas an objective function $f_{xC}(x_C)$ is a function of a flying distance x_C only as shown in Equations (2.4.4.1). An objective function $f_{xC}(x_C)$, an equality constraint $c(x_C, \theta)$, and an inequality constraint $v(x_C, \theta)$ should be analytically derived to formulate Lagrange functions. For a review, kinematic equations of projectile motion are summarized. A speed of a projectile is resolved into horizontal and vertical components as shown in Equations (2.4.4.4-1) and (2.4.4.4-2) so that horizontal and vertical motions can be considered separately. Flying distance is calculated in Equations (2.4.4.5-1) and (2.4.4.5-2) by integrating Equations (2.4.4.4-1) and (2.4.4.4-2).

$$f_{xC}(x_C) = x_C \tag{2.4.4.1}$$

$$c(x_C, \theta) = y_{P_f} - y_A = 0 \tag{2.4.4.2-1}$$

$$c(x_C, \theta) = y_{P_f} - 0.06 = 0 \tag{2.4.4.2-2}$$

$$v(x_C, \theta) = y_{P_t} - y_B \geq 0 \tag{2.4.4.3-1}$$

$$v(x_C, \theta) = y_{P_t} - 0.1 \geq 0 \tag{2.4.4.3-2}$$

$$V_y = V_0 \sin\theta - gt \tag{2.4.4.4-1}$$

$$V_x = V_0 \cos\theta \tag{2.4.4.4-2}$$

$$x = x_C - V_0 t \cos\theta \tag{2.4.4.5-1}$$

$$y = V_0 t \sin\theta - \frac{1}{2} g t^2 \tag{2.4.4.5-2}$$

Altitude trajectory as functions of initial velocity (V_0), flying distance (x_C), and a launching angle (θ) shown in Equation (2.4.4.6) are obtained by substituting time out from Equations (2.4.4.5-1) and (2.4.4.5-2).

$$y = (x_C - x)\tan\theta - \frac{g(x_C - x)^2}{2V_0^2 \cos^2\theta} \tag{2.4.4.6}$$

2.4.4.1.2 Derivation of objective and constraining functions

Flying heights over a flag (y_{P_f}) and tree obstacle (y_{P_t}) are obtained in Equations (2.4.4.7-1) and (2.4.4.7-2) by substituting the positions of a flag and tree into an altitude of a projectile shown in Equation (2.4.4.6). A height of a projectile at a flag target should be equal to a flag elevation, and hence, an equality constraint is defined in Equation (2.4.4.8) and Figure 2.4.4.1 as functions of flying distance (x_C) and a launching angle (θ). A height of a projectile at an obstacle (pine tree) should be greater than or equal to an elevation of a treetop, and hence, an inequality constraint is defined in Equation (2.4.4.9) and Figure 2.4.4.1 as functions of flying distance (x_C) and a launching angle (θ). An optimum launching angle (θ) which is not too high is to be determined to maximize flying distance.

$$y_{P_f} = (x_C - x_A)\tan\theta - \frac{g(x_C - x_A)^2}{2V_0^2\cos^2\theta} \tag{2.4.4.7-1}$$

$$y_{P_t} = (x_C - x_B)\tan\theta - \frac{g(x_C - x_B)^2}{2V_0^2\cos^2\theta} \tag{2.4.4.7-2}$$

$$c(x_C,\ \theta) = y_{P_f} - y_A = 0 \tag{2.4.4.8-1}$$

$$c(x_C,\ \theta) = y_{P_f} - 0.06 = 0 \tag{2.4.4.8-2}$$

$$v(x_C,\ \theta) = y_{P_t} - y_B \geq 0 \tag{2.4.4.9-1}$$

$$v(x_C,\ \theta) = y_{P_t} - 0.1 \geq 0 \tag{2.4.4.9-2}$$

2.4.4.1.3 Derivation of Lagrange function

As shown in Equation (2.4.4.10), Lagrange function based on an analytical function is derived by subtracting an equality constraint $c(x_C,\ \theta)$ shown in Equation(2.4.4.8) and an inequality constraint $v(x_C,\ \theta)$ shown in Equation(2.4.4.9) multiplied by Lagrange multipliers ($\lambda_c,\ \lambda_v$), respectively, from an objective function $f_{xC}(x_C)$ shown in (2.4.4.1). However, an equality constraint $c(x_C,\ \theta)$ shown in Equation (2.4.4.8) does not influence an objective function $f_{xC}(x_C)$ shown in (2.4.4.1) because it is arranged as 0 as shown in Equation (2.4.4.8). It is noted an inequality constraint $v(x_C,\ \theta)$ shown in Equation (2.4.4.9) is either ignored for an inactive KKT condition or an inequality constraint $v(x_C,\ \theta)$ is bound to an equality constraint for an active KKT condition. Thus, an inequality constraint $v(x_C,\ \theta)$ shown in Lagrange function shown in Equation (2.4.4.10) is ignored for an inactive KKT condition or transformed to an equality constraint for an active KKT condition. Readers are recommended to refer to Sections 2.1, 2.2, and 2.3 for Lagrange optimization with inequality constraints. Objective function $f_{xC}(x_C)$ is simply equal to a variable x_C. In this problem, a conventional Lagrange multiplier method is implemented to maximize the distance x_C between a flag target and a cannon. The Lagrange function for this problem becomes Equation (2.4.4.10) in which λ_c and λ_c are Lagrange multipliers of an equality $c(x_C,\ \theta)$ and an inequality constraint $v(x_C,\ \theta)$, respectively. A diagonal matrix S simply becomes a scalar s because S designates only one inequality constraint. S of 1 or 0 is assigned for either active or inactive. Then, first derivative (Jacobi, gradient vector) of Lagrange function is obtained in Equation (2.4.4.11). Jacobi of Lagrange functions based on an explicit analytical function and an ANN is shown in Equations (2.4.4.11-1) and (2.4.4.11-2), respectively.

$$\mathcal{L}(x_C,\theta,\lambda_c,\lambda_v) = f_{xc}(x_C) - \lambda_c c(x_C,\theta) - \lambda_v sv(x_C,\theta)$$
$$= x_C - \lambda_c\left(y_{P_f} - 0.06\right) - \lambda_v s\left(y_{P_t} - 0.1\right) \qquad (2.4.4.10)$$

2.4.4.1.4 Derivation of Jacobi and Hessian matrices

$$\nabla\mathcal{L}(x_C,\ \theta,\ \lambda_c,\ \lambda_v) = \begin{bmatrix} \nabla f_{xc}(x_C) - \left[\mathbf{J}_c(x_C,\ \theta)\right]^T \lambda_c - \left[\mathbf{J}_v(x_C,\ \theta)\right]^T s\lambda_v \\ -c(x_C,\ \theta) \\ -sv(x_C,\ \theta) \end{bmatrix} \qquad (2.4.4.11\text{-}1)$$

$$\nabla\mathcal{L}(x_C,\ \theta,\ \lambda_c,\ \lambda_v) = \begin{bmatrix} \nabla f_{xc}(x_C) - \left[\mathbf{J}_c^{(D)}(x_C,\ \theta)\right]^T \lambda_c - \left[\mathbf{J}_v^{(D)}(x_C,\ \theta)\right]^T s\lambda_v \\ -c(x_C,\ \theta) \\ -sv(x_C,\ \theta) \end{bmatrix} \qquad (2.4.4.11\text{–}2)$$

Jacobi $\nabla f_{xc}(x_C)$ of objective function $f_{xc}(x_C)$ is obtained by partially differentiating objective function with respect to two variables x_C and θ as shown in Equation (2.4.4.12-1). Jacobi $\nabla f_{xc}(x_C)$ and $\mathbf{J}_c(x_C,\ \theta)$ and $\mathbf{J}_v(x_C,\ \theta)$ of an equality function $c(x_C,\ \theta)$ and an inequality function $v(x_C,\ \theta)$ are also derived by partially differentiating them with respect to two variables x_C and θ as shown in Equations (2.4.4.12-2) and (2.4.4.12-3), respectively. It is noted that analytical functions should be at least twice differentiable to obtain Jacobi shown in Equations (2.4.4.12-2) and (2.4.4.12-3) and Hessian Jacobi shown in Equations (2.4.4.14), (2.4.4.15-1), and (2.4.4.16-1) for linearizing Lagrange function. Analytical functions which do not meet these requirements will face some difficulties to establish a linearized Lagrange function, thus Lagrange optimization can be difficult to perform.

$$\nabla f_{xc}(x_C) = \begin{bmatrix} \dfrac{\partial x_C}{\partial x_C} \\ \dfrac{\partial x_C}{\partial \theta} \end{bmatrix} = \begin{bmatrix} 1 \\ 0 \end{bmatrix} \qquad (2.4.4.12\text{-}1)$$

$$\mathbf{J}_c(x_C,\ \theta) = \begin{bmatrix} \dfrac{\partial c(x_C,\ \theta)}{\partial x_C} & \dfrac{\partial c(x_C,\ \theta)}{\partial \theta} \end{bmatrix}$$
$$= \begin{bmatrix} \left(\tan\theta - \dfrac{g(x_C - x_A)}{V_0^2\cos^2\theta}\right) & \left((x_C - x_A)(\tan^2\theta + 1) - \dfrac{g(x_C - x_A)^2\sin\theta}{V_0^2\cos^3\theta}\right) \end{bmatrix} \qquad (2.4.4.12\text{-}2)$$

$$\mathbf{J}_v(x_C,\ \theta) = \begin{bmatrix} \dfrac{\partial v(x_C,\ \theta)}{\partial x_C} & \dfrac{\partial v(x_C,\ \theta)}{\partial \theta} \end{bmatrix}$$

$$= \begin{bmatrix} \left(\tan\theta - \dfrac{g(x_C - x_B)}{V_0^2\cos^2\theta}\right) & \left((x_C - x_B)(\tan^2\theta + 1) - \dfrac{g(x_C - x_B)^2\sin\theta}{V_0^2\cos^3\theta}\right) \end{bmatrix} \quad (2.4.4.12\text{-}3)$$

A solution of a first derivative (Jacobi, gradient vector) of a Lagrange function (or finding stationary or stationary points of Lagrange function) obtained in Equation (2.4.4.11-1) can be obtained based on Newton–Raphson iteration. Lagrange function $\mathcal{L}(x_C, \theta, \lambda_c, \lambda_v)$ has maximum or minimum values at stationary or stationary points. An objective function $f_{xC}(x_C)$ becomes Lagrange function $\mathcal{L}(x_C, \theta, \lambda_c, \lambda_v)$ when an inequality constraint is considered as an inactive KKT condition because an inequality constraint is ignored. However, it is difficult to find solutions when Equation (2.4.4.11-1) is nonlinearly complex. Initial trial input parameters are, then, updated until a first derivative (Jacobi, gradient vector) of a Lagrange function converges to 0. Initial trial input parameters $\left[x_C^{(0)}, \theta^{(0)}, \lambda_c^{(0)}, \lambda_v^{(0)}\right]$ is assumed to begin an iteration to find solutions of a first derivative (Jacobi, gradient vector) of a Lagrange function. Newton–Raphson iteration is implemented after Equation (2.4.4.11-1) is linearized as shown in Equation (2.3.4.1-1). Equation (2.4.4.13-1), which is important in finding stationary points of an objective function, updates initial trial input parameters and Lagrange multipliers to a next step until $\nabla\mathcal{L}(x_C, \theta, \lambda_c, \lambda_v)$ shown in Equation (2.4.4.11-1) converges to 0. Equation (2.3.4.1-1) is constantly checked if a first derivative (Jacobi, gradient vector) of a Lagrange function converges for input parameters updated in Equation (2.4.4.13-1) based on global Hessian matrix of Lagrange function $\mathbf{H}_{\mathcal{L}}(x_C, \theta, \lambda_c, \lambda_v)$ calculated in Equations (2.4.4.13-2) to (2.4.4.16-4). $\mathbf{H}_{\mathcal{L}}(x_C, \theta, \lambda_c, \lambda_v)$ shown in Equation (2.4.4.13-2) is calculated using $\mathbf{H}_{\mathcal{L}}(x_C, \theta)$ shown in Equation(2.4.4.13-3) which is also calculated using Hessian matrix $\mathbf{H}_{f_{xC}}(x_C, \theta)$ of objective function $f_{xC}(x_C)$, Hessian matrix $\mathbf{H}_c(x_C, \theta)$ of equality function $c(x_C, \theta)$, and Hessian matrix $\mathbf{H}_v(x_C, \theta)$ of an inequality function $v(x_C, \theta)$ shown in Equations (2.4.4.14), (2.4.4.15-1), and (2.4.4.16-1), respectively. Hessian matrix $\mathbf{H}_c(x_C, \theta)$ of equality function $c(x_C, \theta)$ shown in Equation (2.4.4.15-1) is obtained using Equations (2.4.4.15-2)–(2.4.4.15-4) while Hessian matrix $\mathbf{H}_v(x_C, \theta)$ of an inequality function $v(x_C, \theta)$ shown in Equation (2.4.4.16-1) is calculated using Equations (2.4.4.16-2)–(2.4.4.16-4). A Newton–Raphson method is employed in finding roots of a first-order partial differential equations $\nabla\mathcal{L}(x_C, \theta, \lambda_c, \lambda_v) = 0$ shown in Equation (2.4.4.11-1), identifying stationary points of the Lagrange function $\mathcal{L}(x_C, \theta, \lambda_c, \lambda_v)$. Initial trial input parameters of [$x_C^{(0)}, \theta^{(0)}, \lambda_c^{(0)}, \lambda_v^{(0)}$] are updated based on first-order approximation of Equation (2.4.4.11-1) using Equation (2.4.4.13-1), in which global Hessian matrix of Lagrange function $\mathbf{H}_{\mathcal{L}}(x_C, \theta, \lambda_c, \lambda_v)$ is calculated as a function of $[x_C, \theta, \lambda_c, \lambda_v]$ whereas $\mathbf{H}_{\mathcal{L}}(x_C, \theta)$ is a function of $[x_C, \theta]$. There are two KKT candidate solutions based on inactive and active conditions for one inequality constraint as shown in Table 2.4.4.1. Final

Table 2.4.4.1 Two KKT Conditions of Active and Inactive Inequalities

Case	*s*	*Comment*
Case 1	1	Inequality $v(x)$ is active
Case 2	0	Inequality $v(x)$ is inactive

stationary points among candidate solutions should be calculated within the ranges defined by inequality constraints $v(x_C, \theta)$.

$$\begin{bmatrix} x_C^{(k+1)} \\ \theta^{(k+1)} \\ \lambda_c^{(k+1)} \\ \lambda_v^{(k+1)} \end{bmatrix} = \begin{bmatrix} x_C^{(k)} \\ \theta^{(k)} \\ \lambda_c^{(k)} \\ \lambda_v^{(k)} \end{bmatrix} - \left[\mathbf{H}_{\mathcal{L}}\left(x_C^{(k)}, \theta^{(k)}, \lambda_c^{(k)}, \lambda_v^{(k)}\right)\right]^{-1} \nabla\mathcal{L}\left(x_C^{(k)}, \theta^{(k)}, \lambda_c^{(k)}, \lambda_v^{(k)}\right) \quad (2.4.4.13\text{-}1)$$

$$\mathbf{H}_{\mathcal{L}}(x_C, \theta, \lambda_c, \lambda_v) = \begin{bmatrix} \mathbf{H}_{\mathcal{L}}(x_C, \theta) & -[J_c(x_C, \theta)]^T & -[sJ_v(x_C, \theta)]^T \\ -J_c(x_C, \theta) & 0 & 0 \\ -sJ_v(x_C, \theta) & 0 & 0 \end{bmatrix} \quad (2.4.4.13\text{-}2)$$

$$\mathbf{H}_{\mathcal{L}}(x_C, \theta) = \mathbf{H}_{f_{xC}}(x_C, \theta) - \lambda_c \mathbf{H}_c(x_C, \theta) - s\lambda_v \mathbf{H}_v(x_C, \theta) \quad (2.4.4.13\text{-}3)$$

$$\mathbf{H}_{f_{xC}}(x_C, \theta) = \begin{bmatrix} \dfrac{\partial^2 x_C}{\partial x_C^2} & \dfrac{\partial^2 x_C}{\partial x_C\, \partial\theta} \\ \dfrac{\partial^2 x_C}{\partial x_C\, \partial\theta} & \dfrac{\partial^2 x_C}{\partial\theta^2} \end{bmatrix} = \begin{bmatrix} 0 & 0 \\ 0 & 0 \end{bmatrix} \quad (2.4.4.14)$$

$$\mathbf{H}_c(x_C, \theta) = \begin{bmatrix} \dfrac{\partial^2 c(x_C, \theta)}{\partial x_C^2} & \dfrac{\partial^2 c(x_C, \theta)}{\partial x_C\, \partial\theta} \\ \dfrac{\partial^2 c(x_C, \theta)}{\partial x_C\, \partial\theta} & \dfrac{\partial^2 c(x_C, \theta)}{\partial\theta^2} \end{bmatrix} \quad (2.4.4.15\text{-}1)$$

$$\frac{\partial^2 c(x_C, \theta)}{\partial x_C^2} = -\frac{g}{V_0^2 \cos^2\theta} \quad (2.4.4.15\text{-}2)$$

$$\frac{\partial^2 c(x_C, \theta)}{\partial x_C\, \partial\theta} = \tan^2\theta + 1 - \frac{2g(x_C - x_A)\sin\theta}{V_0^2 \cos^3\theta} \quad (2.4.4.15\text{-}3)$$

$$\frac{\partial^2 c(x_C, \theta)}{\partial\theta^2} = -2\tan\theta\left(\tan^2\theta + 1\right)(x_C - x_A) - \frac{g(x_C - x_A)^2}{V_0^2 \cos^2\theta} - \frac{3g(x_C - x_A)^2 \sin^2\theta}{V_0^2 \cos^4\theta} \quad (2.4.4.15\text{-}4)$$

$$\mathbf{H}_v(x_C, \theta) = \begin{bmatrix} \dfrac{\partial^2 v(x_C, \theta)}{\partial x_C^2} & \dfrac{\partial^2 v(x_C, \theta)}{\partial x_C\, \partial\theta} \\ \dfrac{\partial^2 v(x_C, \theta)}{\partial x_C\, \partial\theta} & \dfrac{\partial^2 v(x_C, \theta)}{\partial\theta^2} \end{bmatrix} \quad (2.4.4.16\text{-}1)$$

$$\frac{\partial^2 v(x_C, \theta)}{\partial x_C^2} = -\frac{g}{V_0^2 \cos^2\theta} \quad (2.4.4.16\text{-}2)$$

$$\frac{\partial^2 v(x_C,\theta)}{\partial x_C\,\partial\theta} = \tan^2\theta + 1 - \frac{2g(x_C - x_B)\sin\theta}{V_0^2\cos^3\theta} \tag{2.4.4.16-3}$$

$$\frac{\partial^2 v(x_C,\theta)}{\partial\theta^2} = 2\tan\theta(\tan^2\theta+1)(x_C - x_B) - \frac{g(x_C - x_B)^2}{V_0^2\cos^2\theta} - \frac{3g(x_C - x_B)^2\sin^2\theta}{V_0^2\cos^4\theta} \tag{2.4.4.16-4}$$

2.4.4.1.5 Case 1 KKT condition based on active inequality $v(x)$

In this case, inequality $v(x_C,\theta) = y_{P_t} - 0.1 \geq 0$ is assumed to constrain an objective function $f_{xC}(x_C,\theta)$ when reaching a maximum of $f_{xC}(x_C)$ by binding $v(x_C,\theta) = y_{P_t} - 0.1 \geq 0$ to equality constraint $(x_C,\theta) = y_{P_t} - 0.1 = 0$. The Lagrange function $\mathcal{L}(x_C,\theta,\lambda_c,\lambda_v)$ is obtained in Equation (2.4.4.17-1).

$$\mathcal{L}(x_C,\theta,\lambda_c,\lambda_v) = x_C - \lambda_c(y_{P_f} - 0.06) - \lambda_v(y_{P_t} - 0.1) \tag{2.4.4.17-1}$$

An initial trial input parameter of $\mathbf{x}^{(0)} = \left[x_C^{(0)},\theta^{(0)}\right]^T = [0.7, 40]^T$ and initial trial Lagrange multipliers of $\lambda_c^{(0)} = 0$ and $\lambda_v^{(0)} = 0$ are used to solve KKT conditions shown in Equation (2.4.4.11-1) based on iteration, and hence, an initial trial input parameter becomes $\left[x_C^{(0)},\theta^{(0)},\lambda_c^{(0)},\lambda_v^{(0)}\right]^T = [0.7, 40, 0, 0]^T$. Constraints $c(\mathbf{x})$ and $v(\mathbf{x})$ are calculated by substituting $\left[x_C^{(0)},\theta^{(0)},\lambda_c^{(0)},\lambda_v^{(0)}\right]^T = [0.7, 40, 0, 0]^T$ into Equations (2.4.4.18-1) and (2.4.4.18-2) based on Equations (2.4.4.7-1) and (2.4.4.7-2), in which an initial trial input parameter is denoted by red color. Jacobi matrix of constraints $c(\mathbf{x})$ and $v(\mathbf{x})$ are, then, calculated by substituting $\left[x_C^{(0)},\theta^{(0)},\lambda_c^{(0)},\lambda_v^{(0)}\right]^T = [0.7, 40, 0, 0]^T$ into Equations (2.4.4.19-1) and (2.4.4.19-2) based on Equations (2.4.4.12-2) and (2.4.4.12-3). Similarly, Hessian matrices of constraints $c(\mathbf{x})$ and $v(\mathbf{x})$ are calculated by substituting $\left[x_C^{(0)},\theta^{(0)},\lambda_c^{(0)},\lambda_v^{(0)}\right]^T = [0.7, 40, 0, 0]^T$ into Equations (2.4.4.20-1) and (2.4.4.20-2) based on Equations (2.4.4.15-1) and (2.4.4.16-1). Finally, their Jacobi and Hessian matrices are computed as shown in Equations (2.4.5.18)–(2.4.5.20), respectively. A first derivative (Jacobi, gradient vector) of a Lagrange function $\nabla\mathcal{L}\left(x_C^{(0)},\theta^{(0)},\lambda_c^{(0)},\lambda_v^{(0)}\right)$ is calculated in Equation (2.4.4.21-2) by substituting $\nabla f_{xC}(x_C)$ shown in Equation (2.4.4.12-1), $c\left(x_C^{(0)},\theta^{(0)}\right)$ shown in Equation (2.4.4.18-1), $v\left(x_C^{(0)},\theta^{(0)}\right)$ shown in Equation (2.4.4.18-2), $\mathbf{J}_c\left(x_C^{(0)},\theta^{(0)}\right)$ shown in Equation (2.4.4.19-1), $\mathbf{J}_v\left(x_C^{(0)},\theta^{(0)}\right)$ shown in Equation (2.4.4.19-2), and $s = 1$ into Equation (2.4.4.11-1) or Equation (2.4.4.21-1). However, $\nabla\mathcal{L}\left(x_C^{(0)},\theta^{(0)},\lambda_c^{(0)},\lambda_v^{(0)}\right)$ does not converge to 0, indicating $\left[x_C^{(0)},\theta^{(0)},\lambda_c^{(0)},\lambda_v^{(0)}\right]^T = [0.7, 40, 0, 0]^T$ does not calculate a stationary point of the Lagrange function under the Case 1 KKT condition as shown in Equation (2.4.4.21-2). An iteration to update the initial trial input parameter is necessary to further search for a root of KKT condition using Equation (2.4.4.22-1). Initial trial input parameter $[x_C^{(1)},\theta^{(1)},\lambda_c^{(1)},\lambda_v^{(1)}]$ for use in Iteration 1 is calculated in Equation (2.4.4.22-1) based on Equation (2.4.4.13-1) using Hessian matrix $\mathbf{H}_{\mathcal{L}}\left(x_C^{(0)},\theta^{(0)},\lambda_c^{(0)},\lambda_v^{(0)}\right)$ calculated in Equations (2.4.4.22-2), (2.4.4.25-1), and (2.4.4.25-2). For $H_{\mathcal{L}}\left(x_C^{(0)},\theta^{(0)},\lambda_c^{(0)},\lambda_v^{(0)}\right)$, $H_{\mathcal{L}}\left(x_C^{(0)},\theta^{(0)}\right)$ is obtained in Equations (2.4.4.23)–(2.4.4.24) by substituting $H_{f_{xC}}(x_C,\theta)$, $H_c\left(x_C^{(0)},\theta^{(0)}\right)$, $H_v\left(x_C^{(0)},\theta^{(0)}\right)$ obtained in Equations (2.4.4.14), (2.4.4.20-1), (2.4.4.20-2), respectively, into Equation (2.4.4.13-3).

$\mathbf{H}_{\mathcal{L}}\left(x_C^{(0)}, \theta^{(0)}, \lambda_c^{(0)}, \lambda_v^{(0)}\right)$ is, then, obtained in Equations (2.4.4.25-1) and (2.4.4.25-2) by substituting $H_{\mathcal{L}}\left(x_C^{(0)}, \theta^{(0)}\right)$, $J_c\left(x_C^{(0)}, \theta^{(0)}\right)$, $J_v\left(x_C^{(0)}, \theta^{(0)}\right)$ obtained in Equations (2.4.4.24), (2.4.4.19-1), (2.4.4.19-2), respectively, into Equation (2.4.4.22-2). Initial trial input parameter for Iteration 1 $[x_C^{(1)}, \theta^{(1)}, \lambda_c^{(1)}, \lambda_v^{(1)}] = [0.584, 39.98, -1.929, 1.126]$ is obtained in Equation (2.4.4.26) by substituting $\left[\mathbf{H}_{\mathcal{L}}\left(x_C^{(0)}, \theta^{(0)}, \lambda_c^{(0)}, \lambda_v^{(0)}\right)\right]^{-1}$ obtained in Equation (2.4.4.25-3) and $\nabla\mathcal{L}\left(x_C^{(0)}, \theta^{(0)}, \lambda_c^{(0)}, \lambda_v^{(0)}\right)$ obtained in Equation (2.4.4.21-2) into Equation (2.4.4.22-1).

$$c\left(x_C^{(0)}, \theta^{(0)}\right) = \left(x_C^{(0)} - x_A\right)\tan\theta^{(0)} - \frac{g\left(x_C^{(0)} - x_A\right)^2}{2V_0^2\cos^2\theta^{(0)}} - y_A = -0.113 \qquad (2.4.4.18\text{-}1,\ 2.4.5.18a)$$

$$v\left(x_C^{(0)}, \theta^{(0)}\right) = \left(x_C^{(0)} - x_B\right)\tan\theta^{(0)} - \frac{g\left(x_C^{(0)} - x_B\right)^2}{2V_0^2\cos^2\theta^{(0)}} - y_B = -0.090 \qquad (2.4.4.18\text{-}2,\ 2.4.5.18b)$$

$$\mathbf{J}_c\left(x_C^{(0)}, \theta^{(0)}\right) = \begin{bmatrix} -0.989 & 0.119 \end{bmatrix} \qquad (2.4.4.19\text{-}1,\ 2.4.5.19a)$$

$$\mathbf{J}_v\left(x_C^{(0)}, \theta^{(0)}\right) = \begin{bmatrix} -0.806 & 0.204 \end{bmatrix} \qquad (2.4.4.19\text{-}2,\ 2.4.5.19b)$$

$$\mathbf{H}_c\left(x_C^{(0)}, \theta^{(0)}\right) = \begin{bmatrix} -2.612 & -1.364 \\ -1.364 & -1.982 \end{bmatrix} \qquad (2.4.4.20\text{-}1,\ 2.4.5.20b)$$

$$\mathbf{H}_v\left(x_C^{(0)}, \theta^{(0)}\right) = \begin{bmatrix} -2.612 & -1.058 \\ -1.058 & -1.425 \end{bmatrix} \qquad (2.4.4.20\text{-}2,\ 2.4.5.20b)$$

$$\nabla\mathcal{L}\left(x_C^{(0)}, \theta^{(0)}, \lambda_c^{(0)}, \lambda_v^{(0)}\right) = \begin{bmatrix} \nabla f_{xc}\left(x_C^{(0)}\right) - J_c\left(x_C, \theta\right)^T \lambda_c - J_v\left(x_C, \theta\right)^T s\lambda_v \\ -c\left(x_C, \theta\right) \\ -sv\left(x_C, \theta\right) \end{bmatrix}$$

(2.4.4.21-1, 2.4.5.21)

$$\nabla\mathcal{L}\left(x_C^{(0)}, \theta^{(0)}, \lambda_c^{(0)}, \lambda_v^{(0)}\right)$$

$$= \begin{bmatrix} \begin{bmatrix} 1 \\ 0 \end{bmatrix} - \begin{bmatrix} -0.989 \\ 0.119 \end{bmatrix} \times 0 - \begin{bmatrix} -0.806 \\ 0.204 \end{bmatrix} \times 1 \times 0 \\ 0.113 \\ 0.090 \end{bmatrix}$$

$$\rightarrow \nabla\mathcal{L}\left(x_C^{(0)}, \theta^{(0)}, \lambda_c^{(0)}, \lambda_v^{(0)}\right) = \begin{bmatrix} 1 \\ 0 \\ 0.113 \\ 0.090 \end{bmatrix} \neq 0 \qquad (2.4.4.21\text{–}2\ 2.4.5.21b)$$

$$\begin{bmatrix} x_C^{(1)} \\ \theta^{(1)} \\ \lambda_c^{(1)} \\ \lambda_v^{(1)} \end{bmatrix} = \begin{bmatrix} x_C^{(0)} \\ \theta^{(0)} \\ \lambda_c^{(0)} \\ \lambda_v^{(0)} \end{bmatrix} - \left[\mathbf{H}_{\mathcal{L}}\left(x_C^{(0)}, \theta^{(0)}, \lambda_c^{(0)}, \lambda_v^{(0)}\right)\right]^{-1} \nabla\mathcal{L}\left(x_C^{(0)}, \theta^{(0)}, \lambda_c^{(0)}, \lambda_v^{(0)}\right)$$

(2.4.4.22-1, 2.4.5.22)

$$\mathbf{H}_{\mathcal{L}}\left(x_C^{(0)}, \theta^{(0)}, \lambda_c^{(0)}, \lambda_v^{(0)}\right) = \begin{bmatrix} \mathbf{H}_{\mathcal{L}}\left(x_C^{(0)}, \theta^{(0)}\right) & -\left[\mathbf{J}_c\left(x_C^{(0)}, \theta^{(0)}\right)\right]^T & -\left[s\mathbf{J}_v\left(x_C^{(0)}, \theta^{(0)}\right)\right]^T \\ -\mathbf{J}_c\left(x_C^{(0)}, \theta^{(0)}\right) & 0 & 0 \\ -s\mathbf{J}_v\left(x_C^{(0)}, \theta^{(0)}\right) & 0 & 0 \end{bmatrix}$$

(2.4.4.22-2, 2.4.5.23)

Where,

$$\begin{aligned} \mathbf{H}_{\mathcal{L}}\left(\mathbf{x}^{(0)}\right) &= \mathbf{H}_{\mathcal{L}}\left(x_C^{(0)}, \theta^{(0)}\right) \\ &= \mathbf{H}_{f_{xC}}\left(x_C^{(0)}, \theta^{(0)}\right) - \lambda_c \mathbf{H}_c\left(x_C^{(0)}, \theta^{(0)}\right) - s\lambda_v H_v\left(x_C^{(0)}, \theta^{(0)}\right) \end{aligned}$$

(2.4.4.23, 2.4.5.24)

$$\mathbf{H}_{\mathcal{L}}\left(\mathbf{x}^{(0)}\right) = H_{\mathcal{L}}\left(x_C^{(0)}, \theta^{(0)}\right)$$

$$= \begin{bmatrix} 0 & 0 \\ 0 & 0 \end{bmatrix} - 0 \times \begin{bmatrix} -2.612 & -1.364 \\ -1.364 & -1.982 \end{bmatrix} - 1 \times 0 \times \begin{bmatrix} -2.612 & -1.058 \\ -1.058 & -1.425 \end{bmatrix}$$

$$= \begin{bmatrix} 0 & 0 \\ 0 & 0 \end{bmatrix}$$

(2.4.4.24)

$$\mathbf{H}_{\mathcal{L}}\left(x_C^{(0)}, \theta^{(0)}, \lambda_c^{(0)}, \lambda_v^{(0)}\right)$$

$$= \begin{bmatrix} \begin{bmatrix} 0 & 0 \\ 0 & 0 \end{bmatrix} & -\begin{bmatrix} -0.989 \\ 0.119 \end{bmatrix} & -\begin{bmatrix} -0.806 \\ 0.204 \end{bmatrix} \\ -\begin{bmatrix} -0.989 & 0.119 \end{bmatrix} & 0 & 0 \\ -\begin{bmatrix} -0.806 & 0.204 \end{bmatrix} & 0 & 0 \end{bmatrix}$$

(2.4.4.25-1)

$$\mathbf{H}_{\mathcal{L}}\left(x_C^{(0)}, \theta^{(0)}, \lambda_c^{(0)}, \lambda_v^{(0)}\right) = \begin{bmatrix} 0 & 0 & 0.989 & 0.806 \\ 0 & 0 & -0.119 & -0.204 \\ 0.989 & -0.119 & 0 & 0 \\ 0.806 & -0.204 & 0 & 0 \end{bmatrix}$$

(2.4.4.25-2)

$$\left[\mathbf{H}_{\mathcal{L}}\left(x_C^{(0)},\,\theta^{(0)},\,\lambda_c^{(0)},\,\lambda_v^{(0)}\right)\right]^{-1}=\begin{bmatrix}0 & 0 & 1.929 & -1.126\\ 0 & 0 & 7.637 & -9.368\\ 1.929 & 7.637 & 0 & 0\\ -1.126 & -9.368 & 0 & 0\end{bmatrix} \tag{2.4.4.25-3}$$

$$\begin{bmatrix}x_C^{(1)}\\ \theta^{(1)}\\ \lambda_c^{(1)}\\ \lambda_v^{(1)}\end{bmatrix}=\begin{bmatrix}0.7\\ 40\\ 0\\ 0\end{bmatrix}-\begin{bmatrix}0 & 0 & 1.929 & -1.126\\ 0 & 0 & 7.637 & -9.368\\ 1.929 & 7.637 & 0 & 0\\ -1.126 & -9.368 & 0 & 0\end{bmatrix}\begin{bmatrix}1\\ 0\\ 0.113\\ 0.090\end{bmatrix}$$

$$\rightarrow\begin{bmatrix}x_C^{(1)}\\ \theta^{(1)}\\ \lambda_c^{(1)}\\ \lambda_v^{(1)}\end{bmatrix}=\begin{bmatrix}0.584\\ 39.98\\ -1.929\\ 1.126\end{bmatrix} \tag{2.4.4.26}$$

Updated first derivative (Jacobi) of Lagrange function with respect to $\left[x_C^{(1)},\theta^{(1)},\lambda_c^{(1)},\lambda_v^{(1)}\right]=[0.584,\ 39.98,\ -1.929,\ 1.126]$ obtained from Equation (2.4.4.27) does not converge to a stationary point.

$$\nabla\mathcal{L}\left(x_C^{(1)},\,\theta^{(1)},\,\lambda_c^{(1)},\,\lambda_v^{(1)}\right)=\begin{bmatrix}0.5693\\ 40.00\\ -3.703\\ 3.044\end{bmatrix}\neq 0 \tag{2.4.4.27}$$

Initial trial input parameter $\left[x_C^{(0)},\,\theta^{(0)},\,\lambda_c^{(0)},\,\lambda_v^{(0)}\right]=[0.7,\ 40,\ 0,\ 0]$ at Iteration 0, one of candidate solutions for Case 1 KKT condition, converged to $\left[x_C^{(127)},\,\theta^{(127)},\,\lambda_c^{(127)},\,\lambda_v^{(127)}\right]=$ $[0.569,\ 41.12,\ -3.591,\ 2.923]$ with MSE of 9.9E-6 at Iteration 127, resulting in a $f_{xC}\left(x_C,\theta\right)=0.569$ (km) as shown in as shown in Figure 2.4.4.2. Figure 2.4.4.2 also shows that an objective function was maximized as $f_{xC}\left(x_C^{(127)}\right)=0.569$ (km), which is maximized projectile distance with an angle $\theta^{(127)}$ of 41.12°. Three more initial trial input parameters of $\left[x_C^{(0)},\,\theta^{(0)}\right]$ shown in Figure 2.4.4.2 are implemented to find an optimal solution of Case 1. It is noted that trial initial Lagrange multipliers of $\lambda_c^{(0)}$ and $\lambda_v^{(0)}$ are selected as 0. In total, all four selected trial input parameters of $\left[x_C^{(0)},\,\theta^{(0)},\,\lambda_c^{(0)},\,\lambda_v^{(0)}\right]=[0.7,\,40,\,0,\,0]$, $[0.7,\,20,\,0,\,0]$, $[0.6,\,60,\,0,\,0]$, and $[0.2,\,35,\,0,\,0]$ provide the same result of $f_{xC}\left(x_C,\theta\right)=0.569$ (km) at $\left[x_C,\theta\right]=[0.569,\,41.12]$.

In summary, four initial trial input parameters $\left[x_C^{(0)},\,\theta^{(0)},\,\lambda_c^{(0)},\,\lambda_v^{(0)}\right]$ converged to $\left[x_C^{(k)},\,\theta^{(k)},\,\lambda_c^{(k)},\,\lambda_v^{(k)}\right]$ =[0.569, 41.12, −3.591, 2.923], satisfying $\nabla\mathcal{L}\left(x_C,\,\theta\,,\,\lambda_c,\,\lambda_v\right)=0$ as shown in Figure 2.4.4.2. An ordinate on the right side of Figure 2.4.4.2 represents colo r contours for a magnitude of an objective function $f_{xC}\left(x_C^{(k)}\right)$ in which minimized $f_{xC}\left(x_C^{(k)}\right)=0.569$ (km) is shown. Projectile trajectory is shown in Figure 2.4.4.3 for Case 1 KKT condition. A launching angle $\theta^{(k)}$ of 41.12° cannot be read in ordinate on left side Figure 2.4.4.3. A launching angle $\theta^{(k)}$ of 41.12° can also be calculated by equality constraining function shown in Equations (2.4.4.7-1) and (2.4.4.8). A number of iterations for each trial

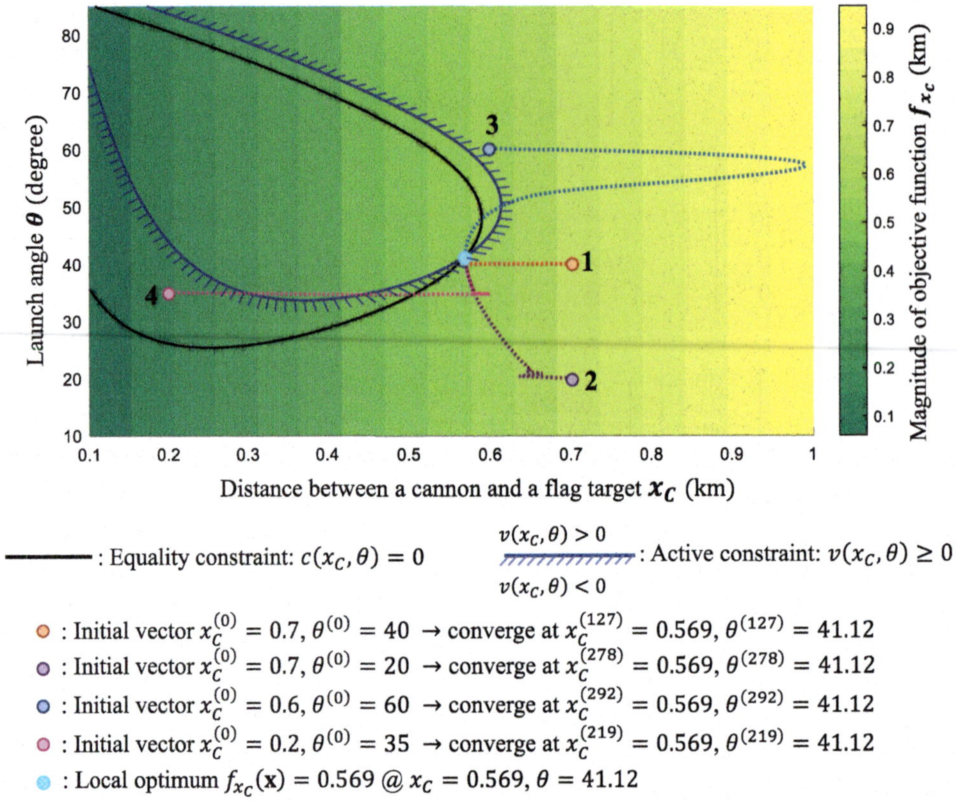

Figure 2.4.4.2 Convergence of initial trial input parameters based on analytical functions for Case 1 KKT; must satisfy $c(x_C, \theta) = y_{P_f} - 0.06 = 0,\ v(x_C, \theta) = y_{P_t} - 0.1 = 0.$

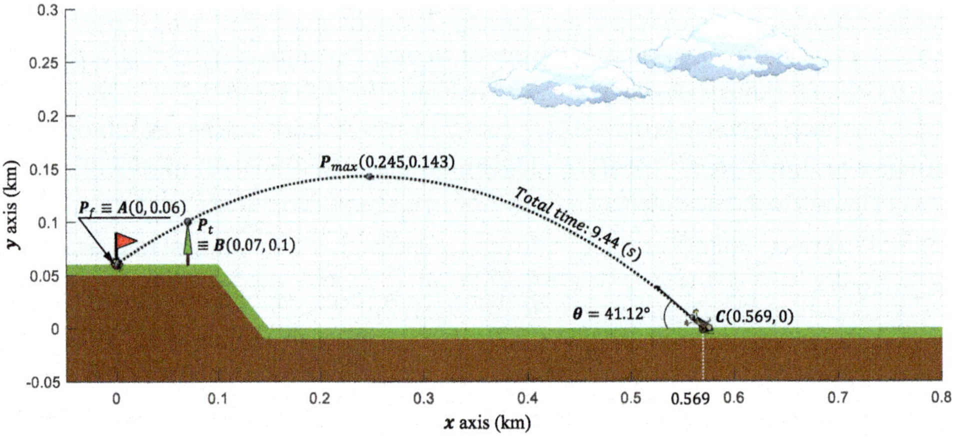

Figure 2.4.4.3 Projectile trajectory for Case 1 KKT; must satisfy $c(x_C, \theta) = y_{P_f} - 0.06 = 0,\ v(x_C, \theta) = y_{P_t} - 0.1 = 0.$

input parameter is indicated in Figure 2.4.4.2. The black line indicates Equality constraint $c\left(x_C^{(k)}, \theta^{(k)}\right)$ for the Case 1 KKT condition whereas the blue line indicates Inequality constraint $v\left(x_C^{(k)}, \theta^{(k)}\right)$ which is bound to equality constraint for the Case 1 KKT condition. In Figure 2.4.4.2, all four initial trial input parameters indicated by 1, 2, 3, 4 converged on sky point$[0.569,\ 41.12,\ -3.591,\ 2.923]$ which is located on both black and blue curves because an inequality condition $v\left(x_C^{(0)},\ \theta^{(0)}\right)$ is bound to an equality Case 1 KKT condition, being transformed to $v(x_C,\ \theta\ ) = y_{P_t} - 0.1 = 0$ when inequality $v(x_C,\ \theta\ ) = y_{P_t} - 0.1 \geq 0$ is active. A maximized distance under Case 1 KKT condition converging $\nabla\mathcal{L}\left(x_C, \theta\ , \lambda_c, \lambda_v\right) = 0$ should be compared with a candidate distance under Case 2 KKT condition to finalize a maximized projectile distance. Convergences of all initial input parameters are traced with a number of repetitions in Figure 2.4.4.2.

2.4.4.1.6 Case 2 KKT condition based on inactive inequality $v(x)$

In this case of Case 2 KKT, inequality $v(x_C,\theta) = y_{P_t} - 0.1 \geq 0$ is assumed be inactive or slack condition, which does not constraint an objective function $f_{xC}(x_C,\theta)$ when reaching a maximum of $f_{xC}(x_C,\theta)$, i.e., inequality $v(x_C,\theta) = y_{P_t} - 0.1 \geq 0$ is not bound to equality constraint, but ignored as shown in Equations (2.4.4.28-1) and (2.4.4.28-2). Roots for Case 2 KKT are obtained similarly to those of Case 1 KKT, however, stationary points of Case 2 KKT should be verified to exist within the region defined by an inequality constraint $v(x_C,\theta) = y_{P_t} - 0.1 \geq 0$, while an inactive inequality constraint is ignored in the formulation of a Lagrange function. A Lagrange function $\mathcal{L}(x_C,\theta,\lambda_c,\lambda_v)$ for Case 2 KKT is found in Equation (2.4.4.27) in which s = 0 is implemented.

$$\mathcal{L}(x_C,\ \theta,\ \lambda_c,\ \lambda_v) = x_C - \lambda_c\left(y_{P_f} - 0.06\right) - 0 \times \lambda_v\left(y_{P_t} - 0.1\right) \tag{2.4.4.27-1}$$

$$\mathcal{L}(x_C,\ \theta,\ \lambda_c,\ \lambda_v) = x_C - \lambda_c\left(y_{P_f} - 0.06\right) \tag{2.4.4.27-2}$$

An initial trial input parameter of $\mathbf{x}^{(0)} = \left[x_C^{(0)},\ \theta^{(0)}\right]^T = [0.7,\ 40]^T$ and initial trial Lagrange multipliers of $\lambda_c^{(0)}$ and $\lambda_v^{(0)}$ are selected as $[-1,\ 0]^T$ are used to find a stationary point under KKT conditions shown in Equation (2.4.4.28-3) using Newton–Raphson method. The initial trial input parameter becomes $\left[x_C^{(0)},\ \theta^{(0)},\ \lambda_c^{(0)},\ \lambda_v^{(0)}\right]^T = [0.7,\ 40,\ -1,\ 0]^T$. Constraints $c\left(x_C^{(0)},\ \theta^{(0)}\right)$ and $v\left(x_C^{(0)},\ \theta^{(0)}\right)$ are, then, calculated in Equations (2.4.4.7) and (2.4.4.8) by substituting $\left[x_C^{(0)},\ \theta^{(0)},\ \lambda_c^{(0)},\ \lambda_v^{(0)}\right]^T = [0.7,\ 40,\ -1,\ 0]^T$ into Equations (2.4.4.7-1) and (2.4.4.7-2), respectively. Jacobi matrix $\nabla f_{xC}(x_C)$ of objective function, Jacobi matrix $\mathbf{J}_c\left(x_C^{(0)},\ \theta^{(0)}\right)$ of constraints $c(\mathbf{x})$, and $\mathbf{J}_c\left(x_C^{(0)},\ \theta^{(0)}\right)$ of $v(\mathbf{x})$ are calculated by substituting $\left[x_C^{(0)},\ \theta^{(0)},\ \lambda_c^{(0)},\ \lambda_v^{(0)}\right]^T = [0.7,\ 40,\ -1,\ 0]^T$ into Equations (2.4.4.12-1)–(2.4.4.12-3), respectively.

A first derivative (Jacobi, gradient vector) of a Lagrange function $\nabla\mathcal{L}\left(x_C^{(0)},\ \theta^{(0)},\ \lambda_c^{(0)},\ \lambda_v^{(0)}\right)$ is, then, calculated in Equation (2.4.4.28-3) by substituting $\nabla f_{xC}(x_C)$ shown in Equation (2.4.4.12-1), $c\left(x_C^{(0)},\ \theta^{(0)}\right)$, $v\left(x_C^{(0)},\ \theta^{(0)}\right)$, $\mathbf{J}_c\left(x_C^{(0)},\ \theta^{(0)}\right)$, $\mathbf{J}_v\left(x_C^{(0)},\ \theta^{(0)}\right)$, and $s = 1$ into Equation (2.4.4.11-1). However, $\nabla\mathcal{L}\left(x_C^{(0)},\ \theta^{(0)},\ \lambda_c^{(0)},\ \lambda_v^{(0)}\right)$ does not converge to 0, indicating $\left[x_C^{(0)},\ \theta^{(0)},\ \lambda_c^{(0)},\ \lambda_v^{(0)}\right]^T = [0.7,\ 40,\ -1,\ 0]^T$ is not a root of Case 1 KKT condition for Case 2 as shown in Equation (2.4.4.28-3). Initial trial input parameter is updated to $\left[x_C^{(1)},\ \theta^{(1)},\ \lambda_c^{(1)},\ \lambda_v^{(1)}\right]$ in Equation (2.4.4.29) and check if a first derivative (Jacobi, gradient vector) of a Lagrange function $\nabla\mathcal{L}\left(x_C^{(1)},\ \theta^{(1)},\ \lambda_c^{(1)},\ \lambda_v^{(1)}\right)$ converges to 0 in Equation

(2.4.4.34) at Iteration 1. A first derivative (Jacobi, gradient vector) of a Lagrange function $\nabla\mathcal{L}\left(x_C^{(0)}, \theta^{(0)}, \lambda_c^{(0)}, \lambda_v^{(0)}\right)$ at Iteration 0 shown in Equation (2.4.4.28-3) and inverse Hessian matrix of Lagrange function $\left[\mathbf{H}_{\mathcal{L}}\left(x_C^{(0)}, \theta^{(0)}, \lambda_c^{(0)}, \lambda_v^{(0)}\right)\right]^{-1}$ shown in Equation (2.4.4.32–4) are substituted into Equation (2.4.4.29) to calculate updated input parameter $\left[x_C^{(1)}, \theta^{(1)}, \lambda_c^{(1)}, \lambda_v^{(1)}\right]$ at Iteration 1. Hessian matrix of Lagrange function $\mathbf{H}_{\mathcal{L}}\left(x_C^{(0)}, \theta^{(0)}\right)$ shown in Equations (2.4.4.31-1) and (2.4.4.31-2) should be obtained to be substituted into $\mathbf{H}_{\mathcal{L}}\left(x_C^{(0)}, \theta^{(0)}, \lambda_c^{(0)}, \lambda_v^{(0)}\right)$ shown in Equation (2.4.4.30). $\mathbf{H}_{\mathcal{L}}\left(x_C^{(0)}, \theta^{(0)}\right)$ is obtained in Equation (2.4.4.31-2) by substituting $\mathbf{H}_{f_{xC}}\left(x_C, \theta\right)$, $\mathbf{H}_c\left(x_C^{(0)}, \theta^{(0)}\right)$, and $\mathbf{H}_v\left(x_C^{(0)}, \theta^{(0)}\right)$ obtained using Equations (2.4.4.14), (2.4.4.15-1), and (2.4.4.16-1) into Equation (2.4.4.31-1). It is noted that Hessian matrices $H_c\left(x_C^{(0)}, \theta^{(0)}\right)$ of constraints $c(\mathbf{x})$ and $H_v\left(x_C^{(0)}, \theta^{(0)}\right)$ of $v(\mathbf{x})$ are calculated by substituting $\left[x_C^{(0)}, \theta^{(0)}, \lambda_c^{(0)}, \lambda_v^{(0)}\right]^T = [0.7, 40, -1, 0]^T$ into Equations (2.4.4.15-1) and (2.4.4.16-1). Hessian matrix of Lagrange function $\mathbf{H}_{\mathcal{L}}\left(x_C^{(0)}, \theta^{(0)}, \lambda_c^{(0)}, \lambda_v^{(0)}\right)$ are, then, obtained in Equations (2.4.4.32-1) and (2.4.4.32-2) by substituting $\mathbf{J}_c\left(x_C^{(0)}, \theta^{(0)}\right)$, $\mathbf{J}_v\left(x_C^{(0)}, \theta^{(0)}\right)$ and $\mathbf{H}_{\mathcal{L}}\left(x_C^{(0)}, \theta^{(0)}\right)$ obtained in Equations (2.4.4.12-2), (2.4.4.12-3), and (2.4.4.31-2) with $s = 0$ into Equation (2.4.4.30). $\mathbf{H}_{\mathcal{L}}\left(x_C^{(0)}, \theta^{(0)}, \lambda_c^{(0)}, \lambda_v^{(0)}\right)$ is inverted in Equations (2.4.4.32-3) and (2.4.4.32-4). Initial trial input parameter is updated in Equation (2.4.4.33) as $\left[x_C^{(1)}, \theta^{(1)}, \lambda_c^{(1)}, \lambda_v^{(1)}\right] = [0.584, 39.98, -1.929, 1.126]$ by substituting $\nabla\mathcal{L}\left(x_C^{(0)}, \theta^{(0)}, \lambda_c^{(0)}, \lambda_v^{(0)}\right)$ calculated in Equation (2.4.4.28-3) and $\left[\mathbf{H}_{\mathcal{L}}\left(x_C^{(0)}, \theta^{(0)}, \lambda_c^{(0)}, \lambda_v^{(0)}\right)\right]^{-1}$ calculated in Equation (2.4.4.32-4) into Equation (2.4.4.29).

A first derivative (Jacobi, gradient vector) of a Lagrange function $\nabla\mathcal{L}\left(x_C^{(1)}, \theta^{(1)}, \lambda_c^{(1)}, \lambda_v^{(1)}\right)$ at Iteration 1 obtained in Equation (2.4.4.34) based on Equation (2.4.4.11-1) did not converge to 0, indicating $\left[x_C^{(0)}, \theta^{(0)}, \lambda_c^{(0)}, \lambda_v^{(0)}\right]^T = [0.7, 40, -1, 0]^T$ is not the root of Case 2 KKT condition, and hence, the same iteration should be repeated until stationary points are identified.

$$\mathcal{L}(x_C, \theta, \lambda_c, \lambda_v) = x_C - \lambda_c\left(y_{P_f} - 0.06\right) - 0 \times \lambda_v\left(y_{P_t} - 0.1\right) \tag{2.4.4.28-1}$$

$$\mathcal{L}(x_C, \theta, \lambda_c, \lambda_v) = x_C - \lambda_c\left(y_{P_f} - 0.06\right) \tag{2.4.4.28-2}$$

$$\nabla\mathcal{L}\left(x_C^{(0)}, \theta^{(0)}, \lambda_c^{(0)}, \lambda_v^{(0)}\right)$$

$$= \begin{bmatrix} \begin{bmatrix} 1 \\ 0 \end{bmatrix} - \begin{bmatrix} -0.989 \\ 0.119 \end{bmatrix} \times (-1) - \begin{bmatrix} -0.806 \\ 0.204 \end{bmatrix} \times 0 \times \lambda_v^{(0)} \\ 0.113 \\ 0 \end{bmatrix}$$

$$\rightarrow \nabla\mathcal{L}\left(x_C^{(0)}, \theta^{(0)}, \lambda_c^{(0)}, \lambda_v^{(0)}\right) = \begin{bmatrix} 0.011 \\ 0.119 \\ 0.113 \\ 0 \end{bmatrix} \neq 0 \tag{2.4.4.28-3}$$

$$\begin{bmatrix} x_C^{(1)} \\ \theta^{(1)} \\ \lambda_c^{(1)} \\ \lambda_v^{(1)} \end{bmatrix} = \begin{bmatrix} x_C^{(0)} \\ \theta^{(0)} \\ \lambda_c^{(0)} \\ \lambda_v^{(0)} \end{bmatrix} - \left[\mathbf{H}_{\mathcal{L}}\left(x_C^{(0)}, \theta^{(0)}, \lambda_c^{(0)}, \lambda_v^{(0)}\right)\right]^{-1} \nabla\mathcal{L}\left(x_C^{(0)}, \theta^{(0)}, \lambda_c^{(0)}, \lambda_v^{(0)}\right) \tag{2.4.4.29}$$

$$\mathbf{H}_{\mathcal{L}}\left(x_C^{(0)}, \theta^{(0)}, \lambda_c^{(0)}, \lambda_v^{(0)}\right) = \begin{bmatrix} \mathbf{H}_{\mathcal{L}}\left(x_C^{(0)}, \theta^{(0)}\right) & -J_c\left(x_C^{(0)}, \theta^{(0)}\right)^T & -\left(sJ_v\left(x_C^{(0)}, \theta^{(0)}\right)\right)^T \\ -J_c\left(x_C^{(0)}, \theta^{(0)}\right) & 0 & 0 \\ -sJ_v\left(x_C^{(0)}, \theta^{(0)}\right) & 0 & 0 \end{bmatrix} \tag{2.4.4.30}$$

$$\begin{aligned} \mathbf{H}_{\mathcal{L}}\left(\mathbf{x}^{(0)}\right) &= \mathbf{H}_{\mathcal{L}}\left(x_C^{(0)}, \theta^{(0)}\right) \\ &= \mathbf{H}_{f_{xC}}\left(x_C^{(0)}, \theta^{(0)}\right) - \lambda_c \mathbf{H}_c\left(x_C^{(0)}, \theta^{(0)}\right) - s\lambda_v \mathbf{H}_v\left(x_C^{(0)}, \theta^{(0)}\right) \end{aligned} \tag{2.4.4.31-1}$$

$$\begin{aligned} \mathbf{H}_{\mathcal{L}}\left(\mathbf{x}^{(0)}\right) &= \mathbf{H}_{\mathcal{L}}\left(x_C^{(0)}, \theta^{(0)}\right) \\ &= \begin{bmatrix} 0 & 0 \\ 0 & 0 \end{bmatrix} - (-1) \times \begin{bmatrix} -2.612 & -1.364 \\ -1.364 & -1.982 \end{bmatrix} \\ &\quad -1 \times 0 \times \begin{bmatrix} -2.612 & -1.058 \\ -1.058 & -1.425 \end{bmatrix} \\ &= \begin{bmatrix} -2.612 & -1.364 \\ -1.364 & -1.982 \end{bmatrix} \end{aligned} \tag{2.4.4.31-2}$$

$$\mathbf{H}_{\mathcal{L}}\left(x_C^{(0)}, \theta^{(0)}, \lambda_c^{(0)}, \lambda_v^{(0)}\right)$$

$$= \begin{bmatrix} \begin{bmatrix} -2.612 & -1.364 \\ -1.364 & -1.982 \end{bmatrix} & -\begin{bmatrix} -0.989 \\ 0.119 \end{bmatrix} & -0\begin{bmatrix} -0.806 \\ 0.204 \end{bmatrix} \\ -\begin{bmatrix} -0.989 & 0.119 \end{bmatrix} & 0 & 0 \\ -0 \times \begin{bmatrix} -0.806 & 0.204 \end{bmatrix} & 0 & 0 \end{bmatrix} \tag{2.4.4.32-1}$$

$$\mathbf{H}_{\mathcal{L}}\left(x_C^{(0)}, \theta^{(0)}, \lambda_c^{(0)}, \lambda_v^{(0)}\right) = \begin{bmatrix} -2.612 & -1.364 & 0.989 & 0 \\ -1.364 & -1.982 & -0.119 & 0 \\ 0.989 & -0.119 & 0 & 0 \\ 0 & 0 & 0 & 0 \end{bmatrix} \tag{2.4.4.32-2}$$

$$\left[\mathbf{H}_{\mathcal{L}}\left(x_C^{(0)},\theta^{(0)},\lambda_c^{(0)},\lambda_v^{(0)}\right)\right]^{-1}=\begin{bmatrix}\begin{bmatrix}-2.612 & -1.364 & 0.989\\ -1.364 & -1.982 & -0.119\\ 0.989 & -0.119 & 0\end{bmatrix}^{-1} & \begin{matrix}0\\0\\0\end{matrix}\\ \begin{matrix}0 & \quad 0 & \quad 0\end{matrix} & 0\end{bmatrix} \tag{2.4.4.32-3}$$

$$\left[\mathbf{H}_{\mathcal{L}}\left(x_C^{(0)},\theta^{(0)},\lambda_c^{(0)},\lambda_v^{(0)}\right)\right]^{-1}=\begin{bmatrix}-0.0062 & -0.0512 & 0.9239 & 0\\ -0.0512 & -0.4260 & -0.7227 & 0\\ 0.9239 & -0.7227 & 1.4426 & 0\\ 0 & 0 & 0 & 0\end{bmatrix} \tag{2.4.4.32-4}$$

$$\begin{bmatrix}x_C^{(1)}\\ \theta^{(1)}\\ \lambda_c^{(1)}\\ \lambda_v^{(1)}\end{bmatrix}=\begin{bmatrix}0.7\\40\\-1\\0\end{bmatrix}-\begin{bmatrix}-0.0062 & -0.0512 & 0.9239 & 0\\ -0.0512 & -0.4260 & -0.7227 & 0\\ 0.9239 & -0.7227 & 1.4426 & 0\\ 0 & 0 & 0 & 0\end{bmatrix}\begin{bmatrix}0.011\\0.119\\0.113\\0\end{bmatrix}$$

$$\rightarrow\begin{bmatrix}x_C^{(1)}\\ \theta^{(1)}\\ \lambda_c^{(1)}\\ \lambda_v^{(1)}\end{bmatrix}=\begin{bmatrix}0.584\\39.98\\-1.929\\1.126\end{bmatrix} \tag{2.4.4.33}$$

$$\nabla\mathcal{L}\left(x_C^{(1)},\theta^{(1)},\lambda_c^{(1)},\lambda_v^{(1)}\right)=\begin{bmatrix}0.6021\\40.13\\-1.086\\0\end{bmatrix}\neq 0 \tag{2.4.4.34}$$

2.4.4.1.7 Summary

Finally, one initial trial input parameter successfully leads $\nabla\mathcal{L}(x_C,\theta,\lambda_c,\lambda_v)$ to converge at $\left[x_C^{(238)},\ \theta^{(238)},\ \lambda_c^{(238)},\ \lambda_v^{(238)}\right]=[0.589,\ 47.81,\ -1.107,\ 0]$ with an MSE of 9.9E-6, resulting in a $f_{xC}(x_C)=0.589$ (km) at Iteration 238 as shown in Figure 2.4.4.4. All four selected initial trial input parameters $\left[x_C^{(0)},\ \theta^{(0)},\ \lambda_c^{(0)},\ \lambda_v^{(0)}\right]=[0.7,40,-1,0],[0.6,30,-1,0],[0.8,66,-1,0]$, and $[0.2,35,-1,0]$ converge to a point indicated by sky color $[0.589,\ 47.81,\ -1.107,\ 0]$ by satisfying $\nabla\mathcal{L}(x_C,\theta,\lambda_c,\lambda_v)=0$ as a candidate solution of a Case 2 KKT condition, resulting in the same result of $f_{xC}(x_C)=0.589$ (km) at $\left[x_C^{(k)},\ \theta^{(k)},\ \lambda_c^{(k)},\ \lambda_v^{(k)}\right]=[0.589,\ 47.81]$. The maximized flying distance and launching angle of the projectile are $\left[x_C^{(k)},\ \theta^{(k)}\right]=[0.589,\ 47.81]$. A number of iterations for each initial trial input parameter is indicated in Figure 2.4.4.4. An ordinate on the right side of Figure 2.4.4.4 represents color contour for a magnitude of an objective function $f_{xC}\left(x_C^{(k)}\right)$ in which minimized $f_{xC}\left(x_C^{(k)}\right)=0.589$ (km) is shown.

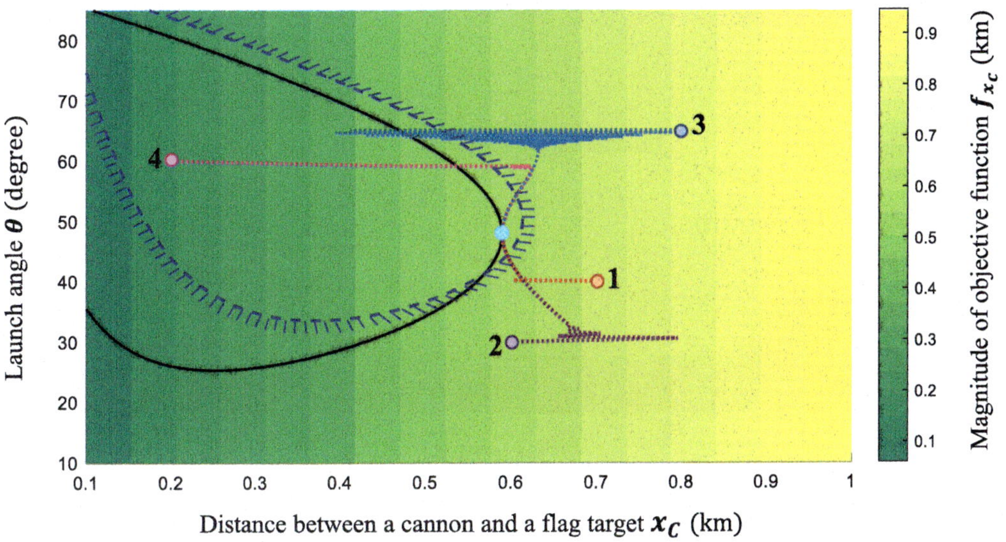

——— : Equality constraint: $c(x_C, \theta) = 0$

$v(x_C, \theta) > 0$ / $v(x_C, \theta) < 0$: Inactive constraint: $v(x_C, \theta) \geq 0$

- : Initial vector $x_C^{(0)} = 0.7, \theta^{(0)} = 40 \rightarrow$ converge at $x_C^{(238)} = 0.589, \theta^{(238)} = 47.81$
- : Initial vector $x_C^{(0)} = 0.6, \theta^{(0)} = 30 \rightarrow$ converge at $x_C^{(276)} = 0.589, \theta^{(276)} = 47.81$
- : Initial vector $x_C^{(0)} = 0.8, \theta^{(0)} = 65 \rightarrow$ converge at $x_C^{(355)} = 0.589, \theta^{(355)} = 47.81$
- : Initial vector $x_C^{(0)} = 0.2, \theta^{(0)} = 60 \rightarrow$ converge at $x_C^{(307)} = 0.589, \theta^{(307)} = 47.81$
- : Local optimum $f_{x_C}(\mathbf{x}) = 0.589$ @ $x_C = 0.589, \theta = 47.81$

Figure 2.4.4.4 Convergence of initial trial input parameters based on analytical functions for Case 2 KKT condition; must satisfy $c(x_C,\theta) = y_{P_f} - 0.06 = 0,\ v(x_C,\theta) = y_{P_t} - 0.1 \geq 0$.

A launching angle $\theta^{(k)}$ of 47.81° cannot be read in the ordinate on the left side Figure 2.4.4.4. A launching angle $\theta^{(k)}$ of 47.81° can be also calculated by equality constraining function shown in Equations (2.4.4.7-1) and (2.4.4.8). The black line indicates Equality constraint $c\left(x_C^{(k)}, \theta^{(k)}\right)$ for the Case 2 KKT condition whereas the blue line indicates Inequality constraint $v\left(x_C^{(k)}, \theta^{(k)}\right)$ which is bound to equality constraint. In Figure 2.4.4.4, all four initial trial input parameters indicated by 1, 2, 3, 4 converge on sky point $[0.589,\ 47.81,\ -1.107,\ 0]$ which are located on a black curve for equality condition $c\left(x_C^{(0)}, \theta^{(0)}\right)$ of Case 2 KKT condition. Sky point is also located in the feasible region of other side the blue shaded area which verifies that the objective function was maximized within inequality region $v(x_C,\theta) = y_{P_t} - 0.1 \geq 0$ as a candidate solution of a Case 2 KKT condition. It is noted that flying distance 0.589 (km) with a Case 2 KKT condition is longer than 0.569 (km) obtained by a Case 1 KKT condition, and hence, the objective function is maximized as 0.589 (km) with a Case 2 KKT condition. Projectile trajectory is shown in Figure 2.4.4.5 for Case 2 KKT condition. Code for Lagrange optimization based on analytical functions is presented in Appendix B3.

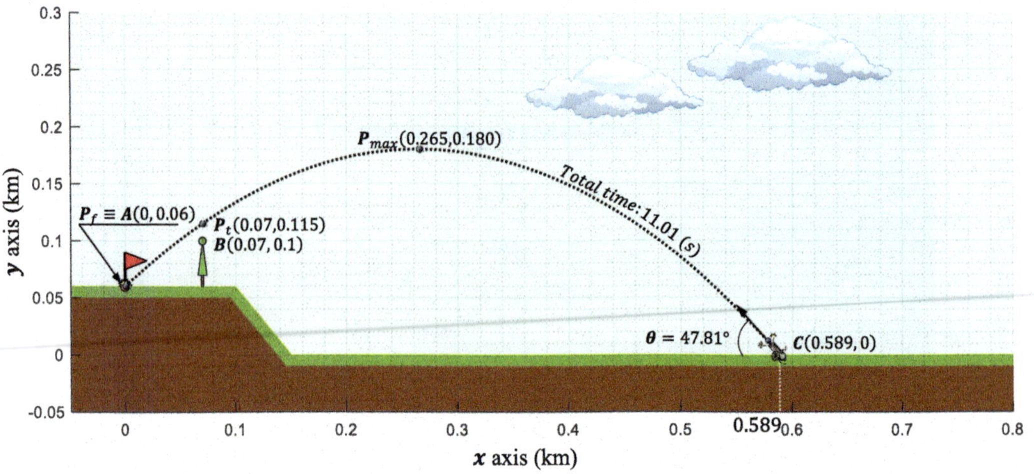

Figure 2.4.4.5 Projectile trajectory for Case 2 KKT condition; must satisfy $c(x_C,\theta)=y_{P_f}-0.06=0,\ v(x_C,\theta)=y_{P_t}-0.1\geq 0$.

2.4.4.2 ANN-based Lagrange optimization

2.4.4.2.1 Step 1: Generation of large datasets for training ANN

In this example, a launching angle θ which maximizes a flying distance of a projectile is determined based on ANN-based Lagrange optimization. Figure 2.4.4.6 illustrates maximized distance of projectile obtained by ANN-based Lagrange optimization. A nonlinear trajectory of a projectile shown in Equations (2.4.4.5-2) and (2.4.4.6) is replaced by an ANN-based differentiable function. As shown in Table 2.4.4.2, a total of 1,000 datasets representing flying distances x_C between 0.1 and 1.2 km and launching angles θ between 10° and 80° is randomly used as inputs to generate the altitudes (y_1–y_{19}) corresponding between $0.95x_C$ and $0.05x_C$ with an interval of $0.05x_C$ to train ANNs. Table 2.4.4.2(a) presents non-normalized large datasets whereas large datasets are normalized in Table 2.4.4.2(b). The X and Y directions of the coordinate shown in Figure 2.4.4.6 are positive to the right and upwards from the origin at O. As shown in Tables 2.4.4.2, 2.4.4.3, and Figure 2.4.4.7, a total of 24 outputs is generated to plot the altitudes of the projectile trajectory for randomly selected 1,000 flying distances x_C and launching angle s θ. Among the generated 24 outputs shown in Tables 2.4.4.2, y_1–y_{19} correspond to the projectile distances x_C between $0.95x_C$ and $0.05x_C$ with an interval of $0.05x_C$ for each randomly selected x_1 (flying distance) and x_2 (launching angle θ). For example, Data 1, one of 1,000 large datasets shown in Table 2.4.4.2, is generated to plot the altitudes of the flying projectile (y_1–y_{19}) corresponding to the projectile distances x_C between $0.95x_C$ and $0.05x_C$ with an interval of $0.05x_C$ for each randomly selected x_1 (flying distance)=0.848 km and x_2 (launching angle θ)=28.15. Big data y_{20} and y_{21} are the flying heights over a flag (y_{P_f}) at $x=x_A=0$ km and tree obstacle (y_{P_t}) at $x=x_B=0.07$ km, respectively, which can be obtained in Equations (2.4.4.7-1) and (2.4.4.7-2) by substituting positions of a flag and tree. y_{22} represents the flying distance at the maximum altitude. whereas $y_{23,}$ and y_{24} represent the maximum altitude and total flying time, respectively. Table 2.4.4.3 and Figure 2.4.4.8 present the ANN-based projectile trajectory (y_1–y_{24}) which was trained by large datasets shown in Table 2.4.4.2 for the eight different locations of the projectile represented by eight flying distances x_c and launching

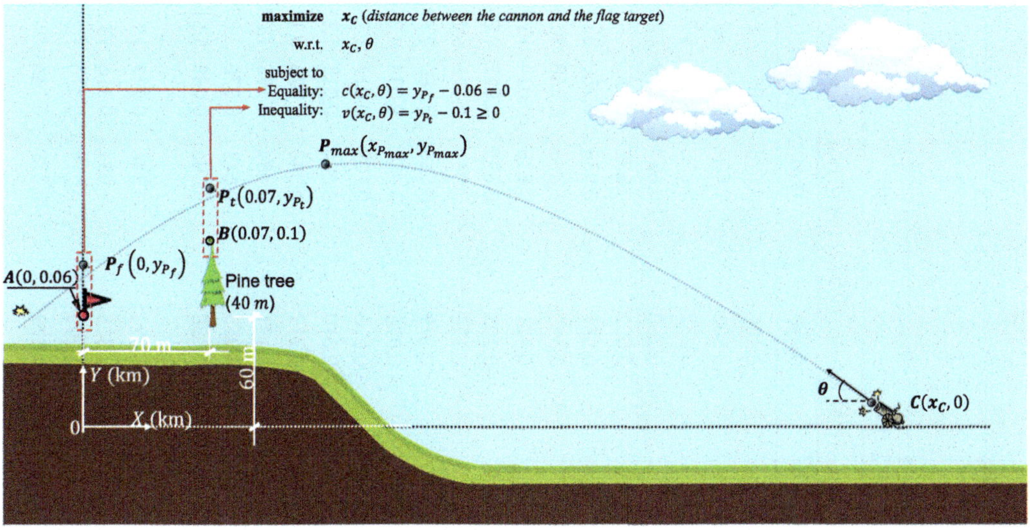

Figure 2.4.4.6 ANN-based Lagrange maximization of flying distance.

angles θ. Projectile trajectory (y_1 to y_{24}) shown in Table 2.4.4.3 and Figure 2.4.4.8 for eight scenarios obtained using ANNs shown in Figure 2.4.4.7 is well compared with one calculated explicitly using Equation (2.4.4.6), demonstrating ANNs shown in Figure 2.4.4.7 were well trained by large datasets given in Table 2.4.4.2.

Big data y_{20} is the flying heights over a flag (y_{P_f}) at $x = x_A = 0$ km which yields positive heights between Scenarios 1 ($x_C = 0.2$ km) and 5 ($x_C = 0.6$ km) whereas yielding negative heights between Scenarios 6 ($x_C = 0.7$ km) and 8 ($x_C = 0.9$ km). Negative heights mean that flying heights fall below the coordinate origin, resulting in the projectile trajectory which did not reach the flag when flying distance and launching angle θ result in negative heights for Scenarios 6 ($x_C = 0.7$ km)–8($x_C = 0.9$ km), as shown in Equation (2.4.4.8). The goal of this example is to calculate the maximum distance while hitting the flag target as illustrated in Figure 2.4.4.8, in which the projectile trajectory indicated by Point O hits the flag target while flies the maximum distance from the cannon. It is noted that Point O is located between Scenarios 4 and 5. The projectile trajectory which flies longer distance while missing the target is not the solution that this example seeks.

2.4.4.2.2 Step 2: ANN training on large datasets

In this step, a proper number of training parameters including hidden layers and neurons needed for training networks should be selected carefully based on a level of complexity for a considered problem. A flying distance of the projectile shown in Figure 2.4.4.6 is maximized based on ANNs. In Figure 2.4.4.7a, ANNs representing a projectile motion shown in Figure 2.4.4.6 have four hidden layers with ten neurons based on Equations (2.4.4.35) and (2.4.4.36). The ANNs are trained on a total of 1,000 datasets generated in Table 2.4.4.2. ANNs shown in Figure 2.4.4.7a map entire input parameters to entire output parameters using large datasets shown in Table 2.4.4.2 based on TED. In Figure 2.4.4.7a, two-input data, x_C (flying distance) and θx_2 (launching angle θ), are mapped to 24 output parameters, simultaneously. A good training accuracy based on 15,000 epochs is obtained in Figure 2.4.4.7b with MSE of test subset (MSE.Tperf) of 9.1E-08.

Table 2.4.4.2 Generation of Large Datasets

(a): Non-normalized

1,000 Datasets (Non-Normalized)

	x_1	x_2	y_1	y_2	y_3		y_{20}	y_{21}	y_{22}	y_{23}	y_{24}
	xC	θ	$y@x=0.95x_C$	$y@x=0.90x_C$	$y@x=0.85x_C$	⋯	$yPf@x=x_A=0$	$yPt@x=x_B=0.07$	$x@y_{max}$	y_{max}	t
Data	(km)	(degree)	(km)	(km)	(km)	...	(km)	(km)	(km)	(km)	(s)
1	0.848	28.15	0.021	0.038	0.052		–0.255	–0.180	0.576	0.073	12.017
2	1.110	31.02	0.030	0.054	0.071	...	–0.618	–0.503	0.822	0.087	16.186
3	0.200	68.23	0.025	0.048	0.070	...	0.278	0.232	–0.025	0.281	6.747
4	0.841	17.22	0.012	0.020	0.026	...	–0.334	–0.260	0.657	0.029	11.006
5	0.495	68.37	0.059	0.111	0.156	...	–0.135	0.052	0.272	0.282	16.795
6	0.798	63.25	0.073	0.134	0.183	...	–0.827	–0.561	0.536	0.260	22.167
7	0.696	65.08	0.070	0.129	0.178	...	–0.592	–0.343	0.446	0.268	20.635
8	1.038	18.90	0.015	0.026	0.033	...	–0.567	–0.471	0.838	0.034	13.716
9	0.757	14.90	0.009	0.015	0.020	...	–0.269	–0.205	0.595	0.022	9.797
10	0.379	38.45	0.015	0.028	0.041	...	0.121	0.126	0.061	0.126	6.042
11	0.283	68.02	0.034	0.066	0.095	...	0.263	0.279	0.056	0.280	9.443
12	0.626	18.76	0.010	0.018	0.024	...	–0.123	–0.076	0.428	0.034	8.268
13	0.255	13.29	0.003	0.006	0.008	...	0.008	0.016	0.109	0.017	3.279
14	0.131	12.24	0.001	0.003	0.004	...	0.015	0.010	–0.004	0.015	1.673
15	0.191	66.97	0.022	0.043	0.063	...	0.267	0.212	–0.044	0.276	6.115
16	0.804	50.64	0.046	0.086	0.119	...	–0.251	–0.131	0.484	0.195	15.839
17	0.202	28.85	0.005	0.011	0.016	...	0.071	0.055	–0.073	0.076	2.889
18	0.376	31.44	0.011	0.022	0.031	...	0.081	0.088	0.086	0.089	5.514
...	...	...	...	...	...	...	...	...	...	...	...
1,000	0.637	58.42	0.049	0.092	0.130	...	–0.098	0.023	0.346	0.237	15.212
x_{max} (y_{max})	1.199	80.00	0.086	0.158	0.216	...	0.309	0.315	1.081	0.316	23.59
x_{min} (y_{min})	0.100	10.00	0.001	0.002	0.003	...	–0.987	–0.799	–0.217	0.010	1.35
x_{mean} (y_{mean})	0.589	40.95	0.027	0.050	0.069	...	–0.146	–0.079	0.329	0.144	10.79

(Continued)

Table 2.4.4.2 (Continued) Generation of Large Datasets

(b): Normalized

1,000 Datasets (Normalized)

	x_1	x_2	y_1	y_2	y_3		y_{20}	y_{21}	y_{22}	y_{23}	y_{24}
	xC	θ	$y@x=0.95x_C$	$y@x=0.90x_C$	$y@x=0.85x_C$	···	$yPf@x=x_A=0$	$yPt@x=x_B=0.07$	$x@y_{\max}$	$y_{\max}$	t
Data	(km)	(degree)	(km)	(km)	(km)	...	(km)	(km)	(km)	(km)	(s)
1	0.361	–0.481	–0.533	–0.537	–0.541		0.130	0.111	0.223	–0.591	–0.041
2	0.838	–0.399	–0.314	–0.336	–0.362	...	–0.430	–0.469	0.600	–0.499	0.334
3	–0.818	0.664	–0.448	–0.413	–0.371	...	0.952	0.850	–0.703	0.771	–0.515
4	0.349	–0.794	–0.754	–0.770	–0.789	...	0.008	–0.033	0.346	–0.878	–0.132
5	–0.281	0.668	0.369	0.400	0.438	...	0.315	0.529	–0.246	0.775	0.389
6	0.271	0.522	0.704	0.699	0.693	...	–0.752	–0.573	0.160	0.633	0.872
7	0.084	0.574	0.621	0.629	0.639	...	–0.390	–0.181	0.022	0.686	0.735
8	0.707	–0.746	–0.662	–0.691	–0.725	...	–0.352	–0.411	0.626	–0.841	0.112
9	0.196	–0.860	–0.817	–0.830	–0.846	...	0.108	0.067	0.252	–0.923	–0.240
10	–0.493	–0.187	–0.682	–0.665	–0.645	...	0.710	0.661	–0.571	–0.241	–0.578
11	–0.667	0.658	–0.224	–0.184	–0.135	...	0.929	0.936	–0.578	0.766	–0.272
12	–0.042	–0.750	–0.796	–0.799	–0.802	...	0.334	0.299	–0.006	–0.844	–0.378
13	–0.717	–0.906	–0.959	–0.958	–0.957	...	0.535	0.463	–0.497	–0.952	–0.827
14	–0.944	–0.936	–0.995	–0.994	–0.994	...	0.545	0.453	–0.672	–0.968	–0.971
15	–0.834	0.628	–0.506	–0.473	–0.435	...	0.934	0.815	–0.732	0.738	–0.572
16	0.281	0.161	0.059	0.074	0.091	...	0.136	0.199	0.080	0.208	0.303
17	–0.814	–0.461	–0.898	–0.891	–0.882	...	0.632	0.534	–0.778	–0.569	–0.862
18	–0.497	–0.387	–0.764	–0.752	–0.738	...	0.648	0.593	–0.533	–0.485	–0.625
...	...	...	...	...	...	...	...	...	...	...	...
1,000	–0.022	0.383	0.132	0.159	0.191	...	0.371	0.477	–0.132	0.480	0.247
$\bar{x}_{max}$ ($\bar{y}_{max}$)	1.000	1.000	1.000	1.000	1.000	...	1.000	1.000	1.000	1.000	1.000
$\bar{x}_{min}$ ($\bar{y}_{min}$)	–1.000	–1.000	–1.000	–1.000	–1.000	...	–1.000	–1.000	–1.000	–1.000	–1.000
$\bar{x}_{mean}$ ($\bar{y}_{mean}$)	–0.111	–0.12	–0.396	–0.389	–0.380	...	0.297	0.293	–0.158	–0.126	–0.151
α_x (α_y)	1.8200	0.0286	23.6770	12.8729	9.4036		1.5427	1.7959	1.5402	6.5247	0.0899

a

1-1 Non-normalized input layer, $\mathbf{x} = [x_C, \theta]^T$

1-2 Normalized input layer, $\mathbf{z}^{(N)} = \mathbf{g}^{(N)}(\mathbf{x})$

2   Four hidden layers with 10 neurons, $\mathbf{z}^{(l)} = f_t^{(l)}(\mathbf{W}^{(l)}\mathbf{z}^{(l-1)} + \mathbf{b}^{(l)})$, $l = \{1, 2, \ldots, 4\}$

3-1 Normalized output layers, $\mathbf{z}^{(L=5)} = f_{lin}^{(5)}(\mathbf{W}^{(5)}\mathbf{z}^{(4)} + \mathbf{b}^{(5)})$

3-2 De-normalized output layer, $\mathbf{y} = \mathbf{z}^{(D)} = \mathbf{g}^{(D)}(\mathbf{z}^{(\text{Output})})$

b

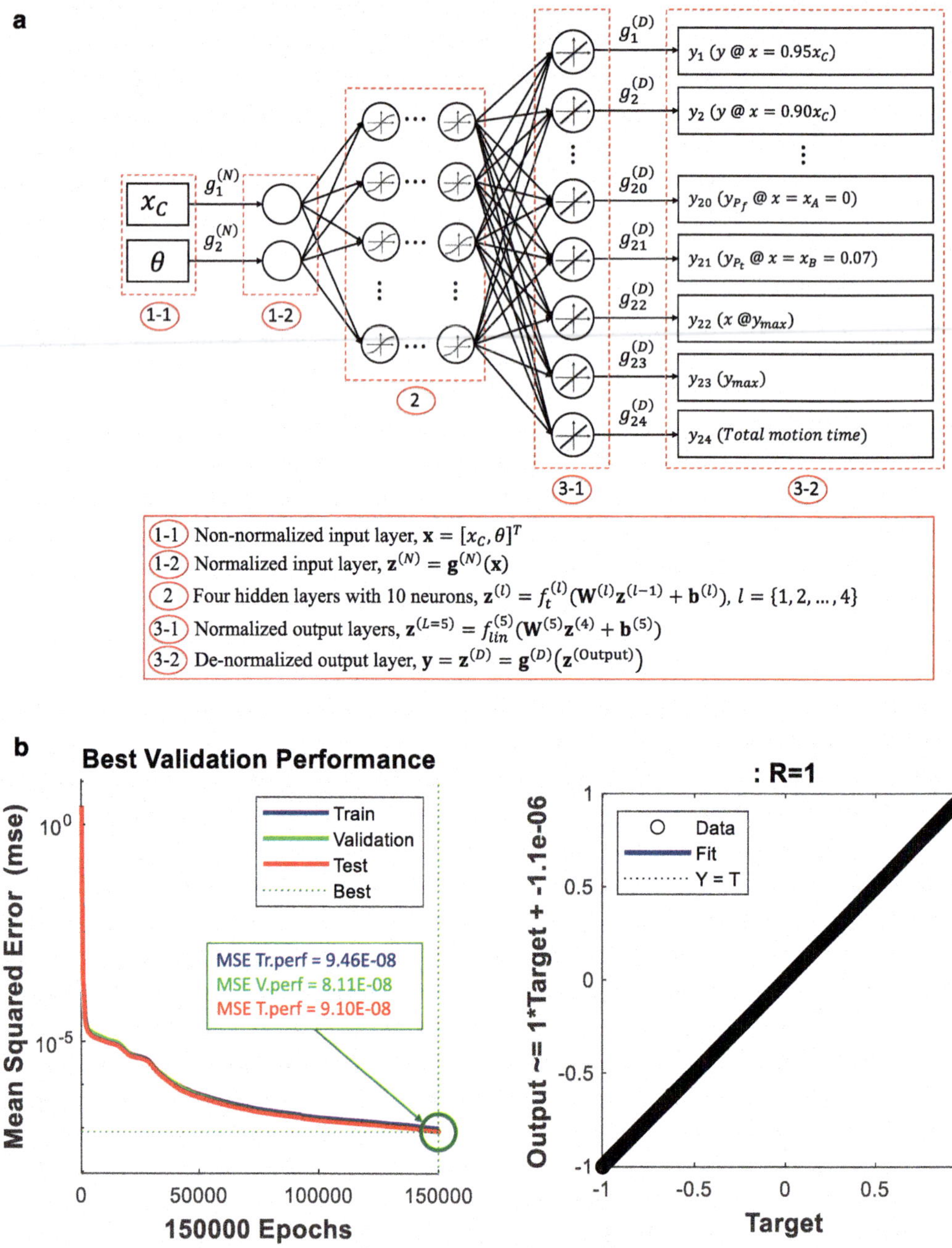

Figure 2.4.4.7 ANNs based on Equations (2.4.4.35) and (2.4.4.36) with four hidden layers and ten neurons. (a) Twenty-four outputs mapped using large datasets of Table 2.4.4.2 based on TED. (b) Training accuracy (MSE.Tperf) with 15,000 epochs 24 obtained based on TED.

Table 2.4.4.3 Comparison of Comparison of mapped projectile trajectories with analytical functions

(a): Scenarios 1 and 2

Forward Design (Based on TED with Four Layers and Ten Neurons)

Parameter		Scenario 1: $x_C = 0.2$, km $\theta = 80°$			Scenario 2: $x_C = 0.3$, km $\theta = 30°$		
		AI Results	Check (Analytical)	Error	AI Results	Check (Analytical)	Error
x_1:	x_C (km)	0.200	0.200	0.00%	0.300	0.300	0.00%
x_2:	θ (degree)	80.00	80.00	0.00%	30.00	30.00	0.00%
y_1:	y @ $x = 0.95x_C$ (km)	0.0541	0.0542	–0.05%	0.0084	0.0084	0.08%
y_2:	y @ $x = 0.90x_C$ (km)	0.1032	0.1033	–0.05%	0.0164	0.0164	0.07%
y_3:	y @ $x = 0.85x_C$ (km)	0.1472	0.1473	–0.05%	0.0239	0.0239	0.06%
y_4:	y @ $x = 0.80x_C$ (km)	0.1861	0.1862	–0.05%	0.0310	0.0310	0.05%
y_5:	y @ $x = 0.75x_C$ (km)	0.2200	0.2200	–0.03%	0.0376	0.0376	0.06%
y_6:	y @ $x = 0.70x_C$ (km)	0.2487	0.2488	–0.03%	0.0437	0.0437	0.06%
y_7:	y @ $x = 0.65x_C$ (km)	0.2724	0.2724	–0.03%	0.0494	0.0494	0.05%
y_8:	y @ $x = 0.60x_C$ (km)	0.2909	0.2910	–0.04%	0.0546	0.0546	0.04%
y_9:	y @ $x = 0.55x_C$ (km)	0.3045	0.3045	–0.03%	0.0593	0.0593	0.05%
y_{10}:	y @ $x = 0.50x_C$ (km)	0.3129	0.3130	–0.02%	0.0636	0.0636	0.05%
y_{11}:	y @ $x = 0.45x_C$ (km)	0.3162	0.3163	–0.02%	0.0675	0.0674	0.04%
y_{12}:	y @ $x = 0.40x_C$ (km)	0.3146	0.3146	0.01%	0.0708	0.0708	0.05%
y_{13}:	y @ $x = 0.35x_C$ (km)	0.3077	0.3077	–0.01%	0.0738	0.0737	0.04%
y_{14}:	y @ $x = 0.30x_C$ (km)	0.2959	0.2958	0.01%	0.0762	0.0762	0.03%
y_{15}:	y @ $x = 0.25x_C$ (km)	0.2789	0.2788	0.02%	0.0782	0.0782	0.03%
y_{16}:	y @ $x = 0.20x_C$ (km)	0.2569	0.2567	0.07%	0.0797	0.0797	0.03%
y_{17}:	y @ $x = 0.15x_C$ (km)	0.2298	0.2296	0.09%	0.0808	0.0808	0.02%
y_{18}:	y @ $x = 0.10x_C$ (km)	0.1976	0.1973	0.12%	0.0814	0.0814	0.03%
y_{19}:	y @ $x = 0.05x_C$ (km)	0.1603	0.1600	0.20%	0.0816	0.0815	0.03%
y_{20}:	y_{Pf} @ $x = x_A = 0$ (km)	0.1180	0.1176	0.32%	0.0813	0.0812	0.02%
y_{21}:	y_{Pt} @ $x = x_B = 0.07$ (km)	0.3079	0.3077	0.05%	0.0785	0.0787	–0.35%
y_{22}:	x @ y_{max} (km)	0.0901	0.0884	1.83%	0.0172	0.0175	–1.98%
y_{23}:	y_{max} (km)	0.3164	0.3164	0.00%	0.0815	0.0815	–0.10%
y_{24}:	Total motion time @ $x = x_A$, (s)	14.375	14.397	–0.16%	4.337	4.330	0.15%

Note: (dotted box) Two inputs for AI design.

(dashed box) Two inputs for Analytical calculation.

(Continued)

Table 2.4.4.3 (Continued) Comparison of Comparison of mapped projectile trajectories with analytical functions

(b): Scenarios 3 and 4

Forward Design (Based on TED with Four Layers and Ten Neurons)

Parameter		*Scenario 3:* $x_C = 0.4$, km $\theta = 70°$			*Scenario 4:* $x_C = 0.5$, km $\theta = 40°$		
		AI Results	*Check (Analytical)*	*Error*	*AI Results*	*Check (Analytical)*	*Error*
x_1:	x_C (km)	0.400	0.400	0.00%	0.500	0.500	0.00%
x_2:	θ (degree)	70.00	70.00	0.00%	40.00	40.00	0.00%
y_1:	y @ $x = 0.95x_C$ (km)	0.0523	0.0523	0.00%	0.0202	0.0202	−0.03%
y_2:	y @ $x = 0.90x_C$ (km)	0.0994	0.0994	0.00%	0.0387	0.0387	−0.02%
y_3:	y @ $x = 0.85x_C$ (km)	0.1413	0.1413	0.00%	0.0556	0.0556	−0.02%
y_4:	y @ $x = 0.80x_C$ (km)	0.1779	0.1779	−0.01%	0.0708	0.0708	−0.02%
y_5:	y @ $x = 0.75x_C$ (km)	0.2092	0.2092	0.00%	0.0845	0.0845	−0.03%
y_6:	y @ $x = 0.70x_C$ (km)	0.2353	0.2354	−0.01%	0.0965	0.0965	−0.02%
y_7:	y @ $x = 0.65x_C$ (km)	0.2562	0.2562	−0.01%	0.1068	0.1068	−0.02%
y_8:	y @ $x = 0.60x_C$ (km)	0.2718	0.2719	−0.01%	0.1156	0.1156	−0.01%
y_9:	y @ $x = 0.55x_C$ (km)	0.2822	0.2823	−0.02%	0.1227	0.1227	−0.01%
y_{10}:	y @ $x = 0.50x_C$ (km)	0.2874	0.2874	−0.03%	0.1281	0.1281	−0.01%
y_{11}:	y @ $x = 0.45x_C$ (km)	0.2873	0.2873	−0.02%	0.1320	0.1320	−0.01%
y_{12}:	y @ $x = 0.40x_C$ (km)	0.2819	0.2820	−0.03%	0.1342	0.1342	−0.01%
y_{13}:	y @ $x = 0.35x_C$ (km)	0.2713	0.2714	−0.04%	0.1348	0.1348	0.00%
y_{14}:	y @ $x = 0.30x_C$ (km)	0.2555	0.2556	−0.05%	0.1337	0.1337	0.00%
y_{15}:	y @ $x = 0.25x_C$ (km)	0.2344	0.2346	−0.06%	0.1310	0.1310	0.00%
y_{16}:	y @ $x = 0.20x_C$ (km)	0.2081	0.2083	−0.09%	0.1267	0.1267	0.01%
y_{17}:	y @ $x = 0.15x_C$ (km)	0.1766	0.1768	−0.11%	0.1207	0.1207	0.00%
y_{18}:	y @ $x = 0.10x_C$ (km)	0.1397	0.1400	−0.19%	0.1132	0.1131	0.04%
y_{19}:	y @ $x = 0.05x_C$ (km)	0.0977	0.0980	−0.26%	0.1039	0.1039	0.03%
y_{20}:	y_{Pf} @ $x = x_A = 0$ (km)	0.0504	0.0507	−0.63%	0.0931	0.0930	0.07%
y_{21}:	y_{Pt} @ $x = x_B = 0.07$(km)	0.1930	0.1932	−0.08%	0.1197	0.1193	0.31%
y_{22}:	x @ y_{max} (km)	0.1903	0.1903	−0.02%	0.1786	0.1788	−0.08%
y_{23}:	y_{max} (km)	0.2881	0.2880	0.03%	0.1348	0.1348	−0.02%
y_{24}:	Total motion time @ $x = x_A$, (s)	14.616	14.619	−0.02%	8.155	8.159	−0.05%

Note: (dotted box) Two inputs for AI design.

(dashed box) Two inputs for Analytical calculation.

(Continued)

Table 2.4.4.3 (Continued) Comparison of Comparison of mapped projectile trajectories with analytical functions

(c): Scenarios 5 and 6

Forward Design (Based on TED with Four Layers and Ten Neurons)

Parameter		*Scenario 5:* $x_C = 0.6$, km $\theta = 55°$			*Scenario 6:* $x_C = 0.7$, km $\theta = 55°$		
		AI Results	*Check (Analytical)*	*Error*	*AI Results*	*Check (Analytical)*	*Error*
x_1:	x_C (km)	0.600	0.600	0.00%	0.700	0.700	0.00%
x_2:	θ (degree)	55.00	55.00	0.00%	55.00	55.00	0.00%
y_1:	y @ $x = 0.95x_C$ (km)	0.0407	0.0407	0.00%	0.0471	0.0471	–0.02%
y_2:	y @ $x = 0.90x_C$ (km)	0.0773	0.0773	0.00%	0.0885	0.0886	–0.02%
y_3:	y @ $x = 0.85x_C$ (km)	0.1097	0.1097	0.01%	0.1243	0.1243	–0.01%
y_4:	y @ $x = 0.80x_C$ (km)	0.1378	0.1378	0.01%	0.1543	0.1543	–0.02%
y_5:	y @ $x = 0.75x_C$ (km)	0.1618	0.1618	0.00%	0.1786	0.1786	–0.01%
y_6:	y @ $x = 0.70x_C$ (km)	0.1816	0.1816	0.00%	0.1972	0.1972	–0.01%
y_7:	y @ $x = 0.65x_C$ (km)	0.1972	0.1972	0.01%	0.2100	0.2101	–0.01%
y_8:	y @ $x = 0.60x_C$ (km)	0.2086	0.2086	0.01%	0.2172	0.2172	–0.01%
y_9:	y @ $x = 0.55x_C$ (km)	0.2158	0.2158	0.01%	0.2187	0.2187	–0.01%
y_{10}:	y @ $x = 0.50x_C$ (km)	0.2188	0.2188	0.01%	0.2145	0.2145	0.00%
y_{11}:	y @ $x = 0.45x_C$ (km)	0.2176	0.2176	0.01%	0.2045	0.2045	0.01%
y_{12}:	y @ $x = 0.40x_C$ (km)	0.2122	0.2122	0.01%	0.1889	0.1889	0.01%
y_{13}:	y @ $x = 0.35x_C$ (km)	0.2027	0.2026	0.02%	0.1676	0.1675	0.02%
y_{14}:	y @ $x = 0.30x_C$ (km)	0.1889	0.1889	0.01%	0.1405	0.1405	0.03%
y_{15}:	y @ $x = 0.25x_C$ (km)	0.1710	0.1709	0.02%	0.1078	0.1077	0.07%
y_{16}:	y @ $x = 0.20x_C$ (km)	0.1488	0.1488	0.01%	0.0693	0.0692	0.10%
y_{17}:	y @ $x = 0.15x_C$ (km)	0.1225	0.1224	0.03%	0.0251	0.0250	0.43%
y_{18}:	y @ $x = 0.10x_C$ (km)	0.0919	0.0919	0.02%	–0.0248	–0.0249	–0.40%
y_{19}:	y @ $x = 0.05x_C$ (km)	0.0572	0.0572	0.06%	–0.0803	–0.0805	–0.19%
y_{20}:	y_{Pf} @ $x = x_A = 0$ (km)	0.0183	0.0182	0.15%	–0.1416	–0.1418	–0.11%
y_{21}:	y_{Pt} @ $x = x_B = 0.07$(km)	0.1023	0.1025	–0.22%	–0.0248	–0.0249	–0.41%
y_{22}:	x @ y_{max} (km)	0.2936	0.2935	0.03%	0.3935	0.3935	0.02%
y_{23}:	y_{max} (km)	0.2189	0.2189	–0.01%	0.2189	0.2189	0.02%
y_{24}:	Total motion time @ $x = x_A$, (s)	13.077	13.076	0.01%	15.255	15.255	0.00%

Note: (dotted box) Two inputs for AI design.

(dashed box) Two inputs for Analytical calculation.

(Continued)

Table 2.4.4.3 (Continued) Comparison of Comparison of mapped projectile trajectories with analytical functions

(d): Scenarios 7 and 8

Forward Design (Based on TED with Four Layers and Ten Neurons)

Parameter		*Scenario 7: $x_C = 0.8$, km $\theta = 45°$*			*Scenario 8: $x_C = 0.9$, km $\theta = 50°$*		
		AI Results	*Check (Analytical)*	*Error*	*AI Results*	*Check (Analytical)*	*Error*
x_1:	x_C (km)	0.800	0.800	0.00%	0.900	0.900	0.00%
x_2:	θ (degree)	45.00	45.00	0.00%	50.00	50.00	0.00%
y_1:	y @ $x = 0.95x_C$ (km)	0.0375	0.0375	−0.01%	0.0499	0.0499	0.00%
y_2:	y @ $x = 0.90x_C$ (km)	0.0702	0.0702	−0.01%	0.0922	0.0922	0.00%
y_3:	y @ $x = 0.85x_C$ (km)	0.0979	0.0979	−0.01%	0.1271	0.1271	0.00%
y_4:	y @ $x = 0.80x_C$ (km)	0.1207	0.1208	−0.01%	0.1544	0.1544	0.00%
y_5:	y @ $x = 0.75x_C$ (km)	0.1387	0.1387	−0.01%	0.1742	0.1742	0.00%
y_6:	y @ $x = 0.70x_C$ (km)	0.1517	0.1517	−0.01%	0.1865	0.1866	0.00%
y_7:	y @ $x = 0.65x_C$ (km)	0.1598	0.1598	−0.01%	0.1913	0.1913	0.00%
y_8:	y @ $x = 0.60x_C$ (km)	0.1630	0.1630	−0.01%	0.1886	0.1886	0.00%
y_9:	y @ $x = 0.55x_C$ (km)	0.1613	0.1613	−0.02%	0.1784	0.1784	−0.01%
y_{10}:	y @ $x = 0.50x_C$ (km)	0.1547	0.1548	−0.02%	0.1607	0.1607	−0.01%
y_{11}:	y @ $x = 0.45x_C$ (km)	0.1432	0.1432	−0.02%	0.1354	0.1354	−0.01%
y_{12}:	y @ $x = 0.40x_C$ (km)	0.1268	0.1268	−0.03%	0.1026	0.1027	−0.01%
y_{13}:	y @ $x = 0.35x_C$ (km)	0.1055	0.1055	−0.04%	0.0624	0.0624	−0.03%
y_{14}:	y @ $x = 0.30x_C$ (km)	0.0793	0.0793	−0.06%	0.0146	0.0146	−0.13%
y_{15}:	y @ $x = 0.25x_C$ (km)	0.0481	0.0482	−0.11%	−0.0407	−0.0407	0.05%
y_{16}:	y @ $x = 0.20x_C$ (km)	0.0121	0.0122	−0.47%	−0.1036	−0.1035	0.03%
y_{17}:	y @ $x = 0.15x_C$ (km)	−0.0288	−0.0288	0.23%	−0.1739	−0.1739	0.01%
y_{18}:	y @ $x = 0.10x_C$ (km)	−0.0747	−0.0746	0.09%	−0.2517	−0.2517	0.02%
y_{19}:	y @ $x = 0.05x_C$ (km)	−0.1254	−0.1254	0.06%	−0.3371	−0.3370	0.01%
y_{20}:	y_{Pf} @ $x = x_A = 0$ (km)	−0.1811	−0.1810	0.04%	−0.4300	−0.4299	0.01%
y_{21}:	y_{Pt} @ $x = x_B = 0.07$(km)	−0.0868	−0.0868	−0.10%	−0.2888	−0.2887	0.02%
y_{22}:	x @ y_{max} (km)	0.4735	0.4738	−0.06%	0.5785	0.5788	−0.04%
y_{23}:	y_{max} (km)	0.1632	0.1631	0.03%	0.1915	0.1914	0.02%
y_{24}:	Total motion time @ $x = x_A$, (s)	14.138	14.142	−0.03%	17.498	17.502	−0.02%

Note: [dotted box] Two inputs for AI design.

[dashed box] Two inputs for Analytical calculation.

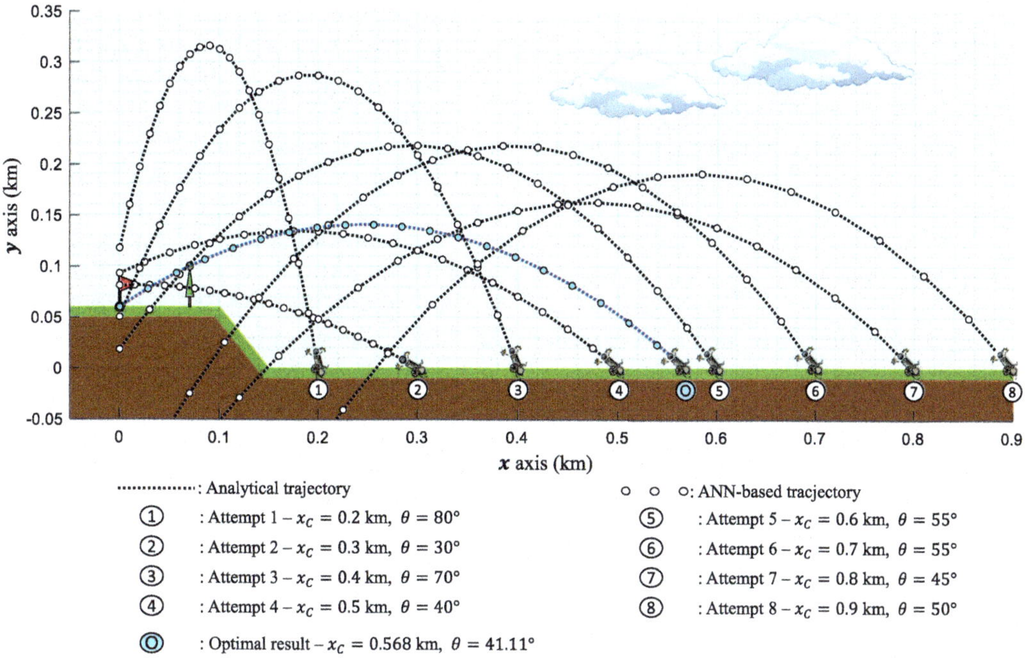

Figure 2.4.4.8 Comparison of ANN-based mapped projectile trajectories with analytical functions.

2.4.4.2.3 Step 3: Formulation of ANN-based objective function

2.4.4.2.3.1 TED-BASED TRAINING TO YIELD 24 OUTPUT PARAMETERS

The aim of an ANN application is to derive a twice differential objective and constraining functions to obtain Jacobian and Hessian matrices, which are needed to solve KKT conditions based on Newton–Raphson method. The goodness of training is verified by eight scenarios presented in Table 2.4.4.3 and Figures 2.4.4.7(a), 2.4.4.8. Figure 2.4.4.7(a) is trained by large datasets generated in Table 2.4.4.2 to obtain weight and bias matrices in Table 2.4.4.4. These matrices are used for Equations (2.4.4.40) to (2.4.4.60) to generalize functions, Jacobi and Hessian matrices.

For given eight heights of the projectile corresponding to different locations represented by a projectile distance x_C and a launching angle θ, ANNs are trained by mapping a projectile distance x_C and a launching angle θ to 24 output parameters using Equation (2.4.4.35) as shown in Table 2.4.4.3. Equation (2.4.4.35) is an objective function with multi hidden layers trained based on TED. Input variables $\mathbf{x}_i = \left[x_C, \theta\right]^T$ represent projectile distance x_C and a launching angle θ. $\mathbf{y}_i$ are the output parameters representing an altitude of a projectile mapped by the corresponding projectile distance x_C and a launching angle θ $\mathbf{x}_i = \left[x_C, \theta\right]^T$. Readers are referred to Section 2.3.3 for an ANN-based function shown in Equation (2.4.4.35). In Table 2.4.4.3, trained results of projectile trajectory between y_1 (y @ $x = 0.95x_C$) and y_{24} (flying time) are presented. The ANN-based projectile trajectories mapped using large datasets shown in Table 2.4.4.2 are similar to those derived analytically as compared in Table 2.4.4.3 and Figure 2.4.4.8. Differentiable functions based on ANNs which yield Jacobian and Hessian matrices will replace analytical objective and constraining functions. As explained in the previous example, Jacobian and Hessian matrices used to approximate Jacobi of Lagrange functions are required to obtain candidate solutions of KKT conditions based on Newton–Raphson iterations.

2.4.4.2.3.2 MULTILAYER PERCEPTRON FUNCTION USING TED

An ANN-based objective function presented in Equation (2.4.4.35) can be expressed as series of composite mathematical operations $f(\mathbf{g}(\mathbf{x})) = (f \circ \mathbf{g} \circ x)$ as shown in Equation (2.4.4.36) based on Equation (2.3.6.3-1). Inputs are relayed to neurons of successive layers fully connected using weights at each neuron and bias in each hidden layer. Layers are then summed at an output layer as a generalized function. Figure 2.4.4.7a shows ANNs derived based on Equations (2.4.4.35) and (2.4.4.36). In Equation (2.4.4.35), red $\mathbf{x}$ is input variables and L is a number of layers including hidden layers and output layer. $\mathbf{z}^{(N)}$ is used as normalized neural input for $\mathbf{z}^{(1)}$ and $\mathbf{z}^{(1)}$ shown in Equation (2.4.4.36) is a neural value at the first layer (1) with weights $\mathbf{W}^{(1)}$ and biases $\mathbf{b}^{(1)}$. $\mathbf{z}^{(4)}$ shown in Equation (2.4.4.36) is a normalized neural value at the last hidden layer (4) whereas $\mathbf{z}^{(L=5=\text{output})}$ shown in Equation (2.4.4.36) is a normalized neural value at the output layer ($L = 5 =$ output) before returning to the original scale by de-normalizing function $\mathbf{g}^D(\bar{x})$. N and D shown in Equation (2.4.4.36) indicate normalizing and de-normalizing steps. $\mathbf{W}^L$ is the weight matrix between layers $L-1$ and L ; and $\mathbf{b}^L$ is the bias matrix of layer L; $\mathbf{g}^N$ and $\mathbf{g}^D$ are normalizing function for inputs based on min–max normalization and de-normalizing function, respectively. The activation functions (*tansig* and *tanh*), f_t^l at layer l, shown in Figure 2.3.3.1 are implemented to formulate nonlinear relations of the networks, whereas a linear activation function, f_{lin}^L, is selected for the output layer because the output values are not bound. Weight and bias matrices which are obtained from an ANN shown in Equations (2.4.4.35), (2.4.4.36), and Figure 2.4.4.7a are presented in Table 2.4.4.4. Large datasets generated in Table 2.4.4.2 were used to train ANN shown in Figure 2.4.4.7a. Subscripts 1 and 2 indicate input parameters x_C and θ.

$$\mathbf{y} = \mathbf{g}^D\left(f_{\text{lin}}^4\left(\mathbf{W}^L f_t^3\left(\mathbf{W}^3 \ldots f_t^1\left(\mathbf{W}^1 \mathbf{g}^N(\mathbf{x}) + \mathbf{b}^1\right)\ldots + \mathbf{b}^3\right) + \mathbf{b}^4\right)\right) \tag{2.4.4.35}$$

$$\mathbf{y} = \mathbf{z}^{(D)} \circ \mathbf{z}^{(L=5=\text{output})} \circ \mathbf{z}^{(4)} \circ \mathbf{z}^{(3)} \circ \mathbf{z}^{(2)} \circ \mathbf{z}^{(1)} \circ \mathbf{z}^{(N)} \tag{2.4.4.36}$$

An ANN-based Lagrange function for a projectile movement is described in Equation (2.4.4.37), similarly to a conventional Lagrange function based on an analytical objective function. First derivative (Jacobi) of Lagrange function is presented in Equation (2.4.4.38-1) based on analytical functions whereas Equation (2.4.4.38-2) is based on ANN-based functions.

$$\begin{aligned}\mathcal{L}(x_C,\theta,\lambda_c,\lambda_v) &= f_{xC}(x_C) - \lambda_c c(x_C,\theta) - \lambda_v sv(x_C,\theta)\\ &= x_C - \lambda_c\left(y_{P_f} - 0.06\right) - \lambda_v s\left(y_{P_t} - 0.1\right)\end{aligned} \tag{2.4.4.37, 2.4.6.3}$$

$$\nabla\mathcal{L}(x_C,\theta,\lambda_c,\lambda_v) = \begin{bmatrix} \nabla f_{xC}(x_C) - \left[J_c(x_C,\theta)\right]^T \lambda_c - \left[J_v(x_C,\theta)\right]^T s\lambda_v \\ -c(x_C,\theta) \\ -sv(x_C,\theta) \end{bmatrix} \tag{2.4.4.38-1, 2.4.6.4a}$$

$$\nabla\mathcal{L}(x_C,\theta,\lambda_c,\lambda_v) = \begin{bmatrix} \nabla f_{xC}(x_C,\theta) - \left[J_c^{(D)}(x_C,\theta)\right]^T \lambda_c - \left[J_v^{(D)}(x_C,\theta)\right]^T s\lambda_v \\ -c(x_C,\theta) \\ -sv(x_C,\theta) \end{bmatrix} \tag{2.4.4.38-2}$$

Table 2.4.4.4 Weight and bias matrices obtained from ANN based on Equations (2.4.4.35), (2.4.4.36), and Figure 2.4.4.7(a).

(a): Hidden Layer #1

$W^{(1)}$		$b^{(1)}$
−0.922	−3.414	3.730
1.224	1.978	−2.059
0.668	−0.814	−2.193
0.576	−0.790	−0.636
0.009	−1.491	−1.214
−0.297	1.238	−0.468
−0.425	−0.428	−0.082
−1.934	−1.265	−2.434
1.024	−0.761	1.807
−3.074	0.977	−5.114
[10×2]		[10×1]

(b): Hidden Layer #2

$W^{(2)}$										$b^{(2)}$
−1.139	1.084	−0.670	0.553	−0.009	−0.319	−0.982	0.155	0.511	0.386	2.376
−0.409	0.507	−0.178	1.290	−0.077	0.427	−0.099	−0.847	0.652	−1.189	1.548
0.889	−0.137	−0.282	−0.051	−0.052	0.547	−0.024	0.062	0.515	0.383	−0.834
0.005	−0.134	−0.474	−0.614	0.066	−0.573	−0.472	0.309	0.588	−0.758	−0.387
−1.988	0.443	−0.463	−0.315	−0.165	−0.711	−0.555	−0.213	0.108	−0.274	−0.028
0.382	0.948	0.930	−0.394	−0.264	−0.161	1.849	0.195	−0.203	−0.634	0.073
−0.417	−0.603	0.545	−0.150	0.534	0.543	0.156	−0.273	0.460	0.911	−0.688
1.031	0.128	−0.426	−0.277	−0.176	−0.766	0.318	0.193	−0.786	1.309	0.013
0.232	−0.278	0.110	0.140	−0.157	0.547	0.294	0.087	−0.493	0.723	1.691
0.043	0.394	0.037	−0.348	0.313	−0.843	1.345	0.048	0.279	0.421	1.935
[10×10]										[10×1]

(Continued)

Table 2.4.4.4 (Continued) Weight and bias matrices obtained from ANN based on Equations (2.4.4.35), (2.4.4.36), and Figure 2.4.4.7(a).

(c): Hidden Layer #3										
$W^{(3)}$										$\mathbf{b}^{(3)}$
–0.222	0.239	0.165	–1.013	0.886	–0.703	–0.697	–0.432	0.795	–0.491	–1.474
1.194	0.708	–1.276	–0.365	0.482	0.370	0.527	–0.969	0.496	–0.535	–1.208
–0.453	–0.098	–0.215	–0.563	–1.049	–0.167	1.103	0.559	–0.413	0.482	0.989
–0.370	0.190	–0.349	0.016	–0.663	–0.339	0.862	0.541	0.494	1.052	0.605
0.399	–0.227	–0.496	0.843	–0.538	0.111	–0.141	–0.359	–0.914	–0.123	–0.180
0.358	0.596	0.546	–0.774	0.246	0.053	–0.752	–0.440	–0.549	–0.370	0.498
–0.505	–0.702	0.052	–0.978	–0.220	0.694	0.065	–0.411	0.251	–1.013	–0.807
–1.816	–0.225	1.074	0.520	–0.656	–0.783	–1.171	0.824	–0.739	–0.331	1.138
0.767	–0.270	0.686	0.455	0.162	0.350	–0.883	–0.516	1.049	0.177	1.872
–1.966	0.774	0.179	–1.018	0.070	–1.225	–0.566	1.079	0.190	0.075	–1.602
[10×10]										[10×1]
(d): Hidden Layer #4 (4)										
$W^{(4)}$										$\mathbf{b}^{(4)}$
0.600	0.301	–0.431	–0.330	0.326	1.707	0.214	–0.988	1.196	–0.100	–1.172
0.334	–0.440	–0.621	0.360	–0.276	1.130	–0.128	0.512	–0.141	0.009	–1.887
–0.175	–0.910	0.714	0.474	0.891	–0.145	–0.238	0.287	0.974	–0.507	1.241
0.394	–0.526	0.244	–1.073	–1.076	–0.369	–0.188	1.445	–0.570	–0.460	0.717
–0.011	0.607	–0.566	–0.277	0.034	–0.835	1.156	–0.971	–0.412	0.315	–0.094
–0.833	0.886	–0.432	0.024	1.208	0.023	0.184	–0.500	–1.143	–1.631	–1.028
–0.334	–0.834	–0.073	0.027	–0.219	1.032	0.418	1.210	0.816	–0.245	0.949
1.293	–0.797	–0.250	0.784	0.775	0.135	0.106	–0.148	–0.697	0.003	0.496
0.140	–0.269	–0.178	1.004	0.781	0.437	–0.465	–0.380	1.105	0.260	1.810
0.552	–0.640	1.275	–0.509	0.902	0.242	0.413	–0.222	0.362	–0.171	1.229
[10×10]										[10×1]

(Continued)

Table 2.4.4.4 (Continued) Weight and bias matrices obtained from ANN based on Equations (2.4.4.35), (2.4.4.36), and Figure 2.4.4.7(a).

(e): Normalized Output Layer (5 = Output)										
$W^{(5)}$										$\mathbf{b}^{(5)}$
1.219	0.596	−0.805	0.957	−0.872	0.285	−0.363	0.424	0.537	−0.162	0.124
1.166	0.650	−0.879	0.976	0.547	0.251	0.570	0.467	0.584	−0.167	0.599
1.106	0.715	0.090	0.998	0.374	0.210	−0.534	0.517	0.781	−0.170	0.318
1.007	0.774	0.019	1.002	0.872	0.158	0.609	0.566	0.385	−0.173	0.104
0.877	0.830	−0.407	0.987	−0.970	0.095	−0.072	0.607	−0.790	−0.161	0.513
0.726	0.892	−0.981	0.969	−0.863	0.023	−0.003	0.655	−0.506	−0.152	0.806
0.551	0.962	−0.258	0.947	0.043	−0.061	−0.224	0.712	−0.265	−0.148	0.933
0.337	1.009	0.712	0.893	1.029	−0.153	0.374	0.755	0.482	−0.145	−0.468
0.102	1.046	−0.099	0.820	0.845	−0.249	0.023	0.783	0.396	−0.123	0.490
−0.145	1.066	−0.569	0.731	0.620	−0.344	0.247	0.802	0.401	−0.103	0.382
−0.396	1.076	0.327	0.632	0.447	−0.440	−0.367	0.815	0.266	−0.085	−0.069
−0.636	1.070	−0.730	0.525	−0.911	−0.526	−0.480	0.811	−0.747	−0.056	0.608
−0.873	1.058	0.072	0.416	0.427	−0.611	−0.526	0.808	0.684	−0.041	−0.380
−1.095	1.040	0.038	0.310	0.401	−0.686	0.425	0.800	−0.409	−0.023	−0.369
−1.295	1.016	0.176	0.207	−0.016	−0.754	−0.301	0.784	−0.042	−0.002	−0.702
−1.484	0.991	−0.439	0.109	0.548	−0.815	0.183	0.769	−0.773	0.020	0.594
−1.660	0.967	−0.215	0.017	−0.748	−0.874	−0.139	0.754	−0.773	0.037	−0.722
−1.827	0.944	−1.033	−0.071	0.829	−0.928	0.359	0.738	0.261	0.054	0.023
−1.960	0.913	−0.867	−0.151	−0.790	−0.969	−0.510	0.716	−0.316	0.073	−0.408
−1.993	0.843	−0.827	−0.216	0.484	−0.962	0.046	0.665	−0.098	0.080	0.073
−0.720	−0.703	−0.363	−1.991	−0.329	−3.849	−1.012	2.319	1.175	−1.894	−0.601
−0.645	0.699	−0.772	−0.942	−0.583	1.204	−0.063	0.099	0.479	−0.587	1.057
−0.675	2.184	0.738	0.550	−1.058	0.813	−0.576	−1.978	−0.554	−1.023	0.755
2.105	0.784	−0.014	−1.230	−0.845	−1.448	−0.383	0.086	−0.038	1.422	−0.200
[24×10]										[24×1]

Like conventional Lagrange, there are two cases of an active inequality status as shown in Table 2.4.4.1. Analytical Jacobi and Hessian matrices are explicitly calculated by directly differentiating functions as shown in Equations (2.4.4.7-1) and (2.4.4.7-2) whereas those based on ANNs with four hidden layers and ten neurons shown in Equations (2.4.4.35), (2.4.4.36) and Figure 2.4.4.7a are derived using weight and bias matrices shown in Table 2.4.4.4 to solve Equation (2.4.4.38-2). A difference between ANN-based and conventional Lagrange function is that a conventional Lagrange calculates y_{P_f} and y_{P_t} based on explicit equations shown in Equations (2.4.4.7-1) and (2.4.4.7-2), whereas an ANN-based Lagrange uses an ANN shown in Equations (2.4.4.35), (2.4.4.36) and Figure 2.4.4.7a to derive y_{P_f} and y_{P_t} and their derivatives. Formulation procedures of ANN-based functions to obtain functions including objective, equality, inequality functions, and their Jacobian and Hessian matrices are illustrated in Figure 2.3.6.2. A total of 24 normalized output parameters $\mathbf{y} = \left[y_1, \ldots, \ y_{24}\right]^T$ are calculated at 24 neurons of the output layer (5=out) as shown in Equation (2.4.4.36), Table 2.4.4.3, and Figure 2.4.4.7a. However, two equality y_{20} $(y_{P_f}) = z^{(5)}_{y20}$ and inequality y_{21} $(y_{P_t}) = z^{(5)}_{y21}$ functions are only used based on ANN because these are needed to optimize an objective function indicated by red color as shown in Equation (2.4.4.39). Weight and bias matrices which are used to calculate Jacobian and Hessian matrices of y_{20} $(y_{P_f}) = z^{(5)}_{y20}$ and y_{21} $(y_{P_t}) = z^{(5)}_{y21}$ are presented at the normalized output layer (5=output) shown in Table 2.4.4.4(e). Output parameters including y_{P_f} (y_{20}) and y_{P_t} (y_{21}) are calculated in Box 3-2 of Figure 2.4.4.7a and Jacobian and Hessian matrices of y_{20} $(y_{P_f}) = z^{(5)}_{y20}$ and y_{21} $(y_{P_t}) = z^{(5)}_{y21}$ are calculated in Step 4 of the next section (calculation of candidate solutions for KKT conditions based on ANN). It is noted that an objective function $f_{xC}(x_C)$ shown in Equation (2.4.4.1) is constant, and hence, Jacobian $\nabla f_{xC}(x_C)$, Hessian $\mathbf{H}_{fxC}(x_C, \ \theta)$ matrices can be simply obtained by differentiating an objective function once and twice as shown in Equations (2.4.4.12-1) and (2.4.4.14). Otherwise, an objective function should be also added in one of the neurons at an output layer to obtain weight and bias matrices.

2.4.4.2.4 Step 4: Calculation of candidate solutions for KKT conditions based on ANN

Table 2.4.4.1 presents two candidate KKT conditions depending on active and inactive inequality functions for an ANN-based optimization which is similar to candidate KKT conditions applied to an analytical function-based optimization.

2.4.4.2.4.1 CASE I KKT CONDITION BASED ON AN ACTIVE EQUALITY $v(x)$

In this case, inequality $v(x_C, \theta) = y_{P_t} - 0.1 \geq 0$ is assumed to constrain an objective function $f_{xC}(x_C)$ when reaching a minimum of $\nabla f_{xC}(x_C)$ by binding inequality $v(x_C, \theta) = y_{P_t} - 0.1 \geq 0$ to equality constraint. Lagrange function is derived in Equation (2.4.4.39) when inequality $v(x_C, \theta) = y_{P_t} - 0.1 \geq 0$ to equality constraint. An initial trial input parameter of $\mathbf{x}^{(0)} = \left[x_C^{(0)}, \ \theta^{(0)}\right]^T = [0.7, 40]^T$ and initial trial Lagrange multipliers of $\lambda_c^{(0)} = 0$ and $\lambda_v^{(0)} = 0$ are used to solve KKT conditions shown in Equation (2.4.4.39) using Newton–Raphson method. The total initial trial input parameter becomes $\left[x_C^{(0)}, \ \theta^{(0)}, \ \lambda_c^{(0)}, \ \lambda_v^{(0)}\right]^T = [0.7, \ 40, \ 0, \ 0]^T$. Figure 2.3.4.2 summarizes a flow of Newton–Raphson iteration to find stationary points of the Lagrange function.

$$\mathcal{L}(x_C, \theta, \lambda_c, \lambda_v) = x_C - \lambda_c\left(y_{P_f} - 0.06\right) - \lambda_v\left(y_{P_t} - 0.1\right) \tag{2.4.4.39}$$

A solution for Equation (2.4.4.38-2) is sought based on the initial trial input parameter $\left[x_C^{(0)},\ \theta^{(0)},\ \lambda_c^{(0)},\ \lambda_v^{(0)}\right]^T = [0.7,\ 40,\ 0,\ 0]^T$ consisting of $\mathbf{x}^{(0)} = \left[x_C^{(0)}, \theta^{(0)}\right]^T = [0.7, 40]^T$ and Lagrange multipliers $\lambda_c^{(0)} = 0$ and $\lambda_v^{(0)} = 0$. Objective, equality, and inequality functions and their Jacobian and Hessian matrices are obtained in Equations (2.4.4.40)–(2.4.4.48) at Iteration 0.

a. Iteration 0; Step (N) for normalizing input, $\mathbf{z}^{(N)}$
A flow of equations to obtain inequality functions y_{P_f} (y_{20}), y_{P_t} (y_{21}), and their Jacobian and Hessian matrices is shown in Figure 2.3.6.2. An initial trial input parameter $\left[x_C^{(0)},\ \theta^{(0)},\ \lambda_c^{(0)},\ \lambda_v^{(0)}\right]^T = [0.7, 40, 0, 0]^T$ is used to begin computing y_{P_f} (y_{20}), y_{P_t} (y_{21}), and their derivatives as indicated in red shown in Equation (2.4.4.40-1). Iterations are performed until solutions converge. Normalizing function $\boldsymbol{\alpha}_\mathbf{x}\begin{bmatrix} 1.820 \\ 0.029 \end{bmatrix}$ is obtained for input variables x_C and θ based on Equation (2.3.6.2-2) which is indicated in blue box of Table 2.4.4.2. Equation (2.4.4.40-3) is a Hessian matrix of an objective function, in which upper and lower lows indicate input variables x_C and θ, respectively. Subscripts 1 and 2 indicate input parameters x_C and θ. Table 2.4.4.4 presents weight and bias matrices obtained by training Equations (2.4.4.35), (2.4.4.36), and Figure 2.4.4.7a on large datasets shown in Table 2.4.4.2. Weight matrix **W** and bias matrices b for Hidden layer (1)–(4) and Output layer (5) are obtained from ANNs. Super scripts (N) and (1) indicate normalizing step and Hidden layer (1), respectively, whereas super script (D) represents de-normalizing step. Equality and inequality constraining functions y_{P_f} (y_{20}), y_{P_t} (y_{21}), Jacobi matrix, and Hessian matrix at Iteration 0 are obtained based on Equations (2.4.4.41-1)–(2.4.4.48-4) using weight matrix **W** and bias matrix b shown in Table 2.4.4.4(a) obtained based on Equations (2.4.4.35), (2.4.4.36), and Figure 2.4.4.7a. Step N and Step D are shown in Equation (2.4.4.40-1) and Equations (2.4.4.47), (2.4.4.48), respectively.

$$\mathbf{z}^{(N)} = \mathbf{g}^{(N)}\left(\mathbf{x}^{(0)}\right) = \boldsymbol{\alpha}_\mathbf{x} \odot \left(\mathbf{x} - \mathbf{x}_{\min}\right) + \overline{\mathbf{x}}_{\min}$$

$$= \begin{bmatrix} 1.820 \\ 0.029 \end{bmatrix} \odot \left(\begin{bmatrix} 0.7 \\ 40 \end{bmatrix} - \begin{bmatrix} 0.1 \\ 10 \end{bmatrix} \right) + \begin{bmatrix} -1 \\ -1 \end{bmatrix}$$

$$= \begin{bmatrix} 0.092 \\ -0.143 \end{bmatrix} \tag{2.4.4.40-1}$$

$$\mathbf{J}^{(N)} = I_2 \odot \boldsymbol{\alpha}_\mathbf{x} = \begin{bmatrix} 1.820 & 0 \\ 0 & 0.029 \end{bmatrix} \tag{2.4.4.40-2}$$

$$\mathbf{H}^{(N)} = \begin{bmatrix} 0 & 0 \\ 0 & 0 \end{bmatrix} \tag{2.4.4.40-3}$$

b. Iteration 0; calculations of output functions, Jacobi and Hessian matrices at Normalized hidden layers, $\mathbf{z}^{(l)}\ \forall l \in \{1, 2, 3, 4\}$

a. Normalized hidden layer, $\mathbf{z}^{(1)}$

Neural value $\mathbf{z}^{(1)}$ at Hidden layer (1) is obtained in Equation (2.4.4.41-1) based on Equations (2.4.4.35), (2.4.436), and (2.3.6.3-2b) whereas Jacobi $\mathbf{J}^{(1)}$ with respect to input variables x_C and θ at Hidden layer (1) is obtained in Equation (2.4.4.41-2) based on in Equation (2.3.6.6-7). Slice Hessian matrices $\mathbf{H}_1^{(1)}$ and $\mathbf{H}_2^{(1)}$ with respect to input variables x_C and θ at Hidden layer (1) are obtained in Equations (2.4.4.41-3a) and (2.4.4.41-3b), respectively, based on in Equation (2.3.6.14). A number of rows is ten because ten neurons are implemented. Weight and bias matrices $\mathbf{W}^{(1)}$ and $\mathbf{b}^{(1)}$ at Hidden layer (1) are shown in Table2.4.4.4(a). Dimensions of matrices for an objective function are shown in Equation (2.4.4.41-1). Weight matrix $\mathbf{W}^{(1)}$, $\mathbf{z}^{(N)}$, and $\mathbf{b}^{(1)}$ are $[10\times2]$, $[2\times1]$, and $[10\times1]$ matrices, respectively, resulting in $\mathbf{z}^{(1)}$ which is $[10\times1]$. Similarly, dimensions of matrices for a Jacobi matrix of an objective function are shown in Equation (2.4.4.41-2) in which $\mathbf{z}^{(1)}$, weight matrix $\mathbf{W}^{(1)}$ and $\mathbf{J}^{(N)}$ are $[10\times1]$, $[10\times2]$, and $[2\times2]$ matrices, respectively, resulting in $\mathbf{J}^{(1)}$ which is $[10\times2]$. Similarly, dimensions of matrices for a slice ($1=x_C$) Hessian matrix of an objective function are shown in Equation (2.4.4.41-3a) in which $\mathbf{z}^{(1)}$, $i_1^{(N)}$, weight matrix $\mathbf{W}^{(1)}$, $\mathbf{J}^{(N)}$, and $\mathbf{H}_1^{(N)}$, are $[10\times1]$, $[2\times1]$, $[10\times2]$, $[2\times2]$, and $[2\times2]$ matrices, respectively, resulting in $\mathbf{H}_1^{(1)}$ which is $[10\times2]$. Similarly, dimensions of matrices for a slice ($2=\theta$) Hessian matrix of an objective function are shown in Equation (2.4.4.41-3b) in which $\mathbf{z}^{(1)}$, $i_2^{(N)}$, weight matrix $\mathbf{W}^{(1)}$, $\mathbf{J}^{(N)}$, and $\mathbf{H}_2^{(N)}$ are $[10\times1]$, $[2\times1]$, $[10\times2]$, $[2\times2]$, and $[2\times2]$ matrices, respectively, resulting in $\mathbf{H}_2^{(1)}$ which is $[10\times2]$.

$$\mathbf{z}^{(1)} = f_t^{(1)}\left(\underbrace{\mathbf{W}^{(1)}}_{[10\times2]}\underbrace{\mathbf{z}^{(N)}}_{[2\times1]} + \underbrace{\mathbf{b}^{(1)}}_{[10\times1]}\right) = \begin{bmatrix} 0.9995 \\ -0.9771 \\ -0.9651 \\ -0.4381 \\ -0.7616 \\ -0.5865 \\ -0.0596 \\ -0.9847 \\ 0.9648 \\ 1.0000 \end{bmatrix}_{[10\times1]} \tag{2.4.4.41-1}$$

$$\mathbf{J}^{(1)} = \left(1-\underbrace{\left(\mathbf{z}^{(1)}\right)^2}_{[10\times1]}\right)\odot\underbrace{\mathbf{W}^{(1)}}_{[10\times2]}\underbrace{\mathbf{J}^{(N)}}_{[2\times2]} = \begin{bmatrix} -0.0017 & -0.0001 \\ 0.1009 & -0.0026 \\ 0.0835 & -0.0016 \\ 0.8469 & -0.0182 \\ 0.0072 & -0.0179 \\ -0.3548 & 0.0232 \\ -0.7714 & -0.0122 \\ -0.1071 & -0.0011 \\ 0.1291 & -0.0015 \\ 0.0003 & 0.0000 \end{bmatrix}_{[10\times2]} \tag{2.4.4.41-2}$$

$$\mathbf{H}_1^{(1)} = -\underbrace{2\mathbf{z}^{(1)}}_{[10\times1]} \odot \left(1-\underbrace{\left(\mathbf{z}^{(1)}\right)^2}_{[10\times1]}\right) \odot \underbrace{\boldsymbol{i}_1^{(N)}}_{[2\times1]} \odot \underbrace{\left(\mathbf{W}^{(1)}\right)^2}_{[10\times2]} \underbrace{\mathbf{J}^{(N)}}_{[2\times2]} + \left(1-\underbrace{\left(\mathbf{z}^{(1)}\right)^2}_{[10\times1]}\right) \odot \underbrace{\mathbf{W}^{(1)}}_{[10\times2]} \underbrace{\mathbf{H}_1^{(N)}}_{[2\times2]}$$

$$= \begin{bmatrix} -0.0058 & -0.0003 \\ 0.4390 & -0.0111 \\ 0.1959 & -0.0037 \\ 0.7775 & -0.0167 \\ 0.0002 & -0.0005 \\ 0.2252 & -0.0147 \\ 0.0712 & 0.0011 \\ 0.7426 & 0.0076 \\ -0.4642 & 0.0054 \\ 0.0039 & 0.0000 \end{bmatrix} \qquad (2.4.4.41\text{-}3a)$$

$$\mathbf{H}_2^{(1)} = -\underbrace{2\mathbf{z}^{(1)}}_{[10\times1]} \odot \left(1-\underbrace{\left(\mathbf{z}^{(1)}\right)^2}_{[10\times1]}\right) \odot \underbrace{\boldsymbol{i}_2^{(N)}}_{[2\times1]} \odot \underbrace{\left(\mathbf{W}^{(1)}\right)^2}_{[10\times2]} \underbrace{\mathbf{J}^{(N)}}_{[2\times2]} + \left(1-\underbrace{\left(\mathbf{z}^{(1)}\right)^2}_{[10\times1]}\right) \odot \underbrace{\mathbf{W}^{(1)}}_{[10\times2]} \underbrace{\mathbf{H}_2^{(N)}}_{[2\times2]}$$

$$= \begin{bmatrix} -0.0003 & -0.0000 \\ 0.0111 & 0.0003 \\ -0.0037 & 0.0001 \\ -0.0167 & 0.0004 \\ -0.0005 & 0.0012 \\ -0.0147 & 0.0010 \\ 0.0011 & 0.0000 \\ 0.0076 & 0.0001 \\ 0.0054 & -0.0001 \\ -0.0000 & 0.0000 \end{bmatrix} \qquad (2.4.4.41\text{-}3b)$$

b. Normalized hidden layer, $\mathbf{z}^{(2)}$
Matrices for a function $\mathbf{z}^{(2)}$, a Jacobi $\mathbf{J}^{(2)}$, and slice Hessian matrices $\mathbf{H}_1^{(2)}$ and $\mathbf{H}_2^{(2)}$ for (1= x_C) and (2= θ) are shown at Hidden layer (2) as shown in Equation (2.4.4.42).

$$\mathbf{z}^{(2)} = f_t^{(2)}\left(\mathbf{W}^{(2)}\mathbf{z}^{(1)} + \mathbf{b}^{(2)}\right) = \begin{bmatrix} 0.6585 \\ 0.9913 \\ 0.2501 \\ 0.9479 \\ -0.6039 \\ -0.6445 \\ -0.9569 \\ -0.2387 \\ 0.4811 \\ 0.9337 \end{bmatrix} \qquad (2.4.4.42\text{-}1)$$

$$\mathbf{J}^{(2)} = \left(1-\left(\mathbf{z}^{(2)}\right)^2\right) \odot \mathbf{W}^{(2)}\mathbf{J}^{(1)} = \begin{bmatrix} 0.8175 & -0.0013 \\ 0.0213 & -0.0002 \\ -0.1860 & 0.0131 \\ 0.0039 & 0.0002 \\ 0.2915 & 0.0007 \\ -0.9228 & -0.0078 \\ -0.0306 & 0.0000 \\ -0.3361 & -0.0109 \\ -0.3047 & 0.0070 \\ -0.1226 & -0.0045 \end{bmatrix} \tag{2.4.4.42-2}$$

$$\mathbf{H}_1^{(2)} = -2\mathbf{z}^{(2)} \odot \left(1-\left(\mathbf{z}^{(2)}\right)^2\right) \odot i_1^{(1)} \odot \left(\mathbf{W}^{(2)}\right)^2 \mathbf{J}^{(1)} + \left(1-\left(\mathbf{z}^{(2)}\right)^2\right) \odot \mathbf{W}^{(2)}\mathbf{H}_1^{(1)}$$

$$= \begin{bmatrix} -1.2603 & 0.0100 \\ -0.0460 & 0.0000 \\ -0.2344 & -0.0031 \\ -0.0854 & 0.0024 \\ -0.1810 & 0.0140 \\ 2.2416 & 0.0267 \\ -0.0254 & -0.0012 \\ 0.1660 & 0.0171 \\ 0.2272 & -0.0093 \\ -0.2546 & -0.0047 \end{bmatrix} \tag{2.4.4.42-3a}$$

$$\mathbf{H}_2^{(2)} = -2\mathbf{z}^{(2)} \odot \left(1-\left(\mathbf{z}^{(2)}\right)^2\right) \odot i_2^{(1)} \odot \left(\mathbf{W}^{(2)}\right)^2 \mathbf{J}^{(1)} + \left(1-\left(\mathbf{z}^{(2)}\right)^2\right) \odot \mathbf{W}^{(2)}\mathbf{H}_1^{(1)}$$

$$= \begin{bmatrix} 0.0100 & 0.0001 \\ -0.0000 & 0.0000 \\ -0.0031 & 0.0002 \\ 0.0024 & -0.0001 \\ 0.0140 & -0.0006 \\ 0.0267 & 0.0000 \\ -0.0012 & 0.0001 \\ 0.0171 & -0.0009 \\ -0.0093 & 0.0002 \\ -0.0047 & -0.0003 \end{bmatrix} \tag{2.4.4.42-3b}$$

c. Normalized hidden layer, $\mathbf{z}^{(3)}$
Dimensions of matrices for a function $\mathbf{z}^{(3)}$, a Jacobi $\mathbf{J}^{(3)}$, and slice Hessian matrices $\mathbf{H}_1^{(3)}$ and $\mathbf{H}_2^{(3)}$ for ($1 = x_C$) and ($2 = \theta$) are shown at Hidden layer (3) as shown in Equation (2.4.4.43).

$$\mathbf{z}^{(3)} = f_t^{(3)}\left(\underbrace{\mathbf{W}^{(3)}}_{[10\times10]}\underbrace{\mathbf{z}^{(2)}}_{[10\times1]} + \underbrace{\mathbf{b}^{(3)}}_{[10\times1]}\right) = \begin{bmatrix} -0.9340 \\ -0.8955 \\ -0.1878 \\ 0.8770 \\ 0.4243 \\ 0.6409 \\ -0.9991 \\ 0.9276 \\ 0.9994 \\ -0.9525 \end{bmatrix}_{[10\times1]} \tag{2.4.4.43-1}$$

$$\mathbf{J}^{(3)} = \left(1 - \underbrace{\left(\mathbf{z}^{(3)}\right)^2}_{[10\times1]}\right) \odot \underbrace{\mathbf{W}^{(3)}}_{[10\times10]}\underbrace{\mathbf{J}^{(2)}}_{[10\times2]} = \begin{bmatrix} 0.0870 & 0.0027 \\ 0.2477 & -0.0009 \\ -0.6181 & -0.0123 \\ -0.1388 & -0.0021 \\ 0.4726 & -0.0082 \\ 0.3577 & 0.0052 \\ -0.0017 & 0.0000 \\ -0.1579 & 0.0013 \\ 0.0001 & 0.0000 \\ -0.0826 & 0.0003 \end{bmatrix}_{[10\times2]} \tag{2.4.4.43-2}$$

$$\mathbf{H}_1^{(3)} = -\underbrace{2\mathbf{z}^{(3)}}_{[10\times1]} \odot \left(1 - \underbrace{\left(\mathbf{z}^{(3)}\right)^2}_{[10\times1]}\right) \odot \underbrace{i_1^{(2)}}_{[10\times1]} \odot \underbrace{\left(\mathbf{W}^{(3)}\right)^2}_{[10\times10]} \underbrace{\mathbf{J}^{(2)}}_{[10\times2]} + \left(1 - \underbrace{\left(\mathbf{z}^{(3)}\right)^2}_{[10\times1]}\right) \odot \underbrace{\mathbf{W}^{(3)}}_{[10\times10]}\underbrace{\mathbf{H}_1^{(2)}}_{[10\times2]}$$

$$= \begin{bmatrix} -0.0385 & 0.0004 \\ 0.4775 & 0.0004 \\ 0.4737 & -0.0111 \\ -0.1903 & -0.0073 \\ -0.5058 & 0.0090 \\ -0.6025 & -0.0010 \\ 0.0080 & 0.0000 \\ -0.2685 & -0.0009 \\ -0.0004 & 0.0000 \\ 0.1351 & -0.0040 \end{bmatrix} \tag{2.4.4.43-3a}$$

$$\mathbf{H}_2^{(3)} = -\underbrace{2\mathbf{z}^{(3)}}_{[10\times1]} \odot \left(\underbrace{1-\left(\mathbf{z}^{(3)}\right)^2}_{[10\times1]}\right) \odot \underbrace{i_2^{(2)}}_{[10\times1]} \odot \underbrace{\left(\mathbf{W}^{(3)}\right)^2}_{[10\times10]} \underbrace{\mathbf{J}^{(2)}}_{[10\times2]} + \left(\underbrace{1-\left(\mathbf{z}^{(3)}\right)^2}_{[10\times1]}\right) \odot \underbrace{\mathbf{W}^{(3)}}_{[10\times10]} \underbrace{\mathbf{H}_2^{(2)}}_{[10\times2]}$$

$$= \begin{bmatrix} 0.0004 & 0.0001 \\ 0.0004 & 0.0002 \\ -0.0111 & 0.0000 \\ -0.0073 & -0.0001 \\ 0.0090 & 0.0002 \\ -0.0010 & 0.0002 \\ 0.0000 & 0.0000 \\ -0.0009 & 0.0000 \\ 0.0000 & 0.0000 \\ -0.0040 & -0.0001 \end{bmatrix} \qquad (2.4.4.43\text{-}3b)$$

d. Normalized hidden layer, $\mathbf{z}^{(4)}$

Matrices for a function $\mathbf{z}^{(4)}$, a Jacobi $\mathbf{J}^{(4)}$, and slice Hessian matrices $\mathbf{H}_1^{(4)}$ and $\mathbf{H}_2^{(4)}$ for ($1= x_C$) and ($2= \theta$) are shown at Hidden layer (4) as shown in Equation (2.4.4.44).

$$\mathbf{z}^{(4)} = f_t^{(4)}\left(\mathbf{W}^{(4)}\mathbf{z}^{(3)} + \mathbf{b}^{(4)}\right) = \begin{bmatrix} -0.6736 \\ -0.3030 \\ 0.9998 \\ 0.4901 \\ -0.9994 \\ -0.5722 \\ 0.9997 \\ 0.2056 \\ 0.9997 \\ 0.7795 \end{bmatrix} \qquad (2.4.4.44\text{-}1)$$

$$\mathbf{J}^{(4)} = \left(1-\left(\mathbf{z}^{(4)}\right)^2\right) \odot \mathbf{W}^{(4)}\mathbf{J}^{(3)} = \begin{bmatrix} 0.7470 & 0.0066 \\ 0.4052 & 0.0154 \\ -0.0001 & 0.0000 \\ -0.7052 & 0.0071 \\ 0.0004 & 0.0000 \\ 0.8092 & -0.0059 \\ 0.0000 & 0.0000 \\ 0.3813 & 0.0002 \\ 0.0002 & 0.0000 \\ -0.1046 & -0.0074 \end{bmatrix} \qquad (2.4.4.44\text{-}2)$$

$$\mathbf{H}_1^{(4)} = -2\mathbf{z}^{(4)} \odot \left(1-\left(\mathbf{z}^{(4)}\right)^2\right) \odot i_1^{(3)} \odot \left(\mathbf{W}^{(4)}\right)^2 \mathbf{J}^{(3)} + \left(1-\left(\mathbf{z}^{(4)}\right)^2\right) \odot \mathbf{W}^{(4)}\mathbf{H}_1^{(3)}$$

$$= \begin{bmatrix} 0.8516 & 0.0177 \\ -1.0383 & 0.0042 \\ -0.0002 & 0.0000 \\ -0.3617 & 0.0036 \\ 0.0014 & 0.0000 \\ 0.8029 & 0.0070 \\ -0.0008 & 0.0000 \\ -1.1449 & 0.0041 \\ -0.0008 & 0.0000 \\ -0.1174 & -0.0037 \end{bmatrix} \qquad (2.4.4.44\text{-}3a)$$

$$\mathbf{H}_2^{(4)} = -2\mathbf{z}^{(4)} \odot \left(1-\left(\mathbf{z}^{(4)}\right)^2\right) \odot i_2^{(3)} \odot \left(\mathbf{W}^{(4)}\right)^2 \mathbf{J}^{(3)} + \left(1-\left(\mathbf{z}^{(4)}\right)^2\right) \odot \mathbf{W}^{(4)}\mathbf{H}_2^{(3)}$$

$$= \begin{bmatrix} 0.0177 & 0.0004 \\ 0.0042 & 0.0002 \\ 0.0000 & 0.0000 \\ 0.0036 & -0.0002 \\ 0.0000 & 0.0000 \\ 0.0070 & 0.0003 \\ 0.0000 & 0.0000 \\ 0.0041 & 0.0001 \\ 0.0000 & 0.0000 \\ -0.0037 & -0.0001 \end{bmatrix} \qquad (2.4.4.44\text{-}3b)$$

c. Normalized output layer (L=5=output), $\mathbf{z}^{(L=5=\text{output})}$

There are 24 nodes in normalized output layer, $\mathbf{z}^{(L=5=\text{output})}$, representing for 24 output parameters $\mathbf{y} = [y_1, \ldots, y_{24}]^T$. However, only outputs y_{20} (y_{P_f}) and y_{21} (y_{P_t}) are needed for Lagrange optimization as shown in Equation (2.4.4.39). Therefore, only two nodes of $z_{y20}^{(5)}$ and $z_{y21}^{(5)}$ are calculated to save calculation efforts. Normalized outputs $z_{y20}^{(5)}$, $z_{y21}^{(5)}$ and their derivatives are calculated following Figure 2.4.4.7a. Matrices for a function $z_{y20}^{(L=5=\text{output})}$, a Jacobi, and slice Hessian matrices for $(1 = x_C)$ and $(2 = \theta)$ are shown in Equation (2.4.4.45-1) based on Equation (2.3.6.3-2e), Equation (2.4.4.45-2) based on Equation (2.3.6.6-8), and Equations (2.4.4.45-3a) and (2.4.4.45-3a) based on Equation (2.3.6.15-2), respectively, at a normalized output layer. Matrices for a function $z_{y21}^{(5=\text{output})}$ and its derivatives are shown in Equation (2.4.4.46). $\mathbf{w}_{20}^{(5)}, b_{20}^{(5)}$ and $\mathbf{w}_{21}^{(5)}, b_{21}^{(5)}$ are the 20th and 21st rows of weight and bias matrices of normalized output layer, mapping previous hidden layer $\mathbf{z}^{(4)}$ to normalized outputs $z_{y20}^{(5)}$ and $z_{y21}^{(5)}$. Weight matrices $\mathbf{w}_{20}^{(5\,=\,\text{output})}$ and $\mathbf{w}_{21}^{(5\,=\,\text{output})}$ shown red color of Table 2.4.4.4(e) describing projectile move are used at an output layer shown in Equations (2.4.4.45) and Equation (2.4.4.46), respectively. Weight matrices are (24×10) matrices which has 24 rows because 24 because 24 outputs are considered at an output layer. They are obtained by TED-based ANNs with ten neurons. A total of 24 outputs is yielded based on (24×10) weight matrix and (24×1) bias matrix. However, Outputs y_{20} (y_{P_f}) and y_{21} (y_{P_t}) are only considered for Lagrange optimization, and hence, only Row 20 and 21 underlined in the weight and bias matrices of Table 2.4.4.4(e) are picked up to calculate y_{20} (y_{P_f}) and y_{21} (y_{P_t}), and their derivatives at an output layer. Readers are referred to Figure 2.3.6.2 for the formulation of a function z, a Jacobi, and slice Hessian matrices shown in this section.

$$z_{y20}^{(L=5=\text{output})} = f_{\text{lin}}^{(5)}\left(\underbrace{\mathbf{w}_{20}^{(5)}}_{[1\times10]}\underbrace{\mathbf{z}^{(4)}}_{[10\times1]} + \underbrace{b_{20}^{(5)}}_{[1\times1]}\right) = 0.4418 \tag{2.4.4.45-1}$$

$$\mathbf{J}_{y20}^{(L=5=\text{output})} = \underbrace{\mathbf{w}_{20}^{(5)}}_{[1\times10]}\underbrace{\mathbf{J}^{(4)}}_{[10\times2]} = \begin{bmatrix} -1.5275 & 0.0032 \end{bmatrix} \tag{2.4.4.45-2}$$

$$\mathbf{H}_{y20,\,1}^{(L=5=\text{output})} = \underbrace{\mathbf{w}_{20}^{(5)}}_{[1\times10]}\underbrace{\mathbf{H}_{1}^{(4)}}_{[10\times2]} = \begin{bmatrix} -4.0374 & -0.0368 \end{bmatrix} \tag{2.4.4.45-3a}$$

$$\mathbf{H}_{y20,\,2}^{(L=5=\text{output})} = \underbrace{\mathbf{w}_{20}^{(5)}}_{[1\times10]}\underbrace{\mathbf{H}_{2}^{(4)}}_{[10\times2]} = \begin{bmatrix} -0.0368 & -0.0009 \end{bmatrix} \tag{2.4.4.45-3b}$$

$$z_{y21}^{(L=5=\text{output})} = f_{\text{lin}}^{(5)}\left(\underbrace{\mathbf{w}_{21}^{(5)}}_{[1\times10]}\underbrace{\mathbf{z}^{(4)}}_{[10\times1]} + \underbrace{b_{21}^{(5)}}_{[1\times1]}\right) = 0.4161 \tag{2.4.4.46-1}$$

$$\mathbf{J}_{y21}^{(L=5=\text{output})} = \underbrace{\mathbf{w}_{21}^{(5)}}_{[1\times10]}\underbrace{\mathbf{J}^{(4)}}_{[10\times2]} = \begin{bmatrix} -1.5063 & 0.0050 \end{bmatrix} \tag{2.4.4.46-2}$$

$$\mathbf{H}_{y21,\,1}^{(L=5=\text{output})} = \underbrace{\mathbf{w}_{21}^{(5)}}_{[1\times10]}\underbrace{\mathbf{H}_{1}^{(4)}}_{[10\times2]} = \begin{bmatrix} -4.1680 & -0.0355 \end{bmatrix} \tag{2.4.4.46-3a}$$

$$\mathbf{H}_{y21,\,2}^{(L=5=\text{output})} = \underbrace{\mathbf{w}_{21}^{(5)}}_{[1\times10]}\underbrace{\mathbf{H}_{2}^{(4)}}_{[1\times10]} = \begin{bmatrix} -0.0355 & -0.0009 \end{bmatrix} \tag{2.4.4.46-3b}$$

d. De-normalized output layer (D, 5=output), $\mathbf{y} = \mathbf{z}^{(D)}$; de-normalizing $\mathbf{z}^{(5=\text{output})}$
   Output parameter $y_{20} = z_{y20}^{(D)}$ is de-normalized in Equation (2.4.4.47-1) based on Equation (2.3.6.3-2f) whereas Jacobi matrix is de-normalized in Equation (2.4.4.47-2) based on Equation (2.3.6.6-9) whereas Hessian matrix is de-normalized in Equations (2.4.4.47-3a), (2.4.4.47-3b) based on Equation (2.3.6.16-2). Slice Hessian matrices calculated in Equations (2.4.4.47-3a) and (2.4.4.47-3b) are, then, combined into global Hessian matrix shown in Equation (2.4.4.47-4).

   Jacobi and Hessian matrices of output parameter $y_{21} = z_{y20}^{(D)}$ are de-normalized in Equation (2.4.4.48). Slice Hessian matrices calculated in Equations (2.4.4.48-3a) and (2.4.4.48-3b) are, then, combined into global Hessian matrix shown in Equation (2.4.4.48-4). $z_{y20}^{(L=5=\text{output})} = 0.4418$ obtained in Equation (2.4.4.45-1) at Hidden layer (5 = output) is de-normalized, being returned to the original unit in Equation (2.4.4.47-1).

α_{y20}, $\bar{y}_{20,\min}$, $y_{20,\min}$ are found in red box of Table 2.4.4.2. Similarly, $z_{y21}^{(L=5=\text{output})}=0.4161$ obtained in Equation (2.4.4.46-1) at Hidden layer (5 = output) is de-normalized, being returned to the original unit in Equation (2.4.4.48-1). α_{y21}, $\bar{y}_{21,\min}$, $y_{21,\min}$ are found in Table 2.4.4.2.

$$y_{20} = z_{y20}^{(D)} = g_{20}^{(D)}\left(z_{y20}^{(L=5=\text{output})}\right)$$

$$= \frac{1}{\alpha_{y20}}\left(z_{y20}^{(5)} - \bar{y}_{20,\min}\right) + y_{20,\min} = \frac{1}{1.5427}\left(0.4418-(-1)\right)+(-0.9871)$$

$$= -0.0526 \tag{2.4.4.47-1}$$

$$\mathbf{J}_{y20}^{(D)} = \frac{1}{\alpha_{y20}}\mathbf{J}_{y20}^{(5=\text{output})} = \begin{bmatrix} -0.9902 & 0.0021 \end{bmatrix} \tag{2.4.4.47-2}$$

$$\mathbf{H}_{y20,1}^{(D)} = \frac{1}{\alpha_{y20}}\mathbf{H}_{y20,1}^{(5=\text{output})} = \begin{bmatrix} -2.6171 & -0.0239 \end{bmatrix} \tag{2.4.4.47-3a}$$

$$\mathbf{H}_{y20,2}^{(D)} = \frac{1}{\alpha_{y20}}\mathbf{H}_{y20,2}^{(5=\text{output})} = \begin{bmatrix} -0.0239 & -0.0006 \end{bmatrix} \tag{2.4.4.47-3b}$$

$$\mathbf{H}_{y20}^{(D)} = \begin{bmatrix} \mathbf{H}_{y20,1}^{(D)} \\ \mathbf{H}_{y20,2}^{(D)} \end{bmatrix} = \begin{bmatrix} -2.6171 & -0.0239 \\ -0.0239 & -0.0006 \end{bmatrix} \tag{2.4.4.47-4}$$

$$y_{21} = z_{y21}^{(D)} = g_{21}^{(D)}\left(z_{y21}^{(L=5=\text{output})}\right)$$

$$= \frac{1}{\alpha_{y21}}\left(z_{y21}^{(5)} - \bar{y}_{21,\min}\right) + y_{21,\min} = \frac{1}{1.7958}\left(0.4161-(-1)\right)+(-0.7988)$$

$$= 0.0105 \tag{2.4.4.48-1}$$

$$\mathbf{J}_{y21}^{(D)} = \frac{1}{\alpha_{y21}}\mathbf{J}_{y21}^{(L=5=\text{output})} = \begin{bmatrix} -0.8075 & 0.0036 \end{bmatrix} \tag{2.4.4.48-2}$$

$$\mathbf{H}_{y21,1}^{(D)} = \frac{1}{\alpha_{y21}}\mathbf{H}_{y21,1}^{(L=5=\text{output})} = \begin{bmatrix} -2.6102 & -0.0186 \end{bmatrix} \tag{2.4.4.48-3a}$$

$$\mathbf{H}_{y21,2}^{(D)} = \frac{1}{\alpha_{y21}}\mathbf{H}_{y21,2}^{(L=5=\text{output})} = \begin{bmatrix} -0.0186 & -0.0004 \end{bmatrix} \tag{2.4.4.48-3b}$$

$$\mathbf{H}_{y21}^{(D)} = \begin{bmatrix} \mathbf{H}_{y21,1}^{(D)} \\ \mathbf{H}_{y21,2}^{(D)} \end{bmatrix} = \begin{bmatrix} -2.6102 & -0.0186 \\ -0.0186 & -0.0004 \end{bmatrix} \tag{2.4.4.48-4}$$

e. Convergence verification of Lagrange function based on Jacobi, Hessian matrix at Iteration 0

Equality $c\left(x_C^{(0)}, \theta^{(0)}\right)$ and inequality $v\left(x_C^{(0)}, \theta^{(0)}\right)$ constraining functions should be obtained in Equations (2.4.4.49-3a) and (2.4.4.49-3b), their Jacobi and Hessian matrices should be also obtained in Equations (2.4.4.50) and (2.4.4.51), respectively, to solve a first derivative (Jacobi) of Lagrange function shown in Equation (2.3.4.2-1b) or Equation (2.4.4.52) with respect to initial trial input parameter $[x_C^{(0)}, \theta^{(0)}]$. ANNs with series of composite mathematical notations for objective, equality $c\left(x_C^{(0)}, \theta^{(0)}\right)$, and inequality $v\left(x_C^{(0)}, \theta^{(0)}\right)$ constraining functions are derived in Equation (2.4.4.36) based on Equations (2.3.6.3-1) and (2.3.6.3-2). The ANNs for y_{P_f} (y_{20}) and y_{P_f} (y_{21}) are shown in Equations (2.4.4.49-1) and (2.4.4.49-2). An objective function $f_{xc}(x_C)$ is obtained easily in Equation (2.4.4.12-1). However, equality $c\left(x_C^{(0)}, \theta^{(0)}\right)$ and inequality $v\left(x_C^{(0)}, \theta^{(0)}\right)$ constraining functions are obtained based on ANN in Equations (2.4.4.49-3a) and (2.4.4.49-3b). A [2×1] initial trial input parameter $x^{(0)} = \left[x_C^{(0)}, \theta^{(0)}\right]^T = [0.7,\ 40]^T$ indicated by red color shown in Equation (2.4.4.40-1) is substituted into Equations (2.4.4.49-1) and (2.4.4.49-2) to obtain normalized y_{P_f} (y_{20}) and y_{P_f} (y_{21}), Jacobi and Hessian matrices in Equations (2.4.4.40)–(2.4.4.46). The normalized y_{P_f} (y_{20}) and y_{P_f} (y_{21}) are then de-normalized as −0.0526 and 0.0105 in Equations(2.4.4.47-1) and (2.4.4.48-1) to obtain equality $c\left(x_C^{(0)}, \theta^{(0)}\right)$, and inequality $v\left(x_C^{(0)}, \theta^{(0)}\right)$ constraining functions, y_{P_f} (y_{20}) and y_{P_f} (y_{21}), based on ANN in Equations (2.4.4.49-3a) and (2.4.4.49-3b).

$$\underbrace{y_{20}}_{[1\times1]} = g_{y20}^{D}\left(f_{\text{lin}}^{L}\left(\underbrace{\mathbf{w}_{20}^{L}}_{[1\times10]} f_t^{L-1}\left(\underbrace{\mathbf{W}^{L-1}}_{[10\times10]} \ldots f_t^{1}\left(\underbrace{\mathbf{W}^{1}}_{[10\times2]}\underbrace{\mathbf{g}^{N}(\mathbf{x})}_{[2\times1]} + \underbrace{\mathbf{b}^{1}}_{[10\times1]}\right)\ldots + \underbrace{\mathbf{b}^{L-1}}_{[10\times1]}\right) + \underbrace{b_{20}^{L}}_{[1\times1]}\right)\right) \tag{2.4.4.49-1}$$

$$\underbrace{y_{21}}_{[1\times1]} = g_{y21}^{D}\left(f_{\text{lin}}^{L}\left(\underbrace{\mathbf{w}_{21}^{L}}_{[1\times10]} f_t^{L-1}\left(\underbrace{\mathbf{W}^{L-1}}_{[10\times10]} \ldots f_t^{1}\left(\underbrace{\mathbf{W}^{1}}_{[10\times2]}\underbrace{\mathbf{g}^{N}(\mathbf{x})}_{[2\times1]} + \underbrace{\mathbf{b}^{1}}_{[10\times1]}\right)\ldots + \underbrace{\mathbf{b}^{L-1}}_{[10\times1]}\right) + \underbrace{b_{21}^{L}}_{[1\times1]}\right)\right) \tag{2.4.4.49-2}$$

$$c\left(x_C^{(0)}, \theta^{(0)}\right) = y_{P_f} - 0.06 = y_{20} - 0.06 = -0.0526 - 0.06 = -0.1126 \tag{2.4.4.49-3a}$$

$$v\left(x_C^{(0)}, \theta^{(0)}\right) = y_{P_f} - 0.06 = y_{21} - 0.1 = 0.0105 - 0.1 = -0.0895 \tag{2.4.4.49-3b}$$

Jacobi matrices $\mathbf{J}_c\left(x_C^{(0)}, \theta^{(0)}\right)$ of equality $c\left(x_C^{(0)}, \theta^{(0)}\right)$ and $\mathbf{J}_v\left(x_C^{(0)}, \theta^{(0)}\right)$ of inequality $v\left(x_C^{(0)}, \theta^{(0)}\right)$ constraining functions are calculated in Equation (2.3.6.6-9) based on a calculation $\mathbf{J}^{(N)}\left(x_C^{(0)}, \theta^{(0)}\right)$ shown in Equation (2.3.6.6-1) to $\mathbf{J}^{(D)}\left(x_C^{(0)}, \theta^{(0)}\right)$ shown in Equation (2.3.6.6-8). Ann-based Jacobi matrices, $\mathbf{J}_c\left(x_C^{(0)}, \theta^{(0)}\right)$ and $\mathbf{J}_v\left(x_C^{(0)}, \theta^{(0)}\right)$ are, then, derived in Equations (2.4.4.40)–(2.4.4.48), from which $\mathbf{J}_c\left(x_C^{(0)}, \theta^{(0)}\right)$ for y_{P_f} (y_{20}) is calculated in Equation (2.4.4.50-1) based on Equation (2.4.4.47-2) whereas, $\mathbf{J}_v\left(x_C^{(0)}, \theta^{(0)}\right)$ for y_{P_f} (y_{21}) is calculated in Equation (2.4.4.50-2) based on Equation (2.4.4.48-2), for given initial trial input parameter $x^{(0)} = \left[x_C^{(0)}, \theta^{(0)}\right]^T = [0.7,\ 40]^T$.

$$\mathbf{J}_c\left(x_C^{(0)}, \theta^{(0)}\right) = \mathbf{J}_c^{(D)}\left(x_C^{(0)}, \theta^{(0)}\right) = \mathbf{J}_{\left(y_{P_f}-0.06\right)}^{(D)} = \mathbf{J}_{y_{P_f}}^{(D)} = \mathbf{J}_{y20}^{(D)}$$

$$= \begin{bmatrix} -0.9902 & 0.0021 \end{bmatrix} \quad (2.4.4.50\text{-}1)$$

$$\mathbf{J}_v\left(x_C^{(0)}, \theta^{(0)}\right) = \mathbf{J}_v^{(D)}\left(x_C^{(0)}, \theta^{(0)}\right) = \mathbf{J}_{\left(y_{P_t}-0.1\right)}^{(D)} = \mathbf{J}_{y_{P_t}}^{(D)} = \mathbf{J}_{y21}^{(D)}$$

$$= \begin{bmatrix} -0.8075 & 0.0036 \end{bmatrix} \quad (2.4.4.50\text{-}2)$$

Hessian matrices $\mathbf{H}_i^{(D)}$ of equality $c\left(x_C^{(0)}, \theta^{(0)}\right)$ and inequality $v\left(x_C^{(0)}, \theta^{(0)}\right)$ constraining functions are calculated using Equations (2.3.6.7-1), (2.3.6.14), (2.3.6.15) and (2.3.6.16). Ann-based Hessian matrices of equality $c\left(x_C^{(0)}, \theta^{(0)}\right)$ and inequality $v\left(x_C^{(0)}, \theta^{(0)}\right)$ are derived in Equations (2.4.4.40)–(2.4.4.48), from which $\mathbf{H}_{y20}^{(D)}$ for y_{P_f} (y_{20}) is calculated in Equation (2.4.4.51-1) based on Equations (2.4.4.47-3) and (2.4.4.47-4), whereas $\mathbf{H}_{y21}^{(D)}$ for y_{P_t} (y_{21}) is calculated in Equation (2.4.4.51-2) based on Equations (2.4.4.48-3) and (2.4.4.48-4), for given initial trial input parameter $x^{(0)} = \left[x_C^{(0)}, \theta^{(0)}\right]^T = [0.7,\ 40]^T$. Slice Hessian matrices are combined into global Hessian matrix $\mathbf{H}_{y20}^{(D)}$ and $\mathbf{H}_{y21}^{(D)}$ shown in Equations (2.4.4.51-1) and (2.4.4.51–2), respectively, based on Equation (2.3.6.17).

$$\mathbf{H}_c\left(x_C^{(0)}, \theta^{(0)}\right) = \mathbf{H}_c^{(D)} = \mathbf{H}_{y20}^{(D)} = \begin{bmatrix} -2.6171 & -0.0239 \\ -0.0239 & -0.0006 \end{bmatrix} \quad (2.4.4.51\text{-}1)$$

$$\mathbf{H}_v\left(x_C^{(0)}, \theta^{(0)}\right) = \mathbf{H}_v^{(D)} = \mathbf{H}_{y21}^{(D)} = \begin{bmatrix} -2.6102 & -0.0186 \\ -0.0186 & -0.0004 \end{bmatrix} \quad (2.4.4.51\text{-}2)$$

A first derivative (Jacobi) of Lagrange function is calculated in Equation (2.3.4.2-1b) or Equation (2.4.4.52) by substituting Equations (2.4.4.49-3a), (2.4.4.49-3b), (2.4.4.50-1), (2.4.4.50-2) and $s = 1$ into Equation (2.4.4.52-1b) for a given initial trial input parameter $x^{(0)} = \left[x_C^{(0)}, \theta^{(0)}\right]^T = [0.7,\ 40]^T$. However, $\nabla\mathcal{L}\left(x_C^{(0)}, \theta^{(0)}, \lambda_c^{(0)}, \lambda_v^{(0)}\right)$ shown in Equation (2.4.4.52-2) does not converge to 0, and hence, initial trial input parameter $\left[x_C^{(0)}, \theta^{(0)}\right]^T = \left[x_C^{(0)}, \theta^{(0)}, \lambda_c^{(0)}, \lambda_v^{(0)}\right]^T = [0.7,\ 40,\ 0,\ 0]^T$ cannot be a candidate solution for Case 1 KKT condition. Another iteration is necessary to check if $\nabla\mathcal{L}\left(x_C^{(1)}, \theta^{(1)}, \lambda_c^{(1)}, \lambda_v^{(1)}\right)$ shown in Equation (2.4.4.59) using updated input parameter $\left[x_C^{(1)}, \theta^{(1)}\right]^T = \left[x_C^{(1)}, \theta^{(1)}, \lambda_c^{(1)}, \lambda_v^{(1)}\right]^T$ at Iteration 1 shown in Equation (2.4.4.53) converges. It is noted that Lagrange multipliers $\left[\lambda_c^{(1)}, \lambda_v^{(1)}\right]^T$ is not used in Equation (2.4.4.40-1) but used in Equation (2.4.4.52-1).

Updated input parameter $\left[x_C^{(1)}, \theta^{(1)}\right]^T = \left[x_C^{(1)}, \theta^{(1)}, \lambda_c^{(1)}, \lambda_v^{(1)}\right]^T$ updated in Iteration 1 is calculated in Equation (2.4.4.58-1) by substituting $\left[\mathbf{H}_{\mathcal{L}}\left(x_C^{(0)}, \theta^{(0)}, \lambda_c^{(0)}, \lambda_v^{(0)}\right)\right]^{-1}$ shown in Equation (2.4.4.57-3) and $\nabla\mathcal{L}\left(x_C^{(0)}, \theta^{(0)}, \lambda_c^{(0)}, \lambda_v^{(0)}\right)$ shown in Equation (2.4.4.52-2) into Equation (2.3.4.1-5) or Equation (2.4.4.53). However, $\boldsymbol{H}_{\mathcal{L}}\left(x_C^{(0)}, \theta^{(0)}\right)$ must be first calculated in Equation (2.4.4.56) by substituting $\boldsymbol{H}_{f_{xC}}\left(x_C, \theta\right)$ shown in Equation (2.4.4.14), $\mathbf{H}_c^{(D)}\left(x_C^{(0)}, \theta^{(0)}\right)$ shown in Equation (2.4.4.51-1), and $\mathbf{H}_v^{(D)}$ shown in Equation (2.4.4.51-2) into Equation (2.4.4.55). $\mathbf{H}_{\mathcal{L}}\left(x_C^{(0)}, \theta^{(0)}, \lambda_c^{(0)}, \lambda_v^{(0)}\right)$ is now ready to be calculated in Equation (2.4.4.57-2) by substituting Jacobi matrix for equality $\mathbf{J}_c\left(x_C^{(0)}, \theta^{(0)}\right)$ shown in Equation (2.4.4.50-1), Jacobi matrix for inequality $\mathbf{J}_v\left(x_C^{(0)}, \theta^{(0)}\right)$ shown in Equation (2.4.4.50-2), $H_{\mathcal{L}}\left(x_C^{(0)}, \theta^{(0)}\right)$ shown in Equation (2.4.4.56) and $s=1$ into Equation (2.4.4.54-2). $\left[\mathbf{H}_{\mathcal{L}}\left(x_C^{(0)}, \theta^{(0)}, \lambda_c^{(0)}, \lambda_v^{(0)}\right)\right]^{-1}$ is, then, obtained in Equation (2.4.4.57-3).

$$\nabla\mathcal{L}\left(x_C^{(0)}, \theta^{(0)}, \lambda_c^{(0)}, \lambda_v^{(0)}\right)=\begin{bmatrix} \nabla f_{xC}\left(x_C^{(0)}\right)-\left[\mathbf{J}_c\left(x_C^{(0)}, \theta^{(0)}\right)\right]^T\lambda_c-\left[\mathbf{J}_v\left(x_C^{(0)}, \theta^{(0)}\right)\right]^T s\lambda_v \\ -c\left(x_C^{(0)}, \theta^{(0)}\right) \\ -sv\left(x_C^{(0)}, \theta^{(0)}\right) \end{bmatrix} \tag{2.4.4.52-1a}$$

$$\nabla\mathcal{L}\left(x_C^{(0)}, \theta^{(0)}, \lambda_c^{(0)}, \lambda_v^{(0)}\right)=\begin{bmatrix} \nabla f_{xC}\left(x_C^{(0)}\right)-\left[\mathbf{J}_c^{(D)}\left(x_C^{(0)}, \theta^{(0)}\right)\right]^T\lambda_c-\left[\mathbf{J}_v^{(D)}\left(x_C^{(0)}, \theta^{(0)}\right)\right]^T s\lambda_v \\ -c\left(x_C^{(0)}, \theta^{(0)}\right) \\ -sv\left(x_C^{(0)}, \theta^{(0)}\right) \end{bmatrix} \tag{2.4.4.52-1b}$$

$$\nabla\mathcal{L}\left(x_C^{(0)}, \theta^{(0)}, \lambda_c^{(0)}, \lambda_v^{(0)}\right)$$

$$=\begin{bmatrix} \begin{bmatrix} 1 \\ 0 \end{bmatrix}-\begin{bmatrix} -0.9902 \\ 0.0021 \end{bmatrix}\times 0-\begin{bmatrix} -0.8075 \\ 0.0036 \end{bmatrix}\times 1\times 0 \\ 0.1126 \\ 0.0895 \end{bmatrix}$$

$$\rightarrow\nabla\mathcal{L}\left(x_C^{(0)}, \theta^{(0)}, \lambda_c^{(0)}, \lambda_v^{(0)}\right)=\begin{bmatrix} 1 \\ 0 \\ 0.1126 \\ 0.0895 \end{bmatrix}\neq 0 \tag{2.4.4.52-2}$$

$$\begin{bmatrix} x_C^{(1)} \\ \theta^{(1)} \\ \lambda_c^{(1)} \\ \lambda_v^{(1)} \end{bmatrix}=\begin{bmatrix} x_C^{(0)} \\ \theta^{(0)} \\ \lambda_c^{(0)} \\ \lambda_v^{(0)} \end{bmatrix}-\left[\mathbf{H}_{\mathcal{L}}\left(x_C^{(0)}, \theta^{(0)}, \lambda_c^{(0)}, \lambda_v^{(0)}\right)\right]^{-1}\nabla\mathcal{L}\left(x_C^{(0)}, \theta^{(0)}, \lambda_c^{(0)}, \lambda_v^{(0)}\right) \tag{2.4.4.53}$$

$$\mathbf{H}_{\mathcal{L}}\left(x_C^{(0)}, \theta^{(0)}, \lambda_c^{(0)}, \lambda_v^{(0)}\right) = \begin{bmatrix} \mathbf{H}_{\mathcal{L}}\left(x_C^{(0)}, \theta^{(0)}\right) & -\left[J_c\left(x_C^{(0)}, \theta^{(0)}\right)\right]^T & -\left[sJ_v\left(x_C^{(0)}, \theta^{(0)}\right)\right]^T \\ -J_c\left(x_C^{(0)}, \theta^{(0)}\right) & 0 & 0 \\ -sJ_v\left(x_C^{(0)}, \theta^{(0)}\right) & 0 & 0 \end{bmatrix} \quad (2.4.4.54\text{-}1)$$

$$\mathbf{H}_{\mathcal{L}}\left(x_C^{(0)}, \theta^{(0)}, \lambda_c^{(0)}, \lambda_v^{(0)}\right) = \begin{bmatrix} \mathbf{H}_{\mathcal{L}}\left(x_C^{(0)}, \theta^{(0)}\right) & -\left[\mathbf{J}_c^{(D)}\left(x_C^{(0)}, \theta^{(0)}\right)\right]^T & -\left[s\mathbf{J}_v^{(D)}\left(x_C^{(0)}, \theta^{(0)}\right)\right]^T \\ -\mathbf{J}_c^{(D)}\left(x_C^{(0)}, \theta^{(0)}\right) & 0 & 0 \\ -s\mathbf{J}_v^{(D)}\left(x_C^{(0)}, \theta^{(0)}\right) & 0 & 0 \end{bmatrix} \quad (2.4.4.54\text{-}2)$$

Where,

$$\mathbf{H}_{\mathcal{L}}\left(x_C^{(0)}, \theta^{(0)}\right) = \mathbf{H}_{f_{xC}}\left(x_C^{(0)}, \theta^{(0)}\right) - \lambda_c^{(0)}\mathbf{H}_c\left(x_C^{(0)}, \theta^{(0)}\right) - s\lambda_v^{(0)}\mathbf{H}_v\left(x_C^{(0)}, \theta^{(0)}\right) \quad (2.4.4.55\text{-}1)$$

$$\mathbf{H}_{\mathcal{L}}\left(x_C^{(0)}, \theta^{(0)}\right) = \mathbf{H}_{f_{xC}}\left(x_C^{(0)}, \theta^{(0)}\right) - \lambda_c^{(0)}\mathbf{H}_c^{(D)}\left(x_C^{(0)}, \theta^{(0)}\right) - s\lambda_v^{(0)}\mathbf{H}_v^{(D)}\left(x_C^{(0)}, \theta^{(0)}\right) \quad (2.4.4.55\text{-}2)$$

$$\mathbf{H}_{\mathcal{L}}\left(x_C^{(0)}, \theta^{(0)}\right) = \begin{bmatrix} 0 & 0 \\ 0 & 0 \end{bmatrix} - 0 \times \begin{bmatrix} -2.6171 & -0.0239 \\ -0.0239 & -0.0006 \end{bmatrix} - 1 \times 0 \times \begin{bmatrix} -2.6102 & -0.0186 \\ -0.0186 & -0.0004 \end{bmatrix} = \begin{bmatrix} 0 & 0 \\ 0 & 0 \end{bmatrix} \quad (2.4.4.56)$$

$$\mathbf{H}_{\mathcal{L}}\left(x_C^{(0)}, \theta^{(0)}, \lambda_c^{(0)}, \lambda_v^{(0)}\right) = \begin{bmatrix} \begin{bmatrix} 0 & 0 \\ 0 & 0 \end{bmatrix} & -\begin{bmatrix} -0.9902 \\ 0.0021 \end{bmatrix} & -\begin{bmatrix} -0.8075 \\ 0.0036 \end{bmatrix} \\ -\begin{bmatrix} -0.9902 & 0.0021 \end{bmatrix} & 0 & 0 \\ -\begin{bmatrix} -0.8075 & 0.0036 \end{bmatrix} & 0 & 0 \end{bmatrix} \quad (2.4.4.57\text{-}1)$$

$$= \begin{bmatrix} 0 & 0 & 0.9902 & 0.8075 \\ 0 & 0 & -0.0021 & -0.0036 \\ 0.9902 & -0.0021 & 0 & 0 \\ 0.8075 & -0.0036 & 0 & 0 \end{bmatrix} \quad (2.4.4.57\text{-}2)$$

$$\left[\mathbf{H}_{\mathcal{L}}\left(x_C^{(0)}, \theta^{(0)}, \lambda_c^{(0)}, \lambda_v^{(0)}\right)\right]^{-1} = \begin{bmatrix} 0 & 0 & 1.9032 & -1.0955 \\ 0 & 0 & 430.70 & -528.17 \\ 1.9032 & 430.70 & 0 & 0 \\ -1.0955 & -528.17 & 0 & 0 \end{bmatrix} \quad (2.4.4.57\text{-}3)$$

ANN-based calculations similar to Equations (2.4.4.40)–(2.4.4.48) are repeated until a solution for a first derivative (Jacobi, gradient vector) of a Lagrange function $\nabla\mathcal{L}\left(x_C^{(k)}, \theta^{(k)}, \lambda_c^{(k)}, \lambda_v^{(k)}\right)$ converges. Updated input parameter $\left[x_C^{(1)}, \theta^{(1)}\right]^T = \left[x_C^{(1)}, \theta^{(1)}, \lambda_c^{(1)}, \lambda_v^{(1)}\right]^T$ is calculated at the end of Iteration 0 as shown in Equation (2.4.4.58-1) by substituting $\left[\mathbf{H}_{\mathcal{L}}\left(x_C^{(0)}, \theta^{(0)}, \lambda_c^{(0)}, \lambda_v^{(0)}\right)\right]^{-1}$ shown in Equation (2.4.4.57-3) and $\nabla\mathcal{L}\left(x_C^{(0)}, \theta^{(0)}, \lambda_c^{(0)}, \lambda_v^{(0)}\right)$ shown in Equation (2.4.4.52-2) into Equation (2.3.4.1-5) or Equation (2.4.4.53). In Equation (A1.1-1), Iteration 1 uses $\left[x_C^{(1)}, \theta^{(1)}\right] = [0.584,\ 38.80]$ instead of $\left[x_C^{(0)}, \theta^{(0)}\right]^T = [0.7, 40]^T$ used in Equation (2.4.4.40-1) for Iteration 0. Equations (A1.10-1)–(A1.11-2) are repeated to calculate $c\left(x_C^{(1)}, \theta^{(1)}\right)$, $v\left(x_C^{(1)}, \theta^{(1)}\right)$ and $\mathbf{J}_c\left(x_C^{(1)}, \theta^{(1)}\right)$, $\mathbf{J}_v\left(x_C^{(1)}, \theta^{(1)}\right)$. $\nabla\mathcal{L}\left(x_C^{(1)}, \theta^{(1)}, \lambda_c^{(1)}, \lambda_v^{(1)}\right)$ shown in Equations (2.4.4.59) or (A1.13) is, then, derived by substituting results obtained in Equations (A1.10-1)–(A1.11-2) into Equation (2.3.4.2-1b) or Equation (2.4.4.58-3). However, a first derivative (Jacobi) of the Lagrange function for Iteration 1 shown in Equations (2.4.4.59) or (A1.13) does not converge yet, failing to find a candidate solution for Case 1 KKT condition. Another iteration is required. New input parameter $\left[x_C^{(2)}, \theta^{(2)}, \lambda_c^{(2)}, \lambda_v^{(2)}\right]$ is updated as $= [0.569, 41.07, -3.833, 3.184]$ at the end of Iteration 1 as shown in Equation (A1.19), after which equality and inequality functions $c\left(x_C^{(2)},\theta^{(2)}\right)$, $v\left(x_C^{(2)}, \theta^{(2)}\right)$ and their Jacobi matrixes $\mathbf{J}_c\left(x_C^{(2)}, \theta^{(2)}\right)$, $\mathbf{J}_v\left(x_C^{(2)}, \theta^{(2)}\right)$ are again updated with respect to $\left[x_C^{(2)}, \theta^{(2)}, \lambda_c^{(2)}, \lambda_v^{(2)}\right] = [0.569, 41.07,\ -3.833, 3.184]$ to calculate a first derivative (Jacobi) of Lagrange function for Iteration 2 as shown in Equation (2.4.4.60). Updated first derivative (Jacobi) of Lagrange function $\nabla\mathcal{L}\left(x_C^{(2)},\theta^{(2)},\lambda_c^{(2)},\lambda_v^{(2)}\right)$ with respect to $\left[x_C^{(2)}, \theta^{(2)}, \lambda_c^{(2)}, \lambda_v^{(2)}\right] = [0.569,\ 41.07,\ -3.833,\ 3.184]$ for Iteration 2 shown in Equation (2.4.4.60) does not converge to 0 even though it comes nearer 0 than Iteration 1. Entire derivation to update $\left[x_C^{(2)}, \theta^{(2)}, \lambda_c^{(2)}, \lambda_v^{(2)}\right]$ is presented in Appendix A1, calculating a new updated input parameter of $\left[x_C^{(2)}, \theta^{(2)}, \lambda_c^{(2)}, \lambda_v^{(2)}\right] = [0.569, 41.07, -3.833, 3.184]$ at the end of Iteration 1.

$$\begin{bmatrix} x_C^{(1)} \\ \theta^{(1)} \\ \lambda_c^{(1)} \\ \lambda_v^{(1)} \end{bmatrix} = \begin{bmatrix} 0.7 \\ 40 \\ 0 \\ 0 \end{bmatrix} - \begin{bmatrix} 0 & 0 & 1.9032 & -1.0955 \\ 0 & 0 & 430.70 & -528.17 \\ 1.9032 & 430.70 & 0 & 0 \\ -1.0955 & -528.17 & 0 & 0 \end{bmatrix} \begin{bmatrix} 1 \\ 0 \\ 0.1126 \\ 0.0895 \end{bmatrix}$$

$$\rightarrow \begin{bmatrix} x_C^{(1)} \\ \theta^{(1)} \\ \lambda_c^{(1)} \\ \lambda_v^{(1)} \end{bmatrix} = \begin{bmatrix} 0.584 \\ 38.80 \\ -1.903 \\ 1.096 \end{bmatrix} \quad (2.4.4.58\text{-}1)$$

$$\nabla\mathcal{L}\left(x_C^{(1)}, \theta^{(1)}, \lambda_c^{(1)}, \lambda_v^{(1)}\right)$$

$$= \begin{bmatrix} \nabla f_{xC}\left(x_C^{(1)}\right) - J_c\left(x_C^{(1)}, \theta^{(1)}\right)^T \lambda_c - J_v\left(x_C^{(1)}, \theta^{(1)}\right)^T s\lambda_v \\ -c\left(x_C^{(1)}, \theta^{(1)}\right) \\ -sv\left(x_C^{(1)}, \theta^{(1)}\right) \end{bmatrix} \quad (2.4.4.58\text{-}2)$$

$$= \begin{bmatrix} \nabla f_{xC}\left(x_C^{(1)}\right) - \left[\mathbf{J}_c^{(D)}\left(x_C^{(1)}, \theta^{(1)}\right)\right]^T \lambda_c - \left[\mathbf{J}_v^{(D)}\left(x_C^{(1)}, \theta^{(1)}\right)\right]^T s\lambda_v \\ -c\left(x_C^{(1)}, \theta^{(1)}\right) \\ -sv\left(x_C^{(1)}, \theta^{(1)}\right) \end{bmatrix} \quad (2.4.4.58\text{-}3)$$

$$\nabla\mathcal{L}\left(x_C^{(1)}, \theta^{(1)}, \lambda_c^{(1)}, \lambda_v^{(1)}\right) = \begin{bmatrix} 0.2666 \\ 0.0030 \\ 0.0206 \\ 0.0198 \end{bmatrix} \neq 0 \quad (2.4.4.59)$$

$$\nabla\mathcal{L}\left(x_C^{(2)}, \theta^{(2)}, \lambda_c^{(2)}, \lambda_v^{(2)}\right) = \begin{bmatrix} -0.0289 \\ -0.0003 \\ 0.0006 \\ 0.0005 \end{bmatrix} \neq 0 \quad (2.4.4.60)$$

f. Summary
Finally, a solution of $\nabla\mathcal{L}\left(x_C^{(k)}, \theta^{(k)}, \lambda_c^{(k)}, \lambda_v^{(k)}\right) = 0$ converged at $\left[x_C^{(5)}, \theta^{(5)}, \lambda_c^{(5)}, \lambda_v^{(5)}\right] =$ $[0.568, 41.10, -3.628, 2.958]$ with an MSE of 2.0E-20, resulting in a $f_{xC}(x_C) =$ 0.568 (km) as shown in Figure 2.4.4.10. Three more initial trial input parameters of $\left[x_C^{(0)}, \theta^{(0)}\right]$ shown in Figure 2.4.4.9 are implemented to find an optimal solution of Case 1. It is noted that initial Lagrange multipliers of $\lambda_c^{(0)}$ and $\lambda_v^{(0)}$ are selected as 0. All four selected initial trial input parameters of $\left[x_C^{(0)}, \theta^{(0)}, \lambda_c^{(0)}, \lambda_v^{(0)}\right] = [0.7, 40, 0, 0]$, $[0.7, 20, 0, 0]$, $[0.6, 60, 0, 0]$, and $[0.2, 35, 0, 0]$ converge to maximize an objective function of $f_{xC}(x_C) = 0.568$ (km) at $[x_C, \theta] = [0.568, 41.10]$. Figure 2.4.4.10 shows projectile trajectory which satisfies Case 1 KKT condition.

2.4.4.2.4.1 CASE 2 KKT CONDITION BASED ON AN INACTIVE EQUALITY $v(x)$

a. Convergence verification of Lagrange function based on initial trial input parameters at Iteration 0

In this case of Case 2 KKT, inequality $v(x_C, \theta) = y_{P_t} - 0.1 \geq 0$ is assumed be inactive or slack condition, which does not constrain an objective function $f_{xC}(x_C)$ when

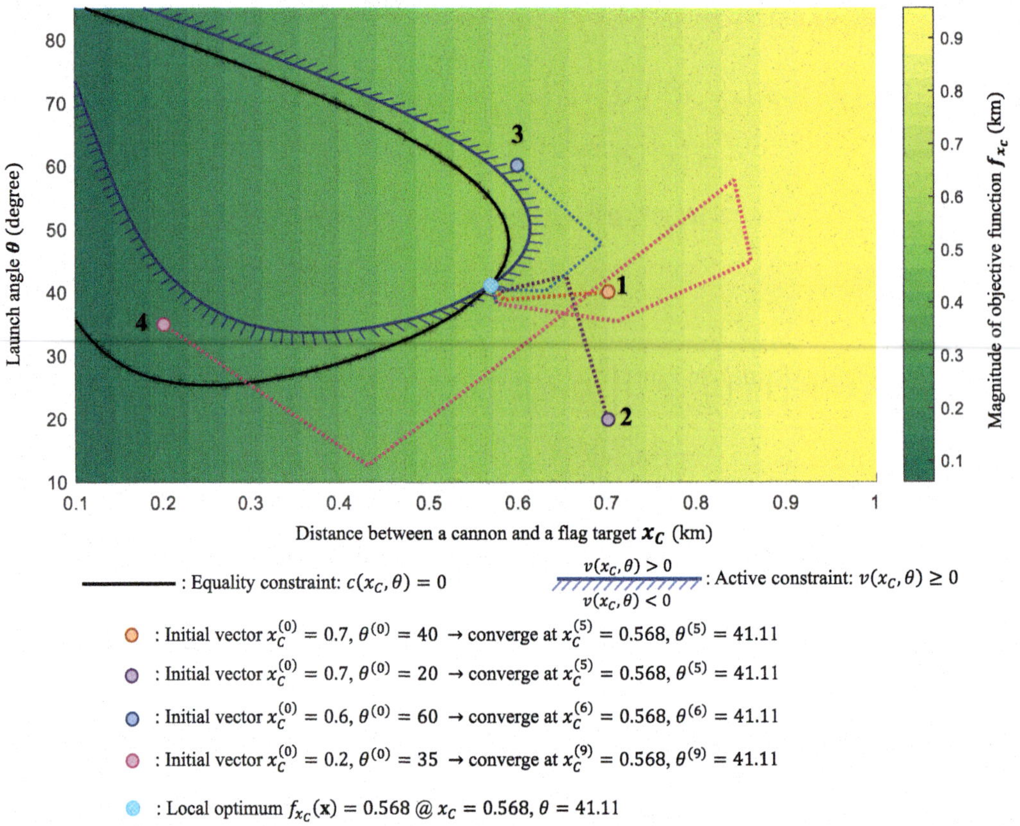

Figure 2.4.4.9 Convergence of initial trial input parameters based on ANN for Case I KKT condition; must satisfy $c(x_C, \theta) = y_{P_f} - 0.06 = 0$, $v(x_C, \theta) = y_{P_t} - 0.1 = 0$.

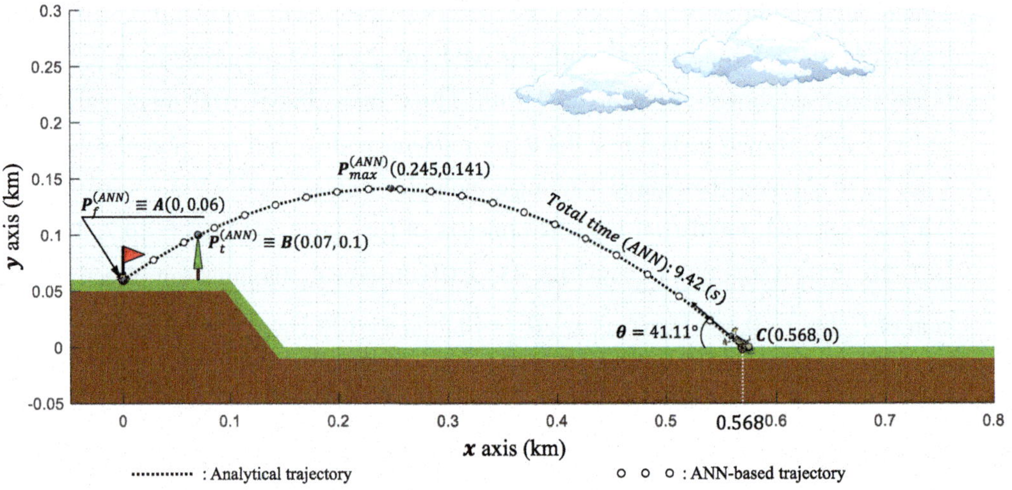

Figure 2.4.4.10 Projectile trajectory for Case I KKT condition; must satisfy $c(x_C, \theta) = y_{P_f} - 0.06 = 0$, $v(x_C, \theta) = y_{P_t} - 0.1 = 0$.

reaching a maximum of $f_{xC}(x_C)$, i.e., inequality $v(x_C, \theta) = y_{P_t} - 0.1 \geq 0$ is not bound to equality constraint, but ignored. Diagonal scalar matrix **S** becomes 0 to inactivate inequality constraint in which inequality constraint are not bound to an equality constraint but considered as slack condition. However, stationary points of Lagrange function of Case 2 KKT condition should be verified to exist within the range defined by an inequality constraint function $v(x_C, \theta) = y_{P_t} - 0.1 \geq 0$. Lagrange function $\mathcal{L}(x_C, \theta, \lambda_c, \lambda_v)$ is obtained in Equation (2.4.4.61), ignoring an inequality constraint, but including an equality constraint.

$$\mathcal{L}(x_C, \theta, \lambda_c, \lambda_v) = x_C - \lambda_c\left(y_{P_f} - 0.06\right) - 0 \times \lambda_v\left(y_{P_t} - 0.1\right) \tag{2.4.4.61}$$

ANNs for an objective function $f_{xC}(x_C)$, equality $c\left(x_C^{(0)}, \theta^{(0)}\right)$, and inequality function $v\left(x_C^{(0)}, \theta^{(0)}\right)$ with a composite notation is shown in Equation (2.4.4.36) based on Equations (2.3.6.3-1) and (2.3.6.3-2). An initial trial input parameter of $\mathbf{x}^{(0)} = \left[x_C^{(0)}, \theta^{(0)}\right]^T = [0.7, 40]^T$ and initial Lagrange multipliers of $\lambda_c^{(0)} = -1$ and $\lambda_v^{(0)} = 0$ shown in red input of Equation (2.4.4.62-2) are used to solve Case 2 KKT condition using Newton–Raphson method. The total initial trial input parameter becomes $\left[x_C^{(0)}, \theta^{(0)}, \lambda_c^{(0)}, \lambda_v^{(0)}\right]^T = [0.7, 40, -1, 0]^T$. Objective function $f_{xC}(x_C)$, equality $c\left(x_C^{(0)}, \theta^{(0)}\right)$, and inequality function $v\left(x_C^{(0)}, \theta^{(0)}\right)$ are calculated in Equations (2.4.4.1), (2.4.4.49-3a), and (2.4.4.49-3b), respectively. For initial trial input parameter $\mathbf{x}^{(0)} = x_C^{(0)}, \theta^{(0)T} = [0.7, 40]^T$, Jacobi matrix $\mathbf{J}_c^{(D)}\left(x_C^{(0)}, \theta^{(0)}\right)$ and $\mathbf{J}_v^{(D)}\left(x_C^{(0)}, \theta^{(0)}\right)$ are calculated in Equations (2.4.4.50-1) and (2.4.4.50-2) whereas Hessian matrix $\mathbf{H}_c^{(D)}\left(x_C^{(0)}, \theta^{(0)}\right)$ and $\mathbf{H}_v^{(D)}\left(x_C^{(0)}, \theta^{(0)}\right)$ are obtained in Equations (2.4.4.51-1) and (2.4.4.51-2), respectively. A first derivative (Jacobi) of Lagrange function is calculated in Equation (2.4.4.62-2) with respect to initial trial input parameter $x^{(0)} = \left[x_C^{(0)}, \theta^{(0)}\right]^T = [0.7,\ 40]^T$ by substituting Equations (2.4.4.49-3a), (2.4.4.49-3b), (2.4.4.50-1), (2.4.4.50-2), and $s = 0$ into Equation (2.4.4.62-1b).

$$\nabla \mathrm{L}\left(x_C^{(0)}, \theta^{(0)}, \lambda_c^{(0)}, \lambda_v^{(0)}\right)$$

$$= \begin{bmatrix} \nabla f_{xC}\left(x_C^{(0)}\right) - \left[\mathbf{J}_c\left(x_C^{(0)}, \theta^{(0)}\right)\right]^T \lambda_c - \left[\mathbf{J}_v\left(x_C^{(0)}, \theta^{(0)}\right)\right]^T s\lambda_v \\ -c\left(x_C^{(0)}, \theta^{(0)}\right) \\ -sv\left(x_C^{(0)}, \theta^{(0)}\right) \end{bmatrix} \tag{2.4.4.62-1a}$$

$$= \begin{bmatrix} \nabla f_{xC}\left(x_C^{(0)}\right) - \left[\mathbf{J}_c^{(D)}\left(x_C^{(0)}, \theta^{(0)}\right)\right]^T \lambda_c - \left[\mathbf{J}_v^{(D)}\left(x_C^{(0)}, \theta^{(0)}\right)\right]^T s\lambda_v \\ -c\left(x_C^{(0)}, \theta^{(0)}\right) \\ -sv\left(x_C^{(0)}, \theta^{(0)}\right) \end{bmatrix} \tag{2.4.4.62-1b}$$

$$\nabla\mathcal{L}\left(x_C^{(0)},\theta^{(0)},\lambda_c^{(0)},\lambda_v^{(0)}\right)=\begin{bmatrix}\begin{bmatrix}1\\0\end{bmatrix}-\begin{bmatrix}-0.9902\\0.0021\end{bmatrix}\times(-1)-\begin{bmatrix}-0.8075\\0.0036\end{bmatrix}\times 0\times 0\\0.1126\\0\times 0.0895\\ \\ \end{bmatrix}$$

$$\rightarrow\nabla\mathcal{L}\left(x_C^{(0)},\theta^{(0)},\lambda_c^{(0)},\lambda_v^{(0)}\right)=\begin{bmatrix}0.0098\\0.0021\\0.1126\\0\end{bmatrix}\neq 0 \tag{2.4.4.62-2}$$

b. Convergence verification of Lagrange function based on updated input parameters at Iteration 1

One of Lagrange multipliers $\boldsymbol{\lambda}_v^{(0)}=\left[\lambda_{v1}^{(0)},\ \lambda_{v2}^{(0)}\right]$ should be non-zero shown in Equations (2.4.4.65) and (2.4.4.66) to avoid a singularity of Hessian matrix $\mathbf{H}_{\mathcal{L}}\left(x_C^{(0)},\theta^{(0)},\lambda_c^{(0)},\lambda_v^{(0)}\right)$ because $\boldsymbol{H}_{\mathcal{L}}\left(x_C^{(0)},\theta^{(0)}\right)$ becomes singular, leading to a singular global Hessian matrix $\mathbf{H}_{\mathcal{L}}\left(x_C^{(0)},\theta^{(0)},\lambda_c^{(0)},\lambda_v^{(0)}\right)$ of Case 2 KKT when all zero Lagrange multipliers $\boldsymbol{\lambda}_v^{(0)}=\left[\lambda_{v1}^{(0)},\ \lambda_{v2}^{(0)}\right]=[0,\ 0]$ are selected with S matrix being 0 for an inactive condition, thus causing iteration diverges from the first iteration regardless a selection of initial trial input parameters $\mathbf{x}^{(0)}=\left[x_C^{(0)},\ \theta^{(0)}\right]$. Hessian matrix of an objective function becomes also zero when being differentiated twice because an objective function is linear, $f_{xC}(x_C)=x_c$, as shown in Equations (2.4.4.1) and (2.4.4.14). Thus, non-zero Lagrange multiplier of $\lambda_c^{(0)}$ shown in red color of Equation (2.4.4.66) should be selected to avoid a singularity of Hessian matrix $\mathbf{H}_{\mathcal{L}}\left(x_C^{(0)},\theta^{(0)},\lambda_c^{(0)},\lambda_v^{(0)}\right)$.

A first derivative (Jacobi) of Lagrange function obtained in Equation (2.4.4.62-2) with respect to initial trial input parameter $x^{(0)}=\left[x_C^{(0)},\ \theta^{(0)}\right]^T=[0.7,\ 40]^T$ does not converge to the solutions, indicating that $\left[x_C^{(0)},\theta^{(0)},\lambda_c^{(0)},\lambda_v^{(0)}\right]^T=[0.7,40,-1,0]^T$ is not the root of Case 2 KKT condition. An initial trial input parameter is updated based on Equation (2.4.4.63) to see if a first derivative (Jacobi) of Lagrange function converges in Iteration 1. It is noted that Lagrange multipliers are substituted in Equation (2.4.4.62-1b), not in Equation (2.4.4.40-1) when updating initial trial input parameter shown in Equation (2.4.4.63). In Equation (2.4.4.67-2), Hessian matrix $\mathbf{H}_{\mathcal{L}}\left(x_C^{(0)},\theta^{(0)},\lambda_c^{(0)},\lambda_v^{(0)}\right)$ is calculated to update initial trial input parameters $\left[x_C^{(0)},\theta^{(0)},\lambda_c^{(0)},\lambda_v^{(0)}\right]=[0.7,\ 40,\ -1,\ 0],[0.6,\ 30,\ -1,\ 0],\text{and}[0.8,\ 66,\ -1,\ 0]$ shown in Equation (2.4.4.63) to $\left[x_C^{(1)},\theta^{(1)},\lambda_c^{(1)},\lambda_v^{(1)}\right]$ shown in Equation (2.4.4.68). Hessian matrix $\mathbf{H}_{\mathcal{L}}\left(x_C^{(0)},\theta^{(0)},\lambda_c^{(0)},\lambda_v^{(0)}\right)$ is calculated in Equation (2.4.4.67-2) by substituting $\boldsymbol{H}_{\mathcal{L}}\left(x_C^{(0)},\theta^{(0)}\right)$, $\mathbf{J}_c^{(D)}\left(x_C^{(0)},\theta^{(0)}\right)$, and $\mathbf{J}_v^{(D)}\left(x_C^{(0)},\theta^{(0)}\right)$ from Equations (2.4.4.66), (2.4.4.50–1), (2.4.4.50–2), respectively, and $s=0$ into Equation (2.4.4.64-2). $\mathbf{H}_{\mathcal{L}}\left(x_C^{(0)},\theta^{(0)}\right)$ is calculated in Equation (2.4.4.66) by substituting $\mathbf{H}_{f_{xC}}(x_C,\theta)$, $\mathbf{H}_c^{(D)}\left(x_C^{(0)},\theta^{(0)}\right)$, and $\mathbf{H}_v^{(D)}\left(x_C^{(0)},\theta^{(0)}\right)$ from Equations (2.4.4.14), (2.4.4.51-1), and (2.4.4.51-2), respectively, into Equation (2.4.4.65-2). Hessian matrix $\mathbf{H}_{\mathcal{L}}\left(x_C^{(0)},\theta^{(0)},\lambda_c^{(0)},\lambda_v^{(0)}\right)$ is, then, inverted as $\left[\mathbf{H}_{\mathcal{L}}\left(x_C^{(0)},\theta^{(0)},\lambda_c^{(0)},\lambda_v^{(0)}\right)\right]^{-1}$ in Equation (2.4.4.67-4).

$$\begin{bmatrix} x_C^{(1)} \\ \theta^{(1)} \\ \lambda_c^{(1)} \\ \lambda_v^{(1)} \end{bmatrix} = \begin{bmatrix} x_C^{(0)} \\ \theta^{(0)} \\ \lambda_c^{(0)} \\ \lambda_v^{(0)} \end{bmatrix} - \left[\mathbf{H}_{\mathcal{L}}\left(x_C^{(0)}, \theta^{(0)}, \lambda_c^{(0)}, \lambda_v^{(0)}\right)\right]^{-1} \nabla\mathcal{L}\left(x_C^{(0)}, \theta^{(0)}, \lambda_c^{(0)}, \lambda_v^{(0)}\right) \quad (2.4.4.63)$$

$$\mathbf{H}_{\mathcal{L}}\left(x_C^{(0)}, \theta^{(0)}, \lambda_c^{(0)}, \lambda_v^{(0)}\right) = \begin{bmatrix} \mathbf{H}_{\mathcal{L}}\left(x_C^{(0)}, \theta^{(0)}\right) & -\left[\mathbf{J}_c\left(x_C^{(0)}, \theta^{(0)}\right)\right]^T & -\left[s\mathbf{J}_v\left(x_C^{(0)}, \theta^{(0)}\right)\right]^T \\ -\mathbf{J}_c\left(x_C^{(0)}, \theta^{(0)}\right) & 0 & 0 \\ -s\mathbf{J}_v\left(x_C^{(0)}, \theta^{(0)}\right) & 0 & 0 \end{bmatrix}$$

(2.4.4.64-1)

$$\mathbf{H}_{\mathcal{L}}\left(x_C^{(0)}, \theta^{(0)}, \lambda_c^{(0)}, \lambda_v^{(0)}\right)$$

$$= \begin{bmatrix} \mathbf{H}_{\mathcal{L}}\left(x_C^{(0)},\theta^{(0)}\right) & -\left[\mathbf{J}_c^{(D)}\left(x_C^{(0)},\theta^{(0)}\right)\right]^T & -\left[s\mathbf{J}_v^{(D)}\left(x_C^{(0)},\theta^{(0)}\right)\right]^T \\ -\mathbf{J}_c^{(D)}\left(x_C^{(0)},\theta^{(0)}\right) & 0 & 0 \\ -s\mathbf{J}_v^{(D)}\left(x_C^{(0)},\theta^{(0)}\right) & 0 & 0 \end{bmatrix} \quad (2.4.4.64-2)$$

$$\mathbf{H}_{\mathcal{L}}\left(x_C^{(0)}, \theta^{(0)}\right) = \mathbf{H}_{f_{xC}}\left(x_C^{(0)}, \theta^{(0)}\right) - \lambda_c^{(0)}\mathbf{H}_c\left(x_C^{(0)}, \theta^{(0)}\right) - s\lambda_v^{(0)}\mathbf{H}_v\left(x_C^{(0)}, \theta^{(0)}\right) \quad (2.4.4.65-1)$$

$$\mathbf{H}_{\mathcal{L}}\left(x_C^{(0)}, \theta^{(0)}\right) = \mathbf{H}_{f_{xC}}\left(x_C^{(0)}, \theta^{(0)}\right) - \lambda_c^{(0)}\mathbf{H}_c^{(D)}\left(x_C^{(0)}, \theta^{(0)}\right) - s\lambda_v^{(0)}\mathbf{H}_v^{(D)}\left(x_C^{(0)}, \theta^{(0)}\right) \quad (2.4.4.65-2)$$

$$\mathbf{H}_{\mathcal{L}}\left(x_C^{(0)}, \theta^{(0)}\right) = \begin{bmatrix} 0 & 0 \\ 0 & 0 \end{bmatrix} - (-1)\times\begin{bmatrix} -2.6171 & -0.0239 \\ -0.0239 & -0.0006 \end{bmatrix}$$

$$-0\times 0\times\begin{bmatrix} -2.6102 & -0.0186 \\ -0.0186 & -0.0004 \end{bmatrix} = \begin{bmatrix} -2.6171 & -0.0239 \\ -0.0239 & -0.0006 \end{bmatrix}$$

(2.4.4.66)

(CASE 1 (Inequality $v(\mathbf{x})$ is active) does not have this problem. Thus, $\mathbf{H}_{\mathcal{L}}\left(x_C^{(0)},\theta^{(0)},\lambda_c^{(0)},\lambda_v^{(0)}\right)$ of Case 1 is non-singular even $\lambda_c^{(0)} = 0$)

$$\mathbf{H}_{\mathcal{L}}\left(x_C^{(0)}, \theta^{(0)}, \lambda_c^{(0)}, \lambda_v^{(0)}\right)$$

$$= \begin{bmatrix} \begin{bmatrix} -2.6171 & -0.0239 \\ -0.0239 & -0.0006 \end{bmatrix} & -\begin{bmatrix} -0.9902 \\ 0.0021 \end{bmatrix} & -0\times\begin{bmatrix} -0.8075 \\ 0.0036 \end{bmatrix} \\ -\begin{bmatrix} -0.9902 & 0.0021 \end{bmatrix} & 0 & 0 \\ -0\times\begin{bmatrix} -0.8075 & 0.0036 \end{bmatrix} & 0 & 0 \end{bmatrix}$$

(2.4.4.67-1)

$$\mathbf{H}_{\mathcal{L}}\left(x_C^{(0)},\ \theta^{(0)},\ \lambda_c^{(0)},\ \lambda_v^{(0)}\right)=\begin{bmatrix} -2.6171 & -0.0239 & 0.9902 & 0 \\ -0.0239 & -0.0006 & -0.0021 & 0 \\ 0.9902 & -0.0021 & 0 & 0 \\ 0 & 0 & 0 & 0 \end{bmatrix} \tag{2.4.4.67-2}$$

$$\left[\mathbf{H}_{\mathcal{L}}\left(x_C^{(0)},\ \theta^{(0)},\ \lambda_c^{(0)},\ \lambda_v^{(0)}\right)\right]^{-1}=\begin{bmatrix} \begin{bmatrix} -2.6171 & -0.0239 & 0.9902 \\ -0.0239 & -0.0006 & -0.0021 \\ 0.9902 & -0.0021 & 0 \end{bmatrix}^{-1} & \begin{matrix} 0 \\ 0 \\ 0 \end{matrix} \\ \begin{matrix} 0 & \quad 0 & \quad 0 \end{matrix} & 0 \end{bmatrix} \tag{2.4.4.67-3}$$

$$\left[\mathbf{H}_{\mathcal{L}}\left(x_C^{(0)},\ \theta^{(0)},\ \lambda_c^{(0)},\ \lambda_v^{(0)}\right)\right]^{-1}=\begin{bmatrix} -0.006 & -2.9038 & 0.9240 & 0 \\ -2.9038 & -1400.1 & -41.431 & 0 \\ 0.9240 & -41.431 & 1.4432 & 0 \\ 0 & 0 & 0 & 0 \end{bmatrix} \tag{2.4.4.67-4}$$

Updated input parameter $\left[x_C^{(1)},\ \theta^{(1)},\ \lambda_c^{(1)},\ \lambda_v^{(1)}\right]=[0.6020,\ 47.57,\ -1.086,\ 0]$ is obtained at the end of Iteration 0 in Equation (2.4.4.68) by substituting Equations (2.4.4.62-2), (2.4.4.67-4) into Equation (2.4.4.63). This input parameter will be used in Iteration 1. A first derivative (Jacobi) of Lagrange function $\nabla\mathcal{L}\left(x_C^{(1)},\ \theta^{(1)},\ \lambda_c^{(1)},\ \lambda_v^{(1)}\right)=[-0.0142,\ -0.0002,\ 0.0116,\ 0]$ with respect to the dated input parameter $\left[x_C^{(1)},\ \theta^{(1)},\ \lambda_c^{(1)},\ \lambda_v^{(1)}\right]$ shown in Equation (2.4.4.68) is obtained in Equations (2.4.4.69) or (A2.13) based on $c\left(x_C^{(1)},\ \theta^{(1)}\right)$, $v\left(x_C^{(1)},\ \theta^{(1)}\right)$, and $\mathbf{J}_c^{(D)}\left(x_C^{(1)},\ \theta^{(1)}\right)$, $\mathbf{J}_v^{(D)}\left(x_C^{(1)},\ \theta^{(1)}\right)$ for Iteration 1 calculated in Equations (A2.10-1)–(A2.11-2).

$$\begin{bmatrix} x_C^{(1)} \\ \theta^{(1)} \\ \lambda_c^{(1)} \\ \lambda_v^{(1)} \end{bmatrix}=\begin{bmatrix} 0.7 \\ 40 \\ -1 \\ 0 \end{bmatrix}-\begin{bmatrix} -0.006 & -2.9038 & 0.9240 & 0 \\ -2.9038 & -1400.1 & -41.431 & 0 \\ 0.9240 & -41.431 & 1.4432 & 0 \\ 0 & 0 & 0 & 0 \end{bmatrix}\begin{bmatrix} 0.0098 \\ 0.0021 \\ 0.1126 \\ 0 \end{bmatrix}$$

$$\rightarrow\begin{bmatrix} x_C^{(1)} \\ \theta^{(1)} \\ \lambda_c^{(1)} \\ \lambda_v^{(1)} \end{bmatrix}=\begin{bmatrix} 0.6020 \\ 47.57 \\ -1.086 \\ 0 \end{bmatrix} \tag{2.4.4.68}$$

$$\nabla\mathcal{L}\left(x_C^{(1)},\ \theta^{(1)},\ \lambda_c^{(1)},\ \lambda_v^{(1)}\right)=\begin{bmatrix} -0.0142 \\ -0.0002 \\ 0.0116 \\ 0 \end{bmatrix}\neq 0 \tag{2.4.4.69, A2.13-2}$$

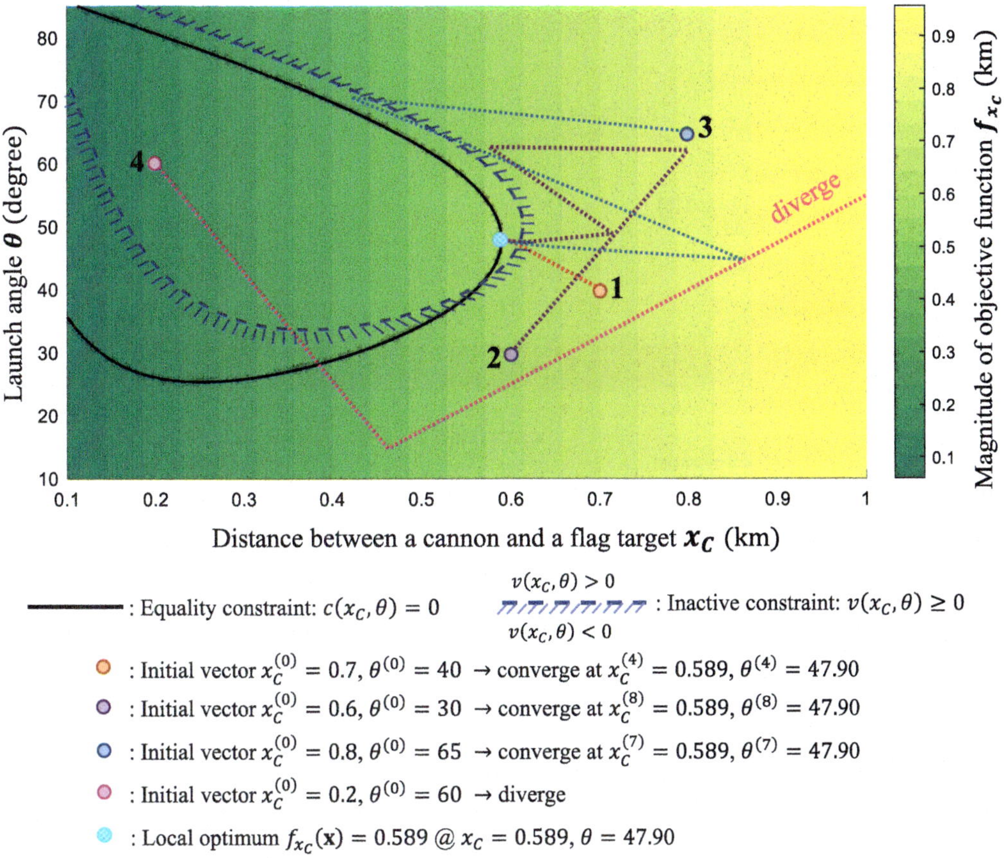

Figure 2.4.4.11 Convergence of initial trial input parameters based on ANN for Case 2 KKT condition; must satisfy $cc(x_C,\theta) = y_{P_f} - 0.06 = 0$, $v(x_C,\theta) = y_{P_t} - 0.1 \geq 0$.

Updated input parameter $\left[x_C^{(1)}, \theta^{(1)}, \lambda_c^{(1)}, \lambda_v^{(1)}\right] = [0.6020,\ 47.57,\ -1.086,\ 0]$ is used for Iteration 1 as shown in red input of Equation (A2.1-1). $\left[x_C^{(1)}, \theta^{(1)}, \lambda_c^{(1)}, \lambda_v^{(1)}\right]$ is updated as $\left[x_C^{(2)}, \theta^{(2)}, \lambda_c^{(2)}, \lambda_v^{(2)}\right]^T = [0.5895, 47.88, -1.106, 0]^T$ at the end of Iteration 1 as shown in Equation (A2.1.19). Updated input parameter $\left[x_C^{(2)}, \theta^{(2)}, \lambda_c^{(2)}, \lambda_v^{(2)}\right]$ will be used for all calculation in Iteration 2. A first derivative (Jacobi) of Lagrange function is updated at Iteration 2 as $[0.0009, 0.000, 0.0002, 0]$ as shown in Equation (2.4.4.70) which almost converges to a solution for Case 2 KKT condition. Calculation for Iteration 1 is presented in Appendix A2.

$$\nabla\mathcal{L}\left(x_C^{(2)}, \theta^{(2)}, \lambda_c^{(2)}, \lambda_v^{(2)}\right) = \begin{bmatrix} 0.0009 \\ 0.0000 \\ 0.0002 \\ 0 \end{bmatrix} \neq 0 \tag{2.4.4.70}$$

c. Convergence verification of Lagrange function at last iteration

Four initial trial input parameters of $\left[x_C^{(0)}, \theta^{(0)}\right]$ shown in Figure 2.4.4.11 are implemented to find optimal solution of Case 2 KKT

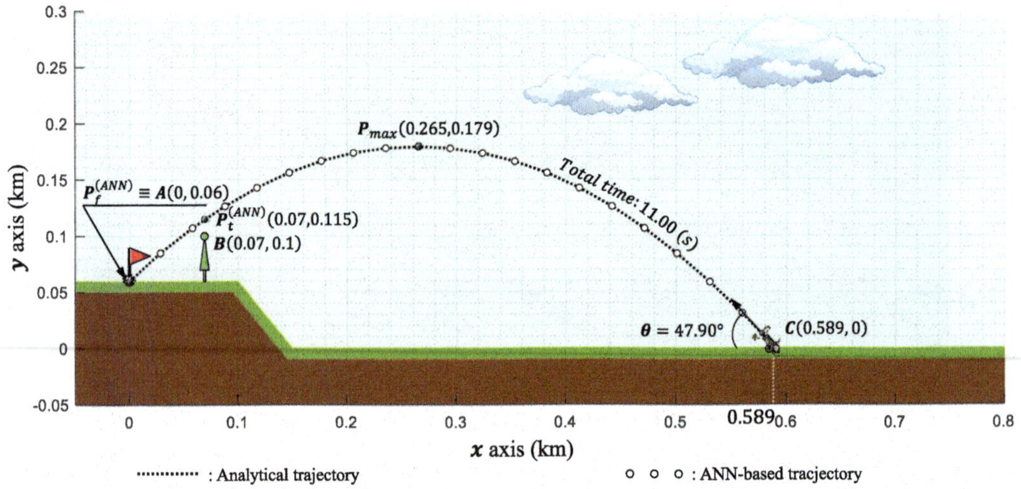

Figure 2.4.4.12 Projectile trajectory for Case 2 KKT condition; must satisfy $c(x_C,\theta) = y_{P_f} - 0.06 = 0$, $v(x_C,\theta) = y_{P_t} - 0.1 \geq 0$.

condition. It is noted that initial Lagrange multipliers of $\lambda_c^{(0)}$ and $\lambda_v^{(0)}$ are selected as −1 and 0, respectively. Finally, $\left[x_C^{(0)},\ \theta^{(0)},\ \lambda_c^{(0)},\ \lambda_v^{(0)}\right] = [0.7,\ 40,\ -1,\ 0]$ converged to $\left[x_C^{(4)}, \theta^{(4)}, \lambda_c^{(4)}, \lambda_v^{(4)}\right] = [0.589,\ 47.90,\ -1.1072,\ 0]$ at Iteration 4 while a first derivative (Jacobi) of Lagrange function $\nabla\mathcal{L}(x_C, \theta, \lambda_c, \lambda_v)$ converged to zero based on with an MSE of 2.0E-20, maximizing an objective function a $f_{xC}(x_C) = 0.589$ (km) as shown in Figures 2.4.4.11 and 2.4.4.12. Three selected initial trial input parameters of $\left[x_C^{(0)},\ \theta^{(0)},\ \lambda_c^{(0)},\ \lambda_v^{(0)}\right] = [0.7,\ 40,\ -1,\ 0]$, $[0.6,\ 30,\ -1,\ 0]$, and $[0.8,\ 66,\ -1,\ 0]$ converge to the same maximized flying distance represented by $f_{xC}(x_C, \theta) = 0.589$ (km) at $[x_C, \theta] = [0.589,\ 47.90]$ at Iterations 4, 8, and 7, respectively, whereas the fourth initial trial input parameter of $\left[x_C^{(0)},\ \theta^{(0)},\ \lambda_c^{(0)},\ \lambda_v^{(0)}\right] = [0.2,\ 60,\ -1,\ 0]$ diverges. An ordinate on the right side of Figure 2.4.4.11 represents color contours for a magnitude of an objective function $f_{xC}\left(x_C^{(k)}\right)$ in which minimized $f_{xC}\left(x_C^{(k)}\right) = 0.589$ (km) with an angle of $\theta^{(k)}$ 47.90° is shown. Figure 2.4.4.12 presents a maximum position of a projectile and its trajectory when an inequality constraint $v(x_C, \theta) = y_{P_t} - 0.1 \geq 0$ is inactive. ANN-based optimized trajectory agrees well with that obtained by an explicit function as shown in Table 2.4.4.5 where convergence history of Case 2 KKT condition similar to that of Case 1 condition is obtained. As shown in Figure 2.4.4.12 and Table 2.4.4.5, ANN-based Lagrange optimization is well compared with that based on analytical functions. Projectile trajectories from the two optimizations are almost identical.

d. Summary

Non-zero Lagrange multiplier of $\lambda_c^{(0)}$ shown in red color of Equation (2.4.4.66) was selected to avoid a singularity Hessian matrix $\mathbf{H}_{\mathcal{L}}\left(x_C^{(0)},\ \theta^{(0)},\ \lambda_c^{(0)},\ \lambda_v^{(0)}\right)$ when S matrix is 0 for an inactive condition. Three initial trial input parameters $\left[x_C^{(0)},\ \theta^{(0)},\ \lambda_c^{(0)},\ \lambda_v^{(0)}\right] = [0.7,\ 40,\ -1,\ 0]$, $[0.6,\ 30,\ -1,\ 0]$, and $[0.8,\ 66,\ -1,\ 0]$ converge on sky point $\left[x_C^{(k)},\ \theta^{(k)},\ \lambda_c^{(k)},\ \lambda_v^{(k)}\right] = [0.589,\ 47.90,\ -1.1072,\ 0]$, satisfying $\nabla\mathcal{L}(x_C, \theta, \lambda_c, \lambda_v) = 0$ with MSE of 2.0E-20. Minimized $f_{xC}\left(x_C^{(k)}\right) = 0.589$ (km) with an angle of $\theta^{(k)}$ 47.90° was obtained. The black line indicates Equality constraint

Table 2.4.4.5 Verification of ANN-Based Lagrange Optimization

ANN-Based Lagrange Optimization (Based on TED with Four Layers and Ten Neurons)							
		Case 1: Inequality $v(x_C,\theta)$ is Active			*Case 2: Inequality $v(x_C,\theta)$ is Inactive*		
Parameter		*Check (Analytical)*	*Error*	*AI Results*	*Check (Analytical)*	*Error*	
x_1:	x_C (km)	0.568	0.568	0.00%	0.589	0.589	0.00%
x_2:	θ (degree)	41.11	41.11	0.00%	47.90	47.90	0.00%
y_1:	y @ $x = 0.95x_C$(km)	0.0237	0.0237	−0.02%	0.0311	0.0311	0.00%
y_2:	y @ $x = 0.90x_C$(km)	0.0452	0.0452	−0.01%	0.0593	0.0593	0.00%
y_3:	y @ $x = 0.85x_C$(km)	0.0646	0.0646	−0.01%	0.0845	0.0845	0.00%
y_4:	y @ $x = 0.80x_C$(km)	0.0817	0.0817	−0.01%	0.1067	0.1067	0.00%
y_5:	y @ $x = 0.75x_C$(km)	0.0967	0.0967	−0.02%	0.1260	0.1260	−0.01%
y_6:	y @ $x = 0.70x_C$(km)	0.1095	0.1095	−0.01%	0.1423	0.1423	−0.01%
y_7:	y @ $x = 0.65x_C$ (km)	0.1201	0.1202	−0.01%	0.1557	0.1557	−0.01%
y_8:	y @ $x = 0.60x_C$ (km)	0.1286	0.1286	0.00%	0.1661	0.1661	0.00%
y_9:	y @ $x = 0.55x_C$ (km)	0.1349	0.1349	0.00%	0.1735	0.1735	0.00%
y_{10}:	y @ $x = 0.50x_C$ (km)	0.1390	0.1390	0.00%	0.1780	0.1780	−0.01%
y_{11}:	y @ $x = 0.45x_C$(km)	0.1409	0.1409	0.00%	0.1795	0.1796	−0.01%
y_{12}:	y @ $x = 0.40x_C$(km)	0.1406	0.1406	−0.01%	0.1781	0.1781	−0.02%
y_{13}:	y @ $x = 0.35x_C$(km)	0.1381	0.1381	0.01%	0.1737	0.1737	−0.01%
y_{14}:	y @ $x = 0.30x_C$(km)	0.1335	0.1335	0.01%	0.1663	0.1664	−0.02%
y_{15}:	y @ $x = 0.25x_C$(km)	0.1267	0.1267	0.01%	0.1560	0.1560	−0.02%
y_{16}:	y @ $x = 0.20x_C$(km)	0.1177	0.1177	0.02%	0.1427	0.1428	−0.03%
y_{17}:	y @ $x = 0.15x_C$(km)	0.1065	0.1065	0.01%	0.1265	0.1265	−0.04%
y_{18}:	y @ $x = 0.10x_C$(km)	0.0932	0.0932	0.06%	0.1073	0.1073	−0.03%
y_{19}:	y @ $x = 0.05x_C$(km)	0.0777	0.0776	0.05%	0.0851	0.0852	−0.07%
y_{20}:	y_{Pf} @ $x = x_A = 0$ (km)	0.0600	0.0599	0.11%	0.0600	0.0601	−0.09%
y_{21}:	y_{Pt} @ $x = x_B = 0.07$(km)	0.1000	0.0996	0.36%	0.1149	0.1149	0.01%
y_{22}:	x @ y_{max} (km)	0.2451	0.2451	−0.01%	0.2651	0.2648	0.13%
y_{23}:	y_{max} (km)	0.1410	0.1410	−0.03%	0.1794	0.1796	−0.06%
y_{24}:	Total motion time @ $x = x_A$, (s)	9.425	9.428	−0.03%	10.988	10.986	0.01%

Note: (dotted box) Two inputs for AI design.

(dashed box) Two inputs for Analytical calculation.

$c\left(x_C^{(k)},\ \theta^{(k)}\right)$ whereas the blue line indicates Inequality constraint $v\left(x_C^{(k)},\ \theta^{(k)}\right)$ which is bound to equality constraint. In Figure 2.4.4.11, three initial trial input parameters indicated by 1, 2, and 3 converged on sky point [0.589, 47.90, −1.1072, 0] which are located on black and the opposite side of the blue curves, indicating an objective function is optimized within the range defined inequality constraint $v(x_C,\ \theta\ ) = y_{P_t} - 0.1 \geq 0$ for Case 2 KKT condition when inequality $v(x_C,\ \theta\ ) = y_{P_t} - 0.1 \geq 0$ is inactive. A root of Case 1 with $0.569(\text{km})$ and Case 2 with $0.589(\text{km})$ for KKT conditions are compared to show that Case 2 KKT condition maximizes projectile distance. Flying distance is maximized with 0.589(km) and an angle of $\theta^{(k)}$ 47.90° under Case 2 KKT condition based on both ANNs and explicit functions.

REFERENCES

Hoffmann, L. D., & G. L. Bradley. (2004). *Calculus for Business, Economics, and the Social and Life Sciences* (8th ed., pp. 575–588). McGraw Hill Education, New York.

Hong, W.K. (2021). *Artificial Intelligence-Based Design of Reinforced Concrete Structures.* Daega Publisher, Gyeonggi, Korea.

Horn, R. A. (1990). The hadamard product. In *Proceedings of Symposia in Applied Mathematics* (Vol. 40, 87–169).

I. Newton. Philosophiae naturalis principia mathematica. London, 1687.

Krenker A, Bešter J, Kos A. Introduction to the artificial neural networks. Artificial Neural Networks: Methodological Advances and Biomedical Applications. InTech, 2011; 1-18. https://doi.org/10.5772/15751.

Kuhn, H. W., & Tucker, A. W. (2014). Traces and emergence of nonlinear programming. In G. Giorgi, & T. Kjeldsen, (Eds.), *Nonlinear Programming.* Birkhäuser, Basel. https://doi.org/10.1007/978-3-0348-0439-4_11

Lagrange, J. L. (1804). *Leçons sur le calcul des fonctions.* Imperiale, Paris, France.

MATLAB. (2022a). *MathWorks Technical Support Department.* The MathWorks Inc., Natick, MA.

MATLAB. (2022b). Hessian matrix of scalar function. The MathWorks Inc., Natick, MA. https://uk.mathworks.com/help/symbolic/sym.hessian.html (accessed May 20, 2022).

MATLAB. (2022c). Jacobian matrix. The MathWorks Inc., Natick, MA. https://uk.mathworks.com/help/symbolic/sym.ja

Protter, M. H., & C. B. Morrey Jr. (1985). *Intermediate Calculus* (2nd ed.). Springer, New York.

Shariat, M., Shariati, M., Madadi, A. & Wakil. K. (2018). Computational Lagrangian multiplier method by using for optimization and sensitivity analysis of rectangular reinforced concrete beams. *Steel & Composite Structures, 29*(2), 243–256. https://doi.org/10.12989/scs.2018.29.2.243.

Silberberg, E., & S. Wing. (2001). *The Structure of Economics: A Mathematical Analysis* (3rd ed., pp. 134–141). Irwin McGraw-Hill, Boston.

Tahouni, S. (2005). *Designing Concrete Structures Based on Iranian Concrete Code.* University of Tehran, Tehran, Iran.

Taylor, B. (1717). *Methodus incrementorum directa et inversa.* Innys, London, UK.

Villarrubia, G., De Paz, J. F., Chamoso, P., & De la Prieta, F. (2018). Artificial neural networks used in optimization problems. *Neurocomputing, 272,* 10–16. https://doi.org/10.1016/j.neucom.2017.04.075.

Chapter 3

Design of reinforced concrete columns using ANN-based Lagrange algorithm

3.1 INTRODUCTION

3.1.1 Overview of Lagrange multiplier method-based KKT conditions

Structural engineers often face several code-restricted design requirements. Codes impose many conditions and requirements for designs of reinforced concrete structures including columns and beams. However, it is difficult to intuitively optimize their designs while satisfying all code requirements simultaneously. Engineers commonly, thus, make design decisions based on their empirical observations. This chapter introduces ANN-based Lagrange algorithm techniques that can be employed to help engineers make more rational engineering decisions, resulting in designs that can meet various code restrictions simultaneously.

Lagrange multiplier method (LMM) optimizes objective functions subject to one or more constraining conditions that have to be met, identifying stationary points of the Lagrange function, as mentioned by (Walsh, 1976, Kalman, 2009). An objective and constraining function must be formed as a function of design input variables and a Lagrange multiplier λ (Protter and Morrey, 1985) to find the stationary points of a Lagrange function. The constrained global or local maxima or minima over the domain of the variables and over the multipliers of the Lagrange function are found at stationary points. Global stationary points are identified among local stationary points based on the Jacobi (which are differentiated once) and Hessian matrix (which are differentiated twice) (Silberberg, 1978) when using Newton–Raphson iteration.

Lagrange optimization techniques subject to inequality and equality constraints are implemented in minimizing or maximizing Lagrange functions (Hoffmann et al., 1989). In this chapter, KKT conditions are used to consider inequality constraints, for designs of reinforced concrete columns that meet various code restrictions at the same time. Candidate stationary points under KKT conditions can be found by solving systems of nonlinear differential equations under strict constraints imposed by design codes. This is why KKT theorem is sometimes referred to as the stationary-point theorem (Hoffmann et al., 1989).

In this chapter, objective functions, such as column cost index CI_c, CO_2 emissions, and weight of column W_c are obtained based on Equations (3.2.2.5-1)–(3.2.2.5-3), respectively, as function of seven design input parameters $\left(b,\ h,\ \rho_s, f_c', f_y,\ P_u,\ M_u\right)$ shown in Figure 3.2.2.1. However, it is difficult to explicitly express objective functions as a function of design variables to use derivative methods, such as Lagrange multipliers. This chapter proposes the use of ANNs to approximate objective functions, constraining functions and other output parameters as a function of input variables for operating Jacobian and Hessian matrices so that stationary points of the Lagrange functions are identified.

DOI: 10.1201/9781003314684-3

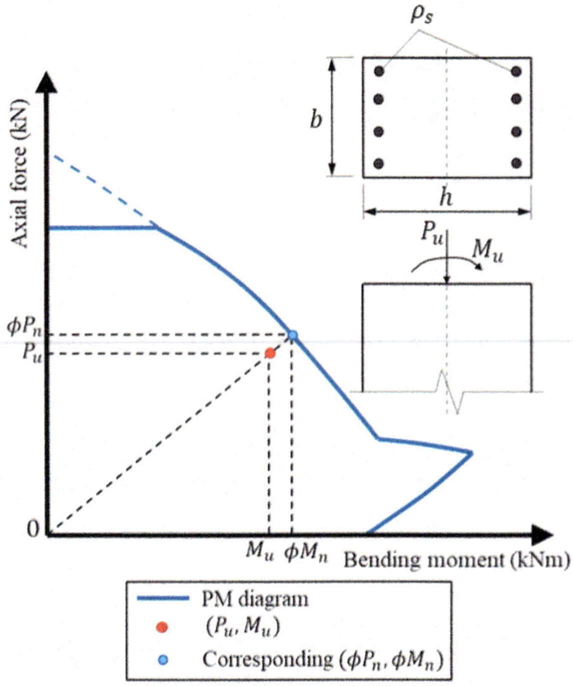

Figure 3.2.2.1 RC column design based on ANNs.

3.1.2 Optimization implemented in structural engineering

Li and Zeng (2008) proposed a neural network algorithm for solving sets of nonlinear equations. The computation employs a simple gradient descent rule with a variable step size. However, the algorithm does not provide accurate solutions when inequality was involved, even if acceptable solutions were obtained in this study. Several studies, including (Aghaee et al., 2015, Fanaie et al., 2016, Madadi et al., 2018, Nasrollahi et al., 2018, and Paknahad et al., 2018), have been conducted to optimize reinforced concrete (RC) structures. These studies mainly focused on minimizing the manufacturing and construction costs; only a few considered structural capabilities against external forces, which are influenced by design codes. Studies on the optimization of RC beams have been reported (Shariati et al., 2010, Fanaie et al., 2012, Toghroli et al., 2014, Awal et al., 2015, Kaveh and Shokohi, 2015, Safa et al., 2016, Shah et al., 2016, Korouzhdeh et al., 2017, and Heydari and Shariati, 2018). Shariat et al. (2018) derived an objective function representing an optimum reinforcement ratio and optimum effective depth of RC beams based on BS (Standard, 1985), ACI (ACI Committee, 2014), and ICS (Tahouni, 2005) regulations, which are presented in Tables 2–4 of their paper, respectively. For example, the procedure for calculating Lagrange multipliers for singly and doubly reinforcement beams according to the ACI code is given in Equations (18)–(21) of their paper. Barros et al. (2005) presented stress–strain diagrams using the LMM to develop nominal moment strengths based on the optimal area of the upper and lower sections of steels for four classes of concrete. Shariat et al. (2018) obtained analytical objective functions for the cost of frames as a function of the design parameters for structural systems in a limited circumstance; however, it is generally very difficult to derive analytical objective functions that represent the entire behavior of structural components such as columns and beams. Villarrubia et al. (2018) employed artificial neural networks (ANNs) to approximate objective functions to optimize analytical objective functions. They approximated objective functions using nonlinear regression that can optimize problems and also using a multilayer perceptron when the use of linear programming or Lagrange multipliers was not feasible.

3.1.3 Significance of the chapter

There are numerous available computer-aided engineering tools, including CAD packages, FEM software, and self-written calculation codes, that are used to study the performance of structures. Objective functions, however, could constitute a mixture of numerical simulations, analytical calculations, and catalog selections, which makes it difficult to apply differentiation to derivative optimization methods, such as a method using Lagrange multipliers. Some non-derivative optimization methods such as Genetic algorithms were applied to structural design problems (Pham and Hong, 2022, Coello Coello et al., 1997, Malasri et al., 1994) as they do not require any derivatives to find an optimal solution. However, computing times of non-derivative methods heavily rely on the computational speed because each trial requires one run of software. In this chapter, the use of artificial neural networks is adopted to universally approximate objective functions. New objective functions, hence, not only enhance the computational speed compared to conventional software but also can be differentiated and implemented in Lagrange optimization. In this chapter, the method of Lagrange multipliers is proposed to identify local maxima and minima of objective functions such as column cost index CI_c, CO_2 emissions, and weight of column W for designs of reinforced concrete columns with rectangular shape subject to one or more constraints. Objective and constraining functions shown in concrete column designs are difficult to be explicitly derived, thus, conventional Lagrange optimizations are also difficult to apply to concrete column designs.

ANN-based networks are used to formulate not only objective functions shown in Equations (3.2.2.5-1)–(3.2.2.5-3), but constraining functions shown in Table 3.2.2.5. The design targets called objective functions are, then, optimized while satisfying constraints as specified by design conditions. Column design scenarios based on forward and reverse designs are shown in Table 3.2.2.4. Optimized column designs based on Table 3.2.2.4 are not possible based on the conventional structural designs.

3.2 ANN-BASED ON LAGRANGE NETWORKS

The reverse LMM is convenient and has good computing speed; however, it may pose conflicts among variables when preassigned as inputs. Conversely, the forward LMM seeks optimized solutions in outputs using feed-forward networks; therefore, it does not pose any cross-relationship conflicts with computational speed lower than that of the reverse LMM. This chapter aims to present objective functions obtained based on forward and reverse artificial neural networks. In this chapter, KKT equations (Karush–Kuhn–Tucker conditions) were implemented, which account for inequality constraints, leading to optimized designs of reinforced concrete columns that meet various design requirements, including code restrictions and/or architectural criteria, while optimizing objective parameters such as cost index, CO_2 emissions, or structural weight.

3.2.1 Obtaining minimum design parameters for reinforced concrete columns based on ACI 318-19 and ACI 318-19

In this chapter, ANNs are used to derive objective shown in Equations (3.2.2.5-1) to (3.2.2.5-3) and constraining functions based on large datasets that can be generated from any computer-aided engineering tools. Optimization using ANN-based objective functions based on large datasets can effectively aid the selection of design parameters for the easy, rapid, but fast for the best engineering applications without a need for explicit objective functions. New ANN-based functions, hence, are differentiated for use in Lagrange optimization. Complex derivations for analytical objective functions, such as a cost and emissions of CO_2 of structural frame are replaced by ANNs-based objective functions.

All these engineering tools are based on forward analysis, in which the required input parameters must be provided to yield output parameters. However, optimization and sensitivity analyses using LMM shown in Equation (3.2.2.6-1) introduced in this chapter are based on both forward and reverse ANNs. The optimized designs are performed and verified using both rectangular and round RC columns. The design of reinforced concrete columns is minimized to obtain design parameters as specified by the American Concrete Institute [ACI 318-14 (ACI Committee 2014) and ACI 318-19 (ACI Committee 2019)] regulations. Various failure criterion can be implemented in the design of RC columns. The ANN-based LMM leads to finding the design parameters that will be the stationary points minimizing objective function subject to the constraining functions for specific design restrictions and requirements.

Solutions of the first derivatives of the Lagrange functions shown in Equation (3.2.2.6-1) are stationary points of Lagrange functions. Stationary points are found based on Newton–Raphson iteration shown in Equation (3.2.2.7-3).

Numerical examples are also presented to better illustrate optimizing design steps. Minimized objective functions including cost with respect to effective parameters including rebar ratios and failure criteria of the reinforced concrete columns are verified by large datasets shown in Figures 3.3.1.2, 3.3.2.2, and 3.3.3.2.

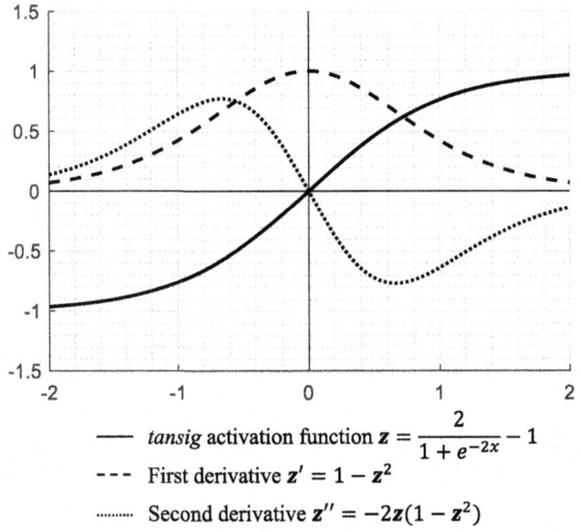

Figure 3.2.2.2 *tansig* activation function and its derivatives.

3.2.2 ANN-based functions including objective functions of RC columns

3.2.2.1 *Weight and bias matrices based on forward ANNs to derive objective functions*

A conventional structural software (*AutoCol*) is used to generate large datasets for training ANNs. ANN-based design capacity for axial force (ϕP_n), moment (ϕM_n) of rectangular RC columns, and rebar strains (ε_s), etc., are evaluated against factored load pair (P_u–M_u) for a given column section ($b \times h$), rebar ratio (ρ_s), and material properties. As shown in Table 3.2.2.1, a total of 100,000 datasets is generated to train ANN for a column design. Tables 3.2.2.1(b-1) and (c-1) present non-normalized large datasets based on ACI 318-14 and ACI 318-19, respectively, whereas large datasets are normalized based on ACI 318-14 and ACI 318-19, respectively, in Tables 3.2.2.1(b-2) and (c-2) where they are normalized between -1 and +1. Minimum and maximum ranges with the means of non-normalized and normalized data are shown below the tables. Large datasets of nine output parameters $\left(\phi P_n, \phi M_n, SF, b/h, \varepsilon_s, CI_c, \mathrm{CO_2}, W_c, \alpha_{e/h}\right)$ are generated for given seven input parameters $\left(b, h, \rho_s, f_c', f_y, P_u, M_u\right)$ as shown in Figure 3.2.2.1 and Table 3.2.2.1.

Nine output parameters including objective functions (CI_c, CO_2, W_c) are functions of seven input variables. The objective functions (CI_c, CO_2, W_c) are the functions that are complex and difficult to derive explicitly. It is also difficult to calculate their Jacobian and Hessian matrices which are necessary to use Newton–Raphson method to identify a stationary point of Lagrange function shown in Equation (3.2.2.1).

- Lagrange function:

$$\mathcal{L}(\mathrm{x}, \lambda_c, \lambda_v) = f(\mathrm{x}) - \lambda_c^T c(\mathrm{x}) - \lambda_v^T v(\mathrm{x}) \tag{3.2.2.1}$$

To overcome difficulties in solving constrained optimization problems for objective functions CI_c, CO_2 emissions, and W_c based on analytically derived functions, objective functions are generalized based on ANN-based forward and reverse neural networks as shown in Equation (2.3.6.3) of Chapter 2 which are repeated in Equations (3.2.2.2)–(3.2.2.4) in which weight and bias matrices should be used. This section is devoted to determining these matrices.

$$f(\mathrm{x}) = \mathbf{g}^D\left(f_{\mathrm{lin}}^L\left(\mathbf{W}^L f_t^{L-1}\left(\mathbf{W}^{L-1} \ldots f_t^1\left(\mathbf{W}^1 \mathbf{g}^N(\mathrm{x}) + \mathbf{b}^1\right) \ldots + \mathbf{b}^{L-1}\right) + b^L\right)\right) \tag{3.2.2.2}$$

Table 3.2.2.1 Summary of Design Parameters and Large Datasets for Reinforced Concrete Columns

(a): Summary of Reinforced Concrete Column Parameters

RC Column Parameters

Seven Inputs for Structural Mechanics		*Nine Corresponding Outputs*	
b	- Column section width	ϕP_n	- Design axial force
h	- Column section height	ϕM_n	- Design moment strength
ρ_s	- Geometric rebar ratio	SF	- Safety factor (ϕM_n / M_u)
f_c'	- Concrete strength	b/h	- Aspect ratio of column section
f_y	- Rebar yield strength	ε_s	- Rebar strain
P_u	- Factored axial force	CI_c	- Cost index of column per height
M_u	- Factored bending moment	CO_2	- CO_2 emission per column height
		W_c	- Weight of column per column height
		$\alpha_{e/h}$	- Angular eccentricity of applied load

(b-1): Non-Normalized Large Dataset Based on ACI 318-14

	Seven Inputs for Engineering Mechanic (AutoCol)							Nine Corresponding Outputs (AutoCol)								
	b (mm)	h (mm)	ρ_s	f_{ck} (MPa)	f_y (MPa)	P_u (kN)	M_u (kN·m)	ϕP_n (kN)	ϕM_n (kN·m)	SF	b/h	ε_s	CI_c (KRW/m)	CO_2 (t-CO_2/m)	W_c (kN/m)	$\alpha_{e/h}$
100,000 datasets	961.0	698	0.050	53	424	11,694.4	6,037.8	7,745.6	3,999.1	0.662	1.377	0.0027	343,383	0.77	1.58	0.934
	716.3	571	0.043	43	422	7,269.5	491.8	1,1054.2	747.9	1.521	1.255	−0.0005	180,498	0.41	0.96	1.453
	630.7	543	0.023	43	515	751.5	986.7	738.2	969.2	0.982	1.162	0.0117	98,151	0.21	0.81	0.392
	…	…	…	…	…	…	…	…	…	…	…	…	…	…	…	…
	1,126.0	1,172	0.029	34	585	7,516.2	4,636.4	14,712.1	9,075.3	1.957	0.961	0.0018	447,115	0.99	3.11	1.086
	634.1	1,250	0.039	56	344	32,695.4	2,549.3	24,393.5	1,902.0	0.746	0.507	−0.0015	332,368	0.74	1.87	1.509
	335.4	388	0.058	56	447	2,614.9	34.6	4,788.8	63.3	1.831	0.865	−0.0022	75,871	0.17	0.31	1.537
Max	2,240.9	1,500	0.080	70	600	218,169.9	112,070.4	141,615.1	66,099.7	2.000	1.500	0.0734	2,543,906	5.78	7.91	1.571
Min	300.0	300	0.010	20	300	0.0	0.1	0.0	0.2	0.500	0.250	−0.0030	15,257	0.03	0.21	0.000
Mean	796.7	898.49	0.045	45.05	450.26	7,757.4	5,702.7	8,414.6	6202.2	1.252	0.900	0.0059	381,430	0.86	1.92	0.782

(c-1): Non-Normalized Large Dataset Based on ACI 318-19

	Seven Inputs for Engineering Mechanic (AutoCol)							Nine Corresponding Outputs (AutoCol)								
	b (mm)	h (mm)	ρ_s	f_{ck} (MPa)	f_y (MPa)	P_u (kN)	M_u (kN·m)	ϕP_n (kN)	ϕM_n (kN·m)	SF	b/h	ε_s	CI_c (KRW/m)	CO_2 (t-CO_2/m)	W_c (kN/m)	$\alpha_{e/h}$
100,000 datasets	1,359.8	1,093	0.035	37	555	4,802.9	19,683.5	3,209.3	13,152.8	0.668	1.244	0.0093	575,628	1.28	3.50	0.261
	1,487	1,242	0.058	25	402	13,765.3	17,683.1	12,982.7	16,677.8	0.943	1.197	0.0024	1,013,257	2.43	4.35	0.769
	300	608	0.038	66	563	295.8	538.9	522.6	952.3	1.767	0.493	0.0089	80,805	0.17	0.43	0.322
	…	…	…	…	…	…	…	…	…	…	…	…	…	…	…	…
	456.4	1,210	0.030	59	474	3,408.1	2,795.5	6,729.3	5,519.7	1.974	0.377	0.0036	197,330	0.42	1.30	0.975
	300	683	0.064	28	415	1,025.9	1,249.7	829.7	1,010.7	0.809	0.439	0.0034	123,652	0.29	0.48	0.511
	620.2	556	0.030	34	427	2,275.4	876.9	2,818.9	1086.4	1.239	1.116	0.0023	114,556	0.26	0.81	0.965
Max	2,247.8	1,500	0.080	70	600	221,251.5	103,019.0	142,073.1	65,425.6	2.000	1.500	0.0691	2,499,403	5.68	7.94	1.571
Min	300	300	0.010	20	300	0.0	0.1	0.0	0.1	0.500	0.250	−0.0030	15,153	0.03	0.21	0.000
Mean	798.3	900.12	0.045	44.899	449.73	7,833.0	5,687.3	8,495.0	6,160.4	1.251	0.900	0.0059	381,632	0.86	1.93	0.785

(b-2): Normalized Large Dataset Based on ACI 318-14

	Seven Inputs for Engineering Mechanic (AutoCol)							Nine Corresponding Outputs (AutoCol)								
	b (mm)	h (mm)	ρ_s	f_{ck} (MPa)	f_y (MPa)	P_u (kN)	M_u (kN·m)	ϕP_n (kN)	ϕM_n (kN·m)	SF	b/h	ε_s	CI_c (KRW/m)	CO_2 (t-CO_2/m)	W_c (kN/m)	$\alpha_{e/h}$
100,000 datasets	−0.319	−0.337	0.142	0.32	−0.173	−0.893	−0.892	−0.891	−0.879	−0.784	0.803	−0.853	−0.74	−0.742	−0.645	0.189
	−0.571	−0.548	−0.071	−0.08	−0.187	−0.933	−0.991	−0.844	−0.977	0.361	0.607	−0.936	−0.869	−0.868	−0.805	0.85
	−0.659	−0.595	−0.634	−0.08	0.433	−0.993	−0.982	−0.99	−0.971	−0.357	0.458	−0.615	−0.934	−0.938	−0.846	−0.501
	…	…	…	…	…	…	…	…	…	…	…	…	…	…	…	…
	−0.839	0.517	−0.437	0.56	0.16	−0.969	−0.946	−0.792	−0.725	0.943	0.137	−0.874	−0.658	−0.668	−0.248	0.383
	−1	−0.362	0.554	−0.68	−0.233	−0.991	−0.976	−0.655	−0.942	−0.672	−0.588	−0.963	−0.749	−0.753	−0.57	0.921
	−0.671	−0.573	−0.428	−0.44	−0.153	−0.979	−0.983	−0.932	−0.998	0.775	−0.017	−0.98	−0.952	−0.952	−0.975	0.957
Max	1	1	1	1	1	1	1	1	1	1	1	1	1	1	1	1
Min	−1	−1	−1	−1	−1	−1	−1	−1	−1	−1	−1	−1	−1	−1	−1	−1
Mean	−0.488	−0.003	0	0.002	0.002	−0.929	−0.898	−0.881	−0.812	0.002	0.04	−0.767	−0.71	−0.712	−0.555	−0.004

(c-2): Normalized Large Dataset Based on ACI 318-19

	Seven Inputs for Engineering Mechanic (AutoCol)							Nine Corresponding Outputs (AutoCol)								
	b (mm)	h (mm)	ρ_s	f_{ck} (MPa)	f_y (MPa)	P_u (kN)	M_u (kN·m)	ϕP_n (kN)	ϕM_n (kN·m)	SF	b/h	ε_s	CI_c (KRW/m)	CO_2 (t-CO_2/m)	W_c (kN/m)	$\alpha_{e/h}$
100,000 datasets	0.088	0.322	−0.280	−0.320	0.700	−0.957	−0.618	−0.955	−0.598	−0.776	0.591	−0.660	−0.549	−0.558	−0.149	−0.668
	0.219	0.570	0.376	−0.800	−0.320	−0.876	−0.657	−0.817	−0.490	−0.409	0.516	−0.851	−0.196	−0.152	0.071	−0.021
	−1.000	−0.487	−0.191	0.840	0.753	−0.997	−0.990	−0.993	−0.971	0.689	−0.611	−0.672	−0.947	−0.952	−0.944	−0.590
	…	…	…	…	…	…	…	…	…	…	…	…	…	…	…	…
	−0.839	0.517	−0.437	0.560	0.160	−0.969	−0.946	−0.905	−0.831	0.966	−0.797	−0.817	−0.853	−0.865	−0.718	0.242
	−1.000	−0.362	0.554	−0.680	−0.233	−0.991	−0.976	−0.988	−0.969	−0.588	−0.697	−0.824	−0.913	−0.908	−0.930	−0.349
	−0.671	−0.573	−0.428	−0.440	−0.153	−0.979	−0.983	−0.960	−0.967	−0.015	0.385	−0.855	−0.920	−0.919	−0.845	0.228
Max	1.000	1.000	1.000	1.000	1.000	1.000	1.000	1.000	1.000	1.000	1.000	1.000	1.000	1.000	1.000	1.000
Min	−1.000	−1.000	−1.000	−1.000	−1.000	−1.000	−1.000	−1.000	−1.000	−1.000	−1.000	−1.000	−1.000	−1.000	−1.000	−1.000
Mean	−0.488	0.000	−0.002	−0.004	−0.002	−0.929	−0.890	−0.880	−0.812	0.002	0.039	−0.754	−0.705	−0.707	−0.555	0.000

$$y = f(\mathbf{x}) = \mathbf{z}^{(D)} \circ \mathbf{z}^{(L)} \circ \mathbf{z}^{(L-1)} \circ \ldots \circ \mathbf{z}^{(l)} \circ \ldots \circ \mathbf{z}^{(1)} \circ \mathbf{z}^{(N)} \quad (3.2.2.3)$$

$$\begin{aligned}
&\mathbf{z}^{(N)} = \mathrm{g}^{(N)}(\mathbf{x}) = \lambda_{\mathrm{x}} \odot (\mathbf{x} - \mathbf{x}_{\mathrm{min}}) + \overline{\mathbf{x}}_{\mathrm{min}} && \text{(a)}\\
&\mathbf{z}^{(1)} = f_t^{(1)}\left(\mathbf{W}^{(1)}\mathbf{z}^{(N)} + \mathbf{b}^{(1)}\right) && \text{(b)}\\
&\ldots\\
&\mathbf{z}^{(l)} = f_t^{(l)}\left(\mathbf{W}^{(l)}\mathbf{z}^{(l-1)} + \mathbf{b}^{(l)}\right) && \text{(c)}\\
&\ldots\\
&\mathbf{z}^{(L-1)} = f_t^{(L-1)}\left(\mathbf{W}^{(L-1)}\mathbf{z}^{(L-2)} + \mathbf{b}^{(L-1)}\right) && \text{(d)}\\
&\mathbf{z}^{(L=\mathrm{output})} = f_{\mathrm{lin}}^{(L)}\left(\mathbf{W}^{(L)}\mathbf{z}^{(L-1)} + b^{(L)}\right) && \text{(e)}\\
&y = \mathbf{z}^{(D)} = \mathrm{g}^{(D)}(y) = \frac{1}{\alpha_y}\left(\mathbf{z}^{(L)} - \overline{\mathbf{z}_{\mathrm{min}}^{(L)}}\right) + \mathbf{z}_{\mathrm{min}}^{(L)} && \text{(f)}
\end{aligned} \quad (3.2.2.4)$$

Objective functions are formulated using ANNs rather than being based on analytical function or structural mechanics (Berrais, 1999). For this, a perceptron with three layers implements an activation function to derive nonlinear objective functions based on ANNs [Kolmogorov's theorem (Kolmogorov, 1957)]. In this chapter, ANNs shown in Equations (3.2.2.5-1)–(3.2.2.5-3) are trained on large structural datasets for obtaining weight and bias matrices which are then used to formulate cost index CI_c, CO_2 emissions, and weight W_c of a RC column as a function of seven input parameters $(b, h, \rho_s, f_c', f_y, P_u, M_u)$. Input design parameters $\mathbf{x} = [b, h, \rho_s, f_c', f_y, P_u, M_u]^T$ shown in red for forward networks of Equations (3.2.2.5-1)–(3.2.2.5-3) are relayed to neurons of fully connected successive layers using weights at each neuron and bias in each hidden layer. The reverse networks change some input and output design parameters between the input and output side. The input parameters of the reverse networks shown in Figure 3.2.2.3 and Table 3.2.2.2 are $\mathbf{x} = [\rho_s, f_c', f_y, P_u, M_u, SF, b/h]^T$ which is different from those of forward network ($\mathbf{x} = [b, h, \rho_s, f_c', f_y, P_u, M_u]^T$. It is noted that SF and b/h appear on the input side, whereas b and h are moved to the output side for the reverse networks. The ANN-based objective functions CI_c, and CO_2 emissions, and W_c shown in Equation (3.2.2.5) are networks with L layers and 80 neurons, which are linked by weighted interconnections and bias through an activation function, thereby performing nonlinear numerical computations (Hong, 2019). ANNs recognized as machine learning are formulated to fit the trends of objective functions between inputs and output parameters, yielding weight and bias matrices. An activation function (*tansig*, *tanh*) shown in Figure 3.2.2.2 is used in Equations (3.2.2.5-1)–(3.2.2.5-3) (Krenker et al., 2011).

$$\underbrace{CI_c}_{[1\times1]} = g_{CI_c}^{D}\left(f_l^{L}\left(\underbrace{\mathbf{W}_{CI_c}^{L}}_{[1\times80]} f_t^{L-1}\left(\underbrace{\mathbf{W}_{CI_c}^{L-1}}_{[80\times80]} \ldots f_t^{1}\left(\underbrace{\mathbf{W}_{CI_c}^{1}}_{[80\times7]} \underbrace{g_{CI_c}^{N}(\mathbf{x})}_{[7\times1]} + \underbrace{\mathbf{b}_{CI_c}^{1}}_{[80\times1]}\right) \ldots + \underbrace{\mathbf{b}_{CI_c}^{L-1}}_{[80\times1]}\right) + \underbrace{b_{CI_c}^{L}}_{[1\times1]}\right)\right) \quad (3.2.2.5\text{-}1)$$

$$\underbrace{CO_2}_{[1\times1]} = g_{CO_2}^{D}\left(f_l^{L}\left(\underbrace{\mathbf{W}_{CO_2}^{L}}_{[1\times80]} f_t^{L-1}\left(\underbrace{\mathbf{W}_{CO_2}^{L-1}}_{[80\times80]} \ldots f_t^{1}\left(\underbrace{\mathbf{W}_{CO_2}^{1}}_{[80\times7]} \underbrace{g_{CO_2}^{N}(\mathbf{x})}_{[7\times1]} + \underbrace{\mathbf{b}_{CO_2}^{1}}_{[80\times1]}\right) \ldots + \underbrace{\mathbf{b}_{CO_2}^{L-1}}_{[80\times1]}\right) + \underbrace{b_{CO_2}^{L}}_{[1\times1]}\right)\right) \quad (3.2.2.5\text{-}2)$$

a

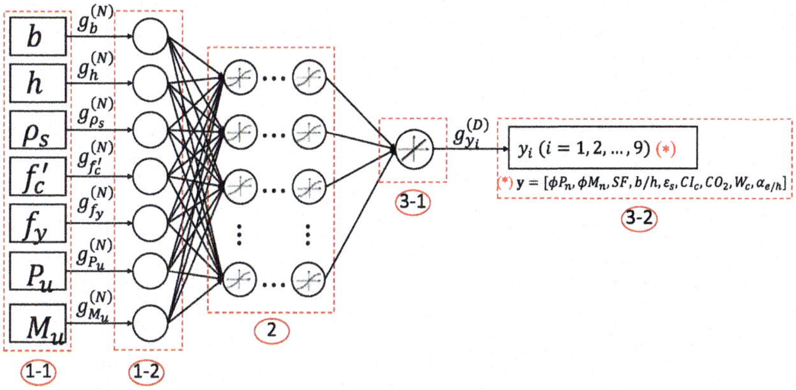

(1-1) Non-normalized input layer, $\mathbf{x} = \left[b, h, \rho_s, f_c', f_y, P_u, M_u\right]^T$

(1-2) Normalized input layer, $\mathbf{z}_{y_i}^{(N)} = \mathbf{g}^{(N)}(\mathbf{x})$

(2) $L-1$ hidden layers with 80 neurons, $\mathbf{z}_{y_i}^{(l)} = f_t^{(l)}(\mathbf{W}_{y_i}^{(l)}\mathbf{z}_{y_i}^{(l-1)} + \mathbf{b}_{y_i}^{(l)}),\ l = \{1, 2, ..., L-1\}$

(3-1) Normalized output layers, $\mathbf{z}_{y_i}^{(L)} = f_{lin}^{(L)}(\mathbf{W}_{y_i}^{(L)}\mathbf{z}_{y_i}^{(L-1)} + b_{y_i}^{(L)})$

(3-2) De-normalized output layer, $y_i = \mathbf{z}_{y_i}^{(D)} = g_{y_i}^{(D)}\left(\mathbf{z}_{y_i}^{(L)}\right)$

b

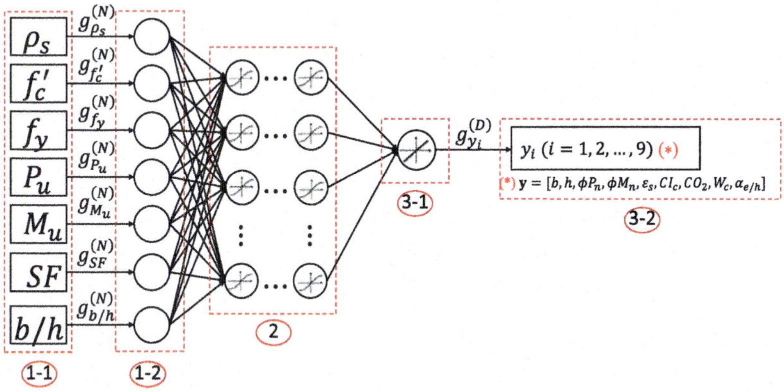

(1-1) Non-normalized input layer, $\mathbf{x} = \left[\rho_s, f_c', f_y, P_u, M_u, SF, b/h\right]^T$

(1-2) Normalized input layer, $\mathbf{z}_{y_i}^{(N)} = \mathbf{g}^{(N)}(\mathbf{x})$

(2) $L-1$ hidden layers with 80 neurons, $\mathbf{z}_{y_i}^{(l)} = f_t^{(l)}(\mathbf{W}_{y_i}^{(l)}\mathbf{z}_{y_i}^{(l-1)} + \mathbf{b}_{y_i}^{(l)}),\ l = \{1, 2, ..., L-1\}$

(3-1) Normalized output layers, $\mathbf{z}_{y_i}^{(L)} = f_{lin}^{(L)}(\mathbf{W}_{y_i}^{(L)}\mathbf{z}_{y_i}^{(L-1)} + b_{y_i}^{(L)})$

(3-2) De-normalized output layer, $y_i = \mathbf{z}_{y_i}^{(D)} = g_{y_i}^{(D)}\left(\mathbf{z}_{y_i}^{(L)}\right)$

Figure 3.2.2.3 ANNs trained based on PTM. (a) Forward network. (b) Reverse network.

$$\underbrace{W_c}_{[1\times1]} = g_{W_c}^{N}\left(f_l^{L}\left(\underbrace{\mathbf{W}_{W_c}^{L}}_{[1\times80]} f_t^{L-1}\left(\underbrace{\mathbf{W}_{W_c}^{L-1}}_{[80\times80]}\cdots f_t^{1}\left(\underbrace{\mathbf{W}_{W_c}^{1}}_{[80\times7]}\underbrace{g_{W_c}^{N}(\mathbf{x})}_{[7\times1]} + \underbrace{\mathbf{b}_{W_c}^{1}}_{[80\times1]}\right)\cdots + \underbrace{\mathbf{b}_{W_c}^{L-1}}_{[80\times1]}\right) + \underbrace{b_{W_c}^{L}}_{[1\times1]}\right)\right) \tag{3.2.2.5-3}$$

Similarly, the rest of the outputs (ϕP_n, ϕM_n, SF, b/h, ε_s, $\alpha_{e/h}$) can also be formulated as functions of seven inputs (b, h, ρ_s, f_c', f_y, P_u, M_u) based on forward ANN as shown in

Table 3.2.2.2 where parallel training method (PTM) is implemented to simultaneously train ANNs shown in Equations (3.2.2.5-1)–(3.2.2.5-3) on one output. PTM requires lesser training time than when using TED where entire input parameters are mapped to entire output parameters, simultaneously. Tables 3.2.2.2(a-1), and (a-2) based on ACI 318-14 and (b-1), (b-2) based on ACI 318-19 present training accuracies of both forward and reverse neural networks, respectively, in which a combination of four types of hidden layers including one, two, five, and ten layers with 80 neurons are implemented. Structural datasets of 100,000 are randomly divided into three subsets: training set, validation set, and test set. According to Brian Ripley (Ripley, 1996), a training set (70% of big datasets) is a set used for learning that is to fit parameters, whereas a validation set is used to tune a neural network to avoid overfitting. A test subset, on the other hand, is independent of the training dataset which does not affect the training procedure. It is, therefore, used only to access the performance of ANNs, and hence, an MSE of test set (*MSE T.perf*) shown in Table 3.2.2.2 is suitable for evaluating the goodness of designs, indicating the capability of ANNs to fit the datasets. Tables 3.2.2.3 shows a sample calculation of weight and bias matrices of the ANN with 1 layer and 80 neurons are obtained based on PTM for CI_c, CO_2, W_c objective function.

Weight and bias matrices shown in Tables 3.2.2.3(a-1) based on ACI 318-14 and (b-1) based on ACI 318-19, respectively, are obtained based on forward ANNs for the three objective functions (CI_c, CO_2, W_c) shown in Equations (3.2.2.5-1)–(3.2.2.5-3) by mapping seven input parameters $\left(b,\ h,\ \rho_s, f_c', f_y,\ P_u,\ M_u\right)$ to each one of the three objective functions (CI_c, CO_2, W_c).

Weight and bias matrices shown in Tables 3.2.2.3(a-2) based on ACI 318-14 and (b-2) based on ACI 318-19, respectively, are obtained from Equations (3.2.2.5-1)–(3.2.2.5-3) based on reverse ANNs by mapping seven input parameters $\left(\rho_s, f_c', f_y, P_u, M_u, SF, b/h\right)$ to each one of output parameters which are objective functions (CI_c,CO_2, W_c). Weight and bias matrices are used to calculate Jacobian and Hessian matrices of the objective and constraining functions to solve KKT solutions. Figure 3.2.2.3a and b are also presented to illustrate forward and reverse networks, respectively, which are based on PTM. As shown in Figure 3.2.2.3, only one weight and bias matrix are obtained to connect the last hidden layer shown in Figure 3.2.2.3-② to (ϕP_n, ϕM_n, SF, b/h, ε_s, $\alpha_{e/h}$ for forward outputs and b, h, ϕP_n, ϕM_n, ε_s, $\alpha_{e/h}$ for reverse outputs) and objective functions (CI_c, CO_2, W_c) of output layer shown in Figure 3.2.2.3-③-2.

Analytical Jacobi and Hessian matrices needed to solve KKT solutions are calculated by directly differentiating analytical functions whereas those based on ANNs are derived using weight and bias matrices. Table 3.2.2.3 shows weight and bias matrices obtained using one hidden layer based on shown in Equations (3.2.2.5-1)–(3.2.2.5-3) and Figure 3.2.2.3a and b. Table 3.2.2.3 lists generated weights and biases of a forward and reverse network for objective functions of CI_c, CO_2, and W_c for both ACI 318-14 and 318-19.

Dimensions of weight matrix (80×7) connecting input layer with seven input variables to the 80 neural values of first hidden layer are determined by a number of rows and columns equal to a number of neurons (nodes) in the successor layer and equal to number of input parameters in predecessor layers, respectively. For example, a 80×7 weight matrix ($\mathbf{W}_{CI_c}^1$) connects seven input variables of the input layer to the 80 neural values of the first hidden layer. Besides, a number of biases (80) equals to a number of neurons (80) since each neuron in the hidden layer and the output layer has only one bias, resulting in a bias matrix by 80×1.

The following relationships are found at the first hidden layer.

At Layer 1: weight matrix ($\mathbf{W}_{CI_c}^1$, 80×7) connecting seven input variables of the input layer to the 80 neural values of the first hidden layer (Hidden layer 1) × seven input parameters (7×1) + bias matrix at Layer 1 (80×1) = Output node of Hidden layer 1 (80×1); a weight matrix ($\mathbf{W}_{CI_c}^l$) connecting the 80 neural values calculated at the predecessor layer (Layer 1) to output layer becomes 1×80 which should be consistent to one output parameter that is an objective function. A bias matrix by 1×1 is resulted.

The following relationships are found at the output layer.

At output layer: weight matrix ($\mathbf{W}^l_{CI_c}$, 1 × 80) connecting the 80 neural values of the predecessor layer (Hidden Layer 1, 80 × 1) to the one output node (1 × 1), resulting in weight matrix ($\mathbf{W}^l_{CI_c}$, 1 × 80) × the predecessor layer (Hidden Layer 1, 80 × 1 + bias matrix at the output layer (1 × 1) = Output layer (1 × 1) = Objective function (1 × 1).

The entire weight and bias matrices are adjusted by back-propagation. Weight and bias matrix are obtained to derive objective functions for cost CI_c, CO_2 emissions and W_c using forward networks with seven forward inputs (b, h, ρ_s, f_c', f_y, P_u, M_u) based on training accuracies as shown in Table 3.2.2.2. The rest of the forward outputs (ϕP_n, ϕM_n, SF, b / h, ε_s, $\alpha_{e/h}$) can be also found as stationary points which are functions of seven forward inputs.

Table 3.2.2.2 Training Accuracies Based on PTM

(a-1): Objective Functions Based on Forward-Lagrange (One, Two, Five, Ten Layers – 80 Neurons); ACI 318-14

Training Table: Forward - Lagrange
7 Inputs ($b, h, \rho_s, f_c', f_y, P_u, M_u$) -
9 Outputs ($\phi P_n, \phi M_n, SF, b/h, \varepsilon_s, CI_c, CO_2, W_c, \alpha_{e/h}$)

No.	*Data*	*Layers*	*Neurons*	*Best Epoch*	*MSE T.perf*	*R at Best Epoch*
7 Inputs ($b, h, \rho_s, f_c', f_y, P_u, M_u$) -1 Output ($\phi P_n$) - PTM						
1a	100,000	1	80	13,229	1.2E-04	0.9982
1b	100,000	2	80	7,799	2.6E-05	0.9996
1c	100,000	5	80	58,582	8.9E-07	1.0000
1d	100,000	10	80	23,743	5.7E-06	0.9999
7 Inputs ($b, h, \rho_s, f_c', f_y, P_u, M_u$) -1 Output ($\phi M_n$) - PTM						
2a	100,000	1	80	23,881	1.4E-04	0.9988
2b	100,000	2	80	24,144	1.0E-05	0.9999
2c	100,000	5	80	16,900	5.4E-07	1.0000
2d	100,000	10	80	35,315	5.2E-06	1.0000
7 Inputs ($b, h, \rho_s, f_c', f_y, P_u, M_u$) -1 Output ($SF$) - PTM						
3a	100,000	1	80	70,810	1.1E-02	0.9829
3b	100,000	2	80	23,782	1.6E-03	0.9978
3c	100,000	5	80	9,616	5.1E-06	1.0000
3d	100,000	10	80	62,772	3.7E-04	0.9995
7 Inputs ($b, h, \rho_s, f_c', f_y, P_u, M_u$) -1 Output ($b/h$) - PTM						
4a	100,000	1	80	68,943	1.1E-07	1.0000
4b	100,000	2	80	44,975	2.0E-08	1.0000
4c	100,000	5	80	5,948	1.4E-05	0.9999
4d	100,000	10	80	23,600	2.0E-06	1.0000
7 Inputs ($b, h, \rho_s, f_c', f_y, P_u, M_u$) -1 Output ($\varepsilon_s$) - PTM						
5a	100,000	1	80	26,197	6.0E-04	0.9921
5b	100,000	2	80	44,138	5.2E-05	0.9994
5c	100,000	5	80	9,734	6.6E-05	0.9992
5d	100,000	10	80	39,003	2.9E-05	0.9997
7 Inputs ($b, h, \rho_s, f_c', f_y, P_u, M_u$) -1 Output ($CI_c$) - PTM						
6a	100,000	1	80	51,325	1.2E-07	1.0000
6b	100,000	2	80	24,824	1.3E-07	1.0000
6c	100,000	5	80	5,948	1.4E-05	0.9999
6d	100,000	10	80	31,260	1.6E-06	1.0000

(Continued)

Table 3.2.2.2 (Continued) Training Accuracies Based on PTM

7 Inputs ($b, h, \rho_s, f_c', f_y, P_u, M_u$) -1 Output ($CO_2$) - PTM						
7a	100,000	1	80	27,427	1.2E-07	1.0000
7b	100,000	2	80	33,193	7.2E-08	1.0000
7c	100,000	5	80	9,616	5.1E-06	1.0000
7d	100,000	10	80	12,221	4.8E-06	1.0000
7 Inputs ($b, h, \rho_s, f_c', f_y, P_u, M_u$) -1 Output ($W_c$) - PTM						
8a	100,000	1	80	62,349	2.1E-08	1.0000
8b	100,000	2	80	29,998	7.5E-08	1.0000
8c	100,000	5	80	16,900	5.4E-07	1.0000
8d	100,000	10	80	25,169	2.3E-06	1.0000
7 Inputs ($b, h, \rho_s, f_c', f_y, P_u, M_u$) -1 Output ($\alpha_{e/h}$) - PTM						
9a	100,000	1	80	31,963	2.6E-03	0.9968
9b	100,000	2	80	13,715	1.7E-04	0.9997
9c	100,000	5	80	58,582	8.9E-07	1.0000
9d	100,000	10	80	138,783	2.4E-07	1.0000

(a-2): Objective Functions Based on Reverse-Lagrange (One, Two, Five, Ten Layers – 80 Neurons); ACI 318-14

Training Table: Reverse - Lagrange
7 Inputs ($\rho_s, f_c', f_y, P_u, M_u, SF, b/h$) -
9 Outputs ($b, h, \phi P_n, \phi M_n, \varepsilon_s, CI_c, CO_2, W_c, \alpha_{e/h}$)

No.	*Data*	*Layers*	*Neurons*	*Best Epoch*	*MSE T.perf*	*R at Best Epoch*
7 Inputs ($\rho_s, f_c', f_y, P_u, M_u, SF, b/h$) -1 Output ($b$) - PTM						
1a	100,000	1	80	55,585	3.5E-04	0.9992
1b	100,000	2	80	67,336	1.5E-05	1.0000
1c	100,000	5	80	63,449	5.4E-06	1.0000
1d	100,000	10	80	72,476	6.2E-06	1.0000
7 Inputs ($\rho_s, f_c', f_y, P_u, M_u, SF, b/h$) -1 Output ($h$) - PTM						
2a	100,000	1	80	36,194	9.0E-04	0.9988
2b	100,000	2	80	44,582	4.6E-05	0.9999
2c	100,000	5	80	36,651	2.7E-05	1.0000
2d	100,000	10	80	100,431	2.4E-05	1.0000
7 Inputs ($\rho_s, f_c', f_y, P_u, M_u, SF, b/h$) -1 Output ($\phi P_n$) - PTM						
3a	100,000	1	80	30,476	1.3E-07	1.0000
3b	100,000	2	80	8,231	1.1E-06	1.0000
3c	100,000	5	80	39,226	2.0E-07	1.0000
3d	100,000	10	80	20,150	2.0E-06	1.0000
7 Inputs ($\rho_s, f_c', f_y, P_u, M_u, SF, b/h$) -1 Output ($\phi M_n$) - PTM						
4a	100,000	1	80	11,649	2.6E-07	1.0000
4b	100,000	2	80	33,103	7.3E-08	1.0000
4c	100,000	5	80	23,665	8.4E-07	1.0000
4d	100,000	10	80	50,481	2.4E-07	1.0000

(Continued)

Table 3.2.2.2 (Continued) Training Accuracies Based on PTM

7 Inputs ($\rho_s, f_c', f_y, P_u, M_u, SF, b/h$) -1 Output ($\varepsilon_s$) - PTM						
5a	100,000	1	80	99,999	3.3E-04	0.9959
5b	100,000	2	80	21,872	7.0E-05	0.9991
5c	100,000	5	80	32,739	2.8E-05	0.9997
5d	100,000	10	80	27,119	2.7E-05	0.9997
7 Inputs ($\rho_s, f_c', f_y, P_u, M_u, SF, b/h$) -1 Output ($CI_c$) - PTM						
6a	100,000	1	80	56,511	7.6E-05	0.9995
6b	100,000	2	80	51,156	1.1E-05	0.9999
6c	100,000	5	80	17,889	1.2E-05	0.9999
6d	100,000	10	80	63,250	5.2E-06	1.0000
7 Inputs ($\rho_s, f_c', f_y, P_u, M_u, SF, b/h$) -1 Output ($CO_2$) - PTM						
7a	100,000	1	80	21,938	1.8E-04	0.9990
7b	100,000	2	80	40,426	1.4E-05	0.9999
7c	100,000	5	80	46,096	4.8E-06	1.0000
7d	100,000	10	80	49,033	5.0E-06	1.0000
7 Inputs ($\rho_s, f_c', f_y, P_u, M_u, SF, b/h$) -1 Output ($W_c$) - PTM						
8a	100,000	1	80	21,920	4.5E-04	0.9988
8b	100,000	2	80	45,884	1.4E-05	1.0000
8c	100,000	5	80	55,162	7.7E-06	1.0000
8d	100,000	10	80	75,074	6.8E-06	1.0000
7 Inputs ($\rho_s, f_c', f_y, P_u, M_u, SF, b/h$) -1 Output ($\alpha_{e/h}$) - PTM						
9a	100,000	1	80	99,808	1.5E-03	0.9978
9b	100,000	2	80	45,009	4.1E-05	0.9999
9c	100,000	5	80	84,980	5.0E-06	1.0000
9d	100,000	10	80	113,526	3.7E-06	1.0000

(b-1): Objective Functions Based on Forward-Lagrange (One, Two, Five, Ten Layers – 80 Neurons); ACI 318-19

Training Table: Forward-Lagrange
7 Inputs ($b, h, \rho_s, f_c', f_y, P_u, M_u$) -
9 Outputs ($\phi P_n, \phi M_n, SF, b/h, \varepsilon_s, CI_c, CO_2, W_c, \alpha_{e/h}$)

No.	*Data*	*Layers*	*Neurons*	*Best Epoch*	*MSE T.perf*	*R at Best Epoch*
7 Inputs ($b, h, \rho_s, f_c', f_y, P_u, M_u$) -1 Output ($\phi P_n$) - PTM						
1a	100,000	1	80	12,345	1.1E-04	0.9983
1b	100,000	2	80	51,637	3.5E-06	1.0000
1c	100,000	5	80	22,694	4.1E-06	0.9999
1d	100,000	10	80	45,002	3.0E-06	1.0000
7 Inputs ($b, h, \rho_s, f_c', f_y, P_u, M_u$) -1 Output ($\phi M_n$) - PTM						
2a	100,000	1	80	18,935	1.4E-04	0.9988
2b	100,000	2	80	22,029	1.0E-05	0.9999
2c	100,000	5	80	33,305	4.6E-06	1.0000
2d	100,000	10	80	60,322	4.2E-06	1.0000
7 Inputs ($b, h, \rho_s, f_c', f_y, P_u, M_u$) -1 Output ($SF$) - PTM						
3a	100,000	1	80	100,000	1.0E-02	0.9842
3b	100,000	2	80	73,171	8.4E-04	0.9988
3c	100,000	5	80	63,672	3.8E-04	0.9995
3d	100,000	10	80	54,463	3.4E-04	0.9995

(Continued)

Table 3.2.2.2 (Continued) Training Accuracies Based on PTM

7 Inputs ($b, h, \rho_s, f_c', f_y, P_u, M_u$) -1 Output ($b/h$) - PTM						
4a	100,000	1	80	99,989	4.2E-08	1.0000
4b	100,000	2	80	50,430	1.6E-08	1.0000
4c	100,000	5	80	47,089	1.1E-07	1.0000
4d	100,000	10	80	80,243	1.0E-07	1.0000
7 Inputs ($b, h, \rho_s, f_c', f_y, P_u, M_u$) -1 Output ($\varepsilon_s$) - PTM						
5a	100,000	1	80	99,818	4.5E-04	0.9947
5b	100,000	2	80	32,551	6.4E-05	0.9993
5c	100,000	5	80	21,597	2.9E-05	0.9997
5d	100,000	10	80	47,824	3.0E-05	0.9997
7 Inputs ($b, h, \rho_s, f_c', f_y, P_u, M_u$) -1 Output ($CI_c$) - PTM						
6a	100,000	1	80	80,194	4.2E-08	1.0000
6b	100,000	2	80	10,276	8.9E-07	1.0000
6c	100,000	5	80	5,765	1.9E-05	0.9999
6d	100,000	10	80	47,041	1.6E-06	1.0000
7 Inputs ($b, h, \rho_s, f_c', f_y, P_u, M_u$) -1 Output ($CO_2$) - PTM						
7a	100,000	1	80	37,649	1.2E-07	1.0000
7b	100,000	2	80	5,495	3.4E-06	1.0000
7c	100,000	5	80	28,549	1.8E-06	1.0000
7d	100,000	10	80	13,174	6.4E-06	1.0000
7 Inputs ($b, h, \rho_s, f_c', f_y, P_u, M_u$) -1 Output ($W_c$) - PTM						
8a	100,000	1	80	94,533	1.4E-08	1.0000
8b	100,000	2	80	56,279	2.2E-08	1.0000
8c	100,000	5	80	15,555	2.3E-06	1.0000
8d	100,000	10	80	40,568	7.1E-07	1.0000
7 Inputs ($b, h, \rho_s, f_c', f_y, P_u, M_u$) -1 Output ($\alpha_{e/h}$) - PTM						
9a	100,000	1	80	41,244	1.8E-03	0.9974
9b	100,000	2	80	54,660	1.8E-05	1.0000
9c	100,000	5	80	94,283	5.8E-07	1.0000
9d	100,000	10	80	121,394	4.3E-07	1.0000

(b-2): Objective Functions Based on Reverse-Lagrange (One, Two, Five, Ten Layers – 80 Neurons); ACI 318-19

Training Table: Reverse-Lagrange
7 Inputs ($\rho_s, f_c', f_y, P_u, M_u, SF, b/h$) -
9 Outputs ($b, h, \phi P_n, \phi M_n, \varepsilon_s, CI_c, CO_2, W_c, \alpha_{e/h}$)

No.	*Data*	*Layers*	*Neurons*	*Best Epoch*	*MSE T.perf*	*R at Best Epoch*
7 Inputs ($\rho_s, f_c', f_y, P_u, M_u, SF, b/h$) -1 Output ($b$) - PTM						
1a	100,000	1	80	29,054	3.7E-04	0.9992
1b	100,000	2	80	77,416	1.5E-05	1.0000
1c	100,000	5	80	54,000	5.7E-06	1.0000
1d	100,000	10	80	61,268	5.7E-06	1.0000

(Continued)

Table 3.2.2.2 (Continued) Training Accuracies Based on PTM

7 Inputs ($\rho_s, f_c', f_y, P_u, M_u, SF, b/h$) -1 Output ($h$) - PTM						
2a	100,000	1	80	70,234	7.1E-04	0.9990
2b	100,000	2	80	51,630	5.8E-05	0.9999
2c	100,000	5	80	49,748	2.3E-05	1.0000
2d	100,000	10	80	57,530	2.6E-05	1.0000
7 Inputs ($\rho_s, f_c', f_y, P_u, M_u, SF, b/h$) -1 Output ($\phi P_n$) - PTM						
3a	100,000	1	80	6,501	1.3E-06	1.0000
3b	100,000	2	80	29,698	2.4E-07	1.0000
3c	100,000	5	80	8,135	5.9E-06	0.9999
3d	100,000	10	80	7,343	3.0E-05	0.9996
7 Inputs ($\rho_s, f_c', f_y, P_u, M_u, SF, b/h$) -1 Output ($\phi M_n$) - PTM						
4a	100,000	1	80	32,566	1.3E-07	1.0000
4b	100,000	2	80	36,280	1.2E-07	1.0000
4c	100,000	5	80	13,394	3.8E-06	1.0000
4d	100,000	10	80	24,401	1.0E-06	1.0000
7 Inputs ($\rho_s, f_c', f_y, P_u, M_u, SF, b/h$) -1 Output ($\varepsilon_s$) - PTM						
5a	100,000	1	80	50,665	4.3E-04	0.9952
5b	100,000	2	80	36,377	6.2E-05	0.9993
5c	100,000	5	80	23,108	3.1E-05	0.9997
5d	100,000	10	80	31,505	3.0E-05	0.9997
7 Inputs ($\rho_s, f_c', f_y, P_u, M_u, SF, b/h$) -1 Output ($CI_c$) - PTM						
6a	100,000	1	80	33,219	1.2E-04	0.9994
6b	100,000	2	80	44,515	9.0E-06	1.0000
6c	100,000	5	80	14,895	8.8E-06	1.0000
6d	100,000	10	80	56,050	5.1E-06	1.0000
7 Inputs ($\rho_s, f_c', f_y, P_u, M_u, SF, b/h$) -1 Output ($CO_2$) - PTM						
7a	100,000	1	80	46,874	9.1E-05	0.9995
7b	100,000	2	80	13,087	2.1E-05	0.9999
7c	100,000	5	80	19,454	1.1E-05	1.0000
7d	100,000	10	80	42,702	6.1E-06	1.0000
7 Inputs ($\rho_s, f_c', f_y, P_u, M_u, SF, b/h$) -1 Output ($W_c$) - PTM						
8a	100,000	1	80	29,072	3.3E-04	0.9991
8b	100,000	2	80	44,270	1.7E-05	1.0000
8c	100,000	5	80	58,614	6.4E-06	1.0000
8d	100,000	10	80	38,101	8.6E-06	1.0000
7 Inputs ($\rho_s, f_c', f_y, P_u, M_u, SF, b/h$) -1 Output ($\alpha_{e/h}$) - PTM						
9a	100,000	1	80	99,971	8.9E-04	0.9986
9b	100,000	2	80	76,044	2.0E-05	1.0000
9c	100,000	5	80	99,912	4.0E-06	1.0000
9d	100,000	10	80	99,924	4.5E-06	1.0000

Table 3.2.2.3 Weight and Bias Matrices for Cl_c, CO_2, W_c objective function obtained by One-Layer-80-Neurons Network Based on PTM for Cl_c, CO_2, W_c Objective Function

			(a-1): Forward Network - ACI 318-14				
			$\mathbf{W}_{Cl_c}{}^{(1)}$				$\mathbf{b}_{Cl_c}{}^{(1)}$
0.493	0.907	0.891	0.125	0.668	0.586	−0.392	−2.658
−0.826	0.865	−0.136	1.200	0.072	−0.591	−1.598	2.856
...	...	...	...	...	...	...	...
0.803	0.412	0.939	−0.600	1.253	1.020	−1.182	2.723
−1.172	−1.052	−0.559	0.725	0.122	−1.542	−1.458	−2.301
			[80×7]				[80×1]
			$\mathbf{W}_{Cl_c}{}^{(2)}$				$\mathbf{b}_{Cl_c}{}^{(2)}$
−0.001	0.000	0.000	...	0.000	0.000	0.000	0.026
			[1×80]				[1×1]
			$\mathbf{W}_{CO_2}{}^{(1)}$				$\mathbf{b}_{CO_2}{}^{(1)}$
0.517	0.332	−0.367	0.001	−0.002	−0.003	−0.029	−1.489
1.264	0.066	0.608	1.062	0.404	−0.675	0.859	−2.678
...	...	...	...	...	...	...	...
0.277	−0.946	−0.653	1.239	−0.545	0.617	1.379	2.627
−0.636	−0.799	−0.141	1.348	1.753	0.674	−0.011	−2.653
			[80×7]				[80×1]
			$\mathbf{W}_{CO_2}{}^{(2}$				$\mathbf{b}_{CO_2}{}^{(2)}$
−0.987	0.000	−0.001	...	−0.001	0.000	0.000	−0.287
			[1×80]				[1×1]
			$\mathbf{W}_{W_c}{}^{(1)}$				$\mathbf{b}_{W_c}{}^{(1)}$
−1.067	−0.472	−0.400	0.245	−0.758	1.083	0.704	2.463
1.115	0.971	1.139	−1.221	−0.405	0.771	0.082	−2.428
...	...	...	...	...	...	...	...
0.762	0.861	1.223	0.391	−0.538	1.885	−0.445	2.339
1.445	1.384	−0.530	0.434	0.019	0.382	−0.781	2.791
			[80×7]				[80×1]
			$\mathbf{W}_{W_c}{}^{(2)}$				$\mathbf{b}_{W_c}{}^{(2)}$
−0.001	0.000	0.000	...	0.000	0.000	0.000	−0.080
			[1×80]				[1×1]

(Continued)

Table 3.2.2.3 (Continued) Weight and Bias Matrices for Cl_c, CO_2, W_c objective function obtained by One-Layer-80-Neurons Network Based on PTM for Cl_c, CO_2, W_c Objective Function

			(b-1): Forward Network - ACI 318-19				
			$\mathbf{W}_{Cl_c}^{(1}$				$\mathbf{b}_{Cl_c}^{(1)}$
1.718	−0.828	0.126	0.646	−0.606	−0.867	−1.013	−2.574
0.861	0.026	0.659	0.864	0.530	−1.539	0.729	−2.532
...	...	...	...	...	...	...	...
1.504	0.698	−0.910	−1.230	0.592	0.432	−0.034	2.586
−1.467	−1.191	−1.119	−0.783	0.081	−0.209	−0.598	−2.696
			[80×7]				[80×1]
			$\mathbf{W}_{Cl_c}^{(2)}$				$\mathbf{b}_{Cl_c}^{(2)}$
0.001	0.000	0.009	...	0.751	0.000	0.000	0.432
			[1×80]				[1×1]
			$\mathbf{W}_{CO_2}^{(1}$				$\mathbf{b}_{CO_2}^{(1)}$
1.357	−1.100	0.258	−0.915	0.037	−0.470	1.457	−2.770
0.354	0.370	0.225	0.014	0.002	−0.189	−0.062	−1.531
...	...	...	...	...	...	...	...
−0.793	−0.576	−0.881	1.429	−1.309	0.486	−1.072	−2.373
1.072	−0.457	0.368	−0.896	−0.316	−1.594	−0.714	2.981
			[80×7]				[80×1]
			$\mathbf{W}_{CO_2}^{(2}$				$\mathbf{b}_{CO_2}^{(2)}$
0.007	0.834	0.001	...	0.000	0.000	0.062	0.156
			[1×80]				[1×1]
			$\mathbf{W}_{W_c}^{(1)}$				$\mathbf{b}_{W_c}^{(1)}$
1.486	−0.013	1.280	−0.824	0.114	1.256	0.217	−2.599
−0.866	−0.166	−1.410	−0.714	0.536	1.232	−1.231	2.411
...	...	...	...	...	...	...	...
1.479	−0.982	−0.330	0.678	−0.819	−0.936	−0.913	2.582
1.552	0.000	−0.970	−0.273	−0.513	1.083	−0.262	2.586
			[80×7]				[80×1]
			$\mathbf{W}_{W_c}^{(2}$				$\mathbf{b}_{W_c}^{(2)}$
0.000	0.000	0.000	...	−0.001	0.000	0.000	0.809
			[1×80]				[1×1]

(Continued)

Table 3.2.2.3 (Continued) Weight and Bias Matrices for Cl_c, CO_2, W_c objective function obtained by One-Layer-80-Neurons Network Based on PTM for Cl_c, CO_2, W_c Objective Function

			(a-2): Reverse Network - ACI 318-14				
			$\mathbf{W}_{Cl_c}^{(1)}$				$\mathbf{b}_{Cl_c}^{(1)}$
−0.038	0.083	−0.026	−0.135	2.825	−0.268	−0.273	2.980
1.096	0.133	−0.203	0.122	−0.645	−1.130	−0.689	−2.920
...	...	...	...	...	...	...	...
1.383	−0.317	0.682	0.742	1.641	−0.622	−0.292	2.231
−0.569	−0.084	0.106	1.457	1.041	0.331	0.951	−3.404
			[80×7]				[80×1]
			$\mathbf{W}_{Cl_c}^{(2)}$				$\mathbf{b}_{Cl_c}^{(2)}$
1.031	0.030	1.244	...	4.953	−0.009	0.553	−0.663
			[1×80]				[1×1]
			$\mathbf{W}_{CO_2}^{(1)}$				$\mathbf{b}_{CO_2}^{(1)}$
−0.711	0.178	−0.629	−1.911	−0.337	−0.219	0.919	2.977
0.371	0.919	−0.388	0.996	−1.376	−0.521	0.382	−2.712
...	...	...	...	...	...	...	...
−1.478	−0.124	0.648	−0.927	−0.775	0.369	1.128	−2.525
−0.713	−1.976	−0.855	0.014	−0.508	−0.213	−0.416	−2.555
			[80×7]				[80×1]
			$\mathbf{W}_{CO_2}^{(2)}$				$\mathbf{b}_{CO_2}^{(2)}$
−0.766	−0.077	0.017	...	0.278	0.001	0.000	−1.523
			[1×80]				[1×1]
			$\mathbf{W}_{W_c}^{(1)}$				$\mathbf{b}_{W_c}^{(1)}$
−0.605	0.920	0.750	−1.878	−0.968	−0.314	−0.457	2.663
0.143	−0.232	0.498	−1.776	−0.256	0.335	−0.107	−3.691
...	...	...	...	...	...	...	...
0.173	−0.575	−0.213	−1.349	−2.247	0.373	0.048	−5.755
0.564	0.039	0.152	7.809	8.568	−0.295	−0.096	18.389
			[80×7]				[80×1]
			$\mathbf{W}_{W_c}^{(2)}$				$\mathbf{b}_{W_c}^{(2)}$
−0.847	−1.253	−0.224	...	1.347	−1.620	8.145	−1.621
			[1×80]				[1×1]

(Continued)

Table 3.2.2.3 (Continued) Weight and Bias Matrices for Cl_c, CO_2, W_c objective function obtained by One-Layer-80-Neurons Network Based on PTM for Cl_c, CO_2, W_c Objective Function

			(b-2): Reverse Network - ACI 318-19				
			$\mathbf{W}_{Cl_c}{}^{(1)}$				$\mathbf{b}_{Cl_c}{}^{(1)}$
−0.006	−0.059	−0.124	−3.475	0.292	0.423	−0.090	−3.318
−0.337	0.886	0.089	−1.803	−0.619	1.273	0.395	2.500
...	...	...	...	...	...	...	...
0.269	1.614	0.880	0.123	−0.723	1.377	−0.698	2.526
0.338	−0.575	0.071	1.771	0.636	1.434	−0.763	2.253
			[80×7]				[80×1]
			$\mathbf{W}_{Cl_c}{}^{(2)}$				$\mathbf{b}_{Cl_c}{}^{(2)}$
−0.380	0.563	−0.496	...	−0.543	−0.002	−0.011	−1.484
			[1×80]				[1×1]
			$\mathbf{W}_{CO_2}{}^{(1)}$				$\mathbf{b}_{CO_2}{}^{(1)}$
−1.062	−0.689	0.400	0.982	−0.911	0.647	−0.605	2.735
−1.209	0.326	−0.241	−0.855	−1.273	−0.681	1.144	2.631
...	...	...	...	...	...	...	...
0.575	0.352	0.000	−1.800	0.394	0.455	0.480	3.347
−0.592	0.354	−0.487	−0.595	−1.166	−0.218	0.016	−3.145
			[80×7]				[80×1]
			$\mathbf{W}_{CO_2}{}^{(2)}$				$\mathbf{b}_{CO_2}{}^{(2)}$
0.033	0.168	1.288	...	−0.226	−0.654	0.561	−0.386
			[1×80]				[1×1]
			$\mathbf{W}_{W_c}{}^{(1)}$				$\mathbf{b}_{W_c}{}^{(1)}$
0.617	0.321	−1.221	0.767	1.058	−0.188	1.271	−2.848
−0.992	0.858	−0.837	0.037	−1.275	−0.425	−0.304	2.731
...	...	...	...	...	...	...	...
0.348	−0.399	0.211	3.009	−0.281	0.144	−0.067	3.566
−0.839	−0.611	−1.214	−0.294	1.720	0.192	−0.777	−2.881
			[80×7]				[80×1]
			$\mathbf{W}_{W_c}{}^{(2)}$				$\mathbf{b}_{W_c}{}^{(2)}$
−0.094	−0.611	−0.291	...	1.151	1.203	−0.142	−1.044
			[1×80]				[1×1]

3.2.2.2 Weight and bias matrices based on reverse ANNs to derive objective functions

Weight and bias matrices are obtained to derive objective functions of CI_c, CO_2 emissions, and W_c under P_u and M_u based on reverse ANNs in Equations (3.2.2.5-1)–(3.2.2.5-3), respectively, for seven reverse inputs $\mathbf{x} = \left[\rho_s, f_c', f_y, P_u, M_u, SF, b/h\right]^T$. A safety factor ($SF$) and aspect ratio ($b/h$) are placed on an input-side in the reverse networks. The rest of the outputs ($\phi P_n, \phi M_n, b, h, \phi P_n, \phi M_n, \varepsilon_s, \alpha_{e/h}$) can also be found as stationary points. Weight and bias matrices shown in Table 3.2.2.3(a-2) based on ACI 318-14 and (b-2) based on ACI 318-19, respectively, are obtained with one hidden layer from Equations (3.2.2.5-1)–(3.2.2.5-3) based on a reverse ANNs for seven reverse inputs $\mathbf{x} = \left[\rho_s, f_c', f_y, P_u, M_u, SF, b/h\right]^T$.

ANNs identical to ones used to derive forward ANNs are used for reverse ANNs except for the exchanged input and output parameters. The reverse network with L layers and 80 neurons are linked using weighted interconnections and bias through an activation function, thereby performing nonlinear numerical computations. Weight and bias matrices are used to calculate Jacobian and Hessian matrices of the objective and constraining functions to solve KKT solutions. As shown in Figure 3.2.2.3 where PTM is used to train ANNs, only weight and bias matrices with one row are obtained to connect the last hidden layer shown in Figure 3.2.2.3-② to ($\phi P_n, \phi M_n, SF, b/h, \varepsilon_s, \alpha_{e/h}$ for forward outputs and $b, h, \phi P_n, \phi M_n, \varepsilon_s, \alpha_{e/h}$ for reverse outputs) and objective functions (CI_c, CO_2, W_c) of output layer shown in Figure 3.2.2.3-③-2. Figure 3.2.2.3a and b are also presented to illustrate forward and reverse networks, respectively. The activation function *tansig* (*tanh*) shown in Figure 3.2.2.2 is used in Equations (3.2.2.5-1)–(3.2.2.5-3) where seven reverse inputs $\mathbf{x} = \left[\rho_s, f_c', f_y, P_u, M_u, SF, b/h\right]^T$ shown in red for reverse networks are relayed to neurons of fully connected successive layers using weights at each neuron and bias in each hidden layer. The layers are then summed up for the outputs, CI_c, and CO_2 emissions, and W_c (Hong, 2019, 2021; Hong and Nguyen, 2021; Hong et al., 2021a, b, c, 2022a, b, c).

3.2.2.3 Jacobian and Hessian matrices derived based on ANNs

Stationary points of Lagrange functions shown in Equation (3.2.2.6-1) are identified when derivatives of Lagrange functions shown in Equations (3.2.2.6-2) and (3.2.2.6-3) converge to zero with respect to input variables $\mathbf{x}$ and Lagrange multipliers $\boldsymbol{\lambda}$. Lagrange functions shown in Equation (3.2.2.6-1) converge, then, to maxima or minima with respect to a stationary point ($\mathbf{x}$ and Lagrange multipliers $\boldsymbol{\lambda}$). Optimization problems are solved under first-order KKT conditions when inequality constraints are imposed as explained in Chapter 2. However, it is difficult to find a stationary point (input variables $\mathbf{x}$ and Lagrange multipliers $\boldsymbol{\lambda}$) at which Equations (3.2.2.6-2) and (3.2.2.6-3) converge to zero when Lagrange functions are complex. A first-order derivative or Jacobi matrix of Lagrange functions shown in Equation (3.2.2.6-2) should be linearized based on Hessian matrix to use Newton–Raphson iteration as shown in Equation (3.2.2.7). Readers are referred to Equations (2.3.4.1-1)–(2.3.4.1–5) of Chapter 2 which are repeated in Equations (3.2.2.7-1)–(3.2.2.7-5).

Jacobi and Hessian matrices of Lagrange function shown in Equation (3.2.2.7) should be obtained to find candidate solutions for the KKT conditions (Villarrubia et al., 2018). Functions based on ANNs must be at least differentiable twice, resulting in the Jacobi and Hessian matrices because first order derivatives or Jacobi matrices of Lagrange functions should be linearized in terms of higher derivatives with respective $\mathbf{x}$, $\boldsymbol{\lambda}_c$, and $\boldsymbol{\lambda}_v$. Jacobi matrix $\nabla\mathcal{L}(\mathbf{x}, \boldsymbol{\lambda}_c, \boldsymbol{\lambda}_v)$ is linearized with respect to $\mathbf{x}^{(0)}$, $\boldsymbol{\lambda}_c^{(0)}$, $\boldsymbol{\lambda}_v^{(0)}$ as shown in Equations (2.3.4.1-1) and (2.3.4.1-2) of Chapter 2 or Equations (3.2.2.7-1) and (3.2.2.7-2) Newton–Raphson method is implemented to find stationary points of Lagrange functions.

$$\mathcal{L}(\mathbf{x}, \boldsymbol{\lambda}_c, \boldsymbol{\lambda}_v) = f(\mathbf{x}) - \boldsymbol{\lambda}_c^T c(\mathbf{x}) - \boldsymbol{\lambda}_v^T v(\mathbf{x}) \quad (3.2.2.6\text{-}1)$$

- First order KKT conditions:

$$\nabla\mathcal{L}\left(\mathbf{x}_i^{(k)},\boldsymbol{\lambda}_c^{(k)},\boldsymbol{\lambda}_v^{(k)}\right)=\nabla f\left(\mathbf{x}_i\right)-\boldsymbol{\lambda}_c^T\nabla\boldsymbol{c}\left(\mathbf{x}_i\right)-\boldsymbol{\lambda}_v^T\mathbf{S}\nabla\boldsymbol{v}\left(\mathbf{x}_i\right)=0 \qquad (3.2.2.6\text{-}2)$$

$$\begin{cases}\nabla_x\mathcal{L}=0\rightarrow\dfrac{\partial\mathcal{L}}{\partial x_i}=0, & i=1,\ldots,n\\ \nabla_{\lambda_c}\mathcal{L}=0\rightarrow\dfrac{\partial\mathcal{L}}{\partial\lambda_{c,j}}=0, & j=1,\ldots,m_1\\ \nabla_{\lambda_v}\mathcal{L}=0\rightarrow\dfrac{\partial\mathcal{L}}{\partial\lambda_{v,k}}=0, & k=1,\ldots,m_2\end{cases} \qquad (3.2.2.6\text{-}3)$$

$$\nabla\mathcal{L}\left(\mathbf{x}^{(0)}+\Delta\mathbf{x},\boldsymbol{\lambda}_c^{(0)}+\Delta\boldsymbol{\lambda}_c,\boldsymbol{\lambda}_v^{(0)}+\Delta\boldsymbol{\lambda}_v\right)\approx\nabla\mathcal{L}\left(\mathbf{x}^{(0)},\boldsymbol{\lambda}_c^{(0)},\boldsymbol{\lambda}_v^{(0)}\right)+\left[\boldsymbol{H}_{\mathcal{L}}\left(\mathbf{x}^{(0)},\boldsymbol{\lambda}_c^{(0)},\boldsymbol{\lambda}_v^{(0)}\right)\right]\begin{bmatrix}\Delta\mathbf{x}\\ \Delta\boldsymbol{\lambda}_c\\ \Delta\boldsymbol{\lambda}_v\end{bmatrix}+\mathbf{0}^+ \qquad (3.2.2.7\text{-}1)$$

$$\begin{bmatrix}\Delta\mathbf{x}\\ \Delta\boldsymbol{\lambda}_c\\ \Delta\boldsymbol{\lambda}_v\end{bmatrix}\approx-\left[\boldsymbol{H}_{\mathcal{L}}\left(\mathbf{x}^{(0)},\boldsymbol{\lambda}_c^{(0)},\boldsymbol{\lambda}_v^{(0)}\right)\right]^{-1}\nabla\mathcal{L}\left(\mathbf{x}^{(0)},\boldsymbol{\lambda}_c^{(0)},\boldsymbol{\lambda}_v^{(0)}\right) \qquad (3.2.2.7\text{-}2)$$

$$\begin{bmatrix}\Delta\mathbf{x}^{(k)}\\ \Delta\boldsymbol{\lambda}_c^{(k)}\\ \Delta\boldsymbol{\lambda}_v^{(k)}\end{bmatrix}=-\left[\boldsymbol{H}_{\mathcal{L}}\left(\mathbf{x}^{(k)},\boldsymbol{\lambda}_c^{(k)},\boldsymbol{\lambda}_v^{(k)}\right)\right]^{-1}\nabla\mathcal{L}\left(\mathbf{x}^{(k)},\boldsymbol{\lambda}_c^{(k)},\boldsymbol{\lambda}_v^{(k)}\right) \qquad (3.2.2.7\text{-}3)$$

$$\begin{bmatrix}\mathbf{x}^{(k+1)}\\ \boldsymbol{\lambda}_c^{(k+1)}\\ \boldsymbol{\lambda}_v^{(k+1)}\end{bmatrix}=\begin{bmatrix}\mathbf{x}^{(k)}\\ \boldsymbol{\lambda}_c^{(k)}\\ \boldsymbol{\lambda}_v^{(k)}\end{bmatrix}+\begin{bmatrix}\Delta\mathbf{x}^{(k)}\\ \Delta\boldsymbol{\lambda}_c^{(k)}\\ \Delta\boldsymbol{\lambda}_v^{(k)}\end{bmatrix} \qquad (3.2.2.7\text{-}4)$$

$$\begin{bmatrix}\mathbf{x}^{(k+1)}\\ \boldsymbol{\lambda}_c^{(k+1)}\\ \boldsymbol{\lambda}_v^{(k+1)}\end{bmatrix}=\begin{bmatrix}\mathbf{x}^{(k)}\\ \boldsymbol{\lambda}_c^{(k)}\\ \boldsymbol{\lambda}_v^{(k)}\end{bmatrix}-\left[\boldsymbol{H}_{\mathcal{L}}\left(\mathbf{x}^{(k)},\boldsymbol{\lambda}_c^{(k)},\boldsymbol{\lambda}_v^{(k)}\right)\right]^{-1}\nabla\mathcal{L}\left(\mathbf{x}^{(k)},\boldsymbol{\lambda}_c^{(k)},\boldsymbol{\lambda}_v^{(k)}\right) \qquad (3.2.2.7\text{-}5)$$

Objective functions shown in Equations (3.2.2.5-1)–(3.2.2.5-3) are trained and optimized based on appropriate training parameters shown in Table 3.2.2.2, such as layers, neurons, and activation functions. Objective functions defined as a function of seven input design parameters $(b, h, \rho_s, f_c', f_y, P_u, M_u)$ for forward networks and $(\rho_s, f_c', f_y, P_u, M_u, SF, b/h)$ for reverse networks are optimized with respect to design parameters under equality and inequality constraints, $\mathbf{c}(\mathbf{x})$ and $\mathbf{v}(\mathbf{x})$ listed in Table 3.2.2.5, respectively. Table 3.2.2.4(a) shows three-column design scenarios with which CI_c, CO_2, and W_c are to be optimized whereas Table 3.2.2.4(b) shows two training scenarios based on a forward and reverse ANN-based Lagrange optimization. In this chapter, an objective function CI_c shown in Equation (3.2.2.5-1) is defined as a function of seven input design parameters $(b, h, \rho_s, f_c', f_y, P_u, M_u)$.

Table 3.2.2.4 Column Design Scenarios for Optimization

(a): Design Scenario

Design Case	*Design Target*	*Design Criteria*
Design 1	Minimized CI_c	Design a **square** column under:
Design 2	Minimized CO_2	P_u = **1,000** kN, & M_u = **3,000** kN·m and Safety factor = **1**.
Design 3	Minimized W_c	Given material properties:
		f_c' = **40** MPa, f_y = **500** MPa.

(b): Training Scenario

	Inputs from Structural Software (AutoCol)							*Outputs from Structural Software (AutoCol)*						*Design Efficiency Indexes*	
Training Network	*bh (mm)*	ρ_s	f_c' *(MPa)*	f_y *(MPa)*	P_u *(kN)*	M_u *(kN×m)*	ϕP_n *(kN)*	ϕM_n *(kN×m)*	*SF*	*b/h*	ε_s	$\alpha_{e/h}$	CI_c *(Won/m)*	CO_2 *(t-CO₂/m)*	W_c *(kN/m)*
1 Forward/ Lagrange	i	i	i	i	i	i	i	o	o	o	o	o	o	o	o
2 Reverse/ Lagrange	i	o	o	i	i	i	i	o	o	i	i	o	o	o	o
Given Input:	i														
Output:	o														

for forward networks and (ρ_s, f_c', f_y, P_u, M_u, SF, b/h) for reverse networks. Equality and inequality constraints are established to meet the requirements of design codes and any restriction imposed by government or architects. Stationary points of Lagrange functions are, then, identified under constraints with one or more equality and inequality constraints based on Newton–Raphson method shown in Figure 3.2.2.4. ANN-based Lagrange optimization is performed following three steps.

- **Step 1:** Structural large datasets are generated from conventional design software, such as *AutoCol*. A proper number of large datasets needed for training ANNs should be selected carefully based on the level of complexity of the considered problem. Acceptable training accuracy based on 100,000 datasets is yielded for an RC column as shown in Table 3.2.2.2.
- **Step 2:** Weight and bias matrices are obtained by training ANNs representing objective functions based on large datasets obtained from Step 1. Accuracies of weight and bias matrices are considerably affected not only by a number of large datasets but also by parameters selected to train ANNs, such as a number of hidden layers, neurons, required epochs, etc. Proper training parameters for ANNs should be implemented to achieve good training results.
- **Step 3:** LMM is applied to optimize ANN-based objective functions which replace analytically obtained analytical objective functions. The aim of ANN-based objective functions is to approximate objective functions and generalize the calculation procedure for Jacobian and Hessian matrices necessary when first order KKT conditions are linearized as shown in Equation (3.2.2.7).

Table 3.2.2.5 Summary of equality and inequality constraints for Designs 1, 2, 3

(a): Forward Network

Equality Constraint	*Inequality Constraints*
$c(x)=\left[c_1(x),c_2(x),\ldots,c_6(x)\right]^T$	$v(x)=\left[v_1(x),v_2(x)\right]^T$
$c_1(x)=b-h=0$	$v_1(x)=\rho_s-0.01\geq 0$
$c_2(x)=f_c'-40=0$	$v_2(x)=-\rho_s+0.08\geq 0$
$c_3(x)=f_y-500=0$	
$c_4(x)=P_u-1,000=0$	
$c_5(x)=M_u-3,000=0$	
$c_6(x)=SF-1=0$	

(b): Reverse Network

Equality Constraint	*Inequality Constraints*
$c(x)=\left[c_1(x),c_2(x),\ldots,c_6(x)\right]^T$	$v(x)=\left[v_1(x),v_2(x)\right]^T$
$c_1(x)=b/h-1=0$	$v_1(x)=\rho_s-0.01\geq 0$
$c_2(x)=f_c'-40=0$	$v_2(x)=-\rho_s+0.08\geq 0$
$c_3(x)=f_y-500=0$	
$c_4(x)=P_u-1,000=0$	
$c_5(x)=M_u-3,000=0$	
$c_6(x)=SF-1=0$	

3.2.2.4 Stationary points of Lagrange functions $\mathcal{L}(\mathbf{x},\lambda_c,\lambda_v)$ subject to constraining conditions based on Newton–Raphson iteration

A Jacobi matrix $\nabla\mathcal{L}(\mathbf{x},\lambda_c,\lambda_v)$ shown in Equation (3.2.2.6-2) of Lagrange functions $\mathcal{L}(\mathbf{x},\lambda_c,\lambda_v)$ shown in Equation (3.2.2.6-1) is derived to find stationary points of Lagrange functions $\mathcal{L}(\mathbf{x},\lambda_c,\lambda_v)$ subject to constraining conditions. Initial trial stationary points $\mathbf{x},\lambda_c,\lambda_v$ are updated using Equation (3.2.2.7-5) based on Hessian matrix which is repeated until a Jacobi matrix $\nabla\mathcal{L}(\mathbf{x},\lambda_c,\lambda_v)$ of Lagrange functions $\mathcal{L}(\mathbf{x},\lambda_c,\lambda_v)$ converges to 0. Iterations for convergence to seek an optimized Lagrange function at maxima or minima based on forward and reverse networks are based on Newton–Raphson iteration as illustrated in Figure 3.2.2.4a and b, respectively. Stationary points of Lagrange functions can be identified by equating Equation (3.2.2.6-2) to 0 with respect to $\mathbf{x}$, λ_c, and λ_v, leading to simultaneous equations shown in Equation (3.2.2.6-3). However, finding $\mathbf{x}$, λ_c, and λ_v to locate stationary points of Lagrange function directly is difficult for complex Lagrange functions, and hence, Equation (3.2.2.6-2) is linearized as shown in Equations (3.2.2.7-1)–(3.2.2.7-5), to find stationary points of Lagrange function. Initial trial input parameters are updated based on Equations (3.2.2.7-1)–(3.2.2.7-5) until $\nabla\mathcal{L}(\mathbf{x},\lambda_c,\lambda_v)$ converges within tolerated error ranges. Hessian matrices shown in Equations (3.2.2.7-1)–(3.2.2.7-5) are calculated using Equations (3.2.2.8-1) to (3.2.2.8-3) or Equation (2.3.4.3-1) in which Hessian matrices

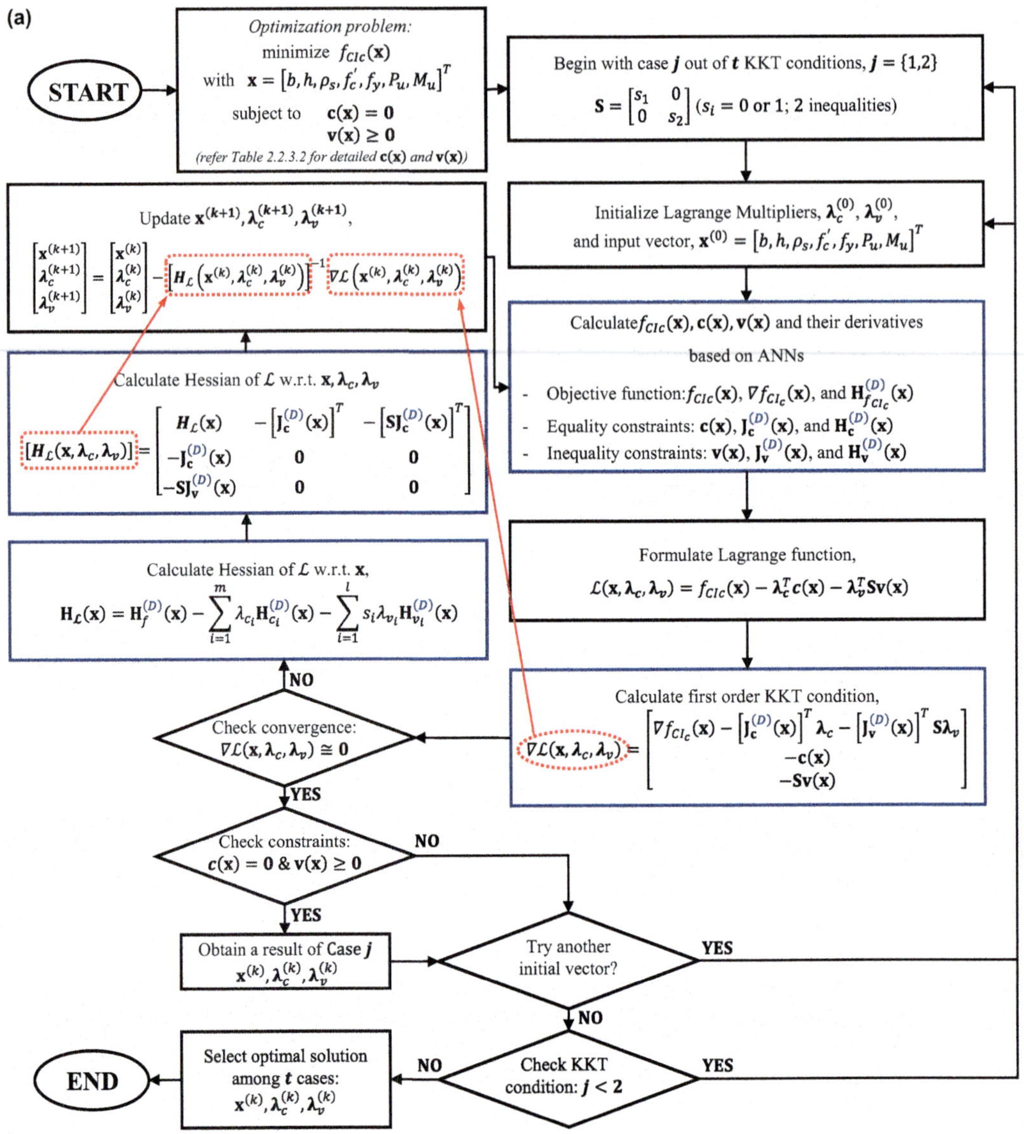

Figure 3.2.2.4 A flow of equations adopted in an iteration based on Newton–Raphson method. (a) Forward network with 2 inequities. (b) Reverse network with 2 inequities.

$\frac{\partial^2 \mathcal{L}}{\partial \mathbf{x}^2} = \mathbf{H}_\mathcal{L}(\mathbf{x})$, $\frac{\partial^2 f(\mathbf{x})}{\partial \mathbf{x}^2} = \mathbf{H}_f(\mathbf{x})$, $\frac{\partial^2 (\boldsymbol{\lambda}_c^T \mathbf{c}(\mathbf{x}))}{\partial \mathbf{x}^2} = \boldsymbol{H}_{c_i}(\mathbf{x})$, and $\frac{\partial^2 (\boldsymbol{\lambda}_v^T \mathbf{S}\mathbf{v}(\mathbf{x}))}{\partial \mathbf{x}^2} = \boldsymbol{H}_{v_i}(\mathbf{x})$ shown in Equations (3.2.2.8-4) to (3.2.2.8-9) or Equations (2.3.4.3-3) to (2.3.4.3-8) are obtained by differentiating a Lagrange functions, objective functions, equality, and inequality functions twice with respect to input variable $\mathbf{x}$, respectively, whereas Hessian matrices $\frac{\partial^2 \mathcal{L}}{\partial \boldsymbol{\lambda}_c^2} = \mathbf{H}_\mathcal{L}(\boldsymbol{\lambda}_c)$ shown in (2.3.4.3–6) and $\frac{\partial^2 \mathcal{L}}{\partial \boldsymbol{\lambda}_v^2} = \mathbf{H}_\mathcal{L}(\boldsymbol{\lambda}_v)$ shown in Equation (2.3.4.3–7) are obtained by differentiating a Lagrange function twice with respect to Lagrange multipliers $\boldsymbol{\lambda}_c$ and $\boldsymbol{\lambda}_v$.

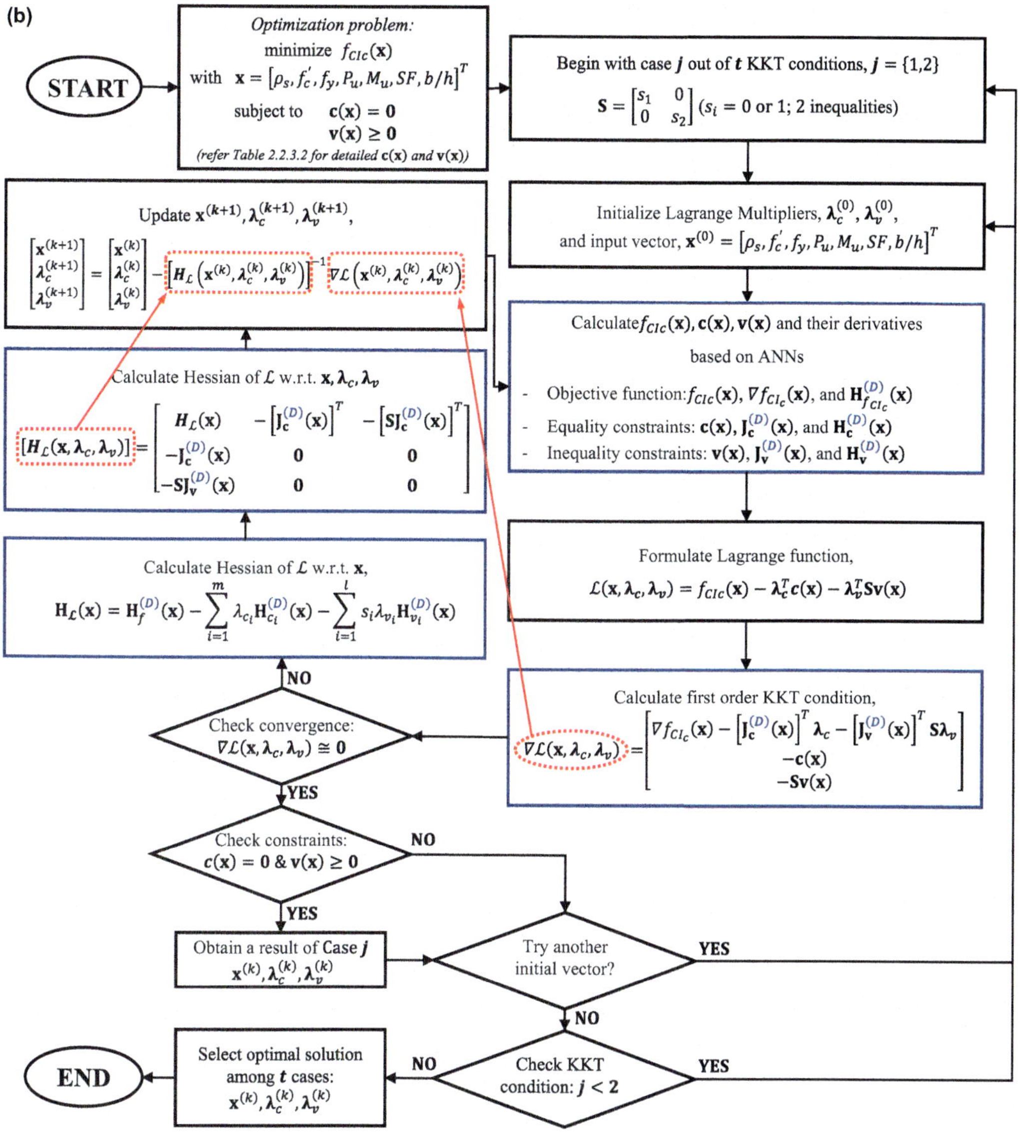

Figure 3.2.2.4 (Continued) A flow of equations adopted in an iteration based on Newton–Raphson method. (a) Forward network with 2 inequities. (b) Reverse network with 2 inequities.

respectively. $\mathrm{J_c}(\mathbf{x})$ and $\mathrm{J_v}(\mathbf{x})$ shown in Equation (3.2.2.8-3) are Jacobi matrices of equality and inequality functions, respectively, with respect to input parameters $\mathbf{x}$ which can be calculated using Equation (2.3.4.2-2b) and Equation (2.3.4.2-2c), respectively. ANN-based functions with their Jacobi and Hessian matrices are obtained based on weight and bias matrices. Table 3.2.2.3 presents an example calculation of weight and bias matrices using an ANN with one hidden layer. A calculation sequence of Jacobi and Hessian matrices such as $\mathbf{z}^{(N)} \Rightarrow \mathbf{J}^{(N)} \Rightarrow \mathbf{H}^{(N)}$, $\mathbf{z}^{(l)} \Rightarrow \mathbf{J}^{(l)} \Rightarrow \mathbf{H}^{(l)}$, $\mathbf{z}^{(L)} \Rightarrow \mathbf{J}^{(L)} \Rightarrow \mathbf{H}^{(L)}$, $\mathbf{z}^{(D)} \Rightarrow \mathbf{J}^{(D)} \Rightarrow \mathbf{H}^{(D)}$ is summarized in Figure 3.2.2.5. Readers are referred to Sections 2.3.4.2 and 2.3.6 for the formulation of ANN-based Jacobian and Hessian matrices.

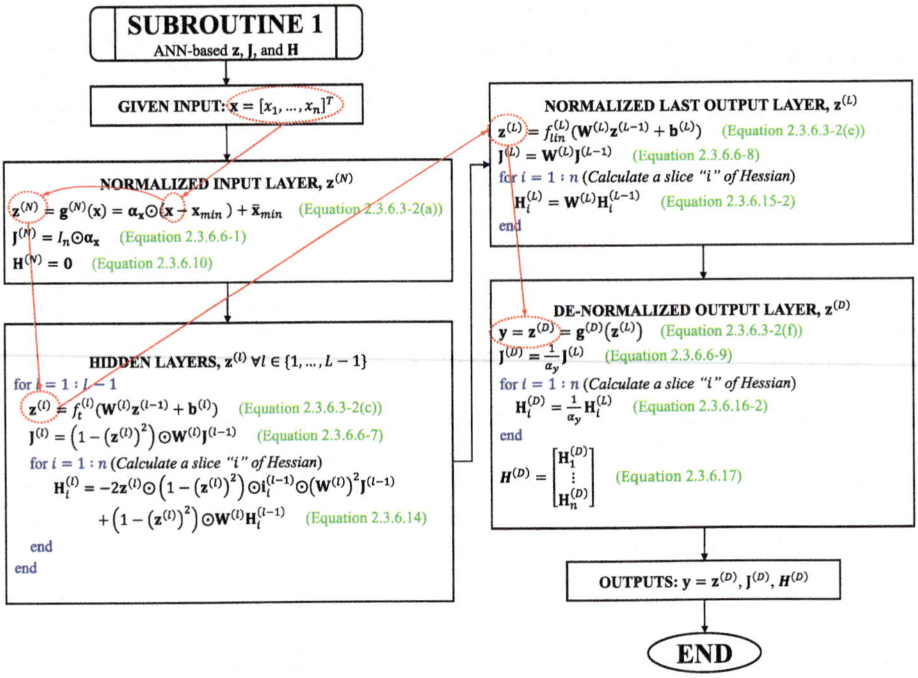

Figure 3.2.2.5 A flow of equations to obtain functions including objective, equality, inequality functions, and their Jacobian and Hessian matrix using weight and bias matrix.

$$\left[\boldsymbol{H}_{\mathcal{L}}\left(\mathbf{x}^{(k)},\boldsymbol{\lambda}_c^{(k)},\boldsymbol{\lambda}_v^{(k)}\right)\right]=\begin{bmatrix}\dfrac{\partial^2\mathcal{L}}{\partial\mathbf{x}^2} & \dfrac{\partial^2\mathcal{L}}{\partial\mathbf{x}\,\partial\boldsymbol{\lambda}_c} & \dfrac{\partial^2\mathcal{L}}{\partial\mathbf{x}\,\partial\boldsymbol{\lambda}_v}\\ \dfrac{\partial^2\mathcal{L}}{\partial\boldsymbol{\lambda}_c\,\partial\mathbf{x}} & \dfrac{\partial^2\mathcal{L}}{\partial\boldsymbol{\lambda}_c^2} & \dfrac{\partial^2\mathcal{L}}{\partial\boldsymbol{\lambda}_c\,\partial\boldsymbol{\lambda}_v}\\ \dfrac{\partial^2\mathcal{L}}{\partial\boldsymbol{\lambda}_v\,\partial\mathbf{x}} & \dfrac{\partial^2\mathcal{L}}{\partial\boldsymbol{\lambda}_v\,\partial\boldsymbol{\lambda}_c} & \dfrac{\partial^2\mathcal{L}}{\partial\boldsymbol{\lambda}_v^2}\end{bmatrix} \qquad (3.2.2.8\text{-}1)$$

$$\left[\boldsymbol{H}_{\mathcal{L}}\left(\mathbf{x}^{(k)},\boldsymbol{\lambda}_c^{(k)},\boldsymbol{\lambda}_v^{(k)}\right)\right]=\begin{bmatrix}\boldsymbol{H}_{\mathcal{L}}\left(\mathbf{x}^{(k)}\right) & -\boldsymbol{J}_c\left(\mathbf{x}^{(k)}\right)^T & -\left(\boldsymbol{S}\boldsymbol{J}_v\left(\mathbf{x}^{(k)}\right)\right)^T\\ -\boldsymbol{J}_c\left(\mathbf{x}^{(k)}\right) & 0 & 0\\ -\boldsymbol{S}\boldsymbol{J}_v\left(\mathbf{x}^{(k)}\right) & 0 & 0\end{bmatrix} \qquad (3.2.2.8\text{-}2)$$

$$\left[\boldsymbol{H}_{\mathcal{L}}\left(\mathbf{x}^{(k)},\boldsymbol{\lambda}_c^{(k)},\boldsymbol{\lambda}_v^{(k)}\right)\right]=\begin{bmatrix}\boldsymbol{H}_{\mathcal{L}}\left(\mathbf{x}^{(k)}\right) & -\left[\mathbf{J}_c^{(D)}\right]^T & -\left[\boldsymbol{S}\mathbf{J}_v^{(D)}\right]^T\\ -\mathbf{J}_c^{(D)} & 0 & 0\\ -\boldsymbol{S}\mathbf{J}_v^{(D)} & 0 & 0\end{bmatrix} \qquad (3.2.2.8\text{-}3)$$

3.3 OPTIMIZATION OF COLUMN DESIGNS BASED ON AN ANN-BASED LAGRANGE ALGORITHM

Objective functions CI_c, CO_2, and W_c based on design criterion shown in Table 3.2.2.4(a) are minimized by ANNs-based forward Lagrange and reverse Lagrange shown in Training scenarios

1 and 2 of Table 3.2.2.4(b). Column designs optimized based on both forward and reverse using PTM training are validated by large datasets. Minimized CI_c, CO_2, and W_c are also identified based on large datasets shown in Figures 3.3.1.2–3.3.3.2 to verify an ANN-based Lagrange optimization. It is challenging to use a conventional method of Lagrange multipliers to optimize column designs similar to those shown in this chapter, but this can be achieved using ANNs.

3.3.1 Column design scenario minimizing CI_c

3.3.1.1 Formulation of Lagrange optimization based on forward network

1. Lagrange multiplier-based optimization imposed by equality constraints $\mathbf{c}(\mathbf{x})$ and inequality constraints $\mathbf{v}(\mathbf{x})$

   Lagrange optimization based on forward network is performed to minimize CI_c in this section. An ANN is derived in Equation (3.3.1.1) to minimize cost index (CI_c) of a RC column shown in Figure 3.2.2.1 under several design requirements shown in Table 3.2.2.5. All design requirements can be expressed in term of equality constraints $\mathbf{c}(\mathbf{x})=\left[c_1(\mathbf{x}),\ldots,c_6(\mathbf{x})\right]^T$ described in Table 3.2.2.5. Besides, rebar ratio ρ_s should be constrained following ACI 318-14 and ACI 318-19 code requirements ($\rho_{s,\min} \le \rho_s \le \rho_{s,\max}$), which are expressed in terms of two inequality constraints: $v_1(\mathbf{x})=\rho_s-\rho_{s,\min}\ge 0$ and $v_2(\mathbf{x})=-\rho_s+\rho_{s,max}\ge 0$ shown in Table 3.2.2.5. An objective function representing CI_c for forward optimization is defined as $CI_c=f_{CI_c}^{FW}(\mathbf{x})$ based on an ANN as shown in Equation (3.3.1.2-1).

$$\underbrace{CI_c}_{[1\times1]}=f_{CI_c}^{RV}(\mathbf{x})$$

$$=g_{CI_c}^{D}\left(f_l^{L}\left(\underbrace{\mathbf{W}_{CI_c}^{L}}_{[1\times80]}f_t^{L-1}\left(\underbrace{\mathbf{W}_{CI_c}^{L-1}}_{[80\times80]}\ldots f_t^{1}\left(\underbrace{\mathbf{W}_{CI_c}^{1}}_{[80\times7]}\underbrace{g_{CI_c}^{N}(\mathbf{x})}_{[7\times1]}+\underbrace{\mathbf{b}_{CI_c}^{1}}_{[80\times1]}\right)\ldots+\underbrace{\mathbf{b}_{CI_c}^{L-1}}_{[80\times1]}\right)+\underbrace{b_{CI_c}^{L}}_{[1\times1]}\right)\right) \tag{3.3.1.1}$$

Lagrange function for cost index $\mathcal{L}_{CI_c}^{FW}$ and its KKT conditions are, then, expressed as a function of input variables $\mathbf{x}=\left[b,\, h,\, \rho_s, f_c', f_y,\, P_u,\, M_u\right]^T$ with Lagrange multiplier of equality and inequality constraints, $\boldsymbol{\lambda}_c=\left[\lambda_1,\, \lambda_2,\ldots,\, \lambda_6\right]^T$ and $\boldsymbol{\lambda}_v=\left[\lambda_1,\, \lambda_2\right]^T$, respectively, as shown in Equation (3.3.1.2-2). Candidate solutions from KKT conditions are obtained from Equation (3.3.1.2-3).

Objective function CI_c:

$$CI_c=f_{CI_c}^{FW}(\mathbf{x}) \tag{3.3.1.2-1}$$

Lagrange function for CI_c:

$$\mathcal{L}_{CI_c}^{FW}(\mathbf{x},\boldsymbol{\lambda}_c,\boldsymbol{\lambda}_v)=f_{CI_c}^{FW}(\mathbf{x})-\boldsymbol{\lambda}_c^T\mathbf{c}(\mathbf{x})-\boldsymbol{\lambda}_v^T\boldsymbol{S}\mathbf{v}(\mathbf{x}) \tag{3.3.1.2-2}$$

Candidate solution from KKT conditions:

$$\nabla\mathcal{L}_{CI_c}^{FW}(\mathbf{x},\boldsymbol{\lambda}_c,\boldsymbol{\lambda}_v)=\begin{bmatrix}\nabla f_{CI_c}^{FW}(\mathbf{x})-\boldsymbol{J}_c(\mathbf{x})^T\boldsymbol{\lambda}_c-\boldsymbol{J}_v(\mathbf{x})^T\mathbf{S}\boldsymbol{\lambda}_v\\ -\mathbf{c}(\mathbf{x})\\ -Sv(\mathbf{x})\end{bmatrix} \tag{3.3.1.2-3}$$

where $f_{CI_c}^{FW}(\mathbf{x})$ is an objective function CI_c based on a forward network shown in Equation (3.3.1.1); $\mathbf{c}(\mathbf{x})=\left[c_1(\mathbf{x}), c_2(\mathbf{x}), \ldots, c_6(\mathbf{x})\right]^T$, $\mathbf{v}(\mathbf{x})=\left[v_1(\mathbf{x}), v_2(\mathbf{x})\right]^T$ are equality and inequality conditions constraining an objective function CI_c, respectively. Table 3.2.2.4(a) describes three-design scenarios to optimize RC columns whereas Table 3.2.2.4(b) shows training scenarios to derive ANNs for the three design scenarios shown in Table 3.2.2.4(a).

2. Formulation of initial trial input parameters to solve KKT equations based on Newton–Raphson method

   It is well-known that Newton–Raphson method relies significantly on good initial trial input parameters assumed as $\mathbf{x}^{(0)}=\left[b, h, \rho_s, f_c', f_y, P_u, M_u\right]^T$ for a forward deign to expedite a run progress as well as enhance accuracy. Good initial trial input parameters are predetermined based on simple equality and inequality which are active or inactive; $c_2(\mathbf{x})$, $c_3(\mathbf{x})$, $c_4(\mathbf{x})$, $c_5(\mathbf{x})$ and $v_1(\mathbf{x})$, $v_2(\mathbf{x})$ shown in Tables 3.2.2.5(a) and 3.2.2.5(b) for forward and reverse networks, respectively.

   1. Case 1 KKT condition for active inequality constraint $\mathbf{v}_1$ and inactive inequality constraint $\mathbf{v}_2$

      Lagrange function is formulated based on an objective function, regular equality constraints, and those obtained with active and inactive inequality conditions. Lagrange multipliers $\boldsymbol{\lambda}_c$ and $\boldsymbol{\lambda}_v$ adopted with active inequality conditions should be greater than 0. In this KKT condition, inequality $v_1(x)$ is assumed to constrain an objective function by binding Inequality $v_1(x)$ to equality constraint. Lagrange function is obtained in Equation (3.3.1.3) by substituting $S=\begin{bmatrix} 1 & 0 \\ 0 & 0 \end{bmatrix}$ into Equation (3.3.1.2-2) when inequality constraint v_1 is activated. However, a stationary point of a Lagrange function of Case 1 KKT condition should be verified to exist within an inequality constraining $v_2(x)$.

$$\mathcal{L}_{CIC}^{FW}(\mathbf{x}, \boldsymbol{\lambda}_c, \boldsymbol{\lambda}_v)=f_{CIC}^{FW}(\mathbf{x})-\boldsymbol{\lambda}_c^T\mathbf{c}(\mathbf{x})-\boldsymbol{\lambda}_v^T\begin{bmatrix} 1 & 0 \\ 0 & 0 \end{bmatrix}\mathbf{v}(\mathbf{x}) \tag{3.3.1.3}$$

      As shown in Equation (3.3.1.4), rebar ratio ρ_s is assumed as 0.01 because inequality $v_1(x)$ is assumed to constrain an objective function by binding Inequality $v_1(x)$ to equality constraint when v_1 is activated. The other four input variables (f_c', f_y, P_u, M_u) are predetermined as 40 MPa, 500 MPa, 1,000 kN, 3,000 kNm based on Case 1 KKT condition based on simple equality constraints $c_2(\mathbf{x})$, $c_3(\mathbf{x})$, $c_4(\mathbf{x})$, and $c_5(\mathbf{x})$ shown in Table 3.2.2.5.

$$\mathbf{x}^{(0)}=\left[b, h, \rho_s, f_c', f_y, P_u, M_u\right]^T=\left[b, h, 0.01, 40, 500, 1{,}000, 3{,}000\right]^T \tag{3.3.1.4}$$

      Beam width b and beam depth h are unknowns to be determined during optimization based on Newton–Raphson iteration. Initial trial input parameters are, then, assumed as $\mathbf{x}^{(0)}=\left[b, h, 0.01, 40, 500, 1{,}000, 3{,}000\right]^T$ to find a stationary point of Lagrange function shown in Equations (3.3.1.2-2) or (3.3.1.3) based on Newton–Raphson iteration. Let's find the unknown initial trial input parameters between $300 \le b \le 2250$ and $300 \le h \le 1500$ in large datasets during the iteration. Newton–Raphson iteration is implemented in solving first order partial differential equations (Jacobi of Lagrange function) shown in Equation (3.3.1.2-3) using one set of initial trial Lagrange multipliers $[\lambda_{c1}, \lambda_{c2}] = [0,\ 0]$ and $5^2 = 25$ initial trial input parameters of $\mathbf{x}^{(0)} = [b, h, 0.01,\ 40,\ 500,\ 1{,}000,\ 3{,}000]^T$ in which b and h are randomly distributed within a training data range. However, a number of initial trial input parameters can be adjusted based on a number of unknowns in Newton–Raphson iterations. A number of initial trial input parameters influences convergence speed. Initial trial Lagrange multipliers are assumed as $\left[\lambda_{c1}^{(0)},\ \lambda_{c2}^{(0)}\right] = [0,\ 0]$.

Initial values of 0 for trial Lagrange multipliers are used because initial trial input parameters can be assumed arbitrarily, and hence, they do not have boundaries; they can be of any number while Newton–Raphson method iterations update to correct Lagrange multipliers. The author tried several initial trial values for Lagrange multipliers and obtained the same stationary points regardless of assumed initial values for Lagrange multipliers. On the other hand, 25 initial trial input parameters of $\mathbf{x}^{(0)}$ are selected being divided each unknown variable (b, h) into five ranges. Final initial trial input parameters including initial trial Lagrange multipliers become $\left[\mathbf{x}^{(0)},\ \lambda_{c1}^{(0)},\ \lambda_{c2}^{(0)}\right]^T = [b, h, 0.01,\ 40,\ 500,\ 1{,}000,\ 3{,}000,\ 0,\ 0]^T$. ANNs yield the best test MSE based on a combination of one, two, five, and ten hidden layers with 80 neurons. Iterations progress to find roots of KKT condition shown in Equation (3.3.1.2-3) to make $[b, h]$ and $[\lambda_{c1}, \lambda_{c2}]$ converge to $[1{,}068.0,\ 1{,}068.0]$ and $[6\times10^{-5},\ 0.0994]$, respectively, for ACI 318-14 and converge to $[1{,}071.1,\ 1{,}071.1]$ and $[6\times10^{-5},\ 0.0984]$, respectively, for ACI 318-19 based on 25 trial initial vectors as shown in Tables 3.3.1.1(e-1) and 3.3.1.1(e-2). As shown in Tables 3.3.1.1(e-1) and 3.3.1.1(e-2) based on best training for the Case 1 KKT condition, the minimized $CI_c = f_{CI_c}^{FW}(\mathbf{x})$ of 202,692.7 based on ACI 318-14 and $CI_c = f_{CI_c}^{FW}(\mathbf{x})$ of 203,141.0 for ACI 318-19 are yielded as roots of the Case 1 KKT condition shown in Equation (3.3.1.2-3) when inequality constraint v_1 is activated. Roots of the Case 1 KKT condition must meet inequality constraint v_2. Tables 3.3.1.2(a-1) and (a-2) based on one hidden layer and 80 neurons for the Case 1 KKT condition show the minimized $CI_c = f_{CI_c}^{FW}(\mathbf{x})$ of 192,921.0 for ACI 318-14 and of 195,308.3 for ACI 318-19, however, errors are relatively large. Minimized CI_c based on two, five, and ten hidden layers with 80 neurons are also shown in Tables 3.3.1.1(b-1)–(d-1) and 3.3.1.1(b-2)–(d-2) for ACI 318-14 and ACI 318-19, respectively. According to Table 3.3.1.1 based on ANN with ten hidden layers and 80 neurons of Case 1 KKT condition, the minimized CI_c is identified as 201,393.5 and as 203,678.7 for ACI 318-14 and ACI 318-19, respectively.

Trainings must be performed which produce best designs, not only depending on training accuracies. Training accuracies with one layer is better than those with ten layers as shown in Table 3.2.2.2, however, design accuracies with ten layers are better than those with one layer as shown in Table 3.3.1.2, indicating that a better training does not always provide a more accurate design as shown in Tables 3.2.2.2 and 3.3.1.2. Choosing training parameters is not like formula, but experiences and intuitions are also needed.

2. Case 2 KKT condition for inactive inequality constraint $\boldsymbol{v}_1$ and active inequality constraint $\boldsymbol{v}_2$

Lagrange function is formulated based on objective functions, regular equality constraints, and those obtained with active and inactive inequality conditions. Lagrange multipliers $\boldsymbol{\lambda}_c$ and $\boldsymbol{\lambda}_v$ adopted with active conditions should be greater than 0. In this KKT condition, inequality $v_2(x)$ is assumed to constrain an objective

Table 3.3.1.1 Lowest Cost Index of Columns (*CIc*) with Corresponding Parameters for KKT Conditions Based on Forward Networks

(a-1): Based on Forward Training with One Layer – 80 Neurons; ACI 318-14 (Refer to Table 3.2.2.2(a-1) for Training Performance)

AI-Based Lagrange Optimization *(One Layer -80 Neurons) (ACI 318-14)*

Parameters		*Case 1*	*Case 2*	*Case 3*
1	b (mm)	1,042.0	665.4	**919.0**
2	h (mm)	1,042.0	665.4	**919.0**
3	ρ_s	0.0100	0.0800	**0.0157**
4	f_c' (MPa)	40	40	40
5	f_y (MPa)	500	500	500
6	P_u (kN)	1,000	1,000	1,000
7	M_u (kN·m)	3,000	3,000	3,000
Objective: CI_c (KRW/m)		192,921.0	334,987.4	**189,969.1**

Note: "Bold values" emphasize the results of optimal case which is the most important case.
Case 1, Inequality $v_1(\boldsymbol{x})$ is active; Case 2, Inequality $v_2(\boldsymbol{x})$ is active; Case 3, None of inequalities are active.

(b-1): Based on Forward Training with Two Layers – 80 Neurons; ACI 318-14 (Refer to Table 3.2.2.2(a-1) for Training Performance)

AI-based Lagrange Optimization *(Two Layers -80 Neurons) (ACI 318-14)*

Parameters		*Case 1*	*Case 2*	*Case 3*
1	b (mm)	1,063.2	662.6	**934.5**
2	h (mm)	1,063.2	662.6	**934.5**
3	ρ_s	0.0100	0.0800	**0.0158**
4	f_c' (MPa)	40	40	40
5	f_y (MPa)	500	500	500
6	P_u (kN)	1,000	1,000	1,000
7	M_u (kN·m)	3,000	3,000	3,000
Objective: CI_c (KRW/m)		200,257.1	332,556.4	**196,623.5**

Note: "Bold values" emphasize the results of optimal case which is the most important case.
Case 1, Inequality $v_1(\boldsymbol{x})$ is active; Case 2, Inequality $v_2(\boldsymbol{x})$ is active; Case 3, None of inequalities are active.

(c-1): Based on Forward Training with Five layers – 80 Neurons; ACI 318-14 (Refer to Table 3.2.2.2(a-1) for Training Performance)

AI-based Lagrange Optimization *(Five Layers -80 Neurons) (ACI 318-14)*

Parameters		*Case 1*	*Case 2*	*Case 3*
1	b (mm)	1,068.5	663.3	**919.5**
2	h (mm)	1,068.5	663.3	**919.5**
3	ρ_s	0.0100	0.0800	**0.0167**
4	f_c' (MPa)	40	40	40
5	f_y (MPa)	500	500	500
6	P_u (kN)	1,000	1,000	1,000
7	M_u (kN·m)	3,000	3,000	3,000
Objective: CI_c (KRW/m)		204,528.9	332,664.1	**193,397.0**

Note: "Bold values" emphasize the results of optimal case which is the most important case.
Case 1, Inequality $v_1(\boldsymbol{x})$ is active; Case 2, Inequality $v_2(\boldsymbol{x})$ is active; Case 3, None of inequalities are active.

(Continued)

Table 3.3.1.1 (Continued) Lowest Cost Index of Columns (*CIc*) with Corresponding Parameters for KKT Conditions Based on Forward Networks

(d-1): Based on Forward Training with Ten Layers – 80 Neurons; ACI 318-14 (Refer to Table 3.2.2.2(a-1) for Training Performance)

AI-Based Lagrange Optimization *(Ten Layers -80 Neurons) (ACI 318-14)*

Parameters		*Case 1*	*Case 2*	*Case 3*
1	b (mm)	1,068.0	661.9	**940.1**
2	h (mm)	1,068.0	661.9	**940.1**
3	ρ_s	0.0100	0.0800	**0.0155**
4	f'_c (MPa)	40	40	40
5	f_y (MPa)	500	500	500
6	P_u (kN)	1,000	1,000	1,000
7	M_u (kN·m)	3,000	3,000	3,000
Objective: CI_c (KRW/m)		201,393.5	329,790.4	**196,049.0**

Note: "Bold values" emphasize the results of optimal case which is the most important case.
Case 1, Inequality $v_1(\mathbf{x})$ is active; Case 2, Inequality $v_2(\mathbf{x})$ is active; Case 3, None of inequalities are active.

(e-1): Based on Best Forward Trainings a Combination of One, Two, Five, and Ten Hidden Layers with 80 Neurons; ACI 318-14 (Refer to Table 3.2.2.2(a-1) for Training Performance)

AI-based Lagrange Optimization *(Best Networks in Term of Test MSE) (ACI 318-14)*

Parameters		*Case 1*	*Case 2*	*Case 3*
1	b (mm)	1,068.0	661.9	**931.8**
2	h (mm)	1,068.0	661.9	**931.8**
3	ρ_s	0.0100	0.0800	**0.0159**
4	f'_c (MPa)	40	40	40
5	f_y (MPa)	500	500	500
6	P_u (kN)	1,000	1,000	1,000
7	M_u (kN·m)	3,000	3,000	3,000
Objective: CI_c (KRW/m)		202,692.7	331,487.6	**196,698.5**

Note: "Bold values" emphasize the results of optimal case which is the most important case.
Case 1, Inequality $v_1(\mathbf{x})$ is active; Case 2, Inequality $v_2(\mathbf{x})$ is active; Case 3, None of inequalities are active.

(a-2): Based on Forward Training with One layer – 80 Neurons; ACI 318-19 (Refer to Table 3.2.2.2(b-1) for Training Performance)

AI-based Lagrange Optimization *(One Layer -80 Neurons) (ACI 318-19)*

Parameters		*Case 1*	*Case 2*	*Case 3*
1	b (mm)	1,050.2	662.9	**956.5**
2	h (mm)	1,050.2	662.9	**956.5**
3	ρ_s	0.0100	0.0800	**0.0142**
4	f'_c (MPa)	40	40	40
5	f_y (MPa)	500	500	500
6	P_u (kN)	1,000	1,000	1,000
7	M_u (kN·m)	3,000	3,000	3,000
Objective: CI_c (KRW/m)		195,308.3	332,604.8	**193,798.6**

Note: "Bold values" emphasize the results of optimal case which is the most important case.
Case 1, Inequality $v_1(\mathbf{x})$ is active; Case 2, Inequality $v_2(\mathbf{x})$ is active; Case 3, None of inequalities are active.

(Continued)

Table 3.3.1.1 (Continued) Lowest Cost Index of Columns (CI_c) with Corresponding Parameters for KKT Conditions Based on Forward Networks

(b-2): Based on Forward Training with Two Layers – 80 Neurons; ACI 318-19 (Refer to Table 3.2.2.2(b-1) for Training Performance)

AI-based Lagrange Optimization *(Two Layers -80 Neurons) (ACI 318-19)*

Parameters		*Case 1*	*Case 2*	*Case 3*
1	*b* (mm)	1,061.9	669.9	**935.4**
2	*h* (mm)	1,061.9	669.9	**935.4**
3	ρ_s	0.0100	0.0800	**0.0157**
4	f_c' (MPa)	40	40	40
5	f_y (MPa)	500	500	500
6	P_u (kN)	1,000	1,000	1,000
7	M_u (kN·m)	3,000	3,000	3,000
Objective: CI_c (KRW/m)		200,078.6	339,157.8	197,238.1

Note: "Bold values" emphasize the results of optimal case which is the most important case.
Case 1, Inequality $v_1(\mathbf{x})$ is active; Case 2, Inequality $v_2(\mathbf{x})$ is active; Case 3, None of inequalities are active.

(c-2): Based on Forward Training with Five Layers – 80 Neurons; ACI 318-19 (Refer to Table 3.2.2.2(b-1) for Training Performance)

AI-based Lagrange Optimization (*Five Layers -80 Neurons*) *(ACI 318-19)*

Parameters		*Case 1*	*Case 2*	*Case 3*
1	*b* (mm)	1,069.4	669.7	**961.2**
2	*h* (mm)	1,069.4	669.7	**961.2**
3	ρ_s	0.0100	0.0800	**0.0144**
4	f_c' (MPa)	40	40	40
5	f_y (MPa)	500	500	500
6	P_u (kN)	1,000	1,000	1,000
7	M_u (kN·m)	3,000	3,000	3,000
Objective: CI_c (KRW/m)		199,538.2	345,367.4	**196,769.0**

Note: "Bold values" emphasize the results of optimal case which is the most important case.
Case 1 – Inequality $v_1(\mathbf{x})$ is active; Case 2 – Inequality $v_2(\mathbf{x})$ is active; Case 3 – None of inequalities are active.

(d-2): Based on Forward Training with Ten Layers – 80 Neurons; ACI 318-19 (Refer to Table 3.2.2.2(b-1) for Training Performance)

AI-based Lagrange Optimization *(Ten Layers -80 Neurons) (ACI 318-19)*

Parameters		*Case 1*	*Case 2*	*Case 3*
1	*b* (mm)	1,071.1	668.6	**927.6**
2	*h* (mm)	1,071.1	668.6	**927.6**
3	ρ_s	0.0100	0.0800	**0.0162**
4	f_c' (MPa)	40	40	40
5	f_y (MPa)	500	500	500
6	P_u (kN)	1,000	1,000	1,000
7	M_u (kN·m)	3,000	3,000	3,000
Objective: CI_c (KRW/m)		203,678.7	338,757.2	**197,370.8**

Note: "Bold values" emphasize the results of optimal case which is the most important case.
Case 1, Inequality $v_1(\mathbf{x})$ is active; Case 2, Inequality $v_2(\mathbf{x})$ is active; Case 3, None of inequalities are active.

(Continued)

Table 3.3.1.1 (Continued) Lowest Cost Index of Columns (CIc) with Corresponding Parameters for KKT Conditions Based on Forward Networks

(e-2): *Based on best forward trainings a combination of 1, 2, 5, and 10 hidden layers with 80 Neurons; ACI 318-19 (Refer to Table 3.2.2.2(b-1) for Training Performance)*				
AI-based Lagrange Optimization *(Best Networks in Term of Test MSE) (ACI 318-19)*				
Parameters		*Case 1*	*Case 2*	*Case 3*
1	b (mm)	1,071.1	668.6	**929.3**
2	h (mm)	1,071.1	668.6	**929.3**
3	ρ_s	0.0100	0.0800	**0.0161**
4	f_c' (MPa)	40	40	40
5	f_y (MPa)	500	500	500
6	P_u (kN)	1,000	1,000	1,000
7	M_u (kN·m)	3,000	3,000	3,000
Objective: CI_c (KRW/m)		203,141.0	338,294.0	**196,562.0**

Note: "Bold values" emphasize the results of optimal case which is the most important case.
Case 1, Inequality $v_1(\mathbf{x})$ is active; Case 2, Inequality $v_2(\mathbf{x})$ is active; Case 3, None of inequalities are active.

function by binding Inequality $v_2(x)$ to equality constraint. However, a stationary point of a Lagrange function of Case 2 KKT condition should be verified to exist within an inequality constraining $v_1(x)$. Lagrange function with $s_2 = 1$ is obtained in Equation (3.3.1.5) by substituting $S_2 = \begin{bmatrix} 0 & 0 \\ 0 & 1 \end{bmatrix}$ in Equation (3.3.1.2-2).

$$\mathcal{L}_{CIC}^{FW}(\boldsymbol{x}, \boldsymbol{\lambda}_c, \boldsymbol{\lambda}_v) = f_{CIC}^{FW}(\boldsymbol{x}) - \boldsymbol{\lambda} \mathbf{c} \mathbf{c}(\boldsymbol{x}) - \boldsymbol{\lambda} \mathbf{v} \begin{bmatrix} 0 & 0 \\ 0 & 1 \end{bmatrix} \mathbf{v}(\boldsymbol{x}) \tag{3.3.1.5}$$

As shown in Equation (3.3.1.6), rebar ratio ρ_s is assumed as 0.08 because inequality $v_2(x)$ is assumed to constrain an objective function by binding Inequality $v_2(x)$ to equality constraint when v_2 is activated. The other four input variables (f_c', f_y, P_u, M_u) are predetermined as 40 MPa, 500 MPa, 1,000 kN, 3,000 kNm based on Case 2 KKT condition based on simple equality constraints $c_2(\boldsymbol{x})$, $c_3(\boldsymbol{x})$, $c_4(\boldsymbol{x})$, and $c_5(\boldsymbol{x})$ shown in Table 3.2.2.5.

$$\mathbf{x}^{(0)} = [b, h, \rho_s, f_c', f_y, P_u, M_u]^T = [b, h, 0.08, 40, 500, 1{,}000, 3{,}000]^T \tag{3.3.1.6}$$

Beam width b and beam depth h are unknowns to be determined during optimization based on Newton–Raphson iteration. Initial trial input parameters are, then, assumed as $\mathbf{x}^{(0)} = [b, h, 0.08, 40, 500, 1{,}000, 3{,}000]^T$ to find a stationary point of Lagrange function shown in Equations (3.3.1.2-2) or (3.3.1.5) based on Newton–Raphson method. Let's find the unknown initial trial input parameters between $300 \le b \le 2250$ and $300 \le h \le 1500$ in large datasets during the iteration. Newton–Raphson method is implemented in solving first order partial differential equations (Jacobi of Lagrange function) shown in Equation (3.3.1.2-3) using one set of initial trial Lagrange multipliers $[\lambda_{c1}, \lambda_{c2}] = [0, 0]$ and $5^2 = 25$ initial trial input parameters of $\mathbf{x}^{(0)} = [b, h, 0.08, 40, 500, 1{,}000, 3{,}000]^T$ shown in Equation

(3.3.1.6) in which b and h are randomly distributed within a training data range. Iterations progress to find the roots of the KKT condition shown in Equation (3.3.1.2-3). As shown in Tables 3.3.1.1(e-1) and 3.1.1.1(e-2) for Case 2 KKT condition, $[b, h] = [661.9, 661.9]$ for ACI 318-14 and $[668.6, 668.6]$ for ACI 318-19 are converged among 25 trials based on the best networks in terms of test MSE among one, two, five, and ten layers with 80 neurons, yielding an optimal value of CI_c = 331,487.6 for ACI 318-14 and 338,294.0 for ACI 318-19 which are roots of Case 2 KKT condition shown in Equation (3.3.1.2-3) when inequality constraint v_2 is activated. Roots of Case 2 KKT condition must meet inequality constraint v_1. Tables 3.3.1.2(a-1) and (a-2) based on one hidden layer and 80 neurons for the Case 2 KKT condition show the minimized $CI_c = f_{CI_c}^{FW}(\boldsymbol{x})$ of 334,987.4 for ACI 318-14 and of 332,604.8 for ACI 318-19, in which errors are relatively large as shown in Table 3.3.1.2. As shown in Tables 3.3.1.1(d-1) and 3.3.1.1(d-2) based on ten hidden layers and 80 neurons for the Case 2 KKT condition, the minimized $CI_c = f_{CI_c}^{FW}(\boldsymbol{x})$ of 339,790.4 based on ACI 318-14 and $CI_c = f_{CI_c}^{FW}(\boldsymbol{x})$ of 338,757.2 based on ACI 318-19 are yielded as roots of the Case 2 KKT condition shown in Equation (3.3.1.2-3). Minimized CI_c based on two, five, and ten hidden layers with 80 neurons are also shown in Tables 3.3.1.1(b-1)–(d-1) and 3.3.1.1(b-2)–(d-2) for ACI 318-14 and ACI 318-19. It is noted that the costs obtained based on rebar ratio ρ_s = 0.08 for the

Table 3.3.1.2 Design Accuracies Minimizing CI_c Based on Forward Lagrange Optimization

(a-1): Based on Forward Training with One Layer – 80 Neurons; ACI 318-14 (Refer to Table 3.2.2.2(a-1) for Training Performance)

Optimal **CI_c** Design *(Based on One Layer With 80 Neurons) (ACI 318-14)*

Parameter		*AI Results*	*Check (AutoCol)*	*Error*
1	b (mm)	**919.00**		
2	h (mm)	**919.00**		
3	ρ_s	**0.0157**		
4	f_c' (MPa)	40		
5	f_y (MPa)	500		
6	P_u (kN)	1,000		
7	M_u (kN·m)	3,000		
8	ϕP_n (kN)	981.9	944.5	3.8%
9	ϕM_n (kN·m)	3,066.7	2,833.6	7.60%
10	SF	1.00	0.94	5.55%
11	b/h	0.9999	1.0000	−0.01%
12	ε_s	0.0211	0.0221	−4.72%
13	CI_c (KRW/m)	**189,969.1**	**189,445.5**	0.28%
14	CO_2 (t-CO_2/m)	0.4043	0.4038	0.12%
15	W_c (kN/m)	19.90	19.90	0.00%
16	$\alpha_{e/h}$	0.317	0.297	6.24%

Note: (dotted box) Seven inputs for AI design.
(dashed box) Structural verification by (*AutoCol*).
Optimal CI_c of **189,969.1** is −2.84% different from **195,515** obtained by large structural datasets.
"Bold values" indicate the important parameters of AI results.

(Continued)

Table 3.3.1.2 (Continued) Design Accuracies Minimizing CI_c Based on Forward Lagrange Optimization

(b-1): Based on Forward Training with Two Layers – 80 Neurons; ACI 318-14 (Refer to Table 3.2.2.2(a-1) for Training Performance)

Optimal $\mathbf{CI_C}$ Design (*Based on Two Layers with 80 Neurons*) (*ACI 318-14*)

Parameter		*AI Results*	*Check (AutoCol)*	*Error*
1	b (mm)	**934.49**		
2	h (mm)	**934.49**		
3	ρ_s	**0.0158**		
4	f_c' (MPa)	40		
5	f_y (MPa)	500		
6	P_u (kN)	1,000		
7	M_u (kN·m)	3,000		
8	ϕP_n (kN)	947.5	1,003.3	−5.9%
9	ϕM_n (kN·m)	3,048.6	3,010.0	1.27%
10	*SF*	1.000	1.003	−0.33%
11	*b/h*	1.000	1.000	0.00%
12	ε_s	0.0227	0.0221	2.89%
13	CI_c (KRW/m)	**196,623.5**	**196,579.0**	0.02%
14	CO_2 (t-CO_2/m)	0.4191	0.4192	−0.02%
15	W_c (kN/m)	20.571	20.574	−0.02%
16	$\alpha_{e/h}$	0.3024	0.3020	0.13%

Note: (dotted box) Seven inputs for AI design.
(dashed box) Structural verification by (*AutoCol*).
Optimal CI_C of **196,623.5** is 0.57% different from **195,515** obtained by large structural datasets.
"Bold values" indicate the important parameters of AI results.

(c-1): Based on Forward Training with Five Layers – 80 Neurons; ACI 318-14 (Refer to Table 3.2.2.2(a-1) for Training Performance)

Optimal $\mathbf{CI_C}$ Design (*Based on Five Layers with 80 Neurons*) (*ACI 318-14*)

Parameter		*AI Results*	*Check (AutoCol)*	*Error*
1	b (mm)	**919.48**		
2	h (mm)	**919.48**		
3	ρ_s	**0.0167**		
4	f_c' (MPa)	40		
5	f_y (MPa)	500		
6	P_u (kN)	1,000		
7	M_u (kN·m)	3,000		
8	ϕP_n (kN)	976.7	1,003.5	−2.7%
9	ϕM_n (kN·m)	3,018.2	3,010.6	0.25%
10	*SF*	1.000	1.004	−0.35%
11	*b/h*	1.000	1.000	0.00%
12	ε_s	0.0215	0.0212	1.12%
13	CI_c (KRW/m)	**193,397.0**	**196,534.9**	−1.62%
14	CO_2 (t-CO_2/m)	0.420	0.421	−0.21%
15	W_c (kN/m)	19.95	19.92	0.14%
16	$\alpha_{e/h}$	0.2973	0.2974	−0.04%

Note: (dotted box) Seven inputs for AI design.
(dashed box) Structural verification by (*AutoCol*).
Optimal CI_c of **193,397.0** is −1.08% different from **195,515** obtained by large structural datasets.
"Bold values" indicate the important parameters of AI results.

(Continued)

Table 3.3.1.2 (Continued) Design Accuracies Minimizing CI_c Based on Forward Lagrange Optimization

(d-1): Based on Forward Training with Ten Layers – 80 Neurons; ACI 318-14 (Refer to Table 3.2.2.2(a-1) for Training Performance)

Optimal $\mathbf{CI_C}$ Design (*Based on Ten Layers with 80 Neurons*) (*ACI 318-14*)

Parameter		*AI Results*	*Check (AutoCol)*	*Error*
1	b (mm)	**940.10**		
2	h (mm)	**940.10**		
3	ρ_s	**0.0155**		
4	f_c' (MPa)	40		
5	f_y (MPa)	500		
6	P_u (kN)	1,000		
7	M_u (kN·m)	3,000		
8	ϕP_n (kN)	986.7	999.4	−1.3%
9	ϕM_n (kN·m)	3,001.3	2,998.1	0.11%
10	SF	1.000	0.999	0.06%
11	b/h	1.0008	1.0000	0.08%
12	ε_s	0.0231	0.0224	2.67%
13	CI_c (KRW/m)	**196,049.0**	**196,172.7**	−0.06%
14	CO_2 (t-CO_2/m)	0.417	0.418	−0.15%
15	W_c (kN/m)	20.737	20.822	−0.41%
16	$\alpha_{e/h}$	0.3034	0.3037	−0.09%

Note: [dotted box] Seven inputs for AI design.
[dashed box] Structural verification by (*AutoCol*).
Optimal CI_c of **196,049.0** is 0.27% different from **195,515** obtained by large structural datasets.
"Bold values" indicate the important parameters of AI results.

(e-1): Based on Best Forward Trainings a Combination of One, Two, Five, and Ten Hidden Layers with 80 Neurons; ACI 318-14 (Refer to Table 3.2.2.2(a-1) for Training Performance)

Optimal $\mathbf{CI_C}$ Design (*Based on Best Networks in Term of MSE*) (*ACI 318-14*)

Parameter		*Best Networks*	*AI Results*	*Check (AutoCol)*	*Error*
1	b (mm)		**931.81**		
2	h (mm)		**931.81**		
3	ρ_s		**0.0159**		
4	f_c' (MPa)		40		
5	f_y (MPa)		500		
6	P_u (kN)		1,000		
7	M_u (kN·m)		3,000		
8	ϕP_n (kN)	5L -80N	(978.2)	999.9	(−2.2%)
9	ϕM_n (kN·m)	10L -80N	(3,002.3)	2,999.8	(0.08%)
10	SF	10L -80N	1.0000	0.9999	0.01%
11	b/h	2L -80N	(1.00)	1.00	(0.00%)
12	ε_s	10L -80N	(0.02254)	0.02196	(2.54%)
13	CI_c (KRW/m)	1L -80N	**196,698.5**	**196,159.3**	0.27%
14	CO_2 (t-CO_2/m)	2L -80N	(0.4183)	0.4185	(−0.03%)
15	W_c (kN/m)	1L -80N	(20.457)	20.456	(0.00%)
16	$\alpha_{e/h}$	10L -80N	(0.3009)	0.3012	(−0.09%)

Note: [dotted box] Seven inputs for AI design.
[dashed box] Structural verification by (*AutoCol*).
Optimal CI_c of **196,698.5** is 0.61% different from **195,515** obtained by large structural datasets.
"Bold values" indicate the important parameters of AI results.

(Continued)

Table 3.3.1.2 (Continued) Design Accuracies Minimizing CI_c Based on Forward Lagrange Optimization

(a-2): Based on Forward Training with One Layer – 80 Neurons; ACI 318-19 (Refer to Table 3.2.2.2(b-1) for Training Performance)

Optimal $\mathbf{CI_C}$ Design (*Based on One Layer with 80 Neurons*) (*ACI 318-19*)

Parameter		*AI Results*	*Check (AutoCol)*	*Error*
1	b (mm)	**956.55**		
2	h (mm)	**956.55**		
3	ρ_s	**0.0142**		
4	f_c' (MPa)	40		
5	f_y (MPa)	500		
6	P_u (kN)	1,000		
7	M_u (kN·m)	3,000		
8	ϕP_n (kN)	1,027.1	974.1	5.2%
9	ϕM_n (kN·m)	3,178.6	2,922.4	8.06%
10	*SF*	1.00	0.97	2.59%
11	*b/h*	1.00	1.00	0.00%
12	ε_s	0.0235	0.0238	−1.48%
13	CI_c (KRW/m)	**193,798.6**	**193,602.9**	0.10%
14	CO_2 (t-CO_2/m)	0.4109	0.4097	0.29%
15	W_c (kN/m)	21.56	21.56	0.00%
16	$\alpha_{e/h}$	0.341	0.309	9.55%

Note: [dotted box] Seven inputs for AI design.
[dashed box] Structural verification by (*AutoCol*).
Optimal CI_c of **193,798.6** is −1.00% different from **195,756** obtained by large structural datasets.
"Bold values" indicate the important parameters of AI results.

(b-2): Based on Forward Training with Two Layers – 80 Neurons; ACI 318-19 (Refer to Table 3.2.2.2(b-1) for Training Performance)

Optimal $\mathbf{CI_C}$ Design (*Based on Two Layers with 80 Neurons*) (*ACI 318-19*)

Parameter		*AI Results*	*Check (AutoCol)*	*Error*
1	b (mm)	935.36		
2	h (mm)	935.36		
3	ρ_s	0.0157		
4	f_c' (MPa)	40		
5	f_y (MPa)	500		
6	P_u (kN)	1,000		
7	M_u (kN·m)	3,000		
8	ϕP_n (kN)	972.4	999.8	−2.8%
9	ϕM_n (kN·m)	3,029.2	2,999.5	0.98%
10	*SF*	1.0000	0.9998	0.02%
11	*b/h*	1.000	1.000	0.00%
12	ε_s	0.0227	0.0222	2.48%
13	CI_c (KRW/m)	**197,238.1**	**196,175.8**	0.54%
14	CO_2 (t-CO_2/m)	0.415	0.418	−0.67%
15	W_c (kN/m)	20.612	20.612	0.00%
16	$\alpha_{e/h}$	0.3029	0.3022	0.22%

Note: [dotted box] Seven inputs for AI design.
[dashed box] Structural verification by (*AutoCol*).
Optimal CI_c of **197,238.1** is 0.76% different from **195,756** obtained by large structural datasets.
"Bold values" indicate the important parameters of AI results.

(Continued)

Table 3.3.1.2 (Continued) Design Accuracies Minimizing CI_c Based on Forward Lagrange Optimization

(c-2): Based on Forward Training with Five Layers – 80 Neurons; ACI 318-19 (Refer to Table 3.2.2.2(b-1) for Training Performance)

Optimal $\mathbf{CI_C}$ Design (*Based on Five Layers with 80 Neurons*) (*ACI 318-19*)

Parameter		*AI Results*	*Check (AutoCol)*	*Error*
1	b (mm)	**961.24**		
2	h (mm)	**961.24**		
3	ρ_s	**0.0144**		
4	f_c' (MPa)	40		
5	f_y (MPa)	500		
6	P_u (kN)	1,000		
7	M_u (kN·m)	3,000		
8	ϕP_n (kN)	990.0	1,000.7	−1.1%
9	ϕM_n (kN·m)	3,001.8	3,002.0	0.00%
10	*SF*	1.000	1.001	−0.07%
11	*b/h*	1.0001	1.0000	0.01%
12	ε_s	0.0244	0.0237	2.96%
13	CI_c (KRW/m)	**196,769.0**	**196,730.6**	0.02%
14	CO_2 (t-CO_2/m)	0.415	0.417	−0.44%
15	W_c (kN/m)	21.744	21.769	−0.12%
16	$\alpha_{e/h}$	0.31014	0.31008	0.02%

Note: (dotted box) Seven inputs for AI design.
(dashed box) Structural verification by (*AutoCol*).
Optimal CI_c of **196,769.0** is 0.52% different from **195,756** obtained by large structural datasets.
"Bold values" indicate the important parameters of AI results.

(d-2): Based on Forward Training with Ten Layers – 80 Neurons; ACI 318-19 (Refer to Table 3.2.2.2(b-1) for Training Performance)

Optimal $\mathbf{CI_C}$ Design (*Based on Ten Layers with 80 Neurons*) (*ACI 318-19*)

Parameter		*AI Results*	*Check (AutoCol)*	*Error*
1	b (mm)	**927.60**		
2	h (mm)	**927.60**		
3	ρ_s	**0.0162**		
4	f_c' (MPa)	40		
5	f_y (MPa)	500		
6	P_u (kN)	1,000		
7	M_u (kN·m)	3,000		
8	ϕP_n (kN)	1,001.2	1,002.0	−0.1%
9	ϕM_n (kN·m)	2,989.2	3,005.9	−0.56%
10	*SF*	1.000	1.002	−0.20%
11	*b/h*	1.0002	1.0000	0.02%
12	ε_s	0.0221	0.0217	2.03%
13	CI_c (KRW/m)	**197,370.8**	**196,368.0**	0.51%
14	CO_2 (t-CO_2/m)	0.417	0.419	−0.56%
15	W_c (kN/m)	20.271	20.272	−0.01%
16	$\alpha_{e/h}$	0.2997	0.2999	−0.07%

Note: (dotted box) Seven inputs for AI design.
(dashed box) Structural verification by (*AutoCol*).
Optimal CI_c of **197,370.8** is 0.82% different from **195,756** obtained by large structural datasets.
"Bold values" indicate the important parameters of AI results.

(*Continued*)

Table 3.3.1.2 (Continued) Design Accuracies Minimizing CI_c Based on Forward Lagrange Optimization

(e-2): Based on Best Forward Trainings a Combination of One, Two, Five, and Ten Hidden Layers with 80 Neurons; ACI 318-19 (Refer to Table 3.2.2.2(b-1) for Training Performance)

Optimal **CI_C** Design (*Based on Best Networks in Term of MSE*) (*ACI 318-19*)

Parameter		*Best Networks*	*AI Results*	*Check (AutoCol)*	*Error*
1	b (mm)		**929.33**		
2	h (mm)		**929.33**		
3	ρ_s		**0.0161**		
4	f_c' (MPa)		40		
5	f_y (MPa)		500		
6	P_u (kN)		1,000		
7	M_u (kN·m)		3,000		
8	ϕP_n (kN)	10L -80N	(1,001.3)	1,001.9	(−0.1%)
9	ϕM_n (kN·m)	10L -80N	(2,989.5)	3,005.6	(−0.54%)
10	SF	10L -80N	1.000	1.002	(−0.19%)
11	b/h	2L-80N	(1.00)	1.00	(0.00%)
12	ε_s	5L -80N	(0.02241)	0.02179	(2.76%)
13	CI_c (KRW/m)	1L -80N	**196,562.0**	**196,367.6**	(0.10%)
14	CO_2 (t-CO_2/m)	1L-80N	(0.420)	0.419	(0.26%)
15	W_c (kN/m)	1L -80N	(20.347)	20.348	(0.00%)
16	$\alpha_{e/h}$	10L -80N	(0.3002)	0.3004	(−0.07%)

Note: (dotted box) Seven inputs for AI design.
(dashed box) Structural verification by (*AutoCol*).
Optimal CI_c of **196,562.0** is 0.41% different from **195,756** obtained by large structural datasets.
"Bold values" indicate the important parameters of AI results.

Case 2 KKT are more expensive than that obtained rebar ratio $\rho_s = 0.01$ for the Case 1 KKT as shown in Table 3.3.1.1.

3. Case 3 KKT condition for both inactive inequality constraint ν_1 and ν_2

   Both inequalities $v_1(x)$ and $v_2(x)$ do not constrain an objective function, whereas a stationary point of a Lagrange function of Case 3 KKT condition should be verified to exist within an inequality constraining set, inequalities $v_1(x)$ and $v_2(x)$. The Lagrange function with s_1 and s_2 equal to 0 is obtained in Equation (3.3.1.7) by substituting $\mathbf{S}_3 = \begin{bmatrix} 0 & 0 \\ 0 & 0 \end{bmatrix}$ into Equation (3.3.1.2-2).

$$\mathcal{L}_{CIC}^{FW}(\boldsymbol{x}, \boldsymbol{\lambda}_c, \boldsymbol{\lambda}_v) = f_{CIC}^{FW}(\boldsymbol{x}) - \boldsymbol{\lambda}_c^T \mathbf{c}(\boldsymbol{x}) - \boldsymbol{\lambda}_v^T \begin{bmatrix} 0 & 0 \\ 0 & 0 \end{bmatrix} \mathbf{v}(\boldsymbol{x}) \tag{3.3.1.7}$$

Lagrange functions contain only equality constraints because inactive inequality constraints are ignored in Lagrange functions when inequalities are non-binding or inactive (refer to Sections 1.3.1. and 1.3.2). However, a stationary point must be identified within the range defined by inactive inequalities $v_1(x)$ and $v_2(x)$, hence, rebar ratio ρ_s are determined within a range of datasets between 0.01 and 0.08 $(0.01 < \rho_s < 0.08)$ when none of the inequality constraints is activated. Initial trial input parameters are assumed as shown in Equation (3.3.1.8).

$$\mathbf{x}^{(0)} = [b, h, \rho_s, f_c', f_y, P_u, M_u]^T = [b, h, \rho_s, 40, 500, 1{,}000, 3{,}000]^T \tag{3.3.1.8}$$

where four input variables (f_c', f_y, P_u, M_u) are predetermined based on simple equality constraints $c_2(\boldsymbol{x})$, $c_3(\boldsymbol{x})$, $c_4(\boldsymbol{x})$, and $c_5(\boldsymbol{x})$. Beam width b, beam depth h, and rebar ratio ρ_s are unknowns to be determined during optimization based on Newton–Raphson iteration.

Initial trial input parameters are, then, assumed as $\mathbf{x}^{(0)} = [b, h, \rho_s, 40, 500, 1{,}000, 3{,}000]^T$ to find a stationary point of a Lagrange function shown in Equations (3.3.1.2-2) or (3.3.1.7) based on Newton–Raphson iteration. Newton–Raphson iteration is implemented to solve partial differential equations (Jacobi of Lagrange function) shown in Equation (3.3.1.2–3) using one set of initial trial Lagrange multipliers $[\lambda_{c1}, \lambda_{c2}] = [0, 0]$ and $5^3 = 125$ initial trial input parameters of $\mathbf{x}^{(0)} = [b, h, \rho_s, 40, 500, 1{,}000, 3{,}000]^T$ in which b, h, and ρ_s are randomly distributed within a training data range. Let's find the unknown input parameters between $300 \le b \le 2{,}250$, $300 \le h \le 1{,}500$, and $0.01 < \rho_s < 0.08$ in large datasets during the iteration. The rebar ratio ρ_s is calculated between 0.01 and 0.08 ($0.01 < \rho_s < 0.08$) when no inequalities are activated. $[b, h, \rho_s] = [931.8, 931.8, 0.0159]$ for ACI 318-14 and $[929.3, 929.3, 0.0161]$ for ACI 318-19 are converged among $5^3 = 125$ trials based on the best networks in term of test MSE among one, two, five, and ten hidden layers with 80 neurons, yielding an optimal value of $f_{CIC}^{FW}(\mathbf{x}) = 196{,}698.5$ for ACI 318-14 and $f_{CIC}^{FW}(\mathbf{x}) = 196{,}562.0$ for ACI 318-19 which are roots of KKT equations shown in Equation (3.3.1.2-3), as shown in Tables 3.3.1.1(e-1) and 3.3.1.1(e-2). Optimized results based on one hidden layer and 80 neurons are also shown in Tables 3.3.1.1(a-1) and (a-2) with minimized $f_{CIC}^{FW}(\mathbf{x}) = 189{,}969.1$ for ACI 318-14 and $f_{CIC}^{FW}(\mathbf{x}) = 193{,}798.6$ for ACI 318-19. Those based on two, five, and ten hidden layers with 80 neurons are also shown in Tables 3.3.1.1(b-1)–(d-1) and (b-2)–(d-2) for ACI 318-14 and ACI 318-19, respectively. As shown in Tables 3.3.1.1 and 3.3.1.2 based on ANN with ten hidden layers and 80 neurons of Case 3 KKT condition, the minimized $f_{CIC}^{FW}(\mathbf{x}) =$ among all KKT conditions with negligible errors is identified as 196,049.0 (–0.06% compared with structural calculation shown in Table 3.3.1.2(d-1)) and as 197,370.8 (0.51% compared with structural calculation shown in Table 3.3.1.2(d-2)) for ACI 318-14 and ACI 318-19, respectively. ANN-based costs obtained using ACI 318-14 and ACI 318-19 are compared in Table 3.3.1.1, whereas accuracies of optimized CI_c based on forward Lagrange optimizations are presented in Table 3.3.1.2.

4. Case 4 KKT condition for both active inequality constraint v_1 and v_2
5. Case 4 KKT condition with all inequality constraints being activated simultaneously will not occur because inequality constraints v_1 and v_2 cannot occur at the same time. Lagrange optimization for this KKT condition dropped.

3. Influence of training parameters (layers and neurons) on KKT equations based on Newton–Raphson iteration

Design parameters representing a stationary point of a Lagrange function, minimizing CI_c are compared with those obtained using a structural software (*AutoCol*) as shown in Table 3.3.1.2 based on forward Lagrange optimization trained with one, two, five, and ten hidden layers based on 80 neurons. In Tables 3.3.1.2(a-1) and (a-2), the largest error of 7.60% (ACI 318-14) and 8.06% (ACI 318-19) for a design moment strength (ϕM_n) is demonstrated with one hidden layer with 80 neurons. For a design moment strength (ϕM_n), largest errors are reduced to 1.27%, 0.25%, and 0.11% for two, five, and ten hidden layers with 80 neurons based on ACI 318-14 as shown in Tables 3.3.1.2(b-1)–(d-1), whereas errors are reduced to 0.98%, 0.00%, and –0.56% for two, five, and ten hidden layers with 80 neurons based on ACI 318-19 as shown in Tables 3.3.1.2(b-2)–(d-2). Figure 3.3.1.2 demonstrates minimum column costs CI_c of 195,515 for ACI 318-14 and 195,756 for ACI 318-19 obtained directly from large datasets, by which column costs CI_c obtained based on forward networks shown in Table 3.3.1.2 are verified. Accuracies of CI_c = 189,969.1 (ACI 318-14) and 193,798.6 (ACI 318-19) are verified with an error of –2.84% (ACI 318-14) and –1.00% (ACI 318-19), respectively, based on one layer shown in Tables 3.3.1.2(a-1) and (a-2) whereas accuracies of CI_c = 196,623.5 (ACI 318-14) and 197,238.1 (ACI 318-19) are verified with 0.57% (ACI 318-14) and 0.76% (ACI 318-19), respectively, by two layers shown in Tables 3.3.1.2(b-1) and (b-2). Accuracies of CI_c = 193,397.0 (ACI 318-14) and 196,769.0 (ACI 318-19) are also verified with –1.08% (ACI 318-14) and 0.52% (ACI 318-19), respectively, by five layers shown in Tables 3.3.1.2(c-1) and (c-2). Accuracies of CI_c = 196,049.0 (ACI 318-14) and 197,370.8 (ACI 318-19) are also verified with –1.08% (ACI 318-14) and 0.82% (ACI 318-19), respectively, by ten layers shown in Tables 3.3.1.2(d-1) and (d-2). Table 3.3.1.2 (e-1) and (e-2) present optimal design based on the best networks in terms of test MSE based on a combination of one, two, five, and ten with 80 neurons, providing an optimal CI_c = 196,698.5 (ACI 318-14) and 196,562.0 (ACI 318-19) with differences of 0.61% and 0.41%, respectively, compared to ones obtained from structural large datasets.

The CI_c obtained by an ANN-based LMM with two, five, and ten hidden layers and that obtained based on large datasets are in good agreement. However, it should be noted that ANN-based minimized CI_c with one hidden layer shown in Tables 3.3.1.1(a) and 3.3.1.2(a) obtained based on forward networks cannot be used because errors associated with design parameters such as design moment strength, ϕM_n (kN·m), are too large even if a network with one layer provides smallest CI_c. Tables 3.3.1.4(a) and 3.3.1.5(a), shown for reverse networks, pose the similar problems. Verification is, therefore, required for all design parameters obtained in design tables before concluding optimized CI_c.

4. Calculating output design parameters minimizing objective functions

   Objective functions CI_c based on a forward network are minimized based on equality constraints for $c_1(\boldsymbol{x}) = b - h = 0$ and $c_6(\boldsymbol{x}) = SF - 1 = 0$ shown in Table 3.2.2.5(a) when Inequality constraints $v_1(\boldsymbol{x}) = \rho_s - 0.01 \geq 0$ and $v_2(\boldsymbol{x}) = -\rho_s + 0.08 \geq 0$ are also implemented in KKT conditions explained in Section 3.3.1.1(2). *Autocol* (conventional structural calculation based on forward design) obtains the output design parameters (ϕP_n, ϕM_n, SF, $b/h\varepsilon_s$, CI_c, CO_2, W_c, $\alpha_{e/h}$) in Boxes 8–16, respectively, using seven forward input design parameters (b, h, ρ_s, f_c', f_y, P_u, M_u) which are 931.81, 931.81, 0.0159, 40, 500, 1,000, 3,000 calculated by ANNs in dotted Boxes 1–7 of Table 3.3.1.2(e), respectively, The *Autocol*-based parameters (ϕP_n, ϕM_n, SF, b/h, ε_s, CI_c, CO_2, W_c, $\alpha_{e/h}$) are then used to verify those calculated based on ANNs.

3.3.1.2 Formulation of Lagrange optimization based on a reverse network

1. Advantages and disadvantages for a reverse network

   One of the advantages of reverse designs is that constrained parameters are simplified by pre-assigning them on an input side with prescribed values (for example, SF = 1 and b/h = 1 on an input side). SF and b/h are reverse input parameters whereas b and h are reverse output parameters. However, conflicts among reverse input parameters may occur when parameters with poor cross-relationships are selected. In Table 3.3.1.3, parameters used for the optimization of RC columns based on forward network are compared with those based on a reverse network. Pre-specified SF and b/h are placed on an input-side for a reverse network. Reverse designs reduce the complexity of constraints associated with SF = 1 and b/h = 1 because a reverse Lagrange network directly assigns b/h ratio to 1, in which b and h are not constrained allowing

Table 3.3.1.3 Comparison of Input and Output Systems Between Forward and Reverse Networks

Forward Network		*Reverse Network*
$\begin{cases} SF = f_{SF}\left(b,\, h,\, \rho_s,\, f'_c, f_y, P_u, M_u\right) \\ b/h = f_{b/h}\left(b,\, h,\, \rho_s,\, f'_c, f_y, P_u, M_u\right) \\ CCM = f_{CBM}\left(b,\, h,\, \rho_s,\, f'_c, f_y, P_u, M_u\right) \end{cases}$	Reverse $\xrightarrow{b,h \text{ and } SF,\, b/h}$	$\begin{cases} b = f_b\left(\rho_s,\, f'_c, f_y, P_u, M_u, SF,\, b/h\right) \\ h = f_h\left(\rho_s,\, f'_c, f_y, P_u, M_u, SF,\, b/h\right) \\ CCM = f_{CBM}\left(\rho_s,\, f'_c, f_y, P_u, M_u, SF,\, b/h\right) \end{cases}$

b and h to be calculated as a function of seven input variables, ρ_s, f'_c, f_y, P_u, M_u, SF, b/h, on an output-side.

2. Formulation of initial trial input parameters and solving KKT equations based on Newton–Raphson iteration

As shown in Equation (3.3.1.9-1), an objective function of CI_c obtained based on a reverse network is a function of seven input parameters $\mathbf{x} = \left[\rho_s, f'_c, f_y,\, P_u,\, M_u, SF, b/h\right]^T$. A reverse network can predetermine six parameters ($f'_c, f_y,\, P_u,\, M_u, SF, b\,/\,h$) out of seven ($\rho_s, f'_c, f_y,\, P_u,\, M_u, SF, b/h$) parameters via equality constraints. A Lagrange function under KKT conditions to minimize an objective function is shown in Equation (3.3.1.9-2) is minimized under KKT conditions as shown in Equation (3.3.1.9-3).

Objective function CI_c:

$$CI_c = f_{CI_c}^{RV}(\mathbf{x}) \tag{3.3.1.9-1}$$

Lagrange function for CI_c:

$$\mathcal{L}_{CI_c}^{RV}(\mathbf{x}, \boldsymbol{\lambda}_c, \boldsymbol{\lambda}_v) = f_{CI_c}^{RV}(\mathbf{x}) - \boldsymbol{\lambda}_c^T \mathbf{c}(\mathbf{x}) - \boldsymbol{\lambda}_v^T \mathbf{S}\mathbf{v}(\mathbf{x}) \tag{3.3.1.9-2}$$

Candidate solution from KKT conditions:

$$\nabla \mathcal{L}_{CI_c}^{RV}(\mathbf{x}, \boldsymbol{\lambda}_c, \boldsymbol{\lambda}_v) = \begin{bmatrix} \nabla f_{CI_c}^{RV}(\mathbf{x}) - \boldsymbol{J}_c(\mathbf{x})^T \boldsymbol{\lambda}_c - \boldsymbol{J}_v(\mathbf{x})^T \mathbf{S}\boldsymbol{\lambda}_v \\ -\mathbf{c}(\mathbf{x}) \\ -Sv(\mathbf{x}) \end{bmatrix} \tag{3.3.1.9-3}$$

1. Case 1 KKT condition for active inequality constraint $\boldsymbol{v}_1$ and inactive inequality constraint $\boldsymbol{v}_2$

Initial trial input parameters are obtained by substituting 0.01 into ρ_s and by substituting 1 into both SF and b/h, respectively, when v_2 is inactivated as shown in Equation (3.3.1.10). Initial trial input parameters are, then, predetermined as ($\rho_s, f'_c, f_y,\, P_u,\, M_u, SF, b\,/\,h$) based on six equality constraints ($c_2(\mathbf{x})$, $c_3(\mathbf{x})$, $c_4(\mathbf{x})$, $c_5(\mathbf{x})$, $c_6(\mathbf{x})$ of Table 3.2.2.5) and active inequality when v_1 is activated. However, a stationary point of a Lagrange function of Case 1 KKT condition should be verified to exist within an inequality constraining $v_2(x)$.

$$\mathbf{x}^{(0)} = \left[\rho_s, f'_c, f_y,\, P_u,\, M_u, SF, b/h\right]^T = [0.01,\, 40,\, 500,\, 1{,}000,\, 3{,}000,\, 1,\, 1]^T \tag{3.3.1.10}$$

Simple equality constraints are established by substituting 0.01, 40, 500, 1,000, 3,000, 1, 1 into $\rho_s, f_c', f_y, P_u, M_u, SF, \frac{b}{h}$, respectively, as shown in Equation (3.3.1.10). Initial trial input parameters $\mathbf{x}^{(0)} = [0.01, 40, 500, 1{,}000, 3{,}000, 1, 1]^T$ are used to find a stationary point of Lagrange function shown in Equation (3.3.1.9-2) based on Newton–Raphson iteration. In this case, a solution converges immediately because all input parameters are predetermined as initial trial input parameters, as shown in Equation (3.3.1.10), resulting in $f_{CIC}^{FW}(\mathbf{x}) = 202{,}254.5$ at Case 1 based on ACI 318-14 and $f_{CIC}^{FW}(\mathbf{x}) = 200{,}787.8$ at Case 1 based on ACI 318-19 as presented in Tables 3.3.1.4(d-1), (e-1) and (d-2), (e-2) obtained using reverse networks, respectively. These costs are roots of KKT equations shown in Equation (3.3.1.9-3) based on the best networks of test MSE (5.2E10-6) for ACI 318-14 and (5.1E10-6) for ACI 318-19 based on ten layers with 80 neurons as shown in 6d of Tables 3.2.2.2(a-2) and (b-2), respectively.

However, the minimized costs $f_{CIC}^{FW}(\mathbf{x}) =$ 196,696.7 for ACI 318-14 and $f_{CIC}^{FW}(\mathbf{x}) = 196{,}453.4$ for ACI 318-19 based on a combination of one, two, five, and ten hidden layers with 80 neurons are obtained for Case 3 KKT condition as shown in Tables 3.3.1.4(e-1) and (e-2) obtained using reverse networks. Tables 3.3.1.5(e-1) and (e-2) summarize design accuracies when minimizing CI_c based on reverse Lagrange optimization.

2. Case 2 KKT condition for inactive inequality constraint ν_1 and active inequality constraint ν_2

   Initial trial input parameters are obtained by substituting 0.08 into ρ_s, and by substituting 1 into both *SF* and *b*/*h*, respectively, when υ_2 is activated as shown in Equation (3.3.1.11). However, a stationary point of a Lagrange function of Case 2 KKT condition should be verified to exist within an inequality constraining $\upsilon_1(x)$.

$$\mathbf{x}^{(0)} = [\rho_s, f_c', f_y, P_u, M_u, SF, b/h]^T = [0.08, 40, 500, 1{,}000, 3{,}000, 1, 1]^T \tag{3.3.1.11}$$

where all input variables $(\rho_s, f_c', f_y, P_u, M_u, SF, \frac{b}{h})$ are predetermined based on simple equality constraints ($c_2(\mathbf{x})$, $c_3(\mathbf{x})$, $c_4(\mathbf{x})$, $c_5(\mathbf{x})$, $c_6(\mathbf{x})$ of Table 3.2.2.5) and active inequality when υ_2 is activated. Initial trial input parameters of $\mathbf{x}^{(0)} = [0.08, 40, 500, 1{,}000, 3{,}000, 1, 1]^T$ are used to find a stationary point of Lagrange function shown in Equation (3.3.1.9-2) based on Newton–Raphson iteration. Solution converges immediately as all input parameters are predetermined as initial trial input parameters, as shown in Equation (3.3.1.11), resulting in $f_{CIC}^{FW}(\mathbf{x}) = 330{,}847.2$ at Case 2 for ACI 318-14 and $f_{CIC}^{FW}(\mathbf{x}) = 338{,}474.8$ at Case 2 for ACI 318-19 as presented in Tables 3.3.1.4(d-1), (e-1) and (d-2), (e-2), respectively. These costs are roots of KKT equations shown in Equation (3.3.1.9-3) based on the best networks of test MSE (5.2E10-6) for ACI 318-14 and (5.1E10-6) for ACI 318-19 based on ten layers with 80 neurons as shown in 6d of Tables 3.2.2.2(a-2) and (b-2), respectively.

3. Case 3 KKT condition for both inactive inequality constraints ν_1 and ν_2

   Inactive inequality constraints υ_1 and υ_2 are ignored in Lagrange function when inequalities υ_1 and υ_2 are inactive. However, roots of KKT equations should converge within the range imposed by inequality constraints υ_1 and υ_2. Initial trial input parameters are presented in Equation (3.3.1.12) in which rebar ratios ρ_s should be determined using Newton–Raphson iteration when inequality constraints for rebar ratios ρ_s are inactivated.

$$\mathbf{x}^{(0)} = \left[\rho_s, f_c', f_y, P_u, M_u, SF, b/h\right]^T = \left[\rho_s, 40, 500, 1{,}000, 3{,}000, 1, 1\right]^T \quad (3.3.1.12)$$

It is noted that rebar ratios ρ_s shown in initial trial input parameters remain unknown because no inequality for rebar ratio ρ_s is active in Case 3 KKT condition, whereas six parameters $(f_c', f_y, P_u, M_u, SF, b/h)$ out of seven $(\rho_s, f_c', f_y, P_u, M_u, SF, b/h)$ input parameters via equality constraints ($c_2(\mathbf{x})$, $c_3(\mathbf{x})$, $c_4(\mathbf{x})$, $c_5(\mathbf{x})$, $c_6(\mathbf{x})$ of Table 3.2.2.5) are predetermined. Initial trial input parameters $\mathbf{x}^{(0)} = \left[\rho_s, 40, 500, 1{,}000, 3{,}000, 1, 1\right]^T$ are used to find a stationary point of Lagrange function shown in Equations (3.3.1.9-2) based on Newton–Raphson iteration. Unknown input parameters ρ_s as one of initial trial input parameters are randomly selected within large datasets $(0.01 < \rho_s < 0.08)$. Rebar ratios ρ_s are, thus, random parameters within the training data range. Newton–Raphson iteration is implemented in solving a system of equations in Equation (3.3.1.9-3) using one pair for an initial trial Lagrange multiplier $\left[\lambda_{c1}, \lambda_{c2}\right] = \left[0, 0\right]$ and 5^1 initial trial input parameters $\mathbf{x}^{(0)} = \left[\rho_s, 40, 500, 1{,}000, 3{,}000, 1, 1\right]^T$. As shown in Tables 3.3.1.4(d-1), (e-1) and (d-2), (e-2), rebar ratios of 0.0159 for ACI 318-14 and ρ_s of 0.0155 for ACI 318-19 are obtained while minimizing column cost based on the best networks trained by a combination of one, two, five, and ten layers with 80 neurons, resulting in minimized costs $f_{CI_C}^{FW}(\mathbf{x}) = 196{,}696.7$ for ACI 318-14 and $f_{CI_C}^{FW}(\mathbf{x}) = 196{,}453.4$ for ACI 318-19. These costs are roots of KKT equations shown in Equation (3.3.1.9-3) based on the best networks of test MSE (5.2E10-6) for ACI 318-14 and (5.1E10-6) for ACI 318-19 based on ten layers with 80 neurons as shown in 6d of Tables 3.2.2.2(a-2) and (b-2), respectively. Tables 3.3.1.5(e-1) and (e-2) summarize design accuracies when minimizing CI_c based on reverse Lagrange optimization. As shown in Table 3.3.1.4, minimized costs and design parameters of Case 3 KKT condition are obtained by Newton–Raphson iteration within the range imposed by inactive (non-binding) constraints $v_1(\mathbf{x}) = \rho_s - 0.01 \ge 0$ and $v_2(\mathbf{x}) = -\rho_s + 0.08 \ge 0$ which is shown in Table 3.2.2.5.

4. Case 4 KKT condition for both active inequality constraint v_1 and v_2

   Case 4 KKT condition with all inequality constraints being activated simultaneously will not occur because inequality constraints $v_1(\mathbf{x}) = \rho_s - 0.01 \ge 0$ and $v_2(\mathbf{x}) = -\rho_s + 0.08 \ge 0$ cannot occur at the same time. Lagrange optimization for this KKT condition drops.

3. Influence of training parameters (layers and neurons) on KKT equations based on Newton–Raphson iteration

   Designs minimized using Lagrange multipliers based on reverse networks are listed in Table 3.3.1.4, in which the proposed optimizations are verified with those obtained using *AutoCol* shown in Table 3.3.1.5. Good accuracies are achieved by the best networks based on a combination of one, two, five, and ten layers with 80 neurons, showing the largest error of 2.03% for ACI 318-14 and 2.17% for ACI 318-19 for rebar strains as shown in Tables 3.3.1.5(e-1) and (e-2). All minimized costs obtained based on roots of KKT equations of the reverse network shown in Equation (3.3.1.9-3) are also consistent with 195,515 for ACI 318-14 and 195,756 for ACI 318-19 shown in Figure 3.3.1.2 which are obtained directly from large datasets. Cost index CI_c of 191,792.6 for ACI 318-14 and 191,683.8 for ACI 318-19 are obtained at Case 3 using ANNs based on one layer with 80 neurons as shown in Tables 3.3.1.4(a-1) and (a-2) (or Tables 3.3.1.5(a-1) and (a-2)), respectively, where differences between CI_c obtained by ANN-based Lagrange and those obtained from large structural datasets shown in Figure 3.3.1.2 are only −1.90% and −2.08% based on ACI 318-14 and ACI 318-19, respectively. Costs of 197,915.2 for ACI 318-14 and 196,510.1 for ACI 318-19 are also obtained at Case 3 using ANNs based on two layers of 80 neurons as shown in Tables 3.3.1.4(b-1) and (b-2) (or Tables 3.3.1.5(b-1) and (b-2)), respectively, where differences between CI_c obtained by ANN-based Lagrange and those obtained from large structural datasets are only 1.23% and 0.39% based on ACI 318-14 and ACI 318-19, respectively shown in Figure 3.3.1.2. Costs of 196,877.4 (ACI 318-14) and 196,012.7 (ACI 318-19) are obtained at Case 3 using ANNs based on five layers with 80 neurons as shown in Tables 3.3.1.4(c-1) and (c-2) (or Tables 3.3.1.5(c-1) and (c-2)), respectively, where differences between CI_c obtained by ANN-based Lagrange and those obtained from large structural datasets are 0.70% and 0.13% based on ACI 318-14 and ACI 318-19, respectively shown in Figure 3.3.1.2. Costs of 196,696.7 for ACI 318-14 and 196,453.4 for ACI 318-19 are obtained at Case 3 using ANN based on ten layers with 80 neurons as shown in Tables 3.3.1.4(d-1), (d-2), and (e-1), (e-2) (or Tables 3.3.1.5(d-1), (d-2), and (e-1), (e-2)) where differences between CI_c obtained by ANN-based Lagrange and those obtained from large structural datasets shown in Figure 3.3.1.2 are 0.60% and 0.36% based on ACI 318-14 and ACI 318-19 and ACI 318-14, respectively. Tables 3.3.1.5(e-1) and (e-2) present an optimized design based on the best networks trained by a combination of one, two, five, and ten with 80 neurons, with minimized costs of 196,696.7 obtained at Case 3 for ACI 318-14 and 196,453.4 obtained at Case 3 for ACI 318-19. Tables 3.3.1.4, 3.3.1.5, and Figure 3.3.1.2 demonstrate that minimized costs obtained by ANNs are almost identical to those obtained directly from large datasets. It is possible for ANN-based Lagrange algorithm to be successfully implemented in optimizing and solving the lowest costs of concrete columns under given constraints.

4. Calculating output design parameters minimizing objective functions

   Objective function CI_c based on a reverse network are minimized based on equality constraints for $c_1(\boldsymbol{x}) = b/h - 1 = 0$ and $c_6(\boldsymbol{x}) = SF - 1 = 0$ shown in Table 3.2.2.5(b) when Inequality constraints $v_1(\boldsymbol{x}) = \rho_s - 0.01 \geq 0$ and $v_2(\boldsymbol{x}) = -\rho_s + 0.08 \geq 0$ are also implemented in KKT conditions explained in Section 3.3.1.1(2). SF and b/h are reverse input parameters whereas b and h are reverse output parameters. *Autocol* (conventional structural calculation based on forward design) obtains the output design parameters

Table 3.3.1.4 Lowest CI_c with Corresponding Parameters for KKT Conditions Base on Reverse Networks

(a-1): Based on Reverse Training with One Layer – 80 Neurons; ACI 318-14 (Refer to Table 3.2.2.2(a-2) for Training Performance)

AI-based Lagrange Optimization *(One Layer -80 Neurons) (ACI 318-14)*				
Parameters		*Case 1*	*Case 2*	*Case 3*
3	ρ_s	0.0100	0.0800	**0.0131**
4	f_c' (MPa)	40	40	40
5	f_y (MPa)	500	500	500
6	P_u (kN)	1,000	1,000	1,000
7	M_u (kN·m)	3,000	3,000	3,000
10	SF	1	1	1
11	b/h	1	1	1
Objective: CI_c (KRW/m)		192,717.7	335,941.7	**191,792.6**

Note: "Bold values" emphasize the results of optimal case which is the most important case.
Case 1, Inequality $v_1(\boldsymbol{x})$ is active; Case 2, Inequality $v_2(\boldsymbol{x})$ is active; Case 3, None of inequalities are active.

Table 3.3.1.4 (Continued) Lowest CI_c with Corresponding Parameters for KKT Conditions Base on Reverse Networks

(b-1): Based on Reverse Training with Two Layers – 80 Neurons; ACI 318-14 (Refer to Table 3.2.2.2(a-2) for Training Performance)

AI-based Lagrange Optimization (*Two Layers -80 Neurons*) (*ACI 318-14*)

Parameters		*Case 1*	*Case 2*	*Case 3*
3	ρ_s	0.0100	0.0800	**0.0153**
4	f_c' (MPa)	40	40	40
5	f_y (MPa)	500	500	500
6	P_u (kN)	1,000	1,000	1,000
7	M_u (kN·m)	3,000	3,000	3,000
10	*SF*	1	1	1
11	*b/h*	1	1	1
Objective: CI_c (KRW/m)		201,370.0	333,063.6	**197,915.2**

Note: "Bold values" emphasize the results of optimal case which is the most important case.
Case 1, Inequality $v_1(\mathbf{x})$ is active; Case 2, Inequality $v_2(\mathbf{x})$ is active; Case 3, None of inequalities are active.

(c-1): Based on Reverse Training with Five layers – 80 Neurons; ACI 318-14 (Refer to Table 3.2.2.2(a-2) for Training Performance)

AI-based Lagrange Optimization (*Five Layers -80 Neurons*) (*ACI 318-14*)

Parameters		*Case 1*	*Case 2*	*Case 3*
3	ρ_s	0.0100	0.0800	**0.0137**
4	f_c' (MPa)	40	40	40
5	f_y (MPa)	500	500	500
6	P_u (kN)	1,000	1,000	1,000
7	M_u (kN·m)	3,000	3,000	3,000
10	*SF*	1	1	1
11	*b/h*	1	1	1
Objective: CI_c (KRW/m)		198,135.8	328,165.7	**196,877.4**

Note: "Bold values" emphasize the results of optimal case which is the most important case.
Case 1, Inequality $v_1(\mathbf{x})$ is active; Case 2, Inequality $v_2(\mathbf{x})$ is active; Case 3, None of inequalities are active.

(d-1): Based on Reverse Training with Ten layers – 80 Neurons; ACI 318-14 (Refer to Table 3.2.2.2(a-2) for Training Performance)

AI-based Lagrange Optimization (*Ten Layers -80 Neurons*) (*ACI 318-14*)

Parameters		*Case 1*	*Case 2*	*Case 3*
3	ρ_s	0.0100	0.0800	**0.0159**
4	f_c' (MPa)	40	40	40
5	f_y (MPa)	500	500	500
6	P_u (kN)	1,000	1,000	1,000
7	M_u (kN·m)	3,000	3,000	3,000
10	*SF*	1	1	1
11	*b/h*	1	1	1
Objective: CI_c (KRW/m)		202,254.5	330,847.2	**196,696.7**

Note: "Bold values" emphasize the results of optimal case which is the most important case.
Case 1, Inequality $v_1(\mathbf{x})$ is active; Case 2, Inequality $v_2(\mathbf{x})$ is active; Case 3, None of inequalities are active.

Table 3.3.1.4 (Continued) Lowest CI_c with Corresponding Parameters for KKT Conditions Base on Reverse Networks

(e-1): Based on Best Reverse Trainings a Combination of One, Two, Five, and Ten Hidden Layers with 80 Neurons; ACI 318-14 (Refer to Table 3.2.2.2(a-2) for Training Performance)

AI-based Lagrange Optimization (*Best Networks in Term of MSE*) (*ACI 318-14*)

Parameters		*Case 1*	*Case 2*	*Case 3*
3	ρ_s	0.0100	0.0800	**0.0159**
4	f_c' (MPa)	40	40	40
5	f_y (MPa)	500	500	500
6	P_u (kN)	1,000	1,000	1,000
7	M_u (kN·m)	3,000	3,000	3,000
10	*SF*	1	1	1
11	*b/h*	1	1	1
Objective: CI_c (KRW/m)		202,254.5	330,847.2	**196,696.7**

Note: "Bold values" emphasize the results of optimal case which is the most important case.
Case 1, Inequality $v_1(\mathbf{x})$ is active; Case 2, Inequality $v_2(\mathbf{x})$ is active; Case 3, None of inequalities are active.

(a-2): Based on Reverse Training with One Layer – 80 Neurons; ACI 318-19 (Refer to Table 3.2.2.2(b-2) for Training Performance)

AI-based Lagrange Optimization (*One Layer -80 Neurons*) (*ACI 318-19*)

Parameters		*Case 1*	*Case 2*	*Case 3*
3	ρ_s	0.0100	0.0800	**0.0114**
4	f_c' (MPa)	40	40	40
5	f_y (MPa)	500	500	500
6	P_u (kN)	1,000	1,000	1,000
7	M_u (kN·m)	3,000	3,000	3,000
10	*SF*	1	1	1
11	*b/h*	1	1	1
Objective: CI_c (KRW/m)		191,793.2	338,530.3	**191,683.8**

Note: "Bold values" emphasize the results of optimal case which is the most important case.
Case 1, Inequality $v_1(\mathbf{x})$ is active; Case 2, Inequality $v_2(\mathbf{x})$ is active; Case 3, None of inequalities are active.

Table 3.3.1.4(b-2) Based on Reverse Training with Two Layers – 80 Neurons; ACI 318-19 (Refer to Table 3.2.2.2(b-2) for Training Performance)

AI-based Lagrange Optimization (*Two Layers -80 Neurons*) (*ACI 318-19*)

Parameters		*Case 1*	*Case 2*	*Case 3*
3	ρ_s	0.0100	0.0800	**0.0150**
4	f_c' (MPa)	40	40	40
5	f_y (MPa)	500	500	500
6	P_u (kN)	1,000	1,000	1,000
7	M_u (kN·m)	3,000	3,000	3,000
10	*SF*	1	1	1
11	*b/h*	1	1	1
Objective: CI_c (KRW/m)		199,846.0	337,425.6	**196,510.1**

Note: "Bold values" emphasize the results of optimal case which is the most important case.
Case 1, Inequality $v_1(\mathbf{x})$ is active; Case 2, Inequality $v_2(\mathbf{x})$ is active; Case 3, None of inequalities are active.

Table 3.3.1.4(b-2) (Continued) Based on Reverse Training with Two Layers – 80 Neurons; ACI 318-19 (Refer to Table 3.2.2.2(b-2) for Training Performance)

(c-2): Based on Reverse Training with Five layers – 80 Neurons; ACI 318-19 (Refer to Table 3.2.2.2(b-2) for Training Performance)

AI-Based Lagrange Optimization (*Five Layers -80 Neurons*) (*ACI 318-19*)

Parameters		*Case 1*	*Case 2*	*Case 3*
3	ρ_s	0.0100	0.0800	**0.0148**
4	f_c' (MPa)	40	40	40
5	f_y (MPa)	500	500	500
6	P_u (kN)	1,000	1,000	1,000
7	M_u (kN·m)	3,000	3,000	3,000
10	*SF*	1	1	1
11	*b/h*	1	1	1
Objective: CI_c (KRW/m)		198,950.9	335,319.7	**196,012.7**

Note: "Bold values" emphasize the results of optimal case which is the most important case.
Case 1, Inequality $v_1(\mathbf{x})$ is active; Case 2, Inequality $v_2(\mathbf{x})$ is active; Case 3, None of inequalities are active.

(d-2): Based on Reverse Training with Ten layers – 80 Neurons; ACI 318-19 (Refer to Table 3.2.2.2(b-2) for Training Performance)

AI-based Lagrange Optimization (*Ten Layers -80 Neurons*) (*ACI 318-19*)

Parameters		*Case 1*	*Case 2*	*Case 3*
3	ρ_s	0.0100	0.0800	**0.0155**
4	f_c' (MPa)	40	40	40
5	f_y (MPa)	500	500	500
6	P_u (kN)	1,000	1,000	1,000
7	M_u (kN·m)	3,000	3,000	3,000
10	*SF*	1	1	1
11	*b/h*	1	1	1
Objective: CI_c (KRW/m)		200,787.8	338,474.8	**196,453.4**

Note: "Bold values" emphasize the results of optimal case which is the most important case.
Case 1, Inequality $v_1(\mathbf{x})$ is active; Case 2, Inequality $v_2(\mathbf{x})$ is active; Case 3, None of inequalities are active.

(e-2): Based on Best Reverse Trainings a Combination of One, Two, Five, and Ten Hidden Layers with 80 Neurons; ACI 318-19 (Refer to Table 3.2.2.2(b-2) for Training Performance)

AI-Based Lagrange Optimization (*Best Networks in Term of MSE*) (*ACI 318-19*)

Parameters		*Case 1*	*Case 2*	*Case 3*
3	ρ_s	0.0100	0.0800	**0.0155**
4	f_c' (MPa)	40	40	40
5	f_y (MPa)	500	500	500
6	P_u (kN)	1,000	1,000	1,000
7	M_u (kN·m)	3,000	3,000	3,000
10	*SF*	1	1	1
11	*b/h*	1	1	1
Objective: CI_c (KRW/m)		200,787.8	338,474.8	**196,453.4**

Note: "Bold values" emphasize the results of optimal case which is the most important case.
Case 1, Inequality $v_1(\mathbf{x})$ is active; Case 2, Inequality $v_2(\mathbf{x})$ is active; Case 3, None of inequalities are active.

(ϕP_n, ϕM_n, SF, b/h, ε_s, CI_c, CO_2, W_c, $\alpha_{e/h}$) in Boxes 8–16, respectively, using seven forward input design parameters (b, h, ρ_s, f_c', f_y, P_u, M_u) which are 933.51, 933.51, 0.0159, 40, 500, 1,000, 3,000 calculated by ANNs in dotted Boxes 1–7 of Table 3.3.1.5(e), respectively. The *Autocol*-based parameters (ϕP_n, ϕM_n, SF, b/h, ε_s, CI_c, CO_2, W_c, $\alpha_{e/h}$) are then used to verify those calculated based on ANNs.

3.3.1.3 Verifications

1. Verification of CI_c based on reverse networks

   An objective function of CI_c obtained by reverse networks shown in Equation (3.2.2.5-1) is a function of initial trial input parameters $\mathbf{x} = \left[\rho_s, f_c', f_y, P_u, M_u, SF, b/h\right]^T$, whereas six of seven input parameters, except ρ_s, can be predetermined based on the design criterion listed in Tables 3.2.2.4(b) and 3.2.2.5(b). Column cost CI_c can be plotted as a function of ρ_s as shown in Figure 3.3.1.1 which is based on reverse networks.

   At minimized column cost CI_c, Figure 3.3.1.1 matches Table 3.3.1.5 prepared based on reverse networks minimizing a design. Figure 3.3.1.1 shows four optimized curves which are generated based on one, two, five, and ten layers with 80 neurons. In general, the minimum column cost obtained based on a reverse network with four types of layers is in good agreement with the valid design range between the minimum and maximum ρ_s of 0.01 and 0.08 for columns. The costs CI_c based on two, five, and ten layers and 80 neurons agree well. Figure 3.3.1.1 uncovers a little difference between the CI_c curve obtained with one layer and that obtained with two, five, and ten layers, indicating that an ANN with one hidden layer does not provide a design accuracy sufficient to minimize an objective function CI_c.

   Charts can be plotted without using Lagrange functions by setting initial trial input parameters as $f_c' = 40$, $f_y = 500$, $P_u = 1{,}000$, $M_u = 3{,}000$, $SF = 1$, $b/h = 1$ while iterating ρ_s from 0 to 0.1, because six input parameters out of seven input parameters are defined by equality conditions. However, this iteration is not practical when multiple design parameters are unknown. For example, many iterations should be needed when column width b and column depth h are unknown, taking too much computational time to optimize designs compared to those based on Lagrange optimization.

   ANN-based Lagrange optimization saves significant computational time when a number of constraints increases. Large datasets verify that four curves converge to present a realistic minimum CI_c of a column if training accuracies are good enough. For ρ_s outside the design range below 0.01 and above 0.08, all four curves do not match because ANNs have difficulties in extrapolating outside a defined domain.
2. Verification of CI_c by large datasets

   The lowest CI_c obtained directly from large datasets is 195,515 for ACI 318-14 and 195,756 for ACI 318-19, as shown in Figure 3.3.1.2, where two million datasets are filtered through $f_c' = 40$ MPa, $f_y = 500$ MPa, $SF = 1$, and $b/h = 1$. The CI_c is consistent with the 193,397.0 for ACI 318-14 shown in Tables 3.3.1.1(c-1), 3.3.1.2(c-1) and 196,769.0 for ACI 318-19 shown in Tables 3.3.1.1(c-2) and 3.3.1.2(c-2) obtained by Lagrange optimization implementing five-layer and 80-neuron based on forward networks trained on 100,000 datasets. Reverse networks with five-layer and 80-neurons yields 196,877.4 for ACI 318-14 shown in Tables 3.3.1.4(c-1), 3.3.1.5(c-1), 196,012.7 for ACI 318-19 shown in Tables 3.3.1.4(c-2) and 3.3.1.5(c-2). Both forward and reverse networks minimize costs which are similar to those minimized by large datasets shown in Figure 3.3.1.2.

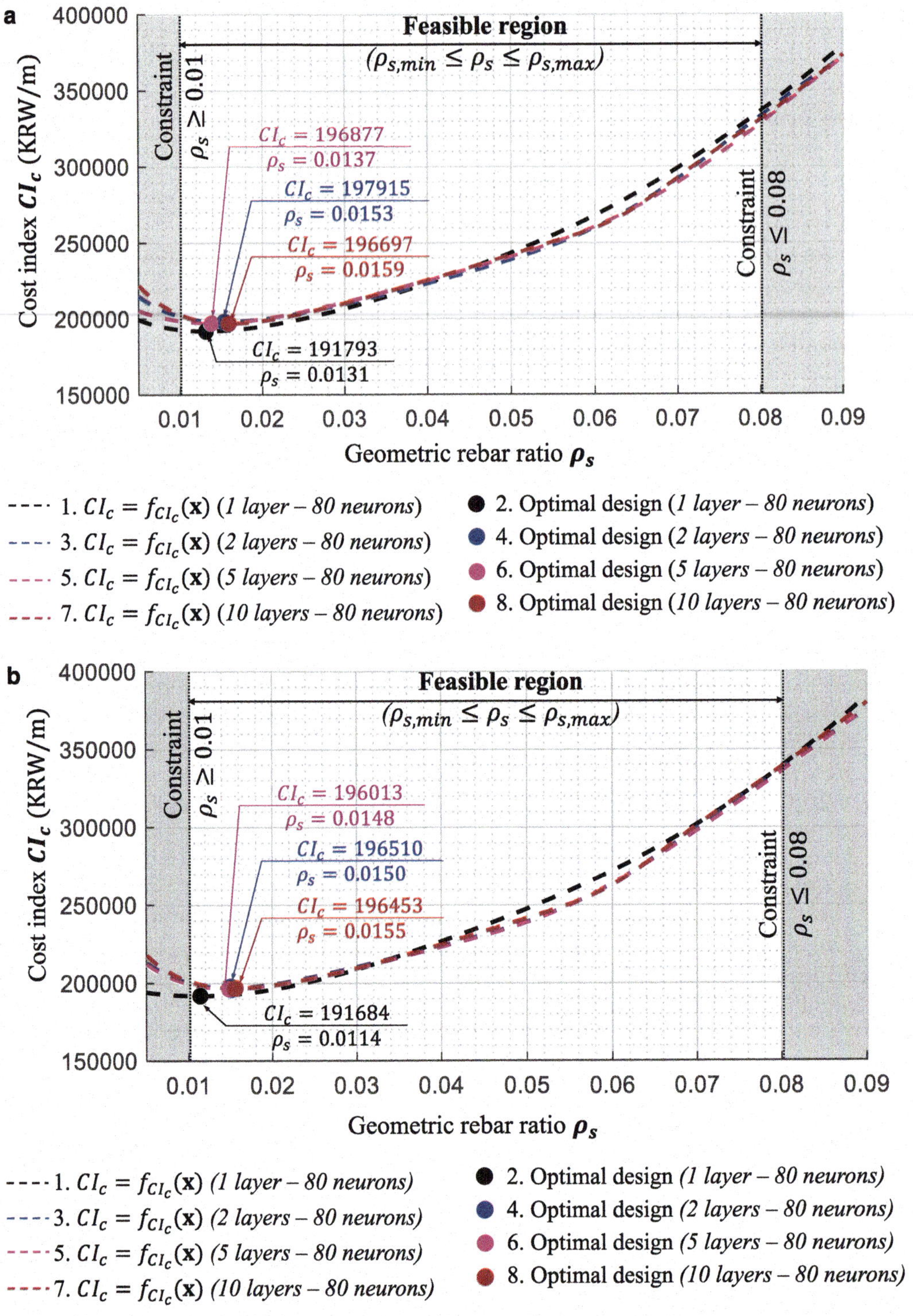

Figure 3.3.1.1 Comparison of CI_c obtained by reverse ANNs based on one, two, five, and ten layers as a function of rebar ratio (ρ_s) between 0.0 and 0.1. (a) ACI 318-14. (b) ACI 318-19.

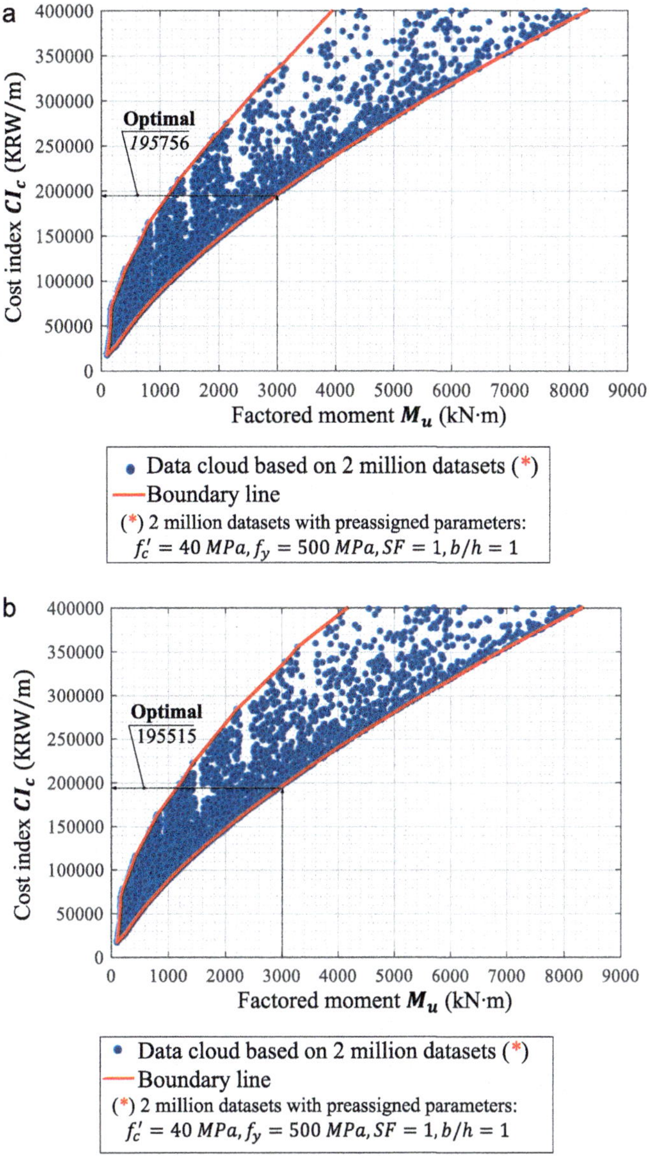

Figure 3.3.1.2 Verification of CI_c based on large datasets. (a) ACI 318-14. (b) ACI 318-19.

3. Accuracies of optimized designs

Accuracies of designs optimized by Lagrange multipliers based on forward networks shown in Table 3.3.1.2 and reverse networks shown in Table 3.3.1.5 are compared with one, two, five, and ten layers. Design errors of ϕP_n (kN) for forward networks with one layer and 80 neurons are 3.8% for ACI 318-14 and 5.2% for ACI 318-19 as shown in Tables 3.3.1.2(a-1) and (a-2), respectively. These errors reduce to –2.7% for ACI 318-14 and –1.1% for ACI 318-19 based on five layers with 80 neurons as shown in Tables 3.3.1.2(c-1) and (c-2), respectively, whereas –1.3% for ACI 318-14 and –0.1% for ACI 318-19 based on ten layers with 80 neurons are shown in Tables 3.3.1.2(d-1) and (d-2), respectively. Similarly, design errors of ϕM_n based on forward networks with one layer and 80 neurons are 7.60% for ACI 318-14 and 8.06% for ACI 318-19 as shown in Tables 3.3.1.2(a-1) and (a-2), respectively. These errors reduce to 1.27% for ACI 318-14 and 0.98% for ACI 318-19 based on two layers with 80 neurons as shown in Tables 3.3.1.2(b-1) and (b-2) whereas errors reduce to 0.25% for ACI 318-14 and 0.00% for ACI 318-19, as shown in Tables 3.3.1.2(c-1) and (c-2) based on five layers with 80 neurons.

Overall design accuracies based on forward networks with two or more layers show better accuracies than those based on one layer.

Optimized designs from networks based on one layer obtained for both forward obtained in Tables 3.3.1.2(a-1) and (a-2) and reverse networks Tables 3.3.1.5(a-1) and (a-2) cannot be used because errors associated with design parameters, such as errors with design moment strength, ϕM_n (kN·m) which are too large, failing to verify all other design parameters obtained in the design tables. It is difficult to explicitly derive objective functions, such as CI_c which is a function of multiple variables, and hence, objective functions as a function of multiple variables are derived based on ANNs and optimized by finding a stationary point of Lagrange functions. Forward and reverse networks with multiple layers and neurons are trained to minimize objective functions. Design accuracies for forward and reverse networks are similar when both trainings are good. However, input conflicts should be avoided when reverse networks are used in training.

Table 3.3.1.5 Design Accuracies Minimizing CI_c Based on Reverse Lagrange Optimization

(a-1): Based on Reverse Training with One Layer – 80 Neurons; ACI 318-14 (refer to Table 3.2.2.2(a-2) for Training Performance)

Optimal $\boldsymbol{CI_C}$ Design (***Based on One Layer with 80 Neurons***) (***ACI 318-14***)

Parameter		*AI Results*	*Check (AutoCol)*	*Error*
1	b (mm)	**959.90**		
2	h (mm)	**978.98**		
3	ρ_s	**0.0131**		
4	f'_c (MPa)	40		
5	f_y (MPa)	500		
6	P_u (kN)	1,000		
7	M_u (kN·m)	3,000		
8	ϕP_n (kN)	992.3	954.6	3.8%
9	ϕM_n (kN·m)	2,999.1	2,863.7	4.51%
10	*SF*	1.00	0.95	4.54%
11	*b/h*	1.00	0.98	1.95%
12	ε_s	0.0254	0.0252	1.01%
13	CI_c (KRW/m)	**191,792.6**	**190,492.9**	0.68%
14	CO_2 (t-CO_2/m)	0.392	0.401	−2.17%
15	W_c (kN/m)	22.278	22.140	0.62%
16	$\alpha_{e/h}$	0.310	0.315	−1.87%

Note: (dotted box) Seven inputs for reverse training.
(dashed box) Structural verification by (AutoCol).
Optimal CI_c of **191,792.6** is −1.90% different from **195,515** obtained by large structural datasets.
"Bold values" indicate the important parameters of AI results.

Table 3.3.1.5 (Continued) Design Accuracies Minimizing CI_c Based on Reverse Lagrange Optimization

(b-1): Based on Reverse Training with Two Layers – 80 Neurons; ACI 318-14 (Refer to Table 3.2.2.2(a-2) for Training Performance)

Optimal CI_C Design (***Based on Two Layers with 80 Neurons***) (***ACI 318-14***)

Parameter		*AI Results*	*Check (AutoCol)*	*Error*
1	b (mm)	**941.58**		
2	h (mm)	**942.08**		
3	ρ_s	**0.0153**		
4	f_c' (MPa)	40		
5	f_y (MPa)	500		
6	P_u (kN)	1,000		
7	M_u (kN·m)	3,000		
8	ϕP_n (kN)	992.6	998.0	−0.5%
9	ϕM_n (kN·m)	2,997.6	2,994.1	0.12%
10	*SF*	1.000	0.998	0.20%
11	*b/h*	1.0000	0.9995	0.05%
12	ε_s	0.0232	0.0226	2.71%
13	CI_c (KRW/m)	**197,915.2**	**196,001.2**	0.97%
14	CO_2 (t-CO_2/m)	0.418	0.417	0.15%
15	W_c (kN/m)	20.828	20.899	−0.34%
16	$\alpha_{e/h}$	0.309	0.304	1.37%

Note: [dotted box] Seven inputs for reverse training.
[dashed box] Structural verification by (AutoCol).
Optimal CI_c of **197,915.2** is 1.23% different from **195,515** obtained by large structural datasets.
"Bold values" indicate the important parameters of AI results.

(c-1): Based on Reverse Training with Five Layers – 80 Neurons; ACI 318-14 (Refer to Table 3.2.2.2(a-2) for Training Performance)

Optimal CI_C Design (***Based on Five Layers with 80 Neurons***) (***ACI 318-14***)

Parameter		*AI Results*	*Check (AutoCol)*	*Error*
1	b (mm)	**973.92**		
2	h (mm)	**974.05**		
3	ρ_s	**0.0137**		
4	f_c' (MPa)	40		
5	f_y (MPa)	500		
6	P_u (kN)	1,000		
7	M_u (kN·m)	3,000		
8	ϕP_n (kN)	998.0	1,000.1	−0.2%
9	ϕM_n (kN·m)	2,972.9	3,000.4	−0.93%
10	*SF*	1.0000	1.0001	−0.01%
11	*b/h*	1.0000	0.9999	0.01%
12	ε_s	0.0250	0.0244	2.29%
13	CI_c (KRW/m)	**196,877.4**	**197,030.9**	−0.08%
14	CO_2 (t-CO_2/m)	0.4157	0.4160	−0.07%
15	W_c (kN/m)	22.353	22.350	0.01%
16	$\alpha_{e/h}$	0.3141	0.3139	0.05%

Note: [dotted box] Seven inputs for reverse training.
[dashed box] Structural verification by (AutoCol).
Optimal CI_c of **196,877.4** is 0.70% different from **195,515** obtained by large structural datasets.
"Bold values" indicate the important parameters of AI results.

Table 3.3.1.5 (Continued) Design Accuracies Minimizing CI_c Based on Reverse Lagrange Optimization

(d-1): Based on Reverse Training with Ten Layers – 80 Neurons; ACI 318-14 (Refer to Table 3.2.2.2(a-2) for Training Performance)

Optimal CI_C Design (***Based on Ten Layers with 80 Neurons***) (***ACI 318-14***)

Parameter		*AI Results*	*Check (AutoCol)*	*Error*
1	b (mm)	**933.28**		
2	h (mm)	**933.17**		
3	ρ_s	**0.0159**		
4	f_c' (MPa)	40		
5	f_y (MPa)	500		
6	P_u (kN)	1,000		
7	M_u (kN·m)	3,000		
8	ϕP_n (kN)	966.7	1,000.5	−3.5%
9	ϕM_n (kN·m)	3,002.7	3,001.5	0.04%
10	SF	1.0000	1.0005	−0.05%
11	b/h	1.0000	1.0001	−0.01%
12	ε_s	0.0225	0.0220	2.03%
13	CI_c (KRW/m)	**196,696.7**	**196,240.7**	0.23%
14	CO_2 (t-CO_2/m)	0.4188	0.4185	0.08%
15	W_c (kN/m)	20.505	20.519	−0.07%
16	$\alpha_{e/h}$	0.3018	0.3016	0.08%

Note: [dotted box] Seven inputs for reverse training.
[dashed box] Structural verification by (AutoCol).
Optimal CI_c of **196,696.7** is 0.60% different from **195,515** obtained by large structural datasets.
"Bold values" indicate the important parameters of AI results.

(e-1): Based on Best Reverse Trainings a Combination of One, Two, Five, and Ten Hidden Layers with 80 Neurons; ACI 318-14 (Refer to Table 3.2.2.2(a-2) for Training Performance)

Optimal CI_C Design (***Based on Best Networks in Term of MSE***) (***ACI 318-14***)

Parameter		*Best Networks*	*AI Results*	*Check (AutoCol)*	*Error*
1	b (mm)	5L -80N	**933.51**		
2	h (mm)	10L -80N	**933.17**		
3	ρ_s		**0.0159**		
4	f_c' (MPa)		40		
5	f_y (MPa)		500		
6	P_u (kN)		1,000		
7	M_u (kN·m)		3,000		
8	ϕP_n (kN)	1L -80N	(991.3)	1,000.7	(−0.9%)
9	ϕM_n (kN·m)	2L -80N	(2,997.8)	3,002.2	(−0.15%)
10	SF		1.0000	1.0007	−0.07%
11	b/h		1.0000	1.0004	−0.04%
12	ε_s	10L -80N	(0.0225)	0.0220	(2.03%)
13	CI_c (KRW/m)	10L -80N	**196,696.7**	**196,287.5**	0.21%
14	CO_2 (t-CO_2/m)	5L -80N	(0.418)	0.419	(−0.03%)
15	W_c (kN/m)	10L -80N	(20.505)	20.524	(−0.09%)
16	$\alpha_{e/h}$	10L -80N	(0.3018)	0.3016	(0.08%)

Note: [dotted box] Seven inputs for reverse training.
[dashed box] Structural verification by (AutoCol).
Optimal CI_c of **196,696.7** is 0.60% different from **195,515** obtained by large structural datasets.
"Bold values" indicate the important parameters of AI results.

Table 3.3.1.5 (Continued) Design Accuracies Minimizing CI_c Based on Reverse Lagrange Optimization

(a-2): Based on Reverse Training with One Layer – 80 Neurons; ACI 318-19 (Refer to Table 3.2.2.2(b-2) for Training Performance)

Optimal CI_C Design (***Based on One Layer with 80 Neurons***) (***ACI 318-19***)

Parameter		*AI Results*	*Check (AutoCol)*	*Error*
1	*b* (mm)	**1,001.71**		
2	*h* (mm)	**1,010.31**		
3	ρ_s	**0.0114**		
4	f_c' (MPa)	40		
5	f_y (MPa)	500		
6	P_u (kN)	1,000		
7	M_u (kN·m)	3,000		
8	ϕP_n (kN)	1,010.6	930.5	7.9%
9	ϕM_n (kN·m)	2,995.3	2,791.4	6.81%
10	*SF*	1.00	0.93	6.95%
11	*b/h*	1.00	0.99	0.85%
12	ε_s	0.0282	0.0278	1.51%
13	CI_c (KRW/m)	**191,683.8**	**190,425.2**	0.66%
14	CO_2 (t-CO_2/m)	0.403	0.397	1.50%
15	W_c (kN/m)	23.997	23.844	0.64%
16	$\alpha_{e/h}$	0.324	0.325	−0.38%

Note: [dotted box] Seven inputs for reverse training.
[dashed box] Structural verification by (*AutoCol*).
Optimal CI_c of **191,683.8** is −2.08% different from **195,756** obtained by large structural datasets.
"Bold values" indicate the important parameters of AI results.

(b-2): Based on Reverse Training with Two Layers – 80 Neurons; ACI 318-19 (Refer to Table 3.2.2.2(b-2) for Training Performance)

Optimal CI_C Design (***Based on Two Layers with 80 Neurons***) (***ACI 318-19***)

Parameter		*AI Results*	*Check (AutoCol)*	*Error*
1	*b* (mm)	**949.08**		
2	*h* (mm)	**947.60**		
3	ρ_s	**0.0150**		
4	f_c' (MPa)	40		
5	f_y (MPa)	500		
6	P_u (kN)	1,000		
7	M_u (kN·m)	3,000		
8	ϕP_n (kN)	998.4	994.9	0.3%
9	ϕM_n (kN·m)	2,996.2	2,984.7	0.38%
10	*SF*	1.000	0.995	0.51%
11	*b/h*	1.0000	1.0016	−0.16%
12	ε_s	0.0237	0.0230	3.20%
13	CI_c (KRW/m)	**196,510.1**	**195,903.6**	0.31%
14	CO_2 (t-CO_2/m)	0.413	0.416	−0.65%
15	W_c (kN/m)	21.238	21.189	0.23%
16	$\alpha_{e/h}$	0.306	0.306	0.16%

Note: [dotted box] Seven inputs for reverse training.
[dashed box] Structural verification by (AutoCol).
Optimal CI_c of **196,510.1** is 0.39% different from **195,756** obtained by large structural datasets.
"Bold values" indicate the important parameters of AI results.

Table 3.3.1.5 (Continued) Design Accuracies Minimizing CI_c Based on Reverse Lagrange Optimization

(c-2): Based on Reverse Training with Five Layers – 80 Neurons; ACI 318-19 (Refer to Table 3.2.2.2(b-2) for Training Performance)

Optimal CI_C Design (***Based on Five Layers with 80 Neurons***) (***ACI 318-19***)

Parameter		*AI results*	*Check (AutoCol)*	*Error*
1	b (mm)	**952.73**		
2	h (mm)	**952.98**		
3	ρ_s	**0.0148**		
4	f_c' (MPa)	40		
5	f_y (MPa)	500		
6	P_u (kN)	1,000		
7	M_u (kN·m)	3,000		
8	ϕP_n (kN)	941.5	998.8	−6.1%
9	ϕM_n (kN·m)	3,024.9	2,996.5	0.94%
10	*SF*	1.00	1.00	0.12%
11	*b/h*	1.00	1.00	0.03%
12	ε_s	0.0237	0.0232	2.21%
13	CI_c (KRW/m)	**196,012.7**	**196,312.5**	−0.15%
14	CO_2 (t-CO_2/m)	0.416	0.417	−0.16%
15	W_c (kN/m)	21.381	21.391	−0.05%
16	$\alpha_{e/h}$	0.308	0.308	−0.01%

Note: (dotted box) Seven inputs for reverse training.
(dashed box) Structural verification by (AutoCol).
Optimal CI_c of **196,012.7** is 0.13% different from **195,756** obtained by large structural datasets.
"Bold values" indicate the important parameters of AI results.

(d-2): Based on Reverse Training with Ten Layers – 80 Neurons; ACI 318-19 (Refer to Table 3.2.2.2(b-2) for Training Performance)

Optimal CI_C Design (***Based on Ten Layers with 80 Neurons***) (***ACI 318-19***)

Parameter		*AI Results*	*Check (AutoCol)*	*Error*
1	b (mm)	**939.00**		
2	h (mm)	**939.22**		
3	ρ_s	**0.0155**		
4	f_c' (MPa)	40		
5	f_y (MPa)	500		
6	P_u (kN)	1,000		
7	M_u (kN·m)	3,000		
8	ϕP_n (kN)	959.9	1,000.3	−4.2%
9	ϕM_n (kN·m)	2,997.2	3,000.8	−0.12%
10	*SF*	1.00	1.00	−0.03%
11	*b/h*	1.00	1.00	0.02%
12	ε_s	0.0229	0.0224	2.17%
13	CI_c (KRW/m)	**196,453.4**	**196,248.7**	0.10%
14	CO_2 (t-CO_2/m)	0.419	0.418	0.18%
15	W_c (kN/m)	20.774	20.778	−0.02%
16	$\alpha_{e/h}$	0.304	0.303	0.08%

Note: (dotted box) Seven inputs for reverse training.
(dashed box) Structural verification by (AutoCol).
Optimal CI_c of **196,453.4** is 0.36% different from **195,756** obtained by large structural datasets.
"Bold values" indicate the important parameters of AI results.

Table 3.3.1.5 (Continued) Design Accuracies Minimizing CI_c Based on Reverse Lagrange Optimization

(e-2): Based on Best Reverse Trainings a Combination of One, Two, Five, and Ten Hidden Layers with 80 Neurons; ACI 318-19 (Refer to Table 3.2.2.2(b-2) for Training Performance)

Optimal CI_c Design (**Based on Best Networks in Term of MSE**) (**ACI 318-19**)

Parameter		*Best Networks*	*AI Results*	*Check (AutoCol)*	*Error*
1	*b* (mm)	5L-80N	**938.93**		
2	*h* (mm)	5L-80N	**939.04**		
3	ρ_s		**0.0155**		
4	f_c' (MPa)		40		
5	f_y (MPa)		500		
6	P_u (kN)		1,000		
7	M_u (kN·m)		3,000		
8	ϕP_n (kN)	2L-80N	(998.4)	999.8	(−0.1%)
9	ϕM_n (kN·m)	2L-80N	(2,996.4)	2,999.3	(−0.10%)
10	*SF*		1.0000	0.9998	0.02%
11	*b/h*		1.0000	1.00	0.03%
12	ε_s	10L-80N	(0.0229)	0.0224	(2.17%)
13	CI_c (KRW/m)	10L-80N	**196,453.4**	**196,196.7**	0.13%
14	CO_2 (t-CO_2/m)	10L-80N	(0.419)	0.418	(0.20%)
15	W_c (kN/m)	5L-80N	(20.763)	20.773	(−0.05%)
16	$\alpha_{e/h}$	5L-80N	(0.3032)	0.3034	(−0.03%)

Note: (dotted box) Seven inputs for reverse training.
(dashed box) Structural verification by (AutoCol).
Optimal CI_c of **196,453.4** is 0.36% different from **195,756** obtained by large structural datasets.
"Bold values" indicate the important parameters of AI results.

3.3.1.4 *P–M diagram*

A minimized CI_c-based *P–M* interaction diagrams for RC columns that satisfy various design criteria presented in Table 3.2.2.4(b) and Table 3.2.2.5 are plotted on a basis of forward and reverse networks with one, two, five, and ten layers with 80 neurons as shown in Figure 3.3.1.3a and b, respectively. In Figure 3.3.1.3, *P–M* diagrams indicated by Legends 1–5 are constructed with parameters shown in the dashed black box of Tables 3.3.1.2 and 3.3.1.5, respectively, obtained by ANN-based Lagrange algorithm. Interaction diagrams minimize CI_c. *P–M* diagrams shown in Figure 3.3.1.3 are plotted using *AutoCol* with input parameters shown in Box 1-7 obtained by ANNs because the accuracies of parameters provided by Tables 3.3.1.2 and 3.3.1.5 are acceptable. All ten *P–M* diagrams (Legends 1–5) shown in Figure 3.3.1.3a and b converge, passing through the given load pair (P_u and M_u) as indicated by a red dot when training accuracies are sufficient.

3.3.2 Column design scenario minimizing CO_2

3.3.2.1 *Formulation of forward network vs. reverse network*

As shown in Equation (3.2.2.5-2), an objective function, CO_2 emission, obtained based on reverse networks is a function of the input parameters $\mathbf{x} = \left[\rho_s, f_c', f_y, P_u, M_u, SF, b/h\right]^T$. According to Design 3, six input parameters, f_c', f_y, P_u, M_u, *SF*, and *b*/*h*, are considered as equality constraints. A reverse network can predetermine six input parameters (f_c', f_y, P_u, M_u, SF, b/h) out of seven input parameters except for ρ_s via equality constraints based on design criteria listed in Tables 3.2.2.4(b) and 3.2.2.5(b). A forward network can predetermine four input parameters $(b, h, \rho_s, f_c', f_y, P_u, M_u)$ out of seven input parameters except for b, h, ρ_s via equality constraints based on design criteria listed in Tables 3.2.2.4(b) and 3.2.2.5(b). CO_2 emission of a square column under P_u = 1,000 kN and M_u = 3,000 kN·m with a safety factor of 1 is minimized for given material properties: f_c' = 40 MPa and f_y = 500 MPa. Rebar ratio is not constrained by this design, but it must comply with ACI-318-14 and ACI-318-19: $\rho_{s,\min} \le \rho_s \le \rho_{s,\max}$. Optimization problems for Design 3 are summarized in Table 3.3.2.1.

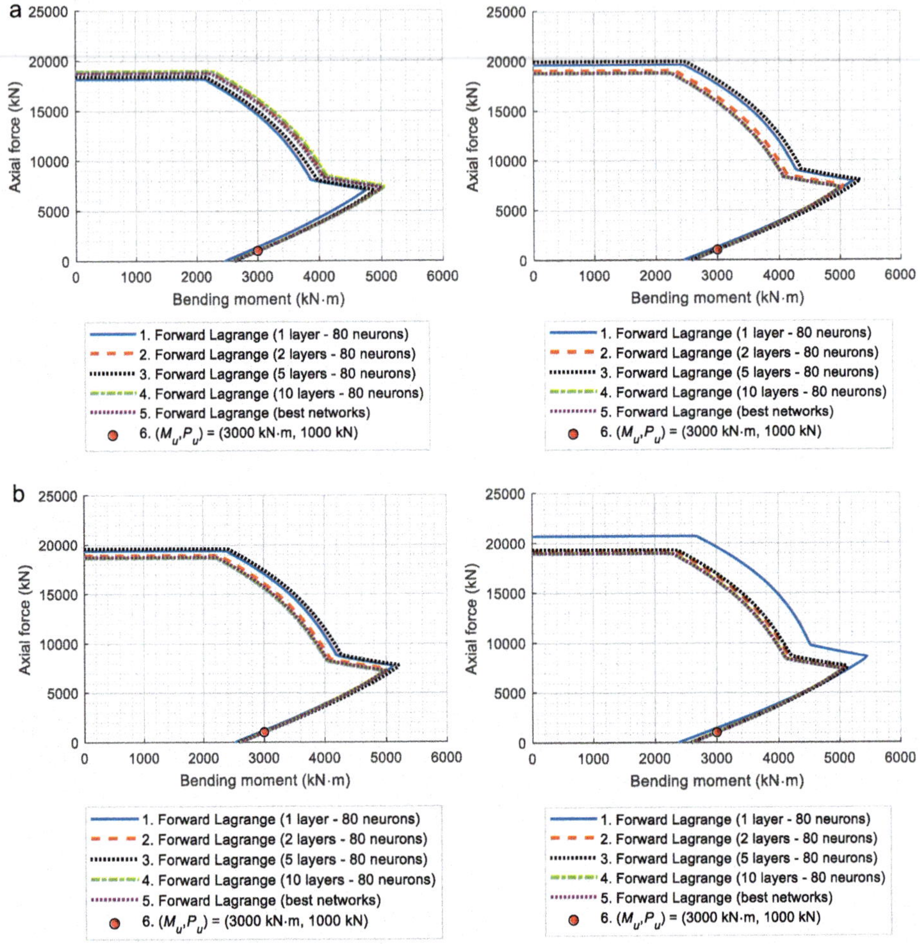

Figure 3.3.1.3 Axial force–moment (*P–M*) interaction diagrams for minimized CI_c. (a) ACI 318-14. (a-1) using forward Lagrange using Table 3.3.1.2. (a-2) using reverse Lagrange based on Table 3.3.1.5. (b) ACI 318-19. (b-1) using forward Lagrange based on Table 3.3.1.2. (b-2) using reverse Lagrange based on Table 3.3.1.5.

Table 3.3.2.1 Design Minimizing CO_2 Emission

Minimize	$CO_2 = f_{CO_2}(x)$
Subject to	
Quality constraints:	$P_u = 1,000$ kN, $M_u = 3,000$ kN·m $f_c' = 40$ MPa, $f_y = 500$ MPa, $SF = 1$, $b/h = 1$
Inequality constraints:	$\rho_{s,\min} \le \rho_s \le \rho_{s,\max}$

3.3.2.2 Solving KKT nonlinear equations based on Newton–Raphson iteration

Newton–Raphson iteration is employed to solve partial differential equations using initial trial Lagrange multipliers and parameters distributed randomly within the training data range, finding a minimized CO_2 emission shown in Tables 3.3.2.2 and 3.3.2.3 similar to minimized CI_c shown in Tables 3.3.1.1 and 3.3.1.4. Case 1 KKT indicates that inequality $v_1(\mathbf{x}) = \rho_s - 0.01 \geq 0$ is active whereas Case 2 KKT indicates that inequality $v_2(\mathbf{x}) = -\rho s + 0.08 \geq 0$ is active. All inequalities are inactive with Case 3 KKT. The minimized CO_2 of 0.415 (t-CO_2/m) at a rebar ratio of 0.0123 for ACI 318-14 (Table 3.3.2.2(e-1)) and 0.417 (t-CO_2/m) at a rebar ratio of 0.0127 for ACI 318-19 (Table 3.3.2.2(e-2)) are identified by Case 3 for forward Lagrange networks based on a combination of one, two, five, and ten layers with 80 neurons which demonstrate the best networks in term of test MSE. The minimized CO_2 of 0.4147 (t-CO_2/m) at a rebar ratio of 0.0116 for ACI 318-14 (Table 3.3.2.3(e-1)) and 0.4140 (t-CO_2/m) at a rebar

Table 3.3.2.2 Lowest CO_2 Emissions of Columns with Corresponding Parameters for KKT Conditions Based on Forward Networks

(a-1): Based on Forward Training with One Layer – 80 Neurons; ACI 318-14 (Refer to Table 3.2.2.2(a-1) for Training Performance)

AI-based Lagrange Optimization *(One Layer -80 Neurons) (ACI 318-14)*

Parameters		*Case 1*	*Case 2*	*Case 3*
1	b (mm)	**1,042.0**	665.4	N/A
2	h (mm)	**1,042.0**	665.4	N/A
3	ρ_s	**0.0100**	0.0800	N/A
4	f_c' (MPa)	40	40	N/A
5	f_y (MPa)	500	500	N/A
6	P_u (kN)	1,000	1,000	N/A
7	M_u (kN·m)	3,000	3,000	N/A
Objective: CO_2 (t-CO_2/m)		**0.397**	0.771	N/A

Note: "Bold values" emphasize the results of optimal case which is the most important case.
Case 1, Inequality $v_1(\mathbf{x})$ is active; Case 2, Inequality $v_2(\mathbf{x})$ is active; Case 3, None of inequalities are active.

(b-1): Based on Forward Training with Two Layers – 80 Neurons; ACI 318-14 (Refer to Table 3.2.2.2(a-1) for Training Performance)

AI-based Lagrange Optimization *(Two Layers -80 Neurons) (ACI 318-14)*

Parameters		*Case 1*	*Case 2*	*Case 3*
1	b (mm)	**1,063.2**	662.6	N/A
2	h (mm)	**1,063.2**	662.6	N/A
3	ρ_s	**0.0100**	0.0800	N/A
4	f_c' (MPa)	40	40	N/A
5	f_y (MPa)	500	500	N/A
6	P_u (kN)	1,000	1,000	N/A
7	M_u (kN·m)	3,000	3,000	N/A
Objective: CO_2 (t-CO_2/m)		**0.413**	0.766	N/A

Note: "Bold values" emphasize the results of optimal case which is the most important case.
Case 1, Inequality $v_1(\mathbf{x})$ is active; Case 2, Inequality $v_2(\mathbf{x})$ is active; Case 3, None of inequalities are active.

Table 3.3.2.2 (Continued) Lowest CO_2 Emissions of Columns with Corresponding Parameters for KKT Conditions Based on Forward Networks

(c-1): Based on Forward Training with Five Layers – 80 Neurons; ACI 318-14 (Refer to Table 3.2.2.2(a-1) for Training Performance)

AI-Based Lagrange Optimization (*Five Layers -80 Neurons*) (*ACI 318-14*)

Parameters		*Case 1*	*Case 2*	*Case 3*
1	b (mm)	1,068.5	663.3	**996.3**
2	h (mm)	1,068.5	663.3	**996.3**
3	ρ_s	0.0100	0.0800	**0.0127**
4	f_c' (MPa)	40	40	40
5	f_y (MPa)	500	500	500
6	P_u (kN)	1,000	1,000	1,000
7	M_u (kN·m)	3,000	3,000	3,000
Objective: CO_2 (t-CO_2/m)		0.418	0.771	**0.415**

Note: "Bold values" emphasize the results of optimal case which is the most important case.
Case 1, Inequality $v_1(\mathbf{x})$ is active; Case 2, Inequality $v_2(\mathbf{x})$ is active; Case 3, None of inequalities are active.

(d-1): Based on Forward Training with Ten Layers – 80 Neurons; ACI 318-14 (Refer to Table 3.2.2.2(a-1) for Training Performance)

AI-based Lagrange Optimization (*Ten Layers -80 Neurons*) (*ACI 318-14*)

Parameters		*Case 1*	*Case 2*	*Case 3*
1	b (mm)	1,068.0	661.9	**957.6**
2	h (mm)	1,068.0	661.9	**957.6**
3	ρ_s	0.0100	0.0800	**0.0145**
4	f_c' (MPa)	40	40	40
5	f_y (MPa)	500	500	500
6	P_u (kN)	1,000	1,000	1,000
7	M_u (kN·m)	3,000	3,000	3,000
Objective: CO_2 (t-CO_2/m)		0.424	0.768	**0.417**

Note: "Bold values" emphasize the results of optimal case which is the most important case.
Case 1, Inequality $v_1(\mathbf{x})$ is active; Case 2, Inequality $v_2(\mathbf{x})$ is active; Case 3, None of inequalities are active.

(e-1): Based on Best Forward Trainings a Combination of One, Two, Five, and Ten Hidden Layers with 80 Neurons; ACI 318-14 (Refer to Table 3.2.2.2(a-1) for Training Performance)

AI-Based Lagrange Optimization (*Best Networks in Term of Test MSE*) (*ACI 318-14*)

Parameters		*Case 1*	*Case 2*	*Case 3*
1	b (mm)	1,068.0	661.9	**1,005.9**
2	h (mm)	1,068.0	661.9	**1,005.9**
3	ρ_s	0.0100	0.0800	**0.0123**
4	f_c' (MPa)	40	40	40
5	f_y (MPa)	500	500	500
6	P_u (kN)	1,000	1,000	1,000
7	M_u (kN·m)	3,000	3,000	3,000
Objective: CO_2 (t-CO_2/m)		0.417	0.765	**0.415**

Note: "Bold values" emphasize the results of optimal case which is the most important case.
Case 1, Inequality $v_1(\mathbf{x})$ is active; Case 2, Inequality $v_2(\mathbf{x})$ is active; Case 3, None of inequalities are active.

Table 3.3.2.2 (Continued) Lowest CO_2 Emissions of Columns with Corresponding Parameters for KKT Conditions Based on Forward Networks

(a-2): Based on Forward Training with One Layer – 80 Neurons; ACI 318-19 (Refer to Table 3.2.2.2(b-1) for Training Performance)

AI-based Lagrange Optimization (*One Layer -80 Neurons*) (*ACI 318-19*)

Parameters		*Case 1*	*Case 2*	*Case 3*
1	b (mm)	**1,050.2**	662.9	N/A
2	h (mm)	**1,050.2**	662.9	N/A
3	ρ_s	**0.0100**	0.0800	N/A
4	f_c' (MPa)	40	40	N/A
5	f_y (MPa)	500	500	N/A
6	P_u (kN)	1,000	1,000	N/A
7	M_u (kN·m)	3,000	3,000	N/A
Objective: CO_2 (t-CO_2/m)		**0.404**	0.767	N/A

Note: "Bold values" emphasize the results of optimal case which is the most important case.
Case 1, Inequality $v_1(\mathbf{x})$ is active; Case 2, Inequality $v_2(\mathbf{x})$ is active; Case 3, None of inequalities are active.

(b-2): Based on Forward Training with Two Layers – 80 Neurons; ACI 318-19 (Refer to Table 3.2.2.2(b-1) for Training Performance)

AI-based Lagrange Optimization (*Two Layers -80 Neurons*) (*ACI 318-19*)

Parameters		*Case 1*	*Case 2*	*Case 3*
1	b (mm)	1,061.9	669.9	**1,006.9**
2	h (mm)	1,061.9	669.9	**1,006.9**
3	ρ_s	0.0100	0.0800	**0.0121**
4	f_c' (MPa)	40	40	40
5	f_y (MPa)	500	500	500
6	P_u (kN)	1,000	1,000	1,000
7	M_u (kN·m)	3,000	3,000	3,000
Objective: CO_2 (t-CO_2/m)		0.413	0.780	**0.411**

Note: "Bold values" emphasize the results of optimal case which is the most important case.
Case 1, Inequality $v_1(\mathbf{x})$ is active; Case 2, Inequality $v_2(\mathbf{x})$ is active; Case 3, None of inequalities are active.

(c-2): Based on Forward Training with Five Layers – 80 Neurons; ACI 318-19 (Refer to Table 3.2.2.2(b-1) for Training Performance)

AI-based Lagrange Optimization (*Five Layers -80 Neurons*) (*ACI 318-19*)

Parameters		*Case 1*	*Case 2*	*Case 3*
1	b (mm)	1,069.4	669.7	**1,014.5**
2	h (mm)	1,069.4	669.7	**1,014.5**
3	ρ_s	0.0100	0.0800	**0.0120**
4	f_c' (MPa)	40	40	40
5	f_y (MPa)	500	500	500
6	P_u (kN)	1,000	1,000	1,000
7	M_u (kN·m)	3,000	3,000	3,000
Objective: CO_2 (t-CO_2/m)		0.415	0.783	**0.413**

Note: "Bold values" emphasize the results of optimal case which is the most important case.
Case 1, Inequality $v_1(\mathbf{x})$ is active; Case 2, Inequality $v_2(\mathbf{x})$ is active; Case 3, None of inequalities are active.

Table 3.3.2.2 (Continued) Lowest CO_2 Emissions of Columns with Corresponding Parameters for KKT Conditions Based on Forward Networks

(d-2): Based on Forward Training with Ten Layers – 80 Neurons; ACI 318-19 (Refer to Table 3.2.2.2(b-1) for Training Performance)

AI-based Lagrange Optimization (*Ten Layers -80 Neurons*) (*ACI 318-19*)				
Parameters		*Case 1*	*Case 2*	*Case 3*
1	b (mm)	1,071.1	668.6	**945.2**
2	h (mm)	1,071.1	668.6	**945.2**
3	ρ_s	0.0100	0.0800	**0.0152**
4	f_c' (MPa)	40	40	40
5	f_y (MPa)	500	500	500
6	P_u (kN)	1,000	1,000	1,000
7	M_u (kN·m)	3,000	3,000	3,000
Objective: CO_2 (t-CO_2/m)		0.427	0.774	**0.417**

Note: "Bold values" emphasize the results of optimal case which is the most important case.
Case 1, Inequality $v_1(\mathbf{x})$ is active; Case 2, Inequality $v_2(\mathbf{x})$ is active; Case 3, None of inequalities are active.

(e-2): Based on Best Forward Trainings a Combination of One, Two, Five, and Ten Hidden Layers with 80 Neurons; ACI 318-19 (Refer to Table 3.2.2.2(b-1) for Training Performance)

AI-based Lagrange Optimization (*Best Networks in Term of Test MSE*) (*ACI 318-19*)				
Parameters		*Case 1*	*Case 2*	*Case 3*
1	b (mm)	**1,071.1**	668.6	997.3
2	h (mm)	**1,071.1**	668.6	997.3
3	ρ_s	**0.0100**	0.0800	0.0127
4	f_c' (MPa)	40	40	40
5	f_y (MPa)	500	500	500
6	P_u (kN)	1,000	1,000	1,000
7	M_u (kN·m)	3,000	3,000	3,000
Objective: CO_2 (t-CO_2/m)		0.420	0.780	0.417

Note: "Bold values" emphasize the results of optimal case which is the most important case.
Case 1, Inequality $v_1(\mathbf{x})$ is active; Case 2, Inequality $v_2(\mathbf{x})$ is active; Case 3, None of inequalities are active.

Table 3.3.2.3 Lowest CO_2 Emissions of Columns with Corresponding Parameters for KKT Conditions Based on Reverse Networks

(a-1): Based on Forward Trainings with One Hidden Layer with 80 Neurons; ACI 318-14 (Refer to Table 3.2.2.2(a-2) for Training Performance)

AI-Based Lagrange Optimization (*One Layer -80 Neurons*) (*ACI 318-14*)				
Parameters		*Case 1*	*Case 2*	*Case 3*
3	ρ_s	**0.0100**	0.0800	N/A
4	f_c' (MPa)	40	40	N/A
5	f_y (MPa)	500	500	N/A
6	P_u (kN)	1,000	1,000	N/A
7	M_u (kN·m)	3,000	3,000	N/A
10	*SF*	1	1	N/A
11	*b/h*	1	1	N/A
Objective: CO_2 (t-CO_2/m)		**0.384**	0.764	N/A

Case 1, Inequality $v_1(\mathbf{x})$ is active; Case 2, Inequality $v_2(\mathbf{x})$ is active; Case 3, None of inequalities are active.

Table 3.3.2.3 (Continued) Lowest CO_2 Emissions of Columns with Corresponding Parameters for KKT Conditions Based on Reverse Networks

(b-1): Based on Reverse Training with Two Layers – 80 Neurons; ACI 318-14 (Refer to Table 3.2.2.2(a-2) for Training Performance)

AI-Based Lagrange Optimization (*Two Layers -80 Neurons*) (*ACI 318-14*)

Parameters		*Case 1*	*Case 2*	*Case 3*
3	ρ_s	0.0100	0.0800	**0.0116**
4	f'_c (MPa)	40	40	40
5	f_y (MPa)	500	500	500
6	P_u (kN)	1,000	1,000	1,000
7	M_u (kN·m)	3,000	3,000	3,000
10	*SF*	1	1	1
11	*b/h*	1	1	1
Objective: CO_2 (t-CO_2/m)		0.4149	0.7670	**0.414**

Case 1, Inequality $v_1(\mathbf{x})$ is active; Case 2, Inequality $v_2(\mathbf{x})$ is active; Case 3, None of inequalities are active.

(c-1): Based on Reverse Training with Five Layers – 80 Neurons; ACI 318-14 (Refer to Table 3.2.2.2(a-2) for Training Performance)

AI-Based Lagrange Optimization (*Five Layers -80 Neurons*) (*ACI 318-14*)

Parameters		*Case 1*	*Case 2*	*Case 3*
3	ρ_s	0.0100	0.0800	**0.0116**
4	f'_c (MPa)	40	40	40
5	f_y (MPa)	500	500	500
6	P_u (kN)	1,000	1,000	1,000
7	M_u (kN·m)	3,000	3,000	3,000
10	*SF*	1	1	1
11	*b/h*	1	1	1
Objective: CO_2 (t-CO_2/m)		0.4154	0.7664	**0.415**

Case 1, Inequality $v_1(\mathbf{x})$ is active; Case 2, Inequality $v_2(\mathbf{x})$ is active; Case 3, None of inequalities are active.

(d-1): Based on Reverse Training with Ten Layers – 80 Neurons; ACI 318-14 (Refer to Table 3.2.2.2(a-2) for Training Performance)

AI-Based Lagrange Optimization (*Ten Layers -80 Neurons*) (*ACI 318-14*)

Parameters		*Case 1*	*Case 2*	*Case 3*
3	ρ_s	0.0100	0.0800	**0.0119**
4	f'_c (MPa)	40	40	40
5	f_y (MPa)	500	500	500
6	P_u (kN)	1,000	1,000	1,000
7	M_u (kN·m)	3,000	3,000	3,000
10	*SF*	1	1	1
11	*b/h*	1	1	1
Objective: CO_2 (t-CO_2/m)		0.4161	0.7651	**0.4150**

Case 1, Inequality $v_1(\mathbf{x})$ is active; Case 2, Inequality $v_2(\mathbf{x})$ is active; Case 3, None of inequalities are active.

Table 3.3.2.3 (Continued) Lowest CO_2 Emissions of Columns with Corresponding Parameters for KKT Conditions Based on Reverse Networks

(e-1): Based on Best Reverse Trainings a Combination of One, Two, Five, and Ten Hidden Layers with 80 Neurons; ACI 318-14 (Refer to Table 3.2.2.2(a-2) for Training Performance)

AI-Based Lagrange Optimization (*Best Networks in Term of MSE*) (*ACI 318-14*)

Parameters		*Case 1*	*Case 2*	*Case 3*
3	ρ_s	0.0100	0.0800	**0.0116**
4	f_c' (MPa)	40	40	40
5	f_y (MPa)	500	500	500
6	P_u (kN)	1,000	1,000	1,000
7	M_u (kN·m)	3,000	3,000	3,000
10	*SF*	1	1	1
11	*b/h*	1	1	1
Objective: CO_2 (t-CO_2/m)		0.4154	0.7664	**0.4147**

Case 1, Inequality $v_1(\mathbf{x})$ is active; Case 2, Inequality $v_2(\mathbf{x})$ is active; Case 3, None of inequalities are active.

(a-2): Based on Reverse Training with One Layer – 80 Neurons; ACI 318-19 (Refer to Table 3.2.2.2(b-2) for Training Performance)

AI-Based Lagrange Optimization (*One Layer -80 Neurons*) (*ACI 318-19*)

Parameters		*Case 1*	*Case 2*	*Case 3*
3	ρ_s	**0.0100**	0.0800	N/A
4	f_c' (MPa)	40	40	N/A
5	f_y (MPa)	500	500	N/A
6	P_u (kN)	1,000	1,000	N/A
7	M_u (kN·m)	3,000	3,000	N/A
10	*SF*	1	1	N/A
11	*b/h*	1	1	N/A
Objective: CO_2 (t-CO_2/m)		**0.400**	0.783	N/A

Case 1, Inequality $v_1(\mathbf{x})$ is active; Case 2, Inequality $v_2(\mathbf{x})$ is active; Case 3, None of inequalities are active.

(b-2): Based on reverse training with Two Layers – 80 Neurons; ACI 318-19 (Refer to Table 3.2.2.2(b-2) for Training Performance)

AI-based Lagrange Optimization (*Two Layers -80 Neurons*) (*ACI 318-19*)

Parameters		*Case 1*	*Case 2*	*Case 3*
3	ρ_s	**0.0100**	0.0800	N/A
4	f_c' (MPa)	40	40	N/A
5	f_y (MPa)	500	500	N/A
6	P_u (kN)	1,000	1,000	N/A
7	M_u (kN·m)	3,000	3,000	N/A
10	*SF*	1	1	N/A
11	*b/h*	1	1	N/A
Objective: CO_2 (t-CO_2/m)		**0.4058**	0.7889	N/A

Case 1, Inequality $v_1(\mathbf{x})$ is active; Case 2, Inequality $v_2(\mathbf{x})$ is active; Case 3, None of inequalities are active.

Table 3.3.2.3 (Continued) Lowest CO_2 Emissions of Columns with Corresponding Parameters for KKT Conditions Based on Reverse Networks

(c-2): Based on Reverse Training with Five Layers – 80 Neurons; ACI 318-19 (Refer to Table 3.2.2.2(b-2) for Training Performance)

AI-based Lagrange Optimization (*Five Layers -80 Neurons*) (*ACI 318-19*)

Parameters		*Case 1*	*Case 2*	*Case 3*
3	ρ_s	**0.0100**	0.0800	N/A
4	f_c' (MPa)	40	40	N/A
5	f_y (MPa)	500	500	N/A
6	P_u (kN)	1,000	1,000	N/A
7	M_u (kN·m)	3,000	3,000	N/A
10	*SF*	1	1	N/A
11	*b/h*	1	1	N/A
Objective: CO_2 (t-CO_2/m)		**0.4107**	0.7780	N/A

Case 1, Inequality $v_1(\mathbf{x})$ is active; Case 2, Inequality $v_2(\mathbf{x})$ is active; Case 3, None of inequalities are active.

(d-2): Based on Reverse Training with Ten Layers – 80 Neurons; ACI 318-19 (Refer to Table 3.2.2.2(b-2) for Training Performance)

AI-based Lagrange Optimization (*Ten Layers -80 Neurons*) (*ACI 318-19*)

Parameters		*Case 1*	*Case 2*	*Case 3*
3	ρ_s	0.0100	0.0800	**0.0105**
4	f_c' (MPa)	40	40	40
5	f_y (MPa)	500	500	500
6	P_u (kN)	1,000	1,000	1,000
7	M_u (kN·m)	3,000	3,000	3,000
10	*SF*	1	1	1
11	*b/h*	1	1	1
Objective: CO_2 (t-CO_2/m)		0.4141	0.7792	**0.4140**

Case 1, Inequality $v_1(\mathbf{x})$ is active; Case 2, Inequality $v_2(\mathbf{x})$ is active; Case 3, None of inequalities are active.

(e-2): Based on Best Reverse Trainings a Combination of One, Two, Five, and Ten Hidden Layers with 80 Neurons; ACI 318-19 (Refer to Table 3.2.2.2(b-2) for Training Performance)

AI-based Lagrange Optimization (*Best Networks in Term of MSE*) (*ACI 318-19*)

Parameters		*Case 1*	*Case 2*	*Case 3*
3	ρ_s	0.0100	0.0800	**0.0105**
4	f_c' (MPa)	40	40	40
5	f_y (MPa)	500	500	500
6	P_u (kN)	1,000	1,000	1,000
7	M_u (kN·m)	3,000	3,000	3,000
10	*SF*	1	1	1
11	*b/h*	1	1	1
Objective: CO_2 (t-CO_2/m)		0.4141	0.7792	**0.4140**

Case 1, Inequality $v_1(\mathbf{x})$ is active; Case 2, Inequality $v_2(\mathbf{x})$ is active; Case 3, None of inequalities are active.

ratio of 0.0105 for ACI 318-19 (Table 3.3.2.3(e-2)) are identified by Case 3 using reverse Lagrange networks based on a combination of one, two, five, and ten layers with 80 neurons which demonstrate the best networks in term of test MSE. It should be noted that minimized CO_2 obtained using a network with one layer shown in Tables 3.3.2.4(a) and 3.3.2.5(a) for both forward and reverse Lagrange networks cannot be adopted because errors associated with design parameters (such as design axial strength ϕP_n (kN), design moment ϕM_n (kN·m)) are too large even if a network with one layer provides least CO_2. Verification is required for all design parameters before concluding minimized CO_2.

Tables 3.3.2.4 and 3.3.2.5 show accuracies of designs which minimize CO_2 emissions by Lagrange multipliers with one, two, five, and ten layers for both forward and reverse networks, respectively. For forward networks trained with two layers and 80 neurons shown in Table 3.3.2.4(b-1) for ACI 318-14 and Table 3.3.2.4 (b-2) for ACI 318-19, the largest errors for ϕP_n are obtained as –10.3% and –3.42% for ACI 318-14 and ACI 318-19, respectively which decrease significantly to 0.7% for ACI 318-14 and 0.98% for ACI 318-19 for five layers as shown in Tables 3.3.2.4(c-1) and (c-2). These errors reduce to –1.1% for ACI 318-14 and 0.03% for ACI 318-19 for ten layers as shown in Tables 3.3.2.4(d-1) and (d-2).

For reverse networks trained with one layer and 80 neurons shown in Table 3.3.2.5(a-1) for ACI 318-14 and Table 3.3.2.5(a-2) for ACI 318-19, the largest errors for ϕP_n are obtained as 8.7% and 10.0% for ACI 318-14 and ACI 318-19, respectively which decrease significantly to 1.2% for ACI 318-14 and 0.7% for ACI 318-19 for two layers as shown in Tables 3.3.2.5(b-1)

Table 3.3.2.4 Design Accuracies Minimizing CO_2 Emissions Based on Forward Lagrange Optimization

(a-1): Based on Forward Training with One Layer – 80 Neurons; ACI 318-14 (Refer to Table 3.2.2.2(a-1) for Training Performance)

Optimal CO_2 Emissions (*Based on One Layer with 80 Neurons*) (*ACI 318-14*)

Parameter		*AI Results*	*Check (AutoCol)*	*Error*
1	*b* (mm)	**1,042.00**		
2	*h* (mm)	**1,042.00**		
3	ρ_s	**0.0100**		
4	f_c' (MPa)	40		
5	f_y (MPa)	500		
6	P_u (kN)	1,000		
7	M_u (kN·m)	3,000		
8	ϕP_n (kN)	984.1	916.5	6.9%
9	ϕM_n (kN·m)	2,938.6	2,749.4	6.44%
10	*SF*	1.00	0.92	8.35%
11	*b/h*	1.00	1.00	0.00%
12	ε_s	0.0297	0.0304	−2.49%
13	CI_c (KRW/m)	192,921.0	191,983.6	0.49%
14	CO_2 (t-CO_2/m)	**0.39650**	**0.39629**	0.05%
15	W_c (kN/m)	25.584	25.581	0.01%
16	$\alpha_{e/h}$	0.351	0.334	4.89%

Note: [dotted box] Seven inputs for forward training.
[dashed box] Structural verification by (AutoCol).
Optimal CO_2 of **0.3965** is −4.34% different from **0.4145** obtained by large structural datasets.
"Bold values" indicate the important parameters of AI results of optimal CO_2 emission.

Table 3.3.2.4 (Continued) Design Accuracies Minimizing CO_2 Emissions Based on Forward Lagrange Optimization

(b-1): Based on Forward Training with Two Layers – 80 Neurons; ACI 318-14 (Refer to Table 3.2.2.2(a-1) for Training Performance)

Optimal CO_2 Emissions (*Based on Two Layers with 80 Neurons*) (*ACI 318-14*)

Parameter		*AI Results*	*Check (AutoCol)*	*Error*
1	b (mm)	**1,063.15**		
2	h (mm)	**1,063.15**		
3	ρ_s	**0.0100**		
4	f_c' (MPa)	40		
5	f_y (MPa)	500		
6	P_u (kN)	1,000		
7	M_u (kN·m)	3,000		
8	ϕP_n (kN)	887.4	978.7	-10.3%
9	ϕM_n (kN·m)	3,058.5	2,936.0	4.00%
10	SF	1.00	0.98	2.13%
11	b/h	0.9999	1.0000	-0.01%
12	ε_s	0.0312	0.0305	2.16%
13	CI_c (KRW/m)	200,257.1	199,856.0	0.20%
14	CO_2 (t-CO_2/m)	**0.4129**	**0.4125**	0.09%
15	W_c (kN/m)	26.630	26.630	0.00%
16	$\alpha_{e/h}$	0.347	0.341	1.98%

Note: [dotted box] Seven inputs for forward training.
[dashed box] Structural verification by (AutoCol).
Optimal CO_2 of **0.4129** is −0.39% different from **0.4145** obtained by large structural datasets.
"Bold values" indicate the important parameters of AI results of optimal CO_2 emission.

(c-1): Based on Forward Training with Five Layers – 80 Neurons; ACI 318-14 (Refer to Table 3.2.2.2(a-1) for Training Performance)

Optimal CO_2 Emissions (*Based on Five Layers with 80 Neurons*) (*ACI 318-14*)

Parameter		*AI Results*	*Check (AutoCol)*	*Error*
1	b (mm)	**996.35**		
2	h (mm)	**996.35**		
3	ρ_s	**0.0127**		
4	f_c' (MPa)	40		
5	f_y (MPa)	500		
6	P_u (kN)	1,000		
7	M_u (kN·m)	3,000		
8	ϕP_n (kN)	1,006.0	999.0	0.7%
9	ϕM_n (kN·m)	3,037.9	2,997.1	1.34%
10	SF	1.000	0.999	0.10%
11	b/h	1.000	1.000	0.00%
12	ε_s	0.0267	0.0258	3.36%
13	CI_c (KRW/m)	196,706.2	197,784.1	−0.55%
14	CO_2 (t-CO_2/m)	**0.415**	**0.415**	−0.02%
15	W_c (kN/m)	23.423	23.388	0.15%
16	$\alpha_{e/h}$	0.3206	0.3207	−0.03%

Note: [dotted box] Seven inputs for forward training.
[dashed box] Structural verification by (AutoCol).
Optimal CO_2 of **0.415** is 0.12% different from **0.4145** obtained by large structural datasets.
"Bold values" indicate the important parameters of AI results of optimal CO_2 emission.

Table 3.3.2.4 (Continued) Design Accuracies Minimizing CO_2 Emissions Based on Forward Lagrange Optimization

(d-1): Based on Forward Training with Ten Layers – 80 Neurons; ACI 318-14 (Refer to Table 3.2.2.2(a-1) for Training Performance)

Optimal CO_2 Emissions (*Based on Ten Layers with 80 Neurons*) (*ACI 318-14*)

Parameter		*AI Results*	*Check (AutoCol)*	*Error*
1	b (mm)	**957.56**		
2	h (mm)	**957.56**		
3	ρ_s	**0.0145**		
4	f'_c (MPa)	40		
5	f_y (MPa)	500		
6	P_u (kN)	1,000		
7	M_u (kN·m)	3,000		
8	ϕP_n (kN)	987.6	998.2	−1.1%
9	ϕM_n (kN·m)	2,998.7	2,994.7	0.14%
10	*SF*	1.000	0.998	0.18%
11	b/h	1.001	1.000	0.08%
12	ε_s	0.0242	0.0235	2.85%
13	CI_c (KRW/m)	196,160.1	196,365.6	−0.10%
14	CO_2 (t-CO_2/m)	**0.417**	**0.416**	0.12%
15	W_c (kN/m)	21.522	21.603	−0.37%
16	$\alpha_{e/h}$	0.3087	0.3090	−0.08%

Note: (dotted box) Seven inputs for forward training.
(dashed box) Structural verification by (AutoCol).
Optimal CO_2 of **0.417** is 0.60% different from **0.4145** obtained by large structural datasets.
"Bold values" indicate the important parameters of AI results of optimal CO_2 emission.

(e-1): Based on Best Forward Trainings a Combination of One, Two, Five, and Ten Hidden Layers with 80 Neurons; ACI 318-14 (Refer to Table 3.2.2.2(a-1) for Training Performance)

Optimal CO_2 Emissions (*Based on Best Networks in Term of MSE*) (*ACI 318-14*)

Parameter		*Best Networks*	*AI Results*	*Check (AutoCol)*	*Error*
1	b (mm)		**1,005.90**		
2	h (mm)		**1,005.90**		
3	ρ_s		**0.0123**		
4	f'_c (MPa)		40		
5	f_y (MPa)		500		
6	P_u (kN)		1,000		
7	M_u (kN·m)		3,000		
8	ϕP_n (kN)	5L -80N	(1,008.5)	996.1	(1.23%)
9	ϕM_n (kN·m)	10L -80N	(2,989.0)	2,988.3	(0.02%)
10	*SF*	10L -80N	1.000	0.996	0.39%
11	b/h	2L -80N	(1.0000)	1.0000	(0.00%)
12	ε_s	10L -80N	(0.02728)	0.02645	(3.07%)
13	CI_c (KRW/m)	1L -80N	(198,661.8)	197,920.9	(0.37%)
14	CO_2 (t-CO_2/m)	2L -80N	**0.4149**	**0.4146**	0.07%
15	W_c (kN/m)	1L -80N	(23.842)	23.839	(0.01%)
16	$\alpha_{e/h}$	10L -80N	(0.3233)	0.3235	(−0.07%)

Note: (dotted box) Seven inputs for forward training.
(dashed box) Structural verification by (AutoCol).
Optimal CO_2 of **0.4149** is 0.10% different from **0.4145** obtained by large structural datasets.
"Bold values" indicate the important parameters of AI results of optimal CO_2 emission.

Table 3.3.2.4 (Continued) Design Accuracies Minimizing CO_2 Emissions Based on Forward Lagrange Optimization

(a-2): Based on Forward Training with One Layer – 80 Neurons; ACI 318-19 (Refer to Table 3.2.2.2(b-1) for Training Performance)

Optimal CO_2 Emissions (*Based on One Layer with 80 Neurons*) (*ACI 318-19*)

Parameter		*AI Results*	*Check (AutoCol)*	*Error*
1	b (mm)	**1,050.21**		
2	h (mm)	**1,050.21**		
3	ρ_s	**0.0100**		
4	f_c' (MPa)	40		
5	f_y (MPa)	500		
6	P_u (kN)	1,000		
7	M_u (kN·m)	3,000		
8	ϕP_n (kN)	1,105.3	940.3	14.9%
9	ϕM_n (kN·m)	3,119.9	2,820.8	9.59%
10	SF	1.00	0.94	5.97%
11	b/h	1.00	1.00	0.00%
12	ε_s	0.0305	0.0304	0.13%
13	CI_c (KRW/m)	195,308.3	195,021.0	0.15%
14	CO_2 (t-CO_2/m)	**0.40379**	**0.40256**	0.30%
15	W_c (kN/m)	25.99	25.99	0.00%
16	$\alpha_{e/h}$	0.364	0.337	7.54%

Note: [dotted box] Seven inputs for forward training.
[dashed box] Structural verification by (AutoCol).
Optimal CO_2 of **0.4038** is −2.58% different from **0.4145** obtained by large structural datasets.
"Bold values" indicate the important parameters of AI results of optimal CO_2 emission.

(b-2): Based on Forward Training with Two Layers – 80 Neurons; ACI 318-19 (Refer to Table 3.2.2.2(b-1) for Training Performance)

Optimal CO_2 Emissions (*Based on Two Layers with 80 Neurons*) (*ACI 318-19*)

Parameter		*AI Results*	*Check (AutoCol)*	*Error*
1	b (mm)	**1,006.86**		
2	h (mm)	**1,006.86**		
3	ρ_s	**0.0121**		
4	f_c' (MPa)	40		
5	f_y (MPa)	500		
6	P_u (kN)	1,000		
7	M_u (kN·m)	3,000		
8	ϕP_n (kN)	955.5	988.1	−3.42%
9	ϕM_n (kN·m)	3,033.6	2,964.4	2.28%
10	SF	1.00	0.99	1.19%
11	b/h	1.0001	1.0000	0.01%
12	ε_s	0.0276	0.0266	3.40%
13	CI_c (KRW/m)	198,278.9	197,124.5	0.58%
14	CO_2 (t-CO_2/m)	**0.411**	**0.413**	−0.31%
15	W_c (kN/m)	23.881	23.884	−0.01%
16	$\alpha_{e/h}$	0.3243	0.3238	0.14%

Note: [dotted box] Seven inputs for forward training.
[dashed box] Structural verification by (AutoCol).
Optimal CO_2 of **0.411** is −0.84% different from **0.4145** obtained by large structural datasets.
"Bold values" indicate the important parameters of AI results of optimal CO_2 emission.

Table 3.3.2.4 (Continued) Design Accuracies Minimizing CO_2 Emissions Based on Forward Lagrange Optimization

(c-2): Based on Forward Training with Five Layers – 80 neurons; ACI 318-19 (Refer to Table 3.2.2.2(b-1) for Training Performance)

Optimal CO_2 Emissions (*Based on Five Layers with 80 Neurons*) (*ACI 318-19*)

Parameter		*AI Results*	*Check (AutoCol)*	*Error*
1	b (mm)	**1,014.50**		
2	h (mm)	**1,014.50**		
3	ρ_s	**0.0120**		
4	f_c' (MPa)	40		
5	f_y (MPa)	500		
6	P_u (kN)	1,000		
7	M_u (kN·m)	3,000		
8	ϕP_n (kN)	1,009.2	999.4	0.98%
9	ϕM_n (kN·m)	3,006.9	2,998.1	0.29%
10	SF	1.000	0.999	0.06%
11	b/h	1.0001	1.0000	0.01%
12	ε_s	0.0276	0.0269	2.34%
13	CI_c (KRW/m)	197,431.6	198,709.6	−0.65%
14	CO_2 (t-CO_2/m)	**0.413**	**0.415**	−0.58%
15	W_c (kN/m)	24.179	24.248	−0.29%
16	$\alpha_{e/h}$	0.326	0.326	0.00%

Note: (dotted box) Seven inputs for forward training.
(dashed box) Structural verification by (AutoCol).
Optimal CO_2 of **0.413** is −0.36% different from **0.4145** obtained by large structural datasets.
"Bold values" indicate the important parameters of AI results of optimal CO_2 emission.

(d-2): Based on Forward Training with Ten Layers – 80 Neurons; ACI 318-19 (Refer to Table 3.2.2.2(b-1) for Training Performance)

Optimal CO_2 Emissions (*Based on Ten Layers with 80 Neurons*) (*ACI 318-19*)

Parameter		*AI Results*	*Check (AutoCol)*	*Error*
1	b (mm)	**945.21**		
2	h (mm)	**945.21**		
3	ρ_s	**0.0152**		
4	f_c' (MPa)	40		
5	f_y (MPa)	500		
6	P_u (kN)	1,000		
7	M_u (kN·m)	3,000		
8	ϕP_n (kN)	1,001.6	1,001.3	0.03%
9	ϕM_n (kN·m)	2,991.8	3,004.0	−0.41%
10	SF	1.000	1.001	−0.13%
11	b/h	1.0002	1.0000	0.02%
12	ε_s	0.0232	0.0227	2.01%
13	CI_c (KRW/m)	197,486.9	196,477.8	0.51%
14	CO_2 (t-CO_2/m)	**0.417**	**0.418**	−0.25%
15	W_c (kN/m)	21.045	21.049	−0.02%
16	$\alpha_{e/h}$	0.3050	0.3052	−0.08%

Note: (dotted box) Seven inputs for forward training.
(dashed box) Structural verification by (AutoCol).
Optimal CO_2 of **0.417** is 0.60% different from **0.4145** obtained by large structural datasets.
"Bold values" indicate the important parameters of AI results of optimal CO_2 emission.

Table 3.3.2.4 (Continued) Design Accuracies Minimizing CO_2 Emissions Based on Forward Lagrange Optimization

(e-2): Based on Best Forward Trainings a Combination of One, Two, Five, and Ten Hidden Layers with 80 Neurons; ACI 318-19 (refer to Table 3.2.2.2(b-1) for Training Performance)

Optimal CO_2 Emissions (*Based on Best Networks in Term of MSE*) (*ACI 318-19*)

Parameter		*Best Networks*	*AI Results*	*Check (AutoCol)*	*Error*
1	b (mm)		**997.33**		
2	h (mm)		**997.33**		
3	ρ_s		**0.0127**		
4	f_c' (MPa)		40		
5	f_y (MPa)		500		
6	P_u (kN)		1,000		
7	M_u (kN·m)		3,000		
8	ϕP_n (kN)	10L -80N	(1,003.4)	1,001.5	(0.19%)
9	ϕM_n (kN·m)	10L -80N	(2,999.1)	3,004.5	(−0.18%)
10	SF	10L -80N	1.000	1.001	−0.15%
11	b/h	2L -80N	(1.0001)	1.0000	(0.01%)
12	ε_s	5L -80N	(0.02652)	0.02582	(2.65%)
13	CI_c (KRW/m)	1L -80N	(198,292.8)	198,092.2	(0.10%)
14	CO_2 (t-CO_2/m)	1L -80N	**0.417**	**0.416**	0.29%
15	W_c (kN/m)	1L -80N	(23.434)	23.435	(0.00%)
16	$\alpha_{e/h}$	10L -80N	(0.3207)	0.3210	(−0.09%)

Note: (dotted box) Seven inputs for forward training.
(dashed box) Structural verification by (AutoCol).
Optimal CO_2 of **0.417** is 0.60% different from **0.4145** obtained by large structural datasets.
"Bold values" indicate the important parameters of AI results of optimal CO_2 emission.

Table 3.3.2.5 Design Accuracies Minimizing CO_2 Emissions Based on Reverse Lagrange Optimization

(a-1): Based on Reverse Training with One Layer – 80 Neurons; ACI 318-14 (Refer to Table 3.2.2.2(a-2) for Training Performance)

Optimal CO_2 Emissions (*Based on One Layer with 80 Neurons*) (*ACI 318-14*)

Parameter		*AI Results*	*Check (AutoCol)*	*Error*
1	b (mm)	**1,023.19**		
2	h (mm)	**1,045.94**		
3	ρ_s	**0.0100**		
4	f_c' (MPa)	40		
5	f_y (MPa)	500		
6	P_u (kN)	1,000		
7	M_u (kN·m)	3,000		
8	ϕP_n (kN)	993.6	907.6	8.7%
9	ϕM_n (kN·m)	3,000.4	2,722.9	9.25%
10	SF	1.00	0.91	9.24%
11	b/h	1.00	2.18%	
12	ε_s	0.0311	0.0304	2.29%
13	CI_c (KRW/m)	192,717.7	189,228.3	1.81%
14	CO_2 (t-CO_2/m)	**0.384**	**0.391**	−1.67%
15	W_c (kN/m)	25.176	25.214	−0.15%
16	$\alpha_{e/h}$	0.330	0.335	−1.60%

Note: (dotted box) Seven inputs for reverse training.
(dashed box) Structural verification by (AutoCol).
Optimal CO_2 of **0.384** is −7.36% different from **0.4145** obtained by large structural datasets.

Table 3.3.2.5 (Continued) Design Accuracies Minimizing CO_2 Emissions Based on Reverse Lagrange Optimization

(b-1): Based on Reverse Training with Two Layers – 80 Neurons; ACI 318-14 (Refer to Table 3.2.2.2(a-2) for Training Performance)

Optimal CO_2 Emissions (*Based on Two Layers with 80 Neurons*) (*ACI 318-14*)

Parameter		*AI Results*	*Check (AutoCol)*	*Error*
1	*b* (mm)	**1,020.93**		
2	*h* (mm)	**1,021.08**		
3	ρ_s	**0.0116**		
4	f_c' (MPa)	40		
5	f_y (MPa)	500		
6	P_u (kN)	1,000		
7	M_u (kN·m)	3,000		
8	ϕP_n (kN)	1,006.2	993.9	1.2%
9	ϕM_n (kN·m)	2,996.0	2,981.6	0.48%
10	*SF*	1.000	0.994	0.61%
11	*b/h*	1.0000	0.9999	0.01%
12	ε_s	0.0288	0.0274	4.59%
13	CI_c (KRW/m)	199,441.7	198,488.6	0.48%
14	CO_2 (t-CO_2/m)	**0.4140**	**0.4142**	−0.05%
15	W_c (kN/m)	24.521	24.560	−0.16%
16	$\alpha_{e/h}$	0.333	0.328	1.61%

Note: (dotted box) Seven inputs for reverse training.
(dashed box) Structural verification by (AutoCol).
Optimal CO_2 of **0.414** is −0.12% different from **0.4145** obtained by large structural datasets.

(c-1): Based on Reverse Training with Five Layers – 80 Neurons; ACI 318-14 (Refer to Table 3.2.2.2(a-2) for Training Performance)

Optimal CO_2 Emissions (*Based on Five Layers with 80 Neurons*) (*ACI 318-14*)

Parameter		*AI Results*	*Check (AutoCol)*	*Error*
1	*b* (mm)	**1,023.36**		
2	*h* (mm)	**1,023.15**		
3	ρ_s	**0.0116**		
4	f_c' (MPa)	40		
5	f_y (MPa)	500		
6	P_u (kN)	1,000		
7	M_u (kN·m)	3,000		
8	ϕP_n (kN)	998.2	999.7	−0.2%
9	ϕM_n (kN·m)	2,969.1	2,999.2	−1.01%
10	*SF*	1.0000	0.9997	0.03%
11	*b/h*	1.0000	1.0002	−0.02%
12	ε_s	0.0283	0.0275	2.84%
13	CI_c (KRW/m)	197,257.3	199,252.1	−1.01%
14	CO_2 (t-CO_2/m)	**0.415**	**0.416**	−0.26%
15	W_c (kN/m)	24.662	24.669	−0.03%
16	$\alpha_{e/h}$	0.3290	0.3287	0.08%

Note: (dotted box) Seven inputs for reverse training.
(dashed box) Structural verification by (AutoCol).
Optimal CO_2 of **0.415** is 0.12% different from **0.4145** obtained by large structural datasets.

Table 3.3.2.5 (Continued) Design Accuracies Minimizing CO_2 Emissions Based on Reverse Lagrange Optimization

(d-1): Based on Reverse Training with Ten Layers – 80 Neurons; ACI 318-14 (Refer to Table 3.2.2.2(a-2) for Training Performance)

Optimal CO_2 Emissions (*Based on Ten Layers with 80 Neurons*) (*ACI 318-14*)

Parameter		*AI Results*	*Check (AutoCol)*	*Error*
1	b (mm)	**1,016.21**		
2	h (mm)	**1,016.51**		
3	ρ_s	**0.0119**		
4	f_c' (MPa)	40		
5	f_y (MPa)	500		
6	P_u (kN)	1,000		
7	M_u (kN·m)	3,000		
8	ϕP_n (kN)	955.7	998.6	−4.5%
9	ϕM_n (kN·m)	3,003.3	2,995.9	0.25%
10	SF	1.000	0.999	0.14%
11	b/h	1.0000	0.9997	0.03%
12	ε_s	0.0279	0.0271	2.86%
13	CI_c (KRW/m)	199,025.5	198,713.9	0.16%
14	CO_2 (t-CO_2/m)	**0.4150**	**0.4153**	−0.06%
15	W_c (kN/m)	24.335	24.337	−0.01%
16	$\alpha_{e/h}$	0.32674	0.32670	0.01%

Note: (dotted box) Seven inputs for reverse training.
(dashed box) Structural verification by (AutoCol).
Optimal CO_2 of **0.415** is 0.12% different from **0.4145** obtained by large structural datasets.

(e-1): Based on Best Reverse Trainings a Combination of One, Two, Five, and Ten Hidden Layers with 80 Neurons; ACI 318-14 (Refer to Table 3.2.2.2(a-2) for Training Performance)

Optimal CO_2 Emissions (*Based on Best Networks in Term of MSE*) (*ACI 318-14*)

Parameter			*AI Results*	*Check (AutoCol)*	*Error*
1	b (mm)	5L -80N	**1,023.36**		
2	h (mm)	10L -80N	**1,023.03**		
3	ρ_s		**0.0116**		
4	f_c' (MPa)		40		
5	f_y (MPa)		500		
6	P_u (kN)		1,000		
7	M_u (kN·m)		3,000		
8	ϕP_n (kN)	1L -80N	(992.9)	999.5	(−0.7%)
9	ϕM_n (kN·m)	2L -80N	(2,996.0)	2,998.4	(−0.08%)
10	SF		1.000	0.999	0.05%
11	b/h		1.0000	−0.03%	
12	ε_s	10L -80N	(0.0283)	0.0275	(2.98%)
13	CI_c (KRW/m)	10L -80N	(199,361.1)	199,228.8	(0.07%)
14	CO_2 (t-CO_2/m)	5L -80N	**0.415**	**0.416**	−0.25%
15	W_c (kN/m)	10L -80N	(24.650)	24.666	(−0.06%)
16	$\alpha_{e/h}$	5L -80N	(0.329)	0.329	(0.00%)

Note: (dotted box) Seven inputs for reverse training.
(dashed box) Structural verification by (AutoCol).
Optimal CO_2 of **0.415** is 0.12% different from **0.4145** obtained by large structural datasets.

Table 3.3.2.5 (Continued) Design Accuracies Minimizing CO_2 Emissions Based on Reverse Lagrange Optimization

(a-2): Based on Reverse Training with One Layer – 80 Neurons; ACI 318-19 (Refer to Table 3.2.2.2(b-2) for Training Performance)

Optimal CO_2 Emissions (*Based on One Layer with 80 Neurons*) (*ACI 318-19*)

Parameter		*AI Results*	*Check (AutoCol)*	*Error*
1	b (mm)	**1,030.17**		
2	h (mm)	**1,042.95**		
3	ρ_s	**0.0100**		
4	f_c' (MPa)	40		
5	f_y (MPa)	500		
6	P_u (kN)	1,000		
7	M_u (kN·m)	3,000		
8	ϕP_n (kN)	1,009.1	907.9	10.0%
9	ϕM_n (kN·m)	2,995.3	2,723.8	9.07%
10	SF	1.00		
11	b/h	1.00		
12	ε_s	0.0308	0.0304	1.47%
13	CI_c (KRW/m)	191,793.2	189,976.4	0.95%
14	CO_2 (t-CO_2/m)	**0.400**	**0.392**	1.93%
15	W_c (kN/m)	25.529	25.313	0.84%
16	$\alpha_{e/h}$	0.333	0.335	−0.43%

Note: [dotted box] Seven inputs for reverse training.
[dashed box] Structural verification by (AutoCol).
Optimal CO_2 of **0.400** is −3.50% different from **0.4145** obtained by large structural datasets.

(b-2): Based on Reverse Training with Two Layers – 80 Neurons; ACI 318-19 (Refer to Table 3.2.2.2(b-2) for Training Performance)

Optimal CO_2 Emissions (*Based on Two Layers with 80 Neurons*) (*ACI 318-19*)

Parameter		*AI Results*	*Check (AutoCol)*	*Error*
1	b (mm)	**1,069.99**		
2	h (mm)	**1,065.97**		
3	ρ_s	**0.0100**		
4	f_c' (MPa)	40		
5	f_y (MPa)	500		
6	P_u (kN)	1,000		
7	M_u (kN·m)	3,000		
8	ϕP_n (kN)	998.3	990.9	0.7%
9	ϕM_n (kN·m)	2,994.9	2,972.7	0.74%
10	SF	1.000	0.991	0.91%
11	b/h	1.000	−0.38%	
12	ε_s	0.0315	0.0305	3.04%
13	CI_c (KRW/m)	199,846.0	201,675.4	−0.92%
14	CO_2 (t-CO_2/m)	**0.406**	**0.416**	−2.59%
15	W_c (kN/m)	26.823	26.872	−0.18%
16	$\alpha_{e/h}$	0.339	0.341	−0.74%

Note: [dotted box] Seven inputs for reverse training.
[dashed box] Structural verification by (AutoCol).
Optimal CO_2 of **0.406** is −2.05% different from **0.4145** obtained by large structural datasets.

Table 3.3.2.5 (Continued) Design Accuracies Minimizing CO_2 Emissions Based on Reverse Lagrange Optimization

(c-2): Based on Reverse Training with Five Layers – 80 Neurons; ACI 318-19 (Refer to Table 3.2.2.2(b-2) for Training Performance)

Optimal CO_2 Emissions (*Based on Five Layers with 80 Neurons*) (*ACI 318-19*)

Parameter		*AI Results*	*Check (AutoCol)*	*Error*
1	b (mm)	**1,067.69**		
2	h (mm)	**1,070.15**		
3	ρ_s	**0.0100**		
4	f_c' (MPa)	40		
5	f_y (MPa)	500		
6	P_u (kN)	1,000		
7	M_u (kN·m)	3,000		
8	ϕP_n (kN)	929.0	997.6	−7.4%
9	ϕM_n (kN·m)	3,048.4	2,992.9	1.82%
10	SF	1.00	1.00	0.24%
11	b/h	1.00	1.00	0.23%
12	ε_s	0.0311	0.0305	1.70%
13	CI_c (KRW/m)	198,950.9	202,029.0	−1.55%
14	CO_2 (t-CO_2/m)	**0.411**	**0.417**	−1.55%
15	W_c (kN/m)	27.058	26.919	0.51%
16	$\alpha_{e/h}$	0.342	0.343	−0.10%

Note: [dotted box] Seven inputs for reverse training.
[dashed box] Structural verification by (AutoCol).
Optimal CO_2 of **0.411** is −0.84% different from **0.4145** obtained by large structural datasets.

(d-2): Based on Reverse Training with Ten Layers – 80 Neurons; ACI 318-19 (Refer to Table 3.2.2.2(b-2) for Training Performance)

Optimal CO_2 Emissions (*Based on Five Layers with 80 Neurons*) (*ACI 318-19*)

Parameter		*AI Results*	*Check (AutoCol)*	*Error*
1	b (mm)	**1,053.99**		
2	h (mm)	**1,053.59**		
3	ρ_s	**0.0105**		
4	f_c' (MPa)	40		
5	f_y (MPa)	500		
6	P_u (kN)	1,000		
7	M_u (kN·m)	3,000		
8	ϕP_n (kN)	1,325.5	997.7	24.7%
9	ϕM_n (kN·m)	3,005.3	2,993.1	0.40%
10	SF	1.00	1.00	0.23%
11	b/h	1.00	1.00	−0.04%
12	ε_s	0.0303	0.0295	2.71%
13	CI_c (KRW/m)	199,923.7	201,054.3	−0.57%
14	CO_2 (t-CO_2/m)	**0.414**	**0.417**	−0.59%
15	W_c (kN/m)	26.159	26.163	−0.01%
16	$\alpha_{e/h}$	0.338	0.338	0.03%

Note: [dotted box] Seven inputs for reverse training.
[dashed box] Structural verification by (AutoCol).
Optimal CO_2 of **0.414** is −0.12% different from **0.4145** obtained by large structural datasets.

Table 3.3.2.5 (Continued) Design Accuracies Minimizing CO_2 Emissions Based on Reverse Lagrange Optimization

(e-2): Based on Best Reverse Trainings a Combination of One, Two, Five, and Ten Hidden Layers with 80 Neurons; ACI 318-19 (Refer to Table 3.2.2.2(b-2) for Training Performance)

Optimal CO_2 Emissions (*Based on Best Networks in Term of MSE*) (*ACI 318-19*)

Parameter			*AI Results*	*Check (AutoCol)*	*Error*
1	b (mm)	5L -80N	**1,052.26**		
2	h (mm)	5L -80N	**1,054.29**		
3	ρ_s		**0.0105**		
4	f_c' (MPa)		40		
5	f_y (MPa)		500		
6	P_u (kN)		1,000		
7	M_u (kN·m)		3,000		
8	ϕP_n (kN)	2L -80N	(998.3)	997.6	(0.1%)
9	ϕM_n (kN·m)	2L -80N	(2,995.1)	2,992.7	(0.08%)
10	SF		1.000	0.998	0.24%
11	b/h		1.000	0.998	0.19%
12	ε_s	10L -80N	(0.0303)	0.0295	(2.69%)
13	CI_c (KRW/m)	10L -80N	(199,923.7)	200,857.4	(−0.47%)
14	CO_2 (t-CO_2/m)	10L -80N	**0.414**	**0.416**	−0.50%
15	W_c (kN/m)	5L -80N	(26.231)	26.137	(0.36%)
16	$\alpha_{e/h}$	5L -80N	(0.3378)	0.3379	(−0.05%)

Note: [dotted box] Seven inputs for reverse training.
[dashed box] Structural verification by (AutoCol).
Optimal CO_2 of **0.414** is −0.12% different from **0.4145** obtained by large structural datasets.

and (b-2). These errors for ϕP_n also decrease to −0.7% for ACI 318-14 and 0.1% for ACI 318-19 for a combination of one, two, five, and ten layers as shown in Tables 3.3.2.5(e-1) and (e-2), respectively. It is noted that uncertainty caused by the randomness of large datasets significantly influences accuracies of optimizations when using one hidden layer whereas adopting more hidden layers removes an influence of uncertainty caused by the randomness of large datasets on optimizations.

3.3.2.3 Verifications

1. Verification of CO_2 emissions based on reverse networks

   As shown in Equation (3.2.2.5-2), an objective function CO_2 emission obtained based on a reverse network is a function of input parameters $\mathbf{x} = \left[\rho_s, f_c', f_y, P_u, M_u, SF, b/h\right]^T$. A reverse network can predetermine six input parameters ($f_c', f_y, P_u, M_u, SF, b/h$) out of seven input parameters except for ρ_s via equality constraints based on design criteria listed in Table 3.2.2.4 (b) and Table 3.2.2.5(b). CO_2 emissions are plotted as a function of ρ_s as shown in Figure 3.3.2.1. Four curves are generated for one, two, five, and ten layers with the same 80 neurons. Large datasets verify that four curves converge to yield a realistic minimum CO_2 emission for a column if training accuracies are sufficient. A CO_2 emission of 0.384 t-CO_2/m for ACI 318-14 (Figure 3.3.2.1a, Table 3.3.2.5(a-1)) and 0.400 t-CO_2/m for ACI 318-19 (Figure 3.3.2.1b, Table 3.3.2.5(a-2)) obtained at ρ_s of 0.01 with one layer and 80 neurons demonstrates a difference of -7.36% (Figure 3.3.2.2a) for

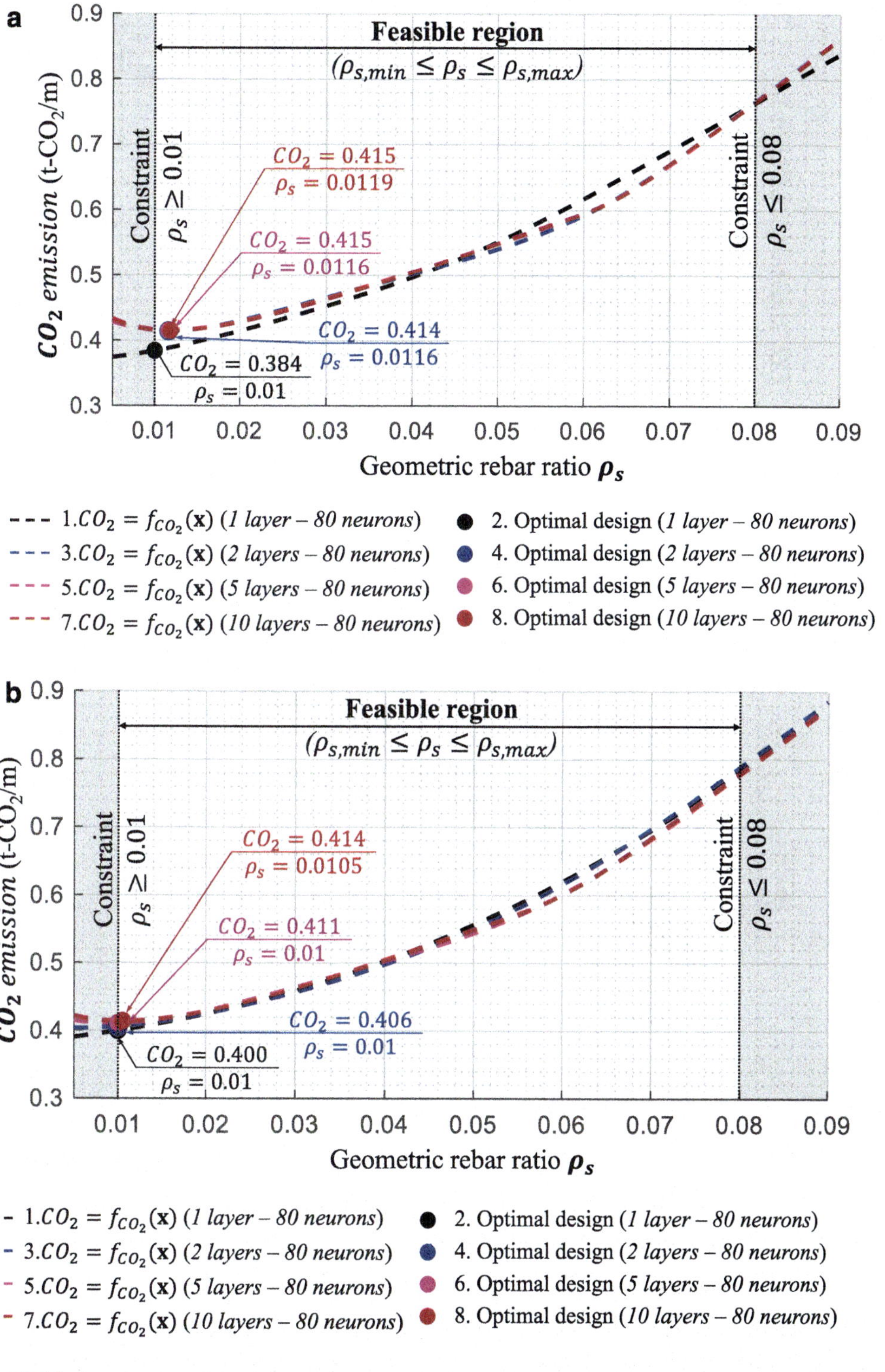

Figure 3.3.2.1 Comparison of the CO_2 emissions obtained by reverse ANNs based on 1, 2, 5, 10 layers as a function of rebar ratio ρ_s between 0.0 and 0.1. (a) ACI 318-14. (b) ACI 318-19.

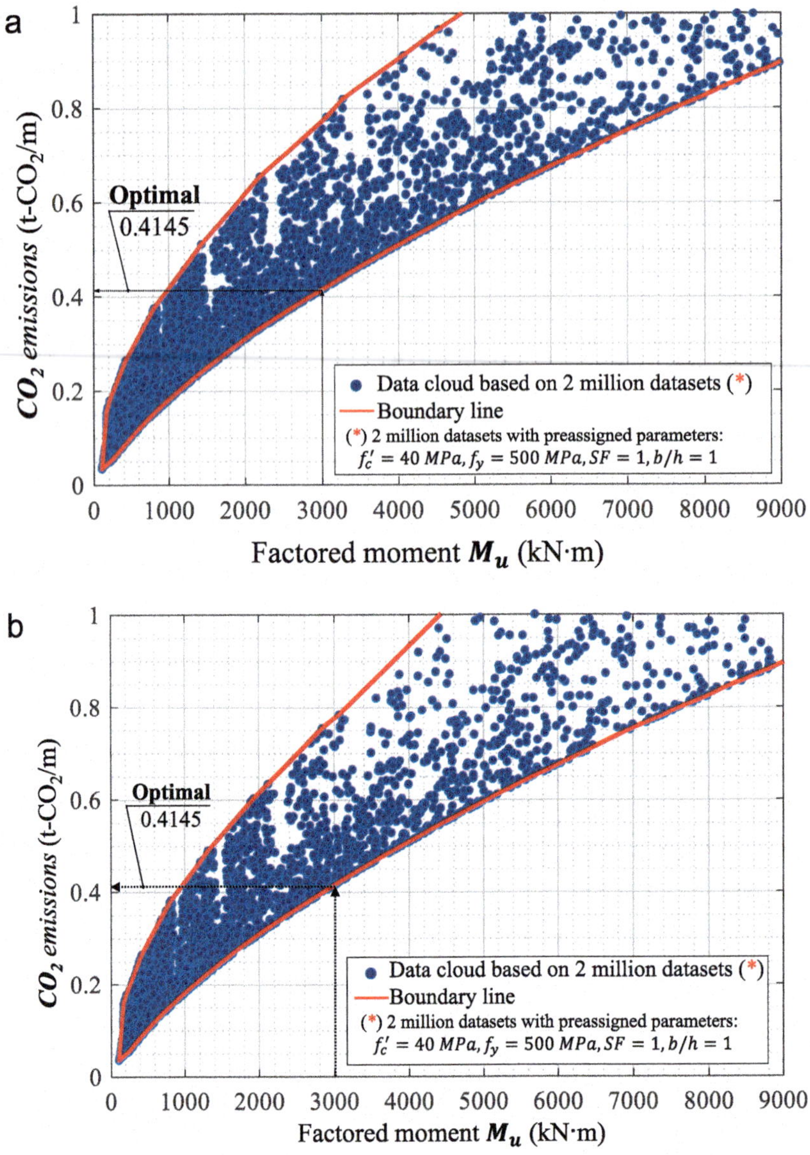

Figure 3.3.2.2 Verification of CO_2 emission based on large datasets. (a) ACI 318-14. (b) ACI 318-19.

ACI 318-14 and –3.50% (Figure 3.3.2.2b) for ACI 318-19 relative to the 0.4145 t-CO_2/m for both ACI 318-14 and ACI 318-19 obtained from large datasets shown in Figure 3.3.2.2. This error can be corrected by using more layers and neurons to improve training accuracy. For example, differences between minimized CO_2 emissions obtained by ANN-based Lagrange optimization and that obtained by large datasets are reduced to –0.12% for ACI 318-14 and –2.05% for ACI 318-19 based on two layers whereas these errors further reduce to 0.12% for ACI 318-14 and –0.84% for ACI 318-19 when five layers are implemented. Four objective functions based on one, two, five, and ten layers with 80 neurons are in good agreement within a valid range of rebar ratios (between a minimum and maximum ρ_s of 0.01 and 0.08). However, all four curves do not match for ρ_s outside the design range (below 0.01 and above 0.08) because datasets outside the design range for training ANNs are not sufficiently generated.

2. Verification of CO_2 using large datasets
   Minimized CO_2 emissions (t-CO_2/m) of 0.4145 for both ACI 318-14 and 318-19 obtained from two million large datasets shown in Figure 3.3.2.2 are consistent with 0.415 for ACI 318-14 and 0.417 for ACI 318-19 shown in Tables 3.3.2.2(e-1) and (e-2), respectively, calculated using forward-Lagrange network. Similarly, 0.4147 for ACI 318-14 and 0.4140 for ACI 318-19 shown in Tables 3.3.2.3(e-1) and (e-2), respectively, calculated using reverse-Lagrange network with the best MSE based on a combination of one, two, five, and ten layers with 80 neurons. However, CO_2 emissions of 0.4145 (t-CO_2/m) for both ACI 318-14 and 318-19 obtained based on large datasets differ from 0.397 for ACI 318-14 and 0.404 for ACI 318-19 shown in Tables 3.3.2.2(a-1) and (a-2), respectively, calculated using forward-Lagrange network based on one layer and 80 neurons. Similarly, CO_2 emissions obtained based on large datasets differ from 0.384 for ACI 318-14 and 0.400 for ACI 318-19 shown in Tables 3.3.2.3(a-1) and (a-2), respectively, calculated using reverse-network based on one layer and 80 neurons. This is because CO_2 emissions minimized from a network with one layer shown in Tables 3.3.2.2(a) and 3.3.2.3(a) cannot be used because errors associated with design parameters, such as an accuracy of design axial strength, ϕP_n (kN) is too large as shown in Tables 3.3.2.4 and 3.3.2.5 even if a network with one layer provides smallest CO_2 emissions. Verification is required for all design parameters before concluding the minimization of CO_2 emissions.

3. Design accuracies
   Tables 3.3.2.4 and 3.3.2.5 demonstrate design accuracies of minimized CO_2 emissions obtained based on the reverse and forward networks, respectively, using one, two, five, and ten layers with 80 neurons. Tables 3.3.2.4(e-1), (e-2) and Tables 3.3.2.5(e-1), (e-2) present acceptable design accuracies for minimizing CO_2 emissions based on a combination of one, two, five, and ten layers with 80 neurons. However, Tables 3.3.2.4(a-1), (a-2) and 3.3.2.5(a-1), (a-2) based on one layer and 80 neurons yield design accuracies insufficient to be adopted in design applications. Designs based on two, five, and ten layers show better accuracies than those based on one layer.

3.3.2.4 *P–M* diagram

P–M interaction diagrams reflecting minimized CO_2 emissions for an RC column are constructed in Figures 3.3.2.3a and b for forward and reverse networks, respectively. Parameters with sufficient accuracies shown in Box 1-7 of Tables 3.3.2.4 and 3.3.2.5 obtained by ANNs are used to plot *P–M* diagrams using *AutoCol*. *P–M* diagrams represented by Legends 1–5 are plotted based on forward and reverse neural networks with one, two, five, and ten layers, and the best network using 80 neurons. All ten optimized *P–M* diagrams represented by Legends 1–5 shown in Figure 3.3.2.3a and b converge to one *P–M* diagram when training accuracies are sufficient whereas *P–M* diagrams pass-through P_u = 1,000 kN and M_u = 3,000 kN·m as indicated by a red dot. However, *P–M* diagram based on both forward and reverse networks with one layer and 80 neurons does not pass-through design points (P_u, M_u) because training accuracies are not good enough. The *P–M* interaction diagrams satisfy all design criteria shown in Table 3.2.2.4(b) and Table 3.2.2.5.

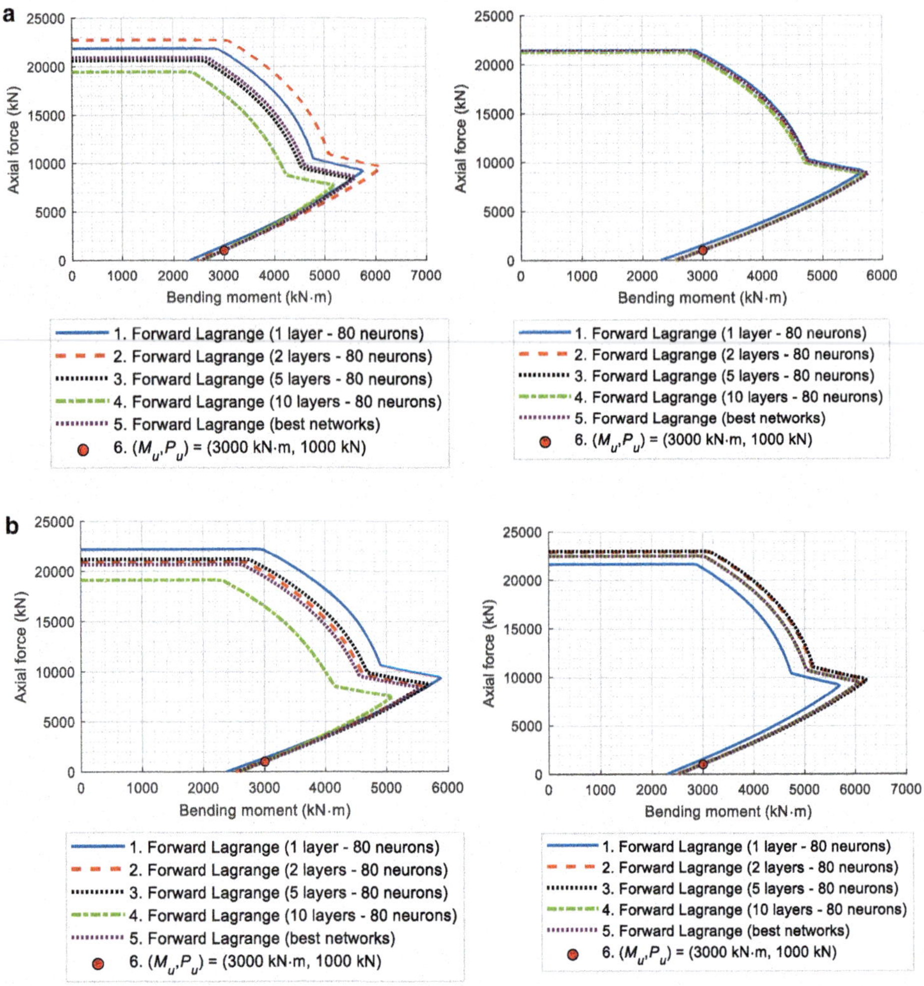

Figure 3.3.2.3 P–M interaction diagrams for minimized CO_2 emissions. (a) ACI 318-14. (a-1) using forward Lagrange based on Table 3.3.2.4. (a-2) using reverse Lagrange based on Table 3.3.2.5. (b) ACI 318-19. (b-1) using forward Lagrange based on Table 3.3.2.4. (b-2) using reverse Lagrange based on Table 3.3.2.5.

3.3.3 Column design scenario minimizing weight

3.3.3.1 Formulation of forward network vs. reverse network

As shown in Equation (3.2.2.5-3), an objective function W_c obtained based on reverse networks is a function of the input parameters $\mathbf{x} = \left[\rho_s, f_c', f_y, P_u, M_u, SF, b/h\right]^T$. According to Design 3, six input parameters, f_c', f_y, P_u, M_u, SF, and b/h, are considered as equality constraints. A reverse network can predetermine six input parameters $(f_c', f_y, P_u, M_u, SF, b/h)$ out of seven input parameters except for ρ_s via equality constraints based on design criteria listed in Table 3.2.2.4(b) and Table 3.2.2.5(b). Weight of a square column W_c subject to $P_u = 1{,}000$ kN and $M_u = 3{,}000$ kN·m with a safety factor of 1 is minimized for given material properties: $f_c' = 40$ MPa and $f_y = 500$ MPa. Rebar ratio is not constrained by this design, but it must comply with ACI-318-19: $\rho_{s,min} \leq \rho_s \leq \rho_{s,max}$. Optimization problems for Design 3 are summarized in Table 3.3.3.1.

Table 3.3.3.1 Design Problem Minimizing Column Weight W_c

Minimize	$W_c = f_{W_c}(x)$
Subject to	
Equality constraints:	$P_u = 1,000$ kN, $M_u = 3,000$ kN·m $f_c' = 40$ MPa, $f_y = 500$ MPa, $SF = 1, b/h = 1$
Inequality constraints:	$\rho_{s,min} \leq \rho_s \leq \rho_{s,max}$

3.3.3.2 *Solving KKT nonlinear equations based on Newton–Raphson method*

Newton–Raphson iteration is employed to solve partial differential equations using initial trial Lagrange multipliers and parameters distributed randomly within the training data range to minimize W_c as shown in Tables 3.3.3.2 and 3.3.3.3 similar to those of CO_2 emissions shown in Tables 3.3.2.2 and 3.3.2.3.

The largest errors for design axial loads ϕP_n (kN) are found as –138% for ACI 318-14 shown in Table 3.3.3.4(a-1) and –127% for ACI 318-19 shown in Table 3.3.3.4(a-2) based on forward network with one layer and 80 neurons, respectively, which, then, decrease to 6.4% for ACI 318-14 shown in Tables 3.3.3.4(b-1) and 1.0% for ACI 318-19 shown in Tables 3.3.3.4(b-2) based on two layers with 80 neurons.

The largest errors for design axial loads ϕP_n (kN) are found as –3.4% for ACI 318-14 shown in Table 3.3.3.5(a-1) and 6.0% for ACI 318-19 shown in Table 3.3.3.5(a-2) based on reverse network with one layer and 80 neurons, respectively, which, then, decrease to 1.3% for ACI 318-14 shown in Tables 3.3.3.5(b-1) and 1.9% for ACI 318-19 shown in Tables 3.3.3.5(b-2) based on two layers with 80 neurons. Designs with one hidden layer shown in Tables 3.3.3.4(a) and 3.3.3.5(a) based on forward and reverse networks, respectively, demonstrate big differences in design accuracies. A minimized W_c obtained by Lagrange functions based on both forward and reverse networks is consistent with that obtained using *AutoCol* when training accuracies are sufficient as shown in Tables 3.3.3.4(e) and 3.3.3.5(e). The rest of design parameters are also accurately predicted while minimizing W_c.

Table 3.3.3.2 Lowest Column Weight (W_c) with Corresponding Parameters for KKT Conditions Based on Forward Networks

(a-1): Based on Forward Training with One Layer – 80 Neurons; ACI 318-14 (Refer to Table 3.2.2.2(a-1) for Training Performance)

AI-based Lagrange Optimization (*One Layer -80 neurons*) (*ACI 318-14*)

Parameters		*Case 1*	*Case 2*	*Case 3*
1	b (mm)	1,042.0	665.4	**664.8**
2	h (mm)	1,042.0	665.4	**664.8**
3	ρ_s	0.0100	0.0800	0.0726
4	f_c' (MPa)	40	40	40
5	f_y (MPa)	500	500	500
6	P_u (kN)	1,000	1,000	1,000
7	M_u (kN·m)	3,000	3,000	3,000
Objective: W_c (kN/m)		25.584	10.437	**10.419**

Note: "Bold values" emphasize the results of optimal case which is the most important case.
Case 1, Inequality $v_1(\mathbf{x})$ is active; Case 2, Inequality $v_2(\mathbf{x})$ is active; Case 3, None of inequalities are active.

Table 3.3.3.2 (Continued) Lowest Column Weight (W_c) with Corresponding Parameters for KKT Conditions Based on Forward Networks

(b-1): Based on Forward Training with Two Layers – 80 Neurons; ACI 318-14 (Refer to Table 3.2.2.2(a-1) for Training Performance)

AI-based Lagrange Optimization (*Two Layers -80 Neurons*) (*ACI 318-14*)

Parameters		*Case 1*	*Case 2*	*Case 3*
1	b (mm)	1,063.2	662.6	**656.8**
2	h (mm)	1,063.2	662.6	**656.8**
3	ρ_s	0.0100	0.0800	**0.0710**
4	f_c' (MPa)	40	40	40
5	f_y (MPa)	500	500	500
6	P_u (kN)	1,000	1,000	1,000
7	M_u (kN·m)	3,000	3,000	3,000
Objective: W_c (kN/m)		26.630	10.344	**10.164**

Note: "Bold values" emphasize the results of optimal case which is the most important case.
Case 1, Inequality $v_1(\mathbf{x})$ is active; Case 2, Inequality $v_2(\mathbf{x})$ is active; Case 3, None of inequalities are active.

(c-1): Based on Forward Training With Five Layers – 80 Neurons; ACI 318-14 (Refer to Table 3.2.2.2(a-1) for Training Performance)

AI-based Lagrange Optimization (*Five Layers -80 Neurons*) (*ACI 318-14*)

Parameters		*Case 1*	*Case 2*	*Case 3*
1	b (mm)	1,068.5	663.3	**656.9**
2	h (mm)	1,068.5	663.3	**656.9**
3	ρ_s	0.0100	0.0800	**0.0654**
4	f_c' (MPa)	40	40	40
5	f_y (MPa)	500	500	500
6	P_u (kN)	1,000	1,000	1,000
7	M_u (kN·m)	3,000	3,000	3,000
Objective: W_c (kN/m)		26.936	10.364	**10.177**

Note: "Bold values" emphasize the results of optimal case which is the most important case.
Case 1, Inequality $v_1(\mathbf{x})$ is active; Case 2, Inequality $v_2(\mathbf{x})$ is active; Case 3, None of inequalities are active.

(d-1): Based on Forward Training with Ten Layers – 80 Neurons; ACI 318-14 (Refer to Table 3.2.2.2(a-1) for Training Performance)

AI-based Lagrange Optimization (*Ten Layers -80 Neurons*) (*ACI 318-14*)

Parameters		*Case 1*	*Case 2*	*Case 3*
1	b (mm)	1,068.5	663.3	**656.9**
2	h (mm)	1,068.5	663.3	**656.9**
3	ρ_s	0.0100	0.0800	**0.0654**
4	f_c' (MPa)	40	40	40
5	f_y (MPa)	500	500	500
6	P_u (kN)	1,000	1,000	1,000
7	M_u (kN·m)	3,000	3,000	3,000
Objective: W_c (kN/m)		26.936	10.364	**10.177**

Note: "Bold values" emphasize the results of optimal case which is the most important case.
Case 1, Inequality $v_1(\mathbf{x})$ is active; Case 2, Inequality $v_2(\mathbf{x})$ is active; Case 3, None of inequalities are active.

Table 3.3.3.2 (Continued) Lowest Column Weight (W_c) with Corresponding Parameters for KKT Conditions Based on Forward Networks

(e-1): Based on Best Forward Trainings a Combination of One, Two, Five, and Ten Hidden Layers with 80 Neurons; ACI 318-14 (Refer to Table 3.2.2.2(a-1) for Training Performance)

AI-based Lagrange Optimization (*Ten Layers -80 Neurons*) (*ACI 318-14*)

Parameters		*Case 1*	*Case 2*	*Case 3*
1	b (mm)	1,068.0	661.9	**656.4**
2	h (mm)	1,068.0	661.9	**656.4**
3	ρ_s	0.0100	0.0800	**0.0651**
4	f_c' (MPa)	40	40	40
5	f_y (MPa)	500	500	500
6	P_u (kN)	1,000	1,000	1,000
7	M_u (kN·m)	3,000	3,000	3,000
Objective: W_c (kN/m)		26.877	10.328	**10.158**

Note: "Bold values" emphasize the results of optimal case which is the most important case.
Case 1, Inequality $v_1(\mathbf{x})$ is active; Case 2, Inequality $v_2(\mathbf{x})$ is active; Case 3, None of inequalities are active.

(a-2): Based on Forward Training with One Layer – 80 Neurons; ACI 318-19 (Refer to Table 3.2.2.2(b-1) for Training Performance)

AI-Based Lagrange Optimization (*One Layer -80 Neurons*) (*ACI 318-19*)

Parameters		*Case 1*	*Case 2*	*Case 3*
1	b (mm)	1,050.2	662.9	**661.0**
2	h (mm)	1,050.2	662.9	**661.0**
3	ρ_s	0.0100	0.0800	0.0690
4	f_c' (MPa)	40	40	40
5	f_y (MPa)	500	500	500
6	P_u (kN)	1,000	1,000	1,000
7	M_u (kN·m)	3,000	3,000	3,000
Objective: W_c (kN/m)		25.986	10.358	**10.297**

Note: "Bold values" emphasize the results of optimal case which is the most important case.
Case 1, Inequality $v_1(\mathbf{x})$ is active; Case 2, Inequality $v_2(\mathbf{x})$ is active; Case 3, None of inequalities are active.

(b-2): Based on forward training with Two Layers – 80 Neurons; ACI 318-19 (Refer to Table 3.2.2.2(b-1) for Training Performance)

AI-based Lagrange Optimization (*Two Layers -80 Neurons*) (*ACI 318-19*)

Parameters		*Case 1*	*Case 2*	*Case 3*
1	b (mm)	1,061.9	669.9	**666.0**
2	h (mm)	1,061.9	669.9	**666.0**
3	ρ_s	0.0100	0.0800	**0.0645**
4	f_c' (MPa)	40	40	40
5	f_y (MPa)	500	500	500
6	P_u (kN)	1,000	1,000	1,000
7	M_u (kN·m)	3,000	3,000	3,000
Objective: W_c (kN/m)		26.561	10.575	**10.452**

Note: "Bold values" emphasize the results of optimal case which is the most important case.
Case 1, Inequality $v_1(\mathbf{x})$ is active; Case 2, Inequality $v_2(\mathbf{x})$ is active; Case 3, None of inequalities are active.

Table 3.3.3.2 (Continued) Lowest Column Weight (W_c) with Corresponding Parameters for KKT Conditions Based on Forward Networks

(c-2): Based on Forward Training with Five Layers – 80 Neurons; ACI 318-19 (Refer to Table 3.2.2.2(b-1) for Training Performance)

AI-based Lagrange Optimization (*Five Layers -80 Neurons*) (*ACI 318-19*)

	Parameters	*Case 1*	*Case 2*	*Case 3*
1	b (mm)	1,069.4	669.7	**664.0**
2	h (mm)	1,069.4	669.7	**664.0**
3	ρ_s	0.0100	0.0800	**0.0625**
4	f_c' (MPa)	40	40	40
5	f_y (MPa)	500	500	500
6	P_u (kN)	1,000	1,000	1,000
7	M_u (kN·m)	3,000	3,000	3,000
Objective: W_c (kN/m)		26.864	10.533	**10.372**

Note: "Bold values" emphasize the results of optimal case which is the most important case.
Case 1, Inequality $v_1(\mathbf{x})$ is active; Case 2, Inequality $v_2(\mathbf{x})$ is active; Case 3, None of inequalities are active.

(d-2): Based on Forward Training with Ten Layers – 80 Neurons; ACI 318-19 (Refer to Table 3.2.2.2(b-1) for Training Performance)

AI-based Lagrange Optimization (*Ten Layers -80 Neurons*) (*ACI 318-19*)

	Parameters	*Case 1*	*Case 2*	*Case 3*
1	b (mm)	1,071.1	668.6	**665.0**
2	h (mm)	1,071.1	668.6	**665.0**
3	ρ_s	0.0100	0.0800	**0.0616**
4	f_c' (MPa)	40	40	40
5	f_y (MPa)	500	500	500
6	P_u (kN)	1,000	1,000	1,000
7	M_u (kN·m)	3,000	3,000	3,000
Objective: W_c (kN/m)		27.010	10.551	**10.439**

Note: "Bold values" emphasize the results of optimal case which is the most important case.
Case 1, Inequality $v_1(\mathbf{x})$ is active; Case 2, Inequality $v_2(\mathbf{x})$ is active; Case 3, None of inequalities are active.

(e-2): Based on Best Forward Trainings a Combination of One, Two, Five, and Ten Hidden Layers with 80 Neurons; ACI 318-19 (Refer to Table 3.2.2.2(b-1) for Training Performance)

AI-based Lagrange Optimization (*Best Networks in Term of Test MSE*) (*ACI 318-19*)

Parameters		*Case 1*	*Case 2*	*Case 3*
1	b (mm)	1,071.1	**668.6**	N/A
2	h (mm)	1,071.1	**668.6**	N/A
3	ρ_s	0.0100	**0.0800**	N/A
4	f_c' (MPa)	40	40	N/A
5	f_y (MPa)	500	500	N/A
6	P_u (kN)	1,000	1,000	N/A
7	M_u (kN·m)	3,000	3,000	N/A
Objective: W_c (kN/m)		27.030	**10.535**	N/A

Note: "Bold values" emphasize the results of optimal case which is the most important case.
Case 1, Inequality $v_1(\mathbf{x})$ is active; Case 2, Inequality $v_2(\mathbf{x})$ is active; Case 3, None of inequalities are active.

Table 3.3.3.3 Lowest W_c with Corresponding Parameters for KKT Conditions Based on Reverse Networks

(a-1): Based on Reverse Training with One Layer – 80 Neurons; ACI 318-14 (Refer to Table 3.2.2.2(a-2) for Training Performance)

AI-based Lagrange Optimization (*One Layer -80 Neurons*) (*ACI 318-14*)

Parameters		*Case 1*	*Case 2*	*Case 3*
3	ρ_s	0.0100	0.0800	**0.0734**
4	f_c' (MPa)	40	40	40
5	f_y (MPa)	500	500	500
6	P_u (kN)	1,000	1,000	1,000
7	M_u (kN·m)	3,000	3,000	3,000
10	*SF*	1	1	1
11	*b/h*	1	1	1
Objective: W_c (kN/m)		25.176	10.511	**10.496**

Note: "Bold values" emphasize the results of optimal case which is the most important case.
Case 1, Inequality $v_1(\mathbf{x})$ is active; Case 2, Inequality $v_2(\mathbf{x})$ is active; Case 3, None of inequalities are active.

(b-1): Based on Reverse Training with Two Layers – 80 Neurons; ACI 318-14 (Refer to Table 3.2.2.2(a-2) for Training Performance)

AI-based Lagrange Optimization (*Two Layers -80 Neurons*) (*ACI 318-14*)

Parameters		*Case 1*	*Case 2*	*Case 3*
3	ρ_s	0.0100	0.0800	**0.0674**
4	f_c' (MPa)	40	40	40
5	f_y (MPa)	500	500	500
6	P_u (kN)	1,000	1,000	1,000
7	M_u (kN·m)	3,000	3,000	3,000
10	*SF*	1	1	1
11	*b/h*	1	1	1
Objective: W_c (kN/m)		26.819	10.336	**10.158**

Note: "Bold values" emphasize the results of optimal case which is the most important case.
Case 1, Inequality $v_1(x)$ is active; Case 2, Inequality $v_2(x)$ is active; Case 3, None of inequalities are active.

(c-1): Based on Reverse Training with Five Layers – 80 Neurons; ACI 318-14 (Refer to Table 3.2.2.2(a-2) for Training Performance)

AI-based Lagrange Optimization (*Five Layers -80 Neurons*) (*ACI 318-14*)

Parameters		*Case 1*	*Case 2*	*Case 3*
3	ρ_s	0.0100	0.0800	**0.0660**
4	f_c' (MPa)	40	40	40
5	f_y (MPa)	500	500	500
6	P_u (kN)	1,000	1,000	1,000
7	M_u (kN·m)	3,000	3,000	3,000
10	*SF*	1	1	1
11	*b/h*	1	1	1
Objective: CI_c (kN/m)		26.993	10.316	**10.177**

Note: "Bold values" emphasize the results of optimal case which is the most important case.
Case 1, Inequality $v_1(x)$ is active; Case 2, Inequality $v_2(x)$ is active; Case 3, None of inequalities are active.

Table 3.3.3.3 (Continued) Lowest W_c with Corresponding Parameters for KKT Conditions Based on Reverse Networks

(d-1): Based on Reverse Training with Ten Layers – 80 Neurons; ACI 318-14 (Refer to Table 3.2.2.2(a-2) for Training Performance)

AI-based Lagrange Optimization (*Ten Layers -80 Neurons*) (*ACI 318-14*)

Parameters		*Case 1*	*Case 2*	*Case 3*
3	ρ_s	0.0100	0.0800	**0.0642**
4	f_c' (MPa)	40	40	40
5	f_y (MPa)	500	500	500
6	P_u (kN)	1,000	1,000	1,000
7	M_u (kN·m)	3,000	3,000	3,000
10	*SF*	1	1	1
11	*b/h*	1	1	1
Objective: CI_c (kN/m)		27.017	10.328	**10.174**

Note: "Bold values" emphasize the results of optimal case which is the most important case.
Case 1, Inequality $v_1(x)$ is active; Case 2, Inequality $v_2(x)$ is active; Case 3, None of inequalities are active.

(e-1): Based on Best Reverse Trainings a Combination of One, Two, Five, and Ten Hidden Layers with 80 Neurons; ACI 318-14 (Refer to Table 3.2.2.2(a-2) for Training Performance)

AI-based Lagrange Optimization (*Best Networks in Term of MSE*) (*ACI 318-14*)

Parameters		*Case 1*	*Case 2*	*Case 3*
3	ρ_s	0.0100	0.0800	**0.0642**
4	f_c' (MPa)	40	40	40
5	f_y (MPa)	500	500	500
6	P_u (kN)	1,000	1,000	1,000
7	M_u (kN·m)	3,000	3,000	3,000
10	*SF*	1	1	1
11	*b/h*	1	1	1
Objective: CI_c (kN/m)		27.017	10.328	**10.174**

Note: "Bold values" emphasize the results of optimal case which is the most important case.
Case 1, Inequality $v_1(x)$ is active; Case 2, Inequality $v_2(x)$ is active; Case 3, None of inequalities are active.

(a-2): Based on Reverse Training with One Layer – 80 Neurons; ACI 318-19 (Refer to Table 3.2.2.2(b-2) for Training Performance)

AI-based Lagrange Optimization (*One Layer -80 Neurons*) (*ACI 318-19*)

Parameters		*Case 1*	*Case 2*	*Case 3*
3	ρ_s	0.0100	**0.0800**	N/A
4	f_c' (MPa)	40	40	N/A
5	f_y (MPa)	500	500	N/A
6	P_u (kN)	1,000	1,000	N/A
7	M_u (kN·m)	3,000	3,000	N/A
10	*SF*	1	1	N/A
11	*b/h*	1	1	N/A
Objective: W_c (kN/m)		25.529	**10.243**	N/A

Note: "Bold values" emphasize the results of optimal case which is the most important case.
Case 1, Inequality $v_1(\mathbf{x})$ is active; Case 2, Inequality $v_2(\mathbf{x})$ is active; Case 3, None of inequalities are active.

Table 3.3.3.3 (Continued) Lowest W_c with Corresponding Parameters for KKT Conditions Based on Reverse Networks

(b-2): Based on Reverse Training with Two Layers – 80 Neurons; ACI 318-19 (Refer to Table 3.2.2.2(b-2) for Training Performance)

AI-based Lagrange Optimization (*Two Layers -80 neurons*) (*ACI 318-19*)

Parameters		*Case 1*	*Case 2*	*Case 3*
3	ρ_s	0.0100	0.0800	**0.0662**
4	f_c' (MPa)	40	40	40
5	f_y (MPa)	500	500	500
6	P_u (kN)	1,000	1,000	1,000
7	M_u (kN·m)	3,000	3,000	3,000
10	*SF*	1	1	1
11	*b/h*	1	1	1
Objective: CI_c (kN/m)		26.823	10.597	**10.390**

Note: "Bold values" emphasize the results of optimal case which is the most important case.
Case 1, Inequality $v_1(x)$ is active; Case 2, Inequality $v_2(x)$ is active; Case 3, None of inequalities are active.

(c-2): Based on Reverse Training with Five Layers – 80 Neurons; ACI 318-19 (Refer to Table 3.2.2.2(b-2) for Training Performance)

AI-based Lagrange Optimization (*Five Layers -80 Neurons*) (*ACI 318-19*)

Parameters		*Case 1*	*Case 2*	*Case 3*
3	ρ_s	0.0100	0.0800	**0.0614**
4	f_c' (MPa)	40	40	40
5	f_y (MPa)	500	500	500
6	P_u (kN)	1,000	1,000	1,000
7	M_u (kN·m)	3,000	3,000	3,000
10	*SF*	1	1	1
11	*b/h*	1	1	1
Objective: CI_c (kN/m)		27.005	9.312	**9.322**

Note: "Bold values" emphasize the results of optimal case which is the most important case.
Case 1, Inequality $v_1(x)$ is active; Case 2, Inequality $v_2(x)$ is active; Case 3, None of inequalities are active.

(d-2): Based on Reverse Training with Ten Layers – 80 Neurons; ACI 318-19 (Refer to Table 3.2.2.2(b-2) for Training Performance)

AI-based Lagrange Optimization (*Ten Layers -80 Neurons*) (*ACI 318-19*)

Parameters		*Case 1*	*Case 2*	*Case 3*
3	ρ_s	0.0100	0.0800	**0.0626**
4	f_c' (MPa)	40	40	40
5	f_y (MPa)	500	500	500
6	P_u (kN)	1,000	1,000	1,000
7	M_u (kN·m)	3,000	3,000	3,000
10	*SF*	1	1	1
11	*b/h*	1	1	1
Objective: CI_c (kN/m)		26.949	10.577	**10.435**

Note: "Bold values" emphasize the results of optimal case which is the most important case.
Case 1, Inequality $v_1(x)$ is active; Case 2, Inequality $v_2(x)$ is active; Case 3, None of inequalities are active.

Table 3.3.3.3 (Continued) Lowest W_c with Corresponding Parameters for KKT Conditions Based on Reverse Networks

(e-2): Based on Best Reverse Trainings a Combination of One, Two, Five, and Ten Hidden Layers with 80 Neurons; ACI 318-19 (Refer to Table 3.2.2.2(b-2) for Training Performance)

AI-based Lagrange Optimization (*Best Networks in Term of MSE*) (*ACI 318-19*)

Parameters		*Case 1*	*Case 2*	*Case 3*
3	ρ_s	0.0100	0.0800	**0.0614**
4	f_c' (MPa)	40	40	40
5	f_y (MPa)	500	500	500
6	P_u (kN)	1,000	1,000	1,000
7	M_u (kN·m)	3,000	3,000	3,000
10	*SF*	1	1	1
11	*b/h*	1	1	1
Objective: CI_c (kN/m)		27.058	10.511	**10.448**

Note: "Bold values" emphasize the results of optimal case which is the most important case.
Case 1, Inequality $v_1(x)$ is active; Case 2, Inequality $v_2(x)$ is active; Case 3, None of inequalities are active.

Table 3.3.3.4 Design Accuracies Minimizing W_c Based on Forward Lagrange Optimization

(a-1): Based on Forward Training with One Layer – 80 Neurons; ACI 318-14 (Refer to Table 3.2.2.2(a-1) for Training Performance)

Optimal W_C Design (*Based on One Layer with 80 Neurons*) (*ACI 318-14*)

Parameter		*AI Results*	*Check (AutoCol)*	*Error*
1	*b* (mm)	**664.82**		
2	*h* (mm)	**664.82**		
3	ρ_s	**0.0726**		
4	f_c' (MPa)	40		
5	f_y (MPa)	500		
6	P_u (kN)	1,000		
7	M_u (kN·m)	3,000		
8	ϕP_n (kN)	426.8	1,016.3	−138%
9	ϕM_n (kN·m)	3,160.4	3,048.9	3.53%
10	*SF*	1.00	1.02	−1.63%
11	*b/h*	1.00	1.00	0.00%
12	ε_s	0.00457	0.00415	9.17%
13	CI_c (KRW/m)	307,516.6	307,380.8	0.04%
14	CO_2 (t-CO_2/m)	0.7067	0.7073	−0.08%
15	W_c (kN/m)	**10.419**	**10.413**	0.06%
16	$\alpha_{e/h}$	0.222	0.218	1.65%

Note: (dotted box) Seven inputs for forward training.
(dashed box) Structural verification by (AutoCol).
Optimal W_c of **10.419** is 1.75% different from **10.240** obtained by large structural datasets.
"Bold values" indicate the important parameters of AI results of optimal Wc.

Table 3.3.3.4 *(Continued)* Design Accuracies Minimizing W_c Based on Forward Lagrange Optimization

(b-1): Based on Forward Training with Two Layers – 80 Neurons; ACI 318-14 (Refer to Table 3.2.2.2(a-1) for Training Performance)

Optimal CI_c Design *(Based on Two Layers with 80 Neurons) (ACI 318-14)*

Parameter		*AI results*	*Check (AutoCol)*	*Error*
1	b (mm)	**656.83**		
2	h (mm)	**656.83**		
3	ρ_s	**0.0710**		
4	f_c' (MPa)	40		
5	f_y (MPa)	500		
6	P_u (kN)	1,000		
7	M_u (kN·m)	3,000		
8	ϕP_n (kN)	1,052.3	985.1	6.4%
9	ϕM_n (kN·m)	3,019.8	2,955.3	2.14%
10	SF	1.00	0.99	1.49%
11	b/h	1.00	1.00	0.00%
12	ε_s	0.00445	0.00426	4.40%
13	CI_c (KRW/m)	294.387.4	294,072.6	0.11%
14	CO_2 (t-CO_2/m)	0.6760	0.6762	−0.03%
15	W_c (kN/m)	**10.164**	**10.164**	0.00%
16	$\alpha_{e/h}$	0.2181	0.2155	1.16%

Note: (dotted box) Seven inputs for forward training.
(dashed box) Structural verification by (AutoCol).
Optimal W_c of **10.164** is −0.74% different from **10.240** obtained by large structural datasets.
"Bold values" indicate the important parameters of AI results of optimal Wc.

(c-1): Based on Forward Training with Five Layers – 80 Neurons; ACI 318-14 (Refer to Table 3.2.2.2(a-1) for Training Performance)

Optimal CI_c Design *(Based on Five Layers with 80 Neurons) (ACI 318-14)*

Parameter		*AI Results*	*Check (AutoCol)*	*Error*
1	b (mm)	**656.87**		
2	h (mm)	**656.87**		
3	ρ_s	**0.0654**		
4	f_c' (MPa)	40		
5	f_y (MPa)	500		
6	P_u (kN)	1,000		
7	M_u (kN·m)	3,000		
8	ϕP_n (kN)	930.6	989.8	−6.4%
9	ϕM_n (kN·m)	2,987.2	2,969.5	0.59%
10	SF	1.00	0.99	1.02%
11	b/h	1.00	1.00	0.00%
12	ε_s	0.00479	0.00470	1.85%
13	CI_c (KRW/m)	275,741.1	274,324.1	0.51%
14	CO_2 (t-CO_2/m)	0.6286	0.6292	−0.09%
15	W_c (kN/m)	**10.177**	**10.166**	0.11%
16	$\alpha_{e/h}$	0.2154	0.2156	−0.08%

Note: (dotted box) Seven inputs for forward training.
(dashed box) Structural verification by (*AutoCol*).
Optimal W_c of **10.177** is −0.62% different from **10.240** obtained by large structural datasets.
"Bold values" indicate the important parameters of AI results of optimal Wc.

Table 3.3.3.4 (Continued) Design Accuracies Minimizing W_c Based on Forward Lagrange Optimization

(d-1): Based on Forward Training with Ten Layers – 80 Neurons; ACI 318-14 (Refer to Table 3.2.2.2(a-1) for Training Performance)

Optimal CI_C Design (*Based on Ten Layers with 80 Neurons*) (*ACI 318-14*)

Parameter		*AI Results*	*Check (AutoCol)*	*Error*
1	b (mm)	**656.42**		
2	h (mm)	**656.42**		
3	ρ_s	**0.0650**		
4	f_c' (MPa)	40		
5	f_y (MPa)	500		
6	P_u (kN)	1,000		
7	M_u (kN·m)	3,000		
8	ϕP_n (kN)	980.0	988.6	−0.9%
9	ϕM_n (kN·m)	2,985.5	2,965.9	0.66%
10	SF	1.00	0.99	1.14%
11	b/h	1.0004	1.0000	0.04%
12	ε_s	0.00489	0.00474	2.97%
13	CI_c (KRW/m)	272,498.6	272,344.9	0.06%
14	CO_2 (t-CO_2/m)	0.6260	0.6245	0.25%
15	W_c (kN/m)	**10.161**	**10.152**	0.09%
16	$\alpha_{e/h}$	0.21543	0.21541	0.01%

Note: [dotted box] Seven inputs for forward training.
[dashed box] Structural verification by (AutoCol).
Optimal W_c of **10.161** is −0.77% different from **10.240** obtained by large structural datasets.
"Bold values" indicate the important parameters of AI results of optimal Wc.

(e-1): Based on Best Forward Trainings a Combination of One, Two, Five, and Ten Hidden Layers with 80 Neurons; ACI 318-14 (Refer to Table 3.2.2.2(a-1) for Training Performance)

Optimal W_c Design (*Based on Best Networks in Term of MSE*) (*ACI 318-14*)

Parameter		*Best Networks*	*AI Results*	*Check (AutoCol)*	*Error*
1	b (mm)		**656.42**		
2	h (mm)		**656.42**		
3	ρ_s		**0.0651**		
4	f_c' (MPa)		40		
5	f_y (MPa)		500		
6	P_u (kN)		1,000		
7	M_u (kN·m)		3,000		
8	ϕP_n (kN)	5L -80N	(927.1)	988.5	(−6.6%)
9	ϕM_n (kN·m)	10L -80N	(2,986.3)	2,965.5	(0.70%)
10	SF	10L -80N	1.00	0.99	1.15%
11	b/h	2L -80N	(1.00)	1.00	(0.00%)
12	ε_s	10L -80N	(0.00488)	0.00474	(2.98%)
13	CI_c (KRW/m)	1L -80N	(272,904.0)	272,651.3	(0.09%)
14	CO_2 (t-CO_2/m)	2L -80N	(0.6250)	0.6252	(−0.04%)
15	W_c (kN/m)	1L -80N	**10.158**	**10.152**	0.06%
16	$\alpha_{e/h}$	10L -80N	(0.21543)	0.21541	(0.01%)

Note: [dotted box] Seven inputs for forward training.
[dashed box] Structural verification by (AutoCol).
Optimal W_c of **10.158** is −0.80% different from **10.240** obtained by large structural datasets.
"Bold values" indicate the important parameters of AI results of optimal Wc.

Table 3.3.3.4 (Continued) Design Accuracies Minimizing W_c Based on Forward Lagrange Optimization

(a-2): Based on Forward Training with One Layer – 80 Neurons; ACI 318-19 (Refer to Table 3.2.2.2(b-1) for Training Performance)

Optimal CI_C Design (*Based on One Layer with 80 Neurons*) (*ACI 318-19*)

Parameter		*AI Results*	*Check (AutoCol)*	*Error*
1	b (mm)	**660.99**		
2	h (mm)	**660.99**		
3	ρ_s	**0.0690**		
4	f'_c (MPa)	40		
5	f_y (MPa)	500		
6	P_u (kN)	1,000		
7	M_u (kN·m)	3,000		
8	ϕP_n (kN)	442.5	1,003.5	−127%
9	ϕM_n (kN·m)	3,023.9	3,010.4	0.44%
10	SF	1.000	1.003	−0.35%
11	b/h	1.00	1.00	0.00%
12	ε_s	0.00474	0.00438	7.55%
13	CI_c (KRW/m)	291,148.3	290,771.1	0.13%
14	CO_2 (t-CO_2/m)	0.667	0.668	−0.09%
15	W_c (kN/m)	**10.297**	**10.294**	0.04%
16	$\alpha_{e/h}$	0.227	0.217	4.60%

Note: (dotted box) Seven inputs for forward training.
(dashed box) Structural verification by (AutoCol).
Optimal W_c of **10.297** is −1.38% different from **10.441** obtained by large structural datasets.
"Bold values" indicate the important parameters of AI results of optimal Wc.

(b-2): Based on Forward Training with Two Layers – 80 Neurons; ACI 318-19 (Refer to Table 3.2.2.2(b-1) for Training Performance)

Optimal W_C Design (*Based on Two Layers with 80 Neurons*) (*ACI 318-19*)

Parameter		*AI Results*	*Check (AutoCol)*	*Error*
1	b (mm)	**666.01**		
2	h (mm)	**666.01**		
3	ρ_s	**0.0645**		
4	f'_c (MPa)	40		
5	f_y (MPa)	500		
6	P_u (kN)	1,000		
7	M_u (kN·m)	3,000		
8	ϕP_n (kN)	995.1	984.7	1.0%
9	ϕM_n (kN·m)	3,021.3	2,954.0	2.23%
10	SF	1.00	0.98	1.53%
11	b/h	1.0000	1.0000	0.00%
12	ε_s	0.00486	0.00477	2.01%
13	CI_c (KRW/m)	278,830.9	278,637.9	0.07%
14	CO_2 (t-CO_2/m)	0.636	0.639	−0.50%
15	W_c (kN/m)	**10.452**	**10.451**	0.02%
16	$\alpha_{e/h}$	0.2180	0.2185	−0.19%

Note: (dotted box) Seven inputs for forward training.
(dashed box) Structural verification by (AutoCol).
Optimal W_c of **10.452** is 0.11% different from **10.441** obtained by large structural datasets.
"Bold values" indicate the important parameters of AI results of optimal Wc.

Table 3.3.3.4 (Continued) Design Accuracies Minimizing W_c Based on Forward Lagrange Optimization

(c-2): Based on Forward Training with Five Layers – 80 Neurons; ACI 318-19 (Refer to Table 3.2.2.2(b-1) for Training Performance)

Optimal CI_C Design (*Based on Five Layers with 80 Neurons*) (*ACI 318-19*)

Parameter		*AI Results*	*Check (AutoCol)*	*Error*
1	*b* (mm)	**664.03**		
2	*h* (mm)	**664.03**		
3	ρ_s	**0.0625**		
4	f_c' (MPa)	40		
5	f_y (MPa)	500		
6	P_u (kN)	1,000		
7	M_u (kN·m)	3,000		
8	ϕP_n (kN)	939.9	976.4	−3.9%
9	ϕM_n (kN·m)	2,973.8	2,929.3	1.50%
10	*SF*	1.00	0.98	2.36%
11	*b/h*	1.000	1.000	0.00%
12	ε_s	0.00510	0.00497	2.58%
13	CI_c (KRW/m)	271,811.1	269,644.0	0.80%
14	CO_2 (t-CO_2/m)	0.6179	0.6175	0.07%
15	W_c (kN/m)	**10.372**	**10.389**	−0.16%
16	$\alpha_{e/h}$	0.2181	0.2178	0.11%

Note: [dotted box] Seven inputs for forward training.
[dashed box] Structural verification by (*AutoCol*).
Optimal W_c of **10.372** is −0.66% different from **10.441** obtained by large structural datasets.
"Bold values" indicate the important parameters of AI results of optimal Wc.

(d-2): Based on Forward Training with Ten Layers – 80 Neurons; ACI 318-19 (Refer to Table 3.2.2.2(b-1) for Training Performance)

Optimal CI_C Design (*Based on Ten Layers with 80 Neurons*) (*ACI 318-19*)

Parameter		AI Results	Check (*AutoCol*)	Error
1	*b* (mm)	**665.04**		
2	*h* (mm)	**665.04**		
3	ρ_s	**0.0616**		
4	f_c' (MPa)	40		
5	f_y (MPa)	500		
6	P_u (kN)	1,000		
7	M_u (kN·m)	3,000		
8	ϕP_n (kN)	989.5	980.4	0.9%
9	ϕM_n (kN·m)	3,000.1	2,941.3	1.96%
10	*SF*	1.00	0.98	1.96%
11	*b/h*	0.9998	1.0000	−0.02%
12	ε_s	0.00519	0.00506	2.56%
13	CI_c (KRW/m)	267,550.9	267,345.1	0.08%
14	CO_2 (t-CO_2/m)	0.6155	0.6119	0.58%
15	W_c (kN/m)	**10.439**	**10.420**	0.18%
16	$\alpha_{e/h}$	0.2179	0.2182	−0.11%

Note: [dotted box] Seven inputs for forward training.
[dashed box] Structural verification by (*AutoCol*).
Optimal W_c of **10.439** is −0.02% different from **10.441** obtained by large structural datasets.
"Bold values" indicate the important parameters of AI results of optimal Wc.

Table 3.3.3.4 (Continued) Design Accuracies Minimizing W_c Based on Forward Lagrange Optimization

(e-2): Based on Best Forward Trainings a Combination of One, Two, Five, and Ten Hidden Layers with 80 Neurons; ACI 318-19 (Refer to Table 3.2.2.2(b-1) for Training Performance)

Optimal CI_C Design (*Based on Best Networks in Term of MSE*) (*ACI 318-19*)

Parameter		*Best Networks*	*AI Results*	*Check (AutoCol)*	*Error*
1	b (mm)		**668.55**		
2	h (mm)		**668.55**		
3	ρ_s		**0.0800**		
4	f_c' (MPa)		40		
5	f_y (MPa)		500		
6	P_u (kN)		1,000		
7	M_u (kN·m)		3,000		
8	ϕP_n (kN)	10L -80N	(1,032.5)	983.4	(4.8%)
9	ϕM_n (kN·m)	10L -80N	(3,018.5)	2,950.2	(2.26%)
10	SF	10L -80N	1.00	0.98	1.66%
11	b/h	2L -80N	(1.00)	1.00	(0.00%)
12	ε_s	5L -80N	(0.00397)	0.00391	(1.55%)
13	CI_c (KRW/m)	1L -80N	(338,294.0)	338,145.7	(0.04%)
14	CO_2 (t-CO_2/m)	1L -80N	(0.7802)	0.7803	(−0.02%)
15	W_c (kN/m)	1L -80N	**10.535**	**10.530**	0.04%
16	$\alpha_{e/h}$	10L -80N	(0.2195)	0.2193	(0.10%)

Note: [dotted box] Seven inputs for forward training.
[dashed box] Structural verification by (*AutoCol*).
Optimal W_c of **10.535** is 0.90% different from **10.441** obtained by large structural datasets.
"Bold values" indicate the important parameters of AI results of optimal Wc.

Table 3.3.3.5 Design Accuracies Minimizing W_c Based on Reverse Lagrange Optimization

(a-1): Based on Reverse Training with One Layer – 80 Neurons; ACI 318-14 (Refer to Table 3.2.2.2(a-2) for Training Performance)

Optimal CI_C Design (*Based on One Layer with 80 Neurons*) (*ACI 318-14*)

Parameter		*AI Results*	*Check (AutoCol)*	*Error*
1	b (mm)	**670.95**		
2	h (mm)	**667.47**		
3	ρ_s	**0.0734**		
4	f_c' (MPa)	40		
5	f_y (MPa)	500		
6	P_u (kN)	1,000		
7	M_u (kN·m)	3,000		
8	ϕP_n (kN)	991.9	1,025.5	−3.4%
9	ϕM_n (kN·m)	2,997.0	3,076.6	−2.66%
10	SF	1.00	1.03	−2.55%
11	b/h	1.00	1.01	−0.52%
12	ε_s	0.0045	0.0041	7.63%
13	CI_c (KRW/m)	310,724.6	314,374.8	−1.17%
14	CO_2 (t-CO_2/m)	0.716	0.724	−1.13%
15	W_c (kN/m)	**10.50**	**10.55**	−0.53%
16	$\alpha_{e/h}$	0.214	0.219	−2.36%

Note: [dotted box] Seven inputs for reverse training.
[dashed box] Structural verification by (*AutoCol*).
Optimal W_c of **10.50** is 2.54% different from **10.240** obtained by large structural datasets.
"Bold values" indicate the important parameters of AI results of optimal Wc.

Table 3.3.3.5 (Continued) Design Accuracies Minimizing W_c Based on Reverse Lagrange Optimization

(b-1): Based on Reverse Training with Two Layers – 80 Neurons; ACI 318-14 (Refer to Table 3.2.2.2(a-2) for Training Performance)

Optimal CI_c Design (*Based on Two Layers with 80 Neurons*) (*ACI 318-14*)

Parameter		*AI Results*	*Check (AutoCol)*	*Error*
1	b (mm)	**656.68**		
2	h (mm)	**655.86**		
3		**0.0674**		
4	f_c' (MPa)	40		
5	f_y (MPa)	500		
6	P_u (kN)	1,000		
7	M_u (kN·m)	3,000		
8	ϕP_n (kN)	997.3	983.9	1.3%
9	ϕM_n (kN·m)	2,996.8	2,951.8	1.50%
10	*SF*	1.00	0.98	1.61%
11	*b/h*	1.000	1.001	−0.13%
12	ε_s	0.00460	0.00452	1.73%
13	CI_c (KRW/m)	282,534.6	281,013.3	0.54%
14	CO_2 (t-CO_2/m)	0.647	0.645	0.36%
15	W_c (kN/m)	**10.158**	**10.147**	0.11%
16	$\alpha_{e/h}$	0.2162	0.2152	0.44%

Note: (dotted box) Seven inputs for reverse training.
(dashed box) Structural verification by (*AutoCol*).
Optimal W_c of **10.158** is −0.80% different from **10.240** obtained by large structural datasets.
"Bold values" indicate the important parameters of AI results of optimal Wc.

(c-1): Based on Reverse Training with Five Layers – 80 Neurons; ACI 318-14 (Refer to Table 3.2.2.2(a-2) for Training Performance)

Optimal CI_c Design (*Based on Five Layers with 80 Neurons*) (*ACI 318-14*)

Parameter		*AI Results*	*Check (AutoCol)*	*Error*
1	b (mm)	**658.22**		
2	h (mm)	**657.88**		
3	ρ_s	**0.0660**		
4	f_c' (MPa)	40		
5	f_y (MPa)	500		
6	P_u (kN)	1,000		
7	M_u (kN·m)	3,000		
8	ϕP_n (kN)	999.9	993.5	0.6%

Table 3.3.3.5 (Continued) Design Accuracies Minimizing W_c Based on Reverse Lagrange Optimization

(c-1): Based on Reverse Training with Five Layers – 80 Neurons; ACI 318-14 (Refer to Table 3.2.2.2(a-2) for Training Performance)

Optimal CI_c Design (*Based on Five Layers with 80 Neurons*) (*ACI 318-14*)

Parameter		*AI Results*	*Check (AutoCol)*	*Error*
9	ϕM_n (kN·m)	2,998.1	2,980.6	0.59%
10	*SF*	1.00	0.99	0.65%
11	*b*/*h*	1.000	1.001	−0.05%
12	ε_s	0.00473	0.00464	1.89%
13	CI_c (KRW/m)	276,340.9	277,357.4	−0.37%
14	CO_2 (t-CO_2/m)	0.634	0.636	−0.36%
15	W_c (kN/m)	**10.177**	**10.202**	−0.25%
16	$\alpha_{e/h}$	0.217	0.216	0.47%

Note: Seven inputs for reverse training.
Structural verification by (*AutoCol*).
Optimal W_c of **10.177** is −0.62% different from **10.240** obtained by large structural datasets.
"Bold values" indicate the important parameters of AI results of optimal Wc.

(d-1): Based on Reverse Training with Ten Layers – 80 Neurons; ACI 318-14 (Refer to Table 3.2.2.2(a-2) for Training Performance)

Optimal CI_c Design (*Based on Ten Layers with 80 Neurons*) (*ACI 318-14*)

Parameter		*AI Results*	*Check (AutoCol)*	*Error*
1	*b* (mm)	**656.99**		
2	*h* (mm)	**655.66**		
3	ρ_s	**0.0642**		
4	f_c' (MPa)	40		
5	f_y (MPa)	500		
6	P_u (kN)	1,000		
7	(kN·m)	3,000		
8	ϕP_n (kN)	961.5	987.5	−2.7%
9	ϕM_n (kN·m)	3,007.6	2,962.6	1.50%
10	*SF*	1.00	0.99	1.25%
11	*b*/*h*	1.0000	−0.20%	
12	ε_s	0.00490	0.00481	1.66%
13	CI_c (KRW/m)	270,757.3	269,612.5	0.42%
14	CO_2 (t-CO_2/m)	0.620	0.618	0.31%
15	W_c (kN/m)	**10.174**	**10.149**	0.25%
16	$\alpha_{e/h}$	0.216	0.215	0.44%

Note: Seven inputs for reverse training.
Structural verification by (*AutoCol*).
Optimal W_c of **10.174** is −0.64% different from **10.240** obtained by large structural datasets.
"Bold values" indicate the important parameters of AI results of optimal Wc.

Table 3.3.3.5 (Continued) Design Accuracies Minimizing W_c Based on Reverse Lagrange Optimization

(e-1): Based on Best Reverse Trainings a Combination of One, Two, Five, and Ten Hidden Layers with 80 Neurons; ACI 318-14 (Refer to Table 3.2.2.2(a-2) for Training Performance)

Optimal CI_c Design (*Based on Best Networks in Term of MSE*) (*ACI 318-14*)

Parameter			*AI Results*	*Check (AutoCol)*	*Error*
1	*b* (mm)	5L -80N	**657.96**		
2	*h* (mm)	10L -80N	**655.66**		
3	ρ_s		**0.0642**		
4	f_c' (MPa)		40		
5	f_y (MPa)		500		
6	P_u (kN)		1,000		
7	M_u (kN·m)		3,000		
8	ϕP_n (kN)	1L -80N	(991.6)	988.2	(0.3%)
9	ϕM_n (kN·m)	2L -80N	(2,996.8)	2,964.6	(1.08%)
10	*SF*		1.00	0.99	1.18%
11	*b/h*		1.000	1.004	(−0.35%)
12	ε_s	10L -80N	(0.0049)	0.0048	(1.73%)
13	CI_c (KRW/m)	10L -80N	(270,757.3)	270,009.3	(0.28%)
14	CO_2 (t-CO_2/m)	5L -80N	(0.621)	0.619	(0.32%)
15	W_c (kN/m)	10L -80N	**10.174**	**10.164**	0.10%
16	$\alpha_{e/h}$	10L -80N	(0.216)	0.215	(0.44%)

Note: [dotted box] Seven inputs for reverse training.
[dashed box] Structural verification by (*AutoCol*).
Optimal W_c of **10.174** is −0.64% different from **10.240** obtained by large structural datasets.
"Bold values" indicate the important parameters of AI results of optimal Wc.

(a-2): Based on Reverse Training with One Layer – 80 Neurons; ACI 318-19 (Refer to Table 3.2.2.2(b-2) for Training Performance)

Optimal CI_c Design (*Based on One Layer with 80 Neurons*) (*ACI 318-19*)

Parameter		*AI Results*	*Check (AutoCol)*	*Error*
1	*b* (mm)	**676.84**		
2	*h* (mm)	**671.16**		
3		**0.0800**		
4	f_c' (MPa)	40		
5	f_y (MPa)	500		
6	P_u (kN)	1,000		
7	M_u (kN·m)	3,000		
8	ϕP_n (kN)	1,059.5	996.0	6.0%
9	ϕM_n (kN·m)	2,995.8	2,988.0	0.26%

Table 3.3.3.5 (Continued) Design Accuracies Minimizing W_c Based on Reverse Lagrange Optimization

(a-2): Based on Reverse Training with One Layer – 80 Neurons; ACI 318-19 (Refer to Table 3.2.2.2(b-2) for Training Performance)

Optimal CI_c Design (*Based on One Layer with 80 Neurons*) (*ACI 318-19*)

Parameter		*AI Results*	*Check (AutoCol)*	*Error*
10	*SF*	1.000	0.996	0.40%
11	*b/h*	1.00	−0.85%	
12	ε_s	0.00396	0.00391	1.42%
13	CI_c (KRW/m)	338,530.3	343,674.0	−1.52%
14	CO_2 (t-CO_2/m)	0.783	0.793	−1.24%
15	W_c (kN/m)	**10.243**	**10.703**	−4.48%
16	$\alpha_{e/h}$	0.229	0.220	4.02%

Note: [dotted box] Seven inputs for reverse training.
[dashed box] Structural verification by (*AutoCol*).
Optimal W_c of **10.243** is −1.90% different from **10.441** obtained by large structural datasets.
"Bold values" indicate the important parameters of AI results of optimal Wc.

(b-2): Based on Reverse Training with Two Layers – 80 Neurons; ACI 318-19 (Refer to Table 3.2.2.2(b-2) for Training Performance)

Optimal CI_c Design (*Based on Two Layers with 80 Neurons*) (*ACI 318-19*)

Parameter		*AI Results*	*Check (AutoCol)*	*Error*
1	*b* (mm)	**663.92**		
2	*h* (mm)	**664.35**		
3		**0.0662**		
4	f'_c (MPa)	40		
5	f_y (MPa)	500		
6	P_u (kN)	1,000		
7	M_u (kN·m)	3,000		
8	ϕP_n (kN)	997.1	978.1	1.9%
9	ϕM_n (kN·m)	2,995.9	2,934.3	2.06%
10	*SF*	1.00	0.98	2.19%
11	*b/h*	1.000	0.07%	
12	ε_s	0.00469	0.00461	1.62%
13	CI_c (KRW/m)	285,044.1	283,420.1	0.57%
14	CO_2 (t-CO_2/m)	0.664	0.650	2.13%
15	W_c (kN/m)	**10.390**	**10.392**	−0.02%
16	$\alpha_{e/h}$	0.2183	0.2179	0.15%

Note: [dotted box] Seven inputs for reverse training.
[dashed box] Structural verification by (*AutoCol*).
Optimal W_c of **10.390** is 0.49% different from **10.441** obtained by large structural datasets.
"Bold values" indicate the important parameters of AI results of optimal Wc.

Table 3.3.3.5 (Continued) Design Accuracies Minimizing W_c Based on Reverse Lagrange Optimization

(c-2): Based on Reverse Training with Five Layers – 80 Neurons; ACI 318-19 (Refer to Table 3.2.2.2(b-2) for Training Performance)

Optimal CI_c Design (*Based on Five Layers with 80 Neurons*) (*ACI 318-19*)

Parameter		*AI Results*	*Check (AutoCol)*	*Error*
1	*b* (mm)	**666.43**		
2	*h* (mm)	**665.76**		
3	ρ_s	**0.0614**		
4	f_c' (MPa)	40		
5	f_y (MPa)	500		
6	P_u (kN)	1,000		
7	(kN·m)	3,000		
8	ϕP_n (kN)	1010.1	983.8	2.6%
9	ϕM_n (kN·m)	3,000.4	2,951.5	1.63%
10	*SF*	1.00	0.98	1.62%
11	*b/h*	1.000	1.001	−0.10%
12	ε_s	0.00527	0.00507	3.67%
13	CI_c (KRW/m)	268,043.9	267,441.4	0.22%
14	CO_2 (t-CO_2/m)	0.611	0.612	−0.19%
15	W_c (kN/m)	**10.448**	**10.453**	−0.05%
16	$\alpha_{e/h}$	0.219	0.218	0.20%

Note: [dotted box] Seven inputs for reverse training.
[dashed box] Structural verification by (*AutoCol*).
Optimal W_c of **10.448** is 0.07% different from **10.441** obtained by large structural datasets.
"Bold values" indicate the important parameters of AI results of optimal Wc.

(d-2): Based on Reverse Training with Ten Layers – 80 Neurons; ACI 318-19 (Refer to Table 3.2.2.3(b-2) for Training Performance)

Optimal CI_c Design (*Based on Ten Layers with 80 Neurons*) (*ACI 318-19*)

Parameter		*AI Results*	*Check (AutoCol)*	*Error*
1	*b* (mm)	**665.60**		
2	*h* (mm)	**665.64**		
3	ρ_s	**0.0626**		
4	f_c' (MPa)	40		
5	f_y (MPa)	500		
6	P_u (kN)	1,000		
7	M_u (kN·m)	3,000		
8	ϕP_n (kN)	881.3	982.8	−11.5%
9	ϕM_n (kN·m)	2,998.0	2,948.5	1.65%

Table 3.3.3.5 (Continued) Design Accuracies Minimizing W_c Based on Reverse Lagrange Optimization

(d-2): Based on Reverse Training with Ten Layers – 80 Neurons; ACI 318-19 (Refer to Table 3.2.2.3(b-2) for Training Performance)

Optimal CI_c Design (*Based on Ten Layers with 80 Neurons*) (*ACI 318-19*)

Parameter		*AI Results*	*Check (AutoCol)*	*Error*
10	*SF*	1.00	0.98	1.72%
11	*b/h*	1.0000	0.9999	0.01%
12	ε_s	0.00511	0.00496	3.09%
13	CI_c (KRW/m)	271,532.4	271,348.5	0.07%
14	CO_2 ($t\text{-}CO_2$/m)	0.619	0.621	−0.37%
15	W_c (kN/m)	**10.435**	**10.438**	−0.03%
16	$\alpha_{e/h}$	0.219	0.218	0.19%

Note: [dotted box] Seven inputs for reverse training.
[dashed box] Structural verification by (*AutoCol*).
Optimal W_c of **10.435** is −0.06% different from **10.441** obtained by large structural datasets.
"Bold values" indicate the important parameters of AI results of optimal Wc.

(e-2): Based on Best Reverse Trainings a Combination of One, Two, Five, and Ten Hidden Layers with 80 Neurons; ACI 318-19 (Refer to Table 3.2.2.2(b-2) for Training Performance)

Optimal CI_C Design (*Based on Best Networks in Term of MSE*) (*ACI 318-19*)

Parameter			*AI Results*	*Check (AutoCol)*	*Error*
1	*b* (mm)	5L -80N	**666.43**		
2	*h* (mm)	5L -80N	**665.76**		
3	ρ_s		**0.0614**		
4	f'_c (MPa)		40		
5	f_y (MPa)		500		
6	P_u (kN)		1,000		
7	M_u (kN·m)		3,000		
8	ϕP_n (kN)	2L -80N	(996.3)	983.8	(1.3%)
9	ϕM_n (kN·m)	2L -80N	(2,996.5)	2,951.5	(1.50%)
10	*SF*		1.00	0.98	1.62%
11	*b/h*		1.000	−0.10%	
12	ε_s	10L -80N	(0.0052)	0.0051	(3.24%)
13	CI_c (KRW/m)	10L -80N	(267,358.9)	267,441.4	(−0.03%)
14	CO_2 ($t\text{-}CO_2$/m)	10L -80N	(0.610)	0.612	(−0.29%)
15	W_c (kN/m)	5L -80N	**10.448**	**10.453**	−0.05%
16	$\alpha_{e/h}$	5L -80N	(0.219)	0.218	(0.20%)

Note: [dotted box] Seven inputs for reverse training.
[dashed box] Structural verification by (*AutoCol*).
Optimal W_c of **10.448** is 0.07% different from **10.441** obtained by large structural datasets.
"Bold values" indicate the important parameters of AI results of optimal Wc.

3.3.3.3 Verifications

1. Verification of a column weight based on reverse networks

   A column weight W_c is plotted as a function of ρ_s as shown in Figure 3.3.3.1 where all four curves, one, two, five, ten layers with the same 80 neurons are identically plotted in the valid design range between the minimum and maximum ρ_s of 0.01 and 0.08. Large datasets verify that four curves also converge to yield a realistic minimum W_c for a column if training accuracies are sufficient. However, all four curves do not match when ρ_s is below 0.01, indicating that neural networks have difficulties in extrapolating beyond a trained region.

2. Verification of weight using large datasets

   The lowest W_c (kN/m) based on a combination of one, two, five, and ten layers with 80 neurons is consistent with 10.158 for ACI 318-14 and 10.535 for ACI 318-19 shown in Tables 3.3.3.4(e-1) and (e-2) calculated using forward-networks based on the best test MSE. Tables 3.3.3.5(e-1) and (e-2) also show 10.174 for ACI 318-14 and 10.448 for ACI 318-19 based on reverse networks, respectively.

   The lowest W_c is 10.240 kN/m for ACI 318-14 and 10.441 kN/m for ACI 318-19 kN/m, corresponding to $f_c' = 40\,\text{MPa}$ obtained from large datasets shown in Figure 3.3.3.2 in which two million datasets are filtered through $f_y = 500\,\text{MPa}$, $SF = 1$, and $b/h = 1$. A minimized W_c (kN/m) obtained from large datasets is similar to a minimized W_c of 10.177 for ACI 318-14 and 10.372 for ACI 318-19 is shown in Tables 3.3.3.4(c-1) and (c-2) obtained based on forward networks trained using five layers with 80 neurons. A minimized W_c (kN/m) obtained from large datasets similar to a minimized W_c *of* 10.177 for ACI 318-14 and 10.448 for ACI 318-19 is shown in Tables 3.3.3.5(c-1) and (c-2) calculated based on reverse networks trained using five layers with 80 neurons.

3. Design accuracies

   Design accuracies obtained in Table 3.3.3.4(a) based on forward networks and Table 3.3.3.5(a) based on reverse networks for one layer are compared with those obtained based on two and five layers with 80 neurons shown in Tables 3.3.3.4(b), (c) and Tables 3.3.3.5(b), (c), respectively. Designs obtained with two and five layers with 80 neurons show better accuracies than those based on one layer with 80 neurons. Design results based on reverse networks are compared with those of structural calculations. Errors found for ϕP_n are −3.4% for ACI 318-14 and 6.0% for ACI 318-19 as shown in Tables 3.3.3.5(a-1) and (a-2) when one layer is implemented. These errors reduce to 1.3% for ACI 318-14 and 1.9% for ACI 318-19 as shown in Tables 3.3.3.5(b-1) and (b-2) when optimization networks are trained with two layers with 80 neurons. A minimized W_c obtained from large structural datasets are 10.240 kN/m for ACI 318-14 and 10.441 kN/m for ACI 318-19 as shown in Figure 3.3.3.2a and b, respectively, demonstrating the biggest difference of −2.54% for ACI 318-14 and −1.90% for ACI 318-19 obtained by a reverse network with one layer. Good design accuracies of design parameters are obtained from networks with good training accuracies (MSE) while minimizing W_c. Training and design accuracies are verified based on forward and reverse networks, showing that ANN-based Lagrange algorithm can be successfully implemented to calculate the lowest W_c for given constraints.

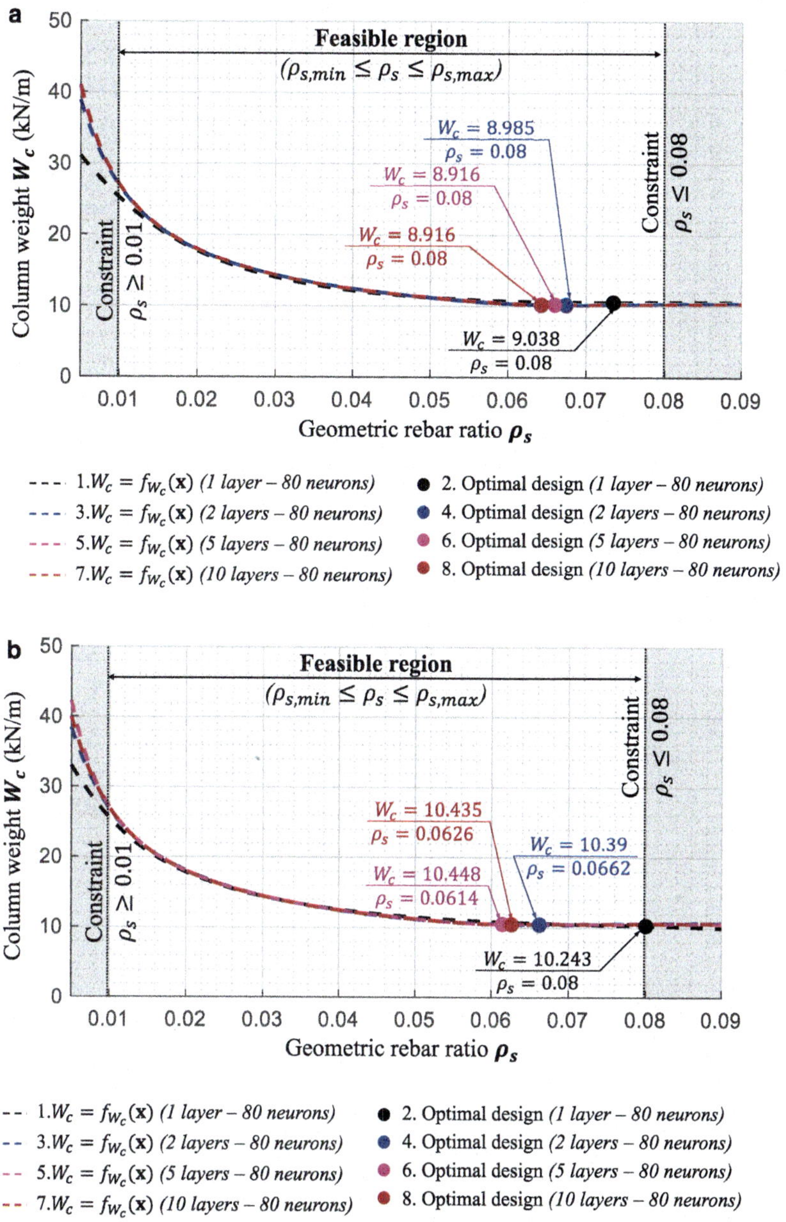

Figure 3.3.3.1 Comparison of W_c obtained by reverse ANNs based on 1, 2, 5, 10 layers as a function of rebar ratio ρ_s between 0 and 0.1. (a) ACI 318-14. (b) ACI 318-19.

3.3.3.4 P–M diagram

P–M interaction diagrams for a square column with $f_c' = 40$ MPa, $f_y = 500$ MPa, and SF of 1 are presented based on a minimized W_c. P–M interaction diagrams indicated by Legends 1–5 shown in Figure 3.3.3.3a based on ACI 318-14 are constructed using *AutoCol* with

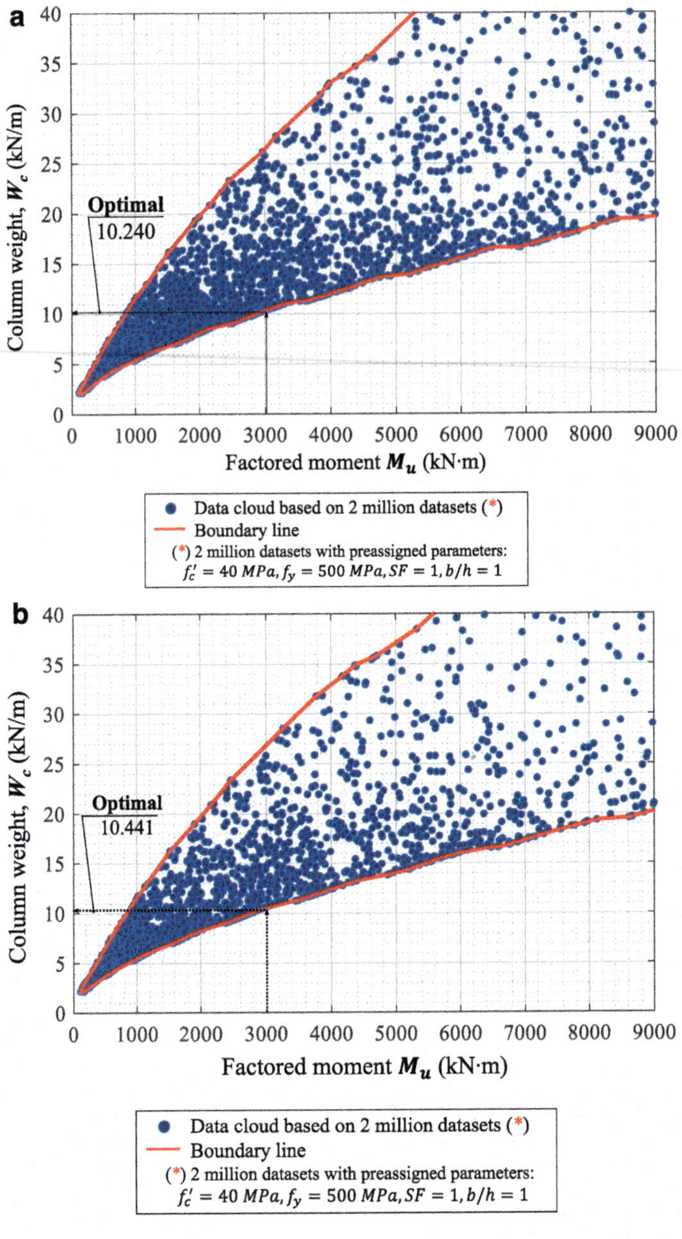

Figure 3.3.3.2 Verification of W_c based on large datasets. (a) ACI 318-14. (b) ACI 318-19.

ANN-based parameters provided in Box 1-7 of Tables 3.3.3.4 for forward Lagrange network and 3.3.3.5 for reverse Lagrange network. *P–M* interaction diagrams are shown in Figure 3.3.3.3b based on ACI 318-19 also satisfy a minimized W_c shown in Table 3.3.3.4 presented for forward Lagrange network and Table 3.3.3.5 presented for reverse Lagrange network. Ten optimized *P–M* diagrams (Legends 1–5) based on both forward and reverse networks shown in Figure 3.3.3.3 converge to one *P-M* diagram whereas curves pass through P_u = 1,000 kN and M_u = 3,000 kN·m as indicated by a red dot when training accuracies are sufficient.

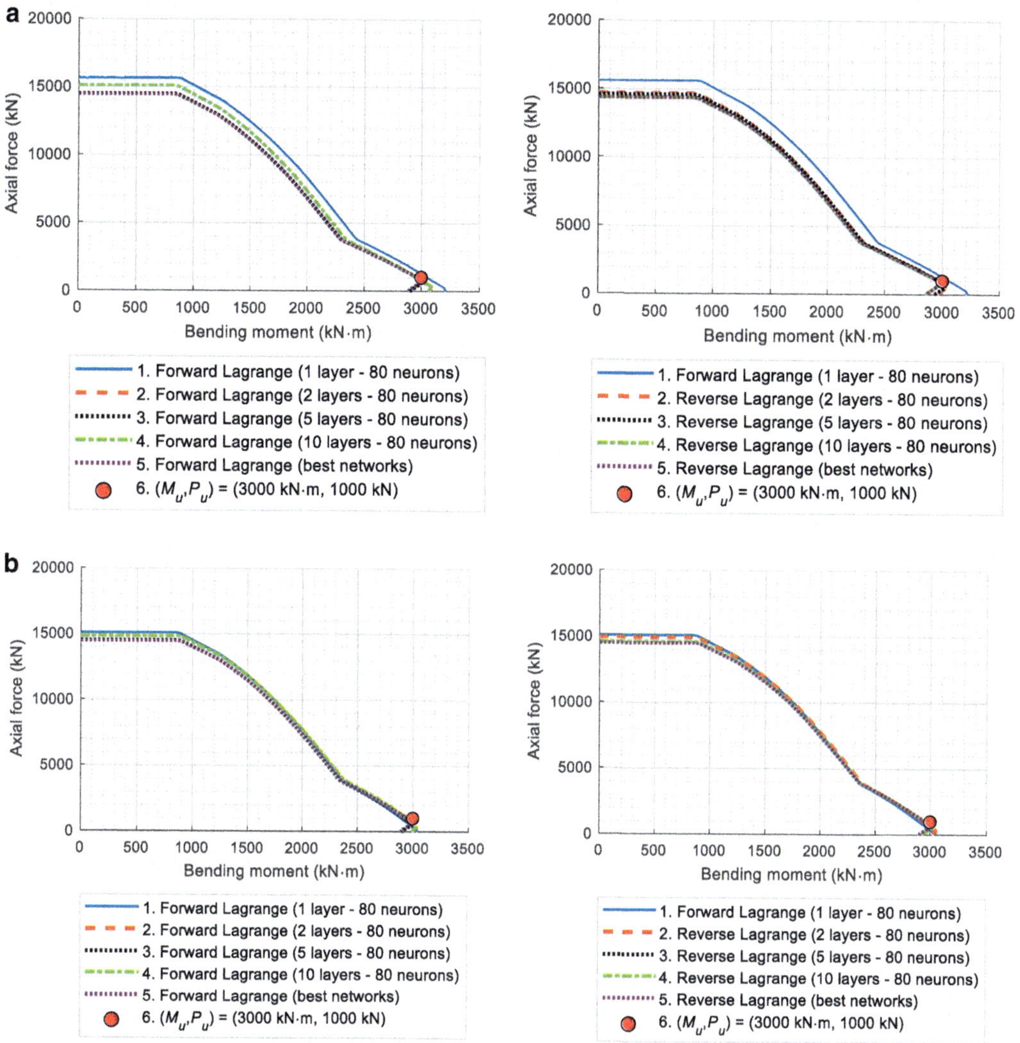

Figure 3.3.3.3 *P–M* interaction diagrams for optimal W_c. (a) ACI 318-14. (a-1) using forward Lagrange based on Tables 3.3.3.4. (a-2) using reverse Lagrange based on Tables 3.3.3.5. (b) ACI 318-19. (b-1) using forward Lagrange based on Tables 3.3.3.4. (b-2) using reverse Lagrange based on Tables 3.3.3.5

3.3.3.5 *Influence of optimization on P–M diagrams*

The shape of *P–M* diagram shown in Figure 3.3.3.3(a) is based on $b \times h$ and ρ_s for forward networks with the best test MSE obtained using a combination of one, two, five, and ten layers with 80 neurons shown in Tables 3.3.3.4(e-1) and (e-2). Weight is minimized by reducing areas of concrete section (weight) to $b \times h = 656.42 \times 656.42$ mm^2 for ACI 318-14 and 668.55×668.55 mm^2 for ACI 318-19, and hence, rebar ratios ρ_s increase up to 0.0651 for ACI 318-14 and 0.08 for ACI 318-19 to compensate for the reduction of a concrete section. However, when cost CI_c is minimized, $b \times h$ are calculated as 931.81×931.81 mm^2 for ACI 318-14 shown in Table 3.3.1.2(e-1) and 929.33×929.33 mm^2 for ACI 318-19 shown in Table 3.3.1.2(e-2). Rebar ratio ρ_s is calculated as 0.0159 for ACI 318-14 and 0.0161 for ACI 318-19, respectively, using forward networks with the best test MSE based on a combination of one, two, five, and ten layers with 80 neurons. Less rebar ratio ρ_s is used whereas large concrete volume is used to reduce the cost. Conclusions for the reverse-based P-M diagrams shown in Figure 3.3.3.3(b) can be drawn. Pure axial load capacities of approximately 14,000–15,000 kN are obtained as shown in Figure 3.3.3.3 while minimizing weight, and these are smaller than those obtained while minimizing CI_c, with which the range of pure axial load capacities are from 18,000 to 20,000 kN as shown in Figure 3.3.1.3. This is because more concrete sections are to be used to minimize CI_c than those used to minimize weight. It is noted that, when minimizing weight, a contribution of concrete section to pure axial load capacity shown in Figure 3.3.3.3 is 0.85 x f_c' x A_c=13,069 kN which is about 86.5% of pure axial load capacity of 15,000 kN, shown in *P–M* diagram. Concrete mostly contributes to pure axial load capacities by about 86.5% in the case shown in Figure 3.3.3.3. The rest is provided by rebars.

3.4 NOTICEABLE UPDATES WITH ACI 318-19 COMPARED WITH 318-14

In a reinforced concrete design, a reduction factor at ultimate limit state ranges from 0.65 to 0.9 based on a strain of reinforcement (ε_s) as shown in Figure 3.4.1. In ACI 318-14, a reduction factor is 0.9 in a tension-controlled section if a net tensile strain (ε_s) in extreme tension reinforcement is sufficiently larger than 0.005, which is established primarily based on Grade 400 (f_y = 420 MPa). Beginning with ACI 318-19, a tension-controlled limit is redefined as ε_{sy} + 0.003 to accommodate reinforcement with higher grades. This update leads to a P-M diagrams of ACI 318-14 which differs from that based on ACI 318-19 at transition region ($\varepsilon_{sy} < \varepsilon_s < \varepsilon_{sy}$ + 0.003 for ACI 318-19). Overall, ACI 318-19 provides a tension-controlled limit at a larger rebar strain than ACI 318-14 when rebar strength is greater than 400MPa (ε_{sy} = 0.002). This difference leads P-M diagram based on ACI 318-19 which is smaller than one based on ACI 318-14 at transition region for rebar strength greater than

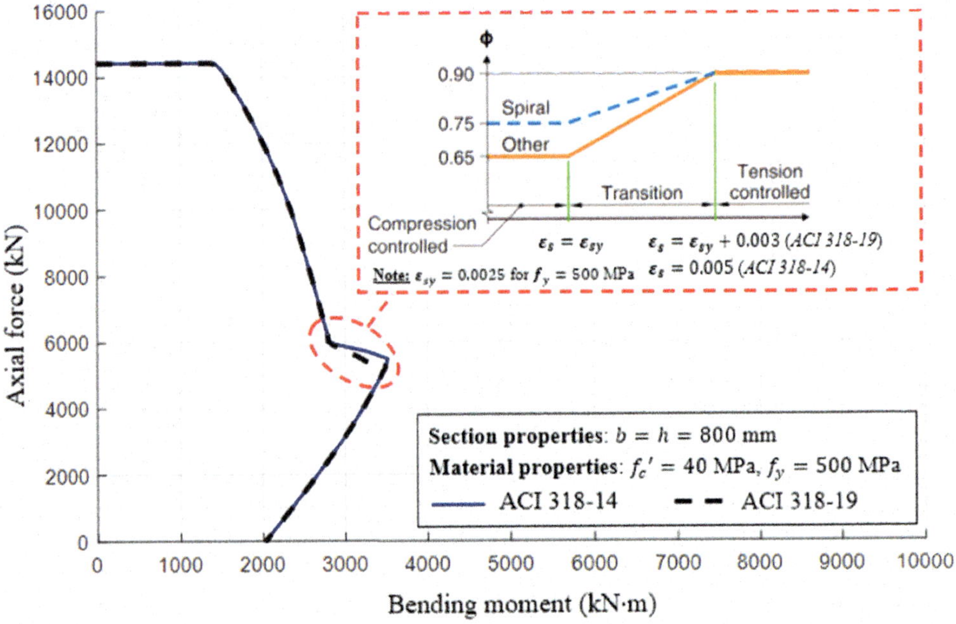

Figure 3.4.1 Noticeable difference of P–M interaction diagrams between ACI 318-19 and ACI 318-14

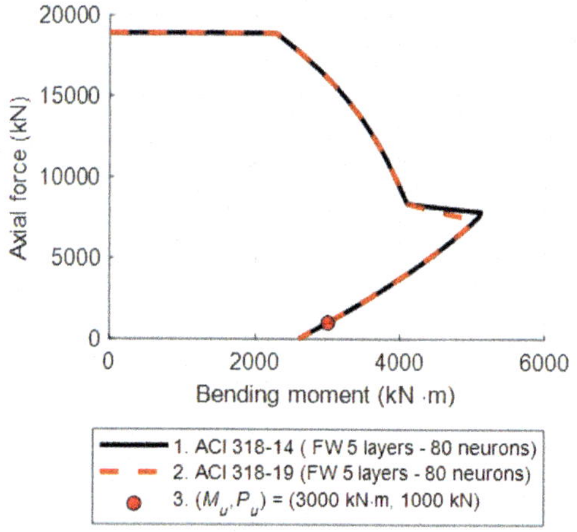

Figure 3.4.2 P–M interaction diagrams minimizing CI_c based on ACI 318-19 and ACI 318-14.

Table 3.4.1 Minimizing CI_c based on ACI 318-14 and ACI 318-19

(a): Based on ACI 318-14					
Optimal CI_c Design (*Based on Best Networks in Term of MSE*) (*ACI 318-14*)					
Parameter		*Best Networks*	*AI Results*	*Check (AutoCol)*	*Error*
1	b (mm)		**931.81**		
2	h (mm)		**931.81**		
3	ρ_s		**0.0159**		
4	f_c' (MPa)		40		
5	f_y (MPa)		500		
6	P_u (kN)		1,000		
7	M_u (kN·m)		3,000		
8	ϕP_n (kN)	5L -80N	978.2	999.9	−2.2%
9	ϕM_n (kN·m)	10L -80N	3,002.3	2,999.8	0.08%
10	SF	10L -80N	1.0000	0.9999	0.01%
11	b/h	2L -80N	1.00	1.00	0.00%
12	ε_s	10L -80N	0.02254	0.02196	2.54%
13	CI_c (KRW/m)	1L -80N	**196,698.5**	**196,159.3**	0.27%
14	CO_2 (t-CO_2/m)	2L -80N	0.4183	0.4185	−0.03%
15	W_c (kN/m)	1L -80N	20.457	20.456	0.00%
16	$\alpha_{e/h}$	10L -80N	0.3009	0.3012	−0.09%

Note: (dotted box) Seven inputs for forward training.
(dashed box) Structural verification by (*AutoCol*).
Optimal CI_c of **196,698.5** is 0.61% different from **195,515** obtained by large structural datasets.
"Bold values" indicate the important parameters of AI results.

400 MPa as shown in Figure 3.4.1. Cost CI_c minimized by ANN-based Lagrange networks based on ACI 318-19 similar to one based on ACI 318-14 is also shown in Tables 3.4.1(a), (b), and Figure 3.4.2 which exhibit negligible differences. Design errors for both ACI 318-14 and 318-19 caused by ANNs are negligible compared with those by structural calculations. In Table 3.4.1, the ANN-based Lagrange networks are trained based on a combination of 1, 2, 5, and 10 layers with 80 neurons, showing a number of layers used to yield best accuracies for the input and output parameters shown in Boxes 1 to 16.

Table 3.4.1 (Continued) Minimizing CI_c based on ACI 318-14 and ACI 318-19

(b): Based on ACI 318-19					
Optimal CI_c Design (Based on Best Networks in Term of MSE) (ACI 318-19)					
Parameter		*Best Networks*	*AI Results*	*Check (AutoCol)*	*Error*
1	b (mm)		**929.33**		
2	h (mm)		**929.33**		
3	ρ_s		**0.0161**		
4	f'_c (MPa)		40		
5	f_y (MPa)		500		
6	P_u (kN)		1,000		
7	M_u (kN·m)		3,000		
8	ϕP_n (kN)	10L -80N	1,001.3	1,001.9	−0.1%
9	ϕM_n (kN·m)	10L -80N	2,989.5	3,005.6	−0.54%
10	SF	10L -80N	1.000	1.002	−0.19%
11	b/h	2L -80N	1.00	1.00	0.00%
12	ε_s	5L -80N	0.02241	0.02179	2.76%
13	CI_c (KRW/m)	1L -80N	**196,562.0**	**196,367.6**	0.10%
14	CO_2 (t-CO_2/m)	1L -80N	0.420	0.419	0.26%
15	W_c (kN/m)	1L -80N	20.347	20.348	0.00%
16	$\alpha_{e/h}$	10L -80N	0.3002	0.3004	−0.07%

Note: (dotted box) Seven inputs for forward training.
(dashed box) Structural verification by (*AutoCol*).
Optimal CI_c of **196,562.0** is 0.41% different from **195,756** obtained by large structural datasets.
"Bold values" indicate the important parameters of AI results.

P–M interaction diagrams reflecting minimized costs for an RC column are constructed in Figures 3.4.2 using forward networks. Parameters with sufficient accuracies shown in dashed black check box of Table 3.4.1 are used to plot *P–M* diagrams using *AutoCol* as shown in Figures 3.4.2. *P–M* diagrams represented by Legends 1–2 are plotted based on forward networks with five layers for 80 neurons for both ACI 318-14 and ACI 318-19, which converge to one P-M diagram when training accuracies are sufficient whereas *P–M* diagrams pass-through P_u = 1,000 kN and M_u = 3,000 kN·m as indicated by a red dot. The *P–M* interaction diagrams satisfy all design criteria shown in Table 3.2.2.4(b) and Table 3.2.2.5.

3.5 CONCLUSIONS

This chapter presents ANN-based Lagrange algorithm techniques, in which objective functions are minimized for rectangular concrete columns based on both forward and reverse networks. A Lagrange optimization with constraints is implemented in achieving rational engineering decisions by finding minimized or maximized design parameters. Differential equations under KKT conditions based on strict constraints, conditions, and requirements imposed by design codes are solved to optimize columns. This chapter helps engineers make final design decisions, not based on the engineers' empirical observations but based on more rational rules to meet various design requirements, including code restrictions and/or architectural criteria while objective functions, such as cost index CI_c,

CO_2 emissions, and structural weight W_c, are optimized. The conclusions drawn from this chapter are as follows:

1. Constructing objective functions and their derivatives analytically for rectangular concrete columns is a challenging task with conventional Lagrange optimization. Objective functions for optimizing rectangular concrete columns using ANN-based Lagrange algorithm techniques can now be generalized, leading to deriving Jacobian and Hessian matrices.
2. This chapter suggests objective functions for CI_c, CO_2 emissions, and W_c using ANN-based forward and reverse networks for Lagrange functions to be minimized. Feed-forward networks provide no input conflict, even though the computation speed is slower than that of the reverse networks.
3. The Karush–Kuhn–Tucker conditions are considered to account for inequality constraints, leading to optimizations of concrete columns that meet various code restrictions simultaneously. Realistic engineering applications are proposed to optimize a design of concrete columns based on all design requirements that must be met simultaneously, thus, achieving design decisions not based on engineers' intuition and empirical observations.
4. Separate optimization processes for CI_c, CO_2 emissions, and weight W_c are proposed, demonstrating negligible errors, which are verified using large structural datasets. Structural mechanics-based calculations also validate the design accuracies of the proposed methods when minimizing CI_c, CO_2 emissions, and weight W_c.
5. Novel and innovative *P–M* diagrams are uniquely constructed to minimize columns in terms of CI_c, CO_2 emissions, and weight W_c. The proposed optimization will offer generic designs for many types of structures, including machinery and structural frames.
6. Lagrange multiplier-based designs of concrete columns are extended to optimize multiple objective functions, such as CI_c, CO_2 emissions, and frame weight W_c, at the same time when columns are subject to multiple load combinations in Chapter 5 where unified objective functions (*UFO*) are proposed.
7. The ANN-based objective functions developed in this chapter can be implemented in broad areas, including engineering, general science, and economics.
8. The goodness of the proposed method is that its performance would be less dependent on problem types such as column, beam, frame, seismic design, etc. but relies on characteristics of big datasets of the considered problem. Once big datasets are good enough to generate objective function as well as other parameters using ANN, an optimization solution proceeds with the Lagrange algorithm. The applications of the proposed method would not be limited to optimizing RC columns but can be extended to other engineering applications.
9. Future studies are as follows.

   A generalizable optimization method based on ANNs proposed herein can be applied to any optimization problem once a sufficient number of datasets can be collected for establishing approximated objective functions and other parameters. Novel objective functions not only enhance a computational speed compared to conventional software but also produce a generalizable calculation method for Jacobian and Hessian matrices for Lagrange optimization. In future work, a robust design for a design of tall buildings would be performed based on ANN-based Lagrange algorithm. One concerning problem is that computational time of Lagrange optimization is heavily dependent on the quantity of inequality constraints because a number of running cases under KKT conditions corresponds to combinations of inequality constraints. Likewise, a building design is composed of many design requirements considered as inequality constraints, such as lateral displacement, story drift, design strength of each component, required total base shears, etc.

REFERENCES

ACI Committee. (2014). *Building Code Requirements For Structural Concrete (ACI 318-14) and Commentary*. American Concrete Institute, Farmington Hills, Michigan, USA.

ACI Committee. (2019). *Building Code Requirements for Structural Concrete (ACI 318–19) and Commentary*. American Concrete Institute, Farmington Hills, Michigan, USA.

Aghaee, K., Yazdi, M. A., & Tsavdaridis, K. D. (2015). Investigation into the mechanical properties of structural lightweight concrete reinforced with waste steel wires. *Magazine of Concrete Research*, *67*(4), 197–205. https://doi.org/10.1680/macr.14.00232

Awal, A. A., Shehu, I. A., & Ismail, M. (2015). Effect of cooling regime on the residual performance of high-volume palm oil fuel ash concrete exposed to high temperatures. *Construction and Building Materials*, *98*, 875–883. https://doi.org/10.1016/j.conbuildmat.2015.09.001.

Barros, M. H. F. M., Martins, R. A. F., & Barros, A. F. M. (2005). Cost optimization of singly and doubly reinforced concrete beams with EC2–2001. *Structural and Multidisciplinary Optimization*, *30*(3), 236–242.

Berrais, A. (1999). Artificial neural networks in structural engineering: Concept and applications. *Engineering Sciences*, 12(1), 53–67.

Coello Coello, C. A., Christiansen, A. D., & Hernandez, F. S. (1997). A simple genetic algorithm for the design of reinforced concrete beams. *Engineering with Computers*, *13*(4), 185–196.

Fanaie, N., Aghajani, S., & Dizaj, E. A. (2016). Theoretical assessment of the behavior of cable bracing system with central steel cylinder. *Advances in Structural Engineering*, *19*(3), 463–472. https://doi.org/10.1177/1369433216630052.

Fanaie, N., Aghajani, S., & Shamloo, S. (2012). Theoretical assessment of wire rope bracing system with soft central cylinder. In *Proceedings of the 15th World Conference on Earthquake Engineering*.

Heydari, A., & Shariati, M. (2018). Buckling analysis of tapered BDFGM nano-beam under variable axial compression resting on elastic medium. *Structural Engineering and Mechanics: An International Journal*, 66(6): 737–748. https://doi.org/10.12989/sem.2018.66.6.737.

Hoffmann, L. D., Bradley, G. L., & Rosen, K. H. (1989). *Calculus for Business, Economics, and the Social and Life Sciences*. McGraw-Hill, New York, USA.

Hong, W. K. (2019). *Hybrid Composite Precast Systems: Numerical Investigation to Construction*. Woodhead Publishing (Elsevier), Sawston, Cambridge, United Kingdom.

Hong, W. K. (2021). *Artificial Intelligence-Based Design of Reinforced Concrete Structures*. Daega Publisher, Gyeonggi, Korea.

Hong, W. K., Nguyen, M.C. (2021). AI-based Lagrange optimization for designing reinforced concrete columns. *Journal of Asian Architecture and Building Engineering TABE*. https://doi.org/10.1080/13467581.2021.1971998

Hong, W. K., Nguyen, M. C., & Pham, T. D. (2022a). Optimized interaction P-M diagram for rectangular reinforced concrete column based on artificial neural networks abstract. *Journal of Asian Architecture and Building Engineering*, 1–25. https://doi.org/10.1080/13467581.2021.2018697

Hong, W. K., Nguyen, V. T., Nguyen, D. H., & Nguyen, M. C. (2022b). Reverse design-based optimizations for reinforced concrete columns encasing H-shaped steel section using ANNs. *Journal of Asian Architecture and Building Engineering*, 1–15. https://doi.org/10.1080/13467581.2022.2047985

Hong, W.-K., Nguyen, V. T., Nguyen, D. H., & Nguyen, M. C. (2022c. An AI-based Lagrange optimization for a design for concrete columns encasing H-shaped steel sections under a biaxial bending. *Journal of Asian Architecture and Building Engineering*. https://doi.org/10.1080/13467581.2022.2060985

Hong, W. K., Nguyen, V. T., & Nguyen, M. C. (2021a). Artificial intelligence-based noble design charts for doubly reinforced concrete beams. *Journal of Asian Architecture and Building Engineering*, *21*(4), 1497–1519. https://doi.org/10.1080/13467581.2021.1928511

Hong, W.-K., Nguyen, V. T., & Nguyen, M. C. (2021b). Optimizing reinforced concrete beams cost based on AI-based Lagrange functions. *Journal of Asian Architecture and Building Engineering*. https://doi.org/10.1080/13467581.2021.2007105

Hong, W. K., Pham, T. D., & Nguyen, V. T. (2021c). Feature selection based reverse design of doubly reinforced concrete beams. *Journal of Asian Architecture and Building Engineering*, 1–25. https://doi.org/10.1080/13467581.2021.1928510

Kalman, D. (2009). Leveling with Lagrange: An alternate view of constrained optimization. *Mathematics Magazine*, *82*(3), 186–196. https://doi.org/10.1080/0025570X.2009.11953617

Kaveh, A., & Shokohi, F. (2015). Optimum design of laterally-supported castellated beams using CBO algorithm. *Steel and Composite Structures*, *18*(2), 305–324. https://doi.org/10.12989/scs.2015.18.2.305.

Kolmogorov, A. N. (1957). On the representation of continuous functions of many variables by superposition of continuous functions of one variable and addition. In *Doklady Akademii* (Vol. *114*, No. 5, pp. 953–956). Russian Academy of Sciences, Saint Petersburg, Russia.

Korouzhdeh, T., Eskandari-Naddaf, H., & Gharouni-Nik, M. (2017). An improved ant colony model for cost optimization of composite beams. *Applied Artificial Intelligence*, *31*(1), 44–63. https://doi.org/10.1080/08839514.2017.1296681.

Krenker A, Bešter J, Kos A. Introduction to the artificial neural networks. Artificial Neural Networks: Methodological Advances and Biomedical Applications. InTech, 2011; 1–18. https://doi.org/10.5772/15751.

KUHN, Harold W.; TUCKER, Albert W. Nonlinear programming. In: Traces and emergence of nonlinear programming. Birkhäuser, Basel, 2014. p. 247–258.

Li, G., & Zeng, Z. (2008). A neural-network algorithm for solving nonlinear equation systems. In *2008 International Conference on Computational Intelligence and Security* (Vol. 1, pp. 20–23). IEEE.

Madadi, A., Eskandari-Naddaf, H., Shadnia, R., & Zhang, L. (2018). Characterization of ferrocement slab panels containing lightweight expanded clay aggregate using digital image correlation technique. *Construction and Building Materials*, *180*: 464–476. https://doi.org/10.1016/j.conbuildmat.2018.06.024.

Malasri, S., Halijan, D. A., & Keough, M. L. (1994). Concrete beam design optimization with genetic algorithms. *Journal of the Arkansas Academy of Science*, *48*(1), 111–115.

Nasrollahi, S., Maleki, S., Shariati, M., Marto, A., & Khorami, M. (2018). Investigation of pipe shear connectors using push out test. *Steel and Composite Structures, An International Journal*, *27*(5), 537–543. https://doi.org/10.12989/scs.2018.27.5.537.

Paknahad, M., Shariati, M., Sedghi, Y., Bazzaz, M., & Khorami, M. (2018). Shear capacity equation for channel shear connectors in steel-concrete composite beams. *Steel and Composite Structures*, *28*(4), 483–494. https://doi.org/10.12989/scs.2018.28.4.483.

Pham, T. D., & Hong, W. K. (2022). Genetic algorithm using probabilistic-based natural selections and dynamic mutation ranges in optimizing precast beams. *Computers & Structures*, *258*, 106681.

Protter, M. H., and C. B. Morrey Jr. (1985). *Intermediate Calculus* (2nd ed., p. 267). Springer, New York.

Ripley, B. D. (1996). *Pattern Classification and Neural Networks.* Cambridge University Press, New York, USA.

Safa, M., Shariati, M., Ibrahim, Z., Toghroli, A., Baharom, S. B., Nor, N. M., & Petković, D. (2016). Potential of adaptive neuro fuzzy inference system for evaluating the factors affecting steel-concrete composite beam's shear strength. *Steel and Composite Structures, An International Journal*, *21*(3), 679–688. https://doi.org/10.12989/scs.2016.21.3.679.

Shah, S. N. R., Sulong, N. R., Khan, R., Jumaat, M. Z., & Shariati, M. (2016). Behavior of industrial steel rack connections. *Mechanical Systems and Signal Processing*, 70, 725–740. https://doi.org/10.12989/scs.2016.21.3.679.

Shariat, M., Shariati, M., Madadi, A., & Wakil, K. (2018). Computational Lagrangian Multiplier Method by using for optimization and sensitivity analysis of rectangular reinforced concrete beams. *Steel and Composite Structures*, 29(2): 243–256. https://doi.org/10.12989/scs.2018.29.2.243.

Shariati, M., Ramli Sulong, N. H., & Arabnejad Khanouki, M. M. (2010). Experimental and analytical study on channel shear connectors in light weight aggregate concrete. In *Proceedings of the 4th International Conference on Steel Composite Structures* (pp. 21–23). https://doi.org/10.3850/978-981-08-6218-3_CC-Fr031.

Silberberg, E. (1978). The structure of economics; A mathematical analysis (No. 04; HB135, S5.).
Standard, B. (1985). *8110: Part 1, Structural Use of Concrete–Code of Practice for Design and Construction* (pp. 3–8), British Standards Institute, London.
Tahouni, S. (2005). *Designing Concrete Structures Based on Iranian Concrete Code*. University of Tehran, Tehran, Iran.
Taylor, B. (1715). *Methodus Incrementorum Directa et Inversa*. Innys, London.
Toghroli, A., Mohammadhassani, M., Suhatril, M., Shariati, M., & Ibrahim, Z. (2014). Prediction of shear capacity of channel shear connectors using the ANFIS model. *Steel and Composite Structures*, *17*(5), 623–639. https://doi.org/10.12989/scs.2014.17.5.623.
Villarrubia, G., De Paz, J. F., Chamoso, P., & De la Prieta, F. (2018). Artificial neural networks used in optimization problems. *Neurocomputing*, 272, 10–16. https://doi.org/10.1016/j.neucom.2017.04.075.
Walsh, G. R. (1976). Methods of optimization. *Bulletin of the American Mathematical Society*, *82*, 540. https://doi.org/10.1090/S0002-9904-1976-14089-X

Chapter 4

Optimization of a reinforced concrete beam design using ANN-based Lagrange algorithm

4.1 SIGNIFICANCE OF THE CHAPTER

4.1.1 Current research

Some optimization studies have been performed widely in civil and architectural engineering, particularly in optimizing reinforced concrete (RC) structures (Aghaee, 2015, Fanaie, 2016, Madadi, 2018, Nasrollahi, 2018, and Paknahad, 2018). Most of these studies only focused on minimizing manufacturing and construction costs. Several optimization methods of RC beams have been performed successfully by Shariati (2010), Fanaie (2012), Toghroli (2014), Awal (2015), Kaveh and Shokohi (2015), Safa (2016), Shah (2016), Korouzhdeh (2017), and Heydari and Shariati (2018). Shariati (2018) also established an analytical objective function for a cost of frames; however, analytical objective functions, which represent an entire behavior of structural components are generally hard to be derived. ANNs are kept being implemented in the field of structural engineering. Rizzo and Caracoglia (2020) proposed a procedure to predict a critical flutter velocity of suspension bridges with closed box deck sections based on an implementation of an ANN. Gomes (2019) proposed to apply an ANN to predict delamination failure in carbon fiber reinforcement polymer (CFRP) plates. A feed-forward-based neural network was used to detect damages based on big data obtained from Finite Element Analysis (FEA), and then results based on an ANN were verified by numerical algorithms. They stated that an ANN was an effective tool for a delamination damage identification problem. Flood and Kartam (1994a, b) presented a concept and an application of neural networks to structural engineering by exploring an influence of number of hidden layers and nodes on accuracies of deep learning networks. They concluded that a success of an ANN implementation is dependent on the quality of data used for training. Hong (2019) also investigated an influence of a number of training datasets on training accuracies. Wu and Jahanshahi (2019) proposed convolutional neural network (CNN) approach to better predict structural responses compared with a multiple layer perceptron (MLP) algorithm against noisy data. They presented a deep convolutional neural network (CNN)-based approach to estimate the dynamic response of a linear single-degree-of-freedom (SDOF) system, a nonlinear SDOF system, and a full-scale 3-story steel frame with multi-degree of freedom (MDOF). Ahmadi (2008) presented back-propagation wavelet neural network (BPW) based on scaled conjugate gradient (SCG) algorithm which replaced sigmoid activation functions of hidden layer neurons, approximating dynamic time history response of frame structures. Fahmy (2016) provided a methodology to use ANNs for the conceptual design of an orthotropic steel deck bridge. They found that ANNs offered a better and cost-effective option compared with international codes or expert opinion for orthotropic deck designs. Adeli (2001) published a large number of articles on structural analysis and design problems since 1989. However, according to Lee (2018), ANNs have shown limitation and numerical instability in the structural engineering area due to a poor

DOI: 10.1201/9781003314684-4

training performance and enormous computational time, especially for complicated problems with multiple hidden layers. Gupta and Sharma (2011) also stated that a use of ANNs in structural engineering application has been significantly decreased over the last decade. In this chapter, a procedure to obtain objective functions for an entire behavior of structural components is proposed based on artificial neural networks (ANNs).

4.1.2 Motivations and objective

This chapter aims to introduce ANN-based forward Lagrange optimizations for ductile doubly reinforced concrete (RC) beams. Structural capabilities which are controlled by design codes are taken into a consideration against external forces. An objective function, the cost (CI_b) of materials and manufacture of RC beams, is minimized considering constraints according to engineer's needs.

Large datasets of 100,000 are used to derive an ANN-based objective function for cost (CI_b) as a function of forward input parameters, leading to minimizing a cost of RC beams. Complex analytical objective functions are replaced by ANN-based objective functions. ANN-based forward Lagrange optimizations are, then, performed to calculate design parameters while minimizing CI_b. In Chapter 4, design charts are proposed to assist engineers to quickly and accurately design a ductile doubly RC beam according to the minimized CI_b. The optimized design including a minimized cost of an RC beam is also validated by structural calculations.

4.1.3 Significance of the proposed methodology

Not many studies were seen to apply ANN-based Lagrange optimization techniques to design structures similar to the ones described in this chapter. Only a few researchers use ANNs for holistic structural design. Chapter 4 aims to develop ANN-based forward Lagrange networks optimizing ductile doubly reinforced concrete (RC) beams. Cost (CI_b) of materials and manufacture is established as an objective function which is minimized based on constraints according to engineer's needs and design codes.

Chapter 4 performs holistic beam designs for engineers who wish to find optimized beams. A resilient design capable of optimizing concrete beams beyond human efficiency is based on equality and inequality constraints which are implanted in ANN-based Lagrange functions as artificial neural genes like human DNAs. This is not a simple task, especially, when multiple constraining conditions are to be considered. Neither is it possible for engineers to pre-assign constraining conditions which is only calculated sequentially on an output-side for a conventional design. Artificial neural genes are not limited to cost of beams, but CO_2 emissions and weight of RC beams can be also constrained as functions of input parameters for structures with regard to codes, serviceability, economies, and environment while minimizing negative influences. A cost of RC beams minimized by forward Lagrange networks is reduced by 18%–26% compared to probable beam designs. ANN-based design charts providing eight forward outputs (ϕM_n, M_u, M_{cr}, $\varepsilon_{rt\_0.003}$, $\varepsilon_{rc\_0.003}$, Δ_{imme}, Δ_{long}, CI_b) based on nine forward inputs (L, h, b, f_y, f_c, ρ_{rt}, ρ_{rc}, M_D, M_L) are proposed to assist engineers to design ductile doubly RC beams while presenting minimized CI_b.

However, solving structural optimization problems by derivative methods such as conventional Lagrange multiplier is challenging because objective functions for structural frames are difficult to express in terms of design variables explicitly. ANN-based studies of optimization grow. Researchers attempt to implement Lagrange optimization in structures with analytical objective functions. Recently, computational powers are significantly enhanced with multiple GPUs in routine ANN training which let researchers attempt to train bigger networks overcoming limitation and numerical instability.

4.2 OPTIMIZATION OF A REINFORCED CONCRETE BEAM DESIGNS BASED ON ANNs

4.2.1 Beam design scenarios

ANN-based forward Lagrange optimizations are formulated based on parameters shown in Table 4.2.1.1(a) that presents forward design scenario based on nine forward input parameters (L, h, b, f_y, f_c', ρ_{rt}, ρ_{rc}, M_D, M_L) and eight forward output parameters (ϕM_n, M_u, M_{cr}, $\varepsilon_{rt\_0.003}$, $\varepsilon_{rc\_0.003}$, Δ_{imme}, Δ_{long}, CI_b). These parameters include material properties and geometries which are predetermined by equality constraints. Table 4.2.1.1(b) lists all input and output parameters with their nomenclatures. Forward Lagrange networks are formulated to avoid input conflicts. Conventional software calculates eight output parameters (ϕM_n, M_u, M_{cr}, $\varepsilon_{rt\_0.003}$, $\varepsilon_{rc\_0.003}$, Δ_{imme}, Δ_{long}, CI_b) based on given nine forward input parameters (L, h, b, f_y, f_c', ρ_{rt}, ρ_{rc}, M_D, M_L) for a design of ductile doubly reinforced concrete beams. In accordance with Section 9.3.3.1 of ACI 318-19, tensile rebar strain ($\varepsilon_{rt\_0.003}$) should be at least ε_{ty} (yield strain of rebars)+0.003 in order to mitigate the brittle flexural behavior of RC beams. This limit for a tensile rebar strain is denoted as Equality v_9 in Table 4.2.1.2. In

Table 4.2.1.1 Beam Design Scenario and Nomenclature

(a): Forward Design Scenario								
Forward Inputs (Nine Parameters)								
L	h	b	f_y	f_c'	ρ_{rt}	ρ_{rc}	M_D	M_L
Forward Outputs (Eight Parameters)								
ϕM_n	M_u	M_{cr}	$\varepsilon_{rt\_0.003}$	$\varepsilon_{rc\_0.003}$	Δ_{imme}	Δ_{long}	CI_b	

No.	*Nomenclature*	
	Forward Input Parameters	
1	L (mm)	Beam span
2	h (mm)	Beam height
3	b (mm)	Beam width
4	f_y (MPa)	Yield strength of rebar
5	f_c' (MPa)	Compressive concrete strength
6	ρ_{rt}	Tensile rebar ratio of A_{rt} to bh; A_{rt} is tensile rebar area
7	ρ_{rc}	Compressive rebar ratio of A_{rc} to bh; A_{rc} is compressive rebar area
8	M_D (kN·m)	Moment due to dead load
9	M_L (kN·m)	Moment due to live load
	Forward Output Parameters	
10	ϕM_n (kN·m)	Design moment strength excluding moment due to self-weight
11	M_u (kN·m)	Factored moment considering moment due to dead and live load
12	M_{cr} (kN·m)	Cracking moment
13	$\varepsilon_{rt\_0.003}$	Tensile rebar strain at concrete strain of 0.003
14	$\varepsilon_{rc\_0.003}$	Compressive rebar strain at concrete strain of 0.003
15	Δ_{imme}	Immediate deflections due to live load
16	Δ_{long}	Long-term deflections
17	CI_b (KRW/m)	Cost index of beam

Table 4.2.1.2 Six Equalities and 11 Inequalities

Equality Constraints	*Inequality Constraints*
$c_1(\mathbf{x}): L = 10{,}000$ mm	$v_1(\mathbf{x}): \phi M_n - M_u \geq 0$
$c_2(\mathbf{x}): f_c' = 30$ MPa	$v_2(\mathbf{x}): \phi M_n - 1.2M_{cr} \geq 0$
$c_3(\mathbf{x}): f_y = 600$ MPa	$v_3(\mathbf{x}): h - 500 \geq 0$
$c_4(\mathbf{x}): M_D = [250 \sim 1500]$ kN·m	$v_4(\mathbf{x}): -h + 1200 \geq 0$
$c_5(\mathbf{x}): M_L = [125 \sim 750]$ kN·m	$v_5(x): b - 0.3h \geq 0$
$c_6(\mathbf{x}): M_u = 1.2M_D + 1.6M_L = [500 \sim 3{,}000$ kN·m$]$	$v_6(\mathbf{x}): -b + 0.8h \geq 0$
	$v_7(\mathbf{x}): -\rho_{rc} + \rho_{rt}/2 \geq 0$
	$v_8(\mathbf{x}): \rho_{rc} - \rho_{rt}/400 \geq 0$
	$v_9(\mathbf{x}): \varepsilon_{rt\_0.003} - 0.006^{a} \geq 0$
	$v_{10}(\mathbf{x}): -\Delta_{imme} + L/360 \geq 0$
	$v_{11}(\mathbf{x}): -\Delta_{long} + L/240 \geq 0$

[a] A limit of tension-controlled sections is $\varepsilon_{ty}\left(\frac{f_y}{E_s} = \frac{600}{200{,}000} = 0.003\right) + 0.003$; ε_{ty} is rebar yield strain, f_y is rebar yield strength, and E_s is young modulus of rebar.

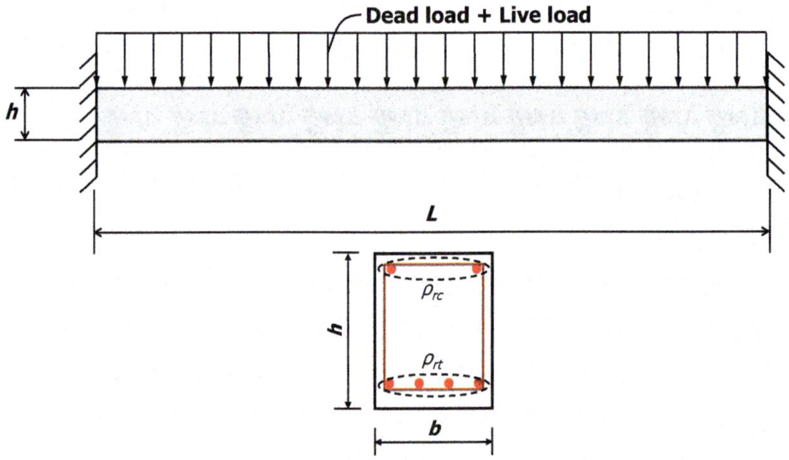

Figure 4.2.1.1 A doubly reinforced concrete beam with fixed-fixed boundary conditions; cost (CI_b) minimized based on ANN-based forward Lagrange network.

this chapter, Lagrange optimization is implemented to design doubly reinforced RC beams with minimized cost CI_b illustrated in Figure 4.2.1.1. Fixed-fixed doubly reinforced concrete (RC) beams are considered in this chapter. Immediate deflections are calculated with only live load being taken into consideration in accordance with Table 24.2.2 of ACI 318-19 (Standard, 2019) as shown in Figure 4.2.1.2. Cost index of beams CI_b is minimized using ANN-based Lagrange algorithm. Besides, design charts identify a minimized CI_b are proposed to assist engineers to design a fixed-fixed ductile doubly RC beam. Cost index of beams CI_b is an objective function as a function of nine forward input parameters (L, h, b, f_y, f_c', ρ_{rt}, ρ_{rc}, M_D, and M_L). In Table 4.2.1.2, six equalities and 11 inequalities are also established to formulate forward Lagrange-based optimization and to construct design

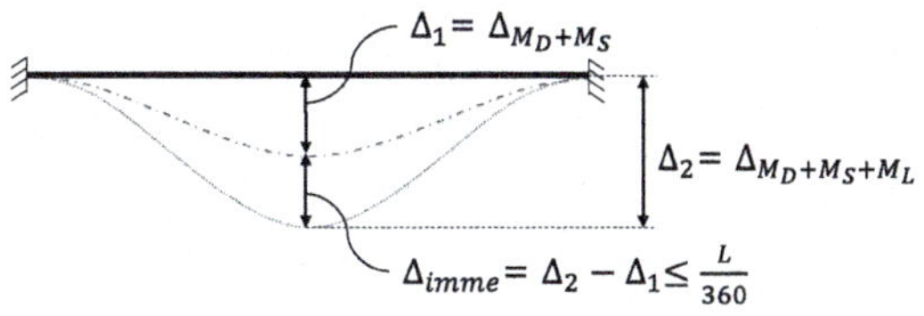

Figure 4.2.1.2 Procedure to calculate immediate deflection.

charts. Self-weight is subtracted from factored moments because the size of a beam size is not known in the beginning of a design. Forward Lagrange-based design charts constructed to minimize cost of beams are functions of factored moment M_u. Fixed support conditions are reflected in the calculation of a factored moment M_u which consists of moment due to dead and live loads (M_u=1.2 M_D +1.6 M_L) as indicated in Equation (4.2.1.1) which is set as equality constraint varying from 500 to 3,000 kN·m.

$$M_u = 1.2M_D + 1.6M_L \tag{4.2.1.1}$$

An objective function, cost of beams CI_b, is minimized by finding a stationary point of a Lagrange function based on ANN-based Lagrange multipliers, during which equality and inequality constraints shown in Table 4.2.1.2 are imposed. Design parameters including h, b, ρ_{rt}, and ρ_{rc} corresponding to minimum CI_b based on equality and inequality constraints shown in Table 4.2.1.2 are also identified while providing output parameters (ϕM_n, M_{cr}, $\varepsilon_{rt\_0.003}$, $\varepsilon_{rc\_0.003}$, Δ_{imme}, Δ_{long}) against a given factored moment M_u.

4.2.2 Formulation of a Lagrange function for optimizing a reinforced concrete beam based on ANNs

Lagrange function is derived for a design of doubly reinforced concrete beams using Lagrange multipliers. ANN-based forward Lagrange functions are formulated based on equality and inequality constraints. A design algorithm shown in Figure 4.2.2.1 minimizes an objective function based on forward ANNs where six steps are presented to solve for KKT conditions when inequality constraints are considered (Hong et al., 2021a, b).

4.2.2.1 Derivation of ANN-based objective functions

ANN-based function is extracted for objective functions based on Equation (4.2.2.1).

$$f(\mathbf{x}) = g^D\left(f_{\text{lin}}^N\left(\mathbf{W}^N f_t^{N-1}\left(\mathbf{W}^{N-1}\ldots f_t^1\left(\mathbf{W}^1 g^N(\mathbf{x}) + \mathbf{b}^1\right)\ldots + \mathbf{b}^{N-1}\right) + b^N\right)\right) \tag{4.2.2.1}$$

(Krenker, 2011)

where $\mathbf{x}$ is an input vector; N is a number of layers including hidden layers and output layer; $\mathbf{W}^n$ is weight matrix between layer $n-1$ and layer n; $\mathbf{b}^n$ is bias matrix of layer n; and g^N, g^D are normalization and de-normalization function, respectively (Hong and Nguyen, 2021, Hong, 2020, 2021).

Activation functions f_t^n at layer n are implemented in formulating non-linear relationships of networks; whereas a linear activation function f_{lin}^n is selected for output layer because output values are not bounded. An ANN-based objection function is used for Lagrange optimization. Type of networks used for Lagrange optimization is determined depending

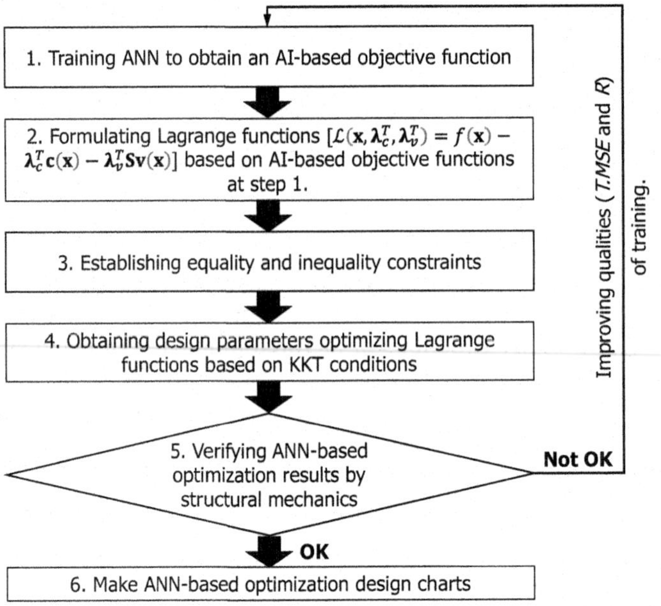

Figure 4.2.2.1 Algorithm for neural networks minimizing designs of a ductile precast beam.

on design natures. Reverse network-based Lagrange optimization may be selected when input conflicts can be obviously avoided. Otherwise, forward neural networks should be employed with an aim of removing any input conflict. An ANN-based objective function shown in Equation (4.2.2.2) is obtained with nine input parameters (L, h, b, f_y, f_c', ρ_{rt}, ρ_{rc}, M_D, M_L) to optimize cost index (CI_b) of a doubly RC beam. In Chapter 4, large datasets are generated for beams with fixed boundary conditions using structural calculation software (*Autobeam*). Output parameters which are influenced mostly by beam length (L) are immediate (Δ_{imme}) and long-term (Δ_{long}) deflections. Immediate (Δ_{imme}) and long-term (Δ_{long}) deflections are constrained not to be greater than $L/360$ and $L/240$ (Table 24.2.2, ACI 318-19 (Standard, 2019)), respectively as indicated as inequality constraints (v_{10} and v_{11} in Table 4.2.1.2). Table 4.2.2.1 presents a summary of training when mapping nine input parameters to each single output parameter based on PTM. As shown in Table 4.2.2.1(a), Network #1 maps nine input parameters (L, h, b, f_y, f_c', ρ_{rt}, ρ_{rc}, M_D, M_L) to eight output parameters (ϕM_n, M_u, M_{cr}, $\varepsilon_{rt\_0.003}$, $\varepsilon_{rc\_0.003}$, Δ_{imme}, Δ_{long}, CI_b) by PTM based on a combination of three hidden layers with three types of neurons (30, 40, and 50), among which a network with the best training accuracies is selected. PTM maps the entire nine input parameters to single output parameter. As shown in Network #2 of Table 4.2.2.1(b), nine input parameters (L, h, b, f_y, f_c', ρ_{rt}, ρ_{rc}, M_D, M_L) are re-trained based on 15 trainings, a combination of three types of hidden layer (3, 4, and 5) with five types of neurons (30, 40, 50, 60, and 70) to improve training accuracies. Testing *MSE* of 1.43E-4 for immediate deflection shown in Table 4.2.2.1(a) with three layers and 40 neurons improved slightly to 1.297E-4 with four layers and 40 neurons shown in Table 4.2.2.1(b). A range of design errors of immediate and long-term deflections are similar as shown in Figure 4.4.1.1c and d.

$$\underbrace{CI_b}_{[1\times1]} = g_{CI_b}^{D}\left(f_l^4\left(\underbrace{\mathbf{W}_{CI_b}^4}_{[1\times30]} f_t^3\left(\underbrace{\mathbf{W}_{CI_b}^3}_{[30\times30]} \cdots f_t^1\left(\underbrace{\mathbf{W}_{CI_b}^1}_{[30\times9]} \underbrace{\mathbf{g}_{CI_b}^{N}(\mathbf{x})}_{[9\times1]} + \underbrace{\mathbf{b}_{CI_b}^1}_{[30\times1]}\right) \cdots + \underbrace{\mathbf{b}_{CI_b}^3}_{[30\times1]}\right) + \underbrace{b_{CI_b}^4}_{[1\times1]}\right)\right) \tag{4.2.2.2}$$

Table 4.2.2.1 Summary of Training When Mapping Nine Input Parameters to Each Single Output Parameter Based on PTM

(a): Network #1

(1) Nine Inputs (L, h, b, f_y, f_c', ρ_{rt}, ρ_{rc}, M_D, M_L) - One Output (ϕM_n)

Datasets	Layers	Neurons	Required Epoch	Best Epoch for Training	Test MSE	R at Best Epoch
100,000	3	40	50,000	19,303	4.23E-07	1.0

(2) Nine Inputs (L, h, b, f_y, f_c', ρ_{rt}, ρ_{rc}, M_D, M_L) - One Output (M_u)

Data	Layers	Neurons	Required Epoch	Best Epoch for Training	Test MSE	R at Best Epoch
100,000	3	50	50,000	12,984	3.74E-07	1.0

(3) Nine Inputs (L, h, b, f_y, f_c', ρ_{rt}, ρ_{rc}, M_D, M_L) - One Output (M_{cr})

Data	Layers	Neurons	Required Epoch	Best Epoch for Training	Test MSE	R at Best Epoch
100,000	3	40	50,000	19,165	7.67E-07	1.0

(4) Nine Inputs (L, h, b, f_y, f_c', ρ_{rt}, ρ_{rc}, M_D, M_L) - One Output ($\varepsilon_{rt\_0.003}$)

Data	Layers	Neurons	Required Epoch	Best Epoch for Training	Test MSE	R at Best Epoch
100,000	3	50	50,000	44,069	3.55E-06	1.0

(5) Nine Inputs (L, h, b, f_y, f_c', ρ_{rt}, ρ_{rc}, M_D, M_L) - 1 Output ($\varepsilon_{rc\_0.003}$)

Data	Layers	Neurons	Required Epoch	Best Epoch for Training	Test MSE	R at Best Epoch
100,000	3	30	50,000	16,335	8.79E-06	1.0

(6) Nine Inputs (L, h, b, f_y, f_c', ρ_{rt}, ρ_{rc}, M_D, M_L) - One Output (Δ_{imme})

Data	Layers	Neurons	Required Epoch	Best Epoch for Training	Test MSE	R at Best Epoch
100,000	3	40	50,000	27,608	1.43E-04	0.999

(7) Nine Inputs (L, h, b, f_y, f_c', ρ_{rt}, ρ_{rc}, M_D, M_L) - One Output (Δ_{long})

Data	Layers	Neurons	Required Epoch	Best Epoch for Training	Test MSE	R at Best Epoch
100,000	3	50	50,000	25,354	4.39E-05	1.000

(8) Nine Inputs (L, h, b, f_y, f_c', ρ_{rt}, ρ_{rc}, M_D, M_L) - One Output (CI_t)

Data	Layers	Neurons	Required Epoch	Best Epoch for Training	Test MSE	R at Best Epoch
100,000	3	30	50,000	30,659	2.44E-08	1.0

(b): Network #2

(1) Nine Inputs (L, h, b, f_y, f_c', ρ_{rt}, ρ_{rc}, M_D, M_L) - One Output (ϕM_n)

Datasets	Layers	Neurons	Required Epoch	Best Epoch for Training	Test MSE	R at Best Epoch
100,000	3	40	50,000	19,303	4.23E-07	1.0

(2) Nine Inputs (L, h, b, f_y, f_c', ρ_{rt}, ρ_{rc}, M_D, M_L) - One Output (M_u)

Data	Layers	Neurons	Required Epoch	Best Epoch for Training	Test MSE	R at Best Epoch
100,000	3	50	50,000	12,984	3.74E-07	1.0

(3) Nine Inputs (L, h, b, f_y, f_c', ρ_{rt}, ρ_{rc}, M_D, M_L) - One Output (M_{cr})

Data	Layers	Neurons	Required Epoch	Best Epoch for Training	Test MSE	R at Best Epoch
100,000	3	40	50,000	19,165	7.67E-07	1.0

(4) Nine Inputs (L, h, b, f_y, f_c', ρ_{rt}, ρ_{rc}, M_D, M_L) - One Output ($\varepsilon_{rt\_0.003}$)

Data	Layers	Neurons	Required Epoch	Best Epoch for Training	Test MSE	R at Best Epoch
100,000	3	50	50,000	44,069	3.55E-06	1.0

(Continued)

Table 4.2.2.1 (Continued) Summary of Training When Mapping Nine Input Parameters to Each Single Output Parameter Based on PTM

(5) Nine Inputs (L, h, b, f_y, f_c', ρ_{rt}, ρ_{rc}, M_D, M_L) - One Output ($\varepsilon_{rc\_0.003}$)						
Data	Layers	Neurons	Required Epoch	Best Epoch for Training	Test MSE	R at Best Epoch
100,000	3	30	50,000	16,335	8.79E-06	1.0
(6) Nine Inputs (L, h, b, f_y, f_c', ρ_{rt}, ρ_{rc}, M_D, M_L) - One Output (Δ_{imme})						
Data	Layers	Neurons	Required Epoch	Best Epoch for Training	Test MSE	R at Best Epoch
100,000	4	40	50,000	34,412	1.29E-04	0.999
(7) Nine Inputs (L, h, b, f_y, f_c', ρ_{rt}, ρ_{rc}, M_D, M_L) - One Output (Δ_{long})						
Data	Layers	Neurons	Required Epoch	Best Epoch for Training	Test MSE	R at Best Epoch
100,000	5	30	50,000	21,010	4.3E-05	1.000
(8) Nine Inputs (L, h, b, f_y, f_c', ρ_{rt}, ρ_{rc}, M_D, M_L) - One Output (CI_b)						
Data	Layers	Neurons	Required Epoch	Best Epoch for Training	Test MSE	R at Best Epoch
100,000	3	30	50,000	30,659	2.44E-08	1.0

4.2.2.2 Derivation of ANN-based Lagrange functions

A Lagrange function $\mathcal{L}$ is derived as a function of input variables $\mathbf{x} = [x_1, x_2,\ldots, x_n]^T$ in which Lagrange multipliers, $\boldsymbol{\lambda}_c = [\lambda_1, \lambda_2,\ldots, \lambda_m]^T$ and $\boldsymbol{\lambda}_v = [\lambda_1, \lambda_2,\ldots, \lambda_l]^T$, are applied to equality $\mathbf{c}(\mathbf{x})$ and inequality $\mathbf{v}(\mathbf{x})$ constraints, respectively, as shown in Equation (4.2.2.3) when an objective function of beam cost (CI_b) is minimized.

$$\mathcal{L}(\mathbf{x}, \boldsymbol{\lambda}_c^T, \boldsymbol{\lambda}_v^T) = f(\mathbf{x}) - \boldsymbol{\lambda}_c^T \mathbf{c}(\mathbf{x}) - \boldsymbol{\lambda}_v^T \mathbf{S}\mathbf{v}(\mathbf{x}) = CI_b - \boldsymbol{\lambda}_c^T \mathbf{c}(\mathbf{x}) - \boldsymbol{\lambda}_v^T \mathbf{S}\mathbf{v}(\mathbf{x}) \tag{4.2.2.3}$$

As shown in Table 4.2.1.2, six simple equalities including $\mathcal{L}$, f_y, f_c', M_D, M_L and M_u can be substituted directly into networks. Eleven inequality functions constraining optimization are shown in Table 4.2.1.2. Equalities are equations, not variables, and hence, they must be written as f_y=600 or f_y − 600=0. Newton-Raphson iteration is applied to solve linearized KKT conditions shown in Equation (4.2.2.4) when inequality constraints are imposed. Jacobi of Lagrange functions $\nabla\mathcal{L}(\mathbf{x}, \boldsymbol{\lambda}_c, \boldsymbol{\lambda}_v)$ shown in Equation (4.2.2.5), which is a set of partially differential equations, derived based on gradient vectors is linearized in terms of Jacobi matrices and Hessian matrices. Hessian matrices are obtained by differentiating Jacobi matrices one more time with respect to input variables.

Newton-Raphson numerical iteration is used to find stationary points of Lagrange function $\mathcal{L}(\mathbf{x}, \boldsymbol{\lambda}_c, \boldsymbol{\lambda}_v)$ with respect to a function of input variables x, Lagrange multiplier of equality constraints $\boldsymbol{\lambda}_c$, and Lagrange multiplier of inequality constraints $\boldsymbol{\lambda}_v$. Jacobian and Hessian matrix are shown in Equations (4.2.2.4)–(4.2.2.6). Input variable $\mathbf{x}$ and Lagrange multipliers $\boldsymbol{\lambda}$ can be updated after every iteration based on Newton-Raphson iteration as indicated in Equation (4.2.2.4). Newton–Raphson approximation is repeated until convergence is achieved. Readers are referred to Section 2.3.4 of Chapter 2 for a detailed explanation for this iteration.

$$\begin{bmatrix} \mathbf{x}^{(k+1)} \\ \boldsymbol{\lambda}_c^{(k+1)} \\ \boldsymbol{\lambda}_v^{(k+1)} \end{bmatrix} = \begin{bmatrix} \mathbf{x}^{(k)} \\ \boldsymbol{\lambda}_c^{(k)} \\ \boldsymbol{\lambda}_v^{(k)} \end{bmatrix} - \left[\mathbf{H}_{\mathcal{L}}\left(\mathbf{x}^{(k)}, \boldsymbol{\lambda}_c^{(k)}, \boldsymbol{\lambda}_v^{(k)}\right)\right]^{-1} \nabla\mathcal{L}\left(\mathbf{x}^{(k)}, \boldsymbol{\lambda}_c^{(k)}, \boldsymbol{\lambda}_v^{(k)}\right) \tag{4.2.2.4}$$

where $\left[\mathbf{H}_{\mathcal{L}}\left(\mathbf{x}^{(k)}, \boldsymbol{\lambda}_c^{(k)}, \boldsymbol{\lambda}_v^{(k)}\right)\right]$ is Hessian matrix of Lagrange function, and $\nabla\mathcal{L}\left(\mathbf{x}^{(k)}, \boldsymbol{\lambda}_c^{(k)}, \boldsymbol{\lambda}_v^{(k)}\right)$ Equation (4.2.2.5) is a Jacobi of Lagrange functions which is a first derivation of Lagrange function. KKT conditions are solved by finding a stationary point based on gradient vectors of objective functions and equality $\mathbf{c}(\mathbf{x})$, inequality $\mathbf{v}(\mathbf{x})$ functions shown in Equation (4.2.2.6) which constrain objective functions.

$$\nabla\mathcal{L}(\mathbf{x},\boldsymbol{\lambda}_c,\boldsymbol{\lambda}_v)=\begin{bmatrix}\nabla f(\mathbf{x})-J_c(\mathbf{x})^T\boldsymbol{\lambda}_c-J_v(\mathbf{x})^T\mathbf{S}\boldsymbol{\lambda}_v\\ -\mathbf{c}(\mathbf{x})\\ -Sv(\mathbf{x})\end{bmatrix} \tag{4.2.2.5}$$

$$J_c(\mathbf{x})=\begin{bmatrix}\nabla c_1(\mathbf{x})\\ \nabla c_2(\mathbf{x})\\ \vdots\\ \nabla c_m(\mathbf{x})\end{bmatrix}$$

and

$$J_v(\mathbf{x})=\begin{bmatrix}\nabla v_1(\mathbf{x})\\ \nabla v_2(\mathbf{x})\\ \vdots\\ \nabla v_l(\mathbf{x})\end{bmatrix} \tag{4.2.2.6}$$

where $J_c(\mathbf{x})$ and $J_v(\mathbf{x})$ are Jacobian matrices of constraining vectors **c** and **v** at **x**, respectively. Lagrange multipliers $\boldsymbol{\lambda}_c$ and $\boldsymbol{\lambda}_v$ for both equality and inequality constraints, respectively, are implemented in converting constrained optimization problems to unconstrained ones.

4.2.2.3 Formulation of KKT conditions based on equality and inequality constraints

Lagrange multiplier method (LMM) finds stationary points at which maxima or minima of Lagrange functions are calculated. A Lagrange function $\mathcal{L}$ is derived as a function of the variables $\mathbf{x}=[x_1, x_2,\ldots, x_n]^T$ whereas equality constrains $c(\mathbf{x}),=[c_1(\mathbf{x}), c_2(\mathbf{x}),\ldots, c_6(\mathbf{x})]$ are shown in Equation (4.2.2.7) and inequality constrains $v(\mathbf{x}),=[v_1(\mathbf{x}), v_2(\mathbf{x}),\ldots, v_{11}(\mathbf{x})]$ are shown in Equation (4.2.2.8). Objective function $f(\mathbf{x})$ is a multivariate function subjected to equality and inequality constraints, $\mathbf{c}(\mathbf{x})=[c_1(\mathbf{x}), c_2(\mathbf{x}), \ldots, c_m(\mathbf{x})]^T=0$ and $\mathbf{v}(\mathbf{x})=[v_1(\mathbf{x}), v_2(\mathbf{x}), \ldots, v_l(\mathbf{x})]^T\geq 0$, respectively. Definition of equality and inequality constraints are presented in Equations (4.2.2.7), (4.2.2.8), and Table 4.2.1.2, whereas Lagrange multipliers of equality constraints $\boldsymbol{\lambda}_c=[\lambda_{c1},\lambda_{c2},\ldots,\lambda_{c6}]^T$ and inequality constraints $\boldsymbol{\lambda}_v=[\lambda_{v1},\lambda_{v2},\ldots,\lambda_{v11}]^T$ are applied to the equality and inequality constraints.

$$\begin{aligned}&\text{Equality constrains}: c_j(\mathbf{x})=0,\quad j=1,\ldots,m_1\\ &\mathbf{c}(\mathbf{x})=[c_1(\mathbf{x}), c_2(\mathbf{x}), \ldots, c_6(\mathbf{x})]^T\end{aligned} \tag{4.2.2.7}$$

$$\text{Inequality constrains}: v_k(\mathbf{x}) \geq 0, \quad k = 1,\ldots,m_2$$
$$\boldsymbol{v}(\mathbf{x}) = \left[v_1(\mathbf{x}), v_2(\mathbf{x}), \ldots, v_{11}(\mathbf{x})\right]^T \tag{4.2.2.8}$$

In Equation (4.2.2.9), inequality matrix **S** is introduced to identify active inequality constraints based on KKT conditions in which active inequality constraints (v_i) are selected with **S** = **1**. Inactive inequality constraints (vi) are selected with **S** = **0**.

$$\mathbf{S} = \begin{bmatrix} s_1 & 0 & \cdots & 0 \\ 0 & s_2 & \cdots & 0 \\ \vdots & \vdots & \ddots & \vdots \\ 0 & 0 & \cdots & s_{11} \end{bmatrix}_{[11\times 11]} \tag{4.2.2.9}$$

1. Formulation of inequality matrix
   KKT conditions assume inequality constraints as either active or inactive conditions, as explained in Section 1.3 of Chapter 1 where active conditions transform inequality constraints to equality constraints to treat Lagrange optimization with only equality constraints. Inequality matrix **S** based on s_i is provided in Equation (4.2.2.9) for inequality constraints **S**. An inequality constraint (v_i) is activated when $s_i = 1$ at i=2 as shown in Equation (4.2.2.10) while inequality constraints (v_i) are not activated when $s_i = 0$. The diagonal matrix of the inequality matrix **S** = **1** of Equation (4.2.2.10) at i=2 activates the inequality constraints to equality constraints by binding inequality conditions to equality conditions. However, inequality conditions are deactivated by setting S = 0 based on $s_i = 0$ (all i except for i=2) as long as stationary points are found within the range defined by ignored inequality constraints.

   In Table 4.2.1.2, code requirements are, then, expressed in terms of inequality constraints based on s_2 = 1; $v_2 = \phi M_n - 1.2M_{cr} \geq 0$ constrained by ACI 318-19 (Section 9.6.1.2) (Standard, 2019) in which a design moment strength ϕM_n is being 1.2 times a cracking moment M_{cr}, leading to Equation (4.2.2.10).

$$\begin{array}{c} \\ \mathbf{S} = \end{array} \begin{array}{c} \begin{array}{cccccc} 1^{\text{st}} & 2^{\text{nd}} & 3^{\text{rd}} & & 10^{\text{th}} & 11^{\text{th}} \end{array} \\ \begin{bmatrix} s_1 = 0 & 0 & 0 & \cdots & 0 & 0 \\ 0 & s_2 = 1 & 0 & \cdots & 0 & 0 \\ 0 & 0 & 0 & \cdots & 0 & 0 \\ \vdots & \vdots & \vdots & \ddots & \vdots & \vdots \\ 0 & 0 & 0 & \cdots & s_{10} = 0 & 0 \\ 0 & 0 & 0 & \cdots & 0 & s_{11} = 0 \end{bmatrix}_{11\times 11} \end{array} \begin{array}{c} 1^{\text{st}} \\ 2^{\text{nd}} \\ 3^{\text{rd}} \\ \vdots \\ 10^{\text{th}} \\ 11^{\text{th}} \end{array} \tag{4.2.2.10}$$

The Lagrange function with $s_2 =$ 1 is obtained in Equation (4.2.2.11) by substituting $s_2 =$ 1 into Equation (4.2.2.10) and Equation (4.2.2.3).

$$\mathcal{L}_{CI_b}(\mathbf{x}, \boldsymbol{\lambda}_c, \boldsymbol{\lambda}_v) = CI_b - \boldsymbol{\lambda}_c^T \boldsymbol{c}(\mathbf{x}) - \boldsymbol{\lambda}_v^T \begin{bmatrix} s_1 = 0 & 0 & \cdots & 0 \\ 0 & s_2 = 1 & \cdots & 0 \\ \vdots & \vdots & \ddots & \vdots \\ 0 & 0 & \cdots & s_{11} = 0 \end{bmatrix}_{[11\times 11]} \mathbf{v}(\mathbf{x}) \tag{4.2.2.11}$$

where $\boldsymbol{\lambda}_c$ and $\boldsymbol{\lambda}_v$ are Lagrange multipliers with respect to $c_i(\mathbf{x})$ and $v_j(\mathbf{x})$, respectively. Section 9.3.3.1 of ACI 318-19 (Standard, 2019) also constrains tensile rebar strain which should be at least $\varepsilon_{rt\_0.003}$ (ε_{ty} + 0.003) in order to mitigate brittle flexural behavior of RC beams. It is noted that yield strain of rebars = f_y / E_s; f_y is rebar yield strength and E_s is young modulus of rebar). Minimum tensile rebar strain can be achieved by limiting a maximum tensile rebar ratio. The minimum limit for tensile rebar strains ($\varepsilon_{rt\_0.003}$) is established as Inequality v_9 to control a maximum limit of tensile rebar ratio as shown in Table 4.2.1.2. The inequality matrix **S** becomes, then, Equation (4.2.2.12) where Inequality constraint (v_9) is activated with s_9 = 1. Lagrange function with s_9 = 1 is obtained in Equation (4.2.2.13) by substituting s_9 equivalent to 1 into Equation (4.2.2.3).

$$S=\begin{bmatrix} s_1=0 & 0 & 0 & \cdots & 0 & 0 & 0 \\ 0 & s_2=0 & 0 & \cdots & 0 & 0 & 0 \\ 0 & 0 & s_3=0 & \cdots & 0 & 0 & 0 \\ \vdots & \vdots & \vdots & \ddots & \vdots & \vdots & \vdots \\ 0 & 0 & 0 & \cdots & s_9=1 & 0 & 0 \\ 0 & 0 & 0 & \cdots & 0 & s_{10}=0 & 0 \\ 0 & 0 & 0 & \cdots & 0 & 0 & s_{11}=0 \end{bmatrix}_{11\times 11} \begin{matrix} 1^{st} \\ 2^{nd} \\ 3^{rd} \\ \vdots \\ 9^{th} \\ 10^{th} \\ 11^{th} \end{matrix} \tag{4.2.2.12}$$

$$\mathcal{L}_{CI_b}(\mathbf{x},\boldsymbol{\lambda}_c,\boldsymbol{\lambda}_v)=CI_b-\boldsymbol{\lambda}_c^T\mathbf{c}(\mathbf{x})-\boldsymbol{\lambda}_v^T$$

$$\times\begin{bmatrix} s_1=0 & 0 & \cdots & \cdots & 0 & 0 \\ 0 & s_2=0 & \cdots & \cdots & 0 & 0 \\ 0 & 0 & \ddots & \cdots & 0 & 0 \\ \vdots & \vdots & \vdots & s_9=1 & \vdots & \vdots \\ 0 & 0 & \cdots & \cdots & s_{10}=0 & 0 \\ 0 & 0 & \cdots & \cdots & 0 & s_{11}=0 \end{bmatrix}_{[11\times 11]} \mathbf{v}(\mathbf{x}) \tag{4.2.2.13}$$

where $\boldsymbol{\lambda}_c$ and $\boldsymbol{\lambda}_v$ are Lagrange multipliers with respect to $c_i(\mathbf{x})$ and $v_j(\mathbf{x})$, respectively.

2. KKT conditions with all equality and inequality constraints

An inequality constraint is said to be active (or tight, binding), otherwise, they are called inactive or slacked. Lagrange function multipliers should be greater than zero for equality and active inequality constraints while they should be zeros for inactive inequality constraints when generating Lagrange functions under KKT conditions, ignoring inactive inequality constraints.

Lagrange functions with multipliers should be tested for candidate solutions based on KKT conditions with all equality and inequality constraints. Lagrange multiplier is going to be strictly positive when a constraint is not slack but active or binding, and hence, imposing constraints is essentially influencing Lagrange function. When there is a slack (inactive) on the constraint, Lagrange multiplier is equal to zero, and hence, Lagrange function is not influenced. Inactive constraints do not cost anything, therefore, Lagrange multipliers (λ_c) are equal to zero when a constraint is a slack or inactive. However, an optimization must occur at stationary points within the ranges defined by inactive and slack constraints.

A design example is shown in Equation (4.2.2.14) where Inequalities $v_2(\mathbf{x})$ and $v_9(\mathbf{x})$ is active with all other s_i being 0.

$$\mathbf{S} = \begin{bmatrix} s_1 = 0 & 0 & \cdots & \cdots & 0 & 0 \\ 0 & s_2 = 1 & \cdots & \cdots & 0 & 0 \\ 0 & 0 & \ddots & \cdots & 0 & 0 \\ \vdots & \vdots & \vdots & s_9 = 1 & \vdots & \vdots \\ 0 & 0 & \cdots & \cdots & s_{10} = 0 & 0 \\ 0 & 0 & \cdots & \cdots & 0 & s_{11} = 0 \end{bmatrix}_{[11\times 11]} \tag{4.2.2.14}$$

Lagrange function is obtained in Equation (4.2.2.15) by substituting Equation (4.2.2.14) into Equation (4.2.2.3).

$$\mathcal{L}_{CI_b}(\mathbf{x}, \boldsymbol{\lambda}_c, \boldsymbol{\lambda}_v) = CI_b - \boldsymbol{\lambda}_c^T \mathbf{c}(\mathbf{x}) - \boldsymbol{\lambda}_v^T$$

$$\times \begin{bmatrix} s_1 = 0 & 0 & \cdots & \cdots & 0 & 0 \\ 0 & s_2 = 1 & \cdots & \cdots & 0 & 0 \\ 0 & 0 & \ddots & \cdots & 0 & 0 \\ \vdots & \vdots & \vdots & s_9 = 1 & \vdots & \vdots \\ 0 & 0 & \cdots & \cdots & s_{10} = 0 & 0 \\ 0 & 0 & \cdots & \cdots & 0 & s_{11} = 0 \end{bmatrix}_{[11\times 11]} \mathbf{v}(\mathbf{x}) \tag{4.2.2.15}$$

where $\boldsymbol{\lambda}_c$ and $\boldsymbol{\lambda}_v$ are Lagrange multipliers with respect to $c_i(\mathbf{x})$ and $v_j(\mathbf{x})$, respectively.

However, S shown in Equation (4.2.2.14) cannot take place at the same time. For example, Inequalities $v_2(\mathbf{x})$ and $v_9(\mathbf{x})$ cannot be activated simultaneously because a tensile rebar ratio cannot reach a maximum and minimum value at the same time; therefore, Equation (4.2.2.14) cannot take place. In particular as shown in Equation (4.2.2.12), design moment strength ϕM_n cannot be equivalent to 1.2 times cracking moment ($1.2M_{cr}$) at the same time when tensile rebar strains $\varepsilon_{rt\_0.003}$ corresponding to a concrete strain of 0.003 reach a limit of controlled-tension section ($\varepsilon_{ty} + 0.003$). This is because a minimum rebar ratio is obtained when design moment strength equals to 1.2 times cracking moment ($\phi M_n = 1.2M_{cr}$), whereas the maximum tensile rebar ratio is reached when tensile rebar strain is equivalent to a limit of controlled-tension section ($\varepsilon_{ty} + 0.003$). ACI 318-19 limits minimum strain in tension-controlled zone ($\varepsilon_{rt\_0.003}$) as shown in $v_9(\mathbf{x})$; $\varepsilon_{rt\_0.003} - 0.006^* \geq 0$ of Table 4.2.1.2 by limiting a maximum tensile rebar ratio whereas a minimum tensile rebar ratio is defined by $v_2(\mathbf{x})$ and $v_9(\mathbf{x})$ cannot be activated at the same time. When both Inequalities $v_2(\mathbf{x})$ and $v_9(\mathbf{x})$ are activated simultaneously, a rebar ratio should be found as both its maximum and minimum ratios simultaneously which does not happen.

Theoretically, there are 2^{11}=2048 KKT conditions for 11 inequalities shown in Table 4.2.1.2. However, many KKT conditions among 2048 KKT conditions are impossible to occur due to conflictions. For another example, two inequalities $v_3(\mathbf{x})$ and $v_4(\mathbf{x})$ in Table 4.2.1.2 cannot be activated at the same time due to conflictions; when $v_3(\mathbf{x})$ is activated, h is bound as 500 mm whereas h is bound as 1,200 mm when $v_4(\mathbf{x})$ is activated, and hence, $v_3(\mathbf{x})$ and $v_4(\mathbf{x})$ cannot be activated simultaneously.

Simple equality and inequality constraints pre-assigned as input parameters shown in Table 4.2.1.2 are substituted into initial input parameters (L, h, b, f_y, f_c', ρ_{rt}, ρ_{rc}, M_D, M_L) for the first iteration based on Newton-Raphson method, reducing the complexity of equality constraints (C) and inequality constraints (V). Newton-Raphson iteration much relies on good initial trial input parameters. Output parameters, such as compressive rebar strain at concrete strain of 0.003 $\varepsilon_{rc\_0.003}$, which is not engaged with any of objective functions, equality, and inequality constraints, not appearing in design scenario, but are obtained as a result (ϕM_n, M_{cr}, $\varepsilon_{rt\_}0.003$, $\varepsilon_{rc\_}0.003$, Δ_{imme}, Δ_{long}) of Lagrange optimization.

4.3 GENERATION OF LARGE STRUCTURAL DATASETS

4.3.1 Input and output parameters selected for large datasets

Nine input and eight output parameters shown in Table 4.2.1.1 are selected to generate large datasets to train ANNs for a design of structural systems in general and a design of doubly reinforced concrete beams, in particular. Nine input parameters include material properties (yield strength of rebar f_y (MPa) and concrete compressive strength f_c' (MPa)), beam geometry (height h (mm), and span length L (m), and moment demand due to dead and live loads (kN·m). Output parameters include design moment strength (ϕM_n), factored moment (M_u), cracking moment (M_{cr}), compressive and tensile trains of rebars ($\varepsilon_{rt\_0.003}$ and $\varepsilon_{rc\_0.003}$), immediate and long-term deflections (Δ_{imme} and Δ_{long}), and cost index (CI_b, KRW/m) for materials and manufacture of an RC beam. However, more network inputs and outputs can be adopted depending on a need of an analysis and design.

4.3.2 Random design ranges

A strain-compatibility-based algorithm (Autobeam) was developed for a design of ductile doubly reinforced concrete beams in a previous study by the author (Nguyen and Hong, 2021). Large structural datasets that are generated based on Autobeam are used to train ANN for a design. Table 4.3.2.1 presents a number of datasets, mean values, variances, standard deviation, and ranges of selected network input and output parameters which are randomly generated for a design of doubly RC beams. Random datasets for rebar yield strengths f_y and for concrete compressive strengths f_c' are generated in the range of 500–600 and 30–50 MPa, respectively, to consider a wide range of beam designs for beam lengths spanning 8–12 m. Beam height and width are randomized in the range of 400–1,500 mm and (0.25–0.8) h.

Table 4.3.2.1 List of Input Variables and Corresponding Ranges for big data generation

	L	h	b	f_c'	f_y	ρ_{rt}	ρ_{rc}	M_D	M_L
	(mm)	(mm)	(mm)	(MPa)	(MPa)			(kN·m)	(kN·m)
Number of datasets	100,000								
Maximum	12,000	1,500	1,200	600	50	0.0475	0.0237	17,407	9,681
Mean	10,003	1,012	525	482	40	0.0109	0.0032	1,276	460
Minimum	8,000	500	125	400	30	0.0025	7.1E-06	10.42	0.26
Variance (V)	1,366,389	82,804	50,865	3,175	37	5.3E-05	1.0E-05	2,153,898	502,367
Standard deviation	1169	288	226	56	6	7.2E-03	3.1E-03	1,468	709

4.3.3 Network training based on parallel training method (PTM) training

Input parameters are mapped to each output parameter separately using parallel training method (PTM), identifying the relationships of entire input parameters with each of output parameters. Less time for training is required for this method than mapping entire input parameters to all output parameters based on TED (Hong, 2020, Hong et al., 2021c), however, feature indexes appearing in a network output cannot be used as input feature indexes to predict the other outputs. In Figure 4.3.3.1, nine input parameters (L, h, b, f_y, f_c', ρ_{rt}, ρ_{rc}, M_D, M_L) of forward network is mapped to eight outputs (ϕM_n, M_u, M_{cr}, $\varepsilon_{rt\_0.003}$, $\varepsilon_{rc\_0.003}$, Δ_{imme}, Δ_{long}, CI_b) using PTM.

4.3.4 Training for forward Lagrange networks

Using PTM, nine input parameters (L, h, b, f_y, f_c', ρ_{rt}, ρ_{rc}, M_D, M_L) of forward Lagrange networks are mapped to eight outputs (ϕM_n, M_u, M_{cr}, $\varepsilon_{rt\_0.003}$, $\varepsilon_{rc\_0.003}$, Δ_{imme}, Δ_{long}, CI_b) with three layers based on three types of neurons (30, 40, and 50) as indicated as Network #1 shown in Table 4.2.2.1(a). Test mean square errors ($T.MSEs$) of immediate and long-term deflections improved when four and five layers based on three types of neurons (30, 40, and 50) are used for Network #2 shown in Table 4.2.2.1(b). In particular, testing MSE of Δ_{imme} (1.43E-4) and Δ_{long} (4.39E-5) with Network #1 decreased slightly to 1.29E-4 and 4.3E-5, respectively, when a greater number of layers is implemented for Network #2. ANNs are trained on Δ_{imme} and Δ_{long} for Network #2 based on four layers with 40 neurons and five layers with 30 neurons, respectively, as shown in Table 4.2.2.1 where the best training results for each parameter are presented. Iterations are performed to calculate stationary points and optimality conditions based on forward Lagrange network as shown in Figure 4.3.4.1.

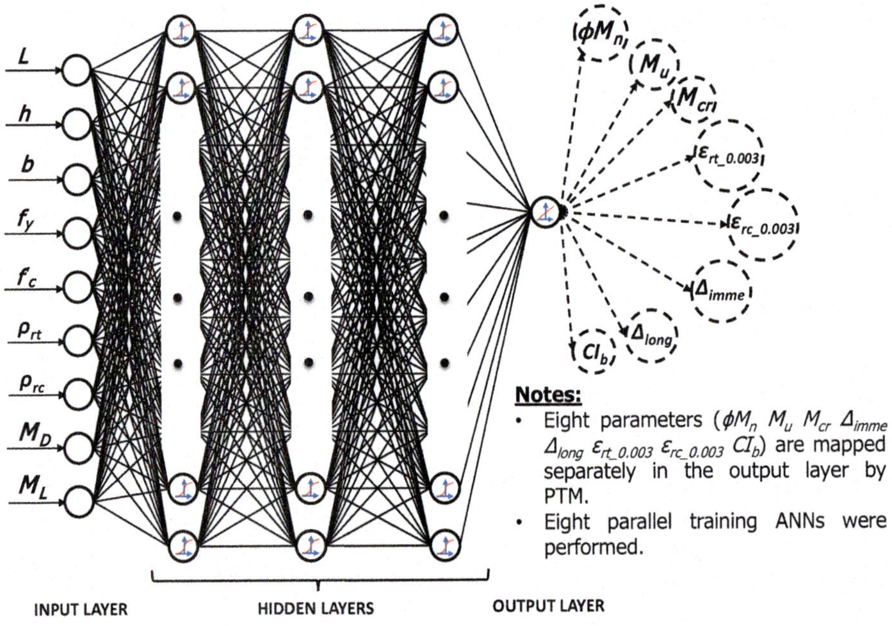

Figure 4.3.3.1 Topology of forward network using PTM.

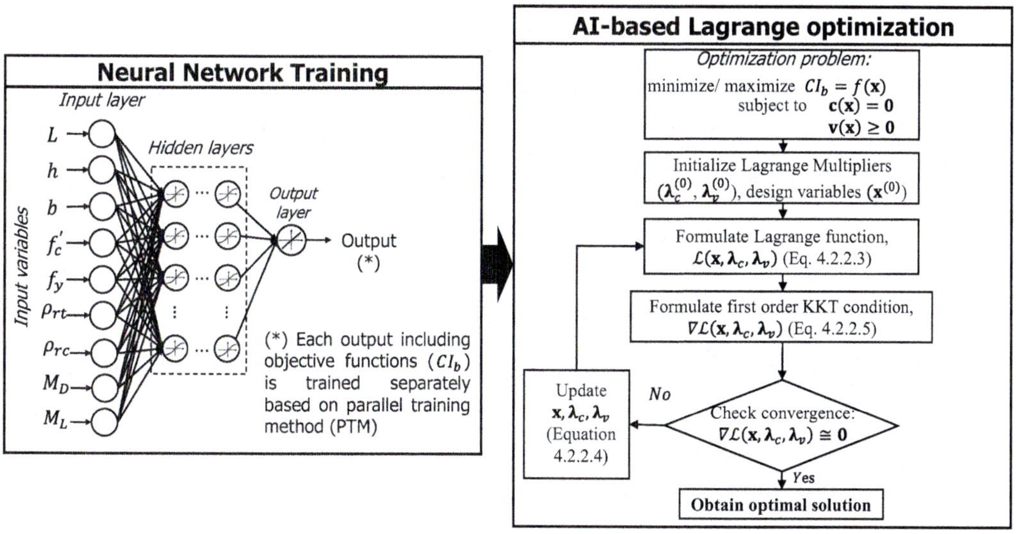

Figure 4.3.4.1 Iterations to calculate stationary points optimizing designs based on forward Lagrange network.

4.3.5 Training for rebar placements with multiple layers

This section describes rebar placements according to ACI 318-14 (Standard, 2014) and ACI 318-19 (Standard, 2019). A total area of reinforcement is calculated based on rebar ratios, in which rebar ratio ρ_r consists of compressive rebar ratio (ρ_{rc}) and tensile rebar ratio (ρ_{rt}). Horizontal spacing of rebars should satisfy a minimum and maximum reinforcement spacings, denoted as $s_{\min}$ and $s_{\max}$, specified in design codes ACI 318-14 (Standard, 2014) and ACI 318-19 (Standard, 2019). Upper and lower limitations of rebar spacings are calculated in Equations (4.3.5.1-1)–(4.3.5.2-2). It is noted that this spacing is based on center-to-center distance of rebars. When more than one rebar layer is required, ACI 318-14 (Standard, 2014) and ACI 318-19 (Standard, 2019) specify a minimum clear space of 25 mm vertically between rebar layers. There are no differences in the two versions of ACI code for rebar placements. Rebar placements in a design of doubly RC beams in this chapter consider a diameter of rebar d_b = 29 mm, clear concrete cover c = 40 mm, tensile rebar strength f_y = 600 MPa, and a diameter of aggregate d_{agg} = 20 mm as shown in Figure 4.3.5.1. Rebar spacings are calculated as follows.

Step 1: Calculation of a total number of rebars.

A total area of rebar is calculated in Equation (4.3.5.3). A number of required rebars having a diameter of d_b = 29 mm is, then, calculated using Equation (4.3.5.4) in which n_{rebars} should be an integer number. For example, when $\rho_r = \rho_{rt} = 0.05$, a total area of rebar is $A_r = \rho_r \times b \times h = 0.05 \times 400 \times 400 = 8{,}000\ (\text{mm}^2)$ if an RC beam has a section of 400×400 mm. A number of rebars should be calculated as $n_{\text{rebars}} = \dfrac{A_{\text{rebars}}}{\frac{\pi \times 29^2}{4}} = \dfrac{8{,}000}{\frac{\pi \times 29^2}{4}} = 12.1$ which becomes $n_{\text{rebars}} = 13$. It is noted that n_{rebars} should be rounded up.

Step 2: Calculation of a horizontal spacing of rebars assuming that all rebars are placed in one layer.

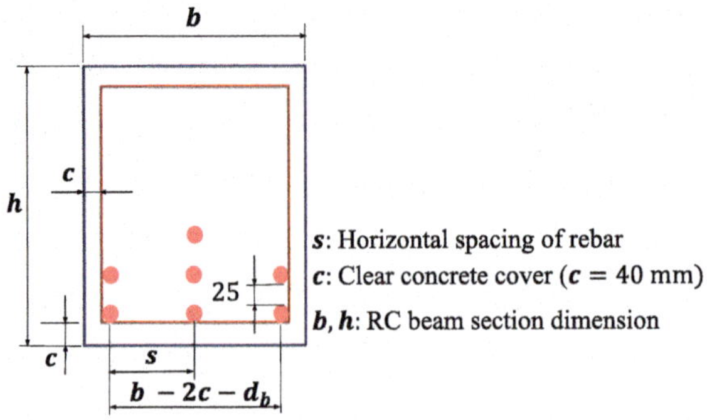

Figure 4.3.5.1 Rebar placements in a design of an RC beam.

A horizontal rebar spacing is calculated in Equation (4.3.5.5). For example, when $n_{\text{rebars}} = 13$, a horizontal spacing is calculated as $s_{\text{horizontal}} = \dfrac{400 - 2\times 40 - 29}{13-1} = 24.25$ mm.

Step 3: Check rebar spacings.

A horizontal spacing of rebars should satisfy a minimum spacing ($s_{\min}$) and a maximum spacing ($s_{\max}$) specified by design codes. If $s_{\min} \le s_{\text{horizontal}} < s_{\max}$, rebars are horizontally placed in one layer. If $s_{\text{horizontal}} < s_{\min}$, more than one layer are required when placing rebars. In this example, $s_{\text{horizontal}} = 24.25$ mm $< s_{\min} = 58$ mm so that rebars should be placed in more than one layer.

Step 4: Calculation of rebar spacings when more than one layer is required.

Step 4.1: Determination of a maximum number of rebars in one layer.

A maximum number of rebars in one layer is obtained when a horizontal rebar spacing satisfies a minimum spacing $s_{\text{horizontal}} = s_{\min}$ presented in Equation (4.3.5.5). A minimum horizontal spacing of rebars in this example is 58 mm, calculated in Equations (4.3.5.1-1) and (4.3.5.1-2). In this example, a maximum number of rebars in one layer is calculated using Equation (4.3.5.6); $n_{r,\max}^{\text{in 1 layer}} = \left(\dfrac{400 - 2\times 40 - 29}{58} + 1\right) = 6.02$ which becomes $n_{r,\max}^{\text{in 1 layer}} = 6$ (rebars).

Step 4.2: Determination a number of rebar layers.

A number of rebar layers is calculated using Equation (4.3.5.7). In this example, a number of rebar layers is calculated as $n_{\text{layers}} = \left(\dfrac{n_{\text{rebars}}}{n_{\max}^{\text{in 1 layer}}}\right) = \dfrac{13}{6} = 2.16$ which becomes $n_{\text{layers}} = 3$ layers. A number of rebar layers should be rounded up.

Step 4.3: Placement of rebars in each rebar layer.

a. Vertical spacing

This example considers a clear vertical spacing between rebar layers which is 25 mm.

b. Horizontal spacing

Except for the last layer, rebars are placed horizontally based on a minimum horizontal spacing $s_{\text{horizontal}} = s_{\min}$ and a number of rebars in each layer is equal to maximum rebars in one layer ($n_{r,\max}^{\text{in 1 layer}}$). In this example, three layers of rebars are required. Rebars in the first and second layers have a horizontal spacing $s_{\text{horizontal}} = s_{\min} = 58$ mm when there are 6 rebars ($n_{r,\max}^{\text{in 1 layer}} = 6$ rebars) in each layer. A number of rebars in the last layer ($n_{\text{rebars}}^{\text{last layer}}$) is calculated using Equation (4.3.5.8); $n_{\text{rebars}}^{\text{last layer}} = n_{\text{rebars}} - n_{r,\max}^{\text{in 1 layer}} \times (n_{\text{layers}} - 1) = 13 - 6\times 2 = 1$ (rebar).

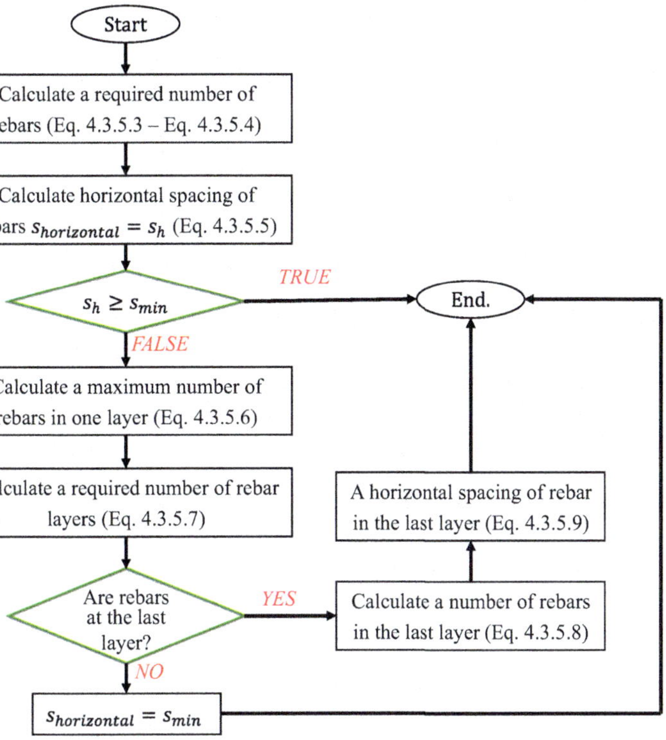

Figure 4.3.5.2 A flowchart to calculate horizontal rebar spacings in a design of an RC beam.

A horizontal spacing of rebar in the last layer is then calculated using Equation (4.3.5.9). Calculations of horizontal center-to-center rebar spacings are illustrated by a flowchart shown in Figure 4.3.5.2. The presented calculations include four steps to obtain rebar spacings based on design codes ACI 318-19 (Standard, 2019) and ACI 318-14 (Standard, 2014). When there are more than one rebar layers, a vertical clear spacing between layers is defined as 25 mm, as shown in Figure 4.3.5.1.

$$s_{\min} - d_b = \max\begin{cases} 25\text{ mm}; \\ d_b; \\ \dfrac{4}{3} d_{\text{agg}} \end{cases} \tag{4.3.5.1-1}$$

$$s_{\min} = \max\left(25 + 29;\ 29 + 29;\ \frac{4}{3} \times 20 + 29\right) = 58\text{ mm} \tag{4.3.5.1-2}$$

$$s_{max} = \min\begin{cases} 380\left(\dfrac{280}{f_s}\right) - 2.5c; \\ 300\left(\dfrac{280}{f_s}\right) \end{cases} \quad \text{where } f_s = 2/3 f_y \tag{4.3.5.2-1}$$

$$s_{\max} = \min\left(380\left(\frac{280}{\frac{2}{3}\times 600}\right) - 2.5\times 40;\ 300\left(\frac{280}{\frac{2}{3}\times 600}\right)\right) = 166 \text{ mm} \quad (4.3.5.2\text{-}2)$$

$$A_r = \rho_s \times b \times h \quad (4.3.5.3)$$

$$n_{rebars} = \frac{A_{rebars}}{\frac{\pi \times 29^2}{4}} \quad (4.3.5.4)$$

$$s_{\text{horizontal}} = \frac{b - 2c - d_b}{n_{\text{rebars}} - 1} = \frac{b - 80 - 29}{n_{\text{rebars}} - 1} \quad (4.3.5.5)$$

$$n_{r,\max}^{\text{in 1 layer}} = \left(\frac{b - 2c - d_b}{s_{\min}} + 1\right) = \left(\frac{b - 80 - 29}{58} + 1\right) \quad (4.3.5.6)$$

$$n_{\text{layers}} = \left(\frac{n_{\text{rebars}}}{n_{r,\max}^{\text{in 1 layer}}}\right) \quad (4.3.5.7)$$

$$n_{\text{rebars}}^{\text{last layer}} = n_{\text{rebars}} - n_{r,\max}^{\text{in 1 layer}} \times \left(n_{\text{layers}} - 1\right) \quad (4.3.5.8)$$

$$s_{\text{horizontal}}^{\text{last layer}} = \frac{b - 2c - d_b}{n_{\text{rebars}}^{\text{last layer}}} = \frac{b - 80 - 29}{n_{\text{rebars}}^{\text{last layer}}} \quad (4.3.5.9)$$

The flow chart shown in Figure 4.3.5.2 is reflected in *Autobeam* which is used to generate big data for training ANNs. A number of rebar layers that is determined based on ANNs is calculated in the output of the ANNs even if it is not printed.

4.4 NETWORK VERIFICATION

4.4.1 Verification of design parameters based on a forward Lagrange network

A forward Lagrange network minimizing cost (CI_b) of material and manufacture for ductile RC beam sections shown in Figure 4.2.1.1 is verified in Figure 4.4.1.1 based on ACI 318-19. ANN-based forward Lagrange optimization is formulated with eight forward output parameters (ϕM_n, M_u, M_{cr}, $\varepsilon_{rt\_0.003}$, $\varepsilon_{rc\_0.003}$, Δ_{imme}, Δ_{long}, CI_b) based on nine forward input parameters (L, h, b, f_y, f_c', ρ_{rt}, ρ_{rc}, M_D, M_L) shown in Table 4.2.1.1. As shown in Figure 4.4.1.1, factored moment (M_u), minimized cost index (CI_b), immediate deflections (Δ_{imme}), and long-term deflections (Δ_{long}) are verified by structural engineering-based software. Figure 4.4.1.1a demonstrates accuracies of Lagrange network in terms of factored moment (including factored moment due to dead and live loads) which is set to between 500 and 3,000 kN·m. Factored moment M_u is established as an equality of a forward Lagrange network. Factored moment M_u preassigned by network minimizing objective function (cost index, CI_b) is verified by structural calculation based on Autobeam software which is used to generate large datasets. Range of errors of calculated M_u based on an ANN is compared to one calculated by structural software is between −0.2% to −1.27% as shown in Figure 4.4.1.1a. Similarly, Figure 4.4.1.1b shows minimized objective function (cost index, CI_b) which is calculated on an output-side of a forward Lagrange network. Cost index (CI_b) obtained by a forward Lagrange network is also verified by that found by a structural calculation using

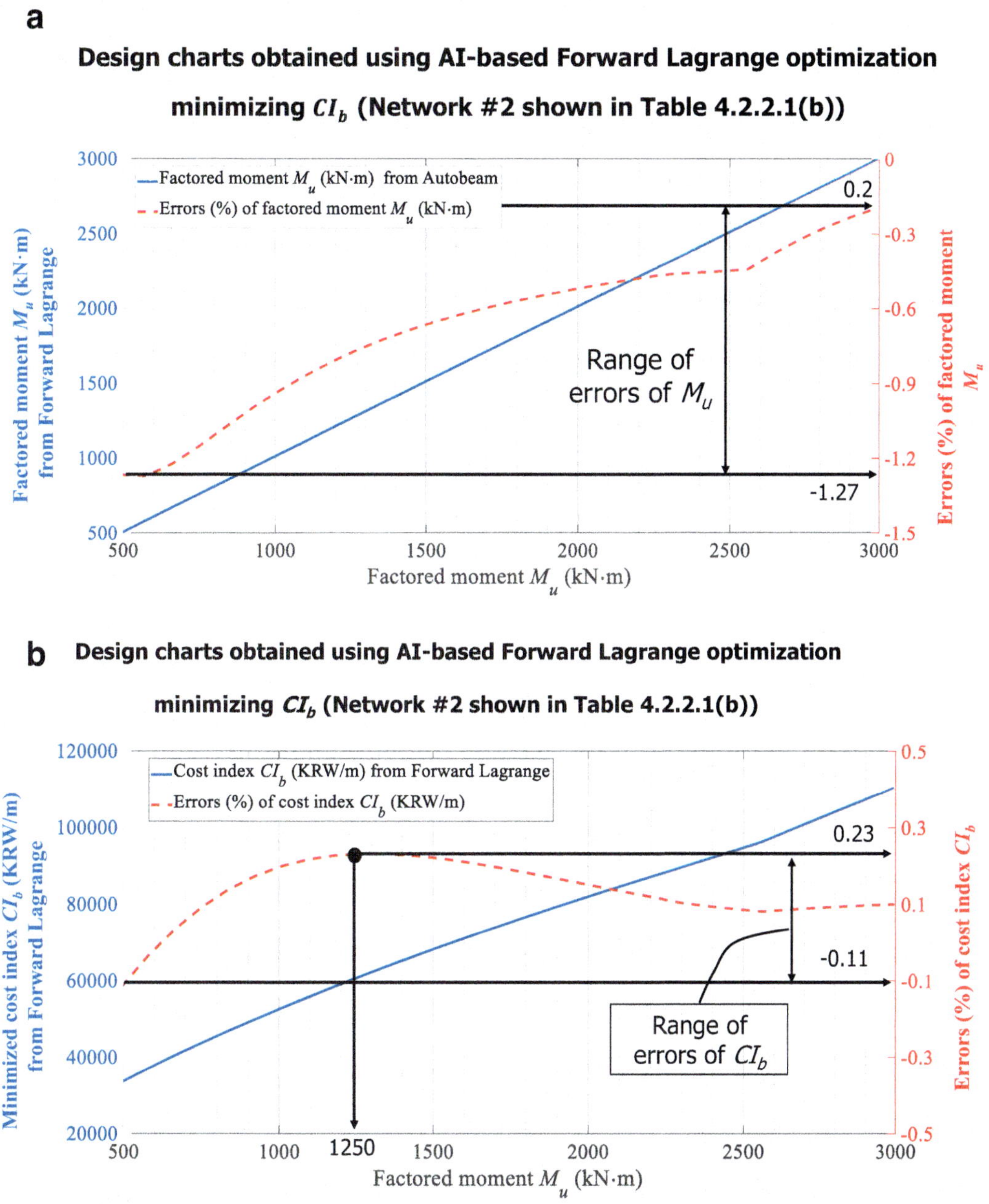

Figure 4.4.1.1 Verification of forward Lagrange network based on ACI 318-19. (a) Factored moment (M_u) (b) Minimized cost index, CI_b. (c) Immediate deflections, Δ_{imme}. (d) Long-term deflections, Δ_{long}.

(Continued)

c

Design charts obtained using AI-based Forward Lagrange optimization minimizing CI_b (Network #1 shown in Table 4.2.2.1(a))

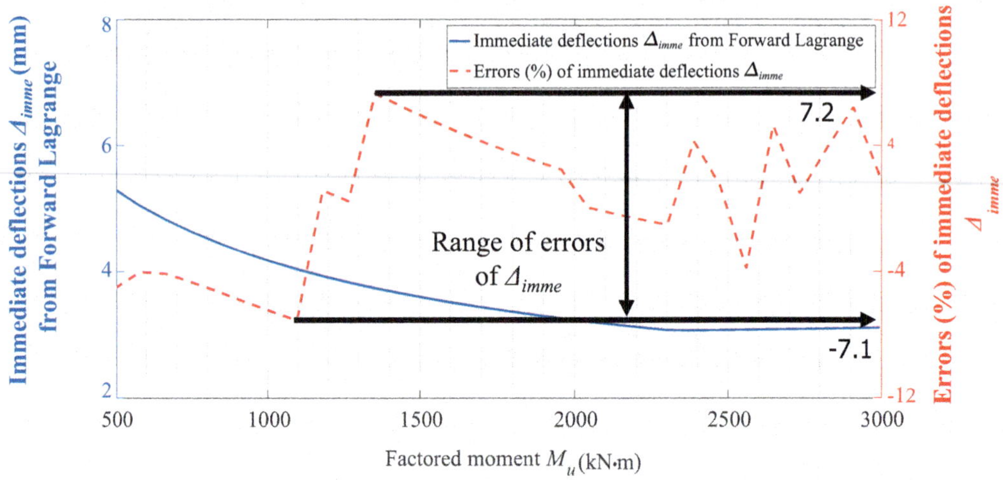

Note: $\Delta_{imme}\ limit = L/360 = 28\ mm$

Design charts obtained using AI-based Forward Lagrange optimization minimizing CI_b (Network #2 shown in Table 4.2.2.1(b))

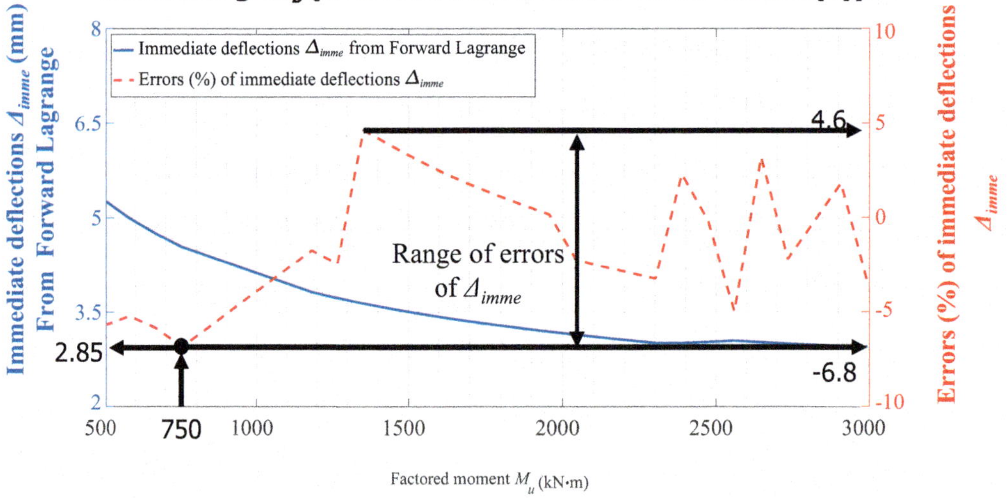

Note: $\Delta_{imme}\ limit = L/360 = 28\ mm$

Figure 4.4.1.1 (Continued).

(Continued)

d

Design charts obtained using AI-based Forward Lagrange optimization minimizing CI_b (Network #1 shown in Table 4.2.2.1(a))

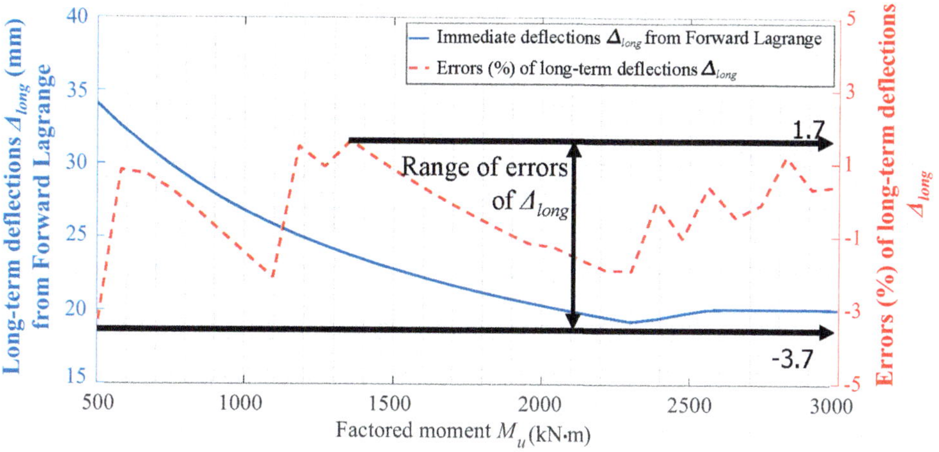

Note: $\Delta_{long}\ limit = L/240 = 42\ mm$

Design charts obtained using AI-based Forward Lagrange optimization minimizing CI_b (Network #2 shown in Table 4.2.2.1(b))

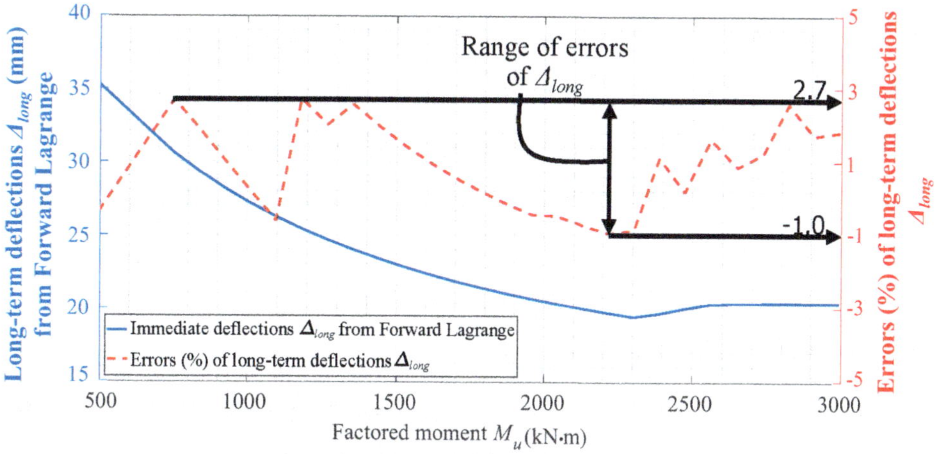

Note: $\Delta_{long}\ limit = L/240 = 42\ mm$

Figure 4.4.1.1 (Continued).

design parameters produced by the network. Maximum error of cost index (CI_b) is 0.23% when M_u is equal to 1,250 kN·m. A Range of errors of cost indexes (CI_b) is between 0.23% and −0.11% as shown in Figure 4.4.1.1b where errors are negligible for the most of cost index (CI_b). As shown in Figure 4.4.1.1c, although large errors from −6.8% to 4.6% compared with those based on structural calculations are found with immediate deflections for Network #2 shown in Table 4.2.2.1(b), an absolute difference between a network prediction and a structural engineering calculation is insignificant. For example, Network #2 predicts an immediate deflection of 2.85 mm for M_u of 750 kN·m when an error is as large as −6.8% whereas immediate deflection calculated based on structural engineering is 3.04 mm, resulting in only a 0.19 mm difference. A range of errors for immediate deflections based on Network #1 is found between −7.1% and 7.2%, which improves slightly to −6.8% to 4.6% based on Network #2 as shown in Figure 4.4.1.1c based on increased layers. On the other hand, a range of errors for long-term deflections based on Network #1 is found between −3.7% and 1.7%, which improves slightly to −1.0% to 2.7% based on Network #2 as shown in Figure 4.4.1.1d when a number of layers increases.

4.4.2 Verification of Selected parameters based on large datasets

A number of 3,000,000 datasets is generated to verify ANN-based Lagrange optimization. Large datasets are created by constraining beam length L as 10 m, compressive concrete strength f_c' as 30 MPa, yield strength of rebar f_y as 600 MPa, beam height h as from 500 to 1,200 mm, beam width b as in the range from 0.3 to 0.5 times beam height, and hence, factored moments M_u are found in the range of 500 ~ 3000 kN·m. In Figure 4.4.2.1, design moment strength ϕM_n is assumed as factored moment M_u between M_u to 1.2 M_u. Design moment strength ϕM_n is constrained to be greater than 1.2 times cracking moment ($1.2M_{cr}$). As shown in Figure 4.4.2.1, cost distributions of three million datasets constrained to L, f_c', f_b, h, b, and M_u are compared with those obtained by forward Lagrange networks which predict a lower bound of a cost index (CI_b). In Figure 4.4.2.1, objective function (cost index, CI_b) minimized using ANN-based Lagrange optimizations match well the lower bound of 112,383 random datasets obtained by constraining 3,000,000 datasets according to Table 4.4.2.1. It is noted that objective function (cost index, CI_b) is well minimized as verified by 112,383 datasets constrained from 3,000,000 datasets as shown in Figure 4.4.2.1.

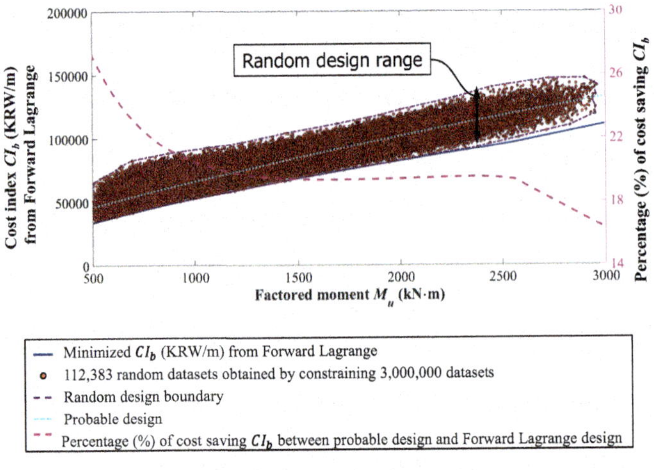

Figure 4.4.2.1 Minimized cost (CI_b) based on 112,383 datasets constrained from 3,000,000 datasets.

Table 4.4.2.1 Constrained Datasets for Verification

No.	Preassigned Constraints
1	$L = 10$ m
2	$f_c' = 30$ MPa
3	$f_y = 600$ MPa
4	$h = [500 \sim 1,200]$mm
5	$b = [0.3 \sim 0.5]h$
No.	Implemented Constraints
1	$M_u = [500 \sim 3,000]kN \cdot m$
2	$\phi M_n = [1 \sim 1.2]M_u$
3	$\phi M_n \geq 1.2M_{cr}$
4	$\varepsilon_{rt\_0.003} \geq 0.006$
5	$\Delta_{imme} \leq L/360 = 28$ mm
6	$\Delta_{long} \leq L/240 = 42$ mm

Percentage (%) of cost saving CI_b between probable design and Forward Lagrange design is also shown in Figure 4.4.2.1.

4.4.3 Cost savings based on Lagrange algorithm

Probable design shown in Figure 4.4.2.1 is established by implementing a trend line function ("polyfit" and "polyval" commands) of MATLAB (MathWorks, 2020). The "polyfit" command returns coefficients for a function that is a best fit for the 112,383 datasets sorted from 3,000,000 datasets. The function of probable design is, then, obtained by "polyval" command based on the coefficients obtained by "polyfit" command. Minimized costs (CI_b) based on ANN-based Lagrange optimizations demonstrate a significant cost saving of beam material and manufacture as shown in Figure 4.4.2.1 where comparison with probable designs is illustrated. A probable design shown in Figure 4.4.2.1 is established by assuming a safety factor of capacity from 1 to 1.2, implying that a range of probable designs for a design moment strength (ϕM_n) is also between factored moment M_u and 1.2 times M_u.

4.5 DESIGN CHARTS BASED ON ANN-BASED LAGRANGE OPTIMIZATIONS MINIMIZING CI_b

4.5.1 Optimization of the cost (CI_b) for material and manufacture for design ductile beam sections based on design charts

In Figure 4.5.1.1, ANN-based design charts for beams are presented using ACI 318-19 with eight forward output parameters (ϕM_n, M_u, M_{cr}, $\varepsilon_{rt\_0.003}$, $\varepsilon_{rc\_0.003}$, Δ_{imme}, Δ_{long}, CI_b) based on nine forward input parameters (L, h, b, f_y, f_c', ρ_{rt}, ρ_{rc}, M_D, M_L) as shown in Table 4.2.1.1. Rebar ratios (ρ_{rt} and ρ_{rc}) prescribed as inequality constraints $v7(\mathbf{x})$ and $v8(\mathbf{x})$ shown in Table 4.2.1.2 are determined to minimize cost index (CI_b), whereas material properties and geometries should be predetermined by engineers via equality constraints $c_1(\mathbf{x})$ and $c_6(\mathbf{x})$. In Figure 4.5.1.1a, cost index CI_b of an RC beam is minimized based on a forward Lagrange network. Stationary points of Lagrange functions derived in Equation (4.2.2.3) are found by solving KKT conditions under inequality constraints shown in Equation (4.2.2.5) based on Newton-Raphson iterations shown in Equation (4.2.2.4).

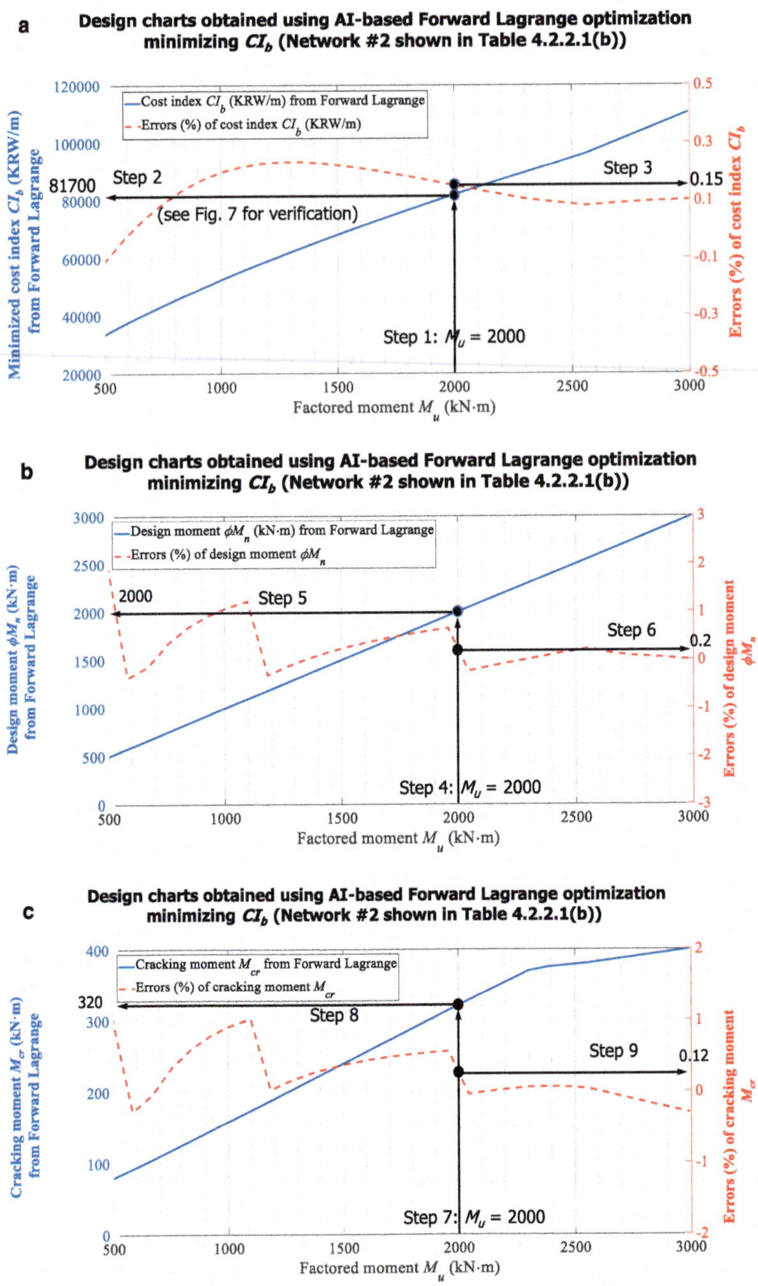

Figure 4.5.1.1 Application of design charts based on ANN-based Lagrange optimizations minimizing CI_b based on ACI 318-19. (a) Cost index (CI_b) corresponding to preassigned M_u of 2,000 kN·m. (b) Design moment ϕM_n corresponding to preassigned M_u of 2,000 kN·m. (c) Cracking moment M_{cr} corresponding to preassigned M_u of 2,000 kN·m. (d) Beam height (h) and beam width (b) corresponding to preassigned M_u of 2,000 kN·m. (e) Tensile rebar (ρ_{rt}) and compressive rebar ratios (ρ_{rc}) corresponding to preassigned M_u of 2,000 kN·m. (f) Tensile rebar strains, $\varepsilon_{rt\_0.003}$ corresponding to preassigned M_u of 2,000 kN·m. (g) Compressive rebar strains, $\varepsilon_{rc\_0.003}$ corresponding to preassigned M_u of 2,000 kN·m. (h) Immediate deflections, Δ_{imme} corresponding to preassigned M_u of 2,000 kN·m. (i) Long-term deflections, Δ_{long} corresponding to preassigned M_u of 2,000 kN·m.

(Continued)

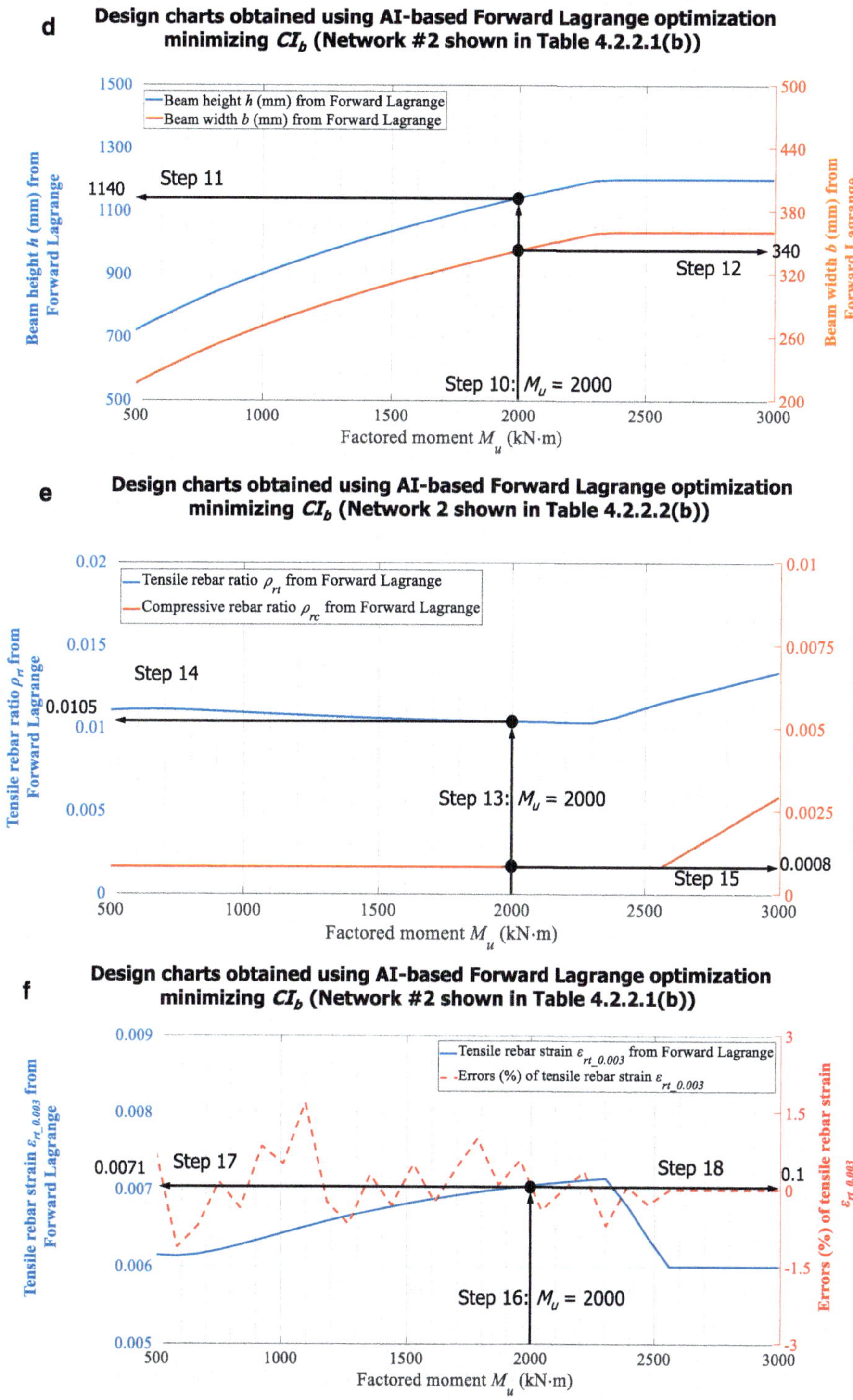

Note: ACI 318-19 requires the minimum $\varepsilon_{rt\_0.003}$ of 0.006 for sufficient ductility in case f_y = 600 MPa

Figure 4.5.1.1 (Continued).

(Continued)

g

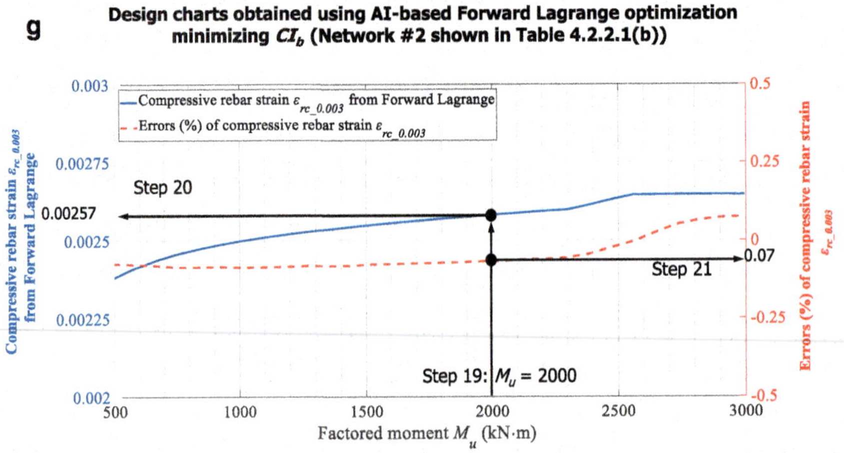

h

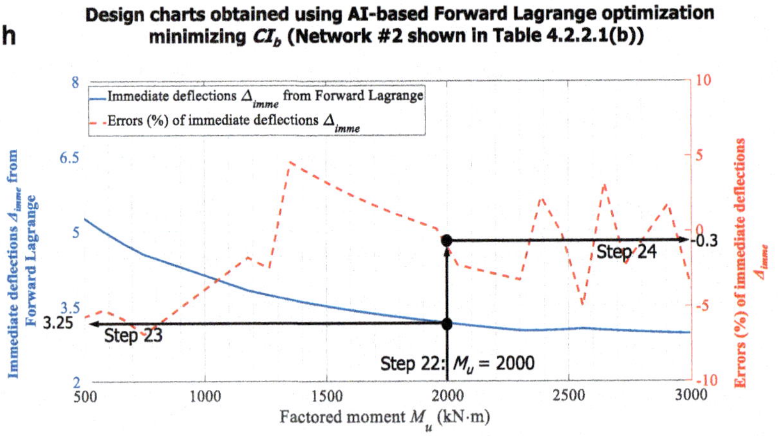

Note: Δ_{imme} $limit = L/360 = 28\ mm$

i

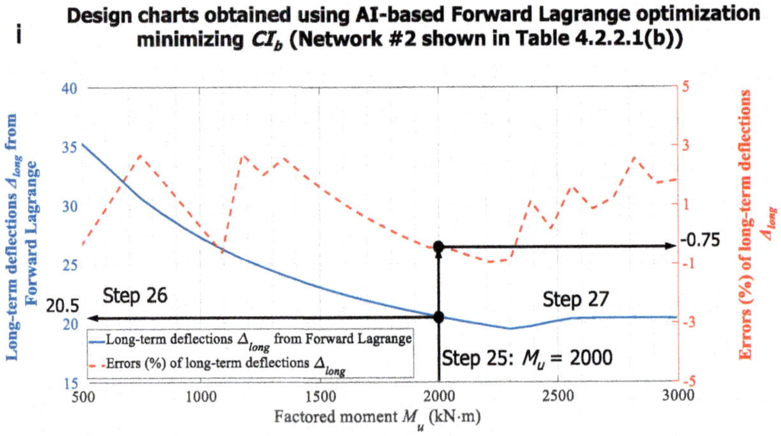

Note: Δ_{long} $limit = L/240 = 42\ mm$

Figure 4.5.1.1 (Continued).

4.5.2 Use of design charts to design ductile beam sections

Design example is shown with preassigned beam dimensions (L=10,000 mm) and beam material properties (f_y=600 MPa and f_c'=30 MPa) as shown in Figure 4.5.2.1. Load demands are also prescribed as M_D=1,000 kN·m, M_L=500 kN·m, and M_u=1.2 M_D+1.6 M_L=2,000 kN·m. Figure 4.5.1.1a shows minimized CI_b can be selected directly from design charts plotted by ANN-based Lagrange algorithm. The rest of design parameters corresponding to a specified factored moment (M_u) of 2,000 kN·m can be found using design charts shown in Figure 4.5.1.1, helping engineers select design parameters with insignificant errors. Final beam design obtained in Figure 4.5.2.1 shows beam section (width b and height h), rebar ratios (ρ_{rt} and ρ_{rc}). ANN-based locations of both tension and compression reinforcements were obtained in Section 4.5.3 to minimize beam cost (CI_b), by which a neutral axis can be calculated as shown in Figure 4.5.2.1.

For Steps 1 to 3 shown in Figure 4.5.1.1a, forward Lagrange-based optimization charts minimizing CI_b are entered with factored moment M_u. The example design shown in Figure 4.5.1.1 begins with M_u of 2,000 kN·m prior to determining beam section (width b and height h), and rebar ratios (ρ_{rt} and ρ_{rc}). A minimized CI_b of 81,700 KRW/m according to preassigned M_u of 2,000 kN·m is selected with a corresponding error of CI_b of 0.15%.

Step 4 shown in Figure 4.5.1.1b supports engineers to move to the axis for determining design moment (ϕM_n) corresponding to preassigned M_u of 2,000 kN·m. Design moment (ϕM_n) and associated error are obtained as 2,000 kN·m and 0.2%, respectively, as shown in Steps 5 and 6 for preassigned M_u of 2,000 kN·m.

Cracking moment (M_{cr}) of 320 kN·m with a corresponding error of 0.12% are obtained in Steps 7–9 of Figure 4.5.1.1c.

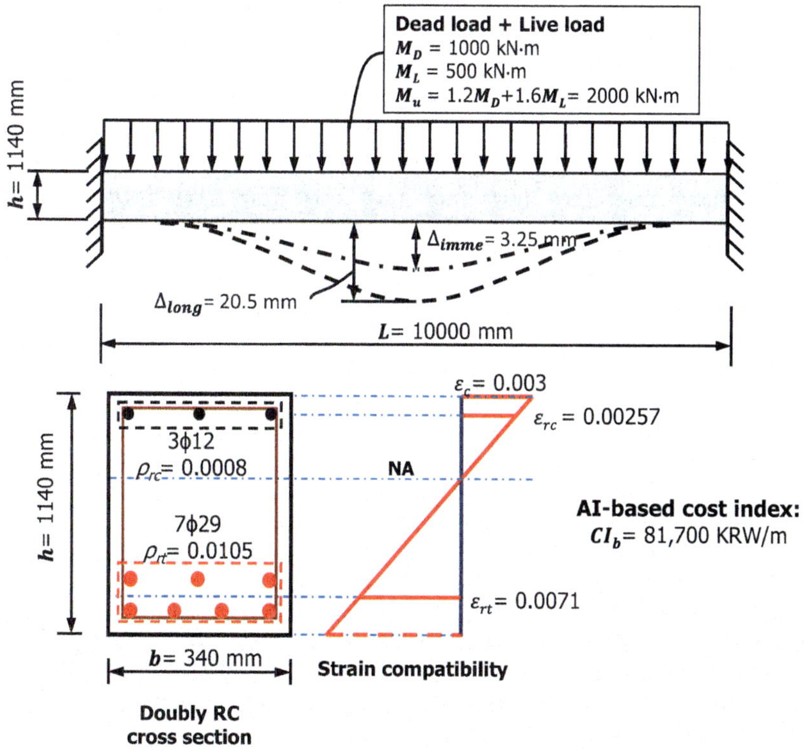

Figure 4.5.2.1 Final beam design corresponding to preassigned M_u of 2,000 kN·m.

Step 10 shown in Figure 4.5.1.1d assists engineers to move to the axes for determining beam height h and beam width b for preassigned factored moment M_u. A beam height h of 1,140 mm and a beam width b of 340 mm are selected in Steps 11 and 12, respectively.

For Steps 13–15 shown in Figure 4.5.1.1e, a compressive rebar ratio (ρ_{rc}) of 0.0008 and a tensile rebar ratio (ρ_{rt}) of 0.0105 are determined, similarly.

A tensile rebar strain ($\varepsilon_{rt\_0.003}$) of 0.0071 is shown in Steps 16–18 of Figure 4.5.1.1f and a compressive rebar strain ($\varepsilon_{rc\_0.003}$) of 0.00257 is shown in Steps 19–21 of Figure 4.5.1.1g for a specified factored moment (M_u) of 2,000 kN·m are identified with corresponding errors where acceptable accuracies are demonstrated as compared with structural calculations. Design charts shown in Figure 4.5.1.1f and g are conveniently adopted in selecting both tensile and compressive rebar strains.

ACI 318-19 (Standard, 2019) recommends that tensile rebar ratios decrease to increase rebar strains when tensile strains of rebars ($\varepsilon_{rt\_0.003}$) are less than a limit of tension-controlled sections which is $\varepsilon_{ty}\left(\varepsilon_{ty}=\frac{f_y}{E_s}=\frac{600}{200{,}000}=0.003\right)+0.003=0.006$ when rebar yield strength f_y is 600 MPa. E_s is young modulus of a rebar. Design requirements imposed by codes (including ACI 318-19, Section 9.3.3.1 (Standard, 2019)) are reflected by inequality constraints as shown in the design charts shown in Figure 4.5.1.1f.

Design parameters meeting this requirement can be selected based on tensile strains of rebars ($\varepsilon_{rt\_0.003}$) greater than 0.006 for $f_y = 600\ MPa$ as shown in Figure 4.5.1.1f with equality constraints including L =10,000 mm, f_y=600 MPa and f_c'=30 MPa. Figure 4.5.1.1f also demonstrates how design charts are used to design tension-controlled ductile beam sections with strains greater than 0.006.

In Figure 4.5.1.1d, beam height reaches its maximum value ($h \le 1{,}200$ mm shown as constrained by Inequality v_4 of Tables 4.2.1.2 or 4.6.1.1(b)) around factored moment (M_u) greater than 2,300 kN·m, and hence, tensile rebars increase rapidly to meet and maintain factored moment (M_u) of 2,300 kN·m as shown in Figure 4.5.1.1e because beam height (h) cannot increase anymore, leading to tensile rebar strains which decrease suddenly in that region as shown in Figure 4.5.1.1f. However, tensile strains of rebars ($\varepsilon_{rt\_0.003}$) do not drop below 0.006 around 2,600 kN·m to design a ductile beam.

Steps 22–27 shown in Figure 4.5.1.1h and i let engineers find an immediate deflection Δ_{imme} of 3.25 mm with corresponding errors of −0.3% shown in Steps 22–24 and a long-term deflection Δ_{long} of 20.5 mm with corresponding errors of −0.75% shown in Steps 25–27 when factored moment (M_u) reaches 2,000 kN·m.

Final beam design corresponding to preassigned factored moment (M_u) of 2,000 kN·m with optimized CI_b is illustrated in Figure 4.5.2.1 which demonstrates an optimized rebars at top (3ϕ12=0.0008) and bottom (7ϕ12=0.0105) of the section with width b=340 mm and depth h=1,140 mm. Seven rebars of diameter=29 mm with two layers are designed via ANNs as per ACI 318-19 requirement as explained in Tables 4.2.1.2 or 4.6.1.1(b).

Neutral axis can be also obtained from strains at top and bottom of the section. Both immediate and long-term deflections calculated from Figures 4.5.1.1h and i based on inequality constraints (v_{10} and v_{11}) shown in Tables 4.2.1.2 or 4.6.1.1(b) are also demonstrated in Figure 4.5.2.1.

Engineers can pre-assign any parameter on an input side as a constraining condition of interests. Notably, this type of technique can be implemented in controlling a behavior of ductility of frames. For example, engineers can pre-determine the performance level of frames by pre-determining ductility (in terms of deformations or strains) for seismic design using the proposed method. Design charts similar to Figures 4.4.1.1 and 4.5.1.1 can be constructed based on any parameter which is pre-assigned and trained on an input side as shown in Table 4.2.2.1. Design sequence of parameters is not important in an ANN-based design because an ANN-based

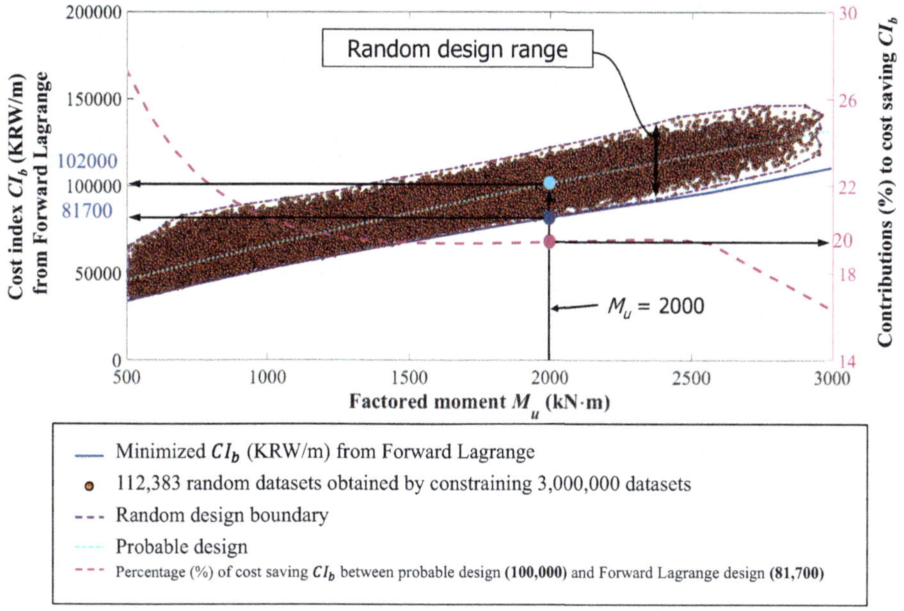

Figure 4.5.2.2 Verification of beam cost (CI_b) for M_u=2,000 kN·m.

design does not depend on structural mechanics but on big datasets that can be collected rather widely. This example demonstrates how AI-based Data-Centric Engineering can contribute to future structural engineering to optimize the interests of design targets. Well-balanced designs are now possible based on the proposed design technologies using ANN-based Lagrange algorithm while optimizing an objective function of any type. It has been challenging to perform beam designs shown in this example based on conventional beam design.

4.5.3 Verification of optimization

Design tables and accuracies of the final design are obtained based on the proposed method as shown in Figure 4.5.2.1 and Table 4.5.3.1 where design errors are negligible for practical applications. An error of 0.15% associated with a Lagrange-based cost (CI_b=81,700 KRW/m) is demonstrated for a beam with factored moment (M_u) of 2,000 kN·m whereas all other errors related with design parameters are not noticeable. As shown in Figure 4.5.2.2 obtained for factored moment (M_u) of 2,000 kN·m, a beam cost CI_b obtained based on the proposed method is well compared with that based on 112,383 observations obtained by constraining 3,000,000 large datasets using Table 4.4.2.1, eliciting that Lagrange functions are well optimized when designing ductile RC beams. As shown in Figure 4.5.2.2, a cost index of a beam CI_b of 81,700 KRW/m based on Lagrange network reduced by 16 – 27% of probable design values (102,000 KRW/m). A Lagrange-based cost CI_b=81,700 KRW/m of a beam for factored moment (M_u) of 2,000 kN·m is compared to the beam cost CI_b=102,000 KRW/m obtained based on a probable design as shown in Figure 4.5.2.2 which shows that contributions (%) to the cost savings increase as design beam capacities increase. Design accuracies are demonstrated as shown in Table 4.5.3.1 where design errors are negligible for practical applications. It is challenging to present design parameters which optimize ductile RC beams by conventional methods. Whereas the ANN-based design shown in Figure 4.5.2.1 is also verified by big datasets.

Table 4.5.3.1 Design Table Corresponding to Preassigned M_u of 2,000 kN·m

AI-based Forward Lagrange Optimization According to Minimized CI_b Nine Inputs (L, h, b, f_y, f_c', ρ_{rt}, ρ_{rc}, M_D, M_L) - Eight Outputs (ϕM_n, M_u, M_{cr}, $\varepsilon_{rt\_0.003}$, $\varepsilon_{rc\_0.003}$, Δ_{imme}, Δ_{long}, CI_b)			
No.	*Parameter*	*Forward Lagrange (PTM)*	*Error (%)*
1	L (mm)	10,000	-
2	h (mm)	1,140	-
3	b (mm)	340	-
4	f_y (MPa)	600	-
5	f_c' (MPa)	30	-
6	ρ_{rt}	0.0105	-
7	ρ_{rc}	0.00080	-
8	M_D (kN·m)	1,000	-
9	M_L (kN·m)	500	-
10	ϕM_n (kN·m)	2,000.0	0.2%
11	M_u (kN·m)	2,000.0	−0.5%
12	M_{cr} (kN·m)	320.0	0.12%
13	$\varepsilon_{rt\_0.003}$	0.0071	0.1%
14	$\varepsilon_{rc\_0.003}$	0.00257	0.07%
15	Δ_{imme} (mm)	3.25	−0.3%
16	Δ_{long} (mm)	20.5	−0.75%
17	CI_b (KRW/m)	81,700	0.15%

Note: Forward inputs parameters: No. 1 ~ 9. Forward outputs parameters: No. 10 ~ 17.

4.6 USE OF ANN-BASED LAGRANGE NETWORKS TO INVESTIGATE CHANGES BETWEEN ACI 318-14 AND ACI 318-19

4.6.1 ACI 318-19

4.6.1.1 Revised limit of tension-controlled sections

A limit of tension-controlled sections is defined when a tensile rebar strain is greater than 0.005 for all rebar yield strengths (f_y) in accordance with ACI 318-14 (Standard, 2014) whereas a limit of tension-controlled sections is defined in Figure R21.2.2b in ACI 318-19 (Standard, 2019) when a tensile rebar strain reaches $\varepsilon_{ty} + 0.003$ based on $\varepsilon_{ty} = f_y / E_s \cdot f_y$ and E_s are yield strength and young modulus of a rebar, respectively. Inequality υ_9 for ACI 318-14 is presented in Table 4.6.1.1(a) where a tension-controlled section is defined when a tensile rebar strain is greater than 0.005 for all rebar yield strengths (f_y). However, for ACI 318-19, a tension-controlled section is defined when a tensile rebar strain is $\varepsilon_{ty} + 0.003$ which is equal to 0.006 when $f_y = 600$ MPa and E_s=200,000 MPa as shown with Inequality υ_9 of Table 4.6.1.1(b) as shown in R21.2.2b in ACI 318-19 (Standard, 2019). ACI 318-19 increases a minimum rebar strain for a rebar yield strength (f_y) greater than 400 MPa. Inequality constraints provided by both design codes are adopted in KKT conditions during Lagrange optimization when plotting the design charts shown in Figures 4.4.1.1 and 4.5.1.1.

4.6.1.2 Reduction in effective moment of inertia for ACI 318-19

During construction periods, a volume of concrete sections is restrained by supporters or by different parts of a component itself, and hence, unexpected tensile, compressive, or flexural

Table 4.6.1.1 Inequality Condition Compared between ACI 318-14 and ACI 318-19

(a):ACI 318-14

Equality Constraints	*Inequality Constraints*
$c_1(\mathbf{x}): L = 10{,}000$ mm	$v_1(\mathbf{x}): \phi M_n - M_u \geq 0$
$c_2(\mathbf{x}): f_c' = 30$ MPa	$v_2(\mathbf{x}): \phi M_n - 1.2M_{cr} \geq 0$
$c_3(\mathbf{x}): f_y = 600$ MPa	$v_3(\mathbf{x}): h - 500 \geq 0$
$c_4(\mathbf{x}): M_D = [250 \sim 1{,}500]$ kN·m	$v_4(\mathbf{x}): -h + 1{,}200 \geq 0$
$c_5(\mathbf{x}): M_L = [125 \sim 750]$ kN·m	$v_5(x): b - 0.3h \geq 0$
$c_6(\mathbf{x}): M_u = 1.2M_D + 1.6M_L = [500 \sim 3{,}000 \text{ kN·m}]$	$v_6(\mathbf{x}): -b + 0.8h \geq 0$
	$v_7(\mathbf{x}): -\rho_{rc} + \rho_{rt}/2 \geq 0$
	$v_8(\mathbf{x}): \rho_{rc} - \rho_{rt}/400 \geq 0$
	$v_9(\mathbf{x}): \varepsilon_{rt\_0.003} - 0.005 \geq 0$
	$v_{10}(\mathbf{x}): -\Delta_{\text{imme}} + L/360 \geq 0$
	$v_{11}(\mathbf{x}): -\Delta_{\text{long}} + L/240 \geq 0$

(b):ACI 318-19

Equality Constraints	*Inequality Constraints*
$c_1(\mathbf{x}): L = 10{,}000$ mm	$v_1(\mathbf{x}): \phi M_n - M_u \geq 0$
$c_2(\mathbf{x}): f_c' = 30$ MPa	$v_2(\mathbf{x}): \phi M_n - 1.2M_{cr} \geq 0$
$c_3(\mathbf{x}): f_y = 600$ MPa	$v_3(\mathbf{x}): h - 500 \geq 0$
$c_4(\mathbf{x}): M_D = [250 \sim 1{,}500]$ kN·m	$v_4(\mathbf{x}): -h + 1{,}200 \geq 0$
$c_5(\mathbf{x}): M_L = [125 \sim 750]$ kN·m	$v_5(x): b - 0.3h \geq 0$
$c_6(\mathbf{x}): M_u = 1.2M_D + 1.6M_L = [500 \sim 3{,}000 \text{ kN·m}]$	$v_6(\mathbf{x}): -b + 0.8h \geq 0$
	$v_7(\mathbf{x}): -\rho_{rc} + \rho_{rt}/2 \geq 0$
	$v_8(\mathbf{x}): \rho_{rc} - \rho_{rt}/400 \geq 0$
	$v_9(\mathbf{x}): \varepsilon_{rt\_0.003} - 0.006^{a} \geq 0$
	$v_{10}(\mathbf{x}): -\Delta_{\text{imme}} + L/360 \geq 0$
	$v_{11}(\mathbf{x}): -\Delta_{\text{long}} + L/240 \geq 0$

[a] A limit of tension-controlled sections is $\varepsilon_{ty}\left(\frac{f_y}{E_s} = \frac{600}{200{,}000} = 0.003\right) + 0.003$; ε_{ty} is rebar yield strain, f_y is rebar yield strength, and E_s is young modulus of rebar.

stresses can occur due to these restraints, causing crack and reducing stiffnesses of elements (Scanlon and Bischoff, 2008). For example, large structural components such as slabs or mat foundations are often constructed part by part, and hence, the adjacent parts constructed earlier have already developed a certain strength to constrain a behavior of structures which is placed later. Structures constructed earlier can constrain shrinkages or expansions of newly poured concrete, causing stresses and cracks. A calculation of effective moment of inertia I_e according to ACI 318-14 is presented in Table 4.6.1.2(a) where M_{cr} is cracking moment, M_a is service moment, I_{cr} is moment of inertia of cracked section transformed to concrete, and I_g is moment of inertia of gross section about centroidal axis neglecting reinforcement. The effective moment of inertia I_e is used to calculate deflections (Δ_{imme} and Δ_{long}) due to service-level gravity loads when designing beam sections for serviceability requirements.

Table 4.6.1.2 Effective moment of inertia Compared between ACI 318-14 and ACI 318-19

(a): Effective Moment of Inertia I_e According to ACI 318-14

Service Moment	*Effective Moment of Inertia, I_e, mm⁴*
$M_a \le M_{cr}$	I_g
$M_a > M_{cr}$	$I_e = \left(\dfrac{M_{cr}}{M_a}\right)^3 I_g + \left[1 - \left(\dfrac{M_{cr}}{M_a}\right)^3\right] I_{cr}$

(b): Effective Moment of Inertia I_e for Non-Prestressed Members According to ACI 318-19

Service Moment	*Effective Moment of Inertia, I_e, mm⁴*
$M_a \le (2/3)M_{cr}$	I_g
$M_a > (2/3)M_{cr}$	$\dfrac{I_{cr}}{1 - \left(\dfrac{(2/3)M_{cr}}{M_a}\right)^2 \left(1 - \dfrac{I_{cr}}{I_g}\right)}$

ACI 318-19 updates this calculation as shown in Table 4.6.1.2(b) because the previous one (ACI 318-14) does not consider restraint so that deflections for beam sections with low reinforcement ratios have been subsequently underestimated according to Section 24.2.3.5 of ACI 318-19. ACI 318-19 revises the cracking moment M_{cr} which is multiplied by two-thirds to consider restraint during construction as well as to account for a reduced tensile strength. An effective cracking moment are reduced to $2/3$ while accounting for a reduced tensile strength of concrete during construction. A cracking resulting in this degradation of concrete tensile strength can increase deflections due to service loads (Reported by ACI Committee 207, n.d.). Updated equations to calculate effective moment of inertia I_e according to ACI 318-19 is presented in Table 4.6.1.2(b).

4.6.2 The Comparisons between ACI 318-14 and ACI 318-19 Based on Conventional Structural Calculations

Changes of the two codes between ACI 318-14 and ACI 318-19 are investigated based on a conventional structural calculation, such as Autobeam. Forward input parameters shown in a Table 4.6.2.1 are implemented in calculating ϕM_n, M_u, M_{cr}, $\varepsilon_{rt\_0.003}$, $\varepsilon_{rc\_0.003}$, Δ_{imme}, Δ_{long}, and CI_b following ACI 318-14 and ACI 318-19. There is no noticeable difference for ϕM_n, M_u, M_{cr}, $\varepsilon_{rt\_0.003}$, $\varepsilon_{rc\_0.003}$, and CI_b, whereas immediate and long-term deflections display a slight difference between ACI 318-14 and ACI 318-19. This is because ACI 318-19 updates an equation to calculate an effective moment inertia as presented in Table 4.6.2.1 4.6.1.2. Immediate deflections are similar based on the two codes as shown with 2.47 and 2.39 mm for ACI 318-14 and 318-19, respectively, however, a long-term deflection obtained based on ACI 318-14 is 7.47% smaller than one based on ACI 318-19, implying that long-term deflections can be underestimated when being calculated based on ACI 318-14.

Table 4.6.2.1 Changes of Codes between ACI 318-14 and ACI 318-19 by Conventional Structural Calculation

Conventional Calculation Based on Autobeam				
		Autobeam		
No.	Parameter	ACI 318-14	ACI 318-19	Difference (%)
1	L (mm)	10,000		-
2	h (mm)	1,200		-
3	b (mm)	400		-
4	f_y (MPa)	600		-
5	f_c' (MPa)	30		-
6	ρ_{rt}	0.01		-
7	ρ_{rc}	0.001		-
8	M_D (kN·m)	1,000		-
9	M_L (kN·m)	500		-
10	ϕM_n (kN·m)	2,521.16	2,521.16	0.00%
11	M_u (kN·m)	2,000	2,000	0.00%
12	M_{cr} (kN·m)	413.38	413.38	0.00%
13	$\varepsilon_{rt\_0.003}$	0.0078	0.0078	0.00%
14	$\varepsilon_{rc\_0.003}$	0.0026	0.0026	0.00%
15	Δ_{imme} (mm)	2.47	2.39	3.24%
16	Δ_{long} (mm)	13.14	14.12	–7.47%
17	CI_b (KRW/m)	98,649	98,649	0.00%

Note: Forward inputs parameters: No. 1 ~ 9. Forward outputs parameters: No. 10 ~ 17.

4.6.3 Changes of Optimized Results between ACI 318-14 and ACI 318-19 Using ANNs

In Table 4.6.3.1, a minimized cost (CI_b) based on the proposed method between ACI 318-14 and ACI 318-19 is compared. Differences of beam height h and beam width b caused by the two codes between ACI 318-14 and ACI 318-19 affect generating large datasets and training results. An optimized cost (CI_b) based on ANN-based Forward Lagrange network between ACI 318-14 (81,500 KRW/m) and ACI 318-19 (81,700 KRW/m) is similar, showing only –0.25% difference even if a large error of –16.39% for a tensile strain ($\varepsilon_{rt\_0.003}$) is observed. However, tensile strains of 0.0061 and 0.0071 for ACI 318-14 and ACI 318-19, respectively, would not present a significant design issue between the two codes. ANN-based investigation shows how design codes evolve in terms of not only a structural behavior, but economy involved with the two designs, which helps code officials understand an influence of the new changes on both behaviors and economy of the structures prior to amendments of codes.

4.7 RESULTS AND DISCUSSIONS

4.7.1 ANN-based formulation of objective functions

1. ANN-based objective function for a cost as a function of nine forward input parameters is formulated to replace complex explicit objective functions. ANN-based networks map network input to output parameters, to obtain weight and bias matrices for a derivation of objective functions constrained by equality and inequality functions.

Table 4.6.3.1 Optimized cost (CI_b) between ACI 318-14 and ACI 318-19

Minimized CI_b Based on AI-based Forward Lagrange Optimization				
		Forward Lagrange (PTM)		
No.	*Parameter*	ACI 318-14	ACI 318-19	*Difference (%)*
1	L (mm)	10,000	10,000	0.00%
2	h (mm)	1110	1140	−2.70%
3	b (mm)	330	340	−3.03%
4	f_y (MPa)	600	600	0.00%
5	f_c' (MPa)	30	30	0.00%
6	ρ_{rt}	0.011	0.0105	4.55%
7	ρ_{rc}	0.0008	0.0008	0.00%
8	M_D (kN·m)	1,000	1,000	0.00%
9	M_L (kN·m)	500	500	0.00%
10	ϕM_n (kN·m)	2,000	2,000	0.00%
11	M_u (kN·m)	2,000	2,000	0.00%
12	M_{cr} (kN·m)	300	320	−6.67%
13	$\varepsilon_{rt\_0.003}$	0.0061	0.0071	−16.39%
14	$\varepsilon_{rc\_0.003}$	0.0026	0.00257	1.15%
15	Δ_{imme} (mm)	3.5	3.25	7.14%
16	Δ_{long} (mm)	21	20.5	2.38%
17	CI_b (KRW/m)	81,500	81,700	−0.25%

Note: Forward inputs parameters: No. 1~9. Forward outputs parameters: No. 10~17.

2. Obtaining analytical objective functions describing structural behavior as a function of multiple design parameters, such as nine forward input parameters ($L, h, b, f_y, f_c', \rho_{rt}, \rho_{rc}, M_D, M_L$) for doubly reinforced concrete beams, is difficult which limits an application of Lagrange optimization.
3. The study of this chapter develops ANN-based forward Lagrange multipliers optimizing ductile doubly reinforced concrete RC beams. An objective function representing cost (CI_b) of materials and manufacture established as one the most favorable design targets is minimized based on six equality and 11 inequality constraints.

4.7.2 Design charts obtained based on Lagrange networks optimizing cost (material and manufacture) of ductile doubly reinforced concrete beams

1. ANN-based design charts for optimizing cost (CI_b) of reinforced concrete beams shown in Figure 4.5.1.1 are developed with eight forward output parameters (ϕM_n, M_u, M_{cr}, $\varepsilon_{rt\_0.003}$, $\varepsilon_{rc\_0.003}$, Δ_{imme}, Δ_{long}, CI_b) based on nine forward input parameters ($L, h, b, f_y, f_c', \rho_{rt}, \rho_{rc}, M_D, M_L$) shown in Table 4.2.1.1.
2. Sequence of selecting parameters including material properties and geometries for a use of design charts are arbitrarily determined by users. A design target represented by an objective function of a beam cost is minimized by equality and inequality conditions which are imposed by design conditions and requirement.

3. Users can impose design conditions and requirement using equality and inequality constraints when constructing design charts minimizing any type of object functions.

4.7.3 Verifying optimized objective functions

1. Proposed neural networks based on Lagrange optimization offer a design for ductile doubly reinforced beams that make cost of beam material and manufacture least. As shown in Figure 4.5.1.1, the most favorable designs for safety and economy are derived based on ANN-based Lagrange optimization, while being verified by large datasets.
2. Predictions of network outputs are verified closely to those calculated by engineering software as shown in Table 4.5.3.1. Close correlations between network outputs and design results based on structural calculations are demonstrated for given beam properties and geometries.
3. ANN-based generalized functions of any type of objective functions such as CO_2 emissions of RC beams or beam weights can be derived by ANNs similar to one developed for beam cost. Minimizing CO_2 emissions of RC beams will contribute to solving for the needs that industries face at the time of climate changes. Optimizations of diverse design problems such as columns, slabs, foundations, and pre-stressed beams are complex and challenging to solve by optimization techniques using analytical Lagrange functions.

4.7.4 ANN-based structural designs beyond human efficiency

1. Generation of large datasets with good quality
   All design information contained in big datasets is delivered to ANNs, encouraging engineers to prepare codes for generating big datasets of good quality. ANNs remember entire data trend including correlations, cross parametric behaviors through recognizing datasets and recalls them for designs beyond human efficiency.
2. This chapter proposes a use of ANNs to derive objective functions and other output parameters as one universal function based on big datasets generated for concrete beams Lagrange function will, then, have to be linearized in terms of Jacobian and Hessian matrices to find stationary points of Lagrange functions.
3. Design efficiencies
   In this chapter, it is shown that ANNs contribute to fast and accurate designs, offering structural designs which are more effective than those designed by human engineers. ANNs replaced explicit Lagrange functions and constraining conditions to automatically determine the most efficient beam designs leading to the cost of beams lower than one achieved by human engineers. The ANN-based Lagrange algorithms laid stepping stone for next step in structural analysis and design research.

4.8 CONCLUSIONS

An ANN-based Lagrange algorithm can provide many creative designs, such as saving endeavor and time for engineers while acquiring design solutions beyond what they intend. The proposed design can replace conventional design methods, eventually making structural designs free from human efforts, judgments, and errors. This chapter demonstrates that analytical Lagrange functions are replaced by those based on ANNs. ANN-based forward Lagrange multipliers are also formulated to optimize ductile doubly reinforced concrete (RC) beams. An ANN-based objective function is derived to minimize Cost (CI_b) of materials and manufacture based on constraints in accordance with both design codes and engineer's needs. Large datasets of 100,000 are generated to formulate an ANN-based objective function for cost CI_b, as a function of forward input parameters, enabling to replace complex analytical objective functions. Structural mechanics-based conventional computations are used to ascertain the network's results. The proposed method can deliver both fast and accurate initial designs to engineering practice. Cost reductions obtained by Lagrange optimization from 18% to 26% are observed compared with probable beam designs obtained from 112,383 observations from 3,000,000 datasets. This chapter proposes design charts obtained while minimizing CI_b based on ANN-based Lagrange algorithm.

REFERENCES

An ACI Standard. (2014). *Building Code Requirements for Structural Concrete (ACI 318-14)*. American Concrete Institute, Farmington Hills, Michigan, USA.

An ACI Standard. (2019). *Building Code Requirements for Structural Concrete (ACI 318–19)*. American Concrete Institute, Farmington Hills, Michigan, USA.

Adeli, H. (2001). Neural networks in civil engineering: 1989–2000. *Computer-Aided Civil and Infrastructure Engineering*, *16*(2), 126–142. https://doi.org/10.1111/0885-9507.00219.

Aghaee, K., Yazdi, M. A., & Tsavdaridis, K. D. (2015). Investigation into the mechanical properties of structural lightweight concrete reinforced with waste steel wires. *Magazine of Concrete Research*, *67*(4), 197–205. https://doi.org/10.1680/macr.14.00232.

Ahmadi, N, Moghadas, R, & Lavaei, A. (2008). Dynamic analysis of structures using neural networks. *American Journal of Applied Sciences*, *5*(9), 1251–1256. https://doi.org/10.3844/ajassp.2008.1251.1256.

Awal, A. A., Shehu, I. A., & Ismail, M. (2015). Effect of cooling regime on the residual performance of high-volume palm oil fuel ash concrete exposed to high temperatures. *Construction and Building Materials*, *98*, 875–883. https://doi.org/10.1016/j.conbuildmat.2015.09.001.

Fahmy, A. S., El-Madawy, M. E. T., & Gobran, Y. A. (2016). Using artificial neural networks in the design of orthotropic bridge decks. *Alexandria Engineering Journal*, *55*(4), 3195–3203. https://doi.org/10.1016/j.aej.2016.06.034.

Fanaie, N, Aghajani, S, & Dizaj, E. A. (2016). Theoretical assessment of the behavior of cable bracing system with central steel cylinder. *Advances in Structural Engineering*, *19*(3), 463–472. https://doi.org/10.1177/1369433216630052.

Fanaie, N., Aghajani, S., & Shamloo, S. (2012). Theoretical assessment of wire rope bracing system with soft central cylinder. In *Proceedings of the 15th World Conference on Earthquake Engineering*.

Flood, I., & Kartam, N. (1994a). Neural networks in civil engineering. I: Principles and understanding. *Journal of Computing in Civil Engineering*, *8*(2), 131–148. https://doi.org/10.1061/(ASCE)0887-3801(1994)8:2(131).

Flood, I., & Kartam N. (1994b). Neural networks in civil engineering. II: Systems and application. *Journal of Computing in Civil Engineering*, *8*(2), 149–162. https://doi.org/10.1061/(ASCE)0887-3801(1994)8:2(149).

Gomes, G. F., de Almeida, F. A., Junqueira, D. M., da Cunha Jr., S. S., & Ancelotti Jr., A. C. (2019). Optimized damage identification in CFRP plates by reduced mode shapes and GA-ANN methods. *Engineering Structures*, *181*, 111–123. https://doi.org/10.1016/j.engstruct.2018.11.081.

Gupta, T, & Sharma, R. K. (2011). Structural analysis and design of buildings using neural network: A review. *International Journal of Engineering and Management Sciences*, 2(4), 216–220.

Heydari, A., & Shariati, M. (2018). Buckling analysis of tapered BDFGM nano-beam under variable axial compression resting on elastic medium. *Structural Engineering & Mechanics*, *66*(6), 737–748. https://doi.org/10.12989/sem.2018.66.6.737.

Hong, W. K. (2019). *Hybrid Composite Precast Systems: Numerical Investigation to Construction* (pp. 427–478). Woodhead Publishing (Elsevier), Sawston, Cambridge, United Kingdom.

Hong, W.-K. (2020). Chapter 10- Artificial-intelligence-based design of the ductile precast concrete beams of Hybrid Composite Precast Systems: Numerical investigation to Construction, *Woodhead Publishing Series in Civil and Structural Engineering* (pp. 427–478). Woodhead Publishing. https://doi.org/10.1016/B978-0-08-102721-9.00010-8.

Hong, W. K. (2021). *Artificial Intelligence-Based Design of Reinforced Concrete Structures.* Daega Publisher, Gyeonggi, Korea.

Hong, W. K., & Nguyen, M. C. (2021). AI-based Lagrange optimization for designing reinforced concrete columns. *Journal of Asian Architecture and Building Engineering TABE.* https://doi.org/10.1080/13467581.2021.1971998.

Hong, W. K., Nguyen, V. T., & Nguyen, M. C. (2021a). Artificial intelligence-based noble design charts for doubly reinforced concrete beams. *Journal of Asian Architecture and Building Engineering*, *21*(4), 1497–1519. https://doi.org/10.1080/13467581.2021.1928511.

Hong, W.-K., Nguyen, V. T., & Nguyen, M. C. (2021b). Optimizing reinforced concrete beams cost based on AI-based Lagrange functions. *Journal of Asian Architecture and Building Engineering.* https://doi.org/10.1080/13467581.2021.2007105.

Hong, W. K., Pham, T. D., & Nguyen, V. T. (2021c). Feature selection based reverse design of doubly reinforced concrete beams. *Journal of Asian Architecture and Building Engineering*, 1–25. https://doi.org/10.1080/13467581.2021.1928510.

Kaveh, A, & Shokohi, F. (2015). Optimum design of laterally supported castellated beams using CBO algorithm. *Steel & Composite Structures*, *18*(2), 305–324. https://doi.org/10.12989/scs.2015.18.2.305.

Korouzhdeh, T., Eskandari-Naddaf, H., & Gharouni-Nik, M. (2017). An improved ant colony model for cost optimization of composite beams. *Applied Artificial Intelligence*, *31*(1), 44–63. https://doi.org/10.1080/08839514.2017.1296681.

Krenker A., Bešter, J., & Kos, A. (2011). Introduction to the artificial neural networks. In *Artificial Neural Networks: Methodological Advances and Biomedical Applications* (pp. 1–18). InTech. https://doi.org/ 10.5772/15751.

Lee, S., Ha, J., Zokhirova, M., Moon, H., & Lee, J. (2018). Background information of deep learning for structural engineering. *Archives of Computational Methods in Engineering*, *25*(1), 121–129. https://doi.org/10.1007/s11831-017-9237-0.

Madadi, A., Eskandari-Naddaf, H., Shadnia, R., & Zhang, L. (2018). Characterization of ferrocement slab panels containing lightweight expanded clay aggregate using digital image correlation technique. *Construction and Building Materials*, *180*, 464–476. https://doi.org/10.1016/j.conbuildmat.2018.06.024.

MathWorks. 2020. *MATLAB R2020b, Version 9.9.0.* The MathWorks Inc., Natick, MA.

Nasrollahi, S., Maleki, S., Shariati, M, Marto, A., & Khorami, M. (2018). Investigation of pipe shear connectors using push out test. *Steel & Composite Structures*, *27*(5): 537–543. https://doi.org/10.12989/scs.2018.27.5.537.

Nguyen, D. H., & Hong, W. K. 2019. Part I: The analytical model predicting post-yield behavior of concrete-encased steel beams considering various confinement effects by transverse reinforcements and steels. *Materials*, *12*(14), 2302. https://doi.org/10.3390/ma12142302.

Paknahad, M., Bazzaz, M., Khorami, & M. (2018). Shear capacity equation for channel shear connectors in steel-concrete composite beams. *Steel & Composite Structures*, *28*(4), 483–494. https://doi.org/10.12989/scs.2018.28.4.483.

Cope, J. L., Cannon, R. W., Abdun-Nur, E. A., Diaz, L. H., Oury, R. F., Anderson, F. A., ... & Bonikowsky, D. A. (2002). Effect of Restraint, Volume Change, and Reinforcement on Cracking of Mass Concrete. Reported by ACI Committee 207. ACI Man. Concr. Pract.

Rizzo F, & Caracoglia, L. (2020). Artificial Neural Network model to predict the flutter velocity of suspension bridges. *Computers & Structures*, *233*, 106236. https://doi.org/10.1016/j.compstruc.2020.106236.

Safa, M., Shariati, M., Ibrahim, Z., Toghroli, A., Baharom, SB., Nor, N. M., & Petkovic, D. (2016). Potential of adaptive neuro fuzzy inference system for evaluating the factors affecting steel-concrete composite beam's shear strength. *Steel & Composite Structures*, 21(3), 679–688.

Scanlon, A., & Bischoff, P. H., 2008, "Shrinkage restraint and loading history effects on deflections of flexural members," *ACI Structural Journal*, *105*(4). https://doi.org/10.14359/19864.

Shah, SNR, Sulong, NR, Khan, R, Jumaat, MZ, & Shariati, M. (2016). Behavior of industrial steel rack connections. *Mechanical Systems and Signal Processing*, *70–71*, 25–740. https://doi.org/10.12989/scs.2016.21.3.679.

Shariati, M., Ramli Sulong, N. H., Maleki, S., & Arabnejad Kh, M. M. (2010). Experimental and analytical study on channel shear connectors in light weight aggregate concrete. In *Proceedings of the 4th International Conference on Steel & Composite Structures* (pp. 21–23). https://doi.org/10.3850/978-981-08-6218-3_CC-Fr031.

Shariat, M., Shariati, M., Madadi, A., & Wakil, K. (2018). Computational Lagrangian Multiplier Method by using for optimization and sensitivity analysis of rectangular reinforced concrete beams. *Steel & Composite Structures*, *29*(2), 243–256. https://doi.org/10.12989/scs.2018.29.2.243.

Toghroli, A, Mohammadhassani, M, Suhatril, M, Shariati, M, & Ibrahim, Z. (2014). Prediction of shear capacity of channel shear connectors using the ANFIS model. *Steel & Composite Structures*, *17*(5), 623–639. https://doi.org/10.12989/scs.2014.17.5.623.

Wu, R. T., & Jahanshahi, M. R. (2019). Deep convolutional neural network for structural dynamic response estimation and system identification. *Journal of Engineering Mechanics*, *145*(1), 04018125. https://doi.org/10.1061/(ASCE)EM.1943-7889.0001556.

Chapter 5

ANN-based structural designs using Lagrange multipliers optimizing multiple objective functions

5.1 INTRODUCTION

5.1.1 Significance of optimizing multiple objective functions

5.1.1.1 Previous studies

Over the last few decades, an optimization in structural design has gained an interest of researchers and engineers. Many previous studies presented applications of MOO in the structural engineering field. MOO algorithms must include good convergence criteria and diverse searching areas to obtain a converged and widespread set of optimal results. However, these aspects are not thoroughly investigated in previous studies and applications of MOO algorithms in structural engineering.

According to Mei and Wang (2021), structural optimization objective functions can be divided into four major categories, including cost minimization, structural performance improvement, environmental impact minimization, and multi-objective optimization. The objective functions are typically conflicting, requiring careful steering to achieve a balance. Multi-objective-based optimization challenges arise in many areas of the structural engineering field. For example, a common task based on multiple objective functions is to minimize cost while reducing an amount of carbon dioxide (CO_2) emissions. However, demands of architects, structural engineers, construction engineers, and governments are conflicting. Some previous studies which are mainly based on metaheuristics have investigated multi-objective optimizations to deal with such conflicts.

Choi (2017) proposed design strategies for reducing high-rise building costs and CO_2 emissions. Martines-Martin (2012) included an economic cost, embedded CO_2 emissions, and reinforcing steel congestion as three objective functions to design bridge piers with hybrid heuristic algorithms. Munk (2015) reviewed studies of bi-objective-based optimization, including topology and shape optimization using evolutionary algorithms. Numerous researchers have investigated and developed optimization algorithms to solve multiple-objective optimization (MOO) challenges.

A prevailing MOO algorithm is based on an evolutionary algorithm such as Nondominated Sorting Genetic Algorithm I and II (NSGA and NSGA II) by Deb (2002), Self-regulated Particle Swarm Multi-task optimization by Zheng (2021) and Mohd Zain (2018), Multi-objective Tabu Search algorithm by Jaeggi (2005), MOO for diversity and performance in a conceptual structural design by Brown (2015, 2016). Zheng and Hu (2018) studied multi-objective optimal design on vibration suppression of building structures with active mass damper. MOO design of truss structures has been studied by Nan (2020) and Kaven and Mahdavi (2018). Afshari (2019) reviewed and compared MOO algorithms with constraints in reinforced concrete (RC) structures. Metaheuristics-based algorithms applied to multi-objective functions optimizing a design of RC frames have been studied by

DOI: 10.1201/9781003314684-5

Babaei and Mollayi (2021), Bekdas and Nigdeli (2017), Kaveh and Sabzi (2012). Barraza et al. (2017) investigated MOO of structural steel buildings. Some other MOO studies of RC structures have been done by Arama (2017), Lee (2020), Park (2013), and Kavabekir (2020). Researchers have investigated populations-based MOO algorithms in a design of RC beams by Shaqfa (2019), Tahmassebi (2020), Zhang (2021), Ferreira (2003), Bekdas and Nigdeli (2013), Coello (1997), Jahjouh (2013). ANNs have been investigated in optimization design of RC members by Yücel (2021a, b), in the design of steel beams by Ferreira (2022), Nguyen (2021), Hosseinpour (2020), and Sharifi (2020).

A Pareto frontier or a Pareto front is regarded as a set of multi-objective optimized results. One of the engineers' interests is to investigate particular trade-off ratios estimating how much sacrifice each objective function makes. The evolutionary-based MOO algorithms such as NSGA and Particle Swarm Optimization require post-processing techniques to show the trade-off ratios, making them known as approaches with a posterior articulation of preferences (Marler, 2004). This seems unwelcome by engineers and decision-makers who are unfamiliar with optimization algorithms while having more interest in trade-off ratios among several objective functions that prevent applications of evolutionary-based MOO algorithms to real-life situations. Furthermore, the methodology of the population-based MOO algorithms is based on biological rules such as crossover and mutation in NSGA I and II that are unfamiliar to structural engineers. When a full understanding of the evolutionary-based optimization algorithms is absent, the evolutionary-based MOO studies can be difficult for engineers to apply them to real-life structural design cases. Few studies have investigated gradient-based MOO algorithms in which the first gradient of objective functions becomes zero at a stationary point, yielding optimized results. Liu and Reynolds (2016) proposed a gradient-based multi-objective-based optimization implementing a weighted sum approach.

5.1.1.2 Problem Descriptions and Motivations of the Chapter

Chapter 5 proposes a novel MOO algorithm and its application in a design of RC rectangular and circular columns. This algorithm is originally developed by Hong and Nguyen (2021) and Hong (2021) for an optimization of single objective function such as a cost of RC columns and RC beams. The proposed algorithm implemented characteristics of artificial neural networks (ANNs) to derive objective functions, which demonstrate a good intercorrelated relationship between input and output design parameters after training (Hong, 2020, 2021, Hong et al., 2021c). Single-objective functions are globalized into a unified function of objective (UFO) by weighting objective functions. A set of MOO results is a Pareto frontier or a Pareto set. These results are solutions of Karush-Kuhn-Turker (KKT) conditions from a Lagrange function, obtained using a Newton–Raphson method. A design example of RC columns is investigated in this chapter, in which three objectives, cost, CO_2 emissions, and column weight, are simultaneously minimized. A Pareto frontier offered by the proposed ANN-based Lagrange algorithm exhibits better convergences compared with that obtained by NSGA-II algorithm (Hong et al., 2022a). Optimal results from the proposed algorithm display a particular trade-off contributed by each objective function, which is beneficial to engineers and decision-makers. Convergence criteria and diversity of searching capability are also discussed.

MOO problems in structural design are subject to mainly bi- and tri-objective functions, in which CO_2 emission and estimated cost index are minimized to attain structural design sustainability. A multi-objective optimization (MOO) problem in structural design provides multiple design goals being optimized simultaneously (Hong et al., 2022a). Meeting multiple objective functions for an engineering design is desired since a structural design must satisfy the interest of contractors who want to minimize construction costs while governmental officials want to achieve sustainability such as by minimizing CO_2 emissions. Trade-off ratios represent contributions between design targets or objective functions, demonstrating that several demands are met simultaneously. However, how to prioritize anyone among those objective functions is often unclear. MOO problem introduces a set of optimized designs which are different from one obtained based on single objective-based optimization to the other. Optimized solutions of structural designs demonstrated in this chapter are based on bi-objective or triple-objective functions, capturing cost, CO_2 emissions, and weight W of structures at the same time, whereas objective functions more than three can be optimized. A general case of MOO problems optimizing m objective functions is defined in Equation (5.1.1.1). $F(\mathbf{x})$ are objective functions containing multiple objective goals $f_1(\mathbf{x}), f_2(\mathbf{x}), ..., f_m(\mathbf{x})$ where $\mathbf{x}$ are multiple input design parameters subjected to equality and in equality constrained conditions $\boldsymbol{c}(\mathbf{x})$ in Equation (5.1.1.2-1) and $\boldsymbol{v}(\mathbf{x})$ in Equation (5.1.1.2-2), respectively.

$$\text{Minimize } \boldsymbol{F}(\mathbf{x}) = \{f_1(\mathbf{x}), f_2(\mathbf{x}), ..., f_m(\mathbf{x})\}^T = \{f_i(\mathbf{x})\}^T \; (i = 1, ..., m) \tag{5.1.1.1}$$

$$\text{subjects to } \boldsymbol{c}(\mathbf{x}) = [c_1(\mathbf{x}), ..., c_n(\mathbf{x})]^T = 0 \tag{5.1.1.2-1}$$

$$\boldsymbol{v}(\mathbf{x}) = [v_1(\mathbf{x}), v_2(\mathbf{x}), ..., v_l(\mathbf{x})]^T \geq 0 \tag{5.1.1.2-2}$$

5.1.1.3 Significance of optimizing UFOs

MOO applications in structural engineering practice are uncommon despite their high demand. Instead, multi-objective population methods have been mainly investigated. This chapter notices ANNs and machine learning (ML) have gained interest from researchers in the structural engineering field. Based on their outstanding learning features, ANNs and ML are favorable to giving predictions of structural behaviors via ANN-based objective functions that are challenging when to be analytically derived.

Hong et al. implemented the ANN-based Lagrange optimization algorithm incorporating one objective functions and its application in a design of RC beams (Hong et al., 2021a,b), in design of RC columns (Hong et al., 2022b and Hong and Nguyen, 2021), and in a design of SRC columns (Hong et al., 2022c, d) following the Building Code Requirements for Structural Concrete such as AISC and ACI 318-19 (ACI Committee, 2019). An application of the ANN-based Lagrange optimization algorithm in a design of RC columns based on three objectives was introduced by Hong et al. (2022a). The ANN-based Lagrange algorithm based on a weighted sum approach (Yang, 2014) to formulate a UFO based on a five-step optimization is used to solve MOO problems in this chapter. A UFO which are globalized multiple objective functions are optimized simultaneously, deriving a Pareto frontier which cannot be found using conventional design methods.

Lagrange multipliers are used with equality and inequality conditions based on Karash-Kuhn-Tucker (KKT) conditions to handle constrained conditions following design requirements (Kuhn and Tucker, 1951, Karush, 1939). Design requirements are imposed by equality and equality constraints. The multiple design variables are obtained using the Newton–Raphson iteration which solves large differential equations represented by Jacobi and Hessian equations of UFO. Stationary points are identified using initial trial input variables which are updated until solutions converge. Thus, the proposed algorithm calculates stationary points at which the first derivative of Lagrange function becomes zero. An application to a design of doubly RC beams minimizing cost, environmental impact, and beam weight is conducted based on the five-step optimization to find optimized points (stationary points) of a Lagrange function using the optimization and training toolbox provided by MATLAB (MathWorks, 2020). Examples of decision-making using optimized results are also introduced in this chapter that can aid engineers for final design decisions. The algorithm has been initially proposed by Hong and Nguyen, 2021 and Hong et al., 2021b, utilizing artificial neural networks (ANN) to generalize objective functions. Lagrange multiplier method and KKT conditions are used to handle constraining conditions (Kuhn and Tucker, 1951, Karush, 1939).

5.1.1.4 Contents of Chapter 5

Chapter 5 describes an ANN-based Lagrange algorithm, leading to a Pareto frontier. The case study validates the design accuracies of a Pareto frontier by mechanics-based large datasets. The important sections include the following sub-sections. Section 5.2 introduces design scenarios of doubly RC structures, defining a MOO problem and describing which design parameters and objective functions are selected. Constraining conditions applied to an optimization problem are also introduced. Section 5.2.3 explains a five-step optimization by which MOO design of doubly RC structures is performed to find optimized design parameters based on multiple objective functions. Sections 5.3 and 5.4 present an optimization based on UFO for circular RC columns sustaining one and multiple loads, respectively, whereas Sections 5.5 and 5.6 present an optimization based on UFO to design uniaxial and biaxial rectangular RC columns, respectively, sustaining multiple loads. Section 5.7 presents an optimization based on UFO to design RC beams. These design examples are based on a set of optimizations minimizing several objective functions simultaneously as decision-making guidance. The optimization results and recommendations are stated while the accuracies of optimized designs (optimal results) are also discussed. Finally, practical conclusions and design recommendations are drawn in Section 5.8. Important contents of Chapter 5 are summarized below.

Section 5.3: Round Column – subject to one load
Section 5.4: Round Column – subject to multi-loads
Section 5.5: Uniaxial rectangular column – subject to multi-loads
Section 5.6: Biaxial rectangular column – subject to multi-loads
Section 5.7: RC beam
Section 5.8: Design recommendations and conclusions

5.1.2 Review of Pareto frontier

Pareto optimality is a concept originally developed by Vilfredo Pareto (1906) in his studies of economic efficiency and income distribution. This concept was later commonly exploited in optimization problems having multiple objective functions by defining their solutions as a set of Pareto optimal points.

A point referred to a Pareto optimal point optimizes multi-objective functions when it is impossible to improve one objective function without worsening at least one of the other objective functions (Peel and Moon, 2020).

Point A dominates Point B if there is at least one objective function of Point A is better optimized than that of point B; and Point A must have the other optimized objective functions equal to or less than in comparison with those belonging to Point B, and hence, Point A is said to be nondominated. This concept is explained in Equations (5.1.2.1-1) and (5.1.2.1-2) where **x** dominates **x*** iff: *all objective functions* for **x** are smaller than or equal to those optimized functions for **x*** and at least one objective function for **x** are smaller than that belonging to **x***.

$$f_i(\mathbf{x}) \leq f_i(\mathbf{x}^*) \text{ for all } i \in \{1, \ldots, m\} \ (m = \text{a number of objective functions}) \tag{5.1.2.1-1}$$

and

$$f_i(\mathbf{x}) < f_i(\mathbf{x}^*) \text{ for some } i \in \{1, \ldots, m\} \ (m = \text{a number of objective functions}) \tag{5.1.2.1-2}$$

Example 1

Considering an example to investigate the concept of "*dominance*" in which two objective functions, cost *CI* and weight W, are minimized. Table 5.1.2.1 presents weight W (Objective function 1) and cost *CI* (Objective function 2) denoted by P1 to P12, each of which represents 12 designs as illustrated in Figure 5.1.2.1.

These designs are performed by minimizing two objective functions (weight W and cost) using a concept of "*dominance*" (Definition 1). Since this example is a bi-objective function-based optimization problem in which **x** and **x*** are the two input variables. Equation (5.1.2.2-1) defines "*dominance*" concept in that **x** dominates **x*** when two objective functions (weight W and cost) for **x** are smaller than or equal to those optimized functions for **x***. In other words, **x** is nondominated by **x*** as shown in Equation (5.1.2.2-2). The two objective functions are compared to decide which one is nondominated. In Example 2 of Section 5.1.4, **x** and **x*** are column dimensions (*D*).

Points P1, ..., P12 denote 12 designs considered in Table 5.1.2.1 where input design parameters are indicated by $\mathbf{x}_k$, k = 1, ..., 12. Design P1 optimizes the two objective functions; weight W is optimized as 19 kN/m and cost is optimized as 200,000 KRW/m, which can be expressed as $f_1(\mathbf{x}_1) = 19\ kN/m$ and $f_2(\mathbf{x}_1) = 200{,}000\ KRW/m$. Design P1 is said to be dominated by Design P2 shown in Equation (5.1.2.2-3) and Design P8 shown in Equation (5.1.2.2-4) because Design P1 has an optimized weight W and cost *CI* which are greater than or equal to those optimized by Designs P2 (weight W and cost *CI* of Designs P2 smaller than Design P1) as shown in Equation (5.1.2.2-3) and P8 (weight W

Table 5.1.2.1 Designs P1 to P12 Optimizing the Two Objective Functions of Weight W and Cost CI

Design	*Weight W (Objective 1)*	*Cost (Objective 2)*
P1	19	200,000
P2	17.5	180,000
P3	15	230,000
P4	11.5	250,000
P5	12.5	300,000
P6	21	230,000
P7	12	230,000
P8	14	200,000
P9	19	270,000
P10	16	310,000
P11	14	290,000
P12	15	400,000

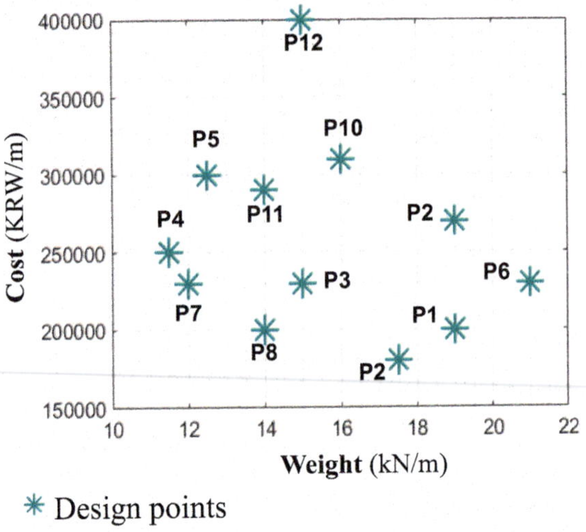

Figure 5.1.2.1 Graphical illustration of design points in terms of weight W and cost *CI*.

of Design P8 smaller than weight W of Design P1 and Cost *CI* of Design P8 equal to Cost *CI* of Design P1) as shown in Equation (5.1.2.2-4).

$$\mathbf{x} \text{ dominates } \mathbf{x}^* \Leftrightarrow \begin{cases} f_1(\mathbf{x}) = f_1(\mathbf{x}^*) \parallel f_1(\mathbf{x}) < f_1(\mathbf{x}^*) \\ f_2(\mathbf{x}) = f_2(\mathbf{x}^*) \parallel f_2(\mathbf{x}) < f_2(\mathbf{x}^*) \end{cases} \tag{5.1.2.2-1}$$

$$\mathbf{x} \text{ is nondominated by } \mathbf{x}^* \Leftrightarrow \begin{cases} f_1(\mathbf{x}) < f_1(\mathbf{x}^*) \\ f_2(\mathbf{x}) = f_2(\mathbf{x}^*) \parallel f_2(\mathbf{x}) > f_2(\mathbf{x}^*) \end{cases} \text{ or } \begin{cases} f_1(\mathbf{x}) = f_1(\mathbf{x}^*) \parallel f_1(\mathbf{x}) > f_1(\mathbf{x}^*) \\ f_2(\mathbf{x}) < f_2(\mathbf{x}^*) \end{cases} \tag{5.1.2.2-2}$$

$$\mathbf{x}_2 \text{ dominates } \mathbf{x}_1 \Leftrightarrow \begin{cases} f_1(\mathbf{x}_2) = 17.5 \text{ kN/m} < f_1(\mathbf{x}_1) = 19 \text{ kN/m} \\ f_2(\mathbf{x}_2) = 180{,}000 \text{ KRW/m} < f_2(\mathbf{x}_1) = 200{,}000 \text{ KRW/m} \end{cases} \tag{5.1.2.2-3}$$

$$\mathbf{x}_8 \text{ dominates } \mathbf{x}_1 \Leftrightarrow \begin{cases} f_1(\mathbf{x}_8) = 14 \text{ kN/m} < f_1(\mathbf{x}_1) = 19 \text{ kN/m} \\ f_2(\mathbf{x}_8) = f_2(\mathbf{x}_1) = 200{,}000 \text{ KRW/m} \end{cases} \tag{5.1.2.2-4}$$

Next, let's evaluate Design P2 optimizing $f_1(\mathbf{x}_2) = 17.5$ kN/m and $f_2(\mathbf{x}_2) = 180{,}000$ KRW/m. There are several designs whose weights W are lighter than weight W of Design P2, such as Designs P3, P4, P5, P7, P8, P10, P11, and P12 shown in Equation (5.1.2.3-1) and Table 5.1.2.2.

However, there are no design points whose costs are less than or equal to that of Design P2 shown in Equation (5.1.2.3-2) and Table 5.1.2.2. In other words, it is impossible to find an improvement in terms of weight W without worsening the cost of Design P2 shown in Equation (5.1.2.4), which indicates that no design points (P*k*) have both the cost and weight less than or equal to cost and weight of

Table 5.1.2.2 Design Points and Their Equivalent Dominance Points

Design	*Weight W (Objective Function 1)*	*Cost CI (Objective Function 2)*	*Equivalent Dominance Points*
P1	19	200,000	P2, P8
P2	17.5	180,000	None
P3	15	230,000	P7
P4	11.5	250,000	None
P5	12.5	300,000	P4, P7
P6	21	230,000	P1, P2, P3, P7, P8
P7	12	230,000	None
P8	14	200,000	None
P9	19	270,000	P1, P2, P3, P4, P7, P8
P10	16	310,000	P3,
P11	14	290,000	P4, P7, P8
P12	15	400,000	P3, P4, P5, P7, P8, P11

Design P2. Therefore, Design P2 is considered as a *nondominated* design point in this case. Similarly, Design P4 ($f_1(\mathbf{x}_4) = 11.5$ kN/m, $f_2(\mathbf{x}_4) = 250{,}000$ KRW/m), Design P7 ($f_1(\mathbf{x}_7) = 12$ kN/m, $f_2(\mathbf{x}_7) = 230{,}000$ KRW/m), and Design P8 ($f_1(\mathbf{x}_8) = 14$ kN/m, $f_2(\mathbf{x}_8) = 200{,}000$ KRW/m) are also *nondominated* design points because weights W of P4, P7, and P8 are less than that of P2 as shown in Equation (5.1.2.3-1). It is noted that even if $f_2(\mathbf{x}_7) = 230{,}000$ KRW/m of Design P7 is larger than $f_2(\mathbf{x}_2) = 180{,}000$ KRW/m of Design P2, $f_1(\mathbf{x}_7) = 12$ kN/m of Design P7 is smaller than $f_1(\mathbf{x}_2) = 17.5$ kN/m of Design P2, as shown in Equation (5.1.2.2-2).

$$\begin{cases} f_1(\mathbf{x}_3) = 15 \text{ kN/m} < f_1(\mathbf{x}_2) = 17.5 \text{ kN/m} \\ f_1(\mathbf{x}_4) = 11.5 \text{ kN/m} < f_1(\mathbf{x}_2) = 17.5 \text{ kN/m} \\ f_1(\mathbf{x}_5) = 12.5 \text{ kN/m} < f_1(\mathbf{x}_2) = 17.5 \text{ kN/m} \\ f_1(\mathbf{x}_7) = 12 \text{ kN/m} < f_1(\mathbf{x}_2) = 17.5 \text{ kN/m} \\ f_1(\mathbf{x}_8) = 14 \text{ kN/m} < f_1(\mathbf{x}_2) = 17.5 \text{ kN/m} \\ f_1(\mathbf{x}_{10}) = 16 \text{ kN/m} < f_1(\mathbf{x}_2) = 17.5 \text{ kN/m} \\ f_1(\mathbf{x}_{11}) = 14 \text{ kN/m} < f_1(\mathbf{x}_2) = 17.5 \text{ kN/m} \\ f_1(\mathbf{x}_{12}) = 15 \text{ kN/m} < f_1(\mathbf{x}_2) = 17.5 \text{ kN/m} \end{cases} \tag{5.1.2.3-1}$$

$$\begin{cases} f_2(\mathbf{x}_1) = 200{,}000 \text{ KRW/m} > f_2(\mathbf{x}_2) = 180{,}000 \text{ KRW/m} \\ f_2(\mathbf{x}_3) = 230{,}000 \text{ KRW/m} > f_2(\mathbf{x}_2) = 180{,}000 \text{ KRW/m} \\ f_2(\mathbf{x}_4) = 250{,}000 \text{ KRW/m} > f_2(\mathbf{x}_2) = 180{,}000 \text{ KRW/m} \\ f_2(\mathbf{x}_5) = 300{,}000 \text{ KRW/m} > f_2(\mathbf{x}_2) = 180{,}000 \text{ KRW/m} \\ f_2(\mathbf{x}_7) = 230{,}000 \text{ KRW/m} > f_2(\mathbf{x}_2) = 180{,}000 \text{ KRW/m} \\ f_2(\mathbf{x}_8) = 200{,}000 \text{ KRW/m} > f_2(\mathbf{x}_2) = 180{,}000 \text{ KRW/m} \\ f_2(\mathbf{x}_9) = 270{,}000 \text{ KRW/m} > f_2(\mathbf{x}_2) = 180{,}000 \text{ KRW/m} \\ f_2(\mathbf{x}_{10}) = 310{,}000 \text{ KRW/m} > f_2(\mathbf{x}_2) = 180{,}000 \text{ KRW/m} \\ f_2(\mathbf{x}_{11}) = 290{,}000 \text{ KRW/m} > f_2(\mathbf{x}_2) = 180{,}000 \text{ KRW/m} \\ f_2(\mathbf{x}_{12}) = 400{,}000 \text{ KRW/m} > f_2(\mathbf{x}_2) = 180{,}000 \text{ KRW/m} \end{cases} \tag{5.1.2.3-2}$$

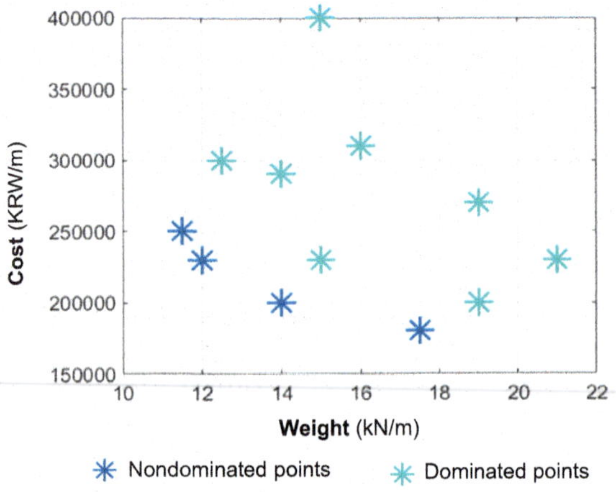

Figure 5.1.2.2 Nondominated design points based on "*dominance*" analysis.

$$\nexists\, k \in \{1, \ldots, 12\}/\{2\}: \begin{cases} f_1(\mathbf{x}_k) = f_1(\mathbf{x}_2) \,||\, f_1(\mathbf{x}_k) < f_1(\mathbf{x}_2) \\ f_1(\mathbf{x}_2) = 17.5 \text{ kN/m} \\ f_2(\mathbf{x}_k) = f_2(\mathbf{x}_2) \,||\, f_2(\mathbf{x}_k) < f_2(\mathbf{x}_2) \\ f_2(\mathbf{x}_2) = 180{,}000 \text{ KRW/m} \end{cases} \tag{5.1.2.4}$$

All 12 design points are evaluated similarly to Equations (5.1.2.2)–(5.1.2.4). Results are presented in Table 5.1.2.2 and Figure 5.1.2.2. Design P2 and Design P4 are minimizing cost *CI* (Objective function 2) and weight W (Objective function 1), respectively. Both points are non-dominated because they optimize at least one objective function.

5.1.3 Criterion space and Pareto frontier

Example 2 shown in Sections 5.1.4 and 5.1.5 discusses how to find nondominated points based on *weighted sum method* (Yang, 2014). Let $\mathbf{x}$ be input parameters of feasible design points, and hence, $\mathbf{x} \in \mathbf{X}$ where $\mathbf{X}$ is a set containing all input parameters of feasible design points. $f_i(\mathbf{x})$ is an objective function i ($i \in \{1, \ldots, m\}$, m = a number of objective functions) with respect to input $\mathbf{x}$. Input parameters of 12 design points P1–P12 of Example 1 shown in Table 5.1.2.1 can be treated as big dataset of a bi-objective function-based optimization problem in which weight W and cost *CI* are the two objective functions. A criterion space $\mathbf{\Gamma}$ is defined as a set containing big data of all 12 points optimizing objective functions ($f_i(\mathbf{x})$, $i \in \{1, \ldots, m\}$) considered in an optimization problem having m objective functions. The criterion space is illustrated in Figure 5.1.3.1. Section 5.1.2 discussed a Pareto frontier to describe nondominated points (Designs P2, P4, P7, and P8) of Example 1 shown in Table 5.1.2.2 and Figure 5.1.2.2 after comparing 12 design points based on "*dominance*" concept. A Pareto frontier is a set of the nondominated points. A blue line connecting nondominated design points of Designs P2, P4, P7, and P8 shown in Figure 5.1.3.1 is a Pareto frontier of a bi-objective function-based optimization problem described by Example 1. All nondominated points lie on a lower boundary of the criterion space $\mathbf{\Gamma}$, however, not all points lying on a boundary of $\mathbf{\Gamma}$ are Pareto optimal. For example, even if Designs P1, P5, P6, P10, and P12 shown in Figure 5.1.2.1 are located on a boundary of $\mathbf{\Gamma}$, these points are dominated points which are not Pareto optimal so that they cannot form a Pareto frontier. A set

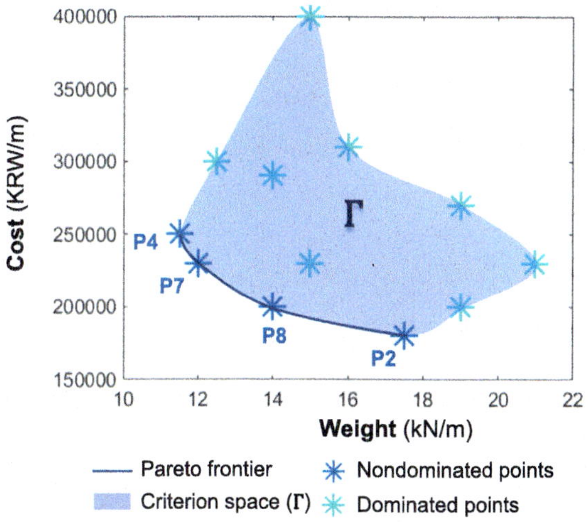

Figure 5.1.3.1 Criterion space and Pareto frontier of design points for Example 1.

of optimal points is the Pareto frontier, which coincides with the lower boundary of the criterion space when minimizing objective functions, or the upper boundary of the criterion space when maximizing objective functions. Searching for a lower boundary of a criterion space is, therefore, more efficient than searching for an upper boundary when minimizing objective functions since they have more chances to represent a Pareto frontier on a lower boundary of a criterion space. Commonly, an upper boundary of criterion space can be neglected in minimizing multiple objective functions to reduce searching time. In general, a Pareto frontier optimizing multi-objective functions based on ANNs can aid engineers and decision-makers to find trade-offs contributed by each objective function. A Pareto frontier is represented by a line identified from a bi-objective function-based optimization problem of Example 1 as illustrated in Figure 5.1.3.1. A Pareto frontier is expressed in m directions when optimizing m objective functions, and hence, optimization algorithms are proposed to generate a Pareto frontier for solving diverse optimization problems optimizing multi-objective functions. This chapter obtains a Pareto frontier in MOO problems by imposing *Weighted sum method*, which is discussed in Section 5.1.4.

5.1.4 Weighted sum method

Weighted sum method is a technique to unify multiple objective functions into one global function by multiplying with fractions or trade-offs contributed by each objective function. A global objective function $F(\mathbf{x})$ is now defined to be a sum of $w_i f_i(\mathbf{x})$ $(i = 1,..., m)$ as shown in Equation (5.1.4.1) where w_i is a fraction (or trade-offs contributed by each objective function) corresponding to objective function f_i and m. is a number of optimized objective functions. The equation shown in Equation (5.1.4.1) is called Unified Functions of Objectives (UFO) in this chapter. All fractions shown in Equation (5.1.4.2) are nonnegative and summed to 1. In this chapter, a Pareto frontier is identified by optimizing UFO shown in Equation (5.1.4.1) in tms of each fraction w_i shown in Equation (5.1.4.2) which varies from 0 to 1. Let's consider an example to apply *weighted sum method*.

$$\text{Minimize } \boldsymbol{F}(\mathbf{x}) = \{w_1, w_2, \ldots, w_m\}\{f_1(\mathbf{x}), f_2(\mathbf{x}), \ldots, f_m(\mathbf{x})\}^T$$
$$= w_1 f_1(\mathbf{x}) + w_2 f_2(\mathbf{x}) + \ldots + w_m\ f_m(\mathbf{x}) = \sum_{i=1}^{m} w_i f_i(\mathbf{x}) \tag{5.1.4.1}$$

$$w_i \in [0, 1];\ \sum_{i=1}^{m} w_i = 1 \tag{5.1.4.2}$$

Example 2

RC circular columns are designed to optimize bi-objective functions, weight W and cost *CI* shown in Equation (5.1.4.3). Objective functions of weight W and cost *CI* are assumed in Equations (5.1.4.4-1) and (5.1.4.4-2) for Example 2. Designs of round and rectangular RC columns based on ANN-based Lagrange optimization using real data are presented in Sections 5.3 and 5.4 and in Sections 5.5 and 5.6, respectively. In fact, a design of RC structures involves many other design parameters, leading to more complex objective functions based on weight W and cost *CI*. Objective functions for weight W and cost *CI* in this example are simplified based on a *weighted sum method* to generate Pareto frontier. An input parameter **x** contains only one variable which is column dimension (*D*) for simplicity as shown in Equation (5.1.4.5).

$$\text{Minimize}: \boldsymbol{F}(\mathbf{x}) = \{f_1(\mathbf{x}), f_2(\mathbf{x})\}^T \tag{5.1.4.3}$$

$$\text{Weight } W: f_1(\mathbf{x}) = 0.02 \times D\ - 0.02 \times D^3 \tag{5.1.4.4-1}$$

$$\text{Cost } CI: f_2(\mathbf{x}) = \left(50 \times D^2\ + \frac{100}{D}\right) \times 10^3 \tag{5.1.4.4-2}$$

$$\text{subjects to } D \in \{1.0 : 0.1 : 2.5\}\ (m) \tag{5.1.4.5}$$

Table 5.1.4.1 presents random big datasets including 16 designs accounting for 16 column dimensions (*D*) varying from 1.0 to 2.5m with a step of 0.1m. Figure 5.1.4.1 graphically visualizes the random bigdata in which the Pareto frontier is naturally demonstrated by five nondominated points P1, P2, P3, P4, and P5 (red points). Two methods are developed to identify Pareto frontier optimizing UFO. The first method considers two objective functions in a separate manner to find nondominated points on the Pareto frontier based on Section 5.1.4.1. In real-world applications, it is difficult to identify the Pareto frontier from the bigdata shown in Table 5.1.4.1 using the first method, and hence, UFO based on weighted sum method is introduced as a more systematic method to identify the Pareto frontier based on the second method described in Section 5.1.4.2.

5.1.4.1 The first method - minimization of bi-objective functions based on a definition of nondominated points

Designs P1 to P16 are evaluated which minimize weight W and cost *CI* simultaneously to identify Pareto frontiers based on definitions of *nondominated points* discussed in Equations (5.1.2.1-1) and (5.1.2.1-2). It is easily seen in Table 5.1.4.1 that Design P1 minimizes weight W (0.18 kgf/m) among 16 designs, and hence, P1 is found as an optimal point on the Pareto frontier even if its cost (150,000 KRW) is higher than the cost of the other designs. A minimum cost (141,429 KRW/m) is provided by Design P5 which is selected as another optimal point (P5) on the Pareto frontier even if the corresponding weight W of 0.225 kgf/m is not minimized.

Table 5.1.4.1 Designs P1–P12 (Random Bigdata) for Example 2

Design	D (m)	Weight W (kgf/m) (Objective Function 1)	Cost (KRW/m) (Objective Function 2)	Dominating Points
P1	1.00	0.180	150,000	None
P2	1.10	0.193	145,909	None
P3	1.20	0.205	143,333	None
P4	1.30	0.216	141,923	None
P5	1.40	0.225	141,429	None
P6	1.50	0.233	141,667	P5
P7	1.60	0.238	142,500	P4, P5, P6
P8	1.70	0.242	143,824	P3, P4, P5, P6
P9	1.80	0.243	145,556	P3, P4, P5, P6
P10	1.90	0.243	147,632	P2–P9
P11	2.00	0.240	150,000	P1–P10
P12	2.10	0.235	152,619	P1–P11
P13	2.20	0.227	155,455	P1–P12
P14	2.30	0.217	158,478	P1–P13
P15	2.40	0.204	161,667	P1–P14
P16	2.50	0.188	165,000	P2–P15

P2, P3, and P4 are also selected as the three nondominated points because there are no other design points that have both lighter weight and less cost than those of P2, P3, and P4. For example, P2 is a nondominated point because its cost (145,909 KRW) is lower than that of P1 even though the weight of P2 (0.193 kgf/m) is greater than that of P1.

Table 5.1.4.1 presents big datasets for Example 2, including five nondominated design points denoted by P1, P2, P3, P4, and P5, and hence, these five design points are the Pareto frontier of Example 2. Connecting these five points offers the Pareto frontier as shown in Figure 5.1.4.1 which illustrates nondominated points identified from big datasets, including 16 design options. A criterion space in this case is presented by a line connecting exterior

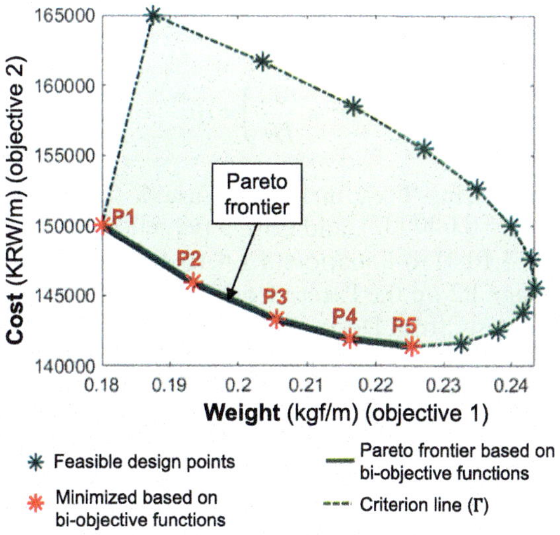

Figure 5.1.4.1 Pareto frontier identified based on non-normalized UFO for Example 2.

design points. Nondominated points (Designs P1–P5) are represented by red marks, forming the Pareto frontier when they are connected as shown in Figure 5.1.4.1. The Pareto frontier is identified with design cases based on big datasets which are not dominated by other design cases.

5.1.4.2 The second method - minimizing bi-objective functions (UFO) based on weighted sum method

Weighted sum method implements fractions in formulating an UFO $F(\mathbf{x})$ shown in Equation (5.1.4.6) in which the two objective functions of weight W and cost CI are linearly combined with a fraction w_1 and w_2, respectively, while their sum is equal to 1 as can be seen in Equation (5.1.4.7). In Example 2, UFO $F(\mathbf{x})$ is minimized with respect to weight W and cost CI simultaneously, calculating stationary points minimizing $F(\mathbf{x})$ while fractions applied to each objective function vary from 0 to 1.

$$\begin{aligned} F(\mathbf{x}) &= \{w_1, w_2\}\{f_1(\mathbf{x}), f_2(\mathbf{x})\}^T = w_1 f_1(\mathbf{x}) + w_2 f_2(\mathbf{x}) \\ &= w_1 \times \left(0.02 \times D - 0.02 \times D^3\right) + w_2 \times \left(50 \times D^2 + \frac{100}{D}\right) \times 10^3 \end{aligned} \tag{5.1.4.6}$$

$$w_1 \in [0, 1];\ w_2 = 1 - w_1 \tag{5.1.4.7}$$

$\{w_1, w_2\}$ denotes a fraction for Objective function 1 (weight, W) and a fraction for Objective function 2 (cost), respectively, to minimize UFO $F(\mathbf{x})$. Case 1, where a fraction combination of $w_1 = 1$ and $w_2 = 0$ are used, leads to $F(\mathbf{x})$ which is equal to $f_1(\mathbf{x})$ based on only Objective function 1 as can be seen in Equation (5.1.4.8-1). Case 1 design minimizes $F(\mathbf{x})$ based on a minimized $f_1(\mathbf{x})$, resulting in Design P1 as one optimal point on the Pareto frontier, as shown in Table 5.1.4.2 and Figure 5.1.4.1. Case 2, where a fraction combination of $w_1 = 0$ and $w_2 = 1$ are used, leads to $F(\mathbf{x})$ which is equal to $f_2(\mathbf{x})$ based on only Objective function 2 as can be seen in Equation (5.1.4.8-2). Case 2 design minimizes $F(\mathbf{x})$ based on a minimized $f_2(\mathbf{x})$, yielding Design P5 as one optimal point on the Pareto frontier as shown in Table 5.1.4.2 and Figure 5.1.4.1.

$$F(\mathbf{x}) = 1 \times f_1(\mathbf{x}) + 0 \times f_2(\mathbf{x}) = 0.02 \times D - 0.02 \times D^3 \tag{5.1.4.8-1}$$

$$F(\mathbf{x}) = 0 \times f_1(\mathbf{x}) + 1 \times f_2(\mathbf{x}) = \left(50 \times D^2 + \frac{100}{D}\right) \times 10^3 \tag{5.1.4.8-2}$$

Other solutions for minimizing $F(\mathbf{x})$, including Case 7 ($w_1 = 0.999990, w_2 = 0.000010$), Case 8 ($w_1 = 0.999993, w_2 = 0.000007$), and Case 9 ($w_1 = 0.999996, w_2 = 0.000004$) are also obtained. Equations (5.1.4.10-1) to Equation (5.1.4.10-16) present step-by-step calculations for Case 9 to obtain Design P2 on the Pareto frontier as an optimal point where W and CI are minimized simultaneously from Designs P1 to P16 when being based on a fraction of $w_1 = 0.999996, w_2 = 0.000004$. Fractions ($w_1, w_2$) of Cases 7 and 8 shown in Table 5.1.4.2 are substituted into Equations (5.1.4.9) to (5.4.1.10-16) based on (1) to (2) to obtain Design P4 and P5 on the Pareto frontier, respectively. The fractions of Cases 3 to 6, w_1 ranging from $0.9 \rightarrow 0.99 \rightarrow 0.999 \rightarrow 0.9999$ and a corresponding fraction w_2 for cost CI ranging from $0.1 \rightarrow 0.01 \rightarrow 0.001 \rightarrow 0.0001$ are also arbitrarily chosen, based on trial-and-error calculations. It is noteworthy that w_2 are relatively small compared to w_1 while a variation of fractions of Cases 7–9 are infinitesimal. For example, a variation of fraction w_1 of Cases 7 and

Table 5.1.4.2 Weighted Sum Method (Yang, 2014) Implementing Fractions in Formulating a Non-Normalized Bi-Objective UFO Functions *F*(*x*) to Minimize Based on Trial-and-Errors

$$F(x) = w_1 \times \left(0.02 \times D - 0.02 \times D^3\right) + w_2 \times \left(50 \times D^2 + \frac{100}{D}\right) \times 10^3$$

Case	*1*	*2*	*3*	*4*	*5*	*6*	*7*	*8*	*9*
w_1	1.0	0.0	0.9	0.99	0.999	0.9999	0.999990	0.999993	0.999996
w_2	0.0	1.0	0.1	0.01	0.001	0.0001	0.000010	0.000007	0.000004
P1	**0.180**	150,000	15,000.2	1,500.18	150.180	15.180	1.680	1.230	0.780
P2	0.193	145,909	14,591.1	1,459.28	146.102	14.784	1.652	1.215	**0.777**
P3	0.205	143,333	14,333.5	1,433.54	143.539	14.539	1.639	**1.209**	0.779
P4	0.216	141,923	14,192.5	1,419.44	142.139	14.408	**1.635**	1.210	0.784
P5	0.225	**141,429**	**14,143.1**	**1,414.51**	**141.653**	**14.368**	1.639	1.215	0.791
P6	0.233	141,667	14,166.9	1,416.90	141.899	14.399	1.649	1.224	0.799
P7	0.238	142,500	14,250.2	1,425.24	142.738	14.488	1.663	1.236	0.808
P8	0.242	143,824	14,382.6	1,438.47	144.065	14.624	1.680	1.249	0.817
P9	0.243	145,556	14,555.8	1,455.80	145.799	14.799	1.699	1.262	0.826
P10	0.243	147,632	14,763.4	1,476.56	147.874	15.006	1.719	1.276	0.833
P11	0.240	150,000	15,000.2	1,500.24	150.240	15.240	1.740	1.290	0.840
P12	0.235	152,619	15,262.1	1,526.42	152.854	15.497	1.761	1.303	0.845
P13	0.227	155,455	15,545.7	1,554.77	155.681	15.772	1.782	1.315	0.849
P14	0.217	158,478	15,848.0	1,585.00	158.695	16.064	1.801	1.326	0.851
P15	0.204	161,667	16,166.8	1,616.87	161.870	16.370	1.820	1.335	0.850
P16	0.188	165,000	16,500.2	1,650.19	165.187	16.687	1.837	1.342	0.847

Red color values indicate Pareto frontier.

8 are $0.999993 - 0.999990 = 0.000003 = 3 \times 10^{-6}$. This is because two objective functions are non-normalized. The UFO $F(\mathbf{x})$ calculated as 14143.1, 1414.51, 141.653, 14.368) yields duplicated P5 when fractions shown in Cases 3–6 ($w_1 = 0.9 \rightarrow 0.99 \rightarrow 0.999 \rightarrow 0.9999$) of Table 5.1.4.2 are used to calculate optimal points. It is noted that UFO used in Table 5.1.4.2 is not normalized, suggesting a small interval be used for a fraction w_1 which is an order of 3×10^{-6} as shown in Cases 7–9 ($w_1 = 0.999990 \rightarrow 0.999993 \rightarrow 0.999996$) to find P4, P3, and P2. A small interval must be used for a fraction w_1 which is an order of 3×10^{-6} as shown in Cases 7–9 ($w_1 = 0.999990 \rightarrow 0.999993 \rightarrow 0.999996$). Five optimal points are found as shown in Table 5.1.4.2, yielding Pareto frontier P1–P5 as illustrated in Figure 5.1.4.1. Therefore, trial-and-error-based fractions based on nine cases are used as shown in Table 5.1.4.2 to find optimal points or nondominated points. UFO $F(\mathbf{x})$ is formulated by applying $w_1 = 0.999996$ and $w_2 = 0.000004$ as shown in Equation (5.1.4.9). In general, when the number of non-dominated points are unknown, a sufficient number of fractions should be used to find the smooth Pareto frontier. In this chapter, the "*linspace*" command defined in MATLAB (MathWorks, 2020) is used to generate the fractions as discussed in Section 5.3.6.

$$F(\mathbf{x}) = 0.999996 \times \left(0.02 \times D - 0.02 \times D^3\right) + 0.000004 \times \left(50 \times D^2 + \frac{100}{D}\right) \times 10^3 \quad (5.1.4.9)$$

1. Designs with 16 column dimensions (Ds) are substituted into Equation (5.1.4.9) to obtain UFO $F(\mathbf{x}_i)$ ($i = 1, 2, \ldots, 16$) for Case 9 in which $\mathbf{x}_i$ is a column dimension with respect to each design (D_i). UFO $F(\mathbf{x}_i)$ are calculated in Equations (5.1.4.10-1)–(5.1.4.10-16).

$$\begin{aligned}&\text{Design P1}: & F(\mathrm{x}_1) &= 0.999996\times\left(0.02\times1.0\ -0.02\times1.0^3\right)\\ &\mathrm{x}_1 = D_1 = 1.0 & &\quad +0.000004\times\left(50\times1.0^2\ +\frac{100}{1.0}\right)\times10^3\\ & & &= 0.780\end{aligned} \tag{5.1.4.10-1}$$

$$\begin{aligned}&\text{Design P2}: & F(\mathrm{x}_2) &= 0.999996\times\left(0.02\times1.1-0.02\times1.1^3\right)\\ &\mathrm{x}_2 = D_2 = 1.1 & &\quad +0.000004\times\left(50\times1.1^2+\frac{100}{1.1}\right)\times10^3\\ & & &= 0.777\end{aligned} \tag{5.1.4.10-2}$$

$$\begin{aligned}&\text{Design P3}: & F(\mathrm{x}_3) &= 0.999996\times\left(0.02\times1.2-0.02\times1.2^3\right)\\ &\mathrm{x}_3 = D_3 = 1.2 & &\quad +0.000004\times\left(50\times1.2^2+\frac{100}{1.2}\right)\times10^3\\ & & &= 0.779\end{aligned} \tag{5.1.4.10-3}$$

$$\begin{aligned}&\text{Design P2}: & F(\mathrm{x}_4) &= 0.999996\times\left(0.02\times1.3-0.02\times1.3^3\right)\\ &\mathrm{x}_4 = D_4 = 1.3 & &\quad +0.000004\times\left(50\times1.3^2+\frac{100}{1.3}\right)\times10^3\\ & & &= 0.784\end{aligned} \tag{5.1.4.10-4}$$

$$\begin{aligned}&\text{Design P5}: & F(\mathrm{x}_5) &= 0.999996\times\left(0.02\times1.4-0.02\times1.4^3\right)\\ &\mathrm{x}_5 = D_5 = 1.4 & &\quad +0.000004\times\left(50\times1.4^2+\frac{100}{1.4}\right)\times10^3\\ & & &= 0.791\end{aligned} \tag{5.1.4.10-5}$$

$$\begin{aligned}&\text{Design P6}: & F(\mathrm{x}_6) &= 0.999996\times\left(0.02\times1.5-0.02\times1.5^3\right)\\ &\mathrm{x}_6 = D_6 = 1.5 & &\quad +0.000004\times\left(50\times1.5^2+\frac{100}{1.5}\right)\times10^3\\ & & &= 0.799\end{aligned} \tag{5.1.4.10-6}$$

$$\begin{aligned}&\text{Design P7}: & F(\mathrm{x}_7) &= 0.999996\times\left(0.02\times1.6-0.02\times1.6^3\right)\\ &\mathrm{x}_7 = D_7 = 1.6 & &\quad +0.000004\times\left(50\times1.6^2+\frac{100}{1.6}\right)\times10^3\\ & & &= 0.808\end{aligned} \tag{5.1.4.10-7}$$

$$\begin{aligned}&\text{Design P8}: & F(\mathrm{x}_8) &= 0.999996\times\left(0.02\times1.7-0.02\times1.7^3\right)\\ &\mathrm{x}_8 = D_8 = 1.7 & &\quad +0.000004\times\left(50\times1.7^2+\frac{100}{1.7}\right)\times10^3\\ & & &= 0.817\end{aligned} \tag{5.1.4.10-8}$$

Design P9: $\quad F(x_9) = 0.999996 \times (0.02 \times 1.8 - 0.02 \times 1.8^3)$

$$x_9 = D_9 = 1.8 \qquad + 0.000004 \times \left(50 \times 1.8^2 + \frac{100}{1.8}\right) \times 10^3 \tag{5.1.4.10-9}$$

$= 0.826$

Design P10: $\quad F(x_{10}) = 0.999996 \times (0.02 \times 1.9 - 0.02 \times 1.9^3)$

$$x_{10} = D_{10} = 1.9 \qquad + 0.000004 \times \left(50 \times 1.9^2 + \frac{100}{1.9}\right) \times 10^3 \tag{5.1.4.10-10}$$

$= 0.833$

Design P11: $\quad F(x_{11}) = 0.999996 \times (0.02 \times 20 - 0.02 \times 2.0^3)$

$$x_{11} = D_{11} = 2.0 \qquad + 0.000004 \times \left(50 \times 2.0^2 + \frac{100}{2.0}\right) \times 10^3 \tag{5.1.4.10-11}$$

$= 0.840$

Design P12: $\quad F(x_{12}) = 0.999996 \times (0.02 \times 2.1 - 0.02 \times 2.1^3)$

$$x_{12} = D_{12} = 2.1 \qquad + 0.000004 \times \left(50 \times 2.1^2 + \frac{100}{2.1}\right) \times 10^3 \tag{5.1.4.10-12}$$

$= 0.845$

Design P13: $\quad F(x_{13}) = 0.999996 \times (0.02 \times 2.2 - 0.02 \times 2.2^3)$

$$x_{13} = D_{13} = 2.2 \qquad + 0.000004 \times \left(50 \times 2.2^2 + \frac{100}{2.2}\right) \times 10^{3'} \tag{5.1.4.10-13}$$

$= 0.849$

Design P14: $\quad F(x_{14}) = 0.999996 \times (0.02 \times 2.3 - 0.02 \times 2.3^3)$

$$x_{14} = D_{14} = 2.3 \qquad + 0.000004 \times \left(50 \times 2.3^2 + \frac{100}{2.3}\right) \times 10^3 \tag{5.1.4.10-14}$$

$= 0.851$

Design P15: $\quad F(x_{15}) = 0.999996 \times (0.02 \times 2.4 - 0.02 \times 2.4^3)$

$$x_{15} = D_{15} = 2.4 \qquad + 0.000004 \times \left(50 \times 2.4^2 + \frac{100}{2.4}\right) \times 10^3 \tag{5.1.4.10-15}$$

$= 0.850$

Design P16: $\quad F(x_{16}) = 0.999996 \times (0.02 \times 2.5 - 0.02 \times 2.5^3)$

$$x_{16} = D_{16} = 2.5 \qquad + 0.000004 \times \left(50 \times 2.5^2 + \frac{100}{2.5}\right) \times 10^3 \tag{5.1.4.10-16}$$

$= 0.847$

2. Note that Table 5.1.4.2 is obtained based on non-normalized UFO $F(\mathbf{x})$ when minimizing bi-objective functions. UFO $F(\mathrm{x}_i)(i=1, 2, \ldots, 16)$ obtained from the above calculations for Case 9 is {0.780, 0.777, 0.779, 0.784, 0.791, 0.799, 0.808, 0.817, 0.826, 0.833, 0.840, 0.845, 0.849, 0.851, 0.850, 0.847}. It is easily seen that Design P2 shown as $F(\mathrm{x}_2)=0.777$ for Case 9 provides a minimum of UFO $F(\mathbf{x})$ among 16 designs, and hence, Design P2 is a minimized UFO $F(\mathbf{x})$ based on a fraction combination of two objective functions, $w_1 = 0.999996$ and $w_2 = 0.000004$. Calculations similar to Case 9 are performed to find Design P3 of Case 6 ($w_1 = 0.9999$, $w_2 = 0.0001$) and Design P4 of Case 7 ($w_1 = 0.99999$, $w_2 = 0.00001$). It is noteworthy that a fraction combination of two objective functions (w_1, w_2) in Case 2 ($w_1 = 0$, $w_2 = 1$), Case 3 ($w_1 = 0.9$, $w_2 = 0.1$), Case 4 (0.99, $w_2 = 0.01$), Case 5 (0.999, $w_2 = 0.001$), and Case 6 (0.9999, $w_2 = 0.0001$), result in Design P5 as a nondominated point.

Weighted sum method is a method where each objective function is weighted with a corresponding ratio in an UFO $F(\mathbf{x})$. Specifically, $w_1, w_2, \ldots, w_m$ represent a fraction based on UFO $F(\mathrm{x}_i)(i=1, 2, \ldots, m)$ in which an optimization problem has m objective functions to minimize simultaneously. Each fraction w_i $(i = 1, 2, \ldots, m)$ is multiplied to its corresponding objective function $f_i(x)$, forming m multiples in UFO. Each fraction w_i varies from 0 to 1 whereas a sum of m fractions $(w_1, w_2, \ldots, w_m)$ is equal to 1. UFO consisting of m objective functions weighted with m fractions $(w_1, w_2, \ldots, w_m)$ is optimized to provide a minimum or maximum of UFO $F(\mathbf{x})$ in a criterion space which is a set of m objective functions. Notably, each combination of m fractions presents a particular contribution made by each objective function so that *weighted sum method* is prioritized based on a contribution made by each objective function. A drawback of this method is time-consuming since it requires calculations to compare UFO $F(\mathbf{x})$ for all feasible points such as calculations shown in Equations (5.1.4.10-1)–(5.1.4.10-16). For example, there are 16 feasible design points in Example 2, requiring 16 calculations of UFO $F(\mathbf{x})$ in each case of fractions. Section 5.2 introduces an optimization algorithm implementing *weighted sum method* and *normalization functions* to overcome this drawback.

5.1.5 Normalized unified function of objectives implementing weighted sum method

5.1.5.1 Normalized UFOs implementing weighted sum method

This chapter introduces a normalized UFO based on the maximum and minimum of each objective function as shown in Equation (5.1.5.1), from which each normalized objective function $f_i^N(\mathbf{x})$ vary from 0 to 1 so that no objective functions outweigh others.

$$f_i^N(\mathbf{x}) = \frac{f_i(\mathbf{x}) - f_i^{\min}(\mathbf{x})}{f_i^{\max}(\mathbf{x}) - f_i^{\min}(\mathbf{x})};\ (i=1, 2) \qquad (5.1.5.1)$$

The following calculations describe a process to optimize Example 2 when objective functions are normalized based on Equation (5.1.5.1). Normalizing functions of two objective functions, weight W and cost CI, are defined as shown in Equations (5.1.5.2-1) and (5.1.5.2-2), respectively. Minimum weight W (Objective 1) and cost CI (Objective 2) are 0.18 kgf/m and 141,429 KRW/m, respectively, while their maximums are 0.243 kgf/m and 165,000 KRW/m, respectively, as can be seen in Table 5.1.4.1.

$$f_1^N(\mathbf{x}) = \frac{f_1(\mathbf{x}) - f_1^{\min}(\mathbf{x})}{f_1^{\max}(\mathbf{x}) - f_1^{\min}(\mathbf{x})} = \frac{f_1(\mathbf{x}) - 0.18}{0.243 - 0.18} \qquad (5.1.5.2\text{-}1)$$

$$f_2^N(\mathbf{x}) = \frac{f_2(\mathbf{x}) - f_2^{\min}(\mathbf{x})}{f_2^{\max}(\mathbf{x}) - f_2^{\min}(\mathbf{x})} = \frac{f_2(\mathbf{x}) - 141{,}429}{160{,}000 - 141{,}429} \tag{5.1.5.2-2}$$

A UFO $\boldsymbol{F}(\mathbf{x})$ consisting of the two objective functions is normalized as shown in Equation (5.1.5.3).

$$\boldsymbol{F}(\mathbf{x}) = w_1 \times \frac{f_1(\mathbf{x}) - 0.18}{0.243 - 0.18} + w_2 \times \frac{f_2(\mathbf{x}) - 141{,}429}{160{,}000 - 141{,}429} \tag{5.1.5.3}$$

Table 5.1.5.1 shows 11 cases representing 11 combinations of fractions $\{w_1, w_2\}$ for the two objective functions, resulting in five optimal designs, Designs P1–P5. Calculations follow a process similar to one shown in Table 5.1.4.2 for Example 2. Substituting column dimensions with respect to each design D_i into a normalized UFO $\boldsymbol{F}(\mathbf{x})$ shown in Equation (5.1.5.3) which is based on objective functions $f_1^N(\mathbf{x})$ and $f_2^N(\mathbf{x})$ shown in Equation (5.1.5.2) yields 16 corresponding UFO $\boldsymbol{F}(\mathrm{x_i})$ to compare. An optimal design is one providing a minimum UFO $\boldsymbol{F}(\mathbf{x})$. Results of these calculations are presented in Table 5.1.5.1. Fractions w_1 and w_2 vary from 0 to 1 with a step of 0.1, leading to 11 calculations required to find five optimal design points P1–P5. It is noted that calculations based on normalized UFO are more efficient compared to those based on non-normalized objective functions.

Table 5.1.5.1 Weighted Sum Method (Yang, 2014) Implementing Fractions for Normalized UFO $\boldsymbol{F}(x)$ to Minimize Bi-Objective Functions

$$\boldsymbol{F}(x) = w_1 \times \frac{f_1(x) - 0.18}{0.243 - 0.18} + w_2 \times \frac{f_2(x) - 141{,}429}{160{,}000 - 141{,}429}$$

Case	*1*	2	3	4	5	6	7	8	9	*10*	*11*
w_1	1.0	0.0	0.9	0.8	0.7	0.6	0.5	0.4	0.3	0.2	0.10
w_2	0.0	1.0	0.1	0.2	0.3	0.4	0.5	0.6	0.7	0.8	0.90
P1	**0.000**	0.364	**0.036**	**0.073**	**0.109**	**0.145**	0.182	0.218	0.255	0.291	0.327
P2	0.211	0.190	0.209	0.207	0.205	0.203	**0.201**	0.199	0.196	0.194	0.192
P3	0.402	0.081	0.369	0.337	0.305	0.273	0.241	**0.209**	0.177	0.145	0.113
P4	0.569	0.021	0.514	0.459	0.405	0.350	0.295	0.240	**0.185**	0.131	0.076
P5	0.712	**0.000**	0.641	0.570	0.498	0.427	0.356	0.285	0.214	**0.142**	**0.071**
P6	0.829	0.010	0.747	0.665	0.583	0.501	0.419	0.338	0.256	0.174	0.092
P7	0.917	0.045	0.830	0.742	0.655	0.568	0.481	0.394	0.307	0.220	0.133
P8	0.974	0.102	0.887	0.800	0.713	0.625	0.538	0.451	0.363	0.276	0.189
P9	1.000	0.175	0.918	0.835	0.753	0.670	0.588	0.505	0.423	0.340	0.258
P10	0.991	0.263	0.919	0.846	0.773	0.700	0.627	0.554	0.482	0.409	0.336
P11	0.947	0.364	0.889	0.830	0.772	0.714	0.655	0.597	0.539	0.480	0.422
P12	0.865	0.475	0.826	0.787	0.748	0.709	0.670	0.631	0.592	0.553	0.514
P13	0.742	0.595	0.728	0.713	0.698	0.683	0.669	0.654	0.639	0.625	0.610
P14	0.579	0.723	0.593	0.608	0.622	0.636	0.651	0.665	0.680	0.694	0.709
P15	0.371	0.859	0.420	0.469	0.517	0.566	0.615	0.664	0.712	0.761	0.810
P16	0.118	1.000	0.207	0.295	0.383	0.471	0.559	0.647	0.736	0.824	0.912

Red color values indicate Pareto frontier.

5.1.5.2 Discussion on normalized objective and nonnormalized functions

This section aims to introduce normalized functions of objectives for a multi-objective optimization problem using *weighted sum method*. Section 5.1.4.2 implements *weighted sum method* to find optimal points based on non-normalized UFO $\mathbf{F}(\mathbf{x})$ leads to a fraction w_1 used for Objective 1 (weight, W) much greater than w_2 used for Objective 2 (cost), as shown in Table 5.1.4.2. This is because two objective functions are non-normalized so that magnitudes of Objective function 1 (weight, W) of 16 designs (P1–P16 of Table 5.1.4.2) are significantly small compared to those of Objective function 2 (cost *CI*), as shown in Example 2. Specifically, weight W ranges from 0.18 to 0.24 kgf/m while cost is in a range of 140,000–165,000 KRW/m as shown in Figure 5.1.4.1 so that cost *CI* is 10^7 times greater than weight W.

Nondominated points or optimal solutions of Example 2 are found based on nine iterations of trial fractions (w_1,w_2) shown in Cases 1 to 9 of Table 5.1.4.2. Figure 5.1.5.1 illustrates these nine iterations based on nine fractions (w_1, w_2) in which the y-axis represents non-normalized UFO $\mathbf{F}(\mathbf{x})$ of 16 design points P1 to P16 whereas the x-axis represents column diameters. UFO $\mathbf{F}(\mathbf{x})$ is calculated based on the formula shown in Equation (5.1.4.6), UFO $F(\mathbf{x}) = w_1 \times \left(0.02 \times D - 0.02 \times D^3\right) + w_2 \times \left(50 \times D^2 + \frac{100}{D}\right) \times 10^3$.

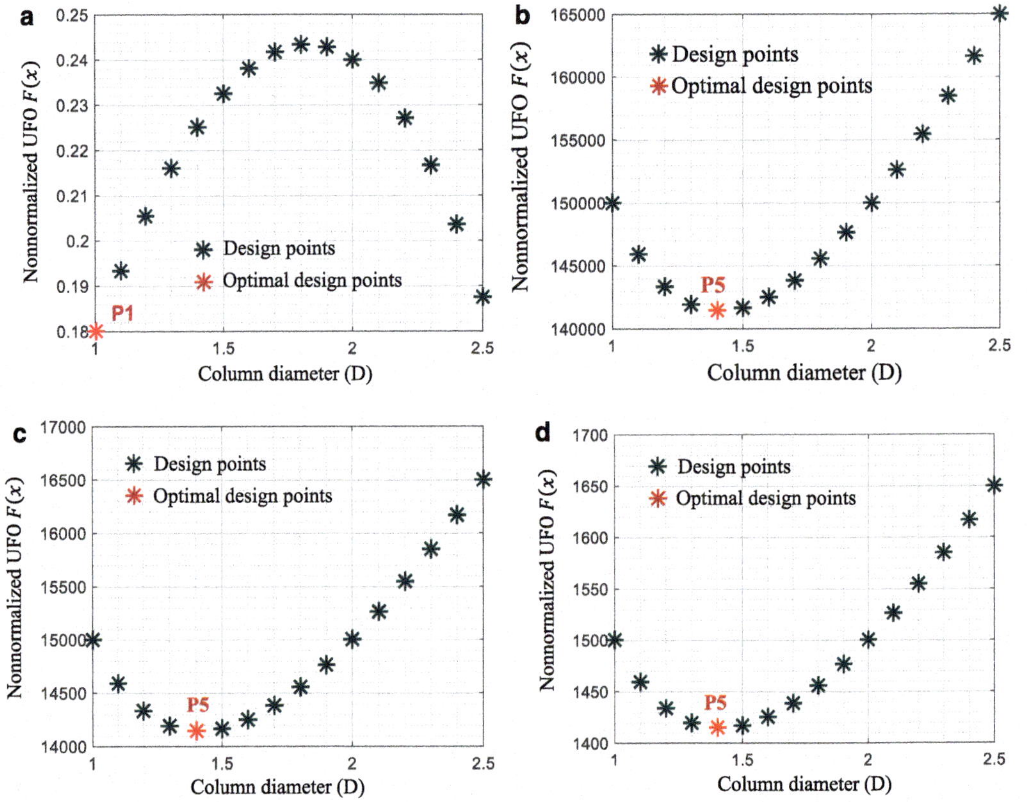

Figure 5.1.5.1 Obtaining optimal point on Pareto frontier with respect to trial fractions w_1, w_2 shown in Table 5.1.4.2. (a) Case 1: $w_1 = 1$, $w_2 = 0$. (b) Case 2: $w_1 = 0$, $w_2 = 1$. (c) Case 3: $w_1 = 0.9$, $w_2 = 0.1$. (d) Case 4: $w_1 = 0.99$, $w_2 = 0.01$. (e) Case 5: $w_1 = 0.999$, $w_2 = 0.001$. (f) Case 6: $w_1 = 0.9999$, $w_2 = 0.0001$. (g) Case 7: $w_1 = 0.999990$, $w_2 = 0.000010$. (h) Case 8: $w_1 = 0.999993$, $w_2 = 0.000007$. (i) Case 9: $w_1 = 0.999996$, $w_2 = 0.000004$.

(Continued)

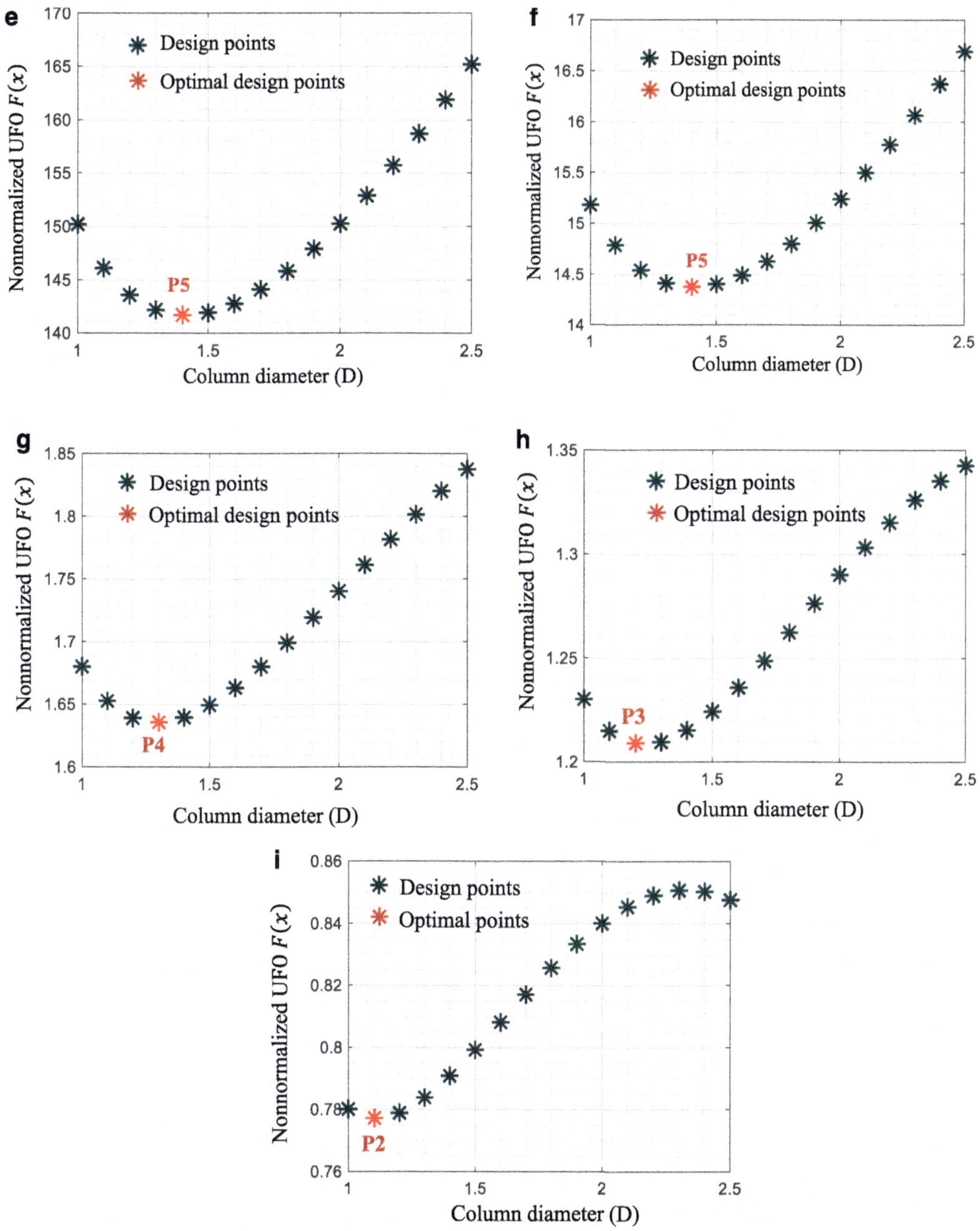

Figure 5.1.5.1 (Continued) Obtaining optimal point on Pareto frontier with respect to trial fractions w_1, w_2 shown in Table 5.1.4.2. (a) Case 1: $w_1 = 1$, $w_2 = 0$. (b) Case 2: $w_1 = 0$, $w_2 = 1$. (c) Case 3: $w_1 = 0.9$, $w_2 = 0.1$. (d) Case 4: $w_1 = 0.99$, $w_2 = 0.01$. (e) Case 5: $w_1 = 0.999$, $w_2 = 0.001$. (f) Case 6: $w_1 = 0.9999$, $w_2 = 0.0001$. (g) Case 7: $w_1 = 0.999990$, $w_2 = 0.000010$. (h) Case 8: $w_1 = 0.999993$, $w_2 = 0.000007$. (i) Case 9: $w_1 = 0.999996$, $w_2 = 0.000004$.

Figure 5.1.5.1c–f of Cases 3–6 show that an optimal point of Design P5 is obtained when using $w_1 = 0.9 \rightarrow 0.99 \rightarrow 0.999 \rightarrow 0.9999$ with respect to $w_2 = 0.1 \rightarrow 0.01 \rightarrow 0.001 \rightarrow 0.0001$. UFO $\boldsymbol{F}(\mathbf{x})$ is noticeably reduced from Cases 3 to 6 as shown in Table 5.1.4.2 because an infinitesimal fraction w_2 reduces $w_2 \times \left(50 \times D^2 + \dfrac{100}{D}\right) \times 10^3$, lessening a contribution of

Objective 2 to a UFO $F(\mathbf{x})$ whereas a contribution of Objective 1 to weight W increases. This is shown in Cases 7–9 illustrated in Figure 5.1.5.1g–i and Table 5.1.4.2, leading to optimal points P4, P3, and P2 where smaller w_2 let optimal points P4, P3, and P2 approach an optimal point P1 (where a fraction $w_1 = 1$ and $w_2 = 0$ are used, leading to $F(\mathbf{x})$ which is equal to $f_1(\mathbf{x})$ based on only Objective function 1; refer to Equation (5.1.4.8-1)), decreasing a contribution of Objective 2.

A small interval must be used for a fraction w_1 which is an order of 3×10^{-6} as shown in Cases 7–9 ($w_1 = 0.999990 \rightarrow 0.999993 \rightarrow 0.999996$) of Table 5.1.4.2 where UFO is not normalized to find P4, P3, and P2. Table 5.1.5.1 shows that normalized UFO (normalized UFOs is explained in Section 5.1.5) avoids the need to select infinitesimal fractions. Table 5.1.5.1 also identifies five nondominated points, P1 to P5, when UFO is normalized. Five nondominated points indicated by red color are found based on 11 trial fractions shown in Table 5.1.5.1 in which UFO consisting of two objective functions are both normalized. This table shows that less duplicated pareto frontiers take place so that an interval of 0.1 can be used to divide fractions ($w_1 = 1.0 \rightarrow 0.9 \rightarrow 0.8 \rightarrow 0.7 \rightarrow 0.6 \rightarrow 0.5 \rightarrow 0.4 \rightarrow 0.3 \rightarrow 0.2 \rightarrow 0.1 \rightarrow 0$). It is also noted that Cases 1, 3, 4, 5, and 6 lead to duplicated P1 whereas Cases 2, 10, and 11 lead to duplicated P5 as shown in Table 5.1.5.1, implying that P1 and P5 are nondominated points in Cases 1, 3, 4, 5, 6 and Cases 2, 10, 11, respectively. A smaller interval of fractions is required when more duplications occur whereas a larger interval of fractions can be used to save time for optimizations if there are less duplications. However, fractions based on intervals close enough are required to provide a smooth Pareto frontier as shown in Sections 5.3.4 and 5.4.2. It concludes that less fractions based on normalized UFO help calculate well-defined smooth nondominated points on a Pareto frontier. Normalization functions helps to in order to find smooth Pareto frontier. In this chapter, the "*linspace*" command defined in MATLAB (MathWorks, 2020) is used to generate the fractions as discussed in Section 5.3.6. The duplication problem is further described in Section 5.3.8. A number of required fractions is adjustable according to the characteristics of optimization problems.

5.2 ANN-BASED LAGRANGE FUNCTIONS OPTIMIZING MULTIPLE OBJECTIVE FUNCTIONS

5.2.1 Significance of considering UFO

Previous chapters discussed an implementation of Lagrange multipliers method in single-objective function-based optimization problems using an ANN. In this chapter, the ANN-based Lagrange optimization algorithm is developed to optimize UFO formulated based on the interests of engineers, contractors, and government officials. Minimizing CI_c, CO_2 emission, and W_c is an interest of structural engineers, governments, and contractors, respectively. However, any design target can be adopted as an objective function other than the cost index, CO_2 emission, and weight if needed. A novel gradient-based algorithm is proposed for MOO problems to optimize multiple objective functions derived for reinforced concrete structures including circular RC columns, uniaxial and biaxial rectangular RC columns and RC beams. Some well-known MOO algorithms have been developed in recent years such as Nondominated Sorting Algorithm – II (NSGA-II) by K. Deb (2002), Particle Swarm optimization (Zheng, 2021 and Mohd Zain, 2018), etc. These methods are referred to as multi-objective evolutionary algorithms (MOEAs) or population methods. Such algorithms, however, are not likely to be conveniently used by engineers and decision-makers since a Pareto frontier they provide does not clearly present trade-off ratios contributed by each objective function. However, ANN-based Lagrange optimization provides a Pareto frontier uncovering the trade-off ratios by implementing *weighted sum method* (Yang, 2014). The proposed algorithm exploits characteristics of ANNs such as learning capability to derive UFO. A learning utility of ANNs is achieved by adjusting and modifying weights and biases, which is referred to as a back-propagation. Good intercorrelated relationships among input and output design parameters are identified by training ANNs on large datasets.

In this chapter, a set of single objective function is unified into a function of objectives (UFO) based on *weighted sum method* which is, then, normalized as discussed in Sections 5.1.4 and 5.1.5. ANN-based Lagrange multiplier methods minimize UFO, leading to a Pareto frontier. Newton-Raphson iterations are used to find stationary inputs for Lagrange function of UFO under KKT conditions. This section describes an algorithm of ANN-based Lagrange multiplier method to optimize UFO in five steps. The algorithms based on three objective functions to optimize structural designs including circular RC columns, uniaxial rectangular RC columns, biaxial rectangular RC columns, and doubly RC beams are introduced in Sections 5.3, 5.5–5.7, respectively. Section 5.4 discusses how to derive ANN-based UFO as a function of input and output parameters to optimize a design of RC columns sustaining multiple load pairs.

5.2.2 Unified function of objectives

UFO is a *unified function of objectives* which is a globalized function utilizing *weighted sum method* and *normalizing functions of objective* to optimize multiple objective functions at the same time. The UFO is treated as an objective function in an optimization. Optimization problems having m numbers of objective functions are described in Equation (5.2.1.1) where $F(\mathbf{x})$ is a global UFO formulated based on all single-objective functions. The UFO shown in Equations (5.2.1.2-1) and (5.2.1.2-2) is implemented in an ANN-based Lagrange optimization algorithm. Each objective function is weighted by non-negative fractions which are summed to 1 as shown in Equation (5.2.1.3). Each fraction varies in a range from 0 to 1 to generate a Pareto frontier. UFO is normalized to become dimensionless prior to optimization. Normalization of UFO is performed based on $f_i^N(\mathbf{x})$ ($i = 1, \ldots, m$) as presented in Equation (5.2.1.4) where $f_i^{min}(\mathbf{x})$ and $f_i^{max}(\mathbf{x})$ are minimum and maximum of a single-objective function, respectively. An optimization of single objective function was investigated in the previous chapters.

$$\text{Minimize}: F(\mathbf{x}) = \{f_1(\mathbf{x}), f_2(\mathbf{x}), \ldots, f_m(\mathbf{x})\}^T = \{f_i(\mathbf{x})\}^T \quad (i = 1, \ldots, m) \tag{5.2.1.1}$$

$$UFO = \{w_1, w_2, \ldots, w_m\}\{f_1^N(\mathbf{x}), f_2^N(\mathbf{x}), \ldots, f_m^N(\mathbf{x})\}^T \tag{5.2.1.2-1}$$

$$UFO = \sum_{i=1}^{m} w_i f_i^N(\mathbf{x}) = w_1 f_1^N(\mathbf{x}) + \ldots + w_m f_m^N(\mathbf{x}) \tag{5.2.1.2-2}$$

$$w_i \in [0, 1]; \ \sum_{i=1}^{m} w_i = 1 \tag{5.2.1.3}$$

$$f_i^N(\mathbf{x}) = \frac{f_i(\mathbf{x}) - f_i^{min}(\mathbf{x})}{f_i^{max}(\mathbf{x}) - f_i^{min}(\mathbf{x})} \tag{5.2.1.4}$$

5.2.3 ANN-based Lagrange optimization algorithm of five steps based on UFO

ANN-based Lagrange optimization of multi-objective functions is performed in five steps as described in Figure 5.2.3.1. ANN-based Lagrange functions for single objective function are obtained in Steps 1–3. An *UFO* is, then, derived in Step 4 to minimize it in Step 5 to identify a Pareto frontier using ANNs optimizing multiple objective functions simultaneously. Design parameters are also obtained during structural optimizations based on an *UFO*.

Step 1: Establishing and training of ANNs.

ANNs are derived based on a topology of networks including a number of hidden layers and neurons, and training method as shown in sub-steps 1.1–1.3. They are trained by large structural datasets including inputs and output parameters selected depending on design natures.

Step 1.1: Selecting design parameters to generate large datasets

Structural mechanics-based software is developed based on conventional structural design to generate large datasets for training ANNs. Inputs and output parameters are selected depending on the types of structures under considerations. Table 5.2.3.1 shows the names of software used to generate large datasets for training ANNs shown in Sections 5.3–5.6. In Section 5.3, "*AutoCC*" is used to generate large datasets for a design of circular RC columns whereas "*AutoCol*" and "*AutoRC2A*" shown in Sections 5.5 and 5.6 are used to generate large datasets for a design of uniaxial and biaxial rectangular RC columns, respectively. Section 5.7 presents "*Autobeam*" to generate large datasets for a design of RC beams.

Step 1.2: Generating large datasets used to train ANNs

Large structural datasets are generated using software shown in Table 5.2.3.1.

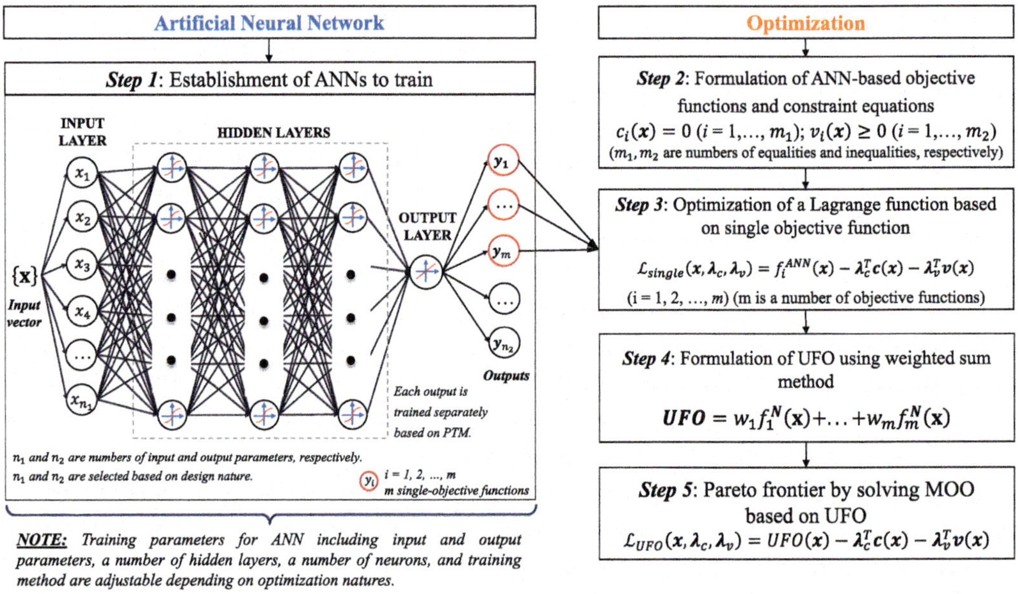

Figure 5.2.3.1 ANN-based Lagrange optimization algorithm following five steps.

Table 5.2.3.1 A Type of Software Used to Generate Large Structural Datasets

Section	*Type of Structure*	*Structural Mechanics-Based Software*
5.3	Circular RC column	*AutoCC*
5.5	Uniaxial Rectangular RC columns	*AutoCol*
5.6	Biaxial Rectangular RC columns	*AutoRC2A*
5.7	RC beams	*Autobeam*

Step 1.3: Training ANNs
Training parameters must be selected to efficiently train ANNs, including the following features.

- A number of hidden layers
- A number of neurons
- A type of activation functions applied to all layers including the output layer
- A number of required epochs and number of validated epochs
- A training method such as training on entire datasets (TED), parallel training method (PTM), chained training scheme with revised sequence (CRS) (Hong, 2020, 2021, Hong et al., 2021c).

MATLAB Training Toolbox (MathWorks, 2020) is used to train ANNs based on training parameters described above.

Step 2: Defining MOO problem
This step initially defines a number of objective functions described in Equation (5.2.3.1) to maximize or minimize m objective functions, where $\mathbf{x}$ is input parameter of ANNs as introduced in Step 1.1.

Step 2.1: Deriving objective functions
$f_1(\mathbf{x}), f_2(\mathbf{x}), \ldots, f_m(\mathbf{x})$ are functions of objective, 1, 2, ..., to m, derived from the trained ANNs in Step 1.3.

Step 2.2: Deriving equalities and inequalities constraints
Constrained conditions for equality constraints ($c_j(\mathbf{x}) = 0;\ j = 1,\ 2,\ \ldots,\ m_1$) shown in Equation (5.2.3.2-1) and for inequality constraints ($v_j(\mathbf{x}) \geq 0;\ j = 1, 2, \ldots, m_2$) shown in Equation (5.2.3.2-2) which must be established prior to optimizing objective functions.

$$\text{maximize / minimize } f_1(\mathbf{x}), f_2(\mathbf{x}), \ldots, f_m(\mathbf{x})\ (m = \text{a number of objectives}) \tag{5.2.3.1}$$

$$\text{subject to}: \boldsymbol{c}(\mathbf{x}) = \left[c_1(\mathbf{x}), c_2(\mathbf{x}), \ldots, c_{m_1}(\mathbf{x})\right]^T = 0 \tag{5.2.3.2-1}$$

$$\boldsymbol{v}(\mathbf{x}) = \left[v_1(\mathbf{x}), v_2(\mathbf{x}), \ldots, v_{m_2}(\mathbf{x})\right]^T \geq 0 \tag{5.2.3.2-2}$$

Step 3: Optimizing single objective function
This step is to optimize each objective function to normalize UFO, which is described in the following steps. The readers can refer to Chapters 1–4 and Hong and Nguyen, 2021 and Hong et al., 2021b for more details.

Step 3.1: Lagrange function based on single-objective function, equalities and inequalities is derived. Lagrange functions are derived by using Lagrange multipliers λ_c and λ_v subject to equalities and inequalities, respectively, shown in Equation 5.2.3.3, and a diagonal matrix $\mathbf{S}$ shown in Equation (5.2.3.4). Components $s_1, s_2, \ldots, s_{m2}$ determines whether inequalities $v_1(\mathbf{x}), v_2(\mathbf{x}), \ldots, v_{m_2}(\mathbf{x})$ are activated or inactivated. For $j \in \{1, 2, ..., m_2\}$ an activated inequality v_j is assumed by setting $s_j = 1$, resulting in a binding condition, whereas s_j is set as 0 when an inequality v_j is inactive for any , resulting in a complementary slackness condition with $\mathrm{S}v(\mathbf{x}) = 0$. Stationary points of UFO are identified to optimize solution by equating first derivative of Lagrange function to 0, which is expressed as Equations (5.2.3.5-1) and (5.2.3.5-2).

$$\mathcal{L}_{single}(\mathbf{x}, \boldsymbol{\lambda}_c, \boldsymbol{\lambda}_v) = f_i(\mathbf{x}) - \boldsymbol{\lambda}_c^{\mathrm{T}} c(\mathbf{x}) - \boldsymbol{\lambda}_v^{\mathrm{T}} \mathrm{S}v(\mathbf{x}); \ (i = 1, ..., m) \tag{5.2.3.3}$$

$$\mathrm{S} = \begin{bmatrix} s_1 & \cdots & 0 \\ \vdots & \ddots & \vdots \\ 0 & \cdots & s_{m_2} \end{bmatrix} \tag{5.2.3.4}$$

$$\nabla\mathcal{L}(\mathbf{x}, \boldsymbol{\lambda}_c, \boldsymbol{\lambda}_v) = \begin{bmatrix} \nabla f(\mathbf{x}) - [\mathrm{J}_c(\mathbf{x})]^{\mathrm{T}} \boldsymbol{\lambda}_c - [\mathrm{J}_v(\mathbf{x})]^{\mathrm{T}} \mathrm{S}\boldsymbol{\lambda}_v \\ -c(\mathbf{x}) \\ -\mathrm{S}v(\mathbf{x}) \end{bmatrix} \tag{5.2.3.5-1}$$

$$\mathrm{J}_c(\mathbf{x}) = \begin{bmatrix} \nabla c_1(\mathbf{x}) \\ \nabla c_2(\mathbf{x}) \\ \vdots \\ \nabla c_{\mathrm{m}_1}(\mathbf{x}) \end{bmatrix} \text{ and } \mathrm{J}_v(\mathbf{x}) = \begin{bmatrix} \nabla v_1(\mathbf{x}) \\ \nabla v_2(\mathbf{x}) \\ \vdots \\ \nabla v_{\mathrm{m}_2}(\mathbf{x}) \end{bmatrix} \tag{5.2.3.5-2}$$

Step 3.2: Lagrange KKT equations by differentiating Lagrange functions for UFO. Active and in active inequalities are selected by setting a diagonal matrix **S** to 1 and 0, respectively.

Step 3.3: Trial input parameters $\mathbf{x}^{(0)}$ and Lagrange multipliers $\boldsymbol{\lambda}_c^{(0)}$, $\boldsymbol{\lambda}_v^{(0)}$ are initialized to begin Newton-Raphson iterations.

Step 3.4: First derivative of Lagrange function is calculated based on initial trial parameters by substituting $\mathbf{x}^{(0)}$, $\boldsymbol{\lambda}_c^{(0)}$ and $\boldsymbol{\lambda}_v^{(0)}$ into Equations (5.2.3.5-1) and (5.2.3.5-2).

Step 3.5: Convergence criteria is checked as shown in Equations (5.2.3.6).

$$\nabla\mathcal{L}(\mathbf{x}, \boldsymbol{\lambda}_c, \boldsymbol{\lambda}_v) \cong 0 \tag{5.2.3.6}$$

Convergence criterion is verified if initial trial parameters ($\mathbf{x}^{(0)}$, $\boldsymbol{\lambda}_c^{(0)}$, $\boldsymbol{\lambda}_v^{(0)}$) satisfy $\nabla\mathcal{L}\left(\mathbf{x}^{(0)}, \boldsymbol{\lambda}_c^{(0)}, \boldsymbol{\lambda}_v^{(0)}\right) \cong 0$, otherwise, next trial parameters should be attempted.

Step 3.6: Lagrange function shown in Equation (5.2.3.12) is linearized based on Jacobi and Hessian matrices as shown in Equation (5.2.3.7). Trial input parameters are updated using Equation (5.2.3.8) to solve for KKT conditions until the solution converges. Trial parameters are updated to ($\mathbf{x}^{(k+1)}$, $\boldsymbol{\lambda}_c^{(k+1)}, \boldsymbol{\lambda}_v^{(k+1)}$) at iteration $(k+1)^{th}$ using Equations (5.2.3.7) and (5.2.3.8) when the previous trial input parameters at k^{th} iteration ($\mathbf{x}^{(k)}$, $\boldsymbol{\lambda}_c^{(k)}, \boldsymbol{\lambda}_v^{(k)}$) do not meet a criterion $\nabla\mathcal{L}\left(\mathbf{x}^{(k)}, \boldsymbol{\lambda}_c^{(k)}, \boldsymbol{\lambda}_v^{(k)}\right) \cong 0$ stated in Equation (5.2.3.6).

$$\begin{bmatrix} \Delta \mathbf{x}^{(k)} \\ \Delta \lambda_c^{(k)} \\ \Delta \lambda_v^{(k)} \end{bmatrix} \approx -\left[\mathbf{H}_{\mathcal{L}}\left(\mathbf{x}^{(k)}, \lambda_c^{(k)}, \lambda_v^{(k)}\right)\right]^{-1} \nabla \mathcal{L}\left(\mathbf{x}^{(k)}, \lambda_c^{(k)}, \lambda_v^{(k)}\right) \tag{5.2.3.7}$$

$$\begin{bmatrix} \mathbf{x}^{(k+1)} \\ \lambda_c^{(k+1)} \\ \lambda_v^{(k+1)} \end{bmatrix} = \begin{bmatrix} \mathbf{x}^{(k)} \\ \lambda_c^{(k)} \\ \lambda_v^{(k)} \end{bmatrix} + \begin{bmatrix} \Delta \mathbf{x}^{(k)} \\ \Delta \lambda_c^{(k)} \\ \Delta \lambda_v^{(k)} \end{bmatrix} \tag{5.2.3.8}$$

Trial parameters $\mathbf{x}^{(k+1)}$, $\lambda_c^{(k+1)}, \lambda_v^{(k+1)}$ at iteration $(k+1)^{th}$ are the stationary point optimizing UFO if convergence $\nabla \mathcal{L}\left(\mathbf{x}^{(k+1)}, \lambda_c^{(k+1)}, \lambda_v^{(k+1)}\right) \cong 0$ is met. This step is repeated until convergence is found.

Step 4: Formulating UFO equation

Step 4.1: Each objective function is normalized as described in Equation (5.2.3.9).

$$f_i^N(\mathbf{x}) = \frac{f_i(\mathbf{x}) - f_i^{min}(\mathbf{x})}{f_i^{max}(\mathbf{x}) - f_i^{min}(\mathbf{x})}; \ (i = 1, 2, \ldots, m) \tag{5.2.3.9}$$

Step 4.2: UFO is formulated in Equation (5.2.3.10) based on fractions w_i of each objective function shown in Equation (5.2.3.11) which varies from 0 to 1.

$$UFO = w_1 f_1^N(\mathbf{x}) + \ldots + w_m f_m^N(\mathbf{x}); \tag{5.2.3.10}$$

$$w_i \in [0, 1]; \ \sum_{i=1}^{m} w_i = 1 \tag{5.2.3.11}$$

Step 5: Optimizing UFO

Lagrange function of UFO (defined in Step 4) constrained by equality and inequality conditions is derived using Lagrange multipliers introduced in Step 2 as presented in Equation (5.2.3.12). Multiple objective functions are simultaneously optimized in this step, yielding a Pareto frontier. Equation 5.2.3.12 shows that contributions of each objective function to the overall optimizations are demonstrated by fractions *wi*.

$$\mathcal{L}_{UFO}(\mathbf{x}, \lambda_c, \lambda_v) = UFO(\mathbf{x}) - \lambda_c^T c(\mathbf{x}) - \lambda_v^T Sv(\mathbf{x}) \tag{5.2.3.12}$$

5.3 ANN-BASED LAGRANGE OPTIMIZATION DESIGN OF RC CIRCULAR COLUMNS HAVING MULTIPLE OBJECTIVE

Section 5.3 discusses an optimization design of a circular RC column in which three objective functions such as cost, estimated CO_2 emission, and column weight, are minimized at the same time. Section 5.3.3 begins with forward design of circular RC columns to generate large datasets, formulating five steps to optimize circular RC column based on three objective functions described in Section 5.2.3. Section 5.3.4 discusses optimal results obtained by ANN-based Lagrange optimization based on multiple objective functions.

5.3.1 Forward design of circular RC columns

A circular RC column section and its corresponding P-M diagram are shown in Figure 5.3.1.1. Conventional design determines column dimensions including diameter (D) and rebar ratio (ρ_s) to sustain given factored axial loads (P_u) and factored bending moments (M_u). Material properties such as specified compressive strength of concrete (f_c') and specified yield strength of reinforcement (f_y) are commonly preassigned in design. A design of RC columns should meet strength requirements specified in ACI 318-19 (ACI Committee, 2019), reflected by safety factor $SF = P_n/P_u = M_n/M_u \geq 1.0$. Rebar ratio ρ_s is also determined between 0.01 and 0.08 as specified by design code (ACI 318-19 (ACI Committee, 2019)). Besides, $\alpha_{e/h}$ is required to construct an axial load-bending moment interaction diagram (P–M diagram) shown in Figure 5.3.1.1 based on a design range between 0 to $\pi/2$. Cost index, CO_2 emission, and column weight denoted as CI_c, CO_2, and W_c, respectively, are determined for RC columns. Columns cost and weight are estimated based on Korean Won indicated in Table 5.3.1.1, while the determination of CO_2 emissions is followed by Hong et al. (2010). Input parameters are selected as $\{D,\ \rho_s,\ f_c', f_y, P_u, M_u\}^T$ to calculate output parameters $\{SF, \varepsilon_s, CI_c, CO_2, W_c, \alpha_{e/h}\}^T$. A selection of design parameters of circular RC columns and their nomenclatures are presented in Table 5.3.1.2.

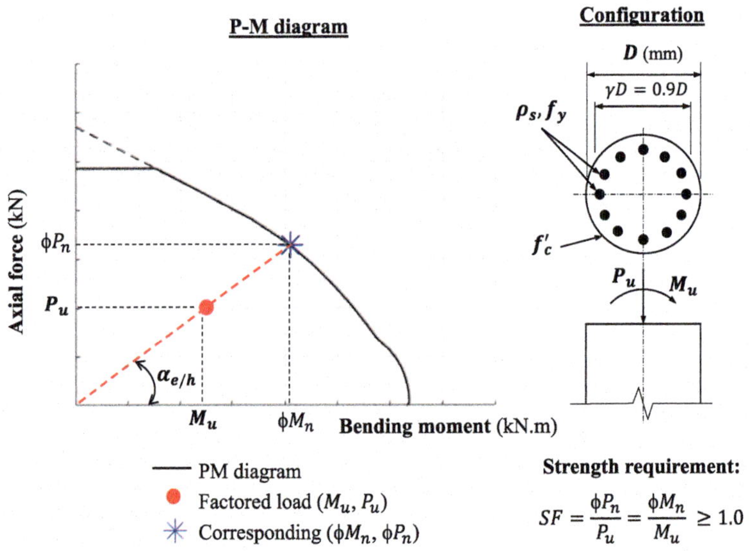

Figure 5.3.1.1 A column section and corresponding P-M diagram.

Table 5.3.1.1 Unit Costs, Weights, and CO_2 Emissions of Materials Based on Korean Won (Hong, 2010)

Material	*Strength*	*Cost (Material and Manufacturing)*	*CO_2 Emission*	*Unit Weight*
Concrete	30 MPa	85,000 (KRW/m^3)	0.1677 (t-CO_2/m^3)	2.356 T/ m^3
	40 MPa	94,000 (KRW/m^3)		
	50 MPa	104,000 (KRW/m^3)		
Rebar	500 MPa	1,055 (KRW/kg)	2.512 (t-CO_2/ton)	7.85 T/ m^3
	600 MPa	1,085 (KRW/kg)		

Table 5.3.1.2 Nomenclature of Design Parameters

Notation	*Nomenclature*
D	Column section diameter
ρ_s	Rebar ratio
f_c'	Concrete strength
f_y	Rebar yield strength
P_u	Factored axial load
M_u	Factored bending moment
SF	Safety factor
s	Rebar strain
CI_c	Cost index of column per height
CO_2	CO_2 emission per column height
W_c	Weight of column per column height
$\alpha_{e/h}$	Angular eccentricity of applied load

5.3.2 Optimization design scenarios

Multi-objective optimization results in a design with a minimized column weight, which is advantageous from both architectural and assembly perspectives, while its cost index (CI_c) and estimated CO_2 emissions are also minimized. A column section and corresponding P-M diagram are shown in Figure 5.3.1.1. An ANN is developed to map five input parameters to six output parameters. Input and output parameters are selected to perform a forward design of RC columns under axial loads and flexural moments. Input parameters, $\mathbf{x} = \left\{D, \rho_s, f_c', f_y, P_u, M_u\right\}^T$, denote column dimension, rebar ratio, material properties (concrete and steel), factored axial load, and factored moment. Output parameters $\mathbf{y} = \left\{SF, \varepsilon_s, CI_c, CO_2, W_c, \alpha_{e/h}\right\}^T$ are selected to represent safety factors, rebar strain, cost index, CO_2 emission, and column weight, in which an output parameter for safety factor, $SF = P_n/P_u = M_n/M_u \geq 1.0$, must reflect a strength requirement for a column design. Constraining conditions, such as the strength requirement of $SF = P_n/P_u = M_n/M_u \geq 1.0$ for a design of RC circular columns, are given in Table 5.3.2.1. Circular RC columns are under a single load combination of factored axial load (P_u = 1,500 kN) and factored moment (M_u = 2,500 kN.m). Material properties, $f_c' = 40$ MPa and $f_y = 500$ MPa are also given.

5.3.3 Five steps to optimize circular RC column based on three-objective functions

Design optimizing UFO of RC columns is a multiple-constrained problem, such that solving for stationary points optimizing multiple objective functions is difficult. The Lagrange multipliers are used to transfer constrained problems into non-boundary problems to optimize

Table 5.3.2.1 Constraining Conditions for a Design of Circular RC Columns

Factored Load	Axial Load	P_u = 1,500 kN
	Moment	M_u = 2,500 kN.m
Material properties	Concrete strength	f_c' = 40 MPa
	Rebar yield strength	f_y = 500 MPa
Strength requirement	$SF \geq 1.0$	

multiple objective functions. A MOO based on UFO proposed in this chapter is described in five steps presented in Section 5.2.3 as follows.

Step 1: ANNs are established based on three sub-steps, which are described as follows.

Step 1.1: Selection of inputs $(D, \rho_s, f_c^{'}, f_y, P_u, M_u)$ and outputs $(SF, \varepsilon_s, CI_c, CO_2, W_c, \alpha_{e/h})$ for ANNs.

An ANN is developed to map five input parameters to six output parameters. Input and output parameters are selected to perform a forward design of RC columns under axial loads and flexural moments. ANN-based Lagrange multiplier method is used to minimize a design of round columns shown in Figure 5.3.1.1. ANN-based input parameters are $\mathbf{x} = \left\{D, \rho_s, f_c^{'}, f_y, P_u, M_u\right\}^T$ including column dimension, rebar ratio, material properties (concrete and steel), factored axial load and factored moment. ANN-based output parameters $\mathbf{y} = \left\{SF, \varepsilon_s, CI_c, CO_2, W_c, \alpha_{e/h}\right\}^T$ are safety factor, rebar strain, cost index, CO_2 emission, and column weight, in which safety factor is to ensure $SF = P_n/P_u = M_n/M_u \geq 1.0$

Step 1.2: Generation of large structural datasets.

Parameters D, ρ_s, $f_c^{'}$, f_y, SF, $\alpha_{e/h}$ are pre-described on an input-side in a range as shown in Table 5.3.3.1 when generating one hundred thousand datasets shown in Table 5.3.3.2. The structural mechanics-based software *AutoCC* is used to generate large datasets by randomly selecting D, ρ_s, $f_c^{'}$, f_y, SF, $\alpha_{e/h}$ within the ranges defined in Table 5.3.3.1 on an input side to calculate $\varepsilon_s, CI_c, CO_2, W_c$, P_u, M_u on an output side. It is noted that P_u and M_u are calculated based on $SF = \frac{\phi P_n}{P_u} = \frac{\phi M_n}{M_u}$ where SF is randomly selected on an input-side and ϕP_n, ϕM_n are the section capacity estimated by *AutoCC* on an output-side (not shown in the large datasets and neural networks) based on randomly selected D and ρ_s. It is noted that $\alpha_{e/h}$ is required to determine P_u and M_u for given design loads (ϕP_n and ϕM_n) when generating large datasets using structural mechanics-based software (AutoCC). ANNs map the input parameters $D, \rho_s, f_c^{'}, f_y, P_u, M_u$ to the output parameter SF during a training, making SF be independent of $\alpha_{e/h}$. Thus, $\alpha_{e/h}$ is neglected when training ANNs to reduce training complexity. Notably, $\alpha_{e/h}$ is neglected in the training. This is because this parameter is only required to construct an interaction P–M diagram of RC columns, which can be determined in design after training using P_u, M_u, and SF. Therefore, $\alpha_{e/h}$ is requisite for data generation but its training is trivial.

Step 1.3: Training ANNs

The parallel training method (PTM) is then used to train ANN using five, ten layers with 20, 50, 80 neurons in each hidden layer. Figure 5.3.3.1 illustrates the

Table 5.3.3.1 Ranges of Input Parameters When Generating Large Datasets Using *AutoCC*

Parameters	*Random Range to Generate Large Datasets*
D	400 ~ 1,200 mm
ρ_s	0.01 ~ 0.08
$f_c^{'}$	20 ~ 70 MPa
f_y	300 ~ 550 MPa
SF	0.1 ~ 3.0
$\alpha_{e/h}$	0 ~ π/2

Table 5.3.3.2 Large Datasets

(a): Non-Normalized												
100,000 datasets	Six Inputs for *AutoCC* to Generate Large Datasets						Six Corresponding Outputs (*AutoCC*) to Generate Large Datasets					
	D[a]	ρ_s[a]	f_c'[a]	f_y[a]	SF[b]	$\alpha_{e/h}$[b]	P_u[a]	M_u[a]	CI_c[b]	CO_2[b]	W_c[b]	ε_s[b]
	mm		MPa	MPa			kN	kN.m∞	KRW/m	t-CO_2/m	kN/m	
	830	0.047	67	469	2.631	1.066	2,817.1	1,290.6	274,563	0.59	12.75	0.0029
	413	0.034	59	437	0.132	0.267	1,291.2	1,949.0	52,112	0.11	3.16	0.0060
	1827	0.057	34	470	2.144	0.945	11,904.4	15,728.7	1,469,495	3.41	61.76	0.0025
	.	.	.	.	.	.	.	.	.	.	.	.
	.	.	.	.	.	.	.	.	.	.	.	.
	.	.	.	.	.	.	.	.	.	.	.	.
	1752	0.040	64	305	1.888	0.928	10,904.1	14,309.0	1,046,932	2.33	56.80	0.0052
	957	0.019	20	411	2.139	1.019	1,594.5	938.5	165,381	0.39	16.95	0.0030
	663	0.054	35	540	2.934	1.303	2,163.2	393.5	188,007	0.43	8.13	0.0010
Max	2,000	0.080	70	600	2.990	1.571	1,352,136.0	543,444.6	2,444,429	5.46	74.02	0.0258
Min	400	0.010	20	300	0.100	0.000	0.0	0.0	21,056	0.05	2.96	–0.0030
Mean	1202	0.045	44.9	450	1.548	0.784	15,214.9	13,513.1	608,982.5	1.38	30.67	0.0040

[a] Parameters on an input-side the ANN.
[b] Parameters on an output-side the ANN.

(*Continued*)

Table 5.3.3.2 (Continued) Large Datasets

(b): *Normalized from −1 to 1*												
100,000 datasets	Six Inputs for *AutoCC* to Generate Large Datasets						Six Corresponding Outputs (*AutoCC*) to Generate Large Datasets					
	D [a]	ρ_s [a]	f_c' [a]	f_y [a]	SF[b]	$\alpha_{e/h}$ [b]	P_u [a]	M_u [a]	CI_c [b]	CO_2 [b]	W_c [b]	ε_s [b]
	mm		MPa	MPa			kN	kN.m∞	KRW/m	t-CO_2/m	kN/m	
	–0.4625	0.0594	0.8800	0.1267	0.7452	0.3578	–0.9958	–0.9953	–0.7908	–0.7982	–0.7245	–0.5935
	–0.9838	–0.3162	0.5600	–0.0867	–0.9780	–0.6600	–0.9981	–0.9928	–0.9744	–0.9758	–0.9945	–0.3773
	0.7838	0.3564	–0.4400	0.1333	0.4100	0.2028	–0.9824	–0.9421	0.1954	0.2427	0.6552	–0.6225
	.	.	.	.	.	.	.	.	.	.	.	.
	.	.	.	.	.	.	.	.	.	.	.	.
	.	.	.	.	.	.	.	.	.	.	.	.
	0.6900	–0.1306	0.7600	–0.9667	0.2331	0.1815	–0.9839	–0.9473	–0.1533	–0.1579	0.5154	–0.4333
	–0.3038	–0.7409	–1.0000	–0.2600	0.4059	0.2979	–0.9976	–0.9965	–0.8809	–0.8728	–0.6063	–0.5875
	–0.6713	0.2666	–0.4000	0.6000	0.9549	0.6591	–0.9968	–0.9986	–0.8622	–0.8592	–0.8544	–0.7216
Max	1.0000	1.0000	1.0000	1.0000	1.0000	1.0000	1.0000	1.0000	1.0000	1.0000	1.0000	1.0000
Min	–1.0000	–1.0000	–1.0000	–1.0000	–1.0000	–1.0000	–1.0000	–1.0000	–1.0000	–1.0000	–1.0000	–1.0000
Mean	0.0021	0.0025	–0.0030	–0.0010	–0.0012	–0.0023	–0.9775	–0.9503	–0.5148	–0.5087	–0.2202	–0.5177

[a] Parameters on an input-side the ANN.
[b] Parameters on an output-side the ANN.

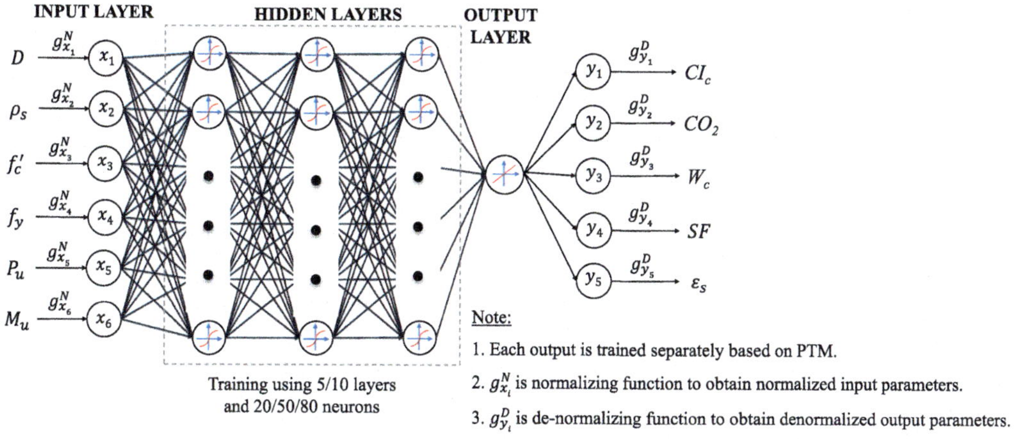

INPUT LAYER
HIDDEN LAYERS
OUTPUT LAYER
D
ρs
f'c
fy
Pu
Mu
CIc
CO2
Wc
SF
εs
Training using 5/10 layers and 20/50/80 neurons
Note:
1. Each output is trained separately based on PTM.
2. $g^N_{x_i}$ is normalizing function to obtain normalized input parameters.
3. $g^D_{y_i}$ is de-normalizing function to obtain denormalized output parameters.

Figure 5.3.3.1 Topology of an ANN.

topology of neural networks used in this chapter. Table 5.3.3.3 presents training results and training accuracies.

Step 2: Defining MOO problem

Three objective functions, cost (CI_c), CO_2 emissions, and column weight (W_c), are minimized simultaneously. Objective functions are obtained by ANNs in STEP 1 as presented by Equations (5.3.3.1-1)–(5.3.3.1-3), in which best training accuracy for objective function CI_c is obtained based on ten hidden layers and 20 neurons, while best training accuracies for CO_2 and W_c are obtained based on five hidden layers with 20 neurons and five hidden layers with 80 neurons, respectively, as shown in Table 5.3.3.3.

$$\underbrace{f^{ANN}_{CI_c}(\mathbf{x})}_{[1\times1]} = g^D_{CI_c}\left(f_{\text{linear}}\left(\underbrace{\omega^{(\text{out})}_{CI_c}}_{[1\times20]} f_{\tanh}\left(\underbrace{\omega^{j}_{CI_c}}_{[20\times20]} \cdots f_{\tanh}\left(\underbrace{\omega^{(1)}_{CI_c}}_{[20\times6]}\underbrace{g^N(X)}_{[6\times1]} + \underbrace{b^{(1)}_{CI_C}}_{[20\times1]}\right)\cdots + \underbrace{b^{(j)}_{CI_C}}_{[20\times1]}\right) + \underbrace{b^{(\text{out})}_{CI_C}}_{[1\times1]}\right)\right) \tag{5.3.3.1-1}$$

Best training with $j = 1, 2, \ldots, 10$ indicating ten hidden layers

Table 5.3.3.3 Training Results and Accuracies

Six Inputs (D, ρ_s, f'_c, f_y, P_u, M_u) - Five Outputs (SF, ε_s, CI_c, CO_2, W_c)						
PTM - 100,000 Data - 100,000 Suggested Epochs – tansig Activation Function						
Output	*Layers*	*Neurons*	*Best Epoch*	*Stopped Epoch*	*Test MSE*	*R at Best Epoch*
SF	10	80	60,372	61,372	4.0.E-05	0.9999
ε_s	10	80	58,913	59,913	2.0.E-04	0.9998
CI_c	10	20	13,224	14,224	7.6.E-07	1.0000
CO_2	5	20	43,211	44,211	6.8.E-08	1.0000
W_c	5	80	40,073	41,073	5.9.E-07	1.0000

$$\underbrace{f_{CO_2}^{ANN}(X)}_{[1\times1]} = g_{CO_2}^{D}\left(f_{linear}\left(\underbrace{\omega_{CO_2}^{(out)}}_{[1\times10]} f_{tanh}\left(\underbrace{\omega_{CO_2}^{j}}_{[10\times10]} \cdots f_{tanh}\left(\underbrace{\omega_{CO_2}^{(1)}}_{[10\times6]} \underbrace{g^N(X)}_{[6\times1]} + \underbrace{b_{CO_2}^{(1)}}_{[20\times1]}\right) \cdots + \underbrace{b_{CO_2}^{(j)}}_{[0\times1]}\right) + \underbrace{b_{CO_2}^{(out)}}_{[1\times1]}\right)\right) \quad (5.3.3.1\text{-}2)$$

Best training with j = 1, 2, ..., 5 indicating five hidden layers

$$\underbrace{f_{W_c}^{ANN}(X)}_{[1\times1]} = g_{W_c}^{D}\left(f_{linear}\left(\underbrace{\omega_{W_c}^{(out)}}_{[1\times10]} f_{tanh}\left(\underbrace{\omega_{W_c}^{j}}_{[10\times10]} \cdots f_{tanh}\left(\underbrace{\omega_{W_c}^{(1)}}_{[10\times6]} \underbrace{g^N(X)}_{[6\times1]} + \underbrace{b_{W_c}^{(1)}}_{[20\times1]}\right) \cdots + \underbrace{b_{W_c}^{(j)}}_{[10\times1]}\right) + \underbrace{b_{W_c}^{(out)}}_{[1\times1]}\right)\right) \quad (5.3.3.1\text{-}3)$$

Best training with j = 1, 2, ..., 5 indicating five hidden layers

Step 2.2: Constrained conditions of design problem are developed based on equality equations $\{c_i(\mathbf{x})=0\}$ and inequality equations $\{v_i(\mathbf{x})\geq 0\}$ shown in Table 5.3.3.4.

Equalities $(c_j(\mathbf{x})=0)$ and inequalities $(v_j(\mathbf{x})\geq 0)$ constraining conditions shown in Table 5.3.3.4 are established. In addition, column dimensions are designed in a range of 400–2,000mm while rebar ratios are limited from 0.01 to 0.08 as specified in design code ACI 318-19 (ACI Committee, 2019).

Step 3: Optimizing based on single-objective function

Step 3 is to calculate the maximum and minimum of each objective function, which are CI_c^{max}, CI_c^{min}, CO_2^{max}, CO_2^{min}, W_c^{max}, W_c^{min} by solving the Lagrange function of each objective L_{CI_c}, L_{CO_2}, L_{W_c} as follows. Steps 3.1–3.3 only minimize CI_c to calculate CI_c^{min}. The rest of maximizing and minimizing CI_c and W_c to calculate CI_c^{max}, CO_2^{max}, CO_2^{min}, W_c^{max}, W_c^{min} follows the similar process shown in Steps 3.1–3.3. Readers are referred to Chapters 1–4 optimizing single objective function including CO_2 and W_c.

Step 3.1: Formulation of Single-aminimize

Lagrange function $L_{CI_c}(\mathbf{x},\boldsymbol{\lambda}_c,\boldsymbol{\lambda}_v)$ is derived in Equation (5.3.3.2-1) where $f_{CI_c}^{ANN}(\mathbf{x})$ is an objective function of CI_c. An ANN for an objective function of CI_c is presented in Equation (5.3.3.1-1) with $\mathbf{x}$ being input parameters $\mathbf{x}=\{x_1, x_2, x_3, x_4, x_5, x_6\}^T$ where x_i $(i = 1, 2, ..., 6)$ is normalized parameters D, ρ_s, f_c', f_y, P_u, M_u in a range of −1 to 1. Lagrange functions are derived by multiplying Lagrange multipliers $\boldsymbol{\lambda}_c=[\lambda_{c1}, \lambda_{c2}, \lambda_{c3}, \lambda_{c4}]^T$ and $\boldsymbol{\lambda}_v=[\lambda_{v1}, \lambda_{v2}, \lambda_{v3}, \lambda_{v4}, \lambda_{v5}]^T$ to four equalities $\boldsymbol{c}(\mathbf{x})=[c_1(\mathbf{x}), c_2(\mathbf{x}), c_3(\mathbf{x}), c_4(\mathbf{x})]^T$ and five inequalities

Table 5.3.3.4 Equality and Inequality Constraints

Equality Constraints $\boldsymbol{c}(\mathbf{x})$	*Inequality Constraints* $\boldsymbol{v}(\mathbf{x})$
$c_1(\mathbf{x})=f_c'-40=0$	$v_1(\mathbf{x})=\rho_s-0.01\geq 0$
$c_2(\mathbf{x})=f_y-500=0$	$v_2(\mathbf{x})=-\rho_s+0.08\geq 0$
$c_3(\mathbf{x})=P_u-1,500=0$	$v_3(\mathbf{x})=D-400\geq 0$
$c_4(\mathbf{x})=M_u-2,500=0$	$v_4(\mathbf{x})=-D+2,000\geq 0$
	$v_5(\mathbf{x})=SF-1.0\geq 0$

$\boldsymbol{v}(\mathbf{x})=\left[v_1(\mathbf{x}),\ v_2(\mathbf{x}),\ v_3(\mathbf{x}),\ v_4(\mathbf{x}),\ v_5(\mathbf{x})\right]^T$, respectively, as shown in Equations (5.3.3.2-1)–(5.3.3.2-3).

$$\mathcal{L}_{CI_c}(\mathbf{x},\boldsymbol{\lambda}_c,\boldsymbol{\lambda}_v)=f_{CI_c}^{ANN}(\mathbf{x})-\boldsymbol{\lambda}_c^T c(\mathbf{x})-\boldsymbol{\lambda}_v^T Sv(\mathbf{x}) \quad (5.3.3.2\text{-}1)$$

$$\mathcal{L}_{CO_2}(\mathbf{x},\boldsymbol{\lambda}_c,\boldsymbol{\lambda}_v)=f_{CO_2}^{ANN}(\mathbf{x})-\boldsymbol{\lambda}_c^T c(\mathbf{x})-\boldsymbol{\lambda}_v^T Sv(\mathbf{x}) \quad (5.3.3.2\text{-}2)$$

$$\mathcal{L}_{W_c}(\mathbf{x},\boldsymbol{\lambda}_c,\boldsymbol{\lambda}_v)=f_{W_c}^{ANN}(\mathbf{x})-\boldsymbol{\lambda}_c^T c(\mathbf{x})-\boldsymbol{\lambda}_v^T Sv(\mathbf{x}) \quad (5.3.3.2\text{-}3)$$

A diagonal matrix S determining whether inequalities are active or inactive is shown in Equation (5.3.3.3).

$$\mathbf{S}=\begin{bmatrix} s_1 & 0 & 0 & 0 & 0 \\ 0 & s_2 & 0 & 0 & 0 \\ 0 & 0 & s_3 & 0 & 0 \\ 0 & 0 & 0 & s_4 & 0 \\ 0 & 0 & 0 & 0 & s_5 \end{bmatrix} \quad (5.3.3.3)$$

KKT conditions are derived based on inequalities which are assumed as active or inactive. Mathematically, there are five cases which are based on one active inequality, ten cases based on two active inequalities, ten cases based on three active inequalities, five cases based on four active inequalities, and one case based on five active inequalities, resulting in total of 31 cases that can occur with five inequalities shown in this example. However, that active Inequality $v_1(\mathbf{x})=\rho_s-0.01=0$ and active Inequality $v_2(\mathbf{x})=-\rho_s+0.08=0$ cannot occur at the same time. Similarly, Inequalities $v_3(\mathbf{x})$ and $v_4(\mathbf{x})$ cannot be active simultaneously, leading to Table 5.3.3.5 which presents 17 possible cases of active inequalities, $v_1(\mathbf{x})$ –$v_5(\mathbf{x})$, including five cases based on one active inequality, eight cases based on two active inequalities, and four cases based on three active inequalities. Stationary points of all 17 cases should be identified to be compared.

Step 3.2: Deriving inequality matrix based on Table 5.3.3.5.

Let Case 5 represent Inequality 5 which is active: $v_5(\mathbf{x})=SF\ -1.0=0$, leading to a diagonal matrix S as shown in Equation (5.3.3.4).

$$\mathbf{S}_{Case\ 5}=\begin{bmatrix} s_1=0 & 0 & 0 & 0 & 0 \\ 0 & s_2=0 & 0 & 0 & 0 \\ 0 & 0 & s_3=0 & 0 & 0 \\ 0 & 0 & 0 & s_4=0 & 0 \\ 0 & 0 & 0 & 0 & s_5=1 \end{bmatrix} \quad (5.3.3.4)$$

Step 3.3: Initializing trial input parameters $\mathbf{x}^{(0)}$ and Lagrange multipliers $\boldsymbol{\lambda}_c^{(0)}$ and $\boldsymbol{\lambda}_v^{(0)}$

A number of initial trial input parameters is determined as 5^n, where n is a number of variables in forward design. Initial trial input parameters $\mathbf{x}=\{D,\rho_s,f_c',f_y,P_u,M_u\}^T$ contain six parameters, four of which are typically preassigned in design: f_c',f_y,P_u,and M_u. Thus, only two parameters (D and ρ_s) vary,

Table 5.3.3.5 Possible Cases Based on Active Inequalities, $v_1(\mathbf{x})$ to $v_5(\mathbf{x})$

Case	*A Number of Active Inequalities that Each Case is Based on*	*Active Inequalities*
1	1	$v_1(x) = \rho_s - 0.01 = 0$
2	1	$v_2(x) = -\rho_s + 0.08 = 0$
3	1	$v_3(x) = D - 400 = 0$
4	1	$v_4(x) = -D + 2{,}000 = 0$
5	1	$v_5(x) = SF - 1.0 = 0$
6	2	$v_1(x) = \rho_s - 0.01 = 0;$ $v_3(x) = D - 400 = 0$
7	2	$v_1(x) = \rho_s - 0.01 = 0;$ $v_4(x) = -D + 2{,}000 = 0$
8	2	$v_1(x) = \rho_s - 0.01 = 0;$ $v_5(x) = SF - 1.0 = 0$
9	2	$v_2(x) = -\rho_s + 0.08 = 0;$ $v_3(x) = D - 400 = 0$
10	2	$v_2(x) = -\rho_s + 0.08 = 0;$ $v_4(x) = -D + 2{,}000 = 0$
11	2	$v_2(x) = -\rho_s + 0.08 = 0;$ $v_5(x) = SF - 1.0 = 0$
12	2	$v_3(x) = D - 400 = 0;$ $v_5(x) = SF - 1.0 = 0$
13	2	$v_4(x) = -D + 2{,}000 = 0;$ $v_5(x) = SF - 1.0 = 0$
14	3	$v_1(x) = \rho_s - 0.01 = 0;$ $v_3(x) = D - 400 = 0;$ $v_5(x) = SF - 1.0 = 0$
15	3	$v_1(x) = \rho_s - 0.01 = 0;$ $v_4(x) = -D + 2{,}000 = 0;$ $v_5(x) = SF - 1.0 = 0$
16	3	$v_2(x) = -\rho_s + 0.08 = 0;$ $v_3(x) = D - 400 = 0;$ $v_5(x) = SF - 1.0 = 0$
17	3	$v_2(x) = -\rho_s + 0.08 = 0;$ $v_4(x) = -D + 2000 = 0;$ $v_5(x) = SF - 1.0 = 0$

Table 5.3.3.6 Optimized Objective Functions

Objectives	*Minimum*	*Maximum*
CI_c (KRW/m)	170,339	2,376,741
CO_2 (t-CO_2/m)	0.3538	5.4846
W_c (kN/m)	9.5566	74.0159

resulting in $n = 2$ and $5^2 = 25$ initial trial input parameters. The readers can refer to Chapters 1 and 2 for more details of Newton-Raphson iterations with initial trial input parameters based on a linearized first derivative of Lagrange function. Table 5.3.3.6 presents optimized (minimum and maximum) CI_c, CO_2, and W_c.

The stationary points or stationary points $\left(\mathbf{x}^*, \lambda_c^*, \text{and } \lambda_v^*\right)$ of Lagrange function shown in Equations (5.3.3.2-1)–(5.3.3.2-3) are the roots of the first-derivative Lagrange function $\nabla\mathcal{L}(\mathbf{x}, \lambda_c, \lambda_v) = 0$. An optimized design of a RC column is obtained at these stationary points for given input parameters, $\mathbf{x} = \left\{D, \rho_s, f_c', f_y, P_u, M_u\right\}^T$, and Lagrange multipliers, λ_c and λ_v. Newton–Raphson iteration is applied to find the stationary points $\left(\mathbf{x}^*, \lambda_c^*, \text{and } \lambda_v^*\right)$ which are the roots of the nonlinear equations (Equations 5.3.3.2-1–5.3.3.2-3), explained in Step 3 of Section 5.2.3. The first derivative of Lagrange function $\nabla\mathcal{L}(\mathbf{x}, \lambda_c, \lambda_v)$ is linearized using Taylor's expansion. Newton-Raphson iteration starts with assumed trial initial input parameters $\left(\mathbf{x}^0, \lambda_c^0, \lambda_v^0\right)$. The gradient of the Lagrange function at the trial initial input parameters is determined as $\nabla\mathcal{L}\left(\mathbf{x}^0, \lambda_c^0, \lambda_v^0\right)$. A convergence criterion is defined as $\mathcal{L}\left(\mathbf{x}^0, \lambda_c^0, \lambda_v^0\right) \le 1e^{-15} (\approx 0)$. Newton-Raphson iteration is employed to determine an incremental initial input parameter, $(\Delta\mathbf{x}^0, \Delta_c^0, \Delta_v^0)$, in multi-variate functions as shown in Equation (5.3.3.5-1) with k being equal to 0, where $\nabla\mathcal{L}\left(\mathbf{x}^{(k=0)}, \lambda_c^{(k=0)}, \lambda_v^{(k=0)}\right)$ and $\mathbf{H}_{\mathcal{L}}\left(\mathbf{x}^{(k=0)}, \lambda_c^{(k=0)}, \lambda_v^{(k=0)}\right)$ are the gradient and Hessian of Lagrange function $\mathcal{L}$, respectively, at the trial initial input parameters $\left(\mathbf{x}^0, \lambda_c^0, \lambda_v^0\right)$ (Hong and Nguyen, 2021; Hong et al., 2021b). The incremental initial input parameters, $(\Delta\mathbf{x}^0, \Delta\lambda_c^0, \lambda)$, are added to update to the next vector, $\left(\mathbf{x}^1, \lambda_c^1, \lambda_v^1\right)$, for a next iteration. The updated input parameters of $\left(\mathbf{x}^1 = \mathbf{x}^0 + \Delta\mathbf{x}^0, \lambda_c^1 = \lambda_c^0 + \Delta_c^0\lambda, \lambda_v^1 = \lambda_v^0 + \Delta\lambda_v^0\right)$ are implemented if the criterion for the first iteration is not met. The convergence criterion is now checked with $\left(\mathbf{x}^1, \lambda_c^1, \lambda_v^1\right)$. The same calculation is repeated until convergence is achieved ($\mathcal{L}\left(\mathbf{x}\ , \lambda_c, \lambda_v\right) \le 1e^{-15}$), using Equations (5.3.3.5-1) and (5.3.3.5-2) for the kth iteration. Another terminating criterion is defined by a maximum number of iterations (as 50 iterations in this study) to save computation effort. Notably, the Newton–Raphson method depends on trial initial input parameters, and hence, the proposed ANN-based Lagrange optimization adopts multiple trial initial input parameters to ensure stationary points are captured. A number of trial initial input parameters is determined as 5^n, where n is a number of unknown variables in the design problem. In RC column designs, trial initial input parameters, $\mathbf{x} = \left\{D, \rho_s, f_c', f_y, P_u, M_u\right\}^T$, contain six parameters, four ($f_c', f_y, P_u,$ and M_u) of which are typically predefined in design problems. Thus, only two unknown parameters, D and ρ_s, vary, resulting in $n = 2$ and 25 initial trial vectors.

$$\begin{bmatrix} \Delta \mathbf{x}^{(k)} \\ \Delta \lambda_c^{(k)} \\ \Delta \lambda_v^{(k)} \end{bmatrix} \approx -\left[\mathbf{H}_{\mathcal{L}}\left(\mathbf{x}^{(k)}, \lambda_c^{(k)}, \lambda_v^{(k)}\right)\right]^{-1} \nabla \mathcal{L}\left(\mathbf{x}^{(k)}, \lambda_c^{(k)}, \lambda_v^{(k)}\right) \tag{5.3.3.5-1}$$

$$\begin{bmatrix} \mathbf{x}^{(k+1)} \\ \lambda_c^{(k+1)} \\ \lambda_v^{(k+1)} \end{bmatrix} = \begin{bmatrix} \mathbf{x}^{(k)} \\ \lambda_c^{(k)} \\ \lambda_v^{(k)} \end{bmatrix} + \begin{bmatrix} \Delta \mathbf{x}^{(k)} \\ \Delta \lambda_c^{(k)} \\ \Delta \lambda_v^{(k)} \end{bmatrix} \tag{5.3.3.5-2}$$

Step 4: Formulating UFO

Step 4.1: Normalizing UFO.

Each objective function $f_{CI_c}^{ANN}(\mathbf{x})$, $f_{CO_2}^{ANN}(\mathbf{x})$, $f_{CO_2}^{ANN}(\mathbf{x})$ shown in Equations (5.3.3.1-1)–(5.3.3.1-3) is normalized as shown in Equations (5.3.3.6-1)–(5.3.3.6-3) that is based on the maximum and minimum of each objective function CI_c^{max}, CI_c^{min}, CO_2^{max}, CO_2^{min}, W_c^{max}, W_c^{min} calculated in Step 3.

$$f_{CI_c}^{N}(x) = \frac{f_{CI_c}^{ANN}(x) - CI_c^{min}}{CI_c^{max} - CI_c^{min}} = \frac{f_{CI_c}^{ANN}(x) - 170{,}339}{2{,}376{,}741 - 170{,}339} \tag{5.3.3.6-1}$$

$$f_{CO_2}^{N}(x) = \frac{f_{CO_2}^{ANN}(x) - CO_2^{min}}{CO_2^{max} - CO_2^{min}} = \frac{f_{CO_2}^{ANN}(x) - 0.3538}{5.4846 - 0.3538} \tag{5.3.3.6-2}$$

$$f_{W_c}^{N}(x) = \frac{f_{W_c}^{ANN}(x) - W_c^{min}}{W_c^{max} - W_c^{min}} = \frac{f_{W_c}^{ANN}(x) - 9.5566}{74.0159 - 9.5566} \tag{5.3.3.6-3}$$

Step 4.2: Deriving UFO

UFO shown in Equation (5.3.3.7-1) is derived based on separate single-objective function $f_{CI_c}^{N}(\mathbf{x})$, $f_{CO_2}^{N}(\mathbf{x})$, $f_{CO_2}^{N}(\mathbf{x})$ shown in Equation (5.3.3.6-1)–(5.3.3.6-3) that are normalized based on their ANN-based functions $f_{CI_c}^{ANN}(\mathbf{x})$, $f_{CO_2}^{ANN}(\mathbf{x})$, $f_{CO_2}^{ANN}(\mathbf{x})$ shown in Equations (5.3.3.1-1)–(5.3.3.1-3) using weighted fractions or trade-offs, $\boldsymbol{w} = \{w_{CI_c}, w_{CO_2}, w_{W_c}\}^T$, to optimize multi-objective functions. All weights are non-negative and summed to one. Each weight varies in a range of 0–1 to generate a Pareto frontier that contains nondominated design points, as shown in Equation (5.3.3.7-2). Cost, CO_2 emissions, and column weight are the objectives, and they are calculated in different units, and hence, all objectives are normalized to be unit-free before being applied into UFO, which are done in Step 4.1.

Normalized UFO function shown in Equation (5.3.3.7-1) in terms of each objective function is calculated by substituting maxima and minima of single-objective function, CI_c^{max}, CI_c^{min}, CO_2^{max}, CO_2^{min}, W_c^{max}, and W_c^{min}, calculated in STEP 3 into Equations (5.3.3.6-1) to (5.3.3.6-3).

$$\text{UFO} = w_{CI_c} f_{CI_c}^{N}(\mathbf{x}) + w_{CO_2} f_{CO_2}^{N}(\mathbf{x}) + w_{W_c} f_{W_c}^{N}(\mathbf{x}) \tag{5.3.3.7-1}$$

$$w_{CI_c}, w_{CO_2}, w_{W_c} \in [0, 1];\ w_{CI_c} + w_{CO_2} + w_{W_c} = 1 \tag{5.3.3.7-2}$$

Step 5: Optimizing *UFO*

Multi-objective functions defined as *UFO* are optimized similar to the one shown in STEP 3.

Lagrange function shown in Equation (5.3.3.8) and its KKT conditions are formulated. An objective function is the UFO equation calculated in Step 4. Stationary points optimizing multi-objective functions are obtained using Newton–Raphson iterations. An appropriate number of fractions of three objectives $\left(w_{CI_c}, w_{CO_2}, w_{W_c}\right)$ are selected in order to generate a Pareto frontier, described in Step 5.2. In this section, 225 fractions and 400 fractions are used.

Step 5.1: Lagrange function based on three objective functions shown in Equation (5.3.3.8) is minimized. Lagrange function is constrained by equalities and inequalities conditions shown in Table 5.3.3.4. Lagrange multipliers are shown in Equations (5.3.3.9) and (5.3.3.10) whereas matrix S defining KKT conditions is also shown in Equation (5.3.3.11).

$$\mathcal{L}_{UFO}\left(\mathbf{x},\lambda_c,\lambda_v\right)=UFO-\lambda_c^{\mathrm{T}}\boldsymbol{c}\left(\mathbf{x}\right)-\lambda_v^{\mathrm{T}}\mathrm{S}\boldsymbol{v}\left(\mathbf{x}\right) \tag{5.3.3.8}$$

$$\lambda_c=\begin{bmatrix}\lambda_{c1}\\ \lambda_{c2}\\ \lambda_{c3}\\ \lambda_{c4}\end{bmatrix};\ \boldsymbol{c}\left(\mathbf{x}\right)=\begin{bmatrix}c_1\left(\mathbf{x}\right)\\ c_2\left(\mathbf{x}\right)\\ c_3\left(\mathbf{x}\right)\\ c_4\left(\mathbf{x}\right)\end{bmatrix} \tag{5.3.3.9}$$

$$\lambda_v=\begin{bmatrix}\lambda_{v1}\\ \lambda_{v2}\\ \lambda_{v3}\\ \lambda_{v4}\\ \lambda_{v5}\end{bmatrix};\ \boldsymbol{c}\left(\mathbf{x}\right)=\begin{bmatrix}v_1\left(\mathbf{x}\right)\\ v_2\left(\mathbf{x}\right)\\ v_3\left(\mathbf{x}\right)\\ v_4\left(\mathbf{x}\right)\\ v_5\left(\mathbf{x}\right)\end{bmatrix} \tag{5.3.3.10}$$

$$\mathrm{S}=\begin{bmatrix}s_1 & 0 & 0 & 0 & 0\\ 0 & s_2 & 0 & 0 & 0\\ 0 & 0 & s_3 & 0 & 0\\ 0 & 0 & 0 & s_4 & 0\\ 0 & 0 & 0 & 0 & s_5\end{bmatrix} \tag{5.3.3.11}$$

Step 5.2: Selecting fractions for three objective functions w_{CI_c}, w_{CO_2}, and w_{W_c}

Lagrange function based on UFO with fractions $\left(w_{CI_c}=w_{CO_2}=w_{W_c}=\frac{1}{3}\right)$ that are applied to three objective functions is presented in Equation (5.3.3.12).

$$\mathcal{L}_{UFO}\left(\mathbf{x},\lambda_c,\lambda_v\right)=\frac{1}{3}f_{CI_c}^{N}\left(\mathbf{x}\right)+\frac{1}{3}f_{CO_2}^{N}\left(\mathbf{x}\right)+\frac{1}{3}f_{W_c}^{N}\left(\mathbf{x}\right)-\lambda_c^{\mathrm{T}}c\left(\mathbf{x}\right)-\lambda_v^{\mathrm{T}}Sv\left(\mathbf{x}\right) \tag{5.3.3.12}$$

Step 5.3: Obtaining matrix S defining KKT conditions based on 17 cases shown in Table 5.3.3.5.

Step 5.4: Finding stationary points of Lagrange functions in each case based on linearized first derivative of Lagrange function $\mathcal{L}_{UFO}\left(\mathbf{x},\lambda_c,\lambda_v\right)$ as shown in Step 3.3.

Step 5.5: Repeating Steps 5.2–5.4 by varying fractions for three objective functions based on Equation (5.3.3.7) to obtain a Pareto frontier.

5.3.4 Discussions on an optimization based on three objective functions

As plotted and compared in Figure 5.3.4.1, two Pareto frontiers (Pareto curves) based on 225 fractions and 400 fractions yield optimal designs of circular RC columns with respect to optimization scenario presented in Section 5.3.2. A circular RC column constrained by given material properties ($f_c' = 40$ MPa, $f_y = 500$ MPa) should sustain factored load ($P_u = 1{,}500$ kN, $M_u = 2{,}500$ kN.m) to satisfy strength requirement ($SF \geq 1.0$).

A number of fractions is selected by "*linspace*" command provided by Matalb which is explained in Section 5.3.6. In the three objective functions of this example, 1500D715 fractions are generated within 0 to 1 based on *linspace* (0,1,15) command whereas 2000D720 fractions are generated when using "*linspace* (0,1,20)" command. A Pareto frontier indicated in yellow color shown in Figure 5.3.4.1 illustrates optimized UFO based on $15 \times 15 = 225$ fractions of three objective functions ($w_{CI_c}, w_{CO_2}, w_{W_c}$), while a Pareto frontier in blue color indicates optimized UFO based on $20 \times 20 = 400$ fractions. It is seen that a Pareto obtained from 400 fractions exhibits better density, indicating more searching areas provide more closely packed points on curve, in comparison with a curve based on 225 fractions. A number of fractions sufficient to provide smooth and dense Pareto frontier are, therefore, recommended based on the trial-and-error method.

5.3.5 Verification to large datasets

Optimized UFO by ANN-based Lagrange optimization is verified based on errors of an ANN and structural mechanics-based software "*AutoCC*". This nonlinear software is developed with MATLAB (MathWorks, 2020) based on strain compatibilities. Parameters D and ρ_s are iterated based on Newton–Raphson method to yield output parameters $y = \{SF, \varepsilon_s, CI_c, CO_2, W_c\}^T$

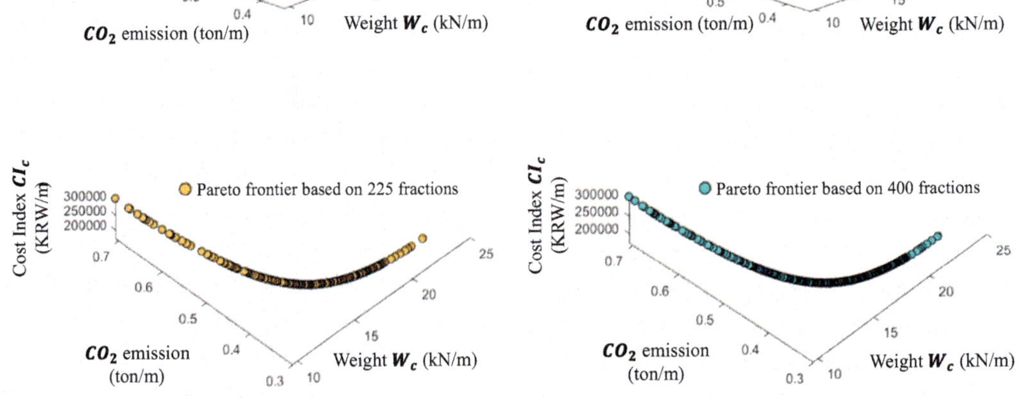

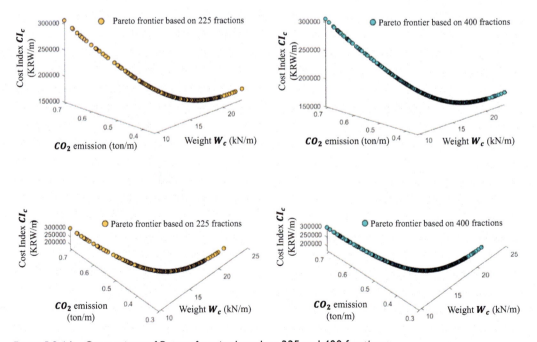

Figure 5.3.4.1 Comparison of Pareto frontier based on 225 and 400 fractions.

whereas the rest of parameters $f_c' = 40$ MPa, $f_y = 500$ MPa, $P_u = 1{,}500$ kN, $M_u = 2{,}500$ kN.m are given as simple equalities. Input parameters, $\mathbf{x} = \{D, \rho_s, f_c' = 40 \text{ MPa}, f_y = 500 \text{ MPa},$ $\mathbf{x} = P_u = 1{,}500 \text{ kN}, M_u = 2{,}500 \text{ kN.m}\}^T$, consisting of optimal values of D and ρ_s, are applied to ANN and *AutoCC* to calculate output parameters, $\mathrm{y} = \{SF, \varepsilon_s, CI_c, \mathrm{CO_2}, W_c\}^T$. Each output calculated using ANN is compared with those estimated using *AutoCC*. Computational errors are calculated as $\text{Error} = \dfrac{\text{ANN} - AutoCC}{\text{ANN}} \times 100\%$. Figure 5.3.5.1 illustrates computational errors of five output parameters $SF, \varepsilon_s, CI_c, \mathrm{CO_2}, W_c$ which are as large as 2% for all 400 fractions applied to three objective functions w_{CI_c}, $w_{\mathrm{CO_2}}, w_{W_c}$.

Two million structural datasets are generated to verify minimized UFO based on ANN-based Lagrange optimization where ranges of design parameters are similar to constraint conditions presented in Table 5.3.3.4 ($400 \text{ mm} \le D \le 2{,}000 \text{ mm}$, $0.01 \le \rho_s \le 0.08$, $f_c' = 40$ MPa, $f_y = 500$ MPa, $P_u = 1{,}500$ kN, $M_u = 2{,}500$ kN.m). From these two million datasets, 140,000 datasets having safety factors from 1.0 to 1.5 $(1.0 \le SF \le 1.5)$ with $f_c' = 40$ MPa, $f_y = 500$ MPa, $P_u = 1{,}500$ kN, $M_u = 2{,}500$ kN.m are sorted. Figure 5.3.5.2 shows that minimized UFO based on ANN-based Lagrange optimization yields a lower boundary of large datasets where three objective functions are minimized simultaneously.

5.3.6 Generation of evenly spaced fractions

This section introduces how fractions indicating contributions of objective functions to the overall optimizations for Pareto frontier can be generated, even if readers can generate their own fractions. However, it is noted there could be some sparse regions on the Pareto frontier when fractions are not appropriately selected as shown in Figure 5.3.6.1. The importance of a good selection of fractions should not be overlooked.

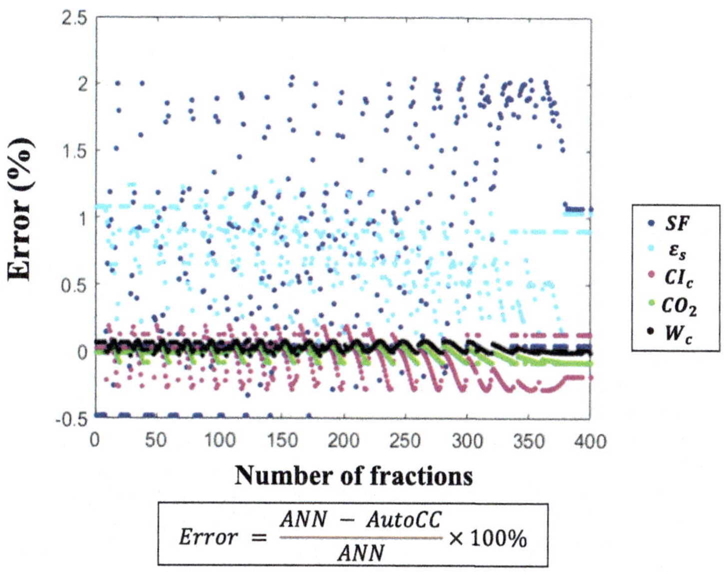

Figure 5.3.5.1 Errors of optimized UFO based on ANN-based Lagrange algorithm compared with structural mechanics-based *AutoCC*.

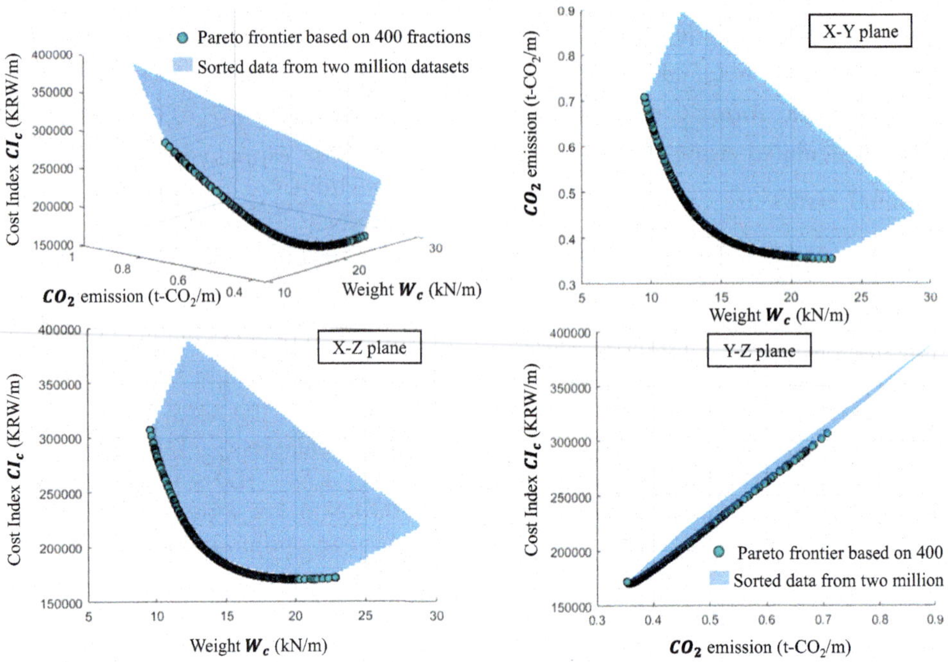

Figure 5.3.5.2 Verification based on 140,000 data having safety factor from 1.0 to 1.5 with $f_c' = 40$ MPa, $f_y = 500$ MPa, $P_u = 1,500$ kN, $M_u = 2,500$ kN.m filtered through two million structural datasets.

Multi-objective MOO optimization provided by ANN-based Lagrange algorithm is obtained using Newton-Raphson iteration under KKT conditions, which depends on the initial trial input parameters. The gradient-based optimization algorithm employs a diverse search of 5^2 initial trial input parameters, in which "2" shown in the power indicates a number of initial trial input parameters (D and ρ_s) which are to converge to stationary (stationary) points to calculate a global optimum which is not trapped in local extrema. Stationary points can be local maximum or local minimum. The proposed algorithm has sufficient searching capability for stationary points because fractions of the three objectives are equally spaced in

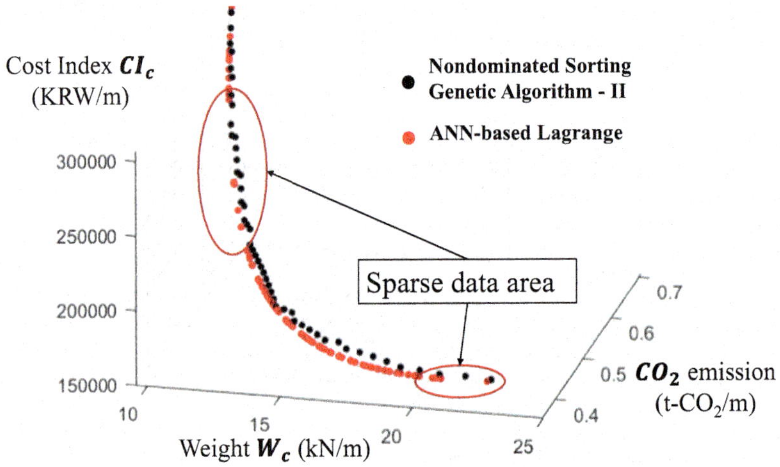

Figure 5.3.6.1 A Pareto frontier having sparse regions when fractions are not appropriately selected.

the searching domains (0–1). A "*linspace*" command provided by MATLAB (MathWorks, 2020) generates equally spaced fractions of the three objective functions as shown in Table 5.3.6.1. Table 5.3.6.2 presents 400 fraction samples calculated based on Table 5.3.6.1. A total of $20 \times 20 = 400$ fractions ($w_{CI_c}^{(i)}$, $w_{CO_2}^{(i,j)}$, and $w_{W_c}^{(i,j)}$) indicating contributions of objective functions to the overall optimizations is generated with an equal spacing as shown in Figure 5.3.6.2a which shows a searching area of the fractions in a range of 0–1 when optimizing RC column sustaining factored axial load of 1,500 kN and factored moment of 2,500 kN.m.

The first fraction $w_{W_c}^{(i=1,j=1)}$ is calculated as $1-0-0=1$ (sum of the three fractions should be 1) when the first fraction of $w_{CO_2}^{(i=1,j=1)} = 0$ based on the first fraction of cost CI_c ($w_{CI_c}^{(i=1)}$) $= 0$ shown in Equation (5.3.6.1-1) and Table 5.3.6.2. Fractions of the three objective functions, in this case, are $w_{CI_c} : w_{CO_2} : w_{W_c} = 0:0:1$. The second fractions of $w_{CO_2}^{(i=1,j=2)}$ and $w_{W_c}^{(i=1,j=2)}$ are calculated using Equations (5.3.6.1-2) and (5.3.6.1-3) corresponding to the first fraction of cost CI_c, $w_{CI_c}^{(i=1)} = 0$, shown in Equation (5.3.6.1-1). The second fraction of CO_2 ($w_{CO_2}^{(i=1,j=2)}$) is, then, calculated as $w_{CO_2}^{(i=1,j=2)} = w_{CO_2}^{(i=1,j=1)} + \frac{1-w_{CI_c}^{(i=1)}}{20-1} = 0 + \frac{1-0}{19} = 0.0526$ using a formula $w_{CO_2} =$ linspace(0,1,20)as shown in Table 5.3.6.2 and Equation (5.3.6.1-2) when a first fraction of cost CI_c $w_{CI_c}^{(i=1)} = 0$. A second fraction $w_{W_c}^{(i=1,j=2)}$ corresponding to CO_2 ($w_{CO_2}^{(i=1,j=2)}$) is, then, calculated as $w_{W_c}^{(i=1,j=2)} = 1-0-0.0526 = 0.9474$ as shown in Equation (5.3.6.1-3). $w_{CI_c}^{(i=1)}$ is the first fraction of cost among 20 fractions. $w_{CO_2}^{(i=1,j=1)} = 0$ and $w_{CO_2}^{(i=1,j=2)} = 0.0526$ are the first and the second fractions of CO_2 corresponding to the first fraction of cost ($w_{CI_c}^{(i=1)} = 0$), respectively. The second fraction $w_{W_c}^{(i=1,j=2)} = 0.9474$ is calculated corresponding to the first fraction of cost ($w_{CI_c}^{(1)} = 0$) and the second fraction of CO_2 ($w_{CO_2}^{(i=1,j=2)} = 0.0526$) as shown in red box of Table 5.3.6.2. This generation is repeated for 20 fractions for cost CI_c. A number of fractions is a number of fraction points plotted in Pareto frontier (curve). Figure 5.3.6.2b illustrates all fractions of CO_2 ($w_{CO_2}^{(i=1,j=1\sim20)}$) and W_c ($w_{W_c}^{(i=1,j=1\sim20)}$) corresponding to the first 20 fractions of $w_{CI_c}^{(i=1)} = 0$.

Table 5.3.6.1 Interpretation of MATLAB Code to Generate Fractions with Equal Spacings; "Linspace" Command Provided by MATLAB (MathWorks, 2020)

MATLAB Code	*Algorithm*
`wCost = linspace(0,1,20);`	Twenty fractions of cost (w_{CI_c}) are generated in a range of 0–1 with an equal spacing of each point calculated as: $\frac{1-0}{20-1} = \frac{1}{19}$

`for i = 1:20`
`w1 = wCost(i);`
`wCO2 = linspace(0,1-w1,20);`
	for each magnitude of fraction of cost ($w_{CI_c}^{(i)}$, $i = 1$ to 20): a corresponding fraction of CO_2 ($w_{CO_2}^{(i,j)}$) are generated in a range of 0–1 with an equal spacing of each point equal to 1/19.

`for j = 1:20`
`w2 = wCO2(j);`
`wWeight = 1-w1-w2;`
`end`
`end`
	for each fraction of CO_2 ($w_{CO_2}^{(i,j)}$; $i, j = 1$ to 20): generating a corresponding fraction of weight ($w_{W_c}^{(i,j)}$) based on an equation: $1 - w_{CI_c}^{(i)} - w_{CO_2}^{(i,j)}$

Table 5.3.6.2 Four hundred Fractions Generated Based on "*Linspace*" Presented in Table 5.3.6.1

i	j	$w_{Cl_b}^{(i)}$	$w_{CO_2}^{(i,i)}$	$w_{W_b}^{(i,i)}$	i	j	$w_{Cl_b}^{(i)}$	$w_{CO_2}^{(i,i)}$	$w_{W_b}^{(i,i)}$		i	j	$w_{Cl_b}^{(i)}$	$w_{CO_2}^{(i,i)}$	$w_{W_b}^{(i,i)}$	i	j	$w_{Cl_b}^{(i)}$	$w_{CO_2}^{(i,i)}$	$w_{W_b}^{(i,i)}$
i=1	j=1	0	0	1	i=4	j=1	0.1579	0	0.8421	...	i=15	j=1	0.6842	0	0.3158	i=18	j=1	0.8421	0	0.1579
	j=2	**0**	**0.0526**	**0.9474**		j=2	0.1579	0.0443	0.7978	...		j=2	0.6842	0.0166	0.2992		j=2	0.8421	0.0083	0.1496
	...	...	...	...		...	...	...	...	...		...	...	...	...		...	...	...	...
	j=19	0	0.9474	0.0526		j=19	0.1579	0.8421	0	...		j=19	0.6842	0.2992	0.01662		j=19	0.8421	0.1496	0.00831
	j=20	0	1	0		j=20	0.1579	0.8421	0	...		j=20	0.6842	0.3158	0		j=20	0.8421	0.1579	0
i=2	j=1	0.0526	0	0.9474	i=5	j=1	0.2105	0	0.7895	...	i=16	j=1	0.7368	0	0.2632	i=19	j=1	0.8947	0	0.1053
	j=2	0.0526	0.0499	0.8975		j=2	0.2105	0.0416	0.7479	...		j=2	0.7368	0.0139	0.2493		j=2	0.8947	0.0055	0.0997
	...	...	...	...		...	...	...	...	...		...	...	...	...		...	...	...	...
	j=19	0.0526	0.8975	0.0499		j=19	0.2105	0.74792	0.04155	...		j=19	0.7368	0.24931	0.01385		j=19	0.8947	0.09972	0.00554
	j=20	0.0526	0.9474	0		j=20	0.2105	0.78947	0	...		j=20	0.7368	0.26316	0		j=20	0.8947	0.10526	0
i=3	j=1	0.1053	0	0.8947	i=6	j=1	0.2632	0	0.7368	...	i=17	j=1	0.7895	0	0.2105	i=20	j=1	0.9474	0	0.0526
	j=2	0.1053	0.0471	0.8476		j=2	0.2632	0.0388	0.6981	...		j=2	0.7895	0.0111	0.1994		j=2	0.9474	0.0028	0.0499
	...	...	...	...		...	...	...	...	...		...	...	...	...		...	...	...	...
	j=19	0.1053	0.8476	0.0471		j=19	0.2632	0.6981	0.0388	...		j=19	0.7895	0.1994	0.0111		j=19	0.9474	0.0526	0.0000
	j=20	0.1053	0.8947	0		j=20	0.2632	0.7368	0	...		j=20	0.7895	0.2105	0		j=20	1	0	0

Bold values indicate Pareto frontier.

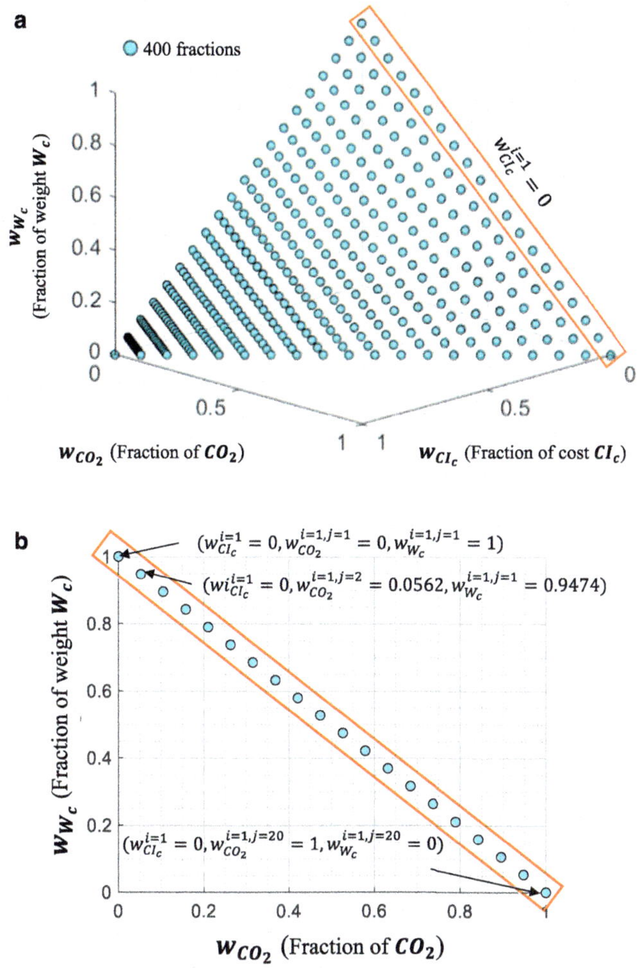

Figure 5.3.6.2 Four hundred equally spaced fractions applied to Pareto frontier based on $f_c' = 40$ MPa, $f_y = 500$ MPa, $P_u = 1,500$ kN, $M_u = 2,500$ kN.m. (a) 400 fractions of three objective functions. (b) Fractions of CO_2 and W_c corresponding to $w_{CI_c}^{(i=1)} = 0$.

$$w_{CI_c}^{(1)} = 0 \tag{5.3.6.1-1}$$

$$w_{CO_2}^{(1,2)} = w_{CO_2}^{(1,1)} + \frac{1}{19} = 0 + 0.0526 = 0.0526 \tag{5.3.6.1-2}$$

$$w_{W_c}^{(1,2)} = 1 - w_{CI_c}^{(1)} - w_{CO_2}^{(1,2)} = 1 - 0 - 0.0526 = 0.9474 \tag{5.3.6.1-3}$$

After 20 fractions of CO_2 ($w_{CO_2}^{(i=1,j=1\sim20)}$) and W_c ($w_{W_c}^{(i=1,j=1\sim20)}$) corresponding to the first 20 fractions of $w_{CI_c}^{(i=1)} = 0$ are generated, the second fraction of $w_{CI_c}^{(i=2)}$ is calculated as $w_{CI_c}^{(i=2)} = w_{CI_c}^{(i=1)} + \frac{1}{19} = 0 + \frac{1}{19} = 0.0526$ because second fraction of cost ($w_{CI_c}^{(i=2)}$) is generated based on a formula `linspace` (0,1,20). If fractions of CO_2 are generated based on an equation similar to one for calculating fractions of cost $CI_c = \texttt{linspace}(0,1,20)$, this will cause fractions of the three objectives to be unequally spaced, resulting in $w_{CI_c} + w_{CO_2} + w_{W_c} > 1.0$. Therefore, the first fraction of CO_2 with respect to the second fraction of $w_{CI_c}^{(i=2)} = 0.0526$ is $w_{CO_2}^{(i=2,j=1)} = 0$ as shown in Table 5.3.6.2. A first fraction of weight $w_{W_c}^{(i=2,j=1)}$ is calculated in accordance with $w_{CI_c}^{(i=2)} = 0.0526$ and $w_{CO_2}^{(i=2,j=1)} = 0$ as: $w_{W_c}^{(i=2,j=1)} = 1 - w_{CI_c}^{(i=2)} - w_{CO_2}^{(i=2,j=1)} = 1 - 0.0526 - 0 = 0.9474$.

Next, the second fraction of CO_2 with respect to the second fraction of $w_{CI_c}^{(i=2)} = 0.0526$ is calculated as: $w_{CO_2}^{(i=2,j=2)} = w_{CO_2}^{(i=2,j=1)} + \dfrac{1 - w_{CI_c}^{(i=2)}}{20 - 1} = 0 + \dfrac{1 - 0.0526}{19} = 0.0499$ (refer to Table 5.3.6.2) because second fraction of CO_2 is generated by a formula $\texttt{linspace}(0, 1 - w_{CI_c}^{(i)}, 20)$. The second fraction of weight $w_{W_c}^{(i=2,j=2)}$ is calculated with respect to the second fraction of $w_{CI_c}^{(i=2)} = 0.0526$ and the second fraction of $w_{CO_2}^{(2,j=2)} = 0.0499$ as: $w_{W_c}^{(2,j=2)} = 1 - w_{CI_c}^{(i=2)} - w_{CO_2}^{(2,j=2)} = 1 - 0.0526 - 0.0499 = 0.8975$.

5.3.7 Interpretation of data trend

5.3.7.1 Relationships among three objective functions

Relationships describing data trends among the three objective functions (CI_c, CO_2 and W_c) are explored in Equations (5.3.7.1-1)–(5.3.7.1-7). Some notations are defined as follows.

W	-	Total weight (kg)
CO_2	-	Total CO_2 emission (t-CO_2)
CI_c	-	Total cost (KRW)
W_R	-	Weight of rebar (kg)
W_c	-	Weight of concrete (kg)
$U_{R\_CO_2}$	-	Unit CO_2 emission of rebar; $U_{R\_CO_2} = 2.512 \times 7.85 = 19.72 \left(\text{ton} - CO_2 / m^3 \right)$
$U_{C\_CO_2}$	-	Unit CO_2 emission of concrete; $U_{C\_CO_2} = 0.1677 \left(\text{ton} - CO_2 / m^3 \right)$
$U_{R\_CI_c}$	-	Unit cost of rebar $f_y = 500$ MPa; $U_{R\_CI_c} = 1,055 \times 7,850 = 8,281,750 \ (\text{KRW}/m^3)$
$U_{C\_CI_c}$	-	Unit cost of concrete $f_c' = 40$ MPa; $U_{C\_CI_c} = 94,000 \left(\text{KRW} / m^3 \right)$

Structural mechanics-based software calculates a total column weight equal to concrete weight, ignoring weight of rebar.

$$W_C = W \tag{5.3.7.1-1}$$

Total CO_2 emission (t-CO_2) is calculated as follows.

$$U_{R\_CO_2} \times W_R + U_{C\_CO_2} \times W_c = CO_2 \tag{5.3.7.1-2}$$

Substituting Equation (5.3.7.1-1) into Equation (5.3.7.1-2) yields Equation (5.3.7.1-3).

$$U_{R\_CO_2} \times W_R + U_{C\_CO_2} \times W = CO_2$$

$$\Rightarrow W_R = \frac{CO_2 - U_{C\_CO_2} \times W}{U_{R\_CO_2}} \tag{5.3.7.1-3}$$

Total cost (KRW) is calculated as follow.

$$U_{R\_CI_c} \times W_R + U_{C\_CI_c} \times W_C = CI_c \tag{5.3.7.1-4}$$

Substituting Equations (5.3.7.1-1) and (5.3.7.1-3) into Equation (5.3.7.1-4) yields Equation (5.3.7.1-5).

$$\begin{aligned} & U_{R\_CI_c} \times \frac{CO_2 - U_{C\_CO_2} \times W}{U_{R\_CO_2}} + U_{C\_CI_c} \times W = CI_c \\ & \Rightarrow CO_2 \times \frac{U_{R\_CI_c}}{U_{R\_CO_2}} + W \times \left(U_{C\_CI_c} - U_{R\_CI_c} \times \frac{U_{C\_CO_2}}{U_{R\_CO_2}} \right) = CI_c \end{aligned} \tag{5.3.7.1-5}$$

Substituting unit values $U_{R\_CO_2}$, $U_{C\_CO_2}$, U_{R_CI}, and $U_{C\_CI_c}$ into Equation (5.3.7.1-5) yields Equation (5.3.7.1-6).

$$\begin{aligned} & CO_2 \times \frac{8{,}281{,}750}{19.72} + W \times \left(94{,}000 - 8{,}281{,}750 \times \frac{0.1677}{19.72} \right) = CI_c \\ & \Rightarrow CO_2 \times 419{,}984 + W \times (23{,}568) = CI_c \\ & \Rightarrow 419{,}984 \times CO_2 + 23{,}568 \times W = CI_c \end{aligned} \tag{5.3.7.1-6}$$

Total CO_2 emission (t-CO_2) is, then, obtained in Equation (5.3.7.1-7) from Equation (5.3.7.1-6).

$$CO_2 = \frac{1}{419{,}984} \times CI_c - \frac{23{,}568}{419{,}984} \times W \tag{5.3.7.1-7}$$

From Equations (5.3.7.1-6) and (5.3.6.1-7), relationships among the three parameters are finally calculated as follows.

Relation 1: From Equation (5.3.7.1-6), CI_c increases when W increases when CO_2 is fixed.
Relation 2: From Equation (5.3.7.1-6), CI_c increases when CO_2 increases when W is fixed.
Relation 3: From Equation (5.3.7.1-7), CO_2 decreases when W increases when CI_c is fixed.

It is noted that these relationships (Relations 1–3) are not correct when one objective function among the three is not fixed. This can be graphically verified by exploring trend of large datasets, which is described in Section 5.3.7.2.

5.3.7.2 Exploring trend of large datasets

Remaining datasets filtered through two million structural datasets based on safety factor from 1.0 to 1.5 with $f_c' = 40$ MPa, $f_y = 500$ MPa, $P_u = 1{,}500$ kN, $M_u = 2{,}500$ kN.m are used to find data trends of the three objective functions. Section 5.3.7.1 explained trends of datasets based on the two objective functions when the third objective function is fixed. Figure 5.3.7.1a graphically shows a proportional trend between cost index (CI_c) and CO_2 emissions of datasets projected to the YZ plane with respect to W_c ranging from 13 to 14 kN/m. Similarly, the data trend between CI_c and W_c is obtained by projecting data into XZ plane as illustrated in Figure 5.3.7.1b when CO_2 ranges from 0.44 to 0.46 ton- CO_2 /m. One of these trends are illustrated by a red line indicating a projected plane having CO_2 in a

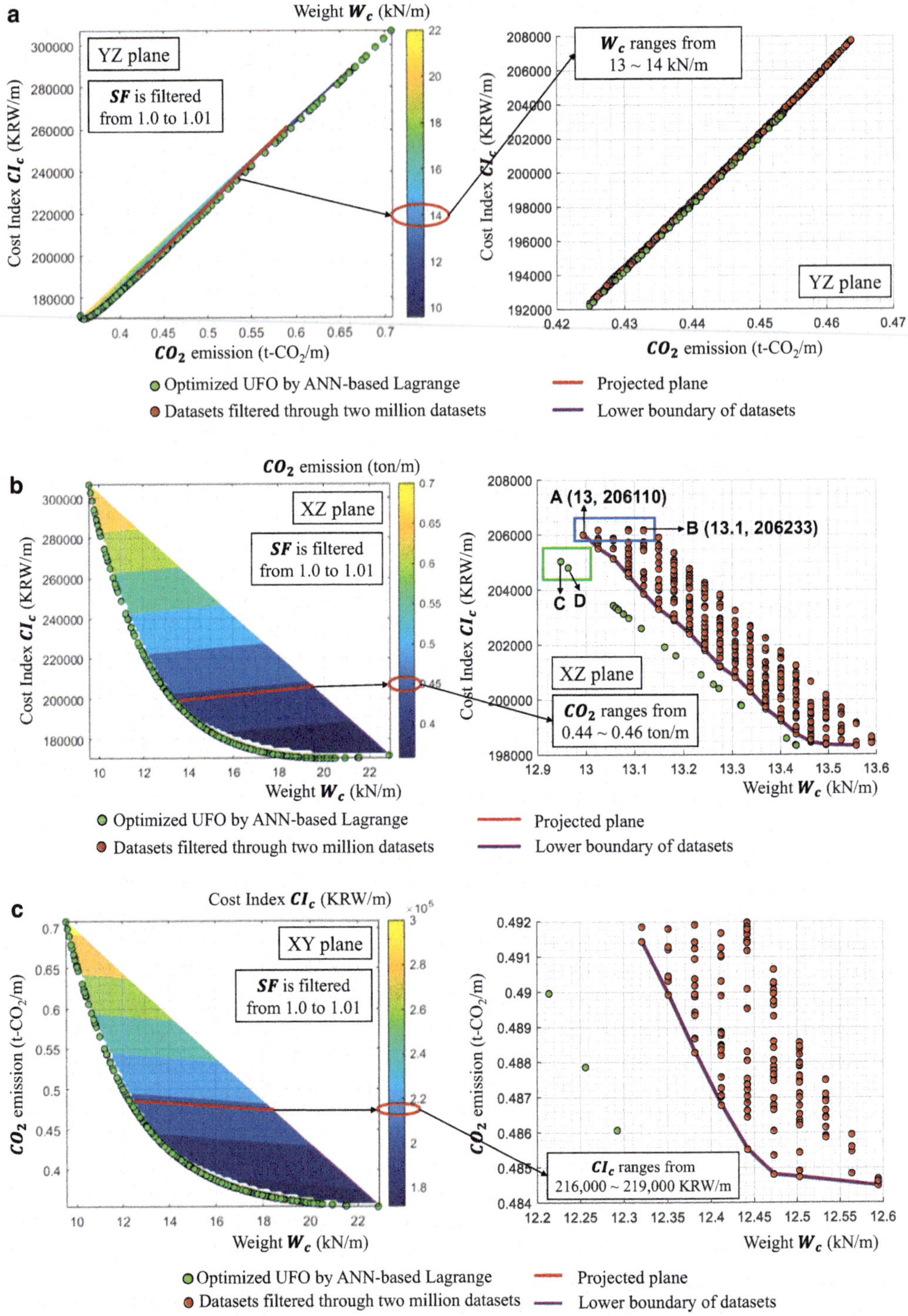

Figure 5.3.7.1 Trends of three objectives subjected to $f_c' = 40$ MPa, $f_y = 500$ MPa, $P_u = 1,500$ kN, $M_u = 2,500$ kN.m with *SF* of 1.0–1.01. (a) Relationships between CI_c and CO_2 for fixed W_c. (b) Relationships between CI_c and W_c for fixed CO_2. (c) Relationships between CO_2 and W_c for fixed CI_c.

range of 0.44 to 0.46 ton- CO_2 /m shown in Figure 5.3.7.1b. The designs illustrated by green dots exhibits an inverse relationship between cost (CI_c) and weight (W_c). The lower bound of the filtered datasets indicated by a brown line are well predicted in Figure 5.3.7.1b. Equation 5.3.7.1.6 and Figure 5.3.7.1b show that CI_c increases when W_c increases if CO_2 is fixed, resulting in an overall cost rise even if concrete weight increases, because an increase rate of concrete cost is greater than decrease rate obtained with less rebars. Commonly, a use of rebars decreases when concrete weight increases. In other words, a decreasing rate of rebars has less impact than that of an increasing rate of concrete on total cost when CO_2 emissions are constant, leading to an increase of cost. An amount of CO_2 emission must be balanced by faster increasing rate of concrete and slower decreasing rate of rebars. Figure 5.3.7.1c and Equation (5.3.7.1-7) show a design case where CO_2 and weight (W_c) exhibit a reverse trend in which an increase in W_c leads to a reduction in CO_2 for fixed CI_c.

There are several datapoints shown in blue box of Figure 5.3.7.1b in which the cost and weight of Points A and B exhibits a proportional trend. Table 5.3.7.1(a) shows the design information of these two points, which indicates that point B presents a design with a higher *SF* value of 1.009 compared with point that of Point A 1.0005, resulting in a 0.06% and 1% increase in cost (CI_c) and in weight (W_c), respectively, in which CI_c trend has been opposite when W_c increases from 12.99 to 13.11 kN/m. However, optimal designs (green dots) have $SF = 1$ so that they have an inverse relationship between cost (CI_c) and weight (W_c), as can be seen in Table 5.3.7.1(b). Two optimal designs (Points C and D) pointed in Figure 5.3.7.1(b) are taken as an example.

5.3.8 Examples of optimal designs based on Pareto frontier

5.3.8.1 Identifying design parameters for a designated fraction

Optimized design parameters, optimized P-M diagram and their corresponding fractions are obtained by three example designs shown in Figure 5.3.8.1 where Pareto frontier is also shown when the three objective functions are minimized at the same time. Design A represents an optimized design where weight W_c is minimized so that fractions of the three objective functions are based on $w_{CI_c} : w_{CO_2} : w_{W_c}$=0:0:1. Similarly, Design B introduces an design minimizing cost CI_c based on fractions=1:0:0 while fractions of Design C minimizing CO_2 is based on 0:1:0. The Design points A, B, and C, therefore, yield optimizations when a single-objective function is minimized separately.

Tables 5.3.8.1–5.3.8.3 present Designs A, B, and C minimizing W_c, ($w_{CI_c} : w_{CO_2} : w_{W_c}$= 0:0:1), CI_c ($w_{CI_c} : w_{CO_2} : w_{W_c} = 1:0:0$), and CO_2 ($w_{CI_c} : w_{CO_2} : w_{W_c} = 0:1:0$), respectively, whereas

Table 5.3.7.1 Pareto Details

(a): Design information of Points A and B shown in Figure 5.3.6.2.1(b) (CO_2 is fixed as 0.46 ton-CO_2/m)

Point	D(mm)	ρ_s	SF	CI_c(KRW/m)	CO_2 (t-CO_2/m)	W_c (kN/m)
A	838	0.0338	1.0005	206,110	0.46	12.99
B	842	0.0334	1.0090	206,233	0.46	13.12

(b): Design Information of Points C and D (Optimal Designs) Shown in Figure 5.3.6.2(b)

Point	D (mm)	ρ_s	SF	CI_c (KRW/m)	CO_2 (t-CO_2/m)	W_c (kN/m)
C	836	0.0337	1.0000	205,029	0.46	12.95
D	837	0.0336	1.0000	204,800	0.46	12.96

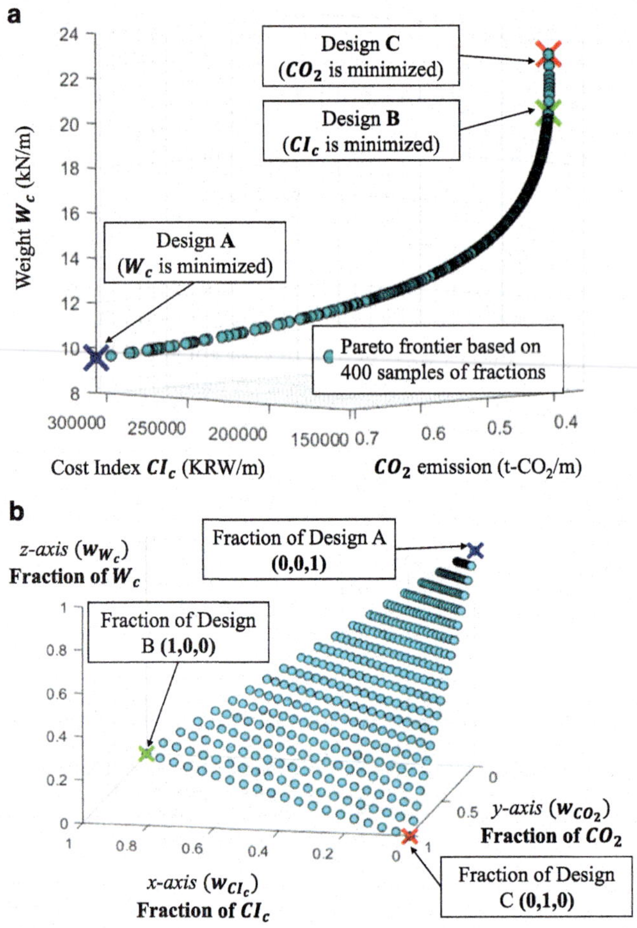

Figure 5.3.8.1 Pareto frontier based on fractions. (a) Designs A, B and C in Pareto frontier. (b) Fractions of Designs A, B and C in Pareto frontier.

Table 5.3.8.4 presents Design D based on $w_{CI_c} : w_{CO_2} : w_{W_c} = 0.32 : 0.32 : 0.36$. Parameters adopted to train ANNs are given in these tables. Evenly spaced fractions are explained in Section 5.3.6. Figure 5.3.8.2 graphically present fractions used in Designs A, B and C in which the same optimizations are calculated by multiple blue and red marks. It is noteworthy that some fractions result in the same optimizations as shown in Figure 5.3.8.2 which illustrates that the same optimizations indicated by Designs A, B, and C are resulted simultaneously by many fractions, not by just one. However, Design B has only one fraction represented by green mark and Table 5.3.8.2(b), indicating that duplication does not happen with Design B. Design A based on $w_{CI_c} : w_{CO_2} : w_{W_c} = 0.0 : 0.0 : 1.0$, shown in Table 5.3.8.1(a) is a design minimizing weight W_c. However, as shown in Figure 5.3.8.2, many fractions indicated by blue marks (x) which are not based on based on $w_{CI_c} : w_{CO_2} : w_{W_c} = 0.0 : 0.0 : 1.0$ can provide an optimization identical to that of Design A which is based on $w_{CI_c} : w_{CO_2} : w_{W_c} = 0.0 : 0.0 : 1.0$. Table 5.3.8.1(b) shows fractions which provide an optimization identical to that of Design A. In other words, many fractions indicated by blue marks lead to the duplicated optimization for Design A. Most of all errors less than 1% are verified based on structural mechanics-based software. Maximum rebar ratio 0.08 is used to reduce column weight as shown in Design A of Table 5.3.8.1(a), whereas

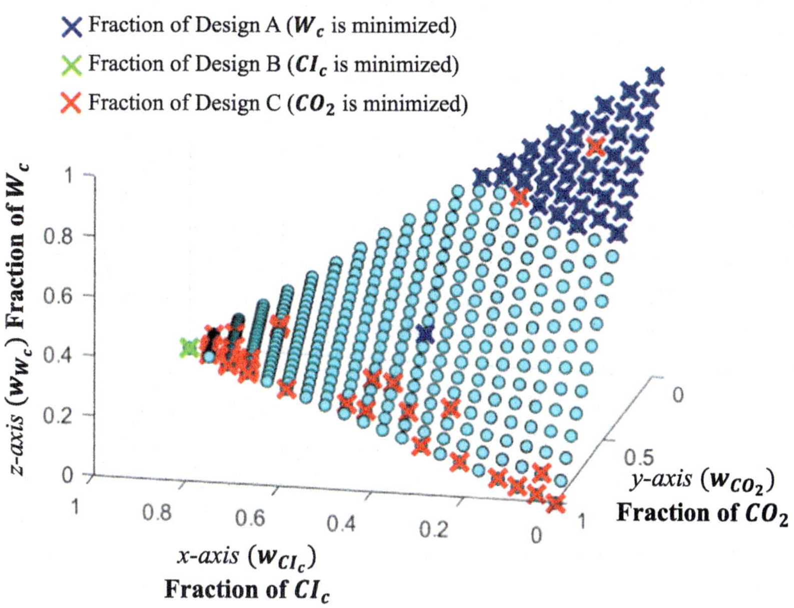

Figure 5.3.8.2 Duplicated fractions of Designs A and C compared to one fraction resulting in Design B.

Table 5.3.8.1 Optimized Parameters for Design A (Minimized W_c Based on $w_{CI_c} : w_{CO_2} : w_{W_c} = 0:0:1$)

(a): Design Parameters Obtained by ANN-Based Lagrange Optimization and Verified by Structural Mechanics-Based Software

	Parameters	*Training Networks*	*AI-Based Lagrange*	*Verification* (**AutoCC**)	*Error*
1	D (mm)		718.4	718.4	-
2	ρ_s		0.0800	0.0800	-
3	f_c' (MPa)		40	40	-
4	f_y (MPa)		500	500	-
5	P_u (kN)		1,500	1,500	-
6	M_u (kN.m)		2,500	2,500	-
7	SF	Ten layers – 80 neurons	1.0000	1.0048	–0.48%
8	ε_s	Ten layers – 80 neurons	0.0032	0.0032	1.07%
9	CI_c (KRW/m)	Ten layers – 20 neurons	**306,731**	**306,679**	0.02%
10	CO_2 (ton-CO_2/m)	Five layers – ten neurons	**0.7076**	**0.7077**	–0.01%
11	W_c (kN/m)	Five layers – ten neurons	**9.5566**	**9.5505**	0.06%
	Inputs for ANN.		$w_{CI_c} : w_{CO_2} : w_{W_c} = 0:0:1$		
	Inputs for structural mechanics-based (*AutoCC*).		$CI_c^{UFO} = 306,679$ KRW/m		
	Outputs for ANN.		$CO_2^{UFO} = 0.7077$ ton – CO_2/m		
	Outputs for structural mechanics-based (*AutoCC*).		$W_c^{UFO} = 9.5505$ kN/m		

Bold values indicate optimized objective functions.

(Continued)

Table 5.3.8.1 (Continued) Optimized Parameters for Design A (Minimized W_c Based on $w_{CI_c} : w_{CO_2} : w_{W_c}$ = 0:0:1)

(b): Fractions Yielding the Same Minimization of W_c Based on Three Objective Functions

Fractions			*Fractions*			*Fractions*		
Cost w_{CI_c}	*CO_2 w_{CO_2}*	*Weight w_{W_c}*	*Cost w_{CI_c}*	*CO_2 w_{CO_2}*	*Weight w_{W_c}*	*Cost w_{CI_c}*	*CO_2 w_{CO_2}*	*Weight w_{W_c}*
0.0000	0.0000	1.0000	0.1053	0.0000	0.8947	0.2105	0.0000	0.7895
0.0000	0.0526	0.9474	0.1053	0.0471	0.8476	0.2105	0.0416	0.7479
0.0000	0.1053	0.8947	0.1053	0.1413	0.7535	0.2105	0.0831	0.7064
0.0000	0.1579	0.8421	0.1053	0.1884	0.7064	0.2105	0.1247	0.6648
0.0000	0.2105	0.7895	0.1053	0.2355	0.6593	0.2105	0.1662	0.6233
0.0000	0.2632	0.7368	0.1053	0.2825	0.6122	0.2632	0.0000	0.7368
0.0000	0.3158	0.6842	0.1579	0.0000	0.8421	0.2632	0.0388	0.6981
0.0000	0.3684	0.6316	0.1579	0.0443	0.7978	0.2632	0.0776	0.6593
0.0526	0.0000	0.9474	0.1579	0.0886	0.7535	0.3158	0.0000	0.6842
0.0526	0.0499	0.8975	0.1579	0.1330	0.7091	0.3158	0.0360	0.6482
0.0526	0.0997	0.8476	0.1579	0.1773	0.6648	0.3684	0.0000	0.6316
0.0526	0.1496	0.7978	0.1579	0.2216	0.6205	0.4211	0.3352	0.2438
0.0526	0.1994	0.7479	0.1053	0.0000	0.8947	0.1053	0.1884	0.7064
0.0526	0.2493	0.6981	0.1053	0.0471	0.8476	0.1053	0.2355	0.6593
0.0526	0.2992	0.6482	0.1053	0.1413	0.7535	0.1053	0.2825	0.6122

Table 5.3.8.2 Optimized parameters for Design B (Minimized CI_c based on $w_{CI_c} : w_{CO_2} : w_{W_c}$ = 1:0:0)

(a): Design Parameters Obtained by ANN-Based Lagrange Optimization and Verified by Structural Mechanics-Based Software

	Parameters	*Training Networks*	*AI-Based Lagrange*	*Verification (**AutoCC**)*	*Error*
1	D (mm)		1,046.7	1,046.7	-
2	CI_c		0.0126	0.0126	-
3	f_c' (MPa)		40	40	-
4	f_y (MPa)		500	500	-
5	P_u (kN)		1,500	1,500	-
6	M_u (kN.m)		2,500	2,500	-
7	SF	Ten layers – 80 neurons	1.0000	0.9893	1.07%
8	ε_s	Ten layers – 80 neurons	0.0090	0.0089	1.03%
9	CI_c (KRW/m)	Ten layers – 20 neurons	**170,339**	**170,658**	–0.19%
10	CO_2 (ton-CO_2/m)	Five layers – ten neurons	**0.3579**	**0.3582**	–0.08%
11	W_c (kN/m)	Five layers – ten neurons	**20.2699**	**20.2714**	–0.01%
	Inputs for ANN.		$w_{CI_c} : w_{CO_2} : w_{W_c} = 1:0:0$		
	Inputs for structural mechanics-based (*AutoCC*).		$CI_c^{UFO} = 170,658$ KRW/m		
	Outputs for ANN.		$CO_2^{UFO} = 0.3582$ ton – CO_2/m		
	Outputs for structural mechanics-based (*AutoCC*).		$W_c^{UFO} = 20.2714$ kN/m		

Bold and red values indicate optimized objective functions.

(Continued)

Table 5.3.8.2 (Continued) Optimized parameters for Design B (Minimized CI_c based on $w_{CI_c} : w_{CO_2} : w_{W_c}$ = 1:0:0)

(b): Fractions Yielding the Minimization of CI_c Based on Three Objective Functions

Fractions

Cost	CO_2	Weight
w_{CI_c}	w_{CO_2}	w_{W_c}
1.0000	0.0000	0.0000

Table 5.3.8.3 Optimized Parameters for Design C (Minimized CO_2 Based on $w_{CI_c} : w_{CO_2} : w_{W_c}$ = 0:1:0)

(a): Design Parameters Obtained by ANN-Based Lagrange Optimization and Verified by Structural Mechanics-Based Software

	Parameters	*Training Networks*	*AI-Based Lagrange*	*Verification (AutoCC)*	*Error*
1	D (mm)		1,111.5	1,111.5	-
2	ρ_s		0.0100	0.0100	-
3	$f_c^{'}$ (MPa)		40	40	-
4	f_y (MPa)		500	500	-
5	P_u (kN)		1,500	1,500	-
6	M_u (kN.m)		2,500	2,500	-
7	SF	Ten layers – 80 neurons	1.0000	0.9996	0.04%
8	ε_s	Ten layers – 80 neurons	0.0102	0.0101	0.90%
9	CI_c (KRW/m)	Ten layers – 20 neurons	**171,779**	**171,567**	0.12%
10	CO_2 (ton-CO_2/m)	Five layers – ten neurons	**0.3538**	**0.3541**	–0.09%
11	W_c (kN/m)	Five layers – ten neurons	**22.8631**	**22.8604**	0.01%
	Inputs for ANN.			$w_{CI_c}:w_{CO_2}:w_{W_c}$ = 0:1:0	
	Inputs for structural mechanics-based (*AutoCC*).			$CI_c^{UFO} = 171,567$ KRW/m	
	Outputs for ANN.			$CO_2^{UFO} = 0.3541$ ton – CO_2/m	
	Outputs for structural mechanics-based (*AutoCC*).			$W_c^{UFO} = 22.8604$ kN/m	

Bold and red color values indicate optimized objective functions.

(b): Fractions Yielding the Same Minimization of CO_2 Based on Three Objective Functions

Fraction Samples			*Fraction Samples*			*Fraction Samples*		
Cost w_{CI_c}	CO_2 w_{CO_2}	Weight w_{W_c}	Cost w_{CI_c}	CO_2 w_{CO_2}	Weight w_{W_c}	Cost w_{CI_c}	CO_2 w_{CO_2}	Weight w_{W_c}
0.0000	1.0000	0.0000	0.4737	0.4155	0.1108	0.8947	0.0332	0.0720
0.0526	0.8975	0.0499	0.5263	0.3740	0.0997	0.8947	0.0665	0.0388
0.0526	0.9474	0.0000	0.5263	0.4488	0.0249	0.8947	0.0831	0.0222
0.1053	0.0942	0.8006	0.5789	0.3989	0.0222	0.8947	0.0886	0.0166
0.1053	0.8947	0.0000	0.7368	0.2632	0.0000	0.8947	0.0997	0.0055
0.1579	0.8421	0.0000	0.7895	0.0776	0.1330	0.8947	0.1053	0.0000
0.2632	0.1163	0.6205	0.7895	0.0886	0.1219	0.9474	0.0028	0.0499
0.2632	0.7368	0.0000	0.8421	0.1247	0.0332	0.9474	0.0277	0.0249
0.3158	0.5762	0.1080	0.8421	0.1413	0.0166	0.9474	0.0332	0.0194
0.3684	0.6316	0.0000	0.8421	0.1496	0.0083	0.9474	0.0388	0.0139
0.4211	0.5180	0.0609	0.8421	0.1579	0.0000	0.9474	0.0471	0.0055

minimum rebar ratio 0.0126 is used to minimize column cost as shown in Design B based on $w_{CI_c} : w_{CO_2} : w_{W_c}$ = 1:0:0 of as shown in Table 5.3.8.2(a). The proposed network is capable of calculating design parameters according to the prescribed fractions when optimizing UFO. It is noted that a structural mechanics-based software can be used to calculate the rest of output parameters using input parameters obtained by ANNs. Design C is a design minimizing W_{CO_2} as shown in Table 5.3.8.3(a) where fraction of Design C is $w_{CI_c} : w_{CO_2} : w_{W_c} = 0.0 : 1.0 : 0.0$.

Another example of an optimized design based on a fraction of $w_{CI_c} : w_{CO_2} : w_{W_c} = 0.32 : 0.32 : 0.36$ is presented in Table 5.3.8.4 and Figure 5.3.8.3 where an optimized design representing a fraction of $w_{CI_c} : w_{CO_2} : w_{W_c} = 0.32 : 0.32 : 0.36$ is indicated with a rebar ratio of 0.0373. Optimized design having an equal trade-off between the three objective functions shown in Equation (5.3.3.12) are presented in Table 5.3.8.4, indicating that contributions of objective functions to the overall optimizations are equal.

5.3.8.2 Optimized P-M diagram

Optimized P-M diagrams of all Designs A, B, C, and D constructed based on parameters in the red box of Tables 5.3.8.1(a), 5.3.8.2(a), 5.3.8.3(a), and 5.3.8.4 are presented in Figure 5.3.8.4 which displays four designs (Design A based on minimized W_c, Design B based on minimized CI_c, Design C based on minimized CO_2, and Design D based equal fraction). Design parameters indicated in the red box of Tables 5.3.8.1–5.3.8.4 are implemented in structural mechanics-based software to plot P-M diagrams. The fractions adopted in Designs A, B, C, and D are reflected in P-M diagrams; small P-M diagram is plotted using less concrete when W_c is minimized whereas larger P-M diagram is plotted using more concrete when CI_c and CO_2 are minimized. The factored load point (P_u = 1,500 kN, M_u = 2,500 kN.m) indicated by red dot passes through all P-M diagrams as shown in Figure 5.3.8.4. Engineers can find all other design parameters including diameters of circular columns, rebar rations, safety factors, rebar strains, and all objective functions minimized based on a Designs

Table 5.3.8.4 Design Parameters of Design D Based on $w_{CI_c} : w_{CO_2} : w_{W_c} = 0.32 : 0.32 : 0.36$

	Parameters	*Training Networks*	*AI-Based Lagrange*	*Verification (AutoCC)*	*Error*
1	D (mm)		820.4	820.4	-
2	ρ_s		0.0373	0.0373	-
3	f'_c (MPa)		40	40	-
4	f_y (MPa)		500	500	-
5	P_u (kN)		1,500	1,500	-
6	M_u (kN.m)		2,500	2,500	-
7	SF	Ten layers – 80 neurons	1.0000	0.9903	0.97%
8	ε_s	Ten layers – 80 neurons	0.0048	0.0047	0.64%
9	CI_c (KRW/m)	Ten layers – 20 neurons	**213,312**	**213,143**	0.08%
10	CO_2 (ton-CO_2/m)	Five layers – ten neurons	**0.4782**	**0.4780**	0.04%
11	W_c (kN/m)	Five layers – ten neurons	**12.4582**	**12.4544**	0.03%
	Inputs for ANN.		$w_{CI_c} : w_{CO_2} : w_{W_c} = 0.32 : 0.32 : 0.36$		
	Inputs for structural mechanics-based (*AutoCC*).		CI_c^{UFO} = 213,143 KRW/m		
	Outputs for ANN.		CO_2^{UFO} = 0.4782 ton – CO_2/m		
	Outputs for structural mechanics-based (*AutoCC*).		W_c^{UFO} = 12.4582 kN/m		

Bold values indicate optimized objective functions.

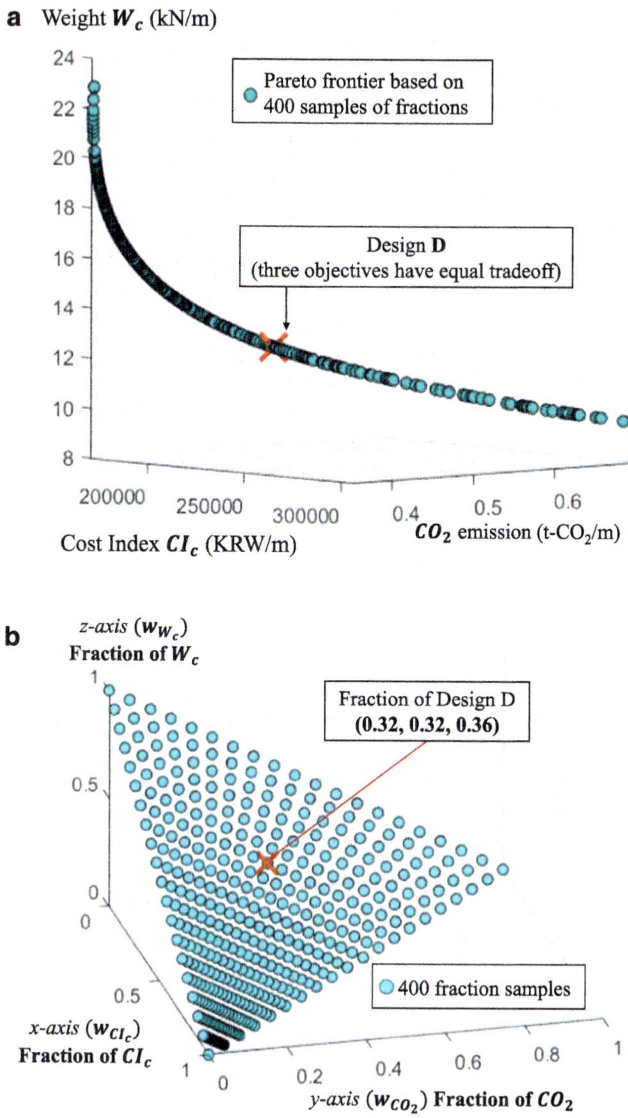

Figure 5.3.8.3 Optimized parameters for Design D based on .$w_{CI_c} : w_{CO_2} : w_{W_c} = 0.32 : 0.32 : 0.36$. (a) Design D in Pareto frontier. (b) Fractions of Design D.

A, B, C, and D shown in Tables 5.3.8.1(a), 5.3.8.2(a), 5.3.8.3(a), and 5.3.8.4, respectively. These figures and tables can serve as preliminary design guides for engineers to identify design parameters for designated trade-offs of design targets.

5.3.9 Decision-making based on Pareto frontier

A Pareto frontier yields designs optimizing multiple objective functions simultaneously. The following design example provides both engineers and decision-makers with preliminary designs minimizing cost, CO_2 emissions and weight, simultaneously. In the previous

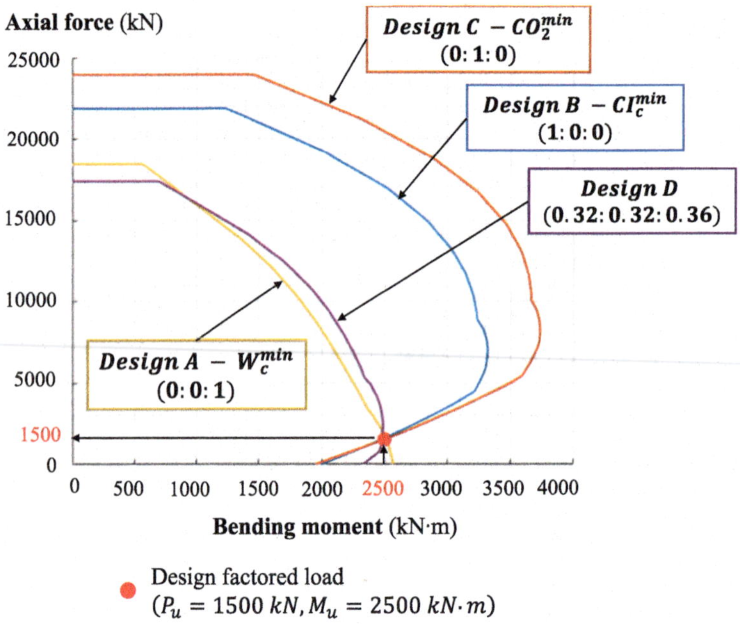

Figure 5.3.8.4 Optimized P-M diagrams of four design examples.

section, a Pareto frontier shown in Figure 5.3.9.1 was constructed based on minimized cost CI_c, CO_2 and W_c for a design of RC circular columns. Design of an RC circular columns must meet factored loads ($P_u = 1{,}500$ kN, $M_u = 2{,}500$ kN.m) and strength requirement ($SF \geq 1.0$) based on given material properties ($f_c' = 40$ MPa, $f_y = 500$ MPa).

Let's assume a project having a budget limited in a range of 200,000~240,000 KRW per meter length for the construction of one RC column. A range of CO_2 emissions and column weight can be optimized as 0.4449 ~ 0.5343 ton- CO_2 /m and 13.29~11.32 kN/m corresponding to budget in a range of 200,000 ~ 240,000 KRW, respectively, based on a Pareto

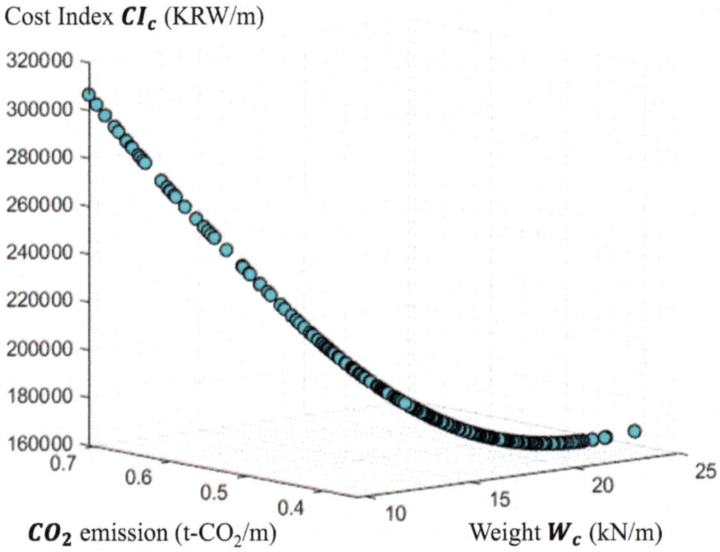

Figure 5.3.9.1 Pareto frontier based on three-objective functions.

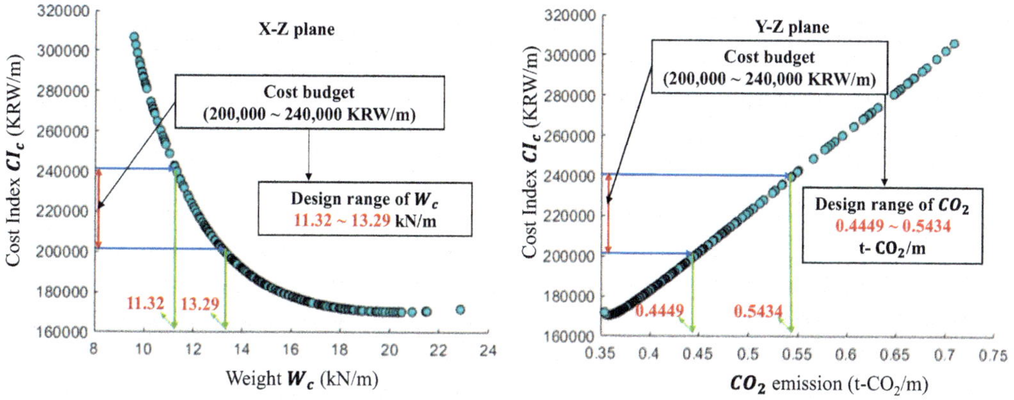

Figure 5.3.9.2 Estimated design range of CO_2 and W_c.

frontier shown in Figure 5.3.9.2 when a design should provide both the least weight and CO_2 emissions. If minimizing CO_2 emission is preferable within the budget between 200,000 and 240,000 KRW, a design having the least CO_2 emissions of 0.4449 ton- CO_2 /m in the range (0.4449 ~ 0.5343 ton- CO_2 /m) should be selected, corresponding to CI_c = 200,050 KRW/m. It is noted that this design gives a weight of W_c = 13.2984 kN/m for a project. It is noted that 50% CI_c (CI_c = 200,050 KRW/m), 50% CO_2 (CO_2 = 0.4449 ton- CO_2 /m), and 29% W_c (W_c = 13.2984 kN/m) contributed to the overall optimizations, respectively. A CO_2 emission of this design is 0.5434 ton- CO_2 /m for a budget in a range of 240,000 KRW while the weight reduces to 11.32 kN/m which is achieved by using more rebars with less concrete. Engineers can freely choose trade-off ratios representing fractions of contributions made by the three objective functions corresponding to a design preference shown in Table 5.3.9.1 where minimized CO_2, weight, and budget CI_c are determined once trade-off ratios are determined by the novel design optimizations (Figures 5.3.9.3 and 5.3.9.4).

5.4 AN ANN-BASED OPTIMIZATION OF UFO FOR CIRCULAR RC COLUMNS SUSTAINING MULTIPLE LOADS

5.4.1 Reusing components of weight matrices subject to one biaxial load pair (ANN-1LP) to derive weight matrices subject to multiple biaxial load pairs (ANN-nLP) load pairs

This section discusses an optimization of circular RC columns subject to multiple loads, in which three-objective functions, *UFO*, including cost, estimated CO_2 emission, and column weight are minimized, simultaneously, by ANN-based Lagrange method. A circular RC column in this design example sustains multiple load pairs $(P_{u,i}, M_{u,i})$.

A weight matrix of an ANN subject to one load pair (P_u, M_u) is explained in Section 5.3. Uniaxial rectangular RC columns sustaining multiple loads and biaxial rectangular RC columns sustaining multiple loads are designed in Sections 5.5 and 5.6 where components of the weight matrix obtained for one load pair (P_u, M_u) are reused in the weight matrix for a design of uniaxial and biaxial rectangular RC columns sustaining multiple loads, respectively. Components of the weight matrix, $\omega_{1,5}$, $\omega_{1,6}$, of $\omega_{SF}^{(1)}$ shown in Equation (5.4.1.2) derived considering one load pair (P_u, M_u) are reused in the weight matrix of $\omega_{SF_i}^{(1)}$ shown in Equations (5.4.1.3-2) and (5.4.1.3-3) when a circular RC column is subject to multiple load pairs (factored multiple load pairs $P_{u,i}$ and $M_{u,i}$). To train ANNs, 100,000

Table 5.3.9.1 Design Options with Trade-off Ratios

Objectives			Fraction Samples			Objectives			Fraction Samples		
Cost	CO_2	Weight	w_{Cl_c}	w_{CO_2}	$c_{14}(x) = M_{u,y}^{(4)} - 8,000$	Cost	CO_2	Weight	w_{Cl_c}	w_{CO_2}	w_{W_c}
200,050	0.4449	13.2984	0.21	0.5	0.29	203,997	0.4549	13.0172	0.49	0.2	0.31
200,497	0.4461	13.2648	0.19	0.51	0.3z	205,712	0.4592	12.9042	0.01	0.66	0.33
200,588	0.4463	13.2581	0.66	0.06	0.28	205,938	0.4598	12.8897	0.37	0.31	0.32
201,318	0.4482	13.2045	0.63	0.08	0.29	207,776	0.4644	12.7749	0.29	0.38	0.33
201,701	0.4491	13.1768	0.7	0.01	0.29	208,133	0.4653	12.7532	0.59	0.08	0.33
201,722	0.4492	13.1753	0.51	0.2	0.29	208,189	0.4654	12.7498	0.29	0.37	0.34
201,868	0.4495	13.1648	0.23	0.47	0.3	209,465	0.4686	12.6738	0.61	0.06	0.33
202,919	0.4522	13.0908	0.08	0.61	0.31	209,870	0.4696	12.6502	0.54	0.12	0.34
203,031	0.4525	13.0831	0.44	0.26	0.3	210,389	0.4709	12.6203	0.3	0.35	0.35
203,352	0.4533	13.0610	0.56	0.14	0.3	211,131	0.4728	12.5781	0.43	0.22	0.35
203,745	0.4543	13.0342	0.58	0.12	0.3	211,409	0.4734	12.5625	0.64	0.02	0.34
203,910	0.4547	13.0230	0.25	0.44	0.31	211,747	0.4743	12.5436	0.47	0.18	0.35
212,349	0.4758	12.5104	0.61	0.04	0.35	214,763	0.4818	12.3815	0.33	0.3	0.37
212,457	0.4760	12.5045	0.2	0.44	0.36	216,421	0.4858	12.2967	0.58	0.05	0.37
213,184	0.4778	12.4651	0.32	0.32	0.36	216,704	0.4865	12.2825	0.36	0.26	0.38
213,362	0.4783	12.4555	0.34	0.3	0.36	217,461	0.4884	12.2450	0.29	0.33	0.38
213,378	0.4783	12.4546	0.4	0.24	0.36	219,468	0.4933	12.1481	0.36	0.25	0.39
214,677	0.4815	12.3860	0.62	0.02	0.36	221,097	0.4973	12.0721	0.23	0.37	0.4
224,024	0.5045	11.9410	0.42	0.17	0.41	222,177	0.5000	12.0229	0.52	0.08	0.4
226,570	0.5107	11.8322	0.19	0.39	0.42	222,316	0.5003	12.0167	0.48	0.12	0.4
228,141	0.5146	11.7673	0.09	0.48	0.43	235,969	0.5337	11.4643	0.12	0.43	0.45
231,829	0.5236	11.6206	0.06	0.5	0.44	239,688	0.5428	11.3304	0.46	0.08	0.46
232,952	0.5263	11.5774	0.06	0.5	0.44	239,927	0.5434	11.3220	0.35	0.19	0.46
235,784	0.5332	11.4712	0.54	0.02	0.44						
170,658	0.3582	20.2714	1	0	0	306,679	0.7077	**9.5505**	0	0	1
171,567	**0.3541**	22.8604	0	1	0						

Bold values indicate optimized objective functions.

datasets are generated based on structural mechanics-based software *AutoCC* as shown in Table 5.3.3.2 of Section 5.3. Training accuracies of ANNs are described in Table 5.3.3.3. Five steps to optimize *UFO* developed in Section 5.2.3 are performed to optimize three objective function, cost, CO_2, and weight, simultaneously, after deriving a weight matrix subject to multiple load pairs.

This section performs STEP 1: An ANN referred to *Model-1LP* is used to design a circular RC column having one load pair as shown in Section 5.3, from which a generalized network defined as *Model-LPs* is derived to consider multiple load pairs. The use of large structural datasets and network training is thus omitted, reducing an overall computation time. The network duplication is generally applied to other types of structures such as rectangular RC columns subject to multiple load pairs. As shown in a flow chart of Figure 5.4.1.1, a generalized ANN subject to multiple load pairs is derived by reusing components of the weight matrix from an ANN established to capture one load pair.

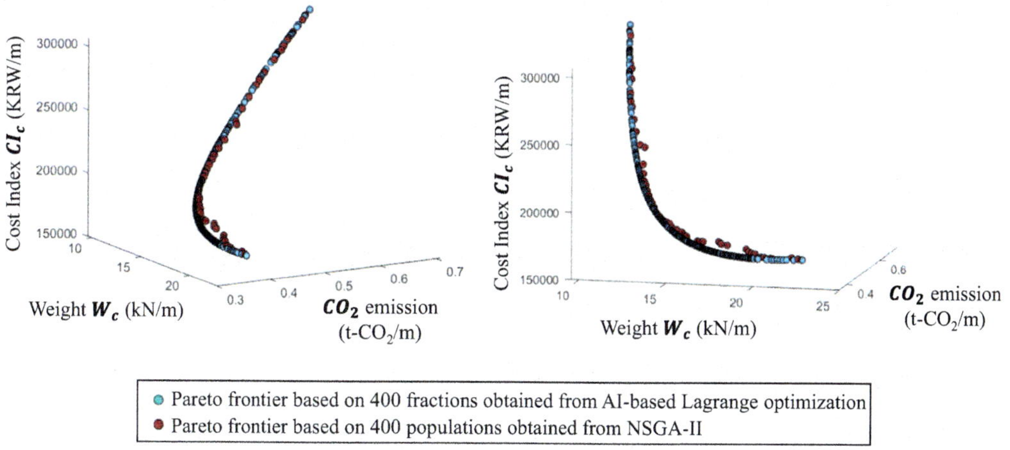

Figure 5.3.10.1 Comparison of ANN-based Pareto frontier and NSGA-II-based Pareto frontier.

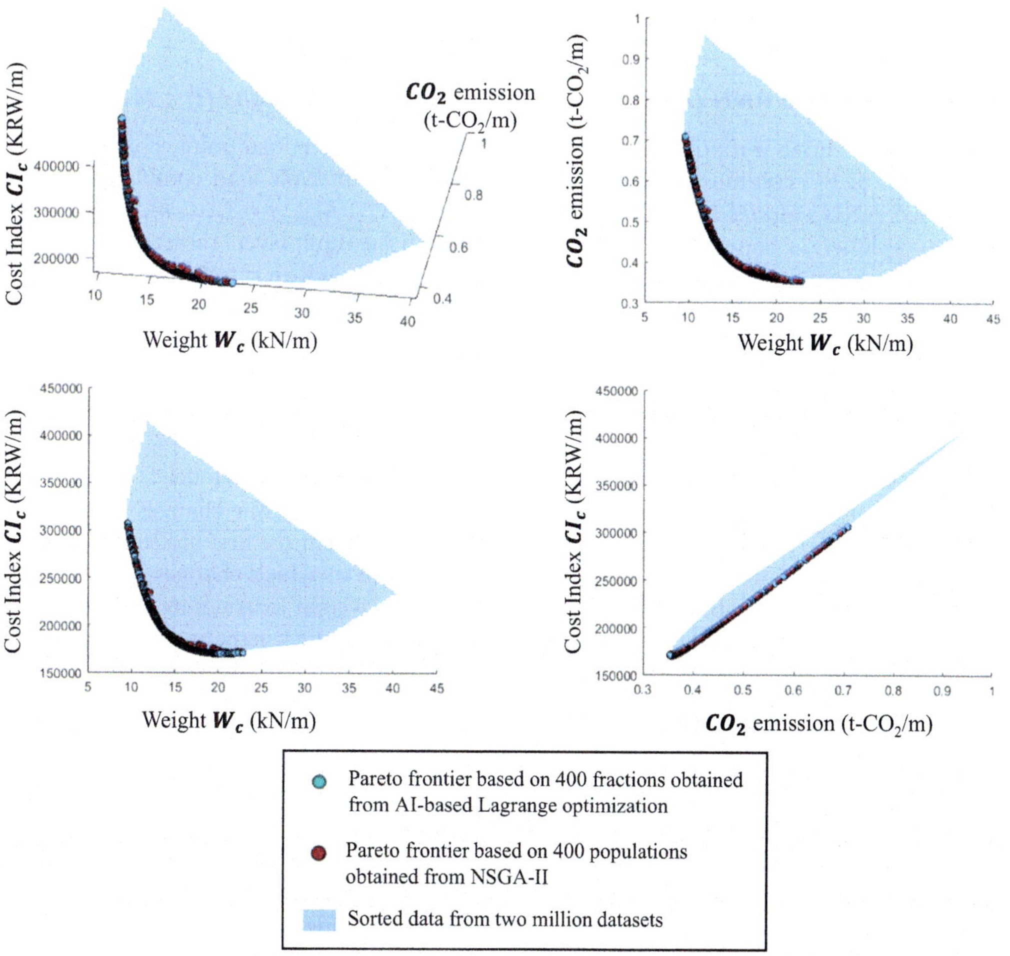

Figure 5.3.10.2 Comparison of ANN-based Pareto frontier and NSGA-II-based Pareto frontier with a large dataset.

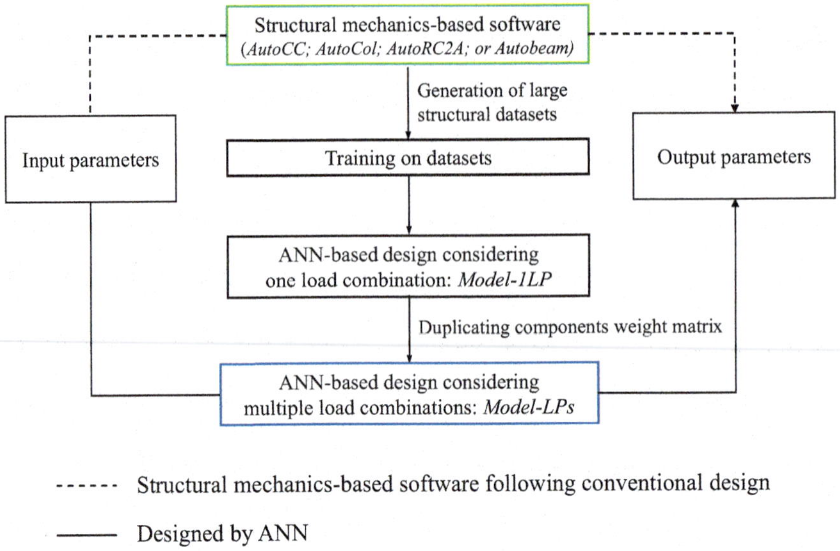

Figure 5.4.1.1 Flow chart to obtain a generalized network capturing multiple load pairs.

5.4.1.1 Generalized ANN (Model-LPs) used to derive n load pairs ($P_{u,i}$, $M_{u,i}$)

Figure 5.4.1.2a shows a topology of *Model-1LP* considering one load point. However, practical designs of RC columns must simultaneously capture multiple load combinations. In a generalized ANN (*Model-LPs*) subject to n load pairs ($P_{u,i}$, $M_{u,i}$, i = 1, ..., n), $2n + 4$ input parameters shown in Figure 5.4.1.2(b) are considered in the input layers rather than six input variables for *Model-1LP* shown in Figure 5.4.1.2(a). It is noted that four input parameters $\{D,\ \rho_s,\ f'_c,\ f_y\}^T$ shown in Figure 5.4.1.2(b) are not influenced by multiple load combinations. Some output parameters appearing on an output-side should be adjusted because multiple load combinations that refer to a pair of SF_i and $\varepsilon_{s,i}$ $(i = 1\ to\ n)$ alter the ANN topology as shown in Figure 5.4.1.2b where n is a number of load pairs, and $i = 1$ indicates Load Pair 1 whereas $i = n$ indicates Load Pair n. Load pairs directly affect mapping of input parameters to Hidden Layer 1, because reusing components of weight matrix of Model-1LP in the first hidden layer $\omega^{(1)}_{input}$ affects weigh matrices. Only the first hidden layer changes as shown in Equations (5.4.1.1-1) and (5.4.1.1-2) because the weight matrix at the first hidden layer is only directly connected to the initial input parameters, the size of which changes due to changing number of load pairs. Based on number of load pairs, weight matrices in the first hidden layer of Model-LPs are modified to by reusing components of the weight and bias matrices of Model-1LP, resulting in the size of weight matrix which changes at the only first hidden layer.

The other hidden layers of Model-LPs use the same weight and bias matrices as those of Model-1LP, indicating that the weight and bias matrices in the other hidden layers of Model-LPs do not change for the rest of the hidden layers of Model-LPs. For example, as shown in the other hidden layers of Equations (5.4.1.1-1) and (5.4.1.1-2), the size of weight matrix depends on the number of neurons of the previous and the next hidden layers which do not change even if a number of load pairs changes, resulting in the size of the weight matrix which does not change. The size of the weight matrix at each hidden layer is [20 00D7 20] which indicates a number of neurons of the previous hidden layer (20 neurons) and the next hidden layer (20 neurons). The number of weight matrices at the other hidden layers depends on the number of neurons.

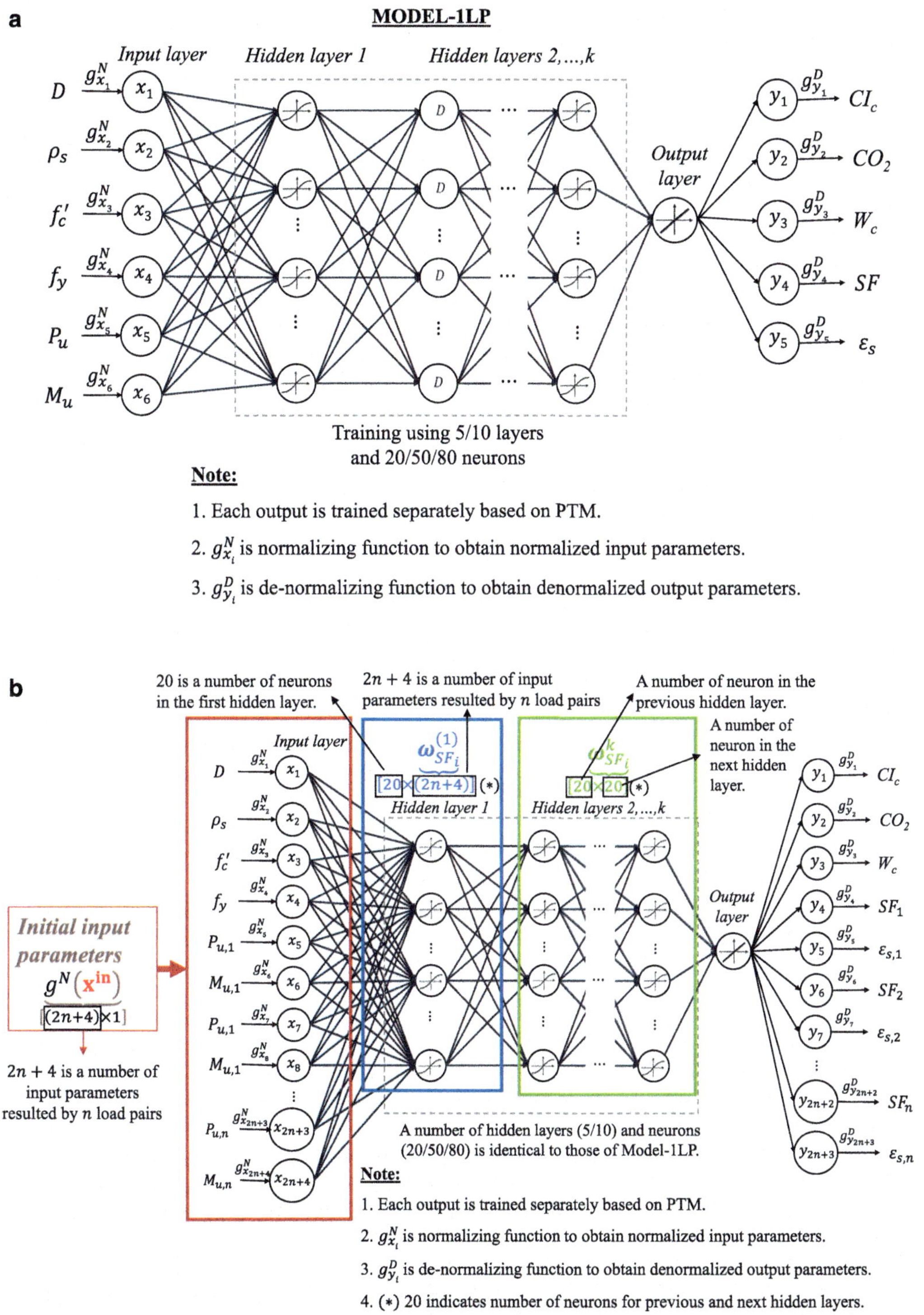

Figure 5.4.1.2 Topology of ANNs with one load pair and multiple load pairs. (a) An ANN subject to one load pair (P_u, M_u). (b) An ANN subject to multiple load pairs $(P_{u,i}, M_{u,i})$.

5.4.1.2 Formulation of the Network subject to multi-load pairs

Equations (5.4.1.1-1) and (5.4.1.1-2) describe ANNs relaying input layer to neurons of successive layers for the output parameter *SF* based on one pair and multi-pair of loads, respectively. As shown in Equations (5.4.1.1-1) and (5.4.1.1-2), the size of initial input parameters $g^N\left(\mathbf{x}^{\text{in}}\right)$ is indicated in red box, while the blue box shows the size of weight matrix at the first hidden layer (Hidden Layer 1) and the green box indicates the size of weight matrix at the rest of the hidden layers. Input parameters of *Model-LPs* shown in Equation (5.4.1.1-2) are mapped to output parameters, similarly to *Model-1LP* shown in Equation (5.4.1.1-1) where input parameters are mapped to output parameters for one load pair (P_u, M_u). Input trial parameters $\mathbf{x}^{\text{in}}$ containing normalized input parameters $\{D,\ \rho_s,\ f_c',\ f_y,\ P_u,\ M_u\}$ which are not influenced is relayed to the first hidden layer (Hidden Layer 1) by the weight $\omega_{SF}^{(1)}$ and bias $\mathbf{b}_{SF}^{(1)}$ matrices. In Model-1LP considering one load pair, the size of $g^N\left(\mathbf{x}^{\text{in}}\right)$ is 6 00D7 1 due to 6 inputs, resulting in a size of 20×6 for the weight matrix connecting initial input parameters to the first hidden layer where 20 is a number of neurons at this layer. Six normalized inputs are relayed to Hidden Layer 1 through weight matrix $\underbrace{\omega_{SF}^{(1)}}_{[20\times6]}$. Here, bias matrix $\underbrace{\mathbf{b}_{SF}^{(1)}}_{[20\times1]}$ is added. It is noted that the best training of the output parameter *SF* is based on ten layers and 80 neurons as shown in shown in Table 5.3.3.3, denoting $\mathbf{b}_{SF}^{(1)}$ as $\underbrace{\mathbf{b}_{SF}^{(1)}}_{[20\times6]}$ where 20 and 6 represent a number of neurons and inputs, respectively, as shown in Equation (5.4.1.1-1). $\mathbf{b}_{SF}^{(1)}$ is also denoted as $\underbrace{\mathbf{b}_{SF}^{(1)}}_{[20\times1]}$ where 20 represent a number of neurons. One bias is calculated for all 20 neurons of one hidden layer. Each neuron in Hidden Layer 1 is, then, relayed to next layers based on the weight matrix $\underbrace{\omega_{SF}^{k}}_{[20\times20]}$ and bias matrix $\underbrace{\mathbf{b}_{SF}^{(k)}}_{[20\times1]}$ for the kth hidden layer. $\underbrace{\omega_{SF}^{k}}_{[20\times20]}$ relays the previous hidden layer having 20 neurons to the next hidden layer having 20 neurons. $\underbrace{\omega_{SF}^{(out)}}_{[1\times20]}$ and $\underbrace{\mathbf{b}_{SF}^{(out)}}_{[1\times1]}$ indicates weight matrix and bias matrix at an output layer where only one neuron exists that is activated by the linear activation function (f_{linear}) before being de-normalized to obtain output y_4 of the original scale of *SF*.

Model-LPs consider n load pairs leading to having $2n+4$ initial input parameters, making the size of $g^N\left(\mathbf{x}^{\text{in}}\right)$ for Model-LPs being $(2n+4)\times1$. The weight matrix at the first hidden layer having a size of $20\times(2n+4)$ are derived in Equations (5.4.1.3-1)–(5.4.1.3-3), (5.4.1.5-1)–(5.4.1.5-3), (5.4.1.6-1)–(5.4.1.6-2), and (5.4.1.7-1)–(5.4.1.7-2). A number of neurons at the rest of the hidden layer of Model-LPs is the same as those of Model-1LP, leading to the size of the weight matrix at k^{th} hidden layer, $\omega_{SF_i}^{k}$, is the same as that of ω_{SF}^{k} as shown in Equations (5.4.1.1-2) and (5.4.1.1-1), respectively. Neurons in Hidden Layer 1 are activated by *tansig* function (f_{tanh}) as shown in Equations (5.4.1.1-1) and (5.4.1.1-2).

$$\underbrace{y_4\ (SF)}_{[1\times1]} = g_{SF}^{D}\left(f_{\text{linear}}\left(\underbrace{\omega_{SF}^{(\text{out})}}_{[1\times10]} f_{\text{tanh}}\left(\underbrace{\omega_{SF}^{k}}_{[20\times20]} \cdots f_{\text{tanh}}\left(\underbrace{\omega_{SF}^{(1)}}_{[10\times6]} \underbrace{g^N\left(X^{\text{in}}\right)}_{[6\times1]} + \underbrace{b_{SF}^{(1)}}_{[20\times1]} \right) \cdots + \underbrace{b_{SF}^{(k)}}_{[20\times1]} \right) + \underbrace{b_{SF}^{(\text{out})}}_{[1\times1]} \right) \right)$$

(5.4.1.1-1)

$$\underbrace{y(SF_i)}_{[1\times1]} = g^D_{SF_i}\left(f_{\text{linear}}\left(\underbrace{\omega^{(\text{out})}_{SF_i}}_{[1\times20]} f_{\tanh}\left(\underbrace{\omega^k_{SF_i}}_{20\times20}\cdots f_{\tanh}\left(\underbrace{\omega^{(1)}_{SF}}_{[20\times(2n+4)]}\underbrace{g^N\left(x^{in}\right)}_{[(2n+4)\times1]} + \underbrace{b^{(1)}_{SF_i}}_{[20\times1]}\right)\cdots + \underbrace{b^{(k)}_{SF_i}}_{[20\times1]}\right) + \underbrace{b^{(\text{out})}_{SF_i}}_{[1\times1]}\right)\right)$$

(5.4.1.1-2)

$i = 1, 2, \ldots, n$ (n = number of load pairs); $\omega^k_{SF_i}$ indicates the weight matrix at kth hidden layer having a size of 20×20 (20 neurons for the previous and the next hidden layers).

An ANN for *Model-LPs* is shown in Figure 5.4.2.1 when three load pairs are considered. Note that SF_i is determined as a ratio of the nominal strength to factored load ($SF_i = P_{n,i} / P_{u,i} = M_{n,i} / M_{u,i}$), indicating that each SF_i is governed by its corresponding load pair ($P_{u,i}$, $M_{u,i}$). An output parameter SF predicted in network *Model-1LP* is equivalent to SF_i in *Model-LPs*, which is governed by the i-th load pair ($P_{u,i}$, $M_{u,i}$). Input layers are relayed to successive layers to obtain SF_i as derived in Equation (5.4.1.1-2). Initial input parameters $\mathbf{x}$ contains $(2n + 4)$ normalized input variables, which increase the size of the weight matrix in the first hidden layer $\omega^{(1)}_{SF_i}$ shown in Equations (5.4.1.3-1)–(5.4.1.3-3) to a $20\times(2n+4)$ matrix when n load pairs are considered. It is noted that the size of the weight matrix in the first hidden layer shown in Equation (5.4.1.1-1) is a 20×6 matrix of $\omega^{(1)}_{SF}$ when one load pair is considered. Other components of the weight and bias matrices in the hidden and output layers shown in Equation (5.4.1.1-2) remain. An output SF_i should have a similar mapping trait to an output SF of *Model-1LP* because the components of weigh matrix, $\omega_{1,5}$, $\omega_{1,6}$, from $\underbrace{\omega^{(1)}_{SF}}_{[20\times6]}$ of *Model-1LP* shown in Equation (5.4.1.2) are reused in the weight matrix $\underbrace{\omega^{(1)}_{SF_i}}_{[20\times(2n+4)]}$ in Hidden Layer 1 shown in Equations (5.4.1.3-1)–(5.4.1.3-3) when n load pairs are considered to modify weight matrix, $\omega^{(1)}_{SF_i}$, in Hidden Layer 1 to predict SF_i of *Model-LPs*. It is noted that the first four input parameters are $\{D,\ \rho_s,\ f_c',\ f_y\}$. Equations (5.4.1.3-1)–(5.4.1.3-3) show that weight matrices of *Model-LPs* that belong to the first four input parameters D, ρ_s, f_c', f_y are unaffected.

The elements of weight matrix $\omega^{(1)}_{SF_1}$ to factored loads $P_{u,1}$ and $M_{u,1}$ shown in Equation (5.4.1.3-1) are $\omega_{1,5}$, $\omega_{1,6}$ which are the same components of the weight matrices referred to P_u and M_u of *Model-1LP* shown in Equation (5.4.1.2). Weight matrix elements $\omega_{1,5}$, $\omega_{1,6}$ are reused in Equation (5.4.1.3-1). The other weight matrices of load pairs ($P_{u,i}$, $M_{u,i}$, $i = 2, \ldots,$ n) are equal to zeros because SF_1 is only controlled by Load Pair 1 ($P_{u,1}$, $M_{u,1}$).

Similarly to weight matrix $\omega^{(1)}_{SF_1}$ subject to factored loads $P_{u,1}$ and $M_{u,1}$ shown in Equation (5.4.1.3-1), new weight matrix elements corresponding to the factored Load Pair ($P_{u,2}$, $M_{u,2}$) of *Model-LPs* that must be added in $\omega^{(1)}_{SF_2}$ are $\omega_{1,7}$, $\omega_{1,8}$ corresponding to SF_2 as shown in Equation (5.4.1.3-2) which are also borrowed from the weight matrices of P_u and M_u of *Model-1LP* shown in Equation (5.4.1.2). The weight elements that are reused in the new weight matrix, $\omega^{(1)}_{SF_2}$, corresponding to SF_2 are the weight components, $\omega_{1,5}$, $\omega_{1,6}$ of the weight matrices of P_u and M_u of *Model-1LP*, respectively, shown in Equation (5.4.1.2). New weight matrix elements $\omega_{1,7}$ and $\omega_{1,8}$ are added in new weight matrix as shown in Equation (5.4.1.3-2). The other weight matrices of load pairs ($P_{u,i}$, $M_{u,i}$, $i=2, \ldots, n$) are equal to zeros because SF_2 is only controlled by Load Pair 1 ($P_{u,2}$, $M_{u,2}$).

Equation (5.4.1.3-3) illustrates the modified weight matrix of output SF_n subject to Load Pairs n ($P_{u,n}$, $M_{u,n}$). The weight matrix of $\varepsilon_{s,i}$ of *Model-LPs* is modularized similarly using Equations (5.4.1.5-1)–(5.4.1.5-3) by reusing the weight matrix components of ε_s of *Model-1LP* shown in Equation (5.4.1.4). The other outputs, such as CI_c, CO_2 and W_c, are not directly affected by multiple load combinations such that their modularized weight matrices are also similarly obtained by adding zero components for the multi-load pairs in the corresponding locations as shown in Equations (5.4.1.6-2), (5.4.1.7-2), and (5.4.1.8-2), respectively.

A generalized network *Model-LPs* is not obtained through training but just modified as explained, and hence, the outputs of a generalized network *Model-LPs* subject to multi-load pairs must inherit the same characteristics as those possessed by *Model-1LP*. In Appendix C, weight and bias matrices for cost (CI_c), CO_2, safety factors (SF), and strains (ε_s) for ANNs with one load pair and multiple load pairs (n load points, $P_{u,1}, M_{u,1}, \ldots, P_{u,n}, M_{u,n}$) shown in Figure 5.4.1.2a and b, respectively, are derived at some selected hidden layers for both *Model-1LP* and *Model-LPs*.

At first hidden layers of network *Model-1LP* for SF:

$$\begin{bmatrix} \text{Neuron 1} \\ \text{Neuron 2} \\ \vdots \\ \text{Neuron 20} \end{bmatrix}_{20\times1}^{(1),SF} = f_{\tanh}\left(\begin{bmatrix} \text{1st} & \text{2nd} & \text{3rd} & \text{4th} & \text{5th} & \text{6th} \\ \omega_{1,1} & \omega_{1,2} & \omega_{1,3} & \omega_{1,4} & \omega_{1,5} & \omega_{1,6} \\ \omega_{2,1} & \omega_{2,2} & \omega_{2,3} & \omega_{2,4} & \omega_{2,5} & \omega_{2,6} \\ \vdots & \vdots & \vdots & \vdots & \vdots & \vdots \\ \omega_{20,1} & \omega_{20,2} & \omega_{20,3} & \omega_{20,4} & \omega_{20,5} & \omega_{20,6} \end{bmatrix}_{20\times6}^{(1),SF} \begin{bmatrix} D & \text{1st} \\ \rho_s & \text{2nd} \\ f_c^{'} & \text{3rd} \\ f_y & \text{4th} \\ P_u & \text{5th} \\ M_u & \text{6th} \end{bmatrix}_{6\times1} + \begin{bmatrix} b_1 \\ b_2 \\ \vdots \\ b_{20} \end{bmatrix}_{20\times1}^{(1),SF} \right) \tag{5.4.1.2}$$

At first hidden layers of network *Model-LPs* for SF_1 $(i = 1)$:

$$\begin{bmatrix} \text{Neuron 1} \\ \text{Neuron 2} \\ \vdots \\ \text{Neuron 20} \end{bmatrix}_{20\times(2n+4)}^{(1),SF_1} = f_{\tanh}\left(\underbrace{\boldsymbol{\omega}_{SF_1}^{(1)}}_{[20\times(2n+4)]} \underbrace{g^N(\mathbf{x})}_{[(2n+4)\times1]} + \underbrace{\mathbf{b}_{SF_1}^{(1)}}_{[20\times1]} \right) \tag{5.4.1.3-1}$$

$$
= f_{\tanh}\left(
\begin{bmatrix}
\text{1st} & \text{2nd} & \text{3rd} & \text{4th} & \text{5th} & \text{6th} & \text{7th} & \text{8th} & \cdots & (2n+3)\,\text{th} & (2n+4)\,\text{th} \\
\omega_{1,1} & \omega_{1,2} & \omega_{1,3} & \omega_{1,4} & \omega_{1,5} & \omega_{1,6} & 0 & 0 & \cdots & 0 & 0 \\
\omega_{2,1} & \omega_{2,2} & \omega_{2,3} & \omega_{2,4} & \omega_{2,5} & \omega_{2,6} & 0 & 0 & \cdots & 0 & 0 \\
\vdots & \vdots & \vdots & \vdots & \vdots & \vdots & \vdots & \vdots & \ddots & \vdots & \vdots \\
\omega_{20,1} & \omega_{20,2} & \omega_{20,3} & \omega_{20,4} & \omega_{20,5} & \omega_{20,6} & 0 & 0 & \cdots & 0 & 0
\end{bmatrix}_{20\times(2n+4)}^{(1),SF_1}
\begin{bmatrix}
D & \text{1st} \\
\rho_s & \text{2nd} \\
f_c' & \text{3rd} \\
f_y & \text{4th} \\
P_{u,1} & \text{5th} \\
M_{u,1} & \text{6th} \\
P_{u,2} & \text{7th} \\
M_{u,2} & \text{8th} \\
\vdots & \vdots \\
P_{u,n} & (2n+3)\,\text{th} \\
M_{u,n} & (2n+4)\,\text{th}
\end{bmatrix}_{(2n+4)\times 1} + +
\begin{bmatrix}
b_1 \\
b_2 \\
\vdots \\
b_{20}
\end{bmatrix}_{20\times 1}^{(1),SF_1}
\right)
$$

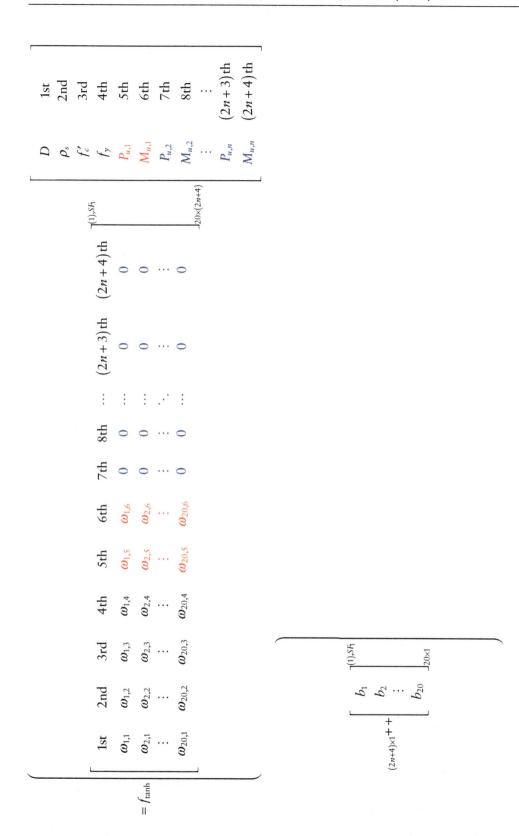

At first hidden layers of network Model-LPs for SF_2 $(i = 2)$:

$$\begin{bmatrix} \text{Neuron 1} \\ \text{Neuron 2} \\ \vdots \\ \text{Neuron 20} \end{bmatrix}_{20\times(2n+4)}^{(1),SF_2} = f_{\tanh}\left(\underbrace{\boldsymbol{\omega}_{SF_2}^{(1)}}_{[20\times(2n+4)]} \underbrace{g^N(\mathbf{x})}_{[(2n+4)\times 1]} + \underbrace{\mathbf{b}_{SF_2}^{(1)}}_{[20\times 1]} \right) \tag{5.4.1.3-2}$$

$$= f_{\tanh}\left(\begin{bmatrix} \text{1st} & \text{2nd} & \text{3rd} & \text{4th} & \text{5th} & \text{6th} & \text{7th} & \text{8th} & \text{9th} & \cdots & (2n+3)\,\text{th} & (2n+4)\,\text{th} \\ \omega_{1,1} & \omega_{1,2} & \omega_{1,3} & \omega_{1,4} & 0 & 0 & \omega_{1,5} & \omega_{1,6} & 0 & \cdots & 0 & 0 \\ \omega_{2,1} & \omega_{2,2} & \omega_{2,3} & \omega_{2,4} & 0 & 0 & \omega_{2,5} & \omega_{2,6} & 0 & \cdots & 0 & 0 \\ \vdots & \vdots & \vdots & \vdots & \vdots & \vdots & \vdots & \vdots & \vdots & \ddots & \vdots & \vdots \\ \omega_{20,1} & \omega_{20,2} & \omega_{20,3} & \omega_{20,4} & 0 & 0 & \omega_{20,5} & \omega_{20,6} & 0 & \cdots & 0 & 0 \end{bmatrix}_{20\times(2n+4)}^{(1),SF_2} \right.$$

$$\left. \begin{bmatrix} D & \text{1st} \\ \rho_s & \text{2nd} \\ f_c' & \text{3rd} \\ f_y & \text{4th} \\ P_{u,1} & \text{5th} \\ M_{u,1} & \text{6th} \\ P_{u,2} & \text{7th} \\ M_{u,2} & \text{8th} \\ P_{u,3} & \text{9th} \\ \vdots & \vdots \\ P_{u,n} & (2n+3)\,\text{th} \\ M_{u,n} & (2n+4)\,\text{th} \end{bmatrix}_{(2n+4)\times 1} + \begin{bmatrix} b_1 \\ b_2 \\ \vdots \\ b_{20} \end{bmatrix}_{20\times 1}^{(1),SF_2} \right)$$

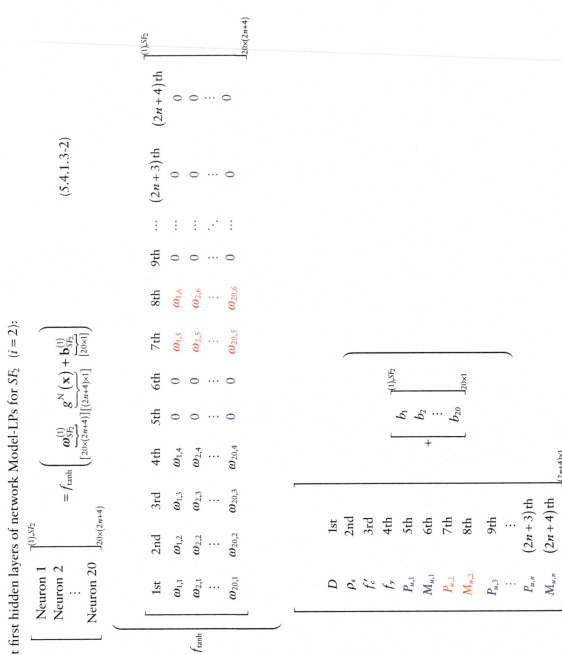

$$\begin{bmatrix} \text{Neuron 1} \\ \text{Neuron 2} \\ \vdots \\ \text{Neuron 20} \end{bmatrix}_{20\times(2n+4)}^{(1),SF_n} = f_{\tanh}\left(\underbrace{\boldsymbol{\omega}_{SF_n}^{(1)}}_{[20\times(2n+4)]} \underbrace{g^N(\mathbf{x})}_{[(2n+4)\times 1]} + \underbrace{\mathbf{b}_{SF_n}^{(1)}}_{[20\times 1]} \right) \tag{5.4.1.3-3}$$

$$= f_{\tanh}\left(\begin{bmatrix} 1\text{th} & 2\text{th} & 3\text{th} & 4\text{th} & 5\text{th} & 6\text{th} & \cdots & (2n+1)\text{th} & (2n+2)\text{th} & (2n+3)\text{th} & (2n+4)\text{th} \\ \omega_{1,1} & \omega_{1,2} & \omega_{1,3} & \omega_{1,4} & 0 & 0 & \cdots & 0 & 0 & \omega_{1,5} & \omega_{1,6} \\ \omega_{2,1} & \omega_{2,2} & \omega_{2,3} & \omega_{2,4} & 0 & 0 & \cdots & 0 & 0 & \omega_{2,5} & \omega_{2,6} \\ \vdots & \vdots & \vdots & \vdots & \vdots & \vdots & \ddots & \vdots & \vdots & \vdots & \vdots \\ \omega_{20,1} & \omega_{20,2} & \omega_{20,3} & \omega_{20,4} & 0 & 0 & \cdots & 0 & 0 & \omega_{20,5} & \omega_{20,6} \end{bmatrix}_{20\times(2n+4)}^{(1),SF_n} \begin{bmatrix} D & 1st \\ \rho_s & 2\text{nd} \\ f_c' & 3\text{rd} \\ f_y & 4\text{th} \\ P_{u,1} & 5\text{th} \\ M_{u,1} & 6\text{th} \\ \vdots & \vdots \\ P_{u,n-1} & (2n+1)\text{th} \\ M_{u,n-1} & (2n+2)\text{th} \\ P_{u,n} & (2n+3)\text{th} \\ M_{u,n} & (2n+4)\text{th} \end{bmatrix}_{(2n+4)\times 1} + \begin{bmatrix} b_1 \\ b_2 \\ \vdots \\ b_{20} \end{bmatrix}_{20\times 1}^{(1),SF_n} \right)$$

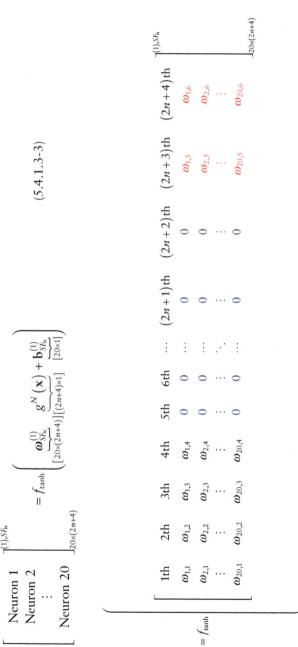

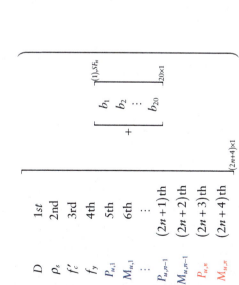

At first hidden layers of network *Model-1LP* for ε_s :

$$\begin{bmatrix} \text{Neuron 1} \\ \text{Neuron 2} \\ \vdots \\ \text{Neuron 80} \end{bmatrix}_{80\times1}^{(1),\varepsilon_s} = f_{\tanh}\left(\begin{bmatrix} \text{1st} & \text{2nd} & \text{3rd} & \text{4th} & \text{5th} & \text{6th} \\ \omega_{1,1} & \omega_{1,2} & \omega_{1,3} & \omega_{1,4} & \omega_{1,5} & \omega_{1,6} \\ \omega_{2,1} & \omega_{2,2} & \omega_{2,3} & \omega_{2,4} & \omega_{2,5} & \omega_{2,6} \\ \vdots & \vdots & \vdots & \vdots & \vdots & \vdots \\ \omega_{80,1} & \omega_{80,2} & \omega_{80,3} & \omega_{80,4} & \omega_{80,5} & \omega_{80,6} \end{bmatrix}_{80\times6}^{(1),\varepsilon_s} \begin{bmatrix} D & \text{1st} \\ \rho_s & \text{2nd} \\ f_c' & \text{3rd} \\ f_y & \text{4th} \\ P_u & \text{5th} \\ M_u & \text{6th} \end{bmatrix}_{6\times1} + \begin{bmatrix} b_1 \\ b_2 \\ \vdots \\ b_{80} \end{bmatrix}_{80\times1}^{(1),\varepsilon_s} \right) \quad (5.4.1.4)$$

At first hidden layers of network *Model-LPs* for $\varepsilon_{s,1}$ $(i=1)$:

$$\begin{bmatrix} \text{Neuron 1} \\ \text{Neuron 2} \\ \vdots \\ \text{Neuron 80} \end{bmatrix}_{80\times(2n+4)}^{(1),\varepsilon_{s,1}} = f_{\tanh}\left(\underbrace{\boldsymbol{\omega}_{\varepsilon_{s,1}}^{(1)}}_{[80\times(2n+4)]} \underbrace{g^N(\mathbf{x})}_{[(2n+4)\times1]} + \underbrace{\mathbf{b}_{\varepsilon_{s,1}}^{(1)}}_{[80\times1]} \right) \quad (5.4.1.5\text{-}1)$$

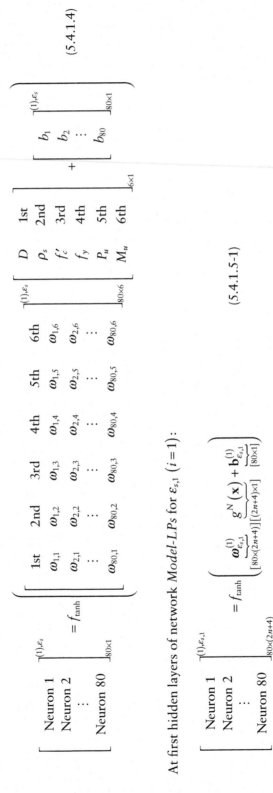

$$= f_{\tanh}\left(\begin{bmatrix} 1\text{th} & 2\text{th} & 3\text{th} & 4\text{th} & 5\text{th} & 6\text{th} & \cdots & (2n+1)\,\text{th} & (2n+2)\,\text{th} & (2n+3)\,\text{th} & (2n+4)\,\text{th} \\ \omega_{1,1} & \omega_{1,2} & \omega_{1,3} & \omega_{1,4} & 0 & 0 & \cdots & 0 & 0 & \omega_{1,5} & \omega_{1,6} \\ \omega_{2,1} & \omega_{2,2} & \omega_{2,3} & \omega_{2,4} & 0 & 0 & \cdots & 0 & 0 & \omega_{2,5} & \omega_{2,6} \\ \vdots & \vdots & \vdots & \vdots & \vdots & \vdots & \ddots & \vdots & \vdots & \vdots & \vdots \\ \omega_{80,1} & \omega_{80,2} & \omega_{80,3} & \omega_{80,4} & 0 & 0 & \cdots & 0 & 0 & \omega_{80,5} & \omega_{80,6} \end{bmatrix}_{80\times(2n+4)}^{(1),\varepsilon_{s,1}} \begin{bmatrix} D & 1st \\ \rho_s & 2\text{nd} \\ f_c' & 3\text{rd} \\ f_y & 4\text{th} \\ P_{u,1} & 5\text{th} \\ M_{u,1} & 6\text{th} \\ \vdots & \vdots \\ P_{u,n-1} & (2n+1)\,\text{th} \\ M_{u,n-1} & (2n+2)\,\text{th} \\ P_{u,n} & (2n+3)\,\text{th} \\ M_{u,n} & (2n+4)\,\text{th} \end{bmatrix}_{(2n+4)\times 1} + \begin{bmatrix} b_1 \\ b_2 \\ \vdots \\ b_{80} \end{bmatrix}_{80\times 1}^{(1),\varepsilon_{s,1}} \right)$$

At first hidden layers of network *Model-LPs* for $\varepsilon_{s,2}$ $(i = 2)$:

$$\begin{bmatrix} \text{Neuron 1} \\ \text{Neuron 2} \\ \vdots \\ \text{Neuron 80} \end{bmatrix}_{80\times(2n+4)}^{(1),\varepsilon_{s,2}} = f_{\tanh}\left(\underbrace{\boldsymbol{\omega}_{\varepsilon_{s,2}}^{(1)}}_{[80\times(2n+4)]} \underbrace{g^N(\mathbf{x})}_{[(2n+4)\times 1]} + \underbrace{\mathbf{b}_{\varepsilon_{s,2}}^{(1)}}_{[80\times 1]} \right) \qquad (5.4.1.5\text{-}2)$$

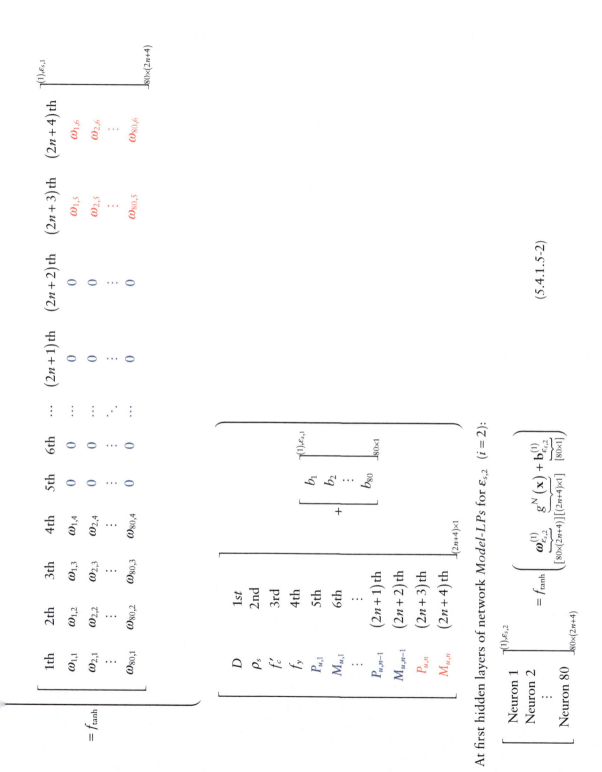

$$
= f_{\tanh}\left(
\begin{bmatrix}
\text{1st} & \text{2nd} & \text{3rd} & \text{4th} & \text{5th} & \text{6th} & \text{7th} & \text{8th} & \text{9th} & \cdots & (2n+3)\,\text{th} & (2n+4)\,\text{th} \\
\omega_{1,1} & \omega_{1,2} & \omega_{1,3} & \omega_{1,4} & 0 & 0 & \omega_{1,5} & \omega_{1,6} & 0 & \cdots & 0 & 0 \\
\omega_{2,1} & \omega_{2,2} & \omega_{2,3} & \omega_{2,4} & 0 & 0 & \omega_{2,5} & \omega_{2,6} & 0 & \cdots & 0 & 0 \\
\vdots & \vdots & \vdots & \vdots & \vdots & \vdots & \vdots & \vdots & \vdots & \ddots & \vdots & \vdots \\
\omega_{80,1} & \omega_{80,2} & \omega_{80,3} & \omega_{80,4} & 0 & 0 & \omega_{80,5} & \omega_{80,6} & 0 & \cdots & 0 & 0
\end{bmatrix}^{(1),\varepsilon_{s,2}}_{80\times(2n+4)}
\begin{bmatrix}
D & \text{1st} \\
\rho_s & \text{2nd} \\
f_c' & \text{3rd} \\
f_y & \text{4th} \\
P_{u,1} & \text{5th} \\
M_{u,1} & \text{6th} \\
P_{u,2} & \text{7th} \\
M_{u,2} & \text{8th} \\
P_{u,3} & \text{9th} \\
\vdots & \vdots \\
P_{u,n} & (2n+3)\,\text{th} \\
M_{u,n} & (2n+4)\,\text{th}
\end{bmatrix}_{(2n+4)\times 1}
+
\begin{bmatrix}
b_1 \\ b_2 \\ \vdots \\ b_{80}
\end{bmatrix}^{(1),\varepsilon_{s,2}}_{80\times 1}
\right)
$$

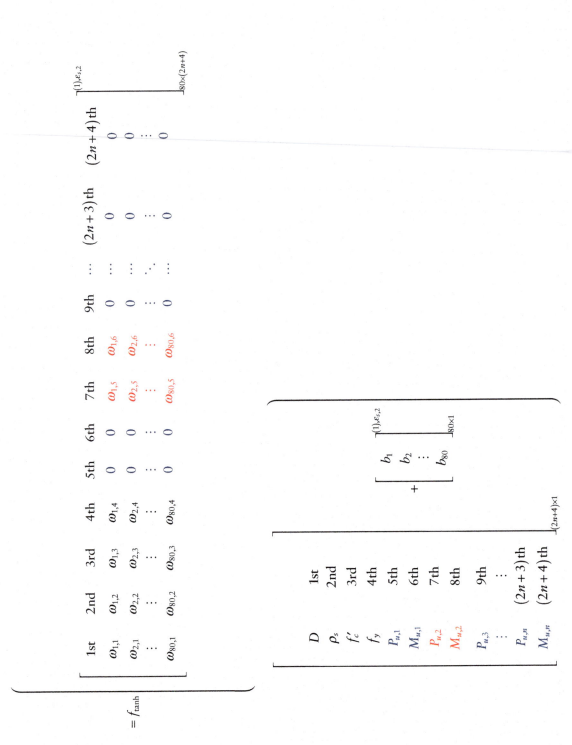

At first hidden layers of network *Model-LPs* for $\varepsilon_{s,n}$ $(i = n)$:

$$\begin{bmatrix} \text{Neuron 1} \\ \text{Neuron 2} \\ \vdots \\ \text{Neuron 80} \end{bmatrix}_{80\times(2n+4)}^{(1),\varepsilon_{s,n}} = f_{\tanh}\left(\underbrace{\boldsymbol{\omega}_{\varepsilon_{s,n}}^{(1)}}_{[80\times(2n+4)]} \underbrace{g^{N}(\mathbf{x})}_{[(2n+4)\times 1]} + \underbrace{\mathbf{b}_{\varepsilon_{s,n}}^{(1)}}_{[80\times 1]} \right) \tag{5.4.1.5-3}$$

$$= f_{\tanh}\left(\begin{bmatrix} \text{1st} & \text{2nd} & \text{3rd} & \text{4th} & \text{5th} & \text{6th} & \cdots & (2n+1)\text{th} & (2n+2)\text{th} & (2n+3)\text{th} & (2n+4)\text{th} \\ \omega_{1,1} & \omega_{1,2} & \omega_{1,3} & \omega_{1,4} & 0 & 0 & \cdots & 0 & 0 & \omega_{1,5} & \omega_{1,6} \\ \omega_{2,1} & \omega_{2,2} & \omega_{2,3} & \omega_{2,4} & 0 & 0 & \cdots & 0 & 0 & \omega_{2,5} & \omega_{2,6} \\ \vdots & \vdots & \vdots & \vdots & \vdots & \vdots & \ddots & \vdots & \vdots & \vdots & \vdots \\ \omega_{80,1} & \omega_{80,2} & \omega_{80,3} & \omega_{80,4} & 0 & 0 & \cdots & 0 & 0 & \omega_{80,5} & \omega_{80,6} \end{bmatrix}_{80\times(2n+4)}^{(1),\varepsilon_{s,n}} \right.$$

$$\left. \begin{bmatrix} D & \text{1st} \\ \rho_s & \text{2nd} \\ f_c' & \text{3rd} \\ f_y & \text{4th} \\ P_{u,1} & \text{5th} \\ M_{u,1} & \text{6th} \\ \vdots & \vdots \\ P_{u,n-1} & (2n+1)\text{th} \\ M_{u,n-1} & (2n+2)\text{th} \\ P_{u,n} & (2n+3)\text{th} \\ M_{u,n} & (2n+4)\text{th} \end{bmatrix}_{(2n+4)\times 1} + \begin{bmatrix} b_1 \\ b_2 \\ \vdots \\ b_{80} \end{bmatrix}_{80\times 1}^{(1),\varepsilon_{s,n}} \right)$$

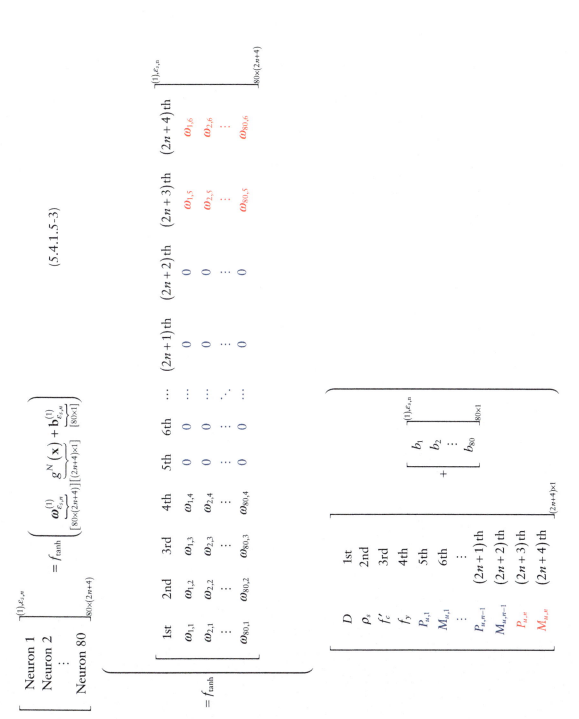

At first hidden layers of network *Model-1LP* for CI_c :

$$\begin{bmatrix} \text{Neuron 1} \\ \text{Neuron 2} \\ \vdots \\ \text{Neuron 20} \end{bmatrix}_{20\times1}^{(1),CI_c} = f_{\text{tanh}}\left(\begin{bmatrix} \text{1st} & \text{2nd} & \text{3rd} & \text{4th} & \text{5th} & \text{6th} \\ \omega_{1,1} & \omega_{1,2} & \omega_{1,3} & \omega_{1,4} & \omega_{1,5} & \omega_{1,6} \\ \omega_{2,1} & \omega_{2,2} & \omega_{2,3} & \omega_{2,4} & \omega_{2,5} & \omega_{2,6} \\ \vdots & \vdots & \vdots & \vdots & \vdots & \vdots \\ \omega_{20,1} & \omega_{20,2} & \omega_{20,3} & \omega_{20,4} & \omega_{20,5} & \omega_{20,6} \end{bmatrix}_{20\times6}^{(1),CI_c} \begin{bmatrix} D & \text{1st} \\ \rho_s & \text{2nd} \\ f_c^{'} & \text{3rd} \\ f_y & \text{4th} \\ P_u & \text{5th} \\ M_u & \text{6th} \end{bmatrix}_{6\times1} + \begin{bmatrix} b_1 \\ b_2 \\ \vdots \\ b_{20} \end{bmatrix}_{20\times1}^{(1),CI_c} \right) \quad (5.4.1.6\text{-}1)$$

At first hidden layers of network *Model-LPs* for CI_c :

$$\begin{bmatrix} \text{Neuron 1} \\ \text{Neuron 2} \\ \vdots \\ \text{Neuron 20} \end{bmatrix}_{20\times(2n+4)}^{(1),CI_c} = f_{\text{tanh}}\left(\underbrace{\boldsymbol{\omega}_{CI_c}^{(1)}}_{[20\times(2n+4)]} \underbrace{g^N(\mathbf{x})}_{[(2n+4)\times1]} + \underbrace{\mathbf{b}_{CI_c}^{(1)}}_{[20\times1]} \right) \quad (5.4.1.6\text{-}2)$$

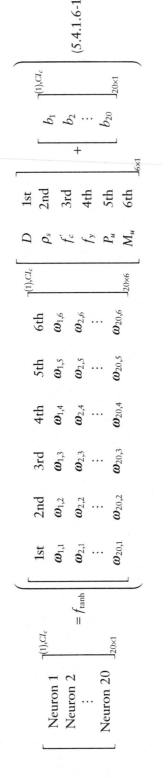

$$= f_{\tanh}\left(\begin{bmatrix} 1\text{st} & 2\text{nd} & 3\text{rd} & 4\text{th} & 5\text{th} & 6\text{th} & 7\text{th} & 8\text{th} & \cdots & (2n+3)\,\text{th} & (2n+4)\,\text{th} \\ \omega_{1,1} & \omega_{1,2} & \omega_{1,3} & \omega_{1,4} & \omega_{1,5} & \omega_{1,6} & 0 & 0 & \cdots & 0 & 0 \\ \omega_{2,1} & \omega_{2,2} & \omega_{2,3} & \omega_{2,4} & \omega_{2,5} & \omega_{2,6} & 0 & 0 & \cdots & 0 & 0 \\ \vdots & \vdots & \vdots & \vdots & \vdots & \vdots & \vdots & \vdots & \ddots & \vdots & \vdots \\ \omega_{20,1} & \omega_{20,2} & \omega_{20,3} & \omega_{20,4} & \omega_{20,5} & \omega_{20,6} & 0 & 0 & \cdots & 0 & 0 \end{bmatrix}_{20\times(2n+4)}^{(1),CI_c} \begin{bmatrix} D & 1st \\ \rho_s & 2\text{nd} \\ f_c^{'} & 3\text{rd} \\ f_y & 4\text{th} \\ P_{u,1} & 5\text{th} \\ M_{u,1} & 6\text{th} \\ P_{u,2} & 7\text{th} \\ M_{u,2} & 8\text{th} \\ \vdots & \vdots \\ P_{u,n} & (2n+3)\,\text{th} \\ M_{u,n} & (2n+4)\,\text{th} \end{bmatrix}_{(2n+4)\times 1} + \begin{bmatrix} b_1 \\ b_2 \\ \vdots \\ b_{20} \end{bmatrix}_{20\times 1}^{(1),CI_c} \right)$$

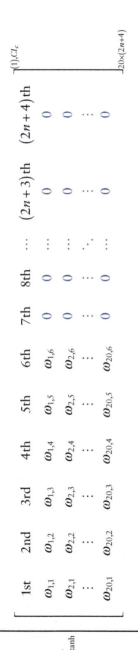

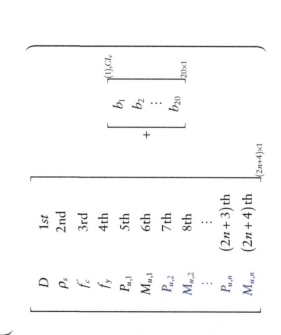

At first hidden layers of network *Model-1LP* for CO_2 :

$$\begin{bmatrix} \text{Neuron 1} \\ \text{Neuron 2} \\ \vdots \\ \text{Neuron 10} \end{bmatrix}_{10\times1}^{(1),CO_2} = f_{\tanh}\left(\begin{bmatrix} \text{1st} & \text{2nd} & \text{3rd} & \text{4th} & \text{5th} & \text{6th} \\ \omega_{1,1} & \omega_{1,2} & \omega_{1,3} & \omega_{1,4} & \omega_{1,5} & \omega_{1,6} \\ \omega_{2,1} & \omega_{2,2} & \omega_{2,3} & \omega_{2,4} & \omega_{2,5} & \omega_{2,6} \\ \vdots & \vdots & \vdots & \vdots & \vdots & \vdots \\ \omega_{10,1} & \omega_{10,2} & \omega_{10,3} & \omega_{10,4} & \omega_{10,5} & \omega_{10,6} \end{bmatrix}_{10\times6}^{(1),CO_2} \begin{bmatrix} D & \text{1st} \\ \rho_s & \text{2nd} \\ f_c' & \text{3rd} \\ f_y & \text{4th} \\ P_u & \text{5th} \\ M_u & \text{6th} \end{bmatrix}_{6\times1} + \begin{bmatrix} b_1 \\ b_2 \\ \vdots \\ b_{10} \end{bmatrix}_{10\times1}^{(1),CO_2} \right) \tag{5.4.1.7-1}$$

At first hidden layers of network *Model-LPs* for CO_2 :

$$\begin{bmatrix} \text{Neuron 1} \\ \text{Neuron 2} \\ \vdots \\ \text{Neuron 10} \end{bmatrix}_{10\times(2n+4)}^{(1),CO_2} = f_{\tanh}\left(\underbrace{\boldsymbol{\omega}_{CO_2}^{(1)}}_{[10\times(2n+4)]} \underbrace{g^N(\mathbf{x})}_{[(2n+4)\times1]} + \underbrace{\mathbf{b}_{CO_2}^{(1)}}_{[10\times1]} \right) \tag{5.4.1.7-2}$$

$$
= f_{\tanh}\left(
\begin{bmatrix}
\text{1st} & \text{2nd} & \text{3rd} & \text{4th} & \text{5th} & \text{6th} & \text{7th} & \text{8th} & \cdots & (2n+3)\text{th} & (2n+4)\text{th} \\
\omega_{1,1} & \omega_{1,2} & \omega_{1,3} & \omega_{1,4} & \omega_{1,5} & \omega_{1,6} & 0 & 0 & \cdots & 0 & 0 \\
\omega_{2,1} & \omega_{2,2} & \omega_{2,3} & \omega_{2,4} & \omega_{2,5} & \omega_{2,6} & 0 & 0 & \cdots & 0 & 0 \\
\vdots & \vdots & \vdots & \vdots & \vdots & \vdots & \vdots & \vdots & \ddots & \vdots & \vdots \\
\omega_{10,1} & \omega_{10,2} & \omega_{10,3} & \omega_{10,4} & \omega_{10,5} & \omega_{10,6} & 0 & 0 & \cdots & 0 & 0
\end{bmatrix}_{10\times(2n+4)}^{(1),Co_2}
\begin{bmatrix}
D & 1st \\
\rho_s & \text{2nd} \\
f_c' & \text{3rd} \\
f_y & \text{4th} \\
P_{u,1} & \text{5th} \\
M_{u,1} & \text{6th} \\
P_{u,2} & \text{7th} \\
M_{u,2} & \text{8th} \\
\vdots & \vdots \\
P_{u,n} & (2n+3)\text{th} \\
M_{u,n} & (2n+4)\text{th}
\end{bmatrix}_{(2n+4)\times 1}
+
\begin{bmatrix}
b_1 \\ b_2 \\ \vdots \\ b_{10}
\end{bmatrix}_{10\times 1}^{(1),Co_2}
\right)
$$

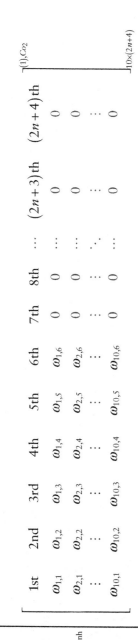

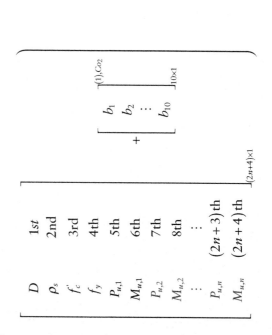

At first hidden layers of network *Model-1LP* for W_c :

$$\begin{bmatrix} \text{Neuron 1} \\ \text{Neuron 2} \\ \vdots \\ \text{Neuron 10} \end{bmatrix}_{10\times1}^{(1),W_c} = f_{\tanh}\left(\begin{bmatrix} \text{1st} & \text{2nd} & \text{3rd} & \text{4th} & \text{5th} & \text{6th} \\ \omega_{1,1} & \omega_{1,2} & \omega_{1,3} & \omega_{1,4} & \omega_{1,5} & \omega_{1,6} \\ \omega_{2,1} & \omega_{2,2} & \omega_{2,3} & \omega_{2,4} & \omega_{2,5} & \omega_{2,6} \\ \vdots & \vdots & \vdots & \vdots & \vdots & \vdots \\ \omega_{10,1} & \omega_{10,2} & \omega_{10,3} & \omega_{10,4} & \omega_{10,5} & \omega_{10,6} \end{bmatrix}_{10\times6}^{(1),W_c} \begin{bmatrix} D & \text{1st} \\ \rho_s & \text{2nd} \\ f_c^{'} & \text{3rd} \\ f_y & \text{4th} \\ P_u & \text{5th} \\ M_u & \text{6th} \end{bmatrix}_{6\times1} + \begin{bmatrix} b_1 \\ b_2 \\ \vdots \\ b_{10} \end{bmatrix}_{10\times1}^{(1),W_c} \right) \quad (5.4.1.8\text{-}1)$$

At first hidden layers of network *Model-LPs* for W_c :

$$\begin{bmatrix} \text{Neuron 1} \\ \text{Neuron 2} \\ \vdots \\ \text{Neuron 10} \end{bmatrix}_{10\times(2n+4)}^{(1),W_c} = f_{\tanh}\left(\underbrace{\boldsymbol{\omega}_{W_c}^{(1)}}_{[10\times(2n+4)]} \underbrace{g^N(\mathbf{x})}_{[(2n+4)\times1]} + \underbrace{\mathbf{b}_{W_c}^{(1)}}_{[10\times1]} \right) \quad (5.4.1.8\text{-}2)$$

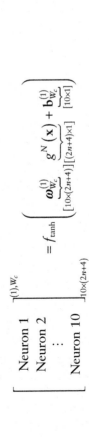

$$
= f_{\tanh}\left(\begin{bmatrix} 1\text{st} & 2\text{nd} & 3\text{rd} & 4\text{th} & 5\text{th} & 6\text{th} & 7\text{th} & 8\text{th} & \cdots & (2n+3)\,\text{th} & (2n+4)\,\text{th} \\ \omega_{1,1} & \omega_{1,2} & \omega_{1,3} & \omega_{1,4} & \omega_{1,5} & \omega_{1,6} & 0 & 0 & \cdots & 0 & 0 \\ \omega_{2,1} & \omega_{2,2} & \omega_{2,3} & \omega_{2,4} & \omega_{2,5} & \omega_{2,6} & 0 & 0 & \cdots & 0 & 0 \\ \vdots & \vdots & \vdots & \vdots & \vdots & \vdots & \vdots & \vdots & \ddots & \vdots & \vdots \\ \omega_{10,1} & \omega_{10,2} & \omega_{10,3} & \omega_{10,4} & \omega_{10,5} & \omega_{10,6} & 0 & 0 & \cdots & 0 & 0 \end{bmatrix}_{10\times(2n+4)}^{(1),\mathrm{W}_c} \begin{bmatrix} D & 1st \\ \rho_s & 2\text{nd} \\ f_c^{'} & 3\text{rd} \\ f_y & 4\text{th} \\ P_{u,1} & 5\text{th} \\ M_{u,1} & 6\text{th} \\ P_{u,2} & 7\text{th} \\ M_{u,2} & 8\text{th} \\ \vdots & \vdots \\ P_{u,n} & (2n+3)\,\text{th} \\ M_{u,n} & (2n+4)\,\text{th} \end{bmatrix}_{(2n+4)\times 1} + \begin{bmatrix} b_1 \\ b_2 \\ \vdots \\ b_{10} \end{bmatrix}_{10\times 1}^{(1),\mathrm{W}_c} \right)
$$

5.4.2 An optimization of a circular RC column sustaining five load pairs based on three-objective functions

5.4.2.1 Optimization design scenario

A design of an RC circular column sustaining five load pairs are optimized based on three-objective functions (CI_c, CO_2 emission, and W_c). Figure 5.4.2.1 shows a circular RC column section sustaining five load pairs and a corresponding P-M diagram. Output parameters $\{SF_i, \ \varepsilon_{s,i}, \ CI_c, \ CO_2, \ W_c\}^T$ are calculated for given input parameters $\{D, \ \rho_s, \ f_c^{'}, f_y, P_{u,i}, M_{u,i}\}^T$ $(i = 1,..., 5)$ for an RC column sustaining five load pairs ($P_{u,i}$, $M_{u,i}$). Strength requirements $SF_i = P_n / P_{u,i} = M_n / M_{u,i} \geq 1.0$ shown in Figure 5.4.2.1 should be met with given material properties such as $f_c^{'} = 40$ MPa and $f_y = 500$ MPa.

5.4.2.2 Five steps to optimize a circular RC column sustaining five load pairs based on three-objective functions

Step 1: Deriving ANNs

A generalized neural network *Model-LPs* is derived by reusing or duplicating weight matrix components from *Model-1LP* as presented in Section 5.4.1.2. A topology of *Model-LPs* to capture five load pairs is illustrated in Figure 5.4.2.2. It is noted that an ANN for *Model-LPs* does not have to be trained because it is obtained by duplicating weight matrix components from a trained neural network (*Model-1LP*) obtained for one load pair. Weight and bias matrices of *Model-LPs* duplicating from *Model-1LP* are presented in Appendix C.

Step 2: Defining MOO problem

Step 2.1: Deriving objective functions

Functions of three objective functions such as cost, CO_2 and weight are derived by duplicating corresponding matrix components from network with one objective function. Specifically, $\underbrace{\boldsymbol{\omega}_{CI_c}^{(1)}}_{[20\times(2n+4)]}$ and $\underbrace{\mathbf{b}_{CI_c}^{(1)}}_{[20\times1]}$ are weight and bias matrices at first hidden layers of networks *Model-LPs* with respect to cost index objective (CI_c) as

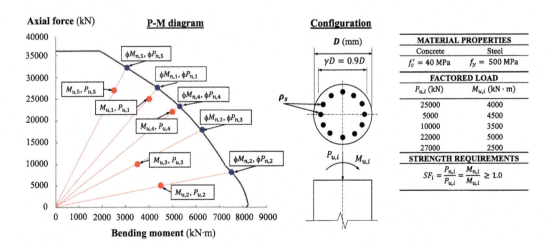

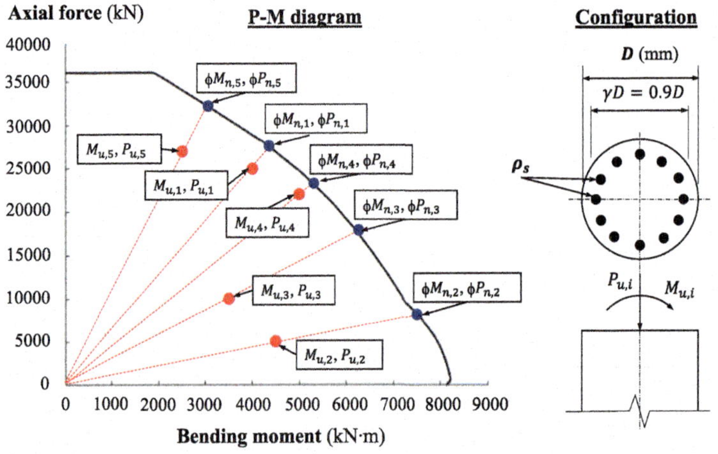

MATERIAL PROPERTIES	
Concrete	Steel
$f_c' = 40$ MPa	$f_y = 500$ MPa
FACTORED LOAD	
$P_{u,i}$ (kN)	$M_{u,i}$ (kN · m)
25000	4000
5000	4500
10000	3500
22000	5000
27000	2500
STRENGTH REQUIREMENTS	
$SF_i = \frac{P_{n,i}}{P_{u,i}} = \frac{M_{n,i}}{M_{u,i}} \geq 1.0$	

Figure 5.4.2.1 A circular RC column section sustaining five load pairs and corresponding P-M diagram.

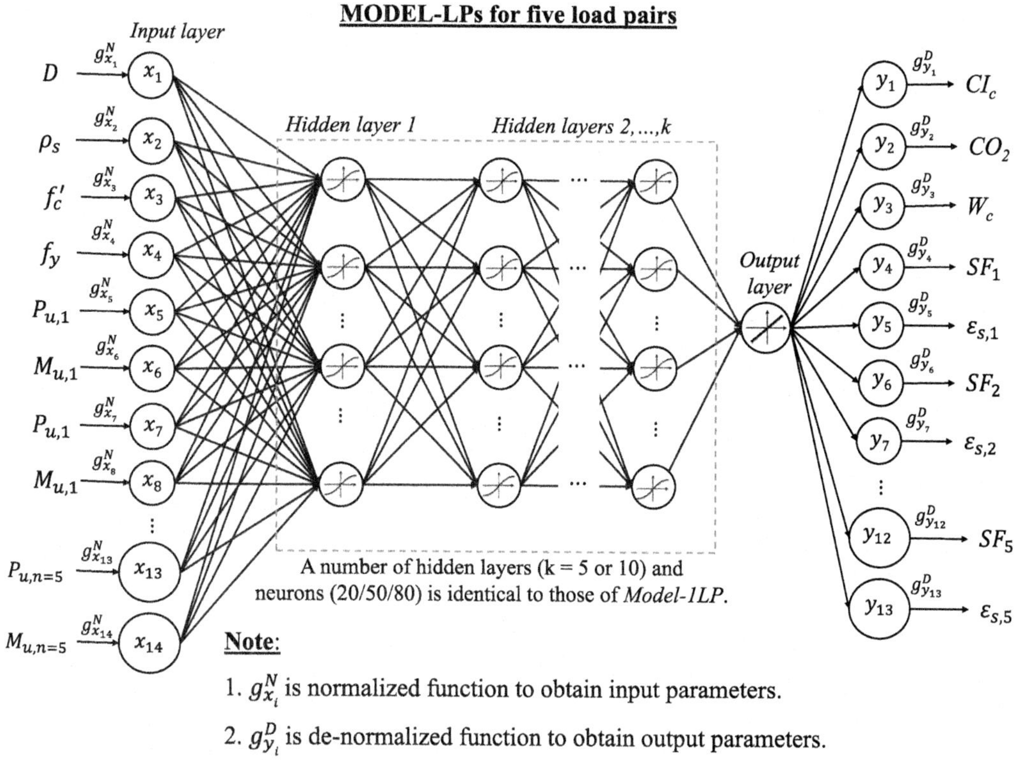

Figure 5.4.2.2 Topology of *Model-LPs* for five load pairs.

provided in Equation (5.4.1.6-2), in which n is equal to 5 load pairs, leading to $2n+4=2\times5+4=14$ input parameters, 20 represents a number of neurons based on the best training of CI_c based on ten hidden layers and 20 neurons shown in Table 5.3.3.3. Objective functions of CI_c, CO_2 and W_c derived from *Model-LPs* are presented in Equations (5.4.2.1-1)–(5.4.2.1-3).

$$\underbrace{y\left(CI_c\right)}_{[1\times1]}=f_{CI_c}^{\text{ANN}}\left(\mathbf{x}\right)$$

$$=g_{CI_c}^{D}\left(f_{\text{linear}}\left(\underbrace{\boldsymbol{\omega}_{CI_c}^{(\text{out})}}_{[1\times20]}f_{\tanh}\left(\underbrace{\boldsymbol{\omega}_{CI_c}^{(k=10)}}_{[20\times20]}\cdots f_{\tanh}\left(\underbrace{\boldsymbol{\omega}_{CI_c}^{(1)}}_{[20\times14]}\underbrace{g^{N}\left(\mathbf{x}\right)}_{[14\times1]}+\underbrace{\mathbf{b}_{CI_c}^{(1)}}_{[20\times1]}\right)\cdots+\underbrace{\mathbf{b}_{CI_c}^{(k=10)}}_{[20\times1]}\right)+\underbrace{\mathbf{b}_{CI_c}^{(\text{out})}}_{[1\times1]}\right)\right) \quad (5.4.2.1\text{-}1)$$

$$\underbrace{y\left(\text{CO}_2\right)}_{[1\times1]}=f_{\text{CO}_2}^{\text{ANN}}\left(\mathbf{x}\right)$$

$$=g_{\text{CO}_2}^{D}\left(f_{\text{linear}}\left(\underbrace{\boldsymbol{\omega}_{\text{CO}_2}^{(\text{out})}}_{[1\times20]}f_{\tanh}\left(\underbrace{\boldsymbol{\omega}_{\text{CO}_2}^{(k=10)}}_{[10\times10]}\cdots f_{\tanh}\left(\underbrace{\boldsymbol{\omega}_{\text{CO}_2}^{(1)}}_{[10\times14]}\underbrace{g^{N}\left(\mathbf{x}\right)}_{[14\times1]}+\underbrace{\mathbf{b}_{\text{CO}_2}^{(1)}}_{[10\times1]}\right)\cdots+\underbrace{\mathbf{b}_{\text{CO}_2}^{(k=5)}}_{[10\times1]}\right)+\underbrace{\mathbf{b}_{\text{CO}_2}^{(\text{out})}}_{[1\times1]}\right)\right) \quad (5.4.2.1\text{-}2)$$

$$\underbrace{y\left(W_c\right)}_{[1\times1]} = f_{W_c}^{ANN}\left(\mathbf{x}\right)$$

$$= g_{W_c}^{D}\left(f_{\text{linear}}\left(\underbrace{\boldsymbol{\omega}_{W_c}^{(out)}}_{[1\times10]} f_{\tanh}\left(\underbrace{\boldsymbol{\omega}_{W_c}^{(k=5)}}_{[10\times10]}\cdots f_{\tanh}\left(\underbrace{\boldsymbol{\omega}_{W_c}^{(1)}}_{[10\times14]}\underbrace{g^{N}\left(\mathbf{x}\right)}_{[14\times1]}+\underbrace{\mathbf{b}_{W_c}^{(1)}}_{[10\times1]}\right)\cdots+\underbrace{\mathbf{b}_{W_c}^{(k=5)}}_{[10\times1]}\right)+\underbrace{\mathbf{b}_{W_c}^{(out)}}_{[1\times1]}\right)\right) \quad (5.4.2.1\text{-}3)$$

Step 2.2: Formulating constrained conditions based on equalities and inequalities

Constrained conditions, equalities $(c_j(\mathbf{x})=0)$ and inequalities $(v_j(\mathbf{x})\geq 0)$ are formulated as shown in Figure 5.4.2.1 and Table 5.4.2.1. Column dimension is restricted to a range of 400–2,000 mm while rebar ratio is limited between 0.01 and 0.08 as specified by the ACI 318-19 (ACI Committee, 2019).

ANN-based Lagrange optimization shown in Steps 3–5 is obtained similarly to those described in previous sections. This section repeats optimization based on UFO.

Step 3: Optimizing single objective

Optimization based on single objective function for a design of circular RC column capturing five load pairs is separately performed in Table 5.4.2.2. The maximum and minimum value of each objective function are obtained by solving the Lagrange function $\mathcal{L}_{CI_c}(\mathbf{x},\boldsymbol{\lambda}_c,\boldsymbol{\lambda}_v)$, $\mathcal{L}_{CO_2}(\mathbf{x},\boldsymbol{\lambda}_c,\boldsymbol{\lambda}_v)$, and $\mathcal{L}_{W_c}(\mathbf{x},\boldsymbol{\lambda}_c,\boldsymbol{\lambda}_v)$ shown in Equations (5.4.2.2-1)–(5.4.2.2-3). The FMINCON function from the MATLAB optimization toolbox (MathWorks, 2020) is used to find stationary points of a Lagrange function that make a first gradient of a Lagrange function become zero.

$$\mathcal{L}_{CI_c}(\mathbf{x},\boldsymbol{\lambda}_c,\boldsymbol{\lambda}_v)=f_{CI_c}^{ANN}(\mathbf{x})-\boldsymbol{\lambda}_c^T c(\mathbf{x})-\boldsymbol{\lambda}_v^T Sv(\mathbf{x}) \quad (5.4.2.2\text{-}1)$$

$$\mathcal{L}_{CO_2}(\mathbf{x},\boldsymbol{\lambda}_c,\boldsymbol{\lambda}_v)=f_{CO_2}^{ANN}(\mathbf{x})-\boldsymbol{\lambda}_c^T c(\mathbf{x})-\boldsymbol{\lambda}_v^T Sv(\mathbf{x}) \quad (5.4.2.2\text{-}2)$$

$$\mathcal{L}_{W_c}(\mathbf{x},\boldsymbol{\lambda}_c,\boldsymbol{\lambda}_v)=f_{W_c}^{ANN}(\mathbf{x})-\boldsymbol{\lambda}_c^T c(\mathbf{x})-\boldsymbol{\lambda}_v^T Sv(\mathbf{x}) \quad (5.4.2.2\text{-}3)$$

Step 4: Formulating UFO

Table 5.4.2.1 Formulation of Equality and Inequality Constraints

Equality Constraint $c(\mathbf{x})$	*Inequality Constraints* $\mathbf{v}(\mathbf{x})$
$c_1(\mathbf{x})=f_c'-40=0$	$v_1(\mathbf{x})=\rho_s-0.01\geq 0$
$c_2(\mathbf{x})=f_y-500=0$	$v_2(\mathbf{x})=-\rho_s+0.08\geq 0$
$c_3(\mathbf{x})=P_{u,1}-25,000=0$	$v_3(\mathbf{x})=D-400\geq 0$
$c_4(\mathbf{x})=M_{u,1}-4,000=0$	$v_4(\mathbf{x})=-D+2,000\geq 0$
$c_5(\mathbf{x})=P_{u,2}-5,000=0$	$v_5(\mathbf{x})=SF_1-1.0\geq 0$
$c_6(\mathbf{x})=M_{u,2}-4,500=0$	$v_6(\mathbf{x})=SF_2-1.0\geq 0$
$c_7(\mathbf{x})=P_{u,3}-10,000=0$	$v_7(\mathbf{x})=SF_3-1.0\geq 0$
$c_8(\mathbf{x})=M_{u,3}-5,500=0$	$v_8(\mathbf{x})=SF_4-1.0\geq 0$
$c_9(\mathbf{x})=P_{u,4}-8,000=0$	$v_9(\mathbf{x})=SF_5-1.0\geq 0$
$c_{10}(\mathbf{x})=M_{u,4}-3,500=0$	
$c_{11}(\mathbf{x})=P_{u,5}-15,000=0$	
$c_{12}(\mathbf{x})=M_{u,5}-5,000=0$	

Table 5.4.2.2 Optimization Based on Separate Objective Function

Objective Function	*Minimum*	*Maximum*	*Normalizing Function*
Cost Index - CI_c (KRW/m)	222,766	2,373,833	$CI_c^{\text{Normalized}} = \dfrac{CI_c - 222,766}{2,373,833 - 222,766}$
CO_2 (ton-CO_2/m)	0.4580	5.4845	$CO_2^{\text{Normalized}} = \dfrac{CO_2 - 0.4580}{5.4845 - 0.4580}$
Weight - W_c (kN/m)	18.1339	74.0674	$W_c^{\text{Normalized}} = \dfrac{W_c - 18.1339}{74.0674 - 18.1339}$

The weighting method is a method to formulate unified functions of objectives (UFO) where multiple objective functions are globalized into one function. Normalized UFO is derived in Equations (5.4.2.3-1) and (5.4.2.3-2) based on normalized objective functions shown in Equations (5.4.2.3-3)–(5.4.2.3-5).

$$\text{UFO} = w_{CI_c} f_{CI_c}^{N}(\mathbf{x}) + w_{\text{CO}_2} f_{\text{CO}_2}^{N}(\mathbf{x}) + w_{\text{W}_c} f_{\text{W}_c}^{N}(\mathbf{x}) \tag{5.4.2.3-1}$$

$$w_{CI_c}, w_{\text{CO}_2}, w_{\text{W}_c} \in [0, 1];\ w_{CI_c} + w_{\text{CO}_2} + w_{\text{W}_c} = 1 \tag{5.4.2.3-2}$$

$$f_{CI_c}^{N}(\mathbf{x}) = \frac{f_{CI_c}^{\text{ANN}}(\mathbf{x}) - 222,766}{2,373,833 - 222,766} \tag{5.4.2.3-3}$$

$$f_{\text{CO}_2}^{N}(\mathbf{x}) = \frac{f_{\text{CO}_2}^{\text{ANN}}(\mathbf{x}) - 0.4580}{5.4845 - 0.4580} \tag{5.4.2.3-4}$$

$$f_{\text{W}_c}^{N}(\mathbf{x}) = \frac{f_{\text{W}_c}^{\text{ANN}}(\mathbf{x}) - 18.1339}{74.0674 - 18.1339} \tag{5.4.2.3-5}$$

Step 5: Optimizing based on UFO

Equation (5.4.2.3) presents Lagrange function $\mathcal{L}_{\text{UFO}}(\mathbf{x}, \boldsymbol{\lambda}_c, \boldsymbol{\lambda}_v)$ based on normalized UFO in Equation (5.4.2.3-1), equalities and inequalities formulated in Step 2.2 as shown in Table 5.4.2.1. Stationary points of a UFO are found by linearization of $\mathcal{L}_{\text{UFO}}(\mathbf{x}, \boldsymbol{\lambda}_c, \boldsymbol{\lambda}_v)$ where multiple objective functions are optimized simultaneously.

$$\mathcal{L}_{\text{UFO}}(\mathbf{x}, \boldsymbol{\lambda}_c, \boldsymbol{\lambda}_v) = \text{UFO}(\mathbf{x}) - \boldsymbol{\lambda}_c^{\text{T}} \boldsymbol{c}(\mathbf{x}) - \boldsymbol{\lambda}_v^{\text{T}} S\boldsymbol{v}(\mathbf{x}) \tag{5.4.2.3}$$

Pareto frontier of a circular RC column sustaining five load pairs based on constrained conditions shown in Table 5.4.2.1 is illustrated in Figure 5.4.2.3, in which 400 and 2,500 fractions are generated. A Pareto frontier based on 400 fractions has some sparse regions because points on a Pareto frontier are overplotted by other fractions as explained in Figure 5.3.8.2 and Table 5.3.8.1 of Section 5.3.8 where multiple fractions yield similar optimizations on a Pareto frontier. More fractions are required to fill Pareto frontier more compact.

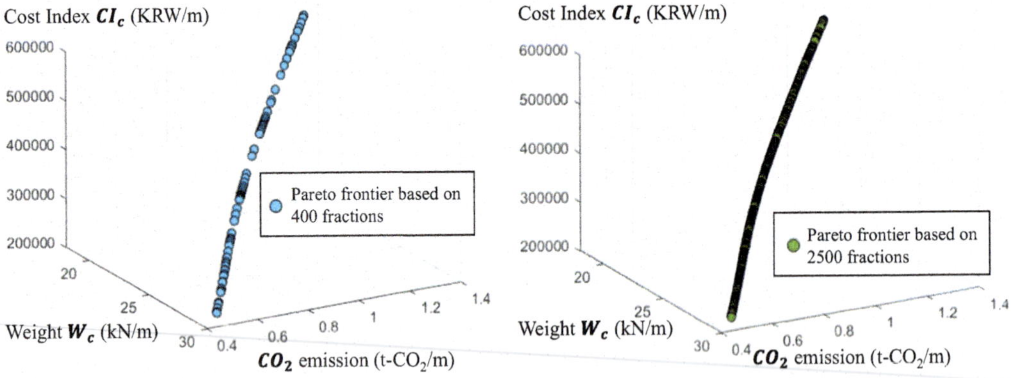

Figure 5.4.2.3 Comparison of Pareto frontier based on 400 and 2,500 fractions.

5.4.3 Verification of Pareto frontier based on large datasets

One hundred thousand structural datasets are generated to verify Pareto frontier obtained by ANN-based Lagrange optimization based on multi-objective functions. Ranges of design parameters and constraint conditions are presented in Table 5.4.2.1 (400 mm≤ D ≤ 2000 mm,0.01 ≤ ρ_s ≤ 0.08, f_c' = 40, f_y = 500, $P_{u,1}$= 25000 kN, $M_{u,1}$ = 4000 kN.m, $P_{u,2}$ = 5000 kN, $M_{u,2}$ = 4500 kN.m, $P_{u,3}$ = 10000 kN, $M_{u,3}$ = 5500 kN.m, $P_{u,4}$)= 8000 kN, $M_{u,4}$ = 3500 kN.m, $P_{u,5}$= 15000 kN, $M_{u,5}$ = 5000 kN.m), having a safety factor varying from 1.0 to 1.5 (1.0 ≤SF≤1.5) having a safety factor varying from 1.0 to 1.5 ($1.0 \le SF \le 1.5$) as shown in Inequality constraints $v_5(\mathbf{x})$–$v_9(\mathbf{x})$. Figure 5.4.2.4 shows that a Pareto frontier calculated

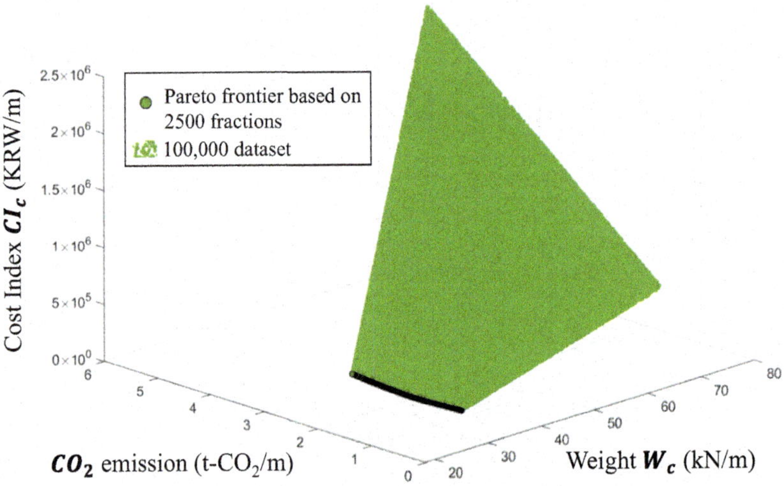

Figure 5.4.2.4 Verification of Pareto frontier based on an ANN-based Lagrange optimization using 100,000 datasets.

by the proposed method predicts lower boundaries of large datasets minimized based on three objective functions.

5.5 ANN-BASED LAGRANGE OPTIMIZATION FOR UFO TO DESIGN UNIAXIAL RECTANGULAR RC COLUMNS SUSTAINING MULTIPLE LOADS

5.5.1 Optimization scenario based on a forward design

A uniaxial rectangular RC column section is illustrated in Figure 5.5.1.1. Design parameters are column section width (b), column section height (h), rebar ratios (ρ_s), material properties of concrete (f_c') and steel (f_y), factored axial loads ($P_{u,i}$), and factored bending moments ($M_{u,i}$). Specifically, b, h and ρ_s are input variables while $f_c', f_y, P_{u,i}, M_{u,i}$ are prescribed design parameters. Design ranges of column dimension b and h are from 300 to 1,500 mm following design requirements while that of rebar ratios ρ_s is 0.01 to 0.08 which is specified by design code, ACI 318-19 (ACI Committee, 2019). An aspect ratio of column section (b/h) is constrained in a range of 0.3–2. A design of RC columns should meet strength

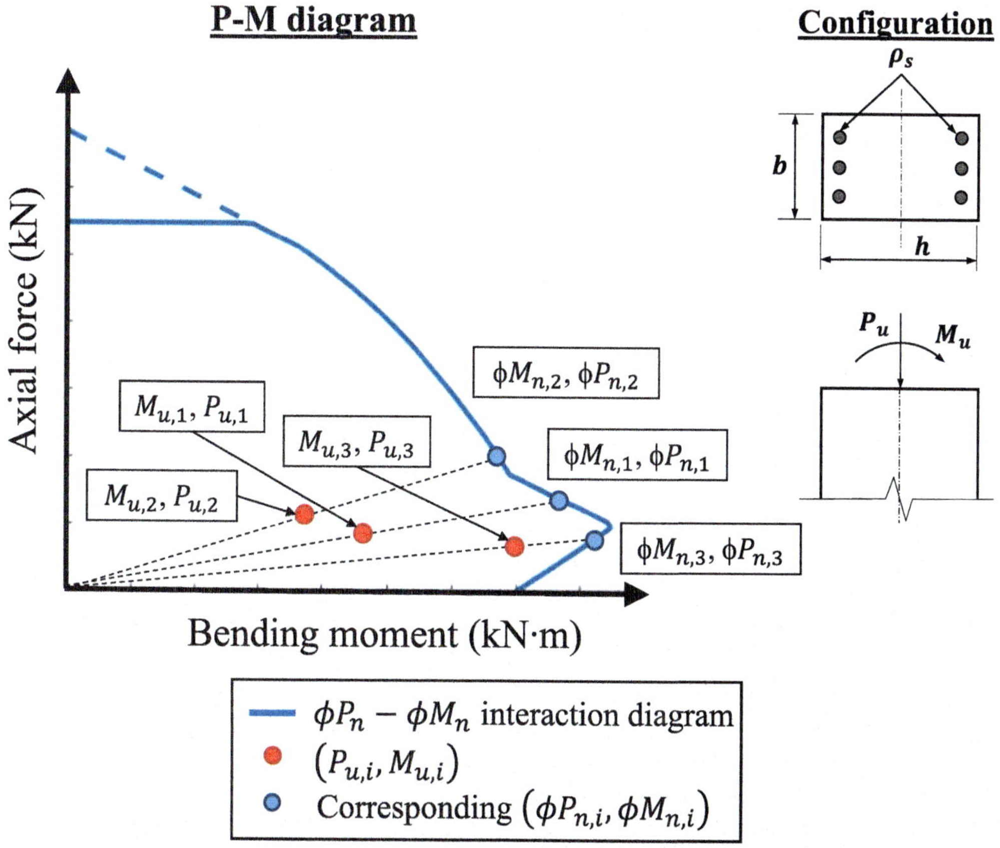

Figure 5.5.1.1 A uniaxial rectangular RC column section and corresponding P-M diagram.

Table 5.5.1.1 Constrained Conditions for Optimization

Factored Loads	Axial Load	$P_{u,1} = 1,500$ kN; $P_{u,2} = 2,000$ kN; $P_{u,3} = 1,200$ kN;
	Moment	$M_{u,1} = 1,000$ kNm; $M_{u,2} = 800$ kNm; $M_{u,3} = 1,500$ kNm
Material properties	Concrete strength	$f_c' = 40$ MPa
	Rebar yield strength	$f_y = 500$ MPa
Strength requirements		$SF_1 \geq 1.0$; $SF_2 \geq 1.0$; $SF_3 \geq 1.0$

requirements specified in ACI 318-19 (ACI Committee, 2019), reflected by safety factor $SF_i = P_n/P_{u,i} = M_n/M_{u,i} \geq 1.0$. Cost index, CO_2 emission, and column weight denoted as CI_c,CO_2, and W_c, respectively, are optimized. An optimization scenario similar to one presented in Section 5.4 is applied to a uniaxial rectangular RC column. A rectangular RC column sustaining three load pairs is simultaneously optimized based on three-objective functions, cost index (CI_c), CO_2 emission, and column weight (W_c). Prescribed material properties, factored loads, constrained conditions of strength, rebar ratios and width-to-height ratio are presented in Table 5.5.1.1.

5.5.2 Five-step optimization based on multiple objective functions

Step 1: Deriving ANNs

An ANN indicated as *Model-1LP* is derived considering one load pair ($i = 1$). Output parameters $\{SF,\ \varepsilon_s,\ CI_c,\ CO_2,\ W_c,\ b/h\}^T$. are calculated for given input parameters $\{b,\ h,\ \rho_s,\ f_c',\ f_y,\ P_u,\ M_u\}^T$ using *Model-1LP*. Table 5.5.2.1 presents ranges of design parameters required to generate large dataset. 150,000 data of 7 inputs ($b,\ h,\ \rho_s,\ f_c', f_y, P_u, M_u$) and 6 outputs ($SF,\ \varepsilon_s,\ CI_c,\ CO_2,\ W_c,\ b/h$) are generated based on structural mechanics-based software (*AutoCol*), shown in Table 5.5.2.2. Large datasets are used to train an ANN for *Model-1LP* which based on 3/4/5/6/7 hidden layers and 30/40/50/60/70 hidden neurons using parallel training method (PTM). Training results are presented in Table 5.5.2.3. The weight matrix components from *Model-1LP* are reused in weight matrix for *Model-LPs* to capture multiple load pairs, similarly to Section 5.4.1.2. A topology of *Model-LPs* in this example to capture three load pairs is illustrated in Figure 5.5.2.1.

Table 5.5.2.1 Ranges to Generate Large Structural Datasets

Parameter		*Design Range*
Column width	b	≥ 300 mm
Column height	h	300 ~ 1,500mm
Aspect ratio of column section	b/h	0.3 ~ 2.0
Rebar ratio	ρ_s	0.01 ~ 0.08
Concrete strength	f_c'	20 ~ 70MPa
Rebar yield strength	f_y	300 ~ 600MPa

Table 5.5.2.2 Large Datasets (150,000 Data) Generated Based on Structural Mechanics-Based Software *AutoCol*

	Seven Inputs for ANN and AutoCol							*Six Outputs for ANN and AutoCol*					
	b	h	ρ_s	f'_c	f_y	P_u	M_u	SF	ε_s	b/h	CI_C	CO_2	W_c
	mm	mm		MPa	MPa	kN	kN·m				KRW/m	ton-CO_2/m	kN/m
Non-normalized	441.9	368	0.055	64	560	3,883.9	550.8	0.82	0.00124	1.201	93,999	0.202	4.319
	300	305	0.025	30	498	154.9	46.4	2.63	0.00323	0.984	26,942	0.061	2.283
	1,422	1,276	0.037	52	500	7,734.0	9,379.5	2.31	0.0045	1.114	744,838	1.620	46.405
	…	…	…	…	…	…	…	…	…	…	…	…	…
	972.0	1,296	0.052	44	396	73,458.5	8,596.2	0.50	–0.0010	0.750	650,787	1.506	33.284
	658.4	397	0.074	57	600	7,645.2	1,698.7	0.50	0.0018	1.658	193,439	0.425	7.219
	436.9	800	0.056	23	443	289.9	4,579.3	0.50	0.0060	0.546	187,984	0.447	9.317
Normalized from –1 to 1	–0.878	–0.887	0.275	0.760	0.733	–0.968	–0.991	–0.857	–0.88316	0.268	–0.946	–0.949	–0.959
	–1.000	–0.992	–0.563	–0.600	0.320	–0.999	–0.999	–0.054	–0.82767	–0.022	–0.992	–0.992	–0.998
	–0.034	0.627	–0.235	0.280	0.333	–0.936	–0.844	–0.194	–0.7925	0.152	–0.5042	–0.520	–0.165
	…	…	…	…	…	…	…	…	…	…	…	…	…
	–1.000	–0.947	0.828	–0.200	0.780	–0.999	–0.999	0.204	–0.850	–0.129	–0.962	–0.961	–0.989
	–0.884	–0.993	0.518	0.120	–0.527	–0.996	–0.999	0.790	–0.884	0.571	–0.956	–0.954	–0.974
	–0.566	–0.633	0.640	–0.280	–0.360	–1.000	–0.992	0.568	–0.732	0.730	–0.830	–0.821	–0.826

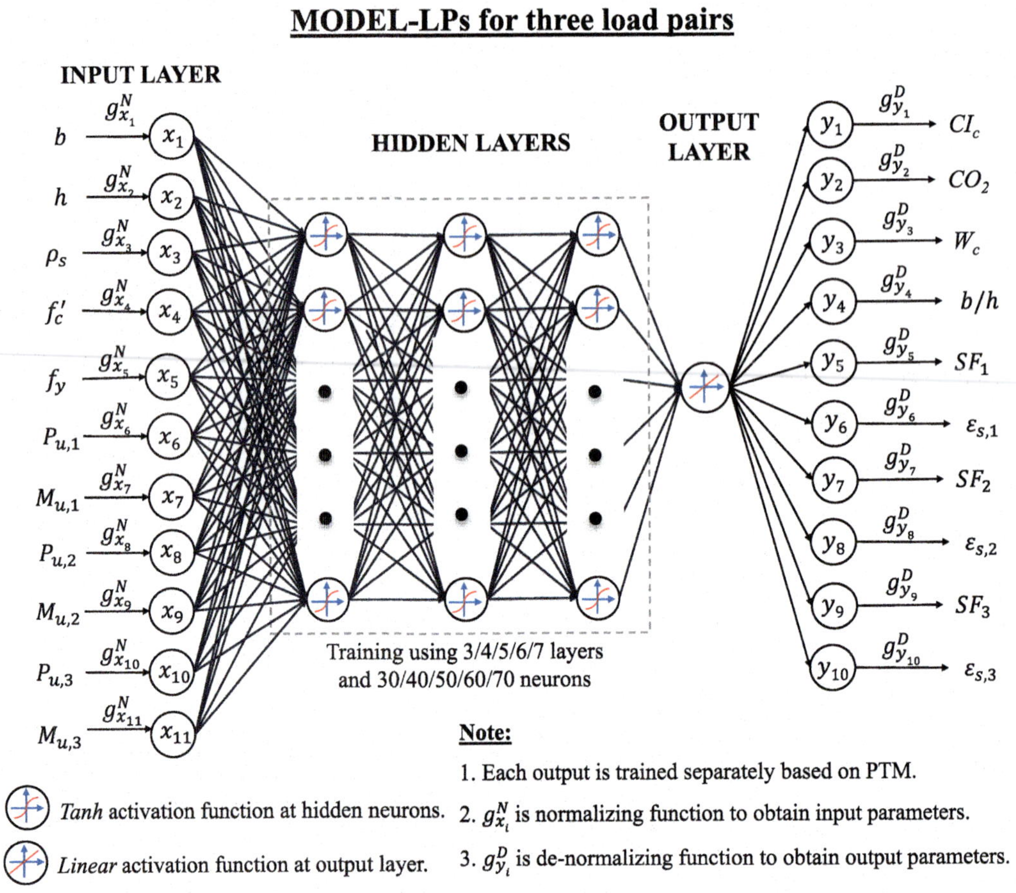

Figure 5.5.2.1 Topology of *Model-LPs* for three load pairs.

Table 5.5.2.3 Training Results and Accuracies

Seven Inputs (b, h, ρ_s, f_c', f_y, P_u, M_u) – Six Outputs (SF, ε_s, b/h, CI_c, CO_2, W_c)						
PTM – 150,000 Data – 100,000 Suggested Epochs – *tansig* Activation Function						
Output	Layers	Neurons	Best Epoch	Stopped Epoch	Test MSE	R at Best Epoch
SF	7	70	99,965	100,000	2.3.E-04	09996
ε_s	5	70	63,922	64,922	2.2.E-05	0.9997
b/h	4	60	81,019	81,183	1.8.E-09	1.0000
CI_c	5	30	34,161	35,161	2.3.E-08	1.0000
CO_2	6	30	72,544	73,544	4.2.E-09	1.0000
W_c	4	30	41,087	41,090	4.0.E-09	1.0000

Step 2: Defining MOO problem

Step 2.1: Deriving objective functions

ANN-based objective functions under three load pairs are derived in Equations (5.5.2.1-1)–(5.5.2.1-3) for CI_c, CO_2, and W_c using weight matrices for *Model-LPs* shown in Equations (5.4.1.6-1)–(5.4.1.8-2) which are modified by duplicating weight matrix components from a *Model-1LP*. *Model-LPs* considers n load pairs leading to having $2n+5$ initial input parameters, making the size of $g^N\left(\mathbf{x}^{\text{in}}\right)$ for *Model-LPs* being $(2n+5)\times 1$, because five input parameters $\left\{b,\ h,\ \rho_s,\ f_c',\ f_y\right\}^T$ are not influenced whereas each load pair has two loads, P_u and M_u. The weight matrix of *Model-LPs* at the first hidden layer, therefore, has a size of $(2n+5) = 2\times 3+5 = 11$ when three load pairs are considered as illustrated in Figure 5.5.1.1. An ANN for *Model-1LP* is derived to design columns sustaining one load pair. The objective functions for *Model-LPs* sustaining three load pairs are derived in Equations (5.5.2.1-1)–(5.5.2.1-3) by duplicating corresponding elements of *Model-1LP* which is derived to design columns sustaining one load pair.

$$\underbrace{f_{CI_c}(\mathbf{x})}_{[1\times1]} = g_{CI_c}^{D}\left(f_{\text{linear}}\left(\underbrace{\omega_{CI_c}^{(\text{out})}}_{[1\times30]} f_{\tanh}\left(\underbrace{\omega_{CI_c}^{(k=5)}}_{[30\times30]} \cdots f_{\tanh}\left(\underbrace{\omega_{CI_c}^{(1)}}_{[30\times11]}\underbrace{g^N(\mathbf{x})}_{[11\times1]} + \underbrace{\mathbf{b}_{CI_c}^{(1)}}_{[30\times1]}\right) \cdots + \underbrace{\mathbf{b}_{CI_c}^{(k=5)}}_{[30\times1]}\right) + \underbrace{\mathbf{b}_{CI_c}^{(\text{out})}}_{[1\times1]}\right)\right) \quad (5.5.2.1\text{-}1)$$

$$\underbrace{f_{CO_2}(\mathbf{x})}_{[1\times1]} = g_{CO_2}^{D}\left(f_{\text{linear}}\left(\underbrace{\omega_{CO_2}^{(\text{out})}}_{[1\times30]} f_{\tanh}\left(\underbrace{\omega_{CO_2}^{(k=6)}}_{[30\times30]} \cdots f_{\tanh}\left(\underbrace{\omega_{CO_2}^{(1)}}_{[30\times11]}\underbrace{g^N(\mathbf{x})}_{[11\times1]} + \underbrace{\mathbf{b}_{CI_c}^{(1)}}_{[30\times1]}\right) \cdots + \underbrace{\mathbf{b}_{CO_2}^{(k=6)}}_{[30\times1]}\right) + \underbrace{\mathbf{b}_{CO_2}^{(\text{out})}}_{[1\times1]}\right)\right) \quad (5.5.2.1\text{-}2)$$

$$\underbrace{f_{W_c}(\mathbf{x})}_{[1\times1]} = g_{W_c}^{D}\left(f_{\text{linear}}\left(\underbrace{\omega_{W_c}^{(\text{out})}}_{[1\times30]} f_{\tanh}\left(\underbrace{\omega_{W_c}^{(k=4)}}_{[30\times30]} \cdots f_{\tanh}\left(\underbrace{\omega_{W_c}^{(1)}}_{[30\times11]}\underbrace{g^N(\mathbf{x})}_{[11\times1]} + \underbrace{\mathbf{b}_{W_c}^{(1)}}_{[30\times1]}\right) \cdots + \underbrace{\mathbf{b}_{W_c}^{(k=4)}}_{[30\times1]}\right) + \underbrace{\mathbf{b}_{W_c}^{(\text{out})}}_{[1\times1]}\right)\right) \quad (5.5.2.1\text{-}3)$$

Step 2.2: Formulating constrained conditions based on equalities and inequalities

Conditions constraining optimization design of uniaxial rectangular RC columns are formulated in Table 5.5.2.4 based on equality and inequality constraints.

Step 3: Optimizing objective function separately

As presented in Table 5.5.2.5, uniaxial rectangular RC columns under three load pairs are separately optimized based on an objective function to normalize the UFO as shown in Equations (5.5.2.2-1)–(5.5.2.2-3).

Step 4: Formulating UFO

Step 4.1: In Equations (5.5.2.2-1)–(5.5.2.2-3), each objectives function is normalized using optimized maxima and minima obtained in Step 3.

$$f_{CI_c}^{N}(\mathbf{x}) = \frac{f_{CI_c}(\mathbf{x}) - 67{,}622}{2{,}976{,}094 - 67{,}622} \quad (5.5.2.2\text{-}1)$$

$$f_{CO_2}^{N}(\mathbf{x}) = \frac{f_{CO_2}(\mathbf{x}) - 0.1398}{6.8643 - 0.1398} \quad (5.5.2.2\text{-}2)$$

Table 5.5.2.4 Equality and Inequality Constraints

Equality Constraint $c(\mathbf{x})$	*Inequality Constraints* $v(\mathbf{x})$
$c_1(\mathbf{x}) = f_c' - 40 = 0$	$v_1(\mathbf{x}) = \rho_s - 0.01 \geq 0$
$c_2(\mathbf{x}) = f_y - 500 = 0$	$v_2(\mathbf{x}) = -\rho_s + 0.08 \geq 0$
$c_3(\mathbf{x}) = P_{u,1} - 1,500 = 0$	$v_3(\mathbf{x}) = h - 300 \geq 0$
$c_4(\mathbf{x}) = M_{u,1} - 1,000 = 0$	$v_4(\mathbf{x}) = -h + 1,500 \geq 0$
$c_5(\mathbf{x}) = P_{u,2} - 2,000 = 0$	$v_5(\mathbf{x}) = b/h - 0.3 \geq 0$
$c_6(\mathbf{x}) = M_{u,2} - 800 = 0$	$v_6(\mathbf{x}) = -b/h - 2.0 \geq 0$
$c_7(\mathbf{x}) = P_{u,3} - 1,200 = 0$	$v_7(\mathbf{x}) = SF_1 - 1.0 \geq 0$
$c_8(\mathbf{x}) = M_{u,3} - 1500 = 0$	$v_8(\mathbf{x}) = SF_2 - 1.0 \geq 0$
	$v_9(\mathbf{x}) = SF_3 - 1.0 \geq 0$

Table 5.5.2.5 Optimization Based on Separate Objective Function

Objectives	*Minimum*	*Maximum*
CI_c (KRW/m)	67,622	2,976,094
CO_2 (t-CO_2/m)	0.1398	6.8643
W_c (kN/m)	5.9600	109.9203

$$f_{W_c}^{N}(\mathbf{x}) = \frac{f_{W_c}(\mathbf{x}) - 5.9600}{109.9203 - 5.9600} \tag{5.5.2.2-3}$$

Step 4.2: UFO is, then, formulated in Equations (5.5.2.3-1) and (5.5.2.3-2) based on fractions applied to each objective function $w_{CI_c}, w_{CO_2},$ *and* w_{W_c} varying from 0 to 1.

$$\text{UFO} = w_{CI_c} f_{CI_c}^{N}(\mathbf{x}) + w_{CO_2} f_{CO_2}^{N}(\mathbf{x}) + w_{W_c} f_{W_c}^{N}(\mathbf{x}) \tag{5.5.2.3-1}$$

$$w_{CI_c}, w_{CO_2}, w_{W_c} \in [0, 1];\ w_{CI_c} + w_{CO_2} + w_{W_c} = 1 \tag{5.5.2.3-2}$$

Step 5: Optimizing based on UFO

Equation (5.5.2.4) presents Lagrange function $\mathcal{L}_{UFO}(\mathbf{x}, \boldsymbol{\lambda}_c, \boldsymbol{\lambda}_v)$ based on normalized UFO in Equation (5.5.2.3-1), equalities and inequalities formulated in Step 2.2 as shown in Table 5.5.2.4. Stationary points of a UFO are found by linearization of $\mathcal{L}_{UFO}(\mathbf{x}, \boldsymbol{\lambda}_c, \boldsymbol{\lambda}_v)$ where multiple objective functions are optimized simultaneously.

$$\mathcal{L}_{UFO}(\mathbf{x}, \boldsymbol{\lambda}_c, \boldsymbol{\lambda}_v) = \text{UFO}(\mathbf{x}) - \boldsymbol{\lambda}_c^{\mathrm{T}} \boldsymbol{c}(\mathbf{x}) - \boldsymbol{\lambda}_v^{\mathrm{T}} \mathbf{S} \boldsymbol{v}(\mathbf{x}) \tag{5.5.2.4}$$

A Pareto frontier of a uniaxial rectangular RC column sustaining three load pairs based on three objective functions is presented in Figure 5.5.2.2 in which constrained conditions shown in Table 5.5.2.4 are applied. A total of 441 fractions are used to obtain a Pareto frontier.

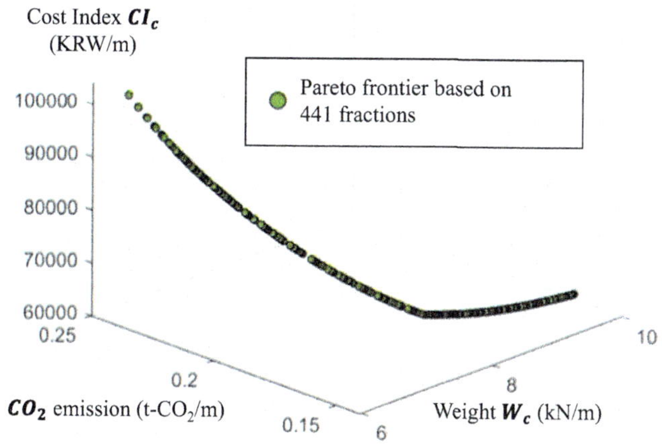

Figure 5.5.2.2 A Pareto frontier with 441 fractions for a uniaxial rectangular RC column sustaining three load pairs based on three objective functions.

5.5.3 Verification of Pareto frontier to large dataset

Fifty thousand (50,000) structural datasets are randomly generated having fixed parameters, such as material properties ($f_c' = 40$ MPa, $f_y = 500$ MPa) and factored loads ($P_{u,1} = 1{,}500$ kN, $M_{u,1} = 1{,}000$ kN.m, $P_{u,2} = 2{,}000$ kN, $M_{u,2} = 800$ kN.m, $P_{u,3} = 1{,}200$ kN, $M_{u,3} = 1{,}500$ kN.m), and three input variables to be determined, such as column width (b), column height (h), and rebar ratios (ρ_s). These input variables are randomly generated within their ranges, specified in Table 5.5.2.1. Specifically, h is determined in a range of 300–1,500 mm, b should be greater than 300 mm. Satisfied b and h should have their aspect ratio b/h in a range of 0.3–2.0. Rebar ratios ρ_s must be within 0.01 and 0.08 as specified in ACI 318-19 (ACI Committee, 2019). All data must have safety factor greater than 1.0 ($SF_i \geq 1.0$). Figure 5.5.3.1 presents a Pareto frontier obtained based on three objective functions (CI_c, CO_2, W_c) minimized simultaneously. Figure 5.5.3.1a shows Pareto frontier in three-dimensions whereas Figure 5.5.3.1b–d project objective functions in two-dimensional plane to illustrate relationships between (CO_2 and W_c), (CI_c and W_c), and (CI_c, and CO_2), respectively. It is verified that a Pareto frontier lies on a lower boundary of 50,000 large structural datasets in Figure 5.5.3.1a–d.

5.5.4 Design parameters corresponding to three fractions of Pareto frontier

Design points corresponding to three fractions ($w_{CI_c} : w_{CO_2} : w_{W_c} = 1:0:0$ and $0:1:0$), ($w_{CI_c} : w_{CO_2} : w_{W_c} = 0:0:1$), and ($w_{CI_c} : w_{CO_2} : w_{W_c} = 0.34:0.33:0.33$) on Pareto frontier are indicated in Figure 5.5.4.1. Design tables obtained for these three fractions are presented in Tables 5.5.4.1–5.5.4.3 which yield design parameters optimized simultaneously based on three objective functions (cost index, CO_2, and weight). Design parameters obtained by minimizing CI_c are identified at Design Point 1 on a Pareto frontier. CI_c based on a fraction of $w_{CI_c} : w_{CO_2} : w_{W_c}$ = 1:0:0 indicates that CI_c is only minimized. CO_2 based on a fraction of $w_{CI_c} : w_{CO_2} : w_{W_c}$ = 0:1:0 indicates that CO_2 is only minimized in a similar manner. Design Point 2 represents an optimized design point where W_c is only minimized based on ($w_{CI_c} : w_{CO_2} : w_{W_c} = 0:0:1$). Design Point 3 identifies optimized design parameters

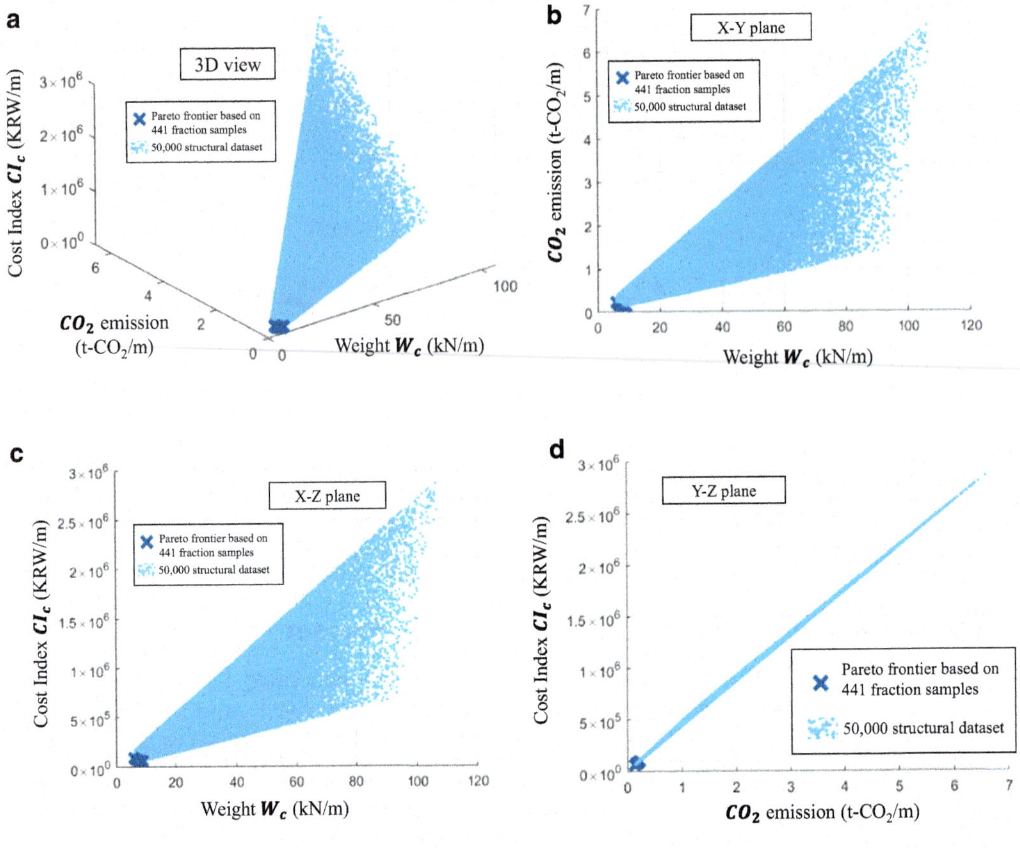

Figure 5.5.3.1 Verification of a Pareto frontier by 50,000 large structural datasets. (a) 3D view. (b) XY plane. (c) XZ plane. (d) YZ plane.

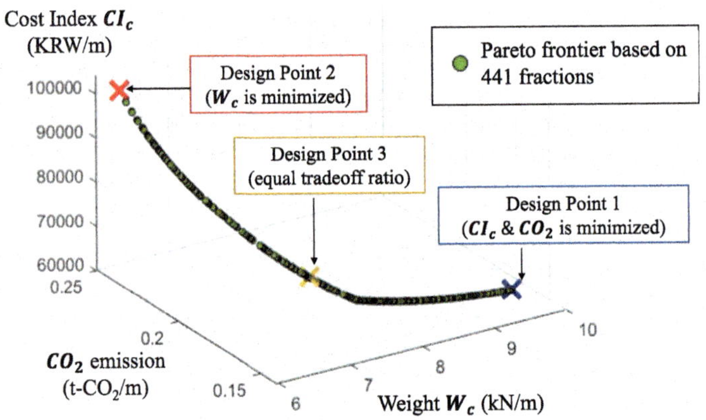

Figure 5.5.4.1 A Pareto frontier obtained by optimizing UFO.

with an equal trade-off among three objective functions based on ($w_{CI_c} : w_{CO_2} : w_{W_c}$ = 0.34:0.33:0.33). Tables 5.5.4.1–5.5.4.3 present design parameters optimized by ANN-based Lagrange algorithm of UFO while insignificant errors are verified by structural mechanics-based *AutoCol*. Design accuracies of ANNs at three fractions are shown in Tables 5.5.4.1–5.5.4.3.

Table 5.5.4.1 Design Table of Design Point 1 in Pareto Frontier ($w_{CI_c} : w_{CO_2} : w_{W_c} = 1:0:0$ and $0:1:0$)

Fraction $CI_c : CO_2 : W_c = 1:0:0; 0:1:0$							
	Parameters	**AI-based Lagrange & *AutoCol***		**Parameters**	**AI-based Lagrange**	**Verificatiown (AutoCol)**	**Error**
Input variables	b (mm)	339.10		SF_1	2.1370	2.1945	2.69%
	h (mm)	1,130.30		SF_2	2.2311	2.2158	0.69%
	ρ_s	0.01		SF_3	1.0000	0.9733	2.67%
Preassigned input parameters	f'_c (MPa)	40	**Output parameters**	$\varepsilon_{s,1}$	0.0054	0.0053	1.85%
	f_y (MPa)	500		$\varepsilon_{s,2}$	0.0018	0.0018	1.67%
	$P_{u,1}$ (kN)	1,500		$\varepsilon_{s,3}$	0.0172	0.0169	1.74%
	$P_{u,2}$ (kN)	2,000		CI_C (KRW/m)	**67,622**	**67,765**	0.21%
	$P_{u,3}$ (kN)	1,200		CO_2 (t-CO_2/m)	**0.1398**	**0.1399**	0.07%
	$M_{u,1}$ (kN·m)	1,000		W_c (kN/m)	**9.2403**	**9.2398**	0.01%
	$M_{u,2}$ (kN·m)	800		b/h	0.0300	0.0300	0.00%
	$M_{u,3}$ (kN·m)	1,500					
	Input variables for ANN (*ModelLPs*) and structural mechanics-based (*AutoCol*).			Outputs for ANN (*ModelLPs*).			
	Preassigned inputs for ANN (*ModelLPs*) and structural mechanics-based (*AutoCol*).			Outputs for structural mechanics-based (*AutoCol*).			

Bold and red color values indicate optimized objective functions.

Table 5.5.4.2 Design Table of Design Point 2 in Pareto Frontier ($w_{Cl_c} : w_{CO_2} : w_{W_c} = 0:0:1$)

Fraction $Cl_c : CO_2 : W_c = 0:0:1$

Parameters		AI-based Lagrange & *AutoCol*	Parameters	AI-based Lagrange	Verification (*AutoCol*)	Error
Input variables	b (mm)	300.00	SF_1	1.2711	1.2690	0.17%
	h (mm)	765.64	SF_2	1.4178	1.4102	0.54%
	ρ_s	0.04	SF_3	1.0000	0.9957	0.43%
Preassigned input parameters	f'_c (MPa)	40	$\varepsilon_{s,1}$	0.0028	0.0028	1.07%
	f_y (MPa)	500	$\varepsilon_{s,2}$	0.0019	0.0019	0.26%
	$P_{u,1}$ (kN)	1,500	$\varepsilon_{s,3}$	0.0050	0.0049	1.60%
	$P_{u,2}$ (kN)	2,000	Cl_c (KRW/m)	**104,170**	**104,142**	0.03%
	$P_{u,3}$ (kN)	1,200	CO_2 (t-CO_2/m)	**0.2352**	**0.2353**	0.04%
	$M_{u,1}$ (kN·m)	1,000	W_c (kN/m)	**5.9600**	**5.9562**	0.06%
	$M_{u,2}$ (kN·m)	800	b/h	0.3918	0.3918	0.00%
	$M_{u,3}$ (kN·m)	1,500				

Input variables for ANN (*ModelLPs*) and structural mechanics-based (*AutoCol*)	Outputs for ANN (*ModelLPs*).
Preassigned inputs for ANN (*ModelLPs*) and structural mechanics-based (*AutoCol*)	Outputs for structural mechanics-based (*AutoCol*).

Red color values indicate optimized objective functions.

Table 5.5.4.3 Design Table of Design Point 3 in Pareto Frontier ($w_{Cl_c} : w_{CO_2} : w_{W_c} = 0.34 : 0.33 : 0.33$)

Fraction $Cl_c : CO_2 : W_c = 0.34 : 0.33 : 0.33$

Parameters		AI-based Lagrange & *AutoCol*	Parameters	AI-based Lagrange	Verification (*AutoCol*)	Error
Input variables	b (mm)	300.00	SF_1	1.5909	1.6607	4.39%
	h (mm)	937.48	SF_2	1.6846	1.6770	0.45%
	ρ_s	0.02	SF_3	1.0000	0.9936	0.64%
Preassigned input parameters	f'_c (MPa)	40	$\varepsilon_{s,1}$	0.0039	0.0038	2.56%
	f_y (MPa)	500	$\varepsilon_{s,2}$	0.0018	0.0018	2.39%
	$P_{u,1}$ (kN)	1,500	$\varepsilon_{s,3}$	0.0100	0.0097	2.90%
	$P_{u,2}$ (kN)	2,000	CI_C (KRW/m)	**75,078**	**74,875**	0.27%
	$P_{u,3}$ (kN)	1,200	CO_2 (t-CO_2/m)	**0.1626**	**0.1625**	0.06%
	$M_{u,1}$ (kN·m)	1,000	W_c (kN/m)	**6.9460**	**6.9474**	0.02%
	$M_{u,2}$ (kN·m)	800	b/h	0.3200	0.3200	0.00%
	$M_{u,3}$ (kN·m)	1,500				
	Input variables for ANN (*ModelLPs*) and structural mechanics-based (*AutoCol*).		Outputs for ANN (ModelLPs).			
	Preassigned inputs for ANN (*ModelLPs*) and structural mechanics-based (*AutoCol*).		Outputs for structural mechanics-based (*AutoCol*).			

Red color values indicate optimized objective functions.

5.6 ANN-BASED LAGRANGE OPTIMIZATION FOR UFO TO DESIGN BIAXIAL RECTANGULAR RC COLUMNS SUSTAINING MULTIPLE LOADS

5.6.1 Optimization scenario based on a forward design subject to multi-loads with small magnitude

A biaxial rectangular RC column section is shown in Figure 5.6.1.1. Columns sustain biaxial moment, including $M_{u,x}$ and $M_{u,y}$, and hence, two parameters, rebar ratios about x-axis ($\rho_{s,x}$) and about y-axis ($\rho_{s,y}$), are to be determined. Rebar ratios ρ_s is, then, summed for $\rho_{s,x}$ and $\rho_{s,y}$. Design parameters for column dimensions, material properties, etc. shown in Section 5.5 are also determined when designing uniaxial rectangular RC columns in this section. A description of notation and nomenclature of design parameters is presented in Table 5.6.1.1. An optimization scenario for a design of biaxial rectangular RC columns is also based on three objective functions, such as cost index (CI_c), CO_2 emission, and column weight (W_c) which are to be minimized simultaneously. In this section, columns sustain ten load pairs of factored axial load and factored bending moment. Prescribed material properties, factored loads, column dimensions, rebar ratio limitations, and conditions constraining strength requirement are presented in Table 5.6.1.2. An aspect ratio of column section b/h varies from 0.25 to 2.0. Input parameters are selected as $\left\{b,\ h,\ \rho_{s,x}\ ,\ \rho_{s,y},\ f_c',\ f_y,\ P_u^{(i)},\ M_{u,x}^{(i)},\ M_{u,y}^{(i)}\right\}^T$ (i = 1,2,..., number of design load pairs) to determine an output of $\left\{SF^{(i)}\ ,\varepsilon_s^{(i)},CI_c,CO_2,W_c,b\,/\,h\right\}^T$. Four parameters such as $b,\ h,\ \rho_{s,x}\ ,\ \rho_{s,y}$ are input variables to be determined by Newton-Raphson iteration while other input parameters $f_c',\ f_y,\ P_u^{(i)},\ M_{u,x}^{(i)},\ M_{u,y}^{(i)}$ are prescribed. It is noteworthy that a number of design load pairs does not affect a number of input variables in an ANN-based optimization problem.

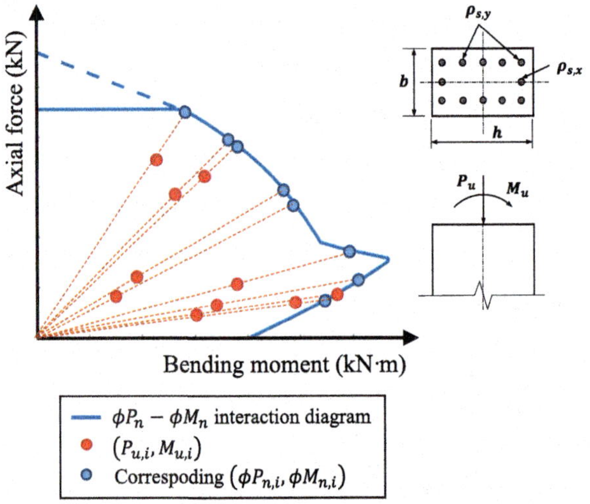

Figure 5.6.1.1 A biaxial rectangular RC column section with multiple load pairs and corresponding P-M diagram.

Table 5.6.1.1 Description of Notations and Nomenclature of Design Parameters

Notation	*Nomenclature*	*Notation*	*Nomenclature*
b	Column width	SF	Safety factor
H	Column height	ε_s	Rebar strain
$\rho_{s,x}$	Rebar ratio about x-axis	CI_c	Cost index of column per height
$\rho_{s,y}$	Rebar ratio about y-axis	CO_2	CO_2 emission per column height
f_c'	Concrete strength	W_c	Weight of column per column height
f_y	Rebar yield strength	$\alpha_{e/h}$	Angular eccentricity of applied load
P_u	Factored axial load	b/h	Aspect ratio of column section
$M_{u,x}$	Factored bending moment about x-axis		
$M_{u,y}$	Factored bending moment about y-axis		

Table 5.6.1.2 Prescribed Parameters and Constrained Conditions

Load Pair (i^{th})		*1*	*2*	*3*	*4*	*5*	*6*	*7*	*8*	*9*	*10*
Axial load	$P_u^{(i)}$ (kN)	1,500	3,000	1,200	500	700	900	2,000	2,500	1,700	3,500
Moment about x-axis	$M_{u,x}^{(i)}$ (kN·m)	500	1200	2,000	1,700	1,300	900	700	800	1,500	1,100
Moment about y-axis	$M_{u,y}^{(i)}$ (kN·m)	1,200	1,000	700	600	1,500	1,300	1,100	900	1.300	500
Strength requirement		$SF^{(i)} \geq 1.0$									
Material properties		Concrete		$f_c' = 40$ MPa			Steel		$f_y = 500$ MPa		
Rebar ratios		$0.01 \leq \rho_{s,x} + \rho_{s,y} \leq 0.08$									
Limits on column sections		$0.01 \leq b,\ h \leq 0.08;\ 0.3 \leq b/h \leq 2.0$									

5.6.2 Five steps optimization based on multiple objective functions

Step 1: Establishing ANNs

An ANN for *Model-1LP* is derived to design columns sustaining one load pair. Nine input parameters $\{b,\ h,\ \rho_{s,x}\ ,\ \rho_{s,y},\ f_c',\ f_y,\ P_u,\ M_{u,x},\ M_{u,y}\}^T$ are mapped to six output parameters $\{SF,\varepsilon_s,CI_c,CO_2,W_c,b/h\}^T$ by training *Model-1LP* based on 200,000 datasets using PTM. Ranges of design parameters for large structural datasets are presented in Table 5.6.2.1. Two hundred thousand datasets are generated based on structural mechanics-based software (*AutoRC2A*) shown in Table 5.6.2.2. *Model-1LP* is trained with four, five, six and seven hidden layers based on 32, 64, and 128 neurons. The weight matrices of *Model-LPs* which captures multiple load pairs are obtained by duplicating corresponding matrix elements from *Model-1LP*. Table 5.6.2.3 presents results and accuracies based on best training. A topology of *Model-LPs* to capture ten load pairs is illustrated in Figure 5.6.2.1.

Table 5.6.2.1 Ranges to Generate Large Structural Dataset

Parameter		*Design Range*
Column width	b	≥300 mm
Column height	h	300 ~ 2,000 mm
Width-to-height ratio	b/h	0.25 ~ 2.0
Total rebar ratio	$\rho_{s,x} + \rho_{s,y}$	0.01 ~ 0.08
Concrete strength	$f_c^{'}$	30 ~ 70 MPa
Rebar yield strength	f_y	300 ~ 600 MPa

Table 5.6.2.2 Large Datasets (200,000 Data) Generated Based on Structural Mechanics-Based Software *AutoRC2A*

	Nine Inputs for ANN and AutoRC2A									*Six Outputs for ANN and AutoRC2A*					
	b (mm)	h (mm)	$\rho_{s,x}$	$\rho_{s,y}$	f'_c (MPa)	f_y (MPa)	P_u (kN)	$M_{u,x}$ (kN·m)	$M_{u,y}$ (kN·m)	*SF*	ε_s	b/h	CI_C (KRW/m)	CO_2 (ton-CO_2/m)	W_c (kN/m)
Non-normalized	858.0	1,075	0.038	0.007	56	556	2,599.1	6,232.8	3,379.1	1.10	0.004	0.798	450,152	0.972	23.954
	1,154	1,122	0.049	0.026	36	573	2,598.7	2,807.4	80.4	4.88	0.002	1.029	939,884	2.137	35.790
	1,612	894	0.012	0.044	40	445	249.5	3,744.2	4,505.4	2.87	0.005	1.803	791,914	1.830	38.295
	...	...	...	...	...	...	...	...	...	...	...	...	...	...	...
	1,680.5	1,193	0.042	0.024	30	430	4,276.8	7,758.1	8,614.1	1.75	0.0031	1.409	124,2391	2.941	54.376
	1,244	998	0.055	0.018	49	469	4,902.5	2,453.2	69.7	4.52	0.001	1.247	870,970	1.994	34.156
	898.7	1,162	0.074	6E-04	64	463	712.9	4,053.1	1,076.2	3.92	0.0055	0.773	761,792	1.712	28.826
Normalized from −1 to 1	−0.344	−0.088	−0.050	−0.826	0.300	0.707	−0.982	−0.931	−0.952	−0.733	−0.798	−0.708	−0.7200	−0.728	−0.604
	0.005	−0.033	0.225	−0.346	−0.700	0.820	−0.982	−0.969	−0.999	0.947	−0.863	−0.585	−0.4051	−0.393	−0.387
	0.544	−0.301	−0.700	0.098	−0.500	−0.033	−0.998	−0.958	−0.936	0.053	−0.779	−0.172	−0.5002	−0.481	−0.341
	...	...		...	...	...	...		...	...	...	...	...	...	...
	0.624	0.051	0.050	−0.402	−1.000	−0.133	−0.971	−0.914	−0.877	−0.444	−0.832	−0.382	−0.210	−0.161	−0.046
	0.111	−0.179	0.375	−0.552	−0.050	0.127	−0.967	−0.973	−0.999	0.787	−0.891	−0.469	−0.449	−0.434	−0.417
	−0.296	0.014	0.850	−0.984	0.700	0.087	−0.995	−0.955	−0.985	0.520	−0.750	−0.721	−0.520	−0.515	−0.514

Table 5.6.2.3 Training Results and Accuracies

Seven Inputs (b, h, $\rho_{s,x}$, $\rho_{s,y}$, $f_c^{'}$, f_y, P_u, $M_{u,x}$, $M_{u,y}$)
Six Outputs (SF, ε_s, b/h, CI_c, CO_2, W_c)

PTM – 200,000 Data – 100,000 Suggested Epochs – *tansig* Activation Function

Output	Layers	Neurons	Best Epoch	Stopped Epoch	Test MSE	R at Best Epoch
SF	5	64	83,538	84,538	3.2.E-04	09954
ε_s	6	64	30,949	31,949	1.9.E-04	0.9860
b/h	5	64	82,102	82,102	4.9.E-09	1.0000
CI_c	3	64	52,532	53,532	2.2.E-08	1.0000
CO_2	7	32	49,280	50,280	1.3.E-09	1.0000
W_c	5	64	99,990	100,000	4.0.E-09	1.0000

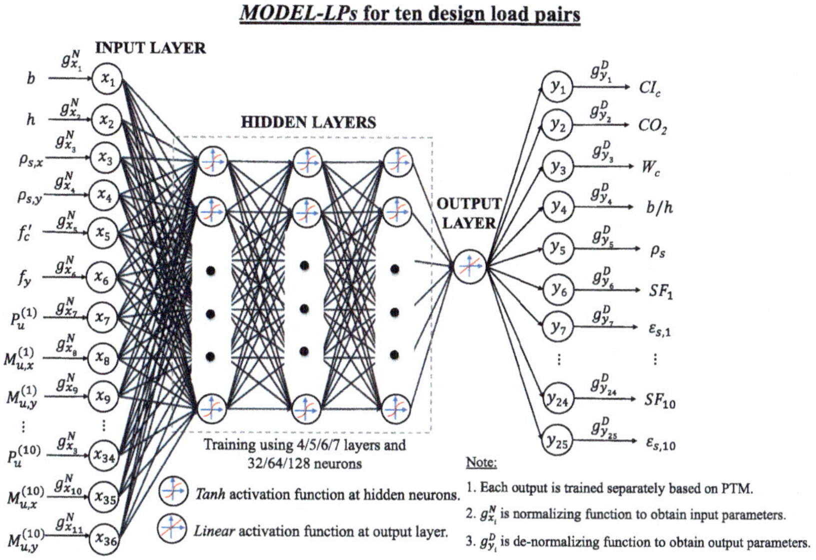
MODEL-LPs for ten design load pairs
INPUT LAYER
HIDDEN LAYERS
OUTPUT LAYER
Training using 4/5/6/7 layers and 32/64/128 neurons
Tanh activation function at hidden neurons.
Linear activation function at output layer.
Note:
1. Each output is trained separately based on PTM.
2. $g_{x_i}^N$ is normalizing function to obtain input parameters.
3. $g_{y_i}^D$ is de-normalizing function to obtain output parameters.

Figure 5.6.2.1 Topology of *Model-LPs* for ten load pairs.

Step 2: Defining MOO problem

Step 2.1: Deriving objective functions

Model-LPs considers n load pairs leading to having $3n+6$ initial input parameters, making the size of $g^N(\mathbf{x}^{\text{in}})$ for Model-LPs being $(3n+6)\times 1$, because six input parameters $\{b,\ h,\ \rho_{s,x},\ \rho_{s,y},\ f_c^{'},\ f_y,\ P_u\}^T$ do not change whereas each load pair has three loads, P_u, $M_{u,x}$, $M_{u,y}$. The weight matrix of *Model-LPs* at the first hidden layer, therefore, has a size of $20\times(3n+6) = 3\times 10+6 = 36$ when ten load pairs are considered as illustrated in Figure 5.6.2.1. An ANN for *Model-1LP* is derived to design columns sustaining one load pair. The objective functions for *Model-LPs* are derived in Equations (5.6.2.1-1)–(5.6.2.1-3) by duplicating corresponding elements of *Model-1LP* which is derived to design columns sustaining one load pair.

$$\underbrace{f_{CI_c}(\mathbf{x})}_{[1\times1]} = g_{CI_c}^{D}\left(f_{\text{linear}}\left(\underbrace{\omega_{CI_c}^{(\text{out})}}_{[1\times64]} f_{\tanh}\left(\underbrace{\omega_{CI_c}^{(k=5)}}_{[64\times64]} \dots f_{\tanh}\left(\underbrace{\omega_{CI_c}^{(1)}}_{[64\times36]} \underbrace{g^{N}(\mathbf{x})}_{[36\times1]} + \underbrace{\mathbf{b}_{CI_c}^{(1)}}_{[64\times1]}\right) \dots + \underbrace{\mathbf{b}_{CI_c}^{(k=5)}}_{[64\times1]}\right) + \underbrace{\mathbf{b}_{CI_c}^{(\text{out})}}_{[1\times1]}\right)\right) \tag{5.6.2.1-1}$$

$$\underbrace{f_{CO_2}(\mathbf{x})}_{[1\times1]} = g_{CO_2}^{D}\left(f_{\text{linear}}\left(\underbrace{\omega_{CO_2}^{(\text{out})}}_{[1\times32]} f_{\tanh}\left(\underbrace{\omega_{CO_2}^{(k=6)}}_{[32\times32]} \dots f_{\tanh}\left(\underbrace{\omega_{CO_2}^{(1)}}_{[32\times36]} \underbrace{g^{N}(\mathbf{x})}_{[36\times1]} + \underbrace{\mathbf{b}_{CI_c}^{(1)}}_{[36\times1]}\right) \dots + \underbrace{\mathbf{b}_{CO_2}^{(k=6)}}_{[32\times1]}\right) + \underbrace{\mathbf{b}_{CO_2}^{(\text{out})}}_{[1\times1]}\right)\right) \tag{5.6.2.1-2}$$

$$\underbrace{f_{W_c}(\mathbf{x})}_{[1\times1]} = g_{W_c}^{D}\left(f_{\text{linear}}\left(\underbrace{\omega_{W_c}^{(\text{out})}}_{[1\times64]} f_{\tanh}\left(\underbrace{\omega_{W_c}^{(k=4)}}_{[64\times64]} \dots f_{\tanh}\left(\underbrace{\omega_{W_c}^{(1)}}_{[64\times36]} \underbrace{g^{N}(\mathbf{x})}_{[36\times1]} + \underbrace{\mathbf{b}_{W_c}^{(1)}}_{[64\times1]}\right) \dots + \underbrace{\mathbf{b}_{W_c}^{(k=4)}}_{[64\times1]}\right) + \underbrace{\mathbf{b}_{W_c}^{(\text{out})}}_{[1\times1]}\right)\right) \tag{5.6.2.1-3}$$

Step 2.2: Formulating constrained conditions based on equalities and inequalities

Conditions constraining optimization for uniaxial rectangular RC columns are formulated based on equality and inequality constraints shown in Table 5.6.2.4.

Step 3: Optimizing single objective

As presented in Table 5.6.2.5, biaxial rectangular RC columns capturing ten load pairs are separately optimized based on an objective function to normalize the UFO as shown in Equations (5.6.2.2-1)–(5.5.2.2-3).

Step 4: Formulating UFO

Step 4.1: Each objective function is normalized as shown in Equations (5.6.2.2-1)–(5.6.2.2-3).

$$f_{CI_c}^{N}(\mathbf{x}) = \frac{f_{CI_c}(\mathbf{x}) - 67{,}622}{2{,}976{,}094 - 67{,}622} \tag{5.6.2.2-1}$$

$$f_{CO_2}^{N}(\mathbf{x}) = \frac{f_{CO_2}(\mathbf{x}) - 0.1398}{6.8643 - 0.1398} \tag{5.6.2.2-2}$$

$$f_{W_c}^{N}(\mathbf{x}) = \frac{f_{W_c}(\mathbf{x}) - 5.9600}{109.9203 - 5.9600} \tag{5.6.2.2-3}$$

Step 4.2: UFO is formulated in Equation (5.6.2.3-1) implementing 121 fractions w_{CI_c}, w_{CO_2}, and w_{W_c} varying from 0 to 1 generated based on Equation (5.6.2.3-2) as illustrated in Figure 5.6.2.2.

$$\text{UFO} = w_{CI_c} f_{CI_c}^{N}(\mathbf{x}) + w_{CO_2} f_{CO_2}^{N}(\mathbf{x}) + w_{W_c} f_{W_c}^{N}(\mathbf{x}) \tag{5.6.2.3-1}$$

$$w_{CI_c}, w_{CO_2}, w_{W_c} = \text{linspace}(0,1,11);\ w_{CI_c} + w_{CO_2} + w_{W_c} = 1 \tag{5.6.2.3-2}$$

Step 5: Optimizing based on UFO

Equation (5.6.2.4) presents Lagrange function $\mathcal{L}_{\text{UFO}}(\mathbf{x}, \boldsymbol{\lambda}_c, \boldsymbol{\lambda}_v)$ based on normalized UFO in Equation (5.6.2.3-1), equalities and inequalities formulated in Step 2.2 as shown in Table 5.6.2.4. Stationary points of a UFO are found by linearization of $\mathcal{L}_{\text{UFO}}(\mathbf{x}, \boldsymbol{\lambda}_c, \boldsymbol{\lambda}_v)$ where multiple objective functions are optimized simultaneously.

Table 5.6.2.4 Equality and Inequality Constraints

Equality Constraint $c_i(\mathbf{x}) = 0$			*Inequality Constraints* $v_i(\mathbf{x}) \geq 0$	
$c_1(\mathbf{x}) = P_u^{(1)} - 1{,}500$	$c_{11}(\mathbf{x}) = M_{u,x}^{(1)} - 500$	$c_{21}(\mathbf{x}) = M_{u,y}^{(1)} - 1{,}800$	$v_1(\mathbf{x}) = SF^{(1)} - 1.0$	$v_{11}(\mathbf{x}) = h - 300$
$c_2(\mathbf{x}) = P_u^{(2)} - 3{,}000$	$c_{12}(\mathbf{x}) = M_{u,x}^{(2)} - 1{,}200$	$c_{22}(\mathbf{x}) = M_{u,y}^{(2)} - 1{,}000$	$v_2(\mathbf{x}) = SF^{(2)} - 1.0$	$v_{12}(\mathbf{x}) = -h + 2{,}000$
$c_3(\mathbf{x}) = P_u^{(3)} - 1{,}200$	$c_{13}(\mathbf{x}) = M_{u,x}^{(3)} - 2{,}000$	$c_{23}(\mathbf{x}) = M_{u,y}^{(3)} - 700$	$v_3(\mathbf{x}) = SF^{(3)} - 1.0$	$v_{13}(\mathbf{x}) = b - 300$
$c_4(\mathbf{x}) = P_u^{(4)} - 500$	$c_{14}(\mathbf{x}) = M_{u,x}^{(4)} - 1{,}700$	$c_{24}(\mathbf{x}) = M_{u,y}^{(41)} - 600$	$v_4(\mathbf{x}) = SF^{(4)} - 1.0$	$v_{14}(\mathbf{x}) = b / h - 0.25$
$c_5(\mathbf{x}) = P_u^{(5)} - 700$	$c_{15}(\mathbf{x}) = M_{u,x}^{(5)} - 1{,}300$	$c_{25}(\mathbf{x}) = M_{u,y}^{(5)} - 1{,}500$	$v_5(\mathbf{x}) = SF^{(5)} - 1.0$	$v_{15}(\mathbf{x}) = -b / h + 2.0$
$c_6(\mathbf{x}) = P_u^{(6)} - 900$	$c_{16}(\mathbf{x}) = M_{u,x}^{(6)} - 900$	$c_{26}(\mathbf{x}) = M_{u,y}^{(6)} - 1{,}400$	$v_6(\mathbf{x}) = SF^{(6)} - 1.0$	$v_{16}(\mathbf{x}) = \rho_{s,x} + \rho_{s,y} - 0.01$
$c_7(\mathbf{x}) = P_u^{(7)} - 1{,}600$	$c_{17}(\mathbf{x}) = M_{u,x}^{(7)} - 700$	$c_{27}(\mathbf{x}) = M_{u,y}^{(7)} - 1{,}100$	$v_7(\mathbf{x}) = SF^{(7)} - 1.0$	$v_{17}(\mathbf{x}) = -\rho_{s,x} - \rho_{s,y} + 0.08$
$c_8(\mathbf{x}) = P_u^{(8)} - 2{,}500$	$c_{18}(\mathbf{x}) = M_{u,x}^{(8)} - 800$	$c_{28}(\mathbf{x}) = M_{u,y}^{(8)} - 900$	$v_8(\mathbf{x}) = SF^{(8)} - 1.0$	$v_{18}(\mathbf{x}) = -b + 2{,}000$
$c_9(\mathbf{x}) = P_u^{(9)} - 1{,}700$	$c_{19}(\mathbf{x}) = M_{u,x}^{(9)} - 1{,}500$	$c_{29}(\mathbf{x}) = M_{u,y}^{(9)} - 1{,}300$	$v_9(\mathbf{x}) = SF^{(9)} - 1.0$	
$c_{10}(\mathbf{x}) = P_u^{(10)} - 2{,}200$	$c_{20}(\mathbf{x}) = M_{u,x}^{(10)} - 1{,}100$	$c_{30}(\mathbf{x}) = M_{u,y}^{(10)} - 500$	$v_{10}(\mathbf{x}) = SF^{(10)} - 1.0$	
$c_{31}(\mathbf{x}) = f_c' - 40$	$c_{32}(\mathbf{x}) = f_y - 500$			

Table 5.6.2.5 Optimization Based on Separate Objective Function

Objectives	*Minimum*	*Maximum*
CI_c(KRW/m)	146,462	2,270,406
$CO_2(t-CO_2/m)$	0.303	5.239
W_c(kN/m))	12.056	83.697

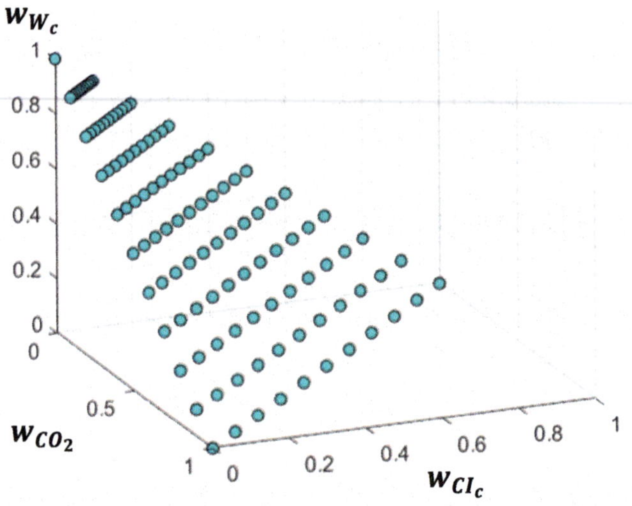

Figure 5.6.2.2 121 fractions of three objective functions ($w_{CI_c} : w_{CO_2} : w_{W_c}$).

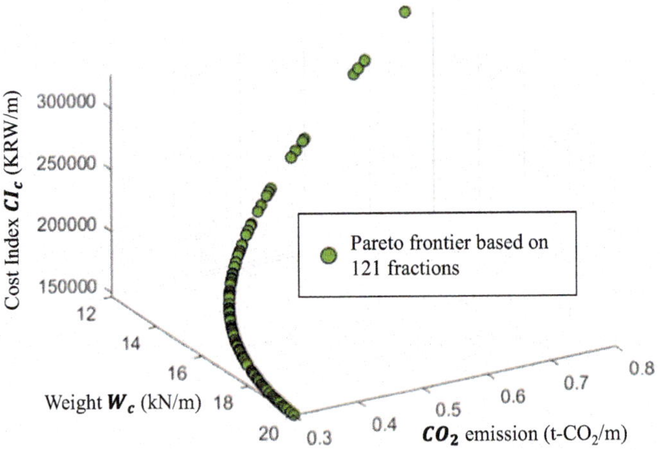

Figure 5.6.2.3 A Pareto frontier based on 121 fractions for biaxial rectangular RC columns sustaining ten load pairs based on three objective functions.

$$\mathcal{L}_{UFO}(\mathbf{x},\boldsymbol{\lambda}_c,\boldsymbol{\lambda}_v) = \text{UFO}(\mathbf{x}) - \boldsymbol{\lambda}_c^{T} c(\mathbf{x}) - \boldsymbol{\lambda}_v^{T} Sv(\mathbf{x}) \tag{5.6.2.4}$$

A Pareto frontier for biaxial rectangular RC column sustaining ten load pairs shown in Table 5.6.1.2 is presented in Figure 5.6.2.3 which employs 121 fractions.

5.6.3 Verification of Pareto frontier based on large datasets

Ten thousand (10,000) structural datasets are generated with prescribed parameters such as material properties and factored loads shown in Table 5.6.1.2. Three input variables such as column width (b), column height (h), and rebar ratio (ρ_s) are calculated based on newton-Raphson iterations using initial trial input variables randomly generated within their design ranges shown in Table 5.6.2.1. For example, b and h are determined in a range of 300–2,000 mm based on a ratio b/h in a range of 0.3–2.0. Rebar ratios ρ_s must be within 0.01 and 0.08 as minimum and maximum rebar ratios as specified in ACI 318-19 (ACI Committee, 2019). All data must have safety factors ($SF_i \geq 1.0$). Figure 5.6.3.1 illustrates a Pareto frontier plotted based on column width (b), column height (h), and rebar ratio (ρ_s) obtained by minimizing objective functions (CI_c, CO_2, W_c) simultaneously. A total of 10,000 datasets are randomly generated for the verifications. It is proven that a Pareto frontier lies on a lower boundary of large structural datasets.

5.6.4 Design parameters corresponding to three points of Pareto frontier

Table 5.6.4.1 presents ten biaxial load pairs that the rectangular RC column must sustain while the column is optimized based on UFO. Design parameters shown in the three design tables corresponding to the three fraction points on Pareto frontier are indicated in Figure 5.6.4.1. These design points are obtained by minimizing three objective functions, cost index, CO_2 and weight, simultaneously. Design parameters obtained by minimizing CI_c which is similar to ones obtained by minimizing CO_2 are yielded at Design Point 1

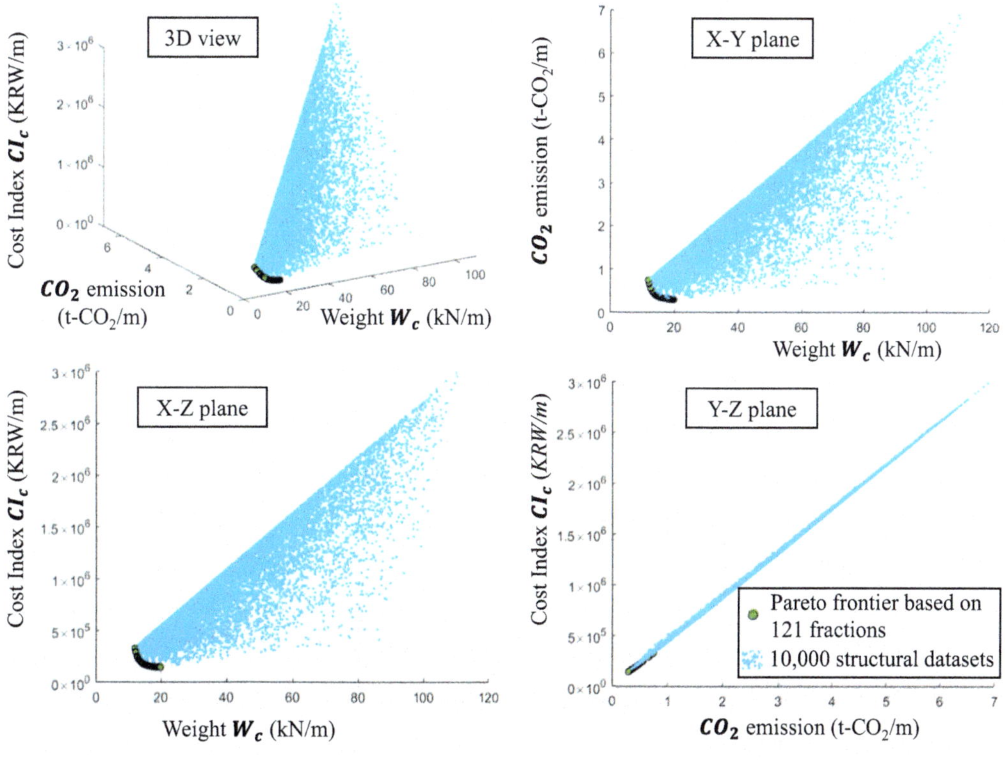

Figure 5.6.3.1 Verification of a Pareto frontier by ten thousand structural datasets.

Table 5.6.4.1 Prescribed Parameters

	Preassigned Parameters		
Factored Load			
	P_u (kN)	$M_{u,x}^{(i)}$ (kN·m)	$M_{u,y}^{(i)}$ (kN·m)
1	1,500	500	1,800
2	3,000	1,200	1,000
3	1,200	2,000	700
4	500	1,700	600
5	700	1,300	1,500
6	900	900	1,400
7	1,600	700	1,100
8	2,500	800	900
9	1,700	1,500	1,300
10	2,200	1,100	500
Material Properties			
12	Concrete	$f_y' = 40\text{MPa}$	
13	Steel	$f_c = 500\text{MPa}$	

presented by a blue point on a Pareto frontier. This is because CI_c based on a fraction of $w_{CI_c} : w_{CO_2} : w_{W_c} = 1:0:0$ (indicating that CI_c is only minimized) and CO_2 based on a fraction of $w_{CI_c} : w_{CO_2} : w_{W_c} = 0:1:0$ (indicating that CO_2 is only minimized) are minimized in a similar manner as illustrated in Figure 5.6.4.1. Design Point 2 represents an optimized design point where W_c is minimized based on $w_{CI_c} : w_{CO_2} : w_{W_c} = 0:0:1$, illustrated by a pink point shown in Figure 5.6.4.1. Design Point 3 represents an optimized design point with an equal trade-off among three objective functions based on $w_{CI_c} : w_{CO_2} : w_{W_c} = 0.30:0.35:0.35$ shown in Table 5.6.4.4. Tables 5.6.4.2–5.6.4.4 present design parameters optimized by ANN-based Lagrange while insignificant errors are verified by structural mechanics-based *AutoRC2A*.

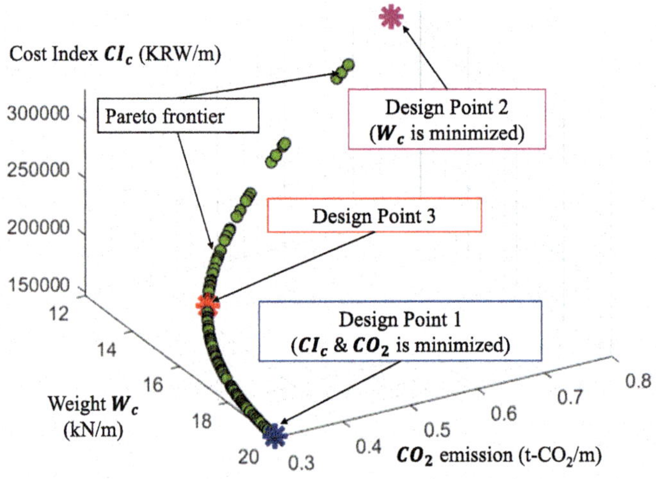

Figure 5.6.4.1 Pareto frontier showing fractions.

Table 5.6.4.2 Design Point I (Cl_c and CO_2 are Minimized; $w_{Cl_c} : w_{CO_2} : w_{W_c} = 1:0:0$ and $0:1:0$)

(a): Input Variables

Fraction $w_{Cl_c} : w_{CO_2} : w_{W_c} = 1:0:0$ & $0:1:0$

Input Variables

Parameters		AI-based Lagrange	Verification (AutoRC2A)
1	b (mm)	780.6	780.6
2	h (mm)	1,062.0	1,062.0
3	$\rho_{s,x}$	0.0056	0.0056
4	$\rho_{s,y}$	0.0044	0.0044

Inputs for ANN (*ModelLPs*).
Inputs for structural mechanics-based software (*AutoRC2A*).

(b): Output Parameters

Fraction: $Cl_c : CO_2 : W_c = 1:0:0$ & $0:1:0$

Output Parameters

Parameters		AI-based Lagrange	Verification (AutoRC2A)	Error	Parameters		AI-based Lagrange	Verification (AutoRC2A)	Error
1	$SF^{(1)}$	1.0000	0.9677	3.23%	14	$\varepsilon_{s,1}$	0.0073	0.0073	0.68%
2	$SF^{(2)}$	1.7139	1.6892	1.44%	15	$\varepsilon_{s,2}$	0.0035	0.0035	1.16%
3	$SF^{(3)}$	1.1302	1.0786	4.56%	16	$\varepsilon_{s,3}$	0.0086	0.0087	0.92%
4	$SF^{(4)}$	1.1652	1.1123	4.54%	17	$\varepsilon_{s,4}$	0.0099	0.0100	1.00%
5	$SF^{(5)}$	1.0000	0.9412	5.88%	18	$\varepsilon_{s,5}$	0.0072	0.0072	0.11%
6	$SF^{(6)}$	1.1576	1.1361	1.86%	19	$\varepsilon_{s,6}$	0.0068	0.0067	1.15%
7	$SF^{(7)}$	1.6696	1.7156	2.75%	20	$\varepsilon_{s,7}$	0.0051	0.0051	0.53%
8	$SF^{(8)}$	2.0364	2.0374	0.05%	21	$\varepsilon_{s,8}$	0.0033	0.0033	0.50%
9	$SF^{(9)}$	1.2161	1.1809	2.90%	22	$\varepsilon_{s,9}$	0.0060	0.0060	0.97%
10	$SF^{(10)}$	2.5919	2.5568	1.35%	23	$\varepsilon_{s,10}$	0.0037	0.0037	1.07%
11	Cl_c (KRW/m)	**146,462**	**146,583**	0.08%	24	b/h	0.7350	0.7350	0.00%
12	CO_2 (t-CO_2/m)	**0.3025**	**0.3026**	0.02%					
13	W_c (kN/m)	**19.9348**	**19.9376**	0.01%					

Outputs for ANN (*ModelLPs*).
Outputs for structural mechanics-based software (*AutoRC2A*).

Outputs for ANN (*ModelLPs*).
Outputs for structural mechanics-based software (*AutoRC2A*).

Bold values indicate optimized objective functions.

5.6.5 An example of ANN-based Lagrange optimization design based on multi-objective functions for biaxial rectangular RC columns sustaining multiple loads with big magnitude

This chapter presents an optimization scenario based on a forward design of biaxial rectangular RC columns sustaining five load pairs ($P_u^{(i)}$, $M_{u,x}^{(i)}$, $M_{u,y}^{(i)}$, $i = 1 \sim 5$) having a big magnitude which are formulated as shown in Table 5.6.5.1. The optimization scenario is similar to one presented in Table 5.6.4.1 of Section 5.6.4 where ten load pairs are considered. Table 5.6.5.2 presents equalities and inequalities for an optimization design of biaxial rectangular RC columns sustaining five load pairs in this section. Section 5.6.5 presents an example

Table 5.6.4.3 Design Point 2 (W_c is Minimized; $w_{CI_c} : w_{CO_2} : w_{W_c} = 0:0:1$)

(a): Input Variables

Parameters		AI-based Lagrange	Verification (AutoRC2A)
		Fraction $w_{CI_c} : w_{CO_2} : w_{W_c} = 0:0:1$	
		Input Variables	
1	b (mm)	564.8	564.8
2	h (mm)	765.1	765.1
3	$\rho_{s,x}$	0.0352	0.0352
4	$\rho_{s,y}$	0.0448	0.0448

Inputs for ANN (*ModelLPs*).
Inputs for structural mechanics-based software (*AutoRC2A*).

(b): Output Parameters

Fraction: $W_{CI_c} : W_{CO_2} : W_{W_c} = 0:0:1$

Output Parameters

Parameters		AI-based Lagrange	Verification (AutoRC2A)	Error	Parameters		AI-based Lagrange	Verification (AutoRC2A)	Error
1	$SF^{(1)}$	1.0000	1.0152	1.52%	14	$\varepsilon_{s,1}$	0.0026	0.0028	6.42%
2	$SF^{(2)}$	1.2613	1.3108	3.92%	15	$\varepsilon_{s,2}$	0.0021	0.0021	2.46%
3	$SF^{(3)}$	1.2628	1.3336	5.61%	16	$\varepsilon_{s,3}$	0.0029	0.0030	3.57%
4	$SF^{(4)}$	1.5337	1.6206	5.66%	17	$\varepsilon_{s,4}$	0.0033	0.0034	2.44%
5	$SF^{(5)}$	1.0000	1.0579	5.79%	18	$\varepsilon_{s,5}$	0.0029	0.0030	3.66%
6	$SF^{(6)}$	1.1325	1.1701	3.31%	19	$\varepsilon_{s,6}$	0.0028	0.0029	4.15%
7	$SF^{(7)}$	1.3757	1.4081	2.36%	20	$\varepsilon_{s,7}$	0.0024	0.0025	5.23%
8	$SF^{(8)}$	1.5207	1.5706	3.28%	21	$\varepsilon_{s,8}$	0.0020	0.0021	4.59%
9	$SF^{(9)}$	1.0269	1.0894	6.08%	22	$\varepsilon_{s,9}$	0.0026	0.0027	3.39%
10	$SF^{(10)}$	1.9130	1.9634	2.64%	23	$\varepsilon_{s,10}$	0.0021	0.0021	1.85%
11	CI_C (KRW/m)	**326,977**	**326,950**	0.01%	24	b/h	0.7383	0.7383	0.01%
12	CO_2 (t-CO_2/m)	**0.7547**	**0.7545**	0.03%					
13	W_C (kN/m)	**12.0563**	**12.0574**	0.01%					
	Outputs for ANN (*ModelLPs*). Outputs for structural mechanics-based software (*AutoRC2A*).					Outputs for ANN (*ModelLPs*). Outputs for structural mechanics-based software (*AutoRC2A*).			

Bold values indicate optimized objective functions.

of an optimization design of biaxial RC columns sustaining five load pairs with big magnitudes that is similar to an example shown in Section 5.6.1. 30 equalities are used with ten load pairs as shown in Table 5.6.2.4 of Section 5.6.1, whereas 15 equalities are used with five load pairs as shown in Table 5.6.5.2 of Section 5.6.5. The magnitudes of the loads shown in Table 5.6.5.2 are around 4~5 times larger compared to those in Table 5.6.4.1 of Section 5.6.4.

5.6.5.1 Identifying design parameters for designated fractions based on two neural networks based on Tables 5.6.2.3 and 5.6.5.6

ANN-based Lagrange optimization minimizing three objectives (CI_c,CO_2, and W_c) at the same time is performed similarly to one described in Section 5.6.2, using an ANN – *ModelLPs* for five load pairs whose best training is based on Table 5.6.2.3. Objective functions

Table 5.6.4.4 Design Point 3 (Fraction: $w_{CI_c} : w_{CO_2} : w_{W_c}$=0.30:0.35:0.35)

(a): Input Variables

Fraction $w_{CI_c} : w_{CO_2} : w_{W_c}$ = 0.30 : 0.35 : 0.35

Parameters		*AI-based Lagrange*	*Verification (AutoRC2A)*
		Input Variables	
1	b (mm)	638.7	638.7
2	h (mm)	946.7	946.7
3	$\rho_{s,x}$	0.0079	0.0079
4	$\rho_{s,y}$	0.0153	0.0153

Inputs for ANN (*ModelLPs*).
Inputs for structural mechanics-based software (*AutoRC2A*).

(b): Output Parameters

Fraction: $W_{CI_c} : W_{CO_2} : W_{W_c}$ = 0.30 : 0.35 : 0.35

Parameters		*AI-based Lagrange*	*Verification (AutoRC2A)*	*Error*	*Parameters*		*AI-based Lagrange*	*Verification (AutoRC2A)*	*Error*
			Output Parameters						
1	$SF^{(1)}$	1.0000	0.9872	1.28%	14	$\varepsilon_{s,1}$	0.0048	0.0048	1.55%
2	$SF^{(2)}$	1.3588	1.3740	1.12%	15	$\varepsilon_{s,2}$	0.0030	0.0029	3.78%
3	$SF^{(3)}$	1.2575	1.3297	5.75%	16	$\varepsilon_{s,3}$	0.0056	0.0057	1.87%
4	$SF^{(4)}$	1.4063	1.4441	2.69%	17	$\varepsilon_{s,4}$	0.0063	0.0065	4.07%
5	$SF^{(5)}$	1.0000	1.0209	2.09%	18	$\varepsilon_{s,5}$	0.0048	0.0049	1.85%
6	$SF^{(6)}$	1.1276	1.1376	0.89%	19	$\varepsilon_{s,6}$	0.0046	0.0046	0.46%
7	$SF^{(7)}$	1.4477	1.4449	0.19%	20	$\varepsilon_{s,7}$	0.0038	0.0037	2.21%
8	$SF^{(8)}$	1.6204	1.6229	0.15%	21	$\varepsilon_{s,8}$	0.0029	0.0028	2.94%
9	$SF^{(9)}$	1.0917	1.1140	2.04%	22	$\varepsilon_{s,9}$	0.0043	0.0042	1.89%
10	$SF^{(10)}$	2.0303	2.0846	2.67%	23	$\varepsilon_{s,10}$	0.0031	0.0029	5.27%
11	CI_C (KRW/m)	**173,274**	**173,274**	0.00%	24	b/h	0.6747	0.6747	0.00%
12	CO_2 (t-CO_2/m)	**0.3786**	**0.3787**	0.04%					
13	W_C (kN/m)	**14.9838**	**14.9841**	0.00%					

Outputs for ANN (*ModelLPs*).
Outputs for structural mechanics-based software (*AutoRC2A*).

Outputs for ANN (*ModelLPs*).
Outputs for structural mechanics-based software (*AutoRC2A*).

Bold values indicate optimized objective functions.

derived based on *Model-LPs* for five load pairs are shown in Equations (5.6.5.1-1)–(5.6.5.1-3). It is noted that, in *Model-LPs,* five load pairs ($P_u^{(i)}$, $M_{u,x}^{(i)}$, $M_{u,y}^{(i)}$) ($i = 1 \sim 5$) lead to 21 inputs including four input variables (b, h, $\rho_{s,x}$, $\rho_{s,y}$), and two prescribed parameters ($f_c^{'}$, f_y) for material properties, 15 prescribed parameters $\left(P_u^{(i)},\ M_{u,x}^{(i)},\ M_{u,y}^{(i)},\ i = 1 \sim 5 \right)$ for three pairs ($P_u^{(i)}$, $M_{u,x}^{(i)}$, $M_{u,y}^{(i)}$) for five loads, and 14 outputs $\left(SF^{(i)}, \varepsilon_{s,i}, CI_c, CO_2, W_c, b/h, i = 1 \sim 5 \right)$.

$$\underbrace{f_{CI_c}(\mathbf{x})}_{[1\times1]} = g_{CI_c}^{D}\left(f_{\text{linear}}\left(\underbrace{\omega_{CI_c}^{(\text{out})}}_{[1\times64]} f_{\tanh}\left(\underbrace{\omega_{CI_c}^{(k=5)}}_{[64\times64]} \dots f_{\tanh}\left(\underbrace{\omega_{CI_c}^{(1)}}_{[64\times21]} \underbrace{g^{N}(\mathbf{x})}_{[21\times1]} + \underbrace{\mathbf{b}_{CI_c}^{(1)}}_{[64\times1]} \right) \dots + \underbrace{\mathbf{b}_{CI_c}^{(k=5)}}_{[64\times1]} \right) + \underbrace{\mathbf{b}_{CI_c}^{(\text{out})}}_{[1\times1]} \right) \right) \quad (5.6.5.1\text{-}1)$$

Table 5.6.5.1 Preassigned Parameters in a Design of Biaxial Rectangular RC Columns Sustaining Five Load Pairs with Big Magnitude

Preassigned Parameters			
Factored Load			
	P_u (kN)	$M_{u,x}^{(i)}$ (kN·m)	$M_{u,y}^{(i)}$ (kN·m)
1	15,000	4,000	1,200
2	5,000	8,000	2,000
3	10,000	2,000	6,000
4	7,000	3,500	8,000
5	12,000	5,000	3,000
Material Properties			
12	Concrete	$f_c' = 40\text{MPa}$	
13	Steel	$f_y = 500\text{MPa}$	

$$\underbrace{f_{\mathrm{CO_2}}(\mathbf{x})}_{[1\times1]} = g_{\mathrm{CO_2}}^{D}\left(f_{\mathrm{linear}}\left(\underbrace{\omega_{\mathrm{CO_2}}^{(\mathrm{out})}}_{[1\times32]} f_{\tanh}\left(\underbrace{\omega_{\mathrm{CO_2}}^{(k=6)}}_{[32\times32]}\cdots f_{\tanh}\left(\underbrace{\omega_{\mathrm{CO_2}}^{(1)}}_{[32\times21]}\underbrace{g^{N}(\mathbf{x})}_{[21\times1]}+\underbrace{\mathbf{b}_{CI_c}^{(1)}}_{[32\times1]}\right)\cdots+\underbrace{\mathbf{b}_{\mathrm{CO_2}}^{(k=6)}}_{[32\times1]}\right)+\underbrace{\mathbf{b}_{\mathrm{CO_2}}^{(\mathrm{out})}}_{[1\times1]}\right)\right) \tag{5.6.5.1-2}$$

$$\underbrace{f_{\mathrm{W}_c}(\mathbf{x})}_{[1\times1]} = g_{\mathrm{W}_c}^{D}\left(f_{\mathrm{linear}}\left(\underbrace{\omega_{\mathrm{W}_c}^{(\mathrm{out})}}_{[1\times64]} f_{\tanh}\left(\underbrace{\omega_{\mathrm{W}_c}^{(k=4)}}_{[64\times64]}\cdots f_{\tanh}\left(\underbrace{\omega_{\mathrm{W}_c}^{(1)}}_{[64\times21]}\underbrace{g^{N}(\mathbf{x})}_{[21\times1]}+\underbrace{\mathbf{b}_{\mathrm{W}_c}^{(1)}}_{[64\times1]}\right)\cdots+\underbrace{\mathbf{b}_{\mathrm{W}_c}^{(k=4)}}_{[64\times1]}\right)+\underbrace{\mathbf{b}_{\mathrm{W}_c}^{(\mathrm{out})}}_{[1\times1]}\right)\right) \tag{5.6.5.1-3}$$

Tables 5.6.5.3–5.6.5.5 based on Table 5.6.2.3 show input and output parameters obtained by minimizing CI_c, CO_2, and W_c separately, based on $\left(w_{CI_c} : w_{\mathrm{CO_2}} : w_{\mathrm{W}_c}\right)$ of 1:0:0, 0:1:0 and 0:0:1. Input and preassigned parameters are 21 inputs $\left(b,\ h,\ \rho_{s,x},\ \rho_{s,y},\ f_c',\ f_y,\ P_u^{(i)},\ M_{u,x}^{(i)},\ M_{u,y}^{(i)},\ i=1\sim5\right)$ where 14 outputs $\left(SF^{(i)}, \varepsilon_{s,i}, CI_c, \mathrm{CO_2}, \mathrm{W}_c, b/h, i=1\sim5\right)$are obtained from 21 inputs based on an ANN (*Model-LPs*) subject to five load pairs. They are verified by structural mechanics-based software *AutoRC2A*. When using five layers and 64 neurons shown in Table 5.6.2.3, Design 1 minimizing CI_c is presented in Table 5.6.5.3 which does not meet the strength requirement because safety factor $SF^{(2)}$ of Load Pair 2 ($P_u^{(2)} = 5{,}000$ kN, $M_{u,x}^{(2)} = 8{,}000$ kN.m, $M_{u,y}^{(2)} = 2{,}000$ kN.m) is equal to 0.8577 based on structural mechanics-based software *AutoRC2A*, leading to 14.23% error between ANN and *AutoRC2A*. Designs 2 and 3 shown in Tables 5.6.5.4 and 5.6.5.5, respectively, meet strength requirements by all safety factors of five load pairs ($SF^{(i)}$) greater than 1.0. However, the biggest error in Design 2 shown in Table 5.6.5.4 is 14.47% when training is based on six layers and 64 neurons shown in Table 5.6.2.3 for a rebar strain of Load Pair 3 ($\varepsilon_{s,3}$). Design 3 shown in Table 5.6.5.5 based on five layers and 64 neurons of Table 5.6.2.3 have all five safety factors $SF^{(1\sim5)}$ from 1.2488 to 3.0695 which is greater than 1.0, implying that this design point was not well optimized.

Best training based on previous ANNs presented in Table 5.6.2.3 shows that training accuracies illustrated by Test MSE of SF and ε_s are relatively low, around an order of 10^{-4},

Table 5.6.5.2 Equality and Inequality Constraints

Equality Constraint $c_i(\boldsymbol{x}) = 0$			*Inequality Constraints* $v_i(\boldsymbol{x}) \geq 0$	
$c_1(\boldsymbol{x}) = P_u^{(1)} - 15{,}000$	$c_6(\boldsymbol{x}) = M_{u,x}^{(1)} - 4{,}000$	$c_{11}(\boldsymbol{x}) = M_{u,y}^{(1)} - 1{,}200$	$v_1(\boldsymbol{x}) = SF^{(1)} - 1.0$	$v_7(\boldsymbol{x}) = h - 300$
$c_2(\boldsymbol{x}) = P_u^{(2)} - 5{,}000$	$c_7(\boldsymbol{x}) = M_{u,x}^{(2)} - 8{,}000$	$c_{12}(\boldsymbol{x}) = M_{u,y}^{(2)} - 2{,}000$	$v_2(\boldsymbol{x}) = SF^{(2)} - 1.0$	$v_8(\boldsymbol{x}) = -h + 2{,}000$
$c_3(\boldsymbol{x}) = P_u^{(3)} - 10{,}000$	$c_8(\boldsymbol{x}) = M_{u,x}^{(3)} - 2{,}000$	$c_{13}(\boldsymbol{x}) = M_{u,y}^{(3)} - 6{,}000$	$v_3(\boldsymbol{x}) = SF^{(3)} - 1.0$	$v_9(\boldsymbol{x}) = b - 300$
$c_4(\boldsymbol{x}) = P_u^{(4)} - 7{,}000$	$c_9(\boldsymbol{x}) = M_{u,x}^{(4)} - 3{,}500$	$c_{14}(\boldsymbol{x}) = M_{u,y}^{(4)} - 8{,}000$	$v_4(\boldsymbol{x}) = SF^{(4)} - 1.0$	$v_{10}(\boldsymbol{x}) = b / h - 0.25$
$c_5(\boldsymbol{x}) = P_u^{(5)} - 12{,}000$	$c_{10}(\boldsymbol{x}) = M_{u,x}^{(5)} - 5{,}000$	$c_{15}(\boldsymbol{x}) = M_{u,y}^{(5)} - 3{,}000$	$v_5(\boldsymbol{x}) = SF^{(5)} - 1.0$	$v_{11}(\boldsymbol{x}) = -b / h + 2.0$
$c_{16}(\boldsymbol{x}) = f_c' - 40$	$c_{17}(\boldsymbol{x}) = f_y - 500$		$v_6(\boldsymbol{x}) = \rho_{s,x} + \rho_{s,y} - 0.01$	$v_{12}(\boldsymbol{x}) = -\rho_{s,x} - \rho_{s,y} + 0.08$

Table 5.6.5.3 Input Parameters and Obtained Output Parameters for Design I for Minimized CI_c Based on Table 5.6.2.3

(a): Input Variables			
Fraction $w_{CI_c} : w_{CO_2} : w_{W_c} = 1:0:0$			
Input Variables			
Parameters		*AI-based Lagrange*	*Verification (AutoRC2A)*
1	*b* (mm)	1,446.3	1,446.3
2	*h* (mm)	1,252.2	1,252.2
3	$\rho_{s,x}$	0.0100	0.0100
4	$\rho_{s,y}$	0.0000	0.0000

Inputs for ANN (*ModelLPs* for five load pairs).
Inputs for structural mechanics-based software (*AutoRC2A*).

(b): Output Parameters					
Fraction: $W_{CI_c} : W_{CO_2} : W_{W_c} = 1:0:0$					
Output Parameters of Design I					
Parameters		*Best Training*	*AI-based Lagrange*	*Verification (AutoRC2A)*	*Error*
1	*SF* (1)	*Five layers*	1.7897	1.8921	5.72%
2	*SF* (2)	*Sixty-four neurons*	1.0000	0.8577	14.23%
3	*SF* (3)		1.4410	1.4196	1.49%
4	*SF* (4)		1.0000	0.9228	7.72%
5	*SF* (5)		1.3890	1.3824	0.48%
6	$\varepsilon_{s,1}$	*Six layers*	0.0008	0.0009	5.22%
7	$\varepsilon_{s,2}$	*Sixty-four neurons*	0.0078	0.0085	8.52%
8	$\varepsilon_{s,3}$		0.0032	0.0034	7.20%
9	$\varepsilon_{s,4}$		0.0057	0.0058	1.80%
10	$\varepsilon_{s,5}$		0.0021	0.0023	8.01%
11	CI_C (KRW/m)	3L - 64N	**320,723**	**320,217**	0.16%
12	CO_2 (t-CO_2/m)	7L - 32N	**0.6607**	**0.6610**	0.04%
13	W_C (kN/m)	5L - 64N	**43.5516**	**43.5546**	0.01%
14	*b/h*	5L - 64N	1.1551	1.1550	0.01%

Outputs for ANN (ModelLPs for five load pairs).
Outputs for structural mechanics-based software (*AutoRC2A*).

Bold and red color values indicate optimized objective functions.

compared to those of other outputs $CI_c, CO_2, W_c, b/h$ which are about an order of 10^{-9}. Therefore, two output parameters SF and ε_s need to be retrained. It is noted that this can be done since ANNs in this chapter is trained by parallel training method (PTM), so that output parameters are trained separately. When outputs are not trained separately (such as training based on TED, CRS (Hong et al., 2021c), entire ANNs should be retrained to improve accuracies of output predictions. Best training based on new training parameters is presented in Table 5.6.5.6, showing that the best new training accuracies of SF is obtained based on six layers and 64 neurons whereas that of ε_s is based on eight layers and 128 neurons. Test MSE of SF and ε_s improves from 3.2×10^{-4} to 1.9×10^{-4} based on previous training parameters shown in Table 5.6.2.3 to 8.1×10^{-5} and 9.3×10^{-5} which is based on new

Table 5.6.5.4 Input Parameters and Obtained Output Parameters for Design 2 for Minimized CO_2 Based on Table 5.6.2.3

(a): Input Variables			
Fraction $w_{CI_c} : w_{CO_2} : w_{W_c} = 0:1:0$			
Input Variables			
Parameters		AI-based Lagrange	Verification (AutoRC2A)
1	b (mm)	1,043.8	1,043.8
2	h (mm)	982.8	982.8
3	$\rho_{s,x}$	0.0293	0.0293
4	$\rho_{s,y}$	0.0507	0.0507

Inputs for ANN (*ModelLPs* for five load pairs).
Inputs for structural mechanics-based software (*AutoRC2A*).

(b): Output Parameters					
Fraction: $W_{CI_c} : W_{CO_2} : W_{W_c} = 0:1:0$					
Output Parameters of Design 2					
Parameters		Best Training	AI-based Lagrange	Verification (AutoRC2A)	Error
1	$SF^{(1)}$	Five layers	1.4947	1.4651	1.98%
2	$SF^{(2)}$	Sixty-four neurons	1.0000	0.9885	1.15%
3	$SF^{(3)}$		1.3734	1.3412	2.35%
4	$SF^{(4)}$		1.0000	1.0037	0.37%
5	$SF^{(5)}$		1.1524	1.1524	0.00%
6	$\varepsilon_{s,1}$	Six layers	0.0008	0.0009	3.92%
7	$\varepsilon_{s,2}$	Sixty-four neurons	0.0030	0.0029	4.26%
8	$\varepsilon_{s,3}$		0.0016	0.0018	14.47%
9	$\varepsilon_{s,4}$		0.0024	0.0025	3.76%
10	$\varepsilon_{s,5}$		0.0016	0.0017	4.63%
11	CI_C (KRW/m)	3L - 64N	**776,337**	**776,082**	0.03%
12	CO_2 (t-CO_2/m)	7L - 32N	**1.7908**	**1.7909**	0.00%
13	W_C (kN/m)	5L - 64N	**28.6229**	**28.6207**	0.01%
14	b/h	5L - 64N	1.0620	1.0620	0.00%

Outputs for ANN (*ModelLPs* for five load pairs).
Outputs for structural mechanics-based software (*AutoRC2A*).

Bold and red color values indicate optimized objective functions.

training parameters, respectively. Regression (R) of *SF* enhances from 0.9954 to 0.9989 at the best epochs, while that of ε_s increases from 0.9860 to 0.9970 as shown in Tables 5.6.2.3 and 5.6.5.6. Readers are referred to Hong (2020, 2021), and Hong et al. (2021c) for further explanations of training parameters including test MSE and R at best epochs. New training results and accuracies shown Table 5.6.5.6 improved compared to previous ones shown in Table 5.6.2.3 in terms of Test MSE and Regression (R) at best epochs.

Calculation errors for optimized parameters of all designs shown in Tables 5.6.5.7–5.6.5.10 decrease by increasing a number of layers and a number of neurons. Designs by ANN-based Lagrange optimization based on previous based on Table 5.6.2.3 and new

Table 5.6.5.5 Input Parameters and Obtained Output Parameters for Design 3 for Minimized W_c Based on Table 5.6.2.3

(a): Input Variables			
Fraction $w_{CI_c} : w_{CO_2} : w_{W_c} = 0:0:1$			
Input Variables			
Parameters		AI-based Lagrange	Verification (AutoRC2A)
1	b (mm)	1,236.7	1,236.7
2	h (mm)	2,000.0	2,000.0
3	$\rho_{s,x}$	0.0062	0.0062
4	$\rho_{s,y}$	0.0038	0.0038

Inputs for ANN (*ModelLPs* for five load pairs).
Inputs for structural mechanics-based software (*AutoRC2A*).

(b): Output Parameters					
Fraction: $W_{CI_c} : W_{CO_2} : W_{W_c} = 0:0:1$					
Output Parameters of Design 3					
Parameters		Best Training	AI-based Lagrange	Verification (AutoRC2A)	Error
1	$SF^{(1)}$	Five layers Sixty-four neurons	3.0695	3.1843	3.74%
2	$SF^{(2)}$		2.3503	2.2170	5.67%
3	$SF^{(3)}$		1.8167	1.7925	1.33%
4	$SF^{(4)}$		1.2488	1.2850	2.90%
5	$SF^{(5)}$		2.6517	2.7556	3.92%
6	$\varepsilon_{s,1}$	Six layers Sixty-four neurons	0.0001	0.0003	355.23%
7	$\varepsilon_{s,2}$		0.0076	0.0075	1.99%
8	$\varepsilon_{s,3}$		0.0033	0.0032	0.84%
9	$\varepsilon_{s,4}$		0.0054	0.0052	3.32%
10	$\varepsilon_{s,5}$		0.0015	0.0014	4.25%
11	CI_C (KRW/m)	3L - 64N	**438,055**	**437,346**	0.16%
12	CO_2 (t-CO_2/m)	7L - 32N	**0.9023**	**0.9028**	0.06%
13	W_c (kN/m)	5L - 64N	**59.4813**	**59.4861**	0.01%
14	b/h	5L - 64N	0.6184	0.6184	0.00%

Outputs for ANN (*ModelLPs* for five load pairs).
Outputs for structural mechanics-based software (*AutoRC2A*)

Bold and red color values indicate optimized objective functions.

training based on Table 5.6.5.6 are also compared. It is noted that the new design based on the new training shown in Table 5.6.5.6 reduce safety factors $SF^{(1\sim5)}$ near 1.0. Designs are performed by ANN-based Lagrange optimization in which their corresponding fractions $\left(w_{CI_c} : w_{CO_2} : w_{W_c}\right)$ are based on 1:0:0, 0:1:0 and 0:0:1 for Designs 1–3, respectively. Optimized parameters of the designs based on previous training shown in Table 5.6.2.3 are presented in Tables 5.6.5.3–5.6.5.5 which improve when new training based on Table 5.6.5.6 are used to design Tables 5.6.5.7–5.6.5.9.

Tables 5.6.5.7–5.6.5.9 compare input and output parameters of designs minimizing CI_c, CO_2, and W_c separately, based on $\left(w_{CI_c} : w_{CO_2} : w_{W_c}\right)$ of 1:0:0, 0:1:0 and 0:0:1.

Table 5.6.5.6 Training Accuracy Based on New Training Parameters for *SF* and ε_s

Nine Inputs (b, h, $\rho_{s,x}$, $\rho_{s,y}$, $f_c^{'}$, f_y, P_u, $M_{u,x}$, $M_{u,y}$) Six Outputs (SF, ε_s, b/h, CI_c, CO_2, W_c)						
PTM – 200,000 Data – 100,000 Suggested Epochs – tansig Activation Function						
Output	*Layers*	*Neurons*	*Best Epoch*	*Stopped Epoch*	*Test MSE*	*R at Best Epoch*
SF	6	64	74,055	75,055	8.1.E-05	0.9989
ε_s	8	128	39,811	40,811	9.3.E-05	0.9970
b/h	5	64	82,102	82,102	4.9.E-09	1.0000
CI_c	3	64	52,532	53,532	2.2.E-08	1.0000
CO_2	7	32	49,280	50,280	1.3.E-09	1.0000
W_c	5	64	99,990	100,000	4.0.E-09	1.0000

Table 5.6.5.7 Comparison of Design 1 (CI_c is Minimized) Based on Tables 5.6.2.3 and 5.6.5.6

(a): Input Variables			
Input Variables of Design 1			
Fraction $w_{CI_c} : w_{CO_2} : w_{W_c} = 1:0:0$			
Parameters		*Results Based on Previous Training (Table 5.6.2.3)*	*Results Based on Revised Training (Table 5.6.5.6)*
1	*b* (mm)	1,446.3	1,364.2
2	*h* (mm)	1,252.2	1,398.8
3	$\rho_{s,x}$	0.0100	0.0079
4	$\rho_{s,y}$	0.0000	0.0021

Note: Inputs for ANN and structural mechanics-based software (*AutoRC2A*).

(b): Output Parameters									
Fraction: $W_{CI_c} : W_{CO_2} : W_{W_c} = 1:0:0$									
Output Parameters of Design 1									
Parameters		*Best Training*	*AI-based Lagrange*	*Verification (AutoRC2A)*	*Error*	*Best Training*	*AI-based Lagrange*	*Verification (AutoRC2A)*	*Error*
1	*SF* (1)	*Five layers Sixty-four neurons*	1.7897	1.8921	5.72%	*Six layers Sixty-four neurons*	2.1062	2.0999	0.30%
2	*SF* (2)		1.0000	0.8577	14.23%		1.0000	1.0202	2.02%
3	*SF* (3)		1.4410	1.4196	1.49%		1.4909	1.4563	2.32%
4	*SF* (4)		1.0000	0.9228	7.72%		1.0000	0.9634	3.66%
5	*SF* (5)		1.3890	1.3824	0.48%		1.5916	1.5703	1.34%
6	$\varepsilon_{s,1}$	*Six layers Sixty-four neurons*	0.0008	0.0009	5.22%	*Eight layers One hundred twenty-eight neurons*	0.0008	0.0007	7.77%
7	$\varepsilon_{s,2}$		0.0078	0.0085	8.52%		0.0081	0.0084	3.43%
8	$\varepsilon_{s,3}$		0.0032	0.0034	7.20%		0.0033	0.0033	1.21%
9	$\varepsilon_{s,4}$		0.0057	0.0058	1.80%		0.0057	0.0057	0.48%
10	$\varepsilon_{s,5}$		0.0021	0.0023	8.01%		0.0020	0.0022	6.28%
11	CI_C (KRW/m)	3L - 64N	**320,723**	**320,217**	0.16%	3L - 64N	**337,992**	**337,408**	0.17%

(Continued)

Table 5.6.5.7 (Continued) Comparison of Design 1 (Cl_c is Minimized) Based on Tables 5.6.2.3 and 5.6.5.6

Parameters	*Best Training*	*AI-based Lagrange*	*Verification (AutoRC2A)*	*Error*	*Best Training*	*AI-based Lagrange*	*Verification (AutoRC2A)*	*Error*
12 CO_2 (t-CO_2/m)	7L - 32N	**0.6607**	**0.6610**	0.04%	7L - 32N	**0.6962**	**0.6965**	0.04%
13 W_C (kN/m)	5L - 64N	**43.5516**	**43.5546**	0.01%	5L - 64N	**45.8899**	**45.8929**	0.01%
14 *b/h*	5L - 64N	1.1551	1.1550	0.01%	5L - 64N	0.9753	0.9752	0.01%
Outputs for ANN (*ModelLPs*).					Best training based on previous training results in Table 5.6.2.3.			
Outputs for structural mechanics-based software (*AutoRC2A*).					Best training based on revised training results in Table 5.6.5.6.			

Bold and red color values indicate optimized objective functions.

Table 5.6.5.8 Comparison of Design 2 (CO_2 is Minimized) Based on Tables 5.6.2.3 and 5.6.5.6

(a): Input Variables

Input Variables of Design 2

Fraction $w_{Cl_c} : w_{CO_2} : w_{W_c} = 0:1:0$

Parameters	*Results Based on Previous Training (Table 5.6.2.3)*	*Results Based on Revised Training (Table 5.6.5.6)*
1 *b* (mm)	1,043.8	1,043.8
2 *h* (mm)	982.8	982.8
3 $\rho_{s,x}$	0.0293	0.0293
4 $\rho_{s,y}$	0.0507	0.0507

Note: Inputs for ANN and structural mechanics-based software (*AutoRC2A*).

(b): Output Parameters

Fraction: $W_{Cl_c} : W_{CO_2} : W_{W_c} = 0:1:0$

Output Parameters of Design 2

Parameters	Best Training	AI-based Lagrange	Verification (*AutoRC2A*)	Error	Best Training	AI-based Lagrange	Verification (*AutoRC2A*)	Error
1 *SF* (1)	*Five layers Sixty-four neurons*	1.4947	1.4651	1.98%	*Six layers Sixty-four neurons*	2.1059	2.0960	0.47%
2 *SF* (2)		1.0000	0.9885	1.15%		1.0000	1.0174	1.74%
3 *SF* (3)		1.3734	1.3412	2.35%		1.4911	1.4562	2.34%
4 *SF* (4)		1.0000	1.0037	0.37%		1.0000	0.9638	3.62%
5 *SF* (5)		1.1524	1.1524	0.00%		1.5914	1.5640	1.72%
6 $\varepsilon_{s,1}$	*Six layers Sixty-four neurons*	0.0008	0.0009	3.92%	*Eight layers One hundred twenty-eight neurons*	0.0008	0.0007	7.21%
7 $\varepsilon_{s,2}$		0.0030	0.0029	4.26%		0.0081	0.0083	2.76%
8 $\varepsilon_{s,3}$		0.0016	0.0018	14.47%		0.0033	0.0033	1.19%
9 $\varepsilon_{s,4}$		0.0024	0.0025	3.76%		0.0057	0.0057	0.50%
10 $\varepsilon_{s,5}$		0.0016	0.0017	4.63%		0.0020	0.0022	6.72%
11 Cl_C (KRW/m)	3L - 64N	**776,337**	**776,082**	0.03%	3L - 64N	**337,993**	**337,408**	0.17%
12 CO_2 (t-CO_2/m)	7L - 32N	**1.7908**	**1.7909**	0.00%	7L - 32N	**0.6962**	**0.6965**	0.04%
13 W_C (kN/m)	5L - 64N	**28.6229**	**28.6207**	0.01%	5L - 64N	**45.8899**	**45.8929**	0.01%
14 *b/h*	5L - 64N	1.0620	1.0620	0.00%	5L - 64N	0.9736	0.9735	0.01%
Outputs for ANN (*ModelLPs*).					Best training based on previous training results in Table 5.6.2.3.			
Outputs for structural mechanics-based software (*AutoRC2A*).					Best training based on revised training results in Table 5.6.5.6.			

Bold and red color values indicate optimized objective functions.

Table 5.6.5.9 Comparison of Design 3 (W_c is Minimized) Based on Tables 5.6.2.3 and 5.6.5.6

(a): Input Variables

Input Variables of Design 3

Fraction $w_{CI_c} : w_{CO_2} : w_{W_c} = 0:0:1$			
Parameters		Results Based on Previous Training (Table 5.6.2.3)	Results Based on Revised Training (Table 5.6.5.6)
1	*b* (mm)	1,236.7	498.8
2	*h* (mm)	2,000.0	1,993.9
3	$\rho_{s,x}$	0.0062	0.0800
4	$\rho_{s,y}$	0.0038	0.0000

Note: Inputs for ANN and structural mechanics-based software (*AutoRC2A*).

(b): Output Parameters

Fraction: $W_{CI_c} : W_{CO_2} : W_{W_c} = 0:0:1$

Output Parameters of Design 3

Parameters	Best Training	AI-based Lagrange	Verification (*AutoRC2A*)	Error	Best Training	AI-based Lagrange	Verification (*AutoRC2A*)	Error
1 $SF^{(1)}$	*Five layers*	3.0695	3.1843	3.74%	*Six layers*	2.3167	2.3752	2.52%
2 $SF^{(2)}$	*Sixty-four*	2.3503	2.2170	5.67%	*Sixty-four*	2.6815	2.8185	5.11%
3 $SF^{(3)}$	neurons	1.8167	1.7925	1.33%	neurons	1.0000	1.0076	0.76%
4 $SF^{(4)}$		1.2488	1.2850	2.90%		1.0009	0.9872	1.37%
5 $SF^{(5)}$		2.6517	2.7556	3.92%		2.2832	2.3258	1.87%
6 $\varepsilon_{s,1}$	*Six layers*	0.0001	0.0003	355.23%	*Eight layers*	–0.0007	0.0003	-136.75%
7 $\varepsilon_{s,2}$	*Sixty-four*	0.0076	0.0075	1.99%	One	0.0014	0.0015	4.48%
8 $\varepsilon_{s,3}$	neurons	0.0033	0.0032	0.84%	hundred twenty-	0.0019	0.0018	1.56%
9 $\varepsilon_{s,4}$		0.0054	0.0052	3.32%	eight	0.0024	0.0023	4.77%
10 $\varepsilon_{s,5}$		0.0015	0.0014	4.25%	neurons	0.0003	0.0003	8.69%
11 CI_C (KRW/m)	3L - 64N	**438,055**	**437,346**	0.16%	3L - 64N	**753,529**	**752,363**	0.15%
12 CO_2 (t-CO_2/m)	7L - 32N	**0.9023**	**0.9028**	0.06%	7L - 32N	**1.7373**	**1.7362**	0.06%
13 W_c (kN/m)	5L - 64N	**59.4813**	**59.4861**	0.01%	5L - 64N	**27.7378**	**27.7460**	0.03%
14 *b/h*	5L - 64N	0.6184	0.6184	0.00%	5L - 64N	0.2500	0.2501	0.06%

Outputs for ANN (*ModelLPs*).

Outputs for structural mechanics-based software (*AutoRC2A*).

Best training based on previous training results in Table 5.6.2.3.

Best training based on revised training results in Table 5.6.5.6.

Bold and red color values indicate optimized objective functions.

Design 1 shown in Table 5.6.5.7(b) based on new training parameters shown in Table 5.6.5.6 satisfies strength requirement although $SF^{(4)}$ is still smaller than 1.0 ($SF^{(4)} = 0.9634$) by *AutoRC2A*. This is considered as acceptable since error of $SF^{(4)}$ is as small as 3.66%. As shown in Table 5.6.5.8(b), large computational errors in Design 2 improve from 14.47% for $\varepsilon_{s,3}$ based on previous training parameters shown in Table 5.6.2.3 to 1.19% based on new training shown in Table 5.6.5.6. A magnitude of $\varepsilon_{s,1}$ obtained by an ANN is 0.0008 which is very close to 0.0007 that is resulted from *AutoRC2A*, even if a relatively large error as 7.21% is seen for $\varepsilon_{s,1}$ in Table 5.6.5.8(b) based on new training of Table 5.6.5.6. This leads to 7.21% error but, acceptable. Table 5.6.5.9(b) proves that W_c are minimized better when using an ANN based on new training parameters than using an ANN based on the previous training parameters. Table 5.6.5.9(a) shows that column sections obtained based on new training parameters shown in Table 5.6.5.6 are much smaller ($b = 498.8$ and $h = 1{,}993.9$ mm) than those ($b = 1{,}236.7$ and $h = 2{,}000.0$ mm) obtained based on previous training parameters shown in Table 5.6.2.3. A minimized W_c using an ANN based on previous training parameters shown in Table 5.6.2.3 is 59.48 kN/m which is compared to 27.7 kN/m based on new training parameters shown in Table 5.6.5.6. As shown in Table 5.6.5.9(b), safety factors, $SF^{(i=1\sim5)}$, obtained based on new training parameters shown in Table 5.6.5.6 are smaller than those based on previous training parameters shown in Table 5.6.2.3, which resulted in smaller column sections. $SF^{(i=1\sim5)}$ of Design 3 obtained between 1.0 and 2.81 based on new training parameters shown in Table 5.6.5.6 are closer to 1.0 than 1.29–3.18 based on previous training parameters shown in Table 5.6.2.3, minimizing W_c better with new training parameters. Tables 5.6.5.9(a) and (b) demonstrate a selection of training parameters which significantly influence not only an accuracy of the optimized design results but also a quality of optimizations.

Table 5.6.5.10 describes the input and output parameters of Design 4 having fractions $w_{CI_c} : w_{CO_2} : w_{W_c} = 0.34 : 0.33 : 0.33$. The largest error of Design 4 based on the previous training parameters is 11.72% for $SF^{(2)}$ whereas that of Design 4 based on the new training parameters is 5.63% for $\varepsilon_{s,1}$, witnessing an improvement in design accuracies.

5.6.5.2 Design accuracies based on the two neural networks based on Tables 5.6.2.3 and 5.6.5.6 for the two Pareto curves

As illustrated in Figure 5.6.5.1, two Pareto curves based on 256 fractions are obtained by ANN-based Lagrange optimization having CI_c, CO_2, and W_c as three objective functions. Pareto curves are projected into X-Y, X-Z, and Y-Z planes to demonstrate the relationships CO_2 - W_c, CI_c - W_c, and CI_c - CO_2, respectively. Pareto curves in red color uses an ANN with previous training parameters whose number of hidden layers and hidden neurons shown in Table 5.6.2.3 whereas those in green color uses an ANN with new training parameters shown in Table 5.6.5.6.

Figure 5.6.5.1 indicates that a Pareto frontier (green dots) shown in blue boxes obtained by an ANN with new training parameters based on a number of hidden layers and neurons shown in Table 5.6.5.6 is scanty compared to one (red dots) obtained by an ANN with previous training parameters based on a number of hidden layers and neurons of Table 5.6.2.3.

Table 5.6.5.10 Comparison of Design 4 ($w_{CI_c} : w_{CO_2} : w_{W_c} = 0.34 : 0.33 : 0.33$) based on Tables 5.6.2.3 and 5.6.5.6

(a): Input Variables		
Input Variables of Design 4		
Fraction $w_{CI_c} : w_{CO_2} : w_{W_c} = 0.34 : 0.33 : 0.33$		
Parameters	Results Based on Previous Training (Table 5.6.2.3)	Results Based on Revised Training (Table 5.6.5.6)
1 b (mm)	1,438.8	1,251.4
2 h (mm)	1,204.7	1,125.0
3 $\rho_{s,x}$	0.0116	0.0114
4 $\rho_{s,y}$	0.0000	0.0131

Note: Inputs for ANN and structural mechanics-based software (*AutoRC2A*).

(*Continued*)

Table 5.6.5.10 (Continued) Comparison of Design 4 ($w_{CI_c} : w_{CO_2} : w_{W_c} = 0.34 : 0.33 : 0.33$) based on Tables 5.6.2.3 and 5.6.5.6

(b): *Output Parameters*

Output Parameters of Design 4 - Fraction: $W_{CI_c} : W_{CO_2} : W_{W_c} = 0.34 : 0.33 : 0.33$

Parameters	Best Training	AI-based Lagrange	Verification (*AutoRC2A*)	Error	Best Training	AI-based Lagrange	Verification (*AutoRC2A*)	Error
1 $SF^{(1)}$	*Five layers Sixty-four neurons*	1.7149	1.8081	5.44%	*six layers Sixty-four neurons*	1.5536	1.5400	0.87%
2 $SF^{(2)}$		1.0000	0.8828	11.72%		1.0000	1.0046	0.46%
3 $SF^{(3)}$		1.4107	1.3824	2.00%		1.3328	1.2903	3.19%
4 $SF^{(4)}$		1.0000	0.9482	5.18%		1.0000	1.0024	0.24%
5 $SF^{(5)}$		1.3373	1.3237	1.02%		1.1819	1.1663	1.31%
6 $\varepsilon_{s,1}$	*Six layers Sixty-four neurons*	0.0009	0.0009	1.87%	*Eight layers One hundred twenty-eight neurons*	0.0010	0.0010	5.63%
7 $\varepsilon_{s,2}$		0.0074	0.0079	6.80%		0.0051	0.0050	0.18%
8 $\varepsilon_{s,3}$		0.0030	0.0033	7.27%		0.0025	0.0025	1.01%
9 $\varepsilon_{s,4}$		0.0055	0.0055	1.04%		0.0041	0.0040	2.11%
10 $\varepsilon_{s,5}$		0.0021	0.0023	7.05%		0.0020	0.0021	5.00%
11 CI_c (KRW/m)	3L - 64N	**329,532**	**329,084**	0.14%	3L - 64N	**417,821**	**417,760**	0.01%
12 CO_2 (t-CO_2/m)	7L - 32N	**0.6862**	**0.6865**	0.04%	7L - 32N	**0.9159**	**0.9160**	0.01%
13 W_c (kN/m)	5L - 64N	**41.8333**	**41.8358**	0.01%	5L - 64N	**34.9776**	**34.9796**	0.01%
14 *b/h*	5L - 64N	1.1943	1.1943	0.00%	5L - 64N	1.1124	1.1123	0.00%

Outputs for ANN (*ModelLPs*).

Outputs for structural mechanics-based software (*AutoRC2A*).

Best training based on previous training results in Table 5.6.2.3.

Best training based on revised training results in Table 5.6.5.6.

Bold and red color values indicate optimized objective functions.

The reason can be the insufficient number of fractions. The scanty Pareto frontier can be improved by increasing a number of fractions.

It is also witnessed in Figure 5.6.5.1 that optimized designs having cost (CI_c) below 420,000 KRW/m and CO_2 below 0.95 $t-CO_2$ /m are noticeably different between the two Pareto curves. As can be seen in yellow box of Figures 5.6.5.1, and 5.6.5.2, two curves are found different, showing that the red Pareto curve based on previous training parameters provides lower cost (CI_c) and CO_2 emissions than the green Pareto curve based on new training parameters. Safety factors corresponding to five load pairs ($SF^{(1)} - SF^{(5)}$) calculated for optimized designs based on two neural networks are presented in y-axis as shown in Figure 5.6.5.3a. However, Figure 5.6.5.3 uncovers that optimized designs based on previous networks shown in Table 5.6.2.3 do not satisfy strength requirements as indicated by safety factors lower than 1.0 which are about 0.85, whereas all optimized designs based on new networks shown in Table 5.6.5.6 meet strength requirements as indicated by all safety factors greater than 1.0 in the specified ranges $CI_c \leq 420{,}000$ KRW / m and $CO_2 \leq 0.95\ t - CO_2\ /\ m$. Errors of safety factors ($SF^{(1)} - SF^{(5)}$) identified based on a formula $\text{Error}\ (\%) = \dfrac{SF^{(i)}_{(\text{ANN})} - SF^{(i)}_{(AutoRC2A)}}{SF^{(i)}_{(\text{ANN})}} \times 100\%; i = 1 \text{ to } 5$ for optimized designs based on new training parameters range from −6% to 6% which are lower than ones based on previous training parameters ranging from −5% to 15%, as shown in Figure 5.6.5.3b.

As illustrated in Figure 5.6.5.4, a Pareto curve with 256 fractions plotted by ANN-based Lagrange optimization based on the new training parameters shown in Table 5.6.2.3 and

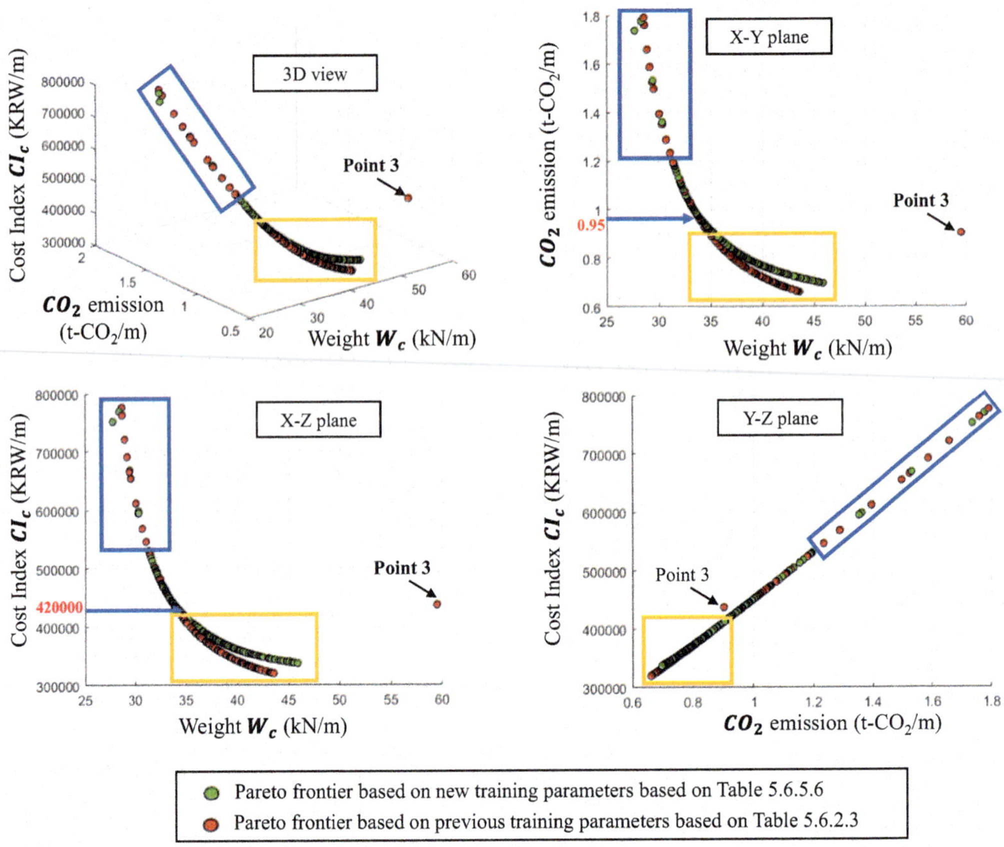

Figure 5.6.5.1 Comparison of Pareto curves based on new and previous training parameters.

previous training parameters shown in Table 5.6.5.6 are projected into X-Y, X-Z, and Y-Z plane to demonstrate the relationships CO_2 - W_c, CI_c - W_c, and CI_c - CO_2, respectively, which are also verified with 20,000 structural datasets. In general, a Pareto curve based on the new training parameters shown in Table 5.6.2.3 demonstrates better accuracies than that obtained based on the previous training parameters shown in Table 5.6.5.6.

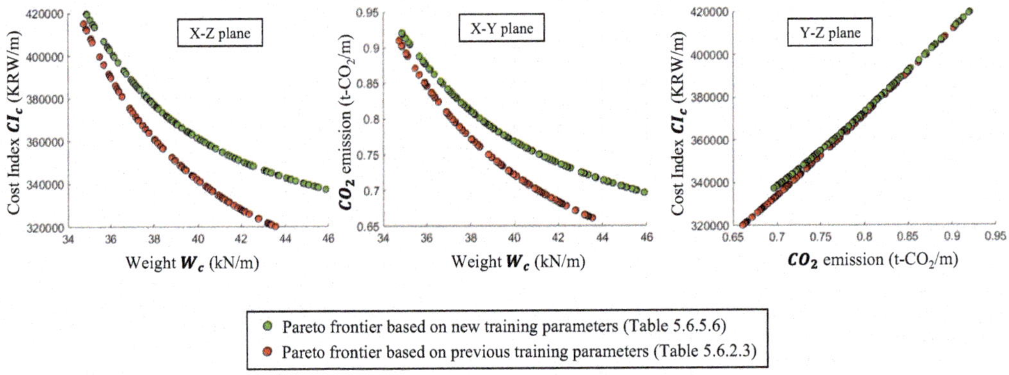

Figure 5.6.5.2 Comparison of Pareto curves based on new and previous training parameters for optimized designs in specified ranges of cost ($CI_c \leq 420,000$ KRW/m) and CO_2 ($CO_2 \leq 0.95\ t - CO_2$/m).

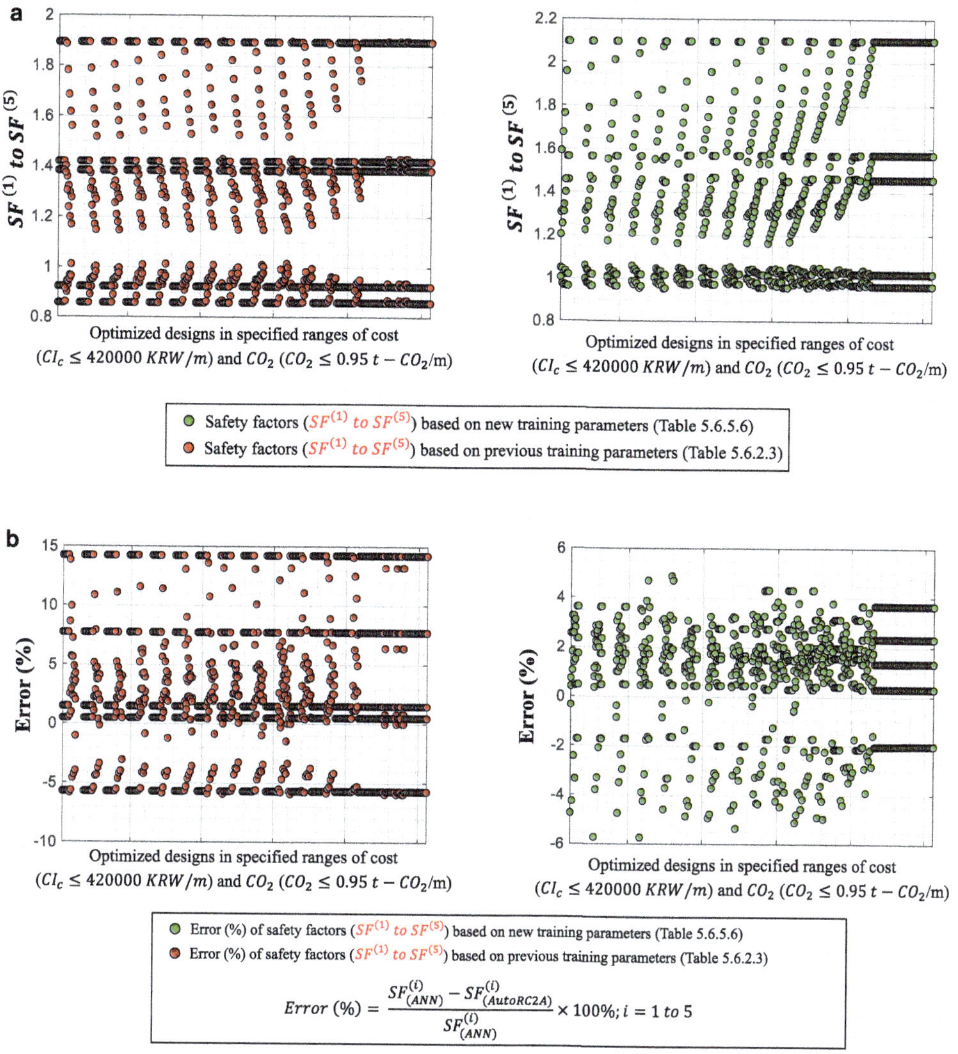

Figure 5.6.5.3 Comparison of safety factors ($SF^{(1)} - SF^{(5)}$) based on new and previous training parameters for optimized designs in specified ranges of cost ($CI_c \leq 420{,}000$ KRW / m) and CO_2 ($CO_2 \leq 0.95\ t - CO_2$/m). (a) Comparison of safety factors ($SF^{(1)} - SF^{(5)}$) calculated by structural mechanics-based software (*AutoRC2A*). (b) Comparison of safety factors ($SF^{(1)} - SF^{(5)}$) based on errors (%).

5.7 ANN-BASED LAGRANGE MULTI-OBJECTIVE OPTIMIZATION DESIGN OF RC BEAMS

This section designs RC beams based on an ANN-based Lagrange method, optimizing three-objective functions, cost (CI_b), estimated CO_2 emission, and beam weight (W_b), simultaneously. Readers are referred to Chapter 4 (Optimizing reinforced concrete beam cost based on ANN-based Lagrange functions) that this section (ANN-based Lagrange optimization for a design of RC beams based on multi-objective functions) is based on.

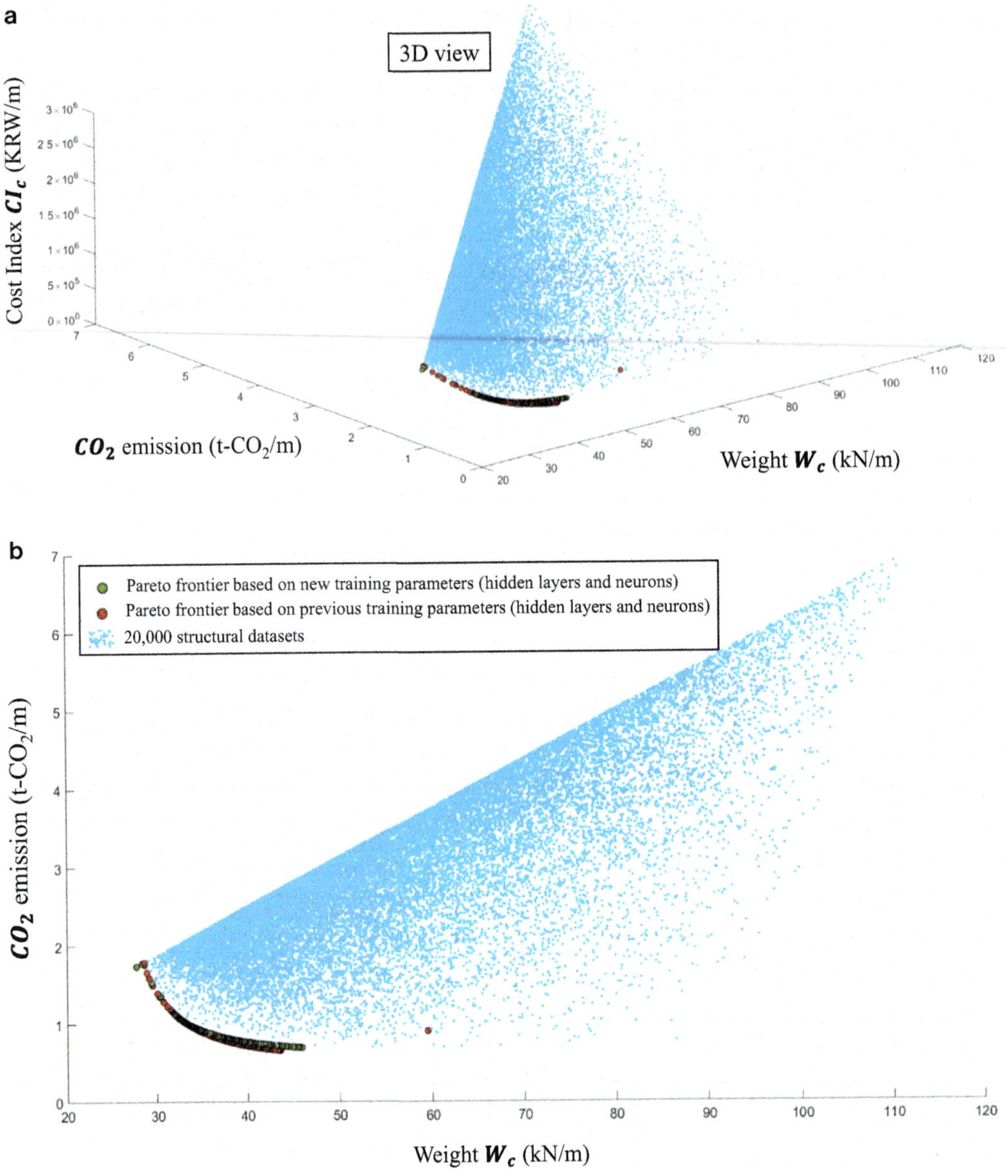

Figure 5.6.5.4 Verification of Pareto frontier based on new training parameters and previous training parameters by 20,000 structural datasets. (a) 3D view. (b) XY plane. (c) XZ plane. (d) YZ plane.

5.7.1 Design scenarios of doubly reinforced concrete beams

5.7.1.1 Selection of design parameters based on design criteria of doubly reinforced concrete beams

This section optimizes doubly reinforced concrete beams with two-end fixed boundary conditions shown in Figure 5.7.1.1 based on an ANN-based Lagrange optimization, where L denotes a span length, cross-sectional dimensions are also denoted by beam width (b) and beam depth (h), and $\rho_{r,c}$ and $\rho_{r,t}$ denote compressive and tensile reinforcement ratio (or rebar ratio),

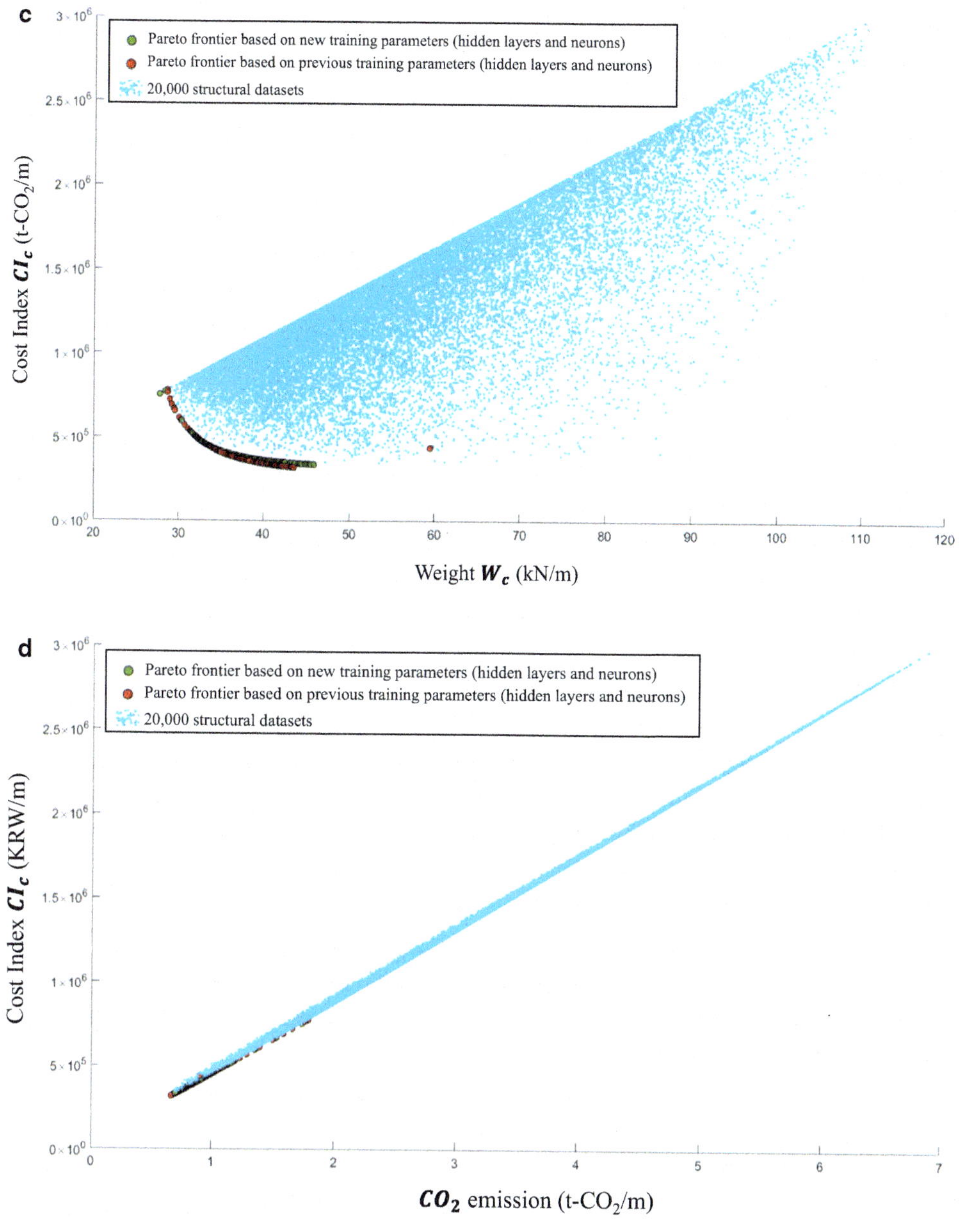

Figure 5.6.5.4 Verification of Pareto frontier based on new training parameters and previous training parameters by 20,000 structural datasets. (a) 3D view. (b) XY plane. (c) XZ plane. (d) YZ plane.

respectively. A beam section (b, h), and reinforcement details ($\rho_{r,c}$ and $\rho_{r,t}$) of an RC beam must be designed to satisfy requirements in terms of strength, serviceability, and reinforcement detailing specified in the design code ACI 318-19 (ACI Committee, 2019).

In this section, design flexural moment strength (ϕM_n) is required to be equal to factored moment (M_u), which is a combination of dead load (M_D) and live load (M_L) meeting strength criteria. To avoid brittle behavior in the case of an overload, a reinforcement strain

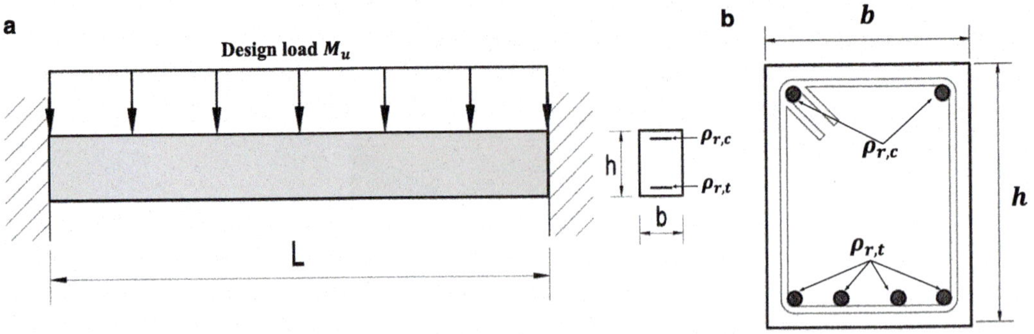

Figure 5.7.1.1 A doubly reinforced concrete beam.

is limited at a design strength. ε_{rt} and ε_{rc} are tensile and compressive rebar strains at a concrete strain of 0.003, respectively. Serviceability is reflected in Δ_{imme} and Δ_{long}, referring to immediate deflections and time-independent deflections, which are limited to $L/360$ and $L/240$, respectively. It is worth noting that the cracking moment M_{cr}, which must be less than $\phi M_n/1.2$ to control a minimum flexural reinforcement according to ACI 318-19 (ACI Committee, 2019), is investigated when calculating Δ_{imme}. A design of a longitudinal compressive rebar and tensile rebar in this study includes a calculation of compressive rebar ratio ($\rho_{r,c}$) and tensile rebar ratio ($\rho_{r,t}$) based on their rebar areas $A_{r,c}$ and $A_{r,t}$, respectively.

For a rectangular cross-section of doubly RC beams, a reinforcement ratio is calculated as the ratio of rebar area to cross-sectional gross area, which is equal to $b \times h$. Rebar details such as maximum and minimum spacing are also determined, where a clear concrete cover c of 40 mm, reinforcement of *03D529*, and aggregates of *03D520* are used in this section. Nine inputs are selected for the design criteria, which include five parameters to define a beam section (L, b, h, $\rho_{r,c}$, $\rho_{r,t}$), two material properties (f'_c,f_y) for concrete strength and rebar yield strength, respectively, and two externally imposed moments (M_D, M_L). Seven output parameters including ϕM_n, M_u, M_{cr}, ε_{rt}, ε_{rc}, Δ_{imme}, and Δ_{long} are calculated from nine inputs. A forward design of doubly RC beams requires determining seven outputs from nine given input parameters. Table 5.7.1.1 describes input and output parameters with their nomenclature while Table 5.7.1.2 presents design criteria according to ACI 318-19 (ACI Committee, 2019).

5.7.1.2 Selection of objective functions

In this design example, a design of doubly RC beams is optimized based on three objective functions including overall construction and material cost (CI_b), beam weight (W_b), and CO_2 emissions, which are selected based on engineers', contractors', and the government's interests. The overall cost or cost index (CI_b) of a doubly RC beam is calculated based on Korean unit cost for concrete and rebar, as defined in Table 5.3.1.1 (Hong et al., 2010) even if readers can use their own units. CO_2 emissions and the weight of concrete, rebar of a reinforced concrete (RC) frame are also calculated using Table 5.3.1.1.

5.7.1.3 An optimization scenario

A design of a doubly RC beam is performed by minimizing CI_b, CO_2, and W_b simultaneously. Beam sections (b, h, $\rho_{r,c}$, $\rho_{r,t}$) are designed for a fix–fix beam having a beam span L = 9,000 mm. Beam depth h is designed in a range of 400 – 1,500 mm, while a beam section

Table 5.7.1.1 Nomenclature of Design Parameters

No.	*Notation*	*Nomenclature*
	Inputs When Generating Large Structural Datasets	
1	L (mm)	Beam span
2	h (mm)	Beam depth
3	b (mm)	Beam width
4	f_c' (MPa)	Rebar yield strength
5	f_y (MPa)	Concrete strength
6	$\rho_{r,t}$	Tensile rebar ratio calculated based on tensile rebar area $A_{r,t}$ $\left(\rho_{r,t} = A_{r,t}/(b \times h)\right)$
7	$\rho_{r,c}$	Tensile rebar ratio calculated based on tensile rebar area $A_{r,c}$ $\left(\rho_{r,c} = A_{r,c}/(b \times h)\right)$
8	M_D (kN.m)	Moment due to dead load
9	M_L (kN.m)	Moment due to live load
	Outputs When Generating Large Structural Datasets	
10	ϕM_n (kN.m)	Nominal flexural strength excluding moment due to self-weight
11	M_u (kN.m)	Factored moment considering as a combination of dead load and live load
12	M_{cr} (kN.m)	Cracking moment
13	$\varepsilon_{r,t}$	Tensile rebar strain at a concrete strain of 0.003
14	$\varepsilon_{r,c}$	Compressive rebar strain at a concrete strain of 0.003
15	Δ_{imme} (mm)	Immediate deflection due to live load
16	Δ_{long} (mm)	Time-dependent (or long-term) deflection
17	CI_b (KRW/m)	Cost index of a beam including material and construction cost
18	CO_2 (t-CO_2/m)	CO_2 emissions
19	W_b (kN/m)	Beam weight

Table 5.7.1.2 Design Criteria for a Design of Ductile RC Beams According to ACI 318-19

Formula	*Notation*	*Referred Section*
Strength Requirements Criteria		
$\phi M_n \geq M_u$	Flexural strength requirement	9.5.1.1(a)
$M_u = 1.2M_D + 1.6M_L$	Load combination and load factors	5.3.1
$\varepsilon_{rt} \geq \left(\frac{f_y}{E_s} + 0.003\right)$	A limitation of reinforcement strain to avoid brittle failure	9.3.3.1
Serviceability Criteria		
$\Delta_{imme} \leq L/360$	A limitation of immediate deflection	24.2.2
$\Delta_{long} \leq L/240$	A limitation of time-dependent deflection	24.2.2
Reinforcement Details Criteria		
$\phi M_n \geq 1.2M_{cr}$	A requirement of minimum flexural reinforcement	9.6.1.2 and R9.6.2.1
$s_{min} \geq \max\left(25;\ 29;\ \frac{4}{3} \times 20\right)$	A requirement of minimum spacing of longitudinal rebar	25.2.1
$s_{max} \leq \min\left(380\frac{280}{f_y} - 2.5c;\ 300\frac{280}{f_y}\right)$	A requirement of maximum spacing of longitudinal rebar	24.3.2

has an aspect ratio b/h within 0.25 – 0.8. The concrete strength and rebar yield strength used in this example are $f_c^{'} = 40$ MPa and $f_y = 550$ MPa, respectively. Dead load M_D and live load M_L are given as 800 and 400 kN.m, respectively. The design of ductile doubly RC beams follows ACI-based requirements in terms of strength, serviceability, and rebar details, as described in Table 5.7.1.2. The conventional design parameters in Table 5.7.1.1 contain nine input parameters and ten output parameters when generating large datasets based on conventional design. Optimized beam parameters (b, h, $\rho_{r,c}$, $\rho_{r,t}$) are then determined on an output-side during an optimization design of a doubly RC beam, resulting in minimizing CI_b, CO_2, and W_b. Input parameters and output parameters in beam optimization will be discussed in Section 5.7.2. Table 5.7.1.3 summarizes an optimization scenario of a design of a doubly RC beam presented in this section.

5.7.2 Five steps to optimize a design of RC beams with which CI_b, CO_2 and W_b are minimized

This section describes five steps minimizing cost, beam weight, and CO_2 emissions simultaneously for doubly RC beams. A Pareto frontier optimizing a doubly RC beam design based on multiple-objective functions is obtained in five steps as shown in Figure 5.2.3.1. In Figure 5.2.3.1, Step 1 describes how to establish an ANN to generate ANN-based objective functions for CI_b, W_b, and CO_2 shown in Equations (5.7.1.1-1)–(5.7.1.1-3). A topology of an ANN used in this example is described in Figure 5.7.2.1. Step 2 introduces an optimization scenario. Equality constraints $c(\mathrm{x})$ and inequality constraints $v(\mathrm{x})$ shown in Table 5.7.2.4 are imposed in Step 2 based on the code requirements and other interests from architects and engineers, for instance. ANN-based objective functions derived in Step 1 and equality constraints $c(\mathrm{x})$, inequality constraints $v(\mathrm{x})$ formulated in Step 2 are used as input parameters of Step 3 for ANN-based Lagrange algorithm of single objective function shown in Equations (5.7.1.2-1)–(5.7.1.2-3). The optimized parameters b, h, ρ_{rt}, ρ_{rc} are obtained on an output-side using the Newton–Raphson iteration, described in Figure 5.7.2.2 which shows a flow of beam designs based on the proposed algorithm established in Figure 5.2.3.1.

The optimized design parameters b, h, ρ_{rt}, ρ_{rc} and prescribed parameters L, $f_c^{'}$, f_y, M_D, M_L are input parameters in the ANN model and structural mechanics-based software to calculate corresponding design parameters ϕM_n, M_u, M_{cr}, ε_{rt}, ε_{rc}, Δ_{imme}, Δ_{long}, CI_b, CO_2, and W_b on an output side. The optimized input parameters b, h, ρ_{rt}, ρ_{rc}, L, $f_c^{'}$, f_y, M_D, M_L and output design parameters ϕM_n, M_u, M_{cr}, ε_{rt}, ε_{rc}, Δ_{imme}, Δ_{long}, CI_b, CO_2, W_b are the set of design stationary points optimizing UFO which yields a Pareto frontier. Step 4 integrates single objective functions (CI_b, W_b, and CO_2 shown in Equations (5.7.1.1-1)–(5.7.1.1-3)

Table 5.7.1.3 An Optimization Design Scenario of a Doubly RC Beam

	(a): Prescribed Parameters	
Load	Dead load	$M_D = 800$ kN.m
	Live load	$M_L = 400$ kN.m
Beam span		$L = 9{,}000$ mm
Material properties	Concrete compressive strength	$f_c^{'} = 40$ MPa
	Rebar yield strength	$f_y = 550$ MPa
	(b): Constrained Conditions	
Beam section	Limitation of beam depth	400 mm $\leq h \leq$ 1,500 mm
	Aspect ratio	$0.25 \leq b/h \leq 0.8$
Design criteria	Specified in Table 2	

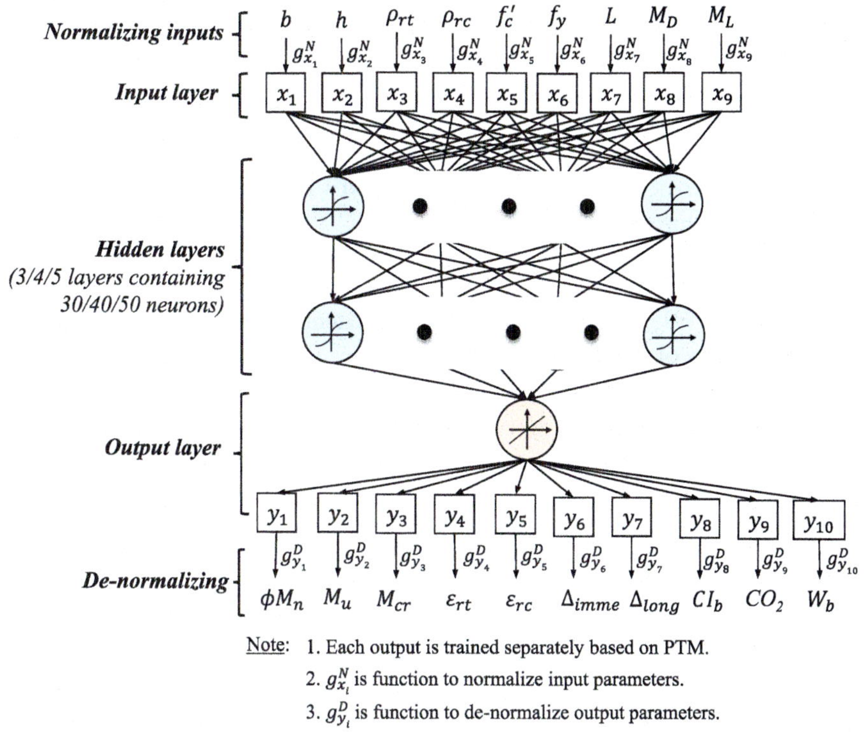

Figure 5.7.2.1 A topology of an ANN for a design of doubly RC beam.

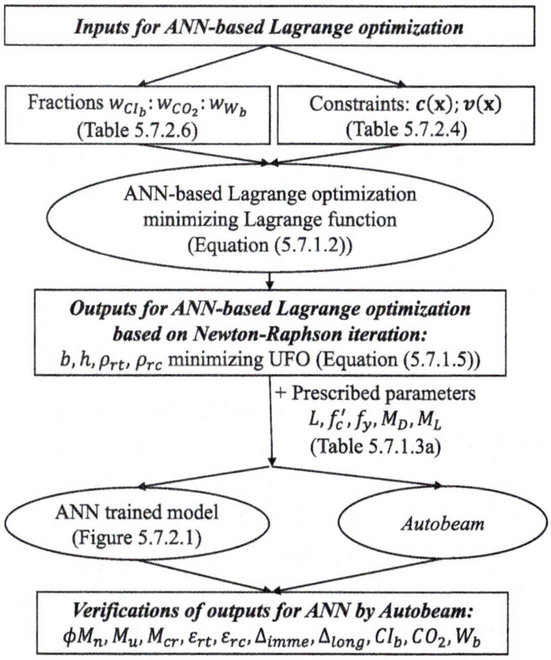

Figure 5.7.2.2 Flow of designs using ANN-based Lagrange optimization algorithm.

derived in Step 1 into multiple objective functions (UFO) shown in Equation (5.7.1.5-1) to formulate Lagrange function of multiple objective functions shown in Equation (5.7.1.5-2). Stationary points of Lagrange function of UFO under KKT conditions are the optimized design parameters to identify a Pareto frontier. The Pareto frontier based on UFO, which is obtained by UFO-based ANN, is verified by structural mechanics-based software *Autobeam* where ϕM_n, M_u, M_{cr}, ε_{rt}, ε_{rc}, Δ_{imme}, Δ_{long}, CI_b, CO_2, and W_b are validated.

$\text{Output}^{\text{ANN}}$ and the structural mechanics-based software denoted as $\text{Output}^{\text{Check}}$ are used to calculate Error $= \dfrac{\text{Output}^{\text{ANN}} - \text{Output}^{\text{Check}}}{\text{Output}^{\text{ANN}}}$. The design accuracies of design points in the Pareto frontier are presented in Section 5.7.4. Three design points in the Pareto frontier are presented in Section 5.7.3, indicating that the design accuracies are as large as 5.5% which validates the accuracies of the proposed ANN.

5.7.2.1 Step 1–Deriving ANNs

Step 1.1: Selecting random design ranges to generate large datasets

This step derives three objective functions CI_b, CO_2, and W_b shown in Equations (5.7.1.1-1) to (5.7.1.1–3) for a design of doubly RC beams based on an optimization scenario described in Section 5.7.1.3.

Table 5.7.1.1 shows the nomenclatures of 19 input and output parameters generated in 100,000 datasets. Large datasets of ductile doubly RC beams are generated by randomly varying input parameters (L, b, h, $\rho_{r,c}$, $\rho_{r,t}$, f_c', f_y, M_D, M_L) within their design ranges described in Table 5.7.2.1. Corresponding outputs (ϕM_n, M_u, M_{cr}, ε_{rt}, ε_{rc}, Δ_{imme}, Δ_{long}, CI_b, CO_2, and W_b) are obtained using *Autobeam*, structural mechanics-based software developed by Nguyen and Hong (2019), which is based on a strain-compatibility-based algorithm. It is worth noting that the design criteria shown in Table 5.7.1.2 are included in *Autobeam*. To consider a wide range of beam designs, a beam length L randomly varies from 8 to 12 m. Beam depth h is randomly selected in a range of 0.4 – 1.5 m, while beam width is generated to satisfy an aspect ratio b/h of 0.25 – 0.8. Load parameters M_D and M_L are generated within 250 – 1,500 kN.m and 125 – 750 kN.m, respectively. Concrete compressive strength and rebar yield strength are also randomly selected in their ranges, such as 30 – 50 MPa and 400 – 600 MPa, respectively.

Step 1.2: Generating large datasets used to train ANNs using structural mechanics-based software "*Autobeam*"

The 100,000 datasets used to train ANNs, derived in Step 1.3, are provided in Table 5.7.2.2, where nonnormalized datasets and datasets normalized between –1 and 1 are shown.

Step 1.3: Training ANNs

Table 5.7.2.1 Design Range to Generate 100,000 Data Used for Training

Parameter		*Lower Boundary*	*Upper Boundary*
Beam length	L (mm)	8,000	12,000
Beam depth	h (mm)	400	1,500
Sectional aspect ratio	b/h	0.25	0.8
Concrete compressive strength	f_c' (MPa)	30	50
Rebar yield strength	f_y (MPa)	500	600
Dead load	M_D (kN·m)	10	20,000
Live load	M_L (kN·m)	0	10,000

Table 5.7.2.2 100,000 Datasets to Train ANNs

(a): Input Parameters When Generating Large Structural Datasets									
	Nine Inputs for ANN and *Autobeam*								
	L (mm)	*b (mm)*	*h (mm)*	$f_c^{'}$ *(MPa)*	f_y *(MPa)*	ρ_{rt}	ρ_{rc}	M_D *(kN.m)*	M_L *(kN.m)*
Non-normalized	8,350	370	580	37	407	0.0083	0.0013	89.1	27.3
	10,250	235	645	38	422	0.0199	0.0072	219.2	127.2
	8,750	365	815	43	462	0.0072	0.0019	143.5	18.0
	…	…	…	…	…	…	…	…	…
	11,150	195	515	48	401	0.0128	0.0033	23.6	17.6
	9,250	485	635	43	428	0.0076	0.0026	200.5	199.3
	8,400	420	1,500	49	488	0.0108	0.0044	1,105.6	30.4
Min	8,000	125	400	30	400	0.0026	0.00001	10.0	0.0
Max	12,000	1,200	1,500	50	600	0.0566	0.0283	20,000	10,000
Mean	10,003	525	1,012	40	482	0.0118	0.0035	1,275.6	460.1
Normalized from −1 to 1	−0.825	−0.088	−0.050	−0.300	−0.930	−0.790	−0.908	−0.991	−0.994
	0.125	−0.033	0.225	−0.200	−0.780	−0.360	−0.494	−0.976	−0.974
	−0.625	−0.301	−0.700	0.300	−0.380	−0.831	−0.864	−0.985	−0.996
	…	…	…	…	…	…	…	…	…
	0.575	0.051	0.050	−0.990	0.800	−0.815	−0.820	−0.998	−0.996
	−0.375	−0.179	0.375	−0.720	0.300	−0.699	−0.687	−0.978	−0.959
	−0.800	0.014	0.850	−0.120	0.900	−0.951	−0.982	−0.874	−0.994
Min	−1	−1	−1	−1	−1	−1	−1	−1	−1
Max	1	1	1	1	1	1	1	1	1
Mean	0.0015	−0.2555	0.0233	−0.0024	−0.1799	−0.6591	−0.7546	−0.8546	−0.9050

(*Continued*)

Table 5.7.2.2 (Continued) 100,000 Datasets to Train ANNs

(b): Output Parameters When Generating Large Structural Datasets										
	Ten Outputs for ANN and *Autobeam*									
	ϕM_n *(kN.m)*	M_u *(kN.m)*	M_{cr} *(kN.m)*	ε_{rt}	ε_{rc}	Δ_{imme} *(mm)*	Δ_{long} *(mm)*	CI_b *(KRW/m)*	CO_2*(t-*CO_2*/m)*	W_b*(kN/m)*
Non-normalized	256.7	150.6	72.2	0.0098	0.0026	23.2	34.8	34,322.9	0.073	4.781
	760.8	466.5	109.8	0.0376	0.0018	28.5	42.7	68,752.1	0.155	5.625
	259.6	201.1	71.6	0.0463	0.0015	24.3	36.5	26,468.3	0.056	3.758
	...	...	...	...	...	...	...	...	...	...
	121.6	56.5	41.8	0.0230	0.0004	31.0	46.5	18,900.4	0.038	2.495
	709.9	559.5	161.8	0.0361	–0.0009	25.7	38.5	69,622.3	0.148	7.776
	1,655.7	1,375.4	741.0	0.0328	–0.0006	23.3	35.0	92,562.8	0.172	15.383
Min	41.4	17.4	19.6	0.0050	–0.0014	22.2	33.3	7,880.2	0.016	1.528
Max	34,555.7	26,657.0	3,328.2	0.0625	0.0027	33.3	50.0	1,059,751.8	2.452	50.990
Mean	3,014.7	2,266.9	550.1	0.0225	0.0017	1.6	13.3	1,339,83.7	0.288	14.553
Normalized from –1 to 1	–0.988	–0.988	–0.968	–0.457	0.044	–0.820	–0.731	–0.9497	–0.953	–0.868
	–0.958	–0.960	–0.946	–0.816	0.577	–0.409	–0.389	–0.8843	–0.886	–0.834
	–0.987	–0.988	–0.969	–0.367	0.179	–0.888	–0.608	–0.9647	–0.968	–0.910
	...	...	...	...	...	...	...	...	...	...
	–0.995	–0.997	–0.987	–0.207	–0.302	–0.684	–0.698	–0.979	–0.982	–0.961
	–0.961	–0.959	–0.914	–0.447	0.246	–0.450	–0.634	–0.883	–0.892	–0.747
	–0.906	–0.898	–0.564	0.510	0.367	–0.996	–0.939	–0.839	–0.872	–0.440
Min	–1	–1	–1	–1	–1	–1	–1	–1	–1	–1
Max	1	1	1	1	1	1	1	1	1	1
Mean	–0.8277	–0.8311	–0.6793	–0.3920	0.5021	–0.7569	–0.6331	–0.7602	–0.7772	–0.4733

Table 5.7.2.3 Training Accuracies of RC Beams Based on Nine Inputs and Ten Outputs

Nine Inputs ($L, b, h, \rho_{st}, \rho_{sc}, f_c', f_y, M_D, M_L$) Ten Outputs ($\phi M_n, M_u, M_{cr}, \varepsilon_{rt}, \varepsilon_{rc}, \Delta_{imme}, \Delta_{long}, CI_b, CO_2, W_b$)					
PTM – 100,000 Data – 50,000 Suggested Epochs – *tanh* Activation Function					
Output	*Layers*	*Neurons*	*Best Epoch*	*Test MSE*	*R at Best Epoch*
ϕM_n	3	40	19,303	4.23E-07	1.000
M_u	3	50	12,984	3.74E-07	1.000
M_{cr}	3	40	19,165	7.67E-07	1.000
ε_{rt}	3	50	44,069	3.55E-06	1.000
ε_{rc}	3	30	16,335	8.79E-06	1.000
Δ_{imme}	3	40	34,412	1.29E-04	0.999
Δ_{long}	3	30	21,010	4.30E-05	1.000
CI_b	3	30	30,659	2.44E-08	1.000
CO_2	3	30	20,792	2.85E-08	1.0000
W_b	3	30	12,031	3.32E-08	1.0000

Training parameters are selected based on the complexity of large datasets and the availability of computing hardware. This section utilizes ANNs similar to ones based on Network #2 presented in Chapter 4 [refer to Section 4.2]. The neural network is trained on 100,000 datasets using three layers with 30, 40, and 50 neurons. Training accuracies of RC beams based on nine inputs and ten outputs are presented in Table 5.7.2.3.

$$\underbrace{f_{CI_2}^{ANN}(\mathbf{x})}_{[1\times1]} = g_{CI_b}^{D}\left(f_{\text{linear}}\left(\underbrace{\boldsymbol{\omega}_{CI_b}^{(\text{out})}}_{[1\times30]} f_{\tanh}\left(\underbrace{\boldsymbol{\omega}_{CI_b}^{(3)}}_{[30\times30]}\cdots f_{\tanh}\left(\underbrace{\boldsymbol{\omega}_{CI_b}^{(1)}}_{[30\times9]}\underbrace{g^{N}(\mathbf{x})}_{[9\times1]} + \underbrace{\mathbf{b}_{CI_b}^{(1)}}_{[30\times1]}\right)\cdots + \underbrace{\mathbf{b}_{CI_b}^{(3)}}_{[30\times1]}\right) + \underbrace{\mathbf{b}_{CI_b}^{(\text{out})}}_{[1\times1]}\right)\right) \tag{5.7.1.1-1}$$

$$\underbrace{f_{CO_2}^{ANN}(\mathbf{x})}_{[1\times1]} = g_{CO_2}^{D}\left(f_{\text{linear}}\left(\underbrace{\boldsymbol{\omega}_{CO_2}^{(\text{out})}}_{[1\times30]} f_{\tanh}\left(\underbrace{\boldsymbol{\omega}_{CO_2}^{(3)}}_{[30\times30]}\cdots f_{\tanh}\left(\underbrace{\boldsymbol{\omega}_{CO_2}^{(1)}}_{[30\times9]}\underbrace{g^{N}(\mathbf{x})}_{[9\times1]} + \underbrace{\mathbf{b}_{CO_3}^{(1)}}_{[30\times1]}\right)\cdots + \underbrace{\mathbf{b}_{CO_2}^{(3)}}_{[30\times1]}\right) + \underbrace{\mathbf{b}_{CO_2}^{(\text{out})}}_{[1\times1]}\right)\right) \tag{5.7.1.1-2}$$

$$\underbrace{f_{W_c}^{ANN}(\mathbf{x})}_{[1\times1]} = g_{W_b}^{D}\left(f_{\text{linear}}\left(\underbrace{\boldsymbol{\omega}_{W_b}^{(\text{out})}}_{[1\times30]} f_{\tanh}\left(\underbrace{\boldsymbol{\omega}_{W_b}^{(3)}}_{[30\times30]}\cdots f_{\tanh}\left(\underbrace{\boldsymbol{\omega}_{W_b}^{(1)}}_{[30\times9]}\underbrace{g^{N}(\mathbf{x})}_{[9\times1]} + \underbrace{\mathbf{b}_{W_b}^{(1)}}_{[30\times1]}\right)\cdots + \underbrace{\mathbf{b}_{W_b}^{(3)}}_{[30\times1]}\right) + \underbrace{\mathbf{b}_{W_b}^{(\text{out})}}_{[1\times1]}\right)\right) \tag{5.7.1.1-3}$$

5.7.2.2 Step 2–Defining MOO problems

Step 2.1: Deriving objective functions

In Step 1, functions of three objectives CI_b, CO_2, and W_b denoted as $f_{CI_b}(\mathbf{x}), f_{CO_2}(\mathbf{x})$, and $f_{W_b}(\mathbf{x})$, respectively, are derived using ANNs as shown in Equations (5.7.1.1-1)–(5.7.1.1-3) where $\mathbf{x}$ consists of five preassigned input parameters (L, f_c', f_y, M_D, and M_L) shown in Table 5.7.1.3(a) and four initial trial input variables ($b, h, p_{r,c}$, and $p_{r,t}$), which are to be determined based on the Newton–Raphson iteration. $g^N(\mathbf{x})$ is used to normalize inputs so that it has a dimension of 9×1. These normalized inputs are then interconnected to hidden layers through interior weight and bias matrices before being sent to an output layer.

$\omega_{CI_b}^{(1)}$ is a weight matrix to transfer input parameters $\mathbf{x}$ to the neurons of first hidden layer, making its dimension 30×9, representing 30 neurons and nine inputs. $\mathbf{b}_{CI_b}^{(1\sim3)}$ is a bias matrix added in hidden Layers 1–3, making their dimensions of 30 neurons × one layer. $\omega_{CI_b}^{(2)}$ and $\omega_{CI_b}^{(3)}$ are weight matrices of interior layers (Hidden layers 2 and 3) so that they are a square matrix of 30×30. $\omega_{CI_b}^{(out)}$ and $\mathbf{b}_{CI_b}^{(out)}$ are matrices of weight and bias at an output layer so that their dimensions are 1×30 and 1×1, respectively.

Normalized output parameters are denormalized based on a de-normalizing function $g_{CI_b}^D$. The matrix dimensions of the objective functions vary because the dimension of training parameters can change to best train objective functions. As shown in Table 5.7.2.3, dimensions of weight and bias matrices in the three objective functions are the same because training parameters of three layers with 30 neurons are used to train three objective functions when yielding the best training accuracies for the three objective functions.

Step 2.2: Formulating constrained conditions based on equalities and inequalities

Six equalities and eleven inequalities are formulated in Table 5.7.2.4 to impose constrained conditions based on design criteria and engineers' needs.

a. Equality constraints: $c(\mathbf{x}) = [c_1(\mathbf{x}), c_2(\mathbf{x}), c_3(\mathbf{x}), c_4(\mathbf{x}), c_5(\mathbf{x}), c_6(\mathbf{x})]^T$

Five equalities $c_{1\sim5}(\mathbf{x})$ represent five preassigned parameters = 9,000 **mm**, $f_c' = 40$ **MPa**, $f_y = 550$ **MPa**, $M_D = 1{,}200$ **kN.m**, $M_L = 600$ **kN.m**. Factored moment $M_U = 1.2M_D + 1.6M_L$ is formulated as another equality $c_6(\mathbf{x})$.

b. Inequality constraints: $v(\mathbf{x}) = [v_1(\mathbf{x}), v_2(\mathbf{x}), \ldots, v_{11}(\mathbf{x})]^T$

Strength requirements $\phi M_n \geq M_u$ and $\phi M_n \geq 1.2M_{cr}$ imposed by the ACI 318-19 (ACI Committee, 2019) are formulated into two inequalities $v_1(\mathbf{x}): \phi M_n - M_u \geq 0$ and $v_2(\mathbf{x}): \phi M_n - 1.2M_{cr} \geq 0$, respectively. A design range of beam depth h that is within 0.4–1.5 m is imposed by $v_3(\mathbf{x})$ and $v_4(\mathbf{x})$. The beam width b is constrained by an aspect ratio $b/h = 0.25 - 0.8$ imposed by $v_5(\mathbf{x})$ and $v_6(\mathbf{x})$. Inequalities $v_{9\sim11}(\mathbf{x})$ also impose serviceability requirements described in Table 5.7.1.2.

Table 5.7.2.4 Equality and Inequality Constraints for Optimization Design of RC Beams (Refer to Table 4.6.1.1(b))

Equality Constraints c(x)	*Inequality Constraints v(x)*	
$c_1(\mathbf{x}): L - 9{,}000 = 0$	$v_1(\mathbf{x}): \phi M_n - M_u \geq 0$	$v_7(\mathbf{x}): -\rho_{rc} + \rho_{rt}/2 \geq 0$
$c_2(\mathbf{x}): f_c' - 40 = 0$	$v_2(\mathbf{x}): \phi M_n - 1.2M_{cr} \geq 0$	$v_8(\mathbf{x}): \rho_{rc} - \rho_{rt}/400 \geq 0$
$c_3(\mathbf{x}): f_y - 550 = 0$	$v_3(\mathbf{x}): h - 400 \geq 0$	$v_9(\mathbf{x}): \varepsilon_{rt} - \left(\frac{f_y}{E_s} + 0.003\right) \geq 0$
$c_4(\mathbf{x}): M_D - 1{,}200 = 0$	$v_4(\mathbf{x}): -h + 1{,}500 \geq 0$	$v_{10}(\mathbf{x}): -\Delta_{imme} + L/360 \geq 0$
$c_5(\mathbf{x}): M_L - 600 = 0$	$v_5(\mathbf{x}): b - 0.25h \geq 0$	$v_{11}(\mathbf{x}): -\Delta_{long} + L/240 \geq 0$
$c_6(\mathbf{x}): M_u = 1.2M_D + 1.6M_L$	$v_6(\mathbf{x}): -b + 0.8h \geq 0$	

5.7.2.3 Step 3–Optimization based on a single-objective function

This step is required to maximize and minimize each objective function to define their boundaries for normalization because each normalized objective function is multiplied by a fraction to derive UFO. Single-objective function (CI_b,CO_2 and W_b) of RC beams is separately minimized and maximized based on the Lagrange functions of each objective function shown in Equations (5.7.1.2-1)–(5.7.1.2-3) which satisfy equalities and inequalities shown in Table 5.7.2.4, yielding Table 5.7.2.5.

How to obtain minimum and maximum CO_2 and W_b is similar to optimizing beam cost (CI_b) described in Chapter 4 [Section 4.2]. Readers also refer to STEP 3 of Section 5.2.3 for the review of ANN-based Lagrange optimization based on single objectives. Maximum and minimum of CI_b, CO_2 and W_b are used to normalize objective functions shown in Equations (5.7.1.3-1)–(5.7.1.3-3), described in Step 4. Equations (5.7.1.2-1)–(5.7.1.2-3) formulate Lagrange functions based on three single-objective functions $f_{CI_b}^{ANN}(\mathbf{x})$, $f_{CO_2}^{ANN}(\mathbf{x})$, and $f_{W_b}^{ANN}(\mathbf{x})$ shown in Equations (5.7.1.1-1)–(5.7.1.1-3).

$$\mathcal{L}_{CI_b}(\mathbf{x},\boldsymbol{\lambda}_c,\boldsymbol{\lambda}_v) = f_{CI_b}^{ANN}(\mathbf{x}) - \boldsymbol{\lambda}_c^T c(\mathbf{x}) - \boldsymbol{\lambda}_v^T v(\mathbf{x}) \quad (5.7.1.2\text{-}1)$$

$$\mathcal{L}_{CO_2}(\mathbf{x},\boldsymbol{\lambda}_c,\boldsymbol{\lambda}_v) = f_{CO_2}^{ANN}(\mathbf{x}) - \boldsymbol{\lambda}_c^T c(\mathbf{x}) - \boldsymbol{\lambda}_v^T v(\mathbf{x}) \quad (5.7.1.2\text{-}2)$$

$$\mathcal{L}_{W_b}(\mathbf{x},\boldsymbol{\lambda}_c,\boldsymbol{\lambda}_v) = f_{W_b}^{ANN}(\mathbf{x}) - \boldsymbol{\lambda}_c^T c(\mathbf{x}) - \boldsymbol{\lambda}_v^T v(\mathbf{x}) \quad (5.7.1.2\text{-}3)$$

5.7.2.4 Step 4–Formulating UFO

Step 4.1: Objective functions shown in Equations (5.7.1.1-1)–(5.7.1.1-3) are normalized as shown in Equations (5.7.1.3-1)–(5.7.1.3-3).

$$f_{CI_b}^{N}(\mathbf{x}) = \frac{f_{CI_b}^{ANN}(\mathbf{x}) - 75{,}359}{552{,}730 - 75{,}359} \quad (5.7.1.3\text{-}1)$$

$$f_{CO_2}^{N}(\mathbf{x}) = \frac{f_{CO_2}^{ANN}(\mathbf{x}) - 0.159}{1.235 - 0.159} \quad (5.7.1.3\text{-}2)$$

$$f_{W_b}^{N}(\mathbf{x}) = \frac{f_{W_b}^{ANN}(\mathbf{x}) - 5.556}{31.721 - 5.556} \quad (5.7.1.3\text{-}3)$$

Step 4.2: Defining fractions and UFO

Fractions of three objective functions w_{CI_b}, w_{CO_2}, and w_{W_b} are generated based on the *linspace* function in MATLAB (MathWorks, 2020), which is a function of linear spacing to assure that they are equally distributed between 0 and 1. Equation (5.7.1.4-1) defines 19

Table 5.7.2.5 Optimized RC Beams Based on Separate Objective Functions

Objectives	*Minimum*	*Maximum*
CI_b (KRW/m)	75,359	552,730
CO_2 (t-CO_2/m)	0.159	1.235
W_b (kN/m)	5.556	31.721

fractions of CI_b $\left(w_{CI_b}\right)$ that are equally spaced between 0 and 1 so that their spacing is equal to 1/18. In Equation (5.7.1.4-2), 19 fractions of CO_2 $\left(w_{CO_2}\right)$ are, then, equally spaced between 0 and w_{CI_b} obtained in Equation (5.7.1.4-1). Fractions of W_b $\left(w_{W_b}\right)$ are calculated with respect to w_{CI_b} and w_{CO_2} using Equation (5.7.1.4-3). A total of 361 fractions applied to the three objective functions are generated based on the formulas presented in Table 5.7.2.6 and Figure 5.7.2.3. A number of fractions is adjustable considering optimization natures.

$$w_{CI_b}^{i} = \text{linspace}(0,1,19);\ i = 1 \text{ to } 19 \qquad (5.7.1.4\text{-}1)$$

$$w_{CO_2}^{i,j} = \text{linspace}\left(0,1-w_{CI_b}^{i},19\right);\ j = 1 \text{ to } 19 \qquad (5.7.1.4\text{-}2)$$

$$w_{W_b}^{i,j} = 1-w_{CI_b}^{i} - w_{CO_2}^{i,j} \qquad (5.7.1.4\text{-}3)$$

UFO and Lagrange function of UFO are formulated as shown in Equations (5.7.1.5-1) and (5.7.1.5-2) implementing 361 fractions for w_{CI_b}, w_{CO_2}, and w_{W_b} obtained according to Equations (5.7.1.4-1)–(5.7.1.4-3).

$$UFO = w_{CI_b} f_{CI_b}^{N}(\mathbf{x}) + w_{CO_2} f_{CO_2}^{N}(\mathbf{x}) + w_{W_b} f_{W_b}^{N}(\mathbf{x}) \qquad (5.7.1.5\text{-}1)$$

$$\mathcal{L}_{\text{UFO}}(\mathbf{x},\lambda_c,\lambda_v) = \text{UFO}(\mathbf{x}) - \lambda_c^T c(\mathbf{x}) - \lambda_v^T v(\mathbf{x}) \qquad (5.7.1.5\text{-}2)$$

5.7.2.5 *Step 5–Optimizing UFO*

A Pareto frontier employing 361 fractions shown in Table 5.7.2.6 for a doubly RC beam is plotted in Figure 5.7.2.4 with three axes representing three-objective functions, cost (CI_b), estimated CO_2 emission, and beam weight (W_b). Designs with any optimization preference can be identified on the Pareto frontier of three-dimensional space. These designs are challenging to obtain based on conventional structural calculation-based methods.

5.7.3 Design parameters corresponding to various fractions on Pareto frontier

There are six design examples indicated by Designs P1–P6 illustrated in Figure 5.7.2.4 where CI_b, CO_2 and W_b are minimized simultaneously. As shown in Figure 5.7.2.4, P1–P3 represent Designs 1–3 where the three objective functions ($w_{CI_b} : w_{CO_2} : w_{W_b}$) are based on fractions with (1.0 : 0.0 : 0.0), (0.0 : 1.0 : 0.0), and (0.0 : 0.0 : 1.0) on Pareto curve, respectively, whereas Design P4 indicates an optimal design based on equal fractions of three objectives ($w_{CI_b} : w_{CO_2} : w_{W_b} = 0.34 : 0.33 : 0.33$) on Pareto curve. Design P5 represents a design that CI_b contributes more than the other two objective functions, CO_2 and W_b ($w_{CI_b} : w_{CO_2} : w_{W_b} = 0.50 : 0.25 : 0.25$), while a fraction of cost in Design P6 is less than those of CO_2 and W_b ($w_{CI_b} : w_{CO_2} : w_{W_b} = 0.12 : 0.44 : 0.44$). The design parameters corresponding to these four fractions for Designs P1–P6 are presented in Table 5.7.2.7. which also identifies entire design parameters corresponding to selected fractions ($w_{CI_b} : w_{CO_2} : w_{W_b}$) such as (1.0 : 0.0 : 0.0), (0.0 : 1.0 : 0.0), (0.0 : 0.0 : 1.0), and (0.34 : 0.33 : 0.33) on the Pareto curve. It is noted, in Table 5.7.2.7, that the tensile rebar ratio (ρ_{st}) of 0.0117 is obtained as shown in Design P1 when a beam cost CI_b is minimized at 75,359 (KRW/m) whereas the rebar ratio (ρ_{st}) of 0.0225 is obtained as shown in Design P3 when a beam weight is minimized to 5.556 (kN/m) from 8.226 (kN/m) of Design P1, resulting increased rebar ratio (ρ_{st})

Table 5.7.2.6 Equally Spaced Fractions Generated Based on MATLAB Function, *Linspace* (MathWorks, 2020)

i	j	$w^{i}_{Cl_b}$	$w^{i,j}_{CO_2}$	$w^{i,j}_{W_b}$	i	j	$w^{i}_{Cl_b}$	$w^{i,j}_{CO_2}$	$w^{i,j}_{BW}$		i	j	$w^{i}_{Cl_b}$	$w^{i,j}_{CO_2}$	$w^{i,j}_{W_b}$
***i* = 1**	***j* = 1**	0	0	1	***i* = 5**	***j* = 1**	0.2222	0	0.7778	…	***i* = 16**	***j* = 1**	0.7778	0	0.2222
	***j* = 2**	0	0.0556	0.9444		***j* = 2**	0.2222	0.0432	0.7346	…		***j* = 2**	0.7778	0.0123	0.2099
	…	…	…	…		…	…	…	…	…		…	…	…	…
	***j* = 19**	0	0.9444	0.0556		***j* = 19**	0.2222	0.7346	0.04321	…		***j* = 19**	0.7778	0.2099	0.0123
	***j* = 20**	0	1	0		***j* = 20**	0.2222	0.7778	0	…		***j* = 20**	0.7778	0.2222	0
***i* = 2**	***j* = 1**	0.0556	0.0000	0.9444	***i* = 6**	***j* = 1**	0.2778	0	0.7222	…	***i* = 17**	***j* = 1**	0.8333	0	0.1667
	***j* = 2**	0.0556	0.0525	0.8920		***j* = 2**	0.2778	0.0401	0.6821	…		***j* = 2**	0.8333	0.0093	0.1574
	…	…	…	…		…	…	…	…	…		…	…	…	…
	***j* = 19**	0.0556	0.8920	0.0525		***j* = 19**	0.2778	0.6821	0.04012	…		***j* = 19**	0.8333	0.1574	0.0093
	***j* = 20**	0.0556	0.9444	0.0000		***j* = 20**	0.2778	0.7222	0	…		***j* = 20**	0.8333	0.1667	0
***i* = 3**	***j* = 1**	0.1111	0.0000	0.8889	***i* = 7**	***j* = 1**	0.3333	0	0.6667	…	***i* = 18**	***j* = 1**	0.8889	0	0.1111
	***j* = 2**	0.1111	0.0494	0.8395		***j* = 2**	0.3333	0.0370	0.6296	…		***j* = 2**	0.8889	0.0062	0.1049
	…	…	…	…		…	…	…	…	…		…	…	…	…
	***j* = 19**	0.1111	0.8395	0.0494		***j* = 19**	0.3333	0.6296	0.0370	…		***j* = 19**	0.8889	0.1049	0.0062
	***j* = 20**	0.1111	0.8889	0.0000		***j* = 20**	0.3333	0.6667	0	…		***j* = 20**	0.8889	0.1111	0
***i* = 4**	***j* = 1**	0.1667	0.0000	0.8333	***i* = 8**	***j* = 1**	0.3889	0	0.6111	…	***i* = 19**	***j* = 1**	0.9444	0	0.0556
	***j* = 2**	0.1667	0.0463	0.7870		***j* = 2**	0.3889	0.0340	0.5772	…		***j* = 2**	0.9444	0.0031	0.0525
	…	…	…	…		…	…	…	…	…		…	…	…	…
	***j* = 19**	0.1667	0.7870	0.0463		***j* = 19**	0.3889	0.5772	0.0340	…		***j* = 19**	0.9444	0.0556	0.0000
	***j* = 20**	0.1667	0.8333	0		***j* = 20**	0.3889	0.6111	0	…		***j* = 20**	1.0000	0.0000	0

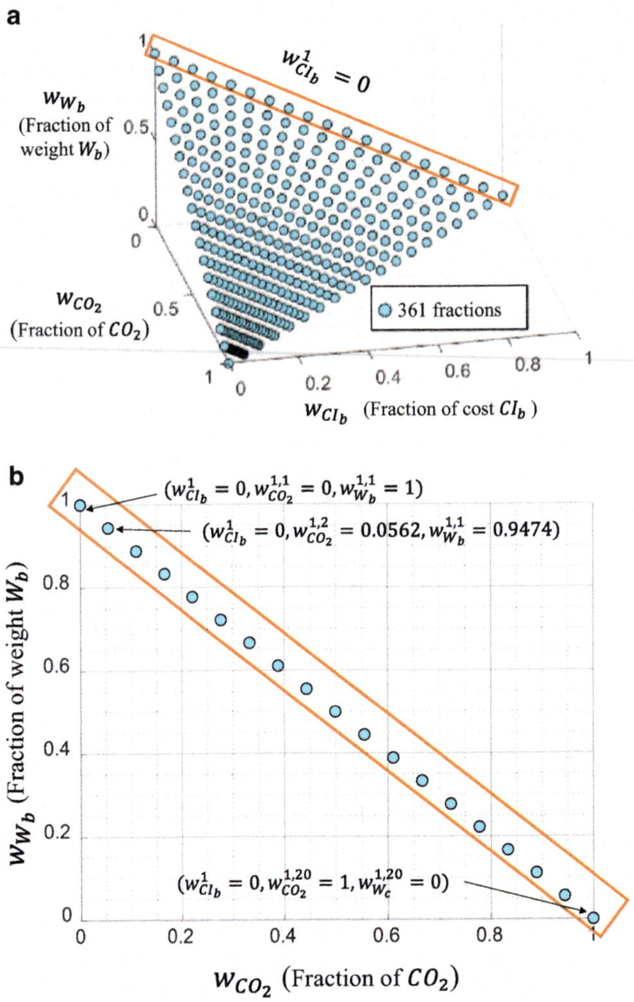

Figure 5.7.2.3 Equally spaced fractions generated based on MATLAB function (MathWorks, 2020). (a) 361 fractions of three objective functions. (b) Fractions of CO_2 and W_b corresponding to $w_{CI_b}^{(1)} = 0$.

by $\dfrac{0.0255-0.0117}{0.0117} \times 100\% = 118\ \%$ to reduce the beam weight, but sacrificing cost which increased from 75,359 to 92,043 (KRW/m) with 22% increase. The rebar ratio $\rho_{st} = 0.0161$ which is mid-range of that obtained between Designs P1 and P3 is obtained as shown in Design P5 when a fraction $(w_{CI_b} : w_{CO_2} : w_{W_b})$ of $(0.50 : 0.25 : 0.25)$ is implemented. The rebar ratio $\rho_{st} = 0.0161$ is greater than the rebar ratio $\rho_{st} = 0.0117$ obtained when minimizing beam cost CI_b as shown in Design P1 and 0099 when minimizing CO_2 emissions as shown in Design P2. Other design parameters than those shown with $(1.0 : 0.0 : 0.0)$, $(0.0 : 1.0 : 0.0)$, $(0.0 : 0.0 : 1.0)$, and $(0.34 : 0.33 : 0.33)$ can also be obtained from the Pareto curve. Nine inputs are used for an ANN and a structural mechanics-based software (*Autobeam*) to obtain ten

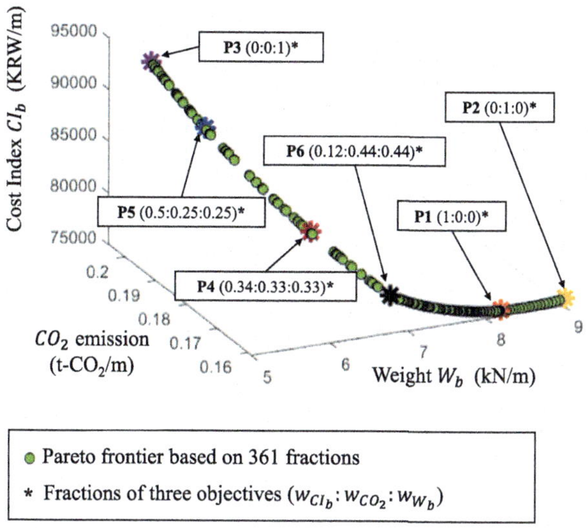

Figure 5.7.2.4 A Pareto frontier based on 361 fractions shown in Figure 5.7.2.3 for RC beams optimizing three objective functions.

corresponding outputs. Design output parameters calculated by an ANN are verified by those resulted by *Autobeam* based on Error $= \frac{\text{Output}^{Autobeam} - \text{Output}^{\text{ANN}}}{\text{Output}^{\text{ANN}}} \times 100\%$. $\text{Output}^{\text{ANN}}$ and $\text{Output}^{Autobeam}$ (output parameters ϕM_n, M_u, M_{cr}, ε_{rt}, ε_{rc}, Δ_{imme}, Δ_{long}, CI_b, CO_2, W_b) are obtained using optimized parameters b, h, ρ_{rt}, ρ_{rc} and prescribed parameters L, f_c', f_y, M_D, M_L based on the ANN and structural mechanics-based software *Autobeam*. Design accuracies obtained using equation Error $= \frac{\text{Output}^{Autobeam} - \text{Output}^{\text{ANN}}}{\text{Output}^{\text{ANN}}} \times 100\%$ based on ANN and structural mechanics-based software *Autobeam* seen in Table 5.7.2.7 are as large as 5.5%.

5.7.4 Verification of Pareto frontier

A Pareto frontier minimizing three objective functions (CI_b,CO_2, and W_b) simultaneously for a doubly RC beam obtained by the ANN-based Lagrange algorithm is verified by a lower boundary generated by 300,000 structural data based on structural mechanics-based software *Autobeam*, as shown in Figures 5.7.2.4 and 5.7.4.1. In Figure 5.7.4.1, optimized designs of RC beams are obtained by a Pareto frontier based on 361 fractions. Pareto frontier in two-dimensional space is also shown for demonstrating the inter-relationships between CO_2 emission beam-weight (W_b) in X-Y plane, cost (CI_b) - weight (W_b) in X-Z plane, and cost (CI_b) - CO_2 emission in Y-Z plane, help understand inter-trends of the three objective functions. Optimized designs based on ANN-based Lagrange methods optimizing UFO form a lower boundary of 300,000 large structural datasets.

Table 5.7.2.7 Optimized Design Parameters on a Pareto Frontier Illustrated in Figure 5.7.2.4

(a): Design Points P1–P3 Based on Fractions $w_{Cl_b} : w_{CO_2} : w_{W_b}$ of 1:0:0, 0:1:0, and 0:0:1										
		Design P1 (Cl_b is Minimized)			Design P2 (CO_2 is Minimized)			Design P3 (W_b is Minimized)		
		Fraction: $w_{Cl_b} : w_{CO_2} : w_{W_b} = 1:0:0$			Fraction: $w_{Cl_b} : w_{CO_2} : w_{W_b} = 0:1:0$			Fraction: $w_{Cl_b} : w_{CO_2} : w_{W_b} = 0:0:1$		
Parameters		AI-based Lagrange	Verification (Autobeam)	Error	AI-based Lagrange	Verification (Autobeam)	Error	AI-based Lagrange	Verification (Autobeam)	Error
1	h (mm)	1,041		-	1,094		-	823		-
2	b (mm)	312		-	328		-	247		-
3	$\rho_{r,t}$	0.0117		-	0.0099		-	0.0255		-
4	$\rho_{r,c}$	0.0008		-	0.0008		-	0.0128		-
5	L (mm)	9,000		-	9,000		-	9,000		-
6	f_c' (MPa)	550		-	550		-	550		-
7	f_y (MPa)	40		-	40		-	40		-
8	M_D (kN.m)	800		-	800		-	800		-
9	M_L (kN.m)	400		-	400		-	400		-
10	ϕM_n (kN.m)	1,597	1,594	0.16%	1,597	1,592	0.31%	1,605	1,576	1.79%
11	M_u (kN.m)	1,597	1,600	0.21%	1,597	1,600	0.16%	1,605	1,600	0.32%
12	M_{cr} (kN.m)	278	277	0.22%	314	313	0.36%	170	169	0.60%
13	$\varepsilon_{r,t}$	0.0094	0.0094	0.43%	0.0119	0.0118	0.74%	0.0057	0.0056	2.61%
14	$\varepsilon_{r,c}$	0.0024	0.0024	0.02%	0.0024	0.0024	0.02%	0.0024	0.0023	1.75%
15	Δ_{imme} (mm)	2.62	2.66	1.73%	2.31	2.26	2.16%	3.73	3.80	1.75%
16	Δ_{long} (mm)	16.11	16.22	0.70%	13.78	13.88	0.77%	21.15	21.44	1.38%
17	Cl_b (KRW/m)	**75,359**	**75,284**	0.10%	**75,755**	**75,670**	0.11%	**92,043**	**91,959**	0.09%
18	CO_2 (t-CO_2/m)	**0.159**	**0.160**	0.03%	**0.159**	**0.159**	0.01%	**0.205**	**0.205**	0.05%
19	W_b (kN/m)	**8.226**	**8.226**	0.00%	**9.003**	**9.003**	0.00%	**5.556**	**5.553**	0.04%

Optimized parameters used as inputs in ANN and structural mechanics-based software (*Autobeam*).
Prescribed parameters used as inputs in ANN and structural mechanics-based software (*Autobeam*).
Outputs for ANN.
Outputs for structural mechanics-based software (*Autobeam*).
Bold and red color values indicate optimized objective functions.

(Continued)

Table 5.7.2.7 (Continued) Optimized Design Parameters on a Pareto Frontier Illustrated in Figure 5.7.2.4

(b): Design Points P4, P5, and P6 Based on Fractions $w_{CI_b} : w_{CO_2} : w_{W_b}$ of 0.34:0.33:0.33, 0.50:0.25:0.25, and 0.12:0.44:0.44

		Design P4			Design P5			Design P6		
		Fraction: $w_{CI_b} : w_{CO_2} : w_{W_b} = 0.34 : 0.33 : 0.33$			Fraction: $w_{CI_b} : w_{CO_2} : w_{W_b} = 0.50 : 0.25 : 0.25$			Fraction: $w_{CI_b} : w_{CO_2} : w_{W_b} = 0.12 : 0.44 : 0.44$		
Parameters		AI-based Lagrange	Verification (Autobeam)	Error	AI-based Lagrange	Verification (Autobeam)	Error	AI-based Lagrange	Verification (Autobeam)	Error
1	h (mm)	823		-	955		-	908		-
2	b (mm)	247		-	286		-	272		-
3	$\rho_{r,t}$	0.0255		-	0.0161		-	0.0187		-
4	$\rho_{r,c}$	0.0128		-	0.0008		-	0.0039		-
5	L (mm)	9,000		-	9,000		-	9,000		-
6	$f_c^{'}$ (MPa)	550		-	550		-	550		-
7	f_y (MPa)	40		-	40		-	40		-
8	M_D (kN.m)	800		-	800		-	800		-
9	M_L (kN.m)	400		-	400		-	400		-
10	ϕM_n (kN.m)	1,605	1,576	1.79%	1,593	1,600	0.42%	1,598	1,569	1.84%
11	M_u (kN.m)	1,605	1,600	0.32%	1,593	1,600	0.42%	1,598	1,600	0.10%
12	M_{cr} (kN.m)	170	169	0.60%	225	226	0.58%	203	202	0.59%
13	$\varepsilon_{r,t}$	0.0058	0.0056	2.61%	0.0058	0.0058	0.87%	0.0058	0.0057	0.87%
14	$\varepsilon_{r,c}$	0.0024	0.0023	1.75%	0.0026	0.0026	0.06%	0.0025	0.0025	0.01%
15	Δ_{imme} (mm)	3.73	3.80	1.75%	3.27	3.09	5.50%	3.27	3.38	3.42%
16	Δ_{long} (mm)	21.15	21.44	1.38%	20.52	20.26	1.28%	21.09	21.53	2.12%
17	CI_b (KRW/m)	92,043	91,959	0.09%	76,767	76,698	0.09%	81,035	80,939	0.12%
18	CO_2 (t-CO_2/m)	0.205	0.205	0.05%	0.166	0.166	0.05%	0.177	0.177	0.08%
19	W_b (kN/m)	5.556	5.553	0.04%	7.041	7.041	0.01%	6.471	6.473	0.03%

Optimized parameters used as inputs in ANN and structural mechanics-bases software (*Autobeam*).
Prescribed parameters used as inputs in ANN and structural mechanics-bases software (*Autobeam*).
Outputs for ANN.
Outputs for structural mechanics-based software (*Autobeam*).

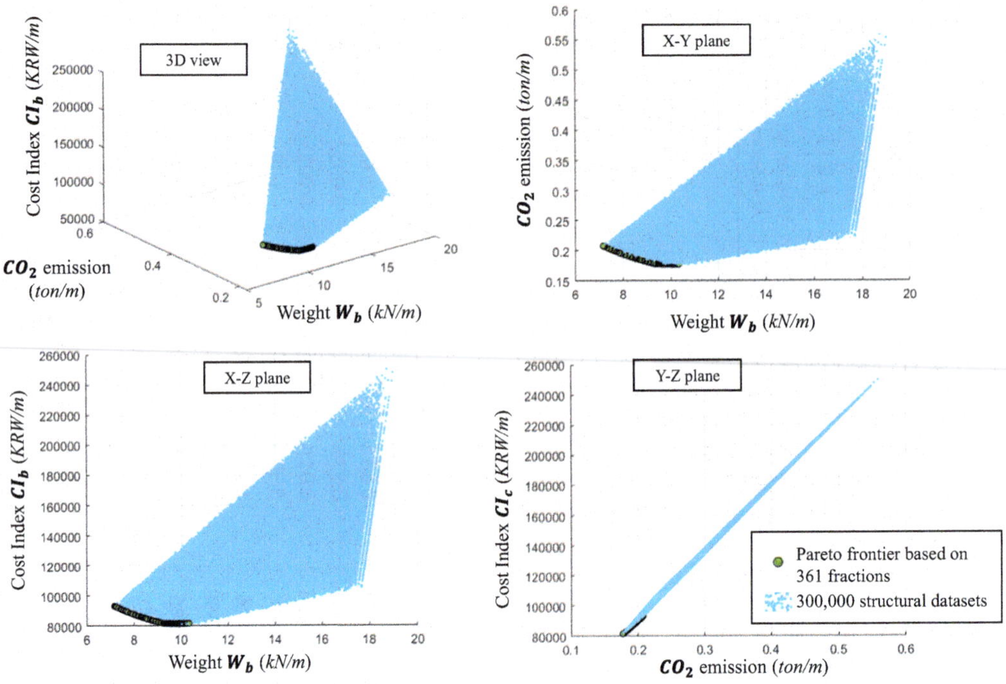

Figure 5.7.4.1 Verification of a Pareto frontier by 300,000 structural datasets.

5.7.5 Decision-making based on the Pareto frontier

As shown in Figure 5.7.2.4, points on the Pareto frontier uncover optimized design parameters by simultaneously minimizing three objective functions (cost CI_b, CO_2, and beam weight W_b), while particular fractions contributed by the three objective functions are represented by a fraction $W_{CIb} : W_{CO2} : W_{Wb}$. The optimized design parameters and their corresponding fractions can aid engineers and decision-makers in a preliminary design stage, offering them optimal designs minimizing CI_b,CO_2, and W_b simultaneously in addition to the fractions contributed by the three objective functions. This section's optimization example presents a design recommendation for a doubly RC beam. When a budget is assumed between 82,000 and 85,000 KRW/m, a doubly RC beam with constraint conditions shown in Table 5.7.1.3 can be optimized. When a design yields the least weight and CO_2 emissions at the same time, the ranges of CO_2 emissions and beam weight corresponding to the budget between 82,000 and 85,000 KRW/m can be estimated as 0.181–0.187 t-CO_2/m and 6.311–6.079 kN/m, respectively, as shown in Figure 5.7.5.1. The green dots in Figure 5.7.5.1 denote a Pareto based on 361 fractions obtained in Figure 5.7.2.3 and Table 5.7.2.6. A Pareto frontier shown in Figure 5.7.5.1 illustrates an estimated design range of CO_2 and W_b when the cost ranges from 82,000 to 85,000 KRW/m. Figure 5.7.5.1 helps determine a design meeting the least W_b of 6.079 kN/m in the range (6.079–6.311 kN/m) that corresponds to the minimized CI_b = 84,662.2 KRW/m and CO_2 = 0.187 t-CO_2/m when minimizing beam weight is preferable. This design corresponds to a fraction of 0.11:0.49:0.40 in which the cost CI_b (CI_b = 84,662.2 KRW/m) contributes 11% to the minimization of entire designs based on multi-objective functions, *UFO*, while contributions of CO_2 (CO_2 = 0.187 t-CO_2/m) and W_b (W_b = 6.079 kN/m) are 49% and 40% to the unified optimization, respectively as shown in Table 5.7.5.1.

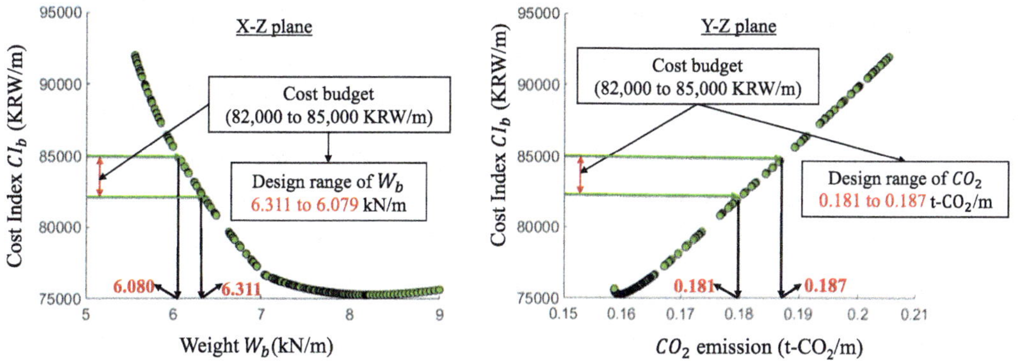

Figure 5.7.5.1 Estimated design ranges for minimized CO_2 and W_b corresponding to a defined cost budget.

Engineers may allow the beam weight W_b to be 6.255 kN/m when lower beam budget with 82,886 KRW/m is preferable by increasing contributions of cost to the minimization of entire designs from 11% to 18.2% whereas contributions of CO_2 (CO_2 = 0.182 t-CO_2/m) and W_b (W_b = 6.255 kN/m) are 36% and 36%, respectively as shown in Table 5.7.5.1.

It is noted that, as shown in Table 5.7.5.1, when cost CI_b and Weight which are more prioritized than contributions of CO_2 based on a fraction of 0.56:0.10:0.35, the cost CI_b = 82,852.3 KRW/m obtained is obtained similar to the CI_b = 82,8886.2 KRW/m obtained based on a fraction of 0.28:0.36:0.36 when three objective functions are roughly equally minimized. These two designs yield an optimized CI_b similar to each other (CI_b = 82,852.3 KRW/m and CI_b = 82,8886.2 KRW/m) because CI_b and CO_2 have a similar trend when a fraction of Weight of the two designs are about the same. This indicates that CI_b has influences similar to that of CO_2 on optimizing UFO. The design parameters of these two designs with two different fractions are presented in Table 5.7.5.2. Parameters shown in Boxes 1 to 4 are design parameters optimized from ANN-based Lagrange networks which are implemented in both ANN and structural mechanics-based software (*Autobeam*). Parameters

Table 5.7.5.1 Design Option with Trade-off Ratios for an Example Shown in Figure 5.7.5.1

Objective Functions			*Fraction Samples*		
Cost (KRW/m)	CO_2 *(t-CO_2/m)*	*Weight (kN/m)*	w_{CI_b}	w_{CO_2}	w_{W_b}
83,811.8	0.184	6.161	0.00	0.61	0.39
82,453.7	0.181	6.301	0.06	0.58	0.37
84,662.2	**0.187**	**6.079**	**0.11**	**0.49**	**0.40**
83,127.9	0.183	6.230	0.17	0.46	0.37
84,653.5	0.187	6.080	0.22	0.39	0.39
82,886.6	**0.182**	**6.255**	**0.28**	**0.36**	**0.36**
83,782.8	0.184	6.164	0.33	0.30	0.37
84,214.4	0.186	6.122	0.39	0.24	0.37
84,209.9	0.186	6.122	0.44	0.19	0.37
83,768.2	0.184	6.165	0.50	0.14	0.36
84,627.3	0.187	6.082	0.56	0.07	0.37
82,852.3	**0.182**	**6.259**	**0.56**	**0.10**	**0.35**
84,623.0	0.187	6.083	0.61	0.02	0.37

Bold and blue color values indicate optimized objective functions.

Table 5.7.5.2 Design Parameters of Two Designs Marked in Blue Color in Table 5.7.5.1

		Fraction: $w_{CI_b} : w_{CO_2} : w_{W_b} = 0.28 : 0.36 : 0.36$			*Fraction:* $w_{CI_b} : w_{CO_2} : w_{W_b} = 0.56 : 0.10 : 0.35$		
Parameters		*AI-based Lagrange*	*Verification (Autobeam)*	*Error*	*AI-based Lagrange*	*Verification (Autobeam)*	*Error*
1	h (mm)	888.9			889.2		-
2	b (mm)	266.7		-	266.8		-
3	$\rho_{r,t}$	0.0200		-	0.0199		-
4	$\rho_{r,c}$	0.0054		-	0.0054		-
5	L (mm)	9,000		-	9,000		-
6	f_c' (MPa)	550		-	550		-
7	f_y (MPa)	40		-	40		-
8	M_D (kN.m)	800		-	800		-
9	M_L (kN.m)	400		-	400		-
10	ϕM_n (kN.m)	1,599.8	1,575.6	1.54%	1,599.8	1,575.5	1.54%
11	M_u (kN.m)	1,599.8	1,600.0	0.01%	1,599.8	1,600.0	0.01%
12	M_{cr} (kN.m)	195.4	194.4	0.49%	195.5	194.6	0.49%
13	$\varepsilon_{r,t}$	0.0057	0.0057	0.88%	0.0057	0.0057	0.88%
14	$\varepsilon_{r,c}$	0.0025	0.0025	0.03%	0.0025	0.0025	0.03%
15	Δ_{imme} (mm)	3.27	3.27	0.03%	3.27	3.27	0.04%
16	Δ_{long} (mm)	21.23	21.09	0.63%	21.23	21.10	0.59%
17	CI_b (KRW/m)	82,974.3	82,886.6	0.11%	82,940.2	82,852.3	0.11%
18	CO_2 (t-CO_2/m)	0.1819	0.1821	0.08%	0.1819	0.1820	0.08%
19	W_b (kN/m)	6.2521	6.2553	0.05%	6.2558	6.2589	0.05%

Optimized parameters used as inputs in ANN and structural mechanics-based software (*Autobeam*).
Prescribed parameters used as inputs in ANN and structural mechanics-based software (*Autobeam*).
Outputs for ANN.
Outputs for structural mechanics-based software (*Autobeam*).

shown in Boxes 5 to 9 are prescribed design parameters used as inputs in ANN and structural mechanics-based software (*Autobeam*). Yellow and green Boxes 10–19 represent output design parameters obtained from ANN and from structural mechanics-based software (*Autobeam*) for input parameters given in Boxes 1–9, indicating the design accuracies are validated by the errors shown in Table 5.7.5.2. Tables 5.7.5.1 and 5.7.5.2 are design tables obtained by optimizing multi-objective functions, UFO, which is very useful for engineers and decision makers in which engineers without any significant efforts can choose design alternatives based on the trade-off ratio or fractions contributed by multi-objective functions according to a project preference, allowing them to calculate minimized CO_2 and W_b for a pre-determined budget CI_b and vice versa. This is a novel method that allows an optimal design to be selected based on any predefined multi-objective function for a project preference. As shown in Table 5.7.5.1, engineers without any computational limitations can choose one of diverse design options based on the trade-off ratios or fractions among the three objective functions corresponding to a project preference, allowing them to calculate minimized CO_2 and W_b for a pre-determined budget CI_b and vice versa. An optimized design can now be selected based on multi-objective functions, assisting engineers to perform efficient designs.

5.7.6 Interpretation of data trend based on relationships among three objective functions

The relationships of three objective functions are presented below, in which unit cost, unit weight and unit CO_2 emissions of reinforcement and concrete are based on Korean unit as shown in Table 5.3.1.1 (Hong et al., 2010). It is noted that the concrete compressive strength of 30 MPa and the steel strength of 500 MPa are used in the case study of this paper, leading to the cost, unit weight and unit CO_2 emissions of rebar and concrete.

Unit cost of concrete:	$U_{\text{concrete}}^{CI_b} = 85,000$ KRW/m^3
Unit cost of rebar:	$U_{\text{rebar}}^{CI_b} = 1055 \times 7850 = 8,281,750$ KRW/m^3
Unit CO_2 emissions of concrete:	$U_{\text{concrete}}^{CO_2} = 0.1677$ ton $- CO_2/m^3$
Unit CO_2 emissions of rebar:	$U_{\text{rebar}}^{CO_2} = 2.512 \times 7.85 = 19.72$ ton $- CO_2/m^3$

Weight of a doubly RC beam is calculated using Equation (1):

$$W_b = W_{\text{concrete}} + W_{\text{rebar}} \tag{5.7.6.1}$$

CO_2 emissions of a beam are calculated in Equation (5.7.6.2) using unit CO_2 emissions of concrete ($U_{\text{concrete}}^{CO_2}$) and unit CO_2 emissions of rebar ($U_{\text{rebar}}^{CO_2}$):

$$CO_2 = U_{\text{concrete}}^{CO_2} \times W_{\text{concrete}} + U_{\text{rebar}}^{CO_2} \times W_{\text{rebar}}$$

$$\Rightarrow CO_2 = U_{\text{concrete}}^{CO_2} \times W_{\text{concrete}} + U_{\text{rebar}}^{CO_2} \times \left(W_b - W_{\text{concrete}}\right) \tag{5.7.6.2}$$

$$CO_2 = U_{\text{rebar}}^{CO_2} \times W_b + W_{\text{concrete}} \times \left(U_{\text{concrete}}^{CO_2} - U_{\text{rebar}}^{CO_2}\right)$$

Using Equation (5.7.6.2), weight of concrete W_{concrete} can be calculated based on total CO_2 emissions as:

$$W_{\text{concrete}} = \frac{CO_2 - U_{\text{rebar}}^{CO_2} \times W_b}{\left(U_{\text{concrete}}^{CO_2} - U_{\text{rebar}}^{CO_2}\right)} \tag{5.7.6.3}$$

Total cost index of a beam is calculated as shown in Equation (5.7.6.4):

$$CI = U_{\text{concrete}}^{CI_b} \times W_{\text{concrete}} + U_{\text{rebar}}^{CI_b} \times W_{\text{rebar}}$$

$$CI = U_{\text{concrete}}^{CI_b} \times W_{\text{concrete}} + U_{\text{rebar}}^{CI_b} \times \left(W_b - W_{\text{concrete}}\right) \tag{5.7.6.4}$$

$$CI = U_{\text{rebar}}^{CI_b} \times W_b + W_{\text{concrete}} \times \left(U_{\text{concrete}}^{CI_b} - U_{\text{rebar}}^{CI_b}\right)$$

Substituting Equation (5.7.6.3) into Equation (5.7.6.4), we have:

$$CI_b = U_{\text{rebar}}^{CI_b} \times W_b + \frac{CO_2 - U_{\text{rebar}}^{CO_2} \times W_b}{\left(U_{\text{concrete}}^{CO_2} - U_{\text{rebar}}^{CO_2}\right)} \times \left(U_{\text{concrete}}^{CI_b} - U_{\text{rebar}}^{CI_b}\right)$$

$$CI_b = \left(U_{\text{rebar}}^{CI_b} - U_{\text{rebar}}^{CO_2} \times \frac{U_{\text{concrete}}^{CI_b} - U_{\text{rebar}}^{CI_b}}{U_{\text{concrete}}^{CO_2} - U_{\text{rebar}}^{CO_2}}\right) \times W_b + \frac{U_{\text{concrete}}^{CI_b} - U_{\text{rebar}}^{CI_b}}{U_{\text{concrete}}^{CO_2} - U_{\text{rebar}}^{CO_2}} \times CO_2$$

$$CI_b = \left(8,281,750 - 19.72 \times \frac{85,000 - 8,281,750}{0.1677 - 19.72}\right) \times W_b + \frac{85,000 - 8,281,750}{0.1677 - 19.72} \times CO_2$$

$$CI_b = -14,696 \times W_b + 419,221 \times CO_2 \tag{5.7.6.5}$$

The relationships of three objective functions are analytically derived as shown in Equation (5.7.6.5), showing that cost index (CI_b) and beam weight (W_b) are conflict objective functions, not proportional. CI_b is proportional to CO_2. However, CO_2 emission is considered as one objective function to reflect engineer's interests.

5.8 DESIGN RECOMMENDATIONS AND CONCLUSIONS

5.8.1 Design recommendations

This chapter proposed an ANN-based Lagrange optimization algorithm to solve MOO problems for structural designs, including circular RC columns, uniaxial and biaxial rectangular RC columns and RC beams. This algorithm uses ANN-trained objective functions. The single-objective function is globalized and normalized into UFO. A Lagrange multiplier method transforms an optimization design problem with multiple constraints into a nonboundary optimization design problem. The proposed method is to perform designs optimizing multiple objective functions. Lagrange multipliers are used with equality and inequality conditions based on KKT conditions. The UFO which is not based on just one objective function is optimized simultaneously to derive a Pareto frontier which cannot be found using conventional design methods. The multiple design variables which are the stationary points are obtained by solving large differential equations represented by Jacobi and Hessian equations of UFO using Newton-Raphson iteration, and hence, stationary points at which the first derivative of Lagrange function becomes zero are calculated by ANN-based Lagrange optimization algorithm.

5.8.2 Conclusions

Previous studies to optimize RC design problems are commonly based on the metaheuristics-based method such as Genetic Algorithm, Nondominated Sorting Genetic Algorithm, and Particle Swarm Optimization, as indicated in the literature review. Trade-off ratios of objective functions are not displayed in the metaheuristics-based optimized results, making it difficult for engineers to apply to real-life structural design cases. The ANN-based Lagrange optimization algorithm proposed in this chapter is a gradient-based method in which trade-off ratios are clearly presented in the Pareto frontier. The algorithm consisting of five steps performed using the MATLAB toolbox (MathWorks, 2020) is introduced for engineers to easily follow. Pareto frontier or Pareto curve obtained by the optimized designs based on ANN-based Lagrange algorithm are placed in the lower boundary obtained by large structural datasets, verifying Pareto curves. This chapter will aid engineers to make efficient decisions by providing them with an overall assessment of a design project, in which three objectives (cost, CO_2, and weight) are optimized simultaneously. Some conclusion drawn from this study are as follows.

1. In this chapter, the ANN-based Lagrange optimization algorithm is proposed to optimize UFO formulated based on interests of engineers, contractors, and government officials. A novel gradient-based algorithm is proposed for MOO problems to optimize multiple objective functions derived for reinforced concrete structures including circular RC columns, uniaxial and biaxial rectangular RC columns and RC beams.
2. Five steps are proposed based on ANN-based Lagrange method to optimize multi-objective functions simultaneously. ANN-based Lagrange functions for single objective function are obtained in Steps 1–3. An UFO is, then, derived in Step 4 to minimize it in Step 5 to identify a Pareto frontier. Design parameters are also obtained during structural optimizations based on an UFO.

3. Using an UFO globalized based on ANNs, MOO is performed by optimizing an UFO function under KKT conditions for given inequality constraints, yielding Pareto curves which are also verified by large structural datasets.
4. Design example with diverse design scenarios are provided to holistically demonstrate that a use of UFO based on ANNs implementing Lagrange optimization yields a Pareto frontier which identifies entire design parameters corresponding to selected contributions $(w_{CI_b} : w_{CO_2} : w_{W_b})$ made by objective functions.
5. This chapter discusses an optimization of round columns subject to single, multiple loads and rectangular RC columns subject to uniaxial and biaxial-single, multiple loads, in which three-objective functions including cost, estimated CO_2 emission, and column weight are minimized, simultaneously. Components of the weight matrix considering one load pair (P_u, M_u) can be used for components of the weight matrix subject to multiple load pairs (factored multiple load pairs $P_{u,i}$ and $M_{u,i}$). P-M interaction diagrams of RC columns are also plotted with multi-load pairs based on selected contributions $(w_{CI_b} : w_{CO_2} : w_{W_b})$. Some weight and bias matrices which are subject to multi-load pairs are provided in appendix to help readers derive their own weight and bias matrices to optimize an UFO of their own RC columns.
6. A Pareto frontier or Pareto curve obtained by the optimized designs based on the ANN-based Lagrange algorithm coincides with the lower boundary obtained by large structural datasets, which verifies Pareto curves. Engineers and decision-makers are able to design structures based on particular contributions made by objective functions, enabling a design meeting holistic design targets. Engineers are now capable of making efficient design decisions with an overall assessment of a design project, in which three objectives (cost, CO_2, and weight) can be optimized simultaneously.
7. Chapter 5 integrates each single-objective function extracted from an artificial neural network (ANN) into a global Lagrange function, a UFO, using trade-off fractions. An ANN-based Lagrange algorithm proposes a novel gradient-based algorithm. Round and rectangular reinforced concrete columns are optimized based on three objectives, such as cost, CO_2 emissions, and column weight, simultaneously. The proposed algorithm brings in a set of optimal results capturing multiple objectives, known as a Pareto frontier. Overall, optimal results determined by the proposed method in this chapter exhibit better convergence than that obtained by NSGA-II (Hong et al., 2022a).

REFERENCES

ACI Committee (2019). *ACI CODE-318-19: Building Code Requirements for Structural Concrete and Commentary*. American Concrete Institute, Farmington Hills, MI.

Afshari, H., Hare, W., & Tesfamariam, S. (2019). "Constrained multi-objective optimization algorithms: Review and comparison with application in reinforced concrete structures". *Applied Soft Computing Journal*, *83*(July). https://doi.org/10.1016/j.asoc.2019.105631

Arama, Z. A., Kayabekir, A. E., Bekdaş, G., & Geem, Z. W. (2020). "CO_2 and cost optimization of reinforced concrete cantilever soldier piles: A parametric study with harmony search algorithm". *Sustainability*, *12*(15). https://doi.org/10.3390/su12155906

Babaei, M., & Mollayi, M. (2021). "Multiobjective optimal design of reinforced concrete frames using two metaheuristic algorithms". *Journal of Engineering Research*, *9*(4 B), 166–192. https://doi.org/10.36909/jer.9973

Barraza, M., Boj00F3rquez, E., Fern00E1ndez-Gonz00E1lez, E., & Reyes-Salazar, A. (2017). "Multi-objective optimization of structural steel buildings under earthquake loads using NSGA-II and PSO". *KSCE Journal of Civil Engineering*, *21*(2), 488–500. https://doi.org/10.1007/s12205-017-1488-7

Bekdas, G., & Nigdeli, S. M. (2013). "Optimization of t-shaped RC flexural members for different compressive strengths of concrete". *International Journal of Mechanics*, 7(2), 109–119.

Bekdaş, G., & Nigdeli, S. M. (2017). "Modified harmony search for optimization of reinforced concrete frames". *Advances in Intelligent Systems and Computing*, *514*, 213–221. https://doi.org/10.1007/978-981-10-3728-3_21

binMohd Zain, M. Z., Kanesan, J., Chuah, J. H., Dhanapal, S., & Kendall, G. (2018). "A multi-objective particle swarm optimization algorithm based on dynamic boundary search for constrained optimization". *Applied Soft Computing Journal*, 70, 680–700. https://doi.org/10.1016/j.asoc.2018.06.022

Brown, N. C. (2016). "Multi-objective optimization conceptual design of structures". Massachusetts Institute of Technology - MSc Thesis, pp. 1–113.

Brown, N., Tseranidis, S., & Mueller, C. (2015). "Multi-objective optimization for diversity and performance in conceptual structural design". In *Proceedings of the International Association for Shell and Spatial Structures (IASS) Symposium "Future Visions,"* (Vol., 20, pp. 1–12). https://www.ingentaconnect.com/content/iass/piass/2015/00002015/00000020/art00012

Choi, S. W., Oh, B. K., & Park, H. S. (2017). "Design technology based on resizing method for reduction of costs and carbon dioxide emissions of high-rise buildings". *Energy and Buildings*, *138*, 612–620. https://doi.org/10.1016/j.enbuild.2016.12.095

Coello, C. C., Hernandez, F. S., & Farrera, F. A. (1997). "Optimal design of reinforced concrete beams using genetic algorithms". *Expert Systems with Applications*, *12*(1), 101–108. https://doi.org/10.1016/S0957-4174(96)00084-X

Deb, K., Pratap, A., Agarwal, S., & Meyarivan, T. (2002). "A fast and elitist multiobjective genetic algorithm: NSGA-II". *IEEE Transactions on Evolutionary Computation*, *6*(2), 182–197. https://doi.org/10.1109/4235.996017

Ferreira, C. C., Barros, M. H. F. M., & Barros, A. F. M. (2003). "Optimal design of reinforced concrete T-sections in bending". *Engineering Structures*, *25*(7), 951–964. https://doi.org/10.1016/S0141-0296(03)00039-7

Ferreira, F. P. V., Shamass, R., Limbachiya, V., Tsavdaridis, K. D., & Martins, C. H. (2022). "Lateral–torsional buckling resistance prediction model for steel cellular beams generated by Artificial Neural Networks (ANN)". *Thin-Walled Structures*, *170*(September 2021), 108592. https://doi.org/10.1016/j.tws.2021.108592

Hong, W.-K. (2020). "Artificial-intelligence-based design of the ductile precast concrete beams". In W.-K. B. T.-H. C. P. S. Hong (Ed.), *Woodhead Publishing Series in Civil and Structural Engineering* (pp. 427–478). Woodhead Publishing. https://doi.org/10.1016/B978-0-08-102721-9.00010-8

Hong, W. K. (2021). "Artificial intelligence-based design of reinforced concrete structures". Daega Publisher, Gyeonggi, Korea.

Hong, W.-K., Kim, J.-M., Park, S.-C., Lee, S.-G., Kim, S.-I., Yoon, K.-J., Kim, H.-C., & Kim, J. T. (2010). A new apartment construction technology with effective CO_2 emission reduction capabilities. *Energy*, *35*(6), 2639–2646. https://doi.org/10.1016/j.energy. .05.036

Hong, W.-K., Le, T.-A., Nguyen, M. C., & Pham, T. D. (2022a). "ANN-based Lagrange optimization for RC circular columns having multi-objective functions". *Journal of Asian Architecture and Building Engineering*. https://doi.org/10.1080/13467581.2022.2064864

Hong, W. K., Nguyen, M.C. (2021). "AI-based Lagrange optimization for designing reinforced concrete columns". *Journal of Asian Architecture and Building Engineering TABE*. https://doi.org/10.1080/13467581.2021.1971998

Hong, W. K., Nguyen, M. C., & Pham, T. D. (2022b). "Optimized interaction P-M diagram for rectangular reinforced concrete column based on artificial neural networks abstract". *Journal of Asian Architecture and Building Engineering*, 1–25. https://doi.org/10.1080/13467581.2021.2018697

Hong, W.-K., Nguyen, V. T., Nguyen, D. H., & Nguyen, M. C. (2022c). "An AI-based Lagrange optimization for a design for concrete columns encasing H-shaped steel sections under a biaxial bending". *Journal of Asian Architecture and Building Engineering*. https://doi.org/10.1080/13467581.2022.2060985

Hong, W. K., Nguyen, V. T., Nguyen, D. H., & Nguyen, M. C. (2022d). "Reverse design-based optimizations for reinforced concrete columns encasing H-shaped steel section using ANNs". *Journal of Asian Architecture and Building Engineering*, 1–15. https://doi.org/10.1080/13467581.2022.2047985

Hong, W. K., Nguyen, V. T., & Nguyen, M. C. (2021a). "Artificial intelligence-based noble design charts for doubly reinforced concrete beams". *Journal of Asian Architecture and Building Engineering, 21*(4), 1497–1519. https://doi.org/10.1080/13467581.2021.1928511
Hong, W.-K., Nguyen, V. T., & Nguyen, M. C. (2021b). "Optimizing reinforced concrete beams cost based on AI-based Lagrange functions". *Journal of Asian Architecture and Building Engineering.* https://doi.org/10.1080/13467581.2021.2007105
Hong, W. K., Pham, T. D., & Nguyen, V. T. (2021c). "Feature selection based reverse design of doubly reinforced concrete beams". *Journal of Asian Architecture and Building Engineering*, 1–25. https://doi.org/10.1080/13467581.2021.1928510
Hosseinpour, M., Sharifi, Y., & Sharifi, H. (2020). "Neural network application for distortional buckling capacity assessment of castellated steel beams". *Structures, 27*, 1174–1183. https://doi.org/10.1016/j.istruc.2020.07.027
Jaeggi, D., Parks, G., Kipouros, T., & Clarkson, J. (2005). "A multi-objective tabu search algorithm for constrained optimisation problems" In C. A. Coello Coello, A. Hernandez Aguirre, & E. Zitzler (Eds.), *BT - Evolutionary Multi-Criterion Optimization* (pp. 490–504). Springer, Berlin, Heidelberg.
Jahjouh, M. M., Arafa, M. H., & Alqedra, M. A. (2013). "Artificial Bee Colony (ABC) algorithm in the design optimization of RC continuous beams". *Structural and Multidisciplinary Optimization, 47*(6), 963–979. https://doi.org/10.1007/s00158-013-0884-y
Karush, W. (1939). "Minima of Functions of Several Variables with Inequalities as Side Constraints". M.Sc. thesis. Department of Mathematics, University of Chicago, Chicago, IL.
Kaveh, A., & Mahdavi, V. R. (2019). "Multi-objective colliding bodies optimization algorithm for design of trusses". *Journal of Computational Design and Engineering, 6*(1), 49–59. https://doi.org/10.1016/j.jcde.2018.04.001
Kaveh, A., & Sabzi, O. (2012). "Optimal design of reinforced concrete frames using big bang-big crunch algorithm". *International Journal of Civil Engineering, 10*(3), 189–200.
Kayabekir, A. E., Arama, Z. A., Bekda015F, G., Nigdeli, S. M., & Geem, Z. W. (2020). "Eco-friendly design of reinforced concrete retaining walls: Multi-objective optimization with harmony search applications". *Sustainability, 12*(15). https://doi.org/10.3390/su12156087
Kuhn, H. W.; Tucker, A. W. (1951). Nonlinear programming. In *Proceedings of 2nd Berkeley Symposium* (pp. 481–492). University of California Press, Berkeley.
Lee, M.-S., Hong, K., & Choi, S.-W. (n.d.). "Genetic algorithm based optimal structural design method for cost and CO_2 emissions of reinforced concrete frames TT - CCA0ADFCCF58D06CB9ACD2B8 BAA8BA58D2B8ACE8C870C758 BE44C6A9 BC0F C774C0B0D654D0C4C18C BC30CD9CB7C9C744 ACE0B824D55C C720C804C790C54CACE0B9ACC998 AE30BC18 AD6CC870CD5CC801D654AE30BC95". *Journal of the Computational Structural Engineering Institute of Korea, 29*(5), 429–436. https://doi.org/10.7734/COSEIK.2016.29.5.429
Liu, X., & Reynolds, A. C. (2016). "Gradient-based multi-objective optimization with applications to waterflooding optimization". *Computational Geosciences, 20*(3), 677–693. https://doi.org/10.1007/s10596-015-9523-6
Marler, R. T., & Arora, J. S. (2004). "Survey of multi-objective optimization methods for engineering". *Structural and Multidisciplinary Optimization, 26*(6), 369–395. https://doi.org/10.1007/s00158-003-0368-6
Martinez-Martin, F. J., Gonzalez-Vidosa, F., Hospitaler, A., & Yepes, V. (2012). "Multi-objective optimization design of bridge piers with hybrid heuristic algorithms". *Journal of Zhejiang University: Science A, 13*(6), 420–432. https://doi.org/10.1631/jzus.A1100304
MathWorks (2020) . *MATLAB R2020b, Version 9.9.0.* The MathWorks Inc., Natick, MA.
Mei, L., & Wang, Q. (2021). "Structural optimization in civil engineering: A literature review". *Buildings, 11*(2), 1–28. https://doi.org/10.3390/buildings11020066
Munk, D. J., Vio, G. A., & Steven, G. P. (2015). "Topology and shape optimization methods using evolutionary algorithms: A review". *Structural and Multidisciplinary Optimization, 52*(3), 613–631. https://doi.org/10.1007/s00158-015-1261-9
Nan, B., Bai, Y., & Wu, Y. (2020). "Multi-objective optimization of spatially truss structures based on node movement". *Applied Sciences, 10*(6). https://doi.org/10.3390/app10061964

Nguyen, D. H., & Hong, W. K. 2019. "Part I: The analytical model predicting post-yield behavior of concrete-encased steel beams considering various confinement effects by transverse reinforcements and steels". *Materials*, *12*(14). https://doi.org/10.3390/ma12142302

Nguyen, T.-A., Ly, H.-B., & Tran, V. Q. (2021). "Investigation of ANN architecture for predicting load-carrying capacity of castellated steel beams". *Complexity*, *2021*, 6697923. https://doi.org/10.1155/2021/6697923

Pareto, V. (1906). Wikipedia, Pareto front, https://en.wikipedia.org/wiki/Pareto_front

Park, H. S., Kwon, B., Shin, Y., Kim, Y., Hong, T., & Choi, S. W. (2013). "Cost and CO_2 emission optimization of steel reinforced concrete columns in high-rise buildings". *Energies*, *6*(11), 5609–5624. https://doi.org/10.3390/en6115609

Peel, C., & Moon, T. K. 2020. "Algorithms for optimization". *IEEE Control Systems*, *40*(2), 92–94. https://doi.org/10.1109/MCS.2019.2961589

Shaqfa, M., & Orb00E1n, Z. (2019). "Modified parameter-setting-free harmony search (PSFHS) algorithm for optimizing the design of reinforced concrete beams". *Structural and Multidisciplinary Optimization*, *60*(3), 999–1019. https://doi.org/10.1007/s00158-019-02252-4

Sharifi, Y., Moghbeli, A., Hosseinpour, M., & Sharifi, H. (2020). "Study of neural network models for the ultimate capacities of cellular steel beams". *Iranian Journal of Science and Technology, Transactions of Civil Engineering*, *44*(2), 579–589. https://doi.org/10.1007/s40996-019-00281-z

Tahmassebi, A., Mohebali, B., Meyer2010Baese, A., & Gandomi, A. H. (2020). "Multiobjective genetic programming for reinforced concrete beam modeling". *Applied AI Letters*, *1*(1), 1–10. https://doi.org/10.1002/ail2.9

Yang, X.-S. (2014). "Multi-objective optimization". In X.-S. Yang (Ed.), *Nature-Inspired Optimization Algorithms* (pp. 197–211). Elsevier. https://doi.org/10.1016/B978-0-12-416743-8.00014-2

Yucel, M., M., Bekdaş, Nigdeli, S. M., & Kayabekir, A. E. (2021a). "An artificial intelligence-based prediction model for optimum design variables of reinforced concrete retaining walls". *International Journal of Geomechanics*, *21*(12), 1–10. https://doi.org/10.1061/(asce)gm.1943-5622.0002234

Yucel, M., Nigdeli, S. M., Kayabekir, A. E., & Bekdaş, G. (2021b). "Optimization and artificial neural network models for reinforced concrete members". In S. Carbas, A. Toktas, & D. Ustun (Eds.), *Nature-Inspired Metaheuristic Algorithms for Engineering Optimization Applications* (pp. 181–199). Springer, Singapore. https://doi.org/10.1007/978-981-33-6773-9_9

Zhang, Z.-Y., Gifari, Z., Ju, Y.-K., & Kim, J. H. (2021). "Multi-objective optimization of the reinforced concrete beam". In S. M. Nigdeli, J. H. Kim, G. Bekda015F, & A. Yadav (Eds.), *Proceedings of 6th International Conference on Harmony Search, Soft Computing and Applications* (pp. 171–178). Springer, Singapore.

Zheng, X., Zhou, D., Li, N., Wu, T., Lei, Y., & Shi, J. (2021). "Self-regulated particle swarm multi-task optimization". *Sensors*, *21*(22). https://doi.org/10.3390/s21227499

Zheng, Y. G., & Hu, X. X. (2018). "Multi-objective optimal design on vibration suppression of building structures with active mass damper based on state difference feedback". In *Chinese Control Conference, CCC, 2018-July* (pp. 1249–1253). https://doi.org/10.23919/ChiCC.2018.8482723

ACKNOWLEDGEMENTS

This work was supported by the National Research Foundation of Korea (NRF) grant funded by the Korean government (MSIT 2019R1A2C2004965).

Appendix A

A1 ITERATION 1 OF CASE 1 KKT FOR ANN-BASED LAGRANGE OPTIMIZATION

At the end of Iteration 0, the initial trial input parameter $\left[x_A^{(0)}, x_b^{(0)}, \lambda_{v1}^{(0)}, \lambda_{v2}^{(0)}\right]^T = [0.7, 40, 0, 0]^T$ shown in Equation (2.4.4.40-1) is updated to $\left[x_C^{(1)}, \theta^{(1)}, \lambda_c^{(1)}, \lambda_v^{(1)}\right] = [0.584,\ 38.80,\ -1.903,\ 1.096]$ shown in Equation (2.4.4.58-1) at the end of Iteration 0. Functions y_{P_f} (y_{20}), y_{P_t} (y_{21}) and their Jacobi, Hessian matrices are derived at Iteration 1 based on $\left[x_C^{(1)}, \theta^{(1)}, \lambda_c^{(1)}, \lambda_v^{(1)}\right]$. The entire formulations are included in Appendix A1. Iteration 1 begins by substituting the red input parameter $\left[x_C^{(1)}, \theta^{(1)}, \lambda_c^{(1)}, \lambda_v^{(1)}\right]$ obtained at the end of Iteration 0 into Equation (A1.1-1). Lagrange multipliers $\lambda_c^{(1)} = -1.903$ and $\lambda_v^{(1)} = 1.096$ are used when formulating Jacobian and Hessian of Lagrange function as shown in red inputs of Equations (A1.13-2) and (A1.17). The updated input parameter $\left[x_C^{(1)}, \theta^{(1)}, \lambda_c^{(1)}, \lambda_v^{(1)}\right]$ is, then, updated to $\left[x_C^{(2)}, \theta^{(2)}, \lambda_c^{(2)}, \lambda_v^{(2)}\right]^T = [0.569, 41.07, -3.833, 3.184]^T$ at the end of Iteration 1 as shown in Equation (A1.19). Weight and bias matrices obtained in Table 2.4.4.4 are used in this appendix. Figure 2.3.6.2 is referred for all procedures.

a. Iteration 1; Step (N) for normalizing input, $\mathbf{z}^{(N)}$

$$\mathbf{z}^{(N)} = \mathbf{g}^{(N)}\left(\mathbf{x}^{(0)}\right) = \boldsymbol{\alpha}_x \odot \left(\mathbf{x} - \mathbf{x}_{min}\right) + \bar{\mathbf{x}}_{min}$$

$$= \begin{bmatrix} 1.820 \\ 0.029 \end{bmatrix} \odot \left(\begin{bmatrix} 0.584 \\ 38.80 \end{bmatrix} - \begin{bmatrix} 0.1 \\ 10 \end{bmatrix} \right) + \begin{bmatrix} -1 \\ -1 \end{bmatrix} \tag{A1.1-1}$$

$$= \begin{bmatrix} -0.1194 \\ -0.1770 \end{bmatrix}$$

$$\mathbf{J}^{(N)} = I_2 \odot \boldsymbol{\alpha}_x = \begin{bmatrix} 1.820 & 0 \\ 0 & 0.029 \end{bmatrix} \tag{A1.1-2}$$

$$\mathbf{H}^{(N)} = \begin{bmatrix} 0 & 0 \\ 0 & 0 \end{bmatrix} \tag{A1.1-3}$$

b. Iteration 1; calculations of output functions, Jacobi and Hessian matrices at Normalized hidden layers, $\mathbf{z}^{(l)}\ \forall l \in \{1, 2, 3, 4\}$

a. Normalized hidden layer, $\mathbf{z}^{(1)}$

$$\mathbf{z}^{(1)} = f_t^{(1)}\left(\mathbf{W}^{(1)}\mathbf{z}^{(N)} + \mathbf{b}^{(1)}\right) = \begin{bmatrix} 0.9997 \\ -0.9880 \\ -0.9721 \\ -0.5114 \\ -0.7402 \\ -0.5729 \\ -0.0449 \\ -0.9625 \\ 0.9488 \\ -0.9999 \end{bmatrix} \tag{A1.2-1}$$

$$\mathbf{J}^{(1)} = \left(1-\left(\mathbf{z}^{(1)}\right)^2\right) \odot \mathbf{W}^{(1)}\mathbf{J}^{(N)} = \begin{bmatrix} -0.009 & -0.0001 \\ 0.0531 & 0.0013 \\ 0.0670 & -0.0013 \\ 0.7739 & -0.0167 \\ 0.0077 & -0.0193 \\ -0.3634 & 0.0238 \\ -0.7726 & -0.0122 \\ -0.2587 & -0.0027 \\ 0.1859 & -0.0022 \\ -0.0012 & 0.0000 \end{bmatrix} \tag{A1.2-2}$$

$$\mathbf{H}_1^{(1)} = -2\mathbf{z}^{(1)} \odot \left(1-\left(\mathbf{z}^{(1)}\right)^2\right) \odot \boldsymbol{i}_1^{(N)} \odot \left(\mathbf{W}^{(1)}\right)^2 \mathbf{J}^{(N)}$$

$$+\left(1-\left(\mathbf{z}^{(1)}\right)^2\right) \odot \mathbf{W}^{(1)}\mathbf{H}_1^{(N)} = \begin{bmatrix} -0.0031 & -0.0002 \\ 0.2336 & 0.0059 \\ 0.1584 & -0.0030 \\ 0.8295 & -0.0179 \\ 0.0002 & -0.0005 \\ 0.2252 & -0.0147 \\ -0.0537 & -0.0008 \\ 1.7535 & 0.0180 \\ -0.6575 & 0.0077 \\ 0.0133 & -0.0001 \end{bmatrix} \tag{A1.2-3a}$$

$$\mathbf{H}_2^{(1)} = -2\mathbf{z}^{(1)} \odot \left(1-\left(\mathbf{z}^{(1)}\right)^2\right) \odot \boldsymbol{i}_2^{(N)} \odot \left(\mathbf{W}^{(1)}\right)^2 \mathbf{J}^{(N)}$$

$$+\left(1-\left(\mathbf{z}^{(1)}\right)^2\right) \odot \mathbf{W}^{(1)}\mathbf{H}_1^{(N)} = \begin{bmatrix} -0.0002 & -0.0000 \\ 0.0059 & 0.0002 \\ -0.0030 & 0.0001 \\ -0.0179 & 0.0004 \\ -0.0005 & 0.0012 \\ -0.0147 & 0.0010 \\ -0.0008 & -0.0000 \\ 0.0180 & 0.0002 \\ 0.0077 & -0.0001 \\ -0.0001 & 0.0000 \end{bmatrix} \quad \text{(A1.2-3a)}$$

b. Normalized hidden layer, $\mathbf{z}^{(2)}$

$$\mathbf{z}^{(2)} = f_t^{(2)}\left(\mathbf{W}^{(2)}\mathbf{z}^{(1)} + \mathbf{b}^{(2)}\right) = \begin{bmatrix} 0.5583 \\ 0.9886 \\ 0.2543 \\ 0.9470 \\ -0.6386 \\ -0.5087 \\ -0.9538 \\ -0.1833 \\ 0.5089 \\ 0.9508 \end{bmatrix} \quad \text{(A1.3-1)}$$

$$\mathbf{J}^{(2)} = \left(1-\left(\mathbf{z}^{(2)}\right)^2\right) \odot \mathbf{W}^{(2)}\mathbf{J}^{(1)} = \begin{bmatrix} 0.9432 & -0.0026 \\ 0.0289 & -0.0002 \\ -0.1571 & 0.0131 \\ 0.0093 & -0.0001 \\ 0.3032 & -0.0001 \\ -1.2243 & -0.0109 \\ -0.0243 & 0.0001 \\ -0.3900 & -0.0118 \\ -0.3274 & 0.0077 \\ -0.0899 & -0.0035 \end{bmatrix} \quad \text{(A1.3-2)}$$

$$\mathbf{H}_1^{(2)} = -2\mathbf{z}^{(2)} \odot \left(1-\left(\mathbf{z}^{(2)}\right)^2\right) \odot \boldsymbol{i}_1^{(1)} \odot \left(\mathbf{W}^{(2)}\right)^2 \mathbf{J}^{(1)}$$

$$+\left(1-\left(\mathbf{z}^{(2)}\right)^2\right) \odot \mathbf{W}^{(2)}\mathbf{H}_1^{(1)} = \begin{bmatrix} -1.0786 & 0.0116 \\ -0.0880 & 0.0004 \\ -0.2215 & -0.0009 \\ -0.0609 & 0.0032 \\ -0.2780 & 0.0107 \\ 2.3340 & 0.0278 \\ -0.0627 & -0.0012 \\ 0.4571 & 0.0165 \\ 0.3422 & -0.0077 \\ -0.2127 & -0.0042 \end{bmatrix} \quad \text{(A1.3-3a)}$$

$$\mathbf{H}_2^{(2)} = -2\mathbf{z}^{(2)} \odot \left(1-\left(\mathbf{z}^{(2)}\right)^2\right) \odot \boldsymbol{i}_2^{(1)} \odot \left(\mathbf{W}^{(2)}\right)^2 \mathbf{J}^{(1)}$$

$$+\left(1-\left(\mathbf{z}^{(2)}\right)^2\right) \odot \mathbf{W}^{(2)}\mathbf{H}_1^{(1)} = \begin{bmatrix} 0.0116 & 0.0000 \\ -0.0004 & 0.0000 \\ -0.0009 & 0.0002 \\ 0.0032 & -0.0001 \\ 0.0107 & -0.0006 \\ 0.0278 & -0.0001 \\ -0.0012 & 0.0001 \\ 0.0165 & -0.0009 \\ -0.0077 & 0.0002 \\ -0.0042 & -0.0003 \end{bmatrix} \quad \text{(A1.3-3b)}$$

c. Normalized hidden layer, $\mathbf{z}^{(3)}$

$$\mathbf{z}^{(3)} = f_t^{(3)}\left(\mathbf{W}^{(3)}\mathbf{z}^{(2)} + \mathbf{b}^{(3)}\right) = \begin{bmatrix} -0.9472 \\ -0.9201 \\ -0.0996 \\ 0.8935 \\ 0.3771 \\ 0.5887 \\ -0.9989 \\ 0.9425 \\ 0.9993 \\ -0.9431 \end{bmatrix} \quad \text{(A1.4-1)}$$

$$\mathbf{J}^{(3)} = \left(1-\left(\mathbf{z}^{(3)}\right)^2\right) \odot \mathbf{W}^{(3)}\mathbf{J}^{(2)} = \begin{bmatrix} 0.0886 & 0.0024 \\ 0.1976 & -0.0010 \\ -0.6610 & -0.0109 \\ -0.1134 & -0.0012 \\ 0.5238 & -0.0095 \\ 0.4407 & 0.0051 \\ -0.0029 & 0.0000 \\ -0.1279 & 0.0013 \\ 0.0001 & 0.0000 \\ -0.0912 & 0.0010 \end{bmatrix} \tag{A1.4-2}$$

$$\mathbf{H}_1^{(3)} = -2\mathbf{z}^{(3)} \odot \left(1-\left(\mathbf{z}^{(3)}\right)^2\right) \odot \boldsymbol{i}_1^{(2)} \odot \left(\mathbf{W}^{(3)}\right)^2 \mathbf{J}^{(2)}$$

$$+\left(1-\left(\mathbf{z}^{(3)}\right)^2\right) \odot \mathbf{W}^{(3)}\mathbf{H}_1^{(2)} = \begin{bmatrix} -0.0015 & 0.0012 \\ 0.3904 & 0.0008 \\ 0.5062 & -0.0119 \\ -0.1161 & -0.0054 \\ -0.5721 & 0.0103 \\ -0.8202 & -0.0011 \\ 0.0129 & -0.0000 \\ -0.2388 & 0.0001 \\ -0.0001 & 0.0000 \\ 0.1172 & -0.0063 \end{bmatrix} \tag{A1.4-3a}$$

$$\mathbf{H}_2^{(3)} = -2\mathbf{z}^{(3)} \odot \left(1-\left(\mathbf{z}^{(3)}\right)^2\right) \odot \boldsymbol{i}_2^{(2)} \odot \left(\mathbf{W}^{(3)}\right)^2 \mathbf{J}^{(2)}$$

$$+\left(1-\left(\mathbf{z}^{(3)}\right)^2\right) \odot \mathbf{W}^{(3)}\mathbf{H}_2^{(2)} = \begin{bmatrix} 0.0012 & 0.0001 \\ -0.0008 & 0.0001 \\ -0.0119 & 0.0000 \\ -0.0054 & -0.0001 \\ 0.0103 & 0.0001 \\ -0.0011 & 0.0002 \\ -0.0000 & 0.0000 \\ 0.0001 & -0.0001 \\ 0.0000 & 0.0000 \\ -0.0063 & -0.0001 \end{bmatrix} \tag{A1.4-3b}$$

d. Normalized hidden layer, $\mathbf{z}^{(4)}$

$$\mathbf{z}^{(4)} = f_t^{(4)}\left(\mathbf{W}^{(4)}\mathbf{z}^{(3)} + \mathbf{b}^{(4)}\right) = \begin{bmatrix} -0.7601 \\ -0.3744 \\ 0.9999 \\ 0.5615 \\ -0.9994 \\ -0.6528 \\ 0.9997 \\ 0.1549 \\ 0.9997 \\ 0.7992 \end{bmatrix} \tag{A1.5-1}$$

$$\mathbf{J}^{(4)} = \left(1-\left(\mathbf{z}^{(4)}\right)^2\right) \odot \mathbf{W}^{(4)}\mathbf{J}^{(3)} = \begin{bmatrix} 0.6305 & 0.0043 \\ 0.5515 & 0.0144 \\ -0.0001 & -0.0000 \\ -0.6692 & 0.0069 \\ 0.0003 & 0.0000 \\ 0.7112 & -0.0069 \\ 0.0000 & 0.0000 \\ 0.5045 & -0.0012 \\ 0.0004 & -0.0000 \\ -0.0870 & -0.0069 \end{bmatrix} \tag{A1.5-2}$$

$$\mathbf{H}_1^{(4)} = -2\mathbf{z}^{(4)} \odot \left(1-\left(\mathbf{z}^{(4)}\right)^2\right) \odot \boldsymbol{i}_1^{(3)} \odot \left(\mathbf{W}^{(4)}\right)^2 \mathbf{J}^{(3)}$$

$$+\left(1-\left(\mathbf{z}^{(4)}\right)^2\right) \odot \mathbf{W}^{(4)}\mathbf{H}_1^{(3)} = \begin{bmatrix} 0.8301 & 0.0138 \\ -0.9895 & 0.0082 \\ -0.0002 & -0.0000 \\ -0.3515 & 0.0050 \\ 0.0012 & 0.0000 \\ 0.7757 & 0.0038 \\ -0.0009 & -0.0000 \\ -1.1035 & 0.0087 \\ -0.0011 & 0.0000 \\ -0.1138 & -0.0031 \end{bmatrix} \tag{A1.5-3a}$$

$$\mathbf{H}_2^{(4)} = -2\mathbf{z}^{(4)} \odot \left(1-\left(\mathbf{z}^{(4)}\right)^2\right) \odot \boldsymbol{i}_2^{(3)} \odot \left(\mathbf{W}^{(4)}\right)^2 \mathbf{J}^{(3)}$$

$$+\left(1-\left(\mathbf{z}^{(4)}\right)^2\right) \odot \mathbf{W}^{(4)}\mathbf{H}_2^{(3)} = \begin{bmatrix} 0.0138 & 0.0003 \\ 0.0082 & 0.0003 \\ -0.0000 & -0.0000 \\ 0.0050 & -0.0002 \\ 0.0000 & -0.0000 \\ 0.0038 & 0.0003 \\ 0.0000 & -0.0000 \\ 0.0087 & 0.0002 \\ 0.0000 & 0.0000 \\ -0.0031 & -0.0001 \end{bmatrix} \tag{A1.5-3b}$$

c. Normalized output layer ($5 =$ output), $\mathbf{z}^{(L=5=\text{output})}$

$$\mathbf{z}_{y20}^{(L=5)} = f_{lin}^{(5)}\left(\mathbf{w}_{20}^{(5)}\mathbf{z}^{(4)} + b_{20}^{(5)}\right) = 0.5836 \tag{A1.6-1}$$

$$\mathbf{J}_{y20}^{(L=5)} = \mathbf{w}_{20}^{(L=5)}\mathbf{J}^{(4)} = \begin{bmatrix} -1.0326 & 0.0073 \end{bmatrix} \tag{A1.6-2}$$

$$\mathbf{H}_{y20,1}^{(L=5)} = \mathbf{w}_{20}^{(L=5)}\mathbf{H}_{1}^{(4)} = \begin{bmatrix} -3.9017 & -0.0197 \end{bmatrix} \tag{A1.6-3a}$$

$$\mathbf{H}_{y20,2}^{(L=5)} = \mathbf{w}_{20}^{(L=5)}\mathbf{H}_{2}^{(4)} = \begin{bmatrix} -0.0197 & -0.0005 \end{bmatrix} \tag{A1.6-3b}$$

$$\mathbf{z}_{y21}^{(L=5)} = f_{lin}^{(5)}\left(\mathbf{w}_{21}^{(5)}\mathbf{z}^{(4)} + b_{21}^{(5)}\right) = 0.5787 \tag{A1.7-1}$$

$$\mathbf{J}_{y21}^{(L=5)} = \mathbf{w}_{21}^{(5)}\mathbf{J}^{(4)} = \begin{bmatrix} -0.8860 & 0.0050 \end{bmatrix} \tag{A1.7-2}$$

$$\mathbf{H}_{y21,1}^{(L=5)} = \mathbf{w}_{21}^{(5)}\mathbf{H}_{1}^{(4)} = \begin{bmatrix} -4.5333 & -0.0142 \end{bmatrix} \tag{A1.7-3a}$$

$$\mathbf{H}_{y21,2}^{(L=5)} = \mathbf{w}_{21}^{(5)}\mathbf{H}_{2}^{(4)} = \begin{bmatrix} -0.0142 & -0.0004 \end{bmatrix} \tag{A1.7-3b}$$

d. De-normalized output layer (D, $5 =$ output), $\mathbf{y} = \mathbf{z}^{(D)}$; de-normalizing $\mathbf{z}^{(5=\text{output})}$

$$\begin{aligned} y_{20} = z_{y20}^{(D)} &= g_{20}^{(D)}\left(z_{y20}^{(L=5)}\right) = \frac{1}{\alpha_{y20}}\left(z_{y20}^{(5)} - \bar{y}_{20,min}\right) + y_{20,min} \\ &= \frac{1}{1.5427}\left(0.5838 - (-1)\right) + (-0.9871) = 0.0394 \end{aligned} \tag{A1.8-1}$$

$$\mathbf{J}_{y20}^{(D)} = \frac{1}{\alpha_{y20}}\mathbf{J}_{y20}^{(L=5)} = \begin{bmatrix} -0.6693 & 0.0047 \end{bmatrix} \tag{A1.8-2}$$

$$\mathbf{H}_{y20,1}^{(D)} = \frac{1}{\alpha_{y20}}\mathbf{H}_{y20,1}^{(L=5)} = \begin{bmatrix} -2.5291 & -0.0128 \end{bmatrix} \tag{A1.8-3a}$$

$$\mathbf{H}_{y20,2}^{(D)} = \frac{1}{\alpha_{y20}}\mathbf{H}_{y20,2}^{(L=5)} = \begin{bmatrix} -0.0128 & -0.0003 \end{bmatrix} \tag{A1.8-3b}$$

$$\mathbf{H}_{y20}^{(D)} = \begin{bmatrix} \mathbf{H}_{y20,1}^{(D)} \\ \mathbf{H}_{y20,2}^{(D)} \end{bmatrix} = \begin{bmatrix} -2.5291 & -0.0128 \\ -0.0128 & -0.0003 \end{bmatrix} \tag{A1.8-4}$$

$$y_{21} = z_{y21}^{(D)} = g_{21}^{(D)}\left(z_{y21}^{(L=5)}\right) = \frac{1}{\alpha_{y21}}\left(z_{y21}^{(5)} - \bar{y}_{21,\min}\right) + y_{21,\min}$$

$$= \frac{1}{1.7958}(0.5787 - (-1)) + (-0.7988) = 0.0802 \quad \text{(A1.9-1)}$$

$$\mathbf{J}_{y21}^{(D)} = \frac{1}{\alpha_{y21}}\mathbf{J}_{y21}^{(L=5)} = \begin{bmatrix} -0.4934 & 0.0055 \end{bmatrix} \quad \text{(A1.9-2)}$$

$$\mathbf{H}_{y21,1}^{(D)} = \frac{1}{\alpha_{y21}}\mathbf{H}_{y21,1}^{(5)} = \begin{bmatrix} -2.5242 & -0.0079 \end{bmatrix} \quad \text{(A1.9-3a)}$$

$$\mathbf{H}_{y21,2}^{(D)} = \frac{1}{\alpha_{y21}}\mathbf{H}_{y21,2}^{(5)} = \begin{bmatrix} -0.0079 & -0.0002 \end{bmatrix} \quad \text{(A1.9-3b}$$

$$\mathbf{H}_{y21}^{(D)} = \begin{bmatrix} \mathbf{H}_{y21,1}^{(D)} \\ \mathbf{H}_{y21,2}^{(D)} \end{bmatrix} = \begin{bmatrix} -2.5242 & -0.0079 \\ -0.0079 & -0.0002 \end{bmatrix} \quad \text{(A1.9-4)}$$

Output variables y_{20} and y_{21}, an equality constraint $c(x_C,\theta)$, and inequality constraints $v(x_C,\theta)$ are obtained in Equations (A1.8-1), (A1.9-1), (A1.10-1), and (A1.10-2), respectively. Jacobi $\mathbf{J}_c\left(x_C^{(1)}, \theta^{(1)}\right)$ and $\mathbf{J}_v\left(x_C^{(1)}, \theta^{(1)}\right)$ are calculated in Equations (A1.11-1) and (A1.11-2), respectively, as functions of $x_C^{(0)}$ and $\theta^{(0)}$ whereas Hessian matrix $\mathbf{H}_c\left(x_C^{(1)}, \theta^{(1)}\right)$ and $\mathbf{H}_v\left(x_C^{(1)}, \theta^{(1)}\right)$ are calculated as functions of $x_C^{(0)}$ and $\theta^{(0)}$ in Equations (A1.12-1) and (A1.12-2). Lagrange multipliers $\lambda_c^{(1)} = -1.903$ and $\lambda_v^{(1)} = 1.096$ obtained in Equation (2.4.4.58-1) are used when formulating Jacobian and Hessian of Lagrange function as shown in red inputs of Equations (A1.13-2) and (A1.17). Weight and bias matrices obtained in Table 2.4.4.4 are used in this appendix. Figure 2.3.6.2 is referred for all procedures. A first derivative (Jacobi) of Lagrange function which is, then, obtained in Equations (A1.13-1) and (A1.13-2) with respect to the updated input parameter $\left[x_C^{(1)}, \theta^{(1)}, \lambda_c^{(1)}, \lambda_v^{(1)}\right] = [0.584,\ 38.80,\ -1.903,\ 1.096]$ obtained in Equation (2.4.4.58-1) and $s = 1$ does not converge to the solutions, indicating that $\left[x_C^{(1)}, \theta^{(1)}, \lambda_c^{(1)}, \lambda_v^{(1)}\right]$ is not the root of Case 1 KKT condition.

$$c\left(x_C^{(1)},\ \theta^{(1)}\right) = y_{P_f} - 0.06 = y_{20} - 0.06 = 0.0394 - 0.06 = -0.0206 \quad \text{(A1.10-1)}$$

$$v\left(x_C^{(1)},\ \theta^{(1)}\right) = y_{P_f} - 0.06 = y_{21} - 0.1 = 0.0802 - 0.1 = -0.0198 \quad \text{(A1.10-2)}$$

$$\mathbf{J}_c\left(x_C^{(1)},\ \theta^{(1)}\right) = \mathbf{J}_c^{(D)}\left(x_C^{(1)},\ \theta^{(1)}\right) = \mathbf{J}_{(yP_f - 0.06)}^{(D)} = \mathbf{J}_{yP_f}^{(D)} = \mathbf{J}_{y20}^{(D)}$$

$$= \begin{bmatrix} -0.6693 & 0.0047 \end{bmatrix} \quad \text{(A1.11-1)}$$

$$\mathbf{J}_v\left(x_C^{(1)},\theta^{(1)}\right)=\mathbf{J}_v^{(D)}\left(x_C^{(1)},\theta^{(1)}\right)=\mathbf{J}_{(y_{P_t}-0.1)}^{(D)}=\mathbf{J}_{y_{P_t}}^{(D)}=\mathbf{J}_{y21}^{(D)}=\begin{bmatrix} -0.4934 & 0.0055 \end{bmatrix} \quad \text{(A1.11–2)}$$

$$\mathbf{H}_c\left(x_C^{(1)},\theta^{(1)}\right)=\mathbf{H}_c^{(D)}\left(x_C^{(1)},\theta^{(1)}\right)=\mathbf{H}_{y20}^{(D)}=\begin{bmatrix} -2.5291 & -0.0128 \\ -0.0128 & -0.0003 \end{bmatrix} \quad \text{(A1.12-1)}$$

$$\mathbf{H}_v\left(x_C^{(1)},\theta^{(1)}\right)=\mathbf{H}_v^{(D)}\left(x_C^{(1)},\theta^{(1)}\right)=\mathbf{H}_{y21}^{(D)}=\begin{bmatrix} -2.5242 & -0.0079 \\ -0.0079 & -0.0002 \end{bmatrix} \quad \text{(A1.12-2)}$$

e. Convergence verification of Lagrange function based on Jacobi, Hessian matrix at Iteration 1

$$\nabla\mathcal{L}\left(x_C^{(1)},\theta^{(1)},\lambda_c^{(1)},\lambda_v^{(1)}\right)$$

$$=\begin{bmatrix} \nabla f_{xc}\left(x_C^{(1)}\right)-\left[J_c\left(x_C^{(1)},\theta^{(1)}\right)\right]^T\lambda_c^{(1)}-\left[J_v\left(x_C^{(1)},\theta^{(1)}\right)\right]^T s\lambda_v^{(1)} \\ -c\left(x_C^{(1)},\theta^{(1)}\right) \\ -sv\left(x_C^{(1)},\theta^{(1)}\right) \end{bmatrix}$$

$$=\begin{bmatrix} \nabla f_{xc}\left(x_C^{(1)}\right)-\left[\mathbf{J}_c^{(D)}\left(x_C^{(1)},\theta^{(1)}\right)\right]^T\lambda_c^{(1)}-\left[\mathbf{J}_v^{(D)}\left(x_C^{(1)},\theta^{(1)}\right)\right]^T s\lambda_v^{(1)} \\ -c\left(x_C^{(1)},\theta^{(1)}\right) \\ -sv\left(x_C^{(1)},\theta^{(1)}\right) \end{bmatrix} \quad \text{(A1.13-1)}$$

$$\nabla\mathcal{L}\left(x_C^{(1)},\theta^{(1)},\lambda_c^{(1)},\lambda_v^{(1)}\right)$$

$$=\begin{bmatrix} \begin{bmatrix} 1 \\ 0 \end{bmatrix}-\begin{bmatrix} -0.6693 \\ 0.0047 \end{bmatrix}\times(-1.903)-\begin{bmatrix} -0.4934 \\ 0.0055 \end{bmatrix}\times 1\times 1.096 \\ 0.0206 \\ 0.0198 \end{bmatrix}$$

$$\rightarrow\nabla\mathcal{L}\left(x_C^{(1)},\theta^{(1)},\lambda_c^{(1)},\lambda_v^{(1)}\right)=\begin{bmatrix} 0.2666 \\ 0.0030 \\ 0.0206 \\ 0.0198 \end{bmatrix}\neq 0 \quad \text{(A1.13-2)}$$

f. Updating the initial trial input parameter for Iteration 2

$\nabla\mathcal{L}\left(x_C^{(1)}, \theta^{(1)}, \lambda_c^{(1)}, \lambda_v^{(1)}\right)$ shown in Equation (A1.13-2) does not converge to zero, indicating $\left[x_C^{(1)}, \theta^{(1)}, \lambda_c^{(1)}, \lambda_v^{(1)}\right]$ is not the root of KKT condition for Case 1. The updated input parameter which will be used for Iteration 2 is to be calculated as Equation (A1.14).

$$\begin{bmatrix} x_C^{(2)} \\ \theta^{(2)} \\ \lambda_c^{(2)} \\ \lambda_v^{(2)} \end{bmatrix} = \begin{bmatrix} x_C^{(1)} \\ \theta^{(1)} \\ \lambda_c^{(1)} \\ \lambda_v^{(1)} \end{bmatrix} - \left[\mathbf{H}_{\mathcal{L}}\left(x_C^{(1)}, \theta^{(1)}, \lambda_c^{(1)}, \lambda_v^{(1)}\right)\right]^{-1} \nabla\mathcal{L}\left(x_C^{(1)}, \theta^{(1)}, \lambda_c^{(1)}, \lambda_v^{(1)}\right) \tag{A1.14}$$

A Hessian of Lagrange function with respect to $\left[x_C^{(1)}, \theta^{(1)}, \lambda_c^{(1)}, \lambda_v^{(1)}\right]$ is calculated based on Equations (A1.15-2) and (A1.18) using $H_{\mathcal{L}}\left(x_C^{(1)}, \theta^{(1)}\right)$ calculated in Equations (A1.16) and (A1.17), $\mathbf{J}_c^{(D)}$ calculated in Equation (A1.11-1), $\mathbf{J}_v^{(D)}$ calculated in Equation (A1.11–2). $H_{\mathcal{L}}\left(x_C^{(1)}, \theta^{(1)}\right)$ is calculated in Equation (A1.17) by substituting $H_{f_{xC}}\left(x_C, \theta\right)$, $\mathbf{H}_c^{(D)}$, and $\mathbf{H}_v^{(D)}$ from Equations (2.4.4.14), (A1.12-1), and (A1.12-2), respectively, into Equation (A1.16). Lagrange multipliers $\left[x_C^{(1)}, \theta^{(1)}, \lambda_c^{(1)}, \lambda_v^{(1)}\right] = [0.584, 38.80, -1.903, 1.096]$ obtained in Equation (2.4.4.58-1) are used in red as shown in Equation (A1.17).

$$\mathbf{H}_{\mathcal{L}}\left(x_C^{(1)}, \theta^{(1)}, \lambda_c^{(1)}, \lambda_v^{(1)}\right)$$

$$= \begin{bmatrix} \mathbf{H}_{\mathcal{L}}\left(x_C^{(1)}, \theta^{(1)}\right) & -\left[J_c\left(x_C^{(1)}, \theta^{(1)}\right)\right]^T & -\left[sJ_v\left(x_C^{(1)}, \theta^{(1)}\right)\right]^T \\ -J_c\left(x_C^{(1)}, \theta^{(1)}\right) & 0 & 0 \\ -sJ_v\left(x_C^{(1)}, \theta^{(1)}\right) & 0 & 0 \end{bmatrix} \tag{A1.15-1}$$

$$\mathbf{H}_{\mathcal{L}}\left(x_C^{(1)}, \theta^{(1)}, \lambda_c^{(1)}, \lambda_v^{(1)}\right)$$

$$= \begin{bmatrix} \mathbf{H}_{\mathcal{L}}\left(x_C^{(1)}, \theta^{(1)}\right) & -\left[\mathbf{J}_c^{(D)}\left(x_C^{(1)}, \theta^{(1)}\right)\right]^T & -\left[s\mathbf{J}_v^{(D)}\left(x_C^{(1)}, \theta^{(1)}\right)\right]^T \\ -\mathbf{J}_c^{(D)}\left(x_C^{(1)}, \theta^{(1)}\right) & 0 & 0 \\ -s\mathbf{J}_v^{(D)}\left(x_C^{(1)}, \theta^{(1)}\right) & 0 & 0 \end{bmatrix} \tag{A1.15-2}$$

where

$$\mathbf{H}_{\mathcal{L}}\left(x_C^{(1)}, \theta^{(1)}\right) = \mathbf{H}_{f_{xC}}\left(x_C^{(1)}, \theta^{(1)}\right) - \lambda_c^{(1)}\mathbf{H}_c\left(x_C^{(1)}, \theta^{(1)}\right) - s\lambda_v^{(1)}\mathbf{H}_v\left(x_C^{(1)}, \theta^{(1)}\right) \tag{A1.16-1}$$

$$\mathbf{H}_{\mathcal{L}}\left(x_C^{(1)}, \theta^{(1)}\right) = \mathbf{H}_{fxC}\left(x_C^{(1)}, \theta^{(1)}\right) - \lambda_c^{(1)}\mathbf{H}_c^{(D)}\left(x_C^{(1)}, \theta^{(1)}\right) - s\lambda_v^{(1)}\mathbf{H}_v^{(D)}\left(x_C^{(1)}, \theta^{(1)}\right) \quad \text{(A1.16-2)}$$

$$\mathbf{H}_{\mathcal{L}}\left(x_C^{(1)}, \theta^{(1)}\right) = \begin{bmatrix} 0 & 0 \\ 0 & 0 \end{bmatrix} - (-1.903) \times \begin{bmatrix} -2.5291 & -0.0128 \\ -0.0128 & -0.0003 \end{bmatrix}$$
$$-1 \times 1.096 \times \begin{bmatrix} -2.5242 & -0.0079 \\ -0.0079 & -0.0002 \end{bmatrix} \quad \text{(A1.17)}$$
$$= \begin{bmatrix} 0.3357 & -0.0285 \\ -0.0285 & -0.0006 \end{bmatrix}$$

$$\mathbf{H}_{\mathcal{L}}\left(x_C^{(1)}, \theta^{(1)}, \lambda_c^{(1)}, \lambda_v^{(1)}\right)$$
$$= \begin{bmatrix} \begin{bmatrix} -2.0483 & -0.0157 \\ -0.0157 & -0.0004 \end{bmatrix} & -\begin{bmatrix} -0.6693 \\ 0.0047 \end{bmatrix} & -\begin{bmatrix} -0.4934 \\ 0.0055 \end{bmatrix} \\ -\begin{bmatrix} -0.6693 & 0.0047 \end{bmatrix} & 0 & 0 \\ -\begin{bmatrix} -0.4934 & 0.0055 \end{bmatrix} & 0 & 0 \end{bmatrix} \quad \text{(A1.18-1)}$$

$$\mathbf{H}_{\mathcal{L}}\left(x_C^{(1)}, \theta^{(1)}, \lambda_c^{(1)}, \lambda_v^{(1)}\right) = \begin{bmatrix} -2.0483 & -0.0157 & 0.6693 & 0.4934 \\ -0.0157 & -0.0004 & -0.0047 & -0.0055 \\ 0.6693 & -0.0047 & 0 & 0 \\ 0.4934 & -0.0055 & 0 & 0 \end{bmatrix} \quad \text{(A1.18-2)}$$

$$\left[\mathbf{H}_{\mathcal{L}}\left(x_C^{(1)}, \theta^{(1)}, \lambda_c^{(1)}, \lambda_v^{(1)}\right)\right]^{-1} = \begin{bmatrix} 0 & 0 & 4.0693 & -3.4938 \\ 0 & 0 & 366.43 & -497.13 \\ 4.0693 & 366.43 & 129.15 & -146.67 \\ -3.4938 & -497.13 & -146.67 & 168.67 \end{bmatrix} \quad \text{(A1.18-3)}$$

Input parameter $\left[x_C^{(1)}, \theta^{(1)}, \lambda_c^{(1)}, \lambda_v^{(1)}\right]$ at Iteration 1 is updated to $\left[x_C^{(2)}, \theta^{(2)}, \lambda_c^{(2)}, \lambda_v^{(2)}\right]$ at Iteration 2 in Equation (A1.19) which will be used in Iteration 2 by substituting Equations (A1.13-2) and (A1.18-3) into Equation (A1.14). The first derivative (Jacobi) of Lagrange function is, then, verified for convergence with respect to the updated input parameter $\left[x_C^{(2)}, \theta^{(2)}, \lambda_c^{(2)}, \lambda_v^{(2)}\right]^T = [0.569,\ 41.07,\ -3.833,\ 3.184]^T$ obtained in Equation (A1.19) and $s = 1$. This procedure will be repeated until a first derivative (Jacobi) of Lagrange function converges.

$$\begin{bmatrix} x_C^{(2)} \\ \theta^{(2)} \\ \lambda_c^{(2)} \\ \lambda_v^{(2)} \end{bmatrix} = \begin{bmatrix} 0.584 \\ 38.80 \\ -1.903 \\ 1.096 \end{bmatrix} - \begin{bmatrix} 0 & 0 & 4.0693 & -3.4938 \\ 0 & 0 & 366.43 & -497.13 \\ 4.0693 & 366.43 & 129.15 & -146.67 \\ -3.4938 & -497.13 & -146.67 & 168.67 \end{bmatrix}$$

$$\begin{bmatrix} 0.2666 \\ 0.0030 \\ 0.0206 \\ 0.0198 \end{bmatrix} \rightarrow \begin{bmatrix} x_C^{(2)} \\ \theta^{(2)} \\ \lambda_c^{(2)} \\ \lambda_v^{(2)} \end{bmatrix} \tag{A1.19}$$

$$= \begin{bmatrix} 0.569 \\ 41.07 \\ -3.833 \\ 3.184 \end{bmatrix}$$

A2 ITERATION 1 OF CASE 2 KKT FOR ANN-BASED LAGRANGE OPTIMIZATION

At the end of Iteration 0, initial trial input parameter $\left[x_A^{(0)}, x_b^{(0)}, \lambda_{v1}^{(0)}, \lambda_{v2}^{(0)}\right]^T = [0.7, 40, -1, 0]^T$ shown in Equations (2.4.4.40-1) and (2.4.4.62-2) is updated to $\left[x_C^{(1)}, \theta^{(1)}, \lambda_c^{(1)}, \lambda_v^{(1)}\right] = [0.6020, 47.57, -1.086, 0]$ shown in Equation (2.4.4.68). Functions y_{P_f} (y_{20}), y_{P_t} (y_{21}) and their Jacobi, Hessian matrices are derived at Iteration 1 based on $\left[x_C^{(1)}, \theta^{(1)}, \lambda_c^{(1)}, \lambda_v^{(1)}\right]$. The entire formulations are included in Appendix A2. Iteration 1 begins by substituting the red input parameter $\left[x_C^{(1)}, \theta^{(1)}, \lambda_c^{(1)}, \lambda_v^{(1)}\right]$ obtained at the end of Iteration 0 into Equation (A2.1-1). Input parameter $\left[x_C^{(1)}, \theta^{(1)}, \lambda_c^{(1)}, \lambda_v^{(1)}\right]$ at Iteration 1 is updated to $\left[x_C^{(2)}, \theta^{(2)}, \lambda_c^{(2)}, \lambda_v^{(2)}\right]^T = [0.5895, 47.88, -1.106, 0]^T$ at the end of Iteration 1 as shown in Equation (A2.19). Lagrange multipliers $\lambda_c^{(1)} = -1.086$ and $\lambda_v^{(1)} = 0$ are used when formulating Jacobian and Hessian of Lagrange function as shown in red inputs of Equations (A2.13-2) and (A2.17). Weight and bias matrices obtained in Table 2.4.4.4 are used in this appendix. Figure 2.3.6.2 is referred for all procedures.

a. Iteration 1; Step (N) for normalizing input, $\mathbf{z}^{(N)}$

$$\mathbf{z}^{(N)} = \mathbf{g}^{(N)}\left(\mathbf{x}^{(0)}\right) = \alpha_x \odot \left(\mathbf{x} - \mathbf{x}_{min}\right) + \bar{\mathbf{x}}_{min}$$

$$= \begin{bmatrix} 1.820 \\ 0.029 \end{bmatrix} \odot \left(\begin{bmatrix} 0.6020 \\ 47.57 \end{bmatrix} - \begin{bmatrix} 0.1 \\ 10 \end{bmatrix} \right) + \begin{bmatrix} -1 \\ -1 \end{bmatrix} \tag{A2.1-1}$$

$$= \begin{bmatrix} -0.0863 \\ 0.0734 \end{bmatrix}$$

$$\mathbf{J}^{(N)} = I_2 \odot \boldsymbol{\alpha}_{\mathrm{x}} = \begin{bmatrix} 1.820 & 0 \\ 0 & 0.029 \end{bmatrix} \tag{A2.1-2}$$

$$\mathbf{H}^{(N)} = \begin{bmatrix} 0 & 0 \\ 0 & 0 \end{bmatrix} \tag{A2.1-3}$$

b. Iteration 1; calculations of output functions, Jacobi and Hessian matrices at Normalized hidden layers, $\mathbf{z}^{(l)}\ \forall l \in \{1, 2, 3, 4\}$

a. Normalized hidden layer, $\mathbf{z}^{(1)}$

$$\mathbf{z}^{(1)} = f_t^{(1)}\left(\mathbf{W}^{(1)}\mathbf{z}^{(N)} + \mathbf{b}^{(1)}\right) = \begin{bmatrix} 0.9984 \\ -0.9654 \\ -0.9805 \\ -0.6311 \\ -0.8678 \\ -0.3378 \\ -0.0762 \\ -0.9823 \\ 0.9323 \\ -0.9999 \end{bmatrix} \tag{A2.2-1}$$

$$\mathbf{J}^{(1)} = \left(1 - \left(\mathbf{z}^{(1)}\right)^2\right) \odot \mathbf{W}^{(1)}\mathbf{J}^{(N)} = \begin{bmatrix} -0.0054 & -0.0003 \\ 0.1515 & 0.0038 \\ 0.0470 & -0.0009 \\ 0.6305 & -0.0136 \\ 0.0042 & -0.0105 \\ -0.4792 & 0.0313 \\ -0.7698 & -0.0122 \\ -0.1233 & -0.0013 \\ 0.2496 & -0.0029 \\ -0.0016 & 0.0000 \end{bmatrix} \tag{A2.2-2}$$

$$\mathbf{H}_1^{(1)} = -2\mathbf{z}^{(1)} \odot \left(1 - \left(\mathbf{z}^{(1)}\right)^2\right) \odot \boldsymbol{i}_1^{(N)} \odot \left(\mathbf{W}^{(1)}\right)^2 \mathbf{J}^{(N)} + \left(1 - \left(\mathbf{z}^{(1)}\right)^2\right) \odot \mathbf{W}^{(1)}\mathbf{H}_1^{(N)}$$

$$= \begin{bmatrix} -0.0182 & -0.0011 \\ 0.6516 & 0.0165 \\ 0.1120 & -0.0021 \\ 0.8341 & -0.0180 \\ 0.0001 & -0.0003 \\ 0.1751 & -0.0115 \\ 0.0908 & 0.0014 \\ 1.8527 & 0.0088 \\ -0.8660 & 0.0101 \\ 0.0178 & -0.0001 \end{bmatrix} \tag{A2.2-3a}$$

$$\mathbf{H}_2^{(1)} = -2\mathbf{z}^{(1)} \odot \left(1-\left(\mathbf{z}^{(1)}\right)^2\right) \odot \boldsymbol{i}_2^{(N)} \odot \left(\mathbf{W}^{(1)}\right)^2 \mathbf{J}^{(N)} + \left(1-\left(\mathbf{z}^{(1)}\right)^2\right) \odot \mathbf{W}^{(1)}\mathbf{H}_1^{(N)}$$

$$= \begin{bmatrix} -0.0011 & -0.0001 \\ 0.0165 & 0.0004 \\ -0.0021 & 0.0000 \\ -0.0180 & 0.0004 \\ -0.0003 & 0.0008 \\ -0.0115 & 0.0007 \\ 0.0014 & 0.0000 \\ 0.0088 & 0.0001 \\ 0.0101 & -0.0001 \\ -0.0001 & 0.0000 \end{bmatrix} \qquad \text{(A2.2–3b)}$$

b. Normalized hidden layer, $\mathbf{z}^{(2)}$

$$\mathbf{z}^{(2)} = f_t^{(2)}\left(\mathbf{W}^{(2)}\mathbf{z}^{(1)} + \mathbf{b}^{(2)}\right) = \begin{bmatrix} 0.5570 \\ 0.9883 \\ 0.3728 \\ 0.9440 \\ -0.6517 \\ -0.6224 \\ -0.9502 \\ -0.3210 \\ 0.5766 \\ 0.9021 \end{bmatrix} \qquad \text{(A2.3-1)}$$

$$\mathbf{J}^{(2)} = \left(1-\left(\mathbf{z}^{(2)}\right)^2\right) \odot \mathbf{W}^{(2)}\mathbf{J}^{(1)} = \begin{bmatrix} 1.0373 & -0.0014 \\ 0.0238 & -0.0000 \\ -0.1677 & 0.0142 \\ 0.0347 & -0.0007 \\ 0.3903 & -0.0039 \\ -0.9089 & -0.0100 \\ -0.0377 & 0.0008 \\ -0.2526 & -0.0176 \\ -0.3834 & 0.0090 \\ -0.1350 & -0.0076 \end{bmatrix} \qquad \text{(A2.3-2)}$$

$$\mathbf{H}_1^{(2)} = -2\mathbf{z}^{(2)} \odot \left(1-\left(\mathbf{z}^{(2)}\right)^2\right) \odot \boldsymbol{i}_1^{(1)} \odot \left(\mathbf{W}^{(2)}\right)^2 \mathbf{J}^{(1)} + \left(1-\left(\mathbf{z}^{(2)}\right)^2\right) \odot \mathbf{W}^{(2)}\mathbf{H}_1^{(1)}$$

$$= \begin{bmatrix} -1.2799 & 0.0157 \\ -0.0447 & -0.0004 \\ -0.4310 & 0.0001 \\ -0.1358 & 0.0031 \\ 0.0890 & 0.0096 \\ 2.2039 & 0.0336 \\ -0.0649 & -0.0017 \\ 0.5387 & 0.0120 \\ 0.1329 & -0.0059 \\ -0.2227 & -0.0048 \end{bmatrix} \quad \text{(A2.3-3a)}$$

$$\mathbf{H}_2^{(2)} = -2\mathbf{z}^{(2)} \odot \left(1-\left(\mathbf{z}^{(2)}\right)^2\right) \odot \boldsymbol{i}_2^{(1)} \odot \left(\mathbf{W}^{(2)}\right)^2 \mathbf{J}^{(1)} + \left(1-\left(\mathbf{z}^{(2)}\right)^2\right) \odot \mathbf{W}^{(2)}\mathbf{H}_1^{(1)}$$

$$= \begin{bmatrix} 0.0157 & 0.0003 \\ -0.0004 & 0.0000 \\ 0.0001 & -0.0000 \\ 0.0031 & -0.0001 \\ 0.0096 & -0.0003 \\ 0.0336 & 0.0002 \\ -0.0017 & 0.0001 \\ 0.0120 & -0.0004 \\ -0.0059 & 0.0001 \\ -0.0048 & -0.0006 \end{bmatrix} \quad \text{(A2.3-3b)}$$

c. Normalized hidden layer, $\mathbf{z}^{(3)}$

$$\mathbf{z}^{(3)} = f_t^{(3)}\left(\mathbf{W}^{(3)}\mathbf{z}^{(2)} + \mathbf{b}^{(3)}\right) = \begin{bmatrix} -0.9183 \\ -0.9209 \\ -0.2115 \\ 0.8755 \\ 0.3117 \\ 0.6478 \\ -0.9987 \\ 0.9502 \\ 0.9995 \\ -0.9405 \end{bmatrix} \quad \text{(A2.4-1)}$$

$$\mathbf{J}^{(3)} = \left(1-\left(\mathbf{z}^{(3)}\right)^2\right) \odot \mathbf{W}^{(3)}\mathbf{J}^{(2)} = \begin{bmatrix} 0.0933 & 0.0039 \\ 0.2151 & 0.0001 \\ -0.7665 & -0.0120 \\ -0.1801 & -0.0025 \\ 0.6078 & -0.0075 \\ 0.4149 & 0.0066 \\ -0.0030 & 0.0000 \\ -0.1390 & 0.0008 \\ 0.0002 & 0.0000 \\ -0.1479 & -0.0000 \end{bmatrix} \tag{A2.4-2}$$

$$\mathbf{H}_1^{(3)} = -2\mathbf{z}^{(3)} \odot \left(1-\left(\mathbf{z}^{(3)}\right)^2\right) \odot \boldsymbol{i}_1^{(2)} \odot \left(\mathbf{W}^{(3)}\right)^2 \mathbf{J}^{(2)}$$

$$+\left(1-\left(\mathbf{z}^{(3)}\right)^2\right) \odot \mathbf{W}^{(3)}\mathbf{H}_1^{(2)} = \begin{bmatrix} -0.0710 & -0.0001 \\ 0.4891 & 0.0036 \\ 0.6025 & -0.0147 \\ -0.2727 & -0.0096 \\ -0.6914 & 0.0116 \\ -0.7643 & -0.0014 \\ 0.0129 & -0.0001 \\ -0.3284 & -0.0018 \\ -0.0007 & 0.0000 \\ 0.4113 & -0.0071 \end{bmatrix} \tag{A2.4-3a}$$

$$\mathbf{H}_2^{(3)} = -2\mathbf{z}^{(3)} \odot \left(1-\left(\mathbf{z}^{(3)}\right)^2\right) \odot \boldsymbol{i}_2^{(2)} \odot \left(\mathbf{W}^{(3)}\right)^2 \mathbf{J}^{(2)}$$

$$+\left(1-\left(\mathbf{z}^{(3)}\right)^2\right) \odot \mathbf{W}^{(3)}\mathbf{H}_2^{(2)} = \begin{bmatrix} -0.0001 & 0.0002 \\ -0.0036 & 0.0002 \\ -0.0147 & -0.0002 \\ -0.0096 & -0.0002 \\ 0.0116 & 0.0003 \\ -0.0014 & 0.0002 \\ -0.0001 & 0.0000 \\ -0.0018 & -0.0001 \\ 0.0000 & -0.0000 \\ -0.0071 & -0.0001 \end{bmatrix} \tag{A2.4-3b}$$

d. Normalized hidden layer, $\mathbf{z}^{(4=L)}$

$$\mathbf{z}^{(4)} = f_t^{(4)}\left(\mathbf{W}^{(4)}\mathbf{z}^{(3)} + \mathbf{b}^{(4)}\right) = \begin{bmatrix} -0.6928 \\ -0.2277 \\ 0.9998 \\ 0.6026 \\ -0.9994 \\ -0.6874 \\ 0.9997 \\ 0.1627 \\ 0.9997 \\ 0.7319 \end{bmatrix} \tag{A2.5-1}$$

$$\mathbf{J}^{(4)} = \left(1-\left(\mathbf{z}^{(4)}\right)^2\right) \odot \mathbf{W}^{(4)}\mathbf{J}^{(3)} = \begin{bmatrix} 0.8159 & 0.0086 \\ 0.5466 & 0.0168 \\ -0.0001 & -0.0000 \\ -0.6431 & 0.0051 \\ 0.0004 & 0.0000 \\ 0.7880 & -0.0038 \\ 0.0000 & 0.0000 \\ 0.5318 & 0.0009 \\ 0.0004 & -0.0000 \\ -0.1246 & -0.0080 \end{bmatrix} \quad \text{(A2.5-2)}$$

$$\mathbf{H}_1^{(4)} = -2\mathbf{z}^{(4)} \odot \left(1-\left(\mathbf{z}^{(4)}\right)^2\right) \odot \boldsymbol{i}_1^{(3)} \odot \left(\mathbf{W}^{(4)}\right)^2 \mathbf{J}^{(3)}$$

$$+\left(1-\left(\mathbf{z}^{(4)}\right)^2\right) \odot \mathbf{W}^{(4)}\mathbf{H}_1^{(3)} = \begin{bmatrix} 1.0919 & 0.0262 \\ -1.3258 & 0.0028 \\ -0.0004 & -0.0000 \\ -0.4548 & 0.0021 \\ 0.0016 & 0.0000 \\ 1.0219 & 0.0111 \\ -0.0009 & -0.0000 \\ -1.4902 & 0.0019 \\ -0.0013 & 0.0000 \\ -0.1633 & -0.0052 \end{bmatrix} \quad \text{(A2.5-3a)}$$

$$\mathbf{H}_2^{(4)} = -2\mathbf{z}^{(4)} \odot \left(1-\left(\mathbf{z}^{(4)}\right)^2\right) \odot \boldsymbol{i}_2^{(3)} \odot \left(\mathbf{W}^{(4)}\right)^2 \mathbf{J}^{(3)}$$

$$+\left(1-\left(\mathbf{z}^{(4)}\right)^2\right) \odot \mathbf{W}^{(4)}\mathbf{H}_2^{(3)} = \begin{bmatrix} 0.0262 & 0.0006 \\ 0.0028 & 0.0002 \\ -0.0000 & -0.0000 \\ 0.0021 & -0.0002 \\ 0.0000 & 0.0000 \\ 0.0111 & 0.0004 \\ -0.0000 & -0.0000 \\ 0.0019 & 0.0003 \\ 0.0000 & 0.0000 \\ -0.0052 & -0.0001 \end{bmatrix} \quad \text{(A2.5-3b)}$$

c. Normalized output layer ($5 = \text{output}$), $\mathbf{z}^{(5=\text{output})}$

$$\mathbf{z}_{y20}^{(L=5)} = f_{lin}^{(5)}\left(\mathbf{w}_{20}^{(5)}\mathbf{z}^{(4)} + b_{20}^{(5)}\right) = 0.5974 \quad \text{(A2.6-1)}$$

$$\mathbf{J}_{y20}^{(L=5)} = \mathbf{w}_{20}^{(L=5)}\mathbf{J}^{(4)} = \begin{bmatrix} -1.4401 & -0.0004 \end{bmatrix} \quad \text{(A2.6-2)}$$

$$\mathbf{H}_{y20,\,1}^{(L=5)} = \mathbf{w}_{20}^{(L=5)}\mathbf{H}_1^{(4)} = \begin{bmatrix} -5.1825 & -0.0600 \end{bmatrix} \quad \text{(A2.6-3a)}$$

$$\mathbf{H}_{y20,\,2}^{(L=5)} = \mathbf{w}_{20}^{(L=5)}\mathbf{H}_2^{(4)} = \begin{bmatrix} -0.0600 & -0.0013 \end{bmatrix} \tag{A2.6-3b}$$

$$\mathbf{z}_{y21}^{(L=5)} = f_{lin}^{(5)}\left(\mathbf{w}_{21}^{(5)}\mathbf{z}^{(4)} + b_{21}^{(5)}\right) = 0.6240 \tag{A2.7-1}$$

$$\mathbf{J}_{y21}^{(L=5)} = \mathbf{w}_{21}^{(5)}\mathbf{J}^{(4)} = \begin{bmatrix} -1.2543 & 0.0038 \end{bmatrix} \tag{A2.7-2}$$

$$\mathbf{H}_{y21,\,1}^{(L=5)} = \mathbf{w}_{21}^{(5)}\mathbf{H}_1^{(4)} = \begin{bmatrix} -6.0311 & -0.0534 \end{bmatrix} \tag{A2.7-3a}$$

$$\mathbf{H}_{y21,\,2}^{(L=5)} = \mathbf{w}_{21}^{(5)}\mathbf{H}_2^{(4)} = \begin{bmatrix} -0.0534 & -0.0010 \end{bmatrix} \tag{A2.7-3b}$$

d. De-normalized output layer (D, $5 = \text{output}$), $\mathbf{y} = \mathbf{z}^{(D)}$; de-normalizing $\mathbf{z}^{(5=\mathbf{output})}$

$$\begin{aligned} y_{20} &= z_{y20}^{(D)} = g_{20}^{(D)}\left(z_{y20}^{(L=5)}\right) = \frac{1}{\alpha_{y20}}\left(z_{y20}^{(5)} - \overline{y}_{20,min}\right) + y_{20,min} \\ &= \frac{1}{1.5427}\left(0.5838 - (-1)\right) + (-0.9871) = 0.0484 \end{aligned} \tag{A2.8-1}$$

$$\mathbf{J}_{y20}^{(D)} = \frac{1}{\alpha_{y20}}\mathbf{J}_{y20}^{(L=5)} = \begin{bmatrix} -0.9335 & -0.0002 \end{bmatrix} \tag{A2.8-2}$$

$$\mathbf{H}_{y20,\,1}^{(D)} = \frac{1}{\alpha_{y20}}\mathbf{H}_{y20,\,1}^{(L=5)} = \begin{bmatrix} -3.3594 & -0.0389 \end{bmatrix} \tag{A2.8-3a}$$

$$\mathbf{H}_{y20,\,2}^{(D)} = \frac{1}{\alpha_{y20}}\mathbf{H}_{y20,\,2}^{(L=5)} = \begin{bmatrix} -0.0389 & -0.0008 \end{bmatrix} \tag{A2.8-3b}$$

$$\mathbf{H}_{y20}^{(D)} = \begin{bmatrix} \mathbf{H}_{y20,\,1}^{(D)} \\ \mathbf{H}_{y20,\,2}^{(D)} \end{bmatrix} = \begin{bmatrix} -3.3594 & -0.0389 \\ -0.0389 & -0.0008 \end{bmatrix} \tag{A2.8-4}$$

$$\begin{aligned} y_{21} &= z_{y21}^{(D)} = g_{21}^{(D)}\left(z_{y21}^{(L=5)}\right) = \frac{1}{\alpha_{y21}}\left(z_{y21}^{(5)} - \overline{y}_{21,min}\right) + y_{21,min} \\ &= \frac{1}{1.7958}\left(0.6240 - (-1)\right) + (-0.7988) = 0.1055 \end{aligned} \tag{A2.9-1}$$

$$\mathbf{J}_{y21}^{(D)} = \frac{1}{\alpha_{y21}}\mathbf{J}_{y21}^{(L=5)} = \begin{bmatrix} -0.6984 & 0.0021 \end{bmatrix} \tag{A2.9-2}$$

$$\mathbf{H}^{(D)}_{y21,1} = \frac{1}{\alpha_{y21}}\mathbf{H}^{(5)}_{y21,1} = \begin{bmatrix} -3.3583 & -0.0297 \end{bmatrix} \quad \text{(A2.9-3a)}$$

$$\mathbf{H}^{(D)}_{y21,2} = \frac{1}{\alpha_{y21}}\mathbf{H}^{(5)}_{y21,2} = \begin{bmatrix} -0.0297 & -0.0006 \end{bmatrix} \quad \text{(A2.9-3b}$$

$$\mathbf{H}^{(D)}_{y21} = \begin{bmatrix} \mathbf{H}^{(D)}_{y21,1} \\ \mathbf{H}^{(D)}_{y21,2} \end{bmatrix} = \begin{bmatrix} -3.3583 & -0.0297 \\ -0.0297 & -0.0006 \end{bmatrix} \quad \text{(A2.9-4)}$$

Output variables y_{20} and y_{21}, an equality constraint $c(x_C,\theta)$, and inequality constraints $v(x_C,\theta)$ are obtained in Equations (A2.8-1), (A2.9-1), (A2.10-1), and (A2.10-2), respectively. Jacobi $\mathbf{J}_c\left(x_C^{(1)}, \theta^{(1)}\right)$ and $\mathbf{J}_v\left(x_C^{(1)}, \theta^{(1)}\right)$ are calculated in Equations (A2.11-1) and (A2.11-2), respectively, as functions of $x_C^{(0)}$ and $\theta^{(0)}$ whereas Hessian matrix $\mathbf{H}_c\left(x_C^{(1)}, \theta^{(1)}\right)$ and $\mathbf{H}_v\left(x_C^{(1)}, \theta^{(1)}\right)$ are calculated as functions of $x_C^{(0)}$ and $\theta^{(0)}$ in Equations (A2.12-1) and (A2.12-2). Lagrange multipliers $\lambda_c^{(1)}=-1.086$ and $\lambda_v^{(1)}=0$ obtained in Equation (2.4.4.68) are used when formulating Jacobian and Hessian of Lagrange function as shown in red inputs of Equations (A2.13-2) and (A2.17), respectively. Weight and bias matrices obtained in Table 2.4.4.4 are used in this appendix. Figure 2.3.6.2 is referred for all procedures. A first derivative (Jacobi) of Lagrange function which is, then, obtained in Equations (A2.13-1) and (A2.13-2) with respect to the updated input parameter $\left[x_C^{(1)}, \theta^{(1)}, \lambda_c^{(1)}, \lambda_v^{(1)}\right] = [0.6020, 47.57, -1.086, 0]$ obtained in Equation (2.4.4.68) and $s = 0$ does not converge to the solutions, indicating that $\left[x_C^{(1)}, \theta^{(1)}, \lambda_c^{(1)}, \lambda_v^{(1)}\right]$ is not the root of Case 2 KKT condition.

$$c\left(x_C^{(1)}, \theta^{(1)}\right) = y_{P_f} - 0.06 = y_{20} - 0.06 = 0.0484 - 0.06 = -0.0116 \quad \text{(A2.10-1)}$$

$$v\left(x_C^{(1)}, \theta^{(1)}\right) = y_{P_f} - 0.06 = y_{21} - 0.1 = 0.1055 - 0.1 = 0.0055 \quad \text{A2.10-2)}$$

$$\mathbf{J}_c\left(x_C^{(1)}, \theta^{(1)}\right) = \mathbf{J}_c^{(D)}\left(x_C^{(1)}, \theta^{(1)}\right) = \mathbf{J}^{(D)}_{\left(y_{P_f}-0.06\right)} = \mathbf{J}^{(D)}_{y_{P_f}} = \mathbf{J}^{(D)}_{y20}$$

$$= \begin{bmatrix} -0.9335 & -0.0002 \end{bmatrix} \quad \text{(A2.11-1)}$$

$$\mathbf{J}_v\left(x_C^{(1)}, \theta^{(1)}\right) = \mathbf{J}_v^{(D)}\left(x_C^{(1)}, \theta^{(1)}\right) = \mathbf{J}^{(D)}_{\left(y_{P_t}-0.1\right)} = \mathbf{J}^{(D)}_{y_{P_t}} = \mathbf{J}^{(D)}_{y21} = \begin{bmatrix} -0.6984 & 0.0021 \end{bmatrix} \quad \text{(A2.11-2)}$$

$$\mathbf{H}_c\left(x_C^{(1)}, \theta^{(1)}\right) = \mathbf{H}_c^{(D)}\left(x_C^{(1)}, \theta^{(1)}\right) = \mathbf{H}^{(D)}_{y20} = \begin{bmatrix} -3.3594 & -0.0389 \\ -0.0389 & -0.0008 \end{bmatrix} \quad \text{(A2.12-1)}$$

$$\mathbf{H}_v\left(x_C^{(1)}, \theta^{(1)}\right) = \mathbf{H}_v^{(D)}\left(x_C^{(1)}, \theta^{(1)}\right) = \mathbf{H}^{(D)}_{y21} = \begin{bmatrix} -3.3583 & -0.0297 \\ -0.0297 & -0.0006 \end{bmatrix} \quad \text{(A2.12-2)}$$

e. Convergence verification of Lagrange function based on Jacobi, Hessian matrix at Iteration 1

$$\nabla\mathcal{L}\left(x_C^{(1)},\theta^{(1)},\lambda_c^{(1)},\lambda_v^{(1)}\right)$$

$$=\begin{bmatrix}\nabla f_{xC}\left(x_C^{(1)}\right)-\left[\mathbf{J}_c\left(x_C^{(1)},\theta^{(1)}\right)\right]^T\lambda_c^{(1)}-\left[\mathbf{J}_v\left(x_C^{(1)},\theta^{(1)}\right)\right]^T s\lambda_v^{(1)}\\-c\left(x_C^{(1)},\theta^{(1)}\right)\\-sv\left(x_C^{(1)},\theta^{(1)}\right)\end{bmatrix}$$

$$=\begin{bmatrix}\nabla f_{xC}\left(x_C^{(1)}\right)-\left[\mathbf{J}_c^{(D)}\left(x_C^{(1)},\theta^{(1)}\right)\right]^T\lambda_c^{(1)}-\left[\mathbf{J}_v^{(D)}\left(x_C^{(1)},\theta^{(1)}\right)\right]^T s\lambda_v^{(1)}\\-c\left(x_C^{(1)},\theta^{(1)}\right)\\-sv\left(x_C^{(1)},\theta^{(1)}\right)\end{bmatrix} \tag{A2.13-1}$$

$$\nabla\mathcal{L}\left(x_C^{(1)},\theta^{(1)},\lambda_c^{(1)},\lambda_v^{(1)}\right)$$

$$=\begin{bmatrix}\begin{bmatrix}1\\0\end{bmatrix}-\begin{bmatrix}-0.9335\\-0.0002\end{bmatrix}\times(-1.086)-\begin{bmatrix}-0.6984\\0.0021\end{bmatrix}\times 0\times 0\\0.0116\\0\end{bmatrix} \tag{A2.13-2}$$

$$\rightarrow\nabla\mathcal{L}\left(x_C^{(1)},\theta^{(1)},\lambda_c^{(1)},\lambda_v^{(1)}\right)=\begin{bmatrix}-0.0142\\-0.0002\\0.0116\\0\end{bmatrix}\neq 0$$

$\nabla\mathcal{L}\left(x_C^{(1)},\theta^{(1)},\lambda_c^{(1)},\lambda_v^{(1)}\right)$. shown in Equation (A2.13-2) does not converge to zeros, indicating $\left[x_C^{(1)},\theta^{(1)},\lambda_c^{(1)},\lambda_v^{(1)}\right]$ is not the root of KKT condition for Case 2. The updated input parameter which will be used for Iteration 2 is to be calculated as Equation (A2.14).

f. Updating input parameter for Iteration 2

$$\begin{bmatrix}x_C^{(2)}\\\theta^{(2)}\\\lambda_c^{(2)}\\\lambda_v^{(2)}\end{bmatrix}=\begin{bmatrix}x_C^{(1)}\\\theta^{(1)}\\\lambda_c^{(1)}\\\lambda_v^{(1)}\end{bmatrix}-\left[\mathbf{H}_{\mathcal{L}}\left(x_C^{(1)},\theta^{(1)},\lambda_c^{(1)},\lambda_v^{(1)}\right)\right]^{-1}\nabla\mathcal{L}\left(x_C^{(1)},\theta^{(1)},\lambda_c^{(1)},\lambda_v^{(1)}\right) \tag{A2.14}$$

A Hessian of Lagrange function with respect to $\left[x_C^{(1)}, \theta^{(1)}, \lambda_c^{(1)}, \lambda_v^{(1)}\right]$ is calculated based on Equations (A2.15-2) and (A2.18) using $\mathbf{H}_{\mathcal{L}}\left(x_C^{(1)}, \theta^{(1)}\right)$ calculated in Equations (A2.16), (A2.17), $\mathbf{J}_c^{(D)}$ calculated in Equation (A2.11-1), $\mathbf{J}_v^{(D)}$ calculated in Equation (A2.11-2). $H_{\mathcal{L}}\left(x_C^{(1)}, \theta^{(1)}\right)$ is calculated in Equation (A2.17) by substituting $\mathbf{H}_{f_{xC}}\left(x_C, \theta\right)$, $\mathbf{H}_c^{(D)}$, and $\mathbf{H}_v^{(D)}$ from Equations (2.4.4.14), (A2.12-1), and (A2.12-2), respectively, into Equation (A2.16). Lagrange multipliers $\left[x_C^{(1)}, \theta^{(1)}, \lambda_c^{(1)}, \lambda_v^{(1)}\right] = [0.6020, 47.57, -1.086, 0]$ obtained in Equation (2.4.4.68) are used in red as shown in Equation (A2.17).

$$\mathbf{H}_{\mathcal{L}}\left(x_C^{(1)}, \theta^{(1)}, \lambda_c^{(1)}, \lambda_v^{(1)}\right) = \begin{bmatrix} \mathbf{H}_{\mathcal{L}}\left(x_C^{(1)}, \theta^{(1)}\right) & -\left[\mathbf{J}_c\left(x_C^{(1)}, \theta^{(1)}\right)\right]^T & -\left[s\mathbf{J}_v\left(x_C^{(1)}, \theta^{(1)}\right)\right]^T \\ -\mathbf{J}_c\left(x_C^{(1)}, \theta^{(1)}\right) & 0 & 0 \\ -s\mathbf{J}_v\left(x_C^{(1)}, \theta^{(1)}\right) & 0 & 0 \end{bmatrix} \tag{A2.15-1}$$

$$\mathbf{H}_{\mathcal{L}}\left(x_C^{(1)}, \theta^{(1)}, \lambda_c^{(1)}, \lambda_v^{(1)}\right) = \begin{bmatrix} \mathbf{H}_{\mathcal{L}}\left(x_C^{(1)}, \theta^{(1)}\right) & -\left[\mathbf{J}_c^{(D)}\left(x_C^{(1)}, \theta^{(1)}\right)\right]^T & -\left[s\mathbf{J}_v^{(D)}\left(x_C^{(1)}, \theta^{(1)}\right)\right]^T \\ -\mathbf{J}_c^{(D)}\left(x_C^{(1)}, \theta^{(1)}\right) & 0 & 0 \\ -s\mathbf{J}_v^{(D)}\left(x_C^{(1)}, \theta^{(1)}\right) & 0 & 0 \end{bmatrix} \tag{A2.15-2}$$

where

$$\mathbf{H}_{\mathcal{L}}\left(x_C^{(1)}, \theta^{(1)}\right) = \mathbf{H}_{f_{xC}}\left(x_C^{(1)}, \theta^{(1)}\right) - \lambda_c^{(1)}\mathbf{H}_c\left(x_C^{(1)}, \theta^{(1)}\right) - s\lambda_v^{(1)}\mathbf{H}_v\left(x_C^{(1)}, \theta^{(1)}\right) \tag{A2.16-1}$$

$$\mathbf{H}_{\mathcal{L}}\left(x_C^{(1)}, \theta^{(1)}\right) = \mathbf{H}_{f_{xC}}\left(x_C^{(1)}, \theta^{(1)}\right) - \lambda_c^{(1)}\mathbf{H}_c^{(D)}\left(x_C^{(1)}, \theta^{(1)}\right) - s\lambda_v^{(1)}\mathbf{H}_v^{(D)}\left(x_C^{(1)}, \theta^{(1)}\right) \tag{A2.16-2}$$

$$\begin{aligned} H_{\mathcal{L}}\left(x_C^{(1)}, \theta^{(1)}\right) &= \begin{bmatrix} 0 & 0 \\ 0 & 0 \end{bmatrix} - (-1.0864) \times \begin{bmatrix} -3.3594 & -0.0389 \\ -0.0389 & -0.0008 \end{bmatrix} \\ &\quad - 0 \times 0 \times \begin{bmatrix} -3.3583 & -0.0297 \\ -0.0297 & -0.0006 \end{bmatrix} \\ &= \begin{bmatrix} -3.6498 & -0.0423 \\ -0.0423 & -0.0009 \end{bmatrix} \end{aligned} \tag{A2.17}$$

$$\mathbf{H}_{\mathcal{L}}\left(x_C^{(1)}, \theta^{(1)}, \lambda_c^{(1)}, \lambda_v^{(1)}\right)$$

$$= \begin{bmatrix} \begin{bmatrix} -3.6498 & -0.0423 \\ -0.0423 & -0.0009 \end{bmatrix} & -\begin{bmatrix} -0.9335 \\ -0.0002 \end{bmatrix} & -0\times\begin{bmatrix} -0.6984 \\ 0.0021 \end{bmatrix} \\ -\begin{bmatrix} -0.9335 & -0.0002 \end{bmatrix} & 0 & 0 \\ -0\times\begin{bmatrix} -0.6984 & 0.0021 \end{bmatrix} & 0 & 0 \end{bmatrix} \quad \text{(A2.18-1)}$$

$$\mathbf{H}_{\mathcal{L}}\left(x_C^{(1)}, \theta^{(1)}, \lambda_c^{(1)}, \lambda_v^{(1)}\right) = \begin{bmatrix} -2.0483 & -0.0157 & 0.9335 & 0 \\ -0.0157 & -0.0004 & 0.0002 & 0 \\ 0.9335 & 0.0002 & 0 & 0 \\ 0 & 0 & 0 & 0 \end{bmatrix} \quad \text{(A2.18-2)}$$

$$\left[\mathbf{H}_{\mathcal{L}}\left(x_C^{(1)}, \theta^{(1)}, \lambda_c^{(1)}, \lambda_v^{(1)}\right)\right]^{-1} = \begin{bmatrix} -0.0001 & 0.2814 & 1.0837 & 0 \\ 0 & -1149.5 & -50.95 & 0 \\ 1.0837 & -50.95 & 1.9297 & 0 \\ 0 & 0 & 0 & 0 \end{bmatrix} \quad \text{(A2.18-3)}$$

Input parameter $\left[x_C^{(1)}, \theta^{(1)}, \lambda_c^{(1)}, \lambda_v^{(1)}\right]$ at Iteration 1 is updated to $\left[x_C^{(2)}, \theta^{(2)}, \lambda_c^{(2)}, \lambda_v^{(2)}\right]$ at Iteration 2 in Equation (A2.19) which will be used in Iteration 2 by substituting Equations (A2.13-2) and (A2.18-3) into Equation (A2.14). The first derivative (Jacobi) of Lagrange function is, then, verified for convergence with respect to the updated input parameter $\left[x_C^{(2)}, \theta^{(2)}, \lambda_c^{(2)}, \lambda_v^{(2)}\right]^T = [0.5895, 47.88, -1.106, 0]^T$ obtained in Equation (A2.19) and $s = 0$. This procedure will be repeated until a first derivative (Jacobi) of Lagrange function converges.

$$\begin{bmatrix} x_C^{(2)} \\ \theta^{(2)} \\ \lambda_c^{(2)} \\ \lambda_v^{(2)} \end{bmatrix} = \begin{bmatrix} 0.6020 \\ 47.57 \\ -1.086 \\ 0 \end{bmatrix} - \begin{bmatrix} -0.0001 & 0.2814 & 1.0837 & 0 \\ 0 & -1149.5 & -50.95 & 0 \\ 1.0837 & -50.95 & 1.9297 & 0 \\ 0 & 0 & 0 & 0 \end{bmatrix} \times$$

$$\begin{bmatrix} -0.0142 \\ -0.0002 \\ 0.0116 \\ 0 \end{bmatrix} \rightarrow \begin{bmatrix} x_C^{(2)} \\ \theta^{(2)} \\ \lambda_c^{(2)} \\ \lambda_v^{(2)} \end{bmatrix} = \begin{bmatrix} 0.5895 \\ 47.88 \\ -1.106 \\ 0 \end{bmatrix} \quad \text{(A2.19)}$$

Appendix B

B1 ANALYTICAL LAGRANGE CODE FOR POLYNOMIAL OPTIMIZATION

```
clc, clear;
format short
%% INPUT
x0 = -3; % initial input x0;
lamda_v0 = 1; % initial Lagrange multiplier, lamda_v

iter_req = 100; % number of required iteration
MSE_target = 1e-10; % Target of mean square error of first derivative of
Lagrange function

fprintf('Initial input vector: [x0, lamda_v0] = [%f, %f]\n',x0, lamda_v0)
fprintf('Number of required iteration: %d\nTarget ofMean square error:
%s\n\n\n',...
                    iter_req,MSE_target)
%% CASE 1: Inequality v(x) = x^2 - 4 >= 0 is inactive
fprintf('...STARTING CASE 1...\n')
x(1) = x0; % initial input x0;
lamda_v(1) = lamda_v0; % initial Lagrange multiplier, lamda_v
for i = 1:iter_req
    % Calculate first derivative of Lagrange function, J, w.r.t. x(i),
lamda_v(i)
    J = 4*x(i)^3-3*x(i)^2-8*x(i)+5;
    % Calculate MSE of J
    MSE(i) = sum(J.^2);
    % Calculate Hessian of Lagrange function, H, w.r.t. x(i), lamda_v(i)
    H = 12*x(i)^2-6*x(i)-8;

    % Calculate objective function, f, w.r.t. x(i), lamda_v(i)
    f(i) = x(i)^4 - x(i)^3 - 4*x(i)^2+5*x(i)+5;

    if MSE(i)>MSE_target % check MSE
        % Update input vector x(i+1) for next iteration
        x(i+1) = x(i) - H^-1*J;
    else
        break
    end
end
table_C1 = array2table([x',MSE',f'],'VariableNames',{'x(i)','MSE','f(x)'});
disp(table_C1)
```

```
% Check result with inactive inequality, v(x)
fprintf('Check final result of CASE 1 with inactive inequality, v(x)\n')
v_x = x(i)^2 - 4;
if v_x >= 0
    check_v = '(Satisfactory)';
else
    check_v = '(Unsatisfactory)';
end
fprintf('x = %f\n',table_C1{end,1})
fprintf('v(x) = x^2 - 4 = %f %s\n\n',v_x,check_v)

%% CASE 2: Inequality v(x) = x^2 - 4 >= 0 is active
fprintf('...STARTING CASE 2...\n')

clear('x','lamda_v','MSE','L','f')
x(1) = x0; % initial input x0;
lamda_v(1) = lamda_v0; % initial Lagrange multiplier, lamda_v
for i = 1:iter_req
    % Calculate first derivative of Lagrange function, J, w.r.t. x(i),
lamda_v(i)
    J = [4*x(i)^3-3*x(i)^2-8*x(i)+5 - 2*lamda_v(i)*x(i);
         4 - x(i)^2];
    % Calculate MSE of J
    MSE(i) = 0.5*sum(J.^2);
    % Calculate Hessian of Lagrange function, H, w.r.t. x(i), lamda_v(i)
    H = [12*x(i)^2-6*x(i)-8-2*lamda_v(i) -2*x(i);
                -2*x(i)          0];
    % Calculate Lagrange function, L, w.r.t. x(i), lamda_v(i)
    L(i) = x(i)^4 - x(i)^3 - 4*x(i)^2+5*x(i)+5 - lamda_v(i)*(x(i)^2 - 4);

    % Calculate objective function, f, w.r.t. x(i), lamda_v(i)
    f(i) = x(i)^4 - x(i)^3 - 4*x(i)^2+5*x(i)+5;

    if MSE(i)>MSE_target % check MSE
        % Update input vector x(i+1), lamda_v(i+1) for next iteration
        temp = [x(i);lamda_v(i)] - H^-1*J;
        x(i+1) = temp(1);
        lamda_v(i+1) = temp(2);
    else
        break
    end
end

table_C2 = array2table([x',lamda_v',MSE',L',f'],'VariableNames',{'x(i)','
lamda_v(i)','MSE','L(x,lamda_v)','f(x)'});
disp(table_C2)

%% Select optimal result
fprintf('...SELECT OPTIMAL RESULT...\n')
fprintf('With an initial input vector: [x0, lamda_v0] = [%f, %f],\n',x0,
lamda_v0)
if v_x >= 0
    fprintf('a local optimum is obtained from CASE 1: Inequality
v(x) = x^2 - 4 >= 0 is inactive \n')
    fprintf('x = %f \nf(x) = %f \n',table_C1{end,1},table_C1{end,3})
else
    fprintf('a local optimum is obtained from CASE 2: Inequality v(x) = x^2
- 4 >= 0 is active \n')
    fprintf('x = %f; lamda_v = %f \nf(x) = %f \nv(x) = %f\n',...
```

```
table_C2{end,1},table_C2{end,2},table_C2{end,5},table_C2{end,1}^2-4)
end

fprintf('Readers need to try several initial input vectors [x0, lamda_v0]
to obtain a global optimum.\n')
```

B2 ANALYTICAL LAGRANGE CODE FOR PROJECTILE OPTIMIZATION

```
clc, clear;
format short
warning('off')
%% INPUT
fy = 200e-3; % yield strength, kN/mm2
x = [2000 20 0 0]'; % initial vector [xC_0, theta_0, lamdaC_0,
lamdaV_0]';
iter_req = 2000; % number of required iteration
MSE_target = 1e-20; % Target of mean square error of first derivative of
Lagrange function

fprintf('Initial input vector: [xA, xh, lamdaV1, lamdaV2] = [%.3f, %.3f,
%.3f, %.3f],\n',...
                    x(1), x(2), x(3), x(4))
fprintf('Number of required iteration: %d\nTarget ofMean square error:
%s\n\n\n',...
                    iter_req,MSE_target)

k = 0;
for case_i = 1:4
    if case_i == 1 % Inequality v(x) = y_Pt - 0.1 >= 0 is active
        fprintf('...STARTING CASE 1: v1(x) is active...\n')
        S = [1 0;
             0 0];
    elseif case_i == 2 % Inequality v(x) = y_Pt - 0.1 >= 0 is inactive
        fprintf('...STARTING CASE 2: v2(x) is active...\n')
        S = [0 0;
             0 1];
    elseif case_i == 3 % Inequality v(x) = y_Pt - 0.1 >= 0 is inactive
        fprintf('...STARTING CASE 3: both v1(x) and v2(x) are
active...\n')
        S = [1 0;
             0 1];
    elseif case_i == 4 % Inequality v(x) = y_Pt - 0.1 >= 0 is inactive
        fprintf('...STARTING CASE 4: both v1(x) and v2(x) are
inactive...\n')
        S = [0 0;
             0 0];
    end
    clear('R')

    R(1).x = x;
```

```
index = 1:4;
 s_index = 1:2;
 status = [1;1;diag(S)];
 index(status==0) = [];
 s_index(diag(S)==0) = [];

 for i = 1:iter_req
     x_A = R(i).x(1); x_h = R(i).x(2);
     lamda = R(i).x(3:4);

     % Calculate objective function, fW, and its derivatives
     R(i).fW = 0.008*abs(x_A)*(2+sqrt(x_h^2+4));
     J_fW = [(0.016+0.008*(x_h^2+4)^0.5);
             0.008*x_A*x_h*(x_h^2+4)^(-0.5)];
     H_fW = [0                            , 0.008*x_h*(x_h^2+4)^(-0.5);
             0.008*x_h*(x_h^2+4)^(-0.5)   , 0.032*x_A/(x_h^2+4)^1.5];

     % Calculate inequalities, c(x), v(x), and their Jacobian, J_c,
J_v, and Hessian, H_c, H_v
     S1 = 55/x_A - 200/(x_h*x_A);
     S2 = 100*sqrt(x_h^2+4) / (x_h*x_A);
     v1 = -abs(S1)+fy;
     v2 = -abs(S2)+fy;
     J_v = [S1/abs(S1)*(55/x_A^2-200/(x_A^2*x_h))   , -S1/abs(S1)*200/
(x_h^2*x_A);
            100*sqrt(x_h^2+4)/(x_A^2*abs(x_h))       , 100*x_h*sqrt(x_
h^2+4)/(x_A*abs(x_h)^3) - 100*x_h/(x_A*abs(x_h)*sqrt(x_h^2+4))];
     H_v1 = [S1/abs(S1)*(400/(x_A^3*x_h)-110/x_A^3)   ,   S1/
abs(S1)*200/(x_h^2*x_A^2);
             S1/abs(S1)*200/(x_h^2*x_A^2)             ,   S1/
abs(S1)*400/(x_h^3*x_A)];

     H_v2 = [-200*sqrt(x_h^2+4)/(x_A^3*abs(x_h)),
(100*x_h/(x_A^2*abs(x_h)*sqrt(x_h^2+4)) - 100*x_h*sqrt(x_h^2+4)/
(x_A^2*abs(x_h)^3));
             (100*x_h/(x_A^2*abs(x_h)*sqrt(x_h^2+4)) - 100*x_h*sqrt(x_
h^2+4)/(x_A^2*abs(x_h)^3))  ,   100*abs(x_h)/(x_A*(x_h^2+4)^1.5) -
200*sqrt(x_h^2+4)/(x_A*abs(x_h)^3)+100/(x_A*abs(x_h)*sqrt(x_h^2+4) )];

     % Calculate Lagrange function, L, w.r.t. to [xA_(i), xh_(i),
lamdaV1_(i), lamdaV2_(i)]'
     R(i).Lagrange = R(i).fW - lamda'*S*[v1;v2];

     % Calculate Jacobian of L w.r.t. to [xA_(i), xh_(i), lamdaV1_(i),
lamdaV2_(i)]'
     R(i).J = [J_fW - J_v'*S*lamda;
                    -S*[v1;v2]   ];

     % Calculate MSE of J
     R(i).E = 1/4*sum(R(i).J .^ 2);

     % Calculate Hessian of L w.r.t. to [xA_(i), xh_(i), lamdaV1_(i),
lamdaV2_(i)]'
     H = H_fW - S(1,1)*lamda(1)*H_v1 - S(2,2)*lamda(2)*H_v2;
     R(i).H = [H          , -(S*J_v)';
               -S*J_v     , zeros(2,2)];

     if R(i).E>MSE_target % Check MSE
         % Update input vector for next iteration
```

```
            R(i+1).x = R(i).x;
            R(i+1).x(index) = R(i).x(index) - R(i).H(index,index) \
R(i).J(index);
        elseif R(i).E <= MSE_target
            fprintf('CASE %d is converged at iteration step #%d with an
MSE of %s, providing a result of:\n',case_i,i,R(i).E)
            fprintf('Input vector: [xA, xh, lamdaV1, lamdaV2] = [%f, %f,
%f, %f]\n',...
                                                R(i).x(1), R(i).x(2),
R(i).x(3), R(i).x(4))
            fprintf('Objective function fW  = %f\n', R(i).fW)
            fprintf('Inequality constraint v1(x) = %f\n', v1)
            fprintf('Inequality constraint v2(x) = %f\n', v2)

            k = k+1;
            summary(k,1) = case_i;
            summary(k,2) = R(end).x(1);
            summary(k,3) = R(end).x(2);
            summary(k,4) = R(end).x(3);
            summary(k,5) = R(end).x(4);
            summary(k,6) = R(end).fW;
            summary(k,7) = v1;
            summary(k,8) = v2;
            break
        elseif isnan(R(i).E)
            fprintf('CASE %d is diverged at iteration step
#%d.\n',case_i,i)
            break
        end
    end
    if i == iter_req && R(i).E>MSE_target % Check MSE
        fprintf('CASE %d is diverged or required iteration number is not
enough\n',case_i)
    end
    fprintf('\n\n')
end
if exist('summary')
    table_sum = array2table(summary,'VariableNames',{'Case','xA','xh','la
mdaV1','lamdaV2','fxC','v1(x)','v2(x)'});
end
%% Select optimal result
fprintf('...SELECT OPTIMAL RESULT...\n')
fprintf('With an initial input vector: [xA, xh, lamdaV1,
lamdaV2] = [%.3f, %.3f, %.3f, %.3f],\n',x(1), x(2), x(3), x(4))
if exist('table_sum')
    disp(table_sum)
end

if exist('table_sum')
    table_sum(table_sum{:,7} <= -1e-6 | table_sum{:,8} <= -1e-6,:) = [];
    [~,index] = min(table_sum{:,6});
    fprintf('a local maximum is obtained from CASE
%d\n',table_sum{index,1});
else
    fprintf('there is no local optimum\n\n')
end
fprintf('Readers need to try several initial input vectors [xA, xh,
lamdaV1, lamdaV2]\nto obtain a global optimum.\n')
```

B3 ANALYTICAL LAGRANGE CODE FOR PROJECTILE OPTIMIZATION

```
clc, clear;
format short

%% INPUT
x = [0.7 40 -1 0]'; % initial vector [xC_0, theta_0, lamdaC_0,
lamdaV_0]';
iter_req = 2000; % number of required iteration
MSE_target = 1e-6; % Target of mean square error of first derivative of
Lagrange function

fprintf('Initial input vector: [xC, theta, lamdaC, lamdaV] = [%.3f, %.3f,
%.3f, %.3f],\n',...
                        x(1), x(2), x(3), x(4))
fprintf('Number of required iteration: %d\nTarget ofMean square error:
%s\n\n\n',...
                        iter_req,MSE_target)

V0 = 0.08; % Projectile velocity, km/s
g = 9.81e-3; % Standard gravity, km.s-2
xA = 0; yA = 0.06; % Position of flag target, km
xB = 0.07; yB = 0.1; % Position of toptree, km

k = 0;
for case_i = 1:2

    if case_i == 1
        fprintf('...STARTING CASE 1: v(x) = y_Pt - 0.1 >= 0 is
active...\n\n')
    elseif case_i == 2
        fprintf('...STARTING CASE 1: v(x) = y_Pt - 0.1 >= 0 is
inactive...\n\n')
    end
    if case_i == 1 % Inequality v(x) = y_Pt - 0.1 >= 0 is active
        S = 1;
    elseif case_i == 2 % Inequality v(x) = y_Pt - 0.1 >= 0 is inactive
        S = 0;
    end
    clear('R')
    R(1).x = x;

    index = 1:4;
    s_index = 4;
    status = [1;1;1;S];
    index(status==0) = [];
    s_index(diag(S)==0) = [];

    for i = 1:iter_req
        xC = R(i).x(1); theta = R(i).x(2);
        lamda_c = R(i).x(3);
        lamda_v = R(i).x(4);

        % Calculate objective function, fxC, and its derivatives
        R(i).xC = xC;
        J_xC = [1;
                0];
        H_xC = [0 , 0;
                0 , 0];
```

```
        % Calculate inequalities, c(x), v(x), and their Jacobian, J_c,
J_v, and Hessian, H_c, H_v
        c = (xC-xA)*tand(theta) - g*(xC-xA)^2/(2*V0^2*cosd(theta)^2)
- yA;
        v = (xC-xB)*tand(theta) - g*(xC-xB)^2/(2*V0^2*cosd(theta)^2)
- yB;
         J_c = [(tand(theta) - g*(xC-xA)/(V0^2*cosd(theta)^2)) ,
((xC-xA)*(tand(theta)^2+1) - g*(xC-xA)^2*sind(theta)/
(V0^2*cosd(theta)^3))];
        J_v = [(tand(theta) - g*(xC-xB)/(V0^2*cosd(theta)^2)) ,
((xC-xB)*(tand(theta)^2+1) - g*(xC-xB)^2*sind(theta)/
(V0^2*cosd(theta)^3))];

        H_c11 = -g/(V0^2*cosd(theta)^2);
        H_c12 = tand(theta)^2+1 - (2*g*(xC-xA)*sind(theta))/
(V0^2*cosd(theta)^3);
        H_c22 = 2*tand(theta)*(tand(theta)^2+1)*(xC-xA) - g*(xC-xA)^2/
(V0^2*cosd(theta)^2) - (3*g*sind(theta)^2*(xC-xA)^2)/
(V0^2*cosd(theta)^4);
        H_c = [H_c11   , H_c12;
               H_c12   , H_c22];

        H_v11 = -g/(V0^2*cosd(theta)^2);
        H_v12 = tand(theta)^2+1 - (2*g*(xC-xB)*sind(theta))/
(V0^2*cosd(theta)^3);
        H_v22 = 2*tand(theta)*(tand(theta)^2+1)*(xC-xB) - g*(xC-xB)^2/
(V0^2*cosd(theta)^2) - (3*g*sind(theta)^2*(xC-xB)^2)/
(V0^2*cosd(theta)^4);
        H_v = [H_v11   , H_v12;
               H_v12   , H_v22];

        % Calculate Lagrange function, L, w.r.t. to [xC_(i), theta_(i),
lamdaC_(i), lamdaV_(i)]'
        R(i).Lagrange = R(i).xC - lamda_c*c - lamda_v*S*v;

        % Calculate Jacobian of L w.r.t. to [xC_(i), theta_(i), lamdaC_
(i), lamdaV_(i)]'
        R(i).J = [J_xC - J_c'*lamda_c - J_v'*S*lamda_v;
                            -c;
                            -S*v];

        % Calculate MSE of J
        R(i).E = 1/4*sum(R(i).J .^ 2);

        % Calculate Hessian of L w.r.t. to [xC_(i), theta_(i), lamdaC_
(i), lamdaV_(i)]'
        H = H_xC - lamda_c*H_c - S*lamda_v*H_v;
        R(i).H = [H       , -J_c'  , -(S*J_v)';
                  -J_c    ,   0    ,     0    ;
                 -S*J_v   ,   0    ,     0    ];

        if R(i).E>MSE_target % Check MSE
            % Update input vector for next iteration
            R(i+1).x = R(i).x;
            R(i+1).x(index) = R(i).x(index) - R(i).H(index,index) \
R(i).J(index);
        elseif R(i).E <= MSE_target
            fprintf('CASE %d is converged at iteration step #%d with an
MSE of %s, providing a result of:\n',case_i,i,R(i).E)
            fprintf('Input vector: [xC, theta, lamdaC, lamdaV] = [%f, %f,
```

```
%f, %f]\n',...
                                    R(i).x(1), R(i).x(2), R(i).x(3), R(i).x(4))
       fprintf('Objective function fxC = xC = %f\n', round(R(i).xC,3))
       fprintf('Equality constraint c(x) = y_Pf - yA = %f\n', round(c,3))
       fprintf('Inequality constraint v(x) = y_Pt - yB = %f\n', round(v,3))

       k = k+1;
       summary(k,1) = case_i;
       summary(k,2) = R(end).x(1);
       summary(k,3) = R(end).x(2);
       summary(k,4) = R(end).x(3);
       summary(k,5) = R(end).x(4);
       summary(k,6) = R(end).xC;
       summary(k,7) = round(c,3);
       summary(k,8) = round(v,3);
       break
        elseif isnan(R(i).E)
            fprintf('CASE %d is diverged at iteration step
#%d.\n',case_i)
            break
        end
    end
    if i == iter_req && R(i).E>MSE_target % Check MSE
        fprintf('CASE %d is diverged or required iteration number is not
enough\n',case_i)
    end
    fprintf('\n\n')
end
table_sum = array2table(summary,'VariableNames',{'Case','xC','theta',
'lamdaC','lamdaV','fxC','c(x) = y_Pf-yA','v(x) = yB'});

%% Select optimal result
fprintf('...SELECT OPTIMAL RESULT...\n')
fprintf('With an initial input vector: [xC, theta, lamdaC,
lamdaV] = [%.3f, %.3f, %.3f, %.3f],\n',x(1), x(2), x(3), x(4))
disp(table_sum)
[~,index] = max(table_sum{:,6});
fprintf('a local maximum is obtained from CASE %d\n',table_sum{index,1});
fprintf('Readers need to try several initial input vectors [xC, theta,
lamdaC, lamdaV] to obtain a global optimum.\n')
```

Appendix C

WEIGHT AND BIAS MATRICES OF MODEL-LPS IN SECTION 5.4

a. Model-1LP:

- Weight and bias matrices of output y_1 (CI_c) that best training results of CI_c are based on ten layers and 20 neurons are as follows. It is noted that k indicates layer number, and input parameters are normalized from –1 to 1 based on normalization functions g^N.

  + At Hidden Layer 1:

$$f_{tanh}\Big(\underbrace{\boldsymbol{\omega}^{(1)}_{CI_c}}_{[20\times6]}\,\underbrace{g^N(\mathbf{x}^{\text{in}})}_{[6\times1]}+\underbrace{\boldsymbol{b}^{(1)}_{CI_c}}_{[20\times1]}\Big)=f_{tanh}\left[\overset{\boldsymbol{\omega}^{(1)}[20\times6]}{\begin{bmatrix}-0.972 & -1.006 & 0.165 & -0.183 & 1.149 & -0.748\\ 1.352 & 0.475 & 0.130 & 0.010 & 0.259 & -0.442\\ 1.308 & -0.696 & -0.623 & 0.237 & 0.464 & 0.624\\ -0.169 & -0.198 & -0.729 & -1.296 & -1.283 & 0.095\\ 0.845 & 0.746 & 0.065 & 0.027 & -0.056 & 0.260\\ & & \cdots & & & \\ 0.782 & -0.650 & -1.172 & 0.467 & -1.501 & -0.439\\ -0.337 & 0.945 & -0.104 & -0.189 & 1.975 & -0.279\end{bmatrix}}\times\overset{\mathbf{x}^{(\text{in})}[6\times1]}{\begin{bmatrix}x_1=g^N(D)\\ x_2=g^N(\rho_s)\\ x_3=g^N(f_c')\\ x_4=g^N(f_y)\\ x_5=g^N(P_u)\\ x_6=g^N(M_u)\end{bmatrix}}+\overset{\mathbf{b}^{(1)}[20\times1]}{\begin{bmatrix}2.202\\ -1.961\\ -1.812\\ 1.823\\ -0.973\\ \cdots\\ 2.160\\ -2.446\end{bmatrix}}\right]=\overset{\mathbf{x}^{(k=1)}[20\times1]}{\begin{bmatrix}\mathrm{x}^{(k=1)}_{1,1}\\ \mathrm{x}^{(k=1)}_{1,2}\\ \mathrm{x}^{(k=1)}_{1,3}\\ \mathrm{x}^{(k=1)}_{1,4}\\ \mathrm{x}^{(k=1)}_{1,5}\\ \cdots\\ \mathrm{x}^{(k=1)}_{1,19}\\ \mathrm{x}^{(k=1)}_{1,20}\end{bmatrix}}=\underbrace{\mathbf{x}^{(1)}}_{[20\times1]} \tag{C1}$$

  + At Hidden Layer 2:

$$f_{tanh}\left[\overset{\boldsymbol{\omega}^{(2)}[20\times20]}{\begin{bmatrix}-0.394 & 0.367 & 0.336 & & 0.358 & -0.294\\ 0.673 & 0.267 & -0.117 & & -0.219 & -0.328\\ -0.327 & -0.267 & -0.442 & \cdots & 0.514 & -0.435\\ -0.447 & -0.386 & -0.246 & & 0.551 & 0.232\\ 0.128 & -0.455 & -0.697 & & -0.327 & -0.511\\ & & \cdots & & & \\ 0.278 & -0.474 & 0.130 & \cdots & 0.175 & 0.463\\ 0.159 & -0.011 & 0.022 & & 0.468 & 0.150\end{bmatrix}}\times\overset{\mathbf{x}^{(k=1)}[20\times1]}{\begin{bmatrix}\mathrm{x}^{(k=1)}_{1,1}\\ \mathrm{x}^{(k=1)}_{1,2}\\ \mathrm{x}^{(k=1)}_{1,3}\\ \mathrm{x}^{(k=1)}_{1,4}\\ \mathrm{x}^{(k=1)}_{1,5}\\ \cdots\\ \mathrm{x}^{(k=1)}_{1,19}\\ \mathrm{x}^{(k=1)}_{1,20}\end{bmatrix}}+\overset{\mathbf{b}^{(2)}[20\times1]}{\begin{bmatrix}1.555\\ 1.461\\ -1.391\\ 1.018\\ 0.828\\ \cdots\\ 1.455\\ -1.672\end{bmatrix}}\right]=\overset{\mathbf{x}^{(k=2)}[20\times1]}{\begin{bmatrix}\mathrm{x}^{(k=2)}_{1,1}\\ \mathrm{x}^{(k=2)}_{1,2}\\ \mathrm{x}^{(k=2)}_{1,3}\\ \mathrm{x}^{(k=2)}_{1,4}\\ \mathrm{x}^{(k=2)}_{1,5}\\ \cdots\\ \mathrm{x}^{(k=2)}_{1,19}\\ \mathrm{x}^{(k=2)}_{1,20}\end{bmatrix}}=\underbrace{\mathbf{x}^{(2)}}_{[20\times1]} \tag{C2}$$

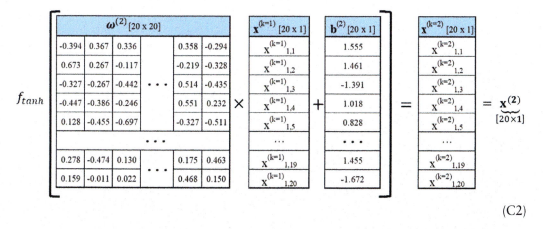

+ At Hidden Layers 3, 4, ..., 10: The structure of weight and bias matrices at Hidden Layers 3–10 similar to that of weight and bias matrices at Hidden Layer 2 is derived in Equation (C2).

+ At output layer: It is noted that y_1 is de-normalized to obtain the original scale of output CI_c.

$$f_{tanh}\left[\underbrace{\begin{pmatrix} -0.304 & 0.987 & 0.153 & \cdots & 0.677 & -0.289 \end{pmatrix}}_{\boldsymbol{\omega}^{(out)}[1 \times 20]} \times \underbrace{\begin{pmatrix} x^{(k=10)}_{1,1} \\ x^{(k=10)}_{1,2} \\ x^{(k=10)}_{1,3} \\ x^{(k=10)}_{1,4} \\ x^{(k=10)}_{1,5} \\ \cdots \\ x^{(k=10)}_{1,19} \\ x^{(k=10)}_{1,20} \end{pmatrix}}_{\mathbf{x}^{(k=10)}[20 \times 1]} + \underbrace{\begin{pmatrix} -0.060 \end{pmatrix}}_{\mathbf{b}^{(out)}[1 \times 1]}\right] = \underbrace{\begin{pmatrix} y_1 \end{pmatrix}}_{\mathbf{y}[1 \times 1]} \quad (C3)$$

- Weight and bias matrices of output y_2 (CO_2) that best training results of CO_2 are based on five layers and 20 neurons are as follows. It is noted that k indicates layer number.

+ At Hidden Layer 1:

$$f_{tanh}\left(\underbrace{\boldsymbol{\omega}^{(1)}_{CO_2}}_{[20\times6]}\underbrace{g^N(\mathbf{x}^{\text{in}})}_{[6\times1]} + \underbrace{\boldsymbol{b}^{(1)}_{CO_2}}_{[20\times1]}\right) = f_{tanh}\left[\underbrace{\begin{pmatrix} -0.597 & -0.185 & -0.714 & -0.077 & -1.113 & -1.128 \\ -1.338 & -0.214 & -0.092 & 0.256 & 0.698 & 0.038 \\ 0.531 & -1.083 & -0.009 & -0.021 & -0.085 & -0.289 \\ 1.291 & -0.980 & 0.029 & 0.023 & 0.333 & 0.950 \\ -0.839 & -0.592 & -0.004 & 0.009 & 0.005 & -0.058 \\ & & \cdots & & & \\ 0.950 & 1.105 & 0.565 & 0.605 & -0.572 & -1.035 \\ -1.688 & -0.164 & 0.005 & 0.005 & -0.292 & -0.504 \end{pmatrix}}_{\boldsymbol{\omega}^{(1)}[20 \times 6]} \times \underbrace{\begin{pmatrix} x_1 = g^N(D) \\ x_2 = g^N(\rho_s) \\ x_3 = g^N(f_c') \\ x_4 = g^N(f_y) \\ x_5 = g^N(P_u) \\ x_6 = g^N(M_u) \end{pmatrix}}_{\mathbf{x}^{(in)}[6 \times 1]} + \underbrace{\begin{pmatrix} 2.825 \\ 2.310 \\ -1.861 \\ -1.781 \\ 1.044 \\ \cdots \\ 2.373 \\ -1.823 \end{pmatrix}}_{\mathbf{b}^{(1)}[20 \times 1]}\right] = \underbrace{\begin{pmatrix} x^{(k=1)}_{2,1} \\ x^{(k=1)}_{2,2} \\ x^{(k=1)}_{2,3} \\ x^{(k=1)}_{2,4} \\ x^{(k=1)}_{2,5} \\ \cdots \\ x^{(k=1)}_{2,19} \\ x^{(k=1)}_{2,20} \end{pmatrix}}_{\mathbf{x}^{(k=1)}[20 \times 1]} = \underbrace{\mathbf{x}^{(1)}}_{[20\times1]} \quad (C4)$$

+ At Hidden Layer 2:

$$f_{tanh}\left[\underbrace{\begin{pmatrix} -0.458 & 0.096 & 0.130 & & -0.483 & 0.488 \\ -0.535 & -0.017 & -0.315 & & 0.653 & -0.241 \\ 0.001 & -0.483 & 0.255 & \cdots & -0.092 & 0.639 \\ -0.233 & 0.045 & 0.215 & & 0.282 & 0.479 \\ -0.535 & 0.138 & -0.357 & & -0.011 & -0.012 \\ & & \cdots & & & \\ 0.335 & 0.420 & 0.049 & \cdots & 0.602 & -0.110 \\ -0.471 & 0.093 & -0.135 & & -0.352 & -0.187 \end{pmatrix}}_{\boldsymbol{\omega}^{(2)}[20 \times 20]} \times \underbrace{\begin{pmatrix} x^{(k=1)}_{2,1} \\ x^{(k=1)}_{2,2} \\ x^{(k=1)}_{2,3} \\ x^{(k=1)}_{2,4} \\ x^{(k=1)}_{2,5} \\ \cdots \\ x^{(k=1)}_{2,19} \\ x^{(k=1)}_{2,20} \end{pmatrix}}_{\mathbf{x}^{(k=1)}[20 \times 1]} + \underbrace{\begin{pmatrix} 1.555 \\ 1.461 \\ -1.391 \\ 1.018 \\ 0.828 \\ \cdots \\ 1.455 \\ -1.672 \end{pmatrix}}_{\mathbf{b}^{(2)}[20 \times 1]}\right] = \underbrace{\begin{pmatrix} x^{(k=2)}_{2,1} \\ x^{(k=2)}_{2,2} \\ x^{(k=2)}_{2,3} \\ x^{(k=2)}_{2,4} \\ x^{(k=2)}_{2,5} \\ \cdots \\ x^{(k=2)}_{2,19} \\ x^{(k=2)}_{2,20} \end{pmatrix}}_{\mathbf{x}^{(k=2)}[20 \times 1]} = \underbrace{\mathbf{x}^{(2)}}_{[20\times1]} \quad (C5)$$

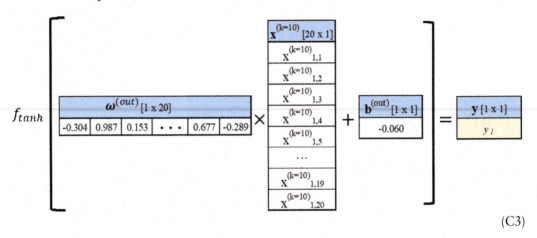

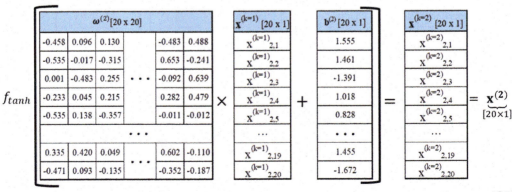

+ At Hidden Layers 3, 4, ..., 10: The structure of weight and bias matrices at Hidden Layers 3–10 similar to that of weight and bias matrices at Hidden Layer 2 is derived in Equation (C5).

+ At output layer: It is noted that y_2 is de-normalized to obtain the original scale of output CO_2.

$$f_{tanh}\left[\underbrace{\begin{bmatrix} 0.729 & 0.166 & 0.405 & \cdots & 0.604 & -0.435 \end{bmatrix}}_{\boldsymbol{\omega}^{(out)}\,[1 \times 20]} \times \underbrace{\begin{bmatrix} x^{(k=5)}_{2,1} \\ x^{(k=5)}_{2,2} \\ x^{(k=5)}_{2,3} \\ x^{(k=5)}_{2,4} \\ x^{(k=5)}_{2,5} \\ \cdots \\ x^{(k=5)}_{2,19} \\ x^{(k=5)}_{2,20} \end{bmatrix}}_{\mathbf{x}^{(k=10)}\,[20 \times 1]} + \underbrace{\begin{bmatrix} 0.492 \end{bmatrix}}_{\mathbf{b}^{(out)}\,[1 \times 1]}\right] = \underbrace{\begin{bmatrix} y_2 \end{bmatrix}}_{\mathbf{y}\,[1 \times 1]} \tag{C6}$$

- Weight and bias matrices of output y_3 (W_C) that best training results of W_C are based on five layers and 80 neurons are as follows. It is noted that k indicates layer number.

  + At Hidden Layer 1:

$$f_{tanh}\Big(\underbrace{\boldsymbol{\omega}^{(1)}_{W_C}}_{[80\times6]} \underbrace{g^N(\mathbf{x}^{\text{in}})}_{[6\times1]} + \underbrace{\boldsymbol{b}^{(1)}_{W_C}}_{[80\times1]}\Big) = f_{tanh}\left[\underbrace{\begin{bmatrix} -1.289 & 0.993 & 0.702 & -0.094 & 1.501 & -1.373 \\ 1.362 & -0.430 & -0.296 & 0.128 & -1.776 & 0.556 \\ -0.143 & 1.315 & 0.542 & 1.101 & 1.735 & -1.195 \\ 1.879 & 0.063 & -0.965 & 0.376 & 0.037 & 1.825 \\ 0.996 & 1.131 & 1.032 & -0.539 & -1.431 & 0.188 \\ & & & \cdots & & \\ 0.046 & -2.111 & -1.542 & 0.066 & -0.782 & -0.530 \\ -0.481 & -1.590 & -1.369 & 0.773 & -0.784 & -0.741 \end{bmatrix}}_{\boldsymbol{\omega}^{(1)}\,[80 \times 6]} \times \underbrace{\begin{bmatrix} x_1 = g^N(D) \\ x_2 = g^N(\rho_s) \\ x_3 = g^N(f_c') \\ x_4 = g^N(f_y) \\ x_5 = g^N(P_u) \\ x_6 = g^N(M_u) \end{bmatrix}}_{\mathbf{x}^{(in)}\,[6 \times 1]} + \underbrace{\begin{bmatrix} 3.016 \\ -2.595 \\ 2.871 \\ -2.714 \\ -2.789 \\ \cdots \\ 2.869 \\ -2.940 \end{bmatrix}}_{\mathbf{b}^{(1)}\,[80 \times 1]}\right] = \underbrace{\begin{bmatrix} x^{(k=1)}_{3,1} \\ x^{(k=1)}_{3,2} \\ x^{(k=1)}_{3,3} \\ x^{(k=1)}_{3,4} \\ x^{(k=1)}_{3,5} \\ \cdots \\ x^{(k=1)}_{3,79} \\ x^{(k=1)}_{3,80} \end{bmatrix}}_{\mathbf{x}^{(k=1)}\,[80 \times 1]} = \underbrace{\mathbf{x}^{(1)}}_{[80\times1]} \tag{C7}$$

  + At Hidden Layer 2:

$$f_{tanh}\left[\underbrace{\begin{bmatrix} -0.127 & 0.255 & -0.176 & & -0.040 & 0.137 \\ 0.050 & 0.031 & -0.173 & & -0.201 & 0.092 \\ 0.162 & -0.070 & -0.219 & \cdots & -0.191 & 0.035 \\ -0.211 & 0.063 & 0.052 & & 0.058 & -0.003 \\ 0.212 & 0.077 & -0.273 & & -0.303 & 0.174 \\ & & & \cdots & & \\ -0.289 & 0.157 & -0.043 & \cdots & -0.125 & 0.174 \\ -0.193 & 0.008 & 0.232 & & -0.224 & 0.067 \end{bmatrix}}_{\boldsymbol{\omega}^{(2)}\,[80 \times 80]} \times \underbrace{\begin{bmatrix} x^{(k=1)}_{3,1} \\ x^{(k=1)}_{3,2} \\ x^{(k=1)}_{3,3} \\ x^{(k=1)}_{3,4} \\ x^{(k=1)}_{3,5} \\ \cdots \\ x^{(k=1)}_{3,79} \\ x^{(k=1)}_{3,80} \end{bmatrix}}_{\mathbf{x}^{(k=1)}\,[80 \times 1]} + \underbrace{\begin{bmatrix} 1.482 \\ -1.446 \\ -1.458 \\ 1.371 \\ -1.353 \\ \cdots \\ -1.438 \\ -1.488 \end{bmatrix}}_{\mathbf{b}^{(2)}\,[80 \times 1]}\right] = \underbrace{\begin{bmatrix} x^{(k=2)}_{3,1} \\ x^{(k=2)}_{3,2} \\ x^{(k=2)}_{3,3} \\ x^{(k=2)}_{3,4} \\ x^{(k=2)}_{3,5} \\ \cdots \\ x^{(k=2)}_{3,79} \\ x^{(k=2)}_{3,80} \end{bmatrix}}_{\mathbf{x}^{(k=2)}\,[80 \times 1]} = \underbrace{\mathbf{x}^{(2)}}_{[80\times1]} \tag{C8}$$

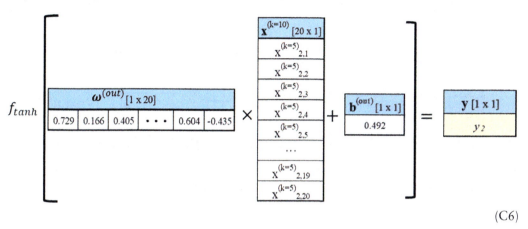

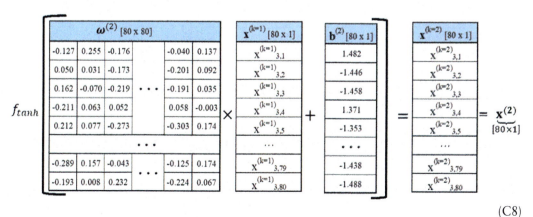

+ At Hidden Layers 3, 4, ..., 10: The structure of weight and bias matrices at Hidden Layers 3–10 similar to that of weight and bias matrices at Hidden Layer 2 is derived in Equation (C8).

+ At output layer: It is noted that y_3 is de-normalized to obtain the original scale of output W_C.

$$f_{tanh}\left[\underbrace{\begin{bmatrix} -0.307 & 0.128 & -0.916 & \cdots & 0.622 & 0.267 \end{bmatrix}}_{\omega^{(out)}\,[1 \times 80]} \times \underbrace{\begin{bmatrix} x^{(k=5)}_{3,1} \\ x^{(k=5)}_{3,2} \\ x^{(k=5)}_{3,3} \\ x^{(k=5)}_{3,4} \\ x^{(k=5)}_{3,5} \\ \cdots \\ x^{(k=5)}_{3,79} \\ x^{(k=5)}_{3,80} \end{bmatrix}}_{\mathbf{x}^{(k=5)}\,[80 \times 1]} + \underbrace{\begin{bmatrix} 0.754 \end{bmatrix}}_{\mathbf{b}^{(out)}\,[1 \times 1]}\right] = \underbrace{\begin{bmatrix} y_3 \end{bmatrix}}_{\mathbf{y}\,[1 \times 1]} \tag{C9}$$

- Weight and bias matrices of output y_4 (SF) that best training results of SF are based on ten layers and 80 neurons are as follows.

  + At Hidden Layer 1:

$$f_{tanh}\left(\underbrace{\boldsymbol{\omega}^{(1)}_{SF}}_{[80\times6]}\underbrace{g^N(\mathbf{x}^{\text{in}})}_{[6\times1]} + \underbrace{\boldsymbol{b}^{(1)}_{SF}}_{[80\times1]}\right) = f_{tanh}\left[\underbrace{\begin{bmatrix} 0.397 & 0.170 & 0.012 & 0.056 & 0.433 & 2.847 \\ -0.374 & -1.097 & -0.025 & -0.522 & -1.591 & 1.845 \\ 0.368 & 1.946 & -0.074 & -0.359 & -0.817 & 1.707 \\ 1.464 & 0.237 & -1.227 & -1.400 & 0.441 & 1.346 \\ 0.479 & -1.116 & -0.795 & -1.180 & 2.045 & -0.569 \\ & & \cdots & & & \\ -0.694 & 0.159 & 0.055 & -0.042 & 0.208 & -2.073 \\ -1.548 & 0.650 & 1.514 & -0.705 & 0.543 & 1.319 \end{bmatrix}}_{\boldsymbol{\omega}^{(1)}\,[80 \times 6]} \times \underbrace{\begin{bmatrix} x_1 = g^N(D) \\ x_2 = g^N(\rho_s) \\ x_3 = g^N(f'_c) \\ x_4 = g^N(f_y) \\ x_5 = g^N(P_u) \\ x_6 = g^N(M_u) \end{bmatrix}}_{\mathbf{x}^{(in)}\,[6 \times 1]} + \underbrace{\begin{bmatrix} 3.651 \\ 2.919 \\ -2.915 \\ -2.817 \\ -2.727 \\ \cdots \\ -2.826 \\ -2.988 \end{bmatrix}}_{\mathbf{b}^{(1)}\,[80 \times 1]}\right] = \underbrace{\begin{bmatrix} x^{(1)}_{4,1} \\ x^{(1)}_{4,2} \\ x^{(1)}_{4,3} \\ x^{(1)}_{4,4} \\ x^{(1)}_{4,5} \\ \cdots \\ x^{(1)}_{4,79} \\ x^{(1)}_{4,80} \end{bmatrix}}_{\mathbf{x}^{(1)}\,[80 \times 1]} = \underbrace{\mathbf{x}^{(1)}}_{[80\times1]} \tag{C10}$$

  + At Hidden Layer 2:

$$f_{tanh}\left[\underbrace{\begin{bmatrix} -0.767 & 0.090 & -0.138 & & 0.423 & -0.211 \\ -0.147 & -0.033 & -0.268 & & 0.139 & -0.265 \\ -0.154 & 0.166 & -0.038 & \cdots & 0.261 & -0.018 \\ 0.272 & 0.149 & 0.210 & & 0.172 & -0.067 \\ 0.051 & -0.183 & -0.172 & & 0.017 & -0.069 \\ & & \cdots & & & \\ 0.178 & 0.159 & 0.171 & \cdots & -0.261 & -0.272 \\ 0.046 & -0.082 & 0.279 & & -0.140 & -0.165 \end{bmatrix}}_{\boldsymbol{\omega}^{(2)}\,[80 \times 80]} \times \underbrace{\begin{bmatrix} x^{(1)}_{4,1} \\ x^{(1)}_{4,2} \\ x^{(1)}_{4,3} \\ x^{(1)}_{4,4} \\ x^{(1)}_{4,5} \\ \cdots \\ x^{(1)}_{4,79} \\ x^{(1)}_{4,80} \end{bmatrix}}_{\mathbf{x}^{(1)}\,[80 \times 1]} + \underbrace{\begin{bmatrix} 1.477 \\ 1.451 \\ 1.452 \\ -1.373 \\ -1.316 \\ \cdots \\ 1.454 \\ 1.486 \end{bmatrix}}_{\mathbf{b}^{(2)}\,[80 \times 1]}\right] = \underbrace{\begin{bmatrix} x^{(2)}_{4,1} \\ x^{(2)}_{4,2} \\ x^{(2)}_{4,3} \\ x^{(2)}_{4,4} \\ x^{(2)}_{4,5} \\ \cdots \\ x^{(2)}_{4,79} \\ x^{(2)}_{4,80} \end{bmatrix}}_{\mathbf{x}^{(2)}\,[80 \times 1]} \tag{C11}$$

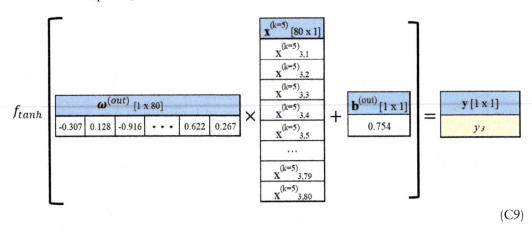

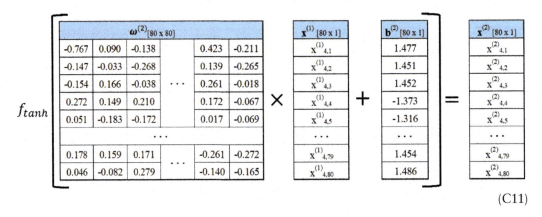

+ At Hidden Layers 3:

$$f_{tanh} \begin{bmatrix} \begin{array}{|ccc|ccc|} \multicolumn{6}{c}{\boldsymbol{\omega}^{(3)}{}_{[80 \times 80]}} \\ \hline 0.070 & -0.142 & 0.294 & & 0.239 & 0.297 \\ 0.001 & -0.106 & 0.025 & & -0.032 & 0.026 \\ -0.312 & -0.013 & 0.172 & \cdots & -0.068 & 0.327 \\ -0.376 & 0.129 & -0.080 & & -0.176 & 0.125 \\ -0.192 & -0.244 & -0.242 & & 0.246 & 0.117 \\ \multicolumn{6}{c}{\cdots} \\ 0.136 & -0.051 & 0.064 & \cdots & -0.078 & 0.073 \\ 0.187 & -0.199 & 0.065 & & -0.218 & 0.148 \\ \end{array} \times \begin{array}{c} \mathbf{x}^{(2)}{}_{[80 \times 1]} \\ \hline x^{(2)}{}_{4,1} \\ x^{(2)}{}_{4,2} \\ x^{(2)}{}_{4,3} \\ x^{(2)}{}_{4,4} \\ x^{(2)}{}_{4,5} \\ \cdots \\ x^{(2)}{}_{4,79} \\ x^{(2)}{}_{4,80} \end{array} + \begin{array}{c} \mathbf{b}^{(3)}{}_{[80 \times 1]} \\ \hline -1.463 \\ -1.343 \\ 1.436 \\ 1.298 \\ 1.346 \\ \cdots \\ 1.425 \\ 1.421 \end{array} \end{bmatrix} = \begin{array}{c} \mathbf{x}^{(3)}{}_{[80 \times 1]} \\ \hline x^{(3)}{}_{4,1} \\ x^{(3)}{}_{4,2} \\ x^{(3)}{}_{4,3} \\ x^{(3)}{}_{4,4} \\ x^{(3)}{}_{4,5} \\ \cdots \\ x^{(3)}{}_{4,79} \\ x^{(3)}{}_{4,80} \end{array}$$

(C12)

+ At Hidden Layers 4:

$$f_{tanh} \begin{bmatrix} \begin{array}{|ccc|ccc|} \multicolumn{6}{c}{\boldsymbol{\omega}^{(4)}{}_{[80 \times 80]}} \\ \hline 0.060 & 0.029 & 0.138 & & 0.113 & -0.201 \\ 0.156 & 0.042 & 0.146 & & 0.035 & -0.127 \\ 0.080 & 0.092 & -0.022 & \cdots & -0.079 & -0.181 \\ 0.060 & 0.058 & -0.153 & & 0.139 & -0.007 \\ 0.106 & 0.122 & 0.043 & & 0.118 & -0.018 \\ \multicolumn{6}{c}{\cdots} \\ -0.059 & 0.292 & 0.137 & \cdots & 0.145 & -0.154 \\ 0.021 & 0.022 & 0.106 & & -0.256 & -0.246 \\ \end{array} \times \begin{array}{c} \mathbf{x}^{(3)}{}_{[80 \times 1]} \\ \hline x^{(3)}{}_{4,1} \\ x^{(3)}{}_{4,2} \\ x^{(3)}{}_{4,3} \\ x^{(3)}{}_{4,4} \\ x^{(3)}{}_{4,5} \\ \cdots \\ x^{(3)}{}_{4,79} \\ x^{(3)}{}_{4,80} \end{array} + \begin{array}{c} \mathbf{b}^{(4)}{}_{[80 \times 1]} \\ \hline -1.451 \\ -1.446 \\ -1.373 \\ -1.324 \\ 1.305 \\ \cdots \\ -1.499 \\ 1.503 \end{array} \end{bmatrix} = \begin{array}{c} \mathbf{x}^{(4)}{}_{[80 \times 1]} \\ \hline x^{(4)}{}_{4,1} \\ x^{(4)}{}_{4,2} \\ x^{(4)}{}_{4,3} \\ x^{(4)}{}_{4,4} \\ x^{(4)}{}_{4,5} \\ \cdots \\ x^{(4)}{}_{4,79} \\ x^{(4)}{}_{4,80} \end{array}$$

(C13)

+ At Hidden Layers 5:

$$f_{tanh} \begin{bmatrix} \begin{array}{|ccc|ccc|} \multicolumn{6}{c}{\boldsymbol{\omega}^{(5)}{}_{[80 \times 80]}} \\ \hline 0.048 & 0.084 & -0.059 & & 0.136 & 0.175 \\ 0.054 & 0.120 & 0.100 & & -0.276 & -0.193 \\ -0.238 & 0.021 & 0.178 & \cdots & -0.034 & -0.072 \\ 0.239 & 0.217 & -0.004 & & 0.037 & -0.071 \\ -0.271 & 0.142 & 0.053 & & -0.170 & 0.050 \\ \multicolumn{6}{c}{\cdots} \\ -0.316 & -0.207 & -0.330 & \cdots & -0.371 & -0.034 \\ 0.029 & 0.081 & 0.277 & & -0.264 & 0.085 \\ \end{array} \times \begin{array}{c} \mathbf{x}^{(4)}{}_{[80 \times 1]} \\ \hline x^{(4)}{}_{4,1} \\ x^{(4)}{}_{4,2} \\ x^{(4)}{}_{4,3} \\ x^{(4)}{}_{4,4} \\ x^{(4)}{}_{4,5} \\ \cdots \\ x^{(4)}{}_{4,79} \\ x^{(4)}{}_{4,80} \end{array} + \begin{array}{c} \mathbf{b}^{(5)}{}_{[80 \times 1]} \\ \hline -1.474 \\ -1.435 \\ 1.446 \\ -1.375 \\ 1.343 \\ \cdots \\ -1.372 \\ 1.473 \end{array} \end{bmatrix} = \begin{array}{c} \mathbf{x}^{(5)}{}_{[80 \times 1]} \\ \hline x^{(5)}{}_{4,1} \\ x^{(5)}{}_{4,2} \\ x^{(5)}{}_{4,3} \\ x^{(5)}{}_{4,4} \\ x^{(5)}{}_{4,5} \\ \cdots \\ x^{(5)}{}_{4,79} \\ x^{(5)}{}_{4,80} \end{array}$$

(C14)

+ At Hidden Layers 6:

$$f_{tanh}\left[\underbrace{\begin{bmatrix} 0.057 & -0.245 & 0.118 & \cdots & -0.133 & 0.218 \\ -0.098 & 0.145 & 0.083 & & 0.155 & 0.194 \\ -0.264 & 0.064 & 0.035 & & -0.151 & 0.248 \\ -0.232 & -0.116 & 0.015 & & 0.061 & -0.140 \\ -0.051 & -0.050 & 0.116 & & -0.039 & 0.020 \\ & & & \cdots & & \\ -0.099 & 0.185 & 0.166 & \cdots & -0.189 & -0.142 \\ -0.027 & 0.158 & 0.084 & & -0.103 & 0.091 \end{bmatrix}}_{\omega^{(6)}_{[80 \times 80]}} \times \underbrace{\begin{bmatrix} x^{(5)}_{4,1} \\ x^{(5)}_{4,2} \\ x^{(5)}_{4,3} \\ x^{(5)}_{4,4} \\ x^{(5)}_{4,5} \\ \cdots \\ x^{(5)}_{4,79} \\ x^{(5)}_{4,80} \end{bmatrix}}_{\mathbf{x}^{(5)}_{[80 \times 1]}} + \underbrace{\begin{bmatrix} -1.484 \\ 1.439 \\ 1.407 \\ 1.375 \\ 1.308 \\ \cdots \\ -1.441 \\ -1.491 \end{bmatrix}}_{\mathbf{b}^{(6)}_{[80 \times 1]}}\right] = \underbrace{\begin{bmatrix} x^{(6)}_{4,1} \\ x^{(6)}_{4,2} \\ x^{(6)}_{4,3} \\ x^{(6)}_{4,4} \\ x^{(6)}_{4,5} \\ \cdots \\ x^{(6)}_{4,79} \\ x^{(6)}_{4,80} \end{bmatrix}}_{\mathbf{x}^{(6)}_{[80 \times 1]}} \tag{C15}$$

+ At Hidden Layers 7:

$$f_{tanh}\left[\underbrace{\begin{bmatrix} -0.177 & 0.026 & 0.259 & \cdots & -0.070 & -0.074 \\ -0.038 & 0.154 & 0.004 & & -0.252 & -0.045 \\ -0.145 & 0.025 & 0.159 & & -0.114 & 0.130 \\ 0.041 & 0.207 & 0.306 & & 0.272 & 0.170 \\ 0.200 & -0.230 & -0.019 & & 0.200 & -0.234 \\ & & & \cdots & & \\ 0.239 & -0.008 & -0.162 & \cdots & 0.012 & 0.082 \\ 0.236 & -0.215 & 0.245 & & 0.045 & 0.254 \end{bmatrix}}_{\omega^{(7)}_{[80 \times 80]}} \times \underbrace{\begin{bmatrix} x^{(6)}_{4,1} \\ x^{(6)}_{4,2} \\ x^{(6)}_{4,3} \\ x^{(6)}_{4,4} \\ x^{(6)}_{4,5} \\ \cdots \\ x^{(6)}_{4,79} \\ x^{(6)}_{4,80} \end{bmatrix}}_{\mathbf{x}^{(6)}_{[80 \times 1]}} + \underbrace{\begin{bmatrix} 1.476 \\ 1.443 \\ 1.390 \\ -1.366 \\ -1.349 \\ \cdots \\ 1.444 \\ 1.423 \end{bmatrix}}_{\mathbf{b}^{(7)}_{[80 \times 1]}}\right] = \underbrace{\begin{bmatrix} x^{(7)}_{4,1} \\ x^{(7)}_{4,2} \\ x^{(7)}_{4,3} \\ x^{(7)}_{4,4} \\ x^{(7)}_{4,5} \\ \cdots \\ x^{(7)}_{4,79} \\ x^{(7)}_{4,80} \end{bmatrix}}_{\mathbf{x}^{(7)}_{[80 \times 1]}} \tag{C16}$$

+ At Hidden Layers 8:

$$f_{tanh}\left[\underbrace{\begin{bmatrix} -0.089 & 0.115 & 0.122 & \cdots & 0.165 & 0.243 \\ 0.057 & -0.237 & 0.086 & & -0.141 & 0.263 \\ -0.118 & -0.278 & 0.123 & & -0.028 & -0.104 \\ -0.103 & 0.150 & -0.253 & & -0.115 & -0.255 \\ -0.090 & 0.244 & -0.043 & & 0.263 & 0.279 \\ & & & \cdots & & \\ 0.102 & -0.107 & 0.061 & \cdots & -0.072 & -0.082 \\ -0.043 & -0.218 & -0.153 & & 0.302 & -0.105 \end{bmatrix}}_{\omega^{(8)}_{[80 \times 80]}} \times \underbrace{\begin{bmatrix} x^{(7)}_{4,1} \\ x^{(7)}_{4,2} \\ x^{(7)}_{4,3} \\ x^{(7)}_{4,4} \\ x^{(7)}_{4,5} \\ \cdots \\ x^{(7)}_{4,79} \\ x^{(7)}_{4,80} \end{bmatrix}}_{\mathbf{x}^{(7)}_{[80 \times 1]}} + \underbrace{\begin{bmatrix} 1.478 \\ -1.442 \\ 1.396 \\ 1.404 \\ 1.341 \\ \cdots \\ 1.447 \\ -1.467 \end{bmatrix}}_{\mathbf{b}^{(8)}_{[80 \times 1]}}\right] = \underbrace{\begin{bmatrix} x^{(8)}_{4,1} \\ x^{(8)}_{4,2} \\ x^{(8)}_{4,3} \\ x^{(8)}_{4,4} \\ x^{(8)}_{4,5} \\ \cdots \\ x^{(8)}_{4,79} \\ x^{(8)}_{4,80} \end{bmatrix}}_{\mathbf{x}^{(8)}_{[80 \times 1]}} \tag{C17}$$

+ At Hidden Layers 9:

$$
f_{tanh}
\begin{bmatrix}
\begin{array}{|ccc c cc|}
\hline
\multicolumn{6}{|c|}{\omega^{(9)}{}_{[80 \times 80]}} \\
\hline
-0.201 & 0.242 & -0.187 & & 0.084 & 0.002 \\
-0.226 & -0.154 & -0.095 & & 0.182 & 0.257 \\
-0.111 & -0.256 & 0.091 & \cdots & -0.094 & -0.243 \\
-0.208 & -0.037 & 0.019 & & -0.254 & 0.176 \\
-0.215 & -0.265 & -0.138 & & 0.022 & 0.129 \\
\multicolumn{6}{|c|}{\cdots} \\
-0.034 & 0.100 & -0.239 & & -0.085 & -0.233 \\
0.137 & -0.266 & -0.228 & \cdots & -0.244 & -0.058 \\
\hline
\end{array}
\times
\begin{array}{|c|}
\hline
\mathbf{x}^{(8)}{}_{[80 \times 1]} \\
\hline
x^{(8)}{}_{4,1} \\
x^{(8)}{}_{4,2} \\
x^{(8)}{}_{4,3} \\
x^{(8)}{}_{4,4} \\
x^{(8)}{}_{4,5} \\
\cdots \\
x^{(8)}{}_{4,79} \\
x^{(8)}{}_{4,80} \\
\hline
\end{array}
+
\begin{array}{|c|}
\hline
\mathbf{b}^{(9)}{}_{[80 \times 1]} \\
\hline
1.476 \\
1.413 \\
1.404 \\
1.359 \\
1.330 \\
\cdots \\
-1.450 \\
1.510 \\
\hline
\end{array}
\end{bmatrix}
=
\begin{array}{|c|}
\hline
\mathbf{x}^{(9)}{}_{[80 \times 1]} \\
\hline
x^{(9)}{}_{4,1} \\
x^{(9)}{}_{4,2} \\
x^{(9)}{}_{4,3} \\
x^{(9)}{}_{4,4} \\
x^{(9)}{}_{4,5} \\
\cdots \\
x^{(9)}{}_{4,79} \\
x^{(9)}{}_{4,80} \\
\hline
\end{array}
\tag{C18}
$$

+ At Hidden Layers 10:

$$
f_{tanh}
\begin{bmatrix}
\begin{array}{|ccc c cc|}
\hline
\multicolumn{6}{|c|}{\omega^{(10)}{}_{[80 \times 80]}} \\
\hline
0.246 & -0.193 & -0.181 & & 0.029 & -0.013 \\
0.145 & 0.289 & 0.045 & & 0.075 & 0.171 \\
-0.174 & 0.256 & 0.033 & \cdots & 0.056 & -0.148 \\
0.171 & 0.057 & 0.039 & & -0.092 & -0.058 \\
-0.076 & 0.259 & -0.174 & & -0.153 & -0.084 \\
\multicolumn{6}{|c|}{\cdots} \\
0.129 & -0.196 & 0.203 & & -0.269 & 0.266 \\
-0.099 & -0.230 & -0.184 & \cdots & -0.087 & 0.092 \\
\hline
\end{array}
\times
\begin{array}{|c|}
\hline
\mathbf{x}^{(9)}{}_{[80 \times 1]} \\
\hline
x^{(9)}{}_{4,1} \\
x^{(9)}{}_{4,2} \\
x^{(9)}{}_{4,3} \\
x^{(9)}{}_{4,4} \\
x^{(9)}{}_{4,5} \\
\cdots \\
x^{(9)}{}_{4,79} \\
x^{(9)}{}_{4,80} \\
\hline
\end{array}
+
\begin{array}{|c|}
\hline
\mathbf{b}^{(10)}{}_{[80 \times 1]} \\
\hline
-1.498 \\
-1.459 \\
1.413 \\
-1.346 \\
1.303 \\
\cdots \\
1.443 \\
-1.480 \\
\hline
\end{array}
\end{bmatrix}
=
\begin{array}{|c|}
\hline
\mathbf{x}^{(10)}{}_{[80 \times 1]} \\
\hline
x^{(10)}{}_{4,1} \\
x^{(10)}{}_{4,2} \\
x^{(10)}{}_{4,3} \\
x^{(10)}{}_{4,4} \\
x^{(10)}{}_{4,5} \\
\cdots \\
x^{(10)}{}_{4,79} \\
x^{(10)}{}_{4,80} \\
\hline
\end{array}
\tag{C19}
$$

+ At output layer: It is noted that y_4 is de-normalized to obtain the original scale of output SF.

$$
f_{linear}
\begin{bmatrix}
\begin{array}{|cccc cc|}
\hline
\multicolumn{6}{|c|}{\omega^{(out)}{}_{[1 \times 80]}} \\
\hline
-0.284 & 0.878 & 0.855 & \cdots & 0.181 & -0.988 \\
\hline
\end{array}
\times
\begin{array}{|c|}
\hline
\mathbf{x}^{(10)}{}_{[80 \times 1]} \\
\hline
x^{(10)}{}_{4,1} \\
x^{(10)}{}_{4,2} \\
x^{(10)}{}_{4,3} \\
x^{(10)}{}_{4,4} \\
x^{(10)}{}_{4,5} \\
\cdots \\
x^{(10)}{}_{4,79} \\
x^{(10)}{}_{4,80} \\
\hline
\end{array}
+
\begin{array}{|c|}
\hline
\mathbf{b}^{(out)}{}_{[1 \times 1]} \\
\hline
-0.819 \\
\hline
\end{array}
\end{bmatrix}
=
\begin{array}{|c|}
\hline
\mathbf{y}\ [1 \times 1] \\
\hline
y_4 \\
\hline
\end{array}
\tag{C20}
$$

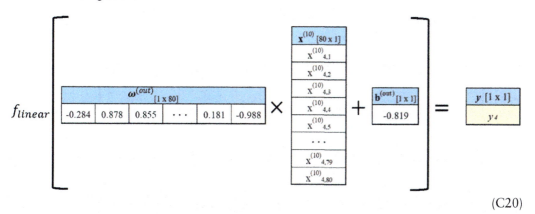

- Weight and bias matrices of output y_5 (ε_s) that best training results of ε_s are based on ten layers and 80 neurons are as follows.

  + At Hidden Layer 1:

$$f_{tanh}\left(\underbrace{\boldsymbol{\omega}_{\varepsilon_s}^{(1)}}_{[80\times6]}\underbrace{g^N(\mathbf{x}^{\text{in}})}_{[6\times1]}+\underbrace{\boldsymbol{b}_{\varepsilon_s}^{(1)}}_{[80\times1]}\right)=f_{tanh}\left[\boldsymbol{\omega}^{(1)}_{[80\times6]}\times\mathbf{x}^{(\text{in})}_{[6\times1]}+\mathbf{b}^{(1)}_{[80\times1]}\right]=\mathbf{x}^{(1)}_{[80\times1]}=\underbrace{\mathbf{x}^{(1)}}_{[80\times1]}$$

$\boldsymbol{\omega}^{(1)}_{[80 \times 6]}$					
-1.263	-1.436	-1.121	-0.036	-0.678	1.533
2.158	0.428	0.584	0.301	-1.010	1.182
-0.085	-0.022	-0.480	0.269	-0.313	-2.228
1.248	-0.420	-0.483	-0.425	1.706	-1.710
0.891	-0.922	0.286	-0.816	-0.732	-1.448
...					
-1.307	1.066	-1.077	-1.136	0.034	-1.074
0.136	-2.292	-0.031	-0.864	-1.087	0.673

×

$\mathbf{x}^{(\text{in})}_{[6 \times 1]}$
$x_1 = g^N(D)$
$x_2 = g^N(\rho_s)$
$x_3 = g^N(f_c')$
$x_4 = g^N(f_y)$
$x_5 = g^N(P_u)$
$x_6 = g^N(M_u)$

+

$\mathbf{b}^{(1)}_{[80 \times 1]}$
2.897
-2.894
-2.890
-2.755
-2.797
...
-2.916
2.998

=

$\mathbf{x}^{(1)}_{[80 \times 1]}$
$x^{(1)}_{5,1}$
$x^{(1)}_{5,2}$
$x^{(1)}_{5,3}$
$x^{(1)}_{5,4}$
$x^{(1)}_{5,5}$
...
$x^{(1)}_{5,79}$
$x^{(1)}_{5,80}$

(C21)

  + At Hidden Layer 2:

$$f_{tanh}\left[\boldsymbol{\omega}^{(2)}_{[80\times80]}\times\mathbf{x}^{(1)}_{[80\times1]}+\mathbf{b}^{(2)}_{[80\times1]}\right]=\mathbf{x}^{(2)}_{[80\times1]}$$

$\boldsymbol{\omega}^{(2)}_{[80 \times 80]}$					
0.041	-0.002	-0.154	...	0.171	0.089
-0.182	-0.127	-0.002		0.112	0.092
-0.208	0.018	0.227		0.040	-0.170
0.171	-0.039	0.157		-0.204	-0.026
0.270	0.003	0.260		-0.158	0.048
...					
-0.253	0.154	0.135	...	0.098	0.236
0.134	0.179	0.088		-0.210	-0.047

×

$\mathbf{x}^{(1)}_{[80 \times 1]}$
$x^{(1)}_{5,1}$
$x^{(1)}_{5,2}$
$x^{(1)}_{5,3}$
$x^{(1)}_{5,4}$
$x^{(1)}_{5,5}$
...
$x^{(1)}_{5,79}$
$x^{(1)}_{5,80}$

+

$\mathbf{b}^{(2)}_{[80 \times 1]}$
1.451
1.466
1.422
-1.365
-1.332
...
-1.451
1.445

=

$\mathbf{x}^{(2)}_{[80 \times 1]}$
$x^{(2)}_{5,1}$
$x^{(2)}_{5,2}$
$x^{(2)}_{5,3}$
$x^{(2)}_{5,4}$
$x^{(2)}_{5,5}$
...
$x^{(2)}_{5,79}$
$x^{(2)}_{5,80}$

(C22)

  + At Hidden Layers 3:

$$f_{tanh}\left[\boldsymbol{\omega}^{(3)}_{[80\times80]}\times\mathbf{x}^{(2)}_{[80\times1]}+\mathbf{b}^{(3)}_{[80\times1]}\right]=\mathbf{x}^{(3)}_{[80\times1]}$$

$\boldsymbol{\omega}^{(3)}_{[80 \times 80]}$					
-0.199	0.174	-0.010	...	0.040	-0.008
0.116	0.092	0.259		0.200	0.143
-0.230	-0.185	0.212		-0.050	-0.186
-0.121	0.020	-0.026		-0.140	-0.274
0.220	0.066	0.276		0.247	0.256
...					
-0.195	0.311	0.133	...	0.027	-0.121
0.204	-0.242	0.242		-0.152	0.054

×

$\mathbf{x}^{(2)}_{[80 \times 1]}$
$x^{(2)}_{5,1}$
$x^{(2)}_{5,2}$
$x^{(2)}_{5,3}$
$x^{(2)}_{5,4}$
$x^{(2)}_{5,5}$
...
$x^{(2)}_{5,79}$
$x^{(2)}_{5,80}$

+

$\mathbf{b}^{(3)}_{[80 \times 1]}$
1.479
-1.422
1.392
1.350
-1.334
...
-1.449
1.488

=

$\mathbf{x}^{(3)}_{[80 \times 1]}$
$x^{(3)}_{5,1}$
$x^{(3)}_{5,2}$
$x^{(3)}_{5,3}$
$x^{(3)}_{5,4}$
$x^{(3)}_{5,5}$
...
$x^{(3)}_{5,79}$
$x^{(3)}_{5,80}$

(C23)

+ At Hidden Layers 4:

$$f_{tanh}\left[\underbrace{\begin{bmatrix} -0.186 & -0.302 & -0.193 & & 0.000 & -0.033 \\ -0.072 & 0.193 & -0.055 & & -0.184 & 0.226 \\ 0.097 & -0.059 & 0.147 & \cdots & 0.267 & 0.035 \\ -0.258 & 0.083 & -0.085 & & 0.247 & -0.036 \\ -0.242 & -0.122 & -0.248 & & 0.119 & 0.125 \\ & & & \cdots & & \\ -0.090 & 0.289 & -0.042 & \cdots & -0.243 & -0.186 \\ 0.222 & 0.155 & -0.143 & & 0.256 & -0.208 \end{bmatrix}}_{\boldsymbol{\omega}^{(4)}[80 \times 80]} \times \underbrace{\begin{bmatrix} x^{(3)}_{5,1} \\ x^{(3)}_{5,2} \\ x^{(3)}_{5,3} \\ x^{(3)}_{5,4} \\ x^{(3)}_{5,5} \\ \cdots \\ x^{(3)}_{5,79} \\ x^{(3)}_{5,80} \end{bmatrix}}_{\mathbf{x}^{(3)}[80 \times 1]} + \underbrace{\begin{bmatrix} 1.470 \\ 1.441 \\ -1.420 \\ 1.365 \\ 1.346 \\ \cdots \\ -1.431 \\ 1.486 \end{bmatrix}}_{\mathbf{b}^{(4)}[80 \times 1]}\right] = \underbrace{\begin{bmatrix} x^{(4)}_{5,1} \\ x^{(4)}_{5,2} \\ x^{(4)}_{5,3} \\ x^{(4)}_{5,4} \\ x^{(4)}_{5,5} \\ \cdots \\ x^{(4)}_{5,79} \\ x^{(4)}_{5,80} \end{bmatrix}}_{\mathbf{x}^{(4)}[80 \times 1]} \tag{C24}$$

+ At Hidden Layers 5:

$$f_{tanh}\left[\underbrace{\begin{bmatrix} -0.021 & 0.114 & -0.031 & & 0.201 & -0.113 \\ -0.272 & -0.216 & 0.210 & & -0.036 & 0.128 \\ 0.014 & -0.148 & -0.227 & \cdots & -0.258 & -0.231 \\ -0.091 & -0.056 & 0.214 & & -0.038 & 0.005 \\ -0.178 & -0.049 & -0.074 & & -0.014 & 0.120 \\ & & & \cdots & & \\ -0.193 & 0.116 & 0.066 & \cdots & 0.279 & 0.285 \\ 0.267 & 0.149 & -0.135 & & -0.261 & -0.239 \end{bmatrix}}_{\boldsymbol{\omega}^{(5)}[80 \times 80]} \times \underbrace{\begin{bmatrix} x^{(4)}_{5,1} \\ x^{(4)}_{5,2} \\ x^{(4)}_{5,3} \\ x^{(4)}_{5,4} \\ x^{(4)}_{5,5} \\ \cdots \\ x^{(4)}_{5,79} \\ x^{(4)}_{5,80} \end{bmatrix}}_{\mathbf{x}^{(4)}[80 \times 1]} + \underbrace{\begin{bmatrix} -1.438 \\ 1.435 \\ -1.402 \\ 1.377 \\ 1.339 \\ \cdots \\ -1.438 \\ 1.477 \end{bmatrix}}_{\mathbf{b}^{(5)}[80 \times 1]}\right] = \underbrace{\begin{bmatrix} x^{(5)}_{5,1} \\ x^{(5)}_{5,2} \\ x^{(5)}_{5,3} \\ x^{(5)}_{5,4} \\ x^{(5)}_{5,5} \\ \cdots \\ x^{(5)}_{5,79} \\ x^{(5)}_{5,80} \end{bmatrix}}_{\mathbf{x}^{(5)}[80 \times 1]} \tag{C25}$$

+ At Hidden Layers 6:

$$f_{tanh}\left[\underbrace{\begin{bmatrix} -0.022 & 0.121 & 0.220 & & -0.128 & 0.264 \\ 0.105 & 0.149 & 0.184 & & -0.040 & 0.135 \\ -0.200 & 0.097 & 0.076 & \cdots & -0.081 & 0.117 \\ -0.188 & 0.255 & -0.126 & & 0.153 & -0.241 \\ -0.021 & 0.227 & -0.052 & & -0.002 & 0.280 \\ & & & \cdots & & \\ 0.070 & -0.174 & 0.080 & \cdots & 0.165 & 0.041 \\ -0.085 & 0.242 & -0.113 & & -0.135 & 0.181 \end{bmatrix}}_{\boldsymbol{\omega}^{(6)}[80 \times 80]} \times \underbrace{\begin{bmatrix} x^{(5)}_{5,1} \\ x^{(5)}_{5,2} \\ x^{(5)}_{5,3} \\ x^{(5)}_{5,4} \\ x^{(5)}_{5,5} \\ \cdots \\ x^{(5)}_{5,79} \\ x^{(5)}_{5,80} \end{bmatrix}}_{\mathbf{x}^{(5)}[80 \times 1]} + \underbrace{\begin{bmatrix} 1.473 \\ -1.428 \\ 1.405 \\ 1.374 \\ 1.341 \\ \cdots \\ 1.438 \\ -1.456 \end{bmatrix}}_{\mathbf{b}^{(6)}[80 \times 1]}\right] = \underbrace{\begin{bmatrix} x^{(6)}_{5,1} \\ x^{(6)}_{5,2} \\ x^{(6)}_{5,3} \\ x^{(6)}_{5,4} \\ x^{(6)}_{5,5} \\ \cdots \\ x^{(6)}_{5,79} \\ x^{(6)}_{5,80} \end{bmatrix}}_{\mathbf{x}^{(6)}[80 \times 1]} \tag{C26}$$

+ At Hidden Layers 7:

	$\omega^{(7)}_{[80 \times 80]}$							$x^{(6)}$ [80 x 1]		$b^{(7)}$ [80 x 1]		$x^{(7)}$ [80 x 1]
	-0.090	-0.053	-0.061		-0.061	-0.038		$x^{(6)}_{5,1}$		1.480		$x^{(7)}_{5,1}$
	-0.068	-0.278	0.038		-0.143	-0.097		$x^{(6)}_{5,2}$		1.463		$x^{(7)}_{5,2}$
	0.157	-0.049	0.195	...	0.140	-0.219		$x^{(6)}_{5,3}$		-1.396		$x^{(7)}_{5,3}$
f_{tanh}	0.300	-0.081	-0.209		0.226	-0.325	×	$x^{(6)}_{5,4}$	+	-1.360	=	$x^{(7)}_{5,4}$
	-0.221	0.264	-0.171		-0.208	-0.119		$x^{(6)}_{5,5}$		1.332		$x^{(7)}_{5,5}$
				...				...		...		...
	0.117	-0.026	-0.115	...	-0.061	0.010		$x^{(6)}_{5,79}$		1.437		$x^{(7)}_{5,79}$
	-0.269	-0.120	0.222		0.111	0.208		$x^{(6)}_{5,80}$		-1.472		$x^{(7)}_{5,80}$

(C27)

+ At Hidden Layers 8:

	$\omega^{(8)}_{[80 \times 80]}$							$x^{(7)}$ [80 x 1]		$b^{(8)}$ [80 x 1]		$x^{(8)}$ [80 x 1]
	0.045	-0.043	-0.051		-0.090	0.033		$x^{(7)}_{5,1}$		1.465		$x^{(8)}_{5,1}$
	-0.023	-0.018	0.278		0.083	-0.149		$x^{(7)}_{5,2}$		1.420		$x^{(8)}_{5,2}$
	0.134	0.127	-0.272	...	0.211	0.011		$x^{(7)}_{5,3}$		-1.418		$x^{(8)}_{5,3}$
f_{tanh}	0.236	-0.162	0.033		-0.209	-0.001	×	$x^{(7)}_{5,4}$	+	-1.365	=	$x^{(8)}_{5,4}$
	-0.276	0.253	-0.046		-0.206	0.157		$x^{(7)}_{5,5}$		1.341		$x^{(8)}_{5,5}$
				...				...		...		...
	0.233	0.046	0.095	...	-0.077	0.139		$x^{(7)}_{5,79}$		1.445		$x^{(8)}_{5,79}$
	0.086	-0.090	-0.175		0.106	-0.242		$x^{(7)}_{5,80}$		1.476		$x^{(8)}_{5,80}$

(C28)

+ At Hidden Layers 9:

	$\omega^{(9)}_{[80 \times 80]}$							$x^{(8)}$ [80 x 1]		$b^{(9)}$ [80 x 1]		$x^{(9)}$ [80 x 1]
	-0.170	0.016	-0.226		0.233	-0.254		$x^{(8)}_{5,1}$		1.484		$x^{(9)}_{5,1}$
	-0.222	-0.128	-0.055		0.101	-0.179		$x^{(8)}_{5,2}$		1.420		$x^{(9)}_{5,2}$
	0.141	-0.266	-0.138	...	0.199	-0.189		$x^{(8)}_{5,3}$		-1.413		$x^{(9)}_{5,3}$
f_{tanh}	-0.282	0.153	-0.237		-0.295	0.247	×	$x^{(8)}_{5,4}$	+	1.345	=	$x^{(9)}_{5,4}$
	-0.181	-0.042	0.319		-0.234	-0.207		$x^{(8)}_{5,5}$		1.301		$x^{(9)}_{5,5}$
				...				...		...		...
	-0.270	0.208	-0.150	...	0.258	-0.151		$x^{(8)}_{5,79}$		-1.434		$x^{(9)}_{5,79}$
	0.010	0.229	0.177		0.113	0.168		$x^{(8)}_{5,80}$		-1.499		$x^{(9)}_{5,80}$

(C29)

+ At Hidden Layers 10:

$$
f_{tanh}\left[
\begin{array}{|c|c|c|c|c|c|}
\hline
\multicolumn{6}{|c|}{\boldsymbol{\omega}^{(10)}{}_{[80 \times 80]}} \\
\hline
-0.052 & 0.167 & -0.230 & & 0.161 & 0.210 \\
-0.226 & 0.254 & -0.082 & & -0.109 & -0.165 \\
0.058 & 0.112 & -0.289 & \cdots & -0.030 & 0.165 \\
-0.230 & 0.015 & -0.011 & & -0.176 & -0.227 \\
0.160 & 0.200 & 0.245 & & -0.240 & -0.057 \\
\multicolumn{6}{|c|}{\cdots} \\
-0.204 & 0.090 & 0.086 & \cdots & -0.146 & 0.154 \\
0.079 & 0.087 & -0.052 & & 0.115 & -0.028 \\
\hline
\end{array}
\times
\begin{array}{|c|}
\hline
\mathbf{b}^{(10)}{}_{[80 \times 1]} \\
\hline
1.483 \\
1.442 \\
-1.386 \\
1.385 \\
-1.339 \\
\cdots \\
-1.437 \\
1.480 \\
\hline
\end{array}
+
\begin{array}{|c|}
\hline
\mathbf{x}^{(9)}{}_{[80 \times 1]} \\
\hline
x^{(9)}_{5,1} \\
x^{(9)}_{5,2} \\
x^{(9)}_{5,3} \\
x^{(9)}_{5,4} \\
x^{(9)}_{5,5} \\
\cdots \\
x^{(9)}_{5,79} \\
x^{(9)}_{5,80} \\
\hline
\end{array}
\right]
=
\begin{array}{|c|}
\hline
\mathbf{x}^{(10)}{}_{[80 \times 1]} \\
\hline
x^{(10)}_{5,1} \\
x^{(10)}_{5,2} \\
x^{(10)}_{5,3} \\
x^{(10)}_{5,4} \\
x^{(10)}_{5,5} \\
\cdots \\
x^{(10)}_{5,79} \\
x^{(10)}_{5,80} \\
\hline
\end{array}
\tag{C30}
$$

+ At output layer: It is noted that y_5 is de-normalized to obtain the original scale of output ε_s.

$$
f_{linear}\left[
\begin{array}{|c|c|c|c|c|c|}
\hline
\multicolumn{6}{|c|}{\boldsymbol{\omega}^{(out)}{}_{[1 \times 80]}} \\
\hline
0.522 & -0.365 & 0.966 & \cdots & -0.152 & 0.036 \\
\hline
\end{array}
\times
\begin{array}{|c|}
\hline
\mathbf{x}^{(10)}{}_{[80 \times 1]} \\
\hline
x^{(10)}_{5,1} \\
x^{(10)}_{5,2} \\
x^{(10)}_{5,3} \\
x^{(10)}_{5,4} \\
x^{(10)}_{5,5} \\
\cdots \\
x^{(10)}_{5,79} \\
x^{(10)}_{5,80} \\
\hline
\end{array}
+
\begin{array}{|c|}
\hline
\mathbf{b}^{(out)}{}_{[1 \times 1]} \\
\hline
-0.794 \\
\hline
\end{array}
\right]
=
\begin{array}{|c|}
\hline
\mathbf{y}_{[1 \times 1]} \\
\hline
y_5 \\
\hline
\end{array}
\tag{C31}
$$

b. ***Model-LPs* considering *n* load points** $(P_{u,1}, M_{u,1}, \ldots, P_{u,n}, M_{u,n})$:

- There are five load points considered in Design Case 1 so that there are $2n+4$ inputs $(D,\ \rho_s, f_c^{'}, f_y,\ P_{u,1}, M_{u,1}, \ldots, P_{u,n}, M_{u,n})$ and $2n+3$ outputs $(CI_c,\ CO_2,\ W_c,\ SF_1,\ \varepsilon_{s,1},\ SF_2,\ \varepsilon_{s,2}, \ldots,\ SF_n,\ \varepsilon_{s,n})$ in *Model-LPs*.

- Weight and bias matrices of output y_1 (CI_c) in *Model-LPs* are presented at hidden layers and at an output layer as follows.

  + At Hidden Layer 1: Weight matrix of output y_1 (CI_c) of *Model-LPs* at Hidden Layer 1 is duplicated based on that of *Model-1LP* and adding zero components that are marked in green color. Bias matrix of output y_1 (CI_c) of *Model-LPs* at Hidden Layer 1 similar to that of CI_c of *Model-1LP* is shown in Equation (C1).

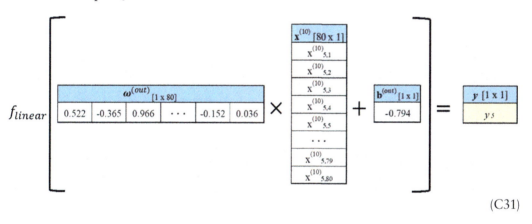

f_{tanh} [

$\boldsymbol{\omega}^{(1)}$ [20 × (2n + 4)]								
-0.972	-1.006	0.165	-0.183	1.149	-0.748	0	...	0
1.352	0.475	0.130	0.010	0.259	-0.442	0		0
1.308	-0.696	-0.623	0.237	0.464	0.624	0		0
-0.169	-0.198	-0.729	-1.296	-1.283	0.095	0		0
0.845	0.746	0.065	0.027	-0.056	0.260	0		0
...			...			...		
0.782	-0.650	-1.172	0.467	-1.501	-0.439	0	...	0
-0.337	0.945	-0.104	-0.189	1.975	-0.279	0		0

$\times$

$\mathbf{x}^{in}$ [(2n + 4) × 1]
$x_1 = g^N(D)$
$x_2 = g^N(\rho_s)$
$x_3 = g^N(f_c')$
$x_4 = g^N(f_y)$
$x_5 = g^N(P_{u,1})$
$x_6 = g^N(M_{u,1})$
$x_7 = g^N(P_{u,2})$
$x_8 = g^N(M_{u,2})$
...
$x_{2i+3} = g^N(P_{u,i})$
$x_{2i+4} = g^N(M_{u,i})$
...
$x_{2n+3} = g^N(P_{u,n})$
$x_{2n+4} = g^N(M_{u,n})$

$+$

$\mathbf{b}^{(1)}$ [20 x 1]
2.202
-1.961
-1.812
1.823
-0.973
...
2.160
-2.446

] $=$

$\mathbf{x}^{(1)}$ [20 x 1]
$x^{(1)}_{1,1}$
$x^{(1)}_{1,2}$
$x^{(1)}_{1,3}$
$x^{(1)}_{1,4}$
$x^{(1)}_{1,5}$
...
$x^{(1)}_{1,19}$
$x^{(1)}_{1,20}$

$= \underbrace{\mathbf{x}^{(1)}}_{[20\times1]}$

(C32)

+ Weight and bias matrices of output y_1 (CI_c) of *Model-LPs* at Hidden Layer 2, 3, ..., 10 and at output layer are similar to those of *Model-1LP*.

- Weight and bias matrices of output y_2 (CO_2) in *Model-LPs* are as follows. It is noted that k indicates layer number.

  + At Hidden Layer 1: Weight matrix of output y_2 (CO_2) of *Model-LPs* at Hidden Layer 1 is duplicated based on that of *Model-1LP* and adding zero components that are marked in green color. Bias matrix of output y_2 (CO_2) of *Model-LPs* at Hidden Layer 1 similar to that of CO_2 of *Model-1LP* is shown in Equation (C4).

f_{tanh} [

$\boldsymbol{\omega}^{(1)}$ [20 × (2n + 4)]								
-0.597	-0.185	-0.714	-0.077	-1.113	-1.128	0	...	0
-1.338	-0.214	-0.092	0.256	0.698	0.038	0		0
0.531	-1.083	-0.009	-0.021	-0.085	-0.289	0		0
1.291	-0.980	0.029	0.023	0.333	0.950	0		0
-0.839	-0.592	-0.004	0.009	0.005	-0.058	0		0
...			...			...		
0.950	1.105	0.565	0.605	-0.572	-1.035	0	...	0
-1.688	-0.164	0.005	0.005	-0.292	-0.504	0		0

$\times$

$\mathbf{x}^{in}$ [(2n + 4) × 1]
$x_1 = g^N(D)$
$x_2 = g^N(\rho_s)$
$x_3 = g^N(f_c')$
$x_4 = g^N(f_y)$
$x_5 = g^N(P_{u,1})$
$x_6 = g^N(M_{u,1})$
$x_7 = g^N(P_{u,2})$
$x_8 = g^N(M_{u,2})$
...
$x_{2i+3} = g^N(P_{u,i})$
$x_{2i+4} = g^N(M_{u,i})$
...
$x_{2n+3} = g^N(P_{u,n})$
$x_{2n+4} = g^N(M_{u,n})$

$+$

$\mathbf{b}^{(1)}$ [20 x 1]
2.825
2.310
-1.861
-1.781
1.044
...
2.373
-1.823

] $=$

$\mathbf{x}^{(1)}$ [20 x 1]
$x^{(1)}_{2,1}$
$x^{(1)}_{2,2}$
$x^{(1)}_{2,3}$
$x^{(1)}_{2,4}$
$x^{(1)}_{2,5}$
...
$x^{(1)}_{2,19}$
$x^{(1)}_{2,20}$

$= \underbrace{\mathbf{x}^{(1)}}_{[20\times1]}$

(C33)

+ Weight and bias matrices of output y_2 (CO_2) of *Model-LPs* at Hidden Layer 2, 3, 4, 5 and at output layer are similar to those of *Model-1LP*.

- Weight and bias matrices of output y_3 (W_C) in *Model-LPs*:

  + At Hidden Layer 1: Weight matrix of output y_3 (W_C) of *Model-LPs* at Hidden Layer 1 is duplicated based on that of *Model-1LP* and adding zero components that are marked in green color. Bias matrix of output y_3 (W_C) of *Model-LPs* at Hidden Layer 1 similar to that of W_C of *Model-1LP* is shown in Equation (C7).

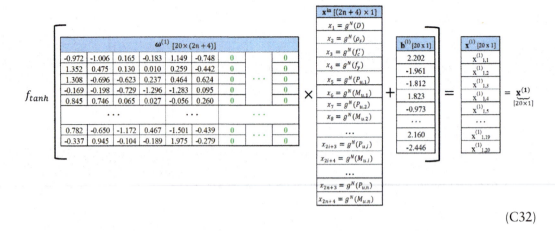

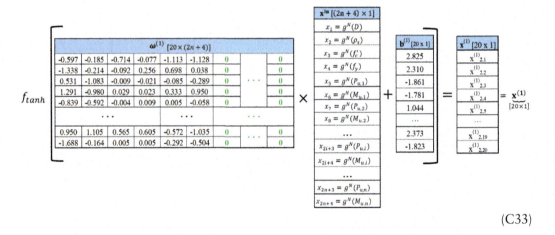

$f_{tanh}\left(\omega^{(1)} \times \mathbf{x}^{in} + \mathbf{b}^{(1)} \right) = \mathbf{x}^{(1)} = \underbrace{\mathbf{x}^{(1)}}_{[80\times1]}$ (C34)

$\omega^{(1)}$ [80 × (2n + 4)]

-1.289	0.993	0.702	-0.094	1.501	-1.373	0		0
1.362	-0.430	-0.296	0.128	-1.776	0.556	0		0
-0.143	1.315	0.542	1.101	1.735	-1.195	0	…	0
1.879	0.063	-0.965	0.376	0.037	1.825	0		0
0.996	1.131	1.032	-0.539	-1.431	0.188	0		0
…			…			…		
0.046	-2.111	-1.542	0.066	-0.782	-0.530	0	…	0
-0.481	-1.590	-1.369	0.773	-0.784	-0.741	0		0

$\mathbf{x}^{in}$ [(2n + 4) × 1]
$x_1 = g^N(D)$
$x_2 = g^N(\rho_s)$
$x_3 = g^N(f_c')$
$x_4 = g^N(f_y)$
$x_5 = g^N(P_{u,1})$
$x_6 = g^N(M_{u,1})$
$x_7 = g^N(P_{u,2})$
$x_8 = g^N(M_{u,2})$
…
$x_{2i+3} = g^N(P_{u,i})$
$x_{2i+4} = g^N(M_{u,i})$
…
$x_{2n+3} = g^N(P_{u,n})$
$x_{2n+4} = g^N(M_{u,n})$

$\mathbf{b}^{(1)}$ [80 x 1]
3.016
-2.595
2.871
-2.714
-2.789
…
2.869
-2.940

$\mathbf{x}^{(1)}$ [80 x 1]
$x^{(1)}_{3,1}$
$x^{(1)}_{3,2}$
$x^{(1)}_{3,3}$
$x^{(1)}_{3,4}$
$x^{(1)}_{3,5}$
…
$x^{(1)}_{3,79}$
$x^{(1)}_{3,80}$

+ Weight and bias matrices of output y_3 (W_C) of *Model-LPs* at Hidden Layer 2, 3, 4, 5 and at output layer are similar to those of *Model-1LP*.

- Weight and bias matrices of output y_4 (SF_1) in *Model-LPs*:

  It is noted that SF_1 is governed by load point $P_{u,1}$, $M_{u,1}$.

  + At Hidden Layer 1: Weight matrix of output y_4 (SF_1) of *Model-LPs* at Hidden Layer 1 is duplicated based on weight matrix of output *SF* at Hidden Layer 1 of *Model-1LP* and adding zero components that are marked in green color. Bias matrix of output y_4 (SF_1) of *Model-LPs* at Hidden Layer 1 similar to that of *SF* of *Model-1LP* is shown in Equation (C10).

$f_{tanh}\left(\omega^{(1)} \times \mathbf{x}^{in} + \mathbf{b}^{(1)} \right) = \mathbf{x}^{(1)} = \underbrace{\mathbf{x}^{(1)}}_{[80\times1]}$ (C35)

$\omega^{(1)}$ [80 × (2n + 4)]

0.397	0.170	0.012	0.056	0.433	2.847	0		0
-0.374	-1.097	-0.025	-0.522	-1.591	1.845	0		0
0.368	1.946	-0.074	-0.359	-0.817	1.707	0	…	0
1.464	0.237	-1.227	-1.400	0.441	1.346	0		0
0.479	-1.116	-0.795	-1.180	2.045	-0.569	0		0
…			…			…		
-0.694	0.159	0.055	-0.042	0.208	-2.073	0	…	0
-1.548	0.650	1.514	-0.705	0.543	1.319	0		0

$\mathbf{x}^{in}$ [(2n + 4) × 1]
$x_1 = g^N(D)$
$x_2 = g^N(\rho_s)$
$x_3 = g^N(f_c')$
$x_4 = g^N(f_y)$
$x_5 = g^N(P_{u,1})$
$x_6 = g^N(M_{u,1})$
$x_7 = g^N(P_{u,2})$
$x_8 = g^N(M_{u,2})$
…
$x_{2i+3} = g^N(P_{u,i})$
$x_{2i+4} = g^N(M_{u,i})$
…
$x_{2n+3} = g^N(P_{u,n})$
$x_{2n+4} = g^N(M_{u,n})$

$\mathbf{b}^{(1)}$ [80 x 1]
3.651
2.919
-2.915
-2.817
-2.727
…
-2.826
-2.988

$\mathbf{x}^{(1)}$ [80 x 1]
$x^{(1)}_{4,1}$
$x^{(1)}_{4,2}$
$x^{(1)}_{4,3}$
$x^{(1)}_{4,4}$
$x^{(1)}_{4,5}$
…
$x^{(1)}_{4,79}$
$x^{(1)}_{4,80}$

+ Weight and bias matrices of output y_4 (SF_1) of *Model-LPs* at Hidden Layer 2, 3, ..., 10 and at output layer similar to those of *SF* of *Model-1LP* are shown in Equations (C11) to (C20).

- Weight and bias matrices of output y_5 ($\varepsilon_{s,1}$) in *Model-LPs* are as follows. It is noted that $\varepsilon_{s,1}$ is governed by load point ($P_{u,1}$, $M_{u,1}$).

  + At Hidden Layer 1: Weight matrix of output y_5 ($\varepsilon_{s,1}$) of *Model-LPs* at Hidden Layer 1 is duplicated based on weight matrix of output ε_s at Hidden Layer 1 of *Model-1LP* and adding zero components that are marked in green color. Bias matrix of output y_5 ($\varepsilon_{s,1}$) of *Model-LPs* at Hidden Layer 1 similar to that of ε_s of *Model-1LP* is shown in Equation (C21).

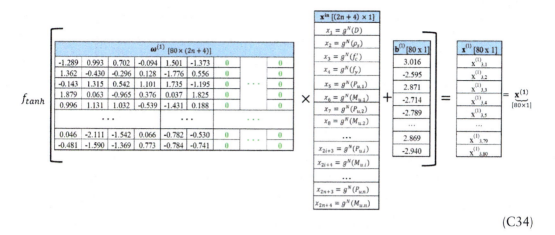

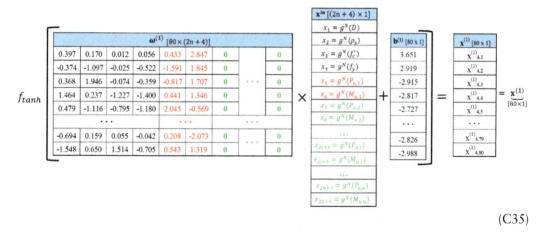

f_{tanh}

$\omega^{(1)}$ [80 × (2n + 4)]								
-1.263	-1.436	-1.121	-0.036	-0.678	1.533	0	...	0
2.158	0.428	0.584	0.301	-1.010	1.182	0		0
-0.085	-0.022	-0.480	0.269	-0.313	-2.228	0		0
1.248	-0.420	-0.483	-0.425	1.706	-1.710	0		0
0.891	-0.922	0.286	-0.816	-0.732	-1.448	0		0
...				...		...		
-1.307	1.066	-1.077	-1.136	0.034	-1.074	0	...	0
0.136	-2.292	-0.031	-0.864	-1.087	0.673	0		0

×

$\mathbf{x}^{in}$ [(2n + 4) × 1]
$x_1 = g^N(D)$
$x_2 = g^N(\rho_s)$
$x_3 = g^N(f_c')$
$x_4 = g^N(f_y)$
$x_5 = g^N(P_{u,1})$
$x_6 = g^N(M_{u,1})$
$x_7 = g^N(P_{u,2})$
$x_8 = g^N(M_{u,2})$
...
$x_{2i+3} = g^N(P_{u,i})$
$x_{2i+4} = g^N(M_{u,i})$
...
$x_{2n+3} = g^N(P_{u,n})$
$x_{2n+4} = g^N(M_{u,n})$

+

$\mathbf{b}^{(1)}$ [80 x 1]
2.897
-2.894
-2.890
-2.755
-2.797
...
-2.916
2.998

=

$\mathbf{x}^{(1)}$ [80 x 1]
$x^{(1)}_{5,1}$
$x^{(1)}_{5,2}$
$x^{(1)}_{5,3}$
$x^{(1)}_{5,4}$
$x^{(1)}_{5,5}$
...
$x^{(1)}_{5,79}$
$x^{(1)}_{5,80}$

(C36)

+ Weight and bias matrices of output y_5 ($\varepsilon_{s,1}$) of *Model-LPs* at Hidden Layer 2, 3, ..., 10 and at output layer similar to those of ε_s of *Model-1LP* are shown in Equations (C22) to (C31).

- Weight matrix of output y_6 (SF_2): (SF_2 is governed by Load Point ($P_{u,2}$, $M_{u,2}$))

  + At Hidden Layer 1: Weight matrix of output y_6 (SF_2) of *Model-LPs* at Hidden Layer 1 is duplicated based on weight matrix of output *SF* at Hidden Layer 1 of *Model-1LP* and adding zero components that are marked in green color. Bias matrix of output y_6 (SF_2) of *Model-LPs* at Hidden Layer 1 similar to that of *SF* of *Model-1LP* is shown in Equation (C10).

f_{tanh}

$\omega^{(1)}$ [80 × (2n + 4)]										
0.397	0.170	0.012	0.056	0	0	0.433	2.847	0	...	0
-0.374	-1.097	-0.025	-0.522	0	0	-1.591	1.845	0		0
0.368	1.946	-0.074	-0.359	0	0	-0.817	1.707	0		0
1.464	0.237	-1.227	-1.400	0	0	0.441	1.346	0		0
0.479	-1.116	-0.795	-1.180	0	0	2.045	-0.569	0		0
...				...		...		...		
-0.694	0.159	0.055	-0.042	0	0	0.208	-2.073	0	...	0
-1.548	0.650	1.514	-0.705	0	0	0.543	1.319	0		0

×

$\mathbf{x}^{in}$ [(2n + 4) × 1]
$x_1 = g^N(D)$
$x_2 = g^N(\rho_s)$
$x_3 = g^N(f_c')$
$x_4 = g^N(f_y)$
$x_5 = g^N(P_{u,1})$
$x_6 = g^N(M_{u,1})$
$x_7 = g^N(P_{u,2})$
$x_8 = g^N(M_{u,2})$
...
$x_{2i+3} = g^N(P_{u,i})$
$x_{2i+4} = g^N(M_{u,i})$
...
$x_{2n+3} = g^N(P_{u,n})$
$x_{2n+4} = g^N(M_{u,n})$

+

$\mathbf{b}^{(1)}$ [80 x 1]
3.651
2.919
-2.915
-2.817
-2.727
...
-2.826
-2.988

=

$\mathbf{x}^{(1)}$ [80 x 1]
$x^{(1)}_{6,1}$
$x^{(1)}_{6,2}$
$x^{(1)}_{6,3}$
$x^{(1)}_{6,4}$
$x^{(1)}_{6,5}$
...
$x^{(1)}_{6,79}$
$x^{(1)}_{6,80}$

$= \underbrace{\mathbf{x}^{(1)}}_{[80\times 1]}$

(C37)

+ Weight and bias matrices of output y_6 (SF_2) of *Model-LPs* at Hidden Layer 2, 3, ..., 10 and at output layer similar to those of *SF* of *Model-1LP* are shown in Equations (C11)–(C20).

- Weight matrix of output y_7 ($\varepsilon_{s,2}$): ($\varepsilon_{s,2}$ is governed by Load Point ($P_{u,2}$, $M_{u,2}$))

+ At Hidden Layer 1:

$$f_{tanh}\left[\underbrace{\begin{bmatrix} -1.263 & -1.436 & -1.121 & -0.036 & 0 & 0 & -0.678 & 1.533 & 0 & \cdots & 0 \\ 2.158 & 0.428 & 0.584 & 0.301 & 0 & 0 & -1.010 & 1.182 & 0 & \cdots & 0 \\ -0.085 & -0.022 & -0.480 & 0.269 & 0 & 0 & -0.313 & -2.228 & 0 & \cdots & 0 \\ 1.248 & -0.420 & -0.483 & -0.425 & 0 & 0 & 1.706 & -1.710 & 0 & \cdots & 0 \\ 0.891 & -0.922 & 0.286 & -0.816 & 0 & 0 & -0.732 & -1.448 & 0 & \cdots & 0 \\ & \cdots & & & \cdots & & \cdots & & & \cdots & \\ -1.307 & 1.066 & -1.077 & -1.136 & 0 & 0 & 0.034 & -1.074 & 0 & \cdots & 0 \\ 0.136 & -2.292 & -0.031 & -0.864 & 0 & 0 & -1.087 & 0.673 & 0 & \cdots & 0 \end{bmatrix}}_{\boldsymbol{\omega}^{(1)}\,[80\times(2n+4)]} \times \underbrace{\begin{bmatrix} x_1 = g^N(D) \\ x_2 = g^N(\rho_s) \\ x_3 = g^N(f_c') \\ x_4 = g^N(f_y) \\ x_5 = g^N(P_{u,1}) \\ x_6 = g^N(M_{u,1}) \\ x_7 = g^N(P_{u,2}) \\ x_8 = g^N(M_{u,2}) \\ \cdots \\ x_{2i+3} = g^N(P_{u,i}) \\ x_{2i+4} = g^N(M_{u,i}) \\ \cdots \\ x_{2n+3} = g^N(P_{u,n}) \\ x_{2n+4} = g^N(M_{u,n}) \end{bmatrix}}_{\mathbf{x}^{in}\,[(2n+4)\times 1]} + \underbrace{\begin{bmatrix} 2.897 \\ -2.894 \\ -2.890 \\ -2.755 \\ -2.797 \\ \cdots \\ -2.916 \\ 2.998 \end{bmatrix}}_{\mathbf{b}^{(1)}\,[80\times 1]}\right] = \underbrace{\begin{bmatrix} \mathbf{x}^{(1)}_{7,1} \\ \mathbf{x}^{(1)}_{7,2} \\ \mathbf{x}^{(1)}_{7,3} \\ \mathbf{x}^{(1)}_{7,4} \\ \mathbf{x}^{(1)}_{7,5} \\ \cdots \\ \mathbf{x}^{(1)}_{7,79} \\ \mathbf{x}^{(1)}_{7,80} \end{bmatrix}}_{\mathbf{x}^{(1)}\,[80\times 1]} \tag{C38}$$

+ Weight and bias matrices of output y_7 ($\varepsilon_{s,2}$) of *Model-LPs* at Hidden Layer 2, 3, …, 10 and at output layer similar to those of ε_s of *Model-1LP* are shown in Equations (C22)–(C31).

- Weight and bias matrices of output SF_i : (SF_i is governed by Load Point i^{th} ($P_{u,i}$, $M_{u,i}$)) (i=1 to n, n is a number of load pair)

+ At Hidden Layer 1:

$$f_{tanh}\left[\underbrace{\begin{bmatrix} 0.397 & 0.170 & 0.012 & 0.056 & 0 & \cdots & 0 & 0.433 & 2.847 & 0 & \cdots & 0 \\ -0.374 & -1.097 & -0.025 & -0.522 & 0 & \cdots & 0 & -1.591 & 1.845 & 0 & \cdots & 0 \\ 0.368 & 1.946 & -0.074 & -0.359 & 0 & \cdots & 0 & -0.817 & 1.707 & 0 & \cdots & 0 \\ 1.464 & 0.237 & -1.227 & -1.400 & 0 & \cdots & 0 & 0.441 & 1.346 & 0 & \cdots & 0 \\ 0.479 & -1.116 & -0.795 & -1.180 & 0 & \cdots & 0 & 2.045 & -0.569 & 0 & \cdots & 0 \\ & \cdots & & & & \cdots & & \cdots & & & \cdots & \\ -0.694 & 0.159 & 0.055 & -0.042 & 0 & \cdots & 0 & 0.208 & -2.073 & 0 & \cdots & 0 \\ -1.548 & 0.650 & 1.514 & -0.705 & 0 & \cdots & 0 & 0.543 & 1.319 & 0 & \cdots & 0 \end{bmatrix}}_{\boldsymbol{\omega}^{(1)}\,[80\times(2n+4)]} \times \underbrace{\begin{bmatrix} x_1 = g^N(D) \\ x_2 = g^N(\rho_s) \\ x_3 = g^N(f_c') \\ x_4 = g^N(f_y) \\ x_5 = g^N(P_{u,1}) \\ x_6 = g^N(M_{u,1}) \\ x_7 = g^N(P_{u,2}) \\ x_8 = g^N(M_{u,2}) \\ \cdots \\ x_{2i+3} = g^N(P_{u,i}) \\ x_{2i+4} = g^N(M_{u,i}) \\ \cdots \\ x_{2n+3} = g^N(P_{u,n}) \\ x_{2n+4} = g^N(M_{u,n}) \end{bmatrix}}_{\mathbf{x}^{in}\,[(2n+4)\times 1]} + \underbrace{\begin{bmatrix} 3.651 \\ 2.919 \\ -2.915 \\ -2.817 \\ -2.727 \\ \cdots \\ -2.826 \\ -2.988 \end{bmatrix}}_{\mathbf{b}^{(1)}\,[80\times 1]}\right] = \underbrace{\begin{bmatrix} \mathbf{x}^{(1)}_{2i+2,1} \\ \mathbf{x}^{(1)}_{2i+2,2} \\ \mathbf{x}^{(1)}_{2i+2,3} \\ \mathbf{x}^{(1)}_{2i+2,4} \\ \mathbf{x}^{(1)}_{2i+2,5} \\ \cdots \\ \mathbf{x}^{(1)}_{2i+2,79} \\ \mathbf{x}^{(1)}_{2i+2,80} \end{bmatrix}}_{\mathbf{x}^{(1)}\,[80\times 1]} \tag{C39}$$

+ Weight and bias matrices of output SF_i of *Model-LPs* at Hidden Layer 2, 3, …, 10 and at output layer similar to those of *SF* of *Model-1LP* are shown in Equations (C11)–(C20).

- Weight and bias matrices of output $\varepsilon_{s,i}$: ($\varepsilon_{s,i}$ is governed by Load Point i^{th} ($P_{u,i}$, $M_{u,i}$)) (i=1 to n, n is a number of load pair)

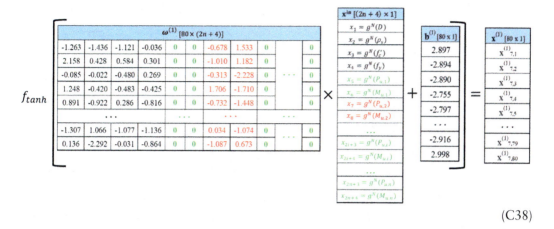

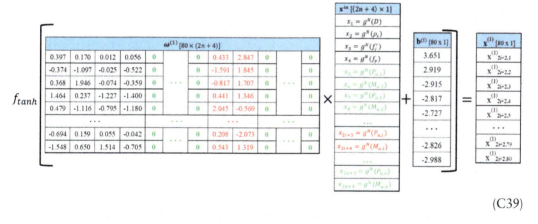

+ At Hidden Layer 1:

$$f_{tanh}\left[\underbrace{\begin{bmatrix} -1.263 & -1.436 & -1.121 & -0.036 & 0 & \cdots & 0 & -0.678 & 1.533 & 0 & \cdots & 0 \\ 2.158 & 0.428 & 0.584 & 0.301 & 0 & & 0 & -1.010 & 1.182 & 0 & & 0 \\ -0.085 & -0.022 & -0.480 & 0.269 & 0 & & 0 & -0.313 & -2.228 & 0 & & 0 \\ 1.248 & -0.420 & -0.483 & -0.425 & 0 & & 0 & 1.706 & -1.710 & 0 & & 0 \\ 0.891 & -0.922 & 0.286 & -0.816 & 0 & & 0 & -0.732 & -1.448 & 0 & & 0 \\ & \cdots & & & & \cdots & & \cdots & & & \cdots & \\ -1.307 & 1.066 & -1.077 & -1.136 & 0 & \cdots & 0 & 0.034 & -1.074 & 0 & \cdots & 0 \\ 0.136 & -2.292 & -0.031 & -0.864 & 0 & & 0 & -1.087 & 0.673 & 0 & & 0 \end{bmatrix}}_{\boldsymbol{\omega}^{(1)}\ [80\times(2n+4)]} \times \underbrace{\begin{bmatrix} x_1 = g^N(D) \\ x_2 = g^N(\rho_s) \\ x_3 = g^N(f_c') \\ x_4 = g^N(f_y) \\ x_5 = g^N(P_{u,1}) \\ x_6 = g^N(M_{u,1}) \\ x_7 = g^N(P_{u,2}) \\ x_8 = g^N(M_{u,2}) \\ \cdots \\ x_{2i+3} = g^N(P_{u,i}) \\ x_{2i+4} = g^N(M_{u,i}) \\ \cdots \\ x_{2n+3} = g^N(P_{u,n}) \\ x_{2n+4} = g^N(M_{u,n}) \end{bmatrix}}_{\mathbf{x}^{in}\ [(2n+4)\times 1]} + \underbrace{\begin{bmatrix} 2.897 \\ -2.894 \\ -2.890 \\ -2.755 \\ -2.797 \\ \cdots \\ -2.916 \\ 2.998 \end{bmatrix}}_{\mathbf{b}^{(1)}\ [80\times 1]}\right] = \underbrace{\begin{bmatrix} x^{(1)}_{2i+3,1} \\ x^{(1)}_{2i+3,2} \\ x^{(1)}_{2i+3,3} \\ x^{(1)}_{2i+3,4} \\ x^{(1)}_{2i+3,5} \\ \cdots \\ x^{(1)}_{2i+3,79} \\ x^{(1)}_{2i+3,80} \end{bmatrix}}_{\mathbf{x}^{(1)}\ [80\times 1]} \tag{C40}$$

+ Weight and bias matrices of output $\varepsilon_{s,i}$ of *Model-LPs* at Hidden Layer 2, 3, ..., 10 and at output layer similar to those of ε_s of *Model-1LP* are shown in Equations (C22)–(C31).

- Weight and bias matrices of output SF_n : (SF_n is governed by Load Point n^{th} ($P_{u.n}$, $M_{u,n}$)) (n is a number of load pair)

+ At Hidden Layer 1:

$$f_{tanh}\left[\underbrace{\begin{bmatrix} 0.397 & 0.170 & 0.012 & 0.056 & 0 & \cdots & 0 & 0.433 & 2.847 \\ -0.374 & -1.097 & -0.025 & -0.522 & 0 & & 0 & -1.591 & 1.845 \\ 0.368 & 1.946 & -0.074 & -0.359 & 0 & & 0 & -0.817 & 1.707 \\ 1.464 & 0.237 & -1.227 & -1.400 & 0 & & 0 & 0.441 & 1.346 \\ 0.479 & -1.116 & -0.795 & -1.180 & 0 & & 0 & 2.045 & -0.569 \\ & \cdots & & & & \cdots & & \cdots & \\ -0.694 & 0.159 & 0.055 & -0.042 & 0 & \cdots & 0 & 0.208 & -2.073 \\ -1.548 & 0.650 & 1.514 & -0.705 & 0 & & 0 & 0.543 & 1.319 \end{bmatrix}}_{\boldsymbol{\omega}^{(1)}\ [80\times(2n+4)]} \times \underbrace{\begin{bmatrix} x_1 = g^N(D) \\ x_2 = g^N(\rho_s) \\ x_3 = g^N(f_c') \\ x_4 = g^N(f_y) \\ x_5 = g^N(P_{u,1}) \\ x_6 = g^N(M_{u,1}) \\ x_7 = g^N(P_{u,2}) \\ x_8 = g^N(M_{u,2}) \\ \cdots \\ x_{2i+3} = g^N(P_{u,i}) \\ x_{2i+4} = g^N(M_{u,n-1}) \\ \cdots \\ x_{2n+3} = g^N(P_{u,n}) \\ x_{2n+4} = g^N(M_{u,n}) \end{bmatrix}}_{\mathbf{x}^{in}\ [(2n+4)\times 1]} + \underbrace{\begin{bmatrix} 3.651 \\ 2.919 \\ -2.915 \\ -2.817 \\ -2.727 \\ \cdots \\ -2.826 \\ -2.988 \end{bmatrix}}_{\mathbf{b}^{(1)}\ [80\times 1]}\right] = \underbrace{\begin{bmatrix} x^{(1)}_{2n+2,1} \\ x^{(1)}_{2n+2,2} \\ x^{(1)}_{2n+2,3} \\ x^{(1)}_{2n+2,4} \\ x^{(1)}_{2n+2,5} \\ \cdots \\ x^{(1)}_{2n+2,79} \\ x^{(1)}_{2n+2,80} \end{bmatrix}}_{\mathbf{x}^{(1)}\ [80\times 1]} = \underbrace{\mathbf{x}^{(1)}}_{[80\times 1]} \tag{C41}$$

+ Weight and bias matrices of output SF_n of *Model-LPs* at Hidden Layer 2, 3, ..., 10 and at output layer similar to those of *SF* of *Model-1LP* are shown in Equations (C11)–(C20).

- Weight and bias matrices of output $\varepsilon_{s,n}$: ($\varepsilon_{s,n}$ is governed by Load Point n^{th} ($P_{u.n}$, $M_{u,n}$)) (n is a number of load pair)

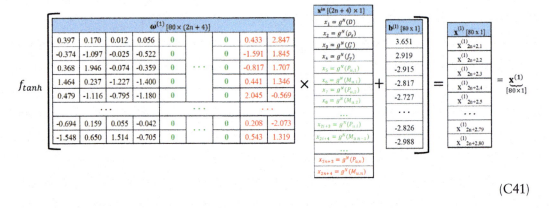

+ At Hidden Layer 1:

f_{tanh} [

$\omega^{(1)}$ [80 × (2n + 4)]								
-1.263	-1.436	-1.121	-0.036	0		0	-0.678	1.533
2.158	0.428	0.584	0.301	0		0	-1.010	1.182
-0.085	-0.022	-0.480	0.269	0	...	0	-0.313	-2.228
1.248	-0.420	-0.483	-0.425	0		0	1.706	-1.710
0.891	-0.922	0.286	-0.816	0		0	-0.732	-1.448
...				...			...	
-1.307	1.066	-1.077	-1.136	0	...	0	0.034	-1.074
0.136	-2.292	-0.031	-0.864	0		0	-1.087	0.673

×

$\mathbf{x}^{in}$ [(2n + 4) × 1]
$x_1 = g^N(D)$
$x_2 = g^N(\rho_s)$
$x_3 = g^N(f_c')$
$x_4 = g^N(f_y)$
$x_5 = g^N(P_{u,1})$
$x_6 = g^N(M_{u,1})$
$x_7 = g^N(P_{u,2})$
$x_8 = g^N(M_{u,2})$
...
$x_{2i+3} = g^N(P_{u,i})$
$x_{2i+4} = g^N(M_{u,n-1})$
...
$x_{2n+3} = g^N(P_{u,n})$
$x_{2n+4} = g^N(M_{u,n})$

+

$\mathbf{b}^{(1)}$ [80 x 1]
2.897
-2.894
-2.890
-2.755
-2.797
...
-2.916
2.998

] =

$\mathbf{x}^{(1)}$ [80 x 1]
$x^{(1)}_{2n+3,1}$
$x^{(1)}_{2n+3,2}$
$x^{(1)}_{2n+3,3}$
$x^{(1)}_{2n+3,4}$
$x^{(1)}_{2n+3,5}$
...
$x^{(1)}_{2n+3,79}$
$x^{(1)}_{2n+3,80}$

(C42)

+ Weight and bias matrices of output $\varepsilon_{s,n}$ of *Model-LPs* at Hidden Layer 2, 3, ..., 10 and at output layer similar to those of ε_s of *Model-1LP* are shown in Equations (C22)–(C31).

Appendix D

D1. ABBA-RC WINDOWS

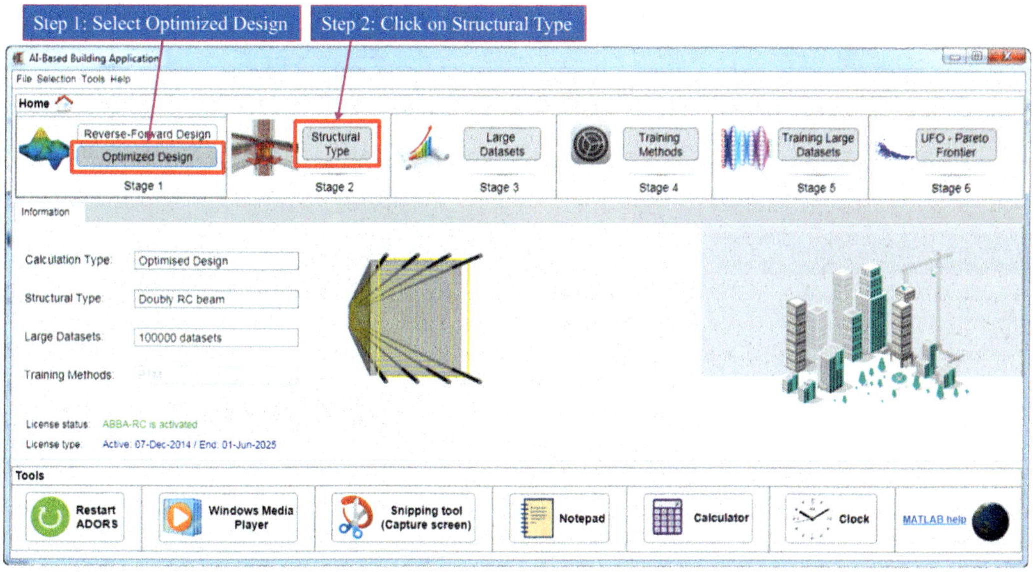

Figure D1 Select Optimized Design and Structural Type from ABBA-RC windows.

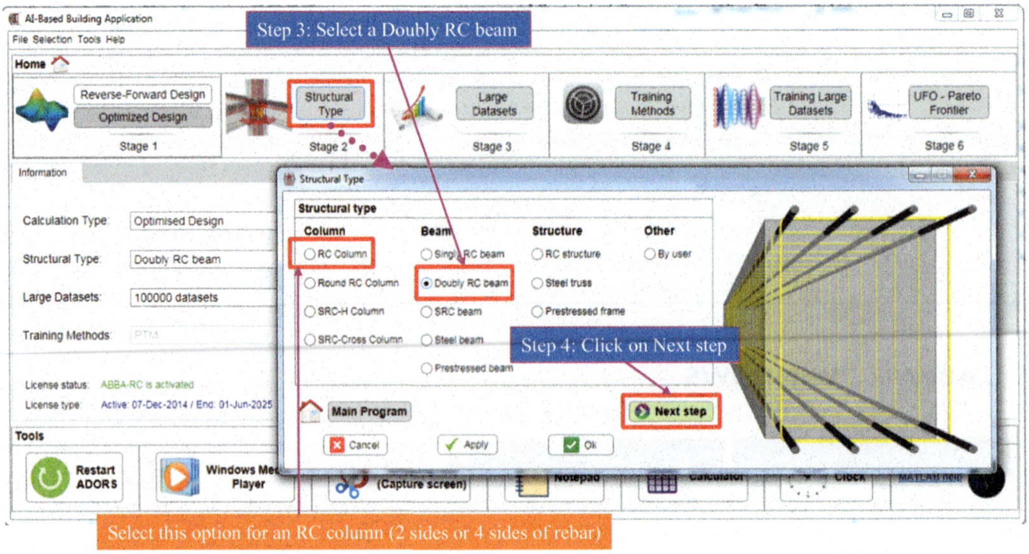

Figure D2 Select a Doubly RC beam or RC column for optimizing an RC beam or a rectangular RC column, respectively.

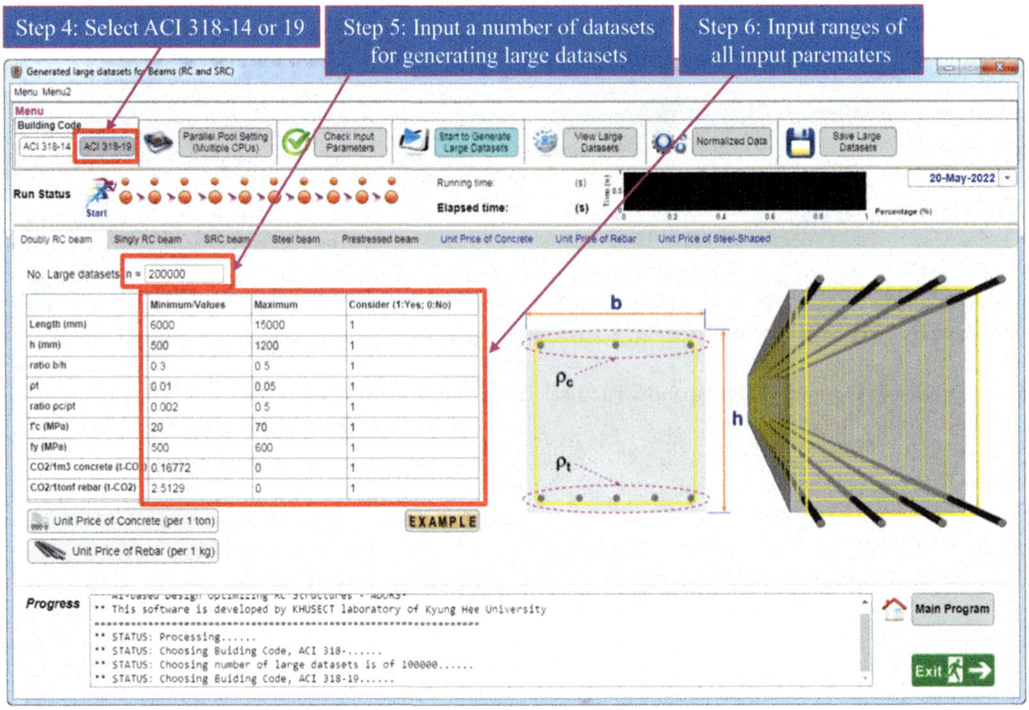

Figure D3 Generate large datasets for a Doubly RC beam.

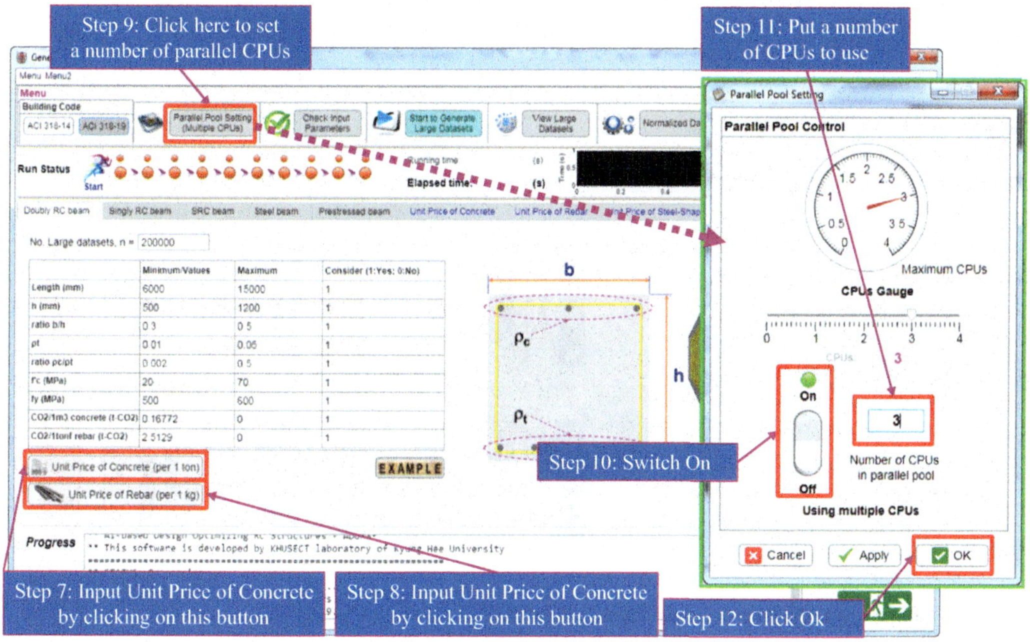

Figure D4 Set a number of CPUs to run large datasets.

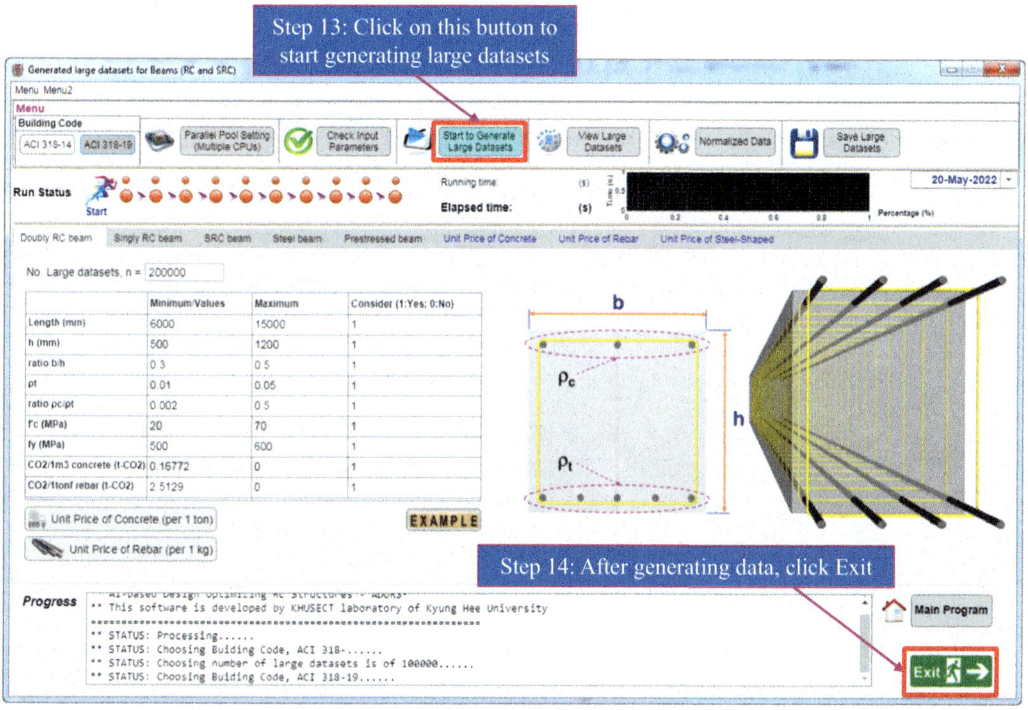

Figure D5 Set a number of CPUs to run large datasets.

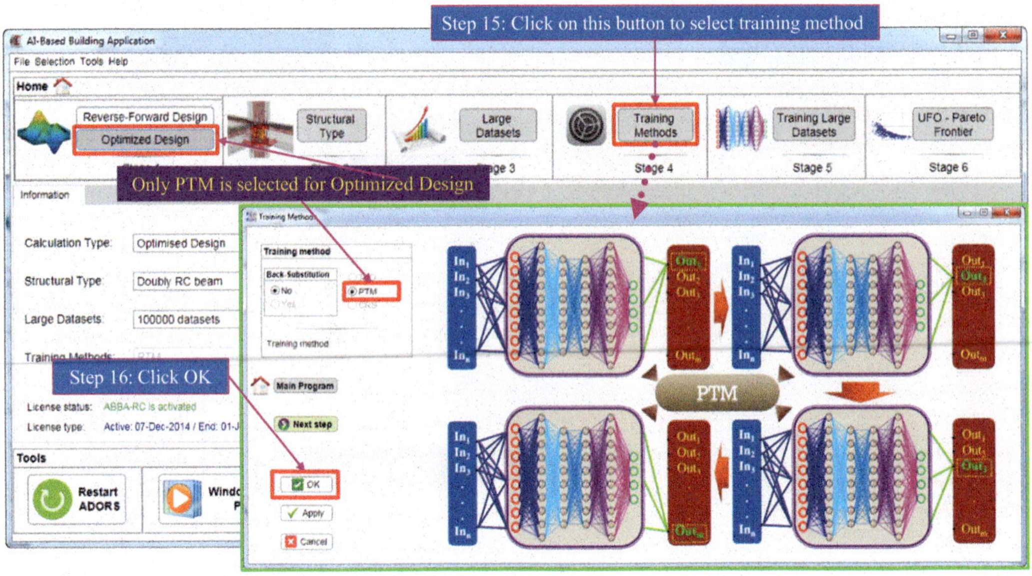

Figure D6 Select a training method for neural networks.

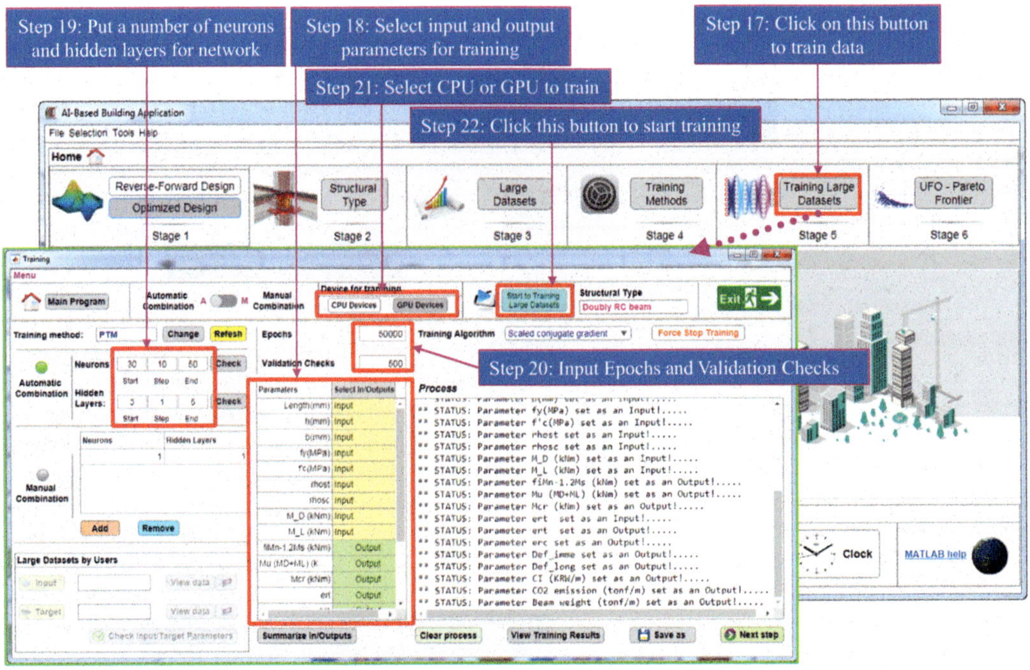

Figure D7 Run artificial neural networks.

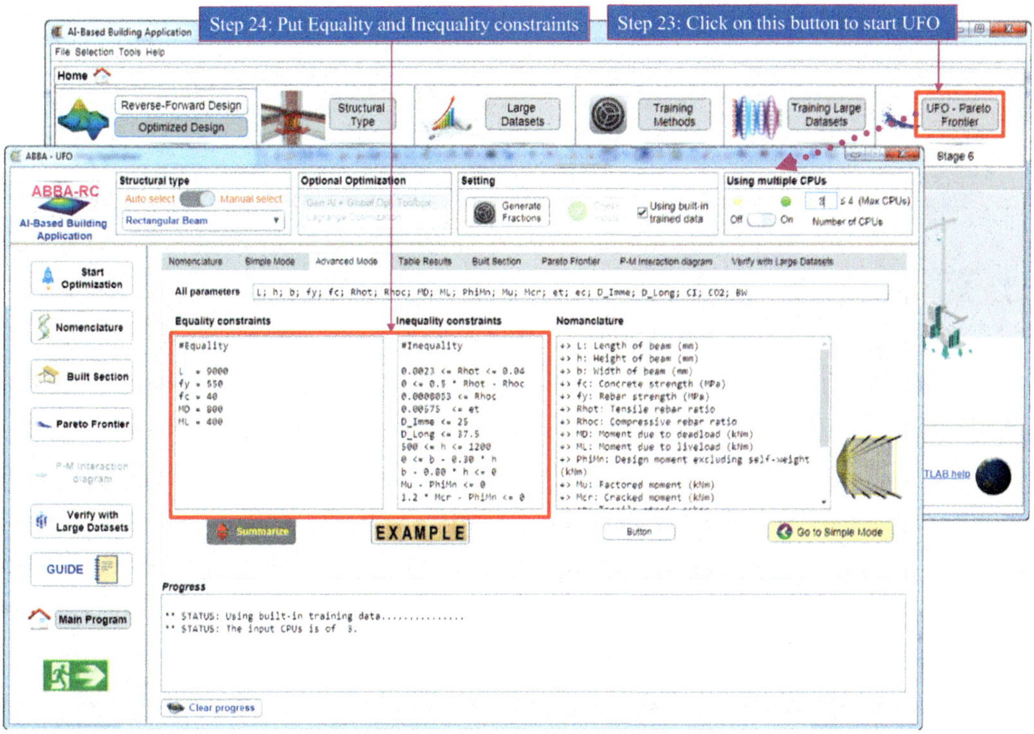

Figure D8 Input Equality and Inequality constraints for UFO of an RC beam.

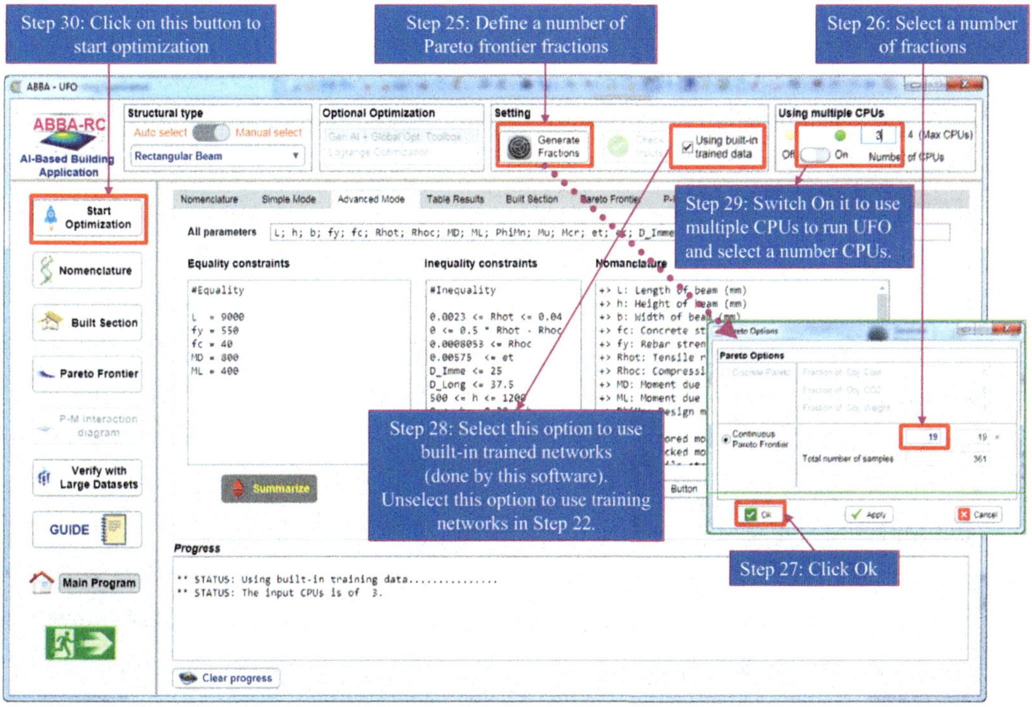

Figure D9 Setup parameters before running UFO.

D2. ABBA-RC UFO FOR AN RC BEAM

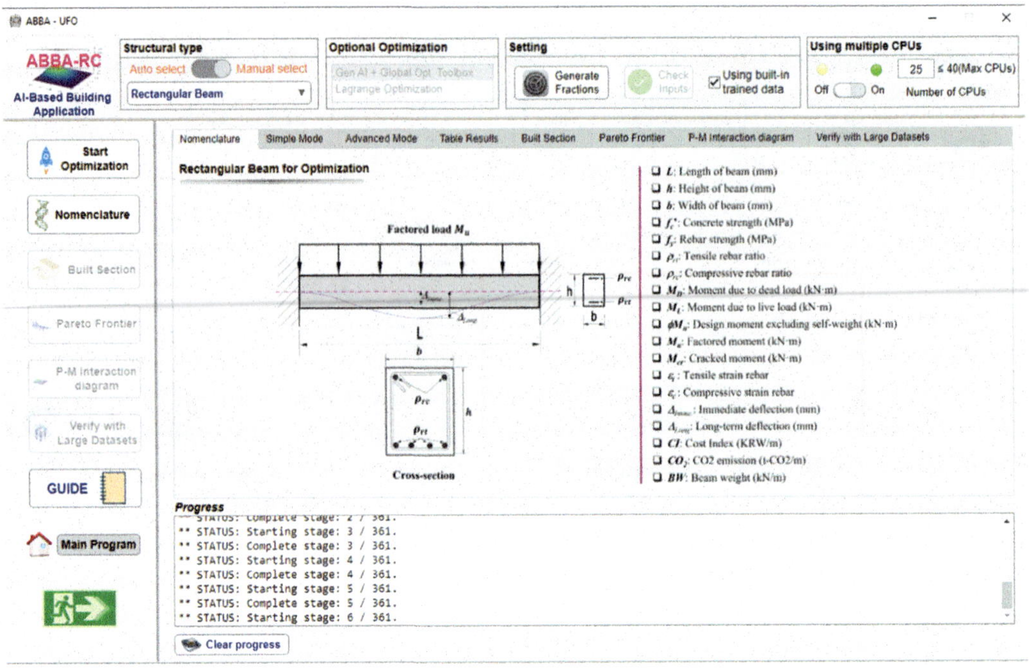

Figure D10 Nomenclature of a Rectangular RC beam.

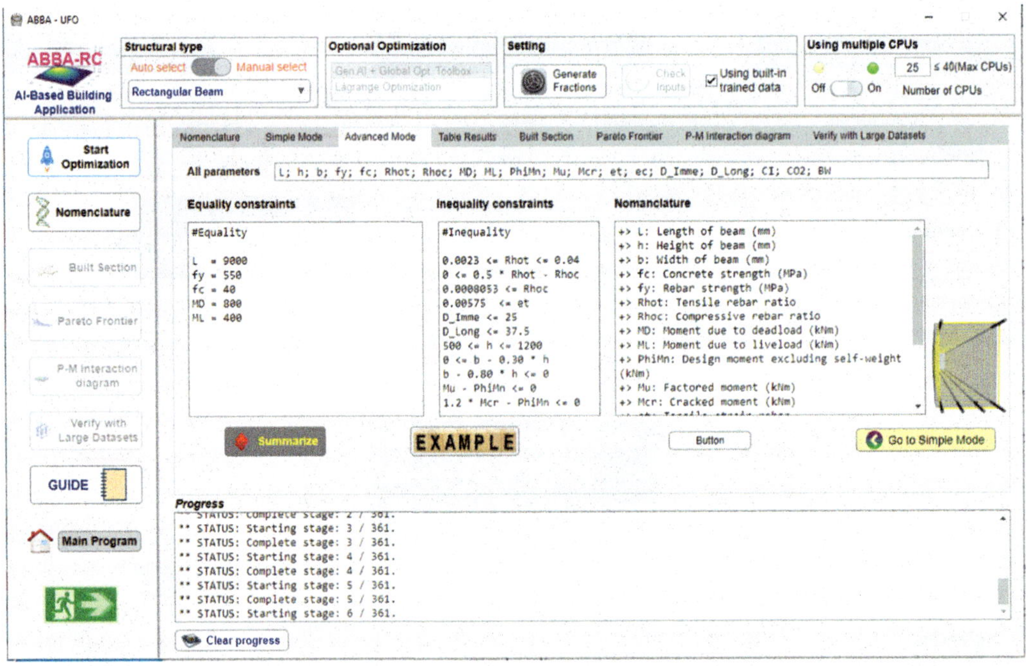

Figure D11 Window shows the running process of UFO for a Rectangular RC beam.

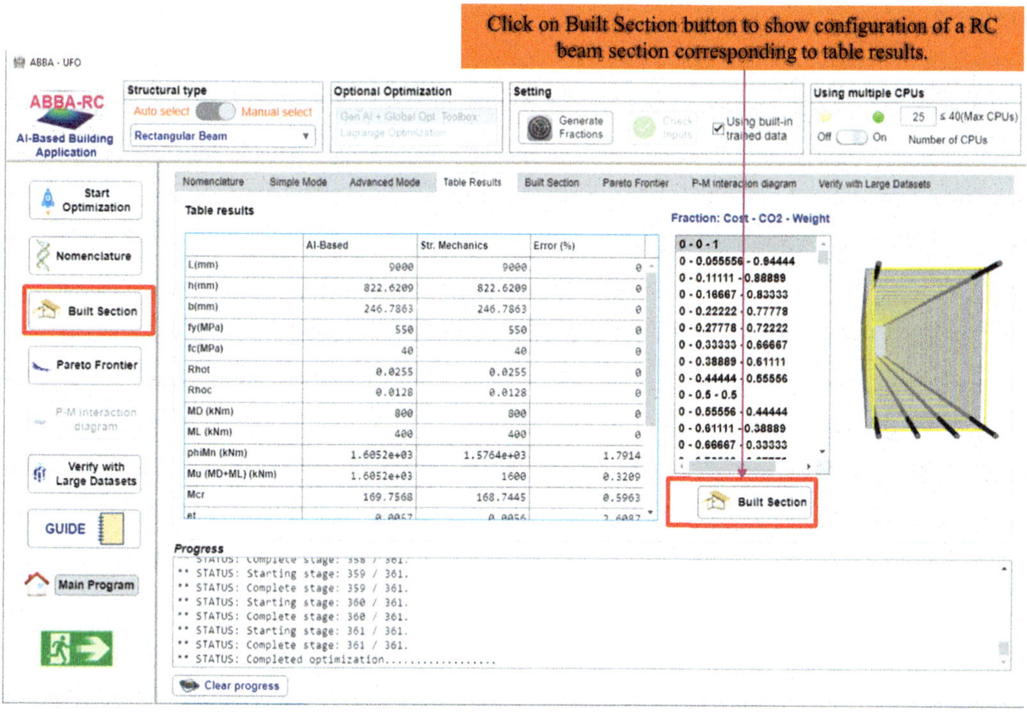

Figure D12 Optimization results of a Rectangular RC beam in case of $W_c : W_{CO_2} : W_{weight} = 0{:}0{:}1$ (corresponding to Table 5.7.2.7(a)).

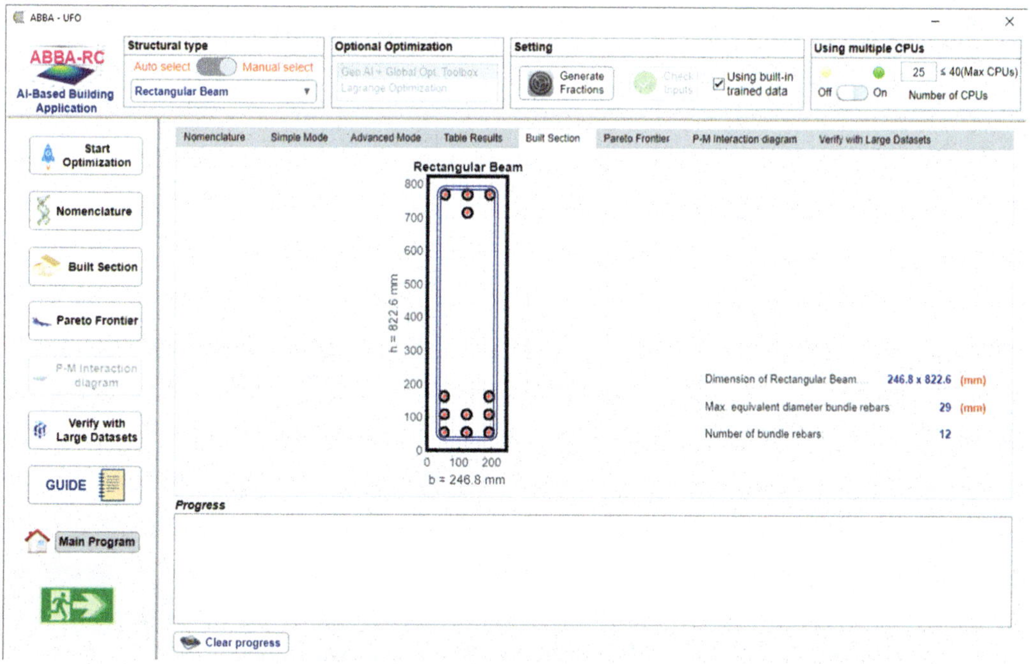

Figure D13 Geometry of a Rectangular RC beam in case of $W_c : W_{CO_2} : W_{weight} = 0{:}0{:}1$.

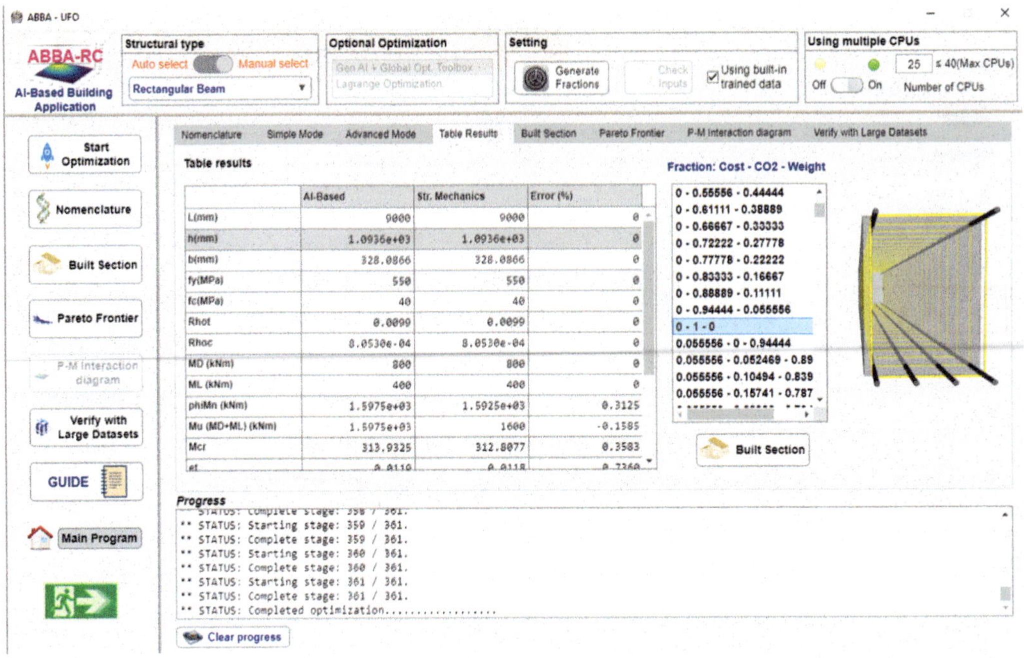

Figure D14 Optimization results of a Rectangular RC beam in case of $W_c : W_{CO_2} : W_{weight} = 0{:}1{:}0$ (corresponding to Table 5.7.2.7(a)).

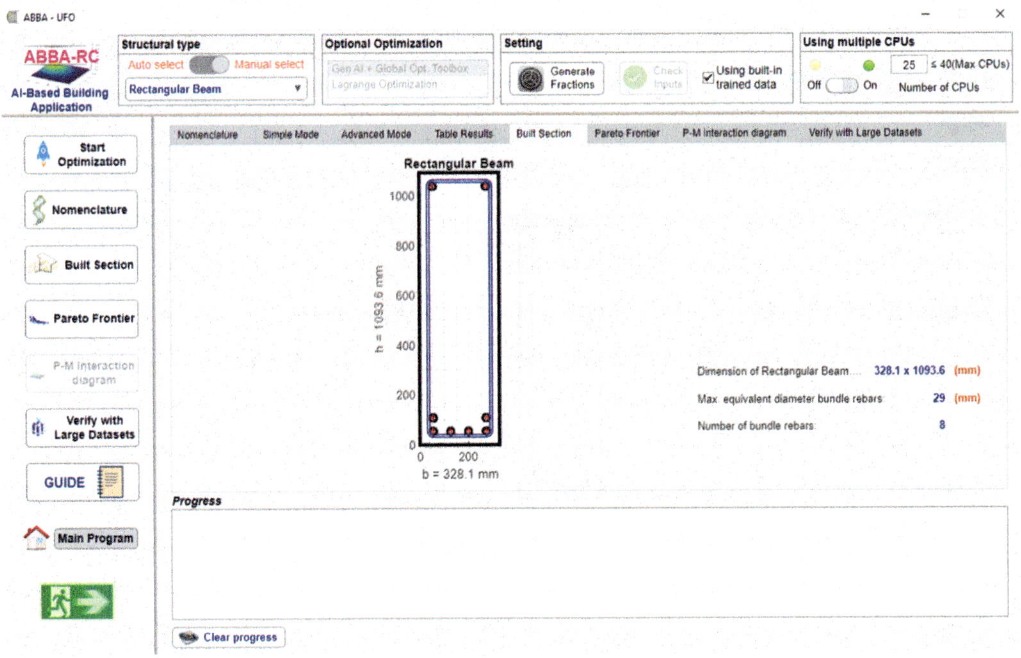

Figure D15 Geometry of a Rectangular RC beam in case of $W_c : W_{CO_2} : W_{weight} = 0{:}1{:}0$.

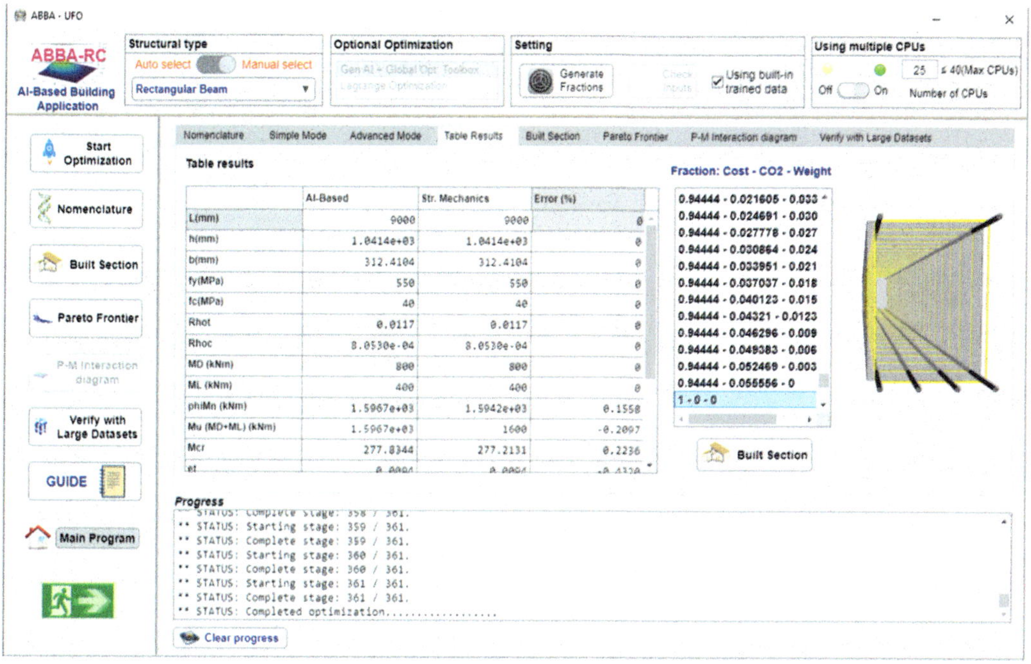

Figure D16 Optimization results of a Rectangular RC beam in case of $W_c : W_{CO_2} : W_{weight} = 1{:}0{:}0$ (corresponding to Table 5.7.2.7(a)).

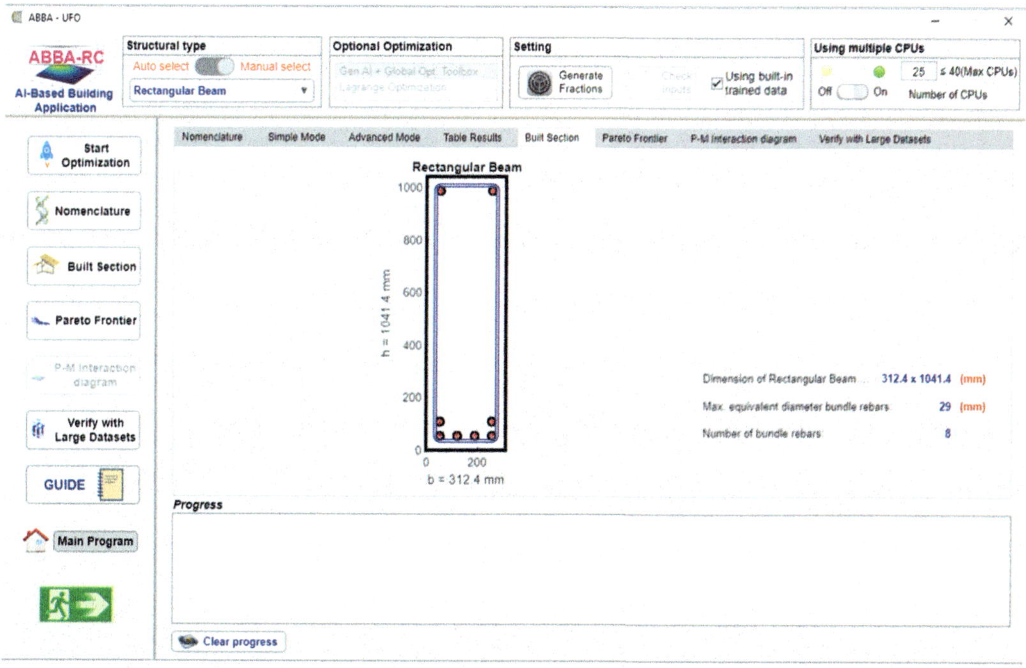

Figure D17 Geometry of a Rectangular RC beam in case of $W_c : W_{CO_2} : W_{weight} = 1{:}0{:}0$.

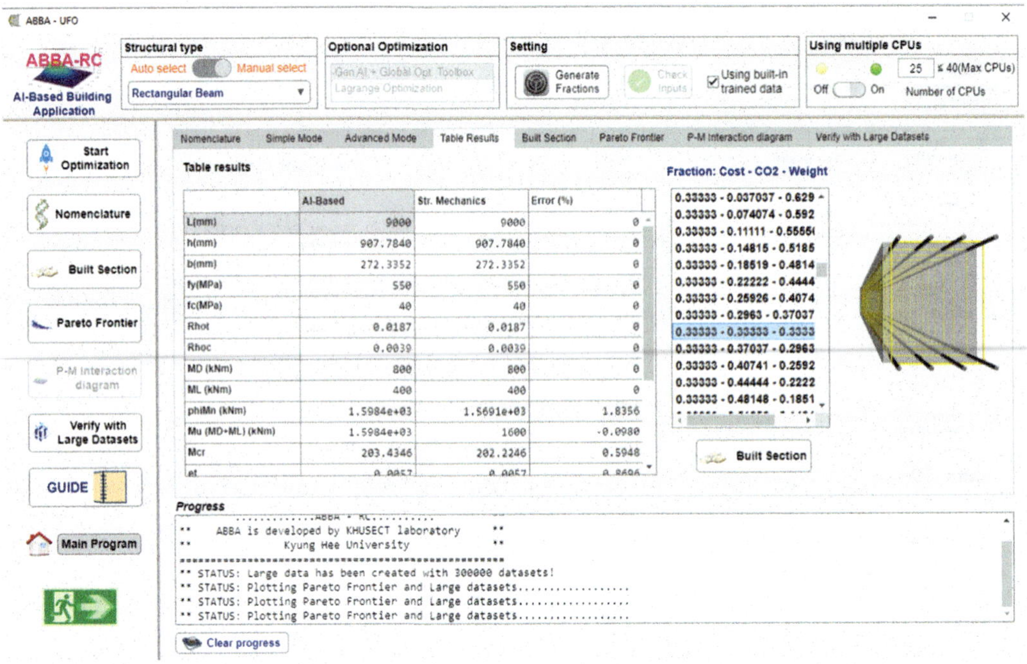

Figure D18 Optimization results of a Rectangular RC beam in case of $W_c : W_{CO_2} : W_{weight} = 1/3{:}1/3{:}1/3$ (corresponding to Table 5.7.2.7(b)).

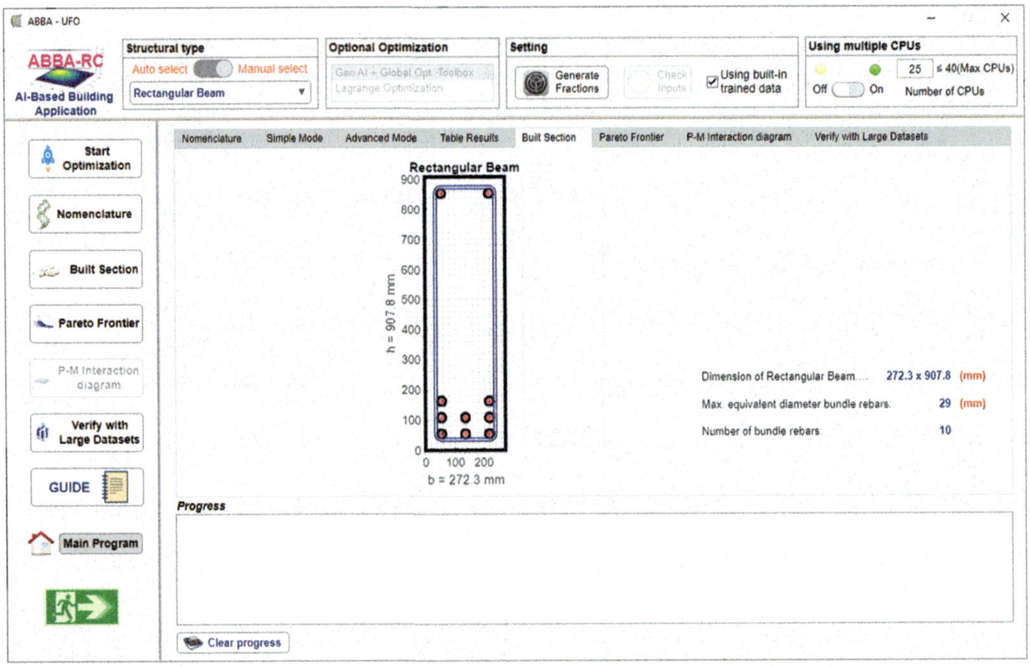

Figure D19 Geometry of a Rectangular RC beam in case of $W_c : W_{CO_2} : W_{weight} = 1/3{:}1/3{:}1/3$.

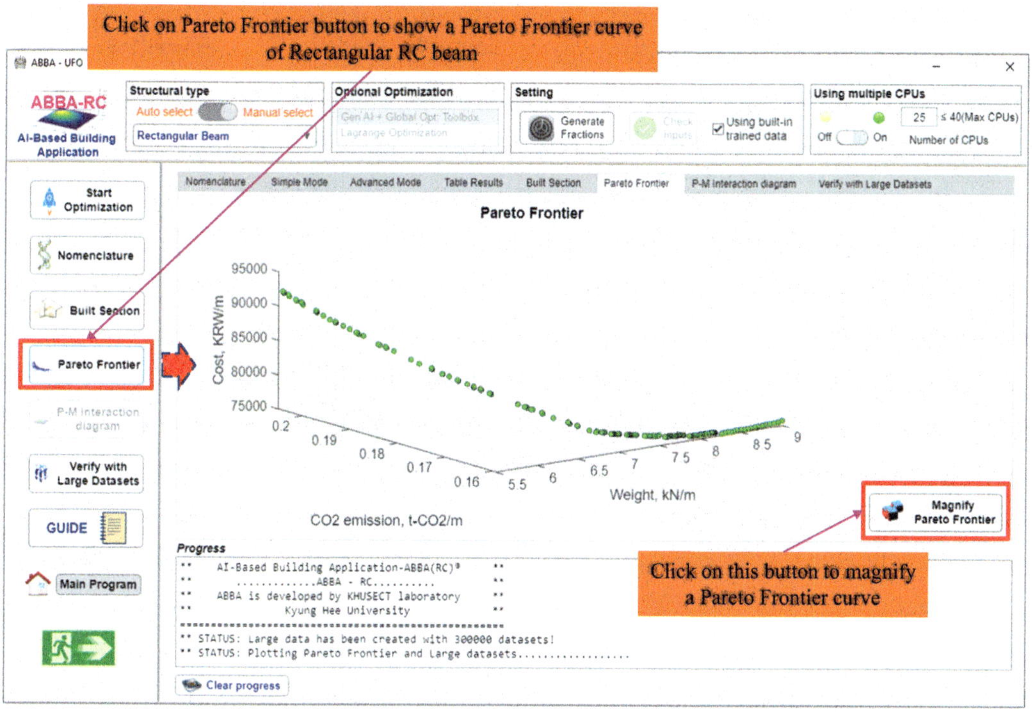

Figure D20 A Pareto Frontier curve of a Rectangular RC beam including 361 points.

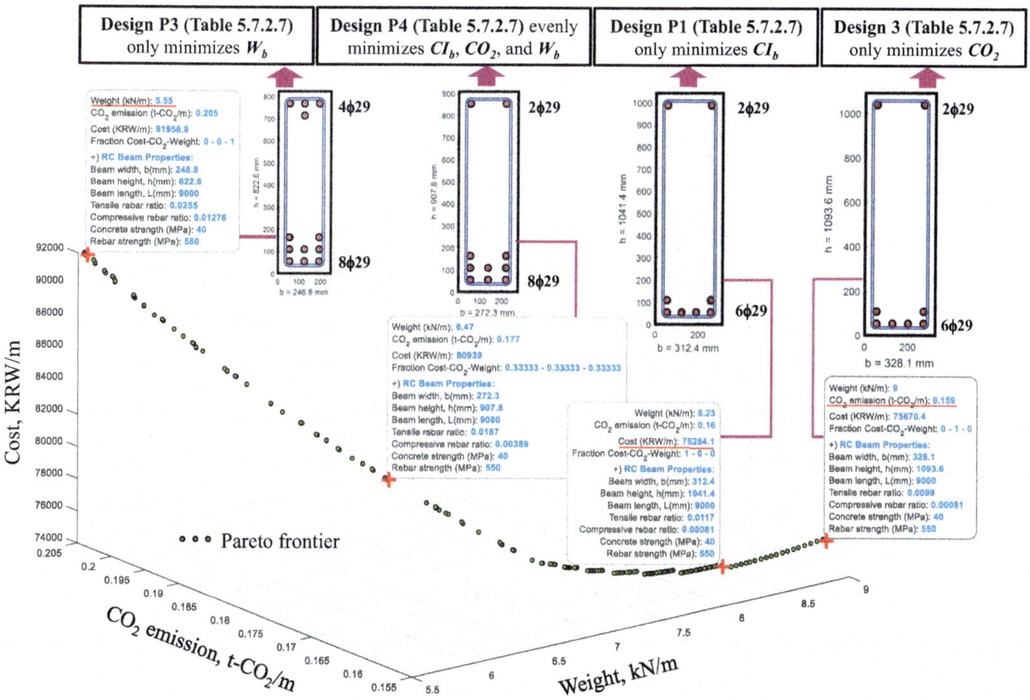

Figure D21 Four optimized designs from a Pareto Frontier curve (corresponding to Figure 5.7.2.4).

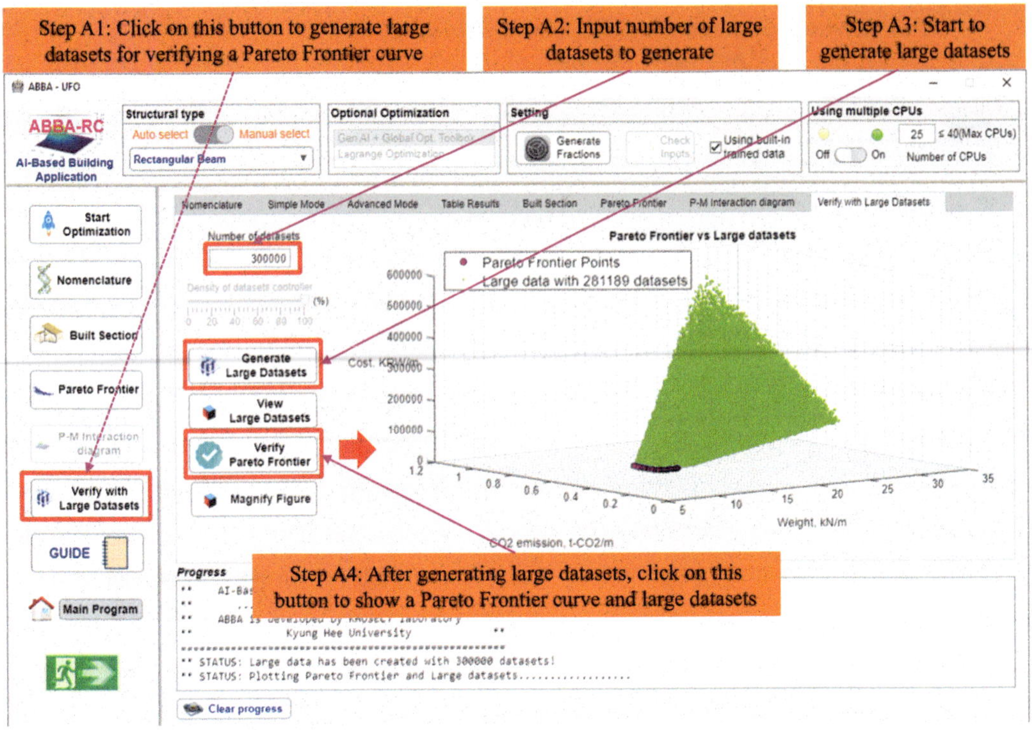

Figure D22 A Pareto Frontier curve and large datasets of an RC beam.

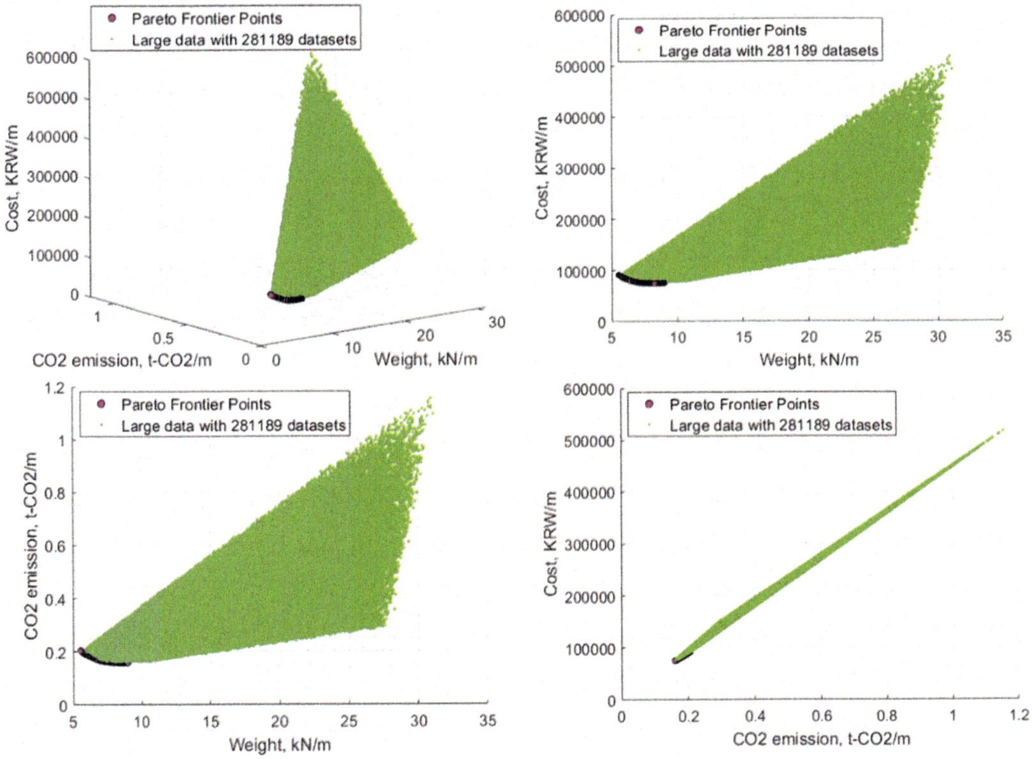

Figure D23 Verification of a Pareto frontier by large structural datasets (corresponding Figure 5.7.4.1).

D3. ABBA-RC UFO FOR A RECTANGULAR BIAXIAL RC COLUMN

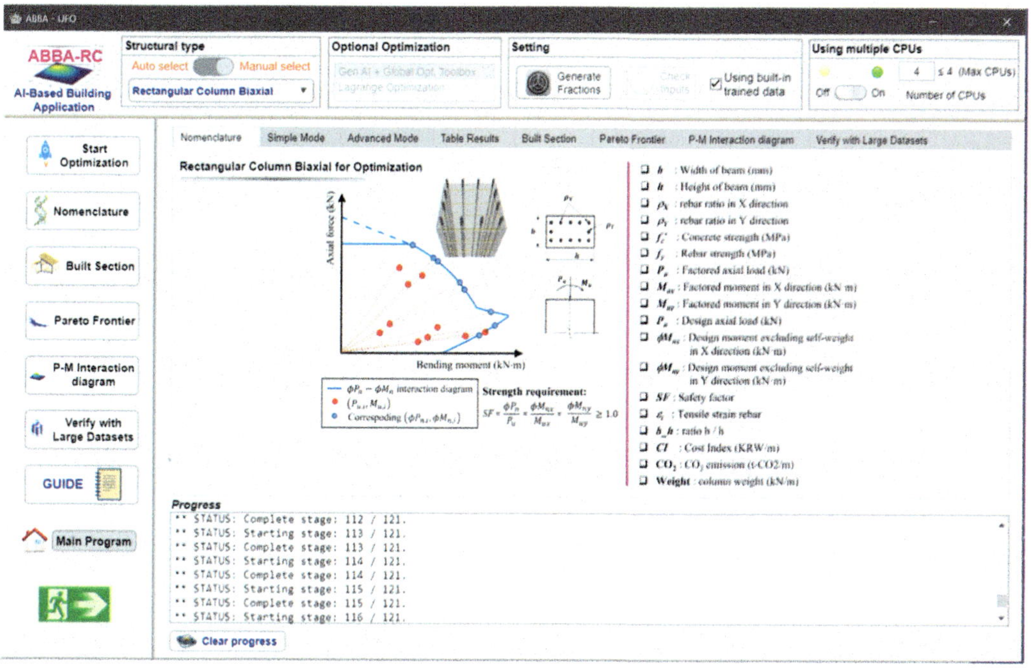

Figure D24 Nomenclature of a Rectangular Biaxial RC Column.

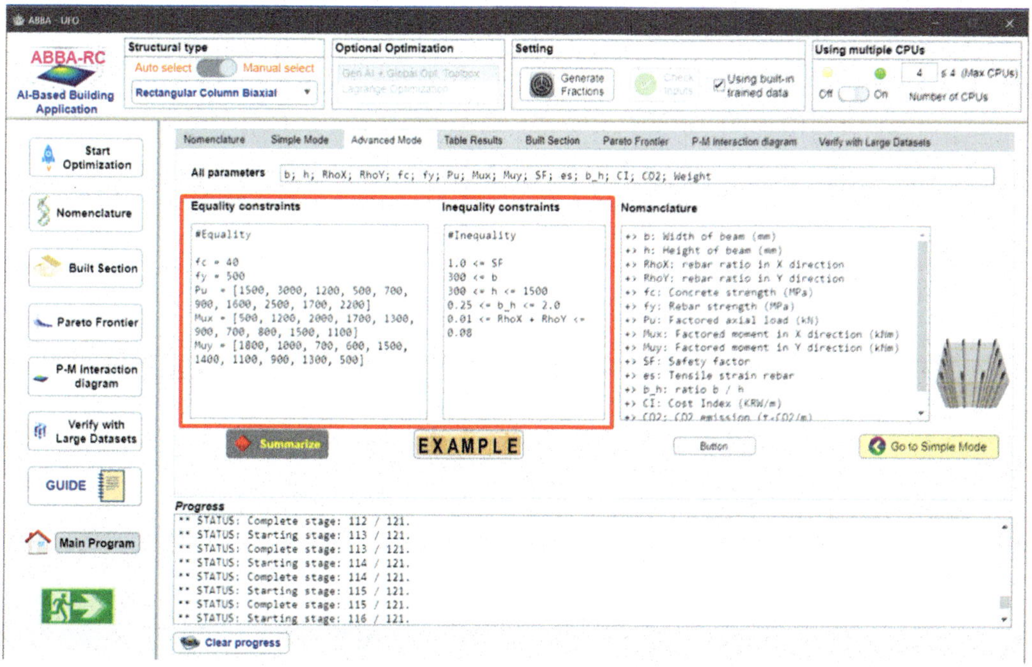

Figure D25 Input Equality and Inequality constraints for UFO of a Rectangular Biaxial RC Column.

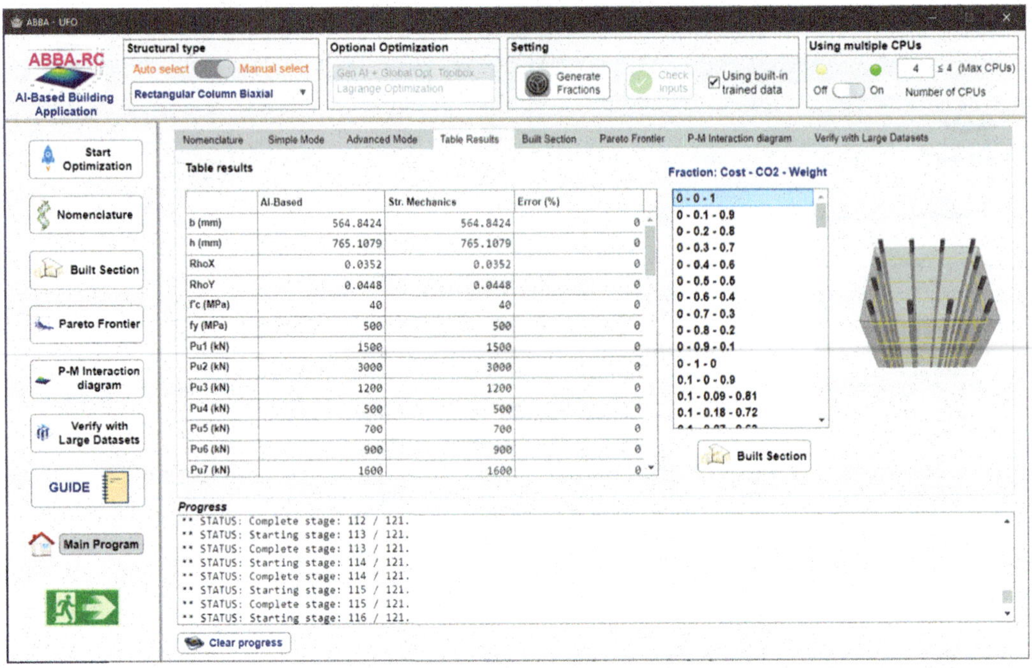

Figure D26 Optimization results of a Rectangular Biaxial RC Column in case of $W_c : W_{CO_2} : W_{weight} = 0{:}0{:}1$ (corresponding to Table 5.6.4.3).

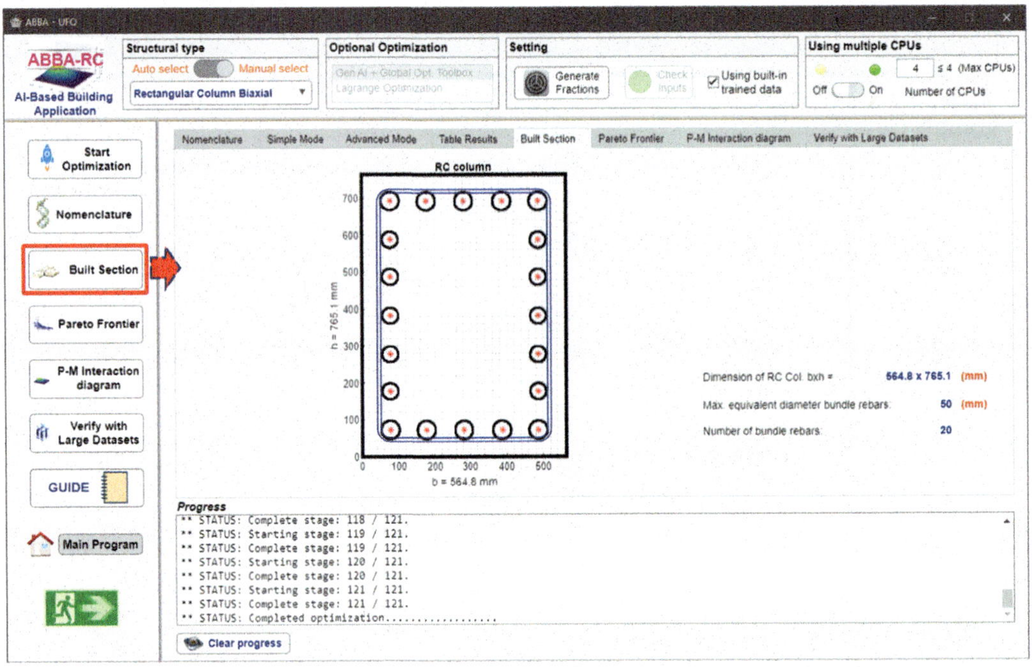

Figure D27 Geometry of a Rectangular Biaxial RC Column in case of $W_c : W_{CO_2} : W_{weight} = 0{:}0{:}1$.

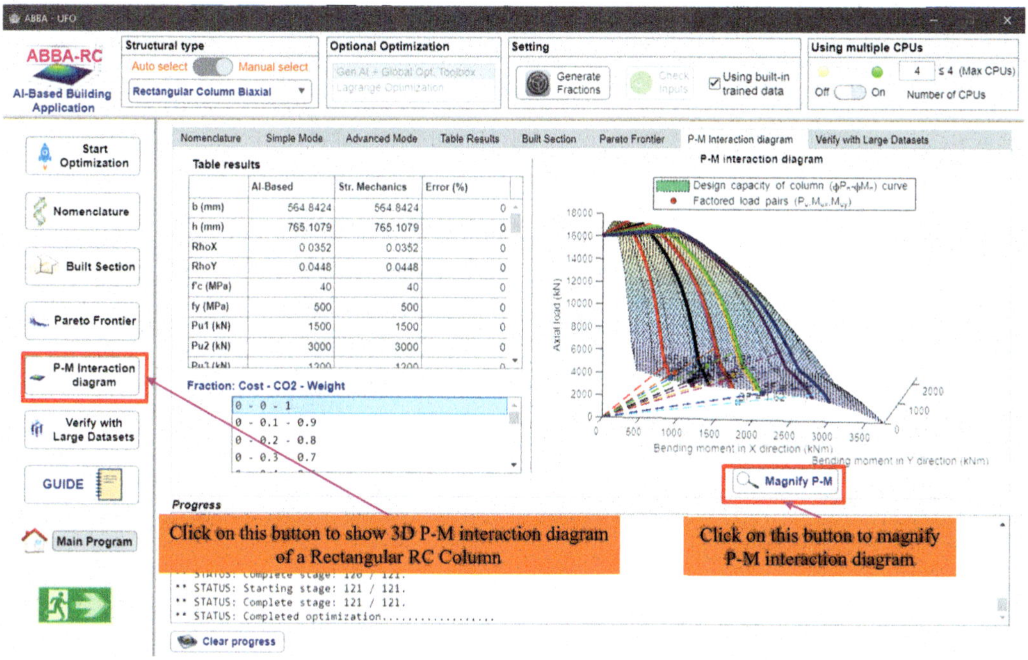

Figure D28 3D P-M interaction diagram of a Rectangular Biaxial RC Column in case of $W_c : W_{CO_2} : W_{weight} = 0{:}0{:}1$.

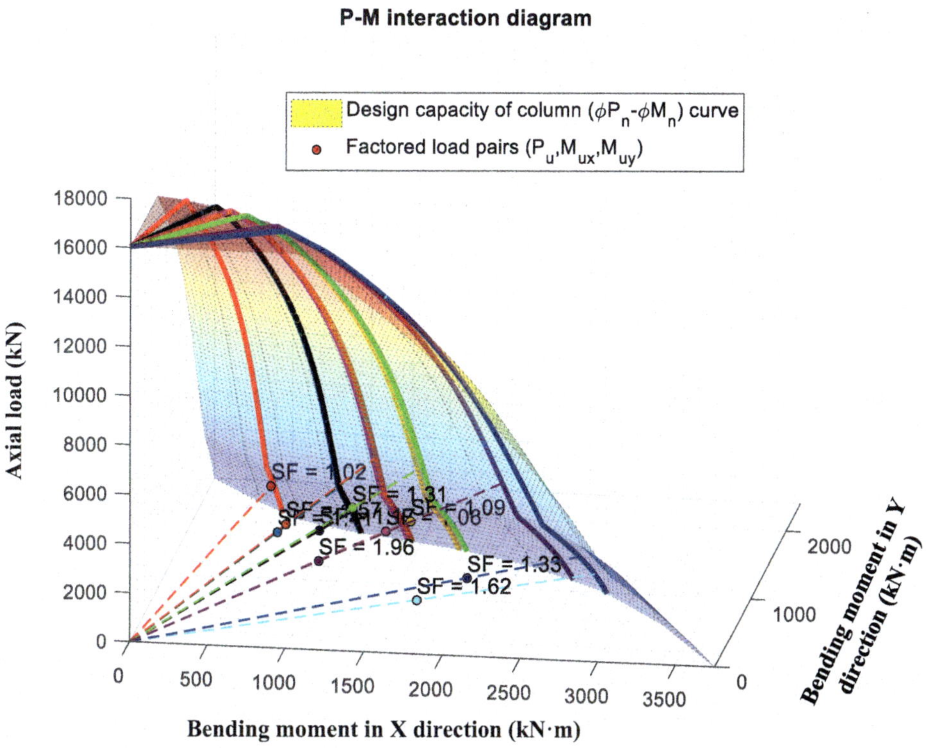

Figure D29 Magnify 3D P-M interaction diagram of a Rectangular Biaxial RC Column in case of $W_c : W_{CO_2} : W_{weight} = 0{:}0{:}1$.

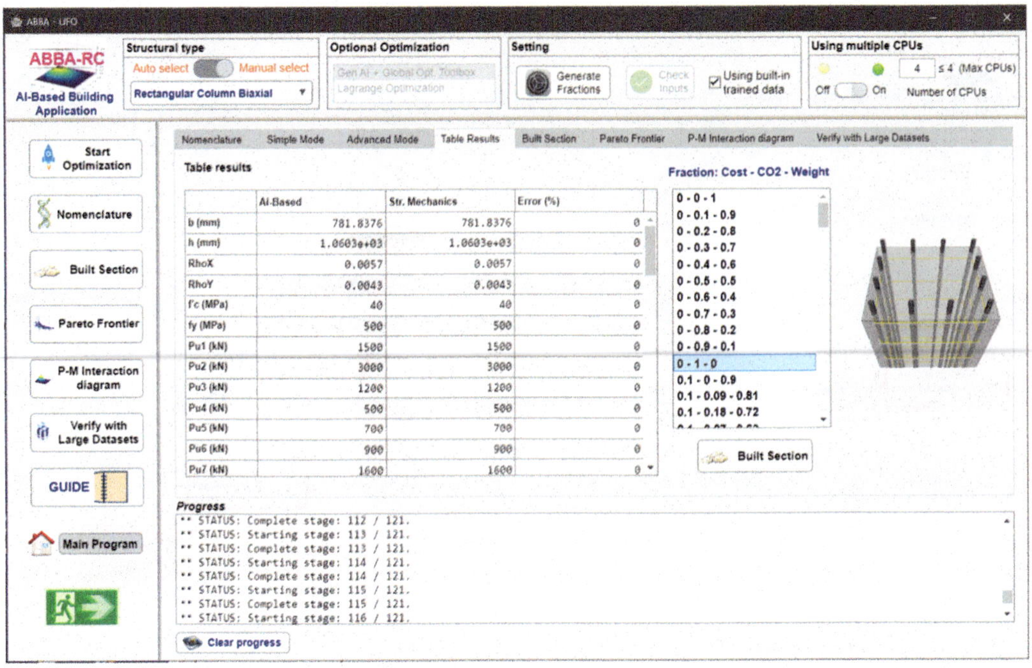

Figure D30 Optimization results of a Rectangular Biaxial RC Column in case of $W_c : W_{CO_2} : W_{weight} = 0{:}1{:}0$ (corresponding to Table 5.6.4.2).

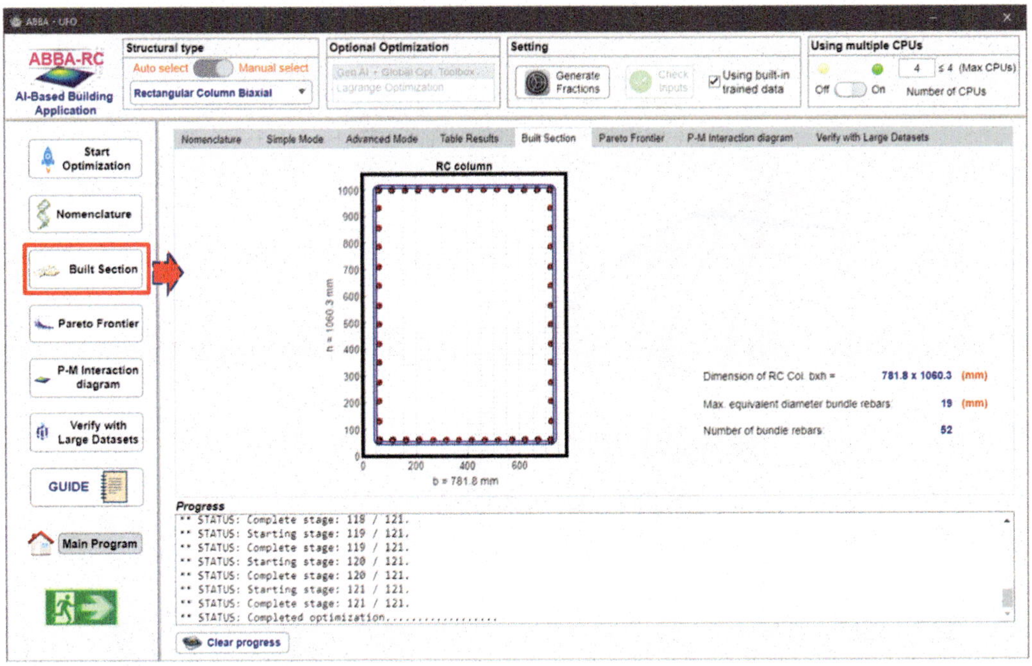

Figure D31 Geometry of a Rectangular Biaxial RC Column in case of $W_c : W_{CO_2} : W_{weight} = 0{:}1{:}0$.

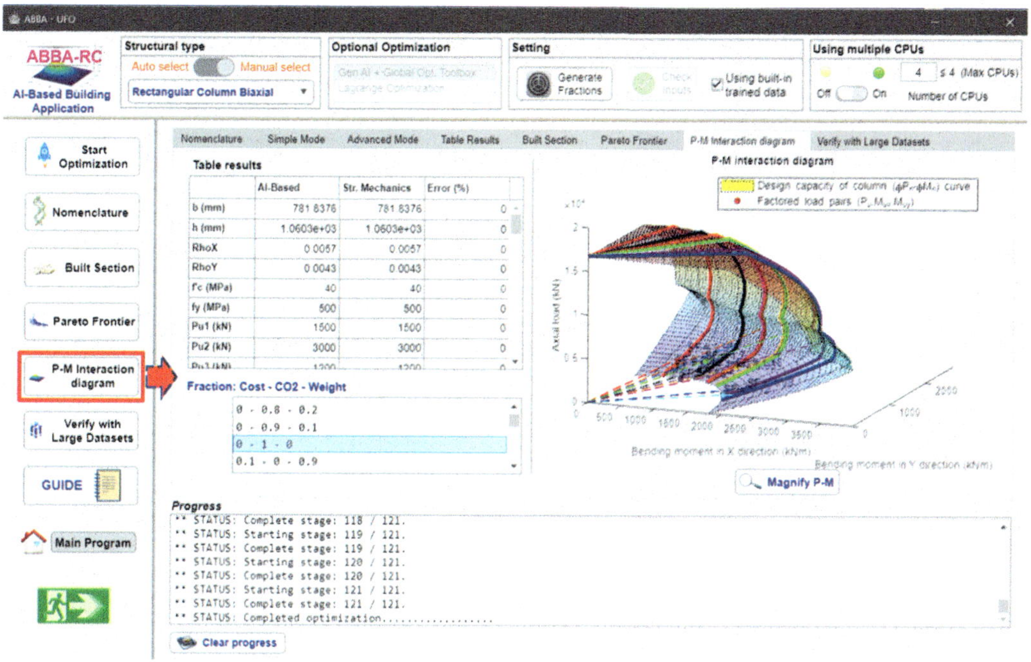

Figure D32 3D P-M interaction diagram of a Rectangular Biaxial RC Column in case of $W_c : W_{CO_2} : W_{weight} = 0{:}1{:}0$.

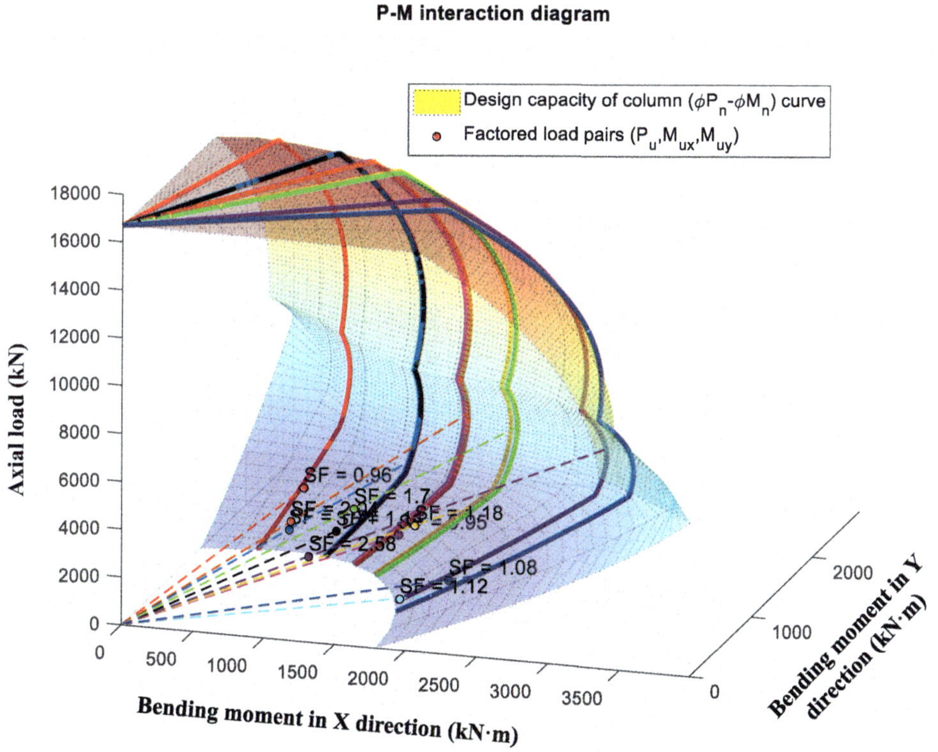

Figure D33 Magnify 3D P-M interaction diagram of a Rectangular Biaxial RC Column in case of $W_c : W_{CO_2} : W_{weight} = 0{:}1{:}0$.

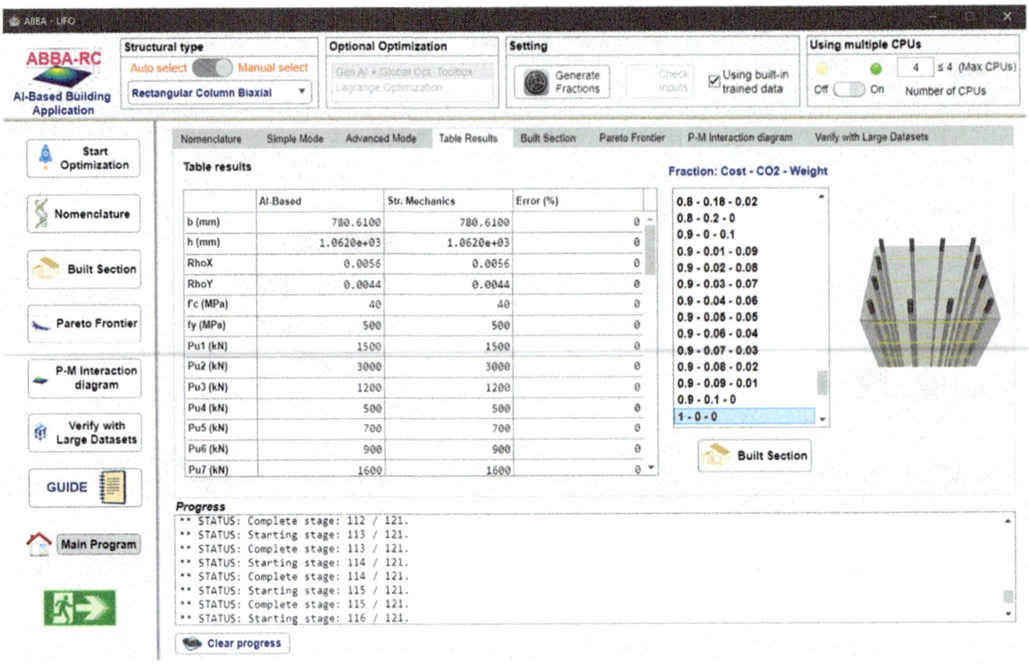

Figure D34 Optimization results of a Rectangular Biaxial RC Column in case of $W_c : W_{CO_2} : W_{weight} = 1{:}0{:}0$ (corresponding to Table 5.6.4.2).

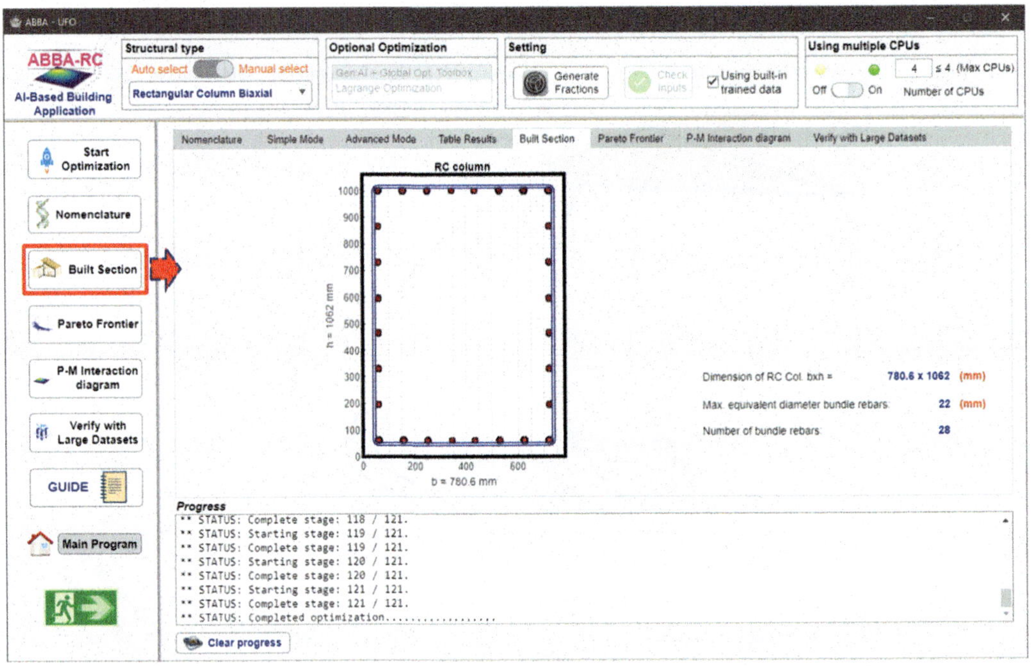

Figure D35 Geometry of a Rectangular Biaxial RC Column in case of $W_c : W_{CO_2} : W_{weight} = 1{:}0{:}0$.

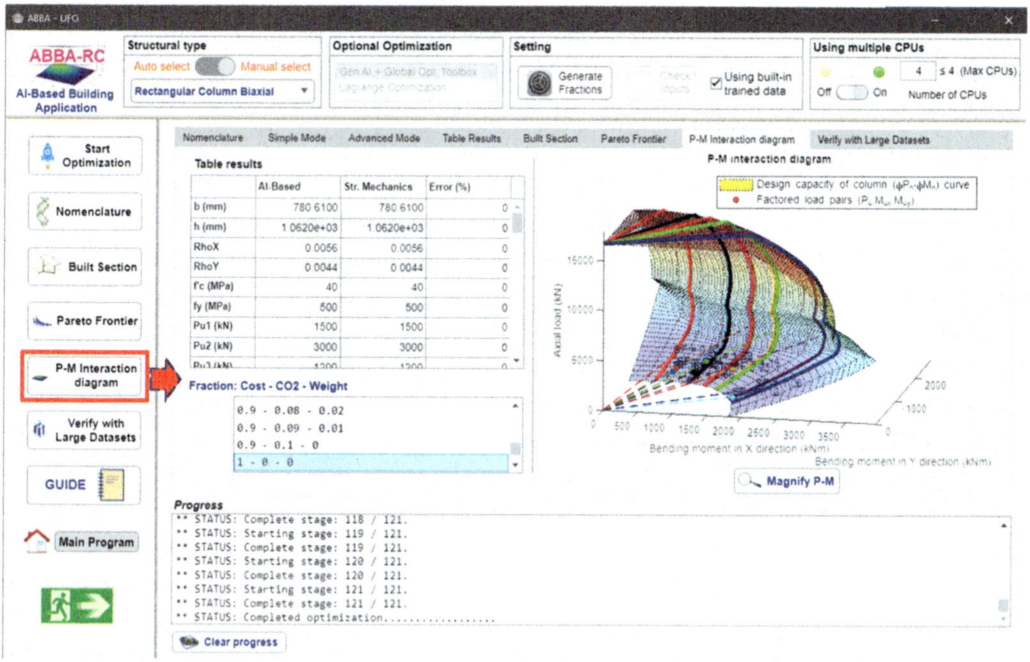

Figure D36 3D P-M interaction diagram of a Rectangular Biaxial RC Column in case of $W_c : W_{CO_2} : W_{weight} = 1:0:0$.

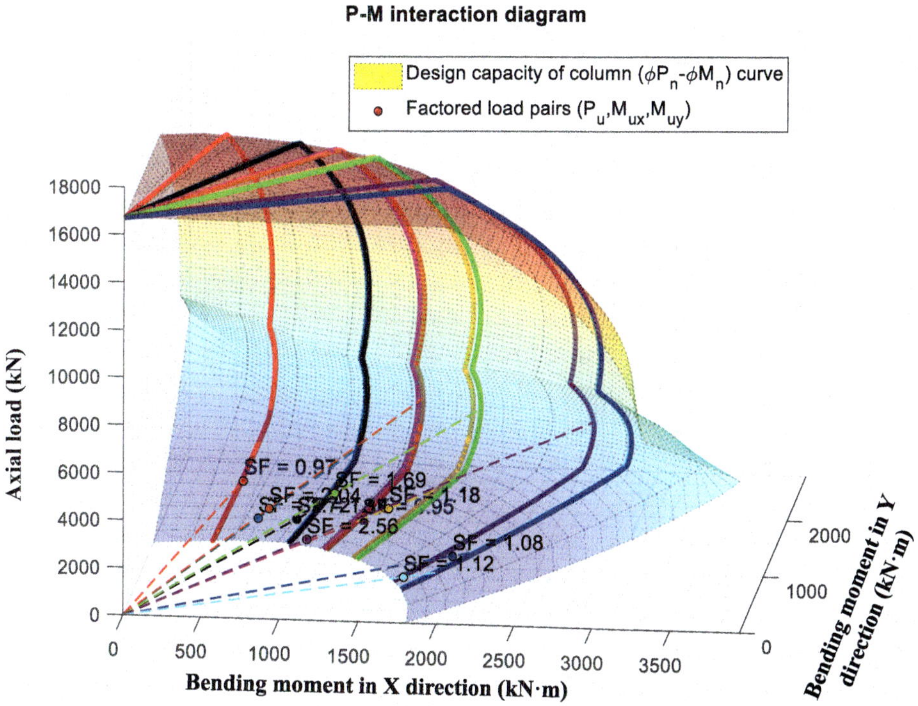

Figure D37 Magnify 3D P-M interaction diagram of a Rectangular Biaxial RC Column in case of $W_c : W_{CO_2} : W_{weight} = 1:0:0$.

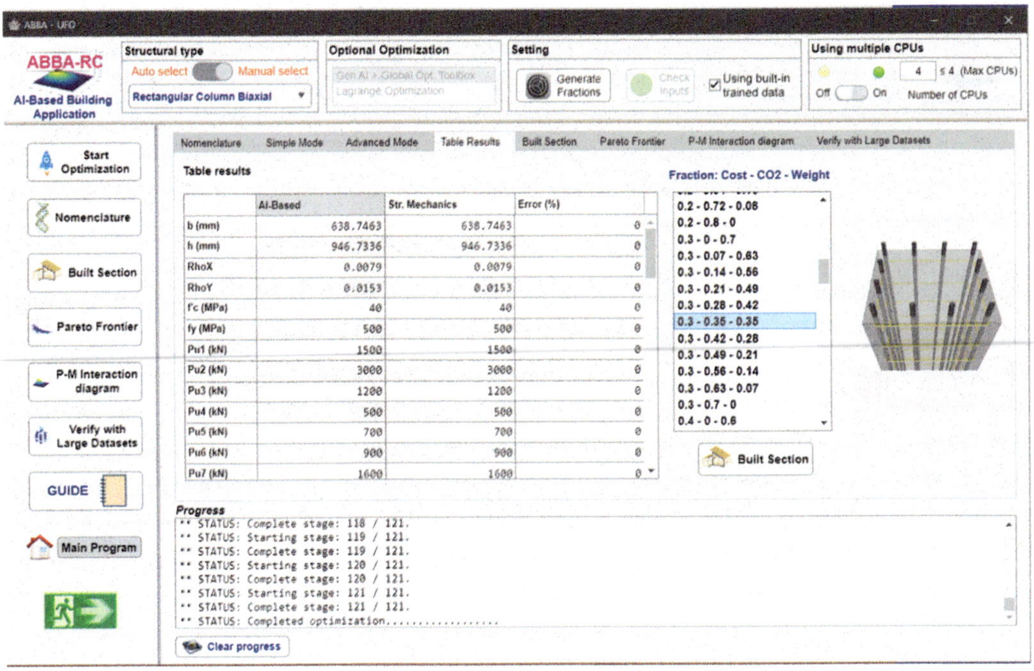

Figure D38 Optimization results of a Rectangular Biaxial RC Column in case of $W_c : W_{CO_2} : W_{weight} = 0.30{:}0.35{:}0.35$ (corresponding to Table 5.6.4.4).

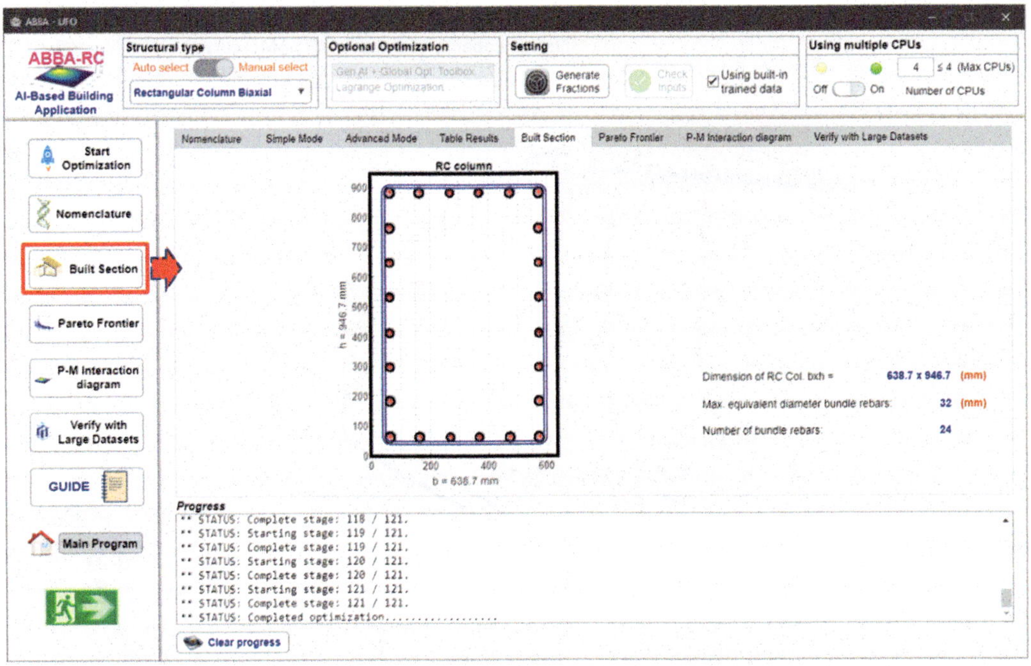

Figure D39 Geometry of a Rectangular Biaxial RC Column in case of $W_c : W_{CO_2} : W_{weight} = 0.30{:}0.35{:}0.35$.

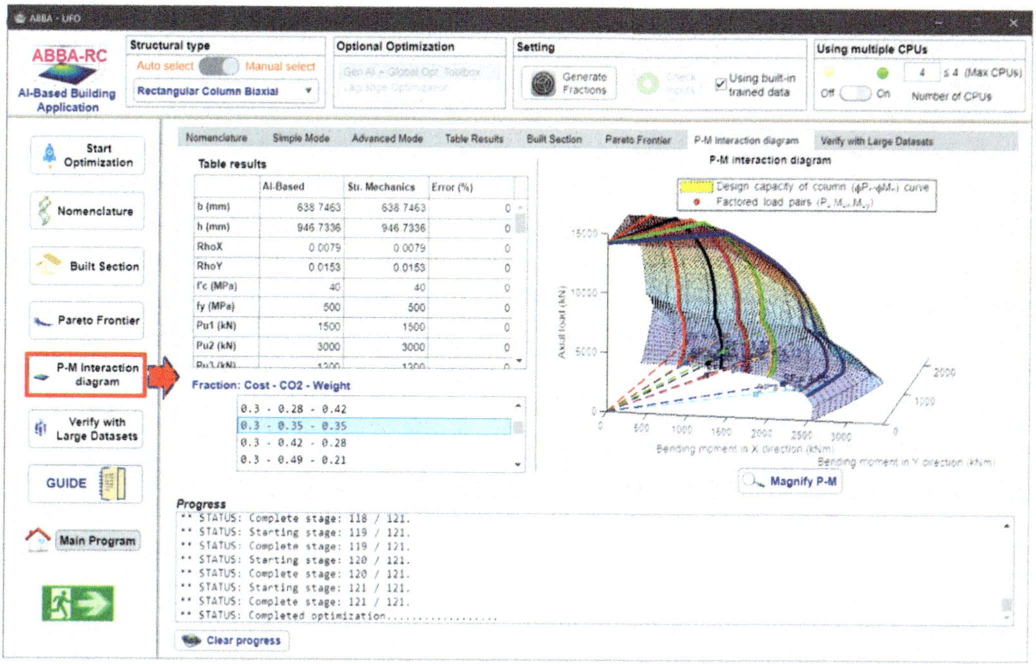

	AI-Based	Str. Mechanics	Error (%)
b (mm)	638.7463	638.7463	0
h (mm)	946.7336	946.7336	0
RhoX	0.0079	0.0079	0
RhoY	0.0153	0.0153	0
f'c (MPa)	40	40	0
fy (MPa)	500	500	0
Pu1 (kN)	1500	1500	0
Pu2 (kN)	3000	3000	0
Pu3 (kN)	1200	1200	0

Figure D40 3D P-M interaction diagram of a Rectangular Biaxial RC Column in case of $W_c : W_{CO_2} : W_{weight} = 0.30:0.35:0.35$.

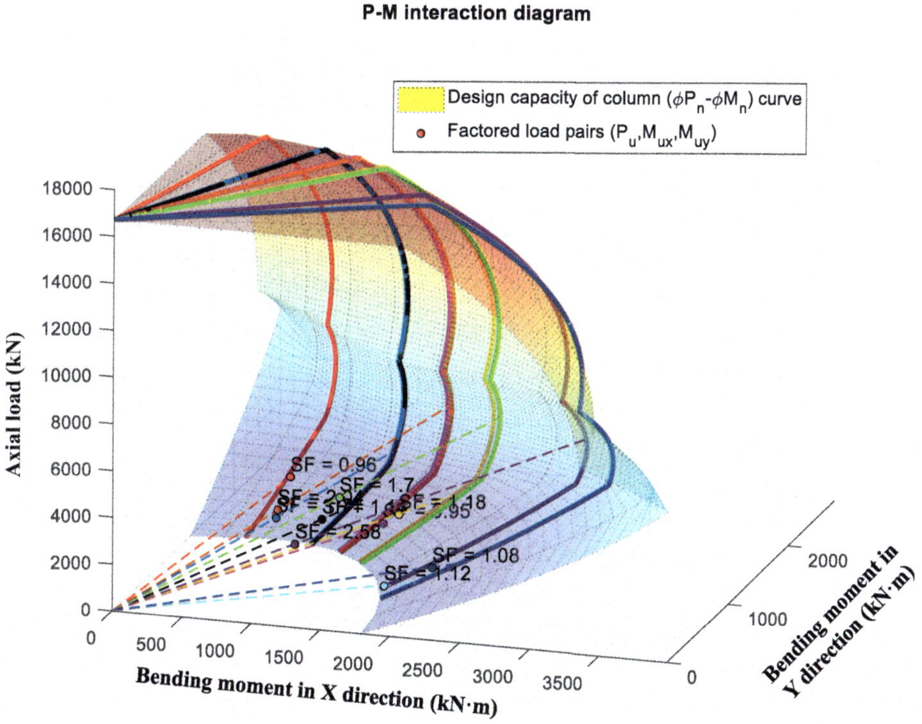

Figure D41 Magnify 3D P-M interaction diagram of a Rectangular Biaxial RC Column in case of $W_c : W_{CO_2} : W_{weight} = 0.30:0.35:0.35$.

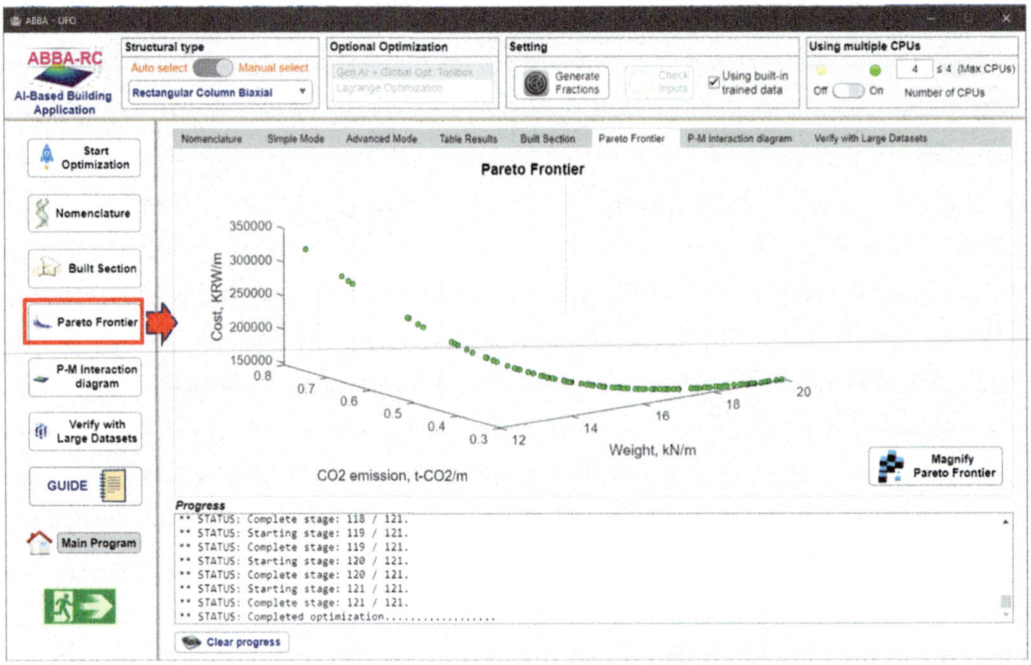

Figure D42 A Pareto Frontier curve of a Rectangular Biaxial RC Column including 121 points (corresponding to Figure 5.6.2.3).

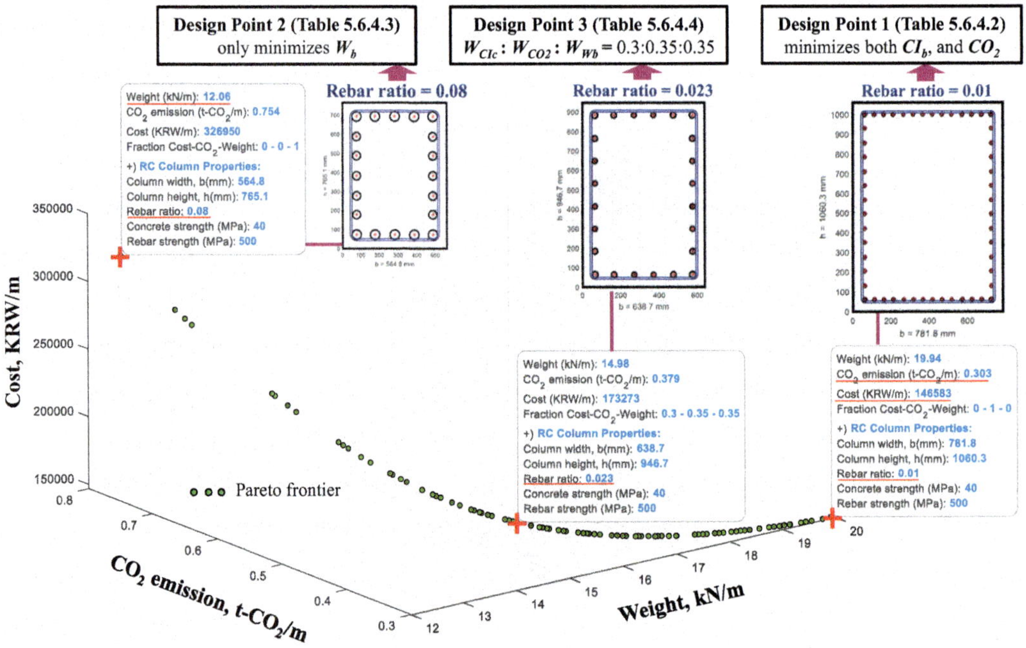

Figure D43 Four optimized designs of a Rectangular Biaxial RC Column from a Pareto Frontier curve (corresponding to Figure 5.6.4.1).

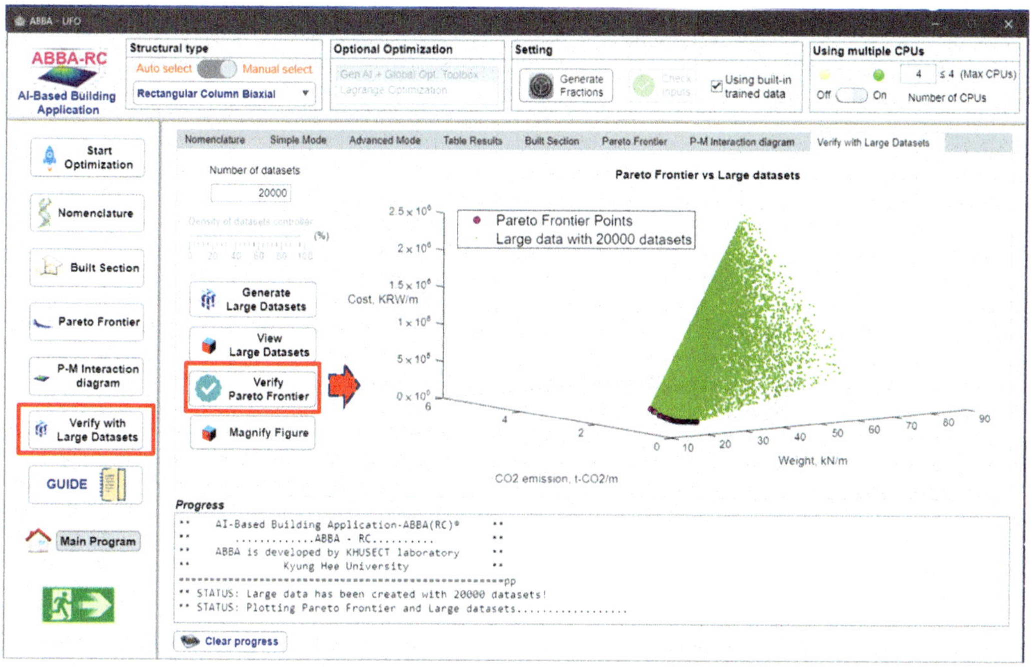

Figure D44 A Pareto Frontier curve and large datasets of a Rectangular Biaxial RC Column.

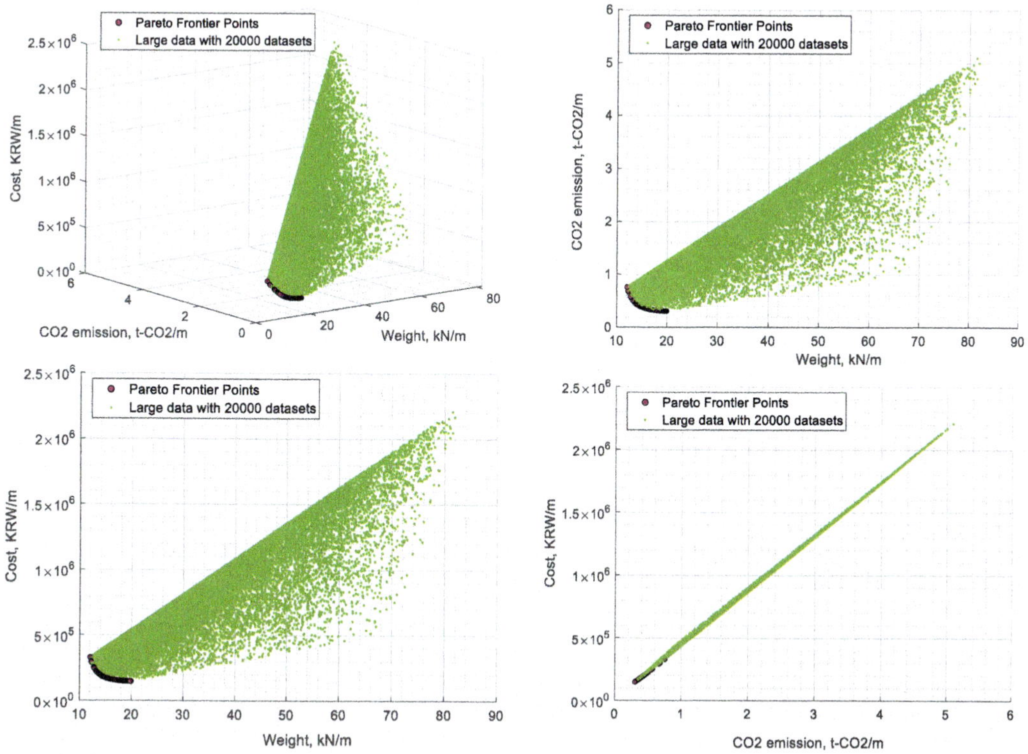

Figure D45 Verification of a Pareto frontier by large structural datasets of a Rectangular RC Column (corresponding Figure 5.6.3.1).

Index

Note: **Bold** page numbers refer to tables; *italic* page numbers refer to figures.